Handbook of Molecular Microbial Ecology I

Handbook of Molecular Microbial Ecology I

Metagenomics and Complementary Approaches

Edited by

Frans J. de Bruijn

WILEY-BLACKWELL

A John Wiley & Sons, Inc., Publication

Published by John Wiley & Sons, Inc., Hoboken, New Jersey
Published simultaneously in Canada

Library of Congress Cataloging-in-Publication Data:

Bruijn, F. J. de (Frans J. de)
Handbook of molecular microbial ecology I : metagenomics and complementary approaches / Frans J. de Bruijn.
p. cm.
Includes index.
ISBN 978-0-470-64479-9 (hardback)
1. Molecular microbiology. 2. Microbial ecology. I. Title.
QR74.B78 2011
576–dc22

2010042169

Printed in Singapore

Set ISBN: 978-0-470-92418-1
oBook ISBN: 978-1-118-01051-8
ePDF ISBN: 978-1-118-01044-0
ePub ISBN: 978-1-118-01049-5

10 9 8 7 6 5 4 3 2 1

To my two daughters, Waverly and Vanessa de Bruijn, for their support even from a distance

Contents

Preface xv

Contributors xvii

1. Introduction 1

Frans J. de Bruijn

Part 1 Background Chapters

2. DNA Reassociation Yields Broad-Scale Information on Metagenome Complexity and Microbial Diversity 5

Vigdis L. Torsvik and Lise Øvreås

3. Diversity of 23S rRNA Genes Within Individual Prokaryotic Genomes 17

Anna Pei, William E. Oberdorf, Carlos W. Nossa, Pooja Chokshi, Martin J. Blaser, Liying Yang, David M. Rosmarin, and Zhiheng Pei

4. Use of the rRNA Operon and Genomic Repetitive Sequences for the Identification of Bacteria 29

Andréa Maria Amaral Nascimento

5. Use of Different PCR Primer-Based Strategies for Characterization of Natural Microbial Communities 41

James I. Prosser, Shahid Mahmood, and Thomas E. Freitag

6. Horizontal Gene Transfer and Recombination Shape Mesorhizobial Populations in the Gene Center of the Host Plants *Astragalus Luteolus* and *Astragalus Ernestii* in Sichuan, China 49

Qiongfang Li, Xiaoping Zhang, Ling Zou, Qiang Chen, David P. Fewer, and Kristina Lindström

7. Amplified rDNA Restriction Analysis (ARDRA) for Identification and Phylogenetic Placement of 16S-rDNA Clones 59

Menachem Y. Sklarz, Roey Angel, Osnat Gillor, and Ines M. Soares

8. Clustering-Based Peak Alignment Algorithm for Objective and Quantitative Analysis of DNA Fingerprinting Data 67

Satoshi Ishii, Koji Kadota, and Keishi Senoo

Part 2 The Species Concept

9. Population Genomics Informs Our Understanding of the Bacterial Species Concept 77

Margaret A. Riley

10. The Microbial Pangenome: Implications for Vaccine Development 83

Annalisa Nuccitelli, Claudio Donati, Michèle A. Barocchi, and Rino Rappuoli

11. Metagenomic Insights into Bacterial Species 89

Konstantinos T. Konstantinidis

12. Reports of Ad Hoc Committees for the Reevaluation of the Species Definition in Bacteriology 99

Erko Stackebrandt

13. Metagenomic Approaches for the Identification of Microbial Species 105

David M. Ward, Melanie C. Melendrez, Eric D. Becraft, Christian G. Klatt, Jason M. Wood, and Frederick M. Cohan

Part 3 Metagenomics

14. Microbial Ecology in the Age of Metagenomics 113

Jianping Xu

15. The Enduring Legacy of Small Subunit rRNA in Microbiology 123

Susannah G. Tringe and Philip Hugenholtz

16. Pitfalls of PCR-Based rRNA Gene Sequence Analysis: An Update on Some Parameters 129

Erko Stackebrandt

17. Empirical Testing of 16S PCR Primer Pairs Reveals Variance in Target Specificity and Efficacy not Suggested by *In Silico* Analysis 135

Sergio E. Morales and William E. Holben

18. The Impact of Next-Generation Sequencing Technologies on Metagenomics 143

George M. Weinstock

19. Accuracy and Quality of Massively Parallel DNA Pyrosequencing 149

Susan M. Huse and David B. Mark Welch

20. Environmental Shotgun Sequencing: Its Potential and Challenges for Studying the Hidden World of Microbes 157

Jonathan A. Eisen

21. A Comparison of Random Sequence Reads Versus 16S rDNA Sequences for Estimating the Biodiversity of a Metagenomic Sample 163

Chaysavanh Manichanh, Charles E. Chapple, Lionel Frangeul, Karine Gloux, Roderic Guigo, and Joel Dore

22. Metagenomic Libraries for Functional Screening 171

Trine Aakvik, Rahmi Lale, Mark Liles, and Svein Valla

23. GC Fractionation Allows Comparative Total Microbial Community Analysis, Enhances Diversity Assessment, and Facilitates Detection of Minority Populations of Bacteria 183

William E. Holben

24. Enriching Plant Microbiota for a Metagenomic Library Construction 197

Ying Zeng, Hao-Xin Wang, Zhao-Liang Geng, and Yue-Mao Shen

25. Towards Automated Phylogenomic Inference 205

Martin Wu and Jonathan A. Eisen

26. Integron First Gene Cassettes: A Target to Find Adaptive Genes in Metagenomes 217

Lionel Huang and Christine Cagnon

27. High-Resolution Metagenomics: Assessing Specific Functional Types in Complex Microbial Communities 225

Ludmila Chistoserdova

28. Gene-Targeted Metagenomics (GT Metagenomics) to Explore the Extensive Diversity of Genes of Interest in Microbial Communities 235

Shoko Iwai, Benli Chai, Ederson da C. Jesus, C. Ryan Penton, Tae Kwon Lee, James R. Cole, and James M. Tiedje

29. Phylogenetic Screening of Metagenomic Libraries Using Homing Endonuclease Restriction and Marker Insertion 245

Torsten Thomas, Staffan Kjelleberg, and Pui Yi Yung

30. ArrayOme- and tRNAcc-Facilitated Mobilome Discovery: Comparative Genomics Approaches for Identifying Rich Veins of Bacterial Novel DNA Sequences 251

Hong-Yu Ou and Kumar Rajakumar

31. Sequence-Based Characterization of Microbiomes by Serial Analysis of Ribosomal Sequence Tags (SARST) 265

Zhongtang Yu and Mark Morrison

Part 4 Consortia and Databases

32. **The Metagenomics of Plant Pathogen-Suppressive Soils** 277
Jan Dirk van Elsas, Anna Maria Kielak, and Mariana Silvia Cretoiu

33. **Soil Metagenomic Exploration of the Rare Biosphere** 287
Tom O. Delmont, Laure Franqueville, Samuel Jacquiod, Pascal Simonet, and Timothy M. Vogel

34. **The BIOSPAS Consortium: Soil Biology and Agricultural Production** 299
Luis Gabriel Wall

35. **The Human Microbiome Project** 307
George M. Weinstock

36. **The Ribosomal Database Project: Sequences and Software for High-Throughput rRNA Analysis** 313
James R. Cole, Qiong Wang, Benli Chai, and James M. Tiedje

37. **The Metagenomics RAST Server: A Public Resource for the Automatic Phylogenetic and Functional Analysis of Metagenomes** 325
Elizabeth M. Glass and Folker Meyer

38. **The EBI Metagenomics Archive, Integration and Analysis Resource** 333
C. Hunter, G. Cochrane, R. Apweiler, S. Hunter

Part 5 Computer-Assisted Analysis

39. **Comparative Metagenome Analysis Using MEGAN** 343
Daniel H. Huson and Suparna Mitra

40. **Phylogenetic Binning of Metagenome Sequence Samples** 353
Alice Carolyn McHardy and Kaustubh Patil

41. **Gene Prediction in Metagenomic Fragments with Orphelia: A Large-Scale Machine Learning Approach** 359
Katharina H. Hoff, Maike Tech, Thomas Lingner, Rolf Daniel, Burkhard Morgenstern, and Peter Meinicke

42. **Binning Metagenomic Sequences Using Seeded GSOM** 369
Ching-Hung Tseng, Chon-Kit Kenneth Chan, Arthur L. Hsu, Saman K. Halgamuge, and Sen-Lin Tang

43. Iterative Read Mapping and Assembly Allows the Use of a More Distant Reference in Metagenome Assembly 379

Bas E. Dutilh, Martijn A. Huynen, Jolein Gloerich, and Marc Strous

44. Ribosomal RNA Identification in Metagenomic and Metatranscriptomic Datasets 387

Ying Huang, Weizhong Li, Patricia W. Finn, and David L. Perkins

45. SILVA: Comprehensive Databases for Quality Checked and Aligned Ribosomal RNA Sequence Data Compatible with ARB 393

Elmar Prüsse, Christian Quast, Pelin Yilmaz, Wolfgang Ludwig, Jörg Peplies, and Frank Oliver Glöckner

46. ARB: A Software Environment for Sequence Data 399

Ralf Westram, Kai Bader, Elmar Prüsse, Yadhu Kumar, Harald Meier, Frank Oliver Glöckner, and Wolfgang Ludwig

47. The Phyloware Project: A Software Framework for Phylogenomic Virtue 407

Daniel N. Frank and Charles E. Robertson

48. MetaSim: A Sequencing Simulator for Genomics and Metagenomics 417

Daniel C. Richter, Felix Ott, Alexander F. Auch, Ramona Schmid, and Daniel H. Huson

49. *ClustScan*: An Integrated Program Package for the Detection and Semiautomatic Annotation of Secondary Metabolite Clusters in Genomic and Metagenomic DNA Datasets 423

John Cullum, Antonio Starcevic, Janko Diminic, Jurica Zucko, Paul F. Long, and Daslav Hranueli

50. MetaGene: Prediction of Prokaryotic and Phage Genes in Metagenomic Sequences 433

Hideki Noguchi

51. Primers4clades: A Web Server to Design Lineage-Specific PCR Primers for Gene-Targeted Metagenomics 441

Bernardo Sachman-Ruiz, Bruno Contreras-Moreira, Enrique Zozaya, Cristina Martínez-Garza, and Pablo Vinuesa

52. A Parsimony Approach to Biological Pathway Reconstruction/Inference for Metagenomes 453

Yuzhen Ye and Thomas G. Doak

53. ESPRIT: Estimating Species Richness Using Large Collections of 16S rRNA Data 461

Yijun Sun, Yunpeng Cai, Li Liu, Fahong Yu, and William Farmerie

Part 6 Complementary Approaches

54. Metagenomic Approaches in Systems Biology **475**

María-Eugenia Guazzaroni and Manuel Ferrer

55. Towards "Focused" Metagenomics: A Case Study Combining DNA Stable-Isotope Probing, Multiple Displacement Amplification, and Metagenomics **491**

Yin Chen, Marc G. Dumont, Joshua D. Neufeld, and J. Colin Murrell

56. Suppressive Subtractive Hybridization Reveals Extensive Horizontal Transfer in the Rumen Metagenome **497**

Elizabeth A. Galbraith, Dionysios A. Antonopoulos, Karen E. Nelson, and Bryan A. White

Part 6A Microarrays

57. GeoChip: A High-Throughput Metagenomics Technology for Dissecting Microbial Community Functional Structure **509**

Joy D. van Nostrand, Zhili He, and Jizhong Zhou

58. Phylogenetic Microarrays (PhyloChips) For Analysis of Complex Microbial Communities **521**

Eoin L. Brodie

59. Phenomics and Phenotype Microarrays: Applications Complementing Metagenomics **533**

Barry R. Bochner

60. Microbial Persistence in Low-Biomass, Extreme Environments: The Great Unknown **541**

Parag Vaishampayan, James N. Benardini, Myron T. La Duc, and Kasthuri Venkateswaran

61. Application of Phylogenetic Oligonucleotide Microarrays in Microbial Analysis **551**

Pankaj Trivedi and Nian Wang

Part 6B Metatranscriptomics

62. Isolation of mRNA From Environmental Microbial Communities for Metatranscriptomic Analyses **569**

Peer M. Schenk

63. Comparative Day/Night Metatranscriptomic Analysis of Microbial Communities in the North Pacific Subtropical Gyre **575**

Rachel S. Poretsky and Mary Ann Moran

64. **The "Double-RNA" Approach to Simultaneously Assess the Structure and Function of a Soil Microbial Community** 587

Tim Urich and Christa Schleper

65. **Soil Eukaryotic Diversity: A Metatranscriptomic Approach** 597

Roland Marmeisse, Julie Bailly, Coralie Damon, Frédéric Lehembre, Marc Lemaire, Micheline Wésolowski-Louvel, and Laurence Fraissinet-Tachet

Part 6C Metaproteomics

66. **Proteomics for the Analysis of Environmental Stress Responses in Prokaryotes** 605

Ksenia J. Groh, Victor J. Nesatyy, and Marc J.-F. Suter

67. **Microbial Community Proteomics** 627

Paul Wilmes

68. **Synchronicity between Population Structure and Proteome Profiles: A Metaproteomic Analysis of Chesapeake Bay Bacterial Communities** 637

Jinjun Kan, Thomas E. Hanson, and Feng Chen

69. **High-Throughput Cyanobacterial Proteomics: Systems-Level Proteome Identification and Quantitation** 645

Saw Yen Ow and Phillip C. Wright

70. **Protein Expression Profile of an Environmentally Important Bacterial Strain: The Chromate Response of *Arthrobacter* Species Strain FB24** 663

Kristene L. Henne, Joshua E. Turse, Cindy H. Nakatsu, and Allan E. Konopka

Part 6D Metabolomics

71. **The Small-Molecule Dimension: Mass-spectrometry-based Metabolomics, Enzyme Assays, and Imaging** 677

Trent R. Northen

72. **Metabolomics: High-Resolution Tools Offer to Follow Bacterial Growth on a Molecular Level** 683

Lucio Marianna, Agnes Fekete, Moritz Frommberger, and Philippe Schmitt-Kopplin

73. **Metabolic Profiling of Plant Tissues by Electrospray Mass Spectrometry** 697

Heather Walker

74. **Metabolite Identification, Pathways, and Omic Integration Using Online Databases and Tools** 709

Matthew P. Davey

Part 6E Single-Cell Analysis

75. Application of Cytomics to Separate Natural Microbial Communities by their Physiological Properties **727**

Susann Müller and David R. Johnson

76. Capturing Microbial Populations for Environmental Genomics **735**

Martha Schattenhofer and Annelie Wendeberg

77. Microscopic Single-Cell Isolation and Multiple Displacement Amplification of Genomes from Uncultured Prokaryotes **741**

Peter Westermann and Thomas Kvist

Index 747

Preface

In the last 25 years, microbiology and molecular microbial ecology have undergone drastic transformations that changed the microbiologist's view of how to study microorganisms. Previously, the main problem was the assumption that microorganisms needed to be culturable, in order to classify them and study their metabolic and organismal diversity. The heart of this transformation was the convincing demonstration that the yet-unculturable world was far greater than the culturable one. In fact, the number of microbial genomes has been estimated from 2000 to 18,000 genomes per gram of soil. In 1985, an experimental advance radically changed our perception of the microbial world. After Carl Woese showed that rRNA genes could be used to derive evolutionary relationships, phylogenetic "trees" and evolutionary chronometers, Norman Pace and colleagues created a new chapter in molecular microbial ecology, using the direct analysis of rRNA sequences in the environment to describe the diversity of microorganisms without culturing (Handelsman, 2004). The next major step forward was the development of the PCR reaction, to amplify rRNA genes for subsequent sequence analysis and classification. The subsequent major advance was the notion that one could extract total DNA or RNA from environmental samples, including culturable and yet unculturable organisms, and clone it into a suitable vector for introduction into a culturable organism, followed by analysis by using high throughput shotgun DNA sequencing of cloned DNA, or by direct sequencing The idea of cloning DNA directly from environmental samples was first proposed by Page; this method was coined "metagenomics" by Handelsman et al. in 1994, and is now used in many laboratories worldwide to study diversity and for the isolation of novel medical and industrial compounds.

These recent studies are reviewed in this book and the companion book, *Handbook of Molecular Microbial Ecology II: Metagenomics in Different Habitats*. Instead of relying only on a limited number of (long) review articles on selected topics, this book provides reviews as well as a large number of case studies, mostly based on original publications and written by expert "at-the-bench" scientists from more than 20 different countries. Both books highlight the databases and computer programs used in each study, by listing them at the end of the chapter, together with their sites. This special feature of both books, facilitates the computer-assisted analysis of the vast amount of data generated by metagenomic studies. In addition, metagenomic studies in a variety of habitats are described, primarily in Volume II, which present a large number of system dependent different approaches in greatly differing habitats. The latter also results in the presentation of multiple biological systems which are interesting to microbial ecologists and microbiologists in their own right. Both books should be of interest to scientists in the fields of soil, water, medicine and industry who are or are contemplating using metagenomics and complementary approaches to address academic, medical, or industrial questions about bacterial communities from varied habitats, but also to those interested in particular biological systems in general.

ACKNOWLEDGMENTS

For their support of this project, I gratefully acknowledge:

The Laboratory for Plant Microbe Interactions (LIPM),

the Institut National de Recherche de Agriculture (INRA), and

the Centre National de Recherche Scientifique (CNRS).

I would like to thank Claude Bruand for his help with the computer work.

Frans J. de Bruijn

Castanet, Tolosan, France
March 2011

Contributors

Editor

Frans J. de Bruijn, Laboratory of Plant Micro-organism Interaction, CNRS-INRA, Castanet Tolosan, France

Authors

Trine Aakvik, Norwegian University of Science and Technology, Trondheim, Norway

Roey Angel, Ben-Gurion University of the Negev, Israel

Dionysios A. Antonopoulos, Argonne National Laboratory, Argonne, Illinois

R. Apweiler, The Wellcome Trust Genome Campus, Hinxton, Cambridgeshire, United Kingdom

Alexander F. Auch, University of Tübingen, Tübingen, Germany

Kai Bader, Technical University of München, Freising, Germany

Julie Bailly, University of Lyon, Villeurbanne Cedex, France

Michèle A. Barocchi, Novartis Vaccines and Diagnostics, Siena, Italy

Eric D. Becraft, Montana State University, Bozeman, Montana

James N. Benardini, California Institute of Technology, Pasadena, California

Martin J. Blaser, New York University School of Medicine, New York, New York

Barry R. Bochner, Biolog, Inc., Hayward, California

Eoin L. Brodie, Lawrence Berkeley National Laboratory, Berkeley, California

Christine Cagnon, Université de Pau et des Pays de l'Adour, Pau, France

Yunpeng Cai, University of Florida, Gainesville, Florida

Benli Chai, Michigan State University, East Lansing, Michigan

Chon-Kit Kenneth Chan, University of Melbourne, Melbourne, Victoria, Australia

Charles E Chapple, Center for Genomic Regulation, Barcelona, Spain

Feng Chen, Biotechnology Institute, University of Maryland, Baltimore, Maryland

Qiang Chen, Sichuan Agricultural University, Ya'an Sichuan, China

Yin Chen, University of Warwick, Coventry, United Kingdom

Ludmila Chistoserdova, University of Washington, Seattle, Washington

Pooja Chokshi, College of Arts and Sciences, Tufts University, Medford, Massachusetts

G. Cochrane, The Wellcome Trust Genome Campus, Hinxton, Cambridgeshire, United Kingdom

Frederick M. Cohan, Wesleyan University, Middletown, Connecticut

James R. Cole, Michigan State University, East Lansing, Michigan
Bruno Contreras-Moreira, Upper Counsel of Scientific Investigations, Zaragoza, Spain
Mariana Silvia Cretoiu, University of Groningen, Haren, The Netherlands
John Cullum, University of Kaiserslautern, Kaiserslautern, Germany
Coralie Damon, University of Lyon, Villeurbanne, Lyon, France
Rolf Daniel, Georg August University at Göttingen, Göttingen, Germany
Matthew P. Davey, University of Cambridge, Cambridge, United Kingdom
Tom O. Delmont, Environmental Microbial Genomics Group, Ecully, France.
Janko Diminic, University of Zagreb, Zagreb, Croatia
Thomas G. Doak, Indiana University, Bloomington, Indiana
Claudio Donati, Novartis Vaccines and Diagnostics, Siena, Italy
Joel Dore, INRA/CNRS, Jouy-en-Josas, France
Marc G. Dumont, University of Warwick, Coventry, United Kingdom
Bas E. Dutilh, Radboud University Nijmegen Medical Center, Nijmegen, The Netherlands
Jonathan A. Eisen, University of California—Davis, Davis, California
Jan Dirk van Elsas, University of Groningen, Haren, The Netherlands
William Farmerie, University of Florida, Gainesville, Florida
Agnes Fekete, Institute of Ecological Chemistry, Neuherberg, Germany
Manuel Ferrer, Institute of Catalysis, Madrid, Spain
David P. Fewer, University of Helsinki, Helsinki, Finland
Patricia W. Finn, University of California—San Diego, La Jolla, California
Laurence Fraissinet-Tachet, University of Lyon, Villeurbanne, Lyon, France
Daniel N. Frank, University of Colorado, Boulder, Colorado
Lionel Frangeul, Genopole, Pasteur Institute, Paris, France
Laure Franqueville, Environmental Microbial Genomics Group, Ecully, France
Thomas E. Freitag, Uppsala BioCenter, Uppsala, Sweden
Moritz Frommberger, Institute of Biological Chemistry, Neuherberg, Germany
Elizabeth A. Galbraith, Agtech Products, USA Inc., Waukesha, Wisconsin
Zhao-Liang Geng, Kunming Institute of Botany, the Chinese Academy of Sciences, Yunnan, China
Osnat Gillor, Ben-Gurion University of the Negev, Beersheba, Israel
Elizabeth M. Glass, The University of Chicago, Chicago, Illinois
Frank Oliver Glöckner, Max Planck Institute for Marine Microbiology, Bremen, Germany; Jacobs University, Bremen, Germany
Jolein Gloerich, Radboud University Nijmegen Medical Center, Nijmegen, The Netherlands
Karine Gloux, INRA-CNRS, Jouy-en-Josas, France
Ksenia J. Groh, Swiss Federal Institute of Science and Technology, Duebendorf, Switzerland
María-Eugenia Guazzaroni, Institute of Catalysis, Madrid, Spain
Roderic Guigo, Center for Genomic Regulation, Barcelona, Spain
Saman K. Halgamuge, The University of Melbourne, Melbourne, Victoria, Australia
Thomas E. Hanson, University of Delaware, Newark, Delaware
Kristene L. Henne, Purdue University, West Lafayette, Indiana
Zhili He, University of Oklahoma, Norman, Oklahoma
Katharina H. Hoff, Medical Center Göttingen, Göttingen, Germany

William E. Holben, The University of Montana, Missoula, Montana
Daslav Hranueli, University of Zagreb, Zagreb, Croatia
Arthur L. Hsu, The University of Melbourne, Melbourne, Victoria, Australia
Lionel Huang, Université de Pau et des Pays de l'Adour, Pau, France
Ying Huang, University of California—San Diego, La Jolla, California
Philip Hugenholtz, Department of Energy Joint Genome Institute, Walnut Creek, California
Daniel H. Huson, University of Tübingen, Tübingen, Germany
C. Hunter, The Wellcome Trust Genome Campus, Hinxton, Cambridgeshire, United Kingdom
S. Hunter, The Wellcome Trust Genome Campus, Hinxton, Cambridgeshire, United Kingdom
Susan M. Huse, Marine Biological Laboratory at Woods Hole, Woods Hole, Massachusetts
Martijn A. Huynen, Radboud University Nijmegen Medical Center, Nijmegen, The Netherlands
Satoshi Ishii, The University of Tokyo, Tokyo, Japan
Shoko Iwai, Michigan State University, East Lansing, Michigan
Samuel Jacquiod, Environmental Microbial Genomics Group, Ecully, France
Ederson da C. Jesus, Michigan State University, East Lansing, Michigan; University of Pará, Belém, Brazil
David R. Johnson, Swiss Federal Institute of Technology Zürich (ETHZ), Zürich, Switzerland
Koji Kadota, The University of Tokyo, Tokyo, Japan
Jinjun Kan, University of Southern California, Los Angeles, California
Anna Maria Kielak, University of Groningen, Haren, The Netherlands
Staffan Kjelleberg, The University of New South Wales, Sydney, Australia
Christian G. Klatt, Montana State University, Bozeman, Montana
Konstantinos T. Konstantinidis, Georgia Institute of Technology, Atlanta, Georgia
Allan E. Konopka, Pacific Northwest National Laboratory, Richland, Washington
Yadhu Kumar, Technical University of München, Freising, Germany
Thomas Kvist, BioGasol ApS, Ballerup, Denmark
Myron T. La Duc, California Institute of Technology, Pasadena, California
Rahmi Lale, Norwegian University of Science and Technology, Trondheim, Norway
Tae Kwon Lee, Yonsei University, Seoul, Republic of Korea
Frédéric Lehembre, University of Lyon, Villeurbanne, Lyon, France
Marc Lemaire, University of Lyon, Villeurbanne, Lyon, France
Qiongfang Li, Sichuan Agricultural University, Ya'an Sichuan, China
Weizhong Li, University of California—San Diego, La Jolla California
Li Liu, University of Florida, Gainesville, Florida
Mark Liles, Auburn University, Auburn, Alabama
Kristina Lindström, University of Helsinki, Helsinki, Finland
Thomas Lingner, Georg August University of Göttingen, Göttingen, Germany
Paul F. Long, University of London, London, United Kingdom
Wolfgang Ludwig, Technical University Munich, Freising, Germany
Shahid Mahmood, Uppsala BioCenter, Uppsala, Sweden
Chaysavanh Manichanh, University Hospital Vall d'Hebron, Barcelona, Spain
Lucio Marianna, Institute of Ecological Chemistry, Neuherberg, Germany
Roland Marmeisse, University of Lyon, Villeurbanne, Lyon, France

Cristina Martínez-Garza, Autonomous University of the State of Morelos, Morelos, Mexico
Alice Carolyn McHardy, Max-Planck Institut for Information, Saarbrücken, Germany
Harald Meier, Technical University of München, Freising, Germany
Peter Meinicke, Georg August University of Göttingen, Göttingen, Germany
Melanie C. Melendrez, Montana State University, Bozeman, Montana
Folker Meyer, The University of Chicago, Chicago, Illinois
Suparna Mitra, Tübingen University, Tübingen, Germany
Mary Ann Moran, University of Georgia, Athens, Georgia
Sergio E. Morales, The University of Montana, Missoula, Montana
Burkhard Morgenstern, Georg August University of Göttingen, Göttingen, Germany
Mark Morrison, The Ohio State University, Columbus, Ohio
Susann Müller, Helmholtz Center for Environmental Research, UFZ, Leipzig, Germany
J. Colin Murrell, University of Warwick, Coventry, United Kingdom
Cindy H. Nakatsu, Purdue University, West Lafayette, Indiana
Andréa Maria Amaral Nascimento, Federal University of General Mines, Minas Gerasis, Brazil
Karen E. Nelson, J. Craig Venter Institute, Rockville, Maryland
Victor J. Nesatyy, EPFL, Lausanne, Switzerland
Joshua D. Neufeld, University of Warwick, Coventry, United Kingdom; University of Waterloo, Ontario, Canada
Hideki Noguchi, Tokyo Institute of Technology, Yokohama, Japan
Trent R. Northen, Lawrence Berkeley National Laboratory, Berkeley, California
Carlos W. Nossa, New York University School of Medicine, New York, New York
Joy D. van Nostrand, University of Oklahoma, Norman, Oklahoma
Annalisa Nuccitelli, Novartis Vaccines and Diagnostics, Siena, Italy
William E. Oberdorf, New York University School of Medicine, New York, New York
Felix Ott, Max-Planck Institute for Developmental Biology, Tübingen, Germany
Hong-Yu Ou, Shanghai Jiaotong University, Shanghai, China
Lise Øvreås, University of Bergen, Bergen, Norway
Saw Yen Ow, The University of Sheffield, Sheffield, United Kingdom
Kaustubh Patil, Max-Planck Institut für Informatik, Saarbrücken, Germany
Anna Pei, Washington University College of Arts and Sciences, St. Louis, Missouri
Zhiheng Pei, New York University School of Medicine, New York, New York
C. Ryan Penton, Michigan State University, East Lansing, Michigan
Jörg Peplies, Ribocon GmbH, Bremen, Germany
David L. Perkins, University of California–San Diego, La Jolla, California
Rachel S. Poretsky, University of Georgia, Athens, Georgia
James I. Prosser, University of Aberdeen, Aberdeen, Scotland
Elmar Prüsse, Max Planck Institute for Marine Microbiology, Bremen, Germany
Christian Quast, Max Planck Institute for Marine Microbiology, Bremen, Germany
Kumar Rajakumar, University of Leicester, Leicester, United Kingdom
Rino Rappuoli, Novartis Vaccines and Diagnostics, Siena, Italy
Daniel C. Richter, University of Tübingen, Tübingen, Germany
Margaret A. Riley, University of Massachusetts, Amherst, Massachusetts

Charles E. Robertson, University of Colorado, Boulder, Colorado
David M. Rosmarin, New York University School of Medicine, New York, New York
Bernardo Sachman-Ruiz, Autonomous National Vniversity of Mexico, Cuernavaca, Morelos, Mexico
Martha Schattenhofer, Helmholtz Centre for Environmental Research, Leipzig, Germany
Peer M. Schenk, The University of Queensland, St. Lucia, Queensland, Australia
Christa Schleper, University of Vienna,Vienna, Austria; University of Bergen, Bergen, Norway
Ramona Schmid, Boehringer Ingelheim Pharma GmbH & Co. KG, Biberach, Germany
Philippe Schmitt-Kopplin, Institute of Ecological Chemistry, Neuherberg, Germany
Keishi Senoo, The University of Tokyo, Tokyo, Japan
Yue-Mao Shen, Kunming Institute of Botany, the Chinese Academy of Sciences, Yunnan, China
Pascal Simonet, Environmental Microbial Genomics Group, Ecully, France
Menachem Y. Sklarz, Ben-Gurion University of the Negev, Beersheba, Israel
Ines M. Soares, Ben-Gurion University of the Negev, Beersheba, Israel
Erko Stackebrandt, German Collection of Microorganisms and Cell Cultures, DSMZ, Braunschweig, Germany
Antonio Starcevic, University of Zagreb, Zagreb, Croatia
Marc Strous, University of Bielefeld, Bielefeld, Germany
Yijun Sun, University of Florida, Gainesville, Florida
Marc J.-F. Suter, Swiss Federal Institute of Science and Technology, Duebendorf, Switzerland
Sen-Lin Tang, Academia Sinica, Taiwan
Maike Tech, Georg August University of Göttingen, Göttingen, Germany
Torsten Thomas, The University of New South Wales, Sydney, Australia
James M. Tiedje, Michigan State University, East Lansing, Michigan
Vigdis L. Torsvik, University of Bergen, Bergen, Norway
Susannah G. Tringe, U. S. Departement of Energy Joint Genome Institute, Walnut Creek, California
Pankaj Trivedi, University of Florida, Lake Alfred, Florida
Ching-Hung Tseng, Academia Sinica, Taiwan
Joshua E. Turse, Pacific Northwest National Laboratory, Richland, Washington
Tim Urich, University of Vienna,Vienna, Austria; University of Bergen, Bergen, Norway
Parag Vaishampayan, California Institute of Technology, Pasadena, California
Svein Valla, Norwegian University of Science and Technology, Trondheim, Norway
Kasthuri Venkateswaran, California Institute of Technology, Pasadena, California
Pablo Vinuesa, Autonomous National University of Mexico, Cuernavaca, Morelos, Mexico
Timothy M. Vogel, Environmental Microbial Genomics Group, Ecully, France
Heather Walker, University of Sheffield, Western Bank, Sheffield, United Kingdom
Luis Gabriel Wall, National University of Quilmes, Bernal, Buenos Aires, Argentina
Hao-Xin Wang, Kunming Institute of Botany, the Chinese Academy of Sciences, Yunnan, China
Nian Wang, University of Florida, Lake Alfred, Florida
Qiong Wang, Center for Microbial Ecology, Michigan State University, Michigan
David M. Ward, Montana State University, Bozeman, Montana
George M. Weinstock, Washington University School of Medicine, St. Louis, Missouri
David B. Mark Welch, Marine Biological Laboratory at Woods Hole, Woods Hole, Massachusetts
Annelie Wendeberg, Helmholtz Center for Environmental Research, Leipzig, Germany

Micheline Wésolowski-Louvel, University of Lyon, Villeurbanne, Lyon, France

Peter Westermann, Aalborg University, Ballerup, Denmark

Ralf Westram, Technical University of München, Freising, Germany

Bryan A. White, University of Illinois at Urbana-Champaign, Urbana, Illinois

Paul Wilmes, Department of Environment and Agro-Biotechnologies; Gabriel Lippmann Public Research Center, Luxembourg, Belgium

Jason M. Wood, Montana State University, Bozeman, Montana

Phillip C. Wright, The University of Sheffield, Sheffield, United Kingdom

Martin Wu, University of Virginia, Charlottesville, Virginia

Jianping Xu, McMaster University, Ontario, Hamilton, Canada

Liying Yang, New York University School of Medicine, New York, New York

Yuzhen Ye, Indiana University, Bloomington, Indiana

Pelin Yilmaz, Max Planck Institute for Marine Microbiology, Bremen, Germany; Jacobs University, Bremen, Germany

Fahong Yu, University of Florida, Gainesville, Florida

Pui Yi Yung, The University of New South Wales, Sydney, Australia

Zhongtang Yu, Ohio State University, Columbus, Ohio

Ying Zeng, Kunming Institute of Botany, the Chinese Academy of Sciences, Yunnan, China

Xiaoping Zhang, Sichuan Agricultural University, Ya'an Sichuan, China

Ling Zou, Sichuan Agricultural University, Ya'an Sichuan, China

Jizhong Zhou, University of Oklahoma, Norman, Oklahoma

Enrique Zozaya, Autonomous National University of Mexico, Cuernavaca, Morelos, Mexico

Jurica Zucko, University of Kaiserslautern, Kaiserslautern, Germany; University of Zagreb, Zagreb, Croatia

Chapter 1

Introduction

Frans J. de Bruijn

In this first volume of the Handbook, metagenomics is introduced, together with computer-assisted analysis, information on consortia and databases, and as a number of complementary methods, such as microarrays, metatranscriptomics, metaproteomics, metabolomics, phenomics (the "omics"), and single-cell analysis.

Part 1, "Background Chapters," contains a number of chapters on nonmetagenomic methods, such as different genomic fingerprinting techniques and their analysis and level of resolution, as well as the first approach to metagenomics (Chapter 2). All these methods are still used today.

In Part 2, "The Species Concept," several experts examine the parameters to call something a new species and provide suggestions to authors when it is proper to call a novel isolate [operating taxonomic unit (OTU)] a new species. The recommendations of two expert meetings on the topic are summarized in another chapter in this part describing the 70% DNA–DNA hybridization level as essential in the species concept. This discussion is very relevant to all phylogenetic studies in both volumes of the Handbook.

In Part 3, metagenomics is introduced and a number of practical parameters of this technique are outlined. An introduction to metagenomics and the other "omics" is presented in Chapter 14. Three subsequent chapters deal with the 16S rRNA gene as phylogenetic marker and also examine the pitfalls of its use. Three chapters describe the impact of next-generation sequencing on metagenomics, examine its accuracy and quality of reads, and review the potential and challenges of environmental shotgun sequences for studying the hidden world of microbes. Metagenomics can involve (a) the generation and analysis of clone libraries which can be screened for particular properties and (b) random sequencing of metagenomic DNA. The former is discussed in an article on vector tools and functional screening of metagenomic libraries (see also Parts 6 and 7, Vol. II). The latter is used in many other articles in the Handbook. The remaining articles in this section introduce various technical aspects of metagenomics, as well as novel approaches such as gene-targeted metagenomics, using homing endonuclease restriction and marker insertion for phylogenetic studies, finding integrons, arrayOme- and tRNAcc-facilitated mobilome discovery, and improved serial analysis of V1 ribosomal sequence tags (SARST-V1) to study bacterial diversity. A plethora of other studies in various habitats are presented in Volume II of this Handbook.

In Part 4, some consortia and databases are discussed, including the Metacontrol consortium focusing on the metagenomics of suppressive soils, the Terragenome consortium to provide a metagenomic shotgun and phosmid sequencing analysis of a "reference" soil, and the Argentinian BIOSPAS consortium aimed at bringing together a group of scientists employing metagenomic and associated approaches. This is followed by a description of the Human Gut Microbiome Initiative (HGMI) and the related Human Microbiome Project (HMP). Chapter 36 in this part describes the Ribosomal Database Project, an irreplaceable source for phylogenetic studies, using the rRNA genes as target (see Chapter 15, Vol. I). The final chapter in this part describes the Metagenomics RAST server a a public resource for automated phylogenetic and functional analysis of Metagenomes.

In Part 5, a smorgasbord of computer programs is presented essential for the analysis of (meta)genomic data. Clearly, computer-assisted analysis is a crucial component of every metagenomic project, and progress in the field is dependent on creating programs and databases for ever-growing datasets and can be the limiting factor for large metagenomic, transcriptomic,

Handbook of Molecular Microbial Ecology, Volume I: Metagenomics and Complementary Approaches, First Edition. Edited by Frans J. de Bruijn.

proteomic, and metabolomic projects. It equals in importance to the development of higher throughput novel sequencing methods (see Chapter 18, Vol. I). The authors in Part 5, as well as all other authors, have been asked to highlight the programs and web sites used in their chapters; therefore in addition to the limited programs highlighted in Part 5, a wealth of further information and other programs can be found in the chapters in Volumes I and II.

In Part 6 a number of complementary approches to metagenomics are presented, including metagenomics approaches in systems biology, the use of stable isotope probing, and subtractive hybridization.

In Part 6A the use of microarrays, including phylochips and geochips and metagenomic arrays, is discussed and examples in different habitats, such as NASA rocket cleanrooms, are given. This part also contains a chapter on phenotypic arrays or "phenomics," another "omic" technique, which can reveal the metabolic capacity of microbes in microplates.

In Part 6B, some examples of metatranscriptomic analysis are presented, which permit a glimpse into the metagene expression profile in various environments, such as the symbiotic protist community in *Reticulitermes* and comparative day and night metatranscriptomics of microbial communities in the North Pacific. In addition a "double RNA" approach is presented to simultaneously assess the structure and function of microbial communities, and one chapter on the metatranscriptomics of eukaryotes is included.

In Part 6C, metaproteomics approaches are highlighted, and examples are presented on the proteomics of microbial stress responses, the metaproteomic analysis of Chesapeake Bay microbial communities, high-throughput proteomics in cyanobacteria, and global proteomic analysis of the chromate response in *Arthrobacter*.

In Part 6D, metabolomics is highlighted, which requires more sophisticated tools such as mass spectrometry. Examples include (a) two chapters that review the small molecule dimension and high-resolution tools to monitor bacterial growth on a molecular level, (b) one chapter on metabolomics in plants, where the metabolomics techniques are well established, and (c) a chapter on metabolite identification, pathways and "omic" integration using databases and other tools.

In Part 6E a highly specialized complementary approach is described, namely the isolation and use of single cells for metagenomic and other analysis.

None of the parts described above are comprehensive. They mainly give a short insight about what one can do in addition to metagenomics to extract more functional data from the system under study to answer the following questions: "Who is there?" and "What are they doing?" An attempt was made to select studies in very different habitats, and a variety of approaches are highlighted. This is continued and expanded upon in Volume II.

Part 1

Background Chapters

Chapter 2

DNA Reassociation Yields Broad-Scale Information on Metagenome Complexity and Microbial Diversity

VIGDIS L. TORSVIK AND LISE ØVREÅS

2.1 INTRODUCTION

2.1.1 Evolution and Development of Diversity

There are close relationships between microbial evolution, diversity, and ecology. Prokaryotic organisms have evolved through 3.8 billion years [Rosing, 1999] in response to varying geological, geochemical, and climatic conditions. For approximately half of their life's history, they resided alone on Earth. Due to their great metabolic flexibility, short generation time, and ability to exchange genes over deep phylogenetic barriers, their ability to adapt and evolve are superior. This means that virtually every (micro) environment on Earth with physical–chemical conditions that can sustain life is occupied by prokaryotic organisms [see Vol. II]. It is therefore not surprising that the biodiversity on Earth is dominated by these organisms, which constitute two of the three primary domains of life, the *Archaea* and *Bacteria* [Woese, 1987; Woese and Fox, 1977]. Their ecological consequences are huge, because ecosystem processes to a large extent are regulated by microbial communities. Important for understanding complex ecosystem functioning is to identify the primary drivers of microbial diversity and community structure. According to ecological theories, relationships between ecosystem functioning and diversity can partly be explained by the resource heterogeneity hypothesis and the "insurance hypothesis" [Yachi and Loreau, 1999]. The insurance hypothesis suggests that high diversity protects communities from unstable environmental conditions because the presence of diverse subpopulations not only increases the range of conditions in which the community as a whole can succeed, but also ensures long-term attainment of the community [Boles et al., 2004].

2.1.2 Methodological Advances, Discoveries, and Issues that Promoted Exploring the Environmental Community DNA

Before the introduction of molecular methods in microbial ecology, it was only possible to study the composition and diversity of microbial communities by investigating cultivated isolates. This traditional reductionist approach has limited our understanding of microbial ecology. In a holistic approach, the microorganisms in a community have been treated as one "black box." The aims were to (a) measure collective variables like biomass, population sizes, process rates, and diversity of cultured microorganisms and (b) integrate these to better understand microbial ecosystems. This approach was hampered by the lack of conceptual models linking biomasses, rate of functions, and diversity to the underlying controlling factors. During the 1970s, methods for direct counts of microorganisms using fluorescence microscopy were developed [Hobbie et al., 1977]. It was then realized

Handbook of Molecular Microbial Ecology, Volume I: Metagenomics and Complementary Approaches, First Edition. Edited by Frans J. de Bruijn.

that the microbial biomass in natural environments was orders of magnitude higher than previously anticipated, one gram of soil and sediment could harbor more than 10^{10} cells. It was demonstrated that there was a factor of 2–3 orders of magnitude between the numbers of microorganisms estimated by direct counts and by colony-forming units (cfu) [Fægri et al., 1977]. A main question was why there was such a discrepancy. One assumption was that the majority of the microorganisms observed in natural environments like soils and sediments were inactive and that those growing in the laboratory represented the active populations. To investigate this, a fractionated centrifugation method for separating the bacteria from soil was developed. By microscopic counts it was estimated that the bacterial fractions contained 50–80% of the bacteria present in the soil samples and that no eukaryotic cells were present. Respiration was used to measure the activity in the bacterial fraction, and the specific oxygen uptake rates (qO_2) calculated on the basis of microscopic counts ranged from 3 to 300 $\mu l\ O_2\ mg^{-1}$ dry weight h^{-1}, indicating that most of the microbial cells observed in the microscope were metabolically active [Fægri et al., 1977]. Furthermore, the amount of DNA in the bacterial fractions (washed with sodium pyrophosphate to remove extracellular DNA) corresponded to an average DNA content per microscopic counted cell of 8.4 fg (10^{-15} g). This is approximately the same as in *Escherichia coli* cells in stationary growth phase [Ritz et al., 1997; Torsvik and Goksoyr, 1978]. It was therefore concluded that virtually all the cells observed in the microscope were viable and belonged to the metabolically active microbial community. A main issue was then whether the cultured bacterial isolates were representative for the total environmental community or whether they constituted a small, exotic subpopulation of microorganisms that could easily be "domesticated" and grown in the laboratory.

Early in the 1980s, ideas emerged that led to a revolution and paradigm shift in microbial ecology. The basic idea was that if it was possible to retrieve DNA from the entire microbial community, this DNA would in principle contain genetic information about nearly all the organisms in the community, including both cultured and uncultured microorganisms. Major problems were (a) the lack of methods for extracting ultrapure DNA from "dirty" samples like soil and sediments and (b) finding tools to analyze and interpret the information harbored in such community metagenomes. During the 1980s, developments of techniques for nucleic acid analyses advanced rapidly. The possibility to study microbial communities at a genomic level led to new avenues of research strategies and made it possible to attach problems that were previously regarded as unsolvable. An advantage of analyzing nucleic acids from microorganisms was that it was a growth-independent approach and that the information could be used to investigate and compare microorganisms at different biological organization levels, from infraspecies and taxon to community level.

2.1.3 Microbial Biodiversity and Metagenome Diversity

Diversity can be defined at different level of biological organization ranging from genomic diversity within an organism, species diversity, and variability within and between species population, to community diversity [Bull, 1992; Harper and Hawksworth, 1994]. Ecological diversity includes community parameters like variability in community structure, the number of guilds (functional diversity), the number of trophic levels, and complexity of interactions. Traditionally, microbial biodiversity has been used to describe the variability among the organisms in an assemblage or a community. Phenotypic diversity is related to the variation in microbial traits, which reflects the expression of genes under a given set of conditions. Genetic diversity measures the total genetic potential in the assemblage or community independent of the environmental conditions.

Commonly, the diversity concept based on taxa includes both the richness (e.g., number of species) and the evenness—that is, how evenly the individuals are distributed among the taxa. The diversity can also be regarded as an expression of the amount of information in a biological assemblage or community [Atlas, 1984]. This definition is adopted from information technology and takes into account both the amount of information and how the information is distributed among the individuals in a community. It can be applied directly to genetic diversity.

Metagenome has been defined as the collection of genomes from the total number of microorganisms in an environmental assemblage or in a whole natural community [Handelsman et al., 1998]. Metagenomics refers to extraction of DNA from natural environmental samples and analyses of this DNA in order to gain information about the organisms the DNA originated from. Our rationale for exploring DNA retrieved from microbial communities in natural environments was that this metagenome, being a mixture of genomes from an unknown number of different microorganisms in amounts corresponding to their relative abundance, ought to provide information about the microbial diversity at the community level. DNA reassociation kinetics was expected to provide such information because it could be used to assess total DNA complexity, and it might therefore be used as a measure of the total genomic diversity in microbial assemblages or communities [Torsvik et al., 1996]. Based on this method, we defined the genetic diversity of microbial communities as the total amount of genetic information in the metagenome, along with the distribution of this information among the different genomic types.

The awareness of an immense microbial diversity in natural environments has evolved rapidly during the last decades. In the first culture-independent analysis of diversity [Torsvik et al., 1990a] we suggested the possibility that there might be as many as 10,000 different taxa in a soil sample of approximately 100 g. At that time, this was a startling discovery that was met with a good deal of skepticism. As the repertoire and improvement of molecular methods evolved dramatically, and the exploration of diversity became more feasible, a consensus has emerged that natural microbial communities are far more diverse than previously recognized.

2.2 BASIC METHODS

2.2.1 Extraction and Purification of DNA from Environmental Samples

Analyses based on DNA melting profile and reassociation kinetics require highly purified DNA, free from extracellular and eukaryotic DNA, humic material, or other contaminants that can interfere with the optical measurements. Great care has to be taken to minimize potential errors caused by impurities by employing thorough cell extraction and DNA purification protocols and applying quantitative and qualitative controls for each step. As a first step the prokaryotic cells are separated from environmental matrixes. The separation protocol is critical for the accuracy of the measurements. For soils and sediments with high organic content, physical disruption of particles combined with fractionated centrifugation proved to give the best results [Fægri et al., 1977; Torsvik, 1980]. The separation method, however, has to be adjusted according to the environmental type under investigation. For cell separation from mineral and clay soils, density gradient centrifugation is regarded as the optimal method [Bakken, 1985]. To ensure that no eukaryotic cells are present and to estimate the fractionation yield, fluorescence microscopy counting is carried out. The cell recovery varied from 50% to more than 80% in high organic soils, and about 60–65% in less organic soils and marine sediments [Fægri et al., 1977; Torsvik et al., 1995]. Other investigators have reported yields of 20–50% [Bakken, 1985, Holben et al., 1988, Steffan et al., 1988]. The cell yields from three different soils calculated both by plate counts and total microscopic counts were very consistent. Thus, there are no indications that the fractionation method is biased, and we assume that virtually all the different microbial types are represented in the bacterial fraction. The prokaryotic fractions were virtually free from fungi and other eukaryotic cells, as confirmed by the fact that 98–100% of the fungal biomass remained in the soil matrix after fractionation [Fægri et al., 1977].

To obtain pure metagenomic DNA, extracellular DNA and humic materials are removed by washing with sodium pyrophosphate (pH 7.0) or sodium hexamethaphosphate (pH 8.5) prior to lysis. The cells are normally lysed by a relatively mild treatment with lysozyme, proteinase K, and sodium dodecyl sulfate (SDS), giving a lysis efficiency of 90–95% [Holben et al., 1988; Steffan et al., 1988; Torsvik et al., 1995]. After bacterial lysis, the DNA is extracted from the cells, and the crude DNA extract is purified two to three times on hydroxyapatite columns [Torsvik et al., 1995]. The DNA extraction and purification causes loss; the highest losses occur during centrifugation (30%) as DNA co-precipitates with cell debris and some of the humic materials, and they also occur during hydroxyapatite purification (50%) [Torsvik, 1980]. It is, however, not likely that these losses are biased toward specific DNA molecules. When starting with 100 g soil wet weight containing in total 4.8×10^{11} microbial cells, typical DNA yields were 350–500 μg. Assuming an average DNA content per cell of 5×10^{-15} [Bak et al., 1970], the theoretical yield would be 2400 μg of DNA. Accordingly, 15–20% of the prokaryotic DNA could be recovered from the soil [Torsvik et al., 1994]. DNA purified twice on hydroxyapatite was very pure, contained $\leq$ 2% RNA, and showed a hyperchromicity higher than 30% upon melting [Torsvik et al., 1990a]. For detailed other protocols to isolate and purify metagenomic DNA, see Chapters 10 and 11 in Volume II.

2.2.2 Melting of Metagenomic DNA to Exhibit Gross Community Profile

Microbial community DNA (metagenome) can provide complementary information about the overall community composition and diversity [Johnsen et al., 2001; Ritz et al., 1997; Torsvik et al., 1990a; Øvreås and Torsvik, 1998]. Gross community composition can be inferred from the base composition (mole % guanine + cytosine; % G+C) in metagenomic DNA. The G+C content in microbial genomes range from 25% to 75% and can be determined by optical measurements of thermal denaturation since single-stranded DNA has approximately 35% higher absorbance than double-stranded DNA at 260 nm. The DNA melting curves are converted to % G+C profiles [Torsvik et al., 1995; Ritz et al., 1997] and provide microbial community profiles. Although such profile analysis is considered to have low resolution, it can be indicative of overall changes in microbial community structure. Single-genome DNA has a steep melting profile with one narrow melting domain, whereas complex community DNA consists of a number of different melting domains as indicated by a much broader profile. The analyses have the limitation that two communities having similar base distributions do not necessarily have similar species composition, since different species often have the same base composition. On the other hand, communities with different base distributions almost certainly have different species composition. We have used % G+C profiles as indicators of changes in microbial community composition along ecological gradients and due to perturbations.

2.2.3 Reassociation of Metagenomic DNA

The complexity in metagenomic DNA can be estimated from measuring the reassociation of single-stranded DNA to double-stranded DNA in solutions when the temperature is lowered to approximately 25°C below its melting point. Britten and Kohne [1968] used this approach to study the size (complexity) of haploid genomes and genome "organization" (repetitive sequences) of eukaryotic organisms. The method has been used to determine genome sizes and phylogenetic relationship within many groups of prokaryotic and eukaryotic organisms.

DNA reassociation follows a second-order kinetics, and the rate is proportional to the square of the concentration of dissociated single-stranded DNA molecules in the solution. The more complex the DNA, the lower the concentration of similar fragments and the lower the renaturation rate. For detailed description of the method, see Torsvik et al. [1995]. The fraction of reassociated DNA is plotted against the C_0t values (C_0 is the initial concentration of dissociated single-stranded DNA in mole nucleotides L^{-1}, and t is time in seconds). $C_0t_{1/2}$, where $t_{1/2}$ denotes the time in seconds for 50% DNA reassociation, is inversely proportional to the rate constant, k, and is proportional to the DNA complexity. Britten and Kohne [1968] used this term as a measure of haploid genome sizes (in base pairs) or the heterogeneity in DNA. Thus, under defined conditions we can use the $C_0t_{1/2}$ to estimate the metagenome size. The $C_0t_{1/2}$ values are determined relative to DNA with known complexity like the *Escherichia coli* B DNA [genome size: 4.6 Mbp (megabase pairs)]. To estimate the size of metagenomic DNA, the $C_0t_{1/2}$ of metagenomic DNA is divided by $C_0t_{1/2}$ of the *E. coli* B genomic DNA, multiplied with the size of the *E. coli* B genome.

2.3 ESTIMATING GENETIC DIVERSITY: MODEL EXPERIMENT WITH CULTURED ISOLATES

The number of prokaryotic "species" in the community is estimated assuming an equal species abundance distribution. Accordingly, the number of species is calculated by dividing the metagenome size by the average genome size of bacterial isolates originating from the same environment. When metagenomes from different environments are compared, we use the *E. coli* B genome as a "standard" genome. Because metagenomic DNA is a mixture of DNA from different bacterial genotypes or haplotypes, which are present in different proportions, reassociation curves for microbial community DNA have flatter slopes than ideal second-order reaction curves. When the number of DNA molecules with different sequences in a mixture increases, the general degree of similarity between them decreases, and the overall reaction deviates from an ideal second-order kinetic. In such cases the rate is a function of an amalgam of reassociation reactions with different reaction constants, and $C_0t_{1/2}$ does not have any precise meaning. Nevertheless, it provides information about the relative DNA complexity of metagenomes from different assemblages or communities.

DNA extracted from natural microbial communities is often extremely complex and the reassociation rate is low; consequently, an experiment can last 1–2 weeks. To increase the rate, the reassociation is measured in solutions with high cation concentrations (6× standard saline citrate) and 30% DMSO [Escara and Hutton, 1980, Torsvik et al., 1995]. To obtain good estimates of DNA complexity, the reaction should reach at least 50% reassociation, but for the most complex DNA this level might never be reached. DNA reassociation is a low-resolution and broad-scale analysis of the sequence complexity in DNA, which can be used as a measure of community diversity. The metagenome complexity as expressed by $C_0t_{1/2}$ is used as an index [Torsvik et al., 1995] of genetic diversity and encompasses both the total range of genetic information in the community (richness component) and the distribution of this information among the different individual genomes (evenness component). This makes it possible for two communities with different structures to have identical $C_0t_{1/2}$ values. The method gives rather conservative (moderate) estimates of microbial diversity. The most abundant DNA molecules reassociate the fastest and contribute most to the metagenome's $C_0t_{1/2}$ value, whereas DNA molecules from rare species are present in low concentrations and may not reassociate during the course of the measurement. If there are a few numerically dominant species in a community, they will therefore tend to lower

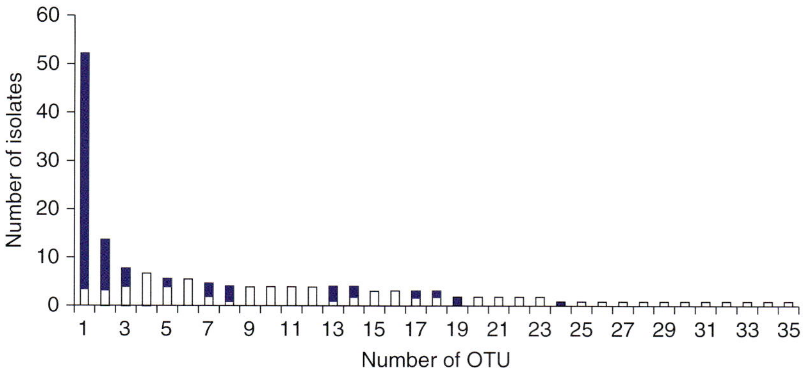

Figure 2.1 Rank abundance distribution of OTUs in assemblages of isolates from soil mesocosms incubated at 4°C (□) and 30°C (■).

the $C_0t_{1/2}$ value, whereas the "rare biosphere" will contribute less to the diversity estimates.

This was demonstrated in a model experiment with soils subjected to stress in the form of elevated temperature, to instigate changes in microbial communities [Torsvik et al., 1994]. Genomic and phenotypic measures of diversity were compared; therefore the investigation was based on assemblages of bacterial isolates. Two parallel soil mesocosms were studied, one incubated at 4°C and the other at 30°C for 3 months, keeping the other environmental parameters constant. For each mesocosm, 80 colonies were picked randomly from standard plating media, isolated, and characterized using a set of 26 morphological and physiological tests (API 20B and API OF; API System S.A., France), followed by cluster analysis of the isolates. When an 80% phenotypic similarity threshold (taxonomic cutoff) was used to delineate OTUs (operating taxonomic units), the 4°C assemblage comprised 33 OTUs, none predominant, and most of them had one to three members. The 30°C assemblage comprised 12 OTUs, and two OTUs were numerically dominant, accounting for 61% (47) and 14% (11) of the isolates, respectively (**Fig. 2.1**).

Genetic diversity was determined by DNA reassociation kinetics. DNA was isolated from mixtures of 10 isolates with equal biomass [Torsvik et al., 1990b]. The reassociation rate was determined in series with equal amounts of DNA from increasing number of groups until DNA from all the 80 isolates in an assemblage had been included. The $C_0t_{1/2}$ for each DNA mixture was plotted against the number of isolates added (**Fig. 2.2**).

The $C_0t_{1/2}$ for the 30°C assemblage leveled off and reached its maximum with 10 isolates. The $C_0t_{1/2}$ for the 4°C assemblage was more than four times higher that that of the 30°C assemblage and it did not reach a maximum level. The two assemblages taken together had higher richness (35 OTUs) than either of the separate assemblages (**Table 2.1**). When equal amounts of DNA from the two assemblages were mixed, however, the mixtures' $C_0t_{1/2}$

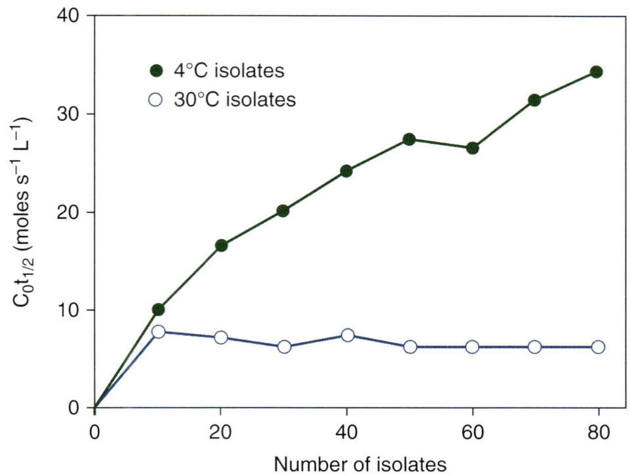

Figure 2.2 $C_0t_{1/2}$ values with increasing number of isolates in assemblages from soil mesocosms incubated at 4°C and 30°C.

value was intermediate between $C_0t_{1/2}$ for the two assemblages reassociated separately (**Fig. 2.3**).

This shows that the numerically dominant isolates in the 30°C assemblage contributed very much to the $C_0t_{1/2}$ by increasing the concentration of DNA molecules with similar sequences in the mixture relative to that of DNA from the more "rare species" of the 4°C assemblage.

This model experiment demonstrates that the reassociation method gives a reliable measure of diversity in microbial assemblages. We did not observe any rapidly reassociating fraction by repetitive DNA in any of the bacterial genomes, which is in agreement with the view that bacterial genomes basically contain a single reassociation kinetic component [Lewin, 2000].

A characteristic feature of $C_0t_{1/2}$ used as a diversity index is that it changes in the same manner as the Shannon–Weaver and the Equitability indices (Table 2.1), which take into account both the species richness and the evenness in a community. Our results were confirmed by Haegeman et al. [2008], who presented theoretical evidence that reassociation kinetics gave accurate diversity

Table 2.1 Numbers of OTUs, Shannon Index (H′ Logaritmic Base), Equitability ($J = H'/H'_{max}$), Simpson's Index of Dominance (D), Genomic Complexity ($C_0t_{1/2}$) for Assemblages of Isolates from Soil Mesocosms at 4°C and 30°C Separately and in Combination

Assemblage:	4°C	30°C	4°C + 30°C
Number of isolates:	80	80	160
Number of OTUs[a]	33	12	35
H':	4.79	2.11	4.07
J:	0.95	0.59	0.79
D:	0.04	0.40	0.13
$C_0t_{1/2}$[b]	34.7	5.8	22.4

[a]OTU, = operational taxonomic unit, delineated at 80% phenotypic similarity level.
[b]DNA reassociation in 4 × standard saline citrate (4 × SSC), 30% dimethylsulfoxide (DMSO).

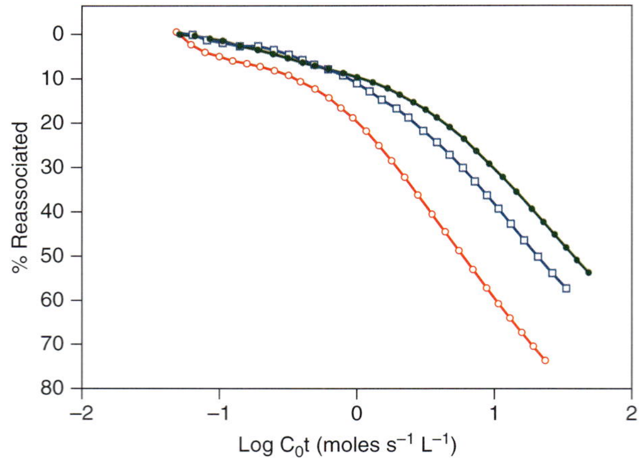

Figure 2.3 Reassociation curves (C_0t plots) in 4 × SSC, (standard saline citrate) 30% DMSO (dimethylsuphoxide) of DNA from isolate assemblages from soil mesocosms incubated at 4°C (•) and 30°C (◦) separately and in mixture (▫).

information that was in accordance with information provided by diversity indices. They argued that the diversity in microbial communities was more properly quantified by diversity indices like the Shannon–Weaver, Simpson, or Rènyi indices than by species numbers.

2.4 EXAMPLES OF APPLICATION OF THE DNA REASSOCIATION METHOD

2.4.1 The First Microbial Metagenomic Analysis Contrasted the High Total Community Diversity as Compared to the Diversity of Cultured Microorganisms

The knowledge and understanding of the dynamics of the structural and functional diversity within microbial communities was hampered by the fact that the majority of the cells as observed in the microscope are recalcitrant to cultivation. Although culture-independent molecular biological techniques provided new valuable insight in microbiology, some fundamental questions arose. One question was whether or not the diversity and composition of microbial isolates from an environment were representative for the total community. If not, what were the extent and patterns of the total diversity in natural microbial communities? To investigate this, we used DNA reassociation to compare the complexity of the metagenome from the total microbial community in a soil sample with that of DNA from an assemblage of 200 randomly picked isolates from the same sample (**Fig. 2.4**).

The $C_0t_{1/2}$ of the soil metagenome was 4500–4700 mol s^{-1} L^{-1}, which was approximately 6000 and 7.7

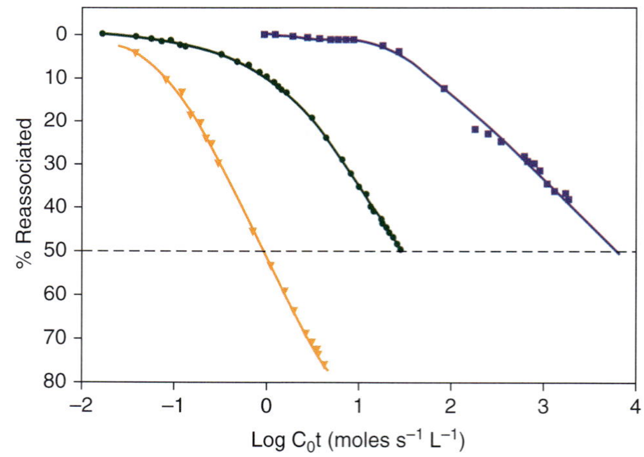

Figure 2.4 Reassociation curves (C_0t plots) in 6 × SSC (standard saline citrate), 30% DMSO (dimethylsuphoxide) of DNA from an assemblage of 206 soil bacterial isolates (•) and from the metagenome of the total soil microbial community (▪). Genomic *Escherichia coli* DNA was used as a control (▾).

Table 2.2 Microbial Diversity in Soils, Marine Sediments, and Solar Salterns Determined by DNA Reassociation Kinetics ($C_0t_{1/2}$; mol s^{-1} L^{-1} at 50% Reassociation) and Estimated Metagenome Complexity in Base Pairs (bp)

Metagenome DNA Source	Abundance Cells cm^{-3})	$C_0t_{1/2}$	Metagenome Complexity (bp)	Genome Equivalents (Relative to *E. coli* Genome; 4.6×10^6 bp)	Reference
Calf thymus nonrepetitive DNA (~ 60%)	—	602[b]	3.4×10^9	760	Torsvik et al. [1990b]
Forest soil; 206 isolates	—	28[a]	1.5×10^8	35	Torsvik et al. [1998]
Forest soil; total community	1.5×10^{10}	4600[b]	3.0×10^{10}	6400	Torsvik et al. [1998]
Agricultural soil, control	—	5700[b]	3.3×10^{10}	7200	Øvreås et al. [1998]
Agricultural soil, methane perturbed	—	270[b]	1.5×10^9	340	Øvreås et al. [1998]
Marine sediment; pristine	3.1×10^9	9000[b]	5.0×10^{10}	11400	Torsvik et al. [1998]
Marine sediment; fish farm	7.7×10^9	40–50[b]	2.3×10^8	50–70	Torsvik et al. [1998]
Marine sediment; abandoned fish farm	1.6×10^9	1300[b]	7.8×10^9	1700	Torsvik et al. [1998]
Salinity pond 22%	7.0×10^7	4.8[b]	3.2×10^7	7	Øvreås et al. [2003]
Salinity pond 32%	9.0×10^7	9.1[b]	6.0×10^7	13	Øvreås et al. [2003]
Salinity pond 37%	6.8×10^7	2.7[b]	1.8×10^7	4	Øvreås et al. [2003]

[a]DNA reassociation in 4 × SSC, 30% DMSO; *E. coli* genome $C_0t_{1/2} = 0.85$.
[b]DNA reassociation in 6 × SSC, 30% DMSO; *E. coli* genome $C_0t_{1/2} = 0.79$.

times higher than $C_0t_{1/2}$ for the *E. coli* B genome and for nonrepetitive (60% of bovine genome with approximately 350 Mbp) calf thymus DNA, respectively. The $C_0t_{1/2}$ value for a mixture of genomes from 200 microbial isolates was 28 mol s^{-1} L^{-1} (**Table 2.2**), which means that the $C_0t_{1/2}$ value for total community DNA was approximately 160 times higher than that of the assemblage of isolates. Therefore we concluded that the isolated microorganisms constituted a minor fraction and were not representative for the total microbial community. We assumed that the diversity was approximately as huge in the uncultured majority as in the 200 cultured isolates. This indicated that the community did not comprise a few numerically dominant haplotypes, but that there was a relatively even distribution of the genetic information among a huge number of haplotypes. The average $C_0t_{1/2}$ for genomes of individual soil bacterial isolates (based on 200 isolates) was approximately 60% higher than $C_0t_{1/2}$ for the *E. coli* B genome. Thus the "standard soil bacterial genome" size was 7.4 Mbp. This is in agreement with Raes et al. [2007], who estimated EGS (effective genomic size) in metagenomes of a complex farm soil sample at about 6.3 Mbp. Bacteria in a nutrient-poor, organism-sparse ocean surface water had EGS values as low as 1.6 Mbp.

In pristine soil and sediments with high organic contents, the DNA diversity encountered in 30- to 100-g samples corresponded to about 3000 to 11,000 different microbial genomes (Table 2.2).

According to the DNA-based species delineation, microbial strains having DNA similarities (reassociation values) of 70% or more belong to the same species [Stackebrandt et al., 2002]. By using this standard delineation, it was estimated that pristine soil samples (30–100 g of soil) contained a minimum of 4000–6000 prokaryotic "species" and that 100-g pristine sediment samples contained a minimum of 12,000–18,000 "species" of equivalent abundances.

2.4.2 DNA Reassociation Assesses the Impact of Perturbation and Pollution on Microbial Diversity

We have used microbial metagenomic diversity and changes in community structure as ecological indicators of perturbations and pollution caused by human activity. Among the environments investigated are perturbed and polluted soils and polluted marine fish farm sediments.

The impact of perturbation and environmental changes on microbial communities was investigated in a model experiment where organic agricultural soil was amended with a sole carbon source (air containing 17% methane) and incubated for 3 weeks at 15°C [Øvreås et al., 1998]. Striking changes in structure and diversity of the soil microbial community was observed after the perturbations. DNA from undisturbed control soil had a $C_0t_{1/2}$ value of 5700 mol s^{-1} L^{-1}, whereas the $C_0t_{1/2}$ of DNA from the methane-amended soil was reduced approximately 20 times (270 mol s^{-1} L^{-1}) (Table 2.2). Community fingerprinting (PCR-DGGE) profiles of the soil metagenome showed that it contained high numbers of different amplicon bands (see Chapter 5, Vol. I). The control soil community profile consisted of weak bands, indicating that there were no predominant populations. In the methane-amended soil community, however, some strong bands appeared on the top of the background of weak bands, indicating that some numerically dominant populations had emerged. Sequencing showed that they were similar to type I methane oxidizing bacteria in the phylum Gammaproteobacteria. Consequently, rather than reduced "species" richness, the reduction in $C_0t_{1/2}$ might reflect reduced evenness because some bacterial types were predominant.

The diversity in polluted environments is often notably reduced as compared to pristine environments. We observed that the metagenome complexity in the top 10-cm sediments under a marine fish farm with accumulated organic wastes was approximately 200 times lower than in pristine sediments (Table 2.2) [Torsvik et al., 1996]. The total number of bacteria in the fish farm and pristine sediment was 7.7×10^9 and 3.1×10^9 per gram, respectively. The organic content in these sediments was similar, 27% and 20%, but the fish farm sediment was heavily polluted with deposits of feed pellets and fecal material. Therefore it is conceivable that the organic matter quality, rather than the quantity, was causing diversity changes. The organic polluted fish farm sediment had an input of a relatively small range of readily available substrates (proteins, carbohydrates, lipids) which sustained a higher bacterial biomass as compared to the natural sediment, where the organic matter mainly was recalcitrant humus. The easily utilized organic substrates exerted a selection pressure, favoring fast-growing microorganisms (r-selection) that became numerically dominant. After the fish farm had been abandoned for 4 years, the microbial diversity had increased again, and the metagenome complexity was 32 times higher than in the operating fish farm sediment, but it was still 7 times lower than in the pristine sediment. Thus after removing the stress factor the community diversity recovered again. These investigations suggest that quantitative measures of microbial diversity and qualitative analysis of community structure can discriminate between environments subjected to different levels of pollution and be useful indicators of stress and perturbation.

2.4.3 Metagenomics Along a Salinity Gradient Indicates Unexpected Diversities and Considerable Changes in Community Composition

Microbial communities in the multi-pond saltern "Bras del Port" in Santa Pola (Alicante, Spain) were investigated using metagenomic approaches. Saltern crystallisers are extreme environments along ecological gradients, which have been studied extensively by molecular methods. They are among the simplest communities known in terms of species richness, as assessed by classical microbiological and molecular methods [Antón et al., 1999; Benlloch et al., 2001, 2002; Martínez-Murcia et al., 1995; Rodriguez-Valera, 1999]. This unique environment represents the only case where a direct comparison of DNA reassociation with deep sequencing of an environmental sample has been performed. The diversity of total metagenomes was analyzed by thermal denaturation (% G+C profiles) and reassociation kinetics and was compared with T-RFLP (terminal restriction fragment length polymorphism) of a small PCR (polymerase chain reaction)-amplified sequence of the conserved 16S rRNA gene [Øvreås et al., 2003; see also Chapter 7, Vol. I]. In addition, 16S rRNA clone library analyses were carried out by Rodriguez-Valera and collaborators [Antón et al., 1999; Benlloch et al., 2001, 2002; Martínez-Murcia et al., 1995; Rodriguez-Valera F, 1999], and an environmental genomics survey was performed by Legault and co-workers [Legault et al., 2006]. All these analyses showed that the diversity was low and that the Archaea was the most important part of the community in terms of numbers, biomass, and genetic heterogeneiety [Antón et al., 2000]. Another notable feature was that the archaeal community was composed of closely related species, and the majority was included in one single genus represented by the square halophilic archaeon *Haloquadratum walsbyi*. Only 18% of the microorganisms were proven to be of bacterial origin. Furthermore, the bacterial community appeared to be even more homogeneous than the archaeal, and it was composed virtually solely of members of the extremely halophilic bacterial genus *Salinibacter* [Antón et al., 2001]. Reassociation rates of the metagenomes showed that the prokaryotic community structure and diversity changed significantly through the salinity gradient of ponds having 22%, 32%, and 37% salinity (**Fig. 2.5**).

Unexpectedly, the total genetic diversity increased from 22% to 32% salinity, although one would expect less species richness in the last one due to more extreme conditions. At 37% salinity the diversity decreased again to nearly half of that at 22% salinity. The complexity of the community genome revealed that there were 7 (22%

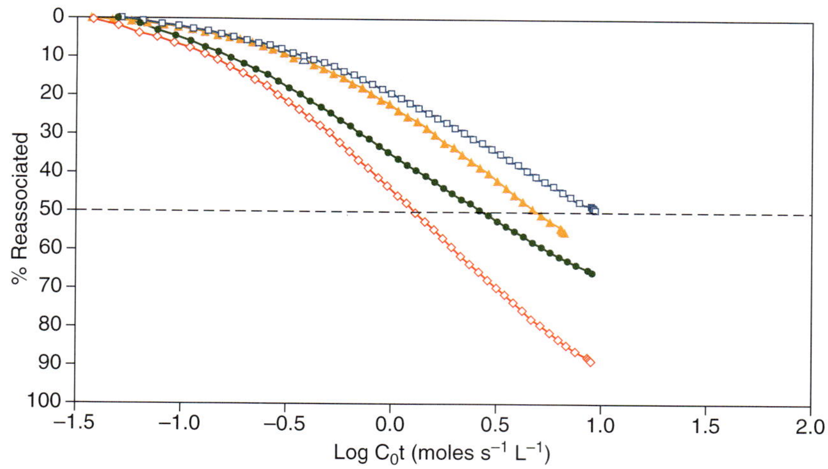

Figure 2.5 Reassociation curves (C_0t plots) for metagenomes from Solar saltern ponds with 22% (▲), 32% (▫), and 36% (•) salinity in 6 × SSC (standard saline citrate), 30% DMSO (dimethylsuphoxide). A mixture of genomic DNA from *Escherichia coli* and *Micrococcus luteus* (◇) was used as a control.

salinity), 13 (32% salinity), and 4 (37% salinity) genome equivalents relative to the *E. coli* genome (Table 2.2) [Øvreås et al., 2003]. These estimates were based on no genome sequence overlap, so they probably underestimate the overall complexity.

Because the reassociation rate depends both on the species richness and evenness, the increased diversity observed does not necessarily mean that there are more species at 32% than at 22% salinity. It may be explained by differences in the population abundance distribution (evenness) in the communities. Percent G+C profiles indicated uneven population distribution in the 22% salinity community, with a more even distribution in the 32% salinity community (**Fig. 2.6**).

The DNA from 22% salinity pond had a major component with 60–65% G+C and a minor component with 45–50% G+C, whereas DNA from 31% salinity pond had two G+C components with nearly equal size. DNA from the 37% salinity pond showed one peak with maximum about 50% G+C. The T-RFLP fingerprinting indicated

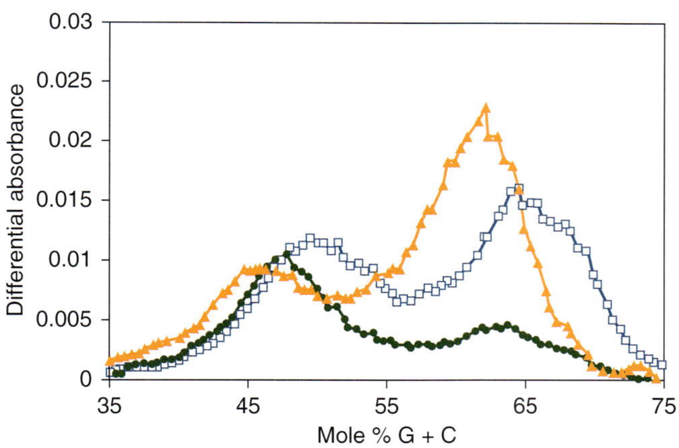

Figure 2.6 Percent G+C profiles of metagenome DNA from Solar saltern ponds with 22% (▲), 32% (▫), and 36% (•) salinity.

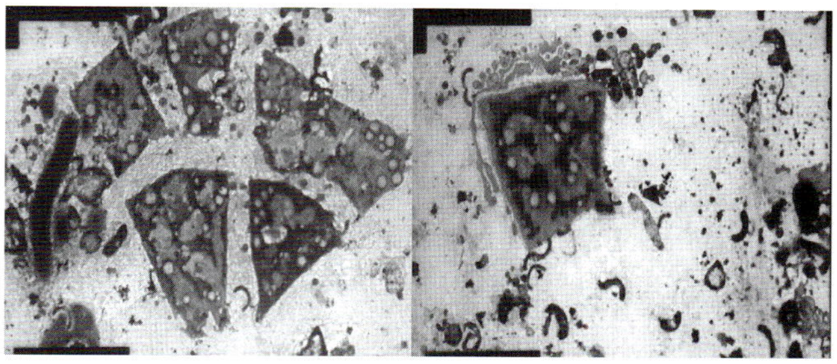

Figure 2.7 Transmission electron micrography of the square bacterium "*Haloquadratum walsbyi*", which predominated in ponds with 37% salinity.

that there was a shift in the microbial community from *Bacteria* dominated community at 22% salinity toward an *Archaea* dominated community at 37% salinity. In the community at 32% salinity, these microbial groups were more equally abundant [Øvreås et al., 2003]. T-RFLP, TEM (**Fig. 2.7**), and fosmid library data confirm the presence of a predominant population at 37% salinity that corresponds to the square halophilic archaeon *Haloquadratum walsbyi*, which has 48% G+C in its genome.

The predominant population in the 22% salinity pond was found to correspond to the bacterium *Salinibacter ruber* (63% G+C) [Øvreås et al., 2003]. Reassociation and T-RFLP indicated that even in the most extreme environments, community genomic complexity and diversity corresponded to 5–10 "species," which is higher than would be expected from monocultures as indicated by the fosmid library. This suggests that there is a potential large gene reservoir in the saltern habitat [Legault et al., 2006] with a considerable degree of microdiversity. The observed diversity may represent ecologically distinct populations or extremely divergent and rearranged genes in many concurrent cell types from the same population due to diversity preservation by phage predation [Rodriguez-Valera et al,. 2009].

As can be seen from the examples given above, the DNA reassociation method has proved useful for assessing the metagenomic complexity and for assessing the overall biodiversity in microbial communities. It has been used to compare the relative diversity in different communities and to study the effects of stress and environmental perturbations on microbial diversity.

2.5 CONCLUSIONS

Direct isolation of DNA from microbial communities and DNA-based analyses of microbial community composition and diversity represented a paradigm shift in microbial ecology. Recent results have confirmed our earlier data that microbial diversity in natural environments is huge and that pristine soil and sediments have among the highest microbial diversity on Earth. Genomic sequencing has revealed that the number of haplotypes present in microbial communities is vast and that there is a high evenness, because even the most common haplotypes have low abundance [Mes, 2008]. Furthermore, 100% identical haplotypes are rare, but there are often many sequence variants with high similarity in natural environments. These observations may reflect the clonal nature of microorganisms, along with the fact that diversification leads to huge numbers of diverging clonal lineages. Ecological theories and empirical information have identified a number of abiotic and biotic forces driving microbial diversity. Among these are: spatial and temporal habitat heterogeneity; low, but qualitative and quantitative, variations in available resources; disturbance and eutrophication; and trophic interactions leading to expansion and reduction of local microbial populations [Torsvik et al., 2002]. In addition, any differences in migration rates and types [Mes, 2008] between local subpopulations may lead to high evenness and can partly explain the differences in genomic diversity between terrestrial and aquatic microbial communities.

A problem when analyzing extremely diverse communities is that due to experimental limitations, only the most abundant DNA sequences reassociate, which means that only a minor fraction of the C_0t curve can be retrieved. Consequently, there are uncertainties about the species abundance distribution derived from reassociation and the diversity of rare species. We therefore choose to use conservative diversity estimates based on relative $C_0t_{1/2}$ value. To comprehend the huge diversity, attempts were made to estimate the "species" richness from reassociation, because estimates of microbial diversity often are based on this parameter. Other investigators have used the information in the reassociation kinetic curves by fitting lines to and extrapolating from such curves to find the best model describing the species abundance distribution (SAD) [Curtis et al., 2002; Gans et al., 2005]. Also, a new nonlinear regression procedure for analysis of C_0t data has recently been presented (C_0tQuest), which generates a variety of qualitative and quantitative model assessments [Bunge et al., 2009]. The SAD can be used to make assumptions on the abundance of rare species and to estimate the total "species" richness. The number of possible SAD is large, and there is no consensus on which are the most appropriate to apply on microbial communities. A problem when trying to estimate the number of species from reassociation curves is that this diversity changes similar to the Shannon–Weaver and Simpson indices and gives stronger weight to numerically dominant haplotypes than to rare haplotypes. Therefore when some haplotypes become numerically dominant, the diversity measure will decrease even if the richness would increase [Torsvik et al., 1994].

Given the average number of 10^{30} bacteria on the planet, combined with the possibility of evolution acting over 3.8 billion years, it is hardly surprising that we are continuously discovering different and new microbial taxa in almost every environment investigated [see Vol. II). Until recently, mapping the extent of microbial diversity was hampered by the discrepancy between sample size and community size, which meant that novel methods of extrapolation were required [Curtis et al., 2002]. The advent of high-throughput sequencing technologies like 454 pyrosequencing allow for larger samples and more robust methods of extrapolation [Quince et al., 2008; see also Vol. II]. Analyses using the new sequencing technology on metagenomic DNA from soil has confirmed that soils are indeed extremely diverse [Roesch et al., 2007], but also other environments have been shown to inhabit a much higher diversity than previously expected [Huse et al., 2007; Quince et al., 2008; Sogin et al., 2006].

Despite the advances in modern molecular tools that have provided huge amounts of new information and knowledge, the link between microbial diversity and ecosystem functions is still a major challenge. To understand the mechanisms and driving forces of microbial diversity and also understand which factors are important in shaping community structure and function, we need theoretical framework and hypothesis, which are still not well developed. Furthermore, good models are needed to predict and possibly control environmental impact, as well as how ecosystems respond to environmental disturbance.

REFERENCES

Antón J, Llobet-Brossa E, Rodríguez-Valera F, Amann R. 1999. Fluorescence in situ hybridization analysis of the prokaryotic community inhabiting crystallizer ponds. *Environ. Microbiol*. **1**:517–523.

Antón J, Rosellò-Mora R, Rodrìguez-Valera F, Amann R. 2000. Extremely halophilic bacteria in crystallizer ponds from solar salterns. *Appl. Environ. Microbiol*. **66**:3052–3057.

Antón J, Oren A, Benlloch S, Rodrìguez-Valera F, Amann R, Rosellò-Mora R. 2001. *Salinibacter ruber* gen.nov., sp.nov., a new species of extremely halophilic bacteria from saltern crystallizer ponds. *Int. J. Syst. Evol. Microbiol*. **52**:485–491.

Atlas RM. 1984. Diversity of microbial communities. In Marshall KC, ed., *Advances in Microbial Ecology*, Vol. 7. New York: Plenum Press, pp. 1–47.

Bak AL, Christiansen C, Stenderup A. 1970. Bacterial genome sizes determined by DNA renaturation studies. *J. Gen. Microbiol*. **64**:377–380.

Bakken LR. 1985. Separation and purification of bacteria from soil. *Appl. Environ. Microbiol*. **49**:1482–1487.

Benlloch S, Acinas SG, Antón J, López-López A, Luz SP, Rodríguez-Valera F. 2001. Archaeal biodiversity in crystallizer ponds from a solar saltern: Culture versus PCR. *Microb. Ecol*. **41**:12–19.

Benlloch S, et al. 2002. Prokaryotic genetic diversity throughout the salinity gradient of a coastal solar saltern. *Environ. Microbiol*. **4**:349–360.

Boles BR, Thoendel M, Singh PK. 2004. Self-generated diversity produces "insurance effects" in biofilm communities. *Proc. Natl. Acad. Sci. USA* **101**:16630–16635.

Britten RJ, Kohne DE. 1968. Repeated sequences in DNA. *Science* **161**:529–540.

Bull AT. 1992. Microbial diversity. *Biodiv. Conserv*. **1**:219–220.

Bunge J, Chouvarine P, Peterson DG. 2009. C_0tQuest: Improved algorithm and software for nonlinear regression analysis of DNA reassociation kinetics data. *Anal. Biochem*. **388**:322–330.

Curtis TP, Sloan WT, Scannell JW. 2002. Estimating prokaryotic diversity and its limits. *Proc. Natl. Acad. Sci. USA* **99**:10494–10499.

Escara JF Hutton, JR. 1980. Thermal stability and renaturation of DNA in dimethyl sulfoxide solutions: Acceleration of the renaturation rate. *Biopolymers* **19**:1315–1328.

Fægri A, Torsvik VL, Goksöyr J. 1977. Bacterial and fungal activities in soil: Separation of bacteria and fungi by a rapid fractionated centrifugation technique. *Soil Biol. Biochem*. **9**:105–112.

Gans J, Wolinsky M, Dunbar J. 2005. Computational improvements reveal great bacterial diversity and high metal toxicity in soil. *Science* **309**:1387–1390.

Haegeman B, Vanpeteghem D, Godon J-J, Hamelin J. 2008. DNA reassociation kinetics and diversity indices: Richness is not rich enough. *Oikos* **117**:177–181.

Handelsman J, Rondon MR, Brady SF, Clardy J, Goodman RM. 1998. Molecular biological access to the chemistry of unknown soil microbes: A new frontier for natural products. *Chem. Biol*. **5**:R245–R249.

Harper JL, Hawksworth DL. 1994. Preface. *Philos. Trans. Royal Soc. London. Series B: Biol. Sci*. **345**:5–12.

Hobbie JE, Daley RJ, Jasper S. 1977. Use of nuclepore filters for counting bacteria by fluorescence microscopy. *Appl. Environ. Microbiol*. **33**:1225–1228.

Holben WE, Jansson JK, Chelm BK, Tiedje JM. 1988. DNA probe method for the detection of specific microorganisms in the soil bacterial community. *Appl. Environ. Microbiol*. **54**:703–711.

Huse S, Huber J, Morrison H, Sogin M, Welch D. 2007. Accuracy and quality of massively parallel DNA pyrosequencing. *Genome Biol*. **8**:R143.

Johnsen K, Jacobsen C, Torsvik V, Sørensen J. 2001. Pesticide effects on bacterial diversity in agricultural soils—A review. *Biol. Fertil. Soils* **33**:443–453.

Legault B, Lopez-Lopez A, Alba-Casado J, Doolittle WF, Bolhuis H, Rodriguez-Valera F, Papke RT. 2006. Environmental genomics of "Haloquadratum walsbyi" in a saltern crystallizer indicates a large pool of accessory genes in an otherwise coherent species. *BMC Genomics*. **7**:171.

Lewin B. 2000. *Genes VII*. New York: Oxford University Press.

Martínez-Murcia AJ, Acinas SG, Rodriguez-Valera F. 1995. Evaluation of prokaryotic diversity by restrictase digestion of 16S rDNA directly amplified from hypersaline environments. *FEMS Microb. Ecol*. **17**:247–255.

Mes T, H. 2008. Microbial diversity—Insights from population genetics. *Environ. Microbiol*. **10**:251–264.

Øvreås L, Torsvik V. 1998. Microbial diversity and community structure in two different agricultural soil communities. *Microb. Ecol*. **36**:303–315.

Øvreås L, Jensen S, Daae FL, Torsvik V. 1998. Microbial community changes in a perturbed agricultural soil investigated by molecular and physiological approaches. *Appl. Environ. Microbiol*. **64**:2739–2742.

Øvreås L, Daae FL, Torsvik V, Rodríguez-Valera F. 2003. Characterization of microbial diversity in hypersaline environments by melting profiles and reassociation kinetics in combination with terminal restriction fragment length polymorphism (T-RFLP). *Microb. Ecol*. **46**:291–301.

Quince C, Curtis TP, Sloan WT. 2008. The rational exploration of microbial diversity. *ISME J*. **2**:997–1006.

Raes J, Korbel J, Lercher M, von Mering C, Bork P. 2007. Prediction of effective genome size in metagenomic samples. *Genome Biol*. **8**:R10.

Ritz K, Griffiths BS, Torsvik VL, Hendriksen NB. 1997. Analysis of soil and bacterioplankton community DNA by melting profiles and reassociation kinetics. *FEMS Microbiol. Lett*. **149**:151–156.

Rodriguez-Valera F. 1999. Contribution of molecular techniques to the study of microbial diversity in hypersaline environments.

Rodriguez-Valera F, Martin-Cuadrado A-B, Rodriguez-Brito B, Pasic L, Thingstad TF, Rohwer F, Mira A. 2009. Explaining microbial population genomics through phage predation. *Nat. Rev. Microbiol*. **7**:828–836.

Roesch L, Fulthorpe R, Riva A, Casella G, Hadwin A, Kent A, Daroub S, Camargo F, Farmerie W, Triplett E. 2007. Pyrosequencing enumerates and contrasts soil microbial diversity. *ISME J*. **1**:283–290.

Rosing MT. 1999. ^{13}C-depleted carbon microparticles in 3700-Ma sea-floor sedimentary rocks from West Greenland. *Science* **283**:674–676.

Sogin ML, Morrison HG, Huber JA, Welch DM, Huse SM, Neal PR, Arrieta JM, Herndl GJ. 2006. Microbial diversity in the deep sea and the underexplored "rare biosphere". *Proc. Natl. Acad. Sci. USA* **103**:12115–12120.

Stackebrandt E, et al. 2002. Report of the ad hoc committee for the re-evaluation of the species definition in bacteriology. *Int. J. Syst. Evol. Microbiol*. **52**:1043–1047.

Steffan RJ, Goksoyr J, Bej AK, Atlas RM. 1988. Recovery of DNA from soils and sediments. *Appl. Environ. Microbiol*. **54**:2908–2915.

Torsvik VL. 1980. Isolation of bacterial DNA from soil. *Soil Biol. Biochem*. **12**:15–21.

Torsvik VL, Goksoyr J. 1978. Determination of bacterial DNA in soil. *Soil Biol. Biochem*. **10**:7–12.

Torsvik V, Goksoyr J, Daae FL. 1990a. High diversity in DNA of soil bacteria. *Appl. Environ. Microbiol*. **56**:782–787.

Torsvik V, Salte K, Sorheim R, Goksoyr J. 1990b. Comparison of phenotypic diversity and DNA heterogeneity in a population of soil bacteria. *Appl. Environ. Microbiol*. **56**:776–781.

Torsvik V, Goksøyr J, Daae FL, Sørheim R, Michalsen J, Salte K. 1994. Use of DNA analysis to determine the diversity of microbial communities. In Ritz K, Dighton J, Giller KE, eds. *Beyond the*

Biomass; Compositional and Functional Analysis of Soil Microbial Communities. New York: John Wiley & Sons, pp. 39–48.

Torsvik V, Daae FL, Goksoyr J. 1995. Extraction, purification, and analysis of DNA from soil bacteria. In Trevors JT, van Elsas JD, eds. *Nucleic Acids in the Environment: Methods and Applications*. Berlin: Springer-Verlag, pp. 29–48.

Torsvik V, Sørheim R, Goksøyr J. 1996. Total bacterial diversity in soil and sediment communities—A review. *J. Ind. Microbiol. Biotechnol*. **17**:170–178.

Torsvik V, Daae FL, Sandaa R-A, Øvreås L. 1998. Novel techniques for analysing microbial diversity in natural and perturbed environments. *J. Biotechnol*. **64**:53–62.

Torsvik V, Øvreås L, Thingstad TF. 2002. Prokaryotic diversity—Magnitude, dynamics, and controlling factors. *Science* **296**:1064–1066.

Woese CR. 1987. Bacterial evolution. *Microbiol. Mol. Biol. Rev*. **51**:221–271.

Woese CR, Fox GE. 1977. Phylogenetic structure of the prokaryotic domain: the primary kingdoms. *Proc. Natl. Acad. Sci. USA* **74**:5088–5090.

Yachi S, Loreau M. 1999. Biodiversity and ecosystem productivity in a fluctuating environment: The insurance hypothesis. *Proc. Natl. Acad. Sci. USA* **96**:1463–1468.

Chapter 3

Diversity of 23S rRNA Genes Within Individual Prokaryotic Genomes

ANNA PEI, WILLIAM E. OBERDORF, CARLOS W. NOSSA, POOJA CHOKSHI, MARTIN J. BLASER, LIYING YANG, DAVID M. ROSMARIN, AND ZHIHENG PEI

3.1 INTRODUCTION

Ribosomes play a vital role in the functioning of all living organisms. They are the ribonucleoprotein machinery in which proteins are synthesized. Prokaryotic ribosomes are composed of a large 50S subunit and a small 30S subunit. The large subunit is composed of a 23S rRNA, a 5S rRNA, and over 30 proteins, while the small subunit contains a 16S rRNA and 20 proteins. The exact spatial arrangement of these components may be critical to proper ribosomal functioning resulting in a constraint on RNA genes from any drastic change [Doolitle, 1999]. Functional constraints dictate which areas of an rRNA sequence must remain conserved and which may be variable without altering integral structural components. Classifying distantly related organisms relies on conserved regions, while the more variable regions separate closely related organisms. It is unlikely that horizontal gene transfer events will falsify the evolutionary history of an organism due to the highly constrained rRNA genes [Eickbush and Eickbush, 2007; Santoyo and Romero, 2005; Gurtler, 1999]. These features make rRNA genes the most suitable molecular chronometer for both phylogenetic analysis and taxonomic classification of cellular organisms [Woese, 1987].

16S and 23S rRNA have both been used to create reliable phylogenetic trees [De Rijk et al., 1995; Cedergren et al., 1988; see also Chapter 15, Vol. I], and the comprehensive phylogenies inferred from sequence comparisons provide three domains, the Bacteria, Archaea, and Eukarya [Woese, 1987; Woese et al., 1990]. However, classification has focused more on 16S rRNA due to a lack of established broad-range sequencing primers for 23S rRNA and early sequencing technologies limitations for sequencing larger genes. Recently, a decrease in sequencing costs with 454 pyro-sequencing, evidence of universally conserved regions within the 23 rRNA gene needed to design broad range primers [Hunt, 2006], and the establishment of the Roadmap Initiative in the Human Microbiome Project (http://nihroadmap.nih.gov/hmp/) have renewed interest in the 23S rRNA gene. Compared to 16S rRNA genes, 23S rRNA genes contain more characteristic sequence stretches due to a greater length, unique insertions and/or deletions, and possibly better phylogenetic resolution because of higher sequence variation [Ludwig and Schleifer, 1994].

One phenomenon that may hinder classification attempts using the 23S rRNA gene is intragenomic heterogeneity. While gene redundancy is uncommon in prokaryotes, rRNA genes may number from 1 to 15 copies in a single genome [Klappenbach et al., 2001]. Divergent evolution between rRNA genes in the same genome may corrupt the record of evolutionary history and obscure the true identity of an organism. When substantial variation occurs, use of rRNA gene sequences may lead to the artificial classification of an organism into more than one species. In this study, we performed a systematic survey of intragenomic variation of 23S rRNA genes in genomes representing 184 prokaryotic species.

Handbook of Molecular Microbial Ecology, Volume I: Metagenomics and Complementary Approaches, First Edition. Edited by Frans J. de Bruijn.

3.2 MATERIALS AND METHODS

3.2.1 Annotation of 23S rRNA Genes

Gene sequences were obtained from the Complete Microbial Genomes database at the NCBI web site (http:// www.ncbi.nlm.nih.gov/genomes/MICROBES/Complete.html). For species with more than one available genome, the most completely annotated was used to avoid duplicates. Lengths of 23S rRNA genes were identified by using experimentally defined 23S rRNA sequences from the closest relatives available and verified by 2° structure analysis based on minimizing free energy, using RNAstructure [Mathews et al., 2004] and Rnaviz [De Rijk and De Wachter, 1997], with experimentally defined 23S rRNA or the consensus 23S rRNA models [Wuyts et al., 2001] used for reference. The number of copies of 23S rRNA genes present in a genome was determined by whole genome BLAST searches based on the known 23S rRNA sequence.

3.2.2 Analysis of Intragenomic Diversity in 23S rRNA Genes

Genomes containing multiple copies of 23S rRNA genes were aligned with Clustalw [Thompson et al., 1994]. To calculate diversity, the number of substitutions including point mutations and insertions/deletions (indels) was divided by the total number of positions, including gaps in the alignment.

3.2.3 Comparison of 2° Structures

To compare two related 2° structures, a mismatch was defined as conserved if located in a loop, or located in a stem but causing GC:GU conversions or covariation resulting in no change in base-pairing. In contrast, a nonconserved mismatch altered base-pairing. Substitutions also were classified by the position-specific relative variability rate calculated from the consensus 23S rRNA model based on an alignment of 187 bacterial 23S rRNA genes [Wuyts et al., 2001]. Positions were classified as variable or nonvariable positions, according to the substitution rate relative to the average substitution rate of all sites [Wuyts et al., 2001]. The relative substitution rate for a variable position $v<1$ indicates a substitution rate higher than that averaged for all sites in the 23S rRNA gene analyzed, while a conserved position had a relative substitution rate $v<1$; uncommon sites are positions occupied in <25% of organisms due to insertions. The expected variability for certain classes of positions was calculated from the consensus models. Differences between expected and observed variability was analyzed by chi-square analysis and considered significant if $p < 0.05$.

3.3 RESULTS

3.3.1 rRNA Gene Database

In total, 342 complete prokaryotic genomes were available in the database of which 184 were unique genomes containing multiple 23 rRNA genes (10 Archaea, 174 Bacteria) and were analyzed. The 184 species represented 11 phyla, of which Proteobacteria was the most abundant phylum (98 species) followed by Firmicutes (43 species), Euryarchaeota (10 species), and Actinobacteria (9 species). The remaining 7 phyla were represented by only 24 species.

3.3.2 Diversity of 23S rRNA Genes in the Primary Structure

The 184 genomes had a median of 4.57 23S rRNA genes/genome (range 1–15). Diversity was found in 113 genomes, with a mean of 0.40% and range from 0.01% to 4.04% (median 0.24%, interquartile range (IQR) 0.10–0.52% sequence variation). The threshold for identification of outliers, calculated as IQR ±1.5 IQR, was 0–1.15%. Using this threshold, 8 genomes were identified as having high intragenomic variation amongst their paralogous 23S rRNA genes (**Table 3.1**). The eight outliers were concentrated in the four most abundant phyla: four in Proteobateria, two in Firmicutes, one in Actinobacteria, and one in Euryarchaeota. The absence of outliers in the remaining seven phyla is likely because they were least represented in the dataset. In 4 genomes, *Carboxydothermus hydrogenoformans, Haloarcula marismortui, Shewanella oneidensis*, and *Streptococcus pyogenes*, there was on average 1.52% diversity due to 44.6 substitutions including 8.8 indels (Table 3.1). In 3 genomes, *Nocardia farcinica, Clostridium perfringens*, and *Salmonella typhimurium*, substitutions tended to concentrate within short segments of 23S rRNA genes causing complex rearrangement of the secondary structure (**Fig. 3.1**), which will be described in detail separately. The greatest diversity was observed in *Thermoanaerobacter tengcongensis* (4.04% diversity), which also contained intervening sequences (IVS) (**Fig. 3.2**); this will be described below.

3.3.3 Effect on Secondary Structure

In *T. tengcongensis*, compared with the consensus 2° structure model of 23S rRNA [Wuyts et al., 2001] (Fig. 3.2), 86 (74.8%) of the 115 mismatch positions occurred at highly variable positions; this is significantly higher than expected (40.4%, $p < 0.0001$) (**Table 3.2**).

Comparing the 2° structures of *rrnB23S* and *rrnC23S*, 105 of the 115 substitutions were conservative

Table 3.1 Genomes with Significant Intragenomic Diversity Among Paralogous 23S rRNA Genes

Organism	Accession Number	Copy	Alignment (nt)	Diversity (%)	Substitution[a]	Covariation	G:U = G:C	In-Loop	Stem–Loop Transition	Indel
Carboxydothermus hydrogenoformans	CP000141	4	2921	1.30	38	4	2	25	7	2
Haloarcula marismortui	AY596297	3	2922	1.54	45	24	2	11	8	0
Shewanella oneidensis	AE014299	9	2894	1.17	34	22	4	5	3	1
Streptococcus pyogenes	AE004092	6	2912	1.65	48	22	8	11	7	8
Nocardia farcinica	AP006618	3	3132	1.18	37					1
Clostridium perfringens	BA000016	10	2915	1.96	57	Complex rearrangement[b]				41
Salmonella typhimurium	AE006468	7	3093	1.27	37					3
Thermoanaerobacter tengcongensis	AE008691	4	2965	4.04	115	24	19	62	10	8

[a] Substitution includes both point mutations and indels.

[b] See Figure 3.1.

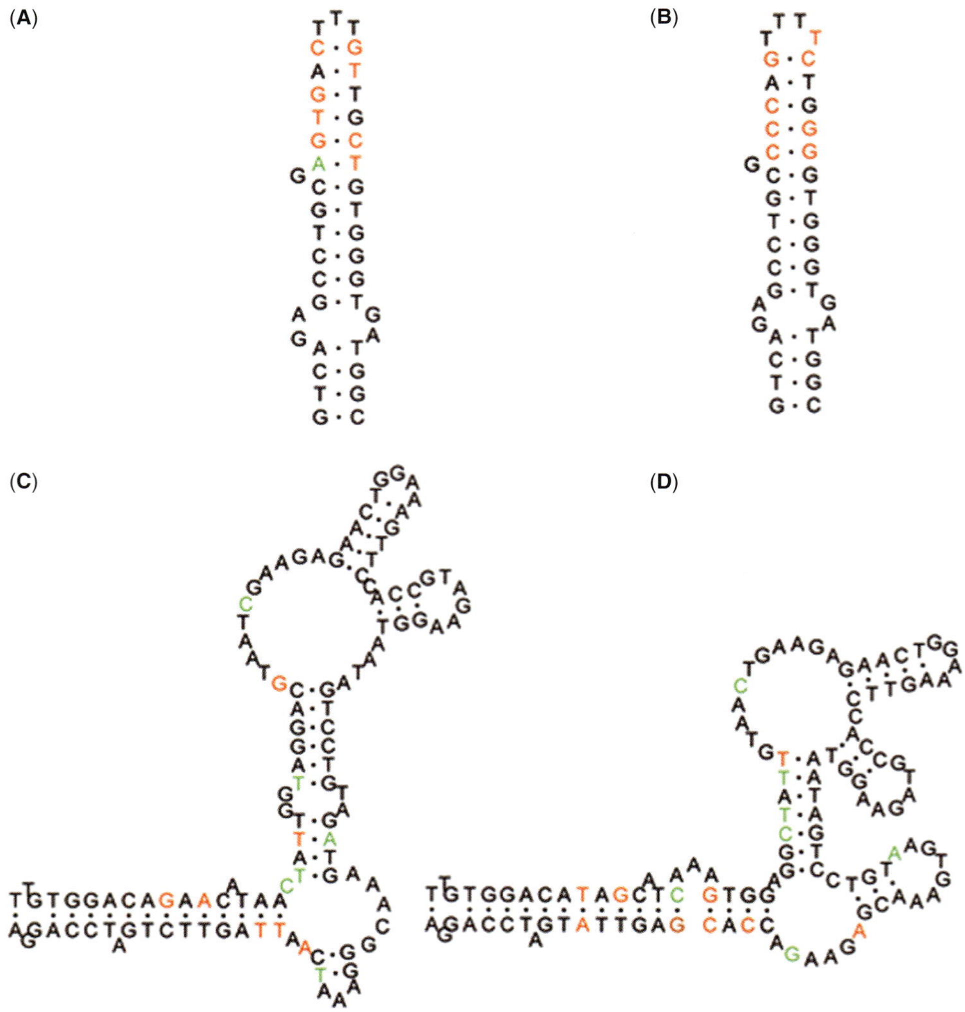

Figure 3.1 Conservation of 2° structure by complex rearrangement of base pairing and substitutions in *rrn23S* of *Nocardia farcinica* (**A**, **B**) and *Clostridium perfringens* (**C**, **D**). Nucleotides related to substitutions are highlighted in red, and indels are highlighted in green. Segments of *rrn23S* of *Nocardia farcinica* shown correspond to positions 620 through 659 of *rrnC23S*. The segments of *rrnC23S* (**A**) and *rrnB23S* (**B**) differ by nine positions, including one indel and eight substitutions. Segments of *rrn23S* of *Clostridium perfringens* shown correspond to positions 292 through 406 of *rrnH23S*. The segments of *rrnH23S* (**C**) and *rrnD23S* (**D**) differ by 22 positions, including 13 indels and 8 substitutions.

(62 in loops [including 4 insertions], 24 covariations, and 19 GU:GC conversions). Only 10 (9 point mutations and 1 single nucleotide insertion) of the 115 substitutions altered the 2° structure by causing stem-loop transition. These features indicate strong structural constraints, suggesting that *rrnC23S* may be functional.

Conservation of the 2° structure despite primary structure alteration is found in other species (Table 3.1). In *C. hydrogenoformans, H. marismortui, S. oneidensis*, and *S. pyogenes*, there are on average 44.6 substitutions per genome, of which only ~20% (8.2 substitutions) may potentially cause stem–loop transition. The effect of other substitutions are minimized by either covariations (14.4 substitutions), G:U to G:C transitions (4 substitutions), or in-loop substitutions (14.4). Complex rearrangements of the 2° structure likely occurred in *N. farcinica, C. perfringens*, and *S. typhimurium* as a result of changes in the primary structure (Fig. 3.1). The complex rearrangement appears to be associated with indels. One example is between *rrnC23S* and *rrnB23S* of *Nocardia farcinica* in which a single deletion accompanied 8 point mutations in a segment of 40 nt. The topographic structure can be conserved by repositioning the bases *rrnB23S*, although the resultant stem is one base pair shorter and the loop is one base larger (Figs. 3.1A and 3.1B). Another example is the *rrnH23S* and *rrnD23S* of *C. perfringens* in which

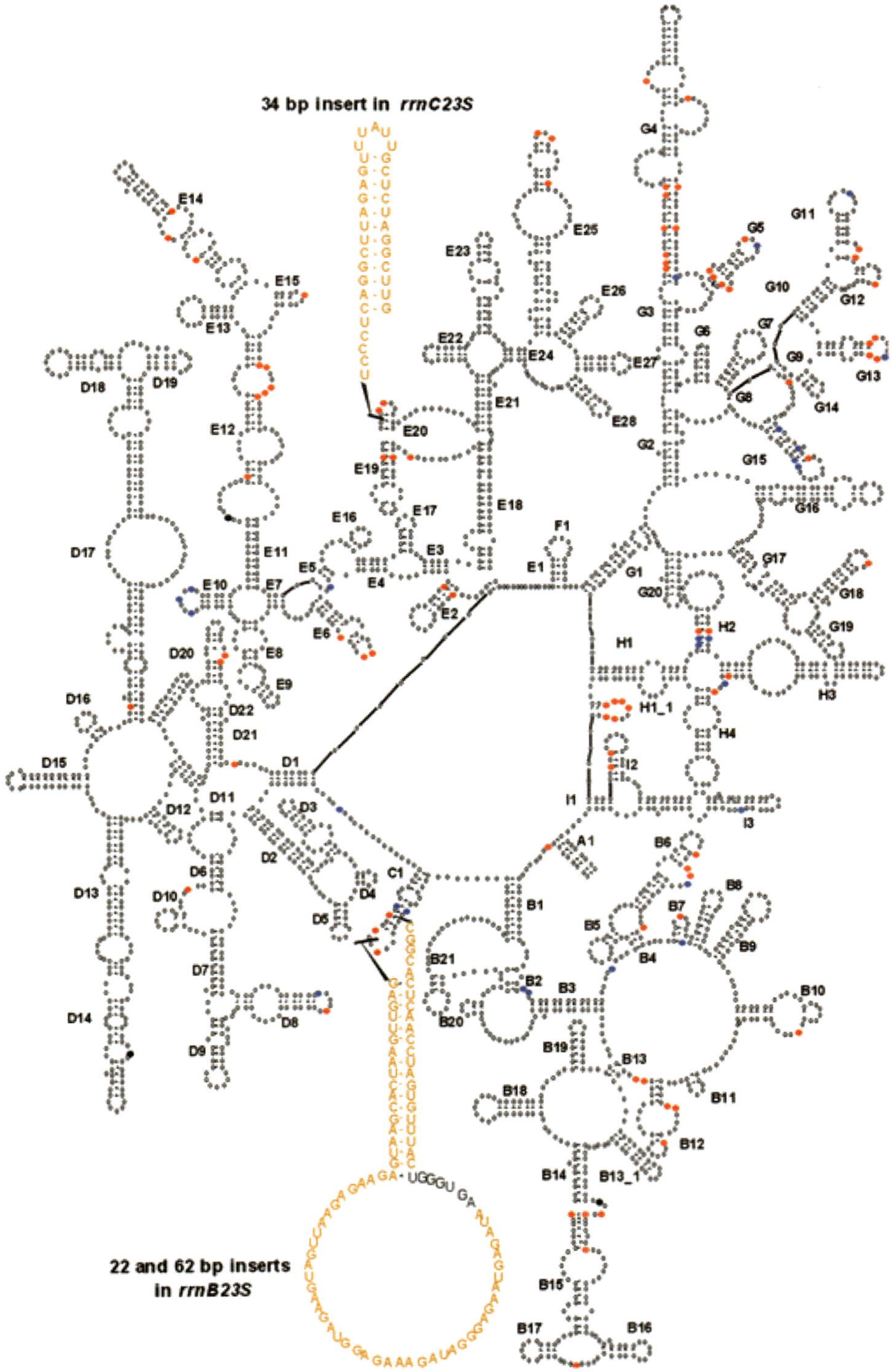

Figure 3.2 Distribution of substituted positions and IVS in 23S rRNA genes of *T. tengcongensis*. Secondary structure of rrnB16S and rrnB23S was based on the 2° structure models for *Thermus thermophilus* [Wuyts et al., 2001]. Substituted positions between rrnB23S and rrnC23S are highlighted in colors according to the position-specific relative variability rate calculated from the consensus rrn23S model based on an alignment of 184 bacterial 23S rRNA genes [Wuyts et al., 2001]. A position with a relative substitution rate $v<1$ (red) implies that it has a substitution rate higher than the average substitution rate of all its sites in the rRNA gene analyzed, while $v < 1$ (blue) indicates that the rate is lower than the average rate. Uncommon sites are positions that are occupied in <25% of organisms because of insertions, which are shown by black dots.

there are 13 indels and 8 point mutations in a segment of 115 nt. Although there was an apparent change in the 2° structure, rearrangement of the positions of the loops and the length of the stems appear to have minimized the effect on the topographic structure (Figs. 3.1C and 3.1D). Variation of rrn23S in *S. typhimurium* has been previously reported [Mattatall and Sanderson, 1996].

3.3.4 Intervening Sequences

Another type of alteration of 23S rRNA genes is insertion of an intervening sequence (**Table 3.3**). IVS was found in 7 genomes (Table 3.3). If the IVS are not included in calculations, the remaining sequences of paralogous 23S rRNA genes have <1% diversity in 6 of the 7 genomes except for *T. tengcongensis*, whose diversity is due to both IVS-like insertions and a high level of substitutions.

3.3.5 Transposon-Mediated Insertion of IVS

Nostoc sp. *PCC7120* contained four 23S rRNA genes; *rrnA23S* is 4299 nt, while each of the other three is 2828 nt [Kaneko et al., 2001]. The variation was due to the presence of a 1471-nt insert in *rrnA23S* (**Fig. 3.3**), including 15-nt imperfect inverted repeats at both ends of the insertion and a 963-nt ORF. The deduced protein is predicted to be a transposase in clusters of orthologous group (COG) COG3547 (expected value = 7e-27). The 303-amino-acid protein of COG3547 consensus consists of both transposases 9 and 20. Blast of the 1471-nt insert against the genome of *Nostoc* sp. *PCC7122* revealed 11 other homologs, each with the terminal inverted repeats. Without the insertion, all four *rrn23S* copies were identical.

The 23S rRNA genes of *Deinococcus radiodurans* were only partially annotated (GenBank accession: AE000513) [White et al., 1999]. The full-length 23S rRNA genes were identified by search of the *D. radiodurans* genome using the unassigned RNA (GenBank accession:30749903) associated with the 50S ribosomal subunit [Schlunzen et al., 2001]. Three putative 23S rRNA genes were identified. The 2° structures of the putative 23S rRNA genes were compared with the consensus prokaryotic 23S rRNA model to confirm identity and to define length [Wuyts et al., 2001]. *rrnC23S* was 3944-nt while *rrnA23S* and *rrnB23S* were only 2877-nt (**Fig. 3.4**). *rrnC23S* contains a 1067-nt insertion including a 519-nt ORF encoding a putative172-amino acid protein. The protein was predicted to be a transposase in COG3335 (expected value = 8e-14), and 27-nt imperfect inverted repeats were present at both ends of the insert. BLAST of the 1067-nt insert against the *D. radiodurans* genome revealed two other nearly identical (>98% homology) copies, including the inserted repeats, located on plasmids MP1 and CP1. *rrnA23S* and *rrnB23S* were nearly identical. Excluding the insertion, only six single-nucleotide substitutions were observed between the three 23S rRNA genes.

3.3.6 Small IVS

Salmonella typhimurium has seven 23S rRNA genes with sizes ranging from 2994 to 3093 bp [McClelland et al., 2001]. Two insertions were found, 83-nt and 99-nt, consistent with previously described IVS in *S. typhimurium* [Mattatall and Sanderson, 1996] (Table 3.3). The largest 23S rRNA gene contains both inserts. The second largest 23S rRNA gene contains the 98-nt insert. The remaining five 23S rRNA genes only contain the 83-nt insert. The largest difference (1.27%, 37/2903) was found between *rrnB23S* and *rrnF23S* with 37 substitutions (Table 3.1).

Photorhabdus luminescens contains seven 23S rRNA genes [Duchaud et al., 2003]. rrnC23S and rrnE23S contain a shared 112-nt insert (Table 3.3). The other five 23S rRNA genes are 2909 nt in length. The insert is similar to those from *S. typhimurium* in size, and thus it does not contain an ORF. No repetitive DNA was present in or near the insertion. Excluding the insert, the largest difference was observed between *rrnD23S* and *rrnE23S*, differing by 28 nt (0.96%). No sequences homologous to the insert were found in *P. luminescens* or GenBank.

Photobacterium profundum contains 15 paralogous 23S rRNA genes ranging from 2893 to 2926 nt in size (GeneBank accession: CR354531.1, unpublished). The size variation was caused by two IVS of 14 and 17 nt, respectively. Maximal diversity was 0.83% (24/2893) between *rrnC23S and rrnJ23S* (Table 3.3). No additional copy of the IVS was found outside of the 23S rRNA genes within the *P. profundum* genome.

Agrobacterium tumefaciens contains four copies of 23S rRNA genes, two on each of its linear and circular chromosomes, respectively [Wood et al., 2001]. The sizes of the 23S rRNA genes vary between 2233 and 2252 nt because the second 23S rRNA gene is shorter, caused by two small IVS of 9 and 13 nt, respectively (Table 3.3). Besides the IVS, 15 mutations were observed (0.67% diversity).

3.3.7 IVS and Point Mutations

T. tengcongensis had four 23S rRNA genes with sizes ranging from 2881 to 2965 nt, reflecting three major alleles [Bao et al., 2002]. The largest difference was between *rrnB23S* and *rrnC23S*, due to inserts of 22, 62, and 34 nt (Figure 3.2, Table 3.1) and 115 substitutions. The 22- and 62-nt inserts were present in *rrnA23S* and *rrnB23S*, respectively, the 34-nt insert was present in *rrnC23S*, and

Table 3.2 Observed and Expected Variability in 23S rRNA Genes in *T. tengcongensis*

Relative Variability(i)	Observed Substitution Rate (%)	Expected Substitution Rate(%)	*P* Value
$i > 1$	78.4 (86/115)	40.4 (1175/2912)	<0.001
$I < 1$	22.6 (26/115)	58.6 (1707/2912)	<0.001
Uncommon sites	2.6 (3/115)	1.0 (30/2912)	0.110

Table 3.3 Genomes with Intervening Sequences in 23S rRNA Genes

Species	Accession #	Number of Copies	Length Range (nt)	IVS (nt)	Diversity (%)	Type of IVS
Agrobacterium tumefaciens	AE007869.2	4	2233–2252	9, 13	0.67	Short IVS
Deinococcus radiodurans	AE000513.1	3	2877–3944	1067	0.07	IVS encoding transposon
Nostoc sp.	BA000019.2	4	2828–4299	1471	0	IVS encoding transposon
Photobacterium profundum	CR354531.1	15	2894–2926	14, 17	0.83	Short IVS
Photorhabdus luminescens	BX470251.1	7	2909–3021	112	0.96	IVS
Salmonella typhimurium	AE006468.1	7	2994–3093	83, 99	1.27	IVS
Thermoanaerobacter tengcongensis	AE008691.1	4	2881–2965	22, 62, 34	4.04	IVS and mutations

there were no inserts present *in rrnD23S*. No sequences homologous to the three inserts were found in GenBank. Excluding the three inserts, there remained differences of 4.0% (115/2884) between *rrnB23S* and *rrnC23S*.

3.4 DISCUSSION

This is the first large-scale survey of intragenomic variation of paralogous 23S rRNA genes. We analyzed 23S rRNA genes from genomes representing 184 prokaryotic species to examine for evidence of ribosomal constraint of rRNA structures at the intragenomic level. Our findings support the hypothesis that individual 23S rRNA genes within a genome are conserved due to such structural constraints.

One level of ribosomal constraint on 23S rRNA genes was observed at the level of primary structure. In the large majority of the genomes (183/184) analyzed, variation of primary structure among paralogous 23S rRNA genes was small (<3%, which is the 16S rrn-based species boundary) [Stackebrandt and Goebel, 1994]. In our dataset, 38% of the 184 genomes containing multiple paralogous 23S rRNA genes were invariable, similar to the finding in 16S rRNA genes [Acinas et al., 2004]. Our findings of 0.4% average diversity for intragenomic variation of rRNA sequences among the 113 diversified genomes is higher than the reported 0.17% diversity among 16S rRNA genes [Stewart and Cavanaugh, 2007], but will be more similar to the 16S rRNA genes diversity (0.25%) if all 184 genomes are included in the calculation. Excluding *T. tengcongensis*, which has unusually high diversity among 23S rRNA genes, the maximal diversity among paralogous 23S rRNA genes is 1.96%, higher than the maximal variation of 1.26% among 16S rRNA genes [Coenye and Vandamme, 2003]. In organisms containing multiple rRNA genes, the homogeneity of primary sequences is believed to be maintained through gene conversion by homologous recombination [Eickbush and Eickbush, 2007; Liao, 2000]. The evolution of these paralogous rRNA genes is concerted in such a fashion that within a species the rRNA genes are nearly identical, whereas orthologous rRNA genes between species can vary considerably [Eickbush and Eickbush, 2007; Liao, 2000]. However, our study suggests that homologous recombination is not always efficient in keeping the primary structure in check. In *T. tengcongensis*, the 4% difference (without considering the IVS) in primary structure between *rrnB23S* and *rrnC23S* is in agreement with the high percent divergence reported between 16S rRNA paralogues [Acinas et al., 2004]. It is tempting to attribute the unusual variation to the presence of IVS in the *T. tengcongensis* 23S rRNA genes, but variation amongst 23S rRNA genes in six other genomes containing IVS is much lower (Table 3.3). It appears that gene conversion or concerted evolution is one, but not the only, way to meet the requirement of ribosomal constraint.

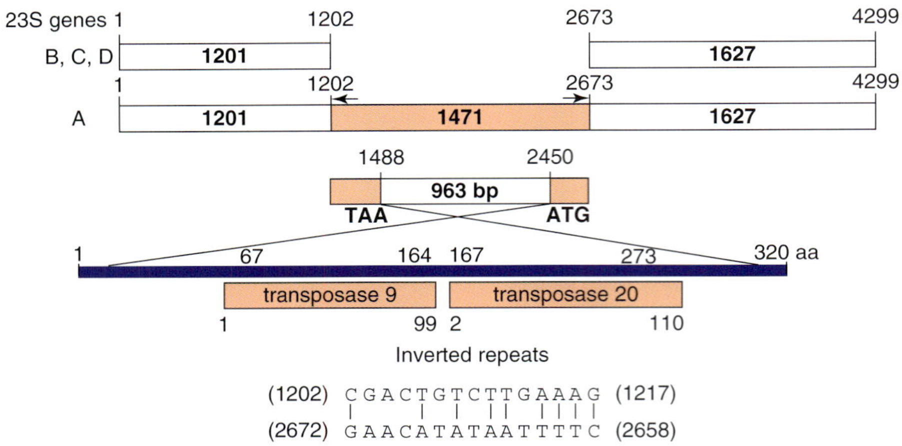

Figure 3.3 IVS in *Nostoc* sp. (strain PCC 7120) 23S rRNA genes. A 1471-bp IVS flanked by partial 16-bp inverted repeats (*arrows*) was found in *rrnA23S*, which contains a 963-bp open reading frame that encodes proteins with strong homology (expected value = $7e^{-27}$) to transposases 9 and 20 in COG3547.

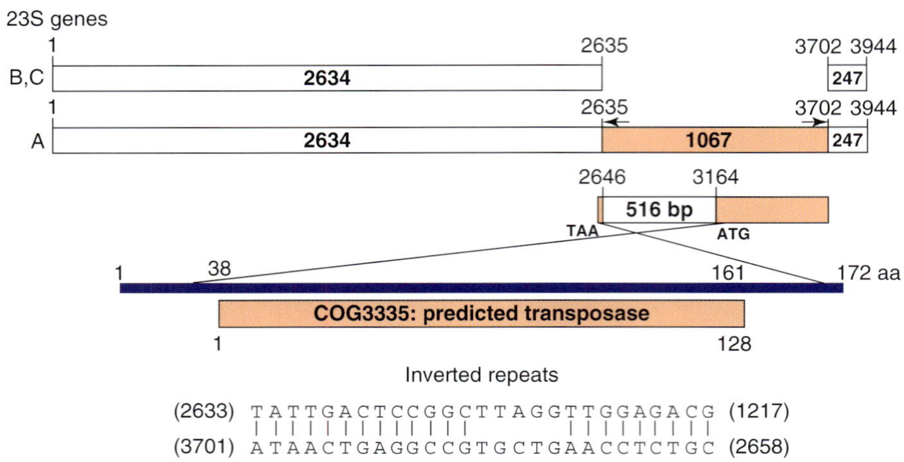

Figure 3.4 IVS in *Deinococcus radiodurans* 23S rRNA genes. *rrnA23S* contains a 1067-bp IVS flanked by 27-bp inverted repeats. The arrows indicate the location of the inverted repeats. The 1067-bp insert contains a 516-bp open reading frame encoding a 172-amino-acid protein with strong homology (expected value = $8e^{-14}$) to predicted transposase COG3335.

Another level of ribosomal constraint was observed at the level of 2° structures. For example, the detailed analysis of the 23S rRNA genes of *T. tengcongensis* showed that only in uncommon instances (10 of 115 point substitutions) a point substitution altered the 2° structure. This observation is in line with the current concept derived from comparing consensus rRNA sequences from closely related species [Zwieb et al., 1981; Stiegler et al., 1981; Curtiss and Vournakis, 1984; Hori and Osawa, 1986; Gutell et al., 1994] in that selection pressure from ribosomal constraints is to preserve 2° structures that are essential for the assembly of functional ribosomes [Higgs, 1998]. In organisms with a single rRNA operon, ribosomal constraints could follow a simple birth-and-death model by linking the fate of a mutation to the viability of the organisms. However, in an organism with multiple rRNA operons, the nature of the selection pressure is unclear. In *Escherichia coli*, which has seven rRNA operons, after deletion of the other 6 operons only a single operon is sufficient to produce >50% of wild-type rRNA levels and support >50% of the wild-type growth rate [Asai et al., 1999]. This experiment suggests that in an organism with multiple rRNA operons, it may not be necessary for survival to keep all rRNA genes under ribosomal constraint. The observed tight constraints on individual 23S rRNA in *T. tengcongensis* needs a new explanation. *T. tengcongensis*, isolated from a freshwater hot spring in Tengchong, China, is a rod-shaped, gram-negative bacterium that grows anaerobically in extreme environments. It propagates at wide ranges of temperatures from 50° to 80°C and pH values between 5.5 and 9 [Bao et al., 2002]. It is possible that all four 23S rRNA genes are essential for survival of *T. tengcongensis* in its natural habitat, though one might be sufficient under laboratory conditions. It also is possible that the divergence of 23S rRNA genes offers competitive advantage in a changing environment in nature and thus are selectively maintained. Ribosomal constraints could be met by maintaining conserved 2° structures through conserved point mutations under sufficient selective pressure to resist the tide of homologous recombination.

Our finding is consistent with the finding by Acinas et al. [2004] in *rrnC16S* of *T. tengcongensis*. Both studies indicate that the 2° structures are conserved despite a high level of change in the primary structures and suggest that the *rrnC* operon is functional. Although Acinas et al. suggested that the *rrnC* operon could be brought into *T. tengcongensis* by horizontal gene transfer, the hypothetic donor has yet to be identified. The ribosomal constraint observed at the level of 2° structures is also evident in 23S rRNA genes in *C. hydrogenoformans, H. marismortui, S. oneidensis*, and *S. pyogenes*.

Complex rearrangement of the position of nucleotides, the position and size of loops, and the length of stems might be another mechanism to conserve the functional structure of an rRNA gene. The examples shown in *N. farcinica* and *C. perfringens* are beyond simple alteration or conservation of the 2° structures since there is no straightforward way to quantify the changes and their effects. Rather, this type of rearrangement seems to aim at conserving the topographic structure in the presence of drastic changes in the primary and secondary structures. However, an experimental approach will be required to address the relationship between the conservation of the topographic structure and function of an rRNA gene.

The concept of rRNA conservation has revolutionized exploration of microbial ecosystems and identifying the causes of diseases of unknown etiologies [Pei et al., 2004; Eckburg et al., 2005; Paster et al., 2001; Relman, 1999]. The most common molecular sequences used for phylogenetic analysis are ribosomal RNAs. The 16S rRNA genes have become the most widely used sequences for phylogenetic analysis as seen by the vast and growing database of 16S rRNA genes [DeSantis et al., 2006; Cole et al., 2009; see also Chapter 36, Vol. I]. The choice of 16S rRNA as the most widely used sequences for phylogenetic analysis has been based on the simple assumption that 16S rRNA is optimal because 5S rRNA is too short while 23S rRNA sequencing requires extra effort without a commensurate increase in phylogenetic signal. However, misidentification has been reported. Strains from different species may have identical 16S rRNA gene sequences, and strains of one species may have 16S rRNA that differ by >4% while 3% is considered as the threshold for within-species dissimilarity [Vandamme et al., 2000; Oberreuter et al., 2002; Harrington and On, 1999]. On the other hand, the main obstacle limiting 23S rRNA from being used in classification of microorganisms has significantly diminished, as the cost for DNA sequencing is becoming less expensive. Evaluation of intragenomic variation of the rRNA genes has become possible since more and more sequences of whole microbial genomes have recently become available. It would be a logical expectation that the intragenomic diversity of a marker for taxonomic classification of microorganisms is below the level of interspecies diversity. We evaluated the suitability of 23S rRNA genes in this regard. Our study indicates that high-level variation among paralogous 23S rRNA genes might exceed the level of variation among 23S rRNA genes from different strains within a species but will not have a significant effect on the integrity of a 23S rRNA-based phylogenetic system, because its occurrence was rare. The insertions encountered when using 23S rRNA genes in phylogenetic analysis can be easily recognized and eliminated in comparisons of homologous sequences. Phylogenetic trees created using similar sets of bacterial 23S and 16S rRNA showed high levels of agreement, and their overall topologies are supported by alternative phylogenetic markers [Ludwig and Schleifer 1994]. 23S rRNA has also been used in correlation with 16S rRNA to confirm specific relations between organisms at multiple different taxonomic levels such as phyla, order, family, and genus [de la Haba et al., 2010a, 2010b; Burggraf et al., 1991; Pilhofer et al., 2008]. Phylogenetic analysis using the 23S rRNA gene has demonstrated a higher resolution in the differentiation of the species of the family Halomonadaceae than analysis using the 16S rRNA gene [de la Haba et al., 2010a].

Intervening sequences (IVS) were frequently observed in 23S rRNA genes. In the seven species in which IVS were identified, five species contain small IVS that did not contain any ORFs; but in *D. radiodurans* and *Nostoc sp*, IVS were large and contained ORFs encoding putative transposases. Experimental studies have demonstrated that IVS in 23S rRNA does not interfere with its functions [Burgin et al., 1990]. The 23S rRNA of *Salmonella typhimurium LT2* are known to carry IVS at two sites, helix-25 and helix-45, which are excised by RNase III during rRNA maturation, resulting in rRNA that is fragmented but nevertheless functional [Burgin et al., 1990]. Fragmentation of the 23S rRNA also has been observed in diverse Eubacteria [Evguenieva-Hackenberg, 2005]. 23S rRNAs with insertions have been shown to rescue an *E. coli* strain in which all chromosomal rRNA operons were deleted [Yokoyama and Suzuki, 2008].

INTERNET RESOURCES

Roadmap Initiative in Human Microbiome Project (http://nihroadmap.nih.gov/hmp/)

Completed Microbial Genomes database (http://www.ncbi.nlm.nih.gov/genomes/MICROBES/Complete.html)

The ARB project (latin, "arbor" = tree) home page (http://rtfm.arb-home.de)

Acknowledgments

We thank Mr. Meisheng Zhou for assistance in verification of sequence diversity.

REFERENCES

Acinas SG, Marcelino LA, Klepac-Ceraj V, Polz MF. 2004. Divergence and redundancy of 16S rRNA sequences in genomes with multiple *rrn* operons. *J. Bacteriol*. **186**:2629–2635.

Asai T, Zaporojets D, Squires C, Squires CL. 1999. An *Escherichia coli* strain with all chromosomal rRNA operons inactivated: complete exchange of rRNA genes between bacteria. *Proc. Natl. Acad. Sci. USA* **96**:1971–1976.

Bao Q, Tian Y, Li W, Xu Z, Xuan Z, et al. 2002. A complete sequence of the *T. tengcongensis* genome. *Genome Res*. **12**:689–700.

Burggraf S, Ching A, Stetter KO, Woese CR. 1991. The sequence of *Methanospirillum hungatei* 23S rRNA confirms the specific relationship between the extreme halophiles and the Methanomicrobiales. *Syst. Appl. Microbiol*. **14**:358–263

Burgin AB, Parodos K, Lane DJ, Pace NR. 1990. The excision of intervening sequences from *Salmonella* 23S ribosomal RNA. *Cell* **60**:405–414.

Cedergren R, Gray MW, Abel Y, Sankoff D. 1988. The evolutionary relationships among known life forms. *J. Mol. Evol*. **28**:98–112.

Coenye T, Vandamme P. 2003. Intragenomic heterogeneity between multiple 16S ribosomal RNA operons in sequenced bacterial genomes. *FEMS Microbiol. Lett*. **228**:45–49.

Cole JR, Wang Q, Cardenas E, Fish J, Chai B, et al. 2009. The Ribosomal Database Project: improved alignments and new tools for rRNA analysis. *Nucleic Acids Res*. **37**: D141–145.

Curtiss WC, Vournakis JN. 1984. Quantitation of base substitutions in eukaryotic 5S rRNA: Selection for the maintenance of RNA secondary structure. *J. Mol. Evol*. **20**:351–361.

de la Haba RR, Arahal M, Marquez MC, Ventosa A. 2010a. Phylogenetic relationships within the family Halomonadaceae based on comparative 23S and 16S rRNA gene sequence analysis. *Int. J. Syst. Evol. Microbiol*. **60**:737–748.

de la Haba RR, Sánchez-Porro C, Marquez MC, Ventosa A. 2010b. Taxonomic study of the genus Salinicola: transfer of *Halomonas salaria* and *Chromohalobacter salarius* to the genus Salinicola as *Salinicola salarius* comb. nov. and *Salinicola halophilus* nom. nov., respectively. *Int. J. Syst. Evol. Microbiol*. **60**:963–971.

DeSantis TZ, Hugenholtz P, Larsen N, Rojas M, Brodie EL, et al. (2006) Greengenes, a chimera-checked 16S rRNA gene database and workbench compatible with ARB. *Appl. Environ. Microbiol*. **72**:5069–5072.

De Rijk P, Van de Peer Y, Van den Broeck I, De Wachter R. 1995. Evolution according to large ribosomal subunit RNA. *J. Mol. Evol*. **41**:366–375.

De Rijk P, De Wachter R. 1997. RnaViz, a program for the visualisation of RNA secondary structure. *Nucleic Acids Res*. **25**:4679–4684.

Duchaud E, Rusniok C, Frangeul L, Buchrieser C, Givaudan A, et al. 2003. The genome sequence of the entomopathogenic bacterium *Photorhabdus luminescens*. *Nat. Biotechnol*. **21**:1307–1313.

Doolittle WF. 1999. Phylogenetic classification and the universal tree. *Science* **284**:2124–2129.

Eckburg PB, Bik EM, Bernstein CN, Purdom E, Dethlefsen L, et al. 2005. Diversity of the human intestinal microbial flora. *Science* **308**:1635–1638.

Eickbush TH, Eickbush DG. 2007. Finely orchestrated movements: evolution of the ribosomal RNA genes. *Genetics* **175**:477–485.

Evguenieva-Hackenberg E. 2005. Bacterial ribosomal RNA in pieces. *Mol. Microbiol*. **57**:318–325.

Gurtler V. 1999. The role of recombination and mutation in 16S–23S rDNA spacer rearrangements. *Gene* **238**:241–252.

Gutell RR, Larsen N, Woese CR. 1994. Lessons from an evolving rRNA: 16S and 23S rRNA structures from a comparative perspective. *Microbiol. Rev*. **58**:10–26.

Harrington CS, On SL. 1999. Extensive 16S rRNA gene sequence diversity in *Campylobacter hyointestinalis* strains: taxonomic and applied implications. *Int. J. Syst. Bacteriol*. 49 Pt 3: 1171–1175.

Higgs PG. 1998. Compensatory neutral mutations and the evolution of RNA. *Genetica* **102–103**:91–101.

Hori H, Osawa S. 1986. Evolutionary change in 5S rRNA secondary structure and a phylogenic tree of 352 5S rRNA species. *Biosystems* **19**:163–172.

Hunt DE, Klepac-Ceraj V, Acinas SG, Gautier C, Bertilsson S, et al. 2006. Evaluation of 23S rRNA PCR primers for use in phylogenetic studies of bacterial diversity. *Appl. Environ. Microbiol*. **72**:2221–2225.

Kaneko T, Nakamura Y, Wolk CP, Kuritz T, Sasamoto S, et al. 2001. Complete genomic sequence of the filamentous nitrogen-fixing cyanobacterium *Anabaena* sp. strain PCC 7120. *DNA Res*. **8**:205–213,227–253.

Klappenbach JA, Saxman PR, Cole JR, Schmidt TM. 2001. rrndb: The ribosomal RNA operon copy number database. *Nucleic Acids Res*. **29**:181–184.

Liao D. 2000. Gene conversion drives within genic sequences: Concerted evolution of ribosomal RNA genes in bacteria and archaea. *J. Mol. Evol*. **51**:305–317.

Ludwig W, Schleifer KH. 1994. Bacterial phylogeny based on 16S and 23S rRNA sequence analysis. *FEMS Microbiol. Rev*. **15**:155–173.

Mathews DH, Disney MD, Childs JL, Schroeder SJ, Zuker M, et al. 2004. Incorporating chemical modification constraints into a dynamic programming algorithm for prediction of RNA secondary structure. *Proc. Natl. Acad. Sci. USA* **101**:7287–7292.

Mattatall NR, Sanderson KE. 1996. *Salmonella typhimurium* LT2 possesses three distinct 23S rRNA intervening sequences. *J. Bacteriol*. **178**:2272–2278.

McClelland M, Sanderson KE, Spieth J, Clifton SW, Latreille P, et al. 2001. Complete genome sequence of *Salmonella enterica* serovar Typhimurium LT2. *Nature* **413**:852–856.

Oberreuter H, Charzinski J, Scherer S. 2002. Intraspecific diversity of *Brevibacterium linens, Corynebacterium glutamicum* and *Rhodococcus erythropolis* based on partial 16S rDNA sequence analysis and Fourier-transform infrared (FT-IR) spectroscopy. *Microbiology* **148**:1523–1532.

Paster BJ, Boches SK, Galvin JL, Ericson RE, Lau CN, et al. 2001. Bacterial diversity in human subgingival plaque. *J. Bacteriol*. **183**:3770–3783.

Pei Z, Bini EJ, Yang L, Zhou M, Francois F, et al. 2004. Bacterial biota in the human distal esophagus. *Proc. Natl. Acad. Sci. USA* **101**:4250–4255.

Pei A, Nossa CW, Chokshi P, Blaser MJ, Yang L, Rosmarin DM, Pei Z. 2009. Diversity of 23S rRNA genes within individual prokaryotic genomes. *PLoS ONE* **4**:e5437.

Pilhofer M, Rappl K, Eckl C, Bauer AP, Ludwig W, Schleifer KH, Petroni G. 2008. Characterization and evolution of cell division and cell wall synthesis genes in the bacterial phyla Verrucomicrobia, Lentisphaerae, Chlamydiae, and Planctomycetes and phylogenetic comparison with rRNA genes. *J Bacteriol* **190**:3192–3202.

Relman DA (1999) The search for unrecognized pathogens. Science **284**:1308–1310.

Santoyo G, Romero D. 2005. Gene conversion and concerted evolution in bacterial genomes. *FEMS Microbiol. Rev*. **29**:169–183.

Schlunzen F, Zarivach R, Harms J, Bashan A, Tocilj A, et al. 2001. Structural basis for the interaction of antibiotics with the peptidyl transferase centre in eubacteria. *Nature* **413**:814–821.

Stackebrandt E, Goebel BM. 1994. Taxonomic note: A place for DNA–DNA reassociation and 16S rRNA sequence analysis in the present species definition in bacteriology. *Int. J. Syst. Bacteriol.* **44**:846–849.

Stewart FJ, Cavanaugh CM. 2007. Intragenomic variation and evolution of the internal transcribed spacer of the rRNA operon in bacteria. *J. Mol. Evol.* **65**:44–67.

Stiegler P, Carbon P, Ebel JP, Ehresmann C. 1981. A general secondary-structure model for procaryotic and eucaryotic RNAs from the small ribosomal subunits. *Eur. J. Biochem.* **120**:487–495.

Thompson JD, Higgins DG, Gibson TJ. 1994. CLUSTAL W: Improving the sensitivity of progressive multiple sequence alignment through sequence weighting, position-specific gap penalties and weight matrix choice. *Nucleic Acids Res.* **22**:4673–4680.

Vandamme P, Harrington CS, Jalava K, On SL. 2000. Misidentifying helicobacters: the *Helicobacter cinaedi* example. *J. Clin. Microbiol.* **38**:2261–2266.

White O, Eisen JA, Heidelberg JF, Hickey EK, Peterson JD, et al. 1999. Genome sequence of the radioresistant bacterium *Deinococcus radiodurans* R1. *Science* **286**:1571–1577.

Woese CR. 1987. Bacterial evolution. *Microbiol. Rev.* **51**: 221–271.

Woese CR, Kandler O, Wheelis ML. 1990. Towards a natural system of organisms: Proposal for the domains Archaea, Bacteria, and Eucarya. *Proc. Natl. Acad. Sci. USA* **87**:4576–4579.

Wood DW, Setubal JC, Kaul R, Monks DE, Kitajima JP, et al. 2001. The genome of the natural genetic engineer *Agrobacterium tumefaciens* C58. *Science* **294**:2317–2323.

Wuyts J, Van de Peer Y, De Wachter R. 2001. Distribution of substitution rates and location of insertion sites in the tertiary structure of ribosomal RNA. *Nucleic Acids Res.* **29**:5017–5028.

Yokoyama T, Suzuki T. 2008. Ribosomal RNAs are tolerant toward genetic insertions: Evolutionary origin of the expansion segments. *Nucleic Acids Res.* **36**:3539–3551.

Zwieb C, Glotz C, Brimacombe R. 1981. Secondary structure comparisons between small subunit ribosomal RNA molecules from six different species. *Nucleic Acids Res.* **9**:3621–3640.

Chapter 4

Use of the rRNA Operon and Genomic Repetitive Sequences for the Identification of Bacteria

ANDRÉA MARIA AMARAL NASCIMENTO

4.1 INTRODUCTION

Historically, scientists have been concerned in arranging organisms into groups based on similarities. Taxonomy is perhaps one of the oldest disciplines of the biological sciences, having begun in Aristotle's time. It provides an accurate communication among scientists facilitating the understanding of organisms and the exchange of knowledge.

Antonie van Leeuwenhoek [1684] was the first person to see bacteria under an amateur microscope. Nevertheless, bacterial taxonomy only had its first impulse in 1880 when [Koch, 1880] described methods for studying bacteria in pure cultures. It began in same way as the taxonomy of animals and plants, and it has been characterized by several milestones that led toward our present insights and ideas about bacteria (**Fig. 4.1**).

The classical methods used in bacterial classification have especially focused on phenotypic comparisons, including morphological, metabolic, physiological, and chemical characteristics of the cells, without considering any phylogenetic relationships. This classification system reflects the technological limitations at the time. The paucity of phenotypic characteristics often causes serious problems in describing or differentiating taxa in several bacterial groups. Nevertheless, the traditional methods guided the bacterial taxonomy for a long time, providing descriptive information of diverse bacterial taxa.

Over the last decades, with the developments in the fields of chemistry, molecular biology, informatics, and statistics, bacterial taxonomy has undergone a deep change [Sneath and Sokal, 1962; Crosa et al., 1973; Sneath, 1995]. Now it employs a molecular approach for the identification of bacteria in a system that relies on the evolutionary relationships among the organisms [Stackebrandt and Goebel, 1987; Vandamme et al., 1996]. In this way, bacterial classification now aims to reflect the biological reality in the best way possible. The introduction of the ribosomal RNA (rRNA) gene sequence and its use as a molecular clock [Woese, 1987] has brought the phylogenetic dimension into bacterial taxonomy. Besides, the polymerase chain reaction (PCR) technology, developed by Kary Mullis [Mullis and Faloona, 1987], and the application of the molecular sequencing method conceived by Fred Sanger have had a strong impact and have been incredibly useful in many areas of science, including, in particular, microbial classification. Both researchers received a Nobel Prize for these achievements. In this chapter I will briefly describe the general concept and applications of 16S rRNA gene sequencing and genomic fingerprinting to assess the taxonomic relationships between prokaryotic species based on these molecular techniques.

Handbook of Molecular Microbial Ecology, Volume I: Metagenomics and Complementary Approaches, First Edition. Edited by Frans J. de Bruijn.

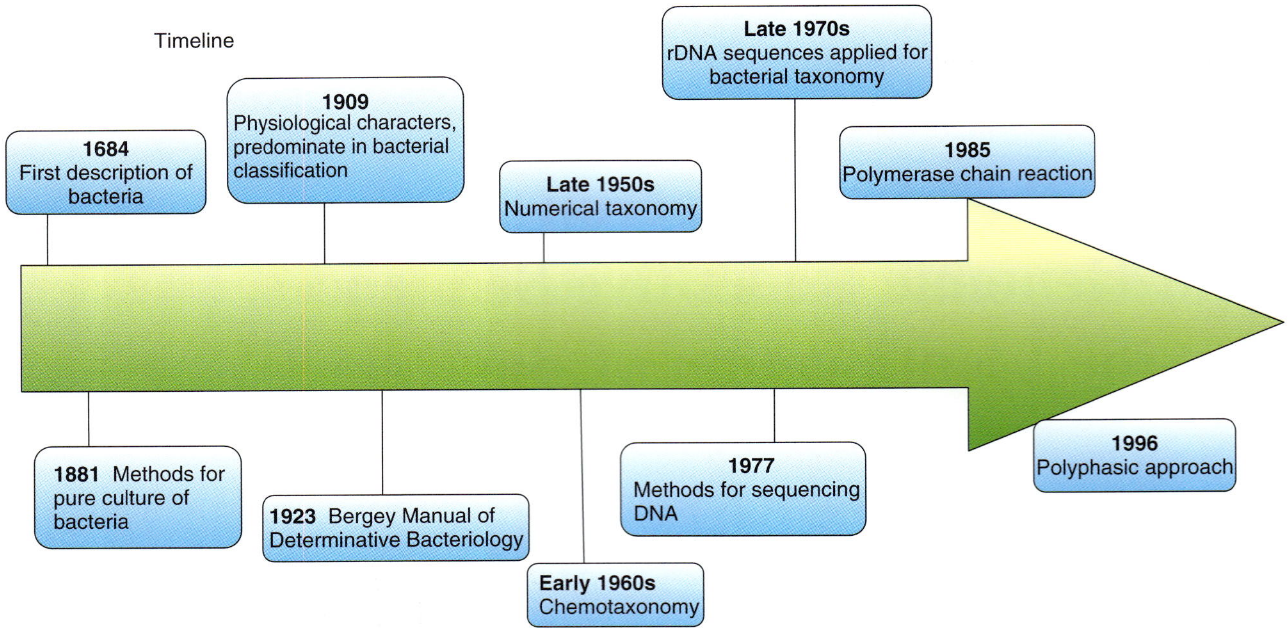

Figure 4.1 Some landmarks of bacterial taxonomy.

4.2 TAXONOMIC RELATIONSHIPS AMONG BACTERIA BASED ON THE rRNA OPERON AND GENOMIC FINGERPRINTING TECHNIQUES

4.2.1 The rRNA Operon

The works of Woese and his colleagues, in which they comparatively analyzed the 16S ribosomal RNA (rRNA) genes sequences from culturable microorganisms, were a landmark in biology. The results of these molecular studies radically transformed our concepts of classification, revolutionizing our insights of bacterial taxonomy and phylogeny. Since then, we have seen an explosion in the applications of molecular tools for elucidating the classification of bacteria, previously dependent on cellular metabolism, and the old determinative classification is rapidly being replaced by a phylogenetics based on nucleotide sequence comparisons. Many of these approaches were predominantly rooted in the use of rRNA genes. Currently, the 16S rRNA approach is one of the most widely used standard techniques in microbial taxonomy (see Chapter 15, Vol. I).

Ribosomal RNAs are ancient molecules, key elements of the protein-synthesizing machinery and of broad evolutionary interest in both prokaryotes and eukaryotes. There are good reasons for rRNA being a useful and reliable molecular marker for phylogenetic analyses: They are universally distributed, functionally constant, sufficiently conserved, and of adequate length, and they are almost unaffected by horizontal genetic transference mechanisms [Woese, 1987]. Additionally, since the 1980s, thousands of rRNA sequences have been added to the databases and are available in widely accessible databases. For example, the Ribosomal Database Project II has gathered a collection of over 1,498,677 sequences (see Chapter 36, Vol. I); and the GenBank, the largest databank of nucleotide sequences, has over 2,818,654 deposited sequences of 16S rRNA. The use of rRNA gene sequences for the classification of bacteria is becoming popular among bacterial taxonomists. Most widely used and useful for defining bacterial identity is the small subunit ribosomal 16S rRNA gene (see Chapter 15, Vol. I). Unfortunately, the resolving power of the 16S rRNA gene is limited when closely related organisms are examined because rRNA sequences are too conserved due to their essential functions and thus not suitable to resolve between similar species [Pontes et. al, 2007]. Nevertheless, despite the intrinsic pitfalls, bacterial taxonomists regard 16S rRNA sequence analysis as suitable for studies of prokaryotic phylogeny and classification. The 23S large subunit rRNA is also phylogenetically highly informative due to its longer sequence, but its length makes its use more costly (see Chapter 3, Vol. I). In contrast, the 5S rRNA gene, which was the first rRNA gene analyzed, is not sufficiently informative for reliable phylogenetic analyses due to its short length.

The rRNA genetic locus *rrn* contains the genes encoding the 16S (*rrs*), 23S (*rrl*) and 5S (*rrf*) rRNAs, usually in this order. Their lengths are about 120, 1650, and 3000 nucleotides, respectively, and they are typically organized into an operon in bacteria (**Fig. 4.2**). The copy number of rRNA operons per bacterial genome varies from 1 to 15 copies [Gürtler and Stanisich, 1996). Slow-growing bacteria have been shown to have fewer copies of this operon than fast-growing bacteria [Krawiec and Riley, 1990]. Thus, the number of copies of the rRNA operon is a component of bacterial life histories, reflecting bacterial ecological strategies [Stevenson and Schmidt, 2004].

The 16S and 23S rRNA genes are separated by a short spacer region, containing tRNA genes and some target sequences for ribonuclease (RNase) III and other recognition signals required for the transcript splicing [Bram et al., 1980]. This region is transcribed together with the ribosomal genes and is called the internal transcribed spacer (ITS) region. It exhibits a significant degree of variation in sequence and length (intercistronic heterogeneity) at the genus and species levels, even among closely related taxa [Barry et al., 1991; Gürtler and Stanisich, 1996; Garcia-Martinez et al., 1999]. These length variations in the ITS region are mainly due to the number and type of tRNA genes that it may contain, which vary considerably among species, among strains of the same species, and even among different operons within a single cell [Condon et al., 1995].

The *rrn* loci have one or two tRNA genes between segments encoding 16S and 23S rRNA and zero to two tRNA genes beyond the segments encoding 5S rRNA; usually no *rrn* locus codes for more than three tRNAs (Fournier and Ozeki, 1985). Recently, a larger ITS containing four tRNA genes coding for $tRNA^{Glu}$, $tRNA^{Lys}$, $tRNA^{Val}$, and $tRNA^{Ala}$ has been described in *Photobacterium damselae* species [Osorio et al., 2005]. Most gram-negative bacteria contain $tRNA^{Ala}$ and $tRNA^{Ile}$ genes, whereas some contain only a $tRNA^{Glu}$ gene. However, new combinations of tRNA genes have been reported [Chun et al., 1999; Lee et al., 2002]. The 16S–23S rDNA spacer region of gram-positive bacteria, in contrast, contains either a $tRNA^{Ala}$, a $tRNA^{Ile}$ gene, or both. In the *Bacillus cereus* group, operons exhibit multiple tRNA genes downstream of the 5S rRNA gene and a putative promoter sequence

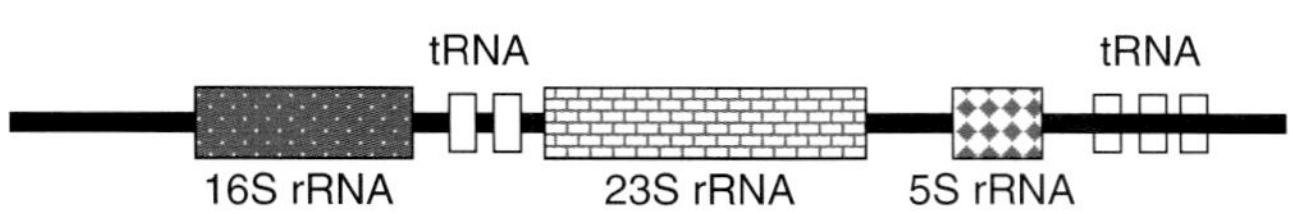

Figure 4.2 Schematic representation of an rRNA operon. The number and type of tRNA genes in the spacer region may vary among species.

in the 23S–5S intergenic region. This suggests that the 5S rRNA and the downstream tRNA genes can be transcribed independently of the 16S and 23S genes [Candelon et al., 2004]. The tRNA genes occur in multiple copies dispersed throughout the genome. The 16S–23S rRNA ITS region exhibits polymorphisms in both length and nucleotide sequence, and this diversity represents a powerful tool to discriminate closely related species or even strains of single species on the basis of the band patterns obtained by polymerase chain reaction (PCR) amplification of the spacer regions, providing higher taxonomic resolution than 16S rRNA genes. Analysis of ITS–PCR patterns has been successfully used to provide adequate discrimination of closely related bacterial species and can reveal intraspecies polymorphisms [Lu et al., 1997; Tilsala-Timisjärvi and Alatossava 1997; Chen et al., 2006; Rocap et al., 2002].

4.2.2 PCR-Based Genetic Fingerprinting Techniques

The 16S rRNA genes sequence analyses have revealed that its resolution, although valuable for classifying bacteria at the genus level, is low for phylogenetic or taxonomic discrimination of bacteria at the species level [Woese, 1987; Fox et al., 1992; Hauben et al., 1997] and thus several alternative or complementary methods have been developed. A better discrimination between species has been achieved by using several highly powerful molecular tools such as PCR-based genetic fingerprinting techniques [see Chapters 3 and 5–7, Vol. I]. These methods have been successfully applied for the differentiation of species of the same genus and are well-suited to distinguish between genotypically related strains of a broad range of bacterial species, detecting the intraspecific diversity. These techniques rely on the electrophoretic separation and subsequent visualization of the differently amplified-sized DNA fragments, thus generating a species-specific or strain-specific genomic fingerprint pattern, which can be analyzed using computer programs and employed to generate phylogenetic trees [Rademaker et al., 2004; see Chapter 8, Vol. I]. One of these fingerprinting methods (rep-PCR), based on the analysis of the whole genome, constitutes a particularly valuable tool. An example of this kind of analysis is illustrated in **Figure 4.3**.

Genomic fingerprints focused on the amplification of highly conserved regions between repetitive DNA sequences randomly interspersed throughout the bacterial genomes, called repetitive sequence-based PCR (rep-PCR) [Versalovic et al., 1994; see Ishii and Sadowsky, 2008], or on amplification of random sequences by arbitrary primers [random amplified polymorphic DNA-PCR (RAPD)] [Williams et al., 1990]. The advantages of repetitive elements sequence-based PCR genomic

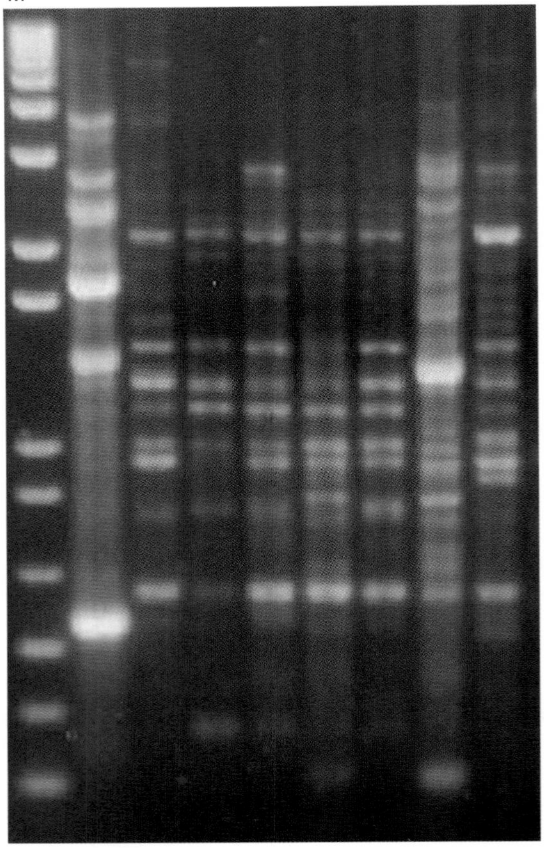

Figure 4.3 DNA fingerprinting with $(GTG)_5$-PCR. Genomic DNAs from eight *Bacillus* isolates of the same species were PCR-amplified using the $(GTG)_5$ primer and the amplicons were separated by gel electrophoresis on the basis of size to generate DNA fingerprints. Lane M, molecular size marker (1 kb Plus-Invitrogen). (Modified from Freitas et al. [2008]).

fingerprinting are its specificity, simplicity, and accuracy for discriminating between a wide range of bacterial genera and species. Moreover, it is relatively less laborious, quicker, and cheaper because it does not require the sequencing and sequence alignment steps used for phylogenetic analysis. These characteristics are ideal for high-throughput analyses. Although the functions of these repetitive sequences remain unclear, they seem to be involved in important RNA and DNA processing.

The intraspecific variation is usually revealed by the different numbers of basic DNA repeat units. The rep-PCR DNA fingerprint is a genotypic method that uses oligonucleotide primers complementary to short intergenic repetitive sequences dispersed throughout the genome of prokaryotes [Versalovic et al., 1991]. Several of these interspersed repetitive elements are conserved in diverse bacteria genera, thereby allowing a single primer set to be used for DNA fingerprinting of many different microorganisms [Versalovic et al., 1994].

The conserved repetitive sequences commonly used for the rep-PCR DNA fingerprinting encompass four types: the repetitive extragenic palindromic (REP) sequences, the enterobacterial repetitive intergenic consensus (ERIC) sequences, the BOX sequences, and the polytrinucleotide $(GTG)_5$ sequences [Versalovic et al., 1994]. Two repetitive families, REP and ERIC, have been found broadly distributed among phylogenetically diverse bacteria [Tobes and Ramos, 2005]. They have been found in many members of the Enterobacteriaceae family and have been extensively used in molecular typing studies on a wide range of gram-negative bacteria [Stern et al., 1984; Hulton et al., 1991; Versalovic et al., 1991; Ishii and Sadowski, 2009; see also Chapter 8, Vol. I]. In contrast, the BOX-repeat motif was first identified in *Streptococcus pneumoniae* and has been used for molecular typing of many gram-positive bacteria [Martin et al., 1992]. Most rep-PCR-based DNA fingerprinting studies have used short polytrinucleotides, such as $(GTG)_5$, 35- to 40-bp REP sequences, 124- to 127-bp ERIC sequences, or the BOX elements (154 bp) composed of the three subunits boxA, boxB, and boxC, which are 59, 45, and 50 nucleotides long, respectively, as priming sites for PCR [Martin et al., 1992; Versalovic et al., 1991, 1994].

Primer sets based on the repetitive elements present in the bacterial genome yield strain-specific genomic fingerprints of bacteria. The primer sets Rep1R-I and Rep2-I ERIC-PCR (primer sets ERIC1R and ERIC2), ERIC2-PCR (primer ERIC2), BOX-PCR (primer BOX A1R), and $(GTG)_5$-PCR [primer $(GTG)_5$] are commonly used for genotyping different bacterial strains [Versalovic et al., 1994]. Amplification of the regions between adjacent repetitive extragenic elements produces strain-specific DNA fingerprints [see also Ishii and Sadowsky, 2009]. In contrast, 16S rRNA gene sequencing is more useful than rep-PCR in distinguishing bacteria at the generic or species level. Rep-PCR fingerprinting with a combination of three primer sets has been shown to generate the same groupings as DNA–DNA reassociation kinetics (Rademaker et al., 2000; 2004).

4.3 DIFFERENT APPROACHES FOR IDENTIFICATION OF *BACILLUS* SPECIES

For a full description of the Materials and Methods, please see the original publication [Freitas et al., 2008].

4.3.1 Material and Methods

The primers and the amplification cycling conditions for the rep-PCR, tDNA, and ITS-PCR are shown in **Table 4.1**.

Table 4.1 Primer Sequences and Amplification Cycling Conditions for the Different PCR-Based Genomic DNA Fingerprints

Primer Set	Primer	Nucleotide Sequence 5′–3′	References	Primer Concentration (μM)	Thermal Cycling Conditions
tDNA	T5A	AGTCCGGTGCTCTAACCAACTGAG	Welsh and McClelland [1991]	0.5	Initial denaturation 94°C for 10 min, 30 cycles of 94°C for 30 s, 50°C for 30 s, 72°C for 1 min, final extension at 72°C for 10 min
	T3B	AGGTCGCGGGTTCGAATCC	Welsh and McClelland [1991]		
ITS	L1	CAAGGCATCCACCGT	Clementino et al. [2001]		
	G1	GAAGTCGTAACAAGG	Clementino et al. [2001]		
ERIC	ERIC2	AAGTAAGTGACTGGGGTGAGCG	Versalovic et al. [1991]	0.5	Initial denaturation 94°C for 5 min, 30 cycles of 94°C for 1 min, 52°C for 1 min, 72°C for 3 min, final extension at 72°C for 10 min
BOX	BOXA1R	CATACGGCAAGGCGACGCT	Versalovic et al. [1991]	0.4	
GTG	GTG_5	GTGGTGGTGGTGGTG	Versalovic et al. [1991]	0.3	

Source: Modified from Freitas et al. [2008].

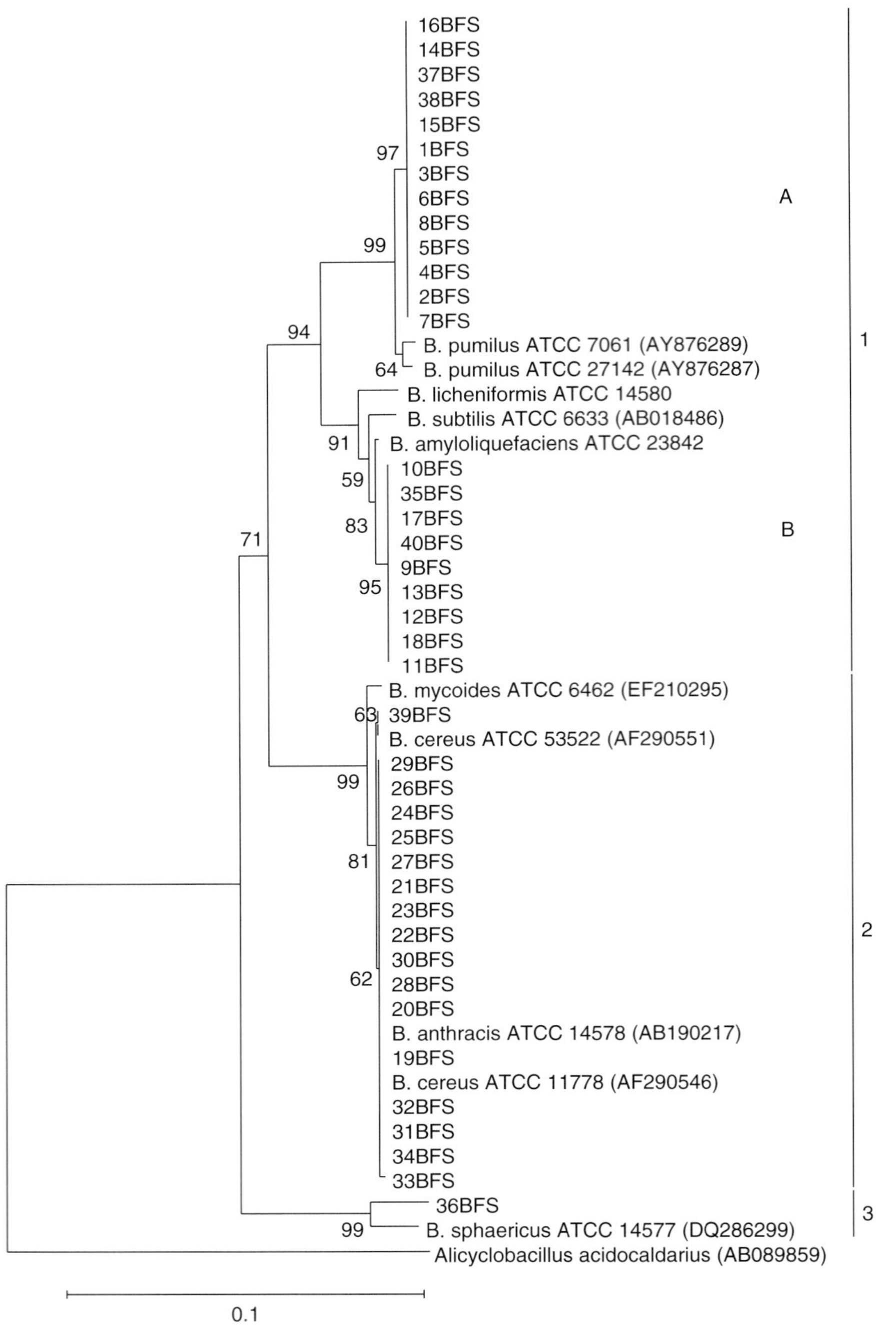

Figure 4.4 Phylogenetic tree of members of the genus *Bacillus*, based on 16S rRNA gene sequences. Numbers at nodes indicate percentages of occurrence in 1000 bootstrapped trees. *Alicyclobacillus acidocaldarius* (AB089859) was used as an outgroup. Reproduced with permission from Freitas et al. [2008].

The phylogenetic tree was constructed using the neighbor-joining method, and genetic distances were computed using Kimura's model [Kimura, 1980].

Cluster analysis of the pairwise similarity values was performed by using the UPGMA (unweighted pair group method using averages) algorithm. Analysis of data was performed using the software PAST 9 (Paleontological Statistics Software Package) [Hammer et al., 2001].

4.4 RESULTS AND DISCUSSION

In the previous section I presented an overview of concepts and applications of the 16S rRNA gene sequencing and fingerprinting techniques, particularly of the ITS-PCR and rep-PCR genomic fingerprinting. In this section I will present some studies that have evaluated the applicability of these techniques for bacterial taxonomy.

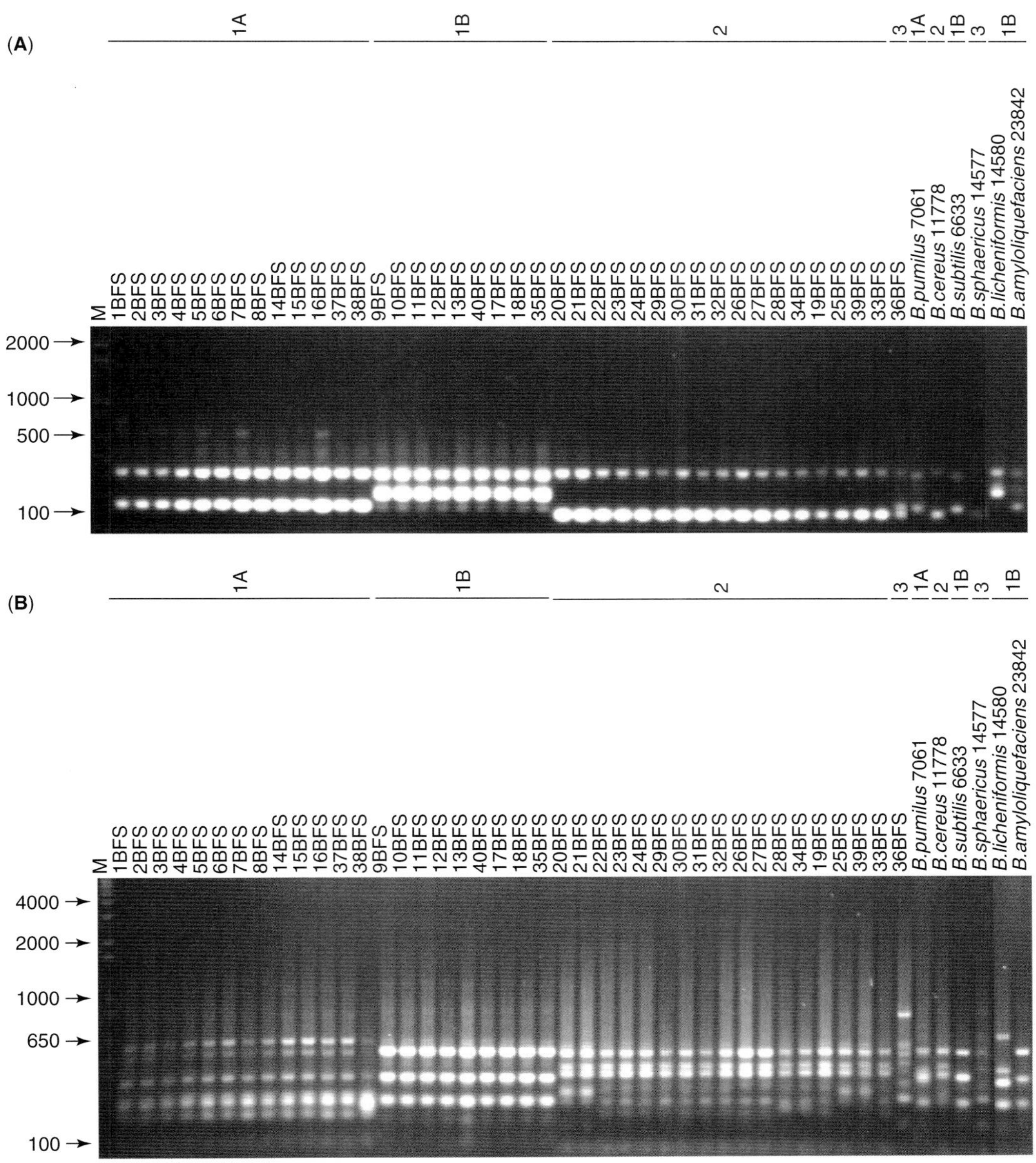

Figure 4.5 (**A**) ITS- and (**B**) tDNA-PCR fingerprinting of isolates and reference strains. Lane M, molecular size marker (1 kb Plus-Invitrogen). The numbers above the figure identify the 16S rDNA-sequence-based phylogeny clusters obtained for the isolates. (Reproduced from Freitas et al. [2008]).

I will especially focus on genera in which the use of 16S rRNA proved less useful, requiring a more comprehensive phylogenetic analysis.

A high similarity has been found among the 16S rRNA sequences of some closely related organisms for which there was no resolution at the species level. One well-characterized example relates to members of the *Bacillus cereus* group (*Bacillus anthracis*, *B. cereus*, *B. thuringiensis*, *B. mycoides*, *B. pseudomycoides*, and *B. weihenstephanensis*) and the *Bacillus subtilis* group (*B. subtilis*, *B. pumilus*, *B. atrophaeus*, *B. licheniformis*, and *B. amyloliquefaciens*). These organisms exhibit almost identical 16S rRNA gene sequences; with intragroup similarities of 99.5% to 100% and 99.2% to 99.6%, respectively [Ash et al., 1991; Fox et al., 1992, Nakamura et al., 1999]. A high similarity of 16S rRNA sequences has also been described in *Enterobacter, Ochrobactrum* [Hoffmann and Roggenkamp, 2003; Lebuhn et al., 2006] and *Chromobacterium* [Martin et al., 2007; Kämpfer et al., 2009].

The use of ITS-PCR, tDNA-PCR and rep-PCR genomic fingerprinting has been particularly useful to

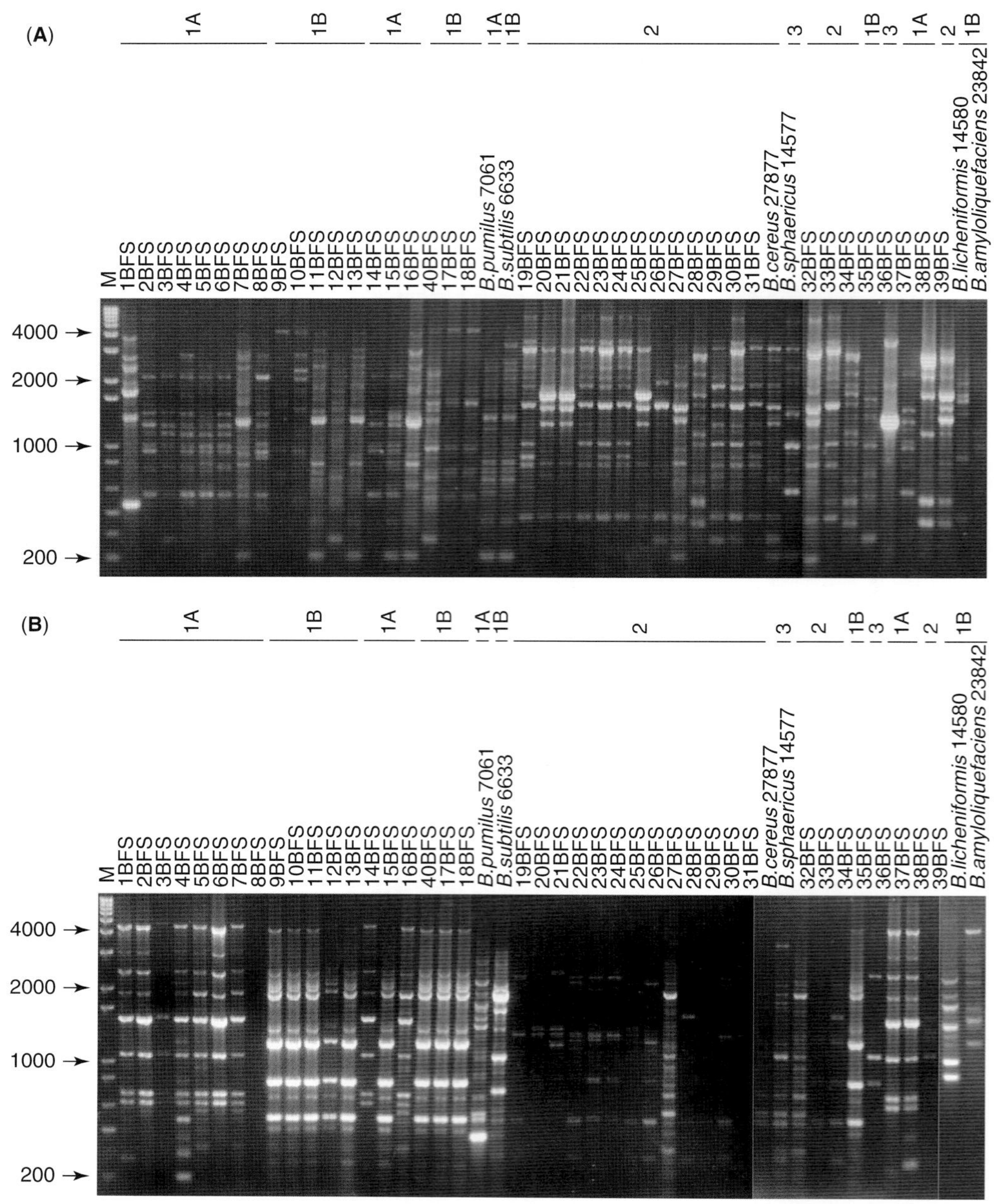

Figure 4.6 (**A**) (GTG)5-PCR. (**B**) BOX-PCR fingerprinting patterns of the isolates and reference strains. Lane M, molecular size marker (1 kb Plus-Invitrogen). The numbers above the figure identify the 16S rDNA-sequence-based phylogeny clusters obtained for the isolates. (Modified from Freitas et al. [2008]).

examine phylogenetic relationships among strains of species that were indistinguishable by their 16S rRNA sequences.

Freitas et al. [2008] used 16S rDNA sequencing as a first identification tool for 40 bacterial isolates from a steel plant waste (**Fig. 4.4**).

The affiliation of species was further compared by three different molecular typing methods: ITS-PCR, tDNA-PCR, and rep-PCR genomic fingerprinting. The *Bacillus* isolates, reunited on the basis of a 16S rRNA phylogeny, had identical ITS-PCR and tDNA-PCR patterns, except for the tDNA-PCR of four isolates of the *B. cereus* group (20BFS, 21BFS, 25BFS, and 39BFS, **Fig. 4.5**). Interestingly, the patterns detected differed from those of all the type strains (*B. subtilis* ATCC 6633^{T}, *B. licheniformis* ATCC 14580^{T}, *B. pumilus* ATCC 7061^{T}, *B. amyloliquefaciens* ATCC 23842, *B.cereus* ATCC 11778, and *B. sphaericus* ATCC 14577^{T}), which were included as references. rep-PCR genomic fingerprinting was applied to discriminate the *Bacilus* isolates that were

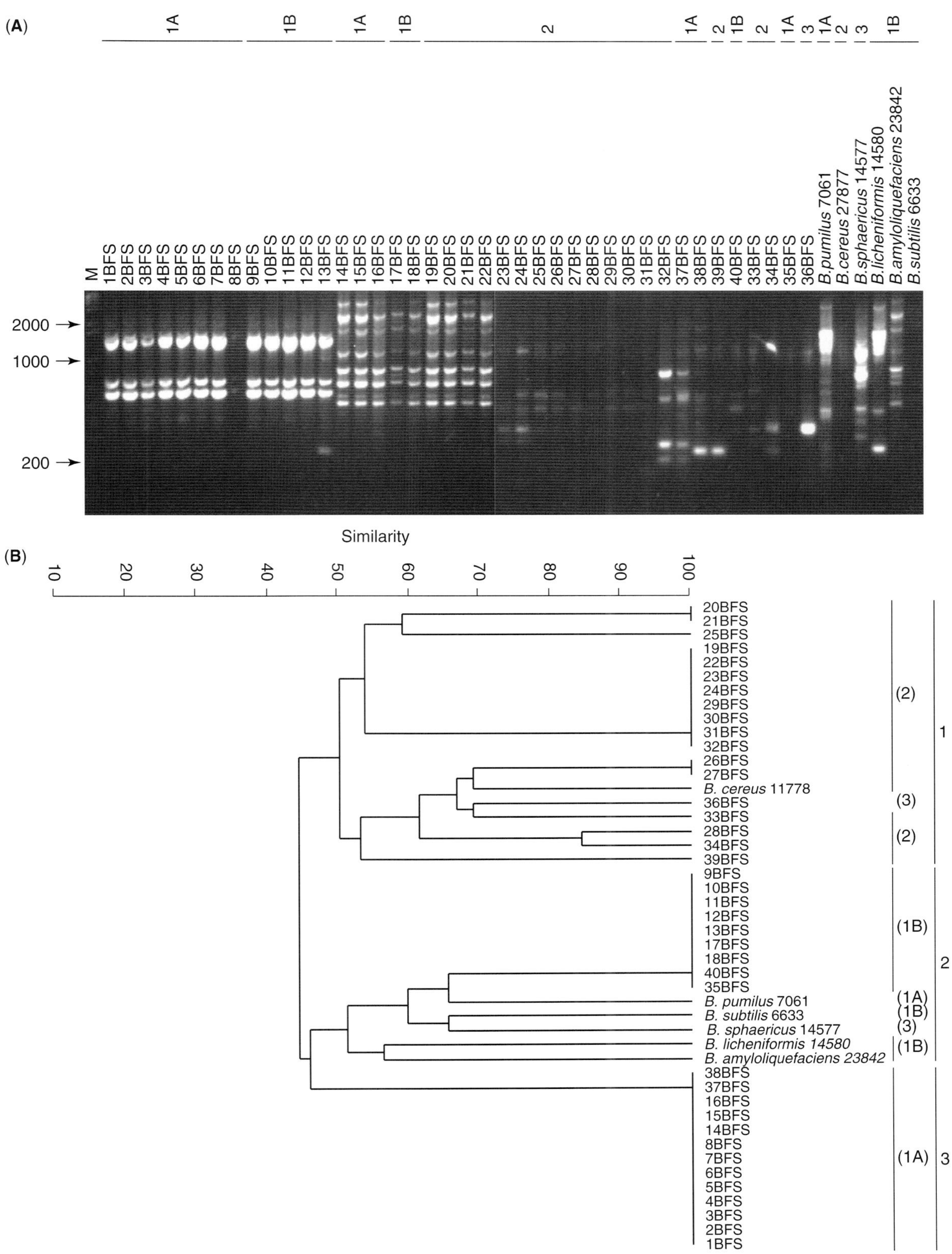

Figure 4.7 (**A**) ERIC-PCR fingerprinting patterns of the isolates and reference strains. Lane M, molecular size marker (1 kb Plus-Invitrogen). The numbers above the figure identify the 16S rDNA-sequence-based phylogeny clusters obtained for the isolates. (**B**) UPGMA cluster analysis of isolates and reference strains based on ERIC-PCR. Numbers in parentheses identify the 16S rDNAsequence-based phylogeny clusters obtained for the isolates. (Modified from Freitas et al. [2008]).

likely to represent the same species. The primer specific for $(GTG)_5$ was less efficient than the BOX and ERIC primer sets for grouping the isolates, since the topology of the dendrogram obtained by $(GTG)_5$-PCR fingerprinting was not congruent with the clusters obtained by the 16S rRNA gene tree. On the other hand, $(GTG)_5$-PCR and BOX-PCR fingerprinting patterns revealed a high genomic diversity among the investigated isolates. The $(GTG)_5$-PCR exhibited a higher resolution (more bands) than the other PCR-based approaches, making it an ideal tool for ecological studies (**Fig. 4.6**).

Interestingly, the analysis by ERIC-PCR genomic fingerprinting revealed a clustering of isolates similar to the one observed in the 16S rDNA-based phylogenetic tree (**Fig. 4.7**). Moreover, ERIC-PCR fingerprinting made it possible to cluster reference and environmental *B. cereus* strains in accordance with the 16S rDNA tree. Similarity analysis of ERIC-PCR profiles grouping is hence an alternative to 16S rRNA gene sequencing as a genomic criterion for species delineation within the genus *Bacillus* and particularly for the *Bacillus cereus* group.

Taken together, ITS-PCR, tDNA-PCR, and ERIC-PCR fingerprinting were as discriminatory as the 16S rRNA gene sequencing analysis. It should be pointed out that all reference strains had unique profiles in all genomic fingerprinting analyses and that their profiles were different from the patterns obtained for the *Bacillus* isolates. Miranda et al. [2008] also observed that the tDNA-PCR approach had the same sensitivity and resolution power of the 16S rRNA sequencing analyses for the identification of the *Bacillus* marine isolates. Differently from the study of Freitas et al. [2008], the reference strains used by Miranda et al. [2008] exhibited the same tDNA-PCR fingerprint pattern of the *Bacillus* marine isolates investigated.

Intraspecific variation in genome size is common in bacteria, as exemplified by *Escherichia coli*, whose genome may differ by over 1 Mb among natural isolates [Parkhill and Thomson, 2003]. The origin of the *Bacillus* isolates could also explain the diversity observed among natural isolates and reference strains in this study.

INTERNET RESOURCES

Ribosomal Database Project II (http://rdp.cme.msu.edu)

GenBank (http://www.ncbi.nlm.nih.gov)

Acknowledgments

I thank the authors from Freitas et al. [2008] and BioMed Central for the permission to use original figures, and I am grateful to Professor Edmar Chartone-Souza and Dr. Marta Svartman for their useful comments. This work was supported by Fundaçao de Amparo à Pesquisa do Estado de Minas Gerais (FAPEMIG) and Conselho Nacional de Desenvolvimento Cientfico e Tecnolgico (CNPq).

REFERENCES

Ash C, Farrow JAE, Wallbanks S, Collins MD. 1991. Phylogenetic heterogeneity of the genus *Bacillus* revealed by comparative analysis of small-subunit-ribosomal RNA sequences. *Lett. Appl. Microbiol.* **13**:202–206.

Barry TG, Colleran G, Glennon M, Dunican LK, Gannon F. 1991. The 16S/23S ribosomal spacer region as a target for DNA probes to identify eubacteria. PCR *Methods Appl.* **1**:51–56.

Bram RJ, Young RA, Steitz JA. 1980. The ribonuclease III site flanking 23S sequences in the 30S ribosomal precursor RNA of *Escherichia coli*. *Cell* **19**:393–401.

Candelon B, Guilloux K, Ehrlich SD, Alexei Sorokin. 2004. Two distinct types of rRNA operons in the *Bacillus cereus* group. *Microbiology* **150**:601–611.

Chen F, Wang K, Kan J, Suzuki MT, Wommack KE. 2006. Diverse and unique picocyanobacteria in Chesapeake Bay, revealed by 16S-23S rRNA internal transcribed spacer sequences. *Appl. Environ. Microbiol*. **72**:2239–2243.

Chun J, A. Huq, and R. R. Colwell. 1999. Analysis of 16S–23S rRNA intergenic spacer regions of *Vibrio cholerae* and *Vibrio mimicus*. *Appl. Environ. Microbiol*. **65**:2202–2208.

Clementino MM, Filippis I, Nascimento CR, Branquinho R, Rocha CL, Martins OB. 2001. PCR analysis of t-RNA intergenic spacer, 16S–23S internal transcribed spacer, and randomly amplified polymorphic DNA reveal inter- and intraspecific relationships of *Enterobacter cloacae* strains. *J. Clin. Microbiol*. **39**:3865–3870.

Condon C, Squires C, Squires CL. 1995. Control of rRNA transcription in *Escherichia coli*. *Microbiol. Rev*. **59**:623–645.

Crosa JH, Brenner DJ, Falkow S. 1973. Use of a single-strand specific nuclease for analysis of bacterial and plasmid deoxyribonucleic acid homo- and heteroduplexes. *J. Bacteriol*. **115**:904–911.

Fournier MJ, Ozeki H. 1985. Structure and organization of the transfer ribonucleic acid genes of *Escherichia coli* K-12. *Microbiol. Rev*. **49**:379–397.

Fox GE, Wisotzkey JD, Jurtshuk PJ. 1992. How close is close: 16S rRNA sequence identity may not be sufficient to guarantee species identity. *Int. J. Syst. Bacteriol*. **42**:166–170.

Freitas DB, Reis MP, Lima-Bittencourt CI, Costa PS, Assis PS, Chartone-Souza E, Nascimento AMA. 2008. Genotypic and phenotypic diversity of Bacillus spp. isolated from steel plant waste. *BMC Res. Notes* **I**: 92.

Garcia-Martinez J, Acinas SG, Anton IA, Rodriguez-Valera F. 1999. Use of the 16S–23S ribosomal genes spacer region in studies ofprokaryotic diversity. *J. Microbiol. Methods* **36**:55–64.

Gürtler V, Stanisich VA. 1996. New approaches to typing and identification of bacteria using the 16S–23S rDNA spacer region. *Microbiology* **142**:3–16.

Hammer Ø, Harper DAT, Ryan PD. 2001 PAST: Paleontological statistics software 20 package for education and data analysis. *Palaeontol. Electronic* **4**:1–9.

Han XY, Han FS, Segal J. 2008. *Chromobacterium haemolyticum* sp. nov., a strongly haemolytic species. *Int. J. Syst. Evol. Microb*. **58**:1398–1403.

Hauben L, Vauterin L, Swings J, Moore ERB. 1997. Comparison of 16S ribosomal DNA sequences of all *Xanthomonas* species. *Int. J. Syst. Bacteriol*. **47**:328–335.

Hoffmann, H., and Roggenkamp, A. 2003. Population genetics of the nomenspecies *Enterobacter cloacae*. *Appl. Environ. Microbiol*. **69**:5306–5318.

Hulton CSJ, Higgins CF, Sharp PM. 1991. ERIC sequences: a novel family of repetitive elements in the genomes of *Escherichia coli, Salmonella typhimurium* and other enteric bacteria. *Mol. Microbiol*. **5**:825–834.

Ishii S, Sadowsky MJ. 2008. *Escherichia coli* in the environment: implications for water quality and human health. *Microbes Environ*. **23**:101–108.

Ishii S, Sadowsky MJ. 2009. Applications of the rep-PCR DNA fingerprinting technique to study microbial diversity, ecology and evolution. *Environ. Microbiol*. **11**:733–740.

Kämpfer P, Busse HJ, Scholz H. 2009. *Chromobacterium piscinae* sp. nov., and *Chromobacterium pseudoviolaceum* sp. nov., two novel *Chromobacterium* species from environmental samples. *Int. J. Syst. Evol. Microbiol*. **59**:2486–2490.

Kimura M. 1980. A simple method for estimating evolutionary rates of base substitutions through comparative studies of nucleotide sequences. *J. Mol. Evol*. **16**:111–120.

Koch, R. 1880. *Investigations into the Etiology of Traumatic Infective Diseases*. London: New Sydenham Society.

Krawiec S, Riley M. 1990. Organization of the bacterial chromosome. *Microbiol. Rev*. **54**:502–539.

Lebuhn M, Bathe S, Achouak W, Hartmann A, Heulin T, Schloter M (2006) Comparative sequence analysis of the internal transcribed spacer of *Ochrobactrum* species. *Syst. Appl. Microbiol*. **29**:265–275

Lee, SKY, Wang HZ, Law SHW, Wu RSS, and Kong RYC. 2002. Analysis of the 16S–23S rDNA intergenic spacers (IGSs) of marine vibrios for species-specific signature DNA sequences. *Marine Pollut. Bull*. **44**:412–420.

Lu W, Evans EH, McColl SM, Saunders VA. 1997. Identification of cyanobacteria by polymorphisms of PCR amplified ribosomal DNA spacer region. *FEMS Microbiol. Lett*. **153**:141–149.

Martin B, Humbert O, Camara M, Guenzi E, Walker J, Mitchell T, Andrew P, Prudhomme M, Alloing G, Hakenbeck R, Morrison DA, Boulnois GJ, Claverys JP 1992. A highly conserved repeated DNA element located in the chromosome of *Streptococcus pneumoniae*. *Nucleic Acids Res*. **20**:3479–3483.

Martin PAW, Gundersen-Rindal D, Blackburn M, Buyer J. 2007. *Chromobacterium subtsugae* sp. nov., a betaproteobacterium toxic to Colorado potato beetle and other insect pests. *Int. J. Syst. Evol. Microbiol*. **57**:993–999.

Miranda CAC, Martins OB, Clementino MM. 2008. Species-level identification of *Bacillus* strains isolates from marine sediments by conventional biochemical, 16S rRNA gene sequencing and inter-tRNA gene sequence lengths analysis. *Antonie van Leeuwenhoek* **93**:297–304.

Mullis KB, Faloona FA. 1987. Specific synthesis of DNA *in vitro* via a polymerase catalyzed chain reaction. *Methods Enzymol*. **155**:335–350.

Nakamura LK, Roberts MS, Cohan FM. 1999. Relationship of *Bacillus subtilis* clades associated with strains 168 and W23: A proposal for *Bacillus subtilis* subsp. subtilis subsp. nov. and *Bacillus subtilis* subsp. spizizenii subsp. nov. *Int. J. Syst. Bacteriol*. **49**:1211–1215.

Osorio CR, Collins MD, Romalde JL, Toranzo AE. 2005. Variation in 16S–23S rRNA intergenic spacer regions in *Photobacterium damselae*: A mosaic-like structure. *Appl. Environ. Microbiol*. **2**:636–645.

Pontes DS, Lima-Bittencourt, C, Chartone-Souza E, Nascimento AMA. 2007. Molecular approaches: Advantages and artifacts in assessing bacterial diversity. *J. Ind. Microbiol. Biotechnol*. **34**:463–473.

Rademaker JL, Hoste B, Louws F, Kersters K, Swings Vauterin L, Vauterin P, de Bruijn FJ. 2000. Comparison of AFLP and rep-PCRgenomic fingerprinting with DNA-DNA homology studies: *Xanthomonas* as a model. *Int. J. Syst. Evol. Microbiol*. **50**:665–667.

Rademaker JL, Louws V, de Bruijn FJ. 2004. Characterization of the diversity of ecologically important microbes by rep-PCR genomic fingerprinting. In Molecular Microbial Ecology Manual, Kowalchuk George A, de Bruijn Frans J, Antoon Ian Head, Akkermans DL, Van Elsas JD, Eds. Environmental Molecular Microbiology. Kluwer Academic Press, pp. 611–643.

Rocap G, Distel DL, Waterbury JB, Chisholm SW. 2002. Resolution of *Prochlorococcus* and *Synechococcus* ecotypes by using 16S–23S ribosomal DNA internal transcribed spacer sequences. *Appl. Environ. Microbiol*. **68**:1180–1191.

Sneath PH, Sokal RR. 1962. Numerical taxonomy. *Nature* **193**:855–860.

Sneath P H. 1995. Thirty years of numerical taxonomy. *Syst. Biol*. **44**:281–298.

Stackebrandt E., Goebel BM. 1987. A place for DNA–DNA reassociation and 16S rRNA sequence analysis in the present species definition in bacteriology. *Int. J. Syst. Bacteriol*. **44**:846–849.

Stern MJ, Ames GFL, Smith NH, Robinson EC, Higgins CF. 1984. Repetitive extragenic palindromic sequences: A major component of the bacterial genome. *Cell* **37**:1015–1026.

Stevenson BS, Schmidt TM. 2004. Life history implications of rRNA gene copy number in *Escherichia coli*. *Appl. Environ. Microbiol*. **70**:6670–667.

Tilsala-Timisjärvi, A, Alatossava T. 1997. Development of oligonucleotide primers from the 16S–23S rRNA intergenic sequences for identifying different dairy and probiotic lactic acid bacteria by PCR. *Int. J. Food Microbiol*. **35**:49–56.

Tobes R, Ramos JL. (2005) REP code: Defining bacterial identity in extragenic space. *Environ. Microbiol*. **7**:225–228.

Vandamme P, Pot B, Gillis M, de Vos P, Kersters K, Swings J. 1996. Polyphasic taxonomy, a consensus approach to bacterial systematics. *Microbiol. Rev*. **60**:407–438.

van Leeuwenhoek A. 1684. An abstract of a Letter from Mr. Anthony Leeuwenhoek at Delft, dated September 17, 1683, containing some Microscopical Observations, about Animals in the scurf of the Teeth, the substance called Worms in the Nose, the Cuticula consisting of Scales. *R. Soc. (London) Philos. Trans*. **14**:568–574.

Versalovic J, Koeuth T, Lupski Jr. 1991. Distribution of repetitive DNA sequences in eubacteria and application to fingerprinting of bacterial genomes. *Nucleic Acids Res*. **19**:6823–6831.

Versalovic J, Schneider M, de Bruijn FJ, Lupski JR. 1994. Genomic fingerprinting of bacteria using repetitive sequence based polymerase chain reaction. *Methods Mol. Cell. Biol*. **5**:25–40.

Welsh J, McClelland M. 1991. Genomic fingerprints produced by PCR with consensus tRNA gene primers. *Nucleic Acids Res*. **19**:861–866.

Williams JGK, Kubelik AR, Livak KJ, Rafalski JA, Tingey SV. 1990. DNA polymorphisms amplified by arbitrary primers are useful as genetic markers. *Nucleic Acids Res*. **18**:6531–6535.

Woese CR. 1987. Bacterial evolution. *Microbiol. Rev*. **51**:221–271.

Chapter 5

Use of Different PCR Primer-Based Strategies for Characterization of Natural Microbial Communities

James I. Prosser, Shahid Mahmood, and Thomas E. Freitag

5.1 INTRODUCTION

Ammonia oxidizers convert ammonia to nitrite in the first step of nitrification, with subsequent oxidation of nitrite to nitrate. The process is usually limited by ammonia oxidation, and ammonia oxidizers therefore control an essential step in the global nitrogen cycle. Ammonia oxidizers exemplify, in many ways, the enormous benefits of molecular techniques in microbial ecology. They grow slowly under laboratory conditions, are difficult to purify, and have few distinguishing phenotypic characteristics. Community or population ecology of ammonia oxidizers is therefore effectively impossible using traditional techniques alone. Quantification of abundance relied on the most probable number technique, which compounds many of the well-accepted limitations of dilution plate counting, including selective media and growth conditions, lack of enumeration of dormant organisms, increases in abundance with incubation time, and additional statistical variability introduced through application of probability distributions. As a consequence, despite their enormous importance in the global nitrogen cycle, traditional techniques provided little reliable information on the ecology, community structure, diversity, and abundance of ammonia oxidizers.

The introduction of 16S rRNA gene-based techniques for analysis of communities of ammonia-oxidizing bacteria (AOB) [Stephen et al., 1996] demonstrated, as for other microbial groups, the enormous diversity within natural communities, links between community composition and environmental conditions, and indications of their activity, through RNA-targeted studies. These studies were complemented by analysis of a key functional gene, *amoA*, encoding subunit A of the ammonia monooxygenase enzyme. This enabled analysis of community structure and transcriptional activity and quantification of 16S rRNA, and *amoA* genes enabled determination of ammonia oxidizer abundance. The importance of molecular techniques was further demonstrated, dramatically, by the discovery of archaeal ammonia oxidizers [Treusch et al., 2005; Venter et al., 2004; Könneke et al., 2005], which are even more difficult to cultivate, leading to research into the respective roles of bacterial and archaeal ammonia oxidisers [Prosser and Nicol, 2008]. Distinction of their activities in natural environments currently relies solely on molecular techniques.

Ammonia oxidizers reach high abundance in wastewater treatment systems, where use of fluorescence in situ hybridization (FISH) of 16S rRNA probes is feasible. Use of FISH is severely limited in soil, because of background fluorescence and high levels of particulate matter; it is also severely limited in both soil and natural aquatic environments, because of low cell abundance and possibly by lower transcriptional activity, reducing signal intensity. FISH also does not capture the high diversity of natural communities. The majority of molecular studies of AOB ecology therefore involve amplification of 16S rRNA or *amoA* genes or gene transcripts from extracted

Handbook of Molecular Microbial Ecology, Volume I: Metagenomics and Complementary Approaches, First Edition. Edited by Frans J. de Bruijn.

environmental DNA or RNA, followed by quantification, using competitive [Phillips et al., 2000] or real-time [e.g., Hermansson and Lindgren, 2001] qPCR, sequencing [e.g., Stephen et al., 1996], or fingerprint analysis [e.g., Stephen et al., 1998].

Molecular analysis of ammonia oxidizers removes the major disadvantages and biases of cultivation-based techniques, associated with selective growth and difficulties in cultivating representative organisms and identification and classification of isolates. Molecular techniques do, however, introduce other biases [Prosser et al., 2010], but increasing acceptance of molecular techniques during the last decade has reduced the degree to which bias is considered, assessed, and controlled for. One potential bias is introduced by analysis of different genes and use of different primers. With the exception of gammaproteobacterial *Nitrosococcus* strains, which are restricted to marine environments, known AOB belong to the betaproteobacteria. Phylogenetic analysis of betaproteobacterial ammonia oxidizer 16S rRNA genes places them within two genera, *Nitrosospira* and *Nitrosomonas*, each containing several subgroups [Purkhold et al., 2000]. This phylogeny is generally congruent with that obtained by analysis of *amoA* genes. Nevertheless, use of 16S rRNA and *amoA* genes is likely to generate different views of bacterial ammonia oxidizer diversity, even when using primers designed to target all betaproteobacterial ammonia oxidizers.

Accumulation of 16S rRNA gene sequences from pure and enrichment cultures and from sequences amplified directly from environmental nucleic acids promoted the design of RNA-targeted probes, specific to all bacterial ammonia oxidizers and subgroups, for FISH studies. A similar range of primers were designed for PCR amplification of bacterial ammonia oxidizer 16S rRNA genes. Purkhold et al. [2000] concluded that βAMO primers [McCaig et al., 1994] were most inclusive of available AOB database sequences, but that specificity was incomplete. The numbers of nontarget sequences amplified by βAMO primers depend on the environment under study and AOB abundance, but their use requires screening of amplicons to eliminate non-AOB sequences. The greatest specificity was found for CTO primers [Kowalchuk et al., 1997], but these did not target several members of the *Nitrosomonas oligotropha* cluster and have some mismatches with the *Nitrosomonas communis* cluster. In addition, in environments with very low AOB abundance, CTO primers amplify many non-AOB sequences [Schmidt et al., 2007].

Fingerprinting analysis of AOB communities has been carried out using denaturing gradient gel electrophoresis (DGGE) [Kowalchuk et al., 1997], temperature gradient gel electrophoresis (TGGE) [Fouratt et al., 2003], terminal-restriction fragment length polymorphism (T-RFLP) [Mintie et al., 2003], and single-strand conformation polymorphism (SSCP) [Backman et al., 2003]. These approaches have, in turn, utilized a range of PCR strategies most frequently amplifying AOB 16S rRNA genes using βAMO and/or CTO primers, sometimes in combination with primers designed to amplify 16S rRNA genes of all bacteria. Each primer set generates products of different size, preventing direct comparison of community structure obtained with different primer strategies and assessment of primer bias. Mahmood et al. [2006] addressed this problem by analysis of natural AOB communities using several commonly used primer strategies, with both direct and nested PCR amplification and with a common final PCR step to enable direct comparison.

5.2 METHODS

Both soil and marine sediment samples were analysed and details of sites and of all methods are provided in Mahmood et al. [2006]. The soil was a managed grassland, subject to fertilizer addition, and marine samples were obtained from sediment of a marine loch (Loch Duich, Scotland). AOB had been characterized previously in several studies of these sites [Webster et al., 2005; Freitag and Prosser, 2003]. DNA was extracted from soil and sediment samples using standard techniques; then it was purified, and 16S rRNA genes were amplified by PCR using five strategies, illustrated in **Figure 5.1**.

a. Direct amplification with βAMO primers (βAMO143f–βAMO1315r [Stephen et al., 1996]).

b. Direct amplification with AOB-specific CTO primers (CTO189f–CTO654r [Kowalchuk et al., 1997]).

c. Nested PCR amplification with first-round amplification using βAMO143f-βAMO1315r and secondary amplification with CTO189f–CTO654r.

d. Nested PCR amplification with first-round amplification using βAMO143f–βAMO1315r and secondary amplification with CTO189f in conjunction with the general bacterial primer Pf1053r [Edwards et al., 1989].

e. Nested PCR amplification with first-round amplification using universal bacterial primers 27f–1492r [Lane, 1991] and secondary amplification with CTO189f–CTO654r.

Amplicons generated by each approach were amplified further using 357f–GC–518r primers [Muyzer et al., 1993], which are considered to be universal bacterial

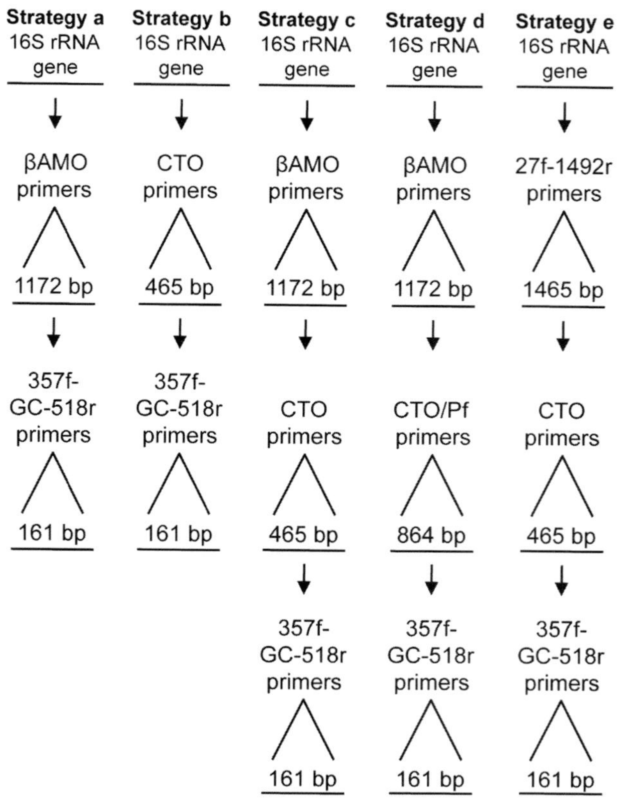

Figure 5.1 Five PCR strategies (a–e) used to amplify ammonia oxidising bacterial 16S rRNA genes from DNA extracted from soil and marine sediment samples. Strategies a and b involve direct amplification with βAMO primers (βAMO143f and βAMO1315r [Stephen et al., 1996]) or CTO primers (CTO189f-CTO654r [Kowalchuk et al., 1997]), which are designed to target ammonia-oxidizing bacteria. Strategies c, d, and e involved nested PCR with primary amplification with either βAMO primers (c and d) or 27f–1492r primers [Lane, 1991], which target all bacteria. Secondary amplification involves either CTO primers (c, e) or a combination of CTO189f [Kowalchuk et al., 1997] and Pf1053r [Edwards et al., 1989], which targets all bacteria. Products from each strategy were amplified using the universal bacterial primer 357f-GC-518r, to generate products of the same size for DGGE analysis. (Modified from Mahmood et al. [2006] with permission.)

primers. They were not selective for AOB and served only to generate products of equal size from each of the five strategies, which could then be characterized and compared using DGGE analysis.

Details of all PCR conditions and DGGE analysis are given in Mahmood et al. [2006]. Gels were silver-stained and scanned, and band intensity was quantified. In addition, DNA from a number of excised DGGE bands was extracted, reamplified, purified, and sequenced to check for bias between gels, to ensure that only AOB sequences had been amplified and to check that bands migrating to the same position had identical sequences. Sequences were obtained for the majority of bands with high relative intensity and for some with low relative intensity.

5.3 RESULTS

5.3.1 Analysis of Soil Ammonia Oxidizer Communities

DGGE patterns from soil samples contained 15 distinct bands, and reproducibility between replicate samples was high, with no detectable difference for any particular PCR strategy. The influence of different PCR strategies can be seen in **Figure 5.2**, with quantitative analysis of DGGE gels in **Table 5.1**. Strategies a and d gave similar profiles that were dominated by three bands consisting of non-AOB sequences related to *Variovorax*. Primary PCR amplification with these two strategies employed βAMO143f–βAMO1315r primers, which are known to amplify non-AOB sequences in some environments, and CTO primers, which amplify non-AOB sequences at low AOB abundance. Although it is not relevant to analysis of AOB communities in this case, the relative intensities of some bands (S5–S8, S15) was greater for strategy d, and this difference may be applicable to communities where AOB sequences only are amplified. Although preferentially amplifying non-AOB sequences, a and d were the only strategies that detected *Nitrosospira*-bands S6 and S7, which were not amplified by CTO-primer-based strategies. Strategy a did not detect a *Nitrosospira briensis*-like band (S15), possibly because of lower abundance than *Variovorax*-like sequences.

Strategies b, c, and e generated similar DGGE profiles with a relatively high number of common bands, most of which consisted of *Nitrosospira*-related sequences, and these were the only strategies to amplify *Nitrosospira*-like bands S9, S10, S13, and S14. Strategies b, c, and e did not amplify two of the *Variovorax*-like bands (S1 and S2) but amplified *Variovorax*-like band S3, although at lower relative intensity than a and d. Strategies c and e generated similar DGGE profiles. Both involved secondary amplification with CTO189f–CTO654r primers, following amplification with either βAMO143f–βAMO1315r or bacterial 27f–1492r primers. All CTO-based approaches led to high relative intensity of *Nitrosospira*-like band S8, which was less intense in strategy a. This may have resulted from relatively high abundance within the AOB community, but lower abundance than *Variovorax*-like sequences. Band S15 (*Nitrosospira briensis*-like sequence) was absent in approach a, probably due to lower copy numbers than *Variovorax*-like sequences types.

Table 5.1 Numbers of Bands and Similarity Coefficients (in parentheses) for DGGE Profiles of Ammonia Oxidizer 16S rRNA Genes Amplified from (a) Soil and (b) Marine Sediment Samples Using PCR Strategies a–e[a]

PCR Strategy	a	b	c	d	e
			(a) Soil		
a	5 (1.00)				
b	2 (0.40)	8 (1.00)			
c	2 (0.40)	8 (1.00)	8 (1.00)		
d	2 (0.44)	4 (0.73)	4 (0.73)	7 (1.00)	
e	2 (0.40)	8 (1.00)	8 (1.00)	4 (0.67)	8 (1.00)
			(b) Marine Sediments		
a	8 (1.00)				
b	8 (0.94)	9 (1.00)			
c	7 (0.82)	8 (0.89)	10 (1.00)		
d	8 (0.89)	8 (0.89)	9 (0.95)	10 (1.00)	
e	7 (0.82)	8 (0.89)	10 (1.00)	9 (0.95)	10 (1.00)

[a] Analysis was restricted to bands that were confirmed, by sequence analysis, to be from ammonia oxidizing bacteria. Values presented are the number of sequenced bands found in each approach (values on diagonal) or shared among different approaches

5.3.2 Analysis of Marine Ammonia Oxidizer Communities

DGGE patterns from marine sediment samples contained 11 distinct bands. Again, reproducibility between replicate samples was high, with no detectable difference for any particular PCR strategy, and the influence of different PCR strategies can be seen in **Figure 5.3**, with quantitative analysis in Table 5.1. All PCR strategies generated similar DGGE profiles for marine sediment samples, and very few non-AOB sequences were amplified. The majority of sequences were related to uncultured *Nitrosospira* and *Nitrosomonas* sequences, in addition to some unidentified betaproteobacterial AOB, and most sequences were closely related to those obtained in cloning studies of sediments from the same site using βAMO143f–βAMO1315r primers [Freitag and Prosser, 2003; Mortimer et al., 2004]. Most major bands were found with all five strategies, except for *Nitrosospira*-like bands M6 and M9, which were only detected by strategies involving the complete CTO primer set (c, e) or, with lower relative intensity, using the CTO189f-primer (strategy d). Some other bands (M1, M4, and M5) detected using strategies a, b, and d were less evident when using strategies c and e. Strategies a and d and c and e generated similar profiles, and UPGMA analysis indicated greater similarities between amplification strategies than in analyzing soils samples.

To assess the influence of the final, tertiary amplification step, both soil and marine sediment samples were analyzed using only primary and secondary amplification steps of strategies c and e. This generated fewer distinct bands (4–5), several of which did not comigrate with AOB cluster controls.

5.4 DISCUSSION

The study of Mahmood et al. [2006] provides important information for those using PCR-based analysis of AOB and also generic information of relevance to any PCR-based study. It does not, however, determine which method is "best," because the exact composition of the soil and marine sediment studies was not known. Determination of the accuracy of analysis would require use of model communities, inoculation of sterile environmental samples with known proportions of different sequence types, or additional, independent techniques that provide greater confidence. The practical difficulties in achieving this, while maintaining relevance to complex soil and sediment communities, preclude such analysis.

The study confirmed the disadvantages of using βAMO143f-βAMO1315r primers for analysis of AOB communities, particularly in the soil. Soil DGGE profiles generated using strategies a and d were dominated by non-AOB (presumably), *Variovorax* sequence types. These organisms were likely more abundant than AOB in this soil and were amplified through close sequence similarity, with only one mismatch against the forward and reverse βAMO primers [Mahmood et al., 2006]. Similar problems have been reported for other soil AOB studies. For example, Stephen et al. [1996] reported amplification of betaproteobacterial sequences related to

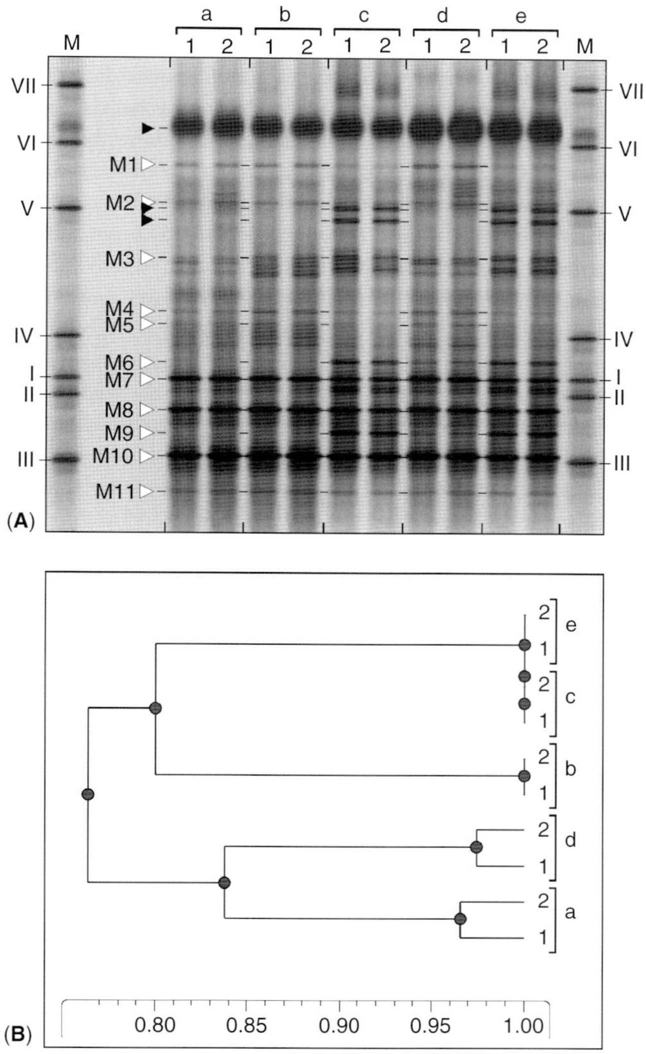

Figure 5.2 DGGE analysis of soil ammonia-oxidizing bacteria (AOB) using PCR strategies a–e to amplify AOB 16S rRNA genes from extracted DNA. (**a**) DGGE profiles of genes amplified from duplicate samples (lanes 1 and 2 for each strategy). White arrowheads indicate major bands whose presence, absence, or relative intensity differed between PCR-amplification strategies. Black arrowheads indicate bands that were PCR artifacts from which a sequence could not be generated. Bands S1–S15 were excised and sequenced.The band above band S8 is an artifact, possibly arising from excessive loading of strong PCR amplification products. Asterisks indicate non-AOB 16S rRNA gene sequences. Markers in lane M consisted of seven AOB cluster controls I–VII: I, EnvB1-8 (*Nitrosospira*); II, pH4.2A/27 (*Nitrosospira*); III, pH4.2A/4 (*Nitrosospira*); IV, pH7B/C3 (*Nitrosospira*); V, EnvA1-21 (*Nitrosomonas*); VI, EnvC1-19 (*Nitrosomonas*); VII, *N. europaea* (*Nitrosomonas*). (**b**) UPGMA (unweighted pairwise grouping method with mathematical averages) dendrogram showing similarities between DGGE banding profiles based on presence or absence of all bands in each lane. (From Mahmood et al. [2006] with permission.)

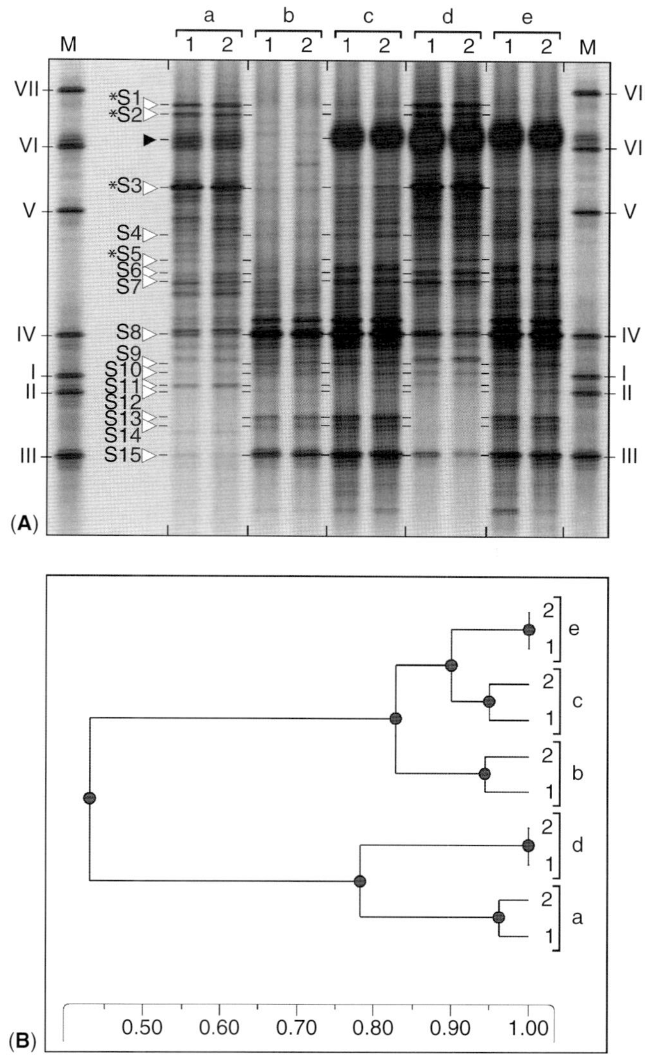

Figure 5.3 DGGE analysis of ammonia oxidising bacteria (AOB) from marine sediment samples using PCR strategies a–e to amplify AOB 16S rRNA genes from extracted DNA. (**a**) DGGE profiles of genes amplified from duplicate samples (lanes 1 and 2 for each strategy). Bands M1–M11 were excised and sequenced, and other information is as described in the legend for Figure 5.2. (**b**) UPGMA (unweighted pairwise grouping method with mathematical averages) dendrogram showing similarities between DGGE banding profiles based on presence or absence of all bands in each lane. (From Mahmood et al. [2006] with permission.)

Comamonas testosteroni and *Rubrivivax gelatinosus*, particularly in soil clone libraries, but also in marine sediments, and Bruns et al. [1999] found nonspecific amplification products in clone libraries generated from several soils. This problem appears to be less important for marine environments [Freitag and Prosser, 2004;

Mahmood et al., 2006], but use of βAMO primers for characterization of AOB communities requires screening of clones to ensure that non-AOB sequences are eliminated from analysis.

CTO primers provided greater specificity, but are less inclusive than βAMO primers and can amplify non-AOB sequences when these are present at low abundance [Schmidt et al., 2007]. This was not the case in the soil and sediment samples investigated by Mahmood et al. [2006] and CTO-primer-based primary amplification strategies generated similar numbers of bands in DGGE profiles. As indicated above, the number of bands generated is not necessarily an indicator of the quality or accuracy of a particular strategy. Many phylotypes will be below the detection limit, and the number of bands detected will depend on evenness rather than richness. Band number will increase with evenness and decrease with dominance, and "accurate" detection of bands requires independent, alternative, and reliable techniques or analysis of model communities with known relative abundances of different phylotypes. It also requires acceptance of relative band intensity as a quantitative measure of relative abundance, which remains a source of debate. Confidence in analysis methods is increased when different strategies generate similar results, reducing concerns about PCR or primer bias. In this respect, marine sediment profiles showed greater consistency with different strategies than soil, but this is mainly a reflection of the increased number of non-AOB sequences amplified from soil.

Nested PCR-based approaches (c, e) are frequently employed, and necessary, for amplification of organisms present at low abundance, particularly in the presence of naturally occurring PCR inhibitors. These are often criticized with respect to their ability to quantify relative abundance (through relative band intensity) because of the increased potential to introduce *Taq*-polymerase reading errors [Barnes, 1992; Tindal and Kunkel, 1988]. However, these criticisms are rarely supported by strong evidence, and the reproducible DGGE banding patterns reported by Mahmood et al. [2006] and high similarities of excised bands sequences and clone library sequences suggest that such errors were not significant. There may, however, be carryover of highly abundant, nontarget sequences from primary to secondary amplifications, possibly explaining detection of *Variovorax*-related sequences (band S3) in strategy c, despite use of AOB-specific CTO primers for secondary amplification. Choice of PCR amplification strategy is also influenced by the length of product generated, and a balance is required between sequence length required for phylogenetic analysis and sequences with melting characteristics suitable for DGGE analysis [Wu et al., 1998]. Thus, βAMO primers amplify 1.1-bp products, which are suitable for phylogenetic analysis but not DGGE. In contrast, CTO189f-CTO654r products are insufficient for phylogenetic analysis, but better for DGGE.

5.5 CONCLUSION

Design and choice of PCR primer strategies when targeting complex communities is often limited by the availability of sequences that are specific for target sequences, that exclude nontarget sequences, that demonstrate efficient PCR reaction kinetics, and that generate products of the required length for sequencing or fingerprinting. The potential for comparison of different primers strategies is therefore limited by lack of alternative primers, but their availability for AOB enabled comparison by Mahmood et al. [2006] of DGGE profiles obtained by different strategies for two environments, facilitated by use of a common primer set for a final amplification stage. Major differences resulted from lack of specificity of one primer set, which was more evident for analysis of soil samples, but otherwise different primers strategies generated similar DGGE profiles and amplified similar AOB sequence types. Potential lack of specificity and potential constraint to known target groups, even if not previously established, requires caution, particularly when initiating study of previously uncharacterised environments. Confidence in results is increased by similarities between sequences of excised DGGE bands, after two or three rounds of amplification, and those obtained from clone libraries employing a single PCR step, emphasizing the desirability of using more than one technique for community analysis.

Acknowledgments

This work was carried out as part of a Marie Curie Individual Fellowship (contract No. QLK3-CT-2000-52161) awarded by the European Commission.

REFERENCES

Backman JSK, Hermansson A, Tebbe CC, Lindgren PE. 2003. Liming induces growth of a diverse flora of ammonia-oxidising bacteria in acid spruce forest soil as determined by SSCP and DGGE. *Soil Biol. Biochem*. **35**:1337–1347.

Barnes WM. 1992. The fidelity of taq polymerase catalyzing PCR is improved by an N-terminal deletion. *Gene* **112**:29–35.

Bruns MA, Stephen JR, Kowalchuk GA, Prosser JI, Paul EA. 1999. Comparative diversity of ammonia oxidizer 16S rRNA gene sequences in native, tilled, and successional soils. *Appl. Environ. Microbiol*. **65**:2994–3000.

Edwards U, Rogall T, Blocker H, Emde M, Bottger EC. 1989. Isolation and direct complete nucleotide determination of entire genes—Characterization of a gene coding for 16S-ribosomal RNA. *Nucleic Acids Res*. **17**:7843–7853.

Fouratt MA, Rhodes JS, Smithers CM, Love NG, Stevens AM. 2003. Application of temperature gradient gel electrophoresis to the characterization of a nitrifying bioaugmentation product. *FEMS Microbiol. Ecol*. **43**:277–286.

Freitag TE, Prosser JI. 2003. Community structure of ammonia-oxidizing bacteria within anoxic marine sediments. *Appl. Environ. Microbiol*. **69**:1359–1371.

Freitag TE, Prosser JI. 2004. Differences between betaproteobacterial ammonia-oxidizing communities in marine sediments and those in overlying water. *Appl. Environ. Microbiol*. **70**:3789–3793.

Hermansson A, Lindgren P-E. 2001. Quantification of ammonia-oxidizing bacteria in arable soil by real-time PCR. *Appl. Environ. Microbiol*. **67**:972–976.

Könneke M, Bernhard AE, de la Torre JR, Walker CB, Waterbury JB, et al. 2005. Isolation of an autotrophic ammonia-oxidizing marine archaeon. *Nature* **437**:543–546.

Kowalchuk GA, Stephen JR, DeBoer W, Prosser JI, Embley TM, Woldendorp JW. 1997. Analysis of ammonia-oxidizing bacteria of the beta subdivision of the class Proteobacteria in coastal sand dunes by denaturing gradient gel electrophoresis and sequencing of PCR-amplified 16S ribosomal DNA fragments. *Appl. Environ. Microbiol*. **63**:1489–1497.

Lane DJ. 1991. 16S/23S rRNA sequencing. In Stackebrandt E, Goodfellow M, eds. *Nucleic Acid Techniques in Bacterial Systematics*. New York: John Wiley & Sons, pp. 115–175.

Mahmood S, Freitag TE, Prosser JI. 2006. Comparison of PCR primer-based strategies for characterization of ammonia oxidizer communities in environmental samples. *FEMS Microbiol. Ecol*. **56**:482–493.

McCaig AE, Embley TM, Prosser JI. 1994. Molecular analysis of enrichment cultures of marine ammonia oxidisers. *FEMS Microbiol. Lett*. **120**:363–367.

Mintie AT, Heichen RS, Cromack K, Myrold DD, Bottomley PJ. 2003. Ammonia-oxidizing bacteria along meadow-to-forest transects in the Oregon cascade mountains. *Appl. Environ. Microbiol*. **69**:3129–3136.

Mortimer RJG, Harris SJ, Krom MD, Freitag TE, Prosser JI, Barnes J, Anschutz P, Hayes PJ, Davies IM. 2004. Anoxic nitrification in marine sediments. *Marine Ecol. Progr. Ser*. **276**:37–51.

Muyzer G, Dewaal EC, Uitterlinden AG. 1993. Profiling of complex microbial populations by denaturing gradient gel electrophoresis analysis of polymerase chain reaction-amplified genes coding for 16S ribosomal RNA. *Appl. Environ. Microbiol*. **59**: 695–700.

Phillips CJ, Paul EA, Prosser JI. 2000. Quantitative analysis of ammonia oxidising bacteria using competitive PCR. *FEMS Microbiol. Ecol*. **32**:167–175.

Prosser JI, Nicol GW. 2008. Relative contributions of archaea and bacteria to aerobic ammonia oxidation in the environment. *Environ. Microbiol*. **10**:2931–2941.

Prosser JI, Jansson JK, Liu W-T. 2010. Nucleic-acid-based characterisation of community structure and function. In Liu W-T, Jansson JK, eds. Environmental Molecular Microbiology. Norwich; Caister Academic Press, pp. 63–86.

Purkhold U, Pommerening-Röser A, Juretschko S, Schmid MC, Koops HP, et al. 2000. Phylogeny of all recognized species of ammonia oxidizers based on comparative 16S rRNA and *amoA* sequence analysis: Implications for molecular diversity surveys. *Appl. Environ. Microbiol*. **66**:5368–5382.

Schmidt CS, Hultman KA, Robinson D, Killham K, Prosser JI. 2007. PCR profiling of ammonia oxidiser communities in acid soils subjected to N and S deposition. *FEMS Microbiol. Ecol*. **61**:305–316.

Stephen JR, McCaig AE, Smith Z, Prosser JI, Embley TM. 1996. Molecular diversity of soil and marine 16S rRNA gene sequences related to beta-subgroup ammonia-oxidizing bacteria. *Appl. Environ. Microbiol*. **62**:4147–4154.

Stephen JR, Kowalchuk GA, Bruns MA, McCaig AE, Phillips CJ, et al. 1998. Analysis of β-subgroup ammonia oxidiser populations in soil by DGGE analysis and hierarchical phylogenetic probing. *Appl. Environ. Microbiol*. **64**:2958–2965.

Tindall KR, Kunkel TA. 1988. Fidelity of DNA synthesis by the *Thermus aquaticus* DNA polymerase. *Biochemistry* **27**:6008–6013.

Treusch AH, Leininger S, Kietzin A, Schuster SC, Klenk H-P, et al. 2005. Novel genes for nitrite reductase and Amo-related proteins indicate a role of uncultivated mesophilic crenarchaeota in nitrogen cycling. *Environ. Microbiol*. **7**:1985–1995.

Venter JC, Remington K, Heidelberg JF, Halpern AL, Rusch D, et al. 2004. Environmental genome shotgun sequencing of the Sargasso Sea. *Science* **304**:66–74.

Webster G, Embley TM, Freitag TE, Smith Z, Prosser JI. 2005. Links between ammonia oxidizer species composition, functional diversity and nitrification kinetics in grassland soils. *Environ. Microbiol*. **7**:676–684.

Wu Y, Hayes VM, Osinga J, Mulder IM, Looman MWG, et al. 1998. Improvement of fragment and primer selection for mutation detection by denaturing gradient gel electrophoresis. *Nucleic Acids Res*. **26**:5432–5440.

Chapter 6

Horizontal Gene Transfer and Recombination Shape Mesorhizobial Populations in the Gene Center of the Host Plants *Astragalus Luteolus* and *Astragalus Ernestii* in Sichuan, China

QIONGFANG LI, XIAOPING ZHANG, LING ZOU, QIANG CHEN, DAVID P. FEWER, AND KRISTINA LINDSTRÖM

6.1 INTRODUCTION

The genus *Astragalus* consists of up to 3000 species, and approximately 1500–2000 recorded species are distributed in the southwestern and the Sino-Himalayan regions of Asia [Sanderson and Wojciechowski, 2000; Osaloo et al., 2005]. *Astragalus* species can form nodules and fix nitrogen in symbiosis with rhizobia, and some representatives of this perennial legume family are used as green manure, Chinese traditional medicine, or bee plants. Since environments for *Astragalus* plants are rapidly being destroyed, characterization and biogeographical studies of rhizobial strains isolated from *Astragalus* species are urgently required. *Astragalus luteolus* and *Astragalus ernestii* are wild plants that grow at altitudes above 3000 m. *Astragalus luteolus* are indigenous plants of the western Sichuan province. *A. ernestii* grows only in western Sichuan, northwest of Yunnan, and in eastern Tibet. *A. ernestii* is also used in Chinese medicine.

The classification of rhizobia into taxa is strongly based on phylogenies of *rrs* genes (encoding 16S rRNA or SSU); but phylogenies of several protein encoding housekeeping genes have recently been used for resolving taxa, especially at the species level [e.g. Gao, et al., 2001; Vinuesa et al., 2005a; Nandasena et al., 2007; Martens et al., 2007; 2008; Chen et al., 2008]. Multilocus sequence analysis is a more reliable classification method than the methodology based solely on ribosomal sequences, for a number of reasons. Firstly, several unlinked genes dispersed in the core genome better represent the true geneaology of the organism than just one single sequence or sequences from a locus that might show within strain variation [Young and Haukka, 1996]. Secondly, especially in rhizobia ribosomal sequences show mosaicism as a consequence of homologous recombination, which interferes with phylogenetic tree construction [Terefework et al., 1998; Van Berkum et al., 2003; Eardley et al., 2005]. Thirdly, *rrs* genes of rhizobia often display low polymorphism in comparison with other taxonomic markers. Thus they are often unreliable for species delineation. This last feature is especially true for the genus *Mesorhizobium*, a prevalent symbiont of *Astragalus* plant species.

Handbook of Molecular Microbial Ecology, Volume I: Metagenomics and Complementary Approaches, First Edition. Edited by Frans J. de Bruijn.

Sampling of Chinese rhizobia has mainly been carried out in provinces Xinjiang, Gansu, Shanxi, Hubei, Hainan, Qinhai, and Tibet, and some novel species have been reported to date [Tan et al., 1997; Wei et al., 2003; Kan et al., 2007; Han et al., 2008a,b; Tian et al., 2008]. In this study we wished to obtain information about hitherto unexplored ecosystems at high elevations on the southeast fringe of the Qinghai–Tibet Plateau, and we also wanted to get more information about rhizobial diversity. We obtained root-nodules from *A. luteolus* and *A. erntstii*, growing in the Aba district on the northwest plateau of the Sichuan provice in southwestern China, which is an important cradleland area for the Yellow River and Yangtze River. The area has a long and cold winter and a cool and moist summer. The mean annual temperature is 7.8°C, the rainfall level is low (average 613.9 mm per year), and the soil is barren. The plants sampled grew at elevations of 3650 m (*A. luteolus)* and 4250 m (*A. ernestii*), respectively. Thirty-three nodule isolates were studied by molecular systematic methods, including ARDRA, AFLP, and sequencing of PCR products corresponding to *rrs, glnA, glnII*, and *nifH* genes, in order to observe diversity, phylogenetic patterns, and speciation in these populations.

6.2 MATERIALS AND METHODS

6.2.1 Bacterial Strains Sampling and Symbiotic Status

Bacterial strains used in this work are listed in **Table 6.1**. Rhizobial strains were isolated using methods described by Weir et al. [2004]. Culture purity was checked by repeated streaking. Strains were stored at −70°C in 20% (v/v) glycerol for long-term preservation. In order to check the capacity of the isolates to nodulate their original hosts, surface sterilized seeds of *A. luteolus* and *A. erntstii* were inoculated and plants were grown in pots filled with vermiculite moistened with N-free plant nutrient solution [Vincent, 1970] and maintained in a growth chamber at 28°C with a photoperiod of 15 h. Nodulation was observed after 1 month. All the strains were able to induce nodules on their original hosts.

6.2.2 PCR Amplification, RFLP, and Nucleotide Sequencing

Complete *rrs* gene fragments were amplified with the universal forward primer P1 and the universal reverse primer P6 [Tan et al., 1997]. The annealing temperature was 56°C. The PCR products of the 16S rRNA genes were digested in separate reactions with the restriction endonucleases *Msp*I, *Hinf*I, *Hae*I, and *Hind*III (Takara) and the fragments separated in 3.0% agarose gels.

Seven representative strains were chosen for partial DNA sequencing of three housekeeping core genes: *rrs, glnA* encoding glutamine synthetase I (GSI), and *glnII* encoding glutamine synthetase II (GSII). The *nifH* gene encoding dinitrogenase reductase was chosen to represent the symbiotic genes. *rrs* gene fragments were amplified as described previously, and purified *rrs* PCR products were directly sequenced. *glnA* and *glnII* gene fragments were amplified with primer pairs GSI-1, GSI-2 and GSII-1, GSII-2, respectively, using an annealing temperature of 55°C [Turner and Young, 2000]. The *nifH* gene was amplified using the primers nifHI and nifHF and an annealing temperature of 57°C [Laguerre et al., 2001]. These gene fragments were cloned into vector pGMT (Takara). Ligated plasmids were transformed into *Escherichia coli* DH5a. Plasmids containing *glnA, glnII*, and *nifH* sequences were extracted and purified as described previously [Tiesman and Rizzino, 1991], and then they were sequenced using an ABI3730 sequencer (ABI, Foster City, CA) using the primers listed in **Table 6.2**.

6.2.3 AFLP Analyses

The AFLP procedure was performed as described [Vos et al., 1995] with 100–500 ng of DNA digested with *EcoR*I and *Mse*I restriction enzymes and, in the same step, ligated with double-stranded adapters specific for each restriction half-site. Two sets of primers were used in separate PCRs: one set with two selective nucleotides [*Mse*I-gc (GAT GAG TCC TGA GTA AGC) and *EcoR*I-gc (GAC GTC GTA CCA ATT CGA GC)] and another set with a mixture of two and three selective nucleotides [*Mse*I-gc and *EcoR*I-gag (GAC GTC GTA CCA ATTCGA GAG)]. The PCR conditions were as described by Terefework et al. [2001]. The PCR products were separated on 5% polyacrylamide gels. Electrophoresis was performed in Tris–borate–EDTA buffer at 120 W and 50°C for 2 h. Silver-stained polyacrylamide gels were scanned and analyzed.

6.2.4 Data Treatment and Sequence Analysis

The ARDRA gels were scored manually, and bands were marked as absent or present. Distances were calculated (sj), and dendrograms were constructed using the UPGMA method in NT-SYS. The AFLP gels were analyzed using the GELCOMPAR 4.0 program (Applied Maths, Kortrijk, Belgium), and UPGMA dendrograms were obtained based on distances calculated with the Dice index. The gene sequences obtained were aligned in CLUSTAL_X (version 1.83) [Thompson et al., 1997] and edited using DNAMAN (version 6.0).

Table 6.1 Strains Used in This Study and Results from ARDRA and *rrs* Sequence Analysis

Strain Code	*rrs* Type	*rrs* Sequence Type (Species)
	***Astragalus luteolus* (3650 m)**	
SCAU 1	A	nd
SCAU2	A	*M. huakuii* clade
SCAU3	A	nd
SCAU4	B	nd
SCAU5	C	nd
SCAU6	D	nd
SCAU7	E	*M. loti/ciceri* clade
SCAU12	G	nd
SCAU13	E	*M. huakuii* clade
SCAU14	H	nd
SCAU15	B	nd
SCAU17	A	nd
SCAU18	A	nd
SCAU 20	H	nd
SCAU 21	I	nd
SCAU 22	C	nd
SCAU 23	E	nd
SCAU 24	D	nd
SCAU 26	F	nd
SCAU 30	E	nd
SCAU 31	E	nd
SCAU 32	B	nd
SCAU 33	A	nd
	***Astragalus ernestii* (4240 m)**	
SCAU8	A	nd
SCAU9	A	*M. huakuii* clade
SCAU10	A	nd
SCAU11	F	*M. loti/ciceri* clade
SCAU16	B	nd
SCAU19	H	nd
SCAU25	I	nd
SCAU27	A	*M. loti/ciceri* clade
SCAU 28	A	nd
SCAU 29	D	nd

[a] Strains used in the phylogenetic studies are bold. M, *Mesorhizobium*; nd, not determined.

6.2.5 Phylogenetic Analysis

Phylogenetic analyses were conducted using Paup* [Swofford, 2001]. Primer sequences and ambiguous regions of the alignments were excluded. Phylogenetic trees were inferred using the neighbor-joining method. One thousand bootstrap replicates were analyzed to test the stability of monophyletic groups in the neighbor-joining analysis. The incongruence of the *glnA* and *glnII* genes was evaluated using the conservative length difference test [Farris et al., 1994] implemented with Paup* [Swofford, 2001]. We used random taxon addition (10 replicates), tree bisection–reconnection branch swapping, and heuristic searches with 10,000 repartitions of the data.

6.2.6 Nucleotide Sequence Accession Numbers

The sequences obtained in this study have been deposited in the GenBank sequence database under accession numbers EU514524–EU514545.

Table 6.2 Oligonucleotides Used as PCR Primers and for Sequencing

Target Gene	Primer Sequence[a]	Position[b]	Size of Amplified Fragment	Reference
rrs	AGAGTTTGATCCTGGCTCAG AACGAACCCT	8–37	~1400 bp	Tan et al. [1997]
	TACGGCTACCTTGTTACGACTTCACCCC	1479–1506		
glnA	AAGGGCGGCTAYTTCCCGGT	532–551	~530 bp	Turner and Young [2000]
	GTCGAGACCGGCCATCAGCA	1143–1124		
glnII	AACGCAGACAAGGAATTCG	69–88	~600 bp	
	ATGCCCGAGCCGTTCCAGTC	686–667		
nifH	TACGGNAARGGSGGNATCGGCAA	25–50	~840 bp	Laguerre et al. [2001]
	AGCATGTCYTCSAGYTCNTCCA	787–808		

[a] Y = C or T; H = A,C, or T; R = A or G; S = C or G; K = G or T; N = A,C,G, or T; I = inosine.

[b] Position of the primers in the corresponding *rrs* sequence of *E. coli, glnA* and *glnII* sequence of *S. meliloti* and *nifH* sequence of *R. leguminosarum* bv. *trifolii*.

6.3 RESULTS

6.3.1 Genetic Diversity

RFLP patterns were produced for *rrs* genes from the 33 Astragalus rhizobia (Table 6.1). Three of the four restriction enzymes, *Msp*I, *Hae*III and *Hind*III, produced two patterns and *Hinf*I produced three. The tested strains were divided into eight *rrs* genotypes A-I (Table 6.1), revealing considerable genetic diversity. AFLP banding patterns were obtained from all strains. An UPGMA dendrogram was constructed from the AFLP fingerprints of the 33 strains isolated from *Astragalus* plants and 18 reference strains (**Fig. 6.1**). Three distinct clades were formed by the novel isolates, and the rest was dispersed among the reference strains. Neither the ARDRA nor the AFLP fingerprinting could differentiate the strains based on host plant or site of isolation.

6.3.2 rrs Phylogeny

A 1400-bp fragment of *rrs* genes was obtained from six selected representative strains: SCAU2, SCAU7, SCAU9, SCAU11, SCAU13, and SCAU27 (Table 6.2). In the phylogenetic tree constructed, genus *Mesorhizobium* was included in a monophyletic group containing all *Astragalus* isolates and *Aminobacter* species (**Fig. 6.2**). The *Astragalus* isolates formed two separate clades. The relationships among the defined reference species were similar to those in previously reported trees [Wei et al. 2003]. SCAU2, SCAU9, and SCAU13 grouped together and formed a sister clade to *Mesorhizobium huakuii, Mesorhizobium plurifarum, Mesorhizobium septentrionale*, and *Mesorhizobium amorphae*. SCAU7, SCAU11, and SCAU27 formed a sister clade to *Mesorhizobium loti* and *Mesorhizobium ciceri*. The novel isolates represented two distinct phylogenetic branches in the *rrs* tree.

6.3.3 Phylogenetic Analysis of *glnA* and *glnII* Sequences

In the phylogenetic tree based on *glnA* gene sequences, genus *Mesorhizobium* formed a well-separated clade (**Fig. 6.3a**). Strains SCAU2 and SCAU9 clustered with *Mesorhizobium* sp. MAFF303099, forming a sister clade to *M. huakuii*. The sequences of strains SCAU5 and SCAU11 were identical to that of SCAU2. Strains SCAU7, SCAU13, and SCAU27 formed a sister clade to *M. ciceri* and *M. loti*. The topology of the *glnA* tree was different from that of the *rrs* tree, with the positions of strains SCAU11 and SCAU13 swapped in relation to the other Astragalus isolates. In the phylogenetic tree based on *glnII*, the genera *Rhizobium* and *Sinorhizobium* separated out as two monophyletic clades.

Genus *Mesorhizobium* was divided into two pleophyletic clades (Fig. 6.3b), with the *Astragalus* strains SCAU2, SCAU9, and SCAU13 and strain MAFF303099 positioned within one of them and strains SCAU7 and SCAU27 within the other, as a sister clade to *M. ciceri*. Unfortunately, no *glnII* sequences were obtained for SCAU5 and SCAU11. The length incongruence test indicated that the two datasets were incongruent with one another ($P = 0.0001$), supporting the topological incongruence we observed. For the *Astragalus* strains, the *glnII* tree was congruent with the *rrs* tree, albeit more deeply branched, the exception being the position of the strains in relation to *M. loti*.

6.3.4 Phylogenetic Analysis of *nifH* Gene Sequences

The *nifH* fragments amplified were approximately 750 bp in size. The *nifH* sequences of SCAU13 and SCAU27

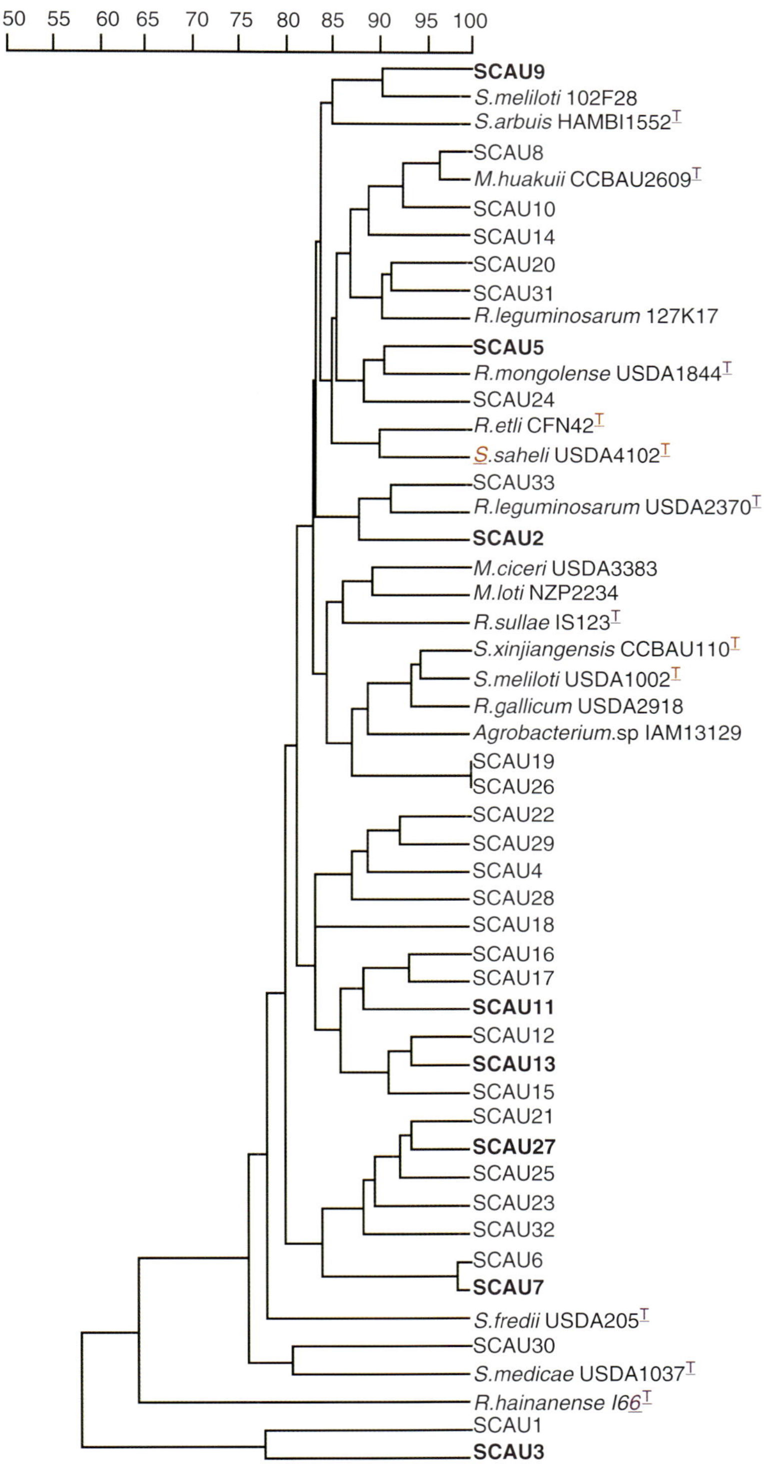

Figure 6.1 Dendrogram generated with the GelCompar 4.0 computer module using Pearson's coefficient clustering by the unweighted pair group method using arithmetic averages. Eighteen reference strains and 33 isolates are included. The scale indicates percent similarity. Abbreviations: *R.*, *Rhizobium*; *M.*, *Mesorhizobium*; *S.*, *Sinorhizobium*.

were not obtained. Sequences of the *Astragalus* strains were all closely related and formed a well-defined, monophyletic clade among sequences from described species of *Mesorhizobium* (**Fig. 6.4**).

6.4 DISCUSSION

We conducted a molecular systematic study of 33 rhizobia isolated from two different *Astragalus* species, *A. luteolus*

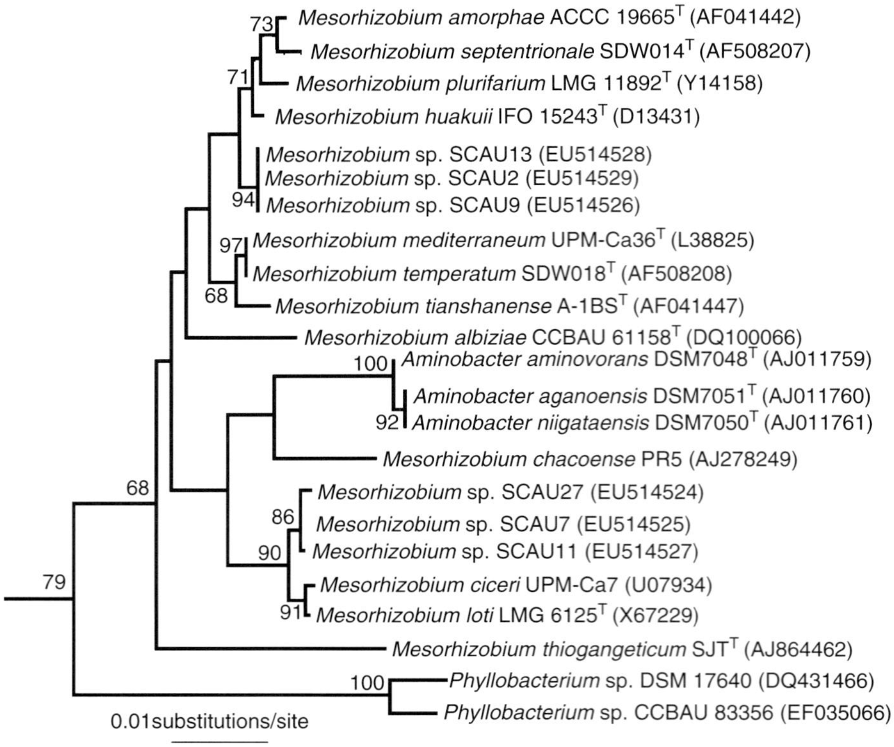

Figure 6.2 Neighbor-joining phylogenetic tree based on 1338 bp aligned bases of *rrs* sequences showing relationship of the novel isolates with reference strains. Bootstrap values above 60 are indicated. A scale bar indicates 0.01 substitutions per site. Strains isolated in this study are in bold. A number of *Rhizobium* and *Sinorhizobium* species were used as the outgroup. But only the *Mesorhizobium*/*Phyllobacterium* clade is shown in this figure.

and *A. ernestii*, growing at two different elevations in the highlands situated on the southeast fringe of the Qinghai-Tibet Plateau in Sichuan, China.

AFLP fingerprinting revealed that the isolates were diverse, but belonged to the same population, because the genomic profiles for isolates from the two plants or sites were intermingled. The AFLP dendrogram did not indicate the actual relationship between the strains recognized because it failed to resolve *Rhizobium* from *Sinorhizobium* and *Mesorhizobium*. Terefework et al. [2001] proposed that AFLP is an excellent technique to differentiate strains or very closely related species, but it is by no means a good phylogenetic tool. Our results support this view.

ARDRA analysis showed that the *Astragalus* isolates were diverse, but belonged to the same *rrs* clade. Sequencing of *rrs* genes of representative strains confirmed that the strains belonged to the *Mesorhizobium* genus and indicated that the novel isolated clades might represent two new species. *Mesorhizobium* sp. MAFF303099, whose genome has been fully sequenced and which is still erroneously called *M. loti* [Turner et al., 2002], is part of the clade of *Astragalus* isolates. This confirmed that the *Astragalus* isolates belonged to the *Mesorhizobium* genus.

The bacterial species has been defined in numerous ways. Considering the biology of the bacteria and species concepts in general, an evolutionary species concept seems desirable (see Chapters 9–13, Vol. I). This primary concept should be followed by secondary concepts that guide the taxonomist in the classification work [Lindström and Gyllenberg, 2006]. So far, DNA:DNA hybridizations have been considered the gold standard in taxonomy and *rrs* phylogeny desirable for species delineation. Considering the nonconserved nature of especially mesorhizobial *rrs* genes [de Lajudie et al., 1998], the species affiliation of the isolates should not be resolved based on the *rrs* phylogeny alone. Also, taking the disadvantages of DNA:DNA hybridizations into account (Lindström and Martínez-Romero, 2007; see Chapter 2, Vol. I), it becomes important to determine the phylogenies of several housekeeping genes for rhizobial species assignments. Glutamine synthetase exists in at least two related forms, GSI and GSII, the sequences of which were used in evolutionary molecular clock studies to analyze the phylogenies of 16 species of rhizobia by Turner and Young [2000]. Their phylogenetic studies indicated that the *glnA* genes was congruent with the *rrs* genes while *glnII* followed the same pattern, but *M. huakuii* showed an aberrant position in the protein phylogeny. Our study confirm the findings of Turner and Young [2000] with respect to the general shapes of the trees obtained with named species. The better resolution of these sequences in comparison with *rrs* suggests that the glutamine synthetase genes are good phylogenetic markers for rhizobial systematics.

The positions of the new isolates from this study, however, showed some aberrant features in the *glnA* tree.

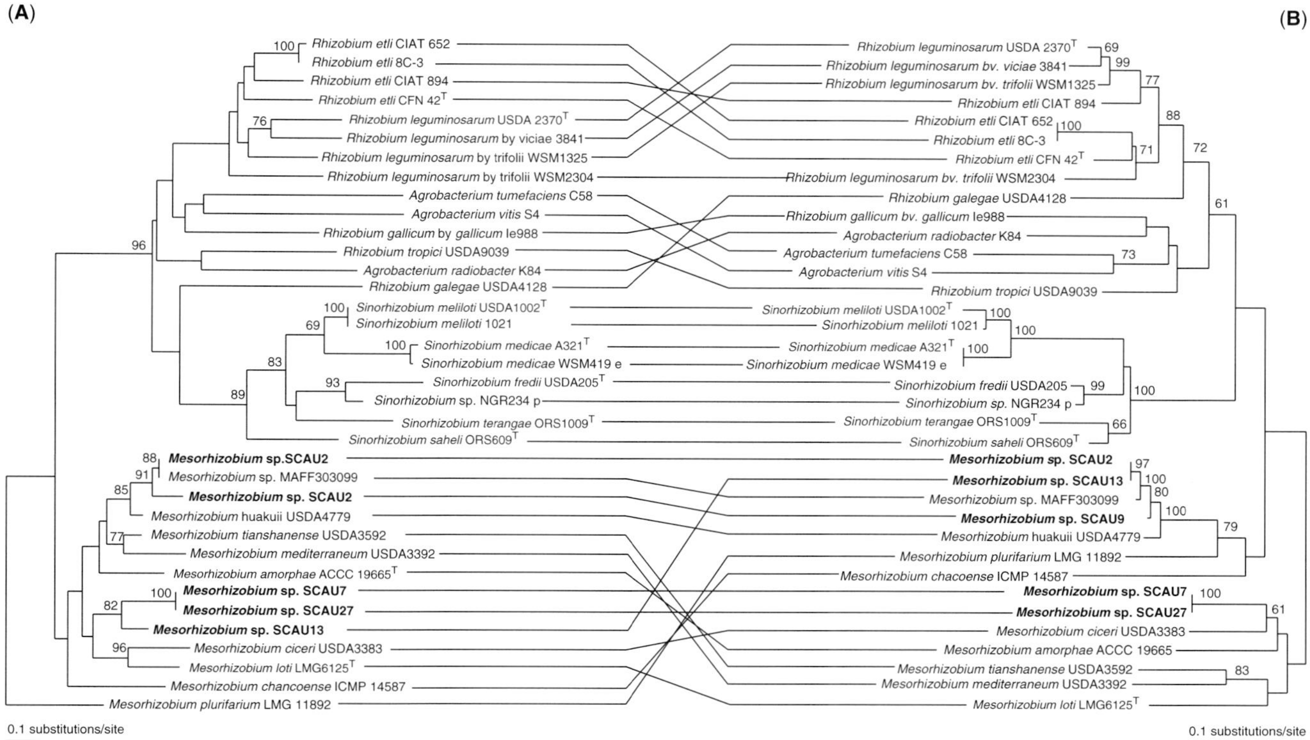

Figure 6.3 Neighbor-joining phylogenetic tree based on 506-bp aligned bases representing glutamine synthetases. (**a**) *gln*A and (**b**) *gln*II. Bootstrap values above 60 are indicated, and scale bars represent 0.01 substitutions per site. The *glnA* and *gln*II trees are rooted at the midpoint. Strains isolated in this study are in bold.

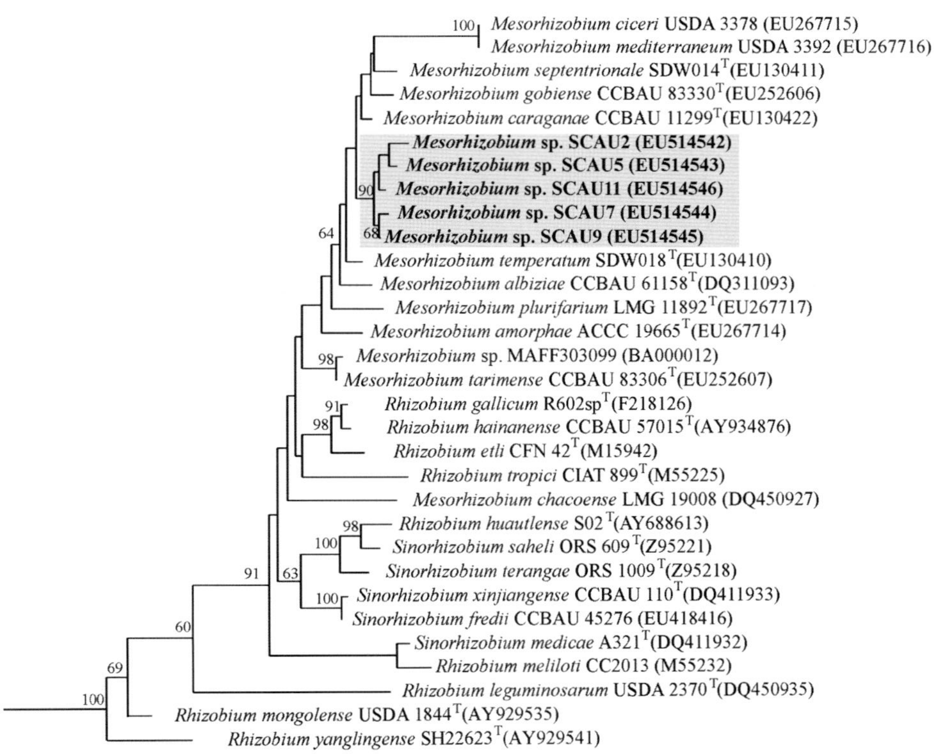

Figure 6.4 Neighbor-joining phylogenetic tree based on 528-bp aligned bases of *nifH*. Only bootstrap probability values greater than 50% are indicated at the branching points. Scale bars represent 0.02 substitution per site. Strains isolated in this study are in bold.

Two of the strains (SCAU2, SCAU9) that grouped as a putative new species in the *rrs* tree maintained their position in both the *glnA* and the *glnII* trees, whereas the third member of this *rrs* clade, SCAU13, had a different *glnA* gene, more closely related to that of strains SCAU7 and SCAU27, which form the second putative new species. Thus, *glnA* seems to have been involved in recombinational events and is not a suitable phylogenetic marker for genus *Mesorhizobium*.

The work by Gao et al. [2001, 2004] with isolates from *A. adsurgens* and that of Zhang et al. [2000] with *A. sinicus* strains demonstrated that populations from *Astragalus* species are genetically diverse, but mainly belong to the *Mesorhizobium* genus. The phylogenies of *glnA* sequences obtained for *A. adsurgens* isolates were congruent with those of *rrs*; and two new species, *M. temperatum* and *M. septentrionale*, were described. The results indicate that the populations representing the species were not undergoing genetic mixing for these housekeeping genes. In this study, we show that if traditional, minimum criteria were to be used, two clades of our *Astragalus* rhizobia in the *rrs* tree would be potentially new species. However, both the AFLP fingerprinting results and the incongruence of the *rrs* and *gln* gene trees indicate that we are dealing with samples from a population that is undergoing genetic exchange and recombination at a core genomic level. Thus, species delineation is not advisable at this stage.

Symbiotic rhizobial genes are commonly found on potentially transmissible elements [MacLean et al., 2007; Nandasena et al., 2007]. Gao et al. [2004] sequenced symbiotic *nodA* genes from *A. adsurgens* rhizobia belonging to the species *M. septentrionale* and *M. temperatum* and found that most of them were closely related, indicating horizontal gene transfer during evolution. We sequenced the symbiotic *nifH* gene encoding one nitrogenase subunit and showed that isolates from the two *Astragalus* species under study were phylogenetically close. They were also part of a large clade harboring *nifH* sequences of bacteria-nodulating plant species that are members of the inverted repeat lacking clade (IRLC) of the Hologalegina in the Papillionideae [Wojciechowski et al., 2004]. The bootstrap values of the branches of this larger part of the mesorhizobial *nifH* tree are, however, low.

According to Polhill [1981a,b] the geographic center of diversity and presumed origin of *Astragalus*, like most of its close relatives in the IRLC clade, is Eurasia and specifically the steppes and mountains of southwestern to south-central Asia and the Himalayan plateau. Sampling of rhizobia in the center of diversity of *Rhizobium galegae* biovar orientalis in the Caucasus yielded a collection with greater diversity than a collection of *R. galegae* biovar officinalis from the same area [Andronov et al., 2003]. Bailly et al. [2007] stressed that empirical studies focusing on the interplay between ecological adaptation and sex in bacteria are needed to gain a better knowledge of the processes leading to bacterial diversification. Vinuesa et al. [2005a] stressed the importance of sufficiently dense sampling of strains in order to achieve reliable gene phylogenies for taxonomic purposes. They showed, using concatenated *atpD-recA* sequences, that three named *Rhizobium* species, *Rhizobium gallicum, Rhizobium mongolense*, and *Rhizobium yanglingense*, should in fact be considered as belonging to the same species. On the other hand, Vinuesa et al. [2005b] demonstrated that strains belonging to the new species *Bradyrhizobium canariense* could be separated from *Bradyrhizobium japonicum* using a population genetics approach.

Several new mesorhizobial species were described recently; but unfortunately, they were not always based on extensive sampling of bacteria and analysis of sufficient housekeeping genes. Thus, the true nature of the population genetics of mesorhizobia is still not obvious. Our results indicate that horizontal gene transfer and recombination shape mesorhizobial populations in the gene center of the host plants *A. luteolus* and *A. ernestii*. We propose that wider sampling of rhizobial populations in the gene centers is likely to reveal interesting evolutionary patterns leading to speciation of symbionts nodulating the large *Astragalus* plant family.

Acknowledgments

This research was supported by the National Natural Science Foundation of China, the Specialized Research Found for the Doctoral Program of Higher Education (20060626006), and the Academy of Finland.

REFERENCES

Andronov EE, Terefework Z, Roumiantseva ML, Dzyubenko NI, Onichtchouk OP, et al. 2003. Symbiotic and genetic diversity of *Rhizobium galegae* isolates collected from the *Galega orientalis* gene center in the Caucasus. *Appl. Environ. Microbiol*. **69**:1067–1074.

Bailly X, Olivieri I, Brunel B, Cleyet-Marel JC, Béna G. 2007. Horizontal gene transfer and homologous recombination drive the evolution of the nitrogen-fixing symbionts of Medicago species. *J. Bacteriol*. **189**:5223–5236.

Chen WF, Guan SH, Zhao CT, Yan XR, Man CX, et al. 2008. Different *Mesorhizobium* species associated with *Caragana* carry similar symbiotic genes and have common host ranges. *FEMS Microbiol. Lett*. **283**:203–209.

de Lajudie P, Willems A, Nick G, Moreira F, Molouba F, et al. 1998. Characterization of tropical tree rhizobia and description of *Mesorhizobium plurifarium* sp. nov. *Int. J. Syst. Bacteriol*. **48**:369–382.

Eardly BD, Nour SM, Van Berkum P, Selander RK. 2005. Rhizobial 16S rRNA and *dnaK* genes: Mosaicism and the uncertain phylogenetic placement of *Rhizobium galegae*. *Appl. Environ. Microbiol*. **71**:1328–1335.

Farris JS, Källersjö M, Kluge AG, Bult C. 1994. Testing significance of congruence. *Cladistics* **10**:315–319.

Gao JL, Terefework Z, Chen WX, Lindström K. 2001. Genetic diversity of rhizobia isolated from *Astragalus adsurgens* growing in different geographical regions of China. *J. Biotechnol.* **91**: 155–168.

Gao JL, Turner SL, Kan FL, Wang ET, Tan ZY, et al. 2004. *Mesorhizobium septentrionale* sp. nov. and *Mesorhizobium temperatum* sp. nov., isolated from *Astragalus adsurgens* growing in the northern regions of China. *Int. J. Syst. Evol*. Microbiol. **54**:2003–2012.

Han TX, Wang ET, Wu LJ, Chen WF, Gu JG, et al. 2008a. *Rhizobium multihospitium* sp. nov., isolated from multiple legume species native of Xinjiang, *China Int. J. Syst. Evol. Microbiol*. **58**:1693–1699.

Han TX, Han LL, Wu LJ, Chen WF, Sui XH, et al. 2008b. *Mesorhizobium gobiense* sp. nov. and *Mesorhizobium tarimense* sp. nov., isolated from wild legumes growing in desert soils of Xinjiang, China. *Int. J. Syst. Evol. Microbiol*. **58**:2610–2618.

Kan FL, Chen ZY, Wang ET, Tian CF, Sui XH, et al. 2007. Characterization of symbiotic and endophytic bacteria isolated from root nodules of herbaceous legumes grown in Qinghai-Tibet plateau and in other zones of China. *Arch. Microbiol*. **188**:103–115.

Laguerre G, Nour SM, Macheret V, Sanjuan J, Drouin P, et al. 2001. Classification of rhizobia based on *nodC* and *nifH* gene analysis reveals a close phylogenetic relationship among *Phaseolus vulgaris* symbionts. *Microbiology* **147**:981–993.

Lindström K, Gyllenberg HG. 2006. A cross-disciplinary species concept. *WFCC Newsletter* July 2006:14–28.

Lindström K, Martínez-Romero E. 2007. International committee on systematics of prokaryotes, subcommittee on the taxonomy of *Agrobacterium* and *Rhizobium*. *Int. J. Syst. Evol. Microbiol*. **57**:1365–1366.

MacLean AM, Finan TM, Sadowsky MJ. 2007. Genomes of the symbiotic nitrogen-fixing bacteria of legumes. *Plant Physiol*. **144**:615–622.

Martens M, Delaere M, Coopman R, De Vos P, Gillis M, et al. 2007. Multilocus sequence analysis of *Ensifer* and related taxa. *Int. J. Syst. Evol. Microbiol*. **57**:489–503.

Martens M, Dawyndt P, Coopman R, Gillis M, De Vos P, et al. 2008. Advantages of multilocus sequence analysis for taxonomic studies: a case study using 10 housekeeping genes in the genus *Ensifer* (including former *Sinorhizobium*). *Int. J. Syst. Evol. Microbiol*. **58**:200–214.

Nandasena KG, O'Hara GW, Tiwari RP, Willlems A, Howieson JG. 2007. *Mesorhizobium ciceri* biovar *biserrulae*, a novel biovar nodulating the pasture legume *Biserrula pelecinus* L. *Int. J. Syst. Evol. Microbiol*. **57**:1041–1045.

Osaloo SK, Maassoumi AA, Murakami N. 2005. Molecular systematics of the Old World *Astragalus* (Fabaceae) as inferred from nrDNA ITS sequence data. *Syst. Brittonia*. **57**:367–381.

Polhill RM. 1981a. *Papilionoideae*. In Polhill RM, Raven PH, eds. *Advances in Legume Systematics*, Part 1. Survey, UK: Royal Botanic Gardens, Kew, pp. 191–198.

Polhill RM. 1981b. *Galegeae*. In Polhill RM, Raven PH, eds. *Advances in Legume Systematics*, Part 1. Survey, UK: Royal Botanic Gardens, Kew, pp. 357–363.

Sanderson MJ, Wojciechowski MF. 2000. Improved bootstrap confidence limits in large-scale phylogenies, with an example from Neo-Astragalus (Leguminosae). *Syst. Biol*. **49**:671–685.

Swofford DL. 2001. *Paup**: Phylogenetic Analysis Using Parsimony (* and Other Methods)*. Sunderland, MA: Sinauer, version 4.0b8.

Tan ZY, Xu XD, Wang ET, Gao JL, Martinez-Romero E, et al. 1997. Phylogenetic and genetic relationships of *Mesorhizobium tianshanense* and related rhizobia. *Int. J. Syst. Bacteriol*. **47**: 874–879.

Terefework Z, Nick G, Suomalainen S, Paulin L, Lindström K. 1998. Phylogeny of *Rhizobium galegae* with respect to other rhizobia and agrobacteria. *Int. J. Syst. Bacteriol*. **48**: 349–356.

Terefework Z, Kaijalainen S, Lindström K. 2001. AFLP fingerprinting as a tool to study the genetic diversity of *Rhizobium galegae* isolated from *Galega orientalis* and *Galega officinalis*. *J. Biotechnol*. **91**:169–180.

Thompson JD, Gibson TJ, Plewniak F, Jeanmougin F, Higgins DG. 1997. The Clustal X windows interface: Flexible strategies for multiple sequence alignment aided by quality analysis tools. *Nucleic Acids Res*. **25**:4876–4882.

Tian CF, Wang ET, Wu LJ, Han TX, Chen WF, et al. 2008. *Rhizobium fabae* sp. nov., a bacterium that nodulates *Vicia faba*. *Int. J. Syst. Evol. Microbiol*. **58**:2871–2875.

Tiesman J, Rizzino A. 1991. A rapid and reliable method for the purification of high-quality plasmid DNA for double-stranded sequencing. *Biotechniques* **10**:326–328.

Turner SL, Young JP. 2000. The glutamine synthetases of rhizobia: phylogenetics and evolutionary implications. *Mol. Biol*. **17**: 309–319.

Turner SL, Zhang XX, Li FD, Young JP. 2002. What does a bacterial genome sequence represent? Misassignment of MAFF 303099 to the genospecies *Mesorhizobium loti. Microbiology* **148**: 3330–3331.

Van Berkum P, Terefework Z, Paulin L, Suomalainen S, Lindström K, et al. 2003. Discordant phylogenies within the *rrn* loci of Rhizobia. *J. Bacteriol*. **185**:2988–2998.

Vincent JM. 1970. *A Manual for the Practical Study of Root Nodule Bacteria*. Oxford: Blackwell Scientific.

Vinuesa P, Silva C, Lorite MJ, Izaguirre-Mayoral ML, Bedmar EJ, et al. 2005a. Molecular systematics of rhizobia based on maximum likelihood and Bayesian phylogenies inferred from *rrs, atpD, recA* and *nifH* sequences, and their use in the classification of *Sesbania* microsymbionts from Venezuelan wetlands. *Syst. Appl. Microbiol*. **28**:702–716.

Vinuesa P, León-Barrios M, Silva C, Willems A, Jarabo-Lorenzo A, et al. 2005b. *Bradyrhizobium canariense* sp.nov., an acid-tolerant endosymbiont that nodulates endemic genistoid legumes (Papilionoideae: Genisteae) growing in the Canary Islands, along with *B. japonicum* bv. *genistearum, Bradyrhizobium* genospecies α and *Bradyrhizobium* genospecies β. *Int. J. Syst. Evol. Microbiol*. **55**:569–575.

Vos PR, Hogers M, Bleeker M, Rejans T, Van de Lee, et al. 1995. AFLP: A new technique for DNA fingerprinting. *Nucleic Acids Res*. **23**:4407–4414.

Wei GH, Tan ZY, Zhu ME, Wang ET, Han SZ, et al. 2003. Characterization of rhizobia isolated from legume species within the genera *Astragalus* and *Lespedeza* grown in the Loess Plateau of China and description of *Rhizobium loessense* sp. nov. *Int. J. Syst. Evol. Microbiol*. **53**:1575–1583.

Weir BS, Turner SJ, Silvester WB, Park DC, Young JM. 2004. Unexpectedly diverse *Mesorhizobium* strains and *Rhizobium leguminosarum* nodulate native legume genera of New Zealand, while introduced legume weeds are nodulated by *Bradyrhizobium* species. *Appl. Environ. Microbiol*. **70**:5980–5987.

Wojciechowski MF, Lavin M, Sanderson MJ. 2004. A phylogeny of legumes (Leguminosae) based on analysis of the plastid *mat*K gene resolves many well-supported subclades within the family. *Am. J. Botany* **91**:1846–1862.

Young JP, Haukka KE. 1996. Diversity and phylogeny of rhizobia. *New Phytol*. **133**:87–94.

Zhang XX, Turner SL, Guo XW, Yang HJ, Debellé F, et al. 2000. The common nodulation genes of *Astragalus sinicus* rhizobia are conserved despite chromosomal diversity. *Appl. Environ. Microbiol*. **66**:2988–2995.

Chapter 7

Amplified rDNA Restriction Analysis (ARDRA) for Identification and Phylogenetic Placement of 16S-rDNA Clones

MENACHEM Y. SKLARZ, ROEY ANGEL, OSNAT GILLOR, AND INES M. SOARES

7.1 INTRODUCTION

Amplified ribosomal DNA restriction analysis (ARDRA), an extension of restriction fragment-length polymorphism (RFLP) analysis, is a tool that relies on DNA polymorphism to study microbial diversity [Deng et al., 2008]. In the ARDRA approach, the 16S rRNA gene is amplified by PCR, using either universal or genus-specific primer sets. The amplicon is digested by restriction endonucleases (REs), followed by separation of the fragments on high-percentage agarose or acrylamide gels. The emerging profiles are then used either to cluster the community into genotypic groups or for strain typing, both in accordance with the digested fragments' profile [Tiedje et al., 1999].

Classically, ARDRA has been employed on either isolated strains or clones (as part of a clone-screening step prior to sequencing), with the aim of providing a level of insight into the phylogeny of unique operational taxonomic units (OTUs) within a community, or as a screening tool to identify clones of interest for sequencing [Nocker et al., 2007]. However, computer analyses have demonstrated that ARDRA clustering of isolated clones requires at least four tetrameric REs and that, even then, sequence similarities >96% may not be distinguished [Moyer et al., 1996], which is less than the threshold required for defining unique species [Gevers et al., 2005]. This observation calls into question the discriminatory power of techniques such as ARDRA, based on restriction of 16S rRNA amplicons.

The ARDRA technique was developed to identify human pathogens [Vaneechoutte et al., 1995] and was therefore focused on specific bacterial groups; later, the technique was used to characterize environmental communities and to monitor the diversity of specific strains in their respective ecosystems [Pukall et al., 1998]. However, studies using ARDRA to predict isolate clustering met with only limited success. A study on the microbial diversity of soil iron–manganese nodules by ARDRA followed by sequencing of the emerging unique profiles revealed inconsistencies between the two analyses: Only 80% of the unique ARDRA profiles resulted in unique isolates [Zhang et al., 2008]. *Rhizobia* species isolated from nodules of *Vicia* were identified using ARDRA, RFLP of the 16S–23S internally transcribed spacer (ITS-RFLP), and sequencing of the 16S rRNA gene: For the most part, the phylogenetic relationships derived from the 16S rRNA sequences supported the relationships estimated from the ARDRA and ITS-RFLP, although some discrepancies were detected [Lei et al., 2008]. Biolog™ and ARDRA profiles detected changes in the structure of a community of cultivable heterotrophic bacteria in Cr-contaminated vs. uncontaminated soils [Viti and Giovannetti, 2005], but these restriction profiles were verified

Handbook of Molecular Microbial Ecology, Volume I: Metagenomics and Complementary Approaches, First Edition. Edited by Frans J. de Bruijn.

by comparison to the profiles of known isolates rather than by sequencing. Attempts to apply ARDRA profiles using a panel of six enzymes to distinguish *Ralstonia* and *Pandoraea* strains isolated from the respiratory tract of cystic fibrosis patients were partially successful: All of the *Ralstonia*, but not *Pandoraea*, strains tested could be differentiated [Segonds et al., 2003].

Attempts to use ARDRA to identify species within a specific genus, based on amplicons generated by group-specific PCR primer pairs, were only partially successful: *Mycoplasma* species isolated from cats [Criado-Fornelio et al., 2003], birds [Kiss et al., 1997], and humans [Stakenborg et al., 2005] were identified using ARDRA, although the isolates' identities were not always verified by sequencing. Similarly, ARDRA fingerprinting has been applied for the identification of *Lactobacillus* strains isolated from grape must and wine [Rodas et al., 2003], dairy products [Giraffa et al., 1998], and fecal isolates [Ventura et al., 2000]. However, the tested isolates were not verified by sequencing but by comparing the emerging ARDRA profiles to predicted profiles of cultured *Lactobacillus* strains.

In-silico digestion of 16S rRNA sequences from species and subspecies of *Mycoplasma*, and in-vitro digestion of isolated 16S rRNA fragment amplicons were shown to match, thus enabling the strain's distinction [Stakenborg et al., 2005]. Identifying species of *Bdellovibrio*-and-like microorganisms using ARDRA groups mostly reflected their diversity and phylogenetic affiliation when compared to the isolates' 16S sequence clusters [Davidov and Jurkevitch, 2004]. In contrast, ARDRA fingerprinting could not distinguish between *Acinetobacter* genomic species, even at a low cutoff level [Koeleman et al., 1998]. Similarly, isolates of nitrogen-fixing *Sphingomonas* in rice plants were sorted by ARDRA and ribosomal rRNA intergenic spacer analysis (RISA) followed by sequencing of the 16S rRNA gene [Videira et al., 2009]. Although four different REs were used for the ARDRA, the obtained clustering was not equivalent to that of the RISA or to the sequencing results, suggesting that identification of this family by ARDRA might be limited. Finally, *Brevibacillus* species isolated from clinical, dairy, and industrial environments were discriminated using ARDRA (the amplicons were digested with five REs), and the emerging profiles revealed inconsistencies when compared to fatty-acid and sequencing analyses [Logan et al., 2002].

In general, ARDRA is a highly useful tool for detecting structural changes in relatively simple microbial communities, but it is not the method of choice to measure diversity or to detect specific phylogenetic groups [Nocker et al., 2007]. The studies cited here could not clearly establish the feasibility of using ARDRA to distinguish species within a given community. In this chapter, we summarize the experimental ARDRA procedure and evaluate its predictive power. We explore two directions in which ARDRA could be used to predict clone identity or phylogeny: (a) environmental clone identification via profile matching to theoretically computed fragmentation profiles and (b) clustering of ARDRA fragmentation profiles in comparison to parallel sequence-based clustering.

7.2 EXPERIMENTAL PROCEDURES

7.2.1 General Considerations

7.2.1.1 Choice of Primers and Restriction Enzymes. For general profiling of clone libraries and unknown isolates, universal 16S rRNA encoding gene primers should be used. Use of primers that yield close to the full length of the 16S rRNA gene (~1500 bp) is recommended and, in any case, they should yield no less than 1000 bp since digestion of shorter amplicons will most likely not produce enough fragments to allow for clear separation between phylotypes. Suggested primers are listed in **Table 7.1**. The most commonly used bacterial primers are the pair 27F and 1492R. However, while these primers yield a long fragment, their coverage of bacterial diversity is somewhat limited. An alternative reverse primer that can be used to screen bacterial strains and can potentially yield better coverage is 1389R. For the Archaea, the primer pair 109F and 1397R yields a sufficiently long fragment and provides near-universal coverage. The reverse primer 1406R is universal and covers both archaeal and bacterial domains, and it can be used as an alternative for the above-mentioned reverse primers.

7.2.1.2 Profiling of Specific Taxonomic Groups. The use of ARDRA is not restricted to profiling microbial communities and isolates using general primers. Indeed, many successful implementations of ARDRA include the screening of specific microbial taxa such as *Bacillus* [Wu et al., 2006], *Mycobacterium* [Vaneechoutte et al., 1993], and *Acinetobacter* [Vaneechoutte et al., 1995]. Limiting ARDRA to screening members of a specific taxon reduces the complexity of the generated data and might therefore be implemented without the need for in-silico analysis. However, due to the limited amount of information available in the rRNA genes and the information-reducing nature of ARDRA, distinguishing between closely related taxa might be difficult, if not impossible. When profiling a narrow range of phylotypes in a community through group-specific primers, or when using universal primers for a low-diversity sample, one can attempt to use ARDRA profiles to obtain the community fingerprint [Arturo et al., 1995].

Table 7.1 Primers Used for ARDRA

Primer Name	Sequence	Domain	Reference
Forward Primers			
27F	AGAGTTTGATCCTGGCTCAG	Bacteria	Lane [1991]
109F	ACKGCTCAGTAACACGT	Archaea	Grosskopf et al. [1998]
Reverse Primers			
1492R	GGTTACCTTGTTACGACTT	Bacteria	Lane [1991]
1389R	ACGGGCGGTGTGTACAAG	Bacteria	Osborn et al. [2000]
1397R	GTGTGCAAGGRGCAGGGA	Archaea	Swan et al. [2010]
1406R	ACGGGCGGTGTGTRC	Universal	Lane [1991]

7.2.1.3 Functional Genes. Functional genes are frequently used in RFLP analysis as powerful markers for the identification of specific microbial groups in complex environments, especially when a specific biochemical reaction is the focus of the study. Functional genes also tend to be less conserved than rRNA genes, and thus the chance of producing identical patterns for different phylotypes is reduced. Nevertheless, the primers most often used for the amplification of functional genes produce only short fragments ($<$1000 bp) and are therefore unsuitable for ARDRA. A promising alternative to the 16S rRNA gene is the large rRNA subunit, namely 23S, in prokaryotes. The 23S rRNA gene is potentially more informative than the 16S rRNA due to its larger size ($\sim$2900 vs. $\sim$1500 bp, respectively) while being as conserved and universal as its smaller counterpart (see Chapter 3, Vol. I.)

7.2.1.4* In-Silico *Analysis of Enzymes and Primers. When designing or implementing a new ARDRA protocol, special attention must be given to the coverage and specificity of the primers, as well as to the resolution power of the REs, with a focus on the restriction sites and the number of the resulting fragments within the PCR amplicon. Because no primer pair is truly universal and virtually no group-specific primer pair can cover an entire phylum or family, excluding species alien to the designated group, one should begin by selecting a primer pair that will sufficiently cover the phylogenetic group or groups of interest. The free online toolkit MICA (http://mica.ibest.uidaho.edu/primer.php) offers a function for the in-silico analysis of primers through which one can test the coverage of new or previously published primers against curated public databases.

Similarly, the resolving power of the selected REs should be tested in silico, to find distinct phylotypes that might yield an identical restriction pattern. This is easily done when specific sequence data are available for the explored microbes. However, predictions can also be made using proxy sequences of the expected community or isolates obtained from public databases. Batch restriction (http://www.mpi-marburg.mpg.de/downloads/conrad/) is an open source tool, written in MATLAB (www.mathworks.com/) but available as an executable file as well, which generates batch in-silico restriction maps. The program can be used to generate full restriction maps using the 220 commercially available REs.

7.2.1.5 PCR Template. As with all PCRs, the preferred template is purified DNA or cDNA. However, PCR amplification using whole cells has been frequently reported and can significantly increase efficiency by reducing the process time and costs. While the preparation of template from colonies can vary with the type of bacteria, the following method is often successful:

1. Pick a colony with a toothpick or pipette tip, carrying as much cell material as possible.
2. Soak the toothpick in a tube containing approximately 10 μl of molecular biology-grade water.
3. Incubate in a thermocycler at 95°C for 10 min (the use of a heat block at this stage is not recommended because the sample might evaporate).
4. Vigorously vortex the solution for a few seconds.
5. Spin down the solution for a few seconds to pellet cells.
6. Use 1 μl of the supernatant as the PCR template.

7.2.1.6 Choice of Electrophoresis Conditions and Reagents. While Tris–acetate–EDTA buffer (TAE) is usually selected for most electrophoresis protocols, ARDRA electrophoresis is normally performed with Tris–borate–EDTA (TBE) buffer since it maintains stable performance during extended electrophoresis. However, TBE buffer tends to precipitate and thus should be prepared at a concentration of no more than 10× and used within a month.

7.2.1.7 *Analysis of Results and Data Storage.*

For experiments involving 100 or less ARDRA profiles, a direct visual comparison of the patterns might suffice. For larger projects or when long-term data storage is required, it is recommended that gel-imaging software be used. Gel-Compar II (Applied Maths, Belgium) is a proprietary software that can digitize, analyze, and store large databases of gel images.

7.2.2 Materials

1. PCR kit
2. Primers and restriction enzymes
3. Electrophoresis-grade agarose
4. 1(X) TBE buffer
5. Gel-imaging device
6. Gel-loading buffer
7. Optional: gel-imaging software

7.2.3 Procedure

7.2.3.1 PCR. It is recommended that the PCR be performed in relatively small volumes, enough to validate the success of the PCR and to digest the amplicon (the enzymatic digestion can later be performed in the same tube, thus saving on time and materials). The following proposed PCR conditions work for most 16S rRNA gene primers and PCR kits. The exact reaction conditions and cycling program should be determined according to the manufacturer's instructions and optimized for the specific sample.

1. Prepare the mixture as described in **Table 7.2** (it is recommended that a batch be prepared in accordance with the number of samples).
2. Run the reaction in a thermocycler (see **Table 7.3** for suggested run conditions).
3. Load 2–3 μl on an agarose gel to validate the success of the reaction.

Table 7.2 Composition of PCR Solution

Reagent	Final Concentration	One Reaction (μl)
PCR H_2O		7.088
PCR buffer 10(X)	1(X)	1.2
dNTP mixture (2.0 mM each)	0.2 mM	1.2
$MgCl_2$ (25 mM)	1.5 mM	0.72
BSA (20 μg/μl)	0.8 μg/μl	0.48
Primer F (25 μM)	0.25 μM/μl	0.12
Primer R (25 μM)	0.25 μM/μl	0.12
Taq polymerase (5 U/μl)	0.36 U	0.072
Template		1
Final volume		**12**

Table 7.3 Suggested Thermocycler Running Conditions for PCR

Step	Temperature	Time	Number of Cycles
Initial denaturation	95°C	4 min	1
Denaturation	95°C	0.5–1 min	25–30
Annealing	42–65°C	0.5–1 min	
Extension	72°C	1 min/kbp	
Final extension	72°C	10 min	1
Hold	4°C	Indefinite	1

7.2.3.2 Digestion with REs. The required amounts of digestion buffer and enzyme should be based on the manufacturer's instructions. Performing the digestion reaction on nonpurified PCR product saves time and reagents, but might lead to inefficient digestion. It is therefore recommended that excess amounts of enzyme be added and the incubation be performed for an extended period of time so as to ensure complete digestion (incomplete digestion may lead to random patterns that cannot be reproduced). Note, however, that some REs should not be used for extended digestion periods because this might lead to nonspecific digestion. If complete digestion is unobtainable under these conditions, purification of the PCR product prior to digestion might be required.

Gel Preparation and Running Conditions

1. Prepare 3% (w/v) agarose gel in 1(X) TBE buffer.
2. Load 5–10 μl of the digested sample (make sure to include a marker lane every 4–5 samples to ensure accurate sizing of the bands).
3. Run the samples for 3–5 h (the extended electrophoresis will yield clear and reproducible patterns). Note that due to the variability in gel and bath sizes, final run length and voltage need to be empirically determined.
4. Following electrophoresis, stain the gel in 1× TBE containing ethidium bromide at a final concentration of 100 μg/ml for 20 min and visualize with UV light.

7.3 IN-SILICO PROCEDURES

The procedures described in this section are similar to those presented in a recent report [Sklarz et al., 2009].

7.3.1 ARDRA-Profile-Based Identification of Clones

All sequences in the Ribosomal Database Project (RDP) version 9.60 [Maidak et al., 2001] were scanned, excluding those shorter than 1400 bp or categorized as unidentified, unclassified, or uncultured. The remaining 48,759 sequences were taken for in-silico ARDRA. The sequences were restricted with the REs reported by Moyer et al. [1996], namely: *Alu*I, *Dde*I, *Hae*III, *Hha*I, *Hinf*I, *Mbo*I, *Msp*I, *Rsa*I, and *Taq*I. The restriction was performed using the `restrict` function of the European Molecular Biology Open Software Suite (EMBOSS) [Rice et al., 2000]. The fragment-length range was divided into discrete segments with sizes given by $s_{i+1} = \lfloor s_i \cdot (1 + \varepsilon) \rfloor$, where s_i is the size of segment i and ε is the error tolerance, to give a list of bins with increasing sizes. Each fragment was changed to the closest value in the list of permitted fragment lengths to produce a new, standardized fragmentation profile. The new fragment profiles were searched for duplicates. This procedure was performed for different RE combinations and for different error values; given that agarose gels cannot be analyzed with absolute precision, identification of bacteria by their fragmentation profiles should be robust to a certain error level.

7.3.2 Phylogenetic Analysis of ARDRA Profiles

Determination of the applicability of ARDRA to phylogenetic analysis of the genus *Mycoplasma* was performed as follows: Fifty species of *Mycoplasma* were selected randomly from the RDP, and dendrograms were constructed based on in-silico ARDRA profiles and on RDP sequence data. ClustalW [Larkin et al., 2007] was used to align the 16S rRNA sequences. Gaps at both ends of the alignment were removed and the alignment was resubmitted to ClustalW to produce neighbor-joining phylogenetic trees. The fragment-based trees were constructed by comparing every pair of profiles in the group and assigning a Jaccard index to each. The Jaccard index between two profiles i and j was calculated based on n REs by

$$J_{i,j} = \sum_{RE=1}^{n} I_{i,j}^{RE} \Big/ \sum_{RE=1}^{n} U_{i,j}^{RE},$$

where $\sum_{RE=1}^{n} I_{i,j}^{RE}$ is the intersection and $\sum_{RE=1}^{n} U_{i,j}^{RE}$ the union of the fragment lengths of profiles i and j, calculated for each of the REs separately and summed.

The matrix of Jaccard distances $(1 - J_{i,j})$ was used to construct a neighbor-joining tree using PHYLIP/neighbor [Felsenstein, 2004]. Trees from both sources were compared with PHYLIP/treedist using the symmetric difference method [Robinson and Foulds, 1981], in which the worst result for a comparison between two trees of a finite number of species (represented by n) is $(2 \times n) - 6$; thus, for trees of 50 species, the maximum distance between two trees is $(2 \times 50) - 6 = 94$. All scripts are available upon request from M. Y. sklarz@bgu.ac.il.

7.4 THEORETICAL CONSIDERATIONS

In the aforementioned study [Sklarz et al., 2009], in which 48,759 sequences from the RDP were subjected to in-silico ARDRA, profiles were found to be unique to genera for two or more REs, even at a high error level of 20% in the sizing of the restriction fragments (as might occur in agarose gels). However, at least three enzymes were necessary to differentiate species. Even then, while most patterns referred to unique sequences, exceptions were observed; however, the use of 10 REs reduced the number of shared patterns. These results were in agreement with the study performed by Moyer et al. [1996], even though a much larger database of sequences was considered (48,759 vs. 106 sequences, respectively). In addition, dendrograms calculated from the theoretical fragmentation profiles of random species collections, both intra- and inter-genus, had little in common with the phylogenetic clustering based on the corresponding 16S rRNA sequence [Sklarz et al., 2009]. However, ARDRA-based trees could distinguish between distant groups, albeit to a lower accuracy than the sequence-based dendrograms. These observations were consistent with a study in which authentic strains of *Brevibacillus* could not be assigned to separate clusters using ARDRA [Logan et al., 2002].

While the global clone identification capacity of ARDRA may be limited, higher resolution can potentially be obtained when amplicons of specific bacterial groups are digested by a small number of REs. For instance, the REs *Alu*I and *Bfu*I (or *Hpy*F10VI) were used to identify 116 species of the genus *Mycoplasma* [Stakenborg et al., 2005]. In our current study, in-silico ARDRA was assessed using 421 sequences from the *Mycoplasma* genus. The sequences were used to analyze both identification and clustering of species based on their theoretical ARDRA profiles. Using eight REs (*Alu*I, *Dde*I, *Hae*III, *Hha*I, *Hinf*I, *Mbo*I, *Rsa*I, and *Taq*I), two patterns were shared by two species each (*M. agalactiae–M. bovis* and *M. collis–M. cricetuli*). Sequence similarity between those pairs was very high (>99%), and hence ARDRA could not resolve them. At the same time, when one RE was used (*Alu*I), five species (*M. anatis, M. columborale, M. gallopavonis, M. mustelae*, and *M. oxoniensis*) shared a common pattern, even though their sequence similarity was 93–97%. When the RE *Dde*I was added to *Alu*I, the group was split in two: *M. columborale,*

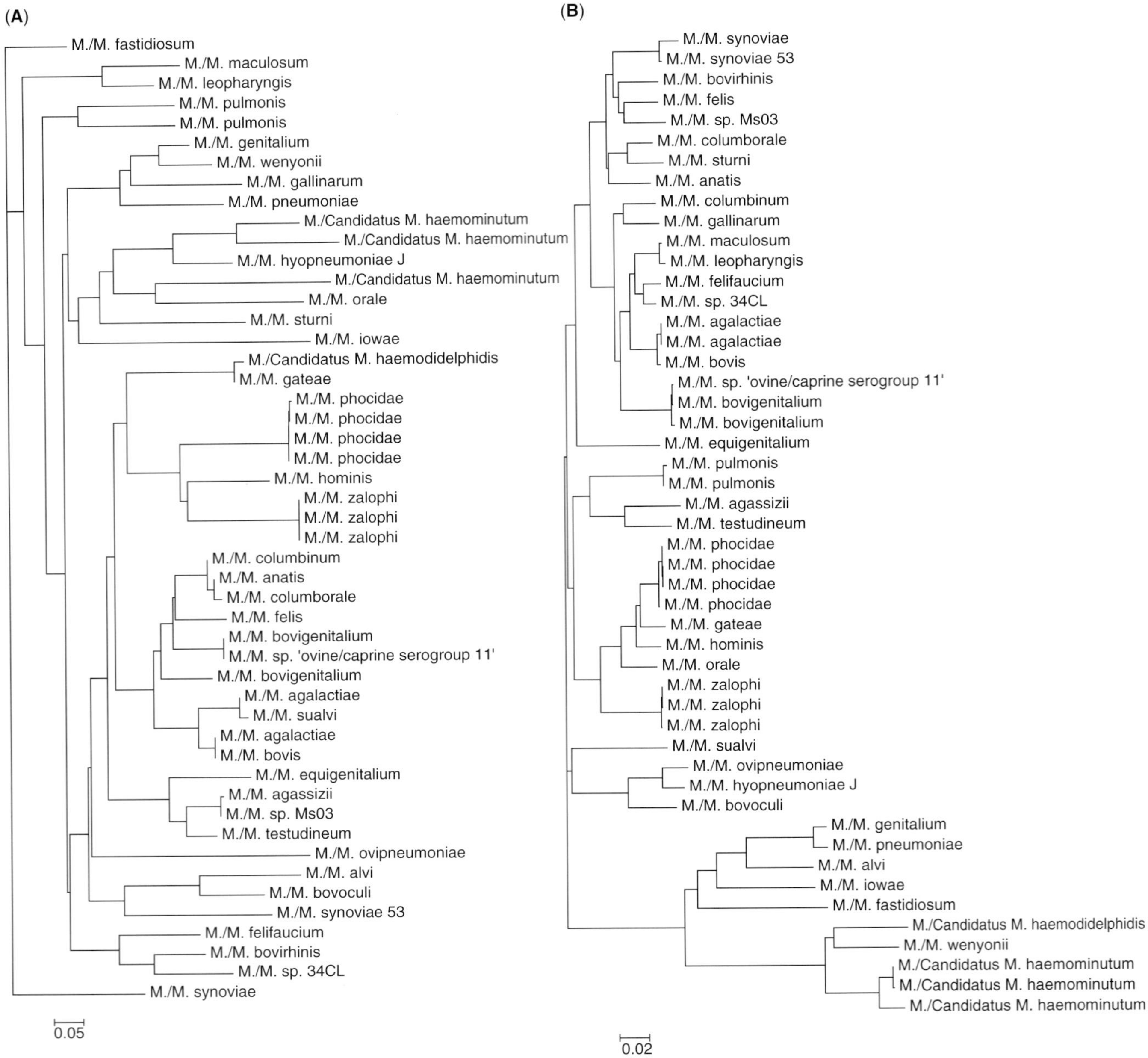

Figure 7.1 ARDRA-based (**A**) and 16S rRNA sequence-based (**B**) phylogenetic tree of 50 representative species of the genus *Mycoplasma*.

M. gallopavonis, and *M. oxoniensis* in one group, and *M. anatis* and *M. mustelae* in the other. If ARDRA-based distinction between these species is required, a search for an appropriate RE set should be conducted as described above.

To compare phylogenetic clustering based on 16S rRNA sequences and ARDRA using *Mycoplasma* sequences, as previously described for random groups from the whole RDP database [Sklarz et al., 2009], 50 *Mycoplasma* species were randomly selected. When the complete set of eight REs was used, a symmetric difference analysis between the two phylograms produced a score of 66 (out of 94), indicating that there is a certain degree of similarity between the trees. However, if only *Alu*I and *Dde*I were used, the score was 82 and the trees varied in the phylogenetic positioning of many of the species **(Fig. 7.1)**.

While it is recognized that the theoretical analysis presented above is not exhaustive, and that there may be sets of REs that are capable of producing high-resolution differentiation and phylogenetic analysis of certain bacterial groups, the resources required to achieve reliable differentiation are often restrictive. In these cases, the option of direct sequencing presents a worthy alternative.

REFERENCES

Arturo A, Odelsor DA, Hichey RF, Tiedje JM. 1995. Bacterial community fingerprint of amplified 16S and 16S–23S ribosomal DNA gene sequences and restriction endonuclease analysis (ARDRA). In Akkermans ADL, Elsas JDV, de Bruijn FJ, eds. *Molecular Microbial Ecology Manual*. Dordrecht: Kluwer Academic Publishers, pp. 1–8.

Criado-Fornelio A, Martinez-Marcos A, Buling-Saraña A, Barba-Carretero JC. 2003. Presence of *Mycoplasma haemofelis, Mycoplasma haemominutum* and piroplasmids in cats from southern Europe: A molecular study. *Vet. Microbiol*. **93**:307–317.

Davidov Y, Jurkevitch E. 2004. Diversity and evolution of *Bdellovibrio*-and-like organisms (BALOs), reclassification of *Bacteriovorax starrii* as *Peredibacter starrii* gen. nov., comb. nov., and description of the *Bacteriovorax-Peredibacter* clade as *Bacteriovoracaceae* fam. nov. *Int. J. Syst. Evol. Microbiol.* **54**:1439–1452.

Deng W, Xi D, Mao H, Wanapat M. 2008. The use of molecular techniques based on ribosomal RNA and DNA for rumen microbial ecosystem studies: A review. *Mol. Bio. Rep*. **35**:265–274.

Felsenstein J. 2004. PHYLIP (phylogeny inference package) version 3.6. Distributed by the author. Seattle, WA: Department of Genome Sciences, University of Washington.

Gevers D, Cohan FM, Lawrence JG, Spratt BG, Coenye T, et al. 2005. Re-evaluating prokaryotic species. *Nat. Rev. Microbiol*. **3**:733–739.

Giraffa G, Vecchi PD, Rossetti L. 1998. Identification of *Lactobacillus delbrueckii* subspecies *bulgaricus* and subspecies *lactis* dairy isolates by amplified rDNA restriction analysis. *J. Appl. Microbiol*. **85**:918–924.

Grosskopf R, Janssen PH, Liesack W. 1998. Diversity and structure of the methanogenic community in anoxic rice paddy soil microcosms as examined by cultivation and direct 16S rRNA gene sequence retrieval. *Appl. Environ. Microbiol*. **64**:960–969.

Kiss I, Matiz K, Kaszanyitzky É, Chávez Y, Johansson K-E. 1997. Detection and identification of avian mycoplasmas by polymerase chain reaction and restriction fragment length polymorphism assay. *Vet. Microbiol*. **58**:23–30.

Koeleman JGM, Stoof J, Biesmans DJ, Savelkoul PHM, Vandenbroucke-Grauls CMJE. 1998. Comparison of amplified ribosomal DNA restriction analysis, random amplified polymorphic DNA analysis, and amplified fragment length polymorphism fingerprinting for identification of *Acinetobacter* genomic species and typing of *Acinetobacter baumannii*. *J. Clin. Microbiol*. **36**: 2522–2529.

Lane DJ. 1991. 16S/23S rRNA sequencing. In Stackebrandt E, Goodfellow M, eds. *Nucleic Acid Techniques in Bacterial Systematics*. New York: John Wiley & Sons, pp. 115–175.

Larkin MA, Blackshields G, Brown NP, Chenna R, McGettigan PA, et al. 2007. Clustal W and clustal X version 2.0. *Bioinformatics* **23**:2947–2948.

Lei X, Wang ET, Chen WF, Sui XH, Chen WX. 2008. Diverse bacteria isolated from root nodules of wild *Vicia* species grown in temperate region of China. *Arch. Microbiol*. **190**:657–671.

Logan NA, Forsyth G, Lebbe L, Goris J, Heyndrickx M, et al. 2002. Polyphasic identification of *Bacillus* and *Brevibacillus* strains from clinical, dairy and industrial specimens and proposal of *Brevibacillus invocatus* sp. nov. *Int. J. Syst. Evol. Microbiol*. **52**: 953–966.

Maidak BL, Cole JR, Lilburn TG, Parker CT Jr, Saxman PR, Farris RJ, Garrity GM, Olsen GJ, Schmidt TM, Tiedje JM. 2001. The RDP-II (ribosomal database project). *Nucleic Acids Res*. **29**:173–174.

Moyer CL, Tiedje JM, Dobbs FC, Karl DM. 1996. A computer-simulated restriction fragment length polymorphism analysis of bacterial small-subunit rRNA genes: Efficacy of selected tetrameric restriction enzymes for studies of microbial diversity in nature. *Appl. Environ. Microbiol*. **62**:2501–2507.

Nocker A, Burr M, Camper A. 2007. Genotypic microbial community profiling: A critical technical review. *Microb. Ecol*. **54**:276–289.

Osborn AM, Moore ERB, Timmis KN, 2000. An evaluation of terminal-restriction fragment length polymorphism (T-RFLP) analysis for the study of microbial community structure and dynamics. *Environ. Microbiol*. **2**:39–50.

Pukall R, Brambilla E, Stackebrandt E. 1998. Automated fragment length analysis of fluorescently-labeled 16S rRNA after digestion with 4-base cutting restriction enzymes. *J. Microbiol. Methods* **32**:55–63.

Rice P, Longden I, Bleasby A. 2000. EMBOSS: The European molecular biology open software suite. *Trends Genet*. **16**:276–277.

Robinson D, Foulds L. 1981. Comparison of phylogenetic trees. *Math. Biosci*. **53**:131–147.

Rodas AM, Ferrer S, Pardo I. 2003. 16S-ARDRA, a tool for identification of lactic acid bacteria isolated from grape must and wine. *Syst. Appl. Microbiol*. **26**:412–422.

Segonds C, Paute S, Chabanon G. 2003. Use of amplified ribosomal DNA restriction analysis for identification of *Ralstonia* and *Pandoraea* species: Interest in determination of the respiratory bacterial flora in patients with cystic fibrosis. *J. Clin. Microbiol*. **41**:3415–3418.

Sklarz MY, Angel R, Gillor O, Soares MIM. 2009. Evaluating amplified rDNA restriction analysis assay for identification of bacterial communities. *A. van Leeuw. J. Microbiol*. **96**:659–664.

Stakenborg T, Vicca J, Butaye P, Maes D, De Baere T, et al. 2005. Evaluation of amplified rDNA restriction analysis (ARDRA) for the identification of *Mycoplasma* species. *BMC Infect. Dis*. **5**: 46.

Swan BK, Ehrhardt CJ, Reifel KM, Moreno LI, Valentine DL. 2010. Archaeal and bacterial communities respond differently to environmental gradients in anoxic sediments of a California hypersaline lake, the Salton sea. *Appl. Environ. Microbiol*. **76**:757–768.

Tiedje JM, Asuming-Brempong S, Nusslein K, Marsh TL, Flynn SJ. 1999. Opening the black box of soil microbial diversity. *Appl. Soil Ecol*. **13**:109–122.

Vaneechoutte M, De Beenhouwer H, Claeys G, Verschraegen G, De Rouck A, Paepe N, Elaichouni A, Portaels F. 1993. Identification of *Mycobacterium* species by using amplified ribosomal DNA restriction analysis. *J. Clin. Microbiol*. **31**:2061–2065.

Vaneechoutte M, Dijkshoorn L, Tjernberg I, Elaichouni A, de Vos P, Claeys G, Verschraegen G. 1995. Identification of *Acinetobacter* genomic species by amplified ribosomal DNA restriction analysis. *J. Clin. Microbiol*. **33**:11–15.

Ventura M, Casas IA, Morelli L, Callegari ML. 2000. Rapid amplified ribosomal DNA restriction analysis (ARDRA) identification of *Lactobacillus* spp. isolated from fecal and vaginal samples. *Syst. Appl. Microbiol*. **23**:504–509.

Videira SS, de Araujo JLS, da Silva Rodrigues L, Baldani VLD, Baldani JI. 2009. Occurrence and diversity of nitrogen-fixing *Sphingomonas* bacteria associated with rice plants grown in Brazil. *FEMS Microbiol. Lett*. **293**:11–19.

Viti C, Giovannetti L. 2005. Characterization of cultivable heterotrophic bacterial communities in Cr-polluted and unpolluted soils using Biolog and ARDRA approaches. *Appl. Soil. Ecol*. **28**:101–112.

Wu XY, Walker MJ, Hornitzky M, Chin J. 2006. Development of a group-specific PCR combined with ARDRA for the identification of *Bacillus* species of environmental significance. *J. Microbiol. Methods*. **64**:107–119.

Zhang L, Liu F, Tan W, Feng X, Zhu Y, He J. 2008. Microbial DNA extraction and analyses of soil iron-manganese nodules. *Soil. Biol. Biochem*. **40**:1364–1369.

Chapter 8

Clustering-Based Peak Alignment Algorithm for Objective and Quantitative Analysis of DNA Fingerprinting Data

SATOSHI ISHII, KOJI KADOTA, AND KEISHI SENOO

8.1 INTRODUCTION

Various DNA fingerprinting techniques have been broadly used in the area of molecular diagnostics, medical microbiology, epidemiological analyses, environmental microbiology, microbial ecology, and so forth. DNA fingerprinting techniques include, but are not limited to, the following: restriction fragment-length polymorphism (RFLP) and its derivative amplified ribosomal DNA restriction analysis (ARDRA), repetitive extragenic palindromic PCR (rep-PCR), ribosomal intergenic spacer analysis (RISA), and denaturing gradient gel electrophoresis (DGGE). These methods are used for different purposes. While ARDRA and rep-PCR can characterize microbes at the genus and subspecies/strain level, respectively [Chèneby et al., 2000; Ishii and Sadowsky, 2008; Rademaker et al., 2008; Sklarz et al., 2009], RISA and DGGE analyses are usually used to compare complex microbial communities [Borneman and Triplett, 1997; Saito et al., 2007; Muyzer et al., 1993]. Please see Chapters 4, 5, and 7 in Volume I for detailed descriptions of each method.

These DNA fingerprinting techniques are still useful in the era of metagenomics, since they are inexpensive to perform. Ishii et al. [2009b] detected specific microbial populations responsive to denitrification-inducing conditions by DGGE analysis, which were also identified by large-scale 16S rRNA gene sequencing. Gilbert et al. [2009] examined the microbial communities by terminal RFLP (T-RFLP) DNA fingerprinting and high-throughput pyrosequencing of the 16S rRNA gene. Similar results were obtained by the two approaches, but the resolution was higher when 16S rRNA pyrosequencing was used [Gilbert et al., 2009]. These results indicate that DNA fingerprint analysis can be used as a "snapshot" of the community, while 16S rRNA pyrosequencing and metagenomics can be used to acquire a "high-resolution picture" of the community.

While many bioinformatics tools are currently available to analyze metagenomic and other biological data [Kunin et al., 2008; see other chapters in this book], these tools have not been used for analysis of DNA fingerprinting data much. Once a DNA fingerprinting image is converted to numerical data (e.g., matrix data), various bioinformatics tools can be applied.

The major difficulty when constructing matrix data from DNA fingerprinting image is the alignment of multiple peak sets (i.e., peak matching). Although it may be possible to align multiple peaks by eye, it is subjective, labor-intensive, and not applicable for between-gel comparisons [Rademaker and de Bruijn, 2008]. Several bioinformatics software applications are commercially available for gel analyses (e.g., BioNumerics and GelCompar, Applied Math, Sint-Martens-Latem, Belgium), but they are usually expensive and the types of analysis included are limited. Therefore, a free and

Handbook of Molecular Microbial Ecology, Volume I: Metagenomics and Complementary Approaches, First Edition. Edited by Frans J. de Bruijn.

flexible program is desirable for the analysis of DNA fingerprinting data.

In this chapter, we report the use of a R program and a clustering-based algorithm to analyze DNA fingerprinting data. The basic concept was originally proposed to analyze mass spectrometry data [Tibshirani et al., 2004; Yu et al., 2006] and differentially expressed fragments from cDNA-based amplified fragment-length polymorphism (AFLP) data [Kadota et al., 2007]. This approach is simple and high-throughput compared to other clustering algorithms. In addition, visual evaluation of peak alignment can be easily performed [Kadota et al., 2007]. Use of R language is also of advantage since many R packages have been established to statistically analyze various biological data. Therefore the output of our R program (i.e., a distance matrix) can be used for subsequent statistical analyses.

In this chapter, we show the basic concept of our clustering-based peak alignment algorithm. We also show several example analyses using this algorithm, and we compare the results obtained by our complete-linkage clustering algorithm with those obtained by BioNumerics software. Applications of the R statistical tools are also used to analyze DNA fingerprinting data.

8.2 MATERIALS AND METHODS

8.2.1 Dataset

Rep-PCR DNA fingerprinting data obtained previously [Ishii et al., 2006] were used in this study. The rep-PCR was performed with ERIC1R and ERIC2 primers [Versalovic et al., 1991] to characterize 38 *Campylobacter* strains isolated from green alga *Cladophora* in Lake Michigan [Ishii et al., 2006]. The original DNA fingerprint patterns were analyzed using BioNumerics software version 2.5 [Ishii et al., 2006]. By analyzing the same data, we could compare the results obtained by our clustering-based peak alignment algorithm with those obtained by BioNumerics software.

DGGE data obtained previously [Ishii et al., 2009b] were also used in this study. The DGGE targeting the V3 region of the 16S rRNA gene [Muyzer et al., 1993] was performed to compare bacterial communities in wetland rice soils amended with or without denitrification stimulants. In TSNS treatment, nitrate and succinate were added to soil as an electron acceptor and donor, respectively, for denitrification. This condition, together with 24-h anoxic incubation at 30°C, was previously shown to enhance soil-denitrifying activity [Ishii et al., 2009b; Saito et al., 2008]. Soil samples supplemented only with nitrate or succinate were also prepared (TSNI and TSSU treatments, respectively). Two negative controls, one without substrate addition (TSCO treatment) and one without 24-h incubation (TSBA treatment), were also included in the experiment. The triplicate samples were prepared for each treatment (total of 15 samples). The samples from each treatment were pooled and used for the large-scale clone library analysis (>1000 clones each) targeting near-full-length 16S rRNA gene [Ishii et al., 2009b]; therefore, we could compare the results obtained by DGGE analysis with those obtained by clone library analysis.

8.2.2 Data Preparation

The gel images were digitized using a CS Analyzer version 3 (ATTO). The molecular weight of each fingerprint band was calculated based on the calibration curve generated using the DNA size standard included in each gel to normalize the banding patterns. For DGGE analysis, physical distance on the gel was used instead of molecular weight. The intensity of each DNA fragment was used to calculate the peak area. The resulting data consisted of sample name, molecular weight (or physical distance), and peak area of each fragment (see Ishii et al. [2009a] for example input data).

All fragment data from multiple samples were combined and sorted by molecular weight (or physical distance). Molecular weight was then $\log_{10}$-transformed. DNA fragments smaller than the minimum size of the DNA size standard or larger than the maximum size of the DNA size standard were removed.

8.2.3 Clustering Algorithm

The clustering algorithm reported here comprises two steps. In the first step, DNA fragments originating from different samples are clustered together if their molecular weights are identical. In the second step, DNA fragments are clustered together if the distance is smaller than the cutoff value and, at the same time, if they have originated from different samples. In this algorithm, the cutoff value greatly influences the clustering pattern; therefore it should be carefully determined and visually evaluated [Kadota et al., 2007]. After optimization and visual evaluation of the resulting clustering patterns, 0.05 and 5 were used as cutoff values for rep-PCR and DGGE analyses, respectively, in this study.

The peak alignment algorithm is explained below in more detail using the example shown in **Fig. 8.1**. First, all fragments, sorted by their molecular weights, are individually numbered (fragment numbers in Fig. 8.1A). In the first step of the peak alignment algorithm, fragments with identical molecular weights are merged and a unique cluster number is assigned (Fig. 8.1A). The second step of the peak alignment algorithm can be divided into two parts. In the first part (step 2-1), distances between fragments

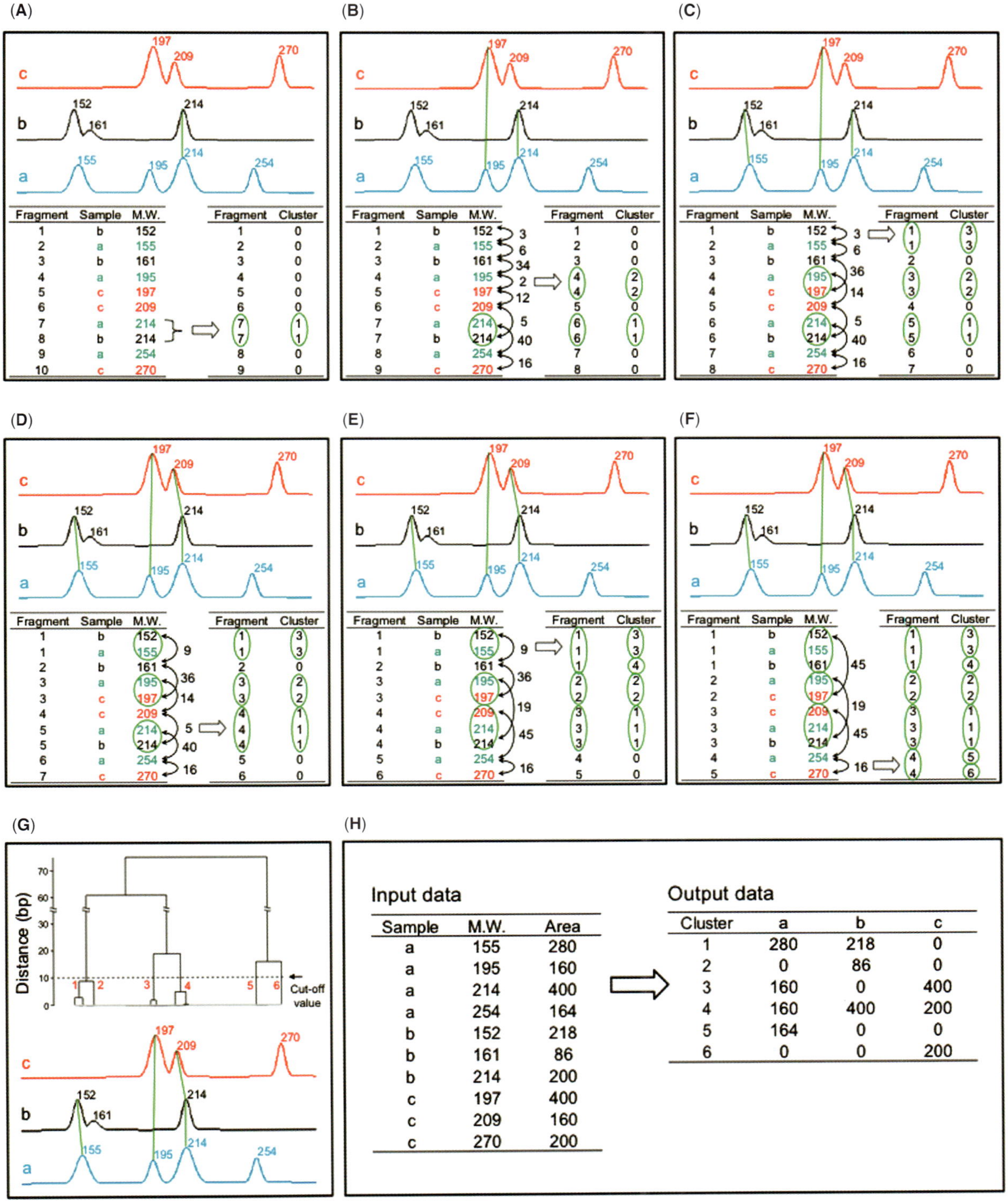

Figure 8.1 Example of the clustering-based peak alignment algorithm used in this study. (**A**) Fragments with identical molecular weights (fragments 7 and 8) were merged, and a unique cluster number "1" was assigned. (**B**) Multiple fragments with the smallest distance (fragments 4 and 5) were merged. Since the merged fragments satisfy the two conditions described in the text (see Section 8.2.3), they were assigned a unique cluster number "2". (**C**) Fragments 1 and 2 were merged, and a cluster number "3" was assigned. (**D**) Fragments 4 and 5 were merged, and a cluster number "1" was assigned. (E) Fragments 1 and 2 were merged, but different cluster numbers "3" and "4," respectively, were assigned (see text for detail). (**F**) Fragments 4 and 5 are merged, but different cluster numbers "5" and "6," respectively, were given (see text for detail). (**G**) As a result, 10 fragments were aligned to form six clusters in this example. (**H**) Input and output data in this example. (Adapted from Ishii et al. [2009a] with permission from the publisher.)

(absolute difference in molecular weights between two fragments) are calculated, based on complete linkage clustering (i.e., furthest distances are considered). In the next part (step 2-2), multiple fragments with the smallest distance (e.g., fragments 4 and 5 in Fig. 8.1B) are merged. A unique cluster number is assigned to the merged fragments if the distance is within the cutoff value (i.e., 10 in this example) and if the merged fragments have originated from different samples (Figs. 8.1B–8.1D). If these two conditions are not satisfied, unique but different cluster numbers are given to each fragment (Figs. 8.1E and 8.1F). As a result, 10 fragments are aligned to form six clusters in this example (Fig. 8.1G).

The resultant output is a matrix of peaks (rows) per sample (columns) (Fig. 8.1H). Each entry can be quantitative data (peak area) or binary data (presence/absence). In this study, quantitative data were used.

8.2.4 Statistical Analyses

The matrix data obtained by the clustering algorithm described above can be used for various statistical tests. The most frequently used analyses are clustering (e.g., construction of a dendrogram) and ordination methods [Fromin et al., 2002; van Verseveld and Röling, 2008]. In this study, a dendrogram was constructed using the pvclust package [Suzuki and Shimodaira, 2006] to calculate the probability values for each cluster by multiscale bootstrap resampling ($n = 1000$). Pearson's product moment correlation coefficient and unweighted pair group method with arithmetic mean (UPGMA, also known as average linkage method) were used to construct a dendrogram. Principal component analysis (PCA) was performed as an example of the ordination methods. In addition, we employed the template match method [Pavlidis and Noble, 2001] using the genefilter package to identify peaks (i.e., DNA bands) specifically appearing in the sample(s) of interest. Peaks showing a similarity index (= 1 − correlation coefficient) of less than 0.2 to the template pattern were defined as specifically enriched peaks in this study. All analyses were performed using R software version 2.8.1.

8.3 RESULTS AND DISCUSSION

8.3.1 Comparison with BioNumerics Software

A total of 565 rep-PCR DNA fragments were obtained from 38 *Campylobacter* strains. These DNA fragments were aligned using our complete linkage clustering algorithm, and the resulting matrix data were used to construct a dendrogram. The same rep-PCR DNA fragment data were also analyzed by using BioNumerics software [Ishii et al., 2006]. Similar groupings were observed between the two dendrograms (**Fig. 8.2**). Since Pearson's product moment correlation coefficient and the UPGMA method was used for both dendrograms, the differences between the two dendrograms should be due to the peak alignment procedure. This suggests that our clustering-based peak alignment algorithm can generate results compatible with those obtained by BioNumerics.

Although bootstrap analysis is widely used to assess uncertainty in phylogenetic trees generated based on nucleic and amino acid sequences [Efron et al. 1996; Felsenstein, 1985], it has not been used much in DNA fingerprinting studies [e.g., Fig. 8.2B; Chèneby et al., 2000; Ishii et al., 2006; Rademaker and de Bruijn, 2008]. The use of R software and pvclust package [Suzuki and Shimodaira, 2006] allows us to evaluate uncertainty in hierarchical cluster by the probability values calculated from bootstrap resampling (Fig. 8.2A).

8.3.2 Identification of Sample-Specific DGGE Bands

DGGE analysis was performed to examine the differences among 15 soil samples (**Fig. 8.3A**). A total of 226 DGGE bands originating from the 15 samples were aligned as described above. A PCA plot generated based on the DGGE profiles showed distinct community structures of TSNS samples from those of the other samples (Fig. 8.3B). Since denitrification stimulant was amended in the TSNS treatment, this may cause a community shift. We then employed a template match method to identify DGGE bands specific to the TSNS-treated samples. A pseudo-heatmap image shows the z-scaled relative intensity of each band among 15 samples (Fig. 8.3C). Bands were sorted by similarity index (= 1 − correlation coefficient) to the TSNS-specific template pattern. Five bands, labeled a, b, c, d, and e in Figures 8.3A and 8.3C, had similarity index of 0.072, 0.045, 0.155, 0.132, and 0.171, respectively, to the template pattern and, therefore, were considered to be TSNS-specific bands.

In many DNA fingerprinting studies, bands specific to samples of interest were identified by eye [e.g., Inaba et al., 2009; Saito et al., 2008]. Using the template match method [Pavlidis and Noble, 2001], we can rank the bands according to their similarity index to the template. This allows us to objectively and quantitatively identify template-specific bands. This method was originally proposed to identify specifically expressed genes in microarray datasets [Pavlidis and Noble, 2001],

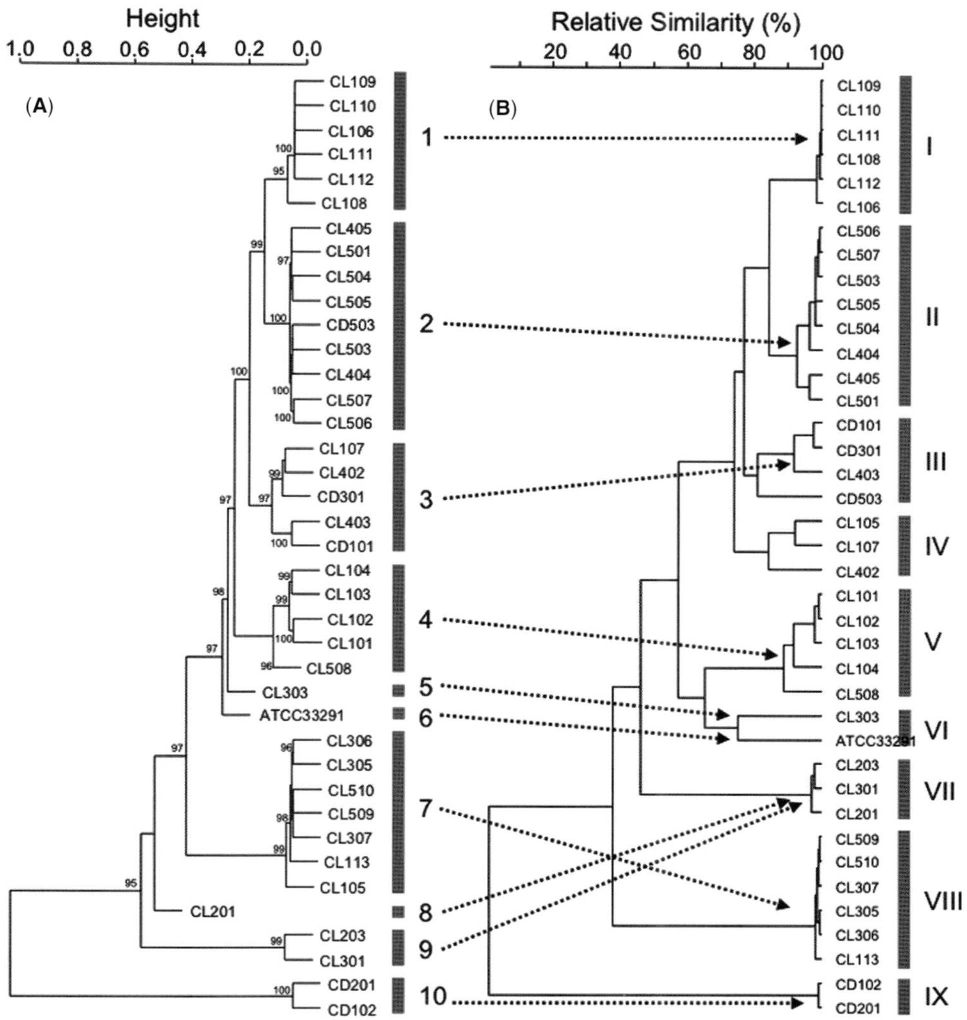

Figure 8.2 Dendrograms based on the rep-PCR DNA fingerprinting of *Campylobacter* strains. Results were analyzed by (**A**) our peak alignment algorithm and (**B**) using BioNumerics software. For both, Pearson's product moment correlation coefficient and UPGMA clustering method were used to construct dendrograms. (Adapted from Ishii et al. 2006, 2009a with permission from the publisher.)

but here we showed that it can also be applied to identify DGGE bands specifically appearing in TSNS samples.

Five bands were identified to be specifically enriched in a denitrification-inductive condition (TSNS samples). Sequence analysis of the excised band suggests that the majority of them originate from the bacteria closely related to the genus *Herbaspirillum* (order Burkholderiales). Similar results were also obtained by comparative, large-scale 16S rRNA gene clone library analysis [Ishii et al., 2009b]. This suggests that DGGE is useful, when analyzed properly, to identify microbial populations responsive to environmental stimuli and can produce results compatible with those obtained by the clone library analysis.

8.3.3 Clustering-Based Algorithm for DNA Fingerprinting and Metagenomic Data Analyses

In this report, we showed the application of complete linkage clustering to align peaks of rep-PCR and DGGE DNA fingerprinting. The resulting distance matrices were used to generate dendrograms and a PCA plot, as well as to identify bands specific to the sample(s) of interest. This approach was also useful to analyze data obtained from mass spectrometry [Tibshirani et al., 2004; Yu et al., 2006], cDNA-AFLP [Kadota et al., 2007], ARDRA, and RISA [Ishii et al., 2009a].

In addition to DNA fingerprinting data analysis, clustering-based algorithm is widely used in various biological data analysis. For example, a clustering-based

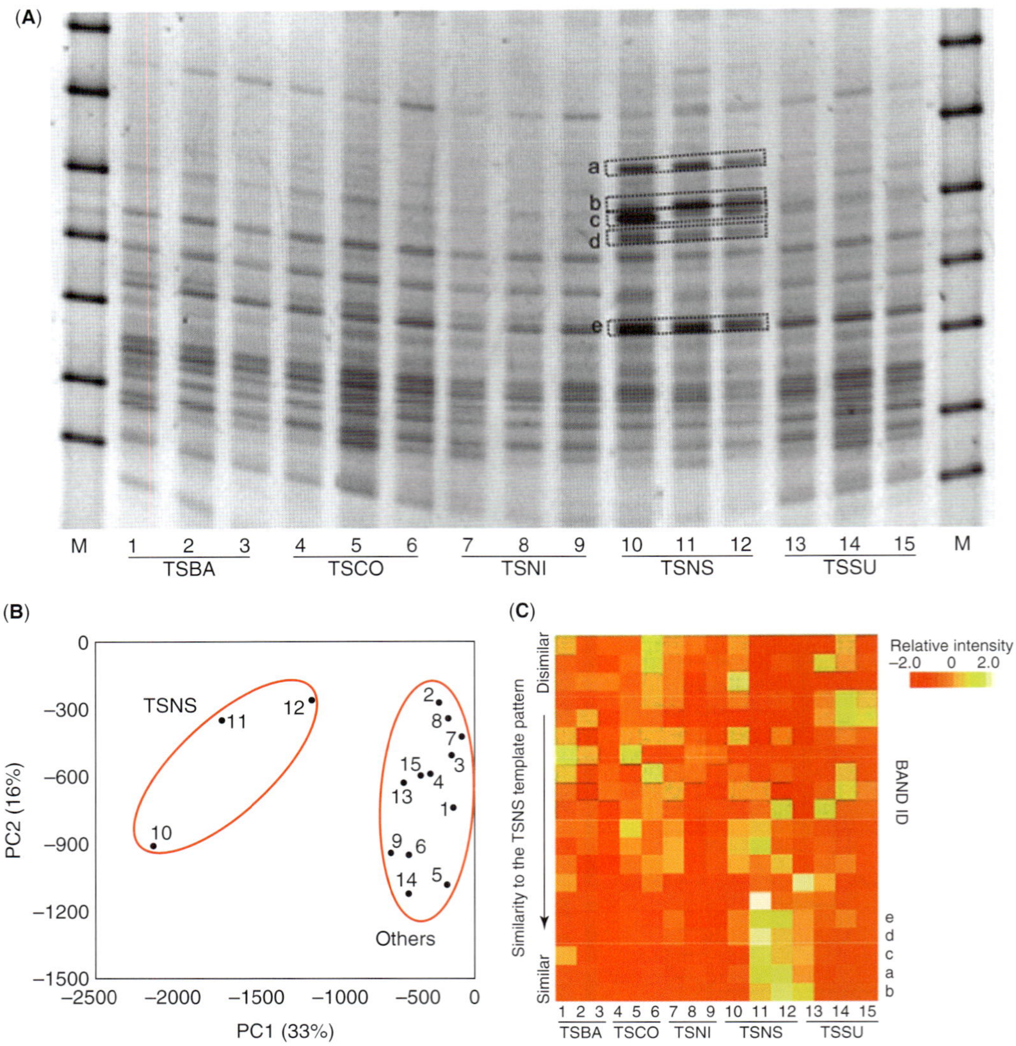

Figure 8.3 DGGE analysis of bacteria community in soil samples amended with different additives. (**A**) DGGE profile. (**B**) PCA plot based on the DGGE profile. The normalized location and intensity of each DGGE band were used in the PCA analysis. Numbers in the plot correspond to the lane numbers in Figure 8.3A. Numbers in parentheses are percentages of variation explained by each component. (**C**) Pseudo-heatmap image generated based on the DGGE profile. Aligned DGGE bands were sorted by similarity index to the TSNS-specific template pattern (TSBA, TSCO, TSNI, TSNS, TSSU = 0, 0, 0, 1, 0, respectively). Five bands, labeled a, b, c, d, and e in Figures 8.3A and 8.3C, were identified as TSNS-specific bands. (Adapted from Ishii et al. [2009b] with permission from the publisher.)

algorithm was used to identify operational taxonomic units (OTUs) in the 16S rRNA gene sequence data [Schloss and Handelsman, 2005] and to identify protein-coding genes in metagenomic sequence data [Yooseph and Sutton, 2008].

8.3.4 DNA Fingerprinting in the Era of Metagenomics

DNA fingerprint analysis can provide a rough overview of the community, while large-scale 16S rRNA clone library, pyrosequencing, and metagenomic analyses allows us to assess detailed structures and potential functions of the community [Gilbert et al., 2009; Ishii et al., 2009b]. Although advancement in technology has dramatically lowered the per-base cost for sequencing [Kunin et al., 2008], metagenomics and other sequence-based approaches are still expensive to perform, especially when multiple samples are analyzed.

DNA fingerprinting patterns can be used to decrease the number of samples to be sequenced, since DNA fingerprinting is relatively inexpensive to perform. For example, based on the DGGE profiles, we can choose distinct microbial community for metagenome sequencing in a time-course perturbation experiment. DNA fingerprinting is also useful to confirm the community evenness among the replicate DNA extracts originating from single environmental samples [e.g., Ishii et al., 2009b].

For any purposes, objective and quantitative data analyses are important (e.g., by using the R program shown here). Such data analyses can increase the reliability of the DNA fingerprinting data and allow us to compare results with those obtained by sequencing-based analyses such as metagenomics.

INTERNET RESOURCES

R program (http://www.r-project.org)

R codes for clustering-based peak alignment, along with example rep-PCR and DGGE input data (linkinghub.elsevier.com/retrieve/pii/S0167701209 00219X)

Acknowledgments

This work was supported by the Program for Promotion of Basic Research Activities for Innovative Biosciences (PROBRAIN) from Bio-oriented Technology Research Advancement Institution, Japan.

REFERENCES

Borneman J, Triplett EW. 1997. Molecular microbial diversity in soils from eastern Amazonia: evidence for unusual microorganisms and microbial population shifts associated with deforestation. *Appl. Environ. Microbiol*. **63**:2647–2653.

Chèneby D, Philippot L, Hartmann A, Henault C, Germon JC. 2000. 16S rDNA analysis for characterization of denitrifying bacteria isolated from three agricultural soils. *FEMS Microbiol. Ecol*. **34**:121–128.

Efron B, Halloran E, Holmes S. 1996. Bootstrap confidence levels for phylogenetic trees. *Proc. Natl. Acad. Sci. USA* **93**:13429–13434.

Felsenstein J. 1985. Confidence limits on phylogenies: An approach using the bootstrap. *Evolution* **39**:783–791.

Fromin N, Hamelin J, Tarnawski S, Roesti D, Jourdain-Miserez K, et al. 2002. Statistical analysis of denaturing gel electrophoresis (DGE) fingerprinting patterns. *Environ. Microbiol*. **4**:634–643.

Gilbert JA, Field D, Swift P, Newbold L, Oliver A, et al. 2009. The seasonal structure of microbial communities in the Western English Channel. *Environ. Microbiol*. **11**:3132–3139.

Inaba S, Tanabe K, Eda S, Ikeda S, Higashitani A, et al. 2009. Nitrous oxide emission and microbial community in the rhizosphere of nodulated soybeans during the late growth period. *Microbes Environ*. **24**:64–67.

Ishii S, Sadowsky MJ. 2008. Applications of the rep-PCR DNA fingerprinting technique to study microbial diversity, ecology and evolution. *Environ. Microbiol*. **11**:733–740.

Ishii S, Yan T, Shively DA, Byappanahalli MN, Whitman RL, Sadowsky MJ. 2006. *Cladophora* (Chlorophyta) spp. harbor human bacterial pathogens in nearshore water of Lake Michigan. *Appl. Environ. Microbiol*. **72**:4545–4553.

Ishii S, Kadota K, Senoo K. 2009a. Application of a clustering-based peak alignment algorithm to analyze various DNA fingerprinting data. *J. Microbiol. Methods* **78**:344–350.

Ishii S, Yamamoto M, Kikuchi M, Oshima K, Hattori M, Otsuka S, Senoo K. 2009b. Microbial populations responsive to denitrification-inducing conditions in rice paddy soil, as revealed by comparative 16S rRNA gene analysis. *Appl. Environ. Microbiol*. **75**:7070–7078.

Kadota K, Araki R, Nakai Y, Abe M. 2007. GOGOT: A method for the identification of differentially expressed fragments from cDNA-AFLP data. *Algorithm. Mol. Biol*. **2**:5.

Kunin V, Copeland A, Lapidus A, Mavromatis K, Hugenholtz P. 2008. A bioinformatician's guide to metagenomics. *Microbiol. Mol. Biol. Rev*. **72**:557–578.

Muyzer G, de Waal EC, Uitterlinden AG. 1993. Profiling of complex microbial populations by denaturing gradient gel electrophoresis analysis of polymerase chain reaction-amplified genes coding for 16S rRNA. *Appl. Environ. Microbiol*. **59**:695–700.

Pavlidis P, Noble W. 2001. Analysis of strain and regional variation in gene expression in mouse brain. *Genome Biol*. **2**: http://genomebiology.com/content/2/10/RESEARCH0042.

Rademaker JLW, de Bruijn FJ. 2008. Computer-assisted analysis of molecular fingerprint profiles and database construction. In Kowalchuk GA, de Bruijn FJ, Head IM, Akkermans ADL, van Elsas JD, eds. *Molecular Microbial Ecology Manual*, 2nd ed. Dordrecht: Springer, pp. 1397–1446.

Rademaker JLW, Louws FJ, Versalovic JV, de Bruijn FJ. 2008. Characterization of the diversity of ecologically important microbes by rep-PCR genomic fingerprinting. In Kowalchuk GA, de Bruijn FJ, Head IM, Akkermans ADL, van Elsas JD, eds. *Molecular Microbial Ecology Manual*, 2nd ed. Dordrecht: Springer, pp. 611–644.

Saito A, Ikeda S, Ezura H, Minamisawa K. 2007. Microbial community analysis of the phytosphere using culture-independent methodologies. *Microbes Environ*. **22**:93–105.

Saito T, Ishii S, Otsuka S, Nishiyama M, Senoo K. 2008. Identification of novel *betaproteobacteria* in a succinate-assimilating population in denitrifying rice paddy soil by using stable isotope probing. *Microbes Environ*. **23**:192–200.

Schloss PD, Handelsman J. 2005. Introducing DOTUR, a computer program for defining operational taxonomic units and estimating species richness. *Appl. Environ. Microbiol*. **71**:1501–1506.

Sklarz MY, Angel R, Gillor O, Soares MIM. 2009. Evaluating amplified rDNA restriction analysis assay for identification of bacterial communities. *Antonie Leeuwenhoek* **96**:659–664.

Suzuki R, Shimodaira H. 2006. pvclust: An R package for assessing the uncertainty in hierarchical clustering. *Bioinformatics* **22**:1540–1542.

Tibshirani R, Hastie T, Narasimhan B, Soltys S, Shi G, Koong A, Le QT. 2004. Sample classification from protein mass spectrometry, by "peak probability contrasts". *Bioinformatics* **20**:3034–3044.

van Verseveld HW, Röling WFM. 2008. Cluster analysis and statistical comparison of molecular community profile data. In Kowalchuk GA, de Bruijn FJ, Head IM, Akkermans ADL, van Elsas JD, eds. *Molecular Microbial Ecology Manual*, 2nd ed. Dordrecht: Springer. pp. 1373–1396.

Versalovic J, Koeuth T, Lupski Jr. 1991. Distribution of repetitive DNA sequences in eubacteria and application to fingerprinting of bacterial genomes. *Nucleic Acids Res*. **19**:6823–6831.

Yooseph S, Li W, Sutton G. 2008. Gene identification and protein classification in microbial metagenomic sequence data via incremental clustering. *BMC Bioinformatics* **9**:182.

Yu W, Wu B, Lin N, Stone K, Williams K, Zhao H. 2006. Detecting and aligning peaks in mass spectrometry data with applications to MALDI. *Comput. Biol. Chem*. **30**:27–38.

Part 2

The Species Concept

Chapter 9

Population Genomics Informs Our Understanding of the Bacterial Species Concept

MARGARET A. RILEY

9.1 INTRODUCTION

The diversity of bacteria is simply staggering. Indeed, according to some, bacterial species diversity is so high that our most powerful methods of estimation provide only the crudest estimate of its magnitude [Torsvik et al., 1990; Torsvik and Ovreas, 2002; see also Chapter 2, Vol. I]. Nonetheless, many such estimates exist, and they suggest that the soil holds the honor of harboring the highest level of species diversity, with potentially over one million species per gram of soil [Gans et al., 2005]. This staggering level of species diversity is even more remarkable when one considers that the number of prokaryotes reported in the National Center for Biotechnology Information molecular database is only about 15,111 [Sayers et al., 2009]. Clearly, there is a vast reservoir of untapped bacterial diversity.

As molecular methods are more widely applied to characterize bacterial diversity, our ability to make taxonomic sense of what we learn is being severely challenged. The focus of this review is to explore how population genomic investigations may provide a rational basis for identifying, and even naming, evolutionary "lineages." In essence, we want to know whether a functional and, better yet, useful bacterial species concept emerges from the burgeoning genomic information overload.

Traditional species designations were based upon extensive phenotypic characterization of a large number of isolates. Now, such designations also require the use of 16s rRNA sequence comparisons to identify the closest relatives of a proposed species [Rossello-Mora and Amann, 2001]. This phylophenetic bacterial species concept posits that a bacterial species is "a monophyletic and genomically coherent cluster of individual organisms that show a high degree of overall similarity with respect to many independent characteristics, and is diagnosable by a discriminative phenotypic property" [Rossello-Mora and Amann, 2001]. Numerous studies have revealed clusters of bacterial isolates that share complex phenotypes, and these clusters are often designated as species [Shute et al., 1984; Sneath and Stevens, 1985; Barrett and Sneath, 1994; Mauchline and Keevil, 1991; Kirschner et al., 2001]. In fact, Cohan [2002] uses the mere existence of these clusters as prima facie evidence of the existence of bacterial species.

Early attempts to use molecular data to delineate bacterial species involved the use of DNA–DNA hybridization, in which bacterial species were defined as those isolated sharing at least 70% hybridization under standard conditions [Stackebrandt et al., 2002]. Given the enormous range of genetic variation detected in different clearly recognized species, it became obvious that a variability cutoff, such as is imposed with hybridization methods, was not appropriate. Levels of variability will vary over the lifetime of a species and will reflect aspects of its life history, particularly the processes by which it adapts to its habitat.

More recent investigations into microbial species distinctions have sought to incorporate more robust estimates of molecular diversity into the process of species

Handbook of Molecular Microbial Ecology, Volume I: Metagenomics and Complementary Approaches, First Edition. Edited by Frans J. de Bruijn.

identification. The assumption is that this molecular diversity will fall into discrete clusters that correspond closely with phenotype-based clusters. In essence, the question has been raised whether DNA sequence variability can be employed to inform the division of a genus into species, to distinguish among similar species, or to address whether bacterial species exist at all [Godoy et al., 2003; Priest et al., 2004; Hanage et al., 2005; Thompson et al., 2005].

9.2 GENOMIC INFORMATION INFORMS OUR SPECIES CONCEPTS

The availability of whole genome sequences provided our first glimpse into the dynamic nature of a species genome. Glasner and Perna [2004] and Mau et al. [2006] compared six complete genomes of *Escherichia coli* and revealed a highly conserved genome backbone with greater than 98% sequence similarity among the isolates. Furthermore, Mau et al. [2006] detected a high level of homologous recombination among these shared genes, confirming earlier studies by Roger Milkman [Milkman et al., 1999, 2003] that the level of within-species recombination for what was then considered a "clonal" species was significant. Even more surprising was the observation that the conserved genome backbone of *E. coli* was interrupted by hundreds of strain specific "sequence islands." Edwards et al. [2002] provided a similar comparison of *Salmonella enterica* strains. As in *E. coli*, a backbone of highly conserved genes was identified, each with an average of greater than 99% sequence similarity and a similar pattern of strain-specific unique sequence islands.

This pattern of shared and unique sequences appears to be common among many bacterial species [Juhas et al., 2007; Coleman et al., 2006; Waterfield et al., 2003]. A comparison of eight genomes of group B *Streptococcus* revealed a core of 1806 genes present in every genome and 907 genes absent in one or more. In contrast, eight genomes of *Bacillus anthracis* revealed very few strain-specific genes. However, the general pattern that emerges is that members of a bacterial species share some large fraction of their genomes, but often carry unique, strain-specific sequences. The fraction of the genome shared versus unique varies greatly from one bacterial species to the next.

The determination of multiple, complete genome sequences of numerous additional bacteria have enabled a new view of the genomic plasticity of several "well-defined" bacterial species. Such an analysis, referred to as genome-tree building, is based on genome-level phylogenetic analysis. This approach involves scaling-up the traditional tree-building approach and analyzing the phylogenetic trees for multiple gene families (in principle, all families represented in many genomes), in an attempt to derive a consensus, "organismal" phylogeny [Brown and Volker, 2004]. It is often not practical to determine the complete genome sequence of multiple isolates within a species. In lieu of complete genomes, a sample of housekeeping genes (referred to as multi-locus sequence typing) (MLST) has become the norm for characterizing genetic diversity within a bacterial species [Maiden et al., 1998; Feil et al., 2003]. This method permits the analysis of large numbers of related bacterial isolates, which is essential to the determination of species designations [Feil et al., 2003; Hanage et al., 2005]. Such studies have confirmed that species designations based upon phenotypic criteria have a corresponding underlying MLST-based genotypic clustering [Woodward et al., 2000; Whitaker et al., 2005; Thompson et al., 2005; Godoy et al., 2003].

9.3 POPULATION GENOMICS DATA PROVIDE THE KEY

One of the first multigene based investigations specifically designed to address the bacterial species concept was conducted by Wertz et al. [2003], who sequenced seven housekeeping genes from a sample of environmental bacteria representing seven species of Enterobacteriaceae [Gordon and FitzGibbon, 1999]. Molecular phylogenies for each of the genes were inferred and the branching patterns of the resulting trees compared. In each case, isolates from a species formed a monophyletic group, which corresponded precisely with the clusters identified by phenotypic data and upon which species distinctions were initially delineated [Holt, 1994; Rossello-Mora and Amann, 2001]. An enteric species phylogeny was inferred from the composite data by concatenating the housekeeping gene sequences [Wertz et al., 2003], which contained enough phylogenetic signal to resolve all of the interspecies nodes and thus provided a robust estimate of the enteric phylogeny corresponding with the existing phenotype-based phylogeny. The authors concluded, at least for this sample of enteric lineages, that bacterial species clearly do exist and the same species emerge from phenotypic and genotypic data. More recently, a highly robust phylogenetic tree was constructed for 13 gamma proteobacteria using a concatenated alignment of several hundred conserved orthologous proteins [Lerat et al., 2003]. Only two of the proteins had incongruent tree topologies in this analysis.

A similar type of investigation was undertaken with *Neisseria*, in which a sample of housekeeping genes was obtained from nearly one thousand strains and 11 named species [Hanage et al., 2005]. Phylogenetic trees were inferred to investigate whether genotypic clusters can be resolved among these highly recombinogenic bacteria and, if so, determine the extent to which they correspond to named species. The analysis suggests that the use of concatenated sequences largely buffered the distorting effect of recombination events and resulted in the resolution of clusters corresponding to the three species most numerous in the sample, *N. meningitidis, N. lactamica* and *N. gonorrhoeae*. Comas et al. [2006] go so far as to propose which genes should be incorporated into these genome-wide surveys for phylogenetic signals. They define "essential" genes, as opposed to "universal" genes, providing the greatest phylogenetic signal-to-noise ratio.

These earlier, human-pathogen focused studies have since been expanded to cover a much broader sampling of bacterial diversity. One such investigation involved an *Agrobacterium* species complex [Popoff et al., 1984; Mougel et al., 2002; Portier et al., 2006]. Genome sequence comparisons among members of the genus highlighted a broad range of intraspecies divergence within very closely related but distinct species of *Agrobacterium*. A subsequent study of the same species compared the sensitivity of recombination to DNA sequence divergence across the species complex. Their data supported earlier claims by Majewski [2001] that "bacterial species experience a degree of sexual isolation from genetically divergent organisms since recombination occurs more frequently within species than between species." A genome-based study of the Actinobacteria, a dominant bacterial phylum in the soil, revealed a surprising level of gene synteny and genome conservation. Prior 16S rRNA gene sequence comparisons had recognized 39 families and 130 genera, all of which share a unique molecular synapomorphy: an insertion in the 23S rRNA gene [Roller et al., 1992]. How this ancient phylum has retained such a high degree of genome conservation remains a mystery. Similar outcomes have been observed for *Haemophilus, Pseudomonas* and *Streptococcus*, where it was shown that as little as 9% divergence at recombination marker genes results in a drop of three orders of magnitude in recombination efficiency.

Konstantinidis and Tiedje [2004; see also Chapter 11, Vol. I] compared the gene content of 70 closely related and fully sequenced bacterial genomes to (a) identify whether species boundaries exist and (b) determine the role of the organism's ecology on its shared gene content. They found the levels of sequence similarity on the order of 94% corresponded to the traditional 70% DNA–DNA reassociation standard of the current species definition. Notably, a large fraction (e.g., up to 65%) of the differences in gene content within species is associated with bacteriophage and transposable elements, revealing an important role of these elements during bacterial speciation. Their findings are consistent with a species definition that would include a more homogeneous set of strains than provided by the current definition and one that considers the ecology of the strains in addition to their evolutionary distance.

9.4 THE CORE GENOME HYPOTHESIS

Lan and Reeves [1996] were the first to recognize the potential link between the *observation* of shared versus unique sequences in bacterial genomes and its *implication* for discriminating bacterial species. They proposed the Core Genome Hypothesis (CGH), which distinguishes between that fraction of the genome (the core) shared by all members of a species and that fraction found in only a subset of the population (the auxiliary). Core genes encode essential metabolic housekeeping and informational processing functions [Feil, 2004]. They are ubiquitous in a species and define the species-specific characteristics. Auxiliary genes may or may not be present in a strain and are generally genes that encode supplementary biochemical pathways, are associated with phage or other mobile elements, or encode products that serve to interact with the external environment. Thus, auxiliary genes serve in the adaptation of strains to local competitive or environmental pressures [Cohan, 2002]. They are likely to encode antibiotic resistance, novel metabolic functions, toxin production, and the like [Dobrindt and Reidl, 2000; Karlin, 2001; White et al., 2001].

The CGH has dramatically influenced how bacteriologists think about the nature of bacterial species. Prior to the CGH, the strongest argument against the recognition of bacterial "species" was the simple observation of horizontal gene transfer (HGT) between bacterial lineages. The fact that bacterial species gene pools may not be tightly closed was enough reason for many bacteriologists to conclude that bacterial species cohesion could not survive such exchanges. This contradicts the clearly demonstrated fact that bacteria exist in phenotypic clusters, which many bacteriologists recognize as species. Even more compelling, it is becoming clear that these well-defined phenotypic clusters correspond to underlying genotype clusters [Woodward et al., 2000; Whitaker et al., 2005; Thompson et al., 2005; Godoy et al., 2003].

Some have argued that it is futile to expect a bacterial species to ever be characterized fully at the genome level, particularly since as more genome sequences are obtained, the pan genomes (i.e., the sum of all genes identified within a species) of numerous species continues to grow [Medini et al., 2005; Tettelin et al., 2005; see also Chapter 10, Vol. I]. In fact, some predict that hundreds of thousands of genome sequences are required to fully define certain bacterial species [Medini et al., 2005]. Others suggest that the wide range of intraspecies variation observed for bacterial species reflects the lack of a universal and meaningful species definition [Feil, 2004].

Many ecological and evolutionary factors will impact how many unique genes a species pan genome may encode and how much genetic variation it harbors. There is no "one size fits all" concept that can, or should, be applied. In fact, no existing species definition requires that either the pan genome or the level of genetic variation be known in order to delineate members of a species.

One of the more commonly accepted species concepts is that proposed by Ernst Mayr, the Biological Species Concept (BSC) [Mayr, 1942]. Mayr proposed that a biological species is comprised of groups of actually or potentially interbreeding natural populations, which are reproductively isolated from other such groups [Mayr, 1942]. Although Mayr developed this definition specifically for eukaryotes, it can easily be modified to apply to bacteria. However, it is important to note that at this juncture, the BSC should not be taken to imply any particular process of speciation, merely that the observation of more gene "sharing" (via recombination and/or lateral gene transfer) is observed within versus between putative bacterial species. The Core Genome Hypothesis provides a perfect backdrop from which to articulate this modification. According to the CGH, a bacterial species is comprised of groups of strains that frequently exchange, or could exchange, core genes, but which are relatively restricted from such exchange with other groups.

The CGH predicts that a subset of genes, the core, is present in all, or nearly all, individuals within the species. These are the genes that provide the defining characteristics of the species and are assumed to experience primarily purifying selection, to remove deleterious mutations and maintain existing functions. As a species evolves, its core genome will evolve as a complex of co-evolved functions. When transferred between species, such genes will most likely experience a selective disadvantage, because this will disrupt co-evolved functions. Such transfer will rarely survive. Thus, core genes will diverge as the species diverge.

In contrast, auxiliary genes will be found in only a subset of individuals within a species. The CGH predicts that these genes experience intermittent positive selection, when their function enhances survival in a varied and ever-changing environment. When such genes are exchanged between species, their functions will often provide a selective advantage to the recipient. Frequent successful transfer between species will serve to limit the divergence of auxiliary genes, relative to the core.

The most specific prediction that emerges from the CGH concerns the rate at which core and auxiliary genes accumulate variability. Core genes will, on average, display a neutral rate of evolution, while auxiliary genes will experience a variety of selective pressures, including diversifying selection (acting to increase levels of variation), directional selection (acting to decrease levels of variation), balancing selection (acting to maintain particular alleles in the species), and purifying selection (acting to weed out deleterious mutations). Thus, the average rate of evolution for auxiliary genes could be just about anything, and the variance around this rate should be extreme. These expectations, based upon the neutral theory [Kimura, 1968], are quite useful for testing predictions from the CGH [Fay et al., 2002]. However, such tests require population-based samples of multiple genomes per species; and, unfortunately, most existing species-based genome samples are chosen to represent the diversity of clinical isolates of human pathogens and thus will often underestimate standing levels of genome diversity. The appropriate data are in the pipeline and should soon be available to permit the sort of population genomics required to address this complex and fascinating matter.

9.5 CONCLUSIONS

We are on the verge of obtaining the type and amount of genomic data required to examine bacterial species definitions in grand detail. It will indeed be exciting to finally understand how we can find such patterns of corresponding phenotypic and genotypic clustering given the extraordinary levels of bacterial diversity, the fluid nature of most genomes, and the unexpected levels and types of horizontal gene transfer that appears to occur between even the most distantly related phyla. A compelling argument can be made that, rather than argue about whether a bacterial species concept exists, we may be better served by focusing on why, in the face of such potentially high levels of HGT, any sign of bacterial species clusters remain at all. Our job remains the development of an understanding of bacterial evolution rich enough to explain this apparent paradox.

REFERENCES

Barrett SJ, Sneath PH. 1994. A numerical phenotypic taxonomic study of the genus *Neisseria*. *Microbiology* **140**(Pt 10):2867–2891.

Brown JR, Volker C. 2004. Phylogeny of gamma-proteobacteria: resolution of one branch of the universal tree? *Bioessays* **26**:463–468.

Cohan FM. 2002. Sexual isolation and speciation in bacteria. *Genetica* **116**:359–370.

Coleman ML, Sullivan MB, Martiny AC, Steglich C, Barry K, et al. 2006. Genomic islands and the ecology and evolution of *Prochlorococcus*. *Science* **311**:1768–1770.

Comas I, Moya A, Azad RK, Lawrence JG, Gonzalez-Candelas F. 2006. The evolutionary origin of Xanthomonadales genomes and the nature of the horizontal gene transfer process. *Mol. Biol. Evol.* **23**:2049–2057.

Dobrindt U, Reidl J. 2000. Pathogenicity islands and phage conversion: evolutionary aspects of bacterial pathogenesis. *Int. J. Med. Microbiol*. **290**:519–527.

Edwards SV, Fertil B, Giron A, Deschavanne PJ. 2002. A genomic schism in birds revealed by phylogenetic analysis of DNA strings. *Syst. Biol*. **51**:599–613.

Fay JC, Wyckoff GJ, Wu CI. 2002. Testing the neutral theory of molecular evolution with genomic data from *Drosophila*. *Nature* **415**:1024–1026.

Feil EJ. 2004. Small change: Keeping pace with microevolution. *Nat. Rev. Microbiol*. **2**:483–495.

Feil EJ, Cooper JE, Grundmann H, Robinson DA, Enright MC, et al. 2003. How clonal is *Staphylococcus aureus*? *J. Bacteriol*. **185**:3307–3316.

Gans J, Wolinsky M, Dunbar J. 2005. Computational improvements reveal great bacterial diversity and high metal toxicity in soil. *Science* **309**:1387–1390.

Glasner JD, Perna NT. (2004). Comparative Genomics of *E. coli, Microbiol. Today* **31**.

Godoy D, Randle G, Simpson AJ, Aanensen DM, Pitt TL, et al. 2003. Multilocus sequence typing and evolutionary relationships among the causative agents of melioidosis and glanders, *Burkholderia pseudomallei* and *Burkholderia mallei*. *J. Clin. Microbiol*. **41**:2068–2079.

Gordon DM, FitzGibbon F. 1999. The distribution of enteric bacteria from Australian mammals: host and geographical effects. *Microbiology* **145**(Pt 10):2663–2671.

Hanage WP, Kaijalainen T, Herva E, Saukkoriipi A, Syrjanen R, et al. 2005. Using multilocus sequence data to define the *Pneumococcus*. *J. Bacteriol*. **187**:6223–6230.

Holt J, Kreig N, Sneath P, Staley J, Williams L, (1994). *Bergey's Manual of Determinative Bacteriology*. 9 edn (Baltimore, Williams Wilkins):

Juhas M, Crook DW, Dimopoulou ID, Lunter G, Harding RM, et al. 2007. Novel type IV secretion system involved in propagation of genomic islands. *J. Bacteriol*. **189**:761–771.

Karlin S. 2001. Detecting anomalous gene clusters and pathogenicity islands in diverse bacterial genomes. *Trends Microbiol*. **9**:335–343.

Kimura M. 1968. Genetic variability maintained in a finite population due to mutational production of neutral and nearly neutral isoalleles. *Genet. Res*. **11**:247–269.

Kirschner C, Maquelin K, Pina P, Ngo Thi NA, Choo-Smith LP, et al. 2001. Classification and identification of enterococci: a comparative phenotypic, genotypic, and vibrational spectroscopic study. *J. Clin. Microbiol*. **39**:1763–1770.

Konstantinidis KT, Tiedje JM. 2004. Trends between gene content and genome size in prokaryotic species with larger genomes. *Proc. Natl. Acad. Sci. USA* **101**:3160–3165.

Lan R, Reeves PR. 1996. Gene transfer is a major factor in bacterial evolution. *Mol. Biol. Evol*. **13**:47–55.

Lerat E, Daubin V, Moran NA. 2003. From gene trees to organismal phylogeny in prokaryotes: The case of the gamma-Proteobacteria. *PLoS Biol*. **1**:E19.

Maiden MC, Bygraves JA, Feil E, Morelli G, Russell JE, et al. 1998. Multilocus sequence typing: a portable approach to the identification of clones within populations of pathogenic microorganisms. *Proc. Natl. Acad. Sci. USA* **95**:3140–3145.

Majewski J. 2001. Sexual isolation in bacteria. *FEMS Microbiol. Lett*. **199**:161–169.

Mau B, Glasner JD, Darling AE, Perna NT. 2006. Genome-wide detection and analysis of homologous recombination among sequenced strains of *Escherichia coli*. *Genome Biol*. **7**:R44.

Mauchline WS, Keevil CW. 1991. Development of the BIOLOG substrate utilization system for identification of *Legionella* spp. *Appl. Environ. Microbiol*. **57**:3345–3349.

Mayr E. 1942. *Systematics and the Origin of Species*. New York: Columbia University Press.

Medini D, Donati C, Tettelin H, Masignani V, Rappuoli R. 2005. The microbial pan-genome. *Curr Opin. Genet. Dev*. **15**:589–594.

Milkman R, Raleigh EA, McKane M, Cryderman D, Bilodeau P, et al. 1999. Molecular evolution of the *Escherichia coli* chromosome. V. Recombination patterns among strains of diverse origin. *Genetics* **153**:539–554.

Milkman R, Jaeger E, McBride RD. 2003. Molecular evolution of the *Escherichia coli* chromosome. VI. Two regions of high effective recombination. *Genetics* **163**:475–483.

Mougel C, Thioulouse J, Perriere G Nesme X. 2002. A mathematical method for determining genome divergence and species delineation using AFLP. *Int. J. Syst. Evol. Microbiol*. **52**: 573–586.

Popoff MY, Kersters K, Kiredjian M, Miras I, Coynault C. 1984. Taxonomic position of *Agrobacterium* strains of hospital origin. *Ann. Microbiol. (Paris)* **135A**:427–442.

Portier P, Fischer-Le Saux M, Mougel C, Lerondelle C, Chapulliot D, et al. 2006. Identification of genomic species in *Agrobacterium* biovar 1 by AFLP genomic markers. *Appl. Environ. Microbiol.* **72**:7123–7131.

Priest FG, Barker M, Baillie LW, Holmes EC, Maiden MC. 2004. Population structure and evolution of the *Bacillus cereus* group. *J. Bacteriol*. **186**:7959–7970.

Roller C, Ludwig W, Schleifer KH. 1992. Gram-positive bacteria with a high DNA G+C content are characterized by a common insertion within their 23S rRNA genes. *J. Gen. Microbiol*. **138**:1167–1175.

Rossello-Mora R, Amann R. 2001. The species concept for prokaryotes. *FEMS Microbiol. Rev.* **25**:39–67.

Sayers EW, Barrett T, Benson DA, Bolton E, Bryant SH, Canese K, Chetvernin V, Church DM, Dicuccio M, Federhen S. 2009. Database resources of the National Center for Biotechnology Information. *Nucleic Acids Res*. 1–12.

Shute LA, Gutteridge CS, Norris JR, Berkeley RC. 1984. Curie-point pyrolysis mass spectrometry applied to characterization and identification of selected *Bacillus* species. *J. Gen. Microbiol*. **130**:343–355.

Sneath PH, Stevens M. 1985. A numerical taxonomic study of *Actinobacillus, Pasteurella* and *Yersinia*. *J. Gen. Microbiol*. **131**:2711–2738.

Stackebrandt E, Frederiksen W, Garrity GM, Grimont PA, Kampfer P, et al. 2002. Report of the ad hoc committee for the re-evaluation of the species definition in bacteriology. *Int. J. Syst. Evol. Microbiol*. **52**:1043–1047.

Tettelin H, Masignani V, Cieslewicz MJ, Donati C, Medini D, et al. 2005. Genome analysis of multiple pathogenic isolates of *Streptococcus agalactiae*: Implications for the microbial "pan-genome". *Proc. Natl. Acad. Sci. USA* **102**:13950–13955.

Thompson JR, Pacocha S, Pharino C, Klepac-Ceraj V, Hunt DE, et al. 2005. Genotypic diversity within a natural coastal bacterioplankton population. *Science* **307**:1311–1313.

Torsvik V, Goksoyr J, Daae FL. 1990. High diversity in DNA of soil bacteria. *Appl. Environ. Microbiol*. **56**:782–787.

Torsvik V, Ovreas L. 2002. Microbial diversity and function in soil: from genes to ecosystems. *Curr. Opin. Microbiol*. **5**:240–245.

Waterfield NR, Daborn PJ, Dowling AJ, Yang G, Hares M, et al. 2003. The insecticidal toxin makes caterpillars floppy 2 (Mcf2) shows similarity to HrmA, an avirulence protein from a plant pathogen. *FEMS Microbiol. Lett*. **229**:265–270.

Wertz JE, Goldstone C, Gordon DM, Riley MA. 2003. A molecular phylogeny of enteric bacteria and implications for a bacterial species concept. *J. Evol. Biol*. **16**:1236–1248.

Whitaker RJ, Grogan DW, Taylor JW. 2005. Recombination shapes the natural population structure of the hyperthermophilic archaeon *Sulfolobus islandicus*. *Mol. Biol. Evol*. **22**:2354–2361.

White PA, McIver CJ, Rawlinson WD. 2001. Integrons and gene cassettes in the enterobacteriaceae. *Antimicrob. Agents Chemother*. **45**:2658–2661.

Woodward MJ, Sojka M, Sprigings KA, Humphrey TJ. 2000. The role of SEF14 and SEF17 fimbriae in the adherence of *Salmonella enterica* serotype Enteritidis to inanimate surfaces. *J. Med. Microbiol*. **49**:481–487.

Chapter 10

The Microbial Pangenome: Implications for Vaccine Development

Annalisa Nuccitelli, Claudio Donati,
Michèle A. Barocchi, and Rino Rappuoli

10.1 CLASSICAL REVERSE VACCINOLOGY FOR VACCINE DEVELOPMENT

Historically, conventional vaccine development relied on the cultivation of microorganisms under laboratory conditions by which potential antigens were isolated through biochemical and microbiological methods. These classical methods were adopted for many years, although limited by cultivability of the bacteria and time.

Nowadays, rapid sequencing and bioinformatic analysis has defined new approaches such as reverse vaccinology (see **Table 10.1** for a glossary of terms used in this chapter), first applied for the development of the *Neisseria meningitis* serogroup B (MenB) vaccine [Pizza et al., 2000]. MenB is the major cause of meningitis and sepsis for children and adults, despite the availability of antibiotics. This gram-negative pathogen is able to asymptomatically colonize the upper nasopharinx of 5–15% of the human population, but in a significant number of cases, it is able to traverse the epithelium and reach the bloodstream, causing septicaemia [Stephens et al., 2007].

In the last 40 years, conventional approaches have failed to provide an efficacious vaccine against *N. meningitides B* that still remains one of the principle causes of bacterial meningitis in developed countries. *M. meningitidis* can be classified by 13 serogroups based on the chemical composition of the capsular polysaccharide. There are five major infective serogroups, A, B, C, Y, and W135, which represent more than 95% of the total cases of invasive disease. The first vaccine against MenB was generated in the 1960s by using the major capsular serotypes purified as antigens. Unfortunately, the chemical composition of this vaccine, resembling a sialic residue present in human tissue, failed in generating a correct immunological response and had the potential to cause autoimmunity [Finne et al., 1987; Nedelec et al., 1990].

In 2000, the whole genome sequence of the MenB strain MC58 was available and opened the way to screen the 2158 predicted ORF for properties that would identify them as novel vaccine candidates [Tettelin et al., 2000]. Briefly, the sequence of the virulent MenB strain MC58 was generated by a shotgun strategy. Based on the assumption that a secreted or a surface-exposed antigen can stimulate the human immunological response and confer protection, the genome of the MenB strain MC58 was screened with bioinformatic tools in order to identify the secreted or surface exposed proteins. Genome mining enabled the identification of 600 potential antigens, 350 of which were successfully expressed in *E. coli*, purified, and used for immunization of mice. Sera obtained from mice immunized with the recombinant proteins were used to evaluate the surface localization of the antigens as well as for their capability to induce protective immune response: 91 novel surface-exposed proteins were identified, 28 of which induced bacteridicidal antibodies. In addition, the new generation antigens were evaluated for gene variability and for

Handbook of Molecular Microbial Ecology, Volume I: Metagenomics and Complementary Approaches, First Edition. Edited by Frans J. de Bruijn.

Table 10.1 Glossary of Terms Used in This Chapter

Reverse Vaccinology: In silico selection of vaccine candidates, using bioinformatics tools, starting from the complete genome sequence of one or more strains of a pathogen.

Pan-genome: Pan, from the Greek word παν, meaning whole. This includes (a) a core genome containing genes present in all strains and (b) a dispensable genome composed of genes absent from one or more strains and genes that are unique to each strain.

Core genome: Those genes present in all strains of a species.

Dispensable genome: Those genes which are missing from at least one strain of a species.

"Open" pan-genome: A pan-genome that grows without limit, increasing in size with the number of independent strains sequenced.

"Closed" pan-genome: A pan-genome that does not grow in size by increasing the number of sequenced strains.

the capacity to confer protection in different strains of *N. meningitidis*.

The vaccine candidates identified by reverse vaccinology were present in the majority of strains tested and shared a conserved sequence. Their capacity to elicit protection was observed by bactericidal activity and passive protection assay in the infant rat model [Serruto et al., 2009]. Five antigens, screened by this approach, were selected to be combined in a universal multicomponent MenB vaccine. In order to increase their immunogenity and facilitate their large-scale production, four were expressed as fusion proteins: GNA2132 with GNA1030, GNA2091 with GNA1870. The NadA antigen, called GNA1994, was included as single antigen [Serruto et al., 2009].

The success of reverse vaccinology in the design of MenB vaccine served as a milestone for this approach and caused a paradigm shift in vaccine design and development which has been applied to other pathogens such as *Streptococcus agalactiae* [Maione et al., 2005], *Streptococcus pneumoniae* [Wizemann et al., 2001], *Porphyromonas gingivalis* [Ross et al., 2001], *Chlamydia pneumoniae* [Montigiani et al., 2002], *Bacillus anthracis* [Ariel et al., 2002, 2008], and *Brucella melitensis* [Yang et al., 2005]. The reverse vaccinology applied to a single genome sequence failed to address how genetic variability drives pathogenesis within a single species and also limited vaccine candidate selection and antibicrobial targets to a single strain. Comparative genome analysis performed on different pathogenic strains allowed the quantitative analysis of species variability.

10.2 THE PAN-GENOMIC ERA

The identification of protein vaccine candidates for a broadly protective vaccine against *Streptococcus agalactiae* (GBS or group B streptococcus) symbolizes how the theory of reverse vaccinology evolved into a new type of analysis known as pan-genomics.

GBS is a gram-positive bacterium that causes life-threatening diseases in newborns and infants. The infection is acquired during delivery by direct mother-to-baby transmission of the pathogen. GBS is able to asymptomatically colonize the anogenital mucosa of 25–40% of healthy women [Gibbs et al., 2004]. GBS is characterized by 10 different capsular serotypes, five of which are known to be the major disease-causing isolates in Europe and the United States: Ia, Ib, II, III, and V. In 2002, two genome sequences of *Streptococcus agalactiae* were generated and were used as the references in subtractive and comparative genome hybridization (CGH) studies. The comparison between the sequenced genomes and 19 GBS isolates revealed genetic heterogeneity, even among those strains sharing the same serotype. In particular, it was found that approximately 18% of the genes present in sequenced strains were absent in at least one of the 19 isolates analyzed. These data suggested that a single genome would not be representative and could not describe the global genetic variability within a single species. In 2005, six additional GBS genome sequence strains, belonging to the five major serotypes, were generated [Tettelin et al., 2005], suggesting that a species could be defined by its pan-genome. Comparative analysis of the eight complete genomes available of *S. agalactiae* revealed an unexpected degree of intraspecies diversity [Maione et al., 2005]. The GBS pan-genome contains 2713 genes: 1806 are present in the "*core genome*" and 907 belong to the "*dispensable genome*"; 358 of the 907 are strain-specific. The *core genome* includes genes conserved in all isolates and is responsible both for the basic aspects of the biological life of a species and for its own phenotypic traits. The *dispensable genome* is divided into genes present in two or more but not all strains and genes unique to each strain, both not essential for bacterial growth. They encode for supplementary biochemical pathways and functions (e.g., adaptation to different niches, antibiotic resistance) and can confer advantages for the infection and colonization mechanism of the pathogen. As expected, the majority of the genes belonging to the "*core genome*" encode for housekeeping functions, such as metabolic pathways, whereas hypothetical genes of unknown functions correspond only to one-third of these genes. On the contrary, the majority of genes included in the "*dispensable genome*" were annotated as genes of hypothetical unknown function. In order to calculate the complete genetic repertoire of the GBS species, the number of new genes added by each genomic sequence was estimated [Tettelin et al., 2005]. Briefly, in order to calculate the number of genes comprising the species pan-genome, an *all-versus-all* comparison of the strains was performed. Every time a new genome sequence is compared to the others, the number of new genes is measured. In the case of GBS, the curve extrapolated from the comparative analysis reached a nonzero asymptotic value, which means that the GBS pan-genome size grows indefinitely with the number of independent strains sequenced. Surprisingly, this model predicts for *S. agalactiae* an *open* pan-genome that grows by an average of 33 new genes every time a new genome sequence is added. Thus, the GBS species will never be fully defined [Tettelin et al., 2005].

10.3 RECENT DEVELOPMENT OF THE PAN-GENOMES

The exponential increase of genome sequences enabled the scientific community to apply the GBS pan-genome approach to many other pathogens [Tettelin et al., 2008].

Based on a methodology similar to that of Tettelin et al., the *open* pan-genome model was described for *E. coli* but not for *Bacillus anthracis*, whose pan-genome is composed only by four genome sequences. In this case, the number of specific genes added dropped to zero after the addition of a fourth genome sequence, meaning that its pan-genome is "*closed*" [Medini et al., 2005]. This probably reflects the fact that *B. anthracis* is extemely clonal and a recent subclone of *Bacillus cereus*, a much more variable species. The diversity in lifestyle between the bacterial species characterized by an open pan-genome indicates that different environments require a high level of compliance and/or promotes the exchange of genetic material [Tettelin et al., 2008].

The pan-genome structure of *Escherichia coli* was determined by comparison of 17 genomes from both commensal and pathogenic isolates, confirming the extensive genome plasticity in *E. coli* previously reported [Pupo et al., 2000; Welch et al., 2002]. The *E. coli* pan-genome is composed by a reservoir of more than 13,000 genes, of which approximately 2200 genes represent its core genome. The pan-genome analysis revealed a number of truly unique genes (TUG) ranging from $\sim$20 in the laboratory adapted strains to more than $\sim$300, depending on the genome used as the reference. The mean number of TUG calculated from the *E. coli* dataset is 176 ± 95, meaning that there is a significant degree of gene variability within this species and that not all isolates share the same clinical characteristics. The application of the exponential decay model to the identification of TUG using the median value revealed that the pan-genome of *Escherichia coli* grew in a number of $\sim$300 TUG per added genome, therefore its pan-genome is "*open*" [Rasko et al., 2008]. Moreover, the complete sequence of the commensal strain, *E. coli* HS, revealed significant genome variability between a pathogen and a commensal strain [Rasko et al., 2008]. This could be important for understanding the evolutionary mechanisms of this species and its relevance for vaccine design.

The comparative analysis of 12 complete genome sequences of *Staphylococcus aureus*, a human pathogen that causes both nocosomial and community-acquired infections, has revealed that most regions are well-conserved, whereas several large genomic islands display high genetic variability. The overall combination of variable elements is strain-specific and appears to reflect the wide spectrum of clinical disease manifestations in humans [Baba et al., 2008].

Questioning the assumption made for the calculation of GBS pan-genome, Hogg et al. [2007] proposed a new pan-genome model, or supragenome, for the analysis of 13 *Haemophilus influenzae* genomes. The Finite Supragenome Model predicts the number of clusters of orthologous genes present in seven population frequencies, varying from rare genes (less than 1% of the strains) to core genes (all the strains). Given that, by definition, this model predicts a finite number of clusters in the species, this model seems to be in contradiction with Tettelin's original model. This approach gave similar results for the pan-genome analysis of 17 *Streptococcus pneumoniae* isolates [Hiller et al., 2007].

Finally, another model for open pan-genomes has been proposed and it is based on an empirical law, called Heaps' law [Heaps, 1978]. This method has been applied for the estimation of a number of species available with annotation in Genbank and in all cases the model fit the data well [Medini et al., 2008].

Irrespective of the pan-genome size and the mathematical calculation used to predict it, it is evident that a species cannot be characterized by a single genome sequence, opening the question of how many genome sequences are needed to fully describe a bacterial species. Unfortunately, the answer varies from species to species. The diversity in the nature of the pan-genome is connected with several factors, such as pathogen lifestyle, capability for acquiring foreign DNA, and the advantages in niche adaptation from the acquisition of laterally transferred DNA. In summary, given that the size of the pan-genome is much larger than the size of a genome sequence of a single isolate, the definition of the ultimate size of a pan-genome species differs slightly, depending on the assumptions that are used for mathematical models. Tettelin et al. predicted that a pan-genome can be large and theoretically unlimited in size, while Hogg et al. assumed that a pan-genome can also be large but finite in size. Determining which of these hypotheses is correct is still a challenge for evolutionary microbiologist, because it requires the analysis of larger datasets to facilitate the construction of more accurate mathematical models [Medini et al., 2008].

10.4 APPLICATION OF PAN-GENOMICS: THE GBS VACCINE

The advantage of comparative analyses of multiple genome isolates of a pathogen in vaccine design is highlighted by the discovery of the universal vaccine candidates for *S. agalactiae*, the group B streptococcus vaccine [Maione et al., 2005]. The pan-genomic approach [Tettelin et al., 2005], applied to the eight genomes available, enabled the identification of 589 genes encoding putative surface-exposed and secreted proteins: 396 belonged to the "*core genome*" and 193 were present in the "*dispensable genome*." Among the putative antigens identified, 312 proteins were produced as soluble recombinant proteins in *E. coli* and tested for their capacity to elicit protection in an active maternal/neonatal challenge model with the use of six GBS strains. This systematic screening enabled the identification of four main antigens: the *Sip* protein, which belongs to the core genome and was previously described to be a potential vaccine candidate [Brodeur et al., 2000], and three other proteins, belonging to the GBS dispensable genome. Each of these antigens elicit protection against more the one strain but not against all strains. Moreover, antigen exposure was evaluated in order to understand if the level of surface exposure was variable and correlated with protective activity. Following the data obtained, they estimated that an antigen was protective if the corresponding serum obtained from immunization of mice was able to determine a greater-than-fivefold increase in fluorescence intensity in pre-immune controls on live bacteria.

The combination of four antigens was tested in the same mouse model with the use of 12 different strains, representing the major pathogenic GBS serotypes belonging to eight multi-locus sequence types (MLST). The combination of those four antigens elicited protection against all 12 strains tested with a percentage of protection ranging from 59% to 100%. Additionally, the sera obtained from immunized mice were also tested in an opsonophagocytic assay with the encapsulated GBS type V strain CJB111, which express all four antigens. The bacterial killing activity was measured in the presence of polymorphonuclear leukocytes (PMNs) and rabbit complement. All sera elicited opsonophagocitic activity of GBS by PMNs and bacterial killing was both complement- and PMNs-dependent.

Therefore, the results obtained in protection and the opsonophagocitic assay suggests that the combination of these antigens could confer protection against GBS strains infection in humans [Maione et al., 2005].

This work demonstrated the importance of multiple genome sequence analyses, rather than single genome analysis, for the development of a broadly protective vaccine against a serious neonatal disease. The latest characterization of two newly identified vaccine antigens revealed that they belong to the dispensable portion of the GBS pan-genome and are involved in the formation of pilus-like structures protruding from the bacterial surface [Lauer et al., 2005].

Nowadays, the estimation of a pathogen's pan-genome is not just a theoretical exercise, but it represents the state-of-the-art method for vaccine design, providing insights into genome evolution and species variability.

REFERENCES

Ariel N, Zvi A, Grosfeld H, Gat O, Inbar Y, et al. 2002. Search for potential vaccine candidate open reading frames in the *Bacillus anthracis* virulence plasmid pXO1: in silico and in vitro screening. *Infect. Immun*. **70**(12):6817–6827.

Ariel N, Zvi A, Makarova KS, Chitlaru T, Elhanany E, et al. 2003. Genome-based bioinformatic selection of chromosomal *Bacillus anthracis* putative vaccine candidates coupled with proteomic identification of surface-associated antigens. *Infect. Immun*. **71**(8):4563–4579.

Baba T, Bae T, Schneewind O, Takeuchi F, Hiramatsu K. 2008. Genome sequence of *Staphylococcus aureus* strain Newman and comparative analysis of staphylococcal genomes: Polymorphism and evolution of two major pathogenicity islands. *J. Bacteriol*. **190**(1):300–310.

Brodeur BR, Boyer M, Charlebois I, Hamel J, Couture F, et al. 2000. Identification of group B streptococcal Sip protein, which elicits cross-protective immunity. *Infect. Immun*. **68**(10):5610–5618.

Finne J, Bitter-Suermann D, Goridis C, Finne U. 1987. An IgG monoclonal antibody to group B meningococci cross-reacts with developmentally regulated polysialic acid units of glycoproteins in neural and extraneural tissues. *J. Immunol*. **138**(12):4402–4407.

Gibbs RS, Schrag S, Schuchat A. 2004. Perinatal infections due to group B streptococci. *Obstet. Gynecol*. **104**(5 Pt 1):1062–1076.

Heaps RA. 1978. Relating physical and psychological fitness: a psychological point of view. *J. Sports Med. Phys. Fitness* **18**(4):399–408.

Hiller NL, Janto B, Hogg JS, Boissy R, Yu S, et al. 2007. Comparative genomic analyses of seventeen *Streptococcus pneumoniae* strains: Insights into the pneumococcal supragenome. *J. Bacteriol*. **189**(22):8186–8195.

Hogg JS, Hu FZ, Janto B, Boissy R, Hayes J, et al. 2007. Characterization and modeling of the *Haemophilus influenzae* core and supragenomes based on the complete genomic sequences of Rd and 12 clinical nontypeable strains. *Genome Biol*. **8**(6):R103.

Lauer P, Rinaudo CD, Soriani M, Margarit I, Maione D, et al. 2005. Genome analysis reveals pili in Group B *Streptococcus*. *Science* **309**(5731):105.

Maione D, Margarit I, Rinaudo CD, Masignani V, Mora M, et al. 2005. Identification of a universal Group B *Streptococcus* vaccine by multiple genome screen. *Science* **309**(5731):148–150.

Medini D, Donati C, Tettelin H, Masignani V, Rappuoli R. 2005. The microbial pan-genome. *Curr. Opin. Genet. Dev*. **15**(6):589–594.

Medini D, Serruto D, Parkhill J, Relman DA, Donati C, et al. 2008. Microbiology in the post-genomic era. *Natl. Rev. Microbiol*. **6**(6):419–430.

Montigiani S, Falugi F, Scarselli M, Finco O, Petracca R, et al. 2002. Genomic approach for analysis of surface proteins in *Chlamydia pneumoniae*. *Infect. Immun*. **70**(1):368–379.

Nedelec JF, Capron-Laudereau M, Adam R, Patry J, Dimicoli JL, et al. 1990. Mouse liver metabolism after 24-hour and 48-hour cold preservation using UW, hydroxyethyl starch-free UW, and Euro-Collins solutions: A 31P, 13C NMR spectroscopy and biochemical analysis. *Transplant. Proc.* **22**(2):492–495.

Pizza M, Scarlato V, Masignani V, Giuliani MM, Arico B, et al. 2000. Identification of vaccine candidates against serogroup B *Meningococcus* by whole-genome sequencing. *Science* **287**(5459): 1816–1820.

Pupo GM, Lan R, Reeves PR, Baverstock PR. 2000. Population genetics of *Escherichia coli* in a natural population of native Australian rats. *Environ. Microbiol.* **2**(6):594–610.

Rasko DA, Rosovitz MJ, Myers GS, Mongodin EF, Fricke WF, et al. 2008. The pangenome structure of *Escherichia coli*: comparative genomic analysis of *E. coli* commensal and pathogenic isolates. *J. Bacteriol.* **190**(20):6881–6893.

Ross BC, Czajkowski L, Hocking D, Margetts M, Webb E, et al. 2001. Identification of vaccine candidate antigens from a genomic analysis of *Porphyromonas gingivalis*. *Vaccine* **19**(30):4135–4142.

Serruto D, Serino L, Masignani V, Pizza M. 2009. Genome-based approaches to develop vaccines against bacterial pathogens. *Vaccine* **27**(25–26):3245–3250.

Stephens DS, Greenwood B, Brandtzaeg P. 2007. Epidemic meningitis, meningococcaemia, and *Neisseria meningitidis*. *Lancet* **369**(9580):2196–210.

Tettelin H, Saunders NJ, Heidelberg J, Jeffries AC, Nelson KE, et al. 2000. Complete genome sequence of *Neisseria meningitidis* serogroup B strain MC58. *Science* **287**(5459):1809–1815.

Tettelin H, Masignani V, Cieslewicz MJ, Donati C, Medini D, et al. 2005. Genome analysis of multiple pathogenic isolates of *Streptococcus agalactiae*: Implications for the microbial "pan-genome". *Proc. Natl. Acad. Sci. USA* **102**(39):13950–13955.

Tettelin H, Riley D, Cattuto C, Medini D. 2008. Comparative genomics: the bacterial pan-genome. *Curr. Opin. Microbiol.* **11**(5):472–477.

Welch RA, Burland V, Plunkett G, 3rd, Redford P, Roesch P, et al. 2002. Extensive mosaic structure revealed by the complete genome sequence of uropathogenic *Escherichia coli*. *Proc. Natl. Acad. Sci. USA* **99**(26):17020–17024.

Wizemann TM, Heinrichs JH, Adamou JE, Erwin AL, Kunsch C, et al. 2001. Use of a whole genome approach to identify vaccine molecules affording protection against *Streptococcus pneumoniae* infection. *Infect. Immun.* **69**(3):1593–1598.

Yang X, Hudson M, Walters N, Bargatze RF, Pascual DW. 2005. Selection of protective epitopes for *Brucella melitensis* by DNA vaccination. *Infect. Immun.* **73**(11):7297–7303.

Chapter 11

Metagenomic Insights into Bacterial Species

KONSTANTINOS T. KONSTANTINIDIS

11.1 INTRODUCTION

The quest for bacterial (and archaeal) species would have been met if bacteria formed monophyletic and genomically coherent clusters of individual organisms (strains) that were diagnosable by at least one discriminative phenotypic property [Rossello-Mora and Amann, 2001]. Bacteria are thought to be primarily asexual organisms; such organisms would commonly form discrete clusters, driven by ecological specialization and population sweeps [Cohan, 2001; Fraser et al., 2007]. However, genomic studies of large collections of bacterial and archaeal strains over the last 10 years have revealed that bacteria are much more amenable to horizontal gene transfer than previously anticipated [Beiko et al., 2005; Konstantinidis and Tiedje, 2005; Tettelin et al., 2005; Sheppard et al., 2008]. These findings have led several scientists to question whether the processes driving the diversification of these microorganisms produce sufficiently discrete groups of individuals (clusters) and, hence, species [Lawrence, 2002; Gevers et al., 2005; Papke et al., 2007; Doolittle and Zhaxybayeva, 2009]. In fact, several studies have recently uncovered indiscrete ("fuzzy") clusters of bacteria and archaea based on the multiple locus sequence (MLSA) or whole-genome analyses of large collections of isolates [Hanage et al., 2005; Mau et al., 2006; Sheppard et al., 2008]. The issue has been extensively discussed previously, and the reader is directed to several excellent reviews [Rossello-Mora and Amann, 2001; Gevers et al., 2005; Konstantinidis et al., 2006; Staley, 2006; Ward, 2006; Fraser et al., 2007] and other chapters of this book for further details.

One problematic aspect of most, if not all, of the previous genomic studies is that they were based on strains isolated in the laboratory. However, lab cultivation is characterized by several known biases [Staley and Konopka, 1985; Amann et al., 1995]. For instance, it remains speculative whether the isolated organism is dormant and/or transient as opposed to being ecologically successful and growing at the place of isolation. Furthermore, the total natural diversity and in-situ abundance of each isolated strain are usually either inaccessible to cultivation-based approaches or not measured. Accordingly, most of the derived conclusions are qualitative rather than quantitative. Finally, several, if not all, of the previous studies analyzed collections of organisms that were recovered from different populations and habitats, further complicating genetic analyses and comparisons.

Metagenomic approaches (discussed in greater detail elsewhere in this book and in Handelsman et al. [2007]) hold great promise for providing new insights into the previous issues because they sidestep several of the limitations of culture-based approaches and can assess quantitatively the total natural diversity within a given habitat. In the remaining text, I will attempt to summarize the emerging findings from the metagenomic surveys of the past five years with respect to the species problem and contrast these findings to those of genome-based surveys of isolates. I will argue that metagenomic approaches reveal that microbial communities are indeed composed of distinct species-like populations and will discuss the implications of these findings for the current species definition.

Handbook of Molecular Microbial Ecology, Volume I: Metagenomics and Complementary Approaches, First Edition. Edited by Frans J. de Bruijn.

11.2 MATERIALS AND METHODS

The data and methods used here are primarily those described by Konstantinidis and DeLong [2008] and Rusch et al. [2007], or slightly modified variations of those. In brief, the fragment recruitment and coverage plots (see examples in **Fig. 11.1**) were obtained as follows: All sequences in a metagenomic dataset (Sanger or 454 sequencing reads) were queried against a reference sequence such as a whole-genome or a fosmid/BAC clone sequence, using the blastn (nucleotide level) algorithm [Altschul et al., 1997] and the following settings and cutoffs: $X = 150$ (drop-off value for gapped alignment), $q = -1$ (penalty for nucleotide mismatch), and $F = -F$ (filter for repeated sequences); the rest of the parameters were at default settings; and a minimum cutoff for a match of at least 70% of the length of the query sequence to be included in the resulting blast alignment. These settings can robustly detect more divergent homologous sequences (e.g., showing 60–80% nucleotide identity) compared to the sequences detected with the default settings (e.g., 90% nucleotide identity or higher). Blastn, even with the previous settings, cannot identify robustly sequences that are related below the 60–70% nucleotide identity level [Konstantinidis and Tiedje, 2005]; thus, such sequences are not considered by the approach described. For the sequences that passed the cutoff above, their nucleotide identity to and orthologous position in the reference sequence were obtained directly from the blastn output and reported in the graphs of **Figures 11.2 and 11.3**.

Evaluating this approach with whole genome sequences of isolates and traditional gene-based phylogenetic analyses showed that it can provide for a robust picture of the microdiversity in the sample (i.e., resolution at the population level or between closely related populations) as long as (i) the reference sequence used originates from within the population or belongs to an organism that is very closely related to the natural population (e.g., shows >95% average nucleotide identity, ANI [Konstantinidis and Tiedje, 2005]), (ii) the reference sequence is at least 40–50 kbp long, and (iii) the genes that are much more conserved at the sequence level compared to the genome average (e.g., the rRNA operon genes), and thus can recruit sequences from distantly related organisms, are not included in the analyses [Konstantinidis and Delong, 2008]. Previous studies have also shown that the latter genes as well as recent horizontally transferred genes are infrequent enough (i.e., comprise less than 5% of the total genes in

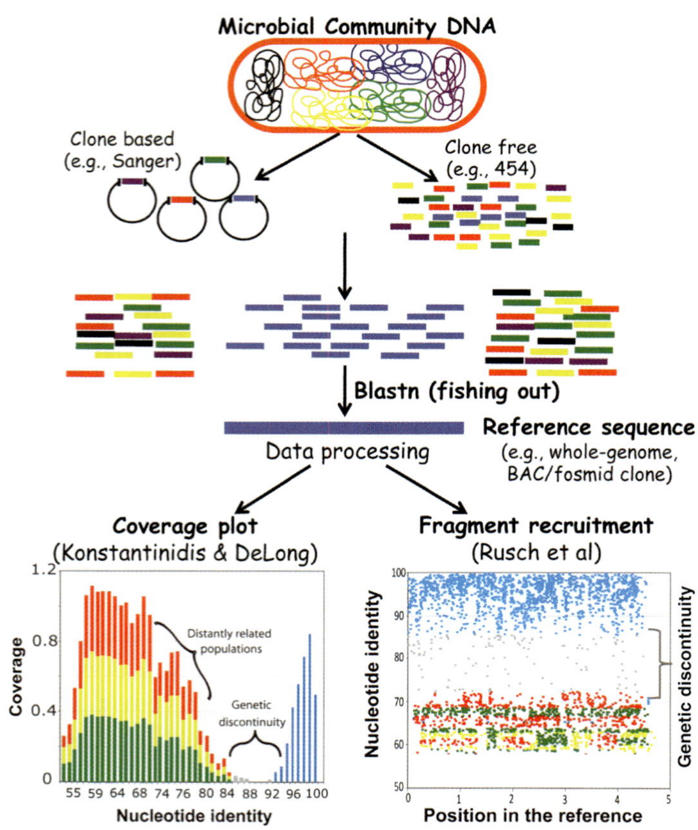

Figure 11.1 Metagenomic approaches for looking into natural microdiversity. A graphical representation of the approaches discussed in the main text is shown. Note that the hypothetical microbial community shown has only six members (populations), each represented by a different color, for simplicity purposes; the communities discussed in the text are typically composed of hundreds or thousands of members. Based on the graphs shown, three abundant populations (red, green, and yellow), closely related to the sequence-discrete blue population represented by the reference sequence, coexist in this community, each showing about 60–80% nucleotide identity to the blue population. This contrasts with the crenarchaeal example shown in Figure 11.2, where no abundant closely related population(s) to the reference crenarchaeal population exists in the community analyzed. The other two populations (black and purple) are too distantly related to the blue population to have any of their sequences recruited by the reference sequence in the 60–100% nucleotide identity range. The gray sequences of intermediate genetic distance between the blue and the remaining populations are discussed in the text. Coverage in the coverage plot is typically measured in "total bases aligned versus total bases in the reference sequence" to provide for a measurement of in-situ abundance that is independent of the length of the reference sequence or the size of the metagenome used, as proposed earlier [Konstantinidis and Delong, 2008].

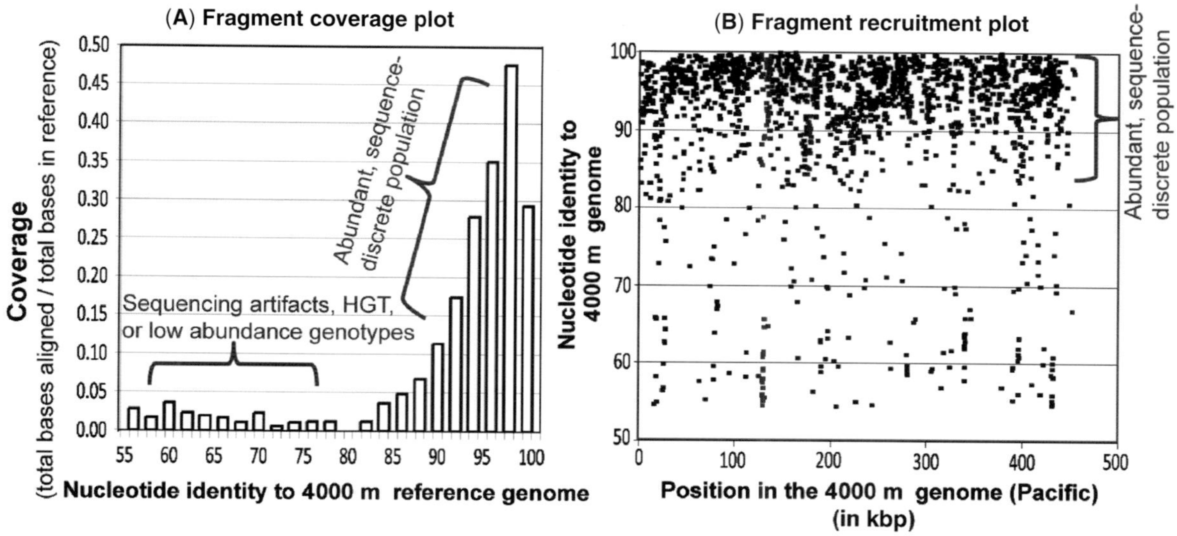

Figure 11.2 *Crenarchaea* form a sequence-discrete population at 4000-m depth in the Pacific Ocean. All short WGS sequences from the 4000-m deep metagenome (about 200,000, 1-kbp-long Sanger sequencing reads) were searched, as described in the materials and methods section, against a crenarchaeal reference genome from the same sample [Konstantinidis and Delong, 2008]. The figure shows two different representations of the results of the search for a 0.5-Mbp-long piece of the genome—that is, a coverage plot (**A**) and a fragment recruitment plot (**B**). (Panel A is adapted from Konstantinidis and Delong [2008].)

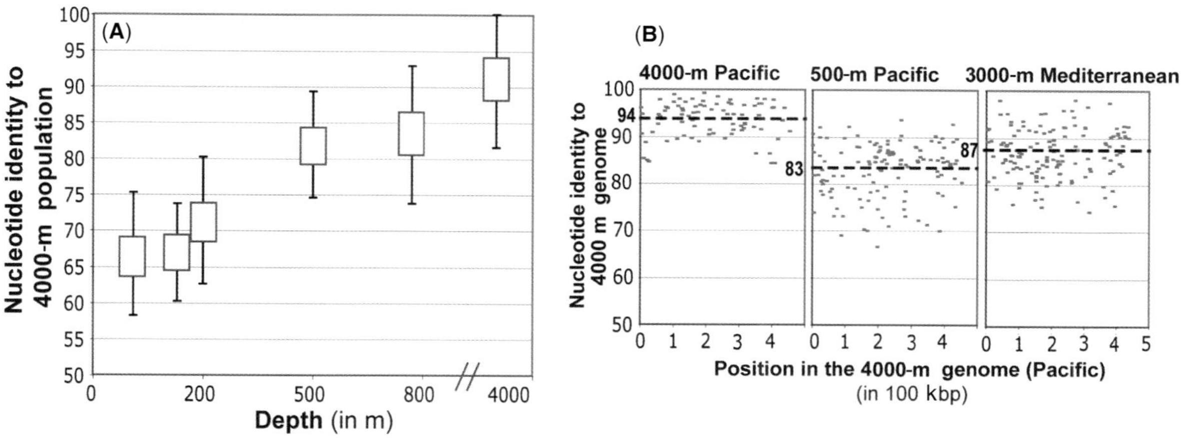

Figure 11.3 Biogeography of sequence-discrete crenarchaeal populations. A reference genome representing the crenarchaeal population at 4000-m depth in the Pacific Ocean was queried against the previously described [Konstantinidis and Delong, 2008] crenarchaeal fosmid end-sequences from seven different depths of the Pacific Ocean [DeLong et al., 2006], the 3000-m depth in the Mediterranean Sea [Martin-Cuadrado et al., 2007], and the crenarchaeal sequences from the WGS metagenome from 4000m in the Pacific Ocean [Konstantinidis and Delong, 2008]. (**A**) Range in nucleotide identities between the end-sequences and the genome (y axis) plotted against the depth that the end-sequences were recovered from (x axis). (**B**) Fragment recruitment representation for selected comparisons performed. Note that *Crenarchaea* are genetically distinct between different depths of the same water column (**A**) but genetically more similar across similar depths in geographically distant locations (**B**).

the genome [Kunin et al., 2005; Caro-Quintero et al., 2009]) to not be problematic. Furthermore, such genes are easily detectable by the approach described above since they recruit many more reads than the remaining genes in the reference sequence [Rusch et al., 2007; Konstantinidis and Delong, 2008]. Finally, assessing the microdiversity at the nucleotide level, as opposed to the amino acid level, offers much higher resolution and sidesteps several limitations such as the confounding effect of convergent evolution at the protein sequence level.

11.3 RESULTS AND DISCUSSION

11.3.1 Complex Microbial Communities Are Organized in Sequence-Discrete, Species-Like Populations

The metagenomic studies that are most relevant for the species issue are those that provided random (unbiased) representations of the total naturally occurring diversity—that is, recovered a large number of whole-genome shotgun (WGS) sequences from community DNA [Tyson et al., 2004; Venter et al., 2004]. Most of such studies to date have been performed in marine environments; accordingly, the main focus of the remaining text will be the diversity within marine microbial communities, with reference to terrestrial communities when appropriate. Furthermore, it is important to note that in these studies the WGS sequences originate typically from different cells in the sample and do not show any obvious strong biases with respect to the genotype or the section of the genome they represent, albeit some rather minor biases specific to the sequencing technology used appear to exist [Gomez-Alvarez et al., 2009; Temperton et al., 2009]. Hence, when these sequences are queried against a reference sequence from within the population, they can uncover the total naturally occurring genetic diversity of the population represented by the reference sequence as well as the genetic relatedness of that (reference) population to other co-occurring populations in the sample (see Fig. 11.1 for an example; also see Section 11.2 for details).

Using this approach, Konstantinidis and DeLong [2008] and others [Rusch et al., 2007] showed that oceanic microbial communities are composed primarily of sequence-discrete populations of bacteria and archaea. The genotypes of a population appear to be genetically homogeneous, showing typically higher than ~90–95% nucleotide identity among themselves (intra-population diversity), while they show substantially lower sequence identity (typically <80%) to their closest co-occurring relatives, if any, in the same sample (Fig. 11.2).

Although these findings were initially [Rusch et al., 2007] based on short (~1 kbp long), error-prone WGS sequences, since they represented single-pass sequencing reads, similar results [Konstantinidis and Delong, 2008] were obtained more recently with longer (~40 kbp long) fosmids clone sequences, which were sequenced at the high draft status (i.e., 10× coverage [Branscomb and Predki, 2002]). In fact, the analyses of the fosmid sequences showed that the intra-population sequence similarity is probably >95% genome-aggregate ANI rather than >90% ANI as indicated by the short WGS sequences, presumably due to the higher sequencing error rate of the latter. This level of intra-population genetic similarity or coherence is comparable to, if not higher than, what corresponds to the frequently used standards for species delineation—that is, 95% ANI, which corresponds tightly to the golden standard of 70% DNA–DNA hybridization [Goris et al., 2007]. Further analyses of the fosmid sequences suggested that most of the (few) short WGS sequences that show nucleotide identities in the 60–90% range to the reference sequence correspond typically to sequencing error artifacts, genes that are more conserved than the genome average (hence, recruit WGS sequences representing distantly related organisms), or recent horizontally transferred genes (gray fragments in Figs. 11.1 and 11.2). Thus, very few, if any, of the latter sequences probably correspond to real genotypes that fall within the gap of genetic relatedness between the reference population and the remaining populations in the community [Konstantinidis and DeLong, 2008]. Even in the latter case, however, these genotypes are most likely ecologically differentiated from the genotypes of the reference population given that, under identical environmental conditions (i.e., they are present in the same sample), they are substantially less abundant (e.g., represented by fewer WGS sequences) and genetically divergent. Therefore, the oceanic microbial communities characterized by the previous metagenomic studies appear to consist quantitatively of sequence-discrete and probably ecologically discrete populations.

Whether or not the sequence-discrete populations are comprised of further ecologically distinct subpopulations is not completely clear. It is likely that some of the sequence variation within the populations is adaptive, as opposed to neutral, and such variation could account for distinct subpopulations. If such subpopulations have emerged only recently from within a population, they would not have been differentiated enough or would not have dominated the intra-population diversity (population sweeps) to be detectable by the metagenomic approaches described above. To detect such subpopulations, whole-genome approaches and time-series data combined with population genetic analyses will be required. Application of the latter approaches in a marine *Vibrio splendidus* population identified recently emerged and ecologically distinct subpopulations [Hunt et al., 2008]. It should be noted, however, that the *V. splendidus* or similar organisms that represent low-abundance (transient?) members of the indigenous community [Sogin et al., 2006] and/or are associated with eukaryotic hosts (like *V. splendidus*) have a fundamentally different ecology and are probably subjected to different selection pressures compared to the free-living abundant planktonic organisms that were primarily assessed by the previous metagenomic surveys.[1] Indeed, it is plausible that such low-abundance organisms form homogeneous sequence-discrete populations when favorable environmental conditions for their growth prevail or when they are transferred to their own favorable habitat (if different from the habitat they were found in). Even though the existence of ecologically distinct subpopulations within the identified sequence-discrete populations cannot be ruled out (e.g., Hunt et al. [2008] and Martiny et al. [2009]), the identification of sequence-discrete populations for almost every known abundant group in the oceans, irrespective of the phylogenetic affiliation of the group or the location of the sample analyzed [Rusch et al., 2007; Konstantinidis and Delong, 2008], indicate that subpopulations may quantitatively be infrequent, at least for the abundant community members.

In summary, it appears that the oceanic microbial communities are composed predominantly of sequence-discrete populations, and these populations possess several important properties expected for species such as genetic and ecological distinctiveness. Metagenomic studies of other environments uncovered that microbial diversity is organized in similar patterns within human-gut associated [Gill et al., 2006], iron-reducing biofilm [Tyson et al., 2004], phosphorus removal bioreactor [Garcia Martin et al., 2006], hot springs [Bhaya et al., 2007; see also Chapter 37, Vol. I], and soil and eutrophic lake microbial communities (Konstantinidis et al., unpublished; see also Chapter 31, Vol. I). These patterns reflect presumably some fundamental properties of the microbial world.

11.3.2 What Are the Forces of Population Cohesion?

Several theories based on recombination frequency[2] [Feil et al., 2001; Fraser et al., 2007] or population sweeps caused by periodic natural selection [Cohan, 2001; Acinas et al., 2004] have been advanced to explain the emergence and maintenance of sequence-discrete populations. An increasing body of evidence indicates that both recombination and periodic selection play a role in driving the diversity patterns observed in the metagenomic datasets, albeit the relative importance of these processes has not been fully determined. For instance, several studies have reported detectable levels of homologous recombination within marine planktonic [Rusch et al., 2007; Vergin et al., 2007; Konstantinidis and Delong, 2008] and terrestrial biofilm natural populations [Tyson et al., 2004; Eppley et al., 2007]. Furthermore, the efficiency of homologous recombination decreases, in general, with increasing divergence of the recombined DNA segments based on laboratory experiments [Vulic et al., 1999; Majewski et al., 2000]. Thus, recombination would have accounted for the sequence-discrete populations if recombination among co-occurring cells was rampant and there was a dramatic decrease in recombination efficiency around the 90–95% nucleotide identity. However, our understanding of the exact relationship between recombination and sequence similarity remains incomplete, while recent metagenomic analyses have implied that this relationship may differ for natural populations compared to laboratory ones [Eppley et al., 2007]. Perhaps more importantly, a precise estimation of the recombination to mutation rate is basically out of reach for current algorithms due to the complications associated with detecting recombination among genetically very similar organisms [Eppley et al., 2007; Konstantinidis and Delong, 2008] despite recent attempts in this area [Kosakovsky Pond et al., 2006; Johnson and Slatkin, 2009]. Thus, the extent to which recombination represents the force of population cohesion and drives the diversity patterns observed remains essentially speculative. The importance of recombination probably depends upon the characteristics of the environment considered as well. For instance, the levels of homologous recombination were found to be about threefold lower in marine oligotrophic planktonic populations compared to populations of terrestrial high-biomass biofilm communities, based on the analyses of comparable WGS metagenomic datasets with the same algorithms [Konstantinidis and Delong, 2008]. However, more populations and habitats must be analyzed before a more complete understanding of the influence of the environment on evolutionary processes such as recombination can emerge.

On the other hand, periodic selection, caused by environmental perturbations, most likely accounts for the fact that the sequence diversity within the identified sequence-discrete populations varied, depending on the organism considered. Some organisms showed considerable intra-population diversity, ranging in the 95–100% ANI level, while others were substantially less diverse—for example, showing 98–100% ANI. For instance, Hallam and colleagues noticed that the

[1] Low-abundance populations cannot be robustly assessed by current metagenomic approaches due to the yet-prohibiting cost of sequencing [Rusch et al., 2007], although the latter is becoming less of an issue with the recent advancements in sequencing technologies [Harris et al., 2010].

[2] Recombination refers to the process that a horizontally transferred piece of DNA is incorporated into the genome of the recipient cell. If the mechanism of incorporation involves the replacement of a homologous and highly identical (at the sequence level) DNA piece, the process is called homologous recombination; otherwise it is called nonhomologous recombination. Ratios of homologous recombination to mutation rate greater than 1.0 are indicative of a sexually reproducing population. Fraser et al. [2007], have calculated this cutoff to be 0.5 in computer simulations but the relevance of these findings to natural populations awaits experimental validation.

population of planktonic *Crenarchaea*, an important nitrifying archaeal group in the oceans [DeLong et al., 1994; Konneke et al., 2005; Karner et al., 2001; Wuchter et al., 2006; Lam et al., 2007], was barely detectable in the summer metagenomic samples of the Sargasso Sea [Venter et al., 2004], while it was substantially more abundant in the winter samples from the same site [Hallam et al., 2006; see also Chapter 29, Vol. II]. The seasonality in detecting the *Crenarchaea*, which primarily reside below the euphotic zone of the oceans (>150 m) [Karner et al., 2001 see also Chapter 28, Vol. II], is presumably linked to the winter upwelling that brings to the surface populations from deeper waters. Importantly, further analyses indicated that the winter surface population of *Crenarchaea* in the Sargasso Sea showed substantially less intra-population diversity compared to its counterpart at 4000-m depth (98–100% vs. 90–100% nucleotide identity based on short sequencing reads) and, in fact, was composed of a mix of (at least) two distinct, clonal and abundant subpopulations [Konstantinidis and Delong, 2008]. These diversity patterns are best explained by lack of or infrequent strong periodic selection in the remarkably stable environment at 4000-m depth [DeLong et al., 2006] and, in contrast, frequent periodic selection in the surface waters due to continuous environmental perturbations.

Regardless of what the actual mechanism(s) of population cohesion are or what their relative importance is, the members of the identified populations appear to somehow cohere together and form sequence-discrete populations.

11.3.3 Biogeography of the Sequence-Discrete Populations

The biogeography of microbial populations remains an important unsettled issue for the species concept and, in general, for microbial ecology [Martiny et al., 2006; Ramette and Tiedje, 2007]. For instance, it is still unclear whether or not the distribution of bacterial populations is constrained by geographic factors and how strong such factors are for different species or environments [Patterson, 2009]. New insights into this issue have started to emerge from the metagenomic analysis of abundant marine populations. A representative example of the latter comes from the nitrifying *Crenarchaea*, which also represents one of the best-sampled groups in several independent metagenomic datasets. The populations of *Crenarchaea* appear to be distinct at different depths of the ocean's water column. For instance, when the genome of the crenarchaeal population living at 4000-m depth, obtained based on a culture-independent approach using fosmid clone sequences, was compared against crenarchaeal sequences from seven different depths at the same sampling station (Station Aloha, North Pacific Ocean), the sequence identity was lower for sequences originating from shallower waters [Konstantinidis and Delong, 2008]. The average nucleotide identity between the population from 4000-m depth and those from 770 m, 500 m, and shallower than 500 m was 83.2%, 81.9%, and <72%, respectively, contrasting with >85% for the intra-population diversity at 4000 m, based on short WGS sequences (Fig. 11.3A).

Further comparisons against datasets recovered from other water basins indicated that the crenarchaeal populations were genetically very similar, if not identical, across the same sampling depths. For instance, the crenarchaeal sequences recovered in the winter samples of the Sargasso Sea were very similar (>90% nucleotide identity) to the crenarchaeal sequences from 200-m depth in the Pacific Ocean (due to water upwelling; see above). Similarly, *Crenarchaea* from 3000 m in the Mediterranean Sea [Martin-Cuadrado et al., 2007] shared 86.8% average nucleotide identity with their counterparts from 4000-m depth in the Pacific Ocean. This level of relatedness was significantly higher ($p < 0.01$) than that observed between the 4000-m-deep and the 500-m-deep populations from the same sampling site in the Pacific Ocean (81.9%), consistent with the idea of depth-stratified crenarchaeal populations and a panmictic population across the seas at the same depth (Fig. 11.3B). The lower genetic similarity between the 3000-m Mediterranean Sea and 4000-m Pacific Ocean populations relatively to the similarity within the latter population may be due to the difference in the depths sampled and the time-separation (on the order of thousands of years) of the two populations due to the global marine water circulation systems [Sexton and Norris, 2008]. Similar patterns, albeit based on a smaller number of sequences available, have been observed for additional important marine groups such as the surface photosynthetic *Prochlorococcus* (*Cyanobacteria*) between the Pacific Ocean and the Sargasso Sea and the deep heterotrophic *Pelagibacter* (α-*Proteobacteria*) between the Mediterranean Sea and the Pacific Ocean [Konstantinidis and Delong, 2008].

Although the *Crenarchaea*, and possibly several other marine groups [Moore et al., 1998], form sequence-distinct populations at different depths, comparisons of the gene content recovered in assembled crenarchaeal contigs from the surface Sargasso Sea and the deep Pacific Ocean metagenomes did not identify any detectable differences in central metabolic pathways present in the surface versus the deep populations [Hallam et al., 2006; see also Chapter 25, Vol. II]. The major crenarchaeal pathways for ammonia oxidation, urea transport, and central metabolism were all present, at the same abundance, in the two datasets. (These conclusions differ from those of Agogue et al., [2008]; based on q-PCR data of the ammonia oxidation genes; however, the difference is most likely due to the q-PCR primers used that had mismatches against the gene homologs from the deep waters and/or the different samples analyzed [Church et al., 2009; Konstantinidis et al., 2009a]). Therefore, it appears as if the depth-specific adaptations of *Crenarchaea* are not related to metabolic properties; rather, these adaptations, if any, are subtle and probably limited to the maintenance of the integrity of the cell membrane and the intracellular proteins [Konstantinidis et al., 2009a].

11.3.4 Implications for the Species Definition

The identification of sequence-discrete clusters that may correspond to genuine species represents a remarkable finding given that, as several scientists believe [Lawrence, 2002; Gevers et al., 2005; Doolittle and Zhaxybayeva, 2009], there is no intrinsic reason why the processes driving diversification and adaptation of bacteria must produce sufficiently coherent groups of individuals (species). The discrepancy appears to be due, at least in part, to the biases introduced by cultivation and the fact that most of the previous studies analyzed collection of isolates recovered from different populations and habitats. For instance, if crenarchaeal isolates from different depths of the water column were available and were compared against each other, a continuous gradient of genetic diversity in the ~80–100% ANI range would have been observed based on the findings from metagenomics (e.g., Fig. 11.3A), instead of the sequence-discrete populations that reside at each depth (e.g., Fig. 11.2). Similarly, if crenarchaeal isolates were available from the winter surface samples of the Sargasso Sea, a heterogeneous, "fuzzy" population would have been revealed, composed of a mix of several distinct subpopulations that originated from different depths of the water column due to winter water upwelling [Konstantinidis and Delong, 2008]. The latter findings also imply that habitats that are characterized by more frequent and intense environmental perturbations compared to the open oceans such as the upper soil layer or the human gut are likely to show higher speciation rates and (perhaps) a higher number of "fuzzy" populations. If, however, genetically homogeneous organisms are maintained within a stable niche for some period of time—in other words, share a similar ecological trajectory, like some of the abundant marine organisms discussed above presumably do—then a sequence-discrete population is expected to emerge.

Thus, these findings reveal that organisms with the same ecological "trajectory" do form species-like populations, although identifying the exact ecological "trajectory" of each organism may be challenging, due mainly to (unpredictable) environmental perturbations and biases introduced by cultivation. Furthermore, most, if not all, of the sequence-discrete populations recovered in the marine or terrestrial metagenomes show an intrapopulation sequence and gene-content diversity that is typically much lower than what corresponds to the current standards for species [Konstantinidis and DeLong, 2008]. These results are therefore consistent with a more ecological and stringent definition for species than the current definition and have important practical implications. For instance, organisms that are recovered from different habitats, even if they share higher than 95% ANI, should not be included in the same species without further evaluation of their metabolic and ecological similarity. In agreement with the latter conclusions, recent work with isolates of comparable genetic relatedness among themselves indicated that gene expression and regulation differences at the whole-genome level were smaller for the isolates with less dramatically different ecological trajectories [Konstantinidis et al., 2009b].

A question of important practical consequences remains: Should each sequence-discrete population of *Crenarchaea* residing at a particular depth of the water column be recognized as a different species, or should all the populations belong to the same species? From the perspective of the microorganism, these populations are clearly genetically distinct and, thus, apparently not interchangeable (a discriminative trait of different species). Their diagnostic properties—that is, the depth that they are adjusted to and their genetic distinctiveness compared to the populations from other depths (e.g., showing <95% ANI to the latter)—might be enough to justify a new species description for each population. From the human perspective and given also their high similarity in metabolic genes, these populations presumably perform the same function or role in the environment and, thus, can be included in the same species, perhaps as different biovars. The former approach could easily result in an intractable number of species described; the latter one may be more pragmatic from the human perspective. The species definition oughts to be pragmatic [Rossello-Mora and Amann, 2001], suggesting that the latter approach may be more appropriate. However, the previous findings also suggest that the decision on which approach is more appropriate may sometimes be in "the eyes of the beholder" and probably depends on how well the overall metabolic similarity of the populations compared is understood or assessed. These finding make it imperative to report any ecological or environmental parameter available when describing a new bacterial isolate or a species (e.g., place and time of isolation, in-situ abundance, etc.) in order to provide a better appreciation of intraspecies diversity.

It should be also pointed out that the 16S rRNA gene sequences among the depth-stratified crenarchaeal populations are almost indistinguishable from each other;

that is, they show higher than 99% nucleotide identity [Pham et al., 2008]. This level of relatedness is substantially higher than the most frequently used standards for environmental species (or operational taxonomic units, OTUs) surveys—that is, 99% or 97% 16S rRNA gene sequence identity. Furthermore, whole genome comparisons of more than 500 complete bacterial and archaeal genomes showed that the 97% 16S rRNA gene identity corresponds typically to about 70% ANI [Konstantinidis and Tiedje, 2007]. Thus, a 97% 16S rRNA gene identity standard would have lumped together, for instance, all crenarchaeal populations from all depths in the Pacific Ocean. These results further corroborate previous findings (e.g., Acinas et al. [2004]) in that the most commonly used 16S rRNA gene standards for environmental surveys cannot frequently resolve the distinct populations of a microbial group (i.e., the microdiversity level). These results also underscore the influence of the standards employed on the level of resolution that can be achieved.

11.4 CONCLUDING REMARKS AND OPEN QUESTIONS

The metagenomic surveys of natural microbial communities clearly indicate that bacteria and archaea do form sequence-discrete clusters (at least) within the same habitat, probably beyond reasonable doubt and the limitations associated with the fragmented nature of the metagenomic datasets. They also provide clues about the biotic and abiotic factors that may affect the detection of the clusters in the environment. While looking for exceptions to these patterns or non-neutral diversity within the identified clusters is valuable and might lead to the discovery of important new underlying principles, the emerging picture is that the immense microbial diversity on the planet may indeed organized predominantly in species-like clusters. Thus, it is time to reevaluate the present way of thinking about bacterial species as well as the methods used to identify and describe species. In particular, it is becoming increasingly more clear that the new omics approaches offer much greater resolving power in unraveling the organism's environmental and ecological adaptations compared to the traditional methods based on cultivation approaches and the 16S rRNA gene sequence identity. It is also likely that many (most?) of the diagnostic adaptations of an organism can be elucidated only when the organism is studied in its natural environment or under in-situ conditions. Such adaptations can be studied using only culture-independent techniques like metagenomics and similar techniques—for example, metatranscriptomics [Shi et al., 2009; see also Chapters 62–65, Vol. II] and metaproteomics [VerBerkmoes et al., 2009; see also Chapters 66–70, Vol. II].

The metagenomic studies to date have also revealed or further corroborated several issues that will be important to advance toward a more educated new definition for bacterial species such as: How fast do microbial populations speciate, and are the rates different for different microbial groups and habitats? What is the relative importance of the drivers of population cohesion or divergence such as homologous recombination and horizontal gene transfer? The sequence-discrete populations recovered in the metagenomic datasets represent a major foundation toward beginning the high-resolution investigations that will provide new insights into these questions and advance our understanding of how populations are organized, interact, and evolve within communities.

Acknowledgments

I thank Professor James Tiedje for inspiring discussions on the species problem over the years, and I thank Professor Edward DeLong for exposing me to the science of metagenomics and to some of the early metagenomic datasets available. My work is supported by the US Department of Energy (award DE-FG02-07ER64389) and the National Science Foundation (award 0919251).

REFERENCES

Acinas, SG, Klepac-Ceraj V, Hunt DE, Pharino C, Ceraj I, et al. 2004. Fine-scale phylogenetic architecture of a complex bacterial community. *Nature* **430**:551–554.

Agogue, H, Brink M, Dinasquet J, Herndl GJ. 2008. Major gradients in putatively nitrifying and non-nitrifying Archaea in the deep North Atlantic. *Nature* **456**:788–791.

Altschul, SF, Madden TL, Schaffer AA, Zhang J, Zhang Z, et al. 1997. Gapped BLAST and PSI-BLAST: A new generation of protein database search programs. *Nucleic Acids Res*. **25**:3389–3402.

Amann, RI, Ludwig W, Schleifer KH. 1995. Phylogenetic identification and *in situ* detection of individual microbial cells without cultivation. *Microbiol. Rev*. **59**:143–169.

Beiko, RG, Harlow TJ, Ragan MA. 2005. Highways of gene sharing in prokaryotes. *Proc. Natl. Acad. Sci. USA* **102**:14332–14337.

Bhaya, D, Grossman AR, Steunou AS, Khuri N, Cohan FM, et al. 2007. Population level functional diversity in a microbial community revealed by comparative genomic and metagenomic analyses. *ISME J*. **1**:703–713.

Branscomb, E, Predki P. 2002. On the high value of low standards. *J. Bacteriol*. **184**:6406–6409; discussion 6409.

Caro-Quintero, A, Rodriguez-Castano GP, Konstantinidis KT. 2009. Genomic insights into the convergence and pathogenicity factors of *Campylobacter jejuni* and *Campylobacter coli* species. *J. Bacteriol*. **191**:5824–5831.

Church, MJ, Wai B, Karl DM, DeLong EF. 2009. Abundances of crenarchaeal amoA genes and transcripts in the Pacific Ocean. *Environ. Microbiol*. 2010 Mar; **12**(3):679–688.

Cohan, FM. 2001. Bacterial species and speciation. *Syst. Biol.* **50**:513–524.

DeLong, EF, Wu KY, Prezelin BB, Jovine RV. 1994. High abundance of Archaea in Antarctic marine picoplankton. *Nature* **371**:695–697.

DeLong, EF, Preston CM, Mincer T, Rich V, Hallam SJ, et al. 2006. Community genomics among stratified microbial assemblages in the ocean's interior. *Science* **311**:496–503.

Doolittle, WF, Zhaxybayeva O. 2009. On the origin of prokaryotic species. *Genome Res.* **19**:744–756.

Eppley, JM, Tyson GW, Getz WM, Banfield JF. 2007. Genetic exchange across a species boundary in the archaeal genus ferroplasma. *Genetics* **177**:407–416.

Feil, EJ, Holmes EC, Bessen DE, Chan MS, Day NP, et al. 2001. Recombination within natural populations of pathogenic bacteria: short-term empirical estimates and long-term phylogenetic consequences. *Proc. Natl. Acad. Sci. USA* **98**:182–187.

Fraser, C, Hanage WP, Spratt BG. 2007. Recombination and the nature of bacterial speciation. *Science* **315**:476–480.

Garcia Martin, H, Ivanova N, Kunin V, Warnecke F, Barry KW, et al. 2006. Metagenomic analysis of two enhanced biological phosphorus removal (EBPR) sludge communities. *Nat. Biotechnol.* **24**:1263–1269.

Gevers, D, Cohan FM, Lawrence JG, Spratt BG, Coenye T, et al. 2005. Opinion: Reevaluating prokaryotic species. *Nat. Rev. Microbiol.* **3**:733–739.

Gill, SR, Pop M, Deboy RT, Eckburg PB, Turnbaugh PJ, et al. 2006. Metagenomic analysis of the human distal gut microbiome. *Science* **312**:1355–1359.

Gomez-Alvarez, V, Teal TK, Schmidt TM. 2009. Systematic artifacts in metagenomes from complex microbial communities. *ISME J.* **3**:1314–1317.

Goris, J, Konstantinidis KT, Klappenbach JA, Coenye T, Vandamme P, et al. 2007. DNA–DNA hybridization values and their relation to whole genome sequence similarities. *Int. J. Syst. Evol. Microbiol.* 81–91.

Hallam, SJ, Konstantinidis KT, Putnam N, Schleper C, Watanabe Y, et al. 2006. Genomic analysis of the uncultivated marine crenarchaeote *Cenarchaeum symbiosum*. *Proc. Natl. Acad. Sci. USA* **103**:18296–18301.

Hanage, WP, Fraser C, Spratt BG. 2005. Fuzzy species among recombinogenic bacteria. *BMC Biol.* **3**:6.

Handelsman, J, Tiedje J, Alvarez-Cohen L, Ashburner M, Cann I, et al. 2007. *The New Science of Metagenomics: Revealing the Secrets of Our Microbial Planet*. Washington, DC: The National Academies Press.

Harris, SR, Feil EJ, Holden MT, Quail MA, Nickerson EK, et al. 2010. Evolution of MRSA during hospital transmission and intercontinental spread. *Science* **327**:469–474.

Hunt, DE, David LA, Gevers D, Preheim SP, Alm EJ, et al. 2008. Resource partitioning and sympatric differentiation among closely related bacterioplankton. *Science* **320**:1081–1085.

Johnson, PL, Slatkin M. 2009. Inference of microbial recombination rates from metagenomic data. *PLoS Genet.* **5**:e1000674.

Karner, MB, DeLong EF, Karl DM. 2001. Archaeal dominance in the mesopelagic zone of the Pacific Ocean. *Nature* **409**:507–510.

Konneke, M, Bernhard AE, de la Torre JR, Walker CB, Waterbury JB, et al. 2005. Isolation of an autotrophic ammonia-oxidizing marine archaeon. *Nature* **437**:543–546.

Konstantinidis, KT, DeLong EF. 2008. Genomic patterns of recombination, clonal divergence and environment in marine microbial populations. *ISME J.* **2**:1052–65.

Konstantinidis, KT, Tiedje JM. 2005. Genomic insights into the species definition for prokaryotes. *Proc. Natl. Acad. Sci. USA* **102**:2567–72

Konstantinidis, KT, Tiedje JM. 2007. Prokaryotic taxonomy and phylogeny in the genomic era: Advancements and challenges ahead. *Curr. Opin. Microbiol.* **10**:504–509.

Konstantinidis, KT, Ramette A, Tiedje JM. 2006. The bacterial species definition in the genomic era. *Philos. Trans. R. Soc. Lond. B Biol. Sci.* **361**:1929–1940.

Konstantinidis, KT, Braff J, Karl DM, DeLong EF. 2009a. Comparative metagenomic analysis of a microbial community residing at a depth of 4,000 meters at station ALOHA in the North Pacific subtropical gyre. *Appl. Environ. Microbiol.* **75**:5345–5355.

Konstantinidis, KT, Serres MH, Romine MF, Rodrigues JL, Auchtung J, et al. 2009b. Comparative systems biology across an evolutionary gradient within the *Shewanella* genus. *Proc. Natl. Acad. Sci. USA* **106**:15909–14.

Kosakovsky Pond, SL, Posada D, Gravenor MB, Woelk CH, Frost SD. 2006. Automated phylogenetic detection of recombination using a genetic algorithm. *Mol. Biol. Evol.* **23**:1891–1901.

Kunin, V, Goldovsky L, Darzentas N, Ouzounis CA. 2005. The net of life: Reconstructing the microbial phylogenetic network. *Genome Res.* **15**:954–959.

Lam, P, Jensen MM, Lavik G, McGinnis DF, Muller B, et al. 2007. Linking crenarchaeal and bacterial nitrification to anammox in the Black Sea. *Proc. Natl. Acad. Sci. USA* **104**:7104–7109.

Lawrence, JG. 2002. Gene transfer in bacteria: speciation without species? *Theor. Popul. Biol.* **61**:449–460.

Majewski, J, Zawadzki P, Pickerill P, Cohan FM, Dowson CG. 2000. Barriers to genetic exchange between bacterial species: *Streptococcus pneumoniae* transformation. *J. Bacteriol.* **182**:1016–1023.

Martin-Cuadrado, AB, Lopez-Garcia P, Alba JC, Moreira D, Monticelli L, et al. 2007. Metagenomics of the deep Mediterranean, a warm bathypelagic habitat. *PLoS One.* **2**:e914.

Martiny, JB, Bohannan BJ, Brown JH, Colwell RK, Fuhrman JA, et al. 2006. Microbial biogeography: Putting microorganisms on the map. *Nat. Rev. Microbiol.* **4**:102–112.

Martiny, AC, Kathuria S, Berube PM. 2009. Widespread metabolic potential for nitrite and nitrate assimilation among *Prochlorococcus ecotypes*. *Proc. Natl. Acad. Sci. USA* **106**:10787–10792.

Mau, B, Glasner JD, Darling AE, Perna NT. 2006. Genome-wide detection and analysis of homologous recombination among sequenced strains of *Escherichia coli*. *Genome Biol.* **7**:R44.

Moore, LR, Rocap G, Chisholm SW. 1998. Physiology and molecular phylogeny of coexisting *Prochlorococcus* ecotypes. *Nature* **393**:464–467.

Papke, RT, Zhaxybayeva O, Feil EJ, Sommerfeld K, Muise D, et al. 2007. Searching for species in haloarchaea. *Proc. Natl. Acad. Sci. USA* **104**:14092–14097.

Patterson, DJ. 2009. Ecology. Seeing the big picture on microbe distribution. *Science* **325**:1506–15067.

Pham, VD, Konstantinidis KT, Palden T, DeLong EF. 2008. Phylogenetic analyses of ribosomal DNA-containing bacterioplankton genome fragments from a 4000 m vertical profile in the North Pacific Subtropical Gyre. *Environ. Microbiol.* **10**:2313–2330.

Ramette, A, Tiedje JM. 2007. Biogeography: An emerging cornerstone for understanding prokaryotic diversity, ecology, and evolution. *Microb. Ecol.* **53**:197–207.

Rossello-Mora, R, Amann R. 2001. The species concept for prokaryotes. *FEMS Microbiol. Rev.* **25**:39–67.

Rusch, DB, Halpern AL, Sutton G, Heidelberg KB, Williamson S, et al. 2007. The Sorcerer II Global Ocean Sampling Expedition: Northwest Atlantic through Eastern Tropical Pacific. *PLoS Biol.* **5**:e77.

Sexton, PF, Norris RD. 2008. Dispersal and biogeography of marine plankton: Long-distance dispersal of the foraminifer *Truncorotalia truncatulinoides*. *Geology* **36**:899–902.

Sheppard, SK, McCarthy ND, Falush D, Maiden MC. 2008. Convergence of *Campylobacter* species: Implications for bacterial evolution. *Science* **320**:237–239.

Shi, Y, Tyson GW, DeLong EF. 2009. Metatranscriptomics reveals unique microbial small RNAs in the ocean's water column. *Nature* **459**:266–269.

Sogin, ML, Morrison HG, Huber JA, Mark Welch D, Huse SM, et al. 2006. Microbial diversity in the deep sea and the underexplored "rare biosphere". *Proc. Natl. Acad. Sci. USA* **103**:12115–12120.

Staley, JT. 2006. The bacterial species dilemma and the genomic–phylogenetic species concept. *Philos. Trans. R. Soc. Lond. B Biol. Sci*. **361**:1899–909.

Staley, JT, Konopka A. 1985. Measurement of in situ activities of nonphotosynthetic microorganisms in aquatic and terrestrial habitats. *Annu. Rev. Microbiol*. **39**:321–346.

Temperton, B, Field D, Oliver A, Tiwari B, Muhling M, et al. 2009. Bias in assessments of marine microbial biodiversity in fosmid libraries as evaluated by pyrosequencing. *ISME J*. **3**:792–796.

Tettelin, H, Masignani V, Cieslewicz MJ, Donati C, Medini D, et al. 2005. Genome analysis of multiple pathogenic isolates of *Streptococcus agalactiae*: Implications for the microbial "pan-genome". *Proc. Natl. Acad. Sci. USA* **102**:13950–13955.

Tyson, GW, Chapman J, Hugenholtz P, Allen EE, Ram RJ, et al. 2004. Community structure and metabolism through reconstruction of microbial genomes from the environment. *Nature* **428**:37–43.

Venter, JC, Remington K, Heidelberg JF, Halpern AL, Rusch D, et al. 2004. Environmental genome shotgun sequencing of the Sargasso Sea. *Science* **304**:66–74.

VerBerkmoes, NC, Denef VJ, Hettich RL, Banfield JF. 2009. Systems biology: Functional analysis of natural microbial consortia using community proteomics. *Nat. Rev. Microbiol*. **7**:196–205.

Vergin, KL, Tripp HJ, Wilhelm LJ, Denver DR, Rappe MS, et al. 2007. High intraspecific recombination rate in a native population of *Candidatus Pelagibacter ubique* (SAR11). *Environ. Microbiol*. **9**:2430–2440.

Vulic, M, Lenski RE, Radman M. 1999. Mutation, recombination, and incipient speciation of bacteria in the laboratory. *Proc. Natl. Acad. Sci. USA* **96**:7348–7351.

Ward, D, A. 2006. A Macrobiological Perspective on Microbial Species. *Microbe Mag*.

Wuchter, C, Abbas B, Coolen MJ, Herfort L, van Bleijswijk J, et al. 2006. Archaeal nitrification in the ocean. *Proc. Natl. Acad. Sci. USA* **103**:12317–12322.

Chapter 12

Reports of Ad Hoc Committees for the Reevaluation of the Species Definition in Bacteriology

Erko Stackebrandt

12.1 INTRODUCTION

The currently applied polyphasic and pragmatic approach to circumscribe a prokaryotic species is the result of a century of struggle to combine two facets of achievements in microbiology: (1) the steadily increasing information on data at the genetic and epigenetic level and (2) the attempt to keep identification as practicable as possible in order to allow recognition of species also by those microbiologists who have no access to the most modern tools of characterization and authentification. The history of species descriptions in the past editions of *Bergey's Manual of Determinative/Systematic Bacteriology* is a vivid example of increasing complexity of characters used—starting with morphology, culture descriptions, and basic metabolic reactions in the mid-1870s; adding physiological properties around 1910; adding serology, ultrastructure, and numerical taxonomy in the 1930s, 1950s, and 1960s, respectively; adding chemotaxonomy, nucleic acid hybridization, and genomic properties in the 1970s; from 1980 on, adding rRNA and DNA gene sequences; and recently adding genomic fingerprinting techniques (see Chapters 2–8, Vol. I) and whole genome sequences [Stackebrandt, 2007]. Hardly any character, once introduced, was considered unsuitable for the purpose of species description, thus making the requirements for the description of this category so much more complex than those used in mycology and for higher evolved eukaryotes. This contribution highlights the influence of two expert committees on the future of bacterial systematics [Wayne et al., 1987; Stackebrandt et al., 2002], especially on the outline of a pragmatic species definition. In the lack of a recognized species concept for microorganisms [Roséllo-Mora and Amann, 2001; Stackebrandt 2004; Staley, 2009], the intuitive process of delineating species from each other defined entities in a pragmatic way, but it did not claim to mirror the products of speciation mechanism(s). Even more, until recently [Gevers et al., 2005; Achtman and Wagner, 2008] it completely ignored the increasing awareness of underlying evolutionary processes that led to genomically coherent biological entities that may indeed exist in the world of prokaryotes. This pragmatism abolishes and voluntarily rejects all properties that may be the result of horizontal gene transfer in order to base characterization on those genes and gene products which, in modern terms, belong to the core genome. In the absence of genomic data, phenotypically instable properties were subsequently deleted from the dataset, making the species circumscription rather robust and identification unambiguous. It is worth noting that the results of recently introduced rRNA and rDNA gene sequencing affected less the phylogenetic position of species at the intrageneric level (except for grossly misclassified species) than that of genera and higher taxa (families, orders, classes).

Handbook of Molecular Microbial Ecology, Volume I: Metagenomics and Complementary Approaches, First Edition. Edited by Frans J. de Bruijn.

12.2 DNA–DNA REASSOCIATION AND GENE SEQUENCING: TWO COMPLEMENTARY APPROACHES

Despite the lack of information on speciation mechanisms and genomic sequence similarities, members of most prokaryotic species do form a genomically coherent entity. This is the result of the application of DNA reassociation techniques that only allow single DNA strands from different strains to form DNA duplices when the different strains are sufficiently highly related to let their DNA reassociate (see Chapter 2, Vol. I). Using conditions under which reassociation occurs only between highly similar sequences, mainly ribosomal rRNA genes and other homologous evolutionary conserved genes match. Deduced from a large dataset of enterobacterial species, for which a substantial set of phenotypic records was available, it was found by Brenner et al. [1969] that at DNA–DNA reassociation values above 70% (not to be confused with genome sequence similarity, which may be as high as 96%), strains share a high degree of phenotypic similarity as well. Thus, the arbitrary cutoff value of 70% DNA–DNA reassociation, determined for a small group of heterotrophic and aerobic bacteria, mostly commensals and pathogens of eukaryotes, became the gold standard for the delineation of species in bacterial systematics. No matter whether Archaea or Bacteria, anaerobes or aerobes, or autotrophs or lithotrophs, differences in the mode and tempo of evolution were sacrificed for an artificial and pragmatic species definition as all natural groups of strain assemblies were tarred with the same brush. However, as compared to previous periods, the introduction of reassociation techniques led to an increase of the taxonomic stability of the construct "species." The standard was accepted worldwide, not only because finally a microbial taxonomist found a common basis to work upon but also because it was accepted and strongly recommended by an ad hoc expert committee representing the International Committee for Systematic Bacteriology (ICSB) (today ICSP [P for Prokaryotes]) (see below).

Application of nucleic acid reassociation in prokaryotic systematics helped to better define the taxon species, but reassociation values were found to drop rapidly even between moderately related species: Because this method already failed to relate species at the intrageneric level, the higher-order structure of taxa could not be unraveled. This taxonomic problem was solved with the introduction of rRNA and rRNA gene sequencing, when Carl Woese provided "experimental grounding to our perspective on evolution" (J. L. Fox [2009] citing N. Pace). From the early to mid-1980s, Woese and his collaborators [Fox et al., 1980; Woese et al., 1985] convincingly demonstrated the failure of previous attempts to show the course of evolution among bacteria by depicting phylogenic relationship among bacterial species. Based upon sequences of rRNA genes, supported by a few homologous housekeeping genes [Gupta 1998; Ludwig et al., 1990], the chapter of bacterial systematics was rewritten and the framework of evolutionary relationships among prokaryotes was placed upon a rational, hence solid, basis (many scientists at least believed in this—until the complexity of genomes of even highly related strains of a species became available in the first decade of the twenty-first century).

12.3 THE AD HOC COMMITTEE ON RECONCILIATION OF APPROACHES TO BACTERIAL SYSTEMATICS

The apparent failure of traditional taxonomy to relate species above the species level and the inability of rRNA gene sequences to solve relationships at the intrageneric level became a problem for taxonomists: The differences between the higher taxa deduced from rRNA gene sequence-based dendrograms and the higher-order structure ad depicted in textbooks of microbiology were too significant to ignore. While some of systematists discussed the possibility of working with two systems—one for practical use (e.g., following the traditional structure as outlined in *Bergey's Manual of Determinative Bacteriology*), the other for scientific achievement—others tried to resolve the apparent differences. In 1987 a group of scientists, representing both schools, met in Paris to discuss the apparent problems. The result of this committee [Wayne et al., 1987] influenced the development of bacterial systematics until today because it found a compromise that calmed the waves down. Committee members scrutinized the state of bacterial systematics from the viewpoint of different perspectives: *phylogenetic* (analyses of evolutionary distances between selected semantides, according to Zuckerkandl and Pauling [1965]), *descriptive* (phenotypic properties help define a taxonomic group that may also be recognized at the phylogenetic level), *diagnostic* (selection of features from phenotypic description or of nucleotide signatures) and *associative* (name to evoke practical information about a strain).

The committee agreed on the following statements:

1. "A single formal overall system appears to be adequate"; thus "an ideal taxonomy would involve one system."
2. "The complete deoxyribonucleic acid (DNA) sequence of the genome would be the reference standard to determine phylogeny, and phylogeny

should determine taxonomy." Considering the year, this statement was rather visionary.

A cautionary note about hierarchical interpretation was expressed with the recommendation that searches should continue actively for additional powerful semantides independent of the ribosomal RNA cistrons as the limitation of the deeper branches to define higher taxa had already been recognized.

With respect to the category species, it was stated that this taxon "is the only taxonomic unit that can be defined in phylogenetic terms." This statement based upon the recognized power of DNA reassociation approaches, representing "the best applicable procedure at the present time." As mentioned above, the committee reiterated that "the phylogenetic definition of a species generally would include strains with approximately 70% or greater DNA–DNA relatedness and with 5°C or less ΔT_m. Both values must be considered." Members of the committee included a cautionary note toward the emerging molecular revolution in that "phenotypic characteristics should agree with this definition and would be allowed to override the phylogenetic concept of species only in a few exceptional cases," and "it is recommended that a distinct genospecies that cannot be differentiated from another genospecies on the basis of any known phenotypic property not be named until they can be differentiated by some phenotypic property."

Most obviously, guidance was given to the taxonomist, and this contribution belongs to one of the most cited papers in bacterial systematics. It acknowledged the merits of gene sequences and positioned their power to define taxa above the category "species." The committee, however, did not recommend that taxonomy should and will be based solely upon molecular data.

12.4 RELATIONSHIP BETWEEN DNA–DNA SIMILARITIES AND rRNA GENE SIMILARITIES

The clear-cut recommendation of the ad-hoc committee resulted in the increased use of DNA–DNA reassociation values to delineate novel species. In 1987, reports on about 60% of all new species described within existing genera or recognized as synonyms in the *International Journal of Systematic Bacteriology* included DNA reassociation studies; the value increased to 75% in 1993. Nevertheless, the gap in the acceptance of the two approaches widened: While rRNA gene sequencing was easy to perform and provided a high-quality product, the DNA pairing approach, though acknowledged, remains the bottleneck in any taxonomic study. Despite its significance, the number of scientists able to perform the complex technical formats is small. Most microbial taxonomists are not in a position to perform these studies and need collaboration with specialized laboratories. Experience is required in isolation and purification of rather large amounts of DNA; and though one can chose from a variety of different hybridization methods [Roséllo-Mora, 2005], none of these are straightforward to apply without thorough training. But the repugnance of this technique is reaching deeper: A significant number of physicochemical parameters—genome size, the presence of large plasmids, DNA purity, and other factors—do influence the hybridization results; reciprocal values may differ up to 15%. Most annoying, the method of reassociation of denatured DNA strands of two different strains does not unfold those homologous genome stretches that are involved in the reassociation process. In times of complete genome sequences and the mediation of sequence techniques to undergraduates, this inability to judge about the process mechanisms appears like a method salved from historical times [Stackebrandt and Ebers, 2006]. Unlike sequences that must be deposited in public databases for inspection of quality, no reviewer of a new species description is in a position to judge upon DNA reassociation values or the conditions used. And, last but not least, because reassociation data are not cumulative, many studies that need to include the type strains of highly related species (e.g., *Aeromonas, Flavobacterium, Streptomyces, Pseudomonas*, or certain groups of *Bacillus*) may turn into a tantalizing occupation.

Already in the 1990s, taxonomists began to realize that the relationship between rRNA gene sequence similarities and DNA–DNA reassociation values is not linear but curvilinear [Stackebrandt and Goebel, 1994]. Strains of species that were clearly separated by reassociation values below 70% may show gene similarity values of higher than 98%, sometimes even 100% [Ash et al., 1991; Roséllo-Mora and Amann, 2001; Fox et al., 1992]. Thus, these studies clearly revealed the limitations of the sequence analysis of this conserved gene and gene product in the determination of relationships at the strain level for which DNA–DNA reassociation experiments still constituted the superior method. The plot of similarity values, generated in 1993, however, clearly indicated that at rRNA gene similarities of 97% and below, the corresponding DNA–DNA similarities were never higher than 70%. This finding led to the recommendation to abolish the need perform DNA–DNA reassociation in those cases where novel strains showed only moderate rRNA similarities ($\leq$70%) with its nearest neighbor [Stackebrandt and Goebel, 1994]. This recommendation, conservative in its interpretation of the threshold value, was well received by authors and reviewers as judged by the high number of citation in subsequent species descriptions (summarized by Stackebrandt [2004] and by Roséllo-Mora and Amann [2001].)

The correlation plot was updated in 2005 by screening all species descriptions published in Volume 55 of the *International Journal of Systematic and Evolutionary Microbiology* [Stackebrandt and Ebers, 2006]. Based upon more than 400 data obtained by all major hybridization techniques performed on representatives of most major phyla of prokaryotes points, the recommendation of 1994 was revised. Rather than 97.0%, a 16S rRNA gene sequence similarity threshold value around 98.8% was now recommended at which DNA–DNA reassociation experiment should be mandatory for testing the genomic uniqueness of novel isolates. This value was already obvious in 1993 but substantiated by the data accumulated in the scientific literature. The increase from 97.0 to 98.8% rRNA gene sequence similarity significantly facilitates the work of taxonomists without sacrificing the quality and precision of a pragmatic "species" description.

Whether scientists decide to put the threshold value for species delineation at 97% or 98.8% or whether they decide to chose a value in between, it is important to stress that the basis for any decision is the availability of a complete or almost complete (>1400 nt) rRNA gene sequence and a proper sequence alignment. The transfer of this delineation value to environmental studies using partial sequences does not reflect the presence of species but only of operational taxonomic unit (OTUs). Partial sequences containing moderately to highly variable regions tend to overemphasize phylogenetic differences and members of different higher taxa of prokaryotes differ in the position of highly variable stretches along the rRNA gene primary structure [Stackebrandt and Rainey, 1995]. Thus, a bacterial community with their myriads of phylogenetically unrelated members cannot be assessed for species numbers when only partial sequences had been generated. As explained above, even above a defined similarity value, a group of genomically closely related organisms may belong to different species as presently defined.

12.5 THE AD HOC COMMITTEE FOR THE REEVALUATION OF THE SPECIES DEFINITION IN BACTERIOLOGY

Because the underlying basis of systematics is evolution and the process of doing systematics requires periodic adjustment to scientific advances, 15 years after the meeting of the ad hoc committee on reconciliation of approaches to bacterial systematics [Wayne et al., 1987], another ad hoc committee was established to reevaluate the previous conclusions and recommendations. This committee met in Ghent, Belgium, February 2002, to specifically reevaluate the pragmatic species definition in bacteriology. The committee made various recommendations regarding the species definition in the light of developments in methodologies available to systematists (for additional references see Stackebrandt et al. [2002]), such as the

- determination of inter- and intraspecies relatedness by rapid DNA typing methods,
- characterization and/or identification of isolates by applying physical methods to prokaryotic cells, such as Fourier-transformed infrared spectroscopy (FTIR), pyrolysis mass spectrometry, and matrix-assisted laser desorption/ionization with time of flight (Maldi-ToF), and
- multilocus sequence typing (MLST) approach that brought a new dimension into the elucidation of genomic relatedness at the inter- and intraspecies level by sequence analyses of housekeeping genes subjected to stabilizing selection [Maiden et al., 1998]. Originally used in epidemiology, MLST and phylogenetic analysis of housekeeping genes (MLSA) allowed first insights into the population structure of the construct "species," leading to the recognition of complex genomic structures ("lumpy diversity" as recognized by Dykhuizen and Green [1991], or "fuzzy species" as defined by Hanage et al. [2005]) and hints for speciation mechanisms.

The committee agreed on the species definition of Roséllo-Mora and Amann [2001] as a "category that circumscribes a (preferably) genomically coherent group of individual isolates/strains sharing a high degree of similarity in (many) independent features, comparatively tested under highly standardized conditions." It also considered the current species definition to be sound, pragmatic, operational, and universally applicable, serving the community well. The importance of the availability of "minimal standards" and the use of standardized phenotypic tests were highlighted. The committee suggested that the technique of DNA–DNA reassociation cannot be improved to maintain (even in the genomic era) the pragmatic delineation of species but instead to integrate into the species definition new techniques and new knowledge as long as there is a sufficient degree of congruence between the technique used and DNA–DNA reassociation. Especially recommended was the MLSA approach of sequencing housekeeping or other genes, in order to genomically circumscribe the taxon "species" and to differentiate it from neighboring "species." Indeed, the availability of housekeeping gene sequences in public databases (as a result of genome sequencing projects and individual MLSA studies) and their use in bacterial systematics is promising. However, DNA profiling methods (such as AFLP, ribotyping, Rep-PCR, PCR-RFLP) were also recommended. In fact a study was carried out by Rademaker et al. [2000] that showed a very high correlation between rep-PCR and AFLP fingerprinting and DNA–DNA similarity values, suggesting that genomic fingerprinting techniques can be used as rapid, highly discriminating screening techniques to determine the taxonomic diversity and phylogenetic structure of bacterial populations (see Chapters 4 and 7, Vol. I). The broad application of recommended DNA arrays is still missing, though the committee believed that this methodology will show great promise.

12.6 OUTLOOK

Sequence analyses of complete genomes has provided scientists with an immeasurable wealth of information, ranging from sequences to chromosome architecture. According to public databases (e.g., http://www.genomesonline.org/gold.cgi), more than 140 archaeal and more than 2400 bacterial genomes have been completed or are under investigation. These genomes, now available for most phyla, not only are an invaluable source of target genes for taxon-specific MLSA but also include those genes the products of which have usually been excluded from the classification process. These data permit the identification of genes that are conserved across a more narrow range of taxa which make use of a specific value for higher discrimination in the classification process. As indicated by Staley [2009], the strength of a future phylogenomic species concept is not restricted to the phylogenetic information but includes the location of genes (synteny), the option to predict gene expression, and the long-awaited possibility to predict DNA–DNA reassociation similarities among complete and selected regions of genomes. In a first attempt, DNA–DNA hybridization values for 28 strains were compared to the extent of average nucleotide identity (ANI) of common genes in fully sequenced genomes of the same strains [Goris et al., 2007; see Chapter 11, Vol. I]: Above the threshold values of 97% DNA–DNA reassociation similarity, the corresponding ANI was 95%; and when restricted to the protein-coding region of the genome, the similarity was 85%. These values support MLSA studies and reinforce the notion about the substantial gene diversity within the pragmatic construct "species."

To conclude this chapter, a visionary view from the reevaluation committee will be repeated: "The dialogue among systematists, population and evolutionary geneticists, ecologists, and microbiologists will be to the benefit of bacterial systematics in general, and of a more transparent species concept in particular" [Stackebrandt et al., 2002].

REFERENCES

Achtman M, Wagner M. 2008. Microbial diversity and the genetic nature of microbial species. *Nat. Rev. Microbiol.* **6**:431–440.

Ash C, Farrow JA, Dorsch M, Stackebrandt E, Collins MD. 1991. Comparative analysis of *Bacillus anthracis, Bacillus cereus*, and related species on the basis of reverse transcriptase sequencing of 16S rRNA. *Int. J. Syst. Bacteriol.* **41**:343–346.

Brenner DJ, Fanning GR, Johnson KE, Citarella RV, Falkow S. 1969. Polynucleotide sequence relationships among members of Enterobacteriaceae. *J. Bacteriol.* **98**:637–650.

Dykhuizen DE, Green L. 1991. Recombination in *Escherichia coli* and the definition of biological species. *J Bacteriol.* **173**:7257–7268.

Fox GE, Stackebrandt E, Hespell RB, Gibson J, Maniloff J, et al. 1980. The phylogeny of prokaryotes. *Science* **209**:457–463.

Fox GE, Wisotzkey JD, Jurtshuk P. 1992. How close Is close: 16S rRNA sequence identity may not be sufficient to guarantee species identity. *Int. J. Syst. Bacteriol.* **42**:166–170.

Fox JL. 2009. Honoring Carl Woese leads to Darwinian adjustments. *Microbe* **4**:354–355.

Gevers D, Cohan FM, Lawrence JG, Spratt BG, Coenye T, et al. 2005. Opinion: Re-evaluating prokaryotic species. *Nat. Rev. Microbiol.* **3**:733–739.

Goris J, Konstantinidis KT, Klappenbach JA, Coenye T, Vandamme P, Tiedje JM. 2007. DNA–DNA hybridization values and their relation to whole genome sequence similarities. *Int. J. Syst. Evol. Microbiol.* **57**:81–91.

Gupta RS. 1998. Protein phylogenies and signature sequences: A reappraisal of evolutionary relationships among Archaea, Eubacteria, and Eukaryotes. *Microbiol. Mol. Biol. Rev.* **62**:1425–1491.

Hanage WP, Fraser C, Spratt BG. 2005. Fuzzy species among recombinogenic bacteria. *BMC Biol.* **3**:6–13.

Ludwig W, Weizenegger M, Betzl D, Leidel E, Lenz T, et al. 1990. Complete nucleotide sequences of seven eubacterial genes coding for the elongation factor Tu: Functional, structural and phylogenetic evaluations. *Arch. Microbiol.* **153**:241–247.

Maiden MCJ, Bygraves JA, Feil E, Morelli G, Russell JE, et al. 1998. Multilocus sequence typing: A portable approach to the identification of clones within populations of pathogenic organisms. *Proc. Nat. Acad. Sci. USA* **95**:3140–3145.

Rademaker JL, Hoste B, Louws FJ, Kersters K, Swings J, Vauterin L, Vauterin P, de Bruijn FJ. 2000. Comparison of AFLP and rep-PCR genomic fingerprinting with DNA–DNA homology studies: *Xanthomonas* as a model system. *Int. J. Syst. Evol. Microbiol.* **50**:665–677.

Rosséllo-Mora R. 2005. DNA–DNA reassociation methods applied to microbial taxonomy and their critical evaluation. In Stackebrandt E, ed. *Molecular Identification, Systematics, and Population Structure of Prokaryotes*. Heidelberg: Springer, pp. 23–50.

Rosséllo -Mora R, Amann R. 2001. The species concept for prokaryotes. *FEMS Microbiol. Rev.* **25**:39–67.

Stackebrandt E. 2007. Forces shaping bacterial systematics. *Microbe* **2**:283–288.

Stackebrandt E. 2004. Defining taxonomic ranks: systematics and classification. In Falkow S, Rosenberg, E, Schleifer KH, Stackebrandt E, Dworkin M, eds. *The Prokaryotes: A Handbook on the Biology of Bacteria*, 3rd ed. New York: Springer, pp. 29–57.

Stackebrandt E, Ebers J. 2006. Taxonomic parameters revisited: tarnished gold standards. *Microbiolol. Today* **33**:152–155.

Stackebrandt E, Goebel BM. 1994. A place for DNA–DNA reassociation and 16S rRNA sequence analysis in the present species definition in bacteriology. *Int. J. Syst. Bacteriol*. **44**:846–849.

Stackebrandt E, Rainey FA. 1995. Partial and complete 16S rDNA sequences, their use in generation of 16S rDNA phylogenetic trees and their implications in molecular ecological studies. In Akkermans AD, van Elsas JD, De Bruijn F, eds. *Molecular Microbial Ecology manual*. Amsterdam: Kluwer Academic Publishers, pp. 1–17.

Stackebrandt E, Frederiksen W, Garrity GM, Grimont PA, Kämpfer P, et al. 2002. Report of the ad hoc committee for the re-evaluation of the species definition in bacteriology. *Int. J. Syst. Evol. Microbiol*. **52**:1043–1047.

Staley JT. 2009. The phylogenomic species concept for Bacteria and Archaea. *Microbe* **4**:361–365

Wayne LG, Brenner DJ, Colwell RR, Grimont PA, Kandler O, et al. 1987. International Committee on Systematic Bacteriology: Report of the ad hoc committee on reconciliation of approaches to bacterial systematics. *Int. J. Syst Bacteriol*. **37**:463–464.

Woese CR, Stackebrandt E, Macke TJ, Fox GE. 1985. A phylogenetic definition of the major eubacterial taxa. *Syst. Appl. Microbiol*. **6**:143–151.

Zuckerkandl E, Pauling L 1965. Molecules as documents of evolutionary history. *J. Theor. Biol*. **8**:357–366.

Chapter 13

Metagenomic Approaches for the Identification of Microbial Species

David M. Ward, Melanie C. Melendrez, Eric D. Becraft, Christian G. Klatt, Jason M. Wood, and Frederick M. Cohan

13.1 INTRODUCTION

In the most recent iteration of our long-term molecular studies of hot spring cyanobacterial mat communities [Ward et al., 1998, 2002, 2006], we have developed and employed several genomics and metagenomics approaches to investigate the existence and character of species populations [Ward et al., 2008, 2011]. The aim of this chapter is to provide a concise overview of these approaches and the inferences to which they have led us, relying on several original papers to report the results in detail [Klatt et al., 2011; Melendrez et al., 2011a,b; Becraft et al., 2011].

We do not believe that there is a single, universal way to conceptualize microbial species. Rather, we believe that, as with plant and animal species, different taxa form species differently depending on how diversity is generated within the taxon and on how nature acts upon this diversity [Cohan and Perry, 2007; Ward et al., 2008]. Our investigations have been of unicellular cyanobacteria (*Synechococcus* spp.), which are predominant members of a mat community living in 50 to ~72°C regions of the effluent channels of alkaline siliceous hot springs in Yellowstone National Park. The evidence in hand suggests that these organisms have undergone speciation through adaptation to parameters that vary in the environments they inhabit [Ward, 1998, 2006]. The Stable Ecotype Model is one of many models created to explain these observations in an evolutionary context [Cohan and Perry, 2007; Ward et al., 2008], and it is currently the best explanation for how these organisms have speciated [Ward and Cohan, 2005; Ward et al., 2006]. The lessons learned in our studies should be broadly applicable, since there is evidence of adaptive speciation in many taxa [Field et al., 1997; Bielawski et al., 2004; Koeppel et al., 2008; Connor et al., 2010; and see Ward, 2006].

To observe the native cyanobacterial inhabitants in this mat community, it was necessary to shift from cultivation-dependent approaches, which led to the false impression that *Synechococcus lividus* was the dominant cyanobacterium in the mat [Ferris et al., 1996], to cultivation-independent molecular approaches, which revealed the genetic signatures of the truly dominant and dramatically unrelated collection of *Synechococcus* spp. (the A/B lineage; **Fig. 13.1A**) [Ward et al., 1990]. The inference that these *Synechococcus* SSU rRNA sequences should be considered to be associated with different ecological species, defined as populations of individuals that occupy the same unique niche [van Valen, 1976; Ward, 1998], was based on observations of the patterning of this molecular diversity relative to natural spatiotemporal gradients of environmental difference. For instance, SSU rRNA-based diversity was found to be distributed along environmental gradients that occur along the flow path away from the hot spring source (e.g., parameters such as changing temperature and nutrients [Ferris and Ward, 1997]). More specifically, *Synechococcus*. SSU rRNA genotypes A″, A′, A, B′, and B were found to be differently distributed along the effluent flow path between the upper temperature limit of mat formation at ~72°C downstream to ~50°C (Fig. 13.1A). Demonstration of adaptations to different temperatures in genetically relevant *Synechococcus* isolates [Allewalt et al., 2006] helped us understand, at

Handbook of Molecular Microbial Ecology, Volume I: Metagenomics and Complementary Approaches, First Edition. Edited by Frans J. de Bruijn.

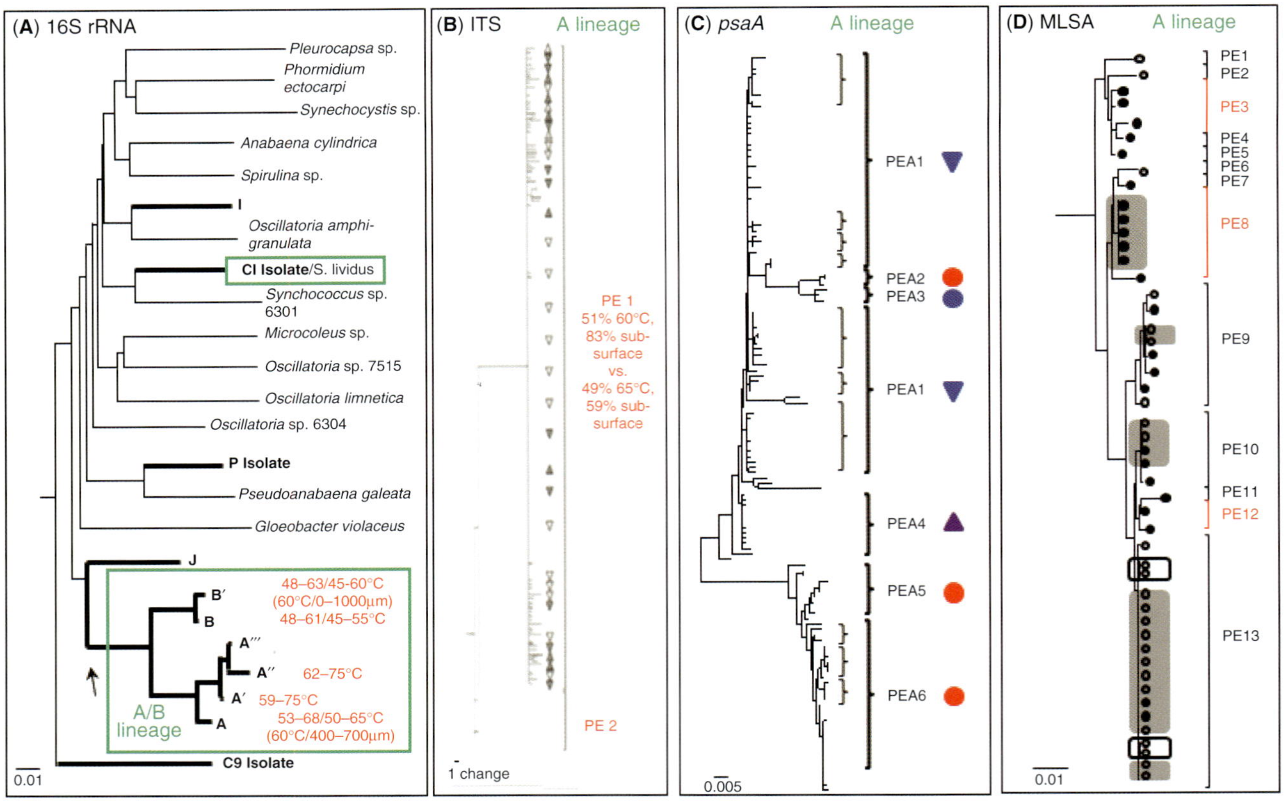

Figure 13.1 Evolutionary ecology of mat *Synechococcus* spp. based on sequence variation in (**A**) SSU rRNA (modified from Ward et al. [1998]), (**B**) 16S–23S rRNA internal transcribed spacer (modified from Ward et al. [2006]), (**C**) *psaA* (modified from Becraft et al. [2011]) and (**D**) multilocus sequence analysis (from [Melendrez et al., 2011a], showing how the number of putative ecotypes (PE) predicted by Evolutionary Simulation analysis rises with the molecular variation offered by these loci or sets of loci. PEs are designated by vertical brackets; red brackets in panel D indicate PEs comprised only of clones collected at 65°C. In panel C, triangles indicate surface and subsurface association as on B; colors indicate predominant collection temperature: blue, 60°C; purple, 63°C; red, 65°C; Brackets within PE brackets represent subclades. In panel A, bold lines indicate sequences retrieved from Octopus and Mushroom Spring mats or isolates therefrom. In panel B, triangles indicate where sequences were collected (upward-pointing, top green layer surface; downward-pointing, subsurface; clear, 60°C; shaded, 65°C). In panel D, the neighbor joining tree shown is based on five concatenated loci (*rbsK*, PK, *lepB, hisF*, CHP and *aroA*) for equal numbers of *Synechococcus* sp. A-like BACs from samples collected at 60°C (open circles) or 65°C (closed circles). Information about temperature and vertical distribution of genotypes and putative ecotypes is summarized in red (in panel A from Ferris and Ward [1997] and Ramsing et al. [2000]; in panel B from Ward et al. [2006]; in panel C from Becraft et al. [2011]). Scale bars indicate the number of fixed point mutations per sequence position (unlabeled) or the number of base differences (labeled changes).

least in part, the basis for this distribution along the flow path. For instance, isolates with SSU rRNA genotypes A, B′, and B were found to have progressively lower growth temperature ranges, consistent with the temperatures at which they are found in situ (Fig. 13.1A). Similarly, differential SSU rRNA genotype distributions within the upper few-millimeter-thick photic zone in the vertical dimension suggested possible adaptations to parameters such as light intensity and quality and/or chemical parameters that vary during the diel light cycle [Ramsing et al., 2000]. More specifically, at ~60°C, *Synechococcus* sp. genotype A occurred as a subsurface population ~400–700 μm beneath the mat surface, whereas B′ was the only *Synechococcus* genotype detected near the mat surface (Fig. 13.1A). We found, however, that it was necessary to use faster-evolving molecular markers, such as the 16S–23S rRNA internal transcribed spacer region, to identify very closely related, but differently distributed, *Synechococcus* spp. [Ferris et al., 2003]. For instance, at 68°C, *Synechococcus* spp. populations with the same SSU rRNA genotype (A′), but with distinct internal transcribed spacer sequences, were found at different mat depths. Because these populations were both genetically and ecologically distinct, we hypothesized that they represent two differently adapted species populations. In general, internal transcribed spacer region analyses led to an increase in species predicted (Fig. 13.1B and see below), making apparent the need for using even faster-evolving, protein-encoding genes to view how variation among individual members of a *Synechococcus* population might be grouped into true ecologically distinct species populations.

13.2 THEORY-BASED PREDICTION OF *SYNECHOCOCCUS* SPECIES POPULATIONS

The grand challenge of species identification using molecular methods is to determine which individual sequence variants belong to which species populations. Various molecular cutoffs have been suggested for species demarcation [Wayne et al., 1987; Goodfellow et al., 1997; Stackebrandt and Ebers, 2006; Konstantinidis and Tiedje, 2005; see also Chapter 11, Vol. I], but these do not account for the reality that species populations may be younger or older, and thus they are cutoffs of convenience, often calibrated to match the variation found among strains of traditional named species [Cohan, 2002], and not natural biological groupings of variants. For the prediction of hot spring *Synechococcus* spp., we used an evolutionary simulation based on the Stable Ecotype Model, called Ecotype Simulation [ES; Koeppel et al., 2008]. This simulation models the evolution of a population into a number of putative ecotypes (i.e., ecological species) by determining the most likely history of (i.e., order and timing of) diversity-purging periodic selection events, adaptive ecotype formation (i.e., ecological speciation) events, and genetic drift that explains the actual phylogeny under investigation. Thus, ES demarcates species based on the evolutionary history of the gene (and taxon) being analyzed.

When used to analyze variation in protein-encoding loci in samples collected at 60°C and 65°C sites in Mushroom Spring (Yellowstone National Park) mat, ES predicted up to 14 times more ecological species populations than were predicted from 16S rRNA variation (i.e., A-like and B′-like species, see Fig. 13.1C], the number depending on the molecular resolution of the gene [Melendrez et al., 2011b,; Becraft et al., 2011]. The most well-sampled populations are dominated by a single sequence variant, which may be of central importance, because this is likely to be the variant that was the driver [and hence survivor] of the most recent periodic selection event in the ecotype population and, thus, the founder of the current species clade. Where a predicted species population has more than one dominant variant, we hypothesize that ES has demarcated species too conservatively.

The ecotypes (i.e., ecological species) predicted by ES are taken as putative until evidence is obtained to confirm their ecological uniqueness as well as the ecological interchangeability of its members. Importantly, the species populations predicted by ES exhibit unique distributions relative to flow and vertical gradients [Melendrez et al., 2011b; Becraft et al., 2011] (Fig. 13.1C). Some have also been found to exhibit unique gene expression patterns and to respond, as populations, to environmental change [Becraft et al., unpublished]. Our expectation is that the genomes of individuals of a true species population, while variable, should contain niche-specifying adaptations, coded by unique alleles and/or genes, which determine the unique ecological character of all members of the population [Ward et al., 2008; Cohan and Koeppel, 2008].

13.3 CULTIVATION-INDEPENDENT MULTILOCUS SEQUENCE ANALYSIS

To better account for the possible influence of recombination, we developed a cultivation-independent, multilocus sequence analysis approach, based on the use of bacterial artificial chromosome cloning to sample multiple loci in the genomic neighborhood of SSU rRNA genes [Melendrez et al., 2011a]. ES also predicted from multiple concatenated loci obtained from the same two mat samples a number of species comparable to that observed with the highest-resolving single loci, even though the recombination/mutation ratio is estimated to be about 2.6–14.4 in these *Synechococcus* populations [Melendrez et al., 2011a]. Many of these predicted species populations contained dominant variants with identical sequences at up to 7 loci and were comprised of variants that came exclusively or predominantly from either a 60°C or a 65°C sample (Fig. 13.1D), suggesting their ecological distinction. Apparently, recombination in the genes analyzed has been unable to erode the clusters of phylogenetically similar variants that represent the ecological species, although it may have displaced some variants in the loci under investigation to positions in the phylogeny outside of this clade in cases where homologous recombination has occurred with a phylogenetically distant individual. An organism containing such a molecular variant may nevertheless belong to the same species population, because it is only excluded from clades that are based on analysis of genes that have been exchanged with distantly related organisms.

13.4 METAGENOMIC ASSEMBLY AND SPECIES POPULATIONS

We have simultaneously examined metagenomic databases to investigate the existence of natural genetically coherent populations, using an assembly approach that is based on molecular sequence similarity, rather than any theoretical basis [Klatt et al., 2011]. Sanger sequences from the ends of 2- to 12-kbp inserts of clones obtained from samples collected at average temperatures of ~60°C and ~65°C in the Mushroom Spring and Octopus Spring mats (same samples as above) were assembled using the Celera assembler. Oligonucleotide frequency analysis of these assemblies, followed by cluster analysis using the K-means algorithm, resulted in a single cluster containing both *Synechococcus* spp. A-like and B′-like populations [Klatt et al., 2011] (**Fig. 13.2**). Thus the current level of species resolution is poor compared to that provided by population genetics studies described above lumping all members of the *Synechococcus* A/B lineage. We are currently investigating databases with much deeper coverage, using alternative clustering algorithms, to test whether it is possible to assemble metagenomic segments of *Synechococcus* spp. populations predicted by single and multi-locus analyses combined with ES.

13.5 CONCLUSIONS

It is abundantly clear that we are approaching a solid understanding of the nature of species populations for hot spring *Synechococcus*. However, it is also clear that, using present approaches, we are unable to resolve these populations through metagenomic assembly. This may never be possible given the many factors that make assembly difficult (e.g., repetitive sequence elements, conserved genes). Nevertheless, comparison of metagenomic clones with genomes of cultivated *Synechococcus* spp. strains A and B′ genomes did reveal genes not found in the isolate, but present in native populations, that might have adaptive character [Bhaya et al., 2007; Klatt et al., 2011], but such analyses depend on the availability of genomes from isolates that are genetically relevant to the community [Ward et al., 2011].

Acknowledgments

This research was supported by long-term support from NSF (currently from the Frontiers in Integrative Biology Research Program, EF-0328698 and the IGERT Program in Geobiological Systems, DGE 0654336), NASA (most recently from the Exobiology Program, NAG5-8824,-8807 and NX09AM87G) and the U.S. Department of Energy (DOE), Office of Biological and Environmental

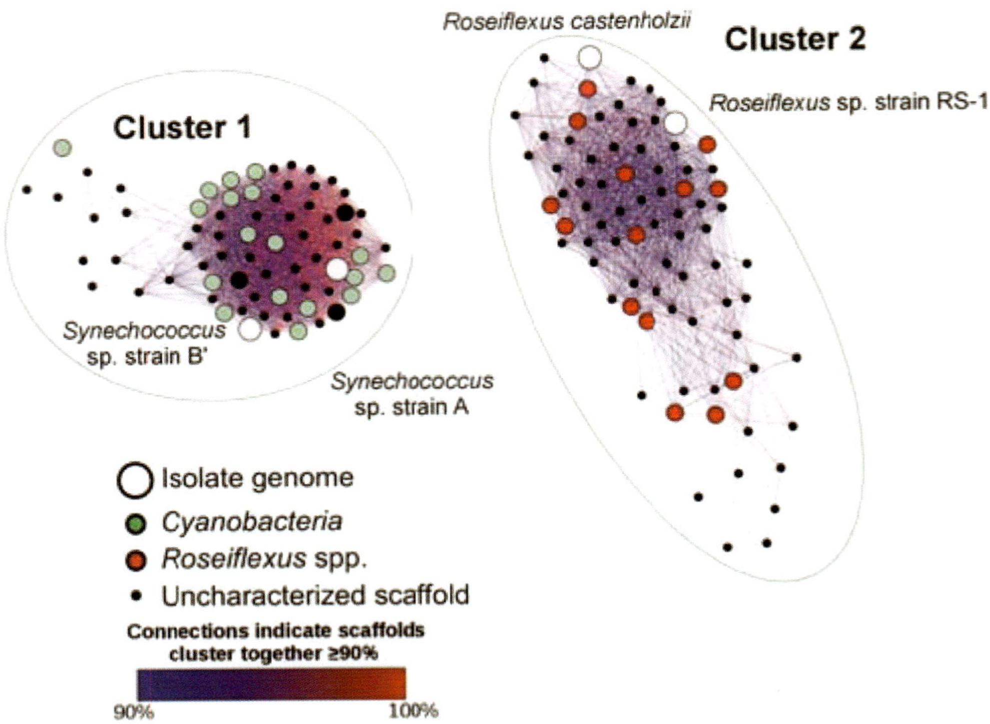

Figure 13.2 Network map showing core cyanobacterial scaffold cluster observed in Celera assemblies of Mushroom Spring and Octopus Spring mat samples collected at 60° and 65°C. The *Roseiflexus* spp. cluster is included to exemplify the discrete separation of cyanobacterial cluster from seven noncyanobacterial clusters observed. Scaffolds with similar oligonucleotide frequency profiles that group together in the same cluster in ≥90% of 100 trials are indicated by connecting lines, whose color reports the percentage that scaffolds group together as defined by the scale bar. Isolate genomes included in this analysis are indicated by large white circles, whereas metagenomic scaffolds that contain characterized phylogenetic marker genes of cyanobacteria are shown as green medium-sized circles. Large color-shaded ovals were drawn by hand to demarcate the different clusters. Genomes and scaffolds that are not linked by lines to other scaffolds formed linkages in <90% of the trials. (from [Klatt et al., 2011]).

Research (BER), as part of BER's Genomic Science Program (GSP). This contribution originates from the GSP Foundational Scientific Focus Area (FSFA) at the Pacific Northwest National Laboratory (PNNL), contract no. 112443. We appreciate the support of the National Park Service personnel at Yellowstone National Park.

REFERENCES

Allewalt JP, Bateson MM, Revsbech NP, Slack K, Ward DM. 2006. Effect of temperature and light on growth and photosynthesis of *Synechococcus* isolates typical of those predominating in the Octopus Spring microbial mat community. *Appl, Environ. Microbiol*. **72**:544–550.

Becraft, ED, Cohan FM, Kühl M Jensen S, Ward DM. 2011. Distribution patterns of ecological species of *Synechococcus* inhabiting the microbial mat of Mushroom Spring. Yellowstone National Park. In preparation.

Bhaya D., Grossman AR, Steunou A-S, Khuri N, Cohan FM, et al. 2007. Population level functional diversity in a microbial community revealed by comparative genomic and metagenomic analyses. *ISME J*. **1**:703–713.

Bielawski JP, Dunn KA, Sabehi G, Béjà O. 2004. Darwinian adaptation of proteorhodopsin to different light intensities in the marine environment. *Proc. Natl. Acad. Sci. USA* **101**:14824–14829.

Cohan FM. 2002. What are bacterial species? *Ann. Rev. Microbiol*. **56**:457–487.

Cohan FM, Koeppel AF. 2008. The origins of ecological diversity in prokaryotes. *Curr. Biol*. **18**:R1024–R1034.

Cohan FM, Perry EB. 2007. A systematics for identifying the fundamental units of bacterial diversity. *Curr. Biol*. **17**:R373–R386.

Connor N, Sikorski J, Rooney AP, Kopac S, Koeppel, et al. 2010. The ecology of speciation in *Bacillus*. *Appl. Environ. Microbiol*. **76**:1349–1358.

Ferris MJ, Ward DM. 1997. Seasonal distributions of dominant 16S rRNA-defined populations in a hot spring microbial mat examined by denaturing gradient gel electrophoresis. *Appl. Environ. Microbiol*. **63**:1375–1381.

Ferris MJ, Ruff-Roberts AL, Kopczynski ED, Bateson MM, Ward DM. 1996. Enrichment culture and microscopy conceal diverse thermophilic *Synechococcus* populations in a single hot spring microbial mat habitat *Appl. Environ. Microbiol*. **62**:1045–1050.

Ferris, MJ, Kühl M, Wieland A, Ward DM. 2003. Different light-adapted ecotypes in a 68°C *Synechococcus* mat community revealed by analysis of 16S–23S intervening transcribed spacer variation. *Appl. Environ. Microbiol*. **69**:2893–2898.

Field KG, Gordon D, Wright T, Rappe, Urbach E, Vergin K, Giovannoni SJ. 1997. Diversity and depth-specific distribution of SAR11 cluster rRNA genes from marine planktonic bacteria. *Appl. Environ.Microbiol*. **63**:63–70.

Goodfellow M, Manfio GP, Chun J. 1997. Towards a practical species concept for cultivable bacteria. In Claridge MF, Dawah HW, Wilson MR, eds. *Species: Units of Biodiversity*. London: Chapman and Hall, pp. 25–59.

Klatt CG, Wood JM, Rusch DB, Bateson MM, Hamamura N, Heidelberg JF, Grossman AR, Bhaya D, Cohan FM, Kühl M, Bryant DA, Ward DM. 2011. Community Ecology of Hot Spring Cyanobacterial Mats: Predominant Populations and their Functional Potential. *ISME J*. (submitted)

Koeppel A, Perry EB, Sikorski J, Krizanc D, et al. 2008. Identifying the fundamental units of bacterial diversity: A paradigm shift to incorporate ecology into bacterial systematics. *Proc. Natl. Acad. Sci. USA* **7**:2504–2509.

Konstantinidis KT, Tiedje JM. 2005. Genomic insights that advance the species definition for prokaryotes. *Proc. Natl. Acad. Sci. USA* **102**:2567–2572.

Melendrez MC, Cohan FM, Ward DM. 2011a. Cultivation-independent multi-locus sequence analysis of *Synechococcus* populations inhabiting a hot spring cyanobacterial mat. In preparation.

Melendrez MC, Lange RK, Cohan FM, Ward DM. 2011b. Influence of molecular resolution on sequence-based discovery of ecological diversity among Synechococcus populations in an alkaline siliceous hot spring microbial mat. *Appl. Environ. Microbiol*. **77**: 1359–1367.

Ramsing NB, Ferris MJ, Ward DM. 2000. Highly ordered vertical structure of *Synechococcus* populations within the one-millimeter thick photic zone of a hot spring cyanobacterial mat. *Appl. Environ. Microbiol*. **66**:1038–1049.

Stackebrandt E, Ebers J. 2006. Taxonomic parameters revisited: Tarnished gold standards. *Microbiol Today* **33**:152–155.

van Valen L. 1976. Ecological species, multispecies, and oaks. *Taxon* **25**:233–239.

Ward DM 1998. A natural species concept for prokaryotes *Curr. Opin. Microbiol*. 1998; **1**:271–277.

Ward DM 2006. A macrobiological perspective on microbial species. *Microbe* **1**:269–278.

Ward DM, Cohan FM. 2005. Microbial diversity in hot spring cyanobacterial mats: Pattern and prediction. In Inskeep B, McDermott T, eds. *Geothermal Biology and Geochemistry in Yellowstone National Park*. Bozeman, MT: Montana State University Thermal Biology Institute, pp. 185–201.

Ward DM, Weller R, Bateson MM. 1990. 16S rRNA sequences reveal numerous uncultured microorganisms in a natural community. *Nature* **345**:63–65.

Ward DM, Bateson MM, Ferris MJ, Nold SC. 1998. A natural view of microbial biodiversity within hot spring cyanobacterial mat communities. *Microbiol. Mol. Biol. Rev*. **62**:1353–1370.

Ward DM, Nübel U, Papke RT, McKitrick MC. 2002. Natural history of microorganisms inhabiting hot spring microbial mats: Clues to the origin of microbial diversity and implications for microbiology and macrobiology. In Staley JT, Reysenbach A-L, eds. *Biodiversity of Microbial Life: Foundation of Earth's Biosphere*. New York: John Wiley & Sons, pp. 25–48.

Ward DM, Bateson MM, Ferris MJ, Kühl M, et al. 2006. Cyanobacterial ecotypes in the microbial mat community of Mushroom Spring [Yellowstone National Park, Wyoming] as species-like units linking microbial community composition, structure and function. *Philos. Trans. R. Soc., London* [Ser. B] **361**:1997–2008.

Ward DM, Cohan FM, Bhaya D, Heidelberg JF, et al. 2008. Genomics, environmental genomics and the issue of microbial species. *Nat. Heredity* **100**:207–219.

Ward DM, Klatt CG, Wood J, Cohan FM, Bryant DA. 2011. Functional genomics in an ecological and evolutionary context: Maximizing the value of genomes in systems biology. In Burnap RL, Vermaas W, eds. *Functional Genomics and Evolution of Photosynthetic Systems*. Dordrecht: Springer; in press.

Wayne LG, Brenner DJ, Colwell RR, Grimont DD, et al. 1987. Report of the ad hoc committee on reconciliation of approaches to bacterial systematics. *Int. J. Syst. Bacteriol*. **37**:463–464.

Part 3

Metagenomics

Chapter 14

Microbial Ecology in the Age of Metagenomics

Jianping Xu

14.1 MICROBIAL ECOLOGY

14.1.1 Introduction and Historical Background

The term *ecology* literally refers to the "law of the household" as it was originally derived from the Greek words *oikos* (household or dwelling) and *logos* (law). It was first defined as a field of scientific investigation by the German biologist Ernst Haeckel [Haeckel, 1866] that explores the interrelationships between organisms and their living (biotic) and nonliving (abiotic) environments. Microbial ecology thus refers to the science that examines the relationships between microorganisms and their biotic and abiotic environments. Compared to the ecological studies of plants and animals, the field of microbial ecology has had a relatively short history. The term *microbial ecology* came into frequent use only in the early 1960s with the first dedicated scientific journal with the same name launched in 1974. The emergence of microbial ecology as an independent field of investigation and its rise in popularity in the 1960s was propelled by both the awakening public interest in environmental issues and the increasing recognition of the essential roles of microbes in the earth's biosphere [e.g., Kasting and Siefert, 2002; Xu, 2006].

Microorganisms have been integral to the history and function of life on earth. They have played central roles in earth's climatic, geochemical, and biological evolution. Although the widespread use of the term microbial ecology is relatively recent, ecologically oriented research on microorganisms was performed as soon as their existence was realized. Notably, in the late 1800s, Russian microbiologist Sergei Winogradsky [Winogradsky, 1887] and Dutch microbiologist Martinus Beijerink [Beijerink, 1888] were among the first to demonstrate the important roles of microorganisms such as iron oxidizers, sulfur oxidizers, and nitrogen-fixing bacteria in the natural environment. However, until very recently, the general importance of microorganisms was appreciated by only a few specialists. Indeed, microorganisms are still most often considered from an anthropocentric perspective, with attention focused on the relatively few species that cause human diseases and the potential of microorganisms to provide useful products and services.

14.1.2 Microbes in Microbial Ecology

The microbial world contains a highly heterogeneous group of organisms sharing only one common characteristic, their small size. Aside from the acellular viruses, cellular microorganisms make up two (out of three) entire domains of life on Earth, the prokaryotic Bacteria and Archaea [Woese, 1987]. Within the third domain, Eukarya, the majority of the evolutionary and metabolic diversity is contained within eukaryotic microorganisms such as protozoa, algae, and fungi. The prokaryotic life emerged about 3.8 billion years ago, about 2 billion years before eukaryotic life arose [Schopf, 1993]. Currently, microbial life forms are found in virtually every imaginable ecological niche on Earth, from the tropics to the Arctic and Antarctica, from underground mines and oil fields to the stratosphere and the top of great mountains, from deserts to the Dead Sea, and from above-ground hot springs to Antarctic lakes and underwater hydrothermal vents (see Vol. II).

Handbook of Molecular Microbial Ecology, Volume I: Metagenomics and Complementary Approaches, First Edition. Edited by Frans J. de Bruijn.

In biological studies, species is the fundamental unit of biological classification and is critical for describing, understanding, and comparing ecological communities at different spatial and temporal levels. However, what constitutes a species remains controversial (see Part 2, "The Species Concept," in Vol. I). For sexually reproducing organisms with the meiotic life cycle (such as the majority of plants, animals, and sexual microbial eukaryotes), although over 20 species concepts exist in the literature [Mayden, 1997], the most widely used is the biological species concept. In this concept, a species consists of individuals capable of interbreeding with each other to produce fertile progeny but are incapable of doing so with members of other species. However, this definition is not applicable to asexual organisms lacking a regular meiotic life cycle. Such organisms include a large proportion of eukaryotic microorganisms as well as all prokaryotes. Because most prokaryotes lack diagnostic morphological characteristics and have no meiotic sexual life cycle, but can exchange genetic materials among each other that are often distantly related and with the exchanged genetic materials localized to only certain parts of the genome [Xu,], the biological species concept is not applicable to prokaryotes.

Instead, the current most widely accepted species concept for prokaryotes is an operational one, rooted in the degree of DNA–DNA reassociation kinetics (see Chapter 2, Vol. I). In this definition, two strains belong to the same species when their purified genomic DNA show at least 70% hybridization [Garrity et al., 2005]. This level of hybridization is equivalent to ~97% nucleotide identity at the 16S ribosomal RNA gene and ~94% average nucleotide identity at the whole genome scale [Konstantinitis and Tiedje, 2005; see also Chapter 11, Vol. I]. It should be noted that this prokaryotic species concept does not translate well to that in plants and animals. For example, using this criterion, all members of primates (e.g., chimpanzees, orangutans, gorillas, gibbons, and humans) would be belonging to the same species [Sibley et al., 1990]. Because of these and other reasons, species concepts for both prokaryotes and eukaryotes are still evolving [e.g., Cohan 2004; Konstantinitis and Tiedje, 2005; Gevers et al., 2005; see also Part 2, Vol. I].

14.1.3 Scales in Microbial Ecology

Microorganisms in the environment can be organized into several levels of hierarchical organizations (from simple to complex): individuals, populations, guilds, communities, and ecosystems. Here, an individual may be defined as a single cell, a single colony consisting of genetically identical cells, or a group of cells derived from a single mating event that are physiologically linked to each other [Xu, 2005]. At the next level, a population is generally referred to a group of individual organisms inhabiting a particular locality, a group of interbreeding individuals that are different from other such groups, or any group of individuals from which samples are take for statistical analyses [Xu, 2005]. A guild refers to groups of microbial populations that are metabolically related to each other. The populations in a guild are often from phylogenetically unrelated and distinct species. A microbial community contains sets of spatially and temporally interacting guilds. At the highest level of ecological organization, a microbial ecosystem consists of both the microbial community and its interacting biotic (macroorganisms such as plants and animals) and abiotic environmental (pH, temperature, inorganic and organic nutrients, etc.) factors. While the spatial scales for microorganisms are typically very small, there is no theoretical limit to the size of a microbial individual, population, guild, or even community.

While we commonly associate microorganisms as decomposers of organic wastes and pathogens of plants, animals, and humans, microorganisms can also form mutualistic associations with each other, as well as be fierce predators of other microorganisms. For example, the minute bacteria *Bdellovibrio* (~0.3 μm in diameter) can quickly destroy an *Escherichia coli* cell many times its own size (1 × 2 μm) [Nunez et al., 2003]. On the other hand, many bacteria and fungi form symbiotic relationships with plants and enhance the growth and reproduction of these plants [e.g., Xu, 2006; Li and Xu, 2009; Wu et al., 2011].

14.1.4 Technical and Methodological Developments Drive Advances in Microbial Ecology

From its inception, advances in microbial ecology have been closely tied to technical and methodological developments. These technical and methodological developments include microscopy (light, phase contrast, dark-field, scanning electron, transmission electron, fluorescent, confocal laser scanning), isotope labeling and tracking, fluorescent in situ hybridization, flow cytometry, enrichment protocols, specialized incubators, and so on. These developments have allowed microbial ecologists, from initially ex situ investigations in the lab to progressively in situ analyses to examine the diversity of microorganisms and how microorganisms interact with each other and with their environment to generate and to maintain such diversities.

Traditional approaches estimating of the diversity of microorganisms include the isolation, identification, and quantification of microorganisms in various habitats. Depending on the objectives, microbial diversity in the environment is typically expressed as a quantitative measure of common indices such as phylogenetic diversity, species diversity, genotype diversity, and gene diversity. Above the species level, microbial diversity is commonly quantified based on evolutionary distances among observed taxonomic groups from a specific environment (e.g., the phylogenetic diversity based on a common chronometer such as the 16S ribosomal RNA subunit) [Faith, 1992; see also Chapter 15 and 17, Vol. I]. Below the species level, microbial diversity is typically described using population genetic parameters such as gene diversity and genotype diversity. Gene diversity and genotype diversity refer respectively to the probability that two randomly drawn genes and genotypes in a population will be different [Xu, 2005]. At the species level, microbial diversity is measured as species diversity. There are various measures of species diversity. One that is commonly used refers to the frequency that two randomly drawn individuals in an environment will be different species. This measure takes into account both the number of species (species richness) and the frequency of each species (species abundance) in the environment. Conceptually, this measure of species diversity is similar to those used for gene diversity and genotype diversity.

14.1.5 Microbial Metabolic Diversity and Distribution Patterns

The spatial and temporal distributions of microbial diversities are often discussed in the context of environmental factors such as temperature, pH, oxygen, salinity, pressure, the availabilities of water and nutrients, and the sources of energy and carbon. The development of a diversity of microelectrodes is allowing unprecedented measurements of many environmental physical and chemical features at increasingly smaller scales [Bond and Scholz, 2005]. All these ecological factors can significantly influence microbial activities and play very important roles in determining the spatial and temporal dynamics of microorganisms in natural environments. Microbial metabolic properties and their spatial and temporal distributions are often independent of their evolutionary relationships.

One of the most common methods to group microorganisms is based on their specific metabolic capacities at using carbon and energy sources [Slonczewski and Foster, 2009]. For example, depending on the energy source, microorganisms are called either phototrophs (obtaining energy from light) or chemotrophs (obtaining energy from chemicals). Among chemotrophs, if the energy sources are from inorganic molecules (such as H_2S, H_2, NH_3, and Fe^{2+}), they are called chemolithotrophs [Slonczewski and Foster, 2009]. In contrast, if their energy sources are from organic compounds, they are called chemoorganotrophs [Slonczewski and Foster, 2009]. Similarly, depending on the carbon source, microorganisms can be either autotrophs (obtaining carbon from inorganic sources such as CO_2 and HCO_3^-) or heterotrophs (obtaining carbon from organic compounds) [Slonczewski and Foster, 2009]. Some microorganisms, either in a free-living state or in association with other organisms, can use atmospheric nitrogen as its nitrogen source [Slonczewski and Foster, 2009]. Indeed, the diversity of microbial metabolisms extends far beyond the typical animal and plant metabolic capabilities. Even more striking are the extreme environmental conditions where many microorganisms are found and thriving. These conditions include extreme high and low pressure, pH, oxygen and metal concentration, salinity, radiation, desiccation, and temperatures [Rothschild and Mancinelli, 2001; see also Vol. II]. For example, the nitrate-reducing chemolithoautotroph *Pyrolobus fumarii* can grow at temperatures of up to 113°C [Blochl et al., 1997].

Until very recently, most of what we know about microbial diversity and microbial activity was derived from cultured microbes and ex situ laboratory experimental investigations. While such studies are essential, recent metagenomic investigations have shown that much of what we thought we knew about our natural microbial world was in fact highly biased.

In the next section, I summarize the main metagenomics approaches and provide examples of how such approaches have impacted microbial ecological research.

14.2 METAGENOMICS AND ITS IMPACT ON MICROBIAL ECOLOGY

The term *metagenomics* was first coined by Jo Handelsman et al. [1998] to describe the emerging field that analyzes genetic material recovered directly from environmental samples. Fundamentally, metagenomics is synonymous with several other terms that have emerged over the last decade including Environmental Genomics, Ecological Genomics, Community Genomics, and, more recently, Megagenomics [Allen and Banfield, 2005; Xu, 2006]. In addition, the broad-sense metagenomics also encompasses metatranscriptomics, metaproteomics, and metabolomics to include any investigation involving the application of modern techniques to study biological molecules directly in their natural environments. At present, due to their small sizes, the enormous gap in our knowledge and the associated opportunities, and the extreme functional diversity, natural microbial communities have been the primary focus of metagenomics studies. Naturally, this is also where most of the surprising discoveries have been made.

Unlike traditional microbiology that relies upon pure clonal cultures, the metagenomic analyses of microbial communities enable studies of organisms that are not easily cultured in the laboratory. In addition, it allows the direct studies of organisms, both culturable and not culturable, in their native environments (see Vol. II). The ability to obtain such information is essential for a comprehensive and realistic understanding of natural biological communities. For example, over 99% of microorgansms in natural environments can't be cultured in the lab [Pace et al., 1985; Xu, 2006]. Even for the ones that we can grow in the laboratory, the artificial environments that we provide are unlikely to be reflective of their natural ecological niches. As a result, for the same organism, their gene expression, protein, and metabolite profiles will likely be different between those observed in the laboratory and those in nature. Therefore, approaches to investigate microbial communities in their native habitats are essential for critical and realistic understanding of microbes in nature.

Early metagenomics research focused on the analyses of single genes (e.g., the small subunit ribosomal RNA gene) or random gene fragments [Pace et al., 1985; Pace, 1991; see also Chapter 15, Vol. I]. In recent years, metagenomics research has developed into analyzing partial or complete genomes as well as other types of molecules in environmental samples, including RNA (transcripts), proteins (translated transcripts), and metabolites (metabolic intermediates and final products). In addition, different approaches (e.g., targeting single cells, size-fractionated cell populations, or the composite community genome) have been developed to analyze different components of the microbial community and the metagenome, depending on the objectives of the research [e.g., Kvist et al., 2007; Venter et al., 2004].

Conceptually, the broad-sense metagenomics can be divided into four complementary fields, depending on the types of molecules (DNA, RNA, protein, and small molecule metabolites) as their primary targets of investigations. Briefly, the conventional, narrow-sense metagenomics targets the analyses of total DNA samples from natural biological communities; metatranscriptomics analyzes the complete RNA samples from natural environments; metaproteomics investigates all the proteins from natural environments; and metabolomics examines the complete collection of metabolites from such samples. It should be noted from the outset that an equally important field, that of bioinformatics, is essential for interpreting the data and for integration of information from the diverse types of analyses (see Part 5, Vol. I). A comprehensive understanding of a natural ecosystem requires the integration of information from all these biological and environmental physical and chemical characteristics.

Below I introduce the main types of metagenomics research. Among the four types of molecules, our analyses of environmental DNA have progressed the most in recent years. As a result, the focus will be on the analyses of the metagenome—the DNA samples from natural microbial communities. While this overview will briefly introduce the other-omics fields, their detailed discussions are presented in Part 6 of this volume.

14.2.1 Narrow-Sense Metagenomics

To study genetic material from natural environments, the first step is to extract the DNA molecules from the environmental sample. Any given environmental sample will likely contain a variety of organisms: bacteria, archaea, viruses, protozoa, algae, fungi, plants, and animals. Different groups of organisms have distinct cell types and structural components. In addition, the same organisms may contain different kinds of cells in the sample—for example, fast-growing and metabolically active vegetative cells vs. resting and metabolically inactive spores or viable but not culturable (VBNC) cells of certain bacteria and fungi. Such structural heterogeneity can differentially influence the efficiency of DNA extraction from these cells. Indeed, different extraction protocols have shown to influence the types of results one would obtain in subsequent metagenomic analyses [e.g., Rondon et al., 2000]. To ensure that what's being extracted is representative of the DNA in the original environmental sample, a combination of extraction protocols is generally recommended [Rondon et al., 2000; see also Chapters 10 and 11, Vol. II]. Once environmental DNA samples are obtained, different approaches can be taken to examine specific issues.

14.2.1.1 The Single-Gene Approach. The analyses of DNA samples directly from the environment started long before the term metagenomics came into use. Early studies examined the distribution and diversity of specific gene fragments in a given environmental sample. Most of the studies have focused on the small subunit of the ribosomal RNA gene (SSU rRNA, commonly known as the 16S rRNA gene in prokaryotes and the 18S rRNA gene in eukaryotes) [Pace et al. 1985; Pace, 1991; Woese, 1987; Slonczewski and Foster, 2009; see also Chapters 15 and 16, Vol. I]. There were several reasons for this choice of marker [Baker et al., 2003; Slonczewski and Foster, 2009]. First, this gene is universally distributed across all cellular organisms. Such a universal distribution permits the potential identification, comparison, and assessment of all cellular organisms in any given environment. Second, this gene is functionally homologous across all groups of cellular life forms. Such functional homogeneity ensures that the comparison performed is meaningful (i.e., apples vs. apples and oranges vs. oranges, but not apples vs. oranges). Third, because of its functional constraint, its secondary and tertiary structures are highly conserved, making its primary sequences relatively stable over evolutionary time scales. This relative evolutionary stability is essential for analyzing taxonomic relationships among organisms and sequences at above the species level. Fourth, this gene has shown to be evolving at a relatively constant rate over time, thus allowing the inferences of relative divergence across broad groups of organisms, from bacteria to archaea to plants and animals. Lastly, both universal and group-specific PCR primers are available that allow the amplification of this specific gene from target organisms [e.g., Baker et al., 2003; Vandenkoornhuyse et al., 2002; see also Chapters 15 and 17, Vol. I].

To study the types and relative abundances of unique SSU rRNA genes in a given environmental sample, after DNA extraction, the most common approach is to use universal (or group-specific) PCR primers to amplify the specific gene from the target organisms. The amplified products would generally contain a mixture of SSU rRNA sequences from diverse groups of organisms in the original sample. The amplified individual gene products can be then cloned and sequenced individually [Pace, 1991]. The more recent, bead-based pyrosequencing technique allows direct sequencing of PCR products, abrogating the cloning and additional amplification steps [Edwards et al., 2006; see Chapter 18, Vol. I]. Once the sequences are obtained, they can be compared to standard databases [e.g., the Ribosomal Database Project (RDP, www.rdp.cme.msu.edu; see Chapter 36, Vol. I), or GenBank] for their identification to various taxonomic levels, from phylum to family, genus, and potentially species levels. The current high-throughput DNA sequencing systems generate tens to hundreds of millions of sequence reads in a single run [see Chapters 18 and 20, Vol. I]. Such data can provide robust estimates of the diversity of microorganisms in specific ecological niches.

Although this approach is limited to exploring universally conserved genes, its application supported early morphology-based observations that microbial diversity was far more complex than was known by culturing methods [e.g., Pace et al. 1985; Pace, 1991]. Many subsequent studies followed, targeting different groups of organisms in different environmental niches. Such work revealed that the vast majority of microbial diversity had been missed by cultivation-based methods. The current consensus is that less than 1% of the Bacteria and Archaea species in a given environmental sample can be grown on existing media and incubation conditions [Daniel, 2005; DeLong, 2005]. The noncultivable groups include some major phyla within all microbial groups (i.e., bacteria, archaea, fungi, viruses, algae, and protozoa) [e.g., Hugenholtz et al., 1998; Dawson and Pace, 2002; Edwards and Rohwer, 2005; Huber et al., 2002; Karner et al., 2001; Schadt et al., 2003; Vandenkoornhuyse et al., 2002].

The rapid accumulation of extensive sequence information for phylogenetic signature molecules has spurred the development of high-throughput microbial detection and monitoring systems for environmental samples. For example, based on DNA sequence information from groups of cultivable and uncultivable organisms, microarrays containing taxa-specific signature sequences have been developed to monitor the distributions of sulfate-reducing prokaryotes [Loy et al., 2002], *Enterococcus* species [Lehner et al., 2005], and other groups of organisms [Wu et al., 2004; Zhou, 2003; see also Chapters 57–61, Vol. I].

14.2.1.2 Targeted Partial Metagenome Sequencing: Linking Phylogeny to Function.

Because sequencing whole metagenomes is very expensive, even with current technology, early efforts to study potential functional genes in natural environments were based on the analyses of partial genomic libraries containing large fragments of environmental DNA [e.g., Beja et al., 2000]. The analyses tended to focus on the clones in these libraries that contained both phylogenetic signature molecules and protein-coding open reading frames. Specifically, the genes close to the SSU rRNA gene cluster or other phylogenetic marker genes were cloned, identified, and sequenced together. To achieve this, relatively high molecular weight DNA needs to be extracted and used to construct clone libraries in Cosmid vectors or bacterial artificial chromosomes (BAC) so that such large fragments could be cloned and analyzed individually [Beja et al., 2000; see also Chapter 22, Vol. I]. Through Southern blotting, phylogenetic probes such as the SSU rRNA gene could be used to identify specific clones in the metagenome library that contain signature gene fragments, allowing further phylogenetic inferences. These clones could then be sequenced at very reasonable cost, through either shotgun sequencing or pyrosequencing. The SSU rRNA gene as well as any other genes on the cloned fragments could be identified and annotated. Together, the information allows the inferences of the organisms and the metabolic processes in the community as well as their potential ecological functions. For example, in a pioneer metagenomic study of genome fragments from a BAC library of marine picoplankton, Beja et al. [2000, 2001] identified a new class of genes of the rhodopsin family, named proteorhodopsin, from an uncultivated alpha-proteobacterium SAR86. At that time, this rhodopsin family was known to exist only in extremely halophilic (salt-loving) archaea and had never before been observed in cultured bacteria. Unlike the archaeal rhodopsin that does not express properly in model laboratory strains, the proteorhodopsin from SAR86 was expressed readily in the laboratory model bacterium *E. coli* and it functioned as a light-driven proton pump. Later studies identified that this new type of light-driven energy generation process is in fact widespread in the ocean and that there are optimized absorption spectra of bacterial rhodopsins at different depths of ocean water [Beja et al., 2001].

14.2.1.3 Random but Comprehensive Metagenome Sequencing. The above two approaches allow specific gene-based investigations of environmental samples. Recent technological advances are allowing the whole or partial metagenomes to be analyzed relatively efficiently [Edwards et al., 2006]. In these analyses, the DNA samples either could be first cloned into a vector and then subjected to sequencing or could be sequenced directly. Because the DNA fragments are randomly sheared, overlapping fragments should exist in the sample. If sufficient numbers of DNA fragments were sequenced using either the shotgun approach or direct sequencing by DNA synthesis, it would be possible to reconstruct large DNA fragments from the randomly sequenced small pieces [Venter et al., 2004; Edwards et al., 2006]. In certain cases, it has allowed the reconstruction of complete or almost complete genomes of several previously unknown bacterial species [e.g., Venter et al., 2004; Tyson et al., 2004; see also Chapter 32, Vol. II]. Because the collection of DNA from an environment is largely unknown and not filtered, the most abundant organisms in an environmental sample should be the most highly represented in the resulting sequence data, thus allowing assembly of large contigs belonging to the dominant species. To achieve the high coverage needed to fully resolve the genomes of underrepresented community members, extremely large number of reads, often prohibitively so, are needed. On the other hand, the random nature of shotgun sequencing ensures that many of the relatively rare organisms will be represented by at least some small sequence segments. Due to the limitations of microbial isolation and observation methods, the vast majority of these organisms would go unnoticed using traditional culturing techniques or microscopy tools.

14.2.1.4 Bioinformatics in Metagenomics. While much progress in metagenomic studies is due to the development of high-throughput sequencing technologies, the development of bioinformatics software has also played a major role and is poised to play a significantly greater role. The initial metagenome assembly and analyses used software developed for genome sequencing projects of cultured or known individual organisms (see Part 5, Vol. I). Huson et al. [2007] developed and published the first stand-alone metagenome analysis tool, MEGAN. This tool was originally developed to analyze the metagenome of a mammoth sample but was later adapted for shotgun reads of metagenome sequences (see Chapter 39, Vol. I). Meyer et al. [2008] released the Metagenomics RAST server (MG-RAST), a community resource for metagenome data analysis. As of April 2009, over 55 giga-base pairs of environmental DNA from over 1700 metagenome data sets have been analyzed via MG-RAST. Other software platforms and databases for analyzing metagenomic data include the IMG/M and CAMERA (see also Parts 4 and 5, Vol. I).

14.2.1.5 Microarrays. Microarray technology is a powerful, high-throughput experimental system that allows the simultaneous analysis of thousands to hundreds of thousands of genes at the same time. Originally developed for monitoring whole-genome gene expressions, microarrays have been developed and evaluated for bacterial detection and microbial community analysis (see Chapters 64–68, Volume I). These arrays include (i) phylogenetic oligonucleotide arrays that contain signature sequences from rRNA of specific groups of organisms, (ii) community genome arrays that contain highly specific signature gene sequences from known cultured microbial species, and (iii) functional gene arrays that contain conserved domains of genes involved in specific metabolic pathways such as the biogeochemical cycling of carbon, nitrogen, sulfate, phosphate, and metals [Zhou, 2003; Chapter 64, Vol. I]. The number of genes and the sizes of arrayed DNA fragments in the functional gene arrays can vary according to analytical purposes [see Chapter 57, Vol. I].

Investigation so far suggested that microarrays have the potential for the detection, identification, and characterization of microorganisms in natural habitats [Wu et al., 2004]. For example, [Loy et al. 2002] constructed a microarray with 132 16S rRNA-targeted oligonucleotide probes (18 nucleotides long) representing all recognized groups of sulfate-reducing prokaryotes and showed that this microarray could be used to distinguish most of the reference strains. Using this array, they determined the diversity of sulfate-reducing prokaryotes in periodontal tooth pockets and a hypersaline cyanobacterial mat [Loy et al., 2002]. Results from the microarray study were similar to those from cloning and sequencing of environmental 16S rRNA. Similar studies have been conducted for other groups of organisms such as the Rhodocyclales in beta-proteobacteria [Loy et al., 2005] and *Enterococcus* species [Lehner et al., 2005].

Despite these successes, significant challenges remain with regard to specificity, sensitivity, and quantification of microbes in natural habitats. This is mainly because microbial communities contain highly heterogeneous groups of organisms with undefined/unknown genomic relationships. The factors that hinder the utility of microarrays in metagenomics research include highly skewed distribution of microbial species, the potential of cross-hybridization between closely related species, the genetic variation among strains within species, and the differential efficiencies of isolating DNA from among the species.

14.2.2 Metatranscriptomics

While the analyses of DNA sequences in environmental samples can provide evidence about the diversity and function of specific groups of organisms in the analyzed ecological niches, the relative importance of those genes in environmental functions cannot be determined from DNA sequences alone. To understand which genes are expressed in a specific environment, the analysis of transcripts from the communities is often needed. The total collection of all expressed genetic information of an ecosystem is called a metatranscriptome. Issues influencing and biasing environmental DNA extraction can also affect RNA samples. Furthermore, RNA samples pose additional technical challenges in the analyses, including (a) the conversion of RNA to cDNA and (b) the overrepresentation of ribosomal RNA over messenger RNA in typical cells. However, despite these difficulties, recent investigations are promising (see Part 6, Vol. I). For example, a recent study by Gilbert et al. [2008] used the GS-FLX pyrosequencing technology to study a complex marine metatranscriptome obtained from random whole-community mRNA. Through analyzing over 323 million base pairs, they found that the highly expressed sequences were frequently novel, consistent with the hypothesis that microorganisms in laboratory environments are often very different in gene expression from those in nature. In addition to sequencing-based approaches for metatranscriptome profiling, microarray platforms have also been used. For example, the comparative metagenome and metatranscriptome analyses of lactic acid bacteria during *kimchi* (a traditional Korean dish of pickled cabbage) fermentation identified relatively limited change in the metagenome of the community. However, the changes in the metatranscriptome profiles were more profound. Some of the microorganisms that were less abundant in the kimchi fermentation process showed relatively high activities [Nam et al., 2009].

14.2.3 Metaproteomics

While the analyses of nucleic acids (both DNA and RNA) give indirect signals about the potential activities of genes in specific environments, proteins are promising alternative molecules that directly reflect the structural components, metabolic reactions, and potentially regulatory cascades within biological communities. The recent growths and interests in metaproteomes have been facilitated by the development of a variety of powerful methods to visualize and identify the proteomes of pure microbial cultures and increasingly the proteomes of microbial communities (see also Chapter 67 and Part 6, Vol. I). The most widely used approach in proteomics is the separation of proteins by two-dimensional gel electrophoresis (2-DE), which combines isoelectric focusing (IEF) in the first dimension and sodium dodecyl sulfate polyacrylamide gel electrophoresis (SDS-PAGE) in the second dimension. This separation is followed by in-gel digestion of selected spots and protein identification by mass spectrometry (MS). In the analyses of proteins from fully sequenced organisms, the identification can be achieved by databank comparison of peptide mass fingerprint [matrix-assisted laser desorption ionization time of flight MS (MALDI-ToF-MS)] or fragmentation data of the peptides (electrospray ionization source tandem mass spectrometry (ESI-MS/MS) or MALDI-ToF/ToF-MS). For the identification of peptides from not yet sequenced species or from unknown organisms in environmental samples, there are two general methods. In cases where proteins share high-level sequence similarity with those already published, one approach is to identify the peptides by analyzing their MS/MS fragments using the approach described above. On the other hand, if the amino acid sequence similarity is low, the peptides can be fully sequenced for their putative identification and characterization (for further details, see Chapters 71–74, Vol. I).

At present, the extraction and analyses of proteomes of cultured microbial samples are relatively well-developed. However, the analyses of metaproteomes from natural environments require more complicated procedures. This is because environmental samples contain mixtures of microorganisms likely in very different proportions, and proteins in some cells will be more readily extracted than others. In addition, other organic and inorganic substances in the samples can affect the quantity and quality of protein extracts. To overcome these problems and to obtain protein samples that are representative of the original community, a variety of sample type-specific extraction protocols are needed, similar to the extractions of DNA and RNA samples. For example, Benndorf et al. [2007] recently developed an extraction method allowing efficient separation of proteins from the inorganic and organic constituents of the soil matrix. Specifically, they used a combination of 0.1 M NaOH treatment and phenol extraction. Incubation with NaOH released humic acids and proteins from soil minerals and, simultaneously, disrupted microorganisms. The subsequent phenol extraction separated the proteins from the humic organic matter. Using spiked samples, their subsequent analyses identified the high-quality protein extracts. Ram et al. [2005] used genomic and mass spectrometry-based proteomic methods to evaluate gene expression, identify key activities, and examine partitioning of metabolic functions in a natural acid mine drainage microbial biofilm community. They detected 2033 proteins from the five most abundant species in the biofilm, including 48% of the predicted proteins from the dominant biofilm organism, *Leptospirillum* group II (also identified directly through metagenome analyses [Ram et al., 2005]). They found that proteins involved in protein refolding and response to oxidative stress were highly expressed, suggesting that damage to biomolecules is a key challenge for survival of microorganisms in such environments [Ram et al., 2005]. Novel proteins of potential economic and environmental importance have been identified through directly analyzing genomes and metagenomes [Voget et al., 2003; Xu, 2006; see also Chapters 47–53, Vol. II].

14.2.4 Metabolomics

A metabolome includes all the naturally occurring metabolites (typically small molecules) in biological samples such as cells, biofluids, and tissues. These small molecules are the products of metabolism and include, for example, carbohydrates, fats (or lipids), amino acids, nucleotides, and their derivatives as well as secondary metabolites such as antibiotics. The complete complement of all metabolites in an environmental sample is called an environmental metabolome. The metabolome and environmental metabolome are the results of interactions between the genes and the environment (see also Chapters 71–74, Vol. I).

Typical biological communities have significant division of labor in metabolism. Waste products of some organisms are substrates for others. Working together, they turn raw resources into fully metabolized products. In these systems, metabolite analyses can greatly accelerate comparative metagenome studies and expression experiments based on microarrays or proteomics. Such analyses would allow researchers to piece together a metabolic network that goes beyond species boundaries, enabling estimates of how metabolites are transferred and transformed through a community.

Similar to metagenomics, metatranscriptomics, and metaproteomics, the development of enviromnetal metabolomics has benefited from the rapid development and popularization of research tools used to measure the metabolites. These tools include nuclear magnetic resonance (NMR) spectroscopy, LC/GC-coupled mass spectrometry, and capillary electrophoresis coupled to mass spectrometry (CE-MS) (see also Part 6, Vol. I). In addition, the fields of metabolomics and environmental metabolomics have their roots on hundreds of years of physiology, biochemistry, and clinical chemistry research.

However, unlike nucleic acids and proteins that are relatively homogeneous in their chemical constituents, metabolites within cells, tissues, and microbial communities are very diverse in both their chemical and physical properties. As a result, different tools targeting different types of molecules may be required to reveal the complete metabolome of a cell population or the environmental metabolome of a microbial community. Because of the heterogeneity in metabolites and the tools used to analyze them, extensive efforts are needed for standardizations. In addition, unlike the widely popular and essential databases for DNA, RNA, and proteins of various types of cells, tissues, and communities, there is no GenBank-like repository for all metabolomics data, although these online databases are under development (see Chapter 74, Vol. I]. Having such a publicly available database would significantly strengthen the analyses of metabolome and environmental metabolome information. It's expected that the next few years will see great improvements in standardization of various analytical platforms, software, and databases. So far, large-scale metabolome surveys have been limited to known cultures, cells, and tissues [Mashego et al., 2007], especially those of model organisms such as *Escherichia coli, Saccharomyces cerevisiae*, plants, and humans.

14.2.4.1 Single-Cell Isolation, Whole Genome Amplification and Sequencing.

While most current studies of uncultivable environmental microorganisms rely on the extraction and sequencing of the whole community genomes, a recent paper reported successful isolation, amplification, and sequencing of the partial genome of a single uncultured archaeon cell from the environment. In this study, Kvist et al. [2007] first mapped the diversity of Archaea in a soil sample by generating a clone library using group-specific primers in combination with a terminal restriction fragment length polymorphism profile. Intact cells were then extracted from the same environmental sample and probed with Cy3-labeled DNA fragment designed from the clone library using fluorescent in situ hybridization. The single cells with a bright fluorescent signal were isolated using a micromanipulator. The genome of the single isolated cells was then amplified through multiple displacement amplification (MDA) using the Phi29 DNA polymerase. The 16S rRNA gene of the single cell showed a >99% sequence identity to a soil crenarchaeotal clone SCA1170. The shotgun sequencing of the generated MDA product found its closest match to a crenarchaeotal BAC clone previously retrieved from a soil sample [Kvist et al.,

2007]. Such an ability to isolate and sequence a microbial cell from the environment (see Chapters 75–77, Vol. I) will bypass many of the obstacles facing metagenomics research and will significantly accelerate our unbiased understanding of natural microbial communities.

14.3 CONCLUSIONS AND PERSPECTIVES

With the development and application of high-throughput "-omics" tools, microbial ecology is undergoing a renaissance. As shown in other chapters of the two volumes, genomics tools have allowed us unprecedented access to natural microbial diversity and their potential activities. These advances are addressing some of the most fundamental issues in microbial ecology such as the number of microbial species on Earth, the diversity and relative importance of specific metabolic pathways in specific ecological niches, and the relationship between microbial diversity and stability in natural environments.

The global environment is facing many significant challenges, rising CO_2 levels, the emergence and reemergence of infectious diseases of plants, animals, and humans, and increasing contaminations on land, in the air, and in our rivers, lakes, seas, and oceans. As it's becoming increasingly evident, microorganisms hold the key for solving these problems. The realistic understanding of microorganisms in their natural environments will be essential for us to meet these challenges. We are just entering the golden years of microbial ecology.

Acknowledgments

Research in my lab on microbial ecology and evolutionary genetics has been supported by the Natural Sciences and Engineering Research Council (NSERC) of Canada, the Ontario Premier's Research Excellence Award, and Genome Canada.

REFERENCES

Allen EE, Banfield JF. 2005. Community genomics in microbial ecology and evolution. *Nat. Rev. Microbiol.* **3**:489–498.

Baker GC, Smith JJ, Cowan DA. 2003. Review and re-analysis of domain-specific 16S primers. *J. Microbiol. Methods* **55**:541–555.

Beijerinck MW. 1888. The root-nodule bacteria. *Bot. Zeitung* **46**: 725–804.

Beja O, Aravind L, Koonin EV, Suzuki MT, Hadd A, et al. 2000. Bacterial rhodopsin: Evidence for a new type of phototrophy in the sea. *Science* **289**:1902–1906.

Beja O, Spudich EN, Spudich JL, Leclerc M, DeLong EF. 2001. Proteorhodopsin phototrophy in the ocean. *Nature* **411**:786–789.

Benndorf D, Balcke GU, Harms H, von Bergen M. 2007. Functional metaproteome analysis of protein extracts from contaminated soil and groundwater. *The ISME J.* **1**:224–234.

Blochl E, Rachel R, Burggraf S, Hafenbradl D, Jannasch HW, Stetter KO. 1997. *Pyrolobus fumarii*, gen. and sp. nov., represents a novel group of archaea, extending the upper temperature limit for life to 113 degrees C. *Extremophiles* **1**:14–21.

Bond AM, Scholz F. 2005. *Electroanalytical Methods: Guide to Experiments and Applications*. Berlin: Springer.

Cohan FM. 2004. Concepts of bacterial biodiversity for the age of genomics. In Fraser CM, Read TD, Nelson KE, eds. *Microbial Genomics*. Totowa, NJ: Humana Press, pp. 175–194.

Daniel R. 2005. The metagenomics of soil. *Nat. Rev. Microbiol.* **3**:470–478.

Dawson SC, Pace NR. 2002. Novel kingdom-level eukaryotic diversity in anoxic environments. *Proc. Natl. Acad. Sci. USA* **99**:8324–8329.

DeLong EF. 2005. Microbial community genomics in the ocean. *Nat. Rev. Microbiol.* **3**:459–469.

Edwards RA, Rohwer F. 2005. Viral metagenomics. *Nat. Rev. Microbiol.* **3**:504–510.

Edwards RA, Rodriguez-Brito B, Wegley L, Haynes M, Breitbart M, et al. 2006. Using pyrosequencing to shed light on deep mine microbial ecology. *BMC Genomics* **7**:e57.

Faith DP. 1992. Conservation evaluation and phylogenetic diversity. *Biol. Conserv.* **61**:1–10.

Garrity GM, Libum TG, Bell JA. 2005. *Bergey's Manual of Systematic Bacteriology*, 2nd ed. New York: Springer-Verlag.

Gevers D, Cohan FM, Lawrence JG, Spratt BG, Coenye T, et al. 2005. Re-evaluating prokaryotic species. *Nat. Rev. Microbiol.* **3**:733–739.

Gilbert JA, Field D, Huang Y, Edwards R, Li W, Gilna P, Joint I. 2008. Detection of large numbers of novel sequences in the metatranscriptomes of complex marine microbial communities. *PLoS ONE* **3**:e3042.

Haeckel E. 1866. *Generelle Morphologie der Organismen*. Berlin: Reimer.

Handelsman J, Rondon MR, Brady SF, Clardy J, and Goodman RM. 1998. Molecular biological access to the chemistry of unknown soil microbes: a new frontier for natural products. *Chem. Biol.* **5**:245–249.

Huber H, Hohn MJ, Rachel R, Fuchs T, Wimmer VC, Stetter KO. 2002. A new phylum of Archaea represented by a nanosized hyperthermophilic symbiont. *Nature* **417**:63–67.

Hugenholtz P, Goebel BM, Pace NR. 1998. Impact of culture-independent studies on the emerging phylogenetic view of bacterial diversity. *J. Bacteriol.* **180**:4765–4774.

Huson DH, Auch A, Qi J, Schuster SC. 2007. MEGAN: Analysis of metagenomic data. *Genome Res.* **17**:377–386.

Karner MB, DeLong EF, Karl DM. 2001. Archaeal dominance in the mesopelagic zone of the Pacific Ocean. *Nature* **409**:507–510.

Kasting JF, Siefert JL. 2002. Life and the evolution of earth's atmosphere. *Science* **296**:1066–1067.

Konstantinidis KT, Tiedje JM. 2005. Genomic insights that advance the species definition for prokaryotes. *Proc. Natl. Acad. Sci. USA* **102**:2567–2572.

Kvist T, Ahring BK, Lasken RS, Westernmann P. 2007. Specific single-cell isolation and genomic amplification of uncultured microorganisms. *Appl. Microbiol. Biotech.* **74**:926–935.

Lehner A, Loy A, Behr T, Gaenge H, Ludwig W, Wagner M, Schleifer KH. 2005. Oligonucleotide microarray for identification of *Enterococcus* species. *FEMS Microbiol. Lett.* **246**:133–142.

Li M, Xu J. 2009. Molecular ecology of ectomycorrhizal fungi: molecular markers, genets and ecological importance. *Acta Botanica Yunnanica* **31**:193–209.

Loy A, Lehner A, Lee N, Adamczyk J, Meier H, Ernst J, Schleifer KH, Wagner M. 2002. Oligonucleotide microarray for 16S rRNA gene-based detection of all recognized lineages of sulfate-reducing prokaryotes in the environment. *Appl. Environ. Microbiol.* **68**:5064–5081.

Loy A, Schulz C, Lucker S, Schopfer-Wendels A, Stoecker K, Baranyi C, Lehner A, Wagner M. 2005. 16S rRNA gene-based oligonucleotide microarray for environmental monitoring of the betaproteobacterial order "Rhodocyclales". *Appl. Environ. Microbiol.* **71**:1373–1386.

Mashego MR, Rumbold K, De Mey M, Vandamme E, Soetaert W, Heijnen JJ. 2007. Microbial metabolomics: Past, present and future methodologies. *Biotech. Lett.* **29**:1–16.

Mayden RL. 1997. A hierarchy of species concepts: The denouement in the saga of the species problem. In Claridge MF, Dawah HA, Wilson MR, eds. *Species: The Unit of Biodiversity*. London: Chapman and Hall, pp. 381–424.

Meyer F, Paarmann D, D'Souza M, Olson R, Glass EM, et al. 2008. The metagenomics RAST server—a public resource for the automatic phylogenetic and functional analysis of metagenomes. *BMC Bioinformatics* **9**:e0.

Nam Y-D, Chang H-W, Kim K-H, Roh S-W, Bae J-W. 2009. Metatranscriptome analysis of lactic acid bacteria during kimchi fermentation with genome-probing microarrays. *Int. J. Food Microbiol.* **130**:140–146.

Nunez ME, Martin MO, Duong LK, Ly E, Spain EM. 2003. Investigations into the life cycle of the bacterial predator *Bdellovibrio bacteriovorus* 109J at an interface by atomic force microscopy. *Biophys. J.* **84**:3379–3388.

Pace NR. 1991. Analysis of a marine picoplankton community by 16S rRNA gene cloning and sequencing. *J. Bacteriol.* **173**:4371–4378.

Pace NR, Stahl DA, Olsen GJ, Lane DJ. 1985. Analyzing natural microbial populations by rRNA sequences. *ASM News* **51**:4–12.

Ram RJ, Verberkmoes NC, Thelen MP, Tyson GW, Baker BJ, Blake RC 2nd, Shah M, Hettich RL, Banfield JF. 2005. Community proteomics of a natural microbial biofilm. *Science* **308**:1915–1920.

Rondon MR, August PR, Bettermann AD, Brady SF, Grossman TH, et al. 2000. Cloning the soil metagenome: A strategy for accessing the genetic and functional diversity of uncultured microorganisms. *Appl. Environ. Microbiol.* **66**:2541–2547.

Rothschild LJ, Mancinelli RL. 2001. Life in extreme environments. *Nature* **409**:1092–1101.

Schadt CW, Martin AP, Lipson DA, Schmidt SK. 2003. Seasonal dynamics of previously unknown fungal lineages in tundra soils. *Science* **301**:1359–1361.

Schopf JW. 1993. Microfossils of the Early Archean Apex chert: New evidence of the antiquity of life. *Science* **260**:640–646.

Sibley CG, Comstock JA, Ahlquist JE. 1990. DNA hybridization evidence of hominoid phylogeny: A reanalysis of the data. *J. Mol. Evol.* **30**:202–236.

Slonczewski JL, Foster JW. 2009. *Microbiology: An Evolving Science*. New York: W.W. Norton & Company.

Tyson GW, Chapman J, Hugenholtz P, Allen EE, Ram RJ, et al. 2004. Community structure and metabolism through reconstruction of microbial genomes from the environment. *Nature* **428**: 37–43.

Vandenkoornhuyse P, Baldauf SL, Leyval C, Straczek J, Young JP. 2002. Extensive fungal diversity in plant roots. *Science* **295**: 2051.

Venter JC, Remington K, Heidelberg JF, Halpern AL, Rusch D, et al. 2004. Environmental genome shotgun sequencing of the Sargasso Sea. *Science* **304**:66–74.

Voget S, Leggewie C, Uesbeck A, Raasch C, Jaeger KE, Streit WR. 2003. Prospecting for novel biocatalysts in a soil metagenome. *Appl. Environ. Microbiol.* **69**:6235–6242.

Winogradsky SN 1887. Concerning sulfur bacteria. *Bot. Zeitung* **45**:489–507.

Woese CR. 1987. Bacterial evolution. *Microbiol. Rev.* **51**:221–271.

Wu JR, Ma HC, Lü M, Han SF, Zhu, YY, Jin H, Liang JF, Liu L, Xu J. 2011. *Rhizoctonia* fungi enhance the growth of the endangered orchid *Cymbidium goeringii. Botany* **88**:20–29.

Wu L, Thompson DK, Liu X, Fields MW, Bagwell CE, Tiedje JM, Zhou J. 2004. Development and evaluation of microarray-based whole-genome hybridization for detection of microorganisms within the context of environmental applications. *Environ. Sci. Technol.* **38**:6775–6782.

Xu J. 2005. *Evolutionary Genetics of Fungi*. UK: Horizon Scientific Press.

Xu J. 2006. Microbial ecology in the age of genomics and metagenomics: concepts, tools, and recent advances. *Mol. Ecol.* **15**:1713–1731.

Xu J. In Press. Reproduction and sex in microorganisms. In Reproduction and Development Biology (Edited by Prof. Andre Pires da Silva). In Encyclopedia of Life Support Systems (EOLSS), Developed under the Auspices of the UNESCO, Eolss Publishers. Oxford: UK. [http://www.eolss.net]

Zhou J. 2003. Microarrays for bacterial detection and microbial community analysis. *Curr. Opin. Microbiol.* **6**:288–294.

Chapter 15

The Enduring Legacy of Small Subunit rRNA in Microbiology

SUSANNAH G. TRINGE AND PHILIP HUGENHOLTZ

15.1 INTRODUCTION

16S ribosomal RNA (16S for short) holds a special place in the study of microbial evolution and ecology. By virtue of a number of uncommon properties (ubiquity, extreme sequence conservation, and a domain structure with variable evolutionary rates [Woese, 1987]), it spearheaded two revolutions in these fields. First, it radically changed our view of evolution from a five-kingdom to three-domain paradigm by providing an objective phylogenetic framework in which to classify cellular life [Woese, 1987]; and second, through the cloning and sequencing of 16S genes directly from the environment using conserved broad-specificity PCR primers (16S surveys), it demonstrated that microbial diversity is far more extensive than we ever imagined from culture-based studies [Pace, 1997].

The application of high-throughput shotgun sequencing to environmental DNA (metagenomics) has exploded in the last several years [Kunin et al., 2008; Rusch et al., 2007; Tringe and Rubin, 2005; Tyson et al., 2004]. Metagenomic sequencing randomly samples all genes present in a habitat rather than just 16S, thereby providing clues to the functional capacity of a community rather than just its phylogenetic composition.

"Classical" community composition profiling by 16S is now often used as a preliminary step prior to metagenomic analysis, and can be of great value in guiding decisions regarding sequencing technology to be used (454 vs. Sanger, shotgun vs. large-insert clones) and amount of sequencing necessary. However, 16S is reemerging as a stand-alone molecular tool due to a confluence of methodological advancements [Tringe and Hugenholtz, 2008].

15.2 MATERIALS AND METHODS

Data were obtained from the greengenes database [DeSantis et al., 2006] and parsed such that each entry for a given study was assumed to be an independent clone, and each unique entry in the "isolation source" field was assumed to be an independent sample. No attempt was made to further manually curate the clone or sample counts based on information in publications or manual examination of clone names or other information, beyond spot-checking the accuracy of the parsing scripts. Submission dates were converted to a numerical format by taking the submission year and adding zero for submissions in January through March, 0.25 for April–June, 0.5 for July-September, and 0.75 for October–December. In some cases the data are unpublished, or the sequences were submitted prior to publication so the date in the figure is earlier than the publication date.

15.3 RESULTS AND DISCUSSION

15.3.1 Data Generation

16S data are being generated at an unprecedented rate due to new and improved sequencing technologies that dramatically increase throughput and decrease cost (see Chapter 18, Vol. I). These include lower Sanger

Handbook of Molecular Microbial Ecology, Volume I: Metagenomics and Complementary Approaches, First Edition. Edited by Frans J. de Bruijn.

sequencing costs as well as inexpensive 454 pyrosequencing and the PhyloChip, a custom microarray for 16S surveys [DeSantis et al., 2007; see also Chapter 58, Vol. I]. The flood of 16S data stemming from these advances has in most cases continued to reveal that most diversity estimates, even those based on culture-independent methodologies, fall far short of reality.

Whereas the typical 16S survey by traditional PCR clone sequencing a decade ago might have included a few dozen sequences, many today encompass thousands [e.g., Eckburg et al., 2005; Elshahed et al., 2008; Grice et al., 2009; Ley et al., 2006, 2008; Turnbaugh et al., 2009]. Indeed, there has been a near-exponential increase in the size of the largest surveys **(Fig. 15.1)**, though these numbers are likely underestimates because many studies only deposit unique phylotypes in the database rather than every clone sequenced. Today's 16S surveys also typically encompass multiple samples, even dozens, rather than targeting a single habitat (Fig. 15.1) [Dunn and Stabb, 2005; Grice et al., 2009; Ley et al., 2006, 2008; Schauer and Hahn, 2005; Turnbaugh et al., 2009]. Despite the expense of the clone-and-sequence approach, it remains the "gold standard" for identifying novel lineages because only full-length or near full-length sequences are adequate for accurate phylogenetic tree building. Such studies continue to expand the known "tree of life" at a steady pace, and they provide a valuable reference base for the high-throughput technologies discussed below.

The widespread availability of 454 pyrosequencing, a technology roughly an order of magnitude less expensive than Sanger sequencing in terms of cost per base, has changed the landscape of genomics [Margulies et al., 2005; see also Chapter 18, Vol. I]. To adapt pyrosequencing technology for 16S analysis, Sogin et al. [2006]

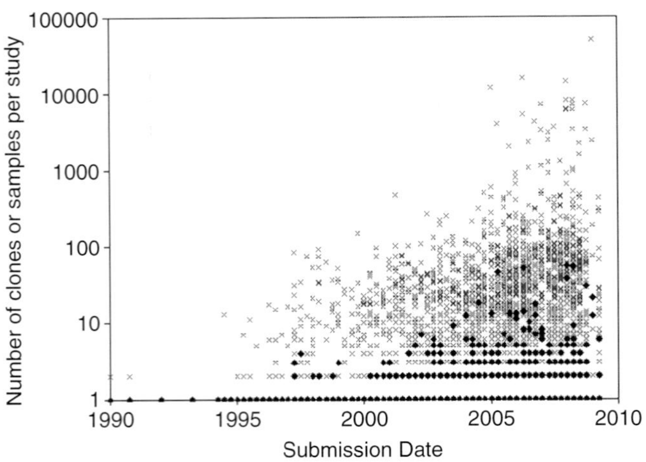

Figure 15.1 Increases in the number of full-length (>1200 nt) Sanger-sequenced 16S clones (gray Xs) and samples (black diamonds) per study since 1990. Note logarithmic scale on *Y* axis.

PCR-amplified the short V6 variable region of the bacterial 16S rRNA gene from eight distinct environments using universal primers and ran them separately within a single 454 run. This single run generated a total of ~118,000 sequence tags ("16S pyrotags"), more than any Sanger-based study to date. A follow-up study, also using GS20 technology, generated more than 900,000 bacterial and archaeal 16S pyrotags [Huber et al., 2007]. Subsequent refinements in the technology have increased the throughput and resolution of 16S pyrotag investigations and will continue to do so, though the increased error rate inherent in the method can both decrease phylogenetic resolution and inflate richness estimates [Kunin et al., 2009; Liu et al., 2007]. Barcoding, in which sequences from particular samples can be identified by unique sequences incorporated into the amplification primers, has enabled multiplexing of samples within runs and has further enhanced the usefulness of this approach [Hamady et al., 2008; Jones et al., 2009; Parameswaran et al., 2007].

Another major development in 16S analysis is not directly dependent on DNA sequencing but involves a high-density microarray of phylogenetically specific probes called the PhyloChip [DeSantis et al., 2007; see also Chapter 58, Vol. I]. Designing such a microarray is nontrivial due to the highly conserved nature of the 16S rRNA gene; however, DeSantis et al. [2007] have been able to use such an array to accurately differentiate among phylotypes in diverse environmental samples, documenting not only the vast majority of taxa identified by traditional cloning and sequencing but also groups not seen in clone libraries that were subsequently confirmed by taxon-specific PCR. Advantages of the PhyloChip are low cost and high speed (facilitated by dedicated software, PhyloTrac, to analyze the output: http://phylotrac.org), and drawbacks include only being able to identify phylotypes targeted on the chip and an inability to determine phylotype abundance distribution in the one sample (although individual phylotypes can be tracked quantitatively across samples).

15.3.2 Analytical Tools

With the ample data produced by these new technologies has come unprecedented statistical power in discerning similarities and differences among communities. The unidimensional diversity indices and total operational taxonomic unit (OTU) estimates commonly used in single-sample studies have given way to tools designed to directly compare the communities found in different samples. Some of these are aimed primarily at discerning overall phylogenetic similarities, while others assess the structure of the communities as well (e.g., abundance information).

Once sequences have been grouped into OTUs based on some set of similarity criteria (e.g., using DOTUR [Schloss and Handelsman, 2005]), similarity indices such as Bray–Curtis can be calculated to estimate the relatedness of different communities. Regression techniques can then be applied to isolate variables that contribute significantly to community composition, as well as correlate the abundances of specific phylogenetic groups with environmental factors [Brodie et al., 2007].

A recent technique more precisely tailored to 16S sequence analysis is UniFrac, a program designed to determine the fraction of unique branch lengths within a phylogenetic tree (comprising sequences from multiple samples) that is attributable to a particular sample [Lozupone and Knight, 2005]. Once this is determined, principal coordinates analysis (PCoA) can be used to identify specific environmental variables that drive differences among communities [Lozupone et al., 2007]. One key advantage of this approach is that it circumvents the controversial and often arbitrary process of assigning sequences to OTUs and deals entirely with tree-based metrics. Thus differences at the species or genus level receive less weight than those at the phylum level, but are still considered in the overall analysis. This method has been applied to data from a spectrum of environments and a variety of studies, often leading to new biological insights [Lozupone and Knight, 2005; Lozupone et al., 2007]. However, the original implementation of UniFrac takes only unique sequences into account, and thus it is insensitive to changes in abundance that may be important to understanding community responses to environmental variation.

A later version of UniFrac, called weighted UniFrac, deals with this weakness by assigning weights to branches of the tree based on the abundance of specific phylotypes. Comparison of the two methods revealed that they measure very different characteristics of the communities, and thus these methods should really be considered complementary approaches rather than different implementations of the same algorithm [Lozupone et al., 2007]. Algorithmic improvements implemented in Fast UniFrac allow the application of either the unweighted or weighted analysis to much larger pyrosequencing datasets, up to 100,000 sequences [Hamady et al., 2009].

A recent rRNA-based study uses a new set of metrics, the phylogenetic species variability (PSV) and phylogenetic species evenness (PSE), to separate out the effects of environmental selection versus interspecies competition [Newton et al., 2007] (discussed in more detail in the next section). These metrics summarize the relatedness of species within communities such that PSV is equal to one if the members of a community are unrelated and approaches zero if all of the members are closely related. PSE incorporates abundance information in addition to prevalence, such that PSE decreases both when community members are closely related and when members are unevenly represented [Helmus et al., 2007a; Helmus et al., 2007b; Newton et al., 2007]. Permutation tests can then be used to indicate whether species in the communities are underdispersed, such that closely related species tend to co-occur, or overdispersed, such that closely related species tend to occur exclusively from one another. Underdispersion may indicate that environmental filtering is an important force in generating community structure, and it supports the use of additional tools to correlate environmental variables with species composition.

Each of these approaches has its strengths and weaknesses, and no one tool can address each individual situation [Schloss, 2008]. But the currently available toolkit of experimental and analytical approaches allows a wide variety of experimental hypotheses to be tested.

15.3.3 Case Studies

These combined experimental and analytical developments are bringing 16S surveys out of the "fishing expedition" stage and producing true hypothesis-driven studies. The ability to take multiple samples over time, space, or other metrics and deeply interrogate each has enabled an entirely new class of studies, in both the environmental and medical arenas, in which 16S presence and abundance are correlated with specific factors.

A recent novel application of 16S rRNA sequencing to medical diagnosis and treatment investigated the microbial diversity of the lung in intubated patients and the effects of antibiotic therapy [Flanagan et al., 2007]. It found that while the lung remained sterile during brief intubation, patients inevitably became colonized during long-term intubation. Intriguingly, though, patients who were culture-positive for *Pseudomonas aeruginosa*, a primary agent in ventilator-associated pneumonia (VAP), were often colonized with a spectrum of other pathogenic and nonpathogenic bacterial species. Paradoxically, when patients were treated with antibiotics targeted to the *Pseudomonas* strains found by culture, the diversity of the flanking populations decreased and *Pseudomonas* became more dominant, potentially as a result of biofilm formation [Tart and Wozniak, 2008]. This finding held true whether the diversity was examined via cloning and sequencing or by PhyloChip, and the decreased diversity correlated with poorer patient outcome in patients with active infections [Flanagan et al., 2007]. This study provided an excellent example of the usefulness of PhyloChip analysis in communities dominated by a single member, because far greater diversity was revealed by microarray than could feasibly have been sampled by a traditional PCR clone library. The availability of multiple samples from individual patients, taken over the course of therapy, greatly increased the ability to correlate community composition with therapeutic intervention and patient outcome. Ribosomal RNA sequencing has also been used to study spatial variability in similar environments, both for complete microbial communities and for specific components of those communities. In one study of insect disease vectors, Jones et al. [2009] compared the microbial communities of two species of prairie dog fleas from multiple animals in six different colonies and two different time periods—a total of 230 samples—using pyrosequencing of variable region 2 (V2). ECOSIM [Gotelli and Entsminger, 2009] analysis of presence/absence matrices of the most common phylotypes revealed patterns of co-occurrence that suggested competitive or cooperative interactions among members of the microbial community. Both weighted and unweighted Unifrac analyses indicated that time was the most significant determinant of microbial community composition, followed by colony of origin, whereas differences between individual host animals or flea species were less significant.

One field in which the relative roles of evolutionary history and environmental selection have been difficult to sort out is the study of mammalian gut microbiota (see Section 3, Vol. II). A number of studies have revealed strong similarities among the gut communities of diverse mammals, but it has been unclear whether this was the result of similarities among the habitats and the nature of the host–microbe symbiosis, or simply the legacy of descent from a common ancestor whose gut community was already established. Ley et al. [2008] recently tackled this question in a study systematically characterizing the fecal communities of 59 distinct mammalian species, from diverse phylogenetic lineages and with widely varying lifestyles, as well as numerous humans. In total, the study encompassed more than 20,000 sequences from 106 samples, including some previously published data. Using UniFrac and PCoA, they found that host phylogeny had a dominant effect on community composition, while diet had a strong secondary role.

15.3.4 Some Remaining Challenges

The increased statistical power that comes with more data, as well as the many tools available to correlate environmental variables with 16S and other molecular data, has highlighted the need for accurate, standardized, and accessible metadata (i.e., nonsequence data associated with the samples being analyzed such as biogeochemical data). Coordinated efforts are now underway to address this need, such as the Genomics Standards Consortium [Field et al., 2008; Garrity et al., 2008; Hirschman et al., 2008; see also Chapter 39, Vol. I].

Almost all of the 16S sequence data in the public repositories to date are the products of PCR amplification of the 16S rRNA gene using 15–25 nucleotide primers broadly targeting bacteria or archaea. However, such primers are known to miss some organisms due to target mismatches [e.g., Baker et al., 2006; Huber et al., 2002], and the recent application of short (10 nucleotide) "miniprimers" suggests that a considerable amount of diversity may be overlooked in environmental samples using standard 16S primers [Isenbarger et al., 2008]. Pyrosequencing of cDNAs prepared from environmental RNAs may be the way of the future. This approach not only bypasses any potential primer bias, but simultaneously provides a community profile of all three domains of life and functional information in the form of expressed messenger RNAs [Urich et al., 2008]. A good-quality reference taxonomy based on phylogenetic inference of full-length 16S sequences is required to classify pyrotag and PhyloChip data. Unfortunately, such a reference tree remains somewhat elusive due to the rate of data accumulation (Fig. 15.1) and difficulties associated with producing and managing trees with hundreds of thousands of taxa. The problem is particularly acute for environmental sequences that are mostly unclassified in the public databases. The issue is currently being addressed through dedicated 16S databases ([e.g., Cole et al., 2007; DeSantis et al., 2006; Pruesse et al., 2007; see also Chapter 45, Vol. I]) and tools developed to handle large sequence datasets ([e.g., Dalevi et al., 2007; Price et al., 2009; Stamatakis, 2006]). It is important to note, however, that a number of the new analytical tools can provide biological insights through correlative analyses without the need to classify the underlying 16S data [Helmus et al., 2007b; Lozupone and Knight, 2005].

While shotgun metagenome sequencing has many benefits, 16S rRNA profiling still has a valuable role in microbial ecology and evolution. Methodological advances in sequencing, far from making ribosomal analysis obsolete, have instead rejuvenated the field.

INTERNET RESOURCES

The ARB project: http://www.arb-home.de/

Greengenes: http://greengenes.lbl.gov/cgi-bin/nph-index.cgi

Acknowledgments

This work was performed under the auspices of the US Department of Energy's Office of Science, Biological, and Environmental Research Program and by the University of California, Lawrence Berkeley National Laboratory under contract No. DE-AC02-05CH11231, Lawrence Livermore National Laboratory under Contract No. DE-AC52-07NA27344, and Los Alamos National Laboratory under contract No. DE-AC02-06NA25396.

REFERENCES

Baker BJ, Tyson GW, Webb RI, Flanagan J, Hugenholtz P et al. 2006. Lineages of acidophilic archaea revealed by community genomic analysis. *Science* **314**:1933–1935.

Brodie EL, DeSantis TZ, Parker JP, Zubietta IX, Piceno YM, et al. 2007. Urban aerosols harbor diverse and dynamic bacterial populations. *Proc. Natl. Acad. Sci. USA* **104**:299–304.

Cole JR, Chai B, Farris RJ, Wang Q, Kulam-Syed-Mohideen AS, et al. 2007. The ribosomal database project (RDP-II): Introducing myRDP space and quality controlled public data. *Nucleic Acids Res.* **35**:D169–D172.

Dalevi D, DeSantis TZ, Fredslund J, Andersen GL, Markowitz VM, et al. 2007. Automated group assignment in large phylogenetic trees using GRUNT: GRouping, Ungrouping, Naming Tool. *BMC Bioinformatics* **8**:402.

DeSantis TZ, Hugenholtz P, Larsen N, Rojas M, Brodie EL, et al. 2006. Greengenes, a chimera-checked 16S rRNA gene database and workbench compatible with ARB. *Appl. Environ. Microbiol.* **72**:5069–5072.

DeSantis TZ, Brodie EL, Moberg JP, Zubieta IX, Piceno YM, et al. 2007. High-density universal 16S rRNA microarray analysis reveals broader diversity than typical clone library when sampling the environment. *Microb. Ecol.* **53**:371–383.

Dunn AK, Stabb EV 2005. Culture-independent characterization of the microbiota of the ant lion *Myrmeleon mobilis* (Neuroptera: Myrmeleontidae). *Appl. Environ. Microbiol.* **71**:8784–8794.

Eckburg PB, Bik EM, Bernstein CN, Purdom E, Dethlefsen L, et al. 2005. Diversity of the human intestinal microbial flora. *Science* **308**:1635–1638.

Elshahed MS, Youssef NH, Spain AM, Sheik C, Najar FZ, et al. 2008. Novelty and uniqueness patterns of rare members of the soil biosphere. *Appl. Environ. Microbiol.* **74**:5422–5428.

Field D, Garrity G, Gray T, Morrison N, Selengut J, et al. 2008. The minimum information about a genome sequence (MIGS) specification. *Nat. Biotechnol.* **26**:541–547.

Flanagan JL, Brodie EL, Weng L, Lynch SV, Garcia O, et al. 2007. Loss of bacterial diversity during antibiotic treatment of intubated patients colonized with *Pseudomonas aeruginosa*. *J. Clin. Microbiol.* **45**:1954–1962.

Garrity GM, Field D, Kyrpides N, Hirschman L, Sansone SA, et al. 2008. Toward a standards-compliant genomic and metagenomic publication record. *Omics* **12**:157–160.

Gotelli NJ, Entsminger GL 2009. *EcoSim: Null Models Software for Ecology*. Jericho, VT: Acquired Intelligence Inc. & Kesey-Bear.

Grice EA, Kong HH, Conlan S, Deming CB, Davis J, et al. 2009. Topographical and temporal diversity of the human skin microbiome. *Science* **324**:1190–1192.

Hamady M, Walker JJ, Harris JK, Gold NJ, Knight R 2008. Error-correcting barcoded primers for pyrosequencing hundreds of samples in multiplex. *Nat. Methods* **5**:235–237.

Hamady M, Lozupone C, Knight R 2009. Fast UniFrac: Facilitating high-throughput phylogenetic analyses of microbial communities including analysis of pyrosequencing and PhyloChip data. *Isme J.* **4**:17–27.

Helmus MR, Bland TJ, Williams CK, Ives AR 2007a. Phylogenetic measures of biodiversity. *Am. Nat.* **169**:E68–E83.

Helmus MR, Savage K, Diebel MW, Maxted JT, Ives AR 2007b. Separating the determinants of phylogenetic community structure. *Ecol. Lett.* **10**:917–925.

Hirschman L, Clark C, Cohen KB, Mardis S, Luciano J, et al. 2008. Habitat-lite: A GSC case study based on free text terms for environmental metadata. *Omics*.

Huber H, Hohn MJ, Rachel R, Fuchs T, Wimmer VC, et al. 2002. A new phylum of Archaea represented by a nanosized hyperthermophilic symbiont. *Nature* **417**:63–67.

Huber JA, Mark Welch DB, Morrison HG, Huse SM, Neal PR, et al. 2007. Microbial population structures in the deep marine biosphere. *Science* **318**:97–100.

Isenbarger TA, Finney M, Rios-Velazquez C, Handelsman J, Ruvkun G. 2008. Miniprimer PCR, a new lens for viewing the microbial world. *Appl. Environ. Microbiol.* **74**:840–849.

Jones RT, Knight R, Martin AP 2009. Bacterial communities of disease vectors sampled across time, space, and species. *Isme J.* **4**:223–231.

Kunin V, Copeland A, Lapidus A, Mavromatis K, Hugenholtz P 2008. A bioinformatician's guide to metagenomics. *Microbiol. Mol. Biol. Rev.* **72**:557–578.

Kunin V, Engelbrektson A, Ochman H, Hugenholtz P 2009. Wrinkles in the rare biosphere: Pyrosequencing errors lead to artificial inflation of diversity estimates. *Environ. Microbiol.* **00**:000–000.

Ley RE, Turnbaugh PJ, Klein S, Gordon JI. 2006. Microbial ecology: Human gut microbes associated with obesity. *Nature* **444**:1022–1023.

Ley RE, Hamady M, Lozupone C, Turnbaugh PJ, Ramey RR, et al. 2008. Evolution of mammals and their gut microbes. *Science* **00**:000–000.

Liu Z, Lozupone C, Hamady M, Bushman FD, Knight R 2007. Short pyrosequencing reads suffice for accurate microbial community analysis. *Nucleic Acids Res.* **35**:e120.

Lozupone C, Knight R 2005. UniFrac: A new phylogenetic method for comparing microbial communities. *Appl. Environ. Microbiol.* **71**:8228–8235.

Lozupone CA, Knight R 2007. Global patterns in bacterial diversity. *Proc. Natl. Acad. Sci. USA* **104**:11436–11440.

Lozupone CA, Hamady M, Kelley ST, Knight R 2007. Quantitative and qualitative beta diversity measures lead to different insights into factors that structure microbial communities. *Appl. Environ. Microbiol.* **73**:1576–1585.

Margulies M, Egholm M, Altman WE, Attiya S, Bader JS, et al. 2005. Genome sequencing in microfabricated high-density picolitre reactors. *Nature* **437**:376–380.

Newton RJ, Jones SE, Helmus MR, McMahon KD 2007. Phylogenetic ecology of the freshwater Actinobacteria acI lineage. *Appl. Environ. Microbiol.* **73**:7169–7176.

Pace NR 1997. A molecular view of microbial diversity and the biosphere. *Science* **276**:734–740.

Parameswaran P, Jalili R, Tao L, Shokralla S, Gharizadeh B, et al. 2007. A pyrosequencing-tailored nucleotide barcode design unveils opportunities for large-scale sample multiplexing. *Nucleic Acids Res.* **35**:e130.

Price MN, Dehal PS, Arkin AP 2009. FastTree: computing large minimum evolution trees with profiles instead of a distance matrix. *Mol. Biol. Evol.* **26**:1641–1650.

Pruesse E, Quast C, Knittel K, Fuchs BM, Ludwig W, et al. 2007. SILVA: A comprehensive online resource for quality checked and aligned ribosomal RNA sequence data compatible with ARB. *Nucleic Acids Res.* **35**:7188–7196.

Rusch DB, Halpern AL, Sutton G, Heidelberg KB, Williamson S, et al. 2007. The Sorcerer II Global Ocean Sampling expedition: Northwest Atlantic through eastern tropical Pacific. *PLoS Biol.* **5**:e77.

Schauer M, Hahn MW 2005. Diversity and phylogenetic affiliations of morphologically conspicuous large filamentous bacteria occurring in the pelagic zones of a broad spectrum of freshwater habitats. *Appl. Environ. Microbiol.* **71**:1931–1940.

Schloss PD 2008. Evaluating different approaches that test whether microbial communities have the same structure. *Isme J.* **2**:265–275.

Schloss PD, Handelsman J 2005. Introducing DOTUR, a computer program for defining operational taxonomic units and estimating species richness. *Appl. Environ. Microbiol*. **71**:1501–1506.

Sogin ML, Morrison HG, Huber JA, Mark Welch D, Huse SM, et al. 2006. Microbial diversity in the deep sea and the underexplored "rare biosphere". *Proc. Natl. Acad. Sci. USA* **103**:12115–12120.

Stamatakis A 2006. RAxML-VI-HPC: maximum likelihood-based phylogenetic analyses with thousands of taxa and mixed models. *Bioinformatics* **22**:2688–2690.

Tart AH, Wozniak DJ 2008. Shifting paradigms in *Pseudomonas aeruginosa* biofilm research. *Curr. Top. Microbiol. Immunol*. **322**:193–206.

Tringe SG, Hugenholtz P 2008. A renaissance for the pioneering 16S rRNA gene. *Curr. Opin. Microbiol*. **11**:442–446.

Tringe SG, Rubin EM 2005. Metagenomics: DNA sequencing of environmental samples. *Nat. Rev. Genet*. **6**:805–814.

Turnbaugh PJ, Hamady M, Yatsunenko T, Cantarel BL, Duncan A, et al. 2009. A core gut microbiome in obese and lean twins. *Nature* **457**:480–484.

Tyson GW, Chapman J, Hugenholtz P, Allen EE, Ram RJ, et al. 2004. Community structure and metabolism through reconstruction of microbial genomes from the environment. *Nature* **428**: 37–43.

Urich T, Lanzen A, Qi J, Huson DH, Schleper C, et al. 2008. Simultaneous assessment of soil microbial community structure and function through analysis of the meta-transcriptome. *PLoS ONE* **3**:e2527.

Woese CR 1987. Bacterial evolution. *Microbiol. Rev*. **51**:221–271.

Chapter 16

Pitfalls of PCR-Based rRNA Gene Sequence Analysis: An Update on Some Parameters

ERKO STACKEBRANDT

16.1 INTRODUCTION

From 1990 to 2000 the search for microbial diversity concentrated on the phylogenetic diversity of environmental samples. The significance of phylogenetic diversity as opposed to functional diversity, however, remained unclear, but became a high-priority research in the past 10 years. The ease with which ribosomal RNA genes could be PCR amplified, cloned, and sequenced, often in concert with the identification of single cells by fluorescent in-situ hybridization [Amann et al., 1995] quickly attracted research groups from academia and the biotec industry all over the world. The analysis of clone libraries and analysis of prokaryotic communities by gradient gel electrophoresis (see Chapter 5, Vol. I and Chapter 12, Vol. II) allowed exciting insides into the unseen cosmos of hitherto uncultured bacteria and archaea and the speed at which habitats were screened were carried out at the expense of thorough evaluation of the various parameters influencing the outcome of such studies. As outlined in the review by von Wintzingerode et al. [1997], almost each methodological step in the complex procedure of diversity assessment was prone to bias the outcome of the study—for example, the influence of sample taking and sample handling, nonquantitative recovery of cells and/or nucleic acids, PCR errors influencing sequence composition, and cloning bias (see also Baker et al. [2003]); but also the interpretation of sequence-based diversity has been—and still is—allowing only a glance at a fraction of the phylogenetic relatedness among community members as it occurs in nature. Here, the number of *rrn* operons, preferential amplification, misprimed elongation (see Chapter 17 and 19, Vol. I), suppression of minority populations, short sequences, sequence alignment, and the quality and the selection of reference sequences have shown to significantly influence the result of any of such studies. The most impressive examples for the inability to assess complex communities is the lack of congruency between the outcome of culture and nonculture studies, demonstrated by the failing of mutual match of sequence identity. The verification of the presence of a single to several types of bacteria in a natural sample by FISH and the identification of small, slow-growing, or starving phylotaxa by its improved version, the "catalyzed reporter deposition fluorescence in situ hybridization" (CARD-FISH) [Pernthaler et al., 2002], was a major breakthrough because it allowed quantification and determination of growth dynamics. Although still today these approaches are often an entry gate to diversity assessments, their results are of restricted significance in answering obvious ecological questions: Who is doing what, when, and under which conditions? One of the first successful approaches to address some of these questions was the visualization of metabolically active cells by microautoradiography (MAR) combined with FISH, obtaining simultaneous information about activity and identity. Because autotrophs and most heterotrophic bacteria assimilate CO_2 in various

Handbook of Molecular Microbial Ecology, Volume I: Metagenomics and Complementary Approaches, First Edition. Edited by Frans J. de Bruijn.

carboxylation reactions during biosynthesis, assimilation of $^{14}CO_2$ or the new $HetCO_2$-MAR approach was used for isotope labeling of active microorganisms in environmental samples [Hesselsoe et al., 2005]. Several isotope-based methods have been introduced since then for cultivation-independent characterization of metabolically active microorganisms in environmental samples. The novel methodologies include direct isotope analysis of extracted biomarkers (including amino acids, fatty acids, and nucleic acids), stable isotope probing (SIP) of DNA or RNA [see Chapter 55, Vol. I], and a isotope microarray [see Hesselsoe et al., 2005] for references. The problems of measurements of microbial metabolism in biofilms, sediments, and soils, especially the ecological physiology of gradient microorganisms, was first tackled by the introduction of microsensors. These devices were able to measure not only O_2, N_2O, NO_3, NH_4, and CH_4 (for example) but also acetate, propionate, isobutyrate, and lactate (see references in Meyer et al. [2002]). The combination of these techniques allowed deep insights into the functioning of an ecosystem, though still restricted to the elucidation of the target organisms. The understanding of microbial processes beyond that of lab-cultured microorganisms was achieved by the new field of metagenomics [Stein et al., 1996] which received a boost by the introduction of high-throuput sequence analysis and advances in bioinformatics. By targeting cDNA, the expressed genes of metabolically active were specifically discovered; with the increasing number of completely sequenced genomes, sequences of the functional gene could often be linked to specific members of the natural community and from there eventually to a cultured strain [see Chapters 20, 27, and 28, Vol. I]. Another functional aspect is based upon comparative gene studies and expression experiments, using microarrays and/or proteomics, which can sketch a metabolic network [see Chapter 58, Vol. I] that reaches beyond species boundaries [see Chapters 10 and 22, Vol. I]. Prerequisite is detailed knowledge about the organisms which requires the skills of experienced microbiologists to provide this information using pure cultures. The deep insight into the microscale level of biodiversity is stunning, as is the the development of techniques that are increasingly refined to scrutinize even smaller niches of an environmental sample within a shorter time. Several of the pitfalls described for the 16S rRNA gene-based studies will also prove true for the other techniques mentioned (e.g., when isolation and manipulation of nucleic acids are involved), while other problems will be related to bottlenecks of statistics and problems connected to quantification of phylogenetic and metabilic diversity. Information of such issues is rare and scattered in individual references [Bent et al., 2007; Robertson et al., 2005; Raes et al., 2007]. In general terms, whenever DNA extraction and PCR techniques are included in an environmental study, its outcome cannot be quantified, neither at the level of genes, organisms or communities. For the latter approach, one has to await large-scale assessments based upon cDNA The following sections will focus on 16S rRNA gene-based studies, updating the progress in some of the methodological downsides reported in the publication of von Wintzingerode et al. [1997].

16.2 UPDATE ON SOME PARAMETERS

16.2.1 Cell Lysis and Extraction of Nucleic Acids

A broad range of protocols are available for the isolation of DNA and RNA (see Web Resources, and see also Chapters 10-11, Vol. II). Among other commercial kits, the MasterPure Gram-Positive DNA Purification Kit (Epicentre MGP04100) and Jetflex Genomic DNA Purification Kit (GENOMED 600100) are of proven reliability and used for providing high-molecular-weight DNA for genome sequence analysis by DSMZ staff. However, the situation is different with environmental nucleic acids when the full range of diversity is to be analyzed. Here, insufficient or preferential disruption of cells is one of the main reasons for bias in the assessment of the composition of microbial diversity. The problem is to achieve lysis of the recalcitrant strains without shearing those cells that lyse more easily. Fragmented nucleic acids are sources of artifacts in reverse transcription or PCR amplification experiments and may contribute to the formation of chimeric PCR products. In addition, various biotic and abiotic components of environmental ecosystems like inorganic particles or organic matter affect lysis efficiency and may interfere with subsequent DNA purification and enzymatic steps (the reader is referred to von Wintzingerode et al. [1997] for references before 1996). More recently, the impact of three different soil DNA extraction methods on bacterial diversity was evaluated using PCR-based 16S ribosomal DNA analysis [Martin-Laurent et al., 2001], confirming that soil DNA extraction methods affect both phylotype abundance and composition of the indigenous bacterial community. While at the genus level, using the ARDRA approach [see Chapters 7, Vol. I], the particular DNA extraction method used does not influence bacterial diversity, the more species-specific ribosomal intergenic spacer analysis (RISA) technique confirmed that the obtained fingerprints depend on the extraction method used. The authors argue that the effect of the extraction method on the efficiency of the 16S rDNA is probably due to differential coextraction of impurities that may affect the activity of the *Taq* polymerase. Using the denaturing gradient gel electrophoresis method, De Lipthay et al. [2004] came to the same conclusion that each of the three different protocols (sonication, grinding–freezing–thawing, and bead beating) used to physically disrupt cells resulted in unique community patterns. While the highest number of DNA bands (phylotypes) was achieved by the bead-beating procedure (which also resulted in more sheared DNA), the bacterial community profile depended on the soil type and less on the extraction protocol. Martin-Laurent et al. [2001], as well as De Lipthay et al. [2004], argue that the choice of DNA extraction protocol will also influence the obtained functional diversity. The influence of the choice of DNA extraction method on bacterial community profiles was also noted by Carrigg et al. [2007]. In their conclusions, two methods, both based on bead beating, were demonstrated to be suitable for comparative studies of a range of soil and sediment types. Krsek and Wellington [1999] investigated the influence of EDTA or monovalent ions on DNA yield and purity, and they found that sonication sheared the DNA more than bead-beating and that lysozyme and SDS lysis without any mechanical treatments allowed isolation of larger. A rapid protocol, including bead-beating, CTAB extraction buffer, and phenol–chloroform–isoamyl alcohol purification, resulted in the co-isolation of DNA and RNA from an established grassland soil [Griffiths et al., 2000]. The method, facilitating concomitant assessment of microbial 16S rRNA diversity by PCR and reverse transcription-PCR amplification from a single extraction, was able to differentiate the active component (rRNA-derived) from the total bacterial diversity (rDNA gene-derived).

16.2.2 Statistical Analysis of Denaturating Gel Electrophoresis Patterns

DGGE and TGGE patterns (see Chapter 5, Vol. I) can be analyzed using different algorithms that are publicly available. Analysis of band migration is often done by the software Bionumerics 4.0 (Applied Maths BVBA, Belgium), followed by the assessment of similarity between DGGE profiles. This is done by calculating similarity indices from the presence or absence of bands [Sorensen, 1948] or is based on the densitometric curves of the profiles compared using the Pearson product–moment correlation [Zoetendal et al., 2001; See also Chapter 8, Vol. I]. Similarities among the communities of different sampling sites can be assessed by normalization of DGGE banding profiles (e.g., Bionumerics), allowing the generation of the absolute intensity value of the bands, which can be imported into the PRIMER 5 software package [Clarke and Gorley, 2001] for further statistical analysis. The UPGMA algorithm or the more elaborate nonmetric multidimensional scaling permutation procedure can be used for the construction of dendrograms and ordination plots. Multivariate statistical analyses packages (e.g., CANOCO software), or PRIMER v5 and PRIMER v6 [Clark and Gorley, 2001, 2006] (see Web Resources) are also applicable for the interpretation of DGGE fingerprints. A discussion of additional statistical approaches, preferrably in joint analysis with environmental datasets, that have been applied to the analysis of microbial community fingerprints is given, for example, by Fromin et al. [2002].

16.2.3 Formation of PCR Chimeric Structures

The formation of PCR artifacts leading to chimeric structures is a potential risk in the PCR-mediated analysis of complex microbiota because it gives the false impression that a previously unknown organism has been discovered when in fact it is merely a combination of sequences of two different taxa. Despite the early notions on artifacts, chimeras and other anomalies are continuing to be generated and submitted without comment to the public repositories [Ashelford et al., 2005]. The average anomaly content per clone library is as high as 9.0% [Ashelford et al., 2006].

The only source of information to decide if a sequence is chimeric or not is to compare it with known, nonchimeric sequences. Several methods have been developed for detecting chimeric sequences, based either on nearest-neighbor methods (e.g., RDP II [Cole et al., 2003; see also Chapter 36 Vol. I]), chimeric alignment methods [Komatsoulis and Waterman, 1997], or partial treeing approaches [Hugenholtz and Huber, 2003]. The latter program, Bellerophon [Huber et al., 2004], detects chimeras based on phylogenetic trees that are inferred from independent regions of a multiple sequence alignment, and the branching patterns are compared for incongruencies that may be indicative of chimeric sequences. A new graphic display, Mallard, to recognize chimera was recently published by Ashelford et al. [2006]. In general, classifying a query sequence as chimeric or nonchimeric is not a simple matter [Gonzalez et al., 2005], and different programs may result in conflicting results. The most widely used programs are listed under Web Resources.

16.2.4 Analysis of 16S rRNA Gene Sequence Data

The quality of results obtained by comparative 16S rRNA sequence analyses strongly depends on the available dataset. Several hundreds of thousands of near-complete and partial 16S rRNA and 16S rDNA sequences of cultivated microorganisms and environmental clones are available in public databases. Though this number appears impressive, the databases contain a substantial though unquantitated percentage of errors, which range from PCR artifacts to reading mistakes and to the presence of chimera. Already published by Clayton et al. [1995], at least 26% of 16S rRNA gene sequence pairs (two sequences deposited for the same species) in GenBank had >1% random sequencing errors; of these, almost half had >2% random sequencing errors. As mentioned above, in a recent multicenter study Ashelford et al. [2005] estimated that at least 5% of the 1399 sequences searched had substantial errors associated with them ranging from chimeras (64%) to sequencing errors or anomalies (35%). Drancourt et al. [2000] made several recommendations concerning proposed criteria for 16S rRNA gene sequencing as a reference method for bacterial identification (mainly for diagnostic purposes), but little progress has been made to assess the quality of such sequences before deposition. Because the assessment of richness in complex communities is futile without extensive sampling, the generation of high-quality sequences and comparison to those of a global well-curated database of 16S rRNA gene sequences is highly desired though not forseeable. A step in the right direction has been published recently which will serve as the backbone of all assessment studies, namely a curated database of 16S rRNA gene sequences of prokaryotic type strains [see Chapter 45, Vol. I]. The living tree database (see Web Resources) [Yarza et al., 2008] provides corrected entries and the best-quality sequences with a manually checked alignment. Among multiple entries for a single type strain, the best-quality sequence was selected for the project. The tree provided in the current release (09.2009) is a result of the calculation of a dataset containing 10,950 single entries, 7710 corresponding to type strain gene sequences, and 3240 additional high-quality sequences to give robustness to the reconstruction. A consortium of laboratories are presently filling the gaps in tree of prokaryotic type strains in order to provide a most comprehensive database.

Various software tools are used to analyze the phylogenetics of rRNA datasets. While the Basic Local Alignment Search Tool (BLAST; see Web Resources) is often the first entry point to find similarity between sequences, a multisequence alignment offers more options. The most widely program is ARB (see Web Resources) [see Chapter 46, Vol. I], which compares nucleotide and/or protein sequences to sequence databases and calculates the statistical significance of matches [Ludwig et al., 2004]. ARB manages, aligns, and annotates sequences and also manages and prints phylogenetic trees. ARB is supplemented with additional software packages such as PHYLIP [Felsenstein, 2004], PAUP [Swofford, 1999], and MrBayes [Huelsenbeck and Ronquist, 2001] (see Web Resources). Robertson et al. [2005] point out some of the limitations of the latter two algorithms which, in combination with parsimony insertion of taxa to a backbone tree, may lead to spurious positioning of sequences in the backbone tree. Due to these computationally intensive algorithms, backbone trees are kept small, but additional sequences can be added using ARB's parsimony feature. However, large phylogenetic trees built by parsimony insertion into small backbone trees run the risk that the more diverse sequences will be positioned incorrectly, which will lead to improper phylogenetic inferences. A recent improvement of the statistically robust but computationally demanding "maximum likelihood" method for phylogenetic inference is RAxML [Stamatakis et al., 2005], apt to analyze a large numbers of rRNA gene sequences.

WEB RESOURCES

Isolation of Nucleic Acids from Pure Bacterial Cultures

http://www.protocol-online.org/prot/Microbiology/Bacteria/Nucleic_Acid_Extraction/index.html

http://www.docstoc.com/docs/640147/DNA-Isolation-Bacterial-CTAB-Protocol—General-Info

Multivariate Statistical Analysis

Canoco: http://www.pri.wur.nl/uk/products/canoco/

Primer v5, v6: http://www.primer-e.com/

Check for Chimeric Sequences

Bellepheron: http://foo.maths.uq.edu.au/huber/bellerophon.pl

Mallard: http://www.bioinformatics-toolkit.org/Web-Pintail

RDPII: http://rdp.cme.msu.edu

Curated 16S rRNA gene sequence database

http://arbsilva.de/livingtree

BLAST Analysis

http://www.ncbi.nlm.nih.gov/BLAST

Multi 16S rRNA Gene (and Other Genes) Sequence Alignment Including Software Packages

http://www.arb-home.de

Software Packages

Phylip: http://evolution.genetics.washington.edu/phylip.html

PAUP: http://paup.csit.fsu.edu

MrBayes: http://mrbayes.csit.fsu.edu

REFERENCES

Amann RI, Ludwig W, Schleifer KH. 1995. Phylogenetic identification and in situ detection of individual microbial cells without cultivation. *Microbiol. Rev*. **59**:143–169.

Ashelford KE, Chuzhanova NA, Fry JC, Jones AJ, Weightman AJ. 2005. At least 1 in 20 16S rRNA sequence records currently held in public repositories is estimated to contain substantial anomalies. *Appl. Environ. Microbiol*. **71**:7724–7736.

Ashelford KE, Chuzhanova NA, Fry JC, Jones AJ, Weightman AJ. 2006. New screening software shows that most recent large 16S rRNA gene clone libraries contain chimeras. *Appl. Environ. Microbiol*. **72**:734–5741.

Baker GC, Smith JJ, Cowan DA. 2003. Review and re-analysis of domain-specific 16S primers. *J. Microbiol. Methods* **55**:541–555.

Bent SJ, Pierson JD, Forney LJ. 2007. Measuring species richness based on microbial community fingerprints: The emperor has no clothes. *Appl. Environ. Microbiol*. **73**:2399–2401.

Carrigg C, Rice O, Kavanagh S, Collins G, O'Flaherty V. 2007. DNA extraction method affects microbial community profiles from soils and sediment. *Appl. Microbiol. Biotechnol*. **77**:955–964.

Clarke KR, Gorley RN. 2001. PRIMER v5: User manual/tutorial. PRIMER-E, Plymouth UK.

Clarke KR, Gorley RN. 2006. PRIMER v6: User manual/tutorial. PRIMER-E, Plymouth UK.

Clayton RA, Sutton G, Hinkle Jr PS, Bult C, Fields C. 1995. Intraspecific variation in small-subunit rRNA sequences in GenBank: Why single sequences may not adequately represent prokaryotic taxa. *Int. J. Syst. Bacteriol*. **45**:595–599.

Cole JR, Chai B, Marsh TL, Farris RJ, Wang Q, et al. 2003. The Ribosomal Database Project (RDP-II): previewing a new autoaligner that allows regular updates and the new prokaryotic taxonomy. *Nucleic Acids Res*. **31**:442–443.

De Lipthay JR, Enzinger C, Johnsen K, Aamand J, Sørensen SJ. 2004. Impact of DNA extraction method on bacterial community composition measured by denaturing gradient gel electrophoresis. *Soil Biol. Biochem*. **36**:1607–1614.

Drancourt M, Bollet C, Carlioz A, Martelin R, Gayral JP, Raoult D. 2000. 16S ribosomal DNA sequence analysis of a large collection of environmental and clinical unidentifiable bacterial isolates. *J. Clin. Microbiol*. **38**:3623–3630.

Felsenstein J. 2004. *Inferring Phylogenies*. Sunderland, MA: Sinauer Associates.

Fromin N, Hamelin J, Tarnawski S, Roesti D, Jourdain-Miserez K, et al. 2002. Statistical analysis of denaturing gel electrophoresis (DGE) fingerprinting patterns. *Environ. Microbiol*. **4**: 634–643.

Gonzalez JM, Zimmermann J, Saiz-Jimenez C. 2005. Evaluating putative chimeric sequences from PCR-amplified products. *Bioinformatics* **21**:333–337.

Griffiths, RI, Whiteley AS, O'Donnell AG, Bailey MJ. 2000. Rapid method for coextraction of DNA and RNA from natural environments for analysis of ribosomal DNA- and rRNA-based microbial community composition. *Appl. Environ. Microbiol*. **66**: 5488–5491.

Hesselsoe M, Nielsen JL, Roslevand P, Nielsen PH. 2005. Isotope labeling and microautoradiography of active heterotrophic bacteria on the basis of assimilation of $^{14}CO_2$. *Appl. Environ. Microbiol*. **71**:646–655.

Huber T, Faulkner G, Hugenholtz P. 2004. Bellerophon; a program to detect chimeric sequences in multiple sequence alignments. *Bioinformatics* **20**:2317–2319.

Huelsenbeck JP, Ronquist F. 2001. MrBAYES: Bayesian inference of phylogenetic trees. *Bioinformatics* **17**:754–755.

Hugenholtz P, Huber T. 2003. Chimeric 16S rDNA sequences of diverse origin are accumulating in the public databases. *Int. J. Syst. Evol. Microbiol*. **53**:289–293.

Komatsoulis G, Waterman M. 1997. A new computational method for detection of chimeric 16S rRNA artifacts generated by PCR amplification from mixed bacterial populations. *Appl. Environ. Microbiol*. **63**:2338–2346.

Krsek M, Wellington EMH. 1999. Comparison of different methods for the isolation and purification of total community DNA from soil. *J. Microbiol. Methods* **39**:1–16.

Ludwig W, Strunk O, Westram R, Richter L, Meier H, Yadhukumar A, et al. 2004. ARB: A software environment for sequence data. *Nucleic Acids Res*. **32**:1363–1371.

Martin-Laurent F, Philippot L, Hallet S, Chaussod R, Germon JC, Soulas G, Catroux G. 2001. DNA extraction from soils: Old bias for new microbial diversity analysis methods. *Appl. Environ. Microbiol*. **67**:2354–2359.

Meyer RL, Larsen LH, Revsbech NP. 2002. Microscale biosensor for measurement of volatile fatty acids in anoxic environments. *Appl. Environ. Microbiol*. **68**:1204–1210.

Morales SE, Holben WE. 2009. Emperical testing of 16S rRNA gene PCR primer pairs reveals variances in target specificity and efficacy not suggested by in silico analysis. *Appl. Environ. Microbiol*. **75**:2677–2683.

Pernthaler A, Pernthaler J, Amann R. 2002. Fluorescence *in situ* hybridization and catalyzed reporter deposition for the identification of marine bacteria. *Appl. Environ. Microbiol*. **68**:3094–3101.

Raes J, Foerstner KU, Bork P. 2007. Get the most out of your metagenome: computational analysis of environmental sequence data. *Curr. Opin. Microbiol*. **10**:490–498.

Robertson CE, Harris JK, Spear JR, Pace NR. 2005. Phylogenetic diversity and ecology of environmental Archaea. *Curr. Opin. Microbiol*. **8**:638–642.

Sorensen T. 1948. A method of establishing groups of equal amplitude in plant sociology based on similarity of species content. *Biol. Skr*. **4**:1–34.

Stamatakis A, Ludwig T, Meier H. 2005. RAxML-III: A fast program for maximum likelihood-based inference of large phylogenetic trees. *Bioinformatics* **21**:456–463.

Stein JL, Marsh TL, Wu KY, Shizuya H, DeLong EF. 1996. Characterization of uncultivated prokaryotes: Isolation and analysis of a 40-kilobase-pair genome fragment from a planktonic marine archaeon. *J. Bacteriol*. **178**:591–599.

Swofford D. 1999. *PAUP: Phylogenetic Analysis Using Parsimony (and Other Methods)*, Sunderland, MA: Sinauer.

von Wintzingerode F, Göbel UB, Stackebrandt E. 1997. Determination of microbial diversity in environmental samples: Pitfalls of PCR-based rRNA analysis. *FEMS Microbiol. Rev*. **21**:213–229.

Yarza P, Richter M, Peplies J, Euzeby J, Amann R, et al. 2008. The all-species living tree project: A 16S rRNA-based phylogenetic tree of all sequenced type strains. *Syst. Appl. Microbiol*. **31**:241–250.

Zoetendal EG, Akkermans ADL, Akkermans-van Vliet WM, de Visser JAGM, de Vos WM. 2001. The host genotype affects the bacterial community in the human gastrointestinal tract. *Microbial Ecol. Health Dis*. **13**:129–134.

Chapter 17

Empirical Testing of 16S PCR Primer Pairs Reveals Variance in Target Specificity and Efficacy not Suggested by *In Silico* Analysis

Sergio E. Morales and William E. Holben

17.1 INTRODUCTION

Molecular (i.e., culture-independent) methods, particularly those based on the 16S rRNA gene, have been a cornerstone in modern microbial ecology [Tringe and Hugenholtz, 2008; see also Chapter 15, Vol. 1]. However, most of these approaches have depended on PCR, which can be affected by a number of underlying biases [Wintzingerode et al., 1997; Polz and Cavanaugh, 1998; Crosby and Criddle, 2003; Ashelford et al., 2005, 2006; Sipos et al., 2007]. These considerations have been studied and mitigating measures developed to allow comparative, but likely not comprehensive, analyses of microbial community and population diversity to be performed [Sipos et al., 2007]. A review of the pitffalls of PCR-mediated 16S rRNA gene analysis is presented in these volumes by Stackebrandt (see Chapter 16, Vol. I). An additional consideration with clear implications in quantitative studies and comprehensive community surveys that has not been directly addressed is that of primer specificity. Methods have been developed to test specificity in silico using freely available software such as PRIMROSE [Ashelford et al., 2002; Cole et al., 2005], but knowledge of how well these simulations correlate to actual PCR reactions is limited [Manz et al., 1996; Meier et al., 1999; Overmann et al., 1999; Buckley and Schmidt, 2001; Reilly et al., 2002; Stach et al., 2003; Fierer et al., 2005; Bathe and Hausner, 2006]. This review provides a look into the incongruence seen between computer simulations and empirical testing of primer sets and the repercussions that these discrepancies likely have in quantitative analyses of microbial populations.

17.1.1 Materials and Methods

For a full outline of the materials and methods employed, please refer to the original publication [Morales and Holben, 2009]. What follows is an abbreviated version.

17.1.1.1 Primer Design and In Silico *Testing.*

Primers targeting specific taxonomic groups were designed based on a 16S rRNA gene sequence library (total 4889 sequences) from soil at the Kellogg Biological Station Long-Term Ecological Research site (KBS-LTER) [Morales et al., 2009] (GenBank accession no. EU352912 to EU357802). Primer design was conducted using the PRIMROSE software [Ashelford et al., 2002] for several major bacterial groups identified in the KBS-specific library based on taxonomic assignments established by ARB [Ludwig et al., 2004; see also Chapter 46, Vol. I] alignments (**Table 17.1**) [Morales et al., 2009]. In silico testing was carried out using PRIMROSE and either the KBS sequence library or an ARB-generated library with over 50,000 16S rRNA gene sequences, including archaeal and eukaryotic sequences not specific to the KBS site. Each primer set was tested against all target and nontarget sequences in the KBS library using PRIMROSE.

Handbook of Molecular Microbial Ecology, Volume I: Metagenomics and Complementary Approaches, First Edition. Edited by Frans J. de Bruijn.

Table 17.1 Primer Targets and Sequence Representation within Libraries

	Number of Sequences[a]	
Taxonomic Level	KBS	ARB
Phylum		
Acidobacteria	955	516
Actinobacteria	491	5,360
Bacteroidetes	453	2,683
CD OD1	68	73
CD OP10	43	55
Chlorobi	22	142
Chloroflexi	27	66
Gemmatimonadetes	251	111
Nitrospira	59	169
Planctomycetes	243	699
Proteobacteria	1,690	20,166
Thermomicrobia	323	361
Verrucomicrobia	103	159
Class		
Alphaproteobacteria	374	5,065
Betaproteobacteria	485	4,527
Deltaproteobacteria	348	1,217
Gammaproteobacteria	478	8,479
OTU[b]		
Genus *Aeromonas*	99	
Acidobacteria grp4	81	
Genus *Lysobacter*	81	
Thermomicrobia#4	63	
Nitrosomonadales	62	
Acidobacteria grp6	61	
Thermomicrobia#7	46	
Genus *Bradyrhizobium*	39	
Genus *Pseudomonas*	38	
Genus *Comamonas*	38	

[a] Total number of sequences found in the database.

[b] Closest classification as determined based on ≥97% sequence similarity at the 16S rRNA gene.

17.1.1.2 Primer Optimization and* In Vitro *Testing. Genomic DNA preparations from *Bradyrhizobium japonicum* USDA 110d, *Streptomyces griseus, Acidovorax facilis, Pseudomonas putida, Acidobacterium capsulatum*, and *Pedobacter heparinus* extracted using the method of Doi [1983] were initially used to assess and optimize amplification from each respective set of phylum- and class-level primers (**Tables 17.2 and 17.3**). Optimization consisted of (a) temperature gradient (±10°C) PCRs using primer 907r as the reverse primer in the pair and (b) the predicted melting temperature (T_m) for the specific primer as the central point in the temperature gradient. Complete reaction conditions are described fully in Morales and Holben [2009].

In order to generate positive controls for phylogenetic groups with no cultured representative available, plasmid DNA encoding the desired target sequence from our library was amplified using the primer pair 536f and 907r as outlined elsewhere [Morales et al., 2009]. Since all specific primers generated in this study are internal to this 16S region, it can be used as template to generate positive controls. Where possible, comparable amplification rates were ensured between cloned inserts and isolated genomic DNA by comparing amplification efficiencies.

Table 17.2 Phylum- and Class-Level Primer Sequences and Target Specificities (% detected of total) as Tested by PRIMROSE

Name	Target	Sequence	Degenerate Bases	KBS		ARB Database			
				[a]Target	[b]Nontarget	Target	Prokaryotes	Archaea	Eucarya
688-706fAB	Acidobacteria	GCGGTGAAAT-GCGTASAT	1	95.29	11.42	83.10	19.40	N.D.	N.D.
691-709fAP	Alphaproteo-bacteria	GTGAAATDCG-TAGAKATT	2	95.19	3.76	87.20	8.70	N.D.	N.D.
697-715fAT	Actinobacteria	TGCGCAGAK-ATCRGGARG	3	94.30	0.09	91.00	0.10	N.D.	N.D.
715-733fBP	Betaproteo-bacteria	AAYACCRATG-GCGAAGGC	2	96.70	1.83	92.60	7.60	N.D.	N.D.
685-703BT	Bacteroidetes	GTAGCGGTGA-AATGCWTA	1	96.25	0.02	90.50	0.20	N.D.	N.D.
554-572fCH	Chlorobi	TCCGGAWTYA-CTGGGTRT	3	95.45	0.08	93.00	3.00	N.D.	N.D.
687-705fCX	Chloroflexi	AGTGGTGAA-ATGCGTWGA	1	96.30	4.19	45.50	0.40	N.D.	N.D.
542-560fDP	Deltaproteo-bacteria	GTGCNARCGT-TGYTCGGA	3	86.21	2.59	55.90	0.60	N.D.	N.D.
680-698fGP	Gammaproteo-bacteria	CMKGTGTAG-CRGTGAAAT	3	93.31	16.85	78.20	10.80	N.D.	N.D.
677-695fGT	Gemmatimon-adetes	TTCCSGGTGT-AGCGGTGG	1	97.21	[c]N.D.	73.90	0.40	N.D.	N.D.
706-724fNP	Nitrospira	ATCGGGAGG-AASRCCKGT	3	93.22	3.94	43.80	0.70	N.D.	N.D.
785-802fOD1	OD1	GGATTAGATA-CYCYWGT	3	85.29	0.70	63.00	4.10	6.60	N.D.
889-907fOP10	OP10	ASTACGGCCG-CAAGGTTG	1	83.72	6.22	78.20	5.60	0.10	N.D.
681-699fPM	Plancto-mycetes	NRGTGRAGC-GGTGAAATG	3	94.24	0.62	83.80	0.60	N.D.	N.D.
767-785fProt	Proteobacteria	AAGCGTGGG-GAGCAAACA	0	89.08	7.55	73.30	23.70	N.D.	N.D.
555-573fTM	Thermo-microbia	CCGGAKTYA-YTGGGCGTA	3	91.02	8.30	74.50	8.20	N.D.	N.D.
562-580fVM	Verrucomi-crobia	YAYTGGGCGT-AAAGGGWG	3	92.23	4.38	80.50	7.50	N.D.	N.D.
853-871fWS3	WS3	GTGCCGCAG-CYAASSCAT	3	94.12	5.42	67.90	5.50	N.D.	N.D.

[a]Target sequences are those belonging to the phylogenetic group intended for amplification.

[b]Nontarget sequences are sequences not belonging to the phylogenetic group intended for amplification.

[c]N.D., not detected.

PCR products were thereafter used for positive controls for all amplification reactions. Minimal DNA concentrations (determined using 10-fold serial dilutions down to 1 pg of control DNA) and optimal empirically determined annealing temperature from prior temperature gradient analyses were used in all reactions. Each optimized primer set was subjected to specificity screening by PCR using negative control DNAs (nontarget sequences) as template as described [Morales et al., 2009].

17.1.1.3 Real-Time PCR Assays. Quantitative real-time PCR (qPCR) was performed on an iCycler iQ thermocycler (Bio-Rad, Hercules, CA) using conditions described in Morales and Holben [2009]. Two independent rounds of triplicate reactions were performed for each target, and the results of at least three qPCRs were analyzed. Abundances for all replicate reactions were related to a standard curve by their respective fluorescence intensity values, giving values of relative concentration.

Table 17.3 Primer Sequences and Target Specificities to the KBS Library as Indicated by PRIMROSE for the Top 10 Most Abundant OTUs (Based on 97% Sequence Similarity)

Name	Target	Sequence	Degenerate Bases	[a]Target	[b]Nontarget
Coma851-869f	Genus *Comamonas*	YCAGTRMCGAAGCTAACG	3	97.37	4.99
Pseudo573-591f	Genus *Pseudomonas*	AAGSGCKCGTAGGYGGTT	3	94.74	2.47
Aero851-869f	Genus *Aeromonas*	GGSTKCCGGMGCTAACGC	3	86.87	1.94
Acido(#4)599-617f	Acidobacteria grp4	CGAYTGTGAAATCTCCGG	1	96.30	0.04
Lyso726-744f	Genus *Lysobacter*	CGAAGGCGGYTSYCTGGA	3	92.59	1.83
Thermo(#4)735-753f	Thermomicrobia#4	CTTCCTGGCCTGTTCTTG	0	95.24	0.14
Nitro813-831f	Nitrosomonadales	TAAACGATGTCGACTGGT	0	95.16	0.10
Acido(#6)654-672f	Acidobacteria grp6	GAGDTYGGGAGAGGGATG	2	93.44	0.50
Thermo(#7)658-676f	Thermomicrobia#7	GCAGGAGAGGGTAGTGGA	0	95.65	0.08
Brady850-868f	Genus *Bradyrhizobium*	CTWGTGGCGMAGCTAACG	2	94.87	0.19

[a]Target sequences are those belonging to the phylogenetic group intended for amplification.

[b]Nontarget sequences are sequences not belonging to the phylogenetic group intended for amplification.

17.1.1.4 Confirmation of Primer Specificity in a Complex Environment Using Clone Libraries.

The specificity of two primer sets was tested by generation of clone libraries that were generated from amplicons produced using total soil DNA from the same KBS-LTER soil and subjected to DNA sequence analysis. PCR reactions were performed in triplicate using primer pairs 688-706fAB plus 907r and Acido (#6)654–672 plus 907r, purified using the Qiaquick PCR purification kit (Qiagen, Valencia, CA) and cloned as previously described [Morales and Holben, 2009]. Taxonomic assignments of the cloned sequences were obtained using the RDP Classifier program [Wang et al., 2007].

17.2 RESULTS AND DISCUSSION

17.2.1 In Silico Versus In Vitro Validation

Of 28 phylogenetic group-specific forward primers targeting the 16S rRNA gene (Tables 17.2 and 17.3), only three (Acido(#4)599-617f, Acido(#6)654-672f, and Nitro813-831f) had sufficient specificity to support their subsequent use in qPCR-based population studies. This was surprising given that all higher-order primers (targeting phylum- and class-level groups) (Table 17.2) were predicted by Primrose to detect about 92.7% (standard error [SE], 1%) of their specific targets, while conversely only predicted to amplify 4.6% (SE, 1.1%) of nontarget sequences. This was also the case with primers targeting the 10 most abundant OTUs (genus-level phylotypes based on 97% sequence similarity) identified from the KBS-LTER dataset (Table 17.3). On average, these primers were predicted to detect 94.2% (SE, 0.9%) of their specific targets, while detecting only 1.2% (SE, 0.5%) of nontarget sequences in the KBS database.

Primer-target mismatch (theoretical number of mismatched base pairs allowed during annealing [**Figs. 17.1 and 17.2**]) is suggested as a potential source of the discrepancy between in silico and in vitro specificity. Single mismatches were sufficient to increase nonspecific target detection six- to ninefold in silico, suggesting that during PCR amplification, primers with slight internal mismatches may bind sufficiently well to enable an initial elongation event (as opposed to 3′ mismatches, which are more effective at preventing nonspecific priming). This initial amplification provides a perfectly matched template allowing for high efficiency amplification of nontargets in subsequent rounds of PCR. We concluded that mismatching between primer and DNA template lends high specificity only where there is virtually no initial elongation taking place, which is unlikely unless multiple, consecutive, or 3′ prime mismatches are present, or when both primers in a reaction have a high degree of specificity.

17.2.2 Effect of Nonspecific Amplification on Real-Time qPCR and Clone Libraries

Unless properly validated, primers used in PCR-based analyses are likely to adversely affect the results and interpretation of an experiment, particularly where quantitative data are desired (e.g., for population studies). The target detection and PCR efficiencies of two primer sets, one that passed validation [Acido (#6)654–672 plus 907r, specific for Acidobacteria group 6 targets] and one that did not (688-706fAB plus 907r, putatively specific for the phylum Acidobacteria), were compared using a real-time qPCR assay to assess the effect of using faulty primers.

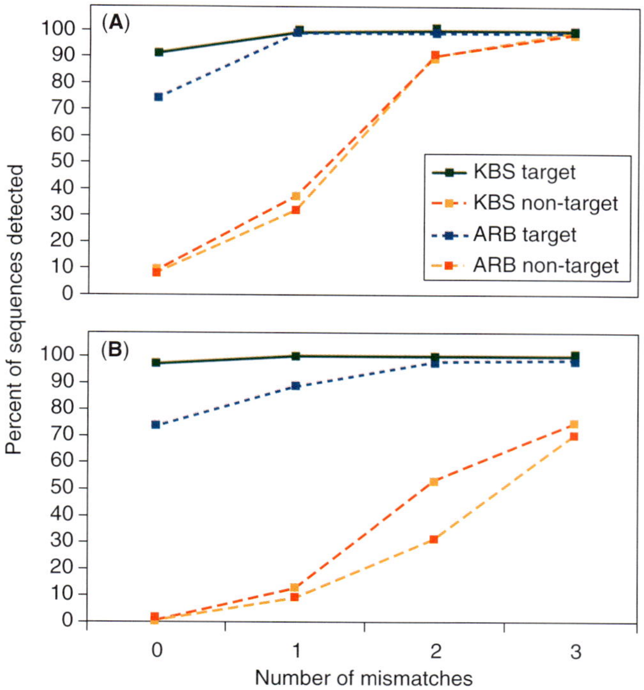

Figure 17.1 Effect of mismatched bases on the recovery of target and nontarget sequences using phylum-level primers. Primers were tested in silico against 4889 sequences from the KBS-LTER library and >50,000 sequences from the ARB database. (**A**) Thermomicrobia-specific primer 555-573fTM; (**B**) Gemmatimonadetes-specific primer 677-695fGT.

A 10% difference in PCR efficiency was observed between the primer pairs, with Acido (#6)654–672 plus 907r exhibiting 103.05% efficiency while 688-706fAB plus 907r showed 92.9% efficiency. Although both primer sets were able to detect and quantify different target concentrations, both alone and in the presence of total community DNA from soil (**Fig. 17.3**), the higher PCR efficiency of the primer pair containing Acido (#6)654–672 resulted in threefold more signal for a given target concentration compared to that with the primer pair containing 688-706fAB. Interestingly, despite the lower efficiency of amplification with the phylum-level primer set, quantification of target sequences from an unspiked soil sample from the KBS-LTER (treatment 1, replicate plot 1) indicated a higher abundance (4 pg, or 10^7 copies) of phylum-level acidobacterial targets than of genus-level Acidobacteria group 6 targets (0.23 pg, or 10^5 copies) per 10 ng of soil extracted DNA, indicating a low proportion of Acidobacteria group 6 within the total Acidobacteria phylum representatives present in that soil. However, any conclusion arising from the primer that did not pass validation would be erroneous due to the high degree of nonspecific amplification as determined by cloning PCR products from the same reactions (**Table 17.4**). Indeed, as

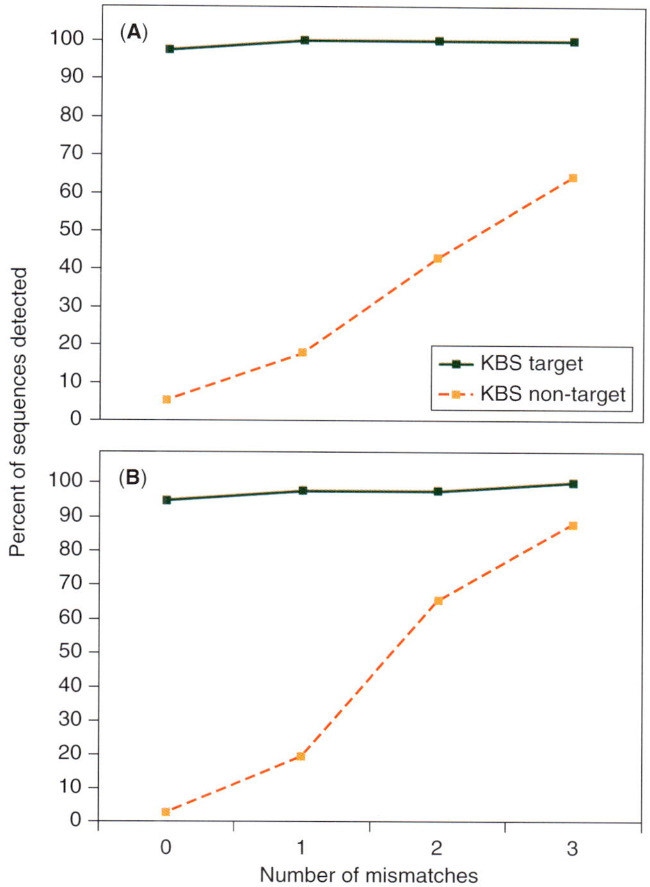

Figure 17.2 Effect of mismatched bases on the recovery of target and nontarget sequences using OTU-level (97% sequence similarity) primers. Primers were tested in silico against 4889 sequences from the KBS-LTER library. (**A**) OTU-specific primer Coma851-869f; (**B**) OTU-specific primer Pseudo573-591f.

expected from the results of the validation experiments, the phylum-level Acidobacteria primer set (which had failed validation) recovered target-specific phylotypes but also resulted in recovery of a large number of nontarget phylotypes from soil community DNA (35% versus 65%, respectively). In contrast, the Acidobacteria group 6-specific primer set Acido (#6)654–672 plus 907r (which passed validation) displayed an extremely high level of specificity, resulting in 96% of all sequences recovered from the soil community being classified as Acidobacteria group 6 and the remaining 4% not being highly associated with any particular phylum (Table 17.4).

17.3 CONCLUSION

Primer design and validation is key to accurate assessment of bacterial community structure and population response.

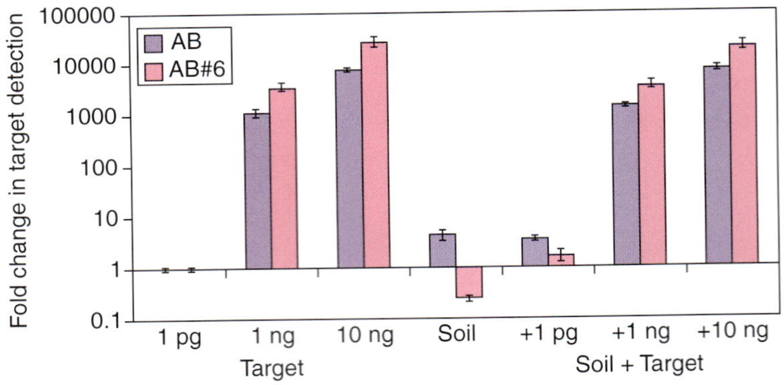

Figure 17.3 Specific detection of Acidobacteria target DNA (Acidobacteria group 6 clone 302.F22 DNA) using the Acido (#6)654–672 plus 907r and 688-706fAB plus 907r primer sets. The given amounts of target DNA were tested alone or after addition to 9 ng of total community DNA isolated from KBS-LTER treatment 1, replicate plot 1. Values indicate the fold change in detection of the target group as a function of the amount of target added. Values for each primer set were normalized to 1 pg of specific target to show fold change in detection. Error bars are one SE of the mean for two rounds of triplicate qPCRs (final $n = 3$). Target, Acidobacteria group 6 clone 302.F22; Soil, 9 ng of total community DNA extracted from soils at the KBS-LTER treatment 1, replicate plot 1.

Table 17.4 Phylogenetic Distribution of Soil Clones Generated Using Primers 688-706fAB and Acido (#6)654–672

Taxonomic Classification[a]	AB	AB & 6
Target[b]	**35% (27)**[c]	**96% (96)**[c]
Nontarget	**65% (50)**	**4% (3)**
Actinobacteria	13% (10)	
Bacteroidetes	3% (2)	
Proteobacteria	9% (7)	
Chloroflexi	1% (1)	
Unclassified	39% (30)	4% (3)

[a] As determined using RDP Classifier with 80% confidence threshhold.
[b] Target for AB is the phylum Acidobacteria, AB & 6 is Acidobacteria Group 6.
[c] Number of sequences in each group is shown in parentheses.

Other studies have highlighted problems in primer design [Baker et al., 2003; Wang and Qian, 2009], but, as demonstrated in this study, empirical testing is indispensable for accurate assessment of efficacy and validation of specificity. This study provides a general strategy for others interested in developing and rigorously testing 16S rRNA gene-based primers for quantitative analysis of specific phyla, classes, or OTUs in environmental samples. Our approach allows accurate validation of primer sets without the need to specifically test each set against all potential targets in a community, an unreasonable feat given the depth of diversity in the studied sample [Morales et al., 2009]. This study also highlights the pitfalls of solely in silico primer design and testing when dealing with complex mixtures of DNA as generally encountered when studying microbial communities.

INTERNET RESOURCES

Primrose (http://www.cardiff.ac.uk/biosi/research/biosoft/)

The KBS LTER Site (http://lter.kbs.msu.edu/)

ARB (http://www.arb-home.de/)

Acknowledgments

The authors would like to thank the *Applied and Environmental Microbiology Journal* for permission to reproduce figures and tables from [Morales and Holben, 2009]. Funding for this project was provided by the U.S. Department of Agriculture National Research Initiative (USDA-CSREES grant 2004–03501). Soil samples for this project were graciously provided by the Kellogg Biological Station Long Term Ecological Research project (KBS-LTER).

REFERENCES

Ashelford KE, Weightman AJ, Fry JC. 2002. PRIMROSE: A computer program for generating and estimating the phylogenetic range of 16S rRNA oligonucleotide probes and primers in conjunction with the RDP-II database. *Nucleic Acids Res*. **30**:3481–3489.

Ashelford KE, Chuzhanova NA, Fry JC, Jones AJ, Weightman AJ. 2005. At least 1 in 20 16S rRNA sequence records currently held in public repositories is estimated to contain substantial anomalies. *Appl. Environ. Microbiol*. **71**:7724–7736.

Ashelford KE, Chuzhanova NA, Fry JC, Jones AJ, Weightman AJ 2006 New screening software shows that most recent large 16S rRNA gene clone libraries contain chimeras. *Appl. Environ. Microbiol*. **72**:5734–5741.

Baker GC, Smith JJ, Cowan DA. 2003. Review and re-analysis of domain-specific 16S primers. *J. Microbiol. Methods* **55**:541–555.

Bathe S, Hausner M. 2006. Design and evaluation of 16S rRNA sequence based oligonucleotideprobes for the detection and quantification of *Comamonas testosteroni* in mixed microbial communities. *BMC Microbiol*. **6**:54.

Buckley DH, Schmidt TM. 2001. Environmental factors influencing the distribution of rRNA from Verrucomicrobia in soil. *FEMS Microbiol. Ecol*. **35**:105–112.

Cole JR, Chai B, Farris RJ, Wang Q, Kulam SA. 2005. The Ribosomal Database Project (RDP-II): sequences and tools for high-throughput rRNA analysis. *Nucleic Acids Res*. **33**:D294–D296.

Crosby LD, Criddle CS. 2003. Understanding bias in microbial community analysis techniques due to rrn operon copy number heterogeneity. *BioTechniques* **34**:790–802.

Doi RH. 1983. In *Recombinant DNA techniques*. Reading, Ma: Addison-Wesley, pp. 162–163.

Fierer N, Jackson JA, Vilgalys R, Jackson RB. 2005. Assessment of soil microbial community structure by use of taxon-specific quantitative PCR assays. *Appl. Environ. Microbiol*. **71**:4117–4120.

Ludwig W, Strunk O, Westram R, Richter L, Meier H, et al. 2004. ARB: A software environment for sequence data. *Nucleic Acids Res*. **32**:1363–1371.

Manz W, Amann R, Ludwig W, Vancanneyt M, Schleifer KH. 1996. Application of a suite of 16S rRNA-specific oligonucleotide probes designed to investigate bacteria of the phylum cytophaga–flavobacter–bacteroides in the natural environment. *Microbiology* **142**:1097–1106.

Meier H, Amann R, Ludwig W, Schleifer KH. 1999. Specific oligonucleotide probes for in situ detection of a major group of gram-positive bacteria with low DNA G+C content. *Syst. Appl. Microbiol*. **22**:186–196.

Morales SE, Holben WE. 2009. Empirical testing of 16S rRNA gene PCR primer pairs reveals variance in target specificity and efficacy not suggested by in silico analysis. *Appl. Environ. Microbiol*. **75**:2677–2683.

Morales SE, Cosart TF, Johnson JV, Holben WE. 2009. Extensive phylogenetic analysis of a soil bacterial community illustrates extreme taxon evenness and the effects of amplicon length, degree of coverage and DNA fractionation on classification and ecological parameters. *Appl. Environ. Microbiol*. **75**:668–675.

Overmann J, Coolen MJL, Tuschak C. 1999. Specific detection of different phylogenetic groups of chemocline bacteria based on PCR and denaturing gradient gel electrophoresis of 16S rRNA gene fragments. *Arch. Microbiol*. **172**:83–94.

Polz MF, Cavanaugh CM. 1998. Bias in template-to-product ratios in multitemplate PCR. *Appl. Environ. Microbiol*. **64**:3724–3730.

Reilly K, Carruthers VR, Attwood GT. 2002. Design and use of 16S ribosomal DNA-directed primers in competitive PCRs to enumerate proteolytic bacteria in the rumen. *Microb. Ecol*. **43**:259–270.

Sipos R, Szekely AJ, Palatinszky M, Revesz S, Marialigeti K, Nikolausz M. 2007. Effect of primer mismatch, annealing temperature and PCR cycle number on 16S rRNA gene-targeting bacterial community analysis. *FEMS Microbiol. Ecol*. **60**:341–350.

Stach JEM, Maldonado LA, Ward AC, Goodfellow M, Bull AT. 2003. New primers for the class Actinobacteria: Application to marine and terrestrial environments. *Environ. Microbiol*. **5**:828–841.

Tringe SG, Hugenholtz P. 2008. A renaissance for the pioneering 16S rRNA gene. *Curr. Opin. Microbiol*. **11**:442–446.

Wang Q, Garrity GM, Tiedje JM, Cole JR. 2007. Naive bayesian classifier for rapid assignment of rRNA sequences into the new bacterial taxonomy. *Appl. Environ. Microbiol*. **73**:5261–5267.

Wang Y, Qian PY. 2009. Conservative fragments in bacterial 16S rRNA genes and primer design for 16S ribosomal DNA amplicons in metagenomic studies. *PLoS One* **4**:e7401.

Wintzingerode F, Gobel UB, Stackebrandt E. 1997. Determination of microbial diversity in environmental samples: Pitfalls of PCR-based rRNA analysis. *FEMS Microbiol. Rev*. **21**:213–229.

Chapter 18

The Impact of Next-Generation Sequencing Technologies on Metagenomics

George M. Weinstock

18.1 INTRODUCTION

The challenge of metagenomics is to sequence an entire microbial community without isolating individual organisms but rather by sequencing the mixture of genomes comprising the community DNA. The term *metagenome* refers to the assemblage of genomes, as if the community were a single organism and all the genomes were its chromosomes, making up its metagenome. Unlike a true genome, with chromosomes in equal stoichiometry (more or less), in a metagenome there can be orders of magnitude differences in stoichiometry: The genomes of the rarer organisms in the community can represent less than 1% of the total, and there may be hundreds of these organisms (genomes). Taken together, this makes the task of the metagenomics approach to describing these communities daunting: One aims to sequence all of the genomes, even the rare ones. Were the organisms in equal stoichiometry, a community of 1000 bacterial species would be of the size of the human genome, which is now routinely sequenced to high coverage. But given that there is such a range of organism abundance, and in fact there may be many more than 1000 species present, the process of metagenomic sequence dwarfs most human/mammalian genome projects.

Most of the classical work in metagenomics focused either on sequencing only the 16S rRNA genes or on selecting for specific clones that expressed a phenotype of interest—for example, antibiotic resistance. In this way, only a very small component of the total metagenome need be sequenced. Moreover, these early studies used Sanger sequencing of clones that were isolated in *E. coli*, and thus they followed familiar protocols for DNA sequencing. This limited the amount of data that could be produced, and thus early studies were limited.

In 2005 the first next-generation sequencing (NGS) instrument, the GS20, was introduced from 454 Life Sciences (now a part of Roche), and this was followed by the Solexa (now Illumina) Genome Analyzer and Applied Biosystems (now Life Technologies) SOLiD instruments. These platforms represented a qualitative change in the amount of data that could be produced at a reduced cost. Consequently, it was now possible to sequence metagenomics samples much more deeply than before, and the details of community structure that could be studied were much enhanced.

This chapter will summarize the pros and cons of these new platforms for metagenomic analysis, as well as introduce the concepts and challenges that they raise for data analysis and computational biology.

18.2 INSTRUMENTATION AND REACTIONS

The area of NGS is in continuous flux with both upgrades on existing instruments and introduction of new instruments. At the time of this writing, the current, the latest versions of instruments are the Roche-454 FLX Titanium and the Illumina GAIIx, although Illumina has recently

Handbook of Molecular Microbial Ecology, Volume I: Metagenomics and Complementary Approaches, First Edition. Edited by Frans J. de Bruijn.

begun marketing the HiSeq2000, the latest in its line of instruments. Life Technologies (Applied Biosystems), with its SOLiD instrument, and Helicos with its instrument round out the main commercially available NGS platforms. However, it is expected that in 2010 there will be instruments available from Pacific Biosciences and Ion Torrent, and possibly other companies. Most experience for metagenomics research has been with the 454 and Illumina platforms, and thus this chapter will focus on those two.

18.2.1 The Roche-454 platform

This system uses "pyrosequencing" in which incorporation of a base into a growing DNA chain is detected by virtue of the pyrophosphate that is released, which is used to drive a chain of other reactions. The pyrophosphate is first used to enzymatically produce ATP, which in turn is used to produce light by luciferase. Thus incorporation of a dNTP into DNA produces light that is detected by a camera.

DNA fragments to be sequenced are first processed by coupling them to beads, one DNA fragment per bead, and then amplified in an oil–water emulsion so that the bead becomes covered with many copies of the same DNA fragment. The beads are deposited in the wells of a picotiter plate, which contains millions of wells, and the sequencing reactions occur by providing each dNTP along with DNA polymerase and the reagents for light generation. Each cycle provides a different dNTP, and a picture is taken of the light produced in each well. These images are processed by software accompanying the instrument to analyze the intensity of light from each well, which is a measure of whether a dNTP was incorporated or, in the case of homopolymer runs, how many molecules of a particular dNTP were incorporated. This information is ultimately converted into a DNA sequence for each well (DNA fragment). In a typical run of the Titanium FLX instrument, one million sequences of 400 bases on average are produced.

18.2.2 The Illumina (Solexa) GAIIx Platform

The Illumina instrument operates by attaching the DNA fragments to be sequenced to a plate by virtue of hybridization of oligonucleotide adaptors that are attached to the fragment to oligonucleotides attached to the plate. Each fragment is then amplified by PCR to create a cluster of identical fragments located near each other on the plate. DNA sequencing is performed by placing the plates in a flow cell that allows sequencing reagents to be plassed over the plate. Key among these are modified dNTPs that have four different fluorescent dyes to distinguish the bases and which also block further incorporation once a modified dNTP has been added to the DNA chain. Following a round of DNA synthesis, a camera records the fluorescent signal from each cluster, and then the dyes are cleaved, freeing the 3′-OH end of the chain and allowing another cycle of reagents to be added and incorporated. The images of the fluorescent signals from the clusters are analyzed to create DNA sequences for each cluster. Typically read lengths of 100 bases are possible, from each end of a DNA fragment, and hundreds of millions of clusters are typically created on a plate. A total yield of 40–50 Gb is typical for an Illumina run.

18.3 FEATURES OF NGS PLATFORMS

Of the four commercially available NGS platforms available (454, Illumina, SOLiD, and Helicos), only 454 and Illumina have been widely used for metagenomic research. The 454 platform differs from the others by its longer read length (400 bases), which makes it more attractive for 16S rRNA and metagenomic whole genome shotgun (mWGS) sequencing. Illumina sequencing, producing 100 base reads, is also valuable for mWGS but has not yet been adapted for 16S rRNA (**Table 18.1**). The other available instruments (SOLiD and Helicos) have shorter read lengths (50 bases), making them less desirable for metagenomic projects. The shorter reads are suitable for projects requiring read mapping—for example, mutation discovery in human genetics or counting transcripts by mapping them to a reference. The instruments expected to be released this year fall into these two categories of read length: Pacific Biosciences is expected to produce longer read lengths and would compete with 454, while Ion Torrent is another short read-length platform not suitable for current metagenomic approaches.

Other characteristics of 454 and Illumina are shown in Table 18.1. The shorter read lengths of Illumina are compensated by the extraordinary amount of data that can be produced, currently about 400 million reads per run, totally 40 Gb of data. These data are produced as read pairs, so 200 million DNA fragments are actually sequenced. These fragments tend to be short (300–500 base pairs), so the two reads are often within a gene. Although Illumina produces two orders of magnitude more data than 454, the cost per run is only slightly higher, and thus the cost per base or per read is significantly reduced.

Enhancements to both of these platforms are being released. The Roche-454 platform is expected to produce longer read lengths, up to 1000 bases, with upgrades that have been under development for the last year or

Table 18.1 Characteristics of NGS Platforms[a]

Platform	Reads per Run	Read Length	Bases per Run	Run Time	Cost/Run (Approximate)	Cost/Read (Approximate)
454 Titanium	1 M	400 bases	400 Mb	0.5 day	$10k	$0.01
Illumina GAIIx	400 M	100 bases	40 Gb	10 days	$25k	$0.00006
3730 (Sanger)	96	750	72 Kb	2 hours	$100	$1.00

[a]The two most commonly used NGS platforms for metagenomics (454 and Illumina) are compared to the traditional sequencing approach: Sanger sequencing using the Applied Biosystems 3730 capillary electrophoresis sequencing instrument. Most values are based on results at the Genome Center at Washington University.

so. A new Illumina instrument, the HiSeq2000, has been released in 2010 that produces twice as much data per run (by doubling the area of the plate/flow cell that is used) and has two flow cells, allowing a total of fourfold more data per run. This instrument also reduces the length of the run by 20%.

The quality of data that is produced is likewise a changing parameter because both hardware and software advances over the years have improved sequence accuracy. At the present time, sequence accuracy of a single read hovers around 99% on both platforms, with higher accuracy early in the read and most errors occurring in the later base calls. Since high sequence coverage is possible with the large data production per run, one often sequences the same region multiple times, and this allows more accurate base calling because most errors are not systematic. When high accuracy is required, it is possible to trim reads aggressively to obtain shorter but more accurate sequences, since most errors occur near the last part of the DNA to be sequenced. When this is not an issue, longer reads can be used—up to 150 bases at present for Illumina—with more errors toward the end of the read (see also Chapter 19, Vol. I).

18.4 TYPES OF NGS APPLICATIONS

As noted earlier, the principal approaches to characterizing metagenomes and describing microbial community structure focus on sequencing 16S rRNA genes or on mWGS. The strategies and limitations for using NGS technology in each of these situations are quite different from the traditional Sanger sequencing approaches.

18.4.1 16S rRNA Sequencing

The 16S rRNA gene is like a bar code for each species, and sequences of 16S genes can uniquely identify which species are present. Often, the abundance of each species is also inferred from the number of sequences for a particular 16S gene. The 16S gene is about 1500 base pairs and has nine variable regions that contribute most of the information to distinguishing the species. The classic approach is to amplify full-length 16S genes with (degenerate) primers, clone them into *E. coli*, and then sequence the clones to obtain the full 16S gene. This method has been extremely valuable and is the only method to date that produces a full-length 16S gene sequence, which allows the highest definition of the source species since all nine variable regions are sequenced.

Despite this significant advantage, there are limitations. First, most studies sequence hundreds of clones (16S genes) per sample, mainly constrained by the need to clone, pick colonies, and make template DNA for sequencing. In addition, the cost limits the number of clones that can be sequenced. Thus the depth of sampling is limited. Other issues are possible biases introduced by the use of degenerate primers as well as the need to clone into *E. coli*, where the cloned 16S genes may be expressed and introduce toxic phenotypes. All of these can lead to undersampling of some organisms.

NGS offers a very different approach to 16S data production. Since none of the platforms make it possible to sequence a full-length 16S gene, the emphasis has been on identifying which variable regions are more useful in species identification. Moreover, since 454 offers the longest read lengths, this has been the platform of choice for 16S sequencing. There are three aspects to the use of NGS that are important: the goal of the 16S sequencing, variable region choice, and cost reduction.

Two main goals in obtaining 16S sequences are defining communities and comparing communities. The first is to obtain a complete census of the organisms present in a community. With this complete description, one can either begin to model community phenotypes based on the known properties of organisms or compare samples to determine their similarity. This latter application is of use in assessing how much variability there is between people in a body site's microbiome, whether people fall into different microbiome types, whether disease and healthy samples group according to phenotypes, and so on. But comparing samples can also be done without an absolute description of the species present: All that is required is that the data be sufficient to group similar samples and distinguish them from less similar samples.

NGS allows sample comparisons to be performed accurately, but is limited in producing a detailed inventory of which species are present. A number of studies have been performed to assess how best to sample the 16S sequence with 454 sequencing. Some focus on a single variable region, and a number of regions have been studied. In the NIH HMP, three different regions within 16S were studied: one encompassing variable regions V1 through V3, one with V3 through V5, and one with V5 through V9. For the purposes of grouping samples, all methods are successful. There are subtle reasons for choosing one region over another, and these reasons may be specific to the system under study; that is, there is no general rule.

At the same time, none of the methods can produce an inventory of the organisms present with high accuracy. Because only a part of the 16S gene is used, often organism identification cannot be made below the genus level and at times it cannot even reach that low a taxonomic classification. Moreover, the somewhat high error rates in NGS data mean that the rarer species (lower coverage) are difficult to distinguish from sequencing errors. Since it is not necessary to clone genes in *E. coli* with NGS, biological biases from this step are removed. But in general, the most useful applications of NGS for 16S analysis are when samples must be compared rather than the absolute community structure determined.

The overall experimental design for NGS 16S analysis is quite different. As mentioned earlier, typical Sanger 16S analyses obtained hundreds of sequences per sample. One desires deeper sequencing to reliably sample rarer organisms, so with 454 16S analysis thousands of sequences per sample are obtained. Thus a typical 454 run will sequence 200 samples, to an average depth of 5000 sequences each. This is achieved by using primers with molecular bar code sequences. In the HMP, each primer pair (for amplifying V1–V3, V3–V5, and V6–V9) exists in 96 versions where each version has a sequence bar code. This allows 96 samples to be independently amplified, then mixed and sequenced together on half of a picotiter plate; two such sets of 96 samples can be sequenced in a single run. The mixture of sequences are then computationally deconvoluted into individual samples based on their bar code. Thus with NGS, 16S sequencing can be performed at a deeper coverage (5000 sequences per sample) for a reduced cost ($0.01 per read for 454 vs. $1.00 per read for Sanger). Although one sacrifices accuracy in absolute species identification with 454, the reduced cost and deeper sampling—allowing more extensive sample comparison—more than compensate.

18.4.2 Metagenomic Whole Genome Shotgun Sequencing

While 16S rRNA sequencing for species identification has been a standard in metagenomics, it does not allow a full picture of a microbial community. To achieve this, one needs a gene list as well as knowledge of the organisms. For example, *E. coli* K12 and *E. coli* O157:H7 both classify as *E. coli* by 16S rRNA sequencing, yet they differ by hundreds of genes, the latter being much more virulent than the former. One wants to provide this type of information on community membership for a full picture.

As mentioned above, comprehensive description of the community gene pool is a sequencing task on the scale of a human genome sequencing project, requiring tens of gigabases of sequence for each sample. Initial analyses of communities by shotgun sequencing used either Sanger or 454 sequencing, but neither of these platforms can produce data on the scale or at a cost that is appropriate. Recently a study [Qin et al., 2010] of the human gut microbiome sampled 124 individuals with Illumina sequencing, producing an average of 4 Gb of sequence per subject. In the NIH HMP, the sequencing of close to 600 samples is underway, aiming for 10 Gb of sequence per sample. Thus it appears that mWGS is moving into a new era of deeper descriptions of the gene content of communities.

This new era for mWGS has a number of challenges. Since many organisms have not previously been sequenced, shotgun reads are not readily attributable to a species. This requires an expansion of reference genome sequencing (see below). Likewise, limitations in the utility of shorter read platforms like Illumina (100 base reads) have not turned out to be a significant factor. Reads of this length seem sufficient for performing Blast searches or comparisons to reference genomes. Another important application of mWGS, for virus discovery, likewise does not appear to be limited as determined in studies by the NIH HMP (unpublished results). One area that does appear to be challenged by read length is assembly of genomes from mWGS data. Since many genomes are from unknown organisms, some of which are not cultured and may only be sampled as part of a mWGS data set, assembling these genomes is an important task. This is an area of ongoing development that may require longer reads to get the best assemblies.

18.4.3 Reference Genome Sequencing

One of the steps in analysis of mWGS data is comparison of reads to reference genome sequences, thus allowing the source species of community sequences to be identified. There are currently about 1000 bacterial reference genomes in GenBank, yet in a typical mWGS dataset only about half the reads align to these sequences (unpublished data). Thus many more reference genomes are needed. With NGS, the cost of sequencing a bacterial genome has dropped remarkably. Sequencing genomes on the 454 platform typically requires at least 15× coverage, so as many as 10 genomes can be sequenced per run, putting the cost around $1000/genome. On the Illumina platform, higher coverage is required, often 75×, so over 150 can be sequenced per run, bringing the cost to $200 or less. These theoretical limits are just starting to be achieved, so NGS is bringing the prospect of a greatly expanded catalog of reference genomes to reality. As for 16S-based analysis, sequencing of this many genomes/run requires bar coding, pooling, and deconvolution of the data before assembly. But this has been achieved (e.g., see Cronn et al. [2008] for an example of genome pooling), and it would seem that the future limitation is in obtaining enough strains to sequence.

18.5 COMPUTATIONAL CHALLENGES

Metagenomic communities are complex collections of genomes, and it is becoming clear that NGS provides the data production capabilities needed to adequately sample the complexity. This leads to new computational challenges that are only beginning to be defined, but have not yet been met. Inexpensive production of 16S sequences at over 10-fold the previous depth means that hundreds of specimens can be sequenced in a single 454 run, leading to millions of sequences in a typical study. For example, the NIH HMP seeks to analyze 12,000 specimens at an average depth of 5000 16S reads, amounting to 60 million sequences. More dramatic is the impact on mWGS analyses. As described earlier, the analysis of gut microbiome with Illumina produced 4 Gb of data for each of 124 subjects, a total of nearly 0.6 Tb of sequence. The NIH HMP aims to do about 10 times this amount of mWGS data production, 6 Tb or 60 billion reads. To analyze these data, one wants to align it to reference genomes (currently 1000), perform Blastx versus GenBank and the KEGG ortholog database, and perform other equally demanding computational exercises. With current methods it will take at least 50 times as long to run these primary analyses as to produce the data. The solution to this computational bottleneck is not simply bigger computer clusters. There will need to be improvements in software as well.

Finally, the challenge of making these data available to the community is also formidable. Will the traditional model of uploading datasets to GenBank or other public databases suffice, given the size of the datasets to upload or download by end users? New research in the use of cloud computing or other distributed methods is testing whether these are adequate solutions or additional technology and strategies are needed.

18.6 CONCLUSIONS

Metagenomic analyses have been limited for many years by sequencing technology. Up until now, only small-scale studies of complex communities have been possible. NGS offers expanded data production capabilities at decreasing costs and is well-matched with metagenomic projects. Considerable progress has been made, and further sequencing technology improvements will keep this momentum going. The new bottleneck that is appearing is in data management and analysis, and this is leading to new research areas designed to provide the computational support for NGS applications.

INTERNET RESOURCES

Many of the protocols the author is familiar with have been developed under the NIH Roadmap Human Microbiome Project (HMP). These are not yet published, but most of this information is publicly available. The reader is referred to the NIH HMP web site at http://nihroadmap.nih.gov/hmp/ and the web site of the Data Analysis and Coordination Center for the HMP at http://www.hmpdacc.org/ where both documents and links to relevant sources of information can be found. In addition, information about the DNA sequencing platforms can be found at the manufacturers' web sites.

Acknowledgments

I would like to thank the extremely talented team at the Genome Center at Washington University for much of the progress in applying NGS to metagenomic problems. In particular, it is a pleasure to acknowledge Dr. Erica Sodergren, leader of the HMP at the Genome Center. Many of the ideas presented here have also been part of the NIH HMP consortium effort, and my thanks go to colleagues at the participating institutions: the Human Genome Sequencing Center at Baylor College of Medicine, the Broad Institute, the J. Craig Venter Institute, and the Data Analysis and Coordination Center at the University of Maryland. I also express my thanks to the NIH for their generous funding.

REFERENCES

Cronn R, Liston A, Parks M, Gernandt DS, Shen R, Mockler T. 2008. Multiplex sequencing of plant chloroplast genomes using Solexa sequencing-by-synthesis technology. *Nucleic Acids Res*. **36**(19):e122.

Qin J, Li R, Raes J, Arumugam M, Burgdorf KS, Manichanh C, Nielsen T, Pons N, Levenez F, Yamada T, et al. 2010. A human gut microbial gene catalogue established by metagenomic sequencing. *Nature* **464**(7285):59–65.

Chapter 19

Accuracy and Quality of Massively Parallel DNA Pyrosequencing

Susan M. Huse and David B. Mark Welch

19.1 INTRODUCTION

Only a very small fraction of the microbial biota is amenable to cultivation. Sequencing of environmental DNA provides a means to assess the diversity and functional repertoire of microbial communities without requiring the cultivation of microbes in the laboratory. However, thorough sampling of metagenomes and the detection of low-abundance taxa require surveys that are many orders of magnitude larger than is financially practical using cloned DNA templates and capillary-based Sanger sequencing. The advent of massively parallel pyrosequencing using platforms such as the Roche Genome Sequencer system developed by 454 Life Sciences [Ronaghi et al., 1998; Margulies et al., 2005; see also Chapter 18, Vol. I] are profoundly changing the field of molecular microbial ecology. It is now possible to generate hundreds of thousands of greater than 400-nucleotide DNA sequence reads in a few hours without requiring the preparation of sequence templates by conventional cloning.

New technology brings new concerns about accuracy and biases that could affect interpretations of environmental sequence data. Sanger (dideoxy) sequencing using dye terminators and capillary separation [Smith et al., 1986], though widely used for only the past 20 years, is considered proven technology with well-characterized error profiles [Ewing and Green, 1998; Ewing et al., 1998].

Pyrosequencing employs a fundamentally different approach, measuring the incorporation of each nucleotide by the chemiluminescence accompanying the release of inorganic pyrophosphate [Margulies et al., 2005]. The protocol for the Roche GS system involves mixing single-stranded template with DNA capture microbeads under conditions in which most beads contain only one template. The beads are encased in an oil emulsion, and the template is amplified to many millions of copies (emPCR). The beads are then distributed on a microtiter plate with wells sized to hold a single bead and additional reagents, including polymerase, luciferase, and ATP sulfurylase [see Chapter 18, Vol. I].

Once in the GS FLX, microfluidics cycle each of the four nucleotide triphosphates through each well, and incorporation of a nucleotide releases pyrophosphate, the substrate for a luminescence reaction. The intensity of each flow of a nucleotide in each well is captured by a cooled CCD camera, producing a flowgram, analogous to a dye-terminator chromatogram, which can be interpreted to produce the order of nucleotides in the template. Each flowgram value corresponds not to the incorporation of a single nucleotide but to the total luminance of that flow, which may correspond to the incorporation of multiple identical nucleotides (a homopolymer). Because the camera records the total intensity in a well for each flow, random polymorphism introduced by emPCR or misincorporation of nucleotides during sequencing is likely to be outweighed by the correct signal.

Thus substitution errors are predicted to be less likely than with dye-terminator Sanger sequencing. However, when a homopolymer is incorporated, the resulting luminance may not be properly interpreted as the integer corresponding to the length of the homopolymer, resulting in an over- or underestimation of homopolymer length. Insufficient flushing between flows can cause single base insertions (carry forward events; usually near but not adjacent to homopolymers). Or, insufficient availability of nucleotides during a flow can cause

Handbook of Molecular Microbial Ecology, Volume I: Metagenomics and Complementary Approaches, First Edition. Edited by Frans J. de Bruijn.

incomplete extension within homopolymers. The GS software makes some corrections for carry forward and incomplete extensions (CAFIE) and truncates the reported read length until fewer than 3% of the remaining flowgram values are of intermediate value. The software identifies ambiguous flow cycles in which no flowgram value was greater than 0.5 and reports these positions as "N".

The original report of pyrosequencing using the 454/Roche technology reported an error rate of nearly 3% per base [Margulies et al., 2005]. The high error rate was overcome by assembling a consensus sequence from many overlapping reads. The high coverage rate provided a consensus with very few errors (~0.07%) [Wicker et al., 2006]. However, this strategy is not viable for metagenomic or tag sequencing surveys, where each read is interpreted to represent a unique gene copy from the environment. Our initial experience with amplicon reads produced by the Roche GS 20 for tag sequencing surveys [Sogin et al., 2006] suggested that the error rate per base per read was substantially lower than 3%, perhaps due to improvements in the protocol, hardware, and software since the publication of the original method. To quantify the types and frequencies of errors present in the pyrosequencing reads generated from tag sequencing amplicons, we sequenced and analyzed more than 340,000 GS 20 reads from an amplicon library of 43 known copies of the V6 hypervariable region of the 16S ribosomal subunit gene [Huse et al., 2007]. We found that the per-base error rate was less than 0.5% and that this could be reduced to 0.2–0.3% through elimination of a small number of sequences that contained a disproportionate number of errors. Here we report our analysis of the error rate of amplicon pyrosequencing on the Roche GS FLX system, using known sequences of the V6 region from *Escherichia coli* and *Staphylococcus epidermidis*.

19.2 METHODS

19.2.1 Generation and Sequencing of V6 Amplicon Libraries

We cloned full-length 16S rRNA coding regions from *Escherichia coli* K-12 ATCC10798 and *Staphylococcus epidermidis* ATCC14990 into plasmids as reported elsewhere [Huse et al., 2010]. We confirmed the sequence of the V6 region of each clone and chose one *E. coli* and one *S. epidermidis* plasmid clone with V6 region lengths of 60 and 62 bases, respectively. We treated each plasmid DNA preparation with plasmid-safe DNase (Epicentre, Madison, WI) to remove *E. coli* genomic DNA and confirmed that each plasmid produced an amplification product of the expected size with primers flanking the V6 region. We then amplified dilutions of each plasmid with primers that flank the V6 region. The forward primers consisted of the Roche GS 20/FLX A sequence, a five-base key for post-sequencing bioinformatics identification [Sogin et al., 2006], and one of five 967F sequences: 5′ CAACGCGAAGAACCTTACC, 5′ CTAACCGANGAACCTYACC, 5′ CNACGCGAAGAACCTTANC, 5′ CAACG-CGAAAAACCTTACC, 5′ CAACGCGCAGAACCTTACC, or ATACGCGARGAACCTTACC. The reverse primers consisted of the Roche GS20/FLX B sequence and one of four 1064R sequences: 5′ CGACAGCCATGCANCACCT, 5′ CGACGGCCATGCANCACCT, 5′ CGACGACCATGCANCACCT, or 5′ CGACAACCATGCANCACCT. To minimize PCR error, we used Platinum Taq polymerase, an excess of dNTPs, and only 30 cycles, as previously described [Huse et al., 2007]. We sequenced in the forward direction only on a Roche GS FLX using standard Roche protocols and supplies and the Roche GS FLX amplicon base-calling pipeline.

19.2.2 Error Rate Calculations

We compared each read to a database of V6 sequences using GAST [Huse et al., 2008]. Reads that had a best match to a nontemplate sequence that was at least 10% better than the best match to a template sequence were removed as contaminants. We considered reads that either did not have a match or did not have a match over at least 80% of their length to be non-V6 DNA or reads with gross errors. Based on our knowledge of the expected sequences, we identified and removed obvious chimeras manually. Some additional chimeras that contain sequencing errors likely remain. We then compared each remaining read to its template sequence using the needledist module of ESPRIT (with options -g 5.75 -e 2.75 -x) [Sun et al., 2009; see also Chapter 53, Vol. I]. The error rate was calculated as the number of insertions and deletions (of any length) plus the number of individual mismatches divided by the length of the template sequence. To evaluate quality score for deletions, we used the quality score at the position where the deleted base would have been and which now also corresponds to the base after the deletion.

19.3 RESULTS

After removing contaminating reads, we obtained 34,790 reads of the *E. coli* template and 44,011 reads of the *S. epidermidis* template. Combined, these two samples represent 8,104,280 sequenced nucleotides. The data included 37,656 errors for an overall error rate of 0.46%. As shown in **Table 19.1**, the number of insertions and deletions were similar and together accounted for 58.5% of the errors; ambiguous base calls (Ns) accounted for 5% of the errors, and 36.5%

Table 19.1 Contribution of Error Types to the Overall Error Rate and Associated Quality Scores

Error Type	Percent of Errors	Per Base Error Rate	Percent of Quality Scores <25	Percent of Quality Scores <30	Percent of Quality Scores <35
N	5.0%	0.00025	100%	100%	100%
Insertions	29.9%	0.00152	66%	79%	90%
Deletions	28.6%	0.00145	53%	64%	73%
Mismatches	36.5%	0.00185	7%	10%	18%

Table 19.2 Rates for Base Mismatches

Mismatch	Percent of Mismatches
A → G	35%
T → C	33%
G → A	9%
C → T	8%
G → T	4%
A → T	3%
T → A	3%
T → G	2%
A → C	1%
C → A	1%
G → C	1%
C → G	1%

were due to mismatches. Of the 12 possible mismatches, transitions accounted for 84% of the mismatches outside of the primer regions, with these heavily biased toward A → G and T → C (see **Table 19.2**). A disproportionate number of mismatches (8% of the total) occurred at the second to last position within the 967F primer sequence, almost three times more than any other position. Because this polymorphism may be caused by an error in primer synthesis, we have not included it in the comparison of base mismatches.

We found no correlation between error rate and distance along the read. Measured with or without the inclusion of the primer sequences, the slope of the cumulative percent of errors over nucleotide position is approximately zero. Instead, positions with high error rates were functions of the sequence bases and were clearly dominated by homopolymer errors. Excluding the primer regions, the most common errors involved a homopolymer run of 5 Gs (G^5) in *S. epidermidis*: an overextension to include a sixth (8.5% of nonprimer errors) or incomplete extension to 4 or fewer Gs (3.8%), a carry forward insertion of a G in advance of the G^5 homopolymer (2.0%), the insertion of a G at the first downstream flow of G after the G^5 homopolymer (1.0%), and the insertion of an extra G at the next instance of a G in the sequence, corresponding to the second flow of Gs after the homopolymer (0.6%). Other common homopolymer-associated errors included an incomplete extension of a C^4 (3.1%) or its overextension (2.0%), and overextension of a T^4 (2.9%).

The errors were not evenly distributed among the reads, with 43% of all errors occurring in 4% of the reads. **(Fig 19.1)**. **(Figure 19.2)** shows that as the number of Ns in a read increases so do the number of other types of errors. Despite the fact that reads with Ns are only 2% of the reads, removing these reads removes 18% of all errors and reduces the error rate from 0.46% to 0.39%,

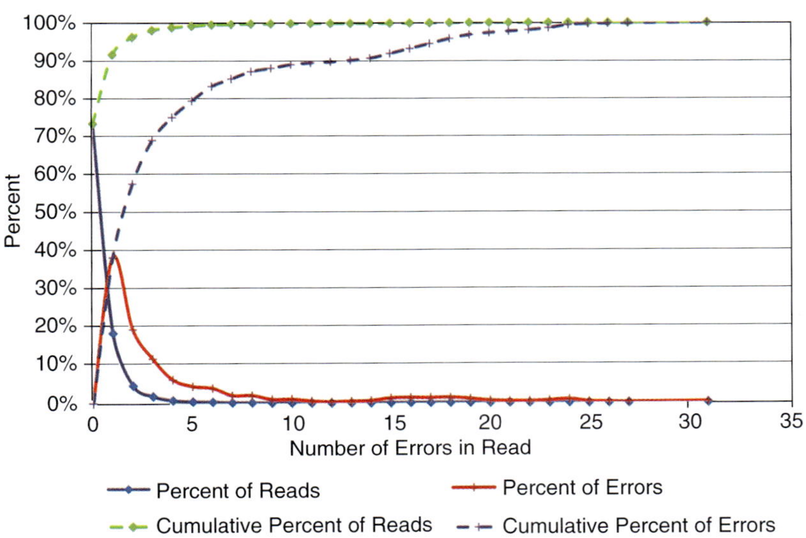

Figure 19.1 Distribution of errors in 454 pyrosequencing reads. The graph shows the proportion of reads that vary from their template by different numbers of errors. Low-quality reads contribute disproportionately to the overall error rate. While 73% of reads contain no errors and 96% of reads contain no more than two errors, 43% of all errors are contributed by the 4% of reads with more than 2 errors.

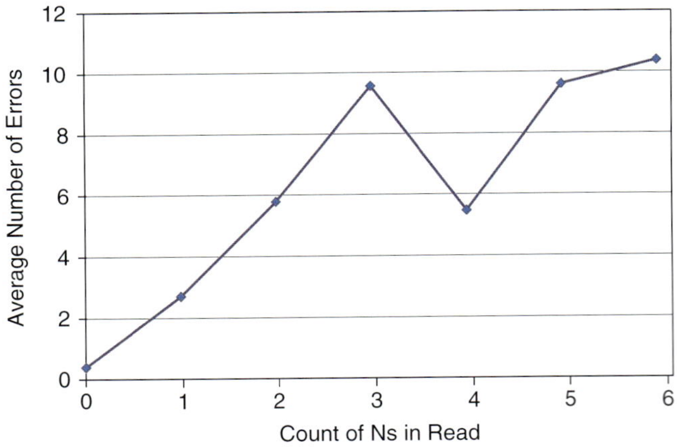

Figure 19.2 Ambiguous bases are associated with increased error rates. As the number of Ns increases within sequence reads, the number of other types of errors also increases.

as compared to the average error rate of reads that contain Ns, which is 5.3%.

Reads that are shorter or longer than expected also account for a disproportionate number of errors **(Fig. 19.3)**. Reads whose length is 10% higher or lower than expected comprise 9% of the total reads, but account for 46% of the errors. The overall per-base error rate of the reads falling outside the expected size range is 2.3%; the error rate of reads within the expected size range is 0.27%.

Another common means used to remove low-quality reads a priori is to remove any read that does not have a perfect match to the proximal amplification primer. The 9% of reads that did not have an exact match to the proximal primer included 38% of the errors and had an error rate of 2.1%. The remaining 91% of reads had an error rate of 0.31%.

The GS FLX image processing software assigns a quality score to every base call. The average quality score of a read has been used as a criterion by which to remove reads that are assumed to have errors [Kunin et al., 2010; Huse et al., 2010]. As can be seen in **(Fig 19.4)**, reads with an average quality score below 25 account for only 0.5% of the total reads, but their averaged error rate is 3.8%. Reads with a quality score below 30 account for 3% of the reads and have an average error rate of 2.8%, while those below 35 account for 15% of reads and have an average error rate of 1.7%. Reads with an average error rate of 30 or greater have an error rate of 0.38%.

We found that 66% of insertions, 53% of deletions, and 7% of mismatches had a quality score less than 25; 79% of insertions, 64% of deletions, and 10% of mismatches had a quality score less than 30; and 90% of insertions, 73% of deletions, and 18% of mismatches had a quality score less than 35 (Table 19.2). Since a deleted base does not directly have a quality score, we used the quality score at the position where the nucleotide would have been. Clearly the quality scoring of the GS FLX is best suited to identifying insertions and deletions and in fact does not provide meaningful predictions about mismatches.

By removing all reads with ambiguous bases (Ns) with mismatches to the proximal primer, which are shorter than expected or have an average quality score less than 30, we removed 12% of the reads and 57% of all errors **(Table 19.3)**. The error rate of the remaining reads was 0.23%. The fraction of reads that contained no errors remained 82%. Raising the average quality score cutoff to 35 reduced the error rate to 0.17%, but at the expense of removing 15% of the reads.

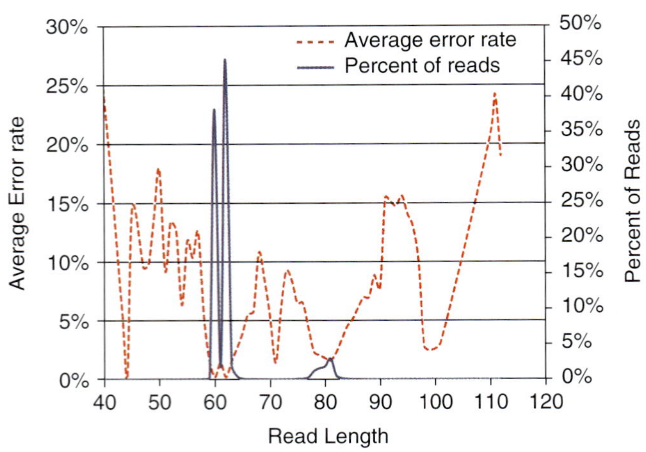

Figure 19.3 Expected length and error rates. As reads diverge from their expected length, their error rates increase. The GS FLX will trim back reads to remove low-quality nucleotides at the distal end, but the remaining nucleotides show a still elevated error rate.

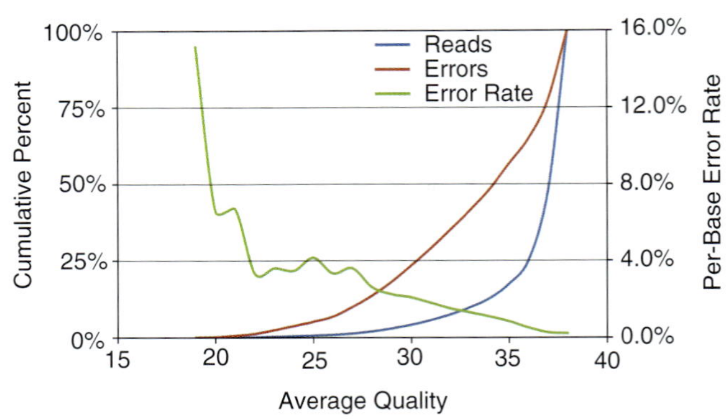

Figure 19.4 Average quality scores and error rates. Although quality scores are biased toward the identification of insertions and deletions rather than mismatches, as the quality score averaged over the read increases, the overall error rate decreases.

Table 19.3 Effect of Quality Filtering Methods

	Reads Retained	Errors Removed	Error Rate, per Base	Reads with No Errors	Reads with At Most 1 Error
All reads	100%	0%	0.46%	73%	91%
Reads with no Ns	98%	18%	0.39%	74%	93%
Reads with length ≥54 nt and ≤68 nt	91%	47%	0.27%	81%	96%
Reads with length ≥50 nt	100%	2%	0.46%	73%	92%
Reads with exact match to proximal primer	91%	38%	0.31%	80%	95%
Reads with average quality ≥30	97%	21%	0.38%	75%	93%
Reads with average quality ≥35	85%	53%	0.26%	80%	96%
Reads with no Ns and average quality ≥30	95%	34%	0.32%	76%	94%
Reads with no Ns, average quality ≥30 and length ≥50	95%	66%	0.32%	76%	94%
Reads with exact primer, length ≥50 nt, no Ns, average quality ≥30	88%	57%	0.23%	82%	97%
Reads with exact primer, length ≥50 nt, no Ns, average quality ≥35	80%	71%	0.17%	85%	98%

19.4 DISCUSSION

The massively parallel pyrosequencing of emulsification PCR-based templates holds great promise for revolutionizing environmental metagenomic and tag sequence surveys. However, even after several years of use, concern remains over the potentially high degree of error for these applications [Reeder and Knight, 2009; Kunin et al., 2010]. Consensus-based projects, such as single-cell genomics, would also benefit from a lower error rate, because fewer sequences would be required to build a reliable consensus. The original description of the 454 Life Sciences system reported an error rate for shotgun library reads of 3–4 bases per hundred nucleotide positions [Margulies et al., 2005]. In addition to the experimental data, that publication presented test fragment data that had much lower error rates despite the fact that the test fragments had extensive homopolymers and were designed to challenge the analysis software. These results were corroborated independently by Huse et al. [2007]. This discrepancy indicates that the basic method of pyrosequencing, luminescence detection, and flow intensity resolution is sound and suggests that the higher error with experimental data may come from the experimental manipulation of the sequences prior to pyrosequencing. Margulies et al. [2005] suggested that this may be due to multiple sequences binding to an individual bead prior to the emPCR amplification, resulting in a heterogeneous amplification pool. The Roche GS analysis software is designed to eliminate flowgrams from beads that contain two highly divergent DNA templates. However, if a bead contains two similar but nonidentical sequences the software may not eliminate the flowgram. Metagenomic and tag survey data may have large numbers of highly similar sequences. It is therefore necessary to determine methods of producing the highest-quality sequence data possible.

We conducted an in-depth analysis of errors in GS FLX reads by sequencing amplicon libraries made from known templates. We found an error rate (incorrect bases/total number of expected nucleotides) of 0.46% (Table 19.3), considerably lower than that reported by Margulies et al. [2005], but similar to the rate we found for the GS20 [Huse et al., 2007]. As with our analysis of GS 20 reads, we found that the errors in GS FLX reads were not randomly distributed across all reads: 73% of the reads contain no errors, while 2% of the reads contain four or more errors each and together account for 43% of the total errors (Fig. 19.1).

The error rate of reads containing Ns (1.6% of all reads) was more than 10 times that of reads without Ns (5.3% vs. 0.39%) (Table 19.3). The presence of even a single ambiguous base (N) is an effective indicator that a sequence is of low quality, and the elimination of reads based on this criterion is universally applicable to any project and is bioinformatically trivial. Reads more than 10% longer or shorter than expected were also poor quality, with an even higher error rate of 2.3%, while those within 10% of the expected length had an error rate of 0.27%. While it is not always possible to identify reads that are longer than expected, it is generally possible to assign a lower expectation to sequence length. For the V6 region of the bacterial rRNA gene, 50 nt is the minimum found in the SILVA database of rRNA sequences [Pruesse et al., 2007; see also Chapter 45, Vol. I]. Selecting a minimum read length of 50 nt did not remove many reads from these two samples (0.11% of reads), but those reads that were removed had an average error rate of 13% and were clearly aberrant.

Removing reads that lack a perfect match to the proximal primer also results in a lower error rate in the remaining sequences. However, the relatively high degree of reproducible mismatches in the primer region suggests that some of this comes from errors in primer synthesis that have not been previously noticed in smaller-scale sequencing efforts (see Chapter 17, Vol. I). One possibility is to record these high-frequency mismatches and treat them as acceptable primer variants. Given the small fraction of reads eliminated, researchers may want to maintain the conservative criterion of a perfect match to the primer. Shotgun metagenomic projects generally lack a known sequence at either end of a read, and therefore they cannot use this filter.

Although we found no evidence for an increased rate of error with increasing read length, we note that our amplicons were only 100 bases in length, considerably shorter than the mean read length possible with the GS FLX and shorter than shotgun metagenomic reads or the amplicons produced by tag surveys using multiple hypervariable regions. Campbell [2008] reported a significant increase in errors beyond position 200 using the GS FLX.

We found that quality scores of individual bases were a good predictor of insertions and deletions caused by inaccurate interpretation of homopolymer length but were a poor predictor of mismatches. The vast majority of mismatches had quality scores of 35–40. This may in part reflect the presence of error introduced during the amplification of the template, but the observed error rate of 3×10^{-4} is several orders of magnitude higher than expected for Platinum Taq polymerase. Despite this bias in quality scoring, culling reads by average quality score did reduce the error rate of the remaining sequences. Removing reads with an average quality score below 35 resulted in an error rate in the remaining reads of 0.26% at a cost of 15% of the total reads. While removing reads with an average quality score below 30 did not remove as many reads (only 3%), it reduced the error rate to 0.39%. A trade-off therefore exists when removing reads with an average quality between 30 and 35. Although the error rate is improved by culling more reads, 43% of the reads that would be lost (those with an average quality score between 30 and 35) have no errors whatsoever.

The approach of removing reads with Ns, are shorter than expected, have inexact matches to the proximal primer, or a low average quality score is an effective means for removing low-quality reads and improving the error rate of the remaining data. For our *E. coli* and *S. epidermidis* reads we removed 12% of the reads, and 57% of the errors using this approach. Of the remaining data, 82% have no errors, 15% have a single error, 2% have two errors, and only 1% of the remaining reads include 3 or more errors, for a final error rate of 0.23%. Thus for the V6 rRNA amplicon sequences, 99% of all reads (82% of all reads with errors) are less than 3% different that their template sequence, and efficient clustering of sequences into operational taxonomic units at the 3% difference level is an accurate means of measuring diversity in a metagenomic sample. We note that the average quality score, absence of Ns, or sequence length range criteria remove an overlapping set of reads; we also note that for some datasets, sequence quality may be the most efficient criterion for improving accuracy. The similarity in the types of error in the GS 20 and GS FLX systems suggest that they are common to the technology employed by the Roche Genome Sequencing system and will be similar in the newer GS Titanium. Thus, massively parallel pyrosequencing data in conjunction with careful quality filtering and trimming provides orders of magnitude more sequence reads, albeit shorter reads, with base-calling as good as or better than traditional Sanger sequencing, at a lower per-base cost.

Acknowledgments

The authors would like to thank their coauthors on Huse et al. [2007]: Julie Huber, Hilary Morrison, and especially Mitchell Sogin for his support and guidance. We also thank Rich Fox for computer systems administration of the W.M. Keck Computational Facility at the Bay Paul Center; and Bette Hecox-Lea, Ekaterina Andreishcheva, and Jessica Mark Welch for laboratory support. This work was supported by a subcontract to Mitchell L. Sogin from the NIH Human Microbiome 1UH2DK083993-01 awarded to the University of Michigan, as well as by foundation awards from the W.M. Keck Foundation to MLS and the Alfred P. Sloan Foundation to MLS and DMW.

REFERENCES

Campbell PJ, Pleasance ED, Stephens PJ, Dicks E, Rance R, Goodhead I, et al. 2008. Subclonal phylogenetic structures in cancer revealed by ultra-deep sequencing. *Proc. Natl. Acad. Sci. USA* **105**:13081–13086.

Ewing B, Green P. 1998. Base-calling of automated sequencer traces using Phred. II. Error probabilities. *Genome Res*. **8**:186–194.

Ewing B, Hillier L, Wendl MC, Green P 1998. Base-calling of automated sequencer traces using Phred I. Accuracy assessment. *Genome Res*. **8**:175–185.

Huse S, Huber J, Morrison H, Sogin M, Welch D 2007. Accuracy and quality of massively parallel DNA pyrosequencing. *Genome Biol*. **8**:R143.

Huse SM, Dethlefsen L, Huber JA, Mark Welch D, Relman DA, Sogin, ML. 2008. Exploring microbial diversity and taxonomy using SSU rRNA hypervariable tag sequencing. *PLoS Genet*. **4**:e1000255.

Huse SM, Mark Welch D, Morrison HG, Sogin, ML. 2010. Ironing out the wrinkles in the rare biosphere. *Environ. Microbiol*. **12**:1889–1898.

Kunin V, Engelbrektson A, Ochman H, Hugenholtz, P. 2010. Wrinkles in the rare biosphere: Pyrosequencing errors lead to artificial inflation of diversity estimates. *Environ. Microbiol*. **12**:118–123.

Margulies M, Egholm M, Altman WE, Attiya S, Bader JS, Bemben LA, et al. 2005. Genome sequencing in microfabricated high-density picolitre reactors. *Nature* **437**:376–380.

Pruesse E, Quast C, Knittel K, Fuchs BM, Ludwig W, Peplies J, Glockner FO. 2007. SILVA: A comprehensive online resource for quality checked and aligned ribosomal RNA sequence data compatible with ARB. *Nucleic Acids Res*. **35**: 7188–7196.

Reeder J, Knight R. 2009. The 'rare biosphere': A reality check. *Nature Methods* **6**:636–637.

Ronaghi M, Uhlén M, Nyrén P. 1998. DNA Sequencing: A sequencing method based on real-time pyrophosphate. *Science* **281**: 363–365.

Smith LM, Sanders JZ, Kaiser RJ, Hughes P, Dodd C, Connell CR, et al. 1986. Fluorescence detection in automated DNA sequence analysis. *Nature* **321**:674–679.

Sogin ML, Morrison HG, Huber JA, Welch DM, Huse SM, Neal PR, et al. 2006. Microbial diversity in the deep sea and the underexplored "rare biosphere". *Proc. Natl. Acad. Sci. USA* **103**:12115–12120.

Sun Y, Cai Y, Liu L, Yu F, Farrell ML, McKendree W, Farmerie W. 2009 ESPRIT: Estimating species richness using large collections of 16S rRNA pyrosequences. *Nucleic Acids Res*. **37**:e76.

Wicker T, Schlagenhauf E, Graner A, Close T, Keller B, Stein N 2006. 454 sequencing put to the test using the complex genome of barley. *BMC Genomics* **7**:275.

Chapter 20

Environmental Shotgun Sequencing: Its Potential and Challenges for Studying the Hidden World of Microbes

JONATHAN A. EISEN

20.1 INTRODUCTION

Since their discovery in the 1670s by Anton van Leeuwenhoek, an incredible amount has been learned about microorganisms and their importance to human health, agriculture, industry, ecosystem functioning, global biogeochemical cycles, and the origin and evolution of life. Nevertheless, it is what is not known that is most astonishing. For example, though there are certainly at least 10 million species of bacteria, only a few thousand have been formally described [Gould, 1996]. This contrasts with the more than 350,000 described species of beetles [Evans and Bellany, 1996]. This is one of many examples indicative of the general difficulties encountered in studying organisms that we cannot readily see or collect in large samples for future analyses. It is thus not surprising that most major advances in microbiology can be traced to methodological advances rather than scientific discoveries per se.

Examples of these key revolutionary methods (**Table 20.1**) include the use of microscopes to view microbial cells, the growth of single types of organisms in the lab in isolation from other types (culturing), the comparison of ribosomal RNA (rRNA) genes to construct the first tree of life that included microbes [Woese and Fox, 1977], the use of the polymerase chain reaction (PCR) [Mullis and Fallona, 1987] to clone rRNA genes from organisms without culturing them [Reysenbach et al., 1992; Medlin et al., 1988; Weisberg et al., 1991], and the use of high-throughput "shotgun" methods to sequence the genomes of cultured species [Fleischmann et al., 1995]. We are now in the midst of another such revolution—this one driven by the use of genome sequencing methods to study microbes directly in their natural habitats, an approach known as metagenomics, environmental genomics, or community genomics [Handelsman, 2004].

In this essay I focus on one particularly promising area of metagenomics—the use of shotgun genome methods to sequence random fragments of DNA from microbes in an environmental sample. The randomness and breadth of this environmental shotgun sequencing (ESS)—first used only a few years ago [Venter et al., 2004; Tyson et al., 2004] and now being used to assay every microbial system imaginable from the human gut [Gill et al., 2006] to waste water sludge [Garcia et al., 2006; see also Vol. II]—has the potential to reveal novel and fundamental insights into the hidden world of microbes and their impact on our world. However, the complexity of analysis required

Reprinted with slight modification from Eisen JA (2007) Environmental Shotgun Sequencing: Its Potential and Challenges for Studying the Hidden World of Microbes. *PLoS Biol* 5(3): e82. doi:10.1371/journal.pbio.0050082.

Handbook of Molecular Microbial Ecology, Volume I: Metagenomics and Complementary Approaches, First Edition. Edited by Frans J. de Bruijn.
Published 2011 by John Wiley & Sons, Inc.

Table 20.1 Some Major Methods for Studying Individual Microbes Found in the Environment

Method	Summary	Comments
Microscopy	Microbial phenotypes can be studied by making them more visible. In conjunction with other methods, such as staining, microscopy can also be used to count taxa and make inferences about biological processes.	The appearance of microbes is not a reliable indicator of what type of microbe one is looking at.
Culturing	Single cells of a particular microbial type are grown in isolation from other organisms. This can be done in liquid or solid growth media.	This is the best way to learn about the biology of a particular organism. However, many microbes are uncultured (i.e., have never been grown in the lab in isolation from other organisms) and may be unculturable (i.e., may not be able to grow without other organisms).
rRNA-PCR	The key aspects of this method are the following: (a) All cell-based organisms possess the same rRNA genes (albeit with different underlying sequences); (b) PCR is used to make billions of copies of basically each and every rRNA gene present in a sample; this amplifies the rRNA signal relative to the noise of thousands of other genes present in each organism's DNA; (c) sequencing and phylogenetic analysis places rRNA genes on the rRNA tree of life; the position on the tree is used to infer what type of organism (a.k.a. phylotype) the gene came from; and (d) the numbers of each microbe type are estimated from the number of times the same rRNA gene is seen.	This method revolutionized microbiology in the 1980s by allowing the types and numbers of microbes present in a sample to be rapidly characterized. However, there are some biases in the process that make it not perfect for all aspects of typing and counting.
Shotgun genome sequencing of cultured species	The DNA from an organism is isolated and broken into small fragments, and then portions of these fragments are sequenced, usually with the aid of sequencing machines. The fragments are then assembled into larger pieces by looking for overlaps in the sequence each possesses. The complete genome can be determined by filling in gaps between the larger pieces.	This has now been applied to over 1000 microbes, as well as some multicellular species, and has provided a much deeper understanding of the biology and evolution of life. One limitation is that each genome sequence is usually a snapshot of one or a few individuals.
Metagenomics	DNA is directly isolated from an environmental sample and then sequenced. One approach to doing this is to select particular pieces of interest (e.g., those containing interesting rRNA genes) and sequence them. An alternative is ESS, which is shotgun genome sequencing as described above, but applied to an environmental sample with multiple organisms, rather than to a single cultured organism.	This method allows one to sample the genomes of microbes without culturing them. It can be used both for typing and counting taxa and for making predictions of their biological functions.

to realize this potential poses unique interdisciplinary challenges, challenges that make the approach both fascinating and frustrating in equal measure.

20.2 DISCUSSION

20.2.1 Who Is Out There? Typing and Counting Microbes in the Environment

One of the most important and conceptually straightforward steps in studying any ecosystem involves cataloging the types of organisms and the numbers of each type. For a long time, such typing and counting was an almost insurmountable problem in microbiology. This is largely because physical appearance does not provide a valid taxonomic picture in microbes. Appearance evolves so rapidly that two closely related taxa may look wildly different and two distantly related taxa may look the same. This vexing problem was partially overcome in the 1980s through the use of rRNA–PCR (Table 20.1). This method allows microorganisms in a sample to be phylogenetically typed and counted based on the sequence of their rRNA genes, genes that are present in all cell-based organisms. In essence, a database of rRNA sequences [Olsen et al., 1991; Cole et al., 2007; see also

Chapter 36, Vol. I] from known organisms functions like a bird field guide, and finding an rRNA–PCR product is akin to seeing a bird through binoculars. Rather than counting species, this approach focuses on "phylotypes," which are defined as organisms whose rRNA sequences are very similar to each other (a cutoff of >97% or >99% identical is frequently used). The ability to use phylotyping to determine who was out there in any microbial sample has revolutionized environmental microbiology [Pace, 1997], has led to many discoveries [e.g., Hugenholtz et al., 1998], and has convinced many people (including me) to become microbiologists.

The selective targeting of a single gene makes rRNA–PCR an efficient method for deep community sampling [Sogin et al., 2006]. However, this efficiency comes with limitations, most of which are complemented or circumvented by the randomness and breadth of ESS. For example, examination of the random samples of rRNA sequences obtained through ESS has already led to the discovery of new taxa—taxa that were completely missed by PCR because of its inability to sample all taxa equally well (e.g., Baker et al. [2006]). In addition, ESS provides the first robust sampling of genes other than rRNA, and many of these genes can be more useful for some aspects of typing and counting. Some universal protein coding genes are better than rRNA both for distinguishing closely related strains (because of third position variation in codons) and for estimating numbers of individuals (because they vary less in copy number between species than do rRNA genes) [Venter et al., 2004]. Perhaps most significantly, ESS is providing groundbreaking insights into the diversity of viruses [Angly et al., 2006; Edwards and Rohwer, 2005], which lack rRNA genes and thus were left out of the previous revolution (see also Section 1, Vol. II).

Certainly, many challenges remain before we can fully realize the potential of ESS for the typing and counting of species, including making automated yet accurate phylogenetic trees of every gene, determining which genes are most useful for which taxa, combining data from different genes even when we do not know if they come from the same organisms, building up databases of genes other than rRNA, and making up for the lack of depth of sampling. If these challenges are met, ESS has the potential to rewrite much of what we thought we knew about the phylogenetic diversity of microbial life.

20.2.2 What Are They Doing? Top Down and Bottom Up Approaches to Understanding Functions in Communities

A community is, of course, more than a list of types of organisms. One approach to understanding the properties and functioning of a microbial community is to start with studies of the different types of organisms and build up from these individuals to the community. Ideally, to do this, one would culture each of the phylotypes and study its properties in the lab. Unfortunately, many, if not most, key microbes have not yet been cultured [Leadbetter, 2003]. Thus, for many years, the only alternative was to make predictions about the biology of particular phylotypes based on what was known about related organisms. Unfortunately, this too does not work well for microbes since very closely related organisms frequently have major biological differences. For example, *Escherichia coli* K12 and *E. coli* O157:H7 are strains of the same species (and considered to be the same phylotype), with genomes containing only about 4000 genes, yet each possesses hundreds of functionally important genes not seen in the other strain [Perna et al., 2001]. Such differences are routine in microbes, and thus one cannot make any useful inferences about what particular phylotypes are doing (e.g., type of metabolism, growth properties, role in nutrient cycling, or pathogenicity) based on the activities of their relatives.

These difficulties—the inability to culture most microbes and the functional disparities between close relatives—led to one of the first kinds of metagenomic analyses, wherein predictions of function were made from analysis of the sequence of large DNA fragments from representatives of known phylotypes. This approach has provided some stunning insights, such as the discovery of a novel form of phototrophy in the oceans [Beja et al., 2000]. However, this large insert approach has the same limitation as predicting properties from characterized relatives—a single cell cannot possibly represent the biological functions of all members of a phylotype.

ESS provides an alternative, more global way of assessing biological functions in microbial communities. As when using the large insert approach, functions can be predicted from sequences. However, in this case the predicted functions represent a random sampling of those encoded in the genomes of all the organisms present. This approach has unquestionably been wildly successful in terms of gene discovery. For example, analysis of ESS data has revealed novel forms of every type of gene family examined, as well as a great number of completely novel families (e.g., Yooseph et al. [2005]). However, there is a major caveat when using ESS data to make community-level inferences. Ecosystems are more than just a bag of genes: They are made up of compartments (e.g., cells, chromosomes, and species), and these compartments matter. The key challenge in analyzing ESS data is to sort the DNA fragments (which are usually less than 1000 base pairs long relative to genome sizes of millions or billions of bases) into bins that correspond to compartments in the system being studied.

Table 20.2 Methods of Binning

Method	Description	Comments
Genome assembly	Identify regions of overlap between different fragments from the same organism to build larger contiguous pieces (contigs).	Getting deep enough sampling for this to work is very expensive except for low diversity systems or for very abundant taxa.
Reference genome alignment	Identify ESS fragments or contigs that are very similar to already assembled sections of the genome of single microbial types.	(a) One of the most effective ways to sort through ESS data, if the reference genome is very closely related to an organism in the sample; (b) the reason why more reference genomes are needed; (c) does not handle regions present in uncultured organisms but not in the reference.
Phylogenetic analysis	Build evolutionary trees of genes encoded by ESS fragments or contigs. Assign fragments or contigs to taxonomic groups based on nearest neighbor(s) in trees.	(a) Very powerful, but level of resolution depends on whether fragments encode useful phylogenetic markers and on how well sampled the database is for the neighbor analysis; (b) would work much better if more genomes were available from across the tree of life.
Word frequency and nucleotide composition analysis	Measure word frequency and composition of each fragment. Group by clustering algorithms or principal component analysis.	(a) Has the potential to work because organisms sometimes have "signatures" of word frequencies that are found throughout the genome and are different between species; (b) very challenging for small fragments.
Population genetics	Build alignments of fragments or contigs with similarity to each other (but not as much as needed for assembly). Examine haplotype structure, predicted effective population size, and synonymous and nonsynonymous substitution patterns.	May be most useful as a way of subdividing bins created by other methods.

A recent study by me and colleagues illustrates the importance of compartments when interpreting ESS data. When we analyzed ESS data from symbionts living inside the gut of the glassy-winged sharpshooter (an insect that has a nutrient-limited diet), we were able to bin the data to two distinct symbionts [Wu et al., 2006]. We then could infer from those data that one of the symbionts synthesizes amino acids for the host while the other synthesizes the needed vitamins and cofactors. Modeling and understanding of this ecosystem are greatly enhanced by the demonstration of this complementary division of labor, in comparison to simply knowing that amino acids, vitamins, and cofactors are made by "symbionts."

How does one go about binning ESS data? A variety of approaches have been developed, some of which are described in **Table 20.2** (see also Chapter 40, 42, Vol. I). In considering the different binning methods and their limitations, the first question one needs to ask is, What are we trying to bin? Is it fragments from the same chromosome from a single cell, which would be useful for studying chromosome structure? If so, then perhaps genome assembly methods are the best. What if instead, as in the sharpshooter example, we are trying to have each bin include every fragment that came from a particular species, knowledge that may be useful for predicting community metabolic potential? If the level of genetic polymorphism among individual cells from the same species is high, then genome assembly methods may not work well (the polymorphisms will break up assemblies). A better approach might be to look for species-specific "word" frequencies in the DNA, such as ones created by patterns in codon usage. The challenge is, How do we tune the methods to find the right target level of resolution? If we are too stringent, most bins will include only a few fragments. But if we are too relaxed, we will create artificial constructs that may prove biologically misleading, such as grouping together sequences from different species. To make matters more complex, most likely the stringency needed will vary for different taxa present in the sample.

Another critical issue is the diversity of the system under study. Generally, binning works better when there are few different phylotypes present, all of which are distantly related and form discrete populations. This is why binning works well for the sharpshooter system and other relatively isolated, low-diversity environments. Binning increases in difficulty exponentially as the number of species increases: The populations and species start to merge together, and the populations get more and more polymorphic and variable in relative abundance (such as in the paper about the Global Ocean Sampling expedition [Rusch et al., 2007]). Further complicating binning is the phenomenon of lateral gene transfer, where genes are exchanged between distantly related lineages at rates that are high enough that random sampling of a genome will frequently include genes with multiple histories.

Despite these challenges, I believe we can develop effective binning methods for complex communities. First, we can combine different approaches together, such as using one method to sort in a relaxed manner and then using another to subdivide the bins provided by the first method. Second, we can incorporate new approaches such as population genetics into the analysis [Johnson and Slatkin, 2006]. In addition, the lessons learned here can be applied to other aspects of metagenomics (e.g., the counting and typing discussed above) and provide insights into the nature of microbial genomes and the structure of microbial populations and communities.

20.3 COMPARATIVE METAGENOMICS

So far, I have discussed issues relating mostly to intrasample analysis of ESS data. However, the area with perhaps the most promise involves the comparative analysis of different samples. This work parallels the comparative analysis of genomes of cultured species. Initial studies of that type compared distantly related taxa with enormous biological differences. What has been learned from these studies pertains mostly to core housekeeping functions, such as translation and DNA metabolism, and to other very ancient processes [Koonin and Mushegian, 1996]. It was not until comparisons were made between closely related organisms that we began to understand events that occurred on shorter time scales, such as selection, gene transfer, and mutation processes [Eisen, 2001]. Similarly, the initial comparisons of ESS data involved comparisons of wildly different environments [Tringe et al., 2005], yielding insights into the general structure of communities. But as more comparisons are made between similar communities [Edwards et al., 2006; Rodriguez-Brito et al., 2006], such as those sampled during vertical and horizontal ocean transects [Rusch et al., 2007; DeLong, 2005; DeLong et al., 2006; Worden et al., 2006], we will begin to learn about shorter time scale processes such as migration, speciation, extinction, responses to disturbance, and succession. It is from a combination of both approaches—comparing both similar and very divergent communities—that we will be able to understand the fundamental rules of microbial ecology and how they relate to ecological principles seen in macroorganisms.

20.4 CONCLUSIONS

In promoting some of the exciting opportunities with ESS, I do not want to give the impression that it is flawless. It is helpful in this respect to compare ESS to the Internet. As with the Internet, ESS is a global portal for looking at what occurs in a previously hidden world. Making sense of it requires one to sort through massive, random, fragmented collections of bits of information. Such searches need to be done with caution because any time you analyze such a large amount of data, patterns can be found. In addition, as with the Internet, there is certainly some hype associated with ESS that gives relatively trivial findings more attention than they deserve. Overall, though, I believe the hype is deserved. As long as we treat ESS as a strong complement to existing methods, and we build the tools and databases necessary for people to use the information, it will live up to its revolutionary potential.

Acknowledgments

I thank Simon Levin, Joshua Weitz, Jonathan Dushoff, Maria-Inés Benito, Doug Rusch, Aaron Halpern, and Shibu Yooseph for helpful discussions, and I am grateful to Melinda Simmons, Merry Youle, and three anonymous reviewers for helpful comments on the manuscript. The writing of this chapter was supported by National Science Foundation Assembling the Tree of Life Grant 0228651 to Jonathan A. Eisen and by the Defense Advanced Research Projects Agency under grants HR0011-05-1-0057 and FA9550-06-1-0478.

REFERENCES

Angly FE, Felts B, Breitbart M, Salamon P, Edwards RA, et al. 2006. The marine viromes of four oceanic regions. *PLoS Biol*. **4**:2121–2131, e368 doi:10.1371/journal.pbio.0040368.

Baker BJ, Tyson GW, Webb RI, Flanagan J, Hugenholtz P, et al. 2006. Lineages of acidophilic archaea revealed by community genomic analysis. *Science* **314**:1933–1935.

Beja O, Aravind L, Koonin EV, Suzuki MT, Hadd A, et al. 2000. Bacterial rhodopsin: Evidence for a new type of phototrophy in the sea. *Science* **289**:1902–1906.

Cole JR, Chai B, Farris RJ, Wang Q, Kulam-Syed-Mohideen AS, et al. 2007. The ribosomal database project (RDP-II): Introducing myRDP space and quality controlled public data. *Nucleic Acids Res*. **35**:D169–D172.

DeLong EF. 2005. Microbial community genomics in the ocean. *Nat. Rev. Microbiol*. **3**:459–469.

DeLong EF, Preston CM, Mincer T, Rich V, Hallam SJ, et al. 2006. Community genomics among stratified microbial assemblages in the ocean's interior. *Science* **311**:496–503.

Edwards RA, Rohwer F. 2005. Viral metagenomics. *Nat. Rev. Microbiol*. **3**:504–510.

Edwards RA, Rodriguez-Brito B, Wegley L, Haynes M, Breitbart M, et al. 2006. Using pyrosequencing to shed light on deep mine microbial ecology. *BMC Genomics* **7**:57.

Eisen JA. 2001. Gastrogenomics. *Nature* **409**:463–466.

Evans AV, Bellamy CL. 1996. An inordinate fondness for beetles. New York: Holt.

Fleischmann RD, Adams MD, White O, Clayton RA, Kirkness EF, et al. 1995. Whole-genome random sequencing and assembly of *Haemophilus influenzae* Rd. *Science* **269**:496–512.

Garcia Martin H, Ivanova N, Kunin V, Warnecke F, Barry KW, et al. 2006. Metagenomic analysis of two enhanced biological phosphorus removal. (EBPR) sludge communities. *Nat. Biotechnol.* **24**:1263–1269.

Gill SR, Pop M, Deboy RT, Eckburg PB, Turnbaugh PJ, et al. 2006. Metagenomic analysis of the human distal gut microbiome. *Science* **312**:1355–1359.

Gould SJ. 1996. *Full house: The Spread of Excellence from Plato to Darwin*. New York: Harmony Books.

Handelsman J. 2004. Metagenomics: Application of genomics to uncultured microorganisms. *Microbiol. Mol. Biol. Rev*. **68**:669–685.

Hugenholtz P, Pitulle C, Hershberger KL, Pace NR. 1998. Novel division level bacterial diversity in a Yellowstone hot spring. *J. Bacteriol*. **180**:366–376.

Johnson PL, Slatkin M. 2006. Inference of population genetic parameters in metagenomics: A clean look at messy data. *Genome Res*. **16**:1320–1327.

Koonin EV, Mushegian AR. 1996. Complete genome sequences of cellular life forms: Glimpses of theoretical evolutionary genomics. *Curr. Opin. Genet. Dev*. **6**:757–762.

Leadbetter Jr. 2003. Cultivation of recalcitrant microbes: Cells are alive, well and revealing their secrets in the 21st century laboratory. *Curr. Opin. Microbiol*. **6**:274–281.

Medlin L, Elwood HJ, Stickel S, Sogin ML. 1988. The characterization of enzymatically amplified eukaryotic 16S-like ribosomal RNA-coding regions. *Gene* **71**:491–500.

Mullis K, Faloona F. 1987. Specific synthesis of DNA *in vitro* via a polymerase-catalyzed chain reaction. *Methods Enzymol*. **155**:335–350.

Olsen GJ, Larsen N, Woese CR. 1991. The ribosomal RNA database project. *Nucleic Acids Res*. **19**:2017–2021.

Pace NR. 1997. A molecular view of microbial diversity and the biosphere. *Science* **276**:734–740.

Perna NT, Plunkett G 3rd, Burland V, Mau B, Glasner JD, et al. 2001. Genome sequence of enterohaemorrhagic *Escherichia coli* O157:H7. *Nature* **409**:529–533.

Reysenbach AL, Giver LJ, Wickham GS, Pace NR. 1992. Differential amplification of rRNA genes by polymerase chain reaction. *Appl. Environ. Microbiol*. **58**:3417–3418.

Rodriguez-Brito B, Rohwer F, Edwards RA. 2006. An application of statistics to comparative metagenomics. *BMC Bioinformatics* **7**:162.

Rusch DB, Halpern AL, Sutton G, Heidelberg KB, Williamson S, et al. 2007. The Sorcerer II Global Ocean Sampling expedition: Northwest Atlantic through Eastern Tropical Pacific. *PLoS Biol* **5**:e77 doi:10.1371/journal.pbio.0050077.

Sogin ML, Morrison HG, Huber JA, Welch DM, Huse SM, et al. 2006. Microbial diversity in the deep sea and the underexplored “rare biosphere”. *Proc. Natl. Acad. Sci. USA* **103**:12115–12120.

Tringe SG, von Mering C, Kobayashi A, Salamov AA, Chen K, et al. 2005. Comparative metagenomics of microbial communities. *Science* **308**:554–557.

Tyson GW, Chapman J, Hugenholtz P, Allen EE, Ram RJ, et al. 2004. Community structure and metabolism through reconstruction of microbial genomes from the environment. *Nature* **428**: 37–43.

Venter JC, Remington K, Heidelberg JF, Halpern AL, Rusch D, et al. 2004. Environmental genome shotgun sequencing of the Sargasso Sea. *Science* **304**:66–74.

Weisburg W, Barns S, Pelletier D, Lane D. 1991. 16S ribosomal DNA amplification for phylogenetic study. *J. Bacteriol*. **173**:697–703.

Woese C, Fox G. 1977. Phylogenetic structure of the prokaryotic domain: The primary kingdoms. *Proc. Natl. Acad. Sci. USA* **74**: 5088–5090.

Worden AZ, Cuvelier ML, Bartlett DH. 2006. In-depth analyses of marine microbial community genomics. *Trends Microbiol.* **14**: 331–336.

Wu D, Daugherty SC, Van Aken SE, Pai GH, Watkins KL, et al. 2006. Metabolic complementarity and genomics of the dual bacterial symbiosis of sharpshooters. *PLoS Biol.* **4**:e188 doi:10.1371/journal.pbio.0040188.

Yooseph S, Sutton G, Rusch DB, Halpern AL, Williamson SJ, et al. 2007. The Sorcerer II Global Ocean Sampling expedition: Expanding the universe of protein families. *PLoS Biol.* **5**:e16 DOI: 10.1371/journal.pbio.0050016.

Chapter 21

A Comparison of Random Sequence Reads Versus 16S rDNA Sequences for Estimating the Biodiversity of a Metagenomic Sample

CHAYSAVANH MANICHANH, CHARLES E. CHAPPLE, LIONEL FRANGEUL, KARINE GLOUX, RODERIC GUIGO, AND JOEL DORE

21.1 INTRODUCTION

The 16S rRNA gene (16S) is a suitable gene for studying bacterial diversity for multiple reasons. It is highly conserved among all microorganisms, and it contains conserved and hypervariable regions suitable for discerning evolutionary relationships among prokaryotic organisms from the phylum to the species level. Furthermore, most of its short hypervariable regions of 2–300 bp are of appropriate length for pyrosequencing [Sogin et al., 2006; McKenna et al., 2008; Claesson et al., 2009; Costello et al., 2009; see also Chapter 15, Vol. I]. For all these reasons, this molecule has given rise to a huge public database (RDPII) containing 1,235,044 16S sequences (November 17, 2009) [Cole et al., 2009; see also Chapter 36, Vol. I]. Although the PCR-based 16S survey is inherently biased by the possible choice of nonuniversal primers [Gibert et al., 2011; see also Chapter 17, Vol. I] and by the existence of multiple copy number of the gene in a given bacterial genome [Acinas et al., 2004], this approach has been extensively applied to all kind of environments [Schmidt et al., 1991; Eckburg et al., 2005; Sogin et al., 2006; see Vol. II]. In a previous study [Manichanh et al., 2006], we built two metagenomic libraries to analyze the intestinal microbiota of healthy volunteers and patients with Crohn's disease. To analyze the diversity of our libraries, we performed a high-throughput extraction of recombinant vector to avoid the *E. coli* genome and we screened the 16S gene by DNA–DNA hybridization on nylon membrane. This technique allowed us to identify 1200 clones containing 16S rDNA sequences from both libraries. The phylogenetic analyses of the 16S sequences showed a reduced diversity in Crohn's disease patients compared to healthy individuals. However, the high cost and time-consuming nature of this technique has encouraged us to find an alternative approach to study the biodiversity of a metagenomic library.

Nowadays, an increasing number of metagenomic projects has generated the release of a huge amount of sequences in databases [Torsvik et al., 1990; Schmidt et al., 1991; Beja et al., 2000; Rondon et al., 2000; Venter et al., 2004]. For instance, the GOLD web site describes 313 completed and ongoing metagenomic projects (February 2011) for different microbial ecosystems. The increasing interest for using metagenomic and high-throughput sequencing approaches has motivated us to use random sequence reads instead of 16S for taxonomic inferences.

The objective of this study was then to demonstrate that the analysis of random sequence reads (RSRs) could serve as a faster and cheaper alternative to 16S screening

Handbook of Molecular Microbial Ecology, Volume I: Metagenomics and Complementary Approaches, First Edition. Edited by Frans J. de Bruijn.

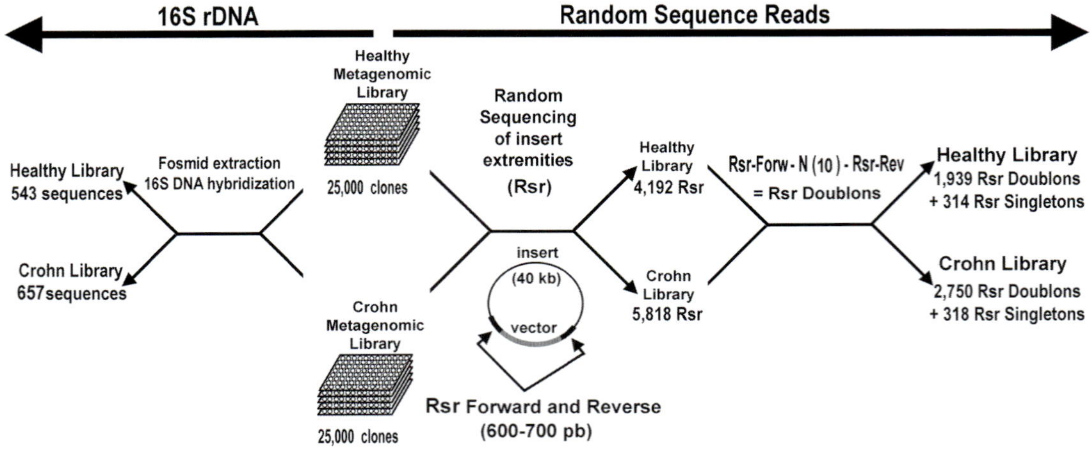

Figure 21.1 About 1200 16S sequences and 10,000 RSRs have been generated from two metagenomic libraries (Healthy and Crohn). The 16S sequences were obtained after screening by DNA–DNA hybridization on macroarray and the RSRs were produced by sequencing the two extremities (forward and reverse sequences) of each insert (Fig. 1 in Manichanh et al. 2008).

and sequencing. For a comparison with the 16S sequences generated in the previous study [Manichanh et al., 2006], we generated a dataset of 10,010 RSRs from the same Crohn and Healthy metagenomic libraries. Then, we applied the same computational pipeline, developed in this study, to both datasets based on the search against a prokaryotic sequence database using BLASTN.

21.2 MATERIALS AND METHODS

21.2.1 Random Sequence Reads (RSRs) (Fig. 21.1)

About 4500 clones were randomly chosen from two metagenomic fosmid libraries (Healthy and Crohn) to perform a one read Sanger sequencing (RSRs) on the two extremities of the insert (5′ and 3′). The sequencing provided 10,010 high-quality sequences of 6–700 bp in average (4192 for healthy and 5818 for Crohn library). To further simplify the parsing, each RSR-5′ was then computationally attached with 10 nucleotides (N) to the RSR-3′ for each clone to form a RSR doublon. In this way, we obtained 1939 RSR doublons and 314 RSR singletons (for which we did not obtain a sequence for the other extremity) for the Healthy library and 2750 RSR doublons and 318 RSR singletons for the Crohn library.

21.2.2 16S rDNA Sequences (Fig. 21.1)

The 1200 16S sequences of around 1200 bp obtained from Manichanh et al. 2006 was used in the present work to validate our computational approach. In the previous work, taxonomic analysis was based on a multiple alignment using the ClustalW program [Thompson et al., 1994] and the clustering in OTUs was based on a 98% cutoff similarity. We then taxonomically assigned each OTU using the online Seqmatch program of the RDPII webpage (Fig. 21.1 in Manichanh et al. 2008; see also Chapter 36, Vol. I).

21.2.3 Sargasso Sea Shotgun Sequences

In order to validate our approach using another bacterial ecosystem, 1.9 millions of sequences from the Sargasso sea shotgun dataset [Venter et al., 2004] were downloaded. From this dataset, a comparable number of RSRs than for the human gut metagenome (5000) was randomly selected to be reanalyzed with our computational pipeline.

21.2.3.1 The Taxonomy Assignment Pipeline (TAP) (Figure 21.2).

We used BLASTN [Altschul et al., 1990] to compare the sequences against a database

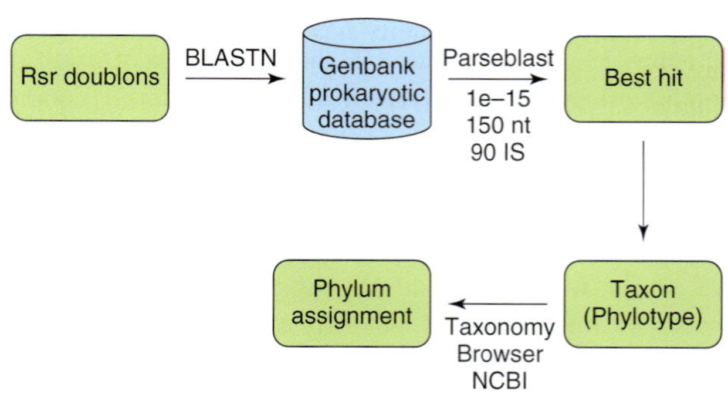

Figure 21.2 Taxonomic assignment pipeline (TAP) (see Section 21.2) (Figure 2 in Manichanh et al. [2008]).

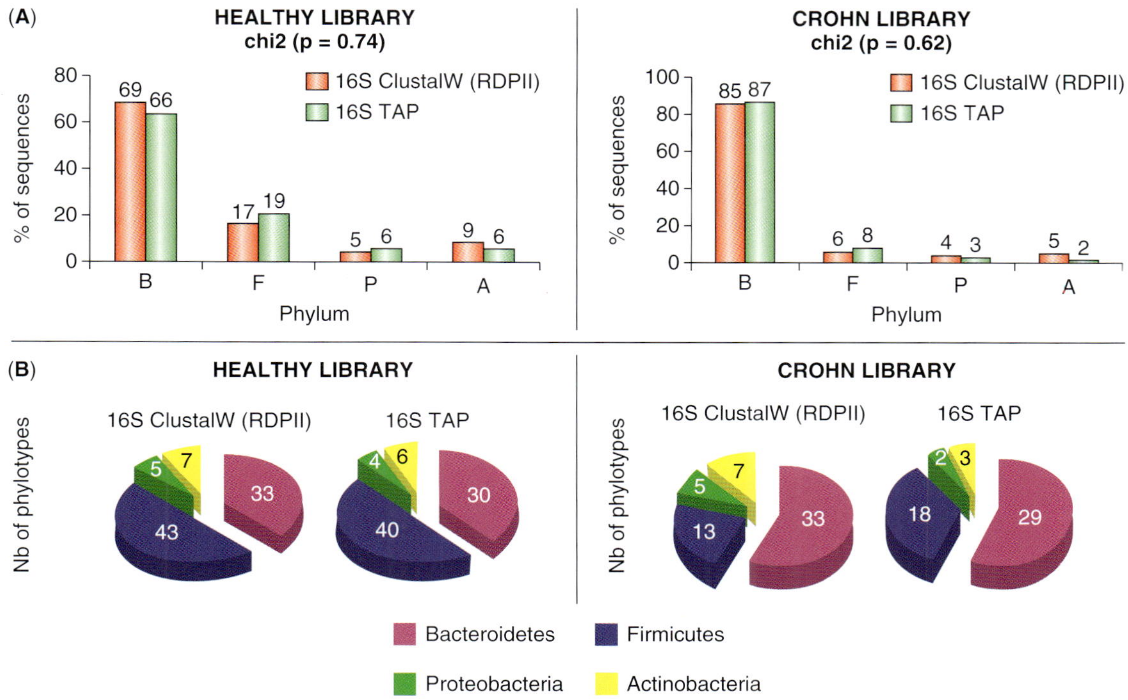

Figure 21.3 Taxonomic assignment of the 16S sequences. Comparison of number of phylotypes obtained from two computational methods: One is based on a multiple alignment to referenced sequences using CLUSTALW, while the other is based on a comparison against a database of GenBank prokaryotic sequence using BLASTN (Fig. 3 in Manichanh et al. [2008]).

of GenBank prokaryotic sequences (GenBank_prok). We applied this local alignment program to find the closest relative for each of the 16S sequence and RSRs against GenBank_prok that contained 516,770 entries. HSPs that met specific cutoff parameters [expected value <e-15, HSP length >150 nt, and IS >90%] were further considered. We then took the top blast hit, identified the species to which the sequence belongs, and used this to perform the taxonomic assignment. Finally, the phylum of each taxon was recovered from the NCBI Taxonomy browser. Bacterial nomenclature used in this study is based on *Bergey's Manual of Systematic Bacteriology*.

21.2.4 Statistical Analyses

A chi-square of homogeneity has been computed to compare the phylotypes frequencies across different datasets. P values smaller than 0.05 were considered enough to reject this hypothesis and to denote therefore significant differences between the compared groups.

21.3 RESULTS AND DISCUSSION

21.3.1 Sequence Datasets

From two Healthy and Crohn metagenomic libraries, two datasets (16S and RSRs) were compared (Fig. 21.1). The 16S dataset contained 1200 sequences obtained from a previous study after a high throughput screening using a DNA–DNA hybridization technique on both libraries and Sanger sequencing. These sequences permitted the phylogenetic analysis and comparison of the two Healthy and Crohn libraries based on a computational method appropriate for 16S sequences in Manichanh et al. [2006]. The results indicated a reduced complexity of the Firmicutes bacterial phylum as a signature of the fecal microbiota in patients with Crohn's disease.

The RSRs dataset was generated for this study and represents a collection of 5818 and 4192 high-quality sequence reads (from Crohn and Healthy libraries, respectively). They were produced by sequencing the two extremities (forward and reverse sequences) of each 40-kb insert. By attaching the two reads of the same insert, we obtained 1939 RSR doublons from the Healthy library and 2750 RSR doublons from the Crohn library.

21.3.2 Choice of the Database

In order to determine the best database to use in TAP, we have tested three different ones for their potential to recover microbial diversity. The first one is GenBank_prok (without environmental sequences), the second is Refseq, and the last one is the 577 complete NCBI microbial genomes. Using the RSRs dataset and

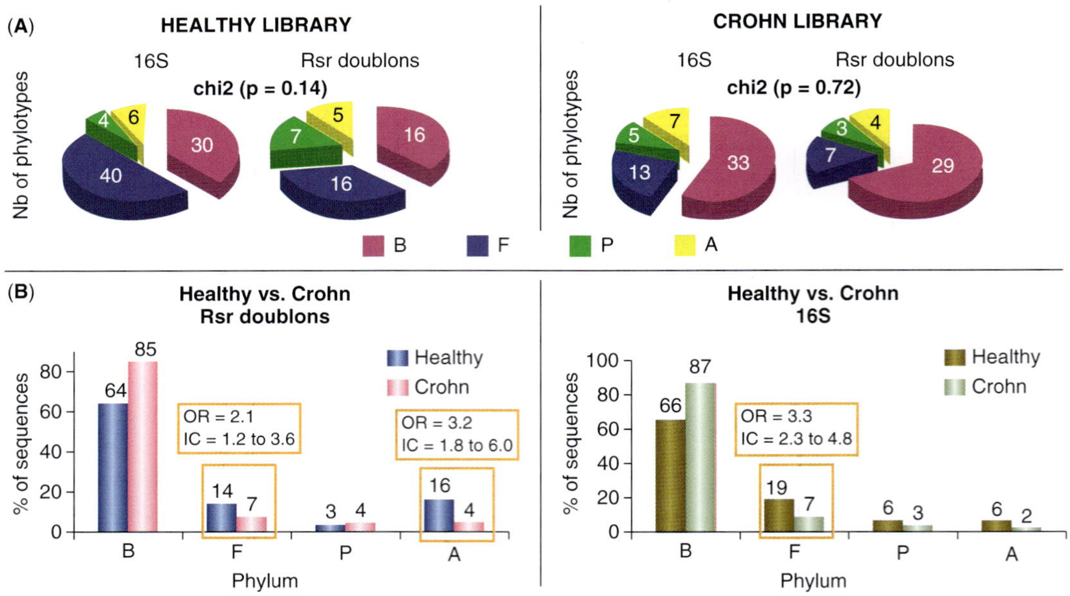

Figure 21.4 Biodiversity with RSRs: (**A**) for the Healthy and Crohn libraries, TAP was applied to the RSRs and we compared the results to those of 16S sequences using the same method. Data were analyzed using the chi-square test for two independent sets of samples. Only a p value <0.05 was considered to denote a significant difference. (**B**) Comparison of two metagenomic libraries using TAP with RSR doublons or 16S. An odds ratio of 1 indicates that the condition or event under study is equally likely in both groups. B, Bacteroidetes; F, Firmicutes; P, Proteobacteria; A, Actinobacteria (Fig. 5 in Manichanh et al. [2008]).

TAP with one the three databases, the results showed that Genbank_prok was the more sensitive in number of hits and phylotypes than both Refseq and the complete genome databases. Therefore, although less well annotated and more redundant, we kept GenBank_prok for further analyses.

21.3.3 Validation of Our Computational pipeline

Using an approach for phylogenetic markers, the analysis of 1200 16S sequences generated from both healthy and Crohn's libraries in a previous study [Manichanh et al., 2006], identified 125 different phylotypes (OTUs) at 98% similarity cutoff. This method was based on the alignment of the sequences (CLUSTLAW program) to referenced sequences of the RDPII, followed by the calculation of distance matrices (DNADIST v3.6 program) and finally by the construction of phylogenetic trees (NEIGHBOR v3.6 program).

On the same data, the analysis, using our computational pipeline TAP (Fig. 21.2), gave similar results to the previous method in terms of phyla, number of phylotypes and percentage of sequences detected in each phylum (**Fig. 21.3**). As expected, we obtained the four dominant bacterial phyla (Bacteroidetes, Firmicutes, Proteobacteria, and Actinobacteria) found in human gut. For the comparison in terms of percentage of sequences in each phylum, a chi-square test was used, which showed no significant differences between the two methods ($P = 0.74$ for Healthy and $P = 0.62$ for Crohn library). However, as also expected, a significant difference between Healthy and Crohn libraries was observed ($P = 0.016$).

21.3.4 Biodiversity of the Gut Microbiota Using RSRs

The use of BLASTN to assign taxonomically the RSRs against a prokaryotic sequence database gave 13% and 11% of hits for the Healthy and Crohn libraries respectively. This very low proportion of match is largely due to the specificity of our BLASTN cut-off parameters. This stringency has been applied to avoid false positive. This method allowed us to detect the four main bacterial phyla found in human gut. Compared to 16S analysis, the proportion of sequences is also comparable for the four bacterial phyla (**Fig. 21.4A**). At the genus (**Table 21.1**) and phylotype levels, RSRs reached 45% and 44% respectively of the diversity detected by 16S. At the phylotype level, the biodiversity is reduced with RSRs (Fig. 21.4A) when compared with that of the 16S. This discrepancy may reflect the limitation in the diversity of genes present in GenBank_prok (see also Chapters 18, 20 and 21, Vol. II).

21.3.5 Comparison of RSRs from Two Human Gut Metagenomic Libraries (Healthy and Crohn)

The comparison between Healthy and Crohn libraries using RSRs and TAP showed a reduced diversity in the

Table 21.1 TAP Analysis of 16S Versus RSRs at the Genus Level

Phylum	Genus	16S		RSR	
		Healthy	Crohn	Healthy	Crohn
Bacteroidetes	**Alistipes**	**0.7**[a]	**1.1**	**ND**	**0.3**
	Bacteroidales	0.7	0.3	ND	ND
	Bacteroides	50.3	43.2	63.7	80.9
	Prevotella	**6.1**	**11.9**	**0.3**	**1.4**
	Tannerella	**ND**	**8.7**	**ND**	**1.4**
Firmicutes	Anaerotruncus	0.6	ND	ND	ND
	Anaerovorax	0.2	ND	ND	ND
	Clostridium	**1.5**	**10.4**	**5.6**	**0.8**
	Dialister	0.2	ND	ND	ND
	Enterococcus	**ND**	**1.1**	**4.0**	**5.5**
	Eubacterium	1.8	ND	ND	ND
	Faecalibacterium	**1.7**	**0.5**	**0.7**	**ND**
	Lachnospira	0.2	ND	ND	ND
	Lachnospiraceae	0.2	ND	ND	ND
	Lactobacillales	0.2	0.3	ND	ND
	Oscillospira	0.2	ND	ND	ND
	Papillibacter	0.6	0.2	ND	ND
	Phascolarctobacterium	0.2	ND	ND	ND
	Roseburia	**1.3**	**0.3**	**0.3**	**ND**
	Ruminococcus	**4.2**	**1.4**	**0.7**	**ND**
	Streptococcus	ND	0.5	ND	0.3
	Subdoligranulum	**2.0**	**0.5**	**0.3**	**ND**
	unclassified_Clostridiaceae	3.5	ND	ND	ND
	unclassified_Lachnospiraceae	0.2	ND	ND	ND
Proteobacteria	Bilophila	0.7	ND	ND	ND
	Desulfovibrio	ND	0.2	ND	ND
	Escherichia	**3.7**	**2.8**	**0.3**	**3.0**
	Shigella	0.2	ND	ND	0.3
	Methylophilus methylotrophus	0.9	ND	ND	ND
Actinobacteria	**Bifidobacterium**	**2.4**	**1.4**	**14.2**	**4.2**
	Brachybacterium	0.2	ND	ND	ND
	Coriobacterium	**3.1**	**0.2**	**1.7**	**ND**
	Corynebacterium	0.6	ND	ND	ND

[a] Values in bold indicate that each genus represented by at least 1% of the clones analyzed by using 16S is also detected when using RSRs. ND, not detected.

Source: Table 1 in Manichanh et al. [2008].

Firmicutes phylum. This reduction was also observed when using 16S sequences (Fig. 21.4A). Regarding the proportion of sequences, the comparison of the two metagenomic libraries showed a significant difference in two bacterial phyla: Firmicutes (as shown with 16S) and Actinobacteria. This difference in the Actinobacteria phylum, not significant when using the 16S sequences, shows an advantage of using RSRs. However, this observation needs to be confirmed by an experimental approach. Here, we show that two metagenomic libraries, built with the same molecular methodology and from the same microbial ecosystem but presenting two different conditions, can be compared in terms of microbial composition using RSRs.

21.3.6 Analyses of the Sargasso Sea Shotgun Dataset

The validation of the BLASTN-based approach was also tested with a different environmental metagenomic library: the Sargasso Sea shotgun dataset [Venter et al., 2004]. Five thousand RSRs (0.25% of the whole dataset) were randomly selected and analyzed with TAP. Twenty

Table 21.2 TAP Analyses of the Sargasso Sea Shotgun Dataset

Phylum	Class	16S (%)[a]	5000 RSRs_AP (%)
Bacteroidetes		0.5	1
Cyanobacteria		5	10.7
Proteobacteria		85	88
	Alphaproteobacteria	41	1.4
	Betaproteobacteria	13	56.6
	Gammaproteobacteria	29	30
	Deltaproteobacteria	1	0
	Epsilonproteobacteria	1	0

[a]Data extracted from Figure 6 of Venter et al. [2004].
Source: Table 2 in Manichanh et al. [2008].

percent of the sequences passed our cutoff parameters. The comparison with the 1164 16S sequences obtained from the same library showed similarity regarding the diversity at the phylum level and the proportion of reads in each phylum (**Table 21.2**).

21.4 CONCLUSION

The use of RSRs will depend on the limitation of the content of the database used in TAP. However, in the coming years, as more species are sequenced, this bias will gradually decrease. So, the use of RSRs to characterize microbial diversity may not be as accurate as the 16S survey, but as demonstrated in this study, it provides a reliable and very quick estimate of the diversity of a metagenomic library.

SOFTWARE/DATABASES

CLUSTLAW, DNADIST, and NEIGHBOR programs (http://bioweb2.pasteur.fr/)

RDPII web site (http://rdp.cme.msu.edu/)

Seqmatch program of the RDPII (http://rdp.cme.msu.edu/seqmatch/seqmatch_intro.jsp)

GOLD web site (http://www.genomesonline.org/gold.cgi?want=Metagenomes)

NCBI Taxonomy browser (http://www.ncbi.nlm.nih.gov/entrez/batchentrez.cgi?db=Taxonomy)

RefSeq (ftp://ftp.ncbi.nih.gov/refseq/release/)

Genbank via FTP (http://www.ncbi.nlm.nih.gov/Genbank/GenBankFtp.html)

Acknowledgments

We acknowledge Dr. Robert Castelo and Anne Bergeron for useful comments and discussions. This work was supported by the Research grant no. BIO2006-03380 from the Spanish Ministry of Education and Science and by a pre-doctoral fellowship from the Spanish Ministry of Education and Science. Funding to pay the Open Access publication charges for this chapter was provided by INRA.

REFERENCES

Acinas SG, Marcelino LA, Klepac-Ceraj V, Polz MF. 2004. Divergence and redundancy of 16S rRNA sequences in genomes with multiple *rrn* operons. *J. Bacteriol*. **186**:2629–2635.

Altschul SF, Gish W, Miller W, Myers EW, et al. 1990. Basic local alignment search tool. *J. Mol. Biol*. **215**:403–410.

Beja O, Suzuki MT, Koonin EV, Aravind L, et al. 2000. Construction and analysis of bacterial artificial chromosome libraries from a marine microbial assemblage. *Environ. Microbiol*. **2**:516–529.

Claesson MJ, O'Sullivan O, Wang Q, Nikkila J, et al. 2009. Comparative analysis of pyrosequencing and a phylogenetic microarray for exploring microbial community structures in the human distal intestine. *PLoS One*. **4**:e6669.

Cole JR, Wang Q, Cardenas E, Fish J, et al. 2009. The Ribosomal Database Project: improved alignments and new tools for rRNA analysis. *Nucleic Acids Res*. **37**:D141–D145.

Costello EK, Lauber CL, Hamady M, Fierer N, et al. 2009. Bacterial community variation in human body habitats across space and time. *Science* **18**:1694–1697.

Eckburg PB, Bik EM, Bernstein CN, Purdom E, et al. 2005. Diversity of the human intestinal microbial flora. *Science* **308**:1635–1638.

Gibert P, Guarner F, Roca J, Manichanh C. 2011. In silico evaluation of universal primers and hypervariable regions of the 16S rRNA gene in microbial ecosystems. Submitted.

Manichanh C, Rigottier-Gois L, Bonnaud E, Gloux K, et al. 2006. Reduced diversity of faecal microbiota in Crohn's disease revealed by a metagenomic approach. *Gut* **55**:205–211.

Manichanh C, Chapple CE, Frangeul L, Gloux K, et al. 2008. A comparison of random sequence reads versus 16S rDNA sequences for estimating the biodiversity of a metagenomic library. *Nucleic Acids Res*. **36**:5180–5188.

McKenna P, Hoffmann C, Minkah N, Aye PP, et al. 2008. The macaque gut microbiome in health, lentiviral infection, and chronic enterocolitis. *PLoS Pathog*. **4**:e20.

Rondon MR, August PR, Bettermann AD, Brady SF, et al. 2000. Cloning the soil metagenome: A strategy for accessing the genetic and functional diversity of uncultured microorganisms. *Appl. Environ. Microbiol*. **66**:2541–2547.

Schmidt TM, DeLong EF, Pace NR. 1991. Analysis of a marine picoplankton community by 16S rRNA gene cloning and sequencing. *J. Bacteriol*. **173**:4371–4378.

Sogin ML, Morrison HG, Huber JA, Mark Welch D, et al. 2006. Microbial diversity in the deep sea and the underexplored "rare biosphere". *Proc. Natl. Acad. Sci. USA* **103**:12115–12120.

Thompson JD, Higgins DG, Gibson TJ. 1994. CLUSTAL W: Improving the sensitivity of progressive multiple sequence alignment through sequence weighting, position-specific gap penalties and weight matrix choice. *Nucleic Acids Res*. **22**: 4673–4680.

Torsvik V, Goksoyr J, Daae FL. 1990. High diversity in DNA of soil bacteria. *Appl. Environ. Microbiol*. **56**:782–787.

Venter JC, Remington K, Heidelberg JF, Halpern AL, et al. 2004. Environmental genome shotgun sequencing of the Sargasso Sea. *Science* **304**:66–74.

Chapter 22

Metagenomic Libraries for Functional Screening

TRINE AAKVIK, RAHMI LALE, MARK LILES, AND SVEIN VALLA

22.1 INTRODUCTION

The genomes of Eubacteria represent the major reservoir of genetic diversity on Earth [Whitman et al., 1998; Ferrer et al., 2009a]. This diversity represents a huge resource that can be exploited for the recovery of novel biomolecules, and its exploration is also important for the understanding of microbial ecology. It is well known that the vast majority of microorganisms (possibly more than 99%) cannot be readily cultured in a laboratory setting [Amann et al., 1995; Rappé and Giovannoni, 2003]. Although the cultivation success rate can certainly be improved [Tyson and Banfield, 2005], the use of a metagenomic approach involving the direct extraction, cloning, and analysis of DNA from its natural environment [Handelsman et al., 1998] is a more inclusive strategy to access microbial genetic reservoirs. The strength of the metagenomics approach for the identification of novel metabolites is reflected in that most of the genes encoding pathways for synthesis of new biomolecules discovered from metagenomic libraries are either weakly related or unrelated to previously known genes [Daniel, 2005]. In contrast, when for instance searching for new/novel antibiotics among cultivable actinobacteria, the rediscovery rate may be as high as 99.9% [Zaehner and Fiedler, 1995].

DNA isolated from the environment can be explored either by direct sequencing, by sequencing of gene libraries prepared from the isolated DNA, or by functional screening of such libraries. Sequence-based methods have resulted in identification of many novel natural products, however, identification of genes of interest by this method depends on the existence of some kind of similarity to already known genes or gene products. Functional screening does not require preexisting sequence information and is thus the best strategy to identify entirely new classes of natural products. A functional metagenomics approach does, however, depend on successful heterologous expression of the gene or genes responsible for their respective function, and it also depends on whether the gene products are active in the host used for expression. In this review we describe some of the tools used to construct metagenomic libraries, as well as the array of different screening technologies for detection of different (and potentially novel) functions that may be expressed from metagenomic clones (see Section 6 and 7, Vol. II).

22.2 ISOLATION OF ENVIRONMENTAL DNA AND ENRICHMENT OF TARGET GENES

The choice of the metagenomic DNA extraction method and the cloning strategy adopted should coincide with the specific genetic sequence and/or functional activity sought within a metagenomic library. Specifically, the criteria that should be considered early on in this process are (1) the likelihood of detecting a particular gene or gene family using a strict sequence-based approach, (2) the availability of a functional screen (or ideally, a selection) for the respective gene product, (3) the predicted size of the genetic loci (or pathway), (4) the previous knowledge of heterologous expression of the targeted gene within an *Escherichia coli* or other host, and (5) the availability of resources and/or time required for

Handbook of Molecular Microbial Ecology, Volume I: Metagenomics and Complementary Approaches, First Edition. Edited by Frans J. de Bruijn.

constructing and screening a large-insert metagenomic library compared to less labor- and resource-intensive strategies. For example, expression of an enzyme for which there is a plate-based screen (i.e., halo formation due to substrate conversion) would indicate a preferred strategy of constructing a small-insert library within an expression vector. On the other hand, screening for the presence of antimicrobial active clones would be dependent in most cases on expression of biosynthetic pathways contained on large-insert clones, preferably with the option of heterologous expression in multiple bacterial hosts. Therefore, careful consideration of metagenomic strategy is critical prior to initiating work in this field. The different steps involved in metagenomics for functional screening approaches are illustrated in **Figure 22.1**.

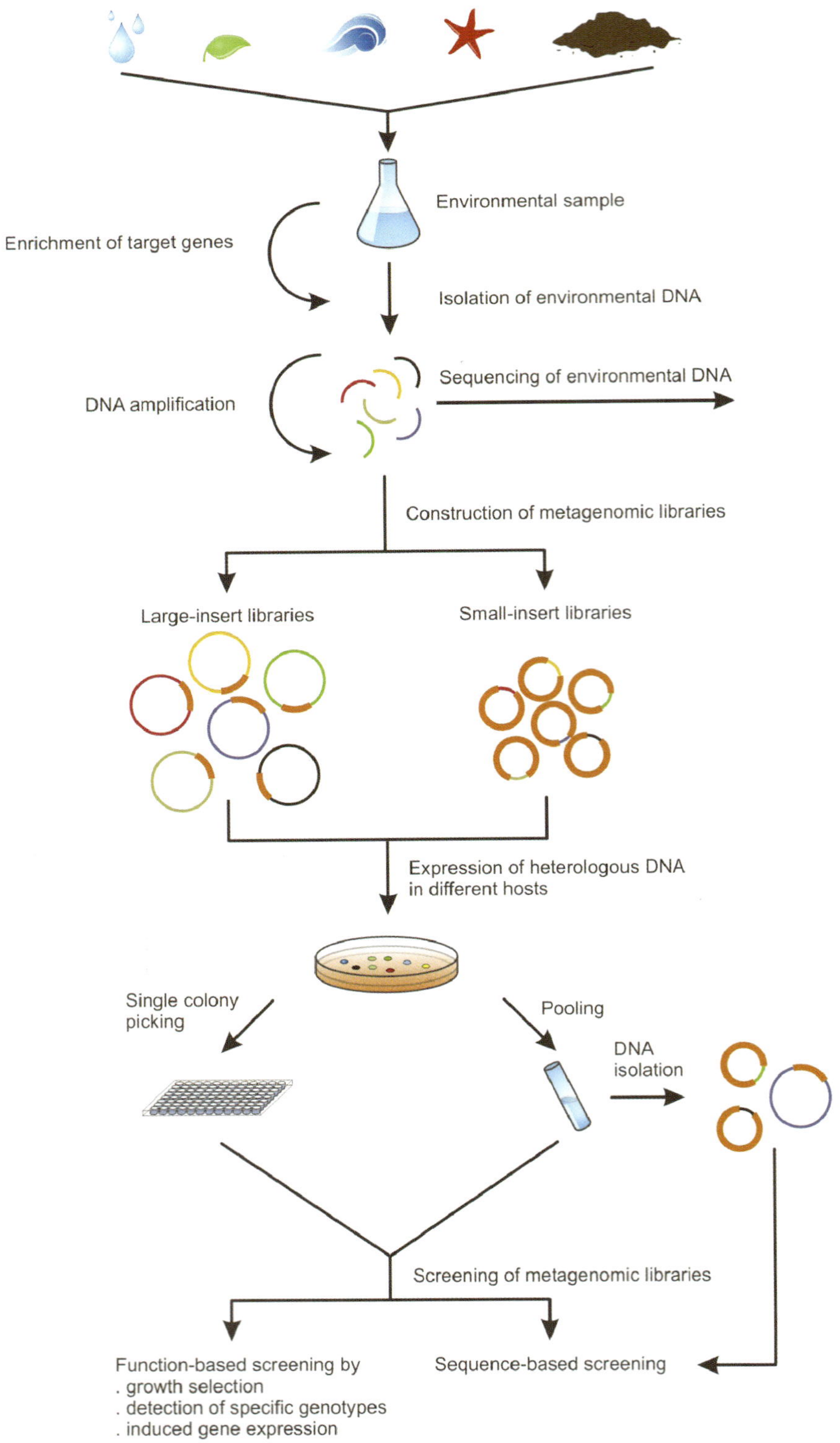

Figure 22.1 Steps involved in metagenomics for functional screening. DNA can be isolated from various sources (water, soil, plants, air, biota, etc.). If required, target genes can be enriched prior to DNA isolation. The isolated DNA can then be subjected to sequencing; or in cases where the quantity of the DNA is low, DNA amplification can be performed. Depending on the goals, either small- or large-insert libraries can be constructed by the use of different vectors. Once the libraries are established in different hosts, they either can be transferred to microtiter plates or can be pooled. Finally, these libraries can be subjected either to function- or sequence-based screenings. For further details, see the text.

For small-insert metagenomic libraries or PCR/pyrosequencing applications, many direct DNA extraction methods are available (e.g., bead-beating) that provide a sufficient degree of cell lysis and subsequent purification [Ogram et al., 1987; Tsai and Olson, 1991; Zhou et al., 1996], and kits are available from many commercial suppliers. The physicochemical characteristics of the environmental sample (i.e., pH, organic content) should be taken into account in order to use the correct sampling scale for DNA extraction. It should also be considered that alternative extraction methods might improve DNA yield and/or purity [Sagova-Mareckova et al., 2008]. It is advisable to use the efficiency of PCR amplification of extracted DNA (e.g., of 16S rRNA genes) as a surrogate for monitoring the suitability of the DNA for cloning. Given that any one method for DNA extraction may have an inherent bias in the representation of specific bacterial taxa as revealed by PCR [Martin-Laurent et al., 2001], there is no single perfectly suitable extraction method for any environmental sample. When using a previously uncharacterized environmental sample, it is therefore wise to use multiple extraction methods and to select a method for metagenomic analysis based on empirical results of DNA yield, ribotype richness, and suitability of the DNA template for PCR amplification and cloning (see also Chapters 10 and 11, Vol. II).

Large-insert cloning requires adoption of DNA extraction methods that are considerably less harsh to prevent DNA shearing in order to procure high-molecular-weight (HMW) metagenomic DNA. The use of indirect DNA extraction methods, in which microbial cells are first recovered and washed prior to cell lysis, have advantages for the removal of the environmental matrix and/or eukaryotic cells and in preserving genomic DNA integrity [Steffan et al., 1988; Torsvik et al., 1990; Krsek and Wellington, 1999]. The primary disadvantage of such methods for extraction of HMW DNA is the inability to use organic extraction and other purification schemes that are highly effective in removing DNA-associated contaminants (e.g., humic acids) due to the resultant DNA fragmentation. Other potential disadvantages in the use of indirect extraction methods include decreased DNA yield and ribotype richness as compared to direct extraction [Steffan et al., 1988; Krsek and Wellington, 1999]. Monitoring of the microbial populations of interest (e.g., Actinobacteria) during the cell recovery and lysis steps is recommended in adopting an indirect extraction method. Purification of HMW metagenomic DNA embedded within agarose plugs may be achieved through use of polyvinylpolypyrrolidone or formamide solutions to remove DNA-associated contaminants and still permit downstream molecular manipulation [Steffan et al., 1988; Liles et al., 2008]. For further discussion of DNA purification strategies for metagenomic applications, please refer to Chapters 10 and 11 in Volume II.

The frequency of metagenomic clones expressing specific traits of interest is low in libraries constructed on the basis of total DNA extraction. This means that in the screening protocols that involve a significant amount of resource input per clone, the vast majority of the clones are expected to give a negative result. One way to reduce this problem is to enrich the environmental sample for the population that carries the genes of interest prior to library construction. This may be achieved by exposing the microorganisms in the sample (in situ) to a selective pressure based on nutritional, physical, or chemical criteria. Such enrichment steps result in loss of diversity in the resulting metagenomic library; however, this has proven to be an advantageous strategy when searching for a specific activity [Entcheva et al., 2001; Knietsch et al., 2003a, b; Rees et al., 2003; Gabor et al., 2004b]. Another possibility is to enrich for high G+C% content in the cloned DNA, since many G+C-rich microorganisms are of particular interest for secondary metabolite expression (see Chapter 23, Vol. I). The DNA from these organisms can be enriched through an ultracentrifugation step after DNA extraction. An even more elegant enrichment strategy is the incorporation of labeled nucleotides into the DNA of metabolically active organisms. This can be accomplished through the use of a stable isotope-labeled substrate (stable isotope probing; see Chapter 55, Vol. I) or a substrate containing bromodeoxyuridine, followed by separation of the labeled DNA through ultracentrifugation or immunocapture, respectively (various enrichment technologies are reviewed in Cowan et al. [2005]).

For identification of enzymatic activities with specific tolerances to environmental extreme conditions, it may be useful to isolate microorganisms living in a respective extreme environment and search for biomolecules that are stable under such hostile conditions. Due to the often low biomass in such environments, it may be challenging to extract sufficient DNA for a metagenomic approach [Ferrer et al., 2009a]. One way to overcome such obstacles is by the use of whole genome amplification, which makes amplification of even very tiny amounts of environmental DNA possible through the use of phi29 DNA polymerase [Abulencia et al., 2006; Yokouchi et al., 2006; Kim et al., 2008; Simon et al., 2009]. Yamada et al. [2008] also demonstrated that this polymerase could successfully be used together with target-specific primers (containing locked nucleic acids) to enrich an environmental sample for target DNA prior to inverse PCR (IPCR) and cloning. Drawbacks of whole genome amplification include amplification biases and formation of chimeric sequences and also include the possible introduction of point mutations that may reduce the chances of cloning a functional gene product.

22.3 VECTORS FOR FUNCTION-BASED METAGENOMIC STUDIES

The ability to detect novel activities in metagenomic libraries is reliant upon successful expression of the vector borne heterologous gene(s) of interest, and this requires several criteria to be fulfilled. First of all, the chosen insert size must be large enough to include the gene or entire cluster of genes needed to express the function of interest. Secondly, a promoter and ribosome binding site that are compatible with the expression machinery of the host are necessary for successful expression of the cloned genes. In addition, several factors may need to be provided by the host cell such as proper transcription factors, inducers, chaperones, cofactors, post-translationally acting factors, and secretion mechanisms. Codon usage may also become a limiting factor for expression because, for example, G + C% content of bacteria can vary significantly. The potential toxicity of the final gene product is yet another possible obstacle to a successful screening result. One potential way to solve these complex problems is to use vectors that can be transferred to and maintained in a variety of different hosts for expression of the metagenomic DNA, such that one can screen a library within a host that is considered likely to express the specific gene product(s) of interest.

22.3.1 Vectors for Small-Insert Metagenomic Libraries

Successful heterologous expression of metagenomic DNA is usually simplest to achieve if only one gene needs to be expressed. In such cases a promoter known to function well in the host used for screening can be a part of the vector such that it transcribes the cloned DNA. One may also use a promoter transcribing in the opposite direction from the other end of the insert and thus become independent of the orientation of the gene searched for—as, for example, described in Lämmle et al. [2007]. Even in such cases, one has to consider prior to library construction at what level one would like the desired gene to be expressed, especially if the gene product has a potential to be toxic to the host. The use of gradually inducible promoters may solve this problem. Vectors with varying copy numbers can also be used to modulate expression levels, but very high copy number vectors may in some cases display toxic levels of expression even in the absence of induction. Instead of using plasmids, one may use λ phage based vectors (phagemids). Since such expression vector systems are accompanied by lysis of the host cell during expression, the problem of gene product toxicity commonly experienced with high copy number vectors is avoided [Ferrer et al., 2005].

22.3.2 Vectors for Large-Insert Metagenomic Libraries

If the product of interest is a compound that requires many gene products, the use of large-insert vectors is essential. The chosen insert size must also be large enough to include all the genes needed to express the function of interest. Since bacterial genetic pathways required for secondary metabolite synthesis (e.g., antibiotics) are very often clustered on a genome, a large-insert metagenomic approach is technically feasible. On the other hand, it should be remembered that fragments are randomly cloned when constructing metagenomic libraries. This means that the insert sizes must be significantly bigger than the expected size of a cluster, because it is otherwise unlikely that an entire cluster will be present in a single clone. Note also that for expression of complex secondary metabolite biosynthetic pathways, the lack of a single essential gene may be sufficient to abolish the synthesis of the final product. For large genetic operons, vector promoters may not achieve expression of all required genes. The transcription of a cloned genetic operon will therefore often depend on native or cryptic promoters already present in the environmental DNA and that are used naturally in the host of origin to express the genes; therefore, heterologous expression of large-insert clones within different hosts may substantially improve the probability that operons may be expressed and their respective natural product(s) detected.

The first large-insert cloning vector described was a cosmid [Collins and Hohn, 1978], which is a multi-copy cloning vector for inserts up to 40 kb that can be packaged into lambda phage heads due to the presence of dual *cos*-sites. The observed instability of genomic libraries constructed in these multi-copy vectors led to the development of the fosmid vector, which is a low copy-number cosmid vector based on the *E. coli* F factor replicon [Kim et al., 1992]. The copy number of this replicon is strictly controlled in *E. coli*, yielding only 1–2 copies per cell. Simultaneously, a bacterial artificial chromosome (BAC) vector based on the same replicon was constructed, but with the capacity of cloning inserts with sizes up to around 300 kb [Shizuya et al., 1992]. The strength of BAC vectors is that large biosynthetic gene clusters can be cloned, such as polyketide or antibiotic synthesis gene clusters which may be over 40 kb in size (e.g., the 55-kb erythromycin clusters [Brikun et al., 2004] and the 124-kb nystatin clusters [Brautaset et al., 2000]). BAC libraries can, however, also contain rather small inserts since there is no selection against small fragments as in the fosmid system. Due to the F factor system, both fosmid and BAC libraries can be established and stably maintained, and these vectors display a low level of chimerism. Wild et al. [2002] combined the stabilizing features of the F factor during the maintenance phase with a possibility of in vivo amplification for achievement of higher levels of DNA recovery and purity upon plasmid isolation (referred to as the copy-control fosmid/BAC vectors, now distributed by Epicentre and Lucigen). Due to the possibility of inducing a higher gene dose of the metagenomic DNA, this copy control system also increases the chances for detection of activities from gene products expressed at low levels.

22.3.3 Host Range of Vectors Useful in Metagenomics

Construction of metagenomic libraries is most easily done in *E. coli* due to its well-established genetics and great transformation competence. Moreover, using *E. coli* as an expression host for function-based screens has also resulted in detection of many novel activities, with a very diverse phylogenetic origin of the DNA sources [Handelsman et al., 1998; Tirawongsaroj et al., 2008; Heath et al., 2009; Simon et al., 2009]. This may be attributable to the fact that *E. coli* has relatively relaxed requirements regarding promoter recognition and translation initiation compared to other well-established expression hosts such as *Bacillus subtilis* and *Streptomyces lividans* [Gabor et al., 2007]. Another advantage of using *E. coli* as host is that it is commonly used in industrial fermentation, thus easing the possible commercial production of the identified product(s). In spite of the benefits from adopting *E. coli* as a heterologous host, it is also clear that many genes will not be expressed from their native promoters in *E. coli* and it is therefore potentially advantageous to expand the library host expression range. This has been demonstrated in several studies, such as for example the one by Martinez et al. [2004], where it was reported that *E. coli, Pseudomonas putida*, and *Streptomyces lividans* differed in their abilities to express heterologous gene clusters. For more examples see Li et al., [2005], Wexler et al., [2005], Angelov et al., [2009], and Craig et al., [2009]. An in silico study by Gabor and co-workers predicted that, on average, only 40% of the genes from 32 different genomes had expression signals that could be recognized by the expression machinery of *E. coli* [Gabor et al., 2004a]. Furthermore, in an experimental study by Warren et al. [2008] the transcription levels of foreign DNA in *E. coli* were compared by RT-PCR, and it was, not surprisingly, found that the degree of expression of genes from phylogenetically diverse organisms was higher the closer the source organism was related to *E. coli*.

The host specificity of the different vectors developed hitherto to improve the chances for heterologous gene expression includes a wide range of both gram-negative and gram-positive hosts (see **Table 22.1**). The vectors are different with regard to how they are transferred to and maintained in the non-*E. coli* host. Some vectors contain an origin for conjugal transfer, usually *oriT* from RK2 (e.g., pRK7813, see Table 22.1), which allows the transfer of constructed libraries from *E. coli* to other hosts with high efficiencies. In the new hosts, either the vectors with inserts are integrated into the chromosome (e.g., pCT3FK and pMBD14; see Table 22.1) or they replicate extra-chromosomally via a vector-provided replicon suitable for the host (e.g., pJWC1 and pUvBBAC; see Table 22.1). The autonomously replicating vectors have the potential for a higher copy number than those integrated into the chromosome, often enhancing the chances for product detection during screening. The pRS44 vector described in Aakvik et al. [2009] is a vector capable of mobilized conjugation and that has both the F-factor replicon and the broad host range RK2 replicon, meaning that a (large-insert) library constructed in this vector can be efficiently transferred to and screened in many different hosts. A particularly important finding in this study was that very large inserts could be established and maintained in species different from *E. coli*.

Expression in bacterial hosts is usually limited to Eubacterial genes, but metagenomic DNA can also contain DNA of archaeal and eukaryotic origin [Gabor et al., 2003]. To circumvent the lack of expression of eukaryotic DNA in bacteria partly due to the presence of noncoding sequences, it is also possible to construct libraries via cDNA [Grant et al., 2006; Bailly et al., 2007. These libraries will, however, only represent expressed genes, and the reverse transcription process will limit the possible insert size. An alternative method to the use of non-*E. coli* hosts is to develop *E. coli* such that its expression capabilities becomes expanded. One such example is given by Bernstein et al. [2007], who reported up to a 12-fold higher expression level from G+C-rich DNA in *E. coli* by directed evolution of the *E. coli* ribosomal protein S1. Other strategies to enhance *E. coli* heterologous expression include introduction of tRNA genes for rare amino acid codons [Rosano and Ceccarelli, 2009] and co-expression of specific chaperone proteins [Wall and Plückthun, 1995; Nishihara et al., 1998; Ferrer et al., 2004].

22.4 CLONING STRATEGIES

The ligation of restriction-digested or blunt-ended metagenomic DNA into vectors and subsequent transformation into a library host strain are performed using

Table 22.1 Vectors Used (or Proposed) for Expression of Metagenomic Libraries in Non-*E. coli* hosts[a]

Vector	Insert Size	Properties/Host Specificities	Reference
Cosmid/Fosmid Vectors			
Cosmid pOS700I	Not explicitly stated	Replicates in *E. coli* and can be transformed to and integrated into the chromosome of *S. lividans*	Courtois et al. [2003]
Cosmid pLAFR3	Ca 25 kb	Contains a broad-host-range RK2-replicon and can be conjugated to a variety of hosts. In this study a library is transferred to *Rhizobium leguminosarum*.	Wexler et al. [2005] Li et al. [2005]
Cosmid pKS13S	Ca 25 kb	Contains a broad-host-range RK2 replicon. In this study a library is transferred to *Pseudomonas putida* by electroporation.	Ono et al. [2007]
Cosmid pRK7813	Ca 33 kb	Contains a broad-host-range RK2 replicon and can be conjugated to a variety of hosts. In this study a library is transferred to *Sinorhizobium meliloti*.	Wang et al. [2006]
Fosmid pCT3FK	Ca 40 kb	Replicates in *E. coli* and can be transformed to and integrated into the chromosome of *Thermus thermophilus*.	Angelov et al. [2009]
Cosmid pJWC1	Not explicitly stated	Contains a broad-host-range RK2 replicon. In this study a library is transferred to *Ralstonia metallidurans* by electroporation.	Craig et al. [2009]
Fosmid and BAC pRS44	Fosmid clones: ca 35 kb BAC clones: up to 195 kb	Contains both an F-replicon and a broad-host-range RK2 replicon. Can be conjugated to a variety of hosts, and in this study a library is transferred to *Pseudomonas fluorescens* and *Xanthomonas campestris*.	Aakvik et al. [2009]
BAC Vectors			
BAC pPAC-S1 ("ESAC")	Up to 140 kb	Replicates in *E. coli* and can be transformed to and integrated into the chromosome of *Streptomyces* hosts. In this study a library is transferred to *S. lividans*.	Sosio et al. [2000]
BAC pMBD14	Up to 85 kb	Replicates in *E. coli* and can be conjugated to and integrated in the chromosome of *S. lividans* and *Pseudomonas putida*.	Martinez et al. [2004]
BAC pUvBBAC	Up to 178 kb	Replicates in *E. coli* and can be transformed to *Listeria* spp. and possibly other G+ bacteria. In this study a library is transferred to *L. innocua*.	Hain et al. [2008]
BAC pBAC-1003	Average:100 kb	A shuttle vector between *E. coli* and *Streptomyces* sp. (In this study a library is not transferred to other hosts.)	Ouyang et al. [2009]
BAC pGNS-BAC	>80 kb	A G-shuttle vector. Contains both an F-replicon and a broad-host-range RK2 replicon.	Kakirde et al., [2010]

[a]The classification of the vectors as fosmid/cosmid or BAC vectors is based on the indicated references.

a variety of different strategies depending on what kind of library is desired. For construction of small-insert libraries the DNA is most often partially restriction-digested and cloned according to standard techniques (see, for example, Henne et al., [1999], Riesenfeld et al., [2004], and Lämmle et al., [2007]). Waschkowitz et al. [2009] compared the use of a T4 DNA ligase-based and a topoisomerase-based cloning method for construction of such small-insert metagenomic libraries, finding that both higher amounts of clones and larger insert sizes were obtained per gram of DNA using the topoisomerase based method. When aiming for large-insert libraries, the cloning strategy depends to a greater extent on the molecular weight of the isolated DNA. It has been demonstrated that the use of optimized extraction methods makes it possible to obtain libraries with average insert sizes up to 100 kb even after restriction treatment of the metagenomic DNA [Liles et al., 2008; Ouyang et al., 2009]. When the isolated DNA is not of high molecular weight, an alternative approach is to employ blunt-end or T-A ligation to avoid further decrease in insert size (see Wilkinson et al., [2002] and Lee et al.

[2004] for examples). In general, however, construction of large libraries with large inserts is often challenging due to problems with the quality of DNA from many environmental sources.

22.5 SCREENING OF METAGENOMIC LIBRARIES

Several different methods exist for functional screening; and because the frequency of the metagenomic clones that express a given activity is low, the method should be either highly sensitive or carried out in a high-throughput manner, or preferably both. Whereas some screening approaches can be carried out on agar plates supplemented with appropriate substrates, other approaches require the use of 96- or 384- (or possibly even higher) well formats, which enable separate cultivation of the individual clones with subsequent assaying. The accomplishment of such high-throughput experiments requires the use of robotic systems, which not only improves the reproducibility and reduces data scattering, but also allows for miniaturization that may reduce costs (due to reduced amounts of both the individual clones and the components of the assay). Handling of clone pools in liquid cultures tends to result in biased libraries, but Hrvatin and Piel [2007] suggested a method for handling primary libraries that reduces this problem, involving semisolid media for generation of growth in three dimensions.

22.5.1 Screening by Growth Selection

The most convenient screening methods are based on growth selection, where the presence of a given activity provides a growth advantage to the organism in which the gene of interest is expressed. Selection for antibiotic or heavy metal resistance is one example [Diaz-Torres et al., 2003; Riesenfeld et al., 2004; Mirete et al., 2007; Mori et al., 2008; Kazimierczak et al., 2009; see also Chapter 35, Vol. II], whereas another approach is the use of mutant host strains that require heterologous complementation for growth under selective conditions. Examples of the latter screening approach are detection of phosphonate utilization pathways [Martinez et al., 2009], DNA polymerase I [Simon et al., 2009], lysine racemases [Chen et al., 2009], naphthalene dioxygenase [Ono et al., 2007], enzymes involved in poly-3-hydroxybutyrate metabolism [Wang et al., 2006], glycerol dehydratases [Knietsch et al., 2003a], operons for biotin biosynthesis [Entcheva et al., 2001], and genes encoding Na^+/H^+ antiporters [Majerník et al., 2001; see also Chapters 47–53, Vol. II].

22.5.2 Screening by Detection of Specific Phenotypes

When the function of interest does not provide the basis for selection, an alternative approach is through detection of specific phenotypes. For this approach the individual clones in the library need to be physically separated, either on agar media or in liquid phase in microtiter plates, and assayed individually for the given trait preferably in a high-throughput manner. To be able to identify enzymatic activities, chemical dyes and insoluble, chromogenic, or fluorogenic substrates can be incorporated into the growth medium [Daniel, 2005]. Recent examples are identification of metalloproteases [Waschkowitz et al., 2009] and esterases [Elend et al., 2006; Chu et al., 2008; Wu and Sun, 2009] that are identified due to formation of clear halos on agar plates containing skimmed milk or tributyrin, respectively, and also extradiol dioxygenases [Suenaga et al., 2007] identified through formation of a yellow colored product (see also Section 6, Vol. II). Antimicrobial or antifungal activity may be detected through growth inhibition of a suitable indicator organism, which either consists of overlaid colonies of the library in a soft agar medium or is grown in microtiter plates in the presence of extracts of the clones [Rondon et al., 2000; Courtois et al., 2003; Brady et al., 2004; Chung et al., 2008; Craig et al., 2009]. Detection assays may also be carried out in mutated strains where heterologous complementation of a certain trait results in a detectable phenotype. In general, accomplishment of detection assays on solid/agar media may be limited in sensitivity because soluble products often diffuse away from the colony, thus leading to detection of only very highly expressing clones.

22.5.3 Screening Based on Induced Gene Expression

Not all activities/functions can be easily linked to a detectable phenotype; for those cases, "substrate-induced gene expression screening" (SIGEX) can be an alternative [Handelsman, 2005; Uchiyama et al., 2005]. SIGEX is a high-throughput screening approach for identification of catabolic genes through the use of an operon trap *gfp*-expression vector. In the vector, metagenomic inserts are cloned upstream of a promoterless *gfp* gene, thus placing expression of green fluorescent protein (GFP) under the control of promoters in the metagenomic DNA. When the clones are incubated in the presence of a target substrate that is acting as an inducer, positive clones are identified by fluorescence cell sorting (FACS). A prescreen without the substrate allows for elimination of false-positive clones. Limitations of the application are that it misses catabolite genes that are not induced upon the substrate or do not have transcriptional regulators localized close to them, and it also misses those genes that are either cloned in opposite direction of the reporter gene or have a transcription terminator between the catabolite genes and the *gfp* gene. Uchiyama et al. [2005] successfully applied SIGEX to isolate 35 aromatic hydrocarbon-induced genes from a groundwater metagenomic library. Another, similar screening technique designated metabolite-related expression (METREX) has been developed by Williamson et al. [2005]. Here, a biosensor that detects small molecules inducing quorum-sensing is inside the same cell as the vector carrying the metagenomic insert. If the clone produces a quorum-sensing inducer, the cell produces GFP, and the fluorescent clone can be identified by fluorescence microscopy or FACS. Detection systems that are induced by the product of a catabolic reaction have also been reported [Mohn et al., 2006; van Sint Fiet et al., 2006]. In these systems a transcriptional regulator that is activated by the product in the reaction of interest is cloned downstream of a reporter gene (*lacZ* or *tetA*). Van Sint Fiet et al. [2006] reported such a system that detects biocatalysts responsible for the formation of benzoate and 2-hydroxybenzoate from their aldehydes, and in the system described by Mohn et al. [2006], biocatalysts converting γ-hexachlorocyclohexan to 1,2,4-trichlorobenzene are detected. This product-sensing reporter system can be modified to extend the range of biocatalysts that is possible to detect.

22.5.4 Sequence-Based Screening

Sequence-based screening approaches include the use of PCR-based or hybridization-based techniques for identification of target genes through the use of primers or probes (respectively) designed from conserved regions of known genes or gene products. A high-throughput alternative to the traditional colony hybridization is the use of microarrays for screening of metagenomic libraries. In such a "metagenome microarray" (MGA) the metagenomic library plasmids are spotted on a slide and specific, labeled gene probes are used for hybridization [Sebat et al., 2003; Park et al., 2008]. Sequence-based approaches may also include direct sequencing of the insert DNA followed by bioinformatics analysis of the obtained sequences [Kunin et al., 2008; Sleator et al., 2008]. In any case, the identification of novel genes are based on predictions made on the basis of already known gene sequences, thus limiting the total potential. Other limitations are that positive clones may not harbor complete genes or pathways, nor give rise to functional gene products. On the other hand, the advantage is that successful expression is not necessary, thus harboring the potential of identifying genes that will never be expressed in heterologous hosts. Jogler et al. [2009] used a hybridization-based method to identify two fosmids containing operons with homology to magnetosome islands of magnetotactic bacteria. Other recent examples reported by Banik and Brady [2008] and Kim et al. [2007] include identification of two novel glycopeptide-encoding gene clusters and a cytochrome P450 monooxygenase gene, respectively, using PCR-based screening methods. There are also examples of using target nonspecific primers, such as those designed for amplification of gene cassettes within integrons [Stokes et al., 2001; Elsaied et al., 2007; Koenig et al., 2009]. Such cassettes harbor open reading frames often encoding important features (e.g., antibiotic resistance determinants [Rowe-Magnus and Mazel, 1999]) that are flanked by *attC* sequences required for the translocation and recombination of integrons [Hall, 1997; see also Chapter 26, Vol I]. Rowe-Magnus [2009] also describes a sequence-independent method for recovery of such gene cassettes through the use of a three-plasmid genetic strategy. Clones containing integron gene cassettes within a metagenomic library are here fused to another plasmid due to recombination at the *attC* site, and selection of fusions is accomplished through subsequent conjugation to a second strain and plating on selective media.

22.6 PERSPECTIVES

Identification of new genes of interest in metagenomic libraries by functional screening has the advantage that completely new traits can be discovered, even if the sequences of the corresponding genes display no similarity to already known genes. The main disadvantages of this approach are that the protocols are often laborious and screening can consequently in some cases be very costly, even though advances in robotics have reduced this problem. Since DNA can now be sequenced efficiently without prior cloning and to a lower and lower price, it is likely that such an approach will become increasingly adopted in the future. The disadvantage of this method is that functions of interest that cannot be detected by bioinformatics analyses will not be identified. Additionally, most biosynthetic pathways will not be assembled from a metagenomic sequence collection into complete contiguous genomic fragments, nor will these pathways be expressed to produce a secondary metabolite. These disadvantages for a purely sequence-based approach will remain an insurmountable obstacle to discovery of much of the genetic diversity present in natural environments.

The metagenomic surveys of microbial and viral communities that are completed so far suggest that the diversity of environmental communities is exceptional and that more than 60% of the sequences are novel sequences with unknown functions [Ferrer et al., 2009b]. This indicates that functional screening may become quite important for many years in order to map more of the gene sequence space that exists in nature. There is little doubt that both direct sequencing and functional screening of libraries will be used in the years to come, and completely new functions and sequences discovered by functional screening will in the long run contribute to improved bioinformatics predictions. We therefore believe that the long-term tendency is likely to move more in the direction of direct sequencing coupled to bioinformatics analyses, but that functional screening will also be important in the foreseeable future. The use of a combined sequence- and function-based metagenomic approach can overcome some of these biases that are inherent in a metagenomic approach, permitting access to the as-yet-uncultured extant functional diversity of microbial life on Earth.

REFERENCES

Aakvik T, Degnes KF, Dahlsrud R, Schmidt F, Dam R, Yu L, Völker U, Ellingsen TE, Valla S. 2009. A plasmid RK2-based broad-host-range cloning vector useful for transfer of metagenomic libraries to a variety of bacterial species. *FEMS Microbiol. Lett*. **296**:149–158.

Abulencia CB, Wyborski DL, Garcia JA, Podar M, Chen W, et al. 2006. Environmental whole-genome amplification to access microbial populations in contaminated sediments. *Appl. Environ. Microbiol*. **72**:3291–3301.

Amann RI, Ludwig W, Schleifer KH. 1995. Phylogenetic identification and in situ detection of individual microbial cells without cultivation. *Microbiol. Rev*. **59**:143–169.

Angelov A, Mientus M, Liebl S, Liebl W. 2009. A two-host fosmid system for functional screening of (meta)genomic libraries from extreme thermophiles. *Syst. Appl. Microbiol*. **32**:177–185.

Bailly J, Fraissinet-Tachet L, Verner M-C, Debaud J-C, Lemaire M, Wésolowski-Louvel M, Marmeisse R. 2007. Soil eukaryotic functional diversity, a metatranscriptomic approach. *ISME J*. **1**:632–642.

Banik JJ, Brady SF. 2008. Cloning and characterization of new glycopeptide gene clusters found in an environmental DNA megalibrary. *Proc. Natl. Acad. Sci. USA* **105**:17273–17277.

Bernstein JR, Bulter T, Shen CR, Liao JC. 2007. Directed evolution of ribosomal protein S1 for enhanced translational efficiency of high GC *Rhodopseudomonas palustris* DNA in *Escherichia coli*. *J. Biol. Chem*. **282**:18929–18936.

Brady SF, Chao CJ, Clardy J. 2004. Long-chain *N*-acyltyrosine synthases from environmental DNA. *Appl. Environ. Microbiol*. **70**:6865–6870.

Brautaset T, Sekurova ON, Sletta H, Ellingsen TE, Strøm AR, Valla S, Zotchev SB. 2000. Biosynthesis of the polyene antifungal antibiotic nystatin in *Streptomyces noursei* ATCC 11455: Analysis of the gene cluster and deduction of the biosynthetic pathway. *Chem. Biol*. **7**:395–403.

Brikun IA, Reeves AR, Cernota WH, Luu MB, Weber JM. 2004. The erythromycin biosynthetic gene cluster of *Aeromicrobium erythreum*. J. Ind. Microbiol. Biotechnol. **31**:335–344.

Chen I-C, Lin W-D, Hsu S-K, Thiruvengadam V, Hsu W-H. 2009. Isolation and characterization of a novel lysine racemase from a soil metagenomic library. *Appl. Environ. Microbiol*. **75**:5161–5166.

Chu X, He H, Guo C, Sun B. 2008. Identification of two novel esterases from a marine metagenomic library derived from South China Sea. *Appl. Microbiol. Biotechnol*. **80**:615–625.

Chung EJ, Lim HK, Kim J-C, Choi GJ, Park EJ, et al. 2008. Forest soil metagenome gene cluster involved in antifungal activity expression in *Escherichia coli*. *Appl. Environ. Microbiol*. **74**:723–730.

Collins J, Hohn B. 1978. Cosmids: a type of plasmid gene-cloning vector that is packageable in vitro in bacteriophage lambda heads. *Proc. Natl. Acad. Sci. USA* **75**:4242–4246.

Courtois S, Cappellano CM, Ball M, Francou F-X, Normand P, et al. 2003. Recombinant environmental libraries provide access to microbial diversity for drug discovery from natural products. *Appl. Environ. Microbiol*. **69**:49–55.

Cowan D, Meyer Q, Stafford W, Muyanga S, Cameron R, Wittwer P. 2005. Metagenomic gene discovery: past, present and future. *Trends Biotechnol*. **23**:321–329.

Craig JW, Chang F-Y, Brady SF. 2009. Natural products from environmental DNA hosted in *Ralstonia metallidurans*. *ACS Chem. Biol*. **4**:23–28.

Daniel R. 2005. The metagenomics of soil. *Nat. Rev. Microbiol*. **3**:470–478.

Diaz-Torres ML, McNab R, Spratt DA, Villedieu A, Hunt N, Wilson M, Mullany P. 2003. Novel tetracycline resistance determinant from the oral metagenome. *Antimicrob. Agents. Chemother*. **47**:1430–1432.

Elend C, Schmeisser C, Leggewie C, Babiak P, Carballeira JD, Steele HL, Reymond J-L, Jaeger K-E, Streit WR. 2006. Isolation and biochemical characterization of two novel metagenome-derived esterases. *Appl. Environ. Microbiol*. **72**:3637–3645.

Elsaied H, Stokes HW, Nakamura T, Kitamura K, Fuse H, Maruyama A. 2007. Novel and diverse integron integrase genes and integron-like gene cassettes are prevalent in deep-sea hydrothermal vents. *Environ. Microbiol*. **9**:2298–2312.

Entcheva P, Liebl W, Johann A, Hartsch T, Streit WR. 2001. Direct cloning from enrichment cultures, a reliable strategy for isolation of complete operons and genes from microbial consortia. *Appl. Environ. Microbiol*. **67**:89–99.

Ferrer M, Chernikova TN, Timmis KN, Golyshin PN. 2004. Expression of a temperature-sensitive esterase in a novel chaperone-based *Escherichia coli* strain. *Appl. Environ. Microbiol*. **70**:4499–4504.

Ferrer M, Golyshina OV, Chernikova TN, Khachane AN, Reyes-Duarte D, Santos VAPMD, Strompl C, Elborough K, Jarvis G, Neef A, et al. 2005. Novel hydrolase diversity retrieved from a metagenome library of bovine rumen microflora. *Environ. Microbiol*. **7**:1996–2010.

Ferrer M, Beloqui A, Timmis KN, Golyshin PN. 2009a. Metagenomics for mining new genetic resources of microbial communities. *J. Mol. Microbiol. Biotechnol*. **16**:109–123.

Ferrer M, Beloqui A, Vieites JM, Guazzaroni ME, Berger I, Aharoni A. 2009b. Interplay of metagenomics and in vitro compartmentalization. *Microbial Biotechnol*. **2**:31–39.

Gabor E, Liebeton K, Niehaus F, Eck J, Lorenz P. 2007. Updating the metagenomics toolbox. Biotechnol. J. **2**:201–206.

Gabor EM, Vries EJ, Janssen DB. 2003. Efficient recovery of environmental DNA for expression cloning by indirect extraction methods. *FEMS Microbiol. Ecol*. **44**:153–163.

Gabor EM, Alkema WBL, Janssen DB. 2004a. Quantifying the accessibility of the metagenome by random expression cloning techniques. *Environ. Microbiol*. **6**:879–886.

Gabor EM, de Vries EJ, Janssen DB. 2004b. Construction, characterization, and use of small-insert gene banks of DNA isolated from soil and enrichment cultures for the recovery of novel amidases. *Environ. Microbiol*. **6**:948–958.

Grant S, Grant WD, Cowan DA, Jones BE, Ma Y, Ventosa A, Heaphy S. 2006. Identification of eukaryotic open reading frames in metagenomic cDNA libraries made from environmental samples. *Appl. Environ. Microbiol.* **72**:135–143.

Hain T, Otten S, von Both U, Chatterjee SS, Technow U, Billion A, Ghai R, Mohamed W, Domann E, Chakraborty T. 2008. Novel bacterial artificial chromosome vector pUvBBAC for use in studies of the functional genomics of *Listeria* spp. *Appl. Environ. Microbiol.* **74**:1892–1901.

Hall RM. 1997. Mobile gene cassettes and integrons: moving antibiotic resistance genes in gram-negative bacteria. *Ciba Found. Symp.* **207**:192–202; discussion 202–205.

Handelsman J. 2005. Sorting out metagenomes. *Nat. Biotechnol.* **23**:38–39.

Handelsman J, Rondon MR, Brady SF, Clardy J, Goodman RM. 1998. Molecular biological access to the chemistry of unknown soil microbes: a new frontier for natural products. *Chem. Biol.* **5**: R245–R249.

Heath C, Hu XP, Cary SC, Cowan D. 2009. Identification of a novel alkaliphilic esterase active at low temperatures by screening a metagenomic library from antarctic desert soil. *Appl. Environ. Microbiol.* **75**:4657–4659.

Henne A, Daniel R, Schmitz RA, Gottschalk G. 1999. Construction of environmental DNA libraries in *Escherichia coli* and screening for the presence of genes conferring utilization of 4-hydroxybutyrate. *Appl. Environ. Microbiol.* **65**:3901–3907.

Hrvatin S, Piel X Jr. 2007. Rapid isolation of rare clones from highly complex DNA libraries by PCR analysis of liquid gel pools. *J. Microbiol. Methods* **68**:434–436.

Jogler C, Lin W, Meyerdierks A, Kube M, Katzmann E, Flies C, Pan Y, Amann R, Reinhardt R, Schüler D. 2009. Toward cloning of the magnetotactic metagenome: Identification of magnetosome island gene clusters in uncultivated magnetotactic bacteria from different aquatic sediments. *Appl. Environ. Microbiol.* **75**:3972–3979.

Kakirde KS, Wild J, Godiska R. Mead DA. Wiggins AG. Goodman RM. Szybalski W. Liles MR . 20106. Gram negative shuttle BAC vector for heterologous expression of metagenomic. *libraries. Gene.*

Kazimierczak KA, Scott KP, Kelly D, Aminov RI. 2009. Tetracycline resistome of the organic pig gut. *Appl. Environ. Microbiol.* **75**:1717–1722.

Kim BS, Kim SY, Park J, Park W, Hwang KY, Yoon, et al. 2007. Sequence-based screening for self-sufficient P450 monooxygenase from a metagenome library. *J. Appl. Microbiol.* **102**:1392–1400.

Kim K-H, Chang H-W, Nam Y-D, Roh SW, Kim M-S, et al. 2008. Amplification of uncultured single-stranded DNA viruses from rice paddy soil. *Appl. Environ. Microbiol.* **74**:5975–5985.

Kim UJ, Shizuya H, de Jong PJ, Birren B, Simon MI. 1992. Stable propagation of cosmid sized human DNA inserts in an F factor based vector. *Nucleic Acids Res.* **20**:1083–1085.

Knietsch A, Bowien S, Whited G, Gottschalk G, Daniel R. 2003a. Identification and characterization of coenzyme B12-dependent glycerol dehydratase- and diol dehydratase-encoding genes from metagenomic DNA libraries derived from enrichment cultures. *Appl. Environ. Microbiol.* **69**:3048–3060.

Knietsch A, Waschkowitz T, Bowien S, Henne A, Daniel R. 2003b. Construction and screening of metagenomic libraries derived from enrichment cultures: Generation of a gene bank for genes conferring alcohol oxidoreductase activity on *Escherichia coli*. *Appl. Environ. Microbiol.* **69**:1408–1416.

Koenig JE, Sharp C, Dlutek M, Curtis B, Joss M, Boucher Y, Doolittle WF. 2009. Integron gene cassettes and degradation of compounds associated with industrial waste: The case of the Sydney tar ponds. *PLoS One* **4**: e5276.

Krsek M, Wellington EM. 1999. Comparison of different methods for the isolation and purification of total community DNA from soil. *J. Microbiol. Methods* **39**:1–16.

Kunin V, Copeland A, Lapidus A, Mavromatis K, Hugenholtz P. 2008. A bioinformatician's guide to metagenomics. *Microbiol. Mol. Biol. Rev.* **72**:557–578.

Lämmle K, Zipper H, Breuer M, Hauer B, Buta C, Brunner H, Rupp S. 2007. Identification of novel enzymes with different hydrolytic activities by metagenome expression cloning. *J. Biotechnol.* **127**:575–592.

Lee S-W, Won K, Lim HK, Kim J-C, Choi GJ, Cho KY. 2004. Screening for novel lipolytic enzymes from uncultured soil microorganisms. *Appl. Microbiol. Biotechnol.* **65**:720–726.

Li Y, Wexler M, Richardson DJ, Bond PL, Johnston AWB. 2005. Screening a wide host-range, waste-water metagenomic library in tryptophan auxotrophs of *Rhizobium leguminosarum* and of *Escherichia coli* reveals different classes of cloned trp genes. *Environ. Microbiol.* **7**:1927–1936.

Liles MR, Williamson LL, Rodbumrer J, Torsvik V, Goodman RM, Handelsman J. 2008. Recovery, purification, and cloning of high-molecular-weight DNA from soil microorganisms. *Appl. Environ. Microbiol.* **74**:3302–3305.

Majerník A, Gottschalk G, Daniel R. 2001. Screening of environmental DNA libraries for the presence of genes conferring Na(+)(Li(+))/H(+) antiporter activity on *Escherichia coli*: characterization of the recovered genes and the corresponding gene products. *J. Bacteriol.* **183**:6645–6653.

Martin-Laurent F, Philippot L, Hallet S, Chaussod R, Germon JC, Soulas G, Catroux G. 2001. DNA extraction from soils: Old bias for new microbial diversity analysis methods. *Appl. Environ. Microbiol.* **67**:2354–2359.

Martinez A, Kolvek SJ, Yip CLT, Hopke J, Brown KA, MacNeil IA, Osburne MS. 2004. Genetically modified bacterial strains and novel bacterial artificial chromosome shuttle vectors for constructing environmental libraries and detecting heterologous natural products in multiple expression hosts. *Appl. Environ. Microbiol.* **70**:2452–2463.

Martinez A, Tyson GW, DeLong EF. 2009. Widespread known and novel phosphonate utilization pathways in marine bacteria revealed by functional screening and metagenomic analyses. *Environ. Microbiol.* **12**:222–238.

Mirete S, de Figueras CG, González-Pastor JE. 2007. Novel nickel resistance genes from the rhizosphere metagenome of plants adapted to acid mine drainage. *Appl. Environ. Microbiol.* **73**:6001–6011.

Mohn WW, Garmendia J, Galvao TC, de Lorenzo Vc. 2006. Surveying biotransformations with Ã la carte genetic traps: Translating dehydrochlorination of lindane (gamma-hexachlorocyclohexane) into lacZ-based phenotypes. *Environ. Microbiol.* **8**:546–555.

Mori T, Mizuta S, Suenaga H, Miyazaki K. 2008. Metagenomic screening for bleomycin resistance genes. *Appl. Environ. Microbiol.* **74**:6803–6805.

Nishihara K, Kanemori M, Kitagawa M, Yanagi H, Yura T. 1998. Chaperone coexpression plasmids: Differential and synergistic roles of DnaK-DnaJ-GrpE and GroEL-GroES in assisting folding of an allergen of Japanese cedar pollen, Cryj2, in *Escherichia coli*. *Appl. Environ. Microbiol.* **64**:1694–1699.

Ogram A, Sayler GS, Barkay T. 1987. The extraction and purification of microbial DNA from sediments. *J. Microbiol. Methods* **7**:57–66.

Ono A, Miyazaki R, Sota M, Ohtsubo Y, Nagata Y, Tsuda M. 2007. Isolation and characterization of naphthalene-catabolic genes and plasmids from oil-contaminated soil by using two cultivation-independent approaches. *Appl. Microbiol. Biotechnol.* **74**:501–510.

Ouyang Y, Dai S, Xie L, Kumar MR, Sun W, Sun H, Tang D, Li X. 2009. Isolation of high molecular weight DNA from marine sponge bacteria for BAC library construction. *Mar. Biotechnol.* **12**:318–325.

PARK S-J, KANG C-H, CHAE J-C, RHEE S-K. 2008. Metagenome microarray for screening of fosmid clones containing specific genes. *FEMS Microbiol. Lett*. **284**:28–34.

RAPPÉ MS, GIOVANNONI SJ. 2003. The uncultured microbial majority. *Annu. Rev. Microbiol*. **57**:369–394.

REES HC, GRANT S, JONES B, GRANT WD, HEAPHY S. 2003. Detecting cellulase and esterase enzyme activities encoded by novel genes present in environmental DNA libraries. *Extremophiles* **7**:415–421.

RIESENFELD CS, GOODMAN RM, HANDELSMAN J. 2004. Uncultured soil bacteria are a reservoir of new antibiotic resistance genes. *Environ. Microbiol*. **6**:981–989.

RONDON MR, AUGUST PR, BETTERMANN AD, BRADY SF, GROSSMAN TH, et al. 2000. Cloning the soil metagenome: A strategy for accessing the genetic and functional diversity of uncultured microorganisms. *Appl. Environ. Microbiol*. **66**:2541–2547.

ROSANO GL, CECCARELLI EA. 2009. Rare codon content affects the solubility of recombinant proteins in a codon bias-adjusted *Escherichia coli* strain. *Microb. Cell Fact*. **8**: 41.

ROWE-MAGNUS DA. 2009. Integrase-directed recovery of functional genes from genomic libraries. *Nucleic Acids Res*. **37**: e118.

ROWE-MAGNUS DA, MAZEL D. 1999. Resistance gene capture. *Curr. Opin. Microbiol*. **2**:483–488.

SAGOVA-MARECKOVA M, CERMAK L, NOVOTNA J, PLHACKOVA K, FORSTOVA J, KOPECKY J. 2008. Innovative methods for soil DNA purification tested in soils with widely differing characteristics. *Appl. Environ. Microbiol*. **74**:2902–2907.

SEBAT JL, COLWELL FS, CRAWFORD RL. 2003. Metagenomic profiling: microarray analysis of an environmental genomic library. *Appl. Environ. Microbiol*. **69**:4927–4934.

SHIZUYA H, BIRREN B, KIM UJ, MANCINO V, SLEPAK T, TACHIIRI Y, SIMON M. 1992. Cloning and stable maintenance of 300-kilobase-pair fragments of human DNA in *Escherichia coli* using an F-factor-based vector. *Proc. Natl. Acad. Sci. USA* **89**:8794–8797.

SIMON C, HERATH J, ROCKSTROH S, DANIEL R. 2009. Rapid identification of genes encoding DNA polymerases by function-based screening of metagenomic libraries derived from glacial ice. *Appl. Environ. Microbiol*. **75**:2964–2968.

SLEATOR RD, SHORTALL C, HILL C. 2008. Metagenomics. *Lett. Appl. Microbiol*. **47**:361–366.

SOSIO M, GIUSINO F, CAPPELLANO C, BOSSI E, PUGLIA AM, DONADIO S. 2000. Artificial chromosomes for antibiotic-producing actinomycetes. *Nat. Biotechnol*. **18**:343–345.

STEFFAN RJ, GOKSØYR J, BEJ AK, ATLAS RM. 1988. Recovery of DNA from soils and sediments. *Appl. Environ. Microbiol*. **54**:2908–2915.

STOKES HW, HOLMES AJ, NIELD BS, HOLLEY MP, NEVALAINEN KM, MABBUTT BC, GILLINGS MR. 2001. Gene cassette PCR: Sequence-independent recovery of entire genes from environmental DNA. *Appl. Environ. Microbiol*. **67**:5240–5246.

SUENAGA H, OHNUKI T, MIYAZAKI K. 2007. Functional screening of a metagenomic library for genes involved in microbial degradation of aromatic compounds. *Environ. Microbiol*. **9**:2289–2297.

TIRAWONGSAROJ P, SRIPRANG R, HARNPICHARNCHAI P, THONGARAM T, CHAMPREDA V, TANAPONGPIPAT S, POOTANAKIT K, EURWILAICHITR L. 2008. Novel thermophilic and thermostable lipolytic enzymes from a Thailand hot spring metagenomic library. *J. Biotechnol*. **133**:42–49.

TORSVIK V, GOKSØYR J, DAAE FL. 1990. High diversity in DNA of soil bacteria. *Appl. Environ. Microbiol*. **56**:782–787.

TSAI YL, OLSON BH. 1991. Rapid method for direct extraction of DNA from soil and sediments. *Appl. Environ. Microbiol*. **57**:1070–1074.

TYSON GW, BANFIELD JF. 2005. Cultivating the uncultivated: A community genomics perspective. *Trends Microbiol*. **13**:411–415.

UCHIYAMA T, ABE T, IKEMURA T, WATANABE K. 2005. Substrate-induced gene-expression screening of environmental metagenome libraries for isolation of catabolic genes. *Nat. Biotechnol*. **23**:88–93.

VAN SINT FIET S, VAN BEILEN JB, WITHOLT B. 2006. Selection of biocatalysts for chemical synthesis. *Proc. Natl. Acad. Sci. USA* **103**:1693–1698.

WALL JG, PLÜCKTHUN A. 1995. Effects of overexpressing folding modulators on the in vivo folding of heterologous proteins in *Escherichia coli*. *Curr. Opin. Biotechnol*. **6**:507–516.

WANG C, MEEK DJ, PANCHAL P, BORUVKA N, ARCHIBALD FS, DRISCOLL BT, CHARLES TC. 2006. Isolation of poly-3-hydroxybutyrate metabolism genes from complex microbial communities by phenotypic complementation of bacterial mutants. *Appl. Environ. Microbiol*. **72**:384–391.

WARREN RL, FREEMAN JD, LEVESQUE RC, SMAILUS DE. FLIBOTTE S, HOLT RA. 2008. Transcription of foreign DNA in *Escherichia coli*. *Genome Res*. **18**:1798–1805.

WASCHKOWITZ T, ROCKSTROH S, DANIEL R. 2009. Isolation and characterization of metalloproteases with a novel domain structure by construction and screening of metagenomic libraries. *Appl. Environ. Microbiol*. **75**:2506–2516.

WEXLER M, BOND PL, RICHARDSON DJ, JOHNSTON AWB. 2005. A wide host-range metagenomic library from a waste water treatment plant yields a novel alcohol/aldehyde dehydrogenase. *Environ. Microbiol*. **7**:1917–1926.

WHITMAN WB, COLEMAN DC, WIEBE WJ. 1998. Prokaryotes: the unseen majority. *Proc. Natl. Acad. Sci. USA* **95**:6578–6583.

WILD J, HRADECNA Z, SZYBALSKI W. 2002. Conditionally amplifiable BACs: switching from single-copy to high-copy vectors and genomic clones. *Genome Res*. **12**:1434–1444.

WILKINSON DE, JEANICKE T, COWAN DA. 2002. Efficient molecular cloning of environmental DNA from geothermal sediments. *Biotechnol. Lett*. **24**:155–161.

WILLIAMSON LL, BORLEE BR, SCHLOSS PD, GUAN C, ALLEN HK, HANDELSMAN J. 2005. Intracellular screen to identify metagenomic clones that induce or inhibit a quorum-sensing biosensor. *Appl. Environ. Microbiol*. **71**:6335–6344.

WU C, SUN B. 2009. Identification of novel esterase from metagenomic library of Yangtze river. *J. Microbiol. Biotechnol*. **19**:187–193.

YAMADA K, TERAHARA T, KURATA S, YOKOMAKU T, TSUNEDA S, HARAYAMA S. 2008. Retrieval of entire genes from environmental DNA by inverse PCR with pre-amplification of target genes using primers containing locked nucleic acids. *Environ. Microbiol*. **10**:978–987.

YOKOUCHI H, FUKUOKA Y, MUKOYAMA D, CALUGAY R, TAKEYAMA H, MATSUNAGA T. 2006. Whole-metagenome amplification of a microbial community associated with scleractinian coral by multiple displacement amplification using phi29 polymerase. *Environ. Microbiol*. **8**:1155–1163.

ZAEHNER H, FIEDLER FP. 1995. Fifty years of antimicrobials: past perspectives and future trends. In HUNTER PA, DARBY GK, RUSSELL NJ, eds. *The Need for New Antibiotics: Possible Ways Forward*. Fifty-Third Symposium of the Society for General Microbiology. Cambridge, UK: Cambridge University Press, pp. 67–84.

ZHOU J, BRUNS MA, TIEDJE JM. 1996. DNA recovery from soils of diverse composition. *Appl. Environ. Microbiol*. **62**:316–322.

Chapter 23

GC Fractionation Allows Comparative Total Microbial Community Analysis, Enhances Diversity Assessment, and Facilitates Detection of Minority Populations of Bacteria

William E. Holben

23.1 INTRODUCTION

A major challenge in modern microbial ecology is to effectively and accurately determine total microbial community diversity, particularly with regard to the detection of (a) unculturable and fastidious bacterial species and (b) those present only in low abundance (i.e., minority populations). A common theme in published studies and textbooks regarding microbial community diversity in most environments is that typically only 0.1–1.0% of bacteria observed by direct microscopic enumeration can be recovered on using general laboratory media (e.g., see Ferguson, et al. [1984], Fliermans and Balkwill [1989], Hazen et al. [1991], Rappel and Giovannoni [2003], Janssen [2006], Jones et al. [2009], and Spain et al. [2009]). As a result, many microbial ecologists are of the opinion that the vast majority of microbial diversity remains uncharacterized, highlighting both potentially huge gaps in our understanding of how microbial communities function in an ecosystem and the large reservoir of possibly useful organisms and genes. This concern has spurred the development of a large number of molecular approaches for microbial community analysis, many of which are based on analysis of nucleic acids extracted directly from environmental samples to enhance our access to, and thus knowledge of, previously unknown microbial populations. The suite of nucleic acid-based community analysis approaches can be organized into two general classes: (1) compilation-based analyses that combine individual bits of data to obtain a sense of community structure (e.g., 16S rRNA gene libraries; metagenomic analyses; and transcriptomic analyses (see Chapter 2, Vol. I)) and (2) total community analyses that attempt to characterize the entire community in a single analysis (e.g., denaturing gradient gel electrophoresis (DGGE, see Chapter 5, Vol. II); terminal restriction fragment length polymorphism (T-RFLP); microarray-based analyses including the GeoChip, PhyloChip (see Chapters 57 and 58, Vol. I), and reverse sample genome probing (RSGP); community DNA reannealing kinetics, see Chapter 2, Vol. I; and GC fractionation).

Compilation-based strategies typically involve a random ("shotgun") sampling approach wherein either related functional gene sequences or ribosomal gene sequences from individual community members are targeted, generally by PCR, and cloned from total community DNA for comparative or phylogenetic analyses. More recently, large-scale, shotgun-based direct sequencing of complex mixtures of community DNA (metagenomic analysis), mRNA (transcriptomic analysis (see Chapters 62–65,

Handbook of Molecular Microbial Ecology, Volume I: Metagenomics and Complementary Approaches, First Edition. Edited by Frans J. de Bruijn.

Vol. I)), or rRNA (representing actively growing bacterial populations) from environmental samples (collectively termed "omic" approaches) is being performed using nascent and still-evolving high-throughput sequencing platforms. Researchers employ these technologies in attempts to develop a view of microbial community composition and function from thousands, millions, or even billions, of snippets of sequence information from individual samples. This has led to a rapidly increasing abundance of microbial community sequence information, although much of it is poorly characterized (i.e., putative genes of unknown function), unannotated and unassembled largely due to the inability of even millions of randomly sampled sequence reads to provide adequate coverage of the numerous individual population genomes comprising the community [Morales and Holben, 2010].

Nonetheless, these compilation-based approaches have proven very powerful and over the last two decades have generated substantial information on microbial community diversity and function in a number of systems. Indeed, research in these areas has been so prevalent that a large number of reviews and opinions on their applications in microbial ecology in a broad variety of systems have been published of late (e.g., see Tiedje et al. [1999], Fuhrman [2002], McCartney [2002], Stahl [2004], Schloss and Handelsman [2005], Spiegelman et al. [2005], Green and Keller [2006], Nagy et al. [2006], Juste et al. [2008], Kunin et al. [2008], Tringe and Hugenholtz [2008], Andoh et al. [2009], and Wooley et al. [2010]). Still, it seems clear that random sampling in support of even high-throughput compilation-based approaches is limited in its ability to afford a comprehensive view of total diversity where communities are complex, leading some researchers to take develop more directed or focused approaches to metagenomic analyses that are explored in other chapters of the current volumes and elsewhere [Morales and Holben, 2011]. As a result, in complex microbial communities comprised of hundreds to thousands of individual taxa present in uneven abundance (e.g., oceans, soils, the gastrointestinal tracts of higher organisms), individual taxa present in low abundance (i.e., minority populations) will continue to go undetected where random sampling approaches are employed [Bent and Forney, 2008; Morales et al., 2009].

This limitation has led some researchers to take a theoretical approach to estimating total community diversity based on mathematical extrapolation from a partial analysis of the total community [Dunbar et al., 2000, 2002; Hughes et al., 2001; McCaig et al., 2001; Curtis et al., 2002; Martin 2002; Torsvik et al., 2002; Curtis and Sloan, 2004; Morales et al., 2009]. However, theoretical approaches provide no specific information regarding the identity or functionality of minority populations since their presence is inferred and no empirical information regarding them is obtained or analyzed.

By contrast, total community analyses generally strive to capture a sense of total community structure or diversity in a single analysis. Total community analyses include, but are not limited to, monitoring DNA reannealing kinetics [Torsvik et al., 1990a,b; see also Chapter 2, Vol. I], T-RFLP [Liu et al., 1997], DGGE [Muyzer et al., 1993; see also Chapter 5, Vol. I], array-based probing using GeoChips, PhyloChips, and RSGP [He et al., 2007; Greene and Voordouw, 2003; Brodie et al., 2006; see also Chapters 57–61, Vol. I], and %G+C-based fractionation of total community DNA (hereafter GC fractionation) [Holben and Harris, 1995]. While these approaches typically probe the entire community in a single analysis of total community DNA, or in some cases rRNA, they generally do not provide direct, high-resolution identification of the populations present without being combined with additional downstream procedures as will be discussed below.

The remainder of this chapter focuses in particular on GC fractionation, with a mechanistic overview of the procedure in the remainder of this Introduction and a detailed description of the approach in the Materials and Methods section. Data illustrating and discussing the outcome of GC fractionation along with representative examples and descriptions of information regarding microbial communities that can be obtained using combined downstream approaches are presented in the Results and Discussion section. Consideration of the strengths and limitations of GC fractionation and the various combined approaches are also provided in the Results and Discussion section.

GC fractionation of total community DNA as originally conceived [Holben and Harris, 1995] is based on two important features of the interaction between the nonintercalating DNA-binding dye bisbenzimidazole (Hoechts reagent No. 33258) and DNA. These are (1) preferential binding to A+T regions of DNA [Weisblum and Haenssler, 1974; Comings, 1975; Muller and Gautier, 1975; Pjura et al., 1987; Searle and Embrey, 1990] and (2) lowering of the buoyant density of DNA proportional to the amount of dye bound [Manuelidis, 1977; Roell and Morse, 1991; Pfeifer et al., 1993]. Accordingly, the higher the %G+C content of a DNA molecule, the less bisbenzimidazole it binds per mole (or unit length) and the greater its buoyant density; this means that, in an equilibrium density gradient, it will band below DNA of lesser %G+C content. These properties of bisbenzimidazole have previously been used in chromosome banding and chromatin analysis experiments, in DNA quantitation, and for the isolation of mitochondrial, chloroplast, and mouse satellite DNA [Manuelidis, 1977; Jorgenson et al., 1978; Brunk et al., 1979; Paul and Myers, 1982; van den Engh et al., 1986; Roell and Morse, 1991; Pfeifer et al., 1993].

These same two binding features of bisbenzimidazole binding also mean that genomic DNA from an individual bacterial population will resolve in an equilibrium gradient at a position corresponding to its characteristic %G+C content and buoyant density when complexed with bisbenzimidazole [Holben and Harris, 1995]. By extension, GC fractionation of total community DNA will result in genomic DNA from the component populations resolving at positions in the gradient corresponding to their individual %G+C contents. The %G+C content of the bacterial genome is characteristic, within a reasonable range, of groups of bacteria at the genus to phylum level [Laskin and Lechevalier, 1973; Holben and Harris, 1995], so this approach resolves complex bacterial community DNA into discrete fractions of similar %G+C content with some element of phylogenetic coherence, although it must be noted that some different taxonomic groups of bacteria can share overlap in %G+C range and thus be contained within the same fraction.

As a result, GC fractionation, of all molecular techniques in the microbial ecologist's toolbox, is probably the only one that is completely independent of any previous knowledge regarding which bacterial populations comprise the community or their genomic content (DNA sequence). In addition, GC fractionation is also independent of PCR-, cloning-, and hybridization-based methods with their inherent biases. Given reliable methods for the recovery of bacterial community DNA from a sample (a topic discussed below), this approach is reliably quantitative, albeit with coarse resolution unless coupled to additional downstream analyses. Furthermore, GC fractionation represents a way to readily access minority populations within the community since it physically separates low biomass fractions of the community from those representing predominant portions of the community.

The output from GC fractionation is a profile of the entire community that indicates relative abundance of DNA as a function of %G+C content, as well as a set of physically separated fractions of total community DNA representing known ranges of %G+C content. These highly purified fractions are of large molecular weight ($\geq$25 kb) and thus suitable for many downstream molecular manipulations including PCR amplification, DGGE analysis, T-RFLP, cloning, and direct sequencing. GC fractionation has been shown to accurately and reproducibly profile and fractionate total microbial community DNA. Indeed, a concerted analysis of the reproducibility of the approach showed that the calculated average standard deviation was 5.0% across six replicate chicken cecal community profiles [Apajalahti et al., 2001]. Furthermore, statistical analysis of GC profiles allows meaningful comparison of microbial communities between samples and has been used to focus downstream analyses on points of interest within and between communities [Nusslein and Tiedje, 1998, 1999; Apajalahti et al., 2002, 2003; Kassinen, et al., 2007; Rodriguez-Minguela, et al., 2009].

GC fractionation of itself or in combination with additional analyses has been successfully employed to study and compare microbial community structure in a variety of environments including soils [Holben et al., 1993; Holben and Harris, 1995; Nusslein and Tiedje, 1998, 1999; Rodriguez-Minguela et al., 2009; Morales et al., 2009], sediments [Gsell et al., 1997; Schleper et al., 1997; Lowell et al., 2009; Morales et al., 2009], bioreactors [Holben et al., 1998; Probert et al., 2004], and the GI tracts of insects [Santo Domingo et al., 1998], animals [Apajalahti et al., 1998, 2001, 2002, 2004; Holben et al., 2004; Peuranen et al., 2004; Apajalahti and Kettunen, 2006; McCracken et al., 2006], and humans [Apajalahti et al., 2003; Kassinen et al., 2007; Dicksved et al., 2008]. The approach can also enable detection and characterization of taxa that are present in low abundance in the community, yet may have important functions such as nitrification or pathogenesis [Holben et al., 2004]. This technique can also be employed to separate the genomic DNA of uncultivated endosymbionts from the genomic DNA of its host [W. E. Holben, unpublished data] and to measure the %G+C content of genomic DNA of individual populations [Wallace et al., 2003]. Furthermore, GC fractionation was adopted by (a) a commercial firm, Diversa, Inc., of San Diego, California, in the 1990s to mine environmental DNA samples for genes encoding enzymes of potential value as originally suggested by W. E. Holben, and (b) another firm, Alimetrics, Inc., of Helsinki, Finland <http://www.alimetrics.com/html/>, to monitor the effects of feed formulations, prebiotics, and probiotics on the gastrointestinal microbiome of host animals.

23.2 MATERIALS AND METHODS

23.2.1 Microbial Community DNA Purification

In principle, microbial community DNA from any type of environmental sample, purified by any of the current plethora of environmental DNA purification protocols, can be used for GC fractionation, provided that it is of sufficient quantity for visualization and detection of DNA in the gradients (35–75 μg) and of sufficiently high molecular weight ($\geq$25-kb fragment length) to avoid effects of localized variance in %G+C content in the genomes of individual taxa. While comparison and discussion of the advantages and disadvantages of various DNA purification approaches is beyond the scope of this chapter (see Chapters 10–11, Vol. II), it can be noted that for our own work we prefer to use large-scale, comprehensive lysis and purification protocols such as those described by Holben

[1994a, b] for soils, sediments, and other environments (*note*: these protocols have been modified in practice so that TE buffer now contains 50 mM EDTA), or Apajalahti et al. [1998] for fecal and gastrointestinal microbiome samples because they minimize within-sample heterogeneity in community composition resulting from small sample size and were specifically developed and demonstrated to provide effective recovery and lysis of the broad array of microbial taxa encountered in such samples.

23.2.1.1 DNA Stock Solutions. Control gradients containing genomic DNA samples of known %G+C composition are prepared using DNA isolated from pure cultures of *Micrococcus lysodeikticus, Escherichia coli*, and *Clostridium perfringens*, which span the breadth of bacterial %G+C contents. These DNAs were obtained from Sigma Chemical Co., St Louis, MO, and were used to prepare 1 mg/ml aqueous stock solutions in TE (10 mM Tris, pH 8.0; 1 mM EDTA) which were stored at 4°C prior to use. Note that other appropriate DNA standards could also be used for GC fractionation experiments, provided that they span the %G+C range of interest for those experiments.

23.2.1.2 GC Fractionation Using Cesium Chloride-Bisbenzimidazole Gradients. GC fractionation employs cesium chloride equilibrium density gradients to fractionate genomic DNA of the component taxa of a microbial community as a function of its characteristic %G+C content. As noted in the introduction, this separation is based on differential density imposed by the AT-dependent DNA-binding dye, bisbenzimidazole. Cesium chloride–bisbenzimidazole equilibrium density gradients are prepared by combining 30 μl of 1 mg/ml bisbenzimidazole stock solution with 15 ml of filtered (0.22 μm) cesium chloride (CsCl) stock solution (refractive index = 1.3980 and containing 5 mM Tris, pH 8.0). DNA samples, typically 50 μg of bacterial community DNA or 15 μg total of control DNA mixture (5 μg each of the three control DNA solutions), are added as 200 μl of aqueous solution, mixed, and then transferred to an 18-ml ultracentrifuge tube, with the remaining tube volume filled with CsCl stock solution. The ultracentrifuge tubes are balanced, sealed, and subjected to centrifugation at 90,000 × *g* for 72 h at 18°C. Note that various vertical and fixed angle ultracentrifuge rotors and corresponding tubes can be used so long as the run conditions described in the previous sentence can be achieved.

Following centrifugation, a syringe pump (or any finely metered pump) is used to inject a higher density displacement solution (typically Fluorinert® FC-40 [Sigma F9755]) into the bottom of the gradient to force it through an ISCO UA-5 (or equivalent) UV absorbance detector set to 280 nm (to effectively detect DNA while minimizing background absorbance due to the cesium chloride gradient) and then to a fraction collector. The density of individual fractions can be determined based on the Rf of individual fractions using a refractometer. More commonly, the %G+C content represented by each fraction is determined by linear regression analysis of data obtained from control gradients containing standard DNA samples of known %G+C composition as described previously [Holben and Harris, 1995].

23.2.1.3 Downstream Analyses of Fractionated Microbial Community DNA. As noted previously, the outcome of GC fractionation is a set of physically separated fractions of high-molecular-weight total community DNA representing known ranges of %G+C content of the component populations. In various studies, these fractions have been effectively used as a template for a number of downstream molecular manipulations including PCR amplification, cloning, direct sequencing, DGGE analysis, and statistically robust comparative analyses of the community profiles obtained. Most of these manipulations are in the realm of general molecular biology and statistical techniques, except that they are applied to fractionated rather than total community DNA or RNA. While examples of the types of data regarding microbial communities that can be obtained using these combined approaches are presented and discussed below, the reader is referred to the cited original publications for procedural details regarding the downstream experimental manipulations performed to obtain the data.

23.3 RESULTS AND DISCUSSION

23.3.1 GC Fractionation

Aqueous mixtures combining 5 μg each of *Micrococcus lysodeikticus* (72% G+C), *Escherichia coli* (50% G+C), and *Clostridium perfringens* (27% G+C) DNA were subjected to equilibrium density gradient centrifugation on cesium chloride–bisbenzimidazole gradients followed by physical fractionation (i.e., GC fractionation). This treatment resolved the DNA mixtures into three distinct bands as visualized under UV illumination (data not shown). Integration of DNA quantification and density data for the fractionated gradients produced a histogram or profile that is presented as a plot of relative abundance versus %G+C content of DNA (**Fig. 23.1A**). There is a linear relationship between the %G+C content of DNA and its buoyant density when complexed with bisbenzimidazole. Since CsCl equilibrium gradients are also linear in form, the position of DNA molecules in gradients is directly and

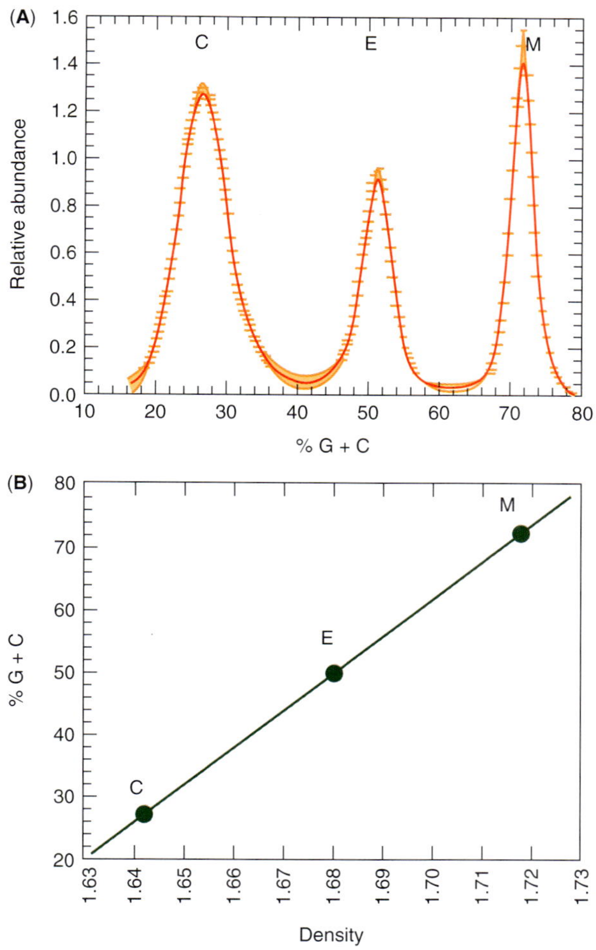

Figure 23.1 Separation of a homogeneous mixture of *Clostridium perfringens* (C), *Escherichia coli* (E), and *Micrococcus lysodeikticus* (M) DNA into discrete bands on cesium chloride–bisbenzimidazole gradients. (**A**) GC profiles of cesium chloride–bisbenzimidazole gradients separating the DNA mixtures indicating relative abundance (as percent of total) versus %G+C content of DNA. DNA amounts were quantified based on A_{260}. The data shown represent the mean of four replicate samples. Error bars indicate one standard deviation in the data and are shown only for every tenth data point for clarity. (**B**) Relationship between %G+C content and buoyant density of DNA derived from panel A; $r^2 \geq 0.99$. Density was determined by measuring R_f values of gradient fractions. (Reprinted with permission from Holben and Harris, 1995. Copyright © 1995 John Wiley & Sons, Inc.)

linearly related to their %G+C content (Fig. 23.1B). Thus, when applied to mixtures of DNA from different bacterial populations, including total bacterial community DNA from environmental samples, this procedure effectively "unscrambles" the mixture to resolve the DNA from each population based on its inherent %G+C content, thereby unraveling the structure of the bacterial community.

Total bacterial community DNA isolated from soil was analyzed using this technique to determine the structure of the native bacterial community. Because bacterial community DNA isolated by our protocol [Holben et al., 1992; Holben, 1994a] is of high molecular weight (average DNA fragment size >25 kb), localized variation in %G+C content within the genome is averaged into its characteristic overall %G+C content, producing a single, discrete DNA band in the gradient for a given bacterial population. Indeed, DNA isolated from pure cultures of bacteria representing several different genera produced a single, discrete band of DNA for each population, whose position in the gradient was determined by its own characteristic %G+C content (data not shown). The bacterial community profile observed in **Figure 23.2** is typical for mid-Michigan Capac agricultural soil maintained at field water moisture content (~24%). As seen in this profile, the majority of DNA (and thus the majority of bacteria in the soil) corresponds to the 55–73% G+C range, which includes bacterial genera known to dominate soil bacterial communities including *Agrobacterium* (59–64% G+C content), *Alcaligenes* (55–61% G+C content), *Arthrobacter* (63–69% G+C content), and *Pseudomonas* (58–66% G+C content). The characteristic %G+C contents of several other common soil bacterial genera are also indicated (Fig. 23.2).

GC fractionation allows monitoring of the effects of perturbation on bacterial community structure and facilitates comparative analyses of bacterial communities from various environmental samples. For example, incubation of the same Capac soil under carbon-amended or water-saturated conditions produced profound effects on bacterial community structure (**Fig. 23.3**). Water saturation and the resulting anaerobiosis produced changes in bacterial community structure composition as an increase in the relative abundance of DNA in the 40–50% G+C range and a concomitant proportional decrease in the 55–70% G+C range (Fig. 23.3A). This response likely resulted from spore germination and outgrowth of *Bacillus* species (38–44% G+C content) in response to the release of carbon and anaerobic conditions produced by water saturation. Anaerobic *Streptococcus* (35–40% G+C content) and *Flavobacterium* species are also found in soil and may have contributed to the increased abundance of 40–50% G+C content DNA, but these genera are less common in soils than *Bacillus*.

Amendment of soil with carbon also produced observable differences in soil bacterial community structure, exhibiting a shift toward higher %G+C content and apparently less complexity than unamended soil, presumably reflecting a net decrease in diversity as one or a few populations best able to utilize the ground leaf amendment proliferated (Fig. 23.3B). Carbon-amended, water-saturated soil had an alternative bacterial community structure with the majority of DNA having 53–63% G+C content (Fig. 23.3C). The genera *Alcaligenes*

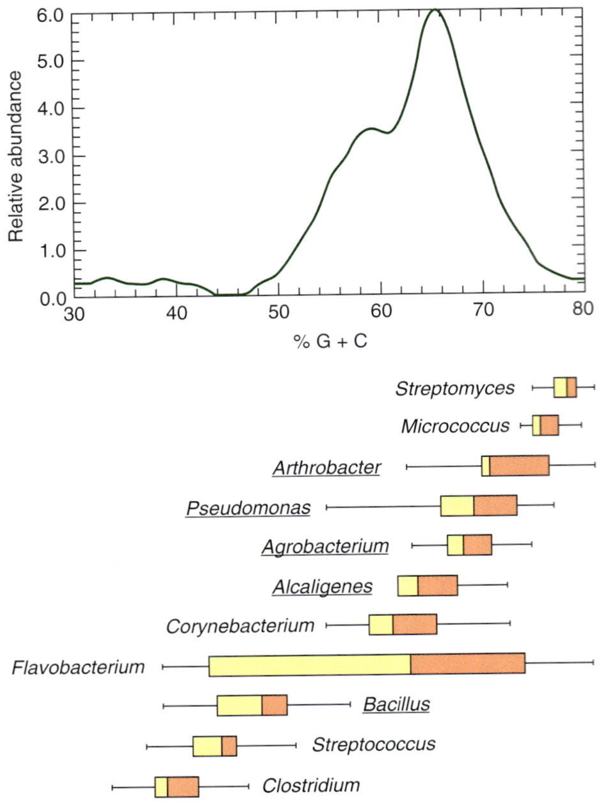

Figure 23.2 Bacterial community profile for mid-Michigan Capac agricultural soil maintained at field water moisture content and box plot showing %G+C content of DNA of soil bacterial genera. Box plots were generated using data from Laskin and Lechevalier [1973]. The vertical line within boxes indicates the median %G+C value for that genus; the left-hand box edge (lower quartile) indicates the value halfway between the lowest value and the median; the right-hand box edge (upper quartile) indicates the value halfway between the highest value and the median; the length of the solid lines to the left and right indicates 1.5× the distance between the lower and upper quartile added to the lower and upper quartile, respectively. Values outside this range are outliers and were deleted for clarity. Underlined names indicate dominant soil bacterial genera (i.e., those that generally comprise >10% of total bacteria in soil). (Reprinted with permission from Holben and Harris, 1995. Copyright © 1995 John Wiley & Sons, Inc.)

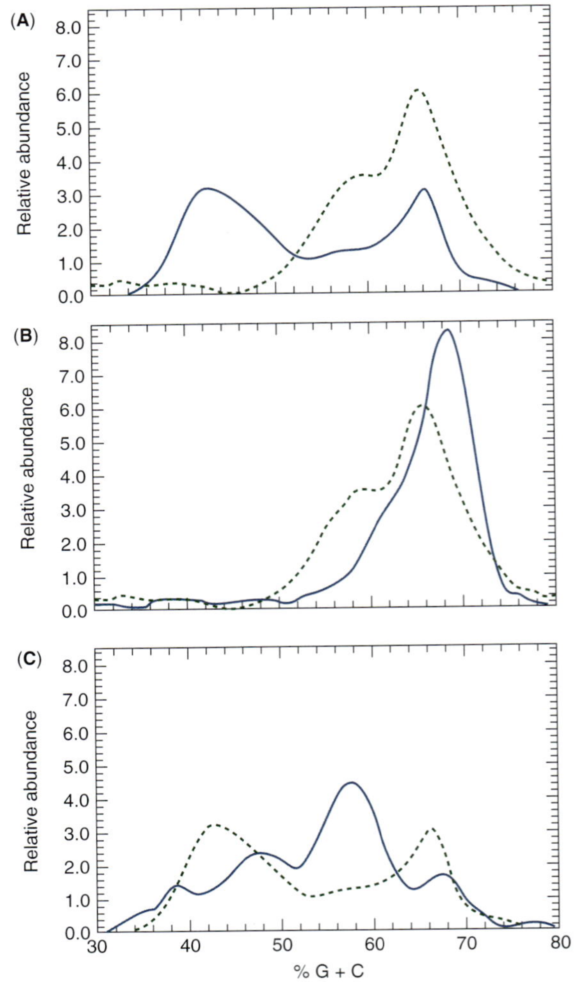

Figure 23.3 Bacterial community profiles showing changes in community structure in response to perturbation. (**A**) Community profiles for soils under water-saturated (——) and field moisture (. . .) conditions. (**B**) Profiles for carbon-amended (——) and unamended (. . .) soil under field moisture conditions. (**C**) Profiles for carbon-amended (——) and unamended (. . .) soil under water-saturated conditions. *x* axes and *y* axes for each panel have the same scale. (Reprinted with permission from Holben and Harris, 1995. Copyright © 1995 John Wiley & Sons, Inc.)

(55–61% G+C), *Corynebacterium* (52–58% G+C), and some species of *Flavobacterium* are facultative anaerobes or denitrifiers capable of growth under anoxic or anaerobic conditions and represent common soil bacteria having this range of G+C content.

Rather than measuring the impact of perturbations or treatments on total processes (i.e., activities), or on one or a few bacterial populations, GC fractionation makes it possible to monitor changes in the composition of the total bacterial community in a single analysis. Conversely, the broad resolution of this approach does not facilitate monitoring of specific bacterial populations in the community since multiple organisms could appear in a single peak or fraction. Determination of the presence or absence of specific populations requires that additional downstream analyses having higher resolution be performed as discussed below.

23.3.2 Creating Sequence Libraries from Fractionated DNA

We and others have previously shown that GC fractionation can be combined with 16S rRNA gene sequencing (GC-16S) to provide broader coverage and/or directed

Table 23.1 Effect of GC Fractionation on 97% Similarity-Based OTU Numbers, Shannon–Weaver Diversity Indices, Evenness, and Richness Estimates

	No. of Unique OTUs	Shannon–Weaver Index	Evenness	Richness Estimate
GC fractionated DNA[a]	335	5.62	0.966	1020
Unfractionated DNA[b]	301	5.45	0.954	780

[a]Based on 487 starting sequences.

[b]Based on 490 starting sequences.

Source: Reprinted with permission from Morales and Holben, 2009. Copyright © ASM.

detection of bacterial populations in the gastrointestinal tract of humans and animals [Apajalahti et al., 2002, 2003; Kassinen et al., 2007] and a volcanic soil [Nusslein and Tiedje, 1998, 1999]. More recently, we have employed the GC-16S approach to develop deep 16S rRNA gene libraries from a midwestern U.S. agricultural soil and a western U.S. river hyporheic sediment [Lowell et al., 2009; Morales et al., 2009]. These libraries were subsequently used to develop and test site-specific primers for quantitative PCR analysis of bacterial populations and groups [Morales and Holben, 2009]. The soil GC-16S analysis also demonstrated that the clone library constructed from GC-fractionated DNA provided deeper coverage and higher values for a suite of commonly employed diversity measures than did a library constructed by random cloning from the same starting DNA sample (**Table 23.1**). The main benefit of GC fractionation prior to sequencing (whether direct or including upstream PCR and cloning steps) is the reduction in DNA complexity within each fraction which allows underrepresented sequences to be detected more readily than in a random survey. This produced a higher recovery rate for minority species, greater detection of unique OTUs, and higher values for community richness, evenness, and diversity from the same-sized random sequence library (Table 23.1). The conclusion was that application of GC fractionation prior to microbial community sequence surveys reduces the required overall survey size needed to reach, or at least approach, complete coverage of the diversity present in the entire bacterial community [Morales et al., 2009].

Efforts are currently underway in collaboration with the Environmental Microbial Genomics Group at the Ecole Centrale de Lyon, Université de Lyon, France to utilize GC fractionated soil community DNA from the Rothamsted long-term agroecosystem experiment in the United Kingdom as template to develop an ultra-deep, directly sequenced, metagenomic library (GC-metagenomics) from the control plots at the site. Similar to the case with the GC-16S rRNA gene library, it is anticipated that creating a metagenomic library from GC fractionated DNA will provide greater depth of coverage of the diversity present than would be obtained from a randomly sequenced metagenomic library from the same starting community DNA sample.

While procedures for the construction of both 16S rRNA gene and metagenomic libraries are well understood and described broadly in the literature (and thus not outlined here), it is worth considering some of the features of utilizing GC fractionated DNA as the starting template that are relevant to experimental objectives and outcomes. According to Fig. 23.2, it is obvious that different fractions contain not just different %G+C content DNA, but also different proportions of the total community DNA. This affords two opportunities for downstream analyses, namely proportional analysis versus normalized analysis.

Using a proportional analysis format, the depth of downstream analysis of individual fractions would be proportional to their relative abundance in the total community DNA to maintain the relative contribution of sequences (or any other data acquired by any downstream technique) from that fraction to its relative abundance in the community (refer to **Table 23.2** for a hypothetical example). In this way, a sequence library constructed from fractionated DNA would have appropriate depth of sampling (numbers of individual sequences) from each fraction such that a study focused on developing a vision of community composition based on composited analysis of all sequences from all samples should approximate their relative abundance in the total community.

It might be argued that this outcome could be obtained more simply by random "shotgun" sampling directly from the total community DNA. However, as we have previously shown [Holben and Harris, 1995; Morales, et al., 2009], using fractionated DNA enhances depth of coverage and detection of minority populations because the reduced complexity and increased relative abundance of templates in individual fractions compared to total DNA allows templates present in low abundance in the total community to be recovered more effectively. Furthermore, in downstream applications where PCR is employed, the limited range of %G+C content within each fraction should facilitate more-even PCR amplification of template mixtures because they have more similar %G+C content than would be the case with total community DNA.

Table 23.2 Hypothetical Distribution of 1500 clones from 15 GC Fractions in a Proportional Analysis

Fraction	Relative Abundance (% of Total)	Number of Clones
1	2	30
2	3	45
3	2	30
4	9	135
5	5	75
6	7	105
7	12	180
8	17	255
9	13	195
10	8	120
11	9	135
12	6	90
13	2	30
14	3	45
15	2	30
Total	100	1500

With the normalized approach, differences in the proportional abundance of DNA in each fraction relative to the total community can be normalized in downstream analyses to facilitate detection of sequences from less abundant taxa. For a rhetorical example, refer to Figure 23.2 where the genomic DNA and the corresponding sequences in the fractions corresponding to 30–35%, 35–40%, 40–45%, and 45–50% G+C content are clearly much less abundant than in the 55–60%, 60–65%, and 65–70% G+C fractions. As a result, sequences in the former would be much less frequently or perhaps not ever encountered in a randomly generated library relative to the latter. However, because there is physical separation of the DNA fractions, analysis of 1000 sequences from each of 10 separate fractions (an arbitrary number since upwards of 50 fractions can readily be collected from a single sample by controlling individual fraction volume) representing a total community would most certainly recover sequences not obtained in a randomly generated 10,000-clone library using total community DNA as template. This approach should be particularly effective for studies aimed at maximal coverage of total diversity in microbial communities.

23.3.3 Functional Gene Probing of Fractionated DNA

GC fractionation can also be paired with functional gene probing (**GC-FGP**) of the DNA in each fraction, which is readily achieved via dot-blotting. This approach was used to compare microbial communities in each compartment of a three-stage, flow-through (i.e., sequential) nitrifying bioreactor system [Holben et al., 1998]. Each stage of triplicate three-stage bioreactors was initially inoculated with identical polyethylene glycol granular activated sludge (GAS) particles and then operated in a single-pass, flow-through format with a feed of 5.0 g of ammonium nitrogen per liter of GAS for 150 days. At that time, a series of sequential nitrifying reactions was observed in each of the three compartments, with half of the ammonium being oxidized to nitrite in compartment 1, the remaining half being oxidized to nitrite in compartment 2, and nitrite oxidation to nitrate being observed only in compartment 3.

Following purification of microbial DNA from each compartment, GC fractionation and community profile analysis clearly demonstrated that compartments 1, 2, and 3 each had a distinctive microbial community composition, despite originally having identical inocula (**Fig. 23.4**). The %G+C contents of the peaks in these community profiles were consistent with the predominance of ammonia oxidizers (46–56% G+C) in compartments 1 and 2, while compartment 3 was dominated by nitrite oxidizers (58–61% G+C) as was suggested by the activity analyses.

Significantly, functional gene probing of individual fractions from the communities in each compartment with *amo* and *hao* gene probes encoding the two sequential functional genes, ammonia monooxygenase and hydroxylamine oxidoreductase, respectively, in the ammonia oxidation pathway of *Nitrosomonas europea*) showed that different ammonia oxidizers were predominant in compartments 1 and 2 and that no ammonia oxidizers were present in compartment 3 (**Fig. 23.5**). The different ammonia oxidizers that predominate in compartments 1 and 2 presumably represent populations having different affinities, sensitivities, or other physiological features related to ammonium and nitrate. The conclusion was that the greater efficiency of ammonia oxidation observed in this three-stage system was achieved via the sequential activities of two different ammonia oxidizers, wherein half of the ammonium was transformed in compartment 1 (i.e., the ammonium concentration went from high to moderate), while residual ammonium was completely oxidized in compartment 2 (i.e., ammonium went from moderate to undetectable). This conclusion is highly consistent with the findings of Suwa et al. [1993], who showed that different ammonia oxidizers have different sensitivities to and affinities for ammonium that confer selective advantage dependent on ammonium concentration, which clearly varied in this sequential, flow-through bioreactor system.

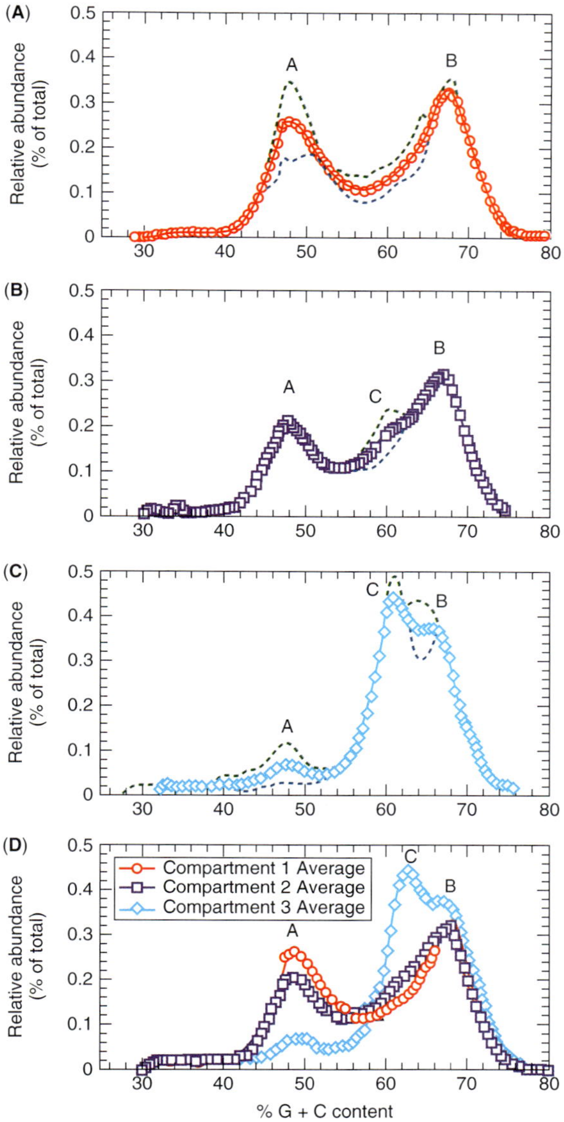

Figure 23.4 Community profile analysis of a three-compartment nitrifying bioreactor system based on GC fractionation of total community DNA. (**A**) Compartment one (○); (**B**) compartment two (□); (**C**) compartment three (◇); and (**D**) overlay of profiles for all three compartments. Dotted lines represent the profiles of individual samples, and dark solid lines represent the average values for two replicate samples. (Reprinted with permission from Holben WE, Suwa Y. 1998. Molecular Analysis of Bacterial Communities in a Three-Compartment Granular Activated Sludge System Indicates Community-Level by Incompatible Nitrification Processes, *Appl. Environ. Microbio.* **64**(7):2528–2532. Copyright © ASM.)

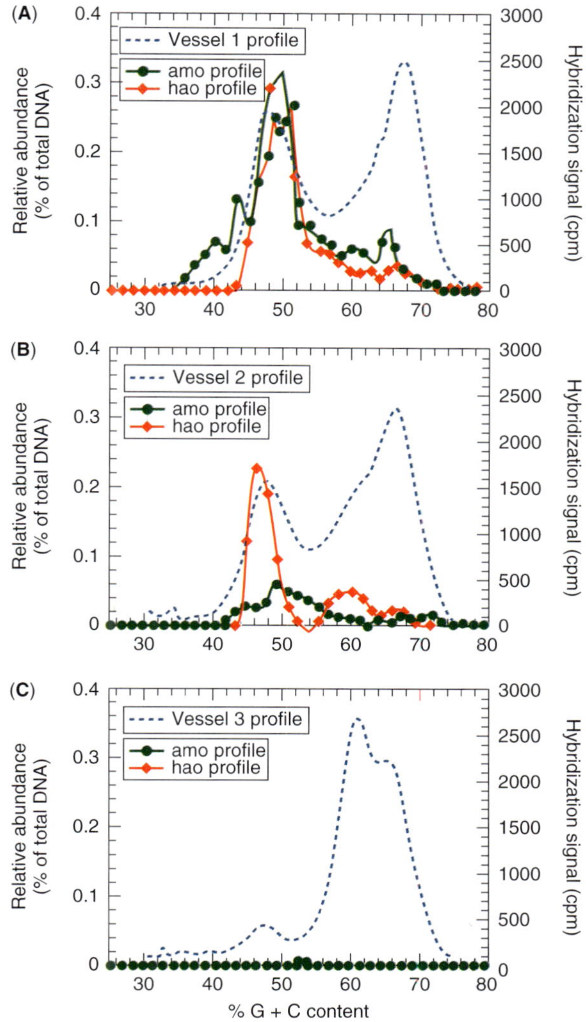

Figure 23.5 Hybridization analysis of fractionated microbial community DNA from the three-compartment bioreactor system in Figure 23.4 using *amo* (●) and *hao* (◆) gene probes from *Nitrosomas europeae*. Hybridization data are shown relative to the community profile data for compartment one (**A**), compartment two (**B**), and compartment three (**C**). Only each third data point is shown for the sake of clarity. (Reprinted with permission from Holben WE, Suwa Y. 1998. Molecular Analysis of Bacterial Communities in a Three-Compartment Granular Activated Sludge System Indicates Community-Level by Incompatible Nitrification Processes, *Appl. Environ. Microbio.* **64**(7):2528–2532. Copyright © ASM.)

23.3.4 DGGE Analysis of Fractionated DNA

GC fractionation is also highly compatible with denaturing gradient gel electrophoresis analysis (GC-DGGE) to provide greater resolution and enhanced access to populations in the community based on the reduced complexity of the mixture of DNA in each fraction relative to unfractionated total community DNA [Holben et al., 2004]. In that study, the DGGE patterns of partial 16s rRNA gene amplicons obtained from individual GC fractions were compared to the pattern obtained by PCR

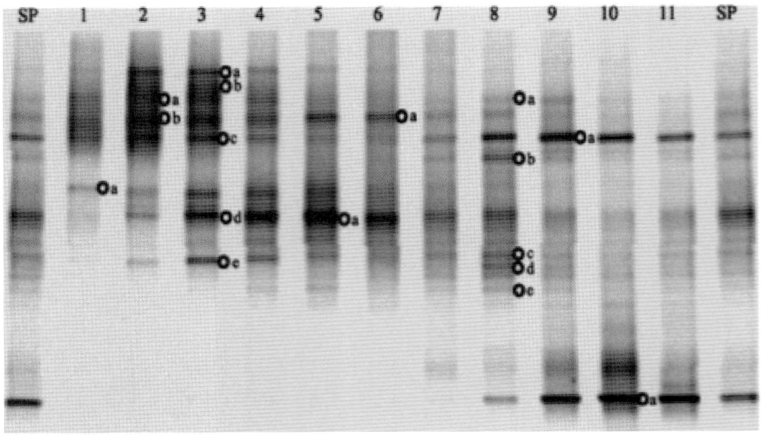

Figure 23.6 DGGE analysis of partial 16S rRNA gene sequences PCR amplified from unfractionated DNA and from individual GC fraction DNA from the same chicken cecum microbiome sample (image normalized to remove smiling, using GelCompar software). SP indicates DGGE patterns from unfractionated super-pool DNA (i.e., pooled DNA from ~500 individual chicken samples worldwide). Lane numbers indicate DGGE patterns from individual GC gradient fractions spanning the breadth of bacterial %G+C content. The lettered circles indicate individual bands excised from the gel for cloning and DNA sequence analysis. (Reprinted with permission from Holben WE, Apajalahti JHA. 2004. GC Fractionation Enhances Microbial Community Diversity Assessment and Detection of Minority Populations of Bacteria by Denaturing Gradient Gel Electrophoresis. *Appl. Environ. Microbio*. **70**(4):2263–2270. Copyright © ASM.)

of unfractionated total community DNA (**Fig. 23.6**). We observed several bands in DGGE lanes from individual GC fractions that were not apparent or only weakly visible in the DGGE lanes corresponding to unfractionated DNA from the same sample. Directed band excision, cloning, and DNA sequencing (refer to Holben, et al. [2004] for a detailed description) were applied to bands selected because they were either abundantly, moderately, or not represented in the unfractionated DNA lanes (refer to Fig. 23.6). Overall, 22 of 28 (79%) phylotypes recovered were not detected by the random cloning approach. More importantly, 20 of 23 (87%) of phylotypes that were abundant in lanes from individual fractions, but not from unfractionated DNA, were not detected in the random cloning analysis. This was because the DNA of these minority populations was localized into one or a few fractions and thus effectively purified relative to the bulk of total community DNA. The combined GC-DGGE approach overcomes (a) the primary limitation of %G+C fractionation; low resolution that does not indicate the number or identity of different taxa in a particular %G+C fraction, and (b) the primary limitation of DGGE; the inability to detect populations present in low abundance, to better assess total community diversity.

23.3.5 Comparative Statistical Analysis of Community Profiles

While some comparative community analyses produce dramatically different %G+C profiles depending on various treatments (refer to Fig. 23.3), others can be much more subtle. To overcome this limitation and to provide a quantitative element of comparative analysis of community GC profiles, robust statistical analyses can be applied to GC profile data (GC-SA). This general approach was first taken and described by Apajalahti et al. [1998] in a study of the chicken gastrointestinal microbiome that simply calculated standard errors of the mean for data from replicate GC fractionation datasets. The data showed, among other things, that birds raised together under the same conditions had highly similar microbial community profiles (data not shown). Subsequent refinements of the GC-SA approach allowed its use to demonstrate that differences in the cecal microbiome community that were smaller than 10% within each 5% increment (fraction) of the %G+C profiles were significant [Apajalahti et al., 2001]. These differences were determined to be most strongly related to the source of the animal feed and to local feed amendment. This fine resolution of differences in total microbial community structure was afforded by mechanical and arithmetic averaging of data from experimental and animal replicate samples (**Fig. 23.7**), *t*-test analysis (**Fig. 23.8**), and multiple linear regression (MLR) modeling and principle component analysis (PCA) (**Fig. 23.9**) of differences between treatments for replicate individual animals from eight different farms in Finland with characteristic feeding regimens.

Differences in the total community profiles of six individual animals could be delineated (Fig. 23.7A), because the experimental variability was so small in these analyses that six replicate gradients from a pooled sample (from the six birds) were virtually indistinguishable from the arithmetic average for those same gradients (Fig. 23.7B).

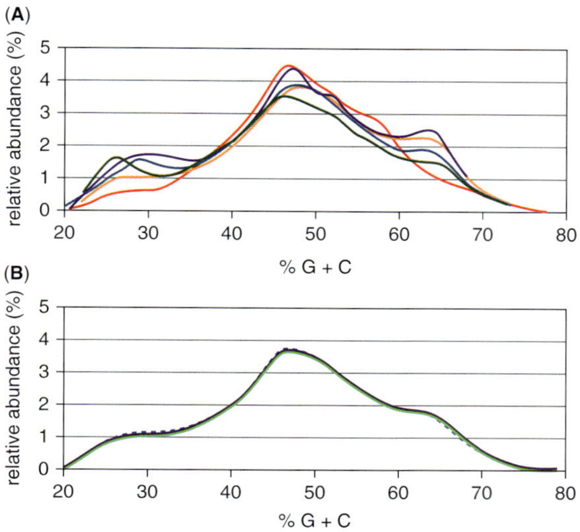

Figure 23.7 GC profiles of cecal bacterial communities in broiler chickens from a commercial Finnish farm. (**A**) Percent G+C profiles of six individual broiler chickens from a farm using commercial feed from feed mill A (farm $A_{Late}1$). (**B**) Six replicate percent G+C profiles of a single digesta sample obtained by pooling the six individual digesta samples of panel A (solid lines) and the arithmetic average of the individual profiles shown in panel A (dotted line). (Reprinted with permission from Apajalhti JH, Holben WE. 2001. Percent GC Profiling Accurately Reveals Diet-Related Differences in the Gastrointestinal Microbial Community of Broiler Chickens. *Appl. Environ. Microbio.* **67**(12):5656–5667. Copyright © ASM.)

Integration of data from t-test analysis of differences in the data for each treatment from the grand average of all data for each 5% increment in %G+C content clearly indicates differences in the microbiome communities as a function of feed type and feed amendment across the entire community (Fig. 23.8), while MLR modeling provides statistically significant determinants ($P < 0.05$) and r^2 values for each 5% increment (not shown). PCA analysis of the GC profiles from eight different farms was used to reveal total bacterial community divergence between individual chickens for each treatment, showing that the structure of the cecal microbiome was strongly feed- and feed-amendment dependent and much less a function of differences between farms or the year the analysis was performed (Fig. 23.9).

Where differences between community profiles are not visually apparent, GC-SA provides a means to identify statistically significant differences between community profiles for more focused examination. Where the community composition of the system of study is at least reasonably well understood (e.g., the chicken cecum), GC-SA can indicate responses to treatments by specific populations or groups of bacteria and their corresponding functions, which can be subsequently confirmed via techniques such as highly specific phylogenetically based or functionally based quantitative PCR. GC-SA continues to

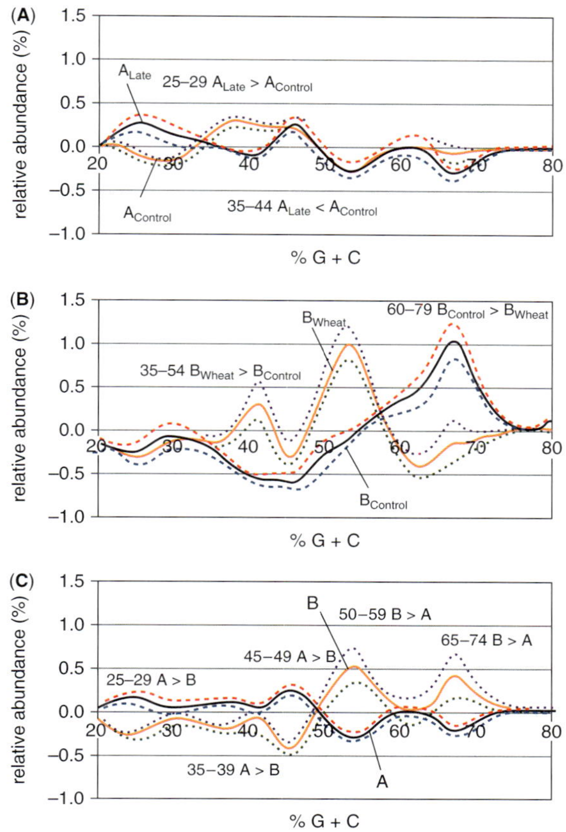

Figure 23.8 Characteristic effects of individual feeding regimens on bacterial community profiles. The average bacterial community profile of all of the broiler chickens analyzed in this study (grand average) was subtracted from the average bacterial community profile of birds on each feeding regimen.Solid lines show the average profile of each feeding regimen, and dotted lines indicate the corresponding standard error of the mean. (**A**) Comparison of birds fed feed A in 1997 and 2000 ($A_{Control}$ versus A_{Late}) with indication of %G+C increments for which feed $A_{Control}$ and A_{Late} differed from each other according to t-test analysis. (**B**) Comparison of birds fed feed B alone and those given feed B amended with whole wheat ($B_{Control}$ versus B_{Wheat}) with indication of %G+C increments for which feed $B_{Control}$ and B_{Wheat} differed from each other according to t-test analysis. (**C**) Comparison of all birds with feed A in their diet to all birds with feed B in their diet ($A_{Control}$ and A_{Late} versus $B_{Control}$ and B_{Wheat}) with indication of %G+C increments for which feeds A and B differed from each other according to t-test analysis. (Reprinted with permission from Apajalhti JH, Holben WE. 2001. Percent GC Profiling Accurately Reveals Diet-Related Differences in the Gastrointestinal Microbial Community of Broiler Chickens. *Appl. Environ. Microbio.* **67**(12):5656–5667. Copyright © ASM.)

be widely applied by Alimetrics, Inc. and collaborators in both publicly reported and proprietary studies to demonstrate diet-, prebiotic-, and probiotic-induced differences in the ileal and cecal microbiomes of humans, dogs, cattle, chickens, swine and rodents [J. Apajalahti, personal communication].

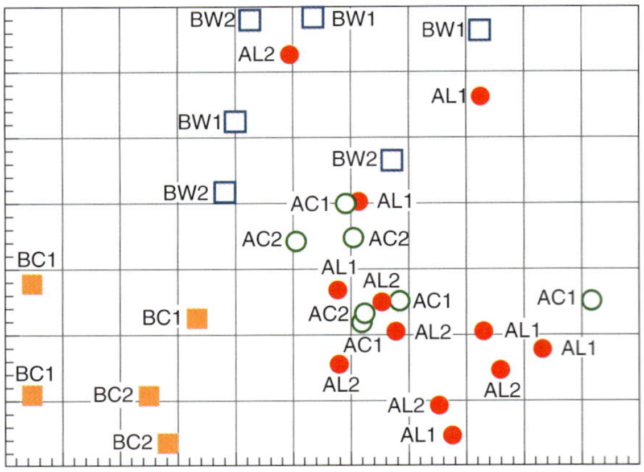

Figure 23.9 PCA of cecal GC profiles of broiler chickens from eight different commercial farms. Positioning of individual broiler chickens is based on the analysis of individual GC profiles using all twelve 5% G+C increments collectively and is indicated by the squares (feed B) and circles (feed A). The label next to each symbol indicates the origin of the sample as follows: AC1 and AC2, farms using feed A in 1997; AL1 and AL2, farms using feed A in 2000; BC1 and BC2, farms using feed B in 1997; BW1 and BW2, farms using feed B amended with whole wheat in 1998. (Reprinted with permission from Apajalhti JH, Holben WE. 2001. Percent GC Profiling Accurately Reveals Diet-Related Differences in the Gastrointestinal Microbial Community of Broiler Chickens. *Appl. Environ. Microbio.* **67**(12):5656–5667. Copyright © ASM.)

23.4 SUMMARY

As mentioned in the introduction, total community analyses, of themselves, typically provide an integrated but low-resolution view of the total bacterial community in a single analysis. However, higher resolution within communities (e.g., greater depth of community sampling), comparative analyses between communities, and, where necessary, taxon identification and characterization are readily achieved through additional downstream analyses. Since each total community or compilation-based approach applied to microbial community analysis has its own inherent strengths and limitations, it is reasonable to assume that combining mechanistically different analyses with GC fractionation would afford better resolution and provide additional information regarding the community being analyzed. Furthermore, GC fractionation can be used to focus downstream analyses on points of interest (e.g., specific fractions) within and between communities, thereby increasing the efficiency of analysis by facilitating very deep analysis of a portion of the community rather than comparatively shallow random sampling of the total community with a given set of resources (e.g., number of clones or direct sequences). Continued development of downstream analytical techniques that complement the GC fractionation approach should further enhance the capabilities and resolution of these combined approaches to levels appropriate to the scientific question being addressed, ranging from the total community structure and function to population dynamics at the subspecies level.

REFERENCES

Andoh A, Benno Y, Kanauchi O, Fujiyama Y. 2009. Recent advances in molecular approaches to gut microbiota in inflammatory bowel disease. *Curr. Pharm. Des*. **15**:2066–2073.

Apajalahti JH, Sarkilahti LK, Maki BR, Heikkinen JP, Nurminen PH, Holben WE. 1998. Effective recovery of bacterial DNA and percent-guanine-plus-cytosine-based analysis of community structure in the gastrointestinal tract of broiler chickens. *Appl. Environ. Microbiol*. **64**:4084–4088.

Apajalahti JH, Kettunen A, Bedford MR, Holben WE. 2001. Percent G+C profiling accurately reveals diet-related differences in the gastrointestinal microbial community of broiler chickens. *Appl. Environ. Microbiol*. **67**:5656–5667.

Apajalahti JH, Kettunen H, Kettunen A, Holben WE, Nurminen PH, Rautonen N, Mutanen M. 2002. Culture-independent microbial community analysis reveals that inulin in the diet primarily affects previously unknown bacteria in the mouse cecum. *Appl. Environ. Microbiol*. **68**:4986–4995.

Apajalahti JH, Kettunen A, Nurminen PH, Jatila H, Holben WE. 2003. Selective plating underestimates abundance and shows differential recovery of bifidobacterial species from human feces. *Appl. Environ. Microbiol*. **69**:5731–5735.

Apajalahti J, Kettunen A, Graham H. 2004. Characteristics of the gastrointestinal microbial communities, with special reference to the chicken. *World Poultry Sci. J*. **60**:223–232.

Apajalahti A, Kettunen A. 2006. Microbes of the chicken gastrointestinal tract. In Perry GC, ed. *Avian Gut Function in Health and Disease*, Vol. 28. United Kingdom: World's Poultry Science Association, pp. 124–137.

Bent SJ, Forney LJ. 2008. The tragedy of the uncommon: Understanding limitations in the analysis of microbial diversity. *ISME J*. **2**:689.

Brodie EL, DeSantis TZ, Joyner DC, et al. 2006. Application of a high-density oligonucleotide microarray approach to study bacterial population dynamics during uranium reduction and reoxidation. *Appl. Environ.Microbiol*. **72**:6288–6298.

Brunk CF, Jones KC, James TW. 1979. Assay for nanogram quantities of DNA in cellular homogenates. *Anal. Biochem*. **92**:497–500.

Comings DE. 1975. Mechanisms of chromosome banding. VIII. Hoechst 33258-DNA interaction. *Chromosoma* **52**:229–243.

Curtis TP, Sloan WT. 2004. Prokaryotic diversity and its limits: microbial community structure in nature and implications for microbial ecology. *Curr. Opin. Microbiol*. **7**:221–226.

Curtis TP, Sloan WT, Scannell JW. 2002. Estimating prokaryotic diversity and its limits. *Proc. Natl. Acad. Sci. USA* **99**:10494–10499.

Dicksved J, Halfvarson J, Rosenquist M, et al. 2008. Molecular analysis of the gut microbiota of identical twins with Crohn's disease. *ISME J*. **2**:716–727.

Dunbar J, Ticknor LO, Kuske CR. 2000. Assessment of microbial diversity in four southwestern United States soils by 16S rRNA gene terminal restriction fragment analysis. *Appl. Environ. Microbiol*. **66**:2943–2950.

Dunbar J, Barns SM, Ticknor LO, Kuske CR. 2002. Empirical and theoretical bacterial diversity in four Arizona soils. *Appl. Environ. Microbiol*. **68**:3035–3045.

Ferguson RL, Buckley EN, Palumbo AV. 1984. Response of marine bacterioplankton to differential filtration and confinement. *Appl. Environ. Microbiol.* **47**:49–55.

Fliermans CB, Balkwill DL. 1989. Microbial life in the deep terrestrial subsurface. *BioScience* **39**:370–377.

Fuhrman JA. 2002. Community structure and function in prokaryotic marine plankton. *A. Van Leeuw. J. Microbiol.* **81**:521–527.

Green BD, Keller M. 2006. Capturing the uncultivated majority. *Curr. Opin. Biotech.* **17**:236–240.

Greene EA, Voordouw G. 2003. Analysis of environmental microbial communities by reverse sample genome probing. *J. Microbiol. Methods* **53**:211–219.

Gsell T, Holben W, Ventullo R. 1997. Characterization of the sediment bacterial community in groundwater discharge zones of an alkaline fen: a seasonal study. *Appl. Environ. Microbiol.* **63**:3111–3118.

Hazen T, Jiménez L, López de Victoria G, Fliermans C. 1991. Comparison of bacteria from deep subsurface sediment and adjacent groundwater. *Microb. Ecol.* **22**:293–304.

He Z, Gentry TJ, Schadt CW, et al. 2007. GeoChip: A comprehensive microarray for investigating biogeochemical, ecological and environmental processes. *ISME J.* **1**:67–77.

Holben WE. 1994a. Isolation and purification of bacterial community DNA from environmental samples. In Hurst GK, Mclnenney M, Stetzenbach LD, Walter M, eds. *Manual of Environmental Microbiology*. Washington, DC: ASM Press, pp. 431–436.

Holben WE. 1994b. Isolation and purification of bacterial DNA from soil. In Wea RW, ed. *Methods in Soil Analysis*, Part 2: *Microbiological and Biochemical Properties*. Madison, WI: Soil Science Society of America, pp. 727–751.

Holben WE, Harris D. 1995. DNA-based monitoring of total bacterial community structure in environmental samples. *Mol. Ecol.* **4**:627–631.

Holben WE, Schroeter BM, Calabrese VGM, Olsen RH, Kukor JK, Biederbeck VO, Smith AE, Tiedje JM. 1992. Gene probe analysis of soil microbial populations selected by amendment with 2,4-dichlorophenoxyacetic acid. *Appl. Environ. Microbiol.* **58**:3941–3948.

Holben WE, Calabrese VG, Harris D, Ka JO, Tiedje JM. 1993. Analysis of structure and selection in microbial communities by molecular methods. In Guerrero R, Pedrós-Alió C, eds. *Trends in Microbial Ecology*. Barcelona, Spain: Spanish Society for Microbiology, pp. 367–370.

Holben WE, Noto K, Sumino T, Suwa Y. 1998. Molecular analysis of bacterial communities in a three-compartment granular activated sludge system indicates community-level control by incompatible nitrification processes. *Appl. Environ. Microbiol.* **64**:2528–2532.

Holben WE, Feris KP, Kettunen A, Apajalahti JH. 2004. GC fractionation enhances microbial community diversity assessment and detection of minority populations of bacteria by denaturing gradient gel electrophoresis. *Appl. Environ. Microbiol.* **70**:2263–2270.

Hughes JB, Hellmann JJ, Ricketts TH, Bohannan BJ. 2001. Counting the uncountable: Statistical approaches to estimating microbial diversity. *Appl. Environ. Microbiol.* **67**:4399–4406.

Janssen PH. 2006. Identifying the dominant soil bacterial taxa in libraries of 16S rRNA and 16S rRNA genes. *Appl. Environ. Microbiol.* **72**:1719–1728.

Jones RT, Robeson MS, Lauber CL, Hamady M, Knight R, Fierer N. 2009. A comprehensive survey of soil acidobacterial diversity using pyrosequencing and clone library analyses. *ISME J.* **3**:442–453.

Jorgenson KF, van de Sande JH, Lin CC. 1978. The use of base pair specific DNA binding agents as affinity labels for the study of mammalian chromosomes. *Chromosoma* **68**:287–302.

Juste A, Thomma BP, Lievens B. 2008. Recent advances in molecular techniques to study microbial communities in food-associated matrices and processes. *Food Microbiol.* **25**:745–761.

Kassinen A, Krogius-Kurikka L, Makivuokko H, et al. 2007. The fecal microbiota of irritable bowel syndrome patients differs significantly from that of healthy subjects. *Gastroenterology* **133**:24–33.

Kunin V, Copeland A, Lapidus A, Mavromatis K, Hugenholtz P. 2008. A bioinformatician's guide to metagenomics. *Microbiol. Mol. Biol. Rev.* **72**:557–578.

Laskin AL, Lechevalier HA. 1973. *Handbook of Microbiology*, Vol. II: *Microbial Structure*. Cleveland, OH: CRC Press.

Liu WT, Marsh TL, Cheng H, Forney LJ. 1997. Characterization of microbial diversity by determining terminal restriction fragment length polymorphisms of genes encoding 16S rRNA. *Appl. Environ. Microbiol.* **63**:4516–4522.

Lowell JL, Gordon N, Engstrom D, Stanford JA, Holben WE, Gannon JE. 2009. Habitat heterogeneity and associated microbial community structure in a small-scale floodplain hyporheic flow path. *Microb. Ecol.* **58**:611–620.

Manuelidis L. 1977. A simplified method for preparation of mouse satellite DNA. *Anal. Biochem.* **78**:561–568.

Martin AP. 2002. Phylogenetic approaches for describing and comparing the diversity of microbial communities. *Appl. Environ. Microbiol.* **68**:3673–3682.

McCaig AE, Glover LA, Prosser JI. 2001. Numerical analysis of grassland bacterial community structure under different land management regimens by using 16S ribosomal DNA sequence data and denaturing gradient gel electrophoresis banding patterns. *Appl. Environ. Microbiol.* **67**:4554–4559.

McCartney AL. 2002. Application of molecular biological methods for studying probiotics and the gut flora. *Br. J. Nutr.* **88**(Suppl 1): S29–S37.

McCracken KJ, Murphy TC, Bedford MR, Apajalahti J. 2006. Chicken caecal microflora correlates with ME:GE using wheat-based diets. In *Proceedings of the XII WPSA European Poultry Conference*, Verona, Italy.

Morales SE, Holben WE. 2009. Empirical testing of 16S rRNA gene PCR primer pairs reveals variance in target specificity and efficacy not suggested by in silico analysis. *Appl. Environ. Microbiol.* **75**:2677–2683.

Morales SE, Holben WE. 2011. Linking bacterial identities and ecosystem processes: can "omic" analyses be more than the sum of their parts?. *FEMS Microbiol. Ecol.* **75**:2–16.

Morales SE, Cosart TF, Johnson JV, Holben WE. 2009. Extensive phylogenetic analysis of a soil bacterial community illustrates extreme taxon evenness and the effects of amplicon length, degree of coverage, and DNA fractionation on classification and ecological parameters. *Appl. Environ. Microbiol.* **75**:668–675.

Muller W, Gautier F. 1975. Interactions of heteroaromatic compounds with nucleic acids. A-T-specific non-intercalating DNA ligands. *Eur. J. Biochem.* **54**:385–394.

Muyzer G, de Waal EC, Uitterlinden AG. 1993. Profiling of complex microbial populations by denaturing gradient gel electrophoresis analysis of polymerase chain reaction-amplified genes coding for 16S rRNA. *Appl. Environ. Microbiol.* **59**:695–700.

Nagy E, Urban E, Soki J, Terhes G, Nagy K. 2006. The place of molecular genetic methods in the diagnostics of human pathogenic anaerobic bacteria. A minireview. *Acta Microbiol. Immunol. Hung.* **53**:183–194.

Nusslein K, Tiedje JM. 1998. Characterization of the dominant and rare members of a young Hawaiian soil bacterial community with small-subunit ribosomal DNA amplified from DNA fractionated on the basis of its guanine and cytosine composition. *Appl. Environ. Microbiol.* **64**:1283–1289.

Nusslein K, Tiedje JM. 1999. Soil bacterial community shift correlated with change from forest to pasture vegetation in a tropical soil. *Appl. Environ. Microbiol*. **65**:3622–3626.

Paul JH, Myers B. 1982. Fluorometric Determination of DNA in Aquatic Microorganisms by Use of Hoechst 33258. *Appl. Environ. Microbiol*. **43**:1393–1399.

Peuranen S, Tiihonen K, Apajalahti J, Kettunen A, Saarinen M, Rautonen N. 2004. Combination of polydextrose and lactitol affects microbial ecosystem and immune responses in rat gastrointestinal tract. *Br. J. Nutr*. **91**:905–914.

Pfeifer TA, Hegedus DD, Khachatourians GG. 1993. The mitochondrial genome of the entomopathogenic fungus *Beauveria bassiana*: Analysis of the ribosomal RNA region. *Can. J. Microbiol*. **39**:25–31.

Pjura PE, Grzeskowiak K, Dickerson RE. 1987. Binding of Hoechst 33258 to the minor groove of B-DNA. *J. Mol. Biol*. **197**:257–271.

Probert HM, Apajalahti JH, Rautonen N, Stowell J, Gibson GR. 2004. Polydextrose, lactitol, and fructo-oligosaccharide fermentation by colonic bacteria in a three-stage continuous culture system. *Appl. Environ. Microbiol*. **70**:4505–4511.

RappeI MS, Giovannoni SJ. 2003. The uncultured microbial majority. *Annu. Rev. Microbiol*. **57**:369–394.

Rodriguez-Minguela CM, Apajalahti JH, Chai B, Cole JR, Tiedje JM. 2009. Worldwide prevalence of class 2 integrases outside the clinical setting is associated with human impact. *Appl. Environ. Microbiol*. **75**:5100–5110.

Roell MK, Morse DE. 1991. Fractionation of nuclear, chloroplast, and mitochondrial DNA from *Polysiphonia boldii* (rhodophyta) using a rapid and simple method for the simultaneous isolation of RNA and DNA. *Phycology* **27**:299–305.

Santo Domingo JW, Kaufman MG, Klug MJ, Holben WE, Harris D, Tiedje JM. 1998. Influence of diet on the structure and function of the bacterial hindgut community of crickets. *Mol. Ecol*. **7**:761–767.

Schleper C, Holben W, Klenk HP. 1997. Recovery of crenarchaeotal ribosomal DNA sequences from freshwater-lake sediments. *Appl. Environ. Microbiol*. **63**:321–323.

Schloss PD, Handelsman J. 2005. Metagenomics for studying unculturable microorganisms: cutting the Gordian knot. *Genome Biol*. **6**:229.

Searle MS, Embrey KJ. 1990. Sequence-specific interaction of Hoechst 33258 with the minor groove of an adenine-tract DNA duplex studied in solution by 1H NMR spectroscopy. *Nucleic Acids Res*. **18**:3753–3762.

Spain AM, Krumholz LR, Elshahed MS. 2009. Abundance, composition, diversity and novelty of soil Proteobacteria. *ISME J*. **3**:992–1000.

Spiegelman D, Whissell G, Greer CW. 2005. A survey of the methods for the characterization of microbial consortia and communities. *Can. J. Microbiol*. **51**:355–386.

Stahl DA. 2004. High-throughput techniques for analyzing complex bacterial communities. *Adv. Exp. Med. Biol*. **547**:5–17.

Suwa YY. Imamura T, Suzuki T, Tashiro, Urushigawa Y. 1993. Ammonia-oxidizing bacteria with different sensitivities to $(NH_4)_2SO_4$ in activated sludges. *Water Resour*. **28**:1523–1532.

Tiedje JM, Asuming-Brempong S, Nusslein K, Marsh TL, Flynn SJ. 1999. Opening the black box of soil microbial diversity. *Appl. Soil Ecol*. **13**:109–122.

Torsvik V, Goksoyr J, Daae FL. 1990a. High diversity in DNA of soil bacteria. *Appl. Environ. Microbiol*. **56**:782–787.

Torsvik V, Salte K, Sorheim R, Goksoyr J. 1990b. Comparison of phenotypic diversity and DNA heterogeneity in a population of soil bacteria. *Appl. Environ. Microbiol*. **56**:776–781.

Torsvik V, Ovreas L, Thingstad TF. 2002. Prokaryotic diversity—magnitude, dynamics, and controlling factors. *Science* **296**:1064–1066.

Tringe SG, Hugenholtz P. 2008. A renaissance for the pioneering 16S rRNA gene. *Curr. Opin. Microbiol*. **11**:442–446.

van den Engh GJ, Trask BJ, Gray JW. 1986. The binding kinetics and interaction of DNA fluorochromes used in the analysis of nuclei and chromosomes by flow cytometry. *Histochemistry* **84**:501–508.

Wallace RJ, McKain N, McEwan NR, et al. 2003. *Eubacterium pyruvativorans* sp. nov., a novel non-saccharolytic anaerobe from the rumen that ferments pyruvate and amino acids, forms caproate and utilizes acetate and propionate. *Int. J. Syst. Evol. Microbiol*. **53**:965–970.

Weisblum B, Haenssler E. 1974. Fluorometric properties of the bibenzimidazole derivative Hoechst 33258, a fluorescent probe specific for AT concentration in chromosomal DNA. *Chromosoma* **46**:255–260.

Wooley JC, Godzik A, Friedberg I. 2010. A primer on metagenomics. *PLoS Comput. Biol*. **6**:e1000667.

Chapter 24

Enriching Plant Microbiota for a Metagenomic Library Construction

YING ZENG, HAO-XIN WANG, ZHAO-LIANG GENG, AND YUE-MAO SHEN

24.1 INTRODUCTION

To date, metagenomics studies have been conducted for a variety of common environments (e.g., soil, water, etc.) and extreme environments (e.g., glaciers, geysers, etc.) as well. Other interesting microbial habitats like marine sponges, insects, and the rumen of ruminants or even the human body have also explored in metagenomics [Piel et al., 2004; Ferrer et al., 2005; Schirmer et al., 2005; Tringe et al., 2005; Turnbaugh et al., 2007; see also Vol. II]. Representing a unique kind of environmental niches, plants serve as complex habitats for colonization by different kinds of microbes. In a few plants, microbes are found inhabiting particular sites such as nodules and leaf galls. For most plants, however, little is known about the diversity and colonization sites of microbes. The microbial biomass may account for only a minute part of all biomass in the host plant. The metagenomics of microorganisms that live in associations with plant tissues (known as plant microbiota) has not been investigated exclusively. This is largely due to a bottleneck generated by the methodology used for microbe enrichment from a host plant. Constructing a metagenomic library for plant microbiota is technically challenging. Without prior enrichment procedures, the metagenomic library could contain an extremely high proportion of plant-derived DNA, doubtless obliterating the microbial contribution and leading to a "waste" library. Therefore, the first and essential step is to enrich for the plant microbiota from plant tissues.

The tropical tree *Mallotus nudiflorus* (also known as *Trewia nudiflora*) was selected for this study because the maytansinoids (19-membered macrocyclic lactams related to ansamycin antibiotics of microbial origin) were isolated from this plant [Yu et al., 2002]. The available evidence suggests production of the core structure of the plant maytansinoids by an associated microbe. However, an intensive study of plant-associated microbial isolates has repeatedly failed to reveal a microbial producer of maytansinoids from maytansinoid-producing plants [Cassady et al., 2004]. We first attempted to enrich for plant-associated microbes of cultured- and uncultured-origin from *M. nudiflorus* stem barks, and then we constructed a metagenomic library from the microbial enrichment to screen genes potentially involved in the biosynthesis of maytansinoids or other polyketides. In our previous study, the constituent microorganisms were enriched by the enzymatic hydrolysis and subsequent differential centrifugation [Jiao et al., 2006]. Whereas this is a fairly crude approach, it increased the representation of prokaryotes in the enrichment from plant leaves or seeds. Our further efforts were focused on how to specifically enrich for plant microbiota—that is, to remain as close as possible to its native community structure while eliminating the plant DNA. Here, a whole new approach creating a predominantly microbial metagenomic library that contained 88% bacterial inserts is described.

24.2 METHODS

For further details about the enriching procedures, please refer to the original publication [Wang et al., 2008]. The general outline of the methods is presented here.

Handbook of Molecular Microbial Ecology, Volume I: Metagenomics and Complementary Approaches, First Edition. Edited by Frans J. de Bruijn.

24.2.1 Enriching Microbial Cells Associated with Stem Barks

Forty grams of fresh stem bark was chopped and homogenized (18,000 rpm × 1 min × 4) in 300 ml of MilliQ water. The suspension was filtered through two layers of bandage and centrifuged at 200 × *g* for 5 min at 4°C. Added to the 260 ml of supernatant was 2.34 g of NaCl (to a final concentration of 0.9%) as well as 1.64 ml of 10% SDS (to a final concentration of 0.063%). The suspension was mixed gently and incubated at 4°C for 1 h, generating a precipitation. The upper phase was carefully transferred to a clean bottle. The upper phase was centrifuged at 5000 × *g* for 10 min (4°C) to collect a pellet. The supernatant was decanted and the pellet was resuspended in 400 ml of MilliQ water. NaCl and SDS were added to the same concentration as described above and the mixture was incubated at 4°C for another 1 h, and subsequently a second pellet was collected by centrifugation at 5000 × *g* for 10 min at 4°C. This pellet was found to be highly enriched for the constituent microorganisms. A 1.6-kg sample of fresh stem bark generated 0.4 g of the final enrichment.

In comparison, enzyme-based enrichment was performed using 40 g of *M. nudiflorus* stem bark exactly according to our previous Method II [Jiao et al., 2006]. As a control, DNA was extracted directly from 2.0 g of stem bark. Meanwhile, stem bark of four different trees grown in Kunming Botanical Garden were collected to verify if our method was effective for microbe enrichment from other plants.

24.2.2 Assessing the Enriching Procedures by 16S rDNA-Based Techniques

DNA was extracted from different pellets as described in the section below. For Amplified Ribosomal DNA Restriction Analysis (ARDRA) of 16S rRNA sequences, the 27F and 1492R primer pair was used to target bacterial genomes [Lane et al., 1985; see also Chapter 7, Vol. I]. The efficiency of microbe enrichment was deduced from the percentage of clones affiliated with bacteria and from the diversity of their ARDRA patterns. In addition to the ARDRA protocol carried out as described in our previous work [Jiao et al., 2006], a simplified version of ARDRA was developed in this study for a fast and efficient assessment of microbial enrichment. Briefly, the PCR-amplified 16S rDNA was first digested by *Pvu*II and then separated on a 1% agarose gel. Two large bands close together could be expected. The undigested band harboring fragments of about 1.5 kb was composed completely of bacterial 16S rDNA. The slightly shorter fragments (~1.3 kb) were usually derived from plastid rDNA digested by *Pvu*II. The ratio of bacterial to plastidial DNA in the sample was easily estimated from the relative intensity between the undigested and digested bands. All the 16S rDNA sequences were checked for chimeras using the program Chimera Check [Cole et al., 2003] at the Ribosomal Database Project (see also Chapter 36, Vol. I), and no obvious chimeras were identified. All 16S rDNA sequences were phylogenetically classified by BLASTN [Altschul et al., 1990] and the RDP classifier. A neighbor-joining phylogenetic tree was generated with AlignX in the Vector NTI Suite.

24.2.3 Metagenomic Library Construction and Evaluation

For metagenomic DNA extraction, 30 mg of enrichment pellet was dispersed in 1 ml of buffer [Kieser et al., 2000] and treated with lysozyme (2 mg ml^{-1}) plus Achromopeptidase (0.5 mg ml^{-1}) for 1 h at 37°C. The pellet was collected by centrifugation at 12,000 × *g* for 10 min and subsequently resuspended in 1 ml proteinase K (0.5 mg ml^{-1}) solution (500 mM Tris [pH 8.0], 10 mM NaCl, 20 mM EDTA, 1% SDS) and incubated for 6 h at 55°C. The lysate was extracted twice with phenol–chloroform, precipitated with isopropanol, and treated with DNase-free RNase A (10 mg ml^{-1}). The DNA was further purified by 0.7 M NaCl containing 1% CTAB.

The final enrichment of 1.6 kg of fresh stem bark yielded a total of 20 μg of metagenomic DNA. The microbial diversity in the metagenomic DNA was analyzed using 16S rDNA-based techniques. From 16 μg of the same source DNA, a fosmid library with a total of 1.37×10^6 clones was generated, using a copy control fosmid library production kit (vector pCC1FOS; Epicentre, Madison, WI), according to the manufacturer's instruction. The average insert size was examined by restriction digest analysis. Fifteen random fosmids were digested by *Bam*, *Pvu*II, or *Xho*I, respectively. To assess the range of prokaryotic DNA inserts in the metagenomic library, the ends (~700 bp) of about 200 random fosmids were sequenced at Invitrogen Biotechnology Co., Ltd (Shanghai, China). All sequences were checked against the NCBI nonredundant databases.

24.3 RESULTS AND DISCUSSION

24.3.1 A Strong Enrichment for Plant Microbiota by Coupling SDS with NaCl

After release of the constituent microorganisms by homogenization, centrifugation at 200 × *g* removed the majority of cell debris and nuclei derived from *M. nudiflorus* stem barks. The deep-green pellet (P1)

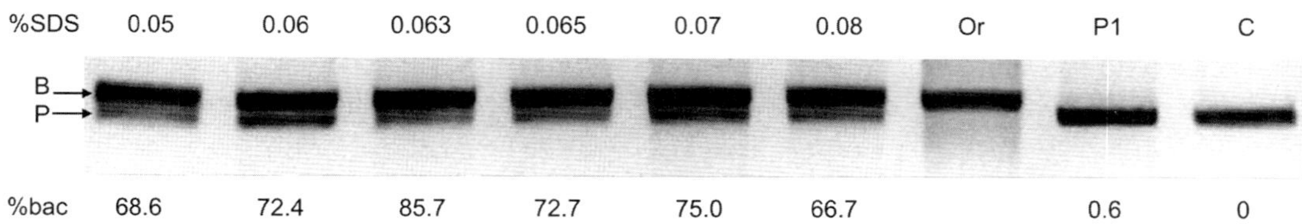

Figure 24.1 Effects of 0.9% NaCl plus SDS on microbe enrichment as detected by ARDRA. The numbers on the top line refer to the concentration (%) of SDS used in the enriching procedures. In the agarose gel the upper band represents the undigested 16S rDNA (~1.5 kb) probably affiliated with bacteria (B-arrow), while the lower band harboring fragments of about 1.3 kb corresponds to the digested 16S rDNA of plastid origin (P-arrow). '% bac' is the percentage of 16S rDNA clones potentially affiliated with bacteria. 'C' is the control and 'Or' is the original 16S rDNA prior to *Pvu*II digestion.

collected by subsequent centrifugation at 5000 × *g* was highly abundant in plant DNA, especially plastid DNA, as evidenced by a complete *Pvu*II digestion of the amplified 16S ribosomal DNA (rDNA) **(Fig. 24.1)**. Extensive studies on disruption of plastids by detergents and isolation of microbial cells from plastid debris were described by Wang et al. [2008]. Enrichments were performed using 0.9% NaCl plus SDS, varying from 0.05% to 0.08% according to the final protocol presented in Section 24.2. 16S rRNA genes were amplified by PCR of DNA extracted from each pellet, and then digested with *Pvu*II or used for a 16S rDNA library construction. As shown in Fig. 24.1, coupling SDS with NaCl conferred a much higher ratio of bacterial to plastidial DNA in the enriched samples. This is supported by the relative intensity between the undigested (B-arrow) and digested (P-arrow) bands of the corresponding 16S rDNA by *Pvu*II. Because most eubacterial 16S rDNA are devoid of the recognition site for *Pvu*II, after digestion they retain the size equal to that of 16S rDNA prior to digestion and form the undigested band upon electrophoresis (Fig. 24.1, B-arrow). However, the *M. nudiflorus* plastid rDNA is sensitive to *Pvu*II [Jiao et al., 2006] and produces a sequence slightly shorter than the original fragment after digestion. The digested band (Fig. 24.1, P-arrow) representing plastid rDNA can be expected to be very close to the undigested band. Given that the PCR-amplified rDNA yield is theoretically proportional to the copies of starting DNA in the PCR reaction [Keohavong et al., 1988], the relative intensity between the undigested and digested bands probably reflects the ratio of bacterial to plastidial DNA in the sample.

The majority (66–85%) of 16S rDNA clones were *Pvu*II-insensitive, potentially affiliated with bacteria. On the contrary, the amplified 16S rDNA of both the control (C) and the direct homogenization (P1) was almost completely digested by *Pvu*II because of its plastidial origin, and very few clones were found to be of bacterial origin (Fig. 24.1). Enrichments were repeatedly performed using 0.9% NaCl plus 0.063% SDS, and these conditions were defined as our final protocol.

A total of 0.4 g of the enriched pellet was obtained from 1.6 kg of fresh stem barks using the final protocol described in Section 24.2. Scanning electron microscopic and fluorescent microscopic analyses, including Acridine Orange staining, found the organisms in the enrichment to be primarily prokaryotic. To verify if our method was effective for microbe enrichment from other plants, we randomly collected stem barks of four different trees grown in the Kunming Botanical Garden. Following the final protocol, NaCl plus SDS resulted in a moderate enrichment in two of those plants as revealed by *Pvu*II-digested patterns of the respective16S rDNA [Wang et al., 2008].

24.3.2 A Considerably Enhanced Microbial Diversity in the Enrichment

For a metagenomic study the microbial diversity in the final enrichment was analyzed using PCR-amplified 16S rDNA libraries generated by using primers specific for Bacteria (see Chapter 15, Vol. I for further information). Phylogenetic diversity of the plant microbiota was surveyed from more than 180 eubacterial 16S rDNA clones. As shown in **Table 24.1,** more than 95% of the 16S rDNA clones were potentially of bacterial origin and 113 ARDRA types were detected in the enrichment-derived

Table 24.1 Microbe Enrichment Assessed by16S rDNA Sequence Analysis

Sources	% bac[a]	ARDRA types	Phyla or Phylum
The final enrichment	96.8	113	*Actinobacteria, Proteobacteria, Firmicutes, Gemmatimonadetes, Bacteroidetes, Planctomycetes, Deinococcus-Thermus*
Enzyme-based	70.1	26	*Proteobacteria*
P1[b]	0.6	NA[a]	NA
Control	0	NA	NA

[a]The percentage of 16S rDNA clones potentially affiliated with bacteria.

[b]P1 refers to the pellet derived from direct homogenization.

[c]NA, not applicable.

16S rDNA library. The metagenomic DNA of final enrichment was demonstrated to originate from very diverse microorganisms. At least 74 distinct ribotypes (at a 97% threshold) from seven different bacterial phyla were identified, mainly distributed among *Actinobacteria* and *Proteobacteria* (**Fig. 24.2**). A few clones were affiliated with uncultured or unclassified bacteria. In addition, half of sequences were singlets, and the most common ribotype accounted for 4.7% of the clones. Based on the sequence analysis, 61.3% of the 16S rDNA had a similarity below 97% with the best hits in the database, indicating that most of the microbes in *M. nudiflorus* microbiota might be previously unknown phylotypes.

To address how representative is the metagenomic library of the actual microbial community on the stem bark, rarefaction curves were built by DOTUR [Schloss and Handelsman, 2005]. As shown in **Figure 24.3**, rarefaction curves failed to reach saturation when using a sequence identity of 97% (species level) or 95% (genus level) as cutoff. However, the curve of the 80% cutoff (phylum level) was more flat. These results demonstrate that the sampling was sufficient at the phylum level and insufficient at the species and genus levels. Coverage estimators such as Chao1 predicted the total number of bacterial OTUs (operational taxonomic units) in our samples to be more than 169 when a 97%

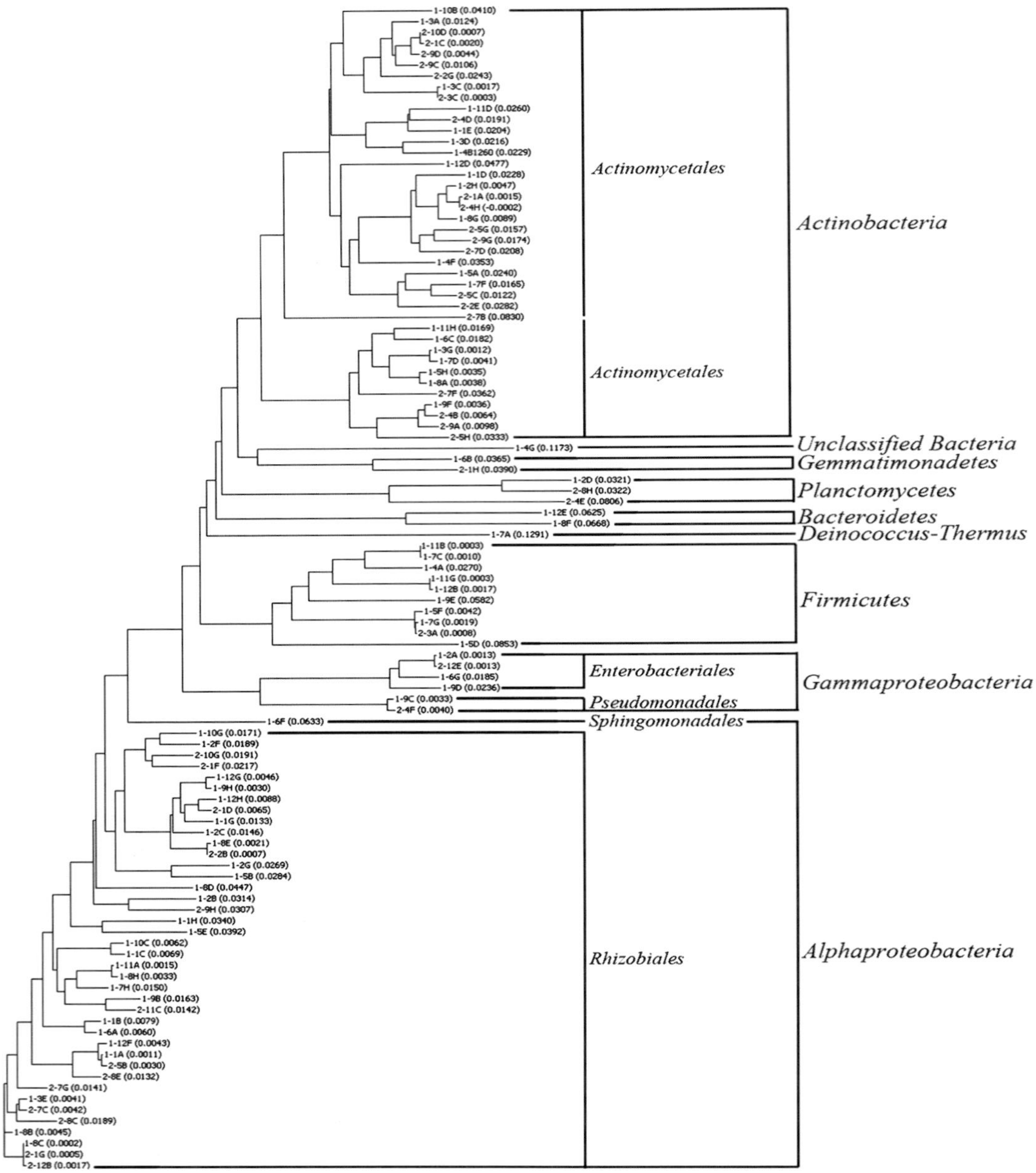

Figure 24.2 A neighbor-joining phylogenetic tree generated with AlignX in the Vector NTI Suite. 16S rDNA sequence analysis revealed that the metagenomic DNA originates from very diverse microorganisms.

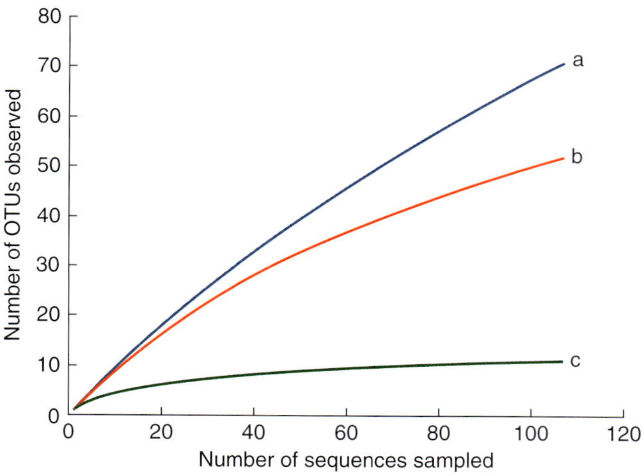

Figure 24.3 Rarefaction curves generated by DOTUR at different levels of sequence identity. Cutoff of 97% identity (a) is typically used to the same species, 95% cutoff (b) to the same genus, and 80% cutoff (c) to the same phylum.

identity cutoff was used and more than 195 when a 98% identity cutoff was used. The Chao1 might underestimate true richness at low sample sizes [Hughes et al., 2001], and the 16S rDNA amplification might also decrease the OTUs observed because not all rRNA genes can be amplified with the same "universal" primers. Thus the "true" richness of the plant microbiota is believed far greater than the OTUs we sampled, as predicted by the Chao1 estimator. On the other hand, 26 unique ARDRA types mapping primarily to only one phylum (*Proteobacteria*) were observed for our previous enzyme-based Method II [Jiao et al., 2006] (Table 24.1). As compared with the control and the direct homogenization (P1), the final enrichment showed a considerably enhanced proportion of bacterium-derived clones and a much wider species and phyla diversity of those clones.

24.3.3 The Metagenomic Library Is Dominated by Prokaryotic Inserts

A fosmid library for the plant microbiota harboring 1.37×10^6 clones was generated from the final enrichments. Based on the restriction digest analysis, the average insert size of the library was found to be 34.5 kb. The overall insert size of the library was, therefore, roughly analogous to 10,000 copies of *E. coli* genomes or 5000 copies of *Streptomyces* genomes. In this respect, the metagenomic library contained an estimated coverage of more than 1000 genomic types when a uniform phylogenetic distribution in the community is assumed.

The end sequences (~700 bp) of 187 random fosmids were determined to assess the range of prokaryotic DNA inserts in the metagenomic library. As expected, 166 of the clones (88.8%) were determined to be prokaryotic at an E value of $\leq 10^{-4}$. A few of the clones were considered not assignable (9.6%) or of eukaryotic origin (1.6%). The putative proteins encoded by end sequences were subjected to assign into specific functional roles using COGnitor [Tatusov et al., 2001]. More than half (68%) of the predicted proteins were assigned into certain COGs (clusters of orthologous groups of proteins) covering all the 18 functional categories in COG database. Estimated from end sequences, the metagenomic library was very rich in genomes of high GC content. The GC content of most inserts (72%) was observed to reach beyond the 70% level (up to 82.3%). This result is consistent with a dominant distribution in the library of *Actinobacteria* and *α-Proteobacteria*, as revealed by 16S rRNA gene analyses. Additionally, both fosmid end and 16S rDNA sequence analyses seemed to achieve similar patterns of phylogenetic diversity contained in the metagenomic library and in its source DNA (**Table 24.2**). The most abundant groups were *Proteobacteria, Actinobacteria, and Firmicutes* (low-GC gram positive), accounting for more than 90% of the OTUs [Wang et al., 2008]. It is very encouraging that the metagenomic library contained a large proportion of prokaryotic inserts and broad microbial diversity, which reinforces the feasibility of our enriching procedures for plant microbiota.

24.4 SUMMARY

In the present work, kilogram quantities of stem barks were thoroughly homogenized in water. The resulting mixture was centrifuged at low speed and the supernatant was collected. Salt and detergent were added to the supernatant to further remove plant DNA, especially plastid DNA. The principle underlying this type of enrichment is probably that the plastids lysed by SDS co-precipitate with plant cell debris during the aggregation stimulated by NaCl or $CaCl_2$. Incubation at 4°C for 1 h appears to be most suitable for a clear precipitation; otherwise, longer incubation time might reduce the microbial yield in the enrichment, presumably by increasing the likelihood of the occurrence of microbe sedimentation and/or lysis by SDS.

For each new plant species being studied, the enrichment procedure for plant microbiota has to be adjusted and optimized. $CaCl_2$ proved to be an alternative when NaCl failed to precipitate the suspension from the stem bark of *Eucommia ulmoides* and *Vernicia montana*. Delicate calculations of SDS concentration and further refinement of the method need to be performed for more efficient microbe enrichment from diverse plant resources. By the enrichment method presented here, the

Table 24.2 Microbial Diversity in the Metagenomic Library and Its Source DNA

Phylum or Subphylum	End Sequence Survey[a]	16S rDNA Survey[b]
α-Proteobacteria	26.5% (44)	38.7% (41)
β-Proteobacteria	3.0% (5)	ND[c]
Proteobacteria		
γ-Proteobacteria	6.0% (10)	5.7% (6)
δ-Proteobacteria	2.4% (4)	ND
Actinobacteria	56.0% (93)	37.7% (40)
Firmicutes	2.4% (4)	9.4% (10)
Deinococcus-Thermus	1.2% (2)	0.9% (1)
Chloroflexi	0.6% (1)	ND
Planctomycetes	0.6% (1)	2.8% (3)
Cyanobacteria	0.6% (1)	ND
Acidobacteria	0.6% (1)	ND
Gemmatimonadetes	ND	0.9% (1)
Bacteroidetes	ND	1.9% (2)
unclassified *Bacteria*	ND	1.9% (2)

[a] End sequence survey in the metagenomic library was carried out using 166 of fosmids harboring prokaryotic DNA inserts based on BLASTX analyses.

[b] 16S rDNA survey of the metagenomic DNA was carried out using 106 of rDNA clones. The figures in parentheses refer to the number of clones belonging to a unique group.

[c] ND, not detectable.

ratio of bacterial to plastidial DNA was considerably enhanced and a much wider species and phyla diversity of 16S rDNA clones were observed. Furthermore, the metagenomic library contained a large proportion of prokaryotic inserts and broad microbial diversity. This method can be applied to study the biology and chemical biology of the uncultivable microorganisms associated with plants, holding potential for drug discovery through a metagenomic strategy. Thus, this work opens further insight into the great biotechnical potential of plant microbiota, paving the way for recovery and biochemical characterization of its functional gene repertoire.

INTERNET RESOURCES

The Chimera Detection program (http://rdp8.cme.msu.edu/docs/chimera_doc.html)

The RDP Classifier (http://rdp.cme.msu.edu/classifier/classifier.jsp)

DOTUR (http://schloss.micro.umass.edu/software/)

Acknowledgments

This work was supported by the National Natural Science Foundation of China (30430020)

REFERENCES

ALTSCHUL SF, GISH W, MILLE, W, MYERS EW, LIPMAN DJ. 1990. Basic local alignment search tool. *J. Mol. Biol*. **215**:403–410.

CASSADY JM, CHAN KK, FLOSS HG, LEISTNER E. 2004. Recent developments in the maytansinoid antitumor agents. *Chem. Pharm. Bull*. **52**:1–26.

COLE JR, CHAI B, MARSH TL, FARRIS RJ, WANG Q, et al. 2003. The Ribosomal Database Project (RDP-II): Previewing a new autoaligner that allows regular updates and the new prokaryotic taxonomy. *Nucleic Acids Res*. **31**:442–443.

FERRER M, GOLYSHINA OV, CHERNIKOVA TN, KHACHANE AN, REYES-DUARTE D, et al. 2005. Novel hydrolase diversity retrieved from a metagenome library of bovine rumen microflora. *Environ. Microbiol*. **7**:1996–2010.

HUGHES JB, HELLMANN JJ, RICKETTS TH, BOHANNAN BJ. 2001. Counting the uncountable: Statistical approaches to estimating microbial diversity. *Appl. Environ. Microbiol*. **67**:4399–4406.

JIAO JY, WANG HX, ZENG Y, SHEN YM. 2006. Enrichment for microbes living in association with plant tissues. *J. Appl. Microbiol*. **100**:830–837.

KEOHAVONG P, WANG CC, CHA RS, THILLY WG. 1988. Enzymatic amplification and characterization of large DNA fragments of genomic DNA. *Gene* **71**:211–216.

KIESER T, BIBB MJ, BUTTNER MJ, CHATER KF, HOPWOOD DA. 2000. *Practical Streptomyces Genetics*. Norwich, England: John Innes Foundation Press.

LANE DJ, PACE B, OLSEN GJ, STAHL DA, SOGIN MI, et al. 1985. Rapid determination of 16S rRNA sequences for phylogenetic analysis. *Proc. Natl. Acad. Sci. USA* **82**:6955–6959.

PIEL J, HUI D, FUSETANI N, MATSUNAGA S. 2004. Targeting modular polyketide synthases with iteratively acting acyltransferases from metagenomes of uncultured bacterial consortia. *Environ. Microbiol*. **6**:921–927.

SCHIRMER A, GADKARI R, REEVES CD, IBRAHIM F, DELONG EF, et al. 2005. Metagenomic analysis reveals diverse polyketide synthase gene clusters in microorganisms associated with the marine sponge *Discodermia dissolute*. *Appl. Environ. Microbiol*. **71**: 4840–4849.

SCHLOSS P D, HANDELSMAN J. 2005. Introducing DOTUR, a computer program for defining operational taxonomic units and estimating species richness. *Appl. Environ. Microbiol*. **71**: 1501–1506.

TATUSOV RL, NATALE DA, GARKAVTSEV IV, TATUSOVA TA, SHANKAVARAM UT, et al. 2001. The COG database: New developments in phylogenetic classification of proteins from complete genomes. *Nucleic Acids Res*. **29**:22–28.

TRINGE SG, MERING C, KOBAYASHI A, SALAMOV AA, CHEN K, et al. 2005. Comparative metagenomics of microbial communities. *Science* **308**:554–557.

TURNBAUGH PJ, LEY RE, HAMADY M, FRASER-LIGGETT CM, KNIGHT R, et al. 2007. The Human Microbiome Project. *Nature* **449**:804–810.

WANG HX, GENG ZL, ZENG Y, SHEN YM. 2008. Enriching plant microbiota for a metagenomic library construction. *Environ. Microbiol*. **10**:2684–2691.

YU TW, BAI L, CLADE D, HOFFMANN D, TOELZER S, et al. 2002. The biosynthetic gene cluster of the maytansinoid antitumor agent ansamitocin from *Actinosynnema pretiosum*. *Proc. Natl. Acad. Sci. USA* **99**:7968–7973.

Chapter 25

Towards Automated Phylogenomic Inference

Martin Wu and Jonathan A. Eisen

25.1 INTRODUCTION

The small subunit ribosomal RNA (SSU rRNA) gene is the current gold standard for microbial classification, systematics, and ecology studies. The benefits of using SSU rRNA are extensive: its universal presence in all unicellular organisms, ease of cloning and sequencing, and the presence of sites with varying rates of evolution within one molecule [see Chapter 15, Vol. I]. Until very recently, the vast majority of microbes have been identified and classified only by recovering and sequencing their SSU rRNA genes. This single sequence of approximately 1.5 kbp is often the only information we have about the organism from which it came, the only way we know it is out there. As a matter of fact, the SSU rRNA gene is the most sequenced gene with hundreds of thousands of its sequences now deposited in public databases [e.g., see Chapter 36, Vol. I].

Despite being extremely valuable for microbial diversity studies, the SSU rRNA has its limitations. For example, it has been well documented that the nucleotide composition bias in the SSU rRNA sequences can mislead phylogenetic methods to incorrectly group two evolutionarily distant SSU rRNA genes together in phylogenetic trees [Woese et al., 1991; Hasegawa and Hashimoto 1993; see also Chapters 16 and 17, Vol. I]. Furthermore, the SSU rRNA gene only represents a tiny fraction of any genome. Inferring the phylogeny of organisms from any single gene bears some risks and needs to be corroborated by the use of other phylogenetic markers. This also has important implications in metagenomic phylotyping, when researchers assign taxonomy to the sequence fragments and use it to learn who is there and their relative abundance in the community. Since SSU rRNA typically makes up less than 1% of the genome, it means that during phylotyping the probability that any given sequence fragment can be anchored to a specific taxonomic clade by using this one gene is small. Thus, phylotyping of metagenomic data can greatly benefit from the use of alternative phylogenetic markers such as the multiple protein markers described below. In addition, the copy number of SSU rRNA gene varies greatly among bacterial species [Rastogi et al., 2009], which complicates the species diversity estimation using metrics such as Chao [Venter et al., 2004].

The recent advent of genomic sequencing greatly expands our ability to use protein-coding genes for microbial molecular systematics. Because protein sequences are constrained at the amino acid level instead of the nucleotide level, phylogenetic analyses of protein sequences are in general more robust to the nucleotide compositional bias seen in SSU rRNA [Loomis and Smith, 1990; Lockhart et al., 1992; Hasegawa and Hashimoto, 1993; Baldauf et al., 2000]. In addition, the less constrained variation at the third codon position allows these genes to be used in studies of more closely related organisms. Now, not only can one build gene trees based on a favorite protein-coding gene (e.g., *EF-Tu, rpoB, recA*, and *HSP70*), but one has the option to concatenate multiple gene sequences to construct trees on the "genome level." Possessing more phylogenetic signals, such "genome trees" are less susceptible to the stochastic errors than those built from a single gene [Jeffroy et al., 2006]. Recent studies attempting to reconstruct the tree of life have demonstrated the power of this approach [Brown et al., 2001;

Handbook of Molecular Microbial Ecology, Volume I: Metagenomics and Complementary Approaches, First Edition. Edited by Frans J. de Bruijn.

Ciccarelli et al., 2006] (for a review, see Delsuc et al. 2005). Likewise, genome trees have also been used successfully to reassess the phylogenetic positions of individual species [Badger et al., 2005; Wu et al., 2005].

The emerging field of metagenomics makes it possible to use protein-coding genes to investigate the microbial diversity in a natural environment. One fundamental goal of metagenomics is to figure out who is present in the community and what they can do. Phylogenetic analysis of markers present in the sequence data can be very informative in revealing who is there. If the marker happens to be part of a larger assembled sequence fragment, then the entire fragment can be anchored by that marker to a specific taxonomic clade. In this way, environmental shotgun sequences can be sorted into taxon-specific "bins" in silico, thereby allowing us to determine who can do what. The more marker genes we use, the better chance a sequence fragment can be anchored. This is critical if our interest is studying the less abundant species in a community (e.g., keystone species). Because they are underrepresented in the metagenomic data, their genomes are usually not well-covered and thus can be easily missed if only one or two marker genes are used during phylotyping. In the Sargasso Sea metagenomic study, multiple protein-coding genes and the SSU rRNA gene were used to estimate the species composition. The protein-coding genes presented a consensus microbial diversity profile that differed from the one estimated using SSU rRNA. This was thought to be due to the large copy number variation of the SSU rRNA between different microbial species. The protein-coding genes used in that study were single-copy genes [Venter et al., 2004].

Despite its demonstrated usefulness, phylogenetic inference based on protein markers has been limited in application, mainly due to the formidable technical difficulties inherent in this approach. Typically, molecular phylogenetic inference involves three steps: retrieval of homologous sequences, creation of multiple sequence alignments, and phylogenetic tree construction. Because only characters of common ancestry can be used to infer the evolutionary history, the most critical step is sequence alignment, when sequences are overlaid horizontally on each other in such a way that, ideally, each column in the alignment would only contain homologous characters (amino acids or nucleotides). To ensure this positional homology, the alignments must be curated—a process that evaluates the probable homology of each column or position in the alignments. Positions for which the assignment of homology is uncertain are then excluded from further analysis by masking [Gatesy et al., 1993]. Judicious masking increases the signal-to-noise ratio and often improves the discriminatory power of the phylogenetic methods [Eisen, 1998]. Unfortunately, curation requires skilled manual intervention, thus making it impractical to suitably process the massive amount of genome sequence data now available. Frequently, the alignment is treated as observation and its uncertainty is simply ignored, which often leads to questionable conclusions [Wong et al., 2008]. In addition, although many programs are available to automate the creation of multiple sequence alignments, using them directly for the *de novo* alignment can be computationally costly for large-scale phylogenetic analyses.

To overcome these problems, we have developed an automated pipeline AMPHORA for large-scale phylogenetic analysis using protein sequences [Wu and Eisen, 2008]. AMPHORA can rapidly and accurately generate highly reproducible multiple sequence alignments for a set of selected phylogenetic markers. More importantly, unlike previous automated methods [Ciccarelli et al., 2006], it can mask the alignments with qualities equivalent to human curations by using masks embedded with the hidden Markov models (HMMs) of the protein families. More recently, we developed a probability-based algorithm named ZORRO to assess the quality of the alignment and use it to mask the regions of uncertainty. Incorporating ZORRO into AMPHORA thus makes it a truly fast and accurate phylogenomics inference tool that should be useful for many applications.

In this chapter, we will demonstrate how AMPHORA can be used to (1) quickly build a genome tree from 578 complete bacterial genomes and (2) identify bacterial phylotypes from the metagenomic data collected from the Global Ocean Survey.

25.2 MATERIALS AND METHODS

25.2.1 Protein Phylogenetic Marker Database

For each marker, we first identified their protein sequences from representative bacterial genomes and used them as the "seed" sequences. The amino acid sequences were aligned using programs such as CLUSTALW [Thompson et al., 1994] or MAFFT [Katoh et al., 2002], and profile HMMs were made using HMMer [Eddy, 1998] (**Fig. 25.1**).

In the original study, we manually embed a mask into the seed alignment using the GDE package [Smith et al., 1994]. The mask was simply a text string of "1" and "0" where reliably aligned columns were labeled "1" and ambiguous columns were labeled "0". The manual masking step has since been replaced by ZORRO. With ZORRO, each column in the alignment is assigned a confidence score between 0 and 1 and the confidence score is then used to mask the alignment.

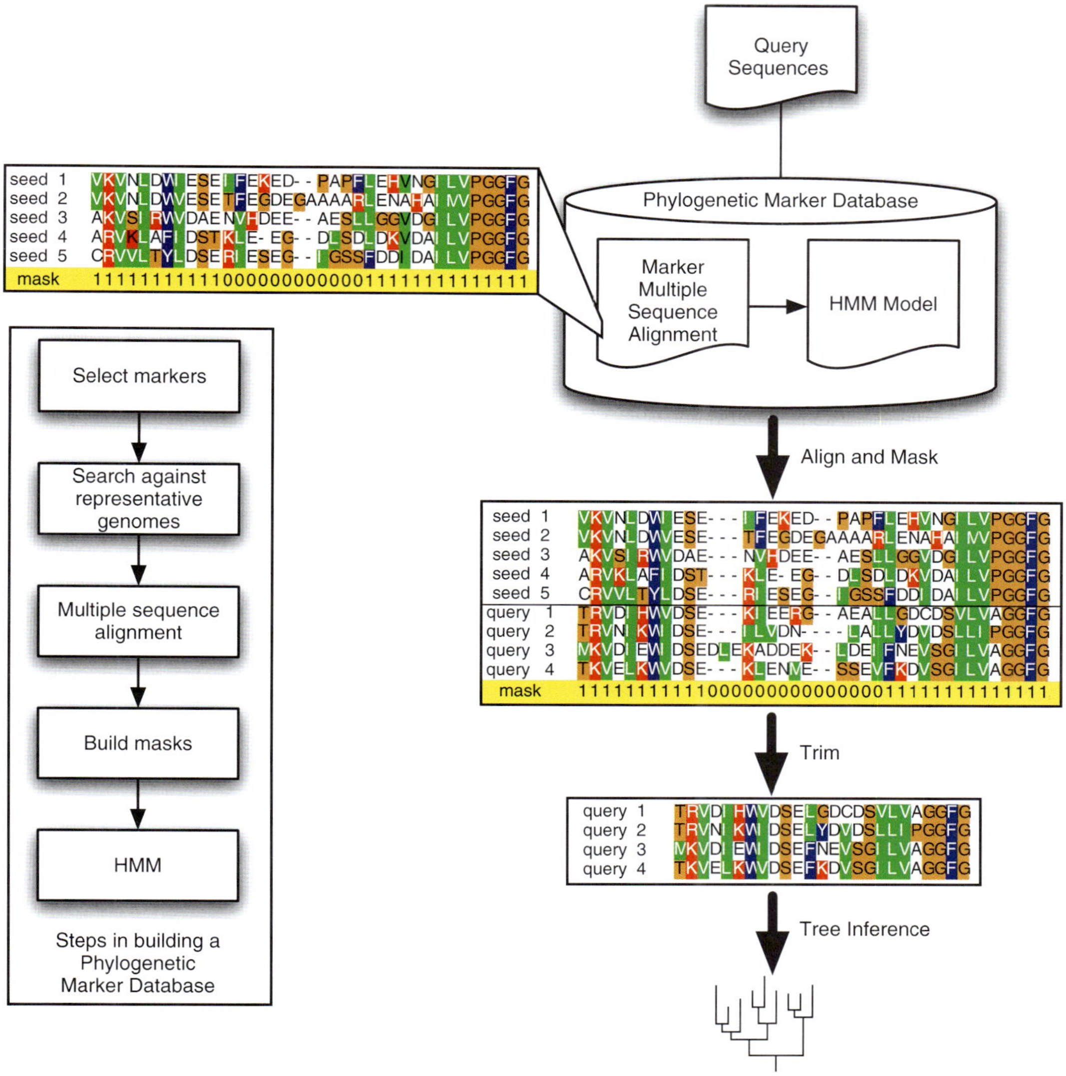

Figure 25.1 The marker protein sequences from representative genomes are retrieved, aligned, and masked. Profile HMMs are then built from those "seed" alignments. New sequences of interest are rapidly and accurately aligned to the trusted seed alignments through HMMs. Predefined masks embedded within the "seed" alignment are then applied to trim off regions of ambiguity prior to phylogenetic inference. Alignment columns marked with "1" or "0" were included or excluded, respectively, during further phylogenetic analysis.

25.2.2 Automated Sequence Alignment and Trimming

Subsequent steps are carried out by a Perl script joining multiple automated processes (Fig. 25.1). First, HMMer efficiently aligns the query amino acid sequences onto the trusted and fixed seed alignments. The Perl script then reads the masks embedded in the seed alignments and automatically trims the query alignments accordingly.

25.2.3 Bacterial Genome Tree Construction

Homologs of each of the 31 phylogenetic marker genes were identified from the 578 complete bacterial genomes, aligned, trimmed using AMPHORA, and concatenated by species into a mega-alignment. A maximum likelihood tree was then constructed from the mega-alignment using PHYML [Guindon and Gascuel, 2003]. The model selected based on the likelihood ratio test was the WAG model of amino acid substitution with gamma-distributed rate variation (five categories) and a proportion of invariable sites. The shape of the gamma-distribution and the proportion of the invariable sites were estimated by the program.

To speed up bootstrapping analyses, very closely related taxa were removed from the original mega-alignment, which left us with 310 taxa. Maximum likelihood trees were made from 100 bootstrapped replicates of this reduced dataset using PHYML with the same parameters described above.

25.2.4 Phylotyping by Phylogenetic Analyses (AMPHORA)

The protein markers used to construct the bacterial genome tree (see above) and the resultant genome tree were used as the reference sequences and the reference tree for phylotyping metagenomic data from the Sargasso Sea, global ocean survey, or the simulated sequences described below. Each marker sequence identified from the metagenomic data or simulated sequences was individually aligned to its corresponding reference sequences and trimmed using the method described earlier. Then it was inserted into the reference tree using a maximum parsimony method of RAXML [Stamatakis, 2006], constraining the topology of the tree to that of the genome tree. This tree-construction procedure was extremely fast, and 100 bootstrap replicates were run for each query sequence to assess the confidence of the branching orders. The trees were rooted arbitrarily using *Deinococcus radiodurans* as the outgroup. Tree branch lengths were calculated using the neighbor-joining algorithm with a fixed tree topology. A tree-based bracketing algorithm was then employed to assign a phylotype to the query sequence. Readers interested in the algorithm are encouraged to refer to the original paper [Wu and Eisen, 2008].

25.2.5 Phylotyping Simulation Study

To assess the performance of the phylotyping methods, a simulation study was carried out. One hundred representative genomes maximizing the phylogenetic diversity of the 578 complete bacterial genomes were selected using the genome tree and an algorithm described in Steel [2005]. From each of the 31 phylogenetic marker genes identified from the 100 bacterial genomes, a DNA sequence fragment of 300–900 bp in length was randomly chosen, which resulted in a total of 3088 simulated shotgun sequences that were used as benchmark query sequences in phylotyping (some markers are missing in some of the genomes). By comparing the predicted taxa with the known taxa, the sensitivity and specificity of phylotyping methods were calculated as described in Krause et al. [2008]. Briefly, for a taxon i, let P_i be the number of query sequences from i, let TP_i be the number of sequences that are correctly assigned to i, and let FP_i be the number of sequences that are incorrectly assigned to i. The sensitivity TP_i/P_i measures the proportion of query sequences that are correctly classified. The specificity $TP_i/(TP_i+FP_i)$ measures the reliability of the phylotyping assignments.

25.3 RESULTS AND DISCUSSION

25.3.1 The AMPHORA Pipeline

With the rapid increase in genomic sequence data, there is an ever-urgent need for automated phylogenetic analyses using protein sequences. However, automation is frequently accompanied by reduced quality. AMPHORA uses a fully automated method that is not only fast but also is of high quality. The main components of AMPHORA are shown in Figure 25.1, and their implementation is briefly described in Section 25.2. Readers interested in the implementation details should refer to the original paper [Wu and Eisen, 2008].

25.3.2 Protein Phylogenetic Marker Database

The core of AMPHORA is a protein phylogenetic marker database containing protein sequence alignments with trimming masks and corresponding profile hidden Markov models (HMMs). Thirty-one protein-coding phylogenetic marker genes (*dnaG, frr, infC, nusA, pgk, pyrG, rplA, rplB, rplC, rplD, rplE, rplF, rplK, rplL, rplM, rplN, rplP, rplS, rplT, rpmA, rpoB, rpsB, rpsC, rpsE, rpsI, rpsJ, rpsK, rpsM, rpsS, smpB, tsf*) are currently included in the database. We selected this initial set of proteins because (1) they are universally distributed in bacteria; (2) the vast majority of them exist as single copy genes within each genome; and (3) they are housekeeping genes involved in information processing (replication, transcription, and translation) or central metabolism and thus are thought to be relatively recalcitrant to lateral gene transfer [Jain et al., 1999].

25.3.3 High-Quality and Highly Reproducible Sequence Alignments

It has been shown that alignment quality can have greater impact on the final tree than does the tree-building method employed [Morrison and Ellis, 1997]. Therefore, preparing high-quality sequence alignments is a most critical part of any molecular phylogenetic analysis. This preparation typically involves careful but tedious manual editing and trimming of the generated alignments and thus remains the biggest challenge to automation.

We overcame this problem by taking advantage of a unique feature of profile HMM-based multiple sequence alignments. When using profile HMMs to align sequences, new sequences can be mapped back, residue by residue, onto the "seed" alignment from which that HMM originated. When the seed alignment includes a high-quality mask, the newly generated alignments can be automatically trimmed accordingly, thus producing high-quality alignments without requiring further human intervention. In addition, the HMM model is the only variable in this automated alignment and trimming. When the same model is used, the alignments generated thereby are completely additive and reproducible, thus enabling meaningful comparison of the results from different phylogenetic studies or different researchers.

In the original AMPHORA, the masks relied on skilled but sometimes arbitrary manual curation, although it needed to be generated only once for each protein family. More recently, we developed a probability-based algorithm named ZORRO to objectively assess the quality of the alignment and use it to mask the regions of uncertainty. ZORRO implements a pair hidden Markov model, calculates the posterior probability that characters in a column should be aligned together, and uses this probability to score the confidence of each aligned position. The probability is evaluated not based on one single alignment but in the context of all probable alignments. Using the Balibase benchmark dataset and a simulation study, we show that masking by ZORRO reduces the alignment uncertainty and thus increases the signal-to-noise ratio of a given alignment, which leads to significant improvements in the accuracy of the phylogenetic analyses of relative divergent sequences (manuscript in preparation).

25.3.4 Speed

Another big advantage of AMPHORA is its speed. It is an order of magnitude faster than methods based on *de novo* pairwise alignment programs such as CLUSTALW and MUSCLE. This is because our HMM-based method aligns sequences by comparing them only once each to the HMM model. As a result, the computational cost increases linearly with the number of sequences to be aligned. In contrast, the computational cost of a pairwise alignment approach increases polynomially and can soon become prohibitively expensive.

25.4 APPLICATION I: BACTERIAL GENOME TREES

25.4.1 Constructing a "Genome Tree"

At the time of this study, 578 complete bacterial genomes were available from the NCBI RefSeq collections. AMPHORA was used to search and identify the 31 protein-coding genes from each genome, generate the sequence alignment for each protein family, and trim the alignment. The alignments were then concatenated by species using a perl script, resulting in a mega-alignment of 5591 good amino acid positions (columns) by 578 species (rows). A maximum likelihood genome tree was constructed from this mega-alignment. A bootstrapped maximum likelihood genome tree of 310 representatives is shown in **Figure 25.2**.

Overall, the genome tree is similar to the SSU rRNA tree. All the major bacterial phyla are well-separated into their own monophyletic groups even though the relationships among some of them remain unclear. However, unlike the SSU rRNA tree, the bushy area (intermediate levels) of the genome tree is highly resolved. In the gamma-proteobacteria, for example, the nodes separating taxa into different orders, families, and genera receive generally excellent bootstrapping support, whereas uncertainty is high in the corresponding regions of the SSU rRNA tree. Highly robust organismal phylogenies of gamma- and alpha-proteobacteria have been inferred previously using hundreds of commonly shared genes [Lerat et al., 2003; Williams et al., 2007] and are congruent to our genome tree. This reflects the much-reduced stochastic noise present in the concatenated protein sequences compared to that of a single, slowly evolving SSU rRNA gene.

25.4.2 Genome-Based Microbial Taxonomy

The uncertainty in the SSU rRNA tree—the backbone of modern microbial systematics—often prevents microbial taxonomists from placing new species or genera within higher taxa, particularly at these intermediate levels [Garrity et al., 2004]. When such assignments were nevertheless made for these problematic taxa, inconsistency was introduced into the taxonomic nomenclature. For example, taxa assigned to the orders *Alteromondales, Pseudomonadales*, and *Oceanospirillales* in *Bergey's Taxonomic Outline of Prokaryotes* [Garrity et al., 2004] are intermingled and paraphyletic in our genome tree. It is our view that the taxonomy needs to be revisited and possibly revised in such cases (see Section 2, Vol. I).

Although use of SSU rRNA was a landmark in microbial systematics, genome trees are more robust in resolving taxonomic relationships below the phylum level and hence provide an excellent alternative phylogenetic framework for microbial systematics. Currently, close to 1000 bacterial genomes have been sequenced and 3500 more are well underway. In particular, there is a systematic effort (*Genomic Encyclopedia of Bacteria and Archaea*, or *GEBA*) to fill in the gaps in genome sequencing along the bacterial and archaeal branches of the tree of life. Although so far we have only scratched the surface in terms of representing the microbial diversity, we expect that the move towards the genome-based microbial systematics will gather speed as more genomes

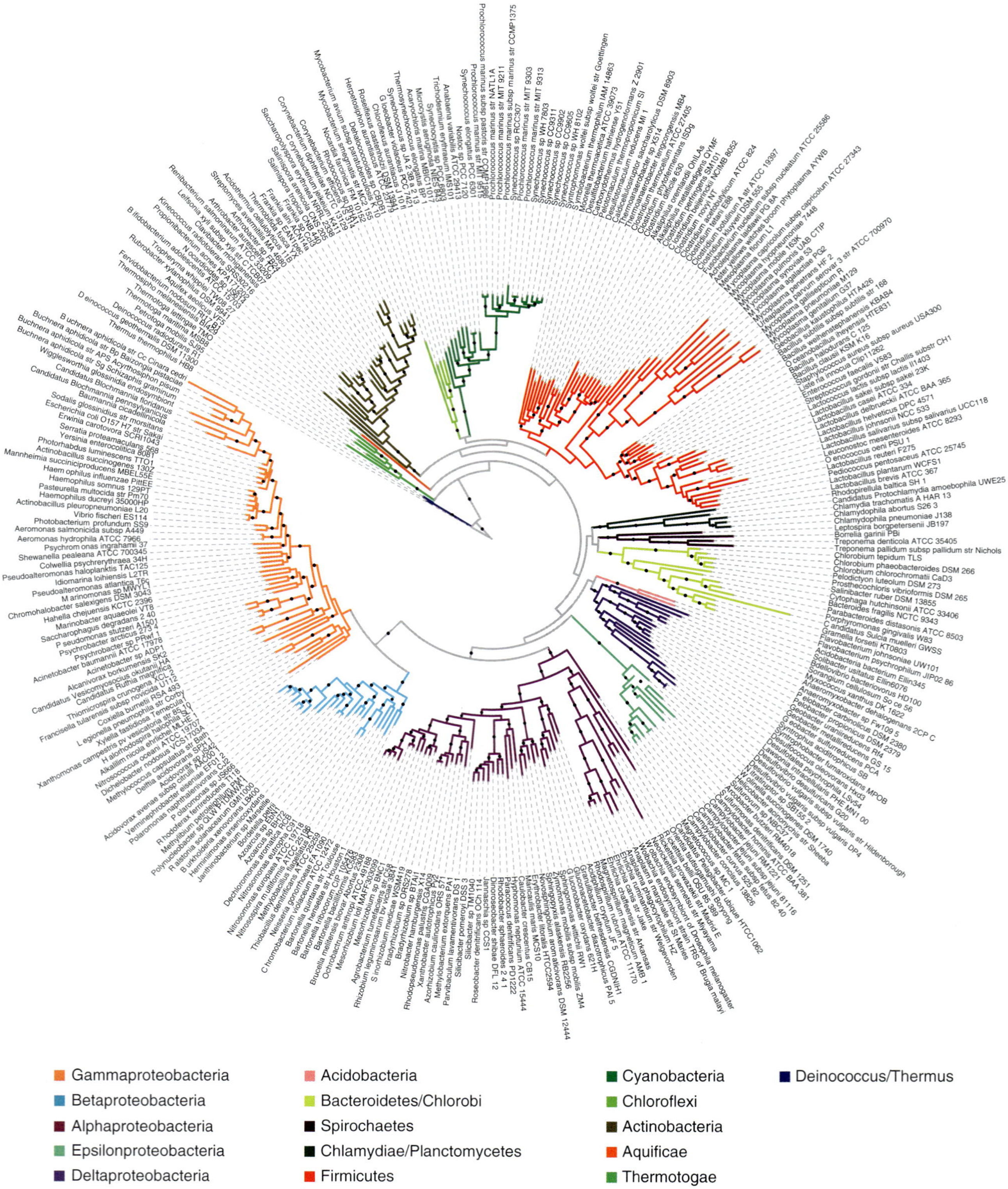

Figure 25.2 The tree was constructed from concatenated protein sequence alignments derived from 31 housekeeping genes. All major phyla are separated into their monophyletic groups and are highlighted by color. The branches with bootstrap support of over 80 (out of 100 replicates) are indicated with black dots. Although the relationships among the phyla are not strongly supported, those below the phylum level show very respectable support. The radial tree was generated using iTOL [Letunic and Bork, 2007].

become available. Until then, a hybrid approach may be most fruitful. A genome tree can be used as a scaffold where species for which we lack full genome sequences can be placed by comparing their SSU rRNA sequences to those of sequenced species.

25.5 APPLICATION II: METAGENOMIC PHYLOTYPING

25.5.1 Global Ocean Survey

In our original paper, we used AMPHORA to reanalyze the environmental shotgun sequencing data collected from the Sargasso Sea and showed that its results were consistent with previous studies [Wu and Eisen, 2008]. We showed that the various individual protein markers present remarkably concordant microbial diversity profiles, thus suggesting that results for different markers may be additive. The global ocean survey represents a much larger dataset and is therefore used to demonstrate the high-throughput capability of AMPHORA for phylotyping.

The global ocean survey collection contains a total of 41,146,566 peptides predicted from the sequence reads [Rusch et al., 2007]. From these peptides, AMPHORA identified 154,102 sequences that belong to one of the 31 protein-coding genes. This corresponds to 0.4% of the whole sequence dataset, which is smaller than the fraction expected assuming an average-sized genome with 3000 genes (31/3000 = 1%). This is not surprising considering that (1) many of the 31 marker genes are ribosomal proteins, which are smaller than an average-sized gene, and (2) sequence reads contain partial genes that might be too short to be recognized or to pass the minimum size limit imposed by AMPHORA. In comparison, 4125 SSU rRNA sequences were identified from the same dataset [Rusch et al., 2007].

Figure 25.3 shows the taxonomic breakdown of the more abundant bacteria species identified from the global ocean survey. It is evident that members of the alpha-proteobacteria dominate at most sites. Different sites exhibit distinct microbial diversity profiles, which we used to generate a heat map (**Fig. 25.4**). Some general patterns are discernible in the map. For example, the habitats seem to cluster by their type and proximity, but there are also notable exceptions. Apparently, more metadata need to be considered to draw meaningful conclusions. However, this is out of the scope of this review.

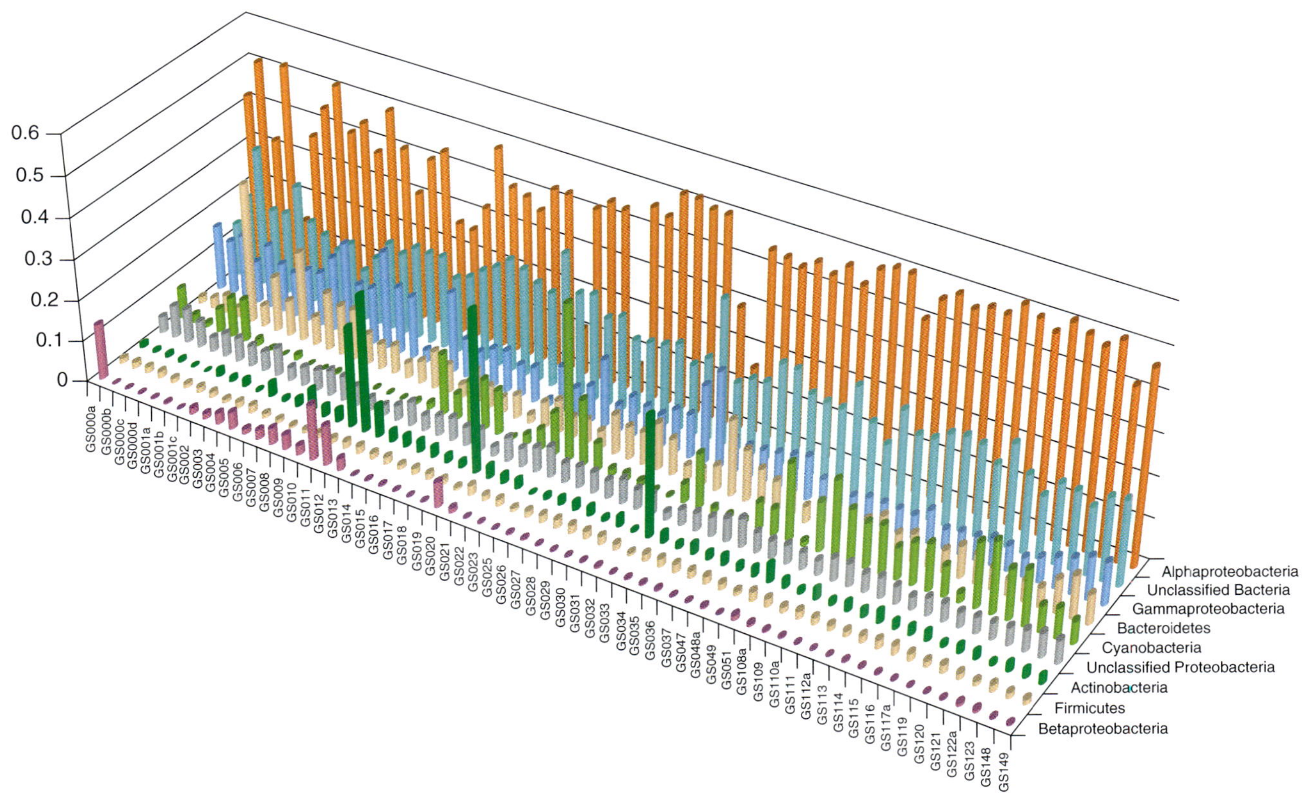

Figure 25.3 The metagenomic data was analyzed using AMPHORA and the 31 protein phylogenetic markers. The microbial diversity profiles obtained from individual markers were combined. The sampling sites listed on the x axis were described in Rusch et al. [2007].

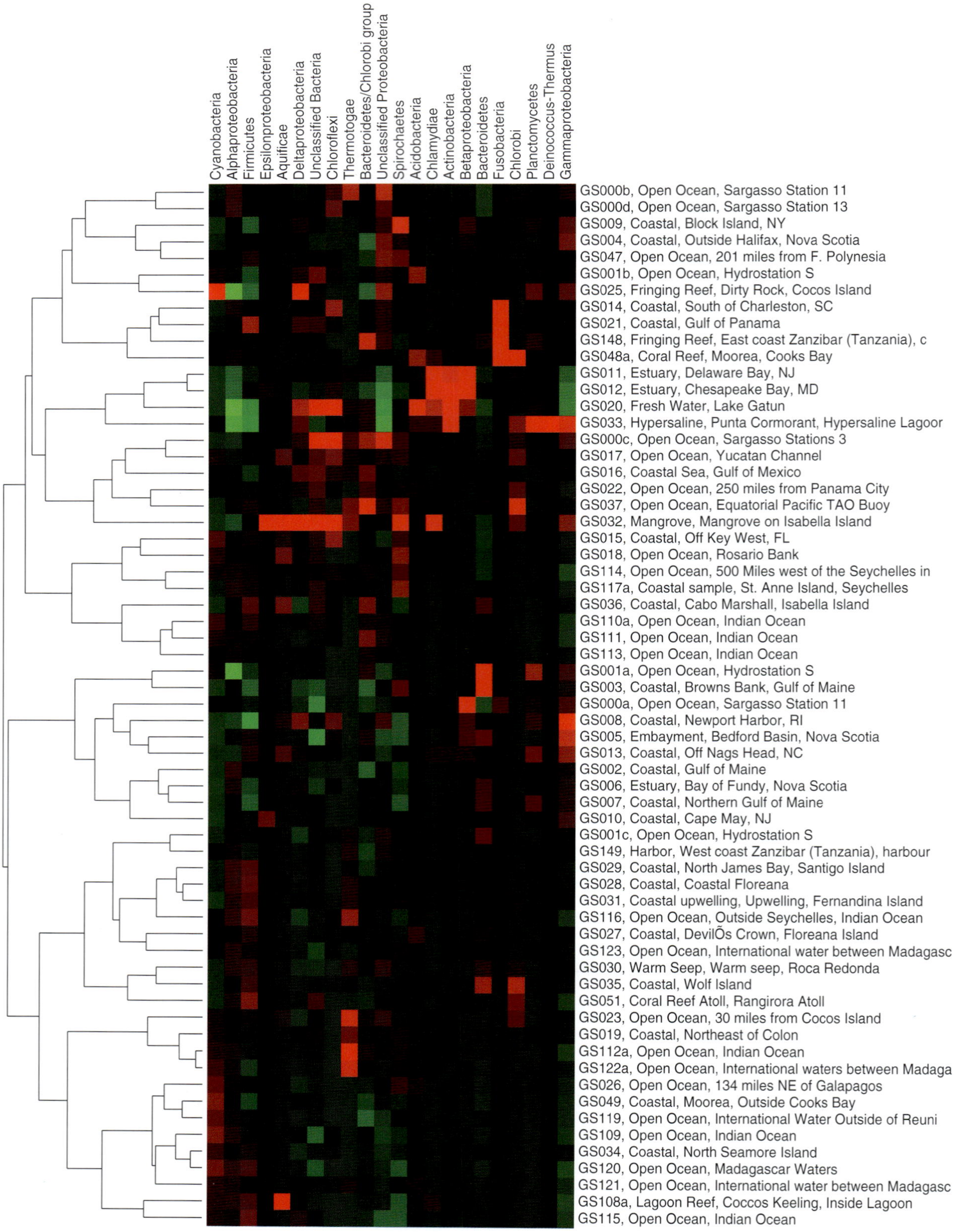

Figure 25.4 Two-way clustering of samples and species based on relative enrichment of bacterial species in the habitats. The over- and underrepresented species are indicated by red and green blocks, respectively.

25.5.2 Comparison of Phylogeny-Based and Similarity-Based Phylotyping

Although our phylogeny-based phylotyping is fully automated, it still requires many more steps than, and is slower than, similarity-based phylotyping methods such as a MEGAN [Huson et al., 2007; see also Chapter 39, Vol. I]. Is it worth the trouble? As previously discussed [Wu and Eisen, 2008], similarity-based phylotyping assigns taxonomy to a query sequence based on its top matches to reference sequences. When species that are closely related to the query sequence exist in the reference database, similarity-based phylotyping can work well. However, if the reference database contains no closely related species to the query, the top hits returned are often not the nearest neighbors and could be misleading [Koski and Golding, 2001]. Furthermore, similarity-based methods require an arbitrary similarity cutoff value to define the top hits. Since species and genes often evolve at very different rates, a universal cutoff that works under all conditions does not exist. As a result, the performance will vary.

In comparison, our tree-based bracketing algorithm places the query sequence within the context of a phylogenetic tree that is more consistent with the taxonomy hierarchy. In addition, it assigns a query to a taxonomic level only if that level has adequate sampling. With the well-sampled species *Prochlorococcus marinus*, for example, our method can distinguish closely related organisms and make taxonomic identifications at the species level. Our reanalysis of the Sargasso Sea data placed 672 sequences (3.6% of the total) within a *Prochlorococcus marinus* clade [Wu and Eisen, 2008]. On the other hand, for sparsely sampled clades such as *Aquifex*, assignments will be made only at the phylum level.

In our original study, we carried out a simulation study to benchmark the performance of the phylogeny-based and the similarity-based phylotyping. We determined the sensitivity and specificity of the taxonomic assignments made by AMPHORA and MEGAN to 3088 simulated shotgun sequences of 31 phylogenetic marker genes identified from 100 known bacterial genomes. To decrease the impact of biased taxon sampling on our results, the 100 genomes were chosen in such a way that maximized their representation of the phylogenetic diversity. **Figure 25.5** compares the sensitivity and specificity of the phylotyping assignments at the phylum, class, order, family, and genus level. The general trend of decreasing sensitivity seen in the figure from the phylum to the species level simply reflects the fact that the amount of reference data available for taxonomic assignment is decreasing. AMPHORA significantly outperformed MEGAN in sensitivity at all taxonomic ranks. Both methods performed extremely well in specificity at all levels (>0.97) except at the species level where AMPHORA (0.63) outperformed MEGAN (0.43) by a large margin.

25.6 FUTURE ISSUES

25.6.1 Additional Markers and Reference Genomes

We are in the process of expanding the phylogenetic marker database by adding more proteins, including the commonly used protein markers RecA, HSP70, and EF-Tu. Major expansion will also require systematic assessment of many other protein families for their suitability as phylogenetic markers. For metagenomic phylotyping, the marker genes do not have to be

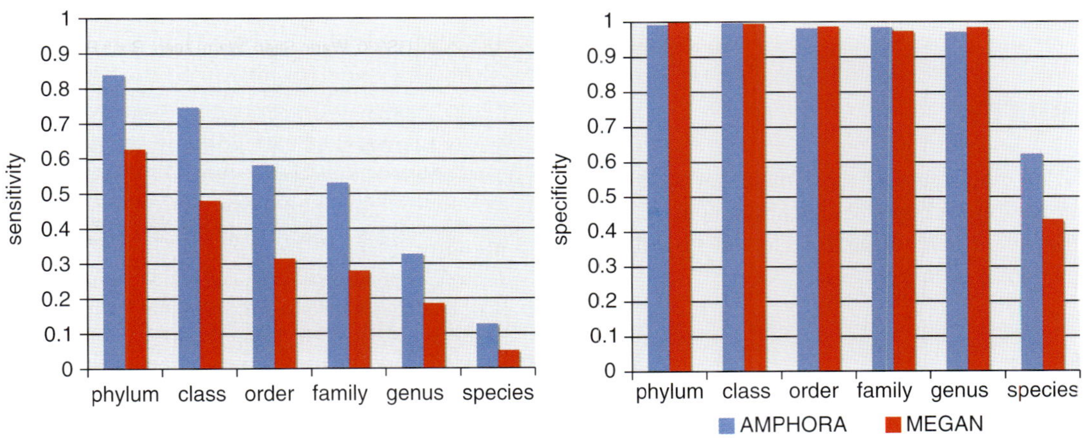

Figure 25.5 The sensitivity and specificity of the phylotyping methods were measured across taxonomic ranks using simulated Sanger shotgun sequences of 31 genes from 100 representative bacterial genomes. It shows that AMPHORA significantly outperforms MEGAN in sensitivity without sacrificing specificity.

single-copy or universal, but they must have sufficient phylogenetic signal and not be frequently exchanged between distantly related lineages. Until we learn more about the extent of lateral gene transfer in natural microbial communities, we would caution against using every protein sequence collected in metagenomics studies for microbial diversity study.

In the original study, we have shown that adding representatives of novel phyla can facilitate metagenomic phylotyping [Wu and Eisen, 2008]. Although the sequencing of thousands of microbial genomes is underway, we have only scratched the surface. More importantly, the organisms are often chosen from an anthropocentric point of view and thus are not truly representative of the total microbial diversity. We see a need to systematically select microbes for sequencing based mainly on their phylogenetic positions, thus maximizing their value for comparative genomics and phylogenomic studies.

25.7 CONCLUSIONS

Currently, SSU rRNA remains the most powerful phylogenetic marker due to the number of sequences available and the scope of taxonomic coverage. However, the imminent arrival of thousands of microbial genome sequences will vastly expand the amount of data available for alternative protein phylogenetic markers, thus presenting us with both a challenge and an opportunity. Designed to rapidly, reliably, and reproducibly align and trim protein sequences, AMPHORA eliminates one of the tightest bottlenecks in large-scale protein phylogenetic inference and can be applied to phylogenetic analyses of single genes, whole genomes, and metagenomes. We believe such a phylogenomic approach will be valuable in helping us make sense of the rapidly accumulating microbial genomic data.

AVAILABILITY

The AMPHORA package and the simulation study data can be downloaded from http://wolbachia.biology.virginia.edu/WuLab/Software.html. ZORRO can be downloaded from http://probmask.sourceforge.net.

ABBREVIATIONS

AMPHORA, automated phylogenomics inference; HMM, hidden Markov model; NCBI, National Center for Biotechnology Information.

INTERNET RESOURCES

Genomes online database http://genomesonline.org

Acknowledgments

The initial development of this work was supported in part by NSF grant DEB-0228651 to JAE. The final development and testing was funded by the Gordon and Betty Moore Foundation (grant #1660 to JAE).

REFERENCES

Badger JH, Eisen JA, Ward NL. 2005. Genomic analysis of *Hyphomonas neptunium* contradicts 16S rRNA gene-based phylogenetic analysis: implications for the taxonomy of the orders 'Rhodobacterales' and Caulobacterales. *Int. J. Syst. Evol. Microbiol.* **55**:1021–1026.

Baldauf SL, Roger AJ, Wenk-Siefert I, Doolittle WF. 2000. A kingdom-level phylogeny of eukaryotes based on combined protein data. *Science* **290**:972–977.

Brown JR, Douady CJ, Italia MJ, Marshall WE, Stanhope MJ. 2001. Universal trees based on large combined protein sequence data sets. *Nat. Genet.* **28**:281–285.

Ciccarelli FD, Doerks T, von Mering C, Creevey CJ, Snel B, et al. 2006. Toward automatic reconstruction of a highly resolved tree of life. *Science* **311**:1283–1287.

Delsuc F, Brinkmann H, Philippe H. 2005. Phylogenomics and the reconstruction of the tree of life. *Nat. Rev. Genet.* **6**:361–375.

Eddy SR. 1998. Profile hidden Markov models. *Bioinformatics* **14**:755–763.

Eisen JA. 1998. Phylogenomics: Improving functional predictions for uncharacterized genes by evolutionary analysis. *Genome Res.* **8**:163–167.

Garrity GM, Bell JA, Lilburn TG. 2004. Taxonomic outline of the prokaryotes. In *Bergey's Manual of Systematic Bacteriology*, 2nd ed. New York: Springer-Verlag. 90–109.

Gatesy J, DeSalle RK, Wheeler W. 1993. Alignment-ambiguous nucleotide sites and the exclusion of systematic data. *Mol. Phylogenet. Evol.* **2**:152–157.

Guindon S, Gascuel O. 2003. A simple, fast, and accurate algorithm to estimate large phylogenies by maximum likelihood. *Syst. Biol.* **52**:696–704.

Hasegawa M, Hashimoto T. 1993. Ribosomal RNA trees misleading? *Nature* **361**:23.

Huson DH, Auch AF, Qi J, Schuster SC. 2007. MEGAN analysis of metagenomic data. *Genome Res.* **17**:377–386.

Jain R, Rivera MC, Lake JA. 1999. Horizontal gene transfer among genomes: The complexity hypothesis. *Proc. Natl. Acad. Sci. USA* **96**:3801–3806.

Jeffroy O, Brinkmann H, Delsuc F, Philippe H. 2006. Phylogenomics: The beginning of incongruence ? *Trends Genet.* **22**:225–231.

Katoh K, Misawa K, Kuma K, Miyata T. 2002. MAFFT: A novel method for rapid multiple sequence alignment based on fast Fourier transform. *Nucleic Acids Res.* **30**:3059–3066.

Koski LB, Golding GB. 2001. The closest BLAST hit is often not the nearest neighbor. *J. Mol. Evol.* **52**:540–542.

Krause L, Diaz NN, Goesmann A, Kelley S, Nattkemper TW, et al. 2008. Phylogenetic classification of short environmental DNA fragments. *Nucleic Acids Res.* **36**:2230–2239.

Lerat E, Daubin V, Moran NA. 2003. From gene trees to organismal phylogeny in prokaryotes: the case of the gamma-proteobacteria. *PLoS Biol*. **1**:19.

Letunic I, Bork P. 2007. Interactive Tree of Life (iTOL): An online tool for phylogenetic tree display and annotation. *Bioinformatics* **23**:127–128.

Lockhart PJ, Howe CJ, Bryant DA, Beanland TJ, Larkum AW. 1992. Substitutional bias confounds inference of cyanelle origins from sequence data. *J. Mol. Evol*. **34**:153–162.

Loomis WF, Smith DW. 1990. Molecular phylogeny of *Dictyostelium discoideum* by protein sequence comparison. *Proc. Natl. Acad. Sci. USA* **87**:9093–9097.

Morrison, DA, Ellis, JT 1997. Effects of nucleotide sequence alignment on phylogeny estimation: A case study of 18S rDNAs of apicomplexa. *Mol. Biol. Evol*. **14**:428–441.

Rastogi R, Wu M, Dasgupta I, Fox GE. 2009. Visualization of ribosomal RNA operon copy number distribution. *BMC Microbiol*. **9**:208.

Rusch DB, Halpern AL, Sutton G, Heidelberg KB, Williamson S, et al. 2007. The Sorcerer II Global Ocean Sampling expedition: northwest Atlantic through eastern tropical Pacific. *PLoS Biol*. **5**:e77.

Smith SW, Overbeek R, Woese CR, Gilbert W, Gillevet PM. 1994. The genetic data environment: An expandable GUI for multiple sequence analysis. *Comput. Appl. Biosci*. **10**:671–675.

Stamatakis A. 2006. RAxML-VI-HPC: Maximum likelihood-based phylogenetic analyses with thousands of taxa and mixed models. *Bioinformatics* **22**:2688–2690.

Steel M. 2005. Phylogenetic diversity and the greedy algorithm. *Syst. Biol*. **54**:527–529.

Thompson JD, Higgins DG, Gibson TJ. 1994. CLUSTAL W: Improving the sensitivity of progressive multiple sequence alignment through sequence weighting, position-specific gap penalties and weight matrix choice. *Nucleic Acids Res*. **22**:4673–4680.

Venter JC, Remington K, Heidelberg JF, Halpern AL, Rusch D, et al. 2004. Environmental genome shotgun sequencing of the Sargasso Sea. *Science* **304**:66–74.

Williams KP, Sobral BW, Dickerman, AW 2007. A robust species tree for the alphaproteobacteria. *J. Bacteriol*. **189**:4578–4586.

Woese CR, Achenbach L, Rouviere P, Mandelco L 1991. Archaeal phylogeny: Reexamination of the phylogenetic position of *Archaeoglobus fulgidus* in light of certain composition-induced artifacts. *Syst. Appl. Microbiol*. **14**:364–371.

Wong KM, Suchard MA, Huelsenbeck JP. 2008. Alignment uncertainty and genomic analysis. *Science* **319**:473–6.

Wu M, Eisen JA. 2008. A simple, fast, and accurate method of phylogenomic inference. *Genome Biol*. **9**:R151.

Wu M, Ren Q, Durkin AS, Daugherty SC, Brinkac LM, et al. 2005. Life in hot carbon monoxide: the complete genome sequence of *Carboxydothermus hydrogenoformans* Z-2901. *PLoS Genet*. **1**:e65.

Chapter 26

Integron First Gene Cassettes: A Target to Find Adaptive Genes in Metagenomes

Lionel Huang and Christine Cagnon

26.1 INTRODUCTION

When a biotope is exposed to conditions of stress, such as the presence of pollutants, bacterial communities are able to adapt to new environmental parameters. Several phenomena lead to this adaptation. The first one is characterized by the evolution of the community structure: The proportion of each bacterial species is correlated to the selective pressure [Bordenave et al., 2004; Païssé et al., 2008]. The second well-studied phenomenon is the modification of bacterial metabolism [Blom et al., 1992] in order to resist [Mazzella et al., 2007] or to catabolize exogenous molecules [Blom et al., 1992; Plaza et al., 2007]. The third, on which we focus in this chapter, concerns the adaptation at the molecular level. It is known that bacteria can acquire new capacities by mutation [Fong et al., 2006] or by gene exchange [Yano et al., 2007]. Gene exchanges imply mobile genetic elements (MGE) [Top and Springael, 2003], such as conjugative plasmids, transposons, some bacteriophages, or integrons. Over the last 10 years, more and more studies have focused on integrons from the environment, because they can possess and exchange many different genes, and seem to be very active in the adaptive response of bacteria. Another factor that encourages scientists to study this subfamily of MGE is that integrons possess conserved structures, facilitating studies in a metagenomic context.

Integrons are genetic elements that were first discovered in clinical environments in 1989 [Stokes and Hall, 1989]. They are known to be involved in bacterial adaptation by spreading antibiotic resistance genes. In 2001, integrons were shown to be also present in environments other than clinical [Nield et al., 2001; see Chapter 31, Vol. II] and able to carry genes other than those conferring antibiotic resistance [Boucher et al., 2007; Nemergut et al., 2004; Rowe-Magnus and Mazel, 2001; Stokes et al., 2001]. Thus integrons are involved in adaptive responses to other agents of stress such as hydrocarbons, metals, and so on. Ten percent of bacteria are estimated to possess an integron [Michael et al., 2004], and thus bacterial adaptation to conditions of stress by gene acquisition should be considered as an important process.

Integrons are characterized by **(Fig. 26.1)** a gene *intI* encoding for an integrase of the tyrosine recombinase family [Collis et al., 2001, 2002; MacDonald et al., 2006], a recombination site *attI* [Partridge et al., 2000], a set of gene cassettes from a few to more than 150 [Heidelberg et al., 2000; Mazel et al., 1998], and one or two promoters allowing the expression of gene cassettes [Collis and Hall, 1995; Kim et al., 2007]. Integrons can be regarded as reservoirs of gene cassettes. Gene cassettes are the smallest MGE, and they are formed by a coding sequence and an *attC* recombination site. The *attC* sites are themselves composed of repeated inverted sequences and their nucleotide sequence is relatively variable according to the cassette [Collis et al., 1998].

Different classes of integrons are defined according to the diversity of *intI* genes. The three first classes described seem to be characteristic of clinical context (classes 1, 2, and 3) [Hall, 1997], but they are also found

Handbook of Molecular Microbial Ecology, Volume I: Metagenomics and Complementary Approaches, First Edition. Edited by Frans J. de Bruijn.

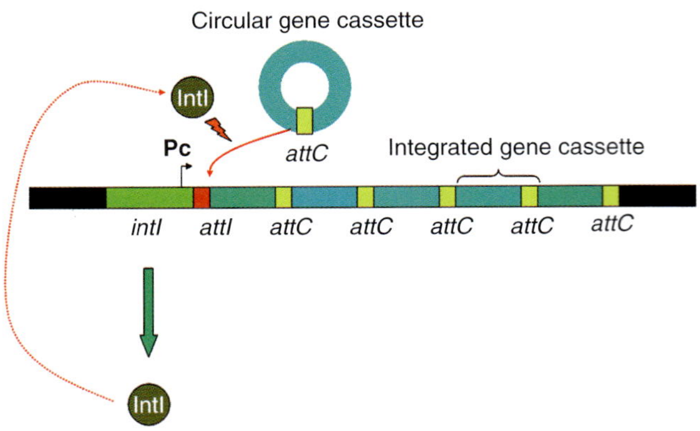

Figure 26.1 Structure of integrons. The *intI* gene encodes an enzyme allowing the integration of new gene cassettes at the *attI* recombination site. Thus, the first gene cassette is the last integrated one. Gene cassettes are formed by a recombination site *attC* and a coding sequence. The Promoter *Pc* permits gene cassette expression.

in studies performed on metagenomes from other environments [Elsaied et al., 2007; Holmes et al., 2003; Nield et al., 2001]. Because new metagenomic studies invariably lead to the discovery of new classes of integrons [see Chapter 31, Vol. II], it is difficult to classify all integrons today. In any case, these data show the importance and the diversity of such genetic elements. However, integrons can be separated in two families: The first one, called "mobile integrons," corresponds to integrons located on another MGE such as a plasmid or a transposon and can be exchanged between bacteria or can be lost [Boucher et al., 2007; Diaz-Mejia et al., 2008; Stokes et al., 2006]. The second family, called "super integrons," concerns integrons located on the bacterial chromosome; this family is usually characterized by integrons with many gene cassettes (up to 200) [Rowe-Magnus et al., 1999].

Since they are involved in bacterial adaptation, integrons could be a great subject of study to find adaptive genes in metagenomes from polluted environments. One simple way is to characterize all gene cassettes associated with integrons. But among the huge abundance of gene cassettes, how is it possible to find those mobilized during the stress considered? That is, how could new adaptive genes be found? It is known that the integration of a new gene cassette, catalyzed by the integrase enzyme, occurs by recombination between the gene cassette *attC* site and the integron *attI* site [Demarre et al., 2007] (Fig. 26.1). Thus, the first gene cassette of an integron is integrated last: If an adaptive gene is newly mobilized in an integron, it will occupy this position. Furthermore, because the first gene cassette is the closest to the integron promoter, its expression level is also the highest in the integron [Collis and Hall, 1995]. These two arguments suggest that the first gene cassette of integrons is a good target to find new adaptive genes in metagenomes. Gene cassettes examined in previous studies from environmental metagenomes did not target first gene cassettes since they were isolated by PCR methods targeting *attC* sites. To amplify the first gene cassettes, a forward primer targeting the *intI* genes or *attI* sites must be used. However, amplification is still unfavorable to isolate first gene cassettes because of the structure of *attC* sites, which contains repeated inverted sequences. The annealing of *attC* primers to both strands allows amplification of gene cassettes other than the first gene cassette. Here we present a method to construct a gene cassette library rich in first gene cassettes, along with an associated screening method to select clones carrying them. The method was developed by using the DNA of a bacterial strain known to possess an integron, and it was validated on the metagenome of a coastal sediment [Huang et al., 2009].

26.2 MATERIAL AND METHODS

26.2.1 DNA Samples

The DNA of the strain *Xanthomonas campestris* ATCC33913^{T} (*X. campestris*) was used to develop the methodology to obtain integron first gene cassettes. DNA was extracted according to Goñi-Urriza et al. [2002].

Mud samples collected in Aber Benoît (Brittany, France) were maintained in microcosms and spiked with oil. Total DNA (the metagenome) was extracted from a sample collected one week after the pollution, using the Ultraclean soil DNA isolation kit (Mo Bio Laboratories) and a previously described protocol with minor modifications [Precigou et al., 2001].

26.2.2 PCR Amplifications of First Gene Cassette

Primers AJH72 targeting *intI* and ICC21 targeting *attC* were used to amplify the integron "first gene cassette" from *X. campestris*. Primers ICC48 and ICC21 were used to amplify integrons first gene cassettes from the metagenome **(Table 26.1)**.

PCR was performed in a final volume of 50 or 100 μl. The reaction mixture contained 1X PCR buffer, 200 μM deoxynucleoside triphosphate, 1.5 mM $MgCl_2$, 0.2 μM primer, and 2.5 or 5 U of *Taq* DNA polymerase (Eurobio). After an initial denaturation step at 95°C for 10 min, we performed 40 cycles of 95°C for 45 s, 52°C when working with primer AJH72 or 51°C when working with ICC48 for 45 s, and 72°C for 1.5 min, followed by a final step at 72°C for 10 min.

Table 26.1 Description of Primers Used for PCR Amplifications

Name	Description	Sequence	Size (bases)	T_m (°C)	Reference
AJH72	*int*I forward	5′-GGATGMCGTTTN-CCGTTGGC-3′	20	52.9	Gillings et al. [2005]
ICC48	*int*I forward	5′-GCAACTGGTCCAGA-ACCTTGAC-3′	22	51	Inverse of intB primer Rosser and Young [1999]
ICC21	*att*C reverse	5′-GTCGGCTTGRAYGA-ATTGTTAGRC-3′	24	54	Huang et al. [2009]
M13 forward(-20)	Cloning vector forward	5′-GTAAAACGACGG-CCAG-3′	16	50.8	Provided with cloning kit
M13 reverse	Cloning vector reverse	5′-CAGGAAACAG-CTATGAC-3′	17	48	Provided with cloning kit

26.2.3 Cloning

PCR products were gel purified by using the GFX DNA and Gel Band Purification kit (GE Healthcare) as recommended by the supplier. Purification products were cloned by using the TOPO TA Cloning® kit (Invitrogen). White transformants were selected on solid Luria medium (10 g/l of bacto-tryptone, 5 g/l of bacto-yeast extract, 10 g/l of NaCl, and 15 g/l of agar) supplemented with 100 mg/l of ampicillin, 40 mg/l of 5-bromo-4-chloro-3-indolyl-β-D-galactopyranoside (X-Gal), and 30 mg/l of IPTG (isopropyl-β-D-thioglucopyranoside).

26.2.4 Clone Screening

Cloned inserts were amplified by PCR performed directly on transformant bacterial cells. Three primers were used concomitantly: M13 Forward (-20) and M13 Reverse primers, specific to the cloning vector surrounding the cloning site, and the primer targeting the *intI* gene previously used to amplify the integrons first gene cassettes (AJH72 or ICC48). This third primer was labeled with the fluorochrome HEX (6-carboxyhexafluoroscein). PCR products were separated by gel electrophoresis without ethidium bromide. The fluorescence was then detected by using a fluorescence scanner Typhoon 9200 (Amersham) to reveal the inserts carrying an integron first gene cassette. Only the amplified fragments carrying the labeled primer were detected.

26.2.5 First Gene Cassettes Analysis

Inserts carrying an integron first gene cassette were sequenced by using a Big Dye Terminator v 1.1 cycle sequencing kit (Applied Biosystem). Electrophoreses were performed at the Genotyping and Sequencing facility of Bordeaux. The sequences were deposited in EMBL database under the following accession numbers: FM210532 to FM210535 and FM210665 to FM210679.

Open reading frames (ORFs) were characterized with ORF finder. Nucleic and amino acid sequences were compared with sequences from Genbank and SwissProt databases, respectively, by using the BLAST algorithm [Altschul et al., 1990]. Protein patterns were determined with ProDom.

26.3 RESULTS

26.3.1 Construction of a Gene Cassette Library Rich in Integron First Gene Cassettes

Our strategy to clone specifically the integron first gene cassettes was developed by using DNA from *Xanthomonas campestris* ATCC33913^T (*X. campestris*), a bacterial strain carrying an integron with 23 gene cassettes [Gillings et al., 2005]. The primers used targeted the integrase genes in forward (since *attI* sites are not conserved, we focused on *intI* gene) and the *attC* sites in reverse (Table 26.1, **Fig. 26.2**). Regardless of the concentration of primers used, several amplified fragments were obtained. The analysis of these fragments, after cloning and sequencing, revealed that most of them (about 75%) were gene cassettes other than the first one. In these cases the reverse primer was used in reverse but also in forward. This kind of amplification could be explained because of the particular structure of *attC* sites with inverted repeat sequences. Thus it is not surprising

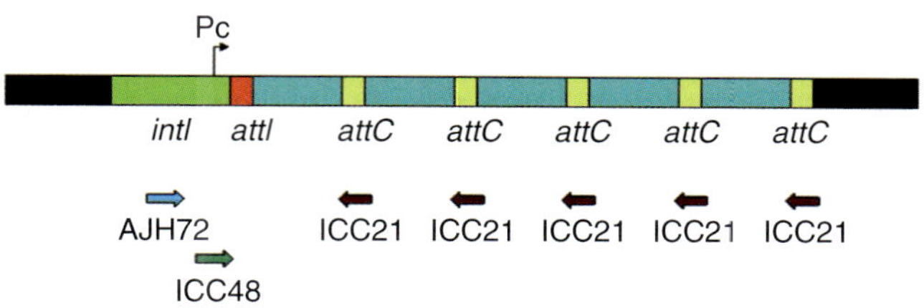

Figure 26.2 Position of the primers used. The primers AJH72, ICC21, and ICC48 used in this study are positioned under the sequences they are able to hybridize with.

that the primer directed to *attC* sites binds in both directions. Then the first gene cassettes selection cannot be performed at the PCR step. Sequencing all clones in order to characterize the first gene cassettes of integrons would be too time-consuming; consequently, a method for screening amplified fragments was developed.

26.3.2 Screening of First Gene Cassettes Amplified Fragments

The screening was performed on the clone library constructed with all PCR products. The inserts carrying a first gene cassette are the unique fragments containing the forward primer sequence, which binds to *intI*. Thus three primers were used for the amplification of the insert of each clone: the two vector-specific primers and the first cassette-specific primer fluorescently labeled (Table 26.1, **Fig. 26.3A**). If the clone does not carry a first gene cassette, a single PCR product is expected. If the clone carries a first gene cassette of the integron, two PCR products are obtained, one of which is labeled. The nonfluorescent fragment is amplified with the two vector primers, and the fluorescent one (slightly smaller) corresponds to the amplification with the first cassette primer and one of the vector primers (Fig. 26.3A). After amplification and electrophoresis, it is easy to detect the labeled fragments (Figs. 26.3B and 26.3C). Sequencing of labeled inserts confirmed the cloning of the integron first gene cassette. As a control, sequencing of nonlabeled inserts showed that such clones carried integron gene cassettes different from the first one.

26.3.3 Application to a Metagenome

Total DNA was extracted from a coastal mud. In order to find newly mobilized gene cassettes, we chose to focus on class 1 integrons known to be the most frequently involved in adaptive responses [Hall, 1997]. Because this class of integrons belongs to the family of mobile integrons, it is likely to be easily propagated throughout the metagenome. A specific integron first gene cassette forward primer was

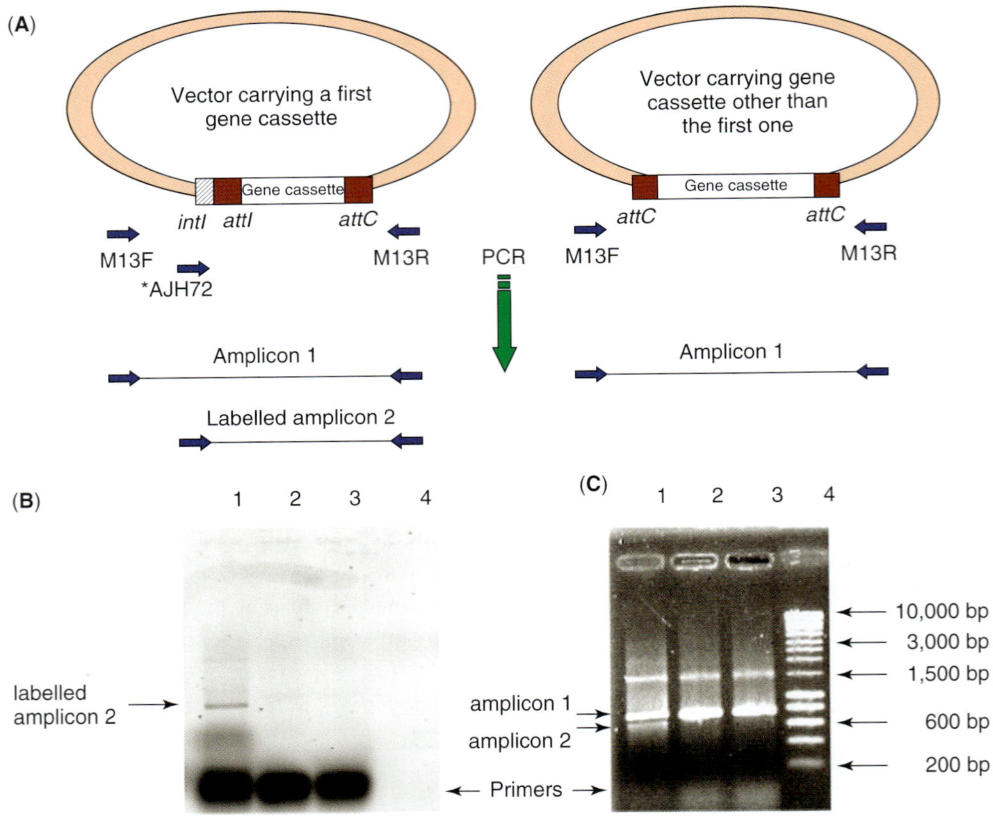

Figure 26.3 Screening strategy of first gene cassette inserts. **(A)** Scheme of the clone library screening PCR strategy. The inserts are amplified by PCR using three primers. Only inserts carrying a first gene cassette lead to labeled amplified fragments. **(B, C)** Gel electrophoresis of insert PCR products from clones of *X. campestris* gene cassettes. 1, Insert carrying the integron first gene cassette of *X. campestris*; 2 and 3, inserts carrying integron gene cassettes of *X. campestris* other than the first one; 4, molecular weight marker (Smart Ladder, Eurogentec). **(B)** Detection of fluorescence. **(C)** Detection of all fragments by ethidium bromide staining. (Reproduced from Huang et al. [2009] with permission from the American Society for Microbiology.)

chosen to target specifically the *intI*-1 gene. The obtained PCR products ranged from 400 bp to 2500 bp. Fragments from 650 bp to 1100 bp were cloned in order to select those that might carry a coding gene cassette. One hundred clones were screened to find those carrying a first gene cassette. About 25% of fluorescent fragments were found and then sequenced.

The primers used for gene cassettes amplification on the metagenome were both found in each sequence. On the side of the primer targeting *intI*, the beginning of this gene was searched for. But, because it was so close to the initiation codon, it was only possible if the *intI* gene had a slightly extended sequence at the 5′ end. Indeed, for one sequence the beginning of an *intI*1 gene was found. On the other side of sequences there should be a part of an *attC* recombination site. Analysis of these sequences showed that some fragments share similarities characteristic of classes 1 or 2 integrons gene cassettes. However, *attC* site could not be found in each case, due to the large variability of this region [Stokes et al., 1997].

Open reading frame (ORF) detection led to the characterization of 29 ORFs potentially transcribed by the integron promoter. Alignments of amino acid sequences corresponding to these ORFs with a protein data bank showed few identities to known proteins. This is not surprising, since previous studies have shown that in environmental integron gene cassettes libraries, few ORFs encode proteins with a characterized function [Koenig et al., 2008].

26.4 DISCUSSION

Integrons are known to carry gene cassettes encoding adaptive proteins in clinic or environmental contexts [Koenig et al., 2008; Michael et al., 2004], and selective pressures may favor the propagation of cassettes conferring a selective advantage [Nemergut et al., 2004; Wright et al., 2008]. To find gene cassettes giving a selective advantage to bacteria in a rapid adaptive response, first gene cassettes of integrons appear to be the best candidate. The following gene cassettes had been integrated previously and are less expressed, depending on the distance from the integron promoter [Collis and Hall, 1995]. This is why we focused on the first gene cassettes in order to detect new adaptive genes from the metagenome. A method was developed to specifically select clones carrying a first gene cassette of integrons. With this protocol, which requires both cloning and screening after PCR amplification, we were able to successfully select clones carrying the first gene cassette of integrons. This strategy was subsequently tested on a metagenome obtained by extracting total DNA from coastal sediment samples contaminated with oil.

There were two limiting steps in this method when working on metagenomes. The first is the choice of primers to cover the greatest number of integrons. For the primer targeting *attC* sites, all primers available from previous studies were unsuccessfully tested with the metagenome analyzed. Thus, we designed a less degenerate primer, targeting class 1 and 2 integrons *attC* sites, ICC21. Concerning the *intI* primer, we chose to focus on class 1 integrons because they are known to be mobile and to carry adaptive genes [Wright et al., 2008]. Because *intI-1* genes from environmental contexts exhibit considerable sequence diversity [Gillings et al., 2008], the *intI* primer used to recover gene cassettes from metagenomes, ICC48 [Rosser and Young, 1999], does not recover all the spectrum of known *intI-1* genes. Thus, to construct a first gene cassette library as complete as possible, several pairs of primers should be used. A preliminary study to characterize the *intI* diversity in the metagenome studied should also help to design specific primers. The second limiting step is cloning. When cloning fragments of large size disparity, small fragments are preferentially cloned. In order to obtain a complete library with a metagenome, the cloning should be performed after separating the fragments by a range of size.

The PCR method we developed is the first that allows the amplification of first gene cassettes from a metagenome, even though all PCR products are not carrying a first gene cassette. The screening method combined with the PCR method lead to 100% of clones carrying a first gene cassette. The clones obtained from polluted environment were analyzed by sequencing. Several ORFs were found inside these cassettes; none for them encoded any known protein, which is in accordance with previous studies [Boucher et al., 2007; Koenig et al., 2008]. Therefore it is difficult to highlight adaptive genes among all the gene cassettes present in a metagenome. With the strategy described, we are now able to reveal last gene acquisition and even to follow the changes of integron first gene cassettes in environmental bacterial communities submitted to conditions of stress. Indeed, such libraries constructed at different times after a stress exposure could reveal gene cassettes spreading in a metagenome. Even if these gene cassettes encode a "hypothetical protein," it would be a good candidate of further analyses. An expression study before, during, and after the stress would confirm the gene implication in adaptive responses to changing conditions.

INTERNET RESOURCES

NCBI blast (http://blast.ncbi.nlm.nih.gov/)

ORF Finder (http://www.ncbi.nlm.nih.gov/gorf/gorf.html)

Prodom (http://prodom.prabi.fr/prodom/current/html/home.php)

Acknowledgments

This work was supported by the Aquitaine Regional Government Council (France) and the ANR/SEST (DHYVA project, No.06SEST09). L.H. was supported partly by a doctoral grant from the *Ministère de l'Enseignement Supérieur et de la Recherche* (France).

We would like to thank all partners of the DHYVA project for their useful discussions. We particularly thank Magalie Stauffert and Laurène Fito-Boncompte for technical support, and we are grateful to Anne Fahy for English correction of this chapter.

REFERENCES

Altschul SF, Gish W, Miller W, Myers EW, Lipman DJ. 1990. Basic local alignment search tool. *J. Mol. Biol*. **215**:403–410.

Blom A, Harder W, Matin A. 1992. Unique and overlapping pollutant stress proteins of *Escherichia coli*. *Appl. Environ. Microbiol*. **58**:331–334.

Bordenave S, Fourçans A, Blanchard S, Goñi MS, Caumette P, Duran R. 2004. Structure and functional analyses of bacterial communities changes in microbial mats following petroleum exposure. *Ophelia* **58**:195–203.

Boucher Y, Labbate M, Koenig JE, Stokes HW. 2007. Integrons: Mobilizable platforms that promote genetic diversity in bacteria. *Trends Microbiol*. **15**:301–309.

Collis CM, Hall RM. 1995. Expression of antibiotic resistance genes in the integrated cassettes of integrons. *Antimicrob. Agents Chemother* **39**:155–162.

Collis CM., Kim MJ, Stokes HW, Hall RM. 1998. Binding of the purified integron DNA integrase IntI1 to integron- and cassette-associated recombination sites. *Mol. Microbiol*. **29**:477–490.

Collis CM, Recchia GD, Kim MJ, Stokes HW, Hall RM. 2001. Efficiency of recombination reactions catalyzed by class 1 integron integrase IntI1. *J. Bacteriol*. **183**:2535–2542.

Collis CM, Kim MJ, Stokes HW, Hall RM. 2002. Integron-encoded IntI integrases preferentially recognize the adjacent cognate *attI* site in recombination with a 59-be site. *Mol. Microbiol*. **46**:1415–1427.

Demarre G, Frumerie C, Gopaul DN, Mazel D. 2007. Identification of key structural determinants of the IntI1 integron integrase that influence *attc* × *attI1* recombination efficiency. *Nucleic Acids Res*. **35**:6475–6489.

Diaz-Mejia JJ, Amabile-Cuevas CF, Rosas I, Souza V. 2008. An analysis of the evolutionary relationships of integron integrases, with emphasis on the prevalence of class 1 integrons in *Escherichia coli* isolates from clinical and environmental origins. *Microbiology* **154**:94–102.

Elsaied H, Stokes HW, Nakamura T, Kitamura K, Fuse H, Maruyama A. 2007. Novel and diverse integron integrase genes and integron-like gene cassettes are prevalent in deep-sea hydrothermal vents. *Environ. Microbiol*. **9**:2298–2312.

Fong SS, Nanchen A, Palsson BO, Sauer U. 2006. Latent pathway activation and increased pathway capacity enable *Escherichia coli* adaptation to loss of key metabolic enzymes. *J. Biol. Chem*. **281**:8024–8033.

Gillings MR, Holley MP, Stokes HW, Holmes AJ. 2005. Integrons in *Xanthomonas*: a source of species genome diversity. *Proc. Natl. Acad. Sci. USA* **102**:4419–4424.

Gillings MR, Krishnan S, Worden PJ, Hardwick SA. 2008. Recovery of diverse genes for class 1 integron-integrases from environmental DNA samples. *FEMS Microbiol. Lett*. **287**:56–62.

Goñi-Urriza M, Arpin C, Capdepuy M, Dubois V, Caumette P, Quentin, C. 2002. Type II topoisomerase quinolone resistance-determining regions of *Aeromonas caviae, A. hydrophila*, and *A. sobria* complexes and mutations associated with quinolone resistance. *Antimicrob. Agents Chemother*. **46**:350–359.

Hall R. 1997. Mobile gene cassettes and integrons: Moving antibiotic resistance genes in gram-negative bacteria. *CIBA Found. Symp*. **207**:192–205.

Heidelberg JF, Elsen JA, Nelson WC, Clayton RA, Gwinn ML, et al. 2000. DNA sequence of both chromosomes of the cholera pathogen *Vibrio cholerae*. *Nature* **406**:477–483.

Holmes AJ, Gillings MR, Nield BS, Mabbutt BC, Nevalainen KMH, Stokes HW. 2003. The gene cassette metagenome is a basic resource for bacterial genome evolution. *Environ. Microbiol*. **5**:383–394.

Huang L, Cagnon C, Caumette P, Duran R. 2009. First gene cassettes of integrons as targets in finding adaptive genes in metagenomes. *Appl. Environ. Microbiol*. **75**:3823–3825.

Kim TE, Kwon HJ, Cho SH, Kim S, Lee BK, Yoo HS, Park YH, Kim SJ. 2007. Molecular differentiation of common promoters in *Salmonella* class 1 integrons. *J. Microbiol. Methods* **68**:453–457.

Koenig JE, Boucher Y, Charlebois RL, Nesbo C, Zhaxybayeva O, et al. 2008. Integron-associated gene cassettes in Halifax Harbour: Assessment of a mobile gene pool in marine sediments. *Environ. Microbiol*. **10**:1024–1038.

MacDonald D, Demarre G, Bouvier M, Mazel D, Gopaul DN. 2006. Structural basis for broad DNA-specificity in integron recombination. *Nature* **440**:1157–1162.

Mazel D, Dychinco B, Webb VA, Davies J. 1998. A distinctive class of integron in the *Vibrio cholerae* genome. *Science* **280**:605–608.

Mazzella N, Molinet J, Syakti AD, Bertrand JC, Doumenq P. 2007. Assessment of the effects of hydrocarbon contamination on the sedimentary bacterial communities and determination of the polar lipid fraction purity: Relevance of intact phospholipid analysis. *Marine Chem*. **103**:304–317.

Michael CA, Gillings MR, Holmes AJ, Hughes L, Andrew NR, Holley MP, Stokes HW. 2004. Mobile gene cassettes: A fundamental resource for bacterial evolution. *Am. Naturalist* **164**:1–12.

Nemergut DR, Martin AP, Schmidt SK. 2004. Integron diversity in heavy-metal-contaminated mine tailings and inferences about integron evolution. *Appl. Environ. Microbiol*. **70**:1160–1168.

Nield BS, Holmes AJ, Gillings MR, Recchia GD, Mabbutt BC, Nevalainen KMH, Stokes HW. 2001. Recovery of new integron classes from environmental DNA. *FEMS Microbiol. Lett*. **195**:59–65.

Païssé S, Coulon F, Goñi-Urriza M, Peperzak L, McGenity T, Duran R. 2008. Structure of bacterial communities along a hydrocarbon contamination gradient in a coastal sediment. *FEMS Microbiol. Ecol*. **66**:295–305.

Partridge SR, Recchia GD, Scaramuzzi C, Collis CM, Stokes HW, Hall RM. 2000. Definition of the *attI1* site of class 1 integrons. *Microbiology* **146**:2855–2864.

Plaza GA, Wypych J, Berry C, Brigmon RL. 2007. Utilization of monocyclic aromatic hydrocarbons individually and in mixture by bacteria isolated from petroleum-contaminated soil. *World J. Microbiol. Biotechnol*. **23**:533–542.

Precigou S, Goulas P, Duran R. 2001. Rapid and specific identification of nitrile hydratase (NHase)-encoding genes in soil samples by polymerase chain reaction. *FEMS Microbiol. Lett*. **204**:155–161.

ROSSER SJ, YOUNG HK. 1999. Identification and characterization of class 1 integrons in bacteria from an aquatic environment. *J. Antimicrob. Chemother*. **44**:11–18.

ROWE-MAGNUS DA, GUÉROUT AM, MAZEL D. 1999. Super-integrons. *Res. Microbiol*. **150**:641–651.

ROWE-MAGNUS DA, MAZEL D. 2001. Integrons: Natural tools for bacterial genome evolution. *Curr. Opin. Microbiol*. **4**:565–569.

STOKES HW, HALL RM. 1989. A novel family of potentially mobile DNA elements encoding site-specific gene-integration functions: Integrons. *Mol. Microbiol*. **3**:1669–1683.

STOKES HW, O'GORMAN DB, RECCHIA GD, PARSEKHIAN M, HALL RM. 1997. Structure and function of 59-base element recombination sites associated with mobile gene cassettes. *Mol. Microbiol*. **26**:731–745.

STOKES HW, HOLMES AJ, NIELD BS, HOLLEY MP, NEVALAINEN KMH, MABBUTT BC, GILLINGS MR. 2001. Gene cassette PCR: Sequence-independent recovery of entire genes from environmental DNA. *Appl. Environ. Microbiol*. **67**:5240–5246.

STOKES HW, NESBO CL, HOLLEY M, BAHL MI, GILLINGS MR, BOUCHER Y. 2006. Class 1 integrons potentially predating the association with Tn402-like transposition genes are present in a sediment microbial community. *J. Bacteriol*. **188**:5722–5730.

TOP EM, SPRINGAEL D. 2003. The role of mobile genetic elements in bacterial adaptation to xenobiotic organic compounds. *Curr. Opin. Biotechnol*. **14**:262–269.

WRIGHT MS, BAKER-AUSTIN C, LINDELL AH, STEPANAUSKAS R, STOKES HW, MCARTHUR JV. 2008. Influence of industrial contamination on mobile genetic elements: class 1 integron abundance and gene cassette structure in aquatic bacterial communities. *ISME J*. **2**:417–428.

YANO H, GARRUTO CE, SOTA M, OHTSUBO Y, NAGATA Y, ZYLSTRA GJ, WILLIAMS PA, TSUDA M. 2007. Complete sequence determination combined with analysis of transposition/site-specific recombination events to explain genetic organization of incp-7 tol plasmid pww53 and related mobile genetic elements. *J. Mol. Biol*. **369**:11–26.

Chapter 27

High-Resolution Metagenomics: Assessing Specific Functional Types in Complex Microbial Communities

LUDMILA CHISTOSERDOVA

27.1 INTRODUCTION

Metagenomics is a fast-growing and diverse field within environmental biology directed at obtaining knowledge on genomes of environmental microbes, without prior cultivation, as well as of entire microbial communities. When applied to communities of low complexity, exemplified by the communities of the acid mine drainage biofilm [Tyson et al., 2004] or the symbionts of a marine oligochaete [Woyke et al., 2006], the metagenomic approach, even at a modest sequencing effort, allows for sequence assembly. Thus, analysis of almost complete genomes of the dominant species in these communities can be carried out, including accurate metabolic reconstruction and even detection of strain-specific genomic variants. However, the situation is quite different when metagenomics is applied to communities of high complexity, such as the communities of marine habitats or soils [Venter et al., 2004; Tringe et al., 2005; Rusch et al., 2007]. In these cases, significantly larger sequencing efforts resulted in very fragmented assemblies even for the most abundant species, with most of the datasets being represented by singleton sequencing reads. While gene-centric analysis [Tringe et al., 2005] can be conducted on the nonassembled metagenomic data and predictions on the major metabolic pathways can be made, the specific metabolic capabilities are hard or impossible to place into the context of individual species.

One way to directly link a function in the environment to a specific guild performing this function is to feed the natural population a substrate of interest, labeled by a heavy isotope, followed by characterization of the heavy fraction of communal DNA that is enriched in DNA of microbes that actively metabolized the labeled substrate. This technique is known as Stable Isotope Probing (DNA-SIP), and it has been effective in identifying microbes involved in specific biogeochemical transformations such as methylotrophy, phenol degradation, glucose metabolism, and so on [Radajewski et al., 2000; Friedrich, 2006; Madsen, 2006; see also Chapter 55, Vol. II]. Typically, small amounts of DNA are isolated in these experiments, and the DNA is used for phylogenetic profiling and detection of key functional genes, after PCR amplification [Friedrich, 2006]. A modification of this approach has been described involving a multiple displacement amplification step, followed by metagenomic library construction and screening for specific marker genes [Chen et al., 2008; see also Chapter 55, Vol. I]. In the work reviewed here, Kalyuzhnaya et al. [2008] scaled up the SIP protocol in order to obtain amounts of DNA enabling the WGS sequencing approach and applied the method to communities of lake sediment involved in utilization of single carbon (C1) compounds (methylotrophs).

Methylotrophy is an important part of the global carbon cycle on this planet [Hanson and Hanson, 1996; Guenter, 2002]. Identities of methylotrophs involved in utilization of specific C1 substrates (such as methane, methanol, methylated amines, etc.) in a variety of environments have previously been assessed by both culture-reliant and culture-independent methods

Handbook of Molecular Microbial Ecology, Volume I: Metagenomics and Complementary Approaches, First Edition. Edited by Frans J. de Bruijn.

[Chistoserdova et al., 2009], the former provides important models for understanding the specific biochemical pathways enabling methylotrophy, and the latter provides insights into species richness within specific functional groups. However, while genomic data for some model methylotrophs are now available [Chistoserdova et al., 2009], these may not represent major players in specific functional guilds. At the same time, current methods for environmental detection provide little insight into the genomic structure of uncultivated methylotrophs. The goal of the targeted metagenomics approach reviewed here has been twofold: to reduce the complexity of the community that has been estimated at approximately 5000 species [Kalyuzhnaya et al., 2008] and to directly link specific substrate repertoires to specific functional guilds.

27.2 METHODS

27.2.1 Sampling and Stable Isotope Probing

Samples of Lake Washington (47°38.075′ N, 122°15.993′ W) sediment were collected using a box core. These were transported to the laboratory on ice and immediately used to set up microcosms. Each microcosm contained 10 ml of sediment, 90 ml of Lake Washington water filtered through 0.22-μm filters (Millipore), and one of the following ^{13}C-labeled substrates: methane (50% of air), methanol (10 mM), methylamine (10 mM), formaldehyde (1 mM), or formate (10 mM). All substrates were 99 atom % ^{13}C and were purchased from Sigma-Aldrich, with the exception of ^{13}C-methanol, which was provided by the National Stable Isotope Resource at Los Alamos National Laboratory. The samples were incubated for 3–5 (methylamine and methane), 5–7 (methanol), or 10–14 (formaldehyde and formate) days at room temperature, with shaking.

27.2.2 DNA Extraction

DNA was extracted using a modified procedure of Sommerville et al. [1989], as follows. Microcosm samples were centrifuged at $5000 \times g$ for 15 min and mixed with freshly prepared lysozyme solution (10 mg ml^{-1} in 10 mM Tris-HCl, pH 8.0, 1 mM EDTA), followed by incubation at 37°C for 1 h. SDS (2% final concentration) and proteinase K (20 mgml^{-1}) were then added, and samples were incubated at 37°C for additional 5 h or overnight. A 5 M NaCl solution was then added to the mixture to a final concentration of 1.25 M. An equal volume of phenol–chloroform–isoamyl alcohol (25:24:1) was added and the mixture was incubated for 30 min at room temperature with horizontal shaking at 150 rpm, followed by centrifugation at $5000 \times g$ for 15 min at 4°C. The aqueous phase was transferred to a clean centrifuge tube and treated again with an equal volume of phenol–chloroform–isoamyl alcohol (25:24:1), as above. An additional purification step, using an equal volume of chloroform, was conducted. The DNA was precipitated from the aqueous phase with 0.7 volume of isopropyl alcohol at room temperature, followed by centrifugation at $5000 \times g$ for 15 min at 4°C. The pellet was washed with 70% ethanol, dried, and resuspended in 1 ml of TE buffer, pH 8.0. DNA concentration was measured spectrophotometrically.

27.2.3 Isopycnic Centrifugation and DNA Recovery

DNA was prepared for CsCl–ethidium bromide density gradient ultracentrifugation as previously described [Nercessian et al., 2005] and centrifuged at $265,000 \times g$ (Beckman VTi 65 rotor) for 16 h at 20°C. DNA fractions were visualized in UV, and the ^{13}C-DNA fraction was collected using a 19-gauge needle. DNA was purified following standard procedures and used in a second round of CsCl–ethidium bromide density gradient ultracentrifugation, as above, and the ^{13}C-DNA fraction was collected using a 19-gauge needle.

27.2.4 DNA Sequencing and Assembly

Five shotgun libraries were constructed, one per each microcosm, in the pUC18 vector (1 to 3-kb inserts). Vector inserts were sequenced with BigDye Terminators v3.1 and resolved with ABI PRISM 3730 (ABI) sequencers at the Joint Genome Institute—Production Genomics Facility (JGI-PGF; Walnut Creek, CA). A total of 344,832 reads comprising 255 megabases (Mb) of Phred Q20 sequence were generated. Sequences were stringently quality- and vector-trimmed using LUCY [Chou and Holmes, 2001], and the trimmed sequences were assembled, both *en masse* and by sample, using the PGA assembler. Assembly statistics are shown in **Table 27.1**. These draft quality assemblies were manually validated and used for all downstream analyses.

27.2.5 Compositional Binning

Assembled metagenomic fragments were binned (classified) using PhyloPythia, a phylogenetic classifier that uses a multi-class Support Vector Machine (SVM) for the composition-based assignment of fragments at different taxonomic ranks, essentially as previously described [McHardy et al., 2007; see also Chapter 40, Vol. I].

Table 27.1 Summary of Sequencing, Assembly, and Gene Prediction Statistics

	Methane	Methanol	Methylamine	Formaldehyde	Formate	Combined	*Methylotenera*
Assembly Statistics							
Number of reads	71,808	67,200	83,712	80,640	41,472	344,832	NA[a]
Average read length (bp)	792	797	709	712	638	741	NA
Trimmed read length (Mbp)	56.85	53.53	59.34	58.91	26.45	255.08	NA
Nonredundant sequence (bp)	52.16	50.25	37.23	57.62	17.57	211.47	11.16
Percent of reads in contigs	10.2	10.0	55.5	7.3	34.3	27.6	100
Total contigs (>2 kb)	2,797	2,871	7,558	2,583	3,618	25,877	4,078
Total singlets	59,417	56,408	29,217	69,104	18,857	215,581	0
Average sequence coverage (x)	1.6	1.6	1.9	1.7	1.9	1.7	2.1
Highest sequence coverage (x)	7.0	4.8	20.4	6.4	4.7	23.1	20.4
Average size of contigs (bp)	1,418	1,288	2,065	1,166	1,265	1,593	2,736
Largest contig (bp)	6,174	5,913	20,771	4,714	6,276	22,407	15,820
GC content (%)	58.9	59.5	53.0	57.9	65.8	58.3	46.2
Gene Predictions							
Protein coding genes	81,076	77,229	54,340	89,729	28,700	321,503	12,719
Genes in COGs	43,456	40,773	33,643	46,032	17,112	174,344	10,082
Genes in Pfams	28,090	26,494	23,586	29,375	10,585	115,228	8,543
Predicted enzymes	3,089	3,047	5,005	3,065	1,417	16,780	3,264
Number of 16S rRNA genes	12	12	10	18	5	61	3
Number of tRNA genes	405	412	376	504	121	1,728	181

[a]NA, not available.

Source: Reproduced from Kalyuzhnaya et al. [2008].

Generic models for the ranks of domain, phylum and class were combined with models for the dominant clades in the sample. The generic models represent all clades covered by three or more species at the corresponding ranks among the sequenced microbial isolates. At the rank of family, a sample-specific model was created with classes for the clades *Methylococcaceae, Burkholderiaceae, Rhodocyclaceae, Methylophilaceae*, and *Comamonadaceae* and a class "other." A sample-specific model for the dominant sample populations was created with classes for the *Methylotenera* and *Methylobacter* populations and a class "other." The sample-specific population model was trained on 138 kb and 141 kb of contigs for the *Methylotenera* and *Methylobacter* populations, respectively, that were identified based on phylogenetic marker genes, as well as sequenced isolates for the class "other." The family-level model was trained using the sample-specific data and additional sequenced isolates available for the corresponding clades. For each model, five sample-specific multi-class SVMs were created using fragments of lengths of 3, 5, 10, 15, and 50 kb, respectively. All input sequences were extended by their reverse complement prior to computation of the compositional feature vectors. The parameters w and l were both set to 5 for the sample-specific models. The final classifier consisting of the sample-specific and generic clade models was applied to assign all fragments >1 kb of the samples. In case of conflicting assignments, preference was given to assignments of the sample-specific models. Data were incorporated into the Integrated Microbial Genomes with Microbiome Samples (IMG/M) system (http://img.jgi.doe.gov/m).

27.2.6 Protein Recruitment

Protein recruitment was carried out essentially as previously described [Rusch et al., 2007] except for protein sequences rather than DNA sequences were used. The Phylogenetic Profiler tool that is part of the IMG/M package was used. In the case of *M. tundripaludum/M. capsulatus* pair, cutoffs of 60% to 80% were used, based on 89% 16S rRNA gene similarity between the two strains. In the case of *R. eutropha/R. eutropha* H-16 pair (99% 16S rRNA gene similarity), a cutoff of 90% was used.

27.3 RESULTS AND DISCUSSION

27.3.1 Specific Shifts in Community Composition and Enrichment for Methylotroph DNA

We selected, as our test model, populations of microbes active in methylotrophy in the sediment of a freshwater lake, Lake Washington in Seattle, WA, an environment known for high rates of methane consumption [Auman et al., 2000]. Sediment samples were exposed separately to ^{13}C-labeled methane, methanol, methylamine, formaldehyde, and formate, to target populations actively utilizing each of the C1 compounds. Total DNA was extracted from each microcosm, and the ^{13}C-labeled fractions were separated from unlabeled DNA by isopycnic centrifugation. ^{13}C-labeled DNA was then used to construct five separate shotgun libraries, and these were sequenced at the JGI-PGF. Between 26 and 59 million base pairs (Mb) of sequence were produced from each microcosm, totaling 255 Mb. Sequences were assembled, automatically annotated, and loaded into the JGI's IMG/M system (Table 27.1), followed by manual analysis. Sequence coverage and degree of assembly depended on the sequencing effort applied and on the species richness and evenness of the enriched communities. Based on analysis of 16S rRNA gene sequences, community complexity was significantly reduced in microcosms exposed to each of the C_1 substrates compared to the complexity of the nonenriched community that we estimate to be over 5000 species [Kalyuzhnaya et al., 2008], and it shifted toward specific functional guilds that included both *bona fide* methylotrophs (*Methylobacter tundripaludum, Methylomonas sp., Methylotenera mobilis, Methyloversatilis universalis, Ralstonia eutropha*) and organisms only distantly related to any cultivated species, implicating the latter in environmental cycling of C1 compounds (**Fig. 27.1**). The closest relatives of these included *Verrucomicrobia, Nitrospirae, Planctomycetes, Acidobacteria, Cyanobacteria*, and *Proteobacteria* (detailed in Supplementary Table 1 of the original publication). The 16S rRNA data were supported by data on phylogenetic profiling of each metagenomic dataset, based on top BLAST hit distribution patterns (data not shown). From these analyses, the methylamine microcosm was one of the least complex in terms of species richness and most enriched in genes diagnostic for C1 transforming capability. It was dominated by a group of closely related strains identified as *M. mobilis*, represented by the only cultivated strain, *M. mobilis* JLW8 previously isolated from Lake Washington [Kalyuzhnaya et al., 2006].

27.3.1.1 The Composite Genome of Methylotenera mobilis. Based on 16S rRNA gene sequence coverage (up to 20×), complete or nearly complete genomes of a few *M. mobilis* strains were predicted to be encoded in the methylamine microcosm metagenomic dataset. A composite genome of *M. mobilis* totaling slightly over 11 Mb was extracted from the methylamine microcosm metagenome using the recently described compositional binning method, PhyloPythia [McHardy et al., 2007; see also Chapter 40, Vol. I] (genome statistics are shown in Table 27.1). Genome completeness was validated by examination of the presence of key metabolic and housekeeping genes (detailed in Supplementary Tables 4 and 5 of the original publication). In terms of central metabolism, we identified a complete set of genes for specific pathways enabling methylamine utilization in *M. mobilis*. Multiple copies for each gene were identified (3–15), consistent with the composite genome being representative of a few closely related strains. In terms of the housekeeping functions, completeness of the genome was demonstrated by the presence of (a) 181 tRNA genes

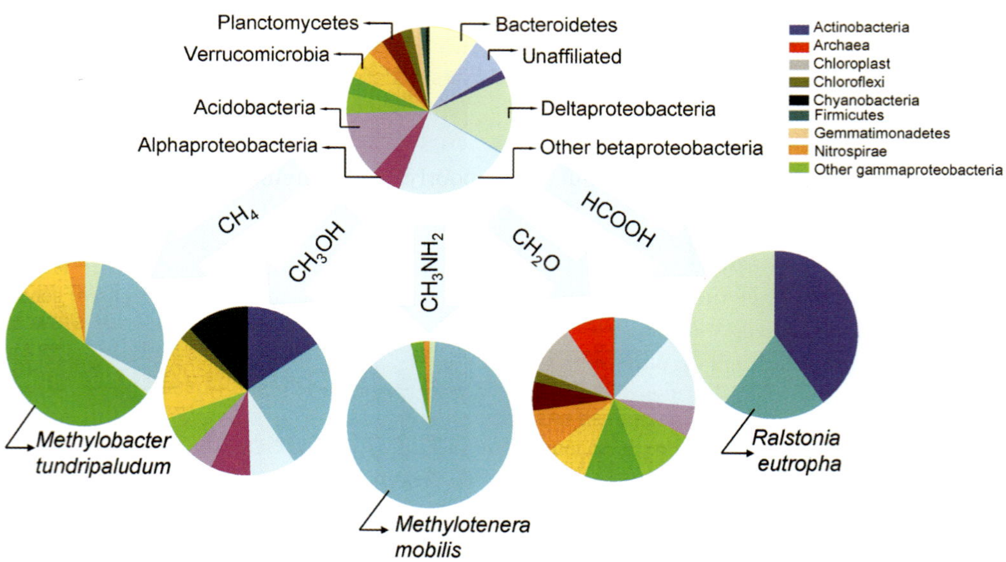

Figure 27.1 Taxonomic distribution of 16S rRNA genes in metagenomic datasets. The size of each wedge reflects the number of 16S rRNA genes multiplied by sequence coverage for each gene. (Reproduced from Kalyuzhnaya et al. [2008].)

corresponding to 36 tRNA acceptors for recognizing all 20 amino acids (not shown) and (b) a complete set of aminoacyl-tRNA transferases. Standard sets of genes for DNA replication, transcription, and translation were identified, and complete pathways were reconstructed for biosynthesis of all the amino acids and nucleotides and all the essential vitamins.

We reconstructed the metabolism of *M. mobilis* and conducted genome-wide comparisons with the genome of *Methylobacillus flagellatus*, a methylotroph closely related to *M. mobilis*, of a similar genome size [Kalyuzhnaya et al., 2006; Chistoserdova et al., 2007]. *M. mobilis* from Lake Washington and *M. flagellatus* are 93–95% similar at the 16S rRNA gene sequence level and share most of the pathways enabling methylotrophy. However, they were found to be quite different in their genomic content, gene synteny and gene conservation. Reciprocal BLAST analyses using the Phylogenetic Profiler tool that is part of the IMG/M interface revealed that only 57% of the proteins translated from the *M. flagellatus* chromosome had homologs in *M. mobilis* at a 50% cutoff, and only 62% of the proteins translated from the composite genome of *M. mobilis* had homologs in *M. flagellatus*. Focusing on some of the highly conserved genomic regions encoding methylotrophy functions, we uncovered examples of nonhomologous replacements in common biochemical pathways as well as examples of homolog recruitment into novel/secondary functions. These examples include proposal of a function for a novel cytochrome as an alternative electron acceptor from methylamine dehydrogenase and proposal of a novel role for Fae (formaldehyde-activating enzyme), as a sensor component of regulatory and/or signal transduction systems [Kalyuzhnaya et al., 2008].

Global genome–genome comparisons between *M. mobilis* and *M. flagellatus* revealed that the conserved parts of the genomes encode central metabolism and housekeeping functions (methylotrophy, energy transduction, replication, transcription, translation, amino acid, and vitamin biosynthesis), while the variable parts of the genomes encode auxiliary functions (transport, regulation, electron transfer, CRISPR, prophage, nonessential biochemical pathways). We were able to precisely map 63 indels of more than two genes on the chromosome of *M. flagellatus*, totaling approximately 1070 kb, not present in the composite genome of *M. mobilis*. The number and the size of indels could not be estimated with such precision for the composite genome of *M. mobilis*, but we were able to calculate that approximately 600 kb of sequences per genome were unique, estimating the genome size for *M. mobilis* at approximately 2.5 Mb. One notable element missing from the composite genome of *M. mobilis* was the methanol dehydrogenase-encoding gene cluster, thought to be highly conserved in most methylotrophs [Lidstrom, 2006]. Conversely, some enzymes and pathways not present in *M. flagellatus* were identified in *M. mobilis*, such as the methylcitric acid cycle. Comparisons of energy-generating electron transfer pathways encoded in the two genomes showed little overlap, suggesting adaptation to significantly different life styles (**Table 27.2**). For example, an incomplete denitrification pathway was reconstructed from the *M. mobilis* composite genome, suggesting a potential role in reduction of nitrate to nitrous oxide, the ability subsequently proven in experiments with cultivated *M. mobilis* [Kalyuzhnaya et al., 2009], while the *M. flagellatus* genome lacks genes for denitrification.

Sequences of *M. mobilis* were also present in the metagenomes of microcosms incubated with methane, methanol, and formaldehyde. To test whether *M. mobilis* strains labeled by these substrates were metabolically different from *M. mobilis* strains in the methylamine microcosm, we conducted substrate-specific genome–genome comparisons, interrogating each dataset with the *M. mobilis* composite genome. In this way we detected a number of genes that were not present in the combined dataset for methane, methanol, and formaldehyde microcosms, but were unique to the methylamine microcosm. Remarkably, the entire gene cluster encoding methylamine oxidation (*mauFBEAGLMNO*) was missing from the former, suggesting that methylamine-oxidizing capability is "an acquired taste" and not an attribute of *M. mobilis* as a species and suggesting alternative primary substrates for *M. mobilis*. In contrast, in all microcosms, hits were found for the entire set of genes involved in the methylcitric acid cycle, pointing to its potential role as a central metabolic pathway. One proposed function for this cycle could be in utilization of propionate that is a product of demethylation of a compound(s) typical of aquatic environments (such as dimethylsulfoniopropionate) [Ginzburg et al., 1998]. Indeed, we later demonstrated that type strain *M. mobilis* JLW8 is capable of nitrate-dependent growth on methanol [Kalyuzhnaya et al., 2009] and that it also grows, albeit poorly, on a lignin-derived compound methoxyphenol [Kalyuzhnaya et al., 2010].

More recently we obtained an ultimate proof of the precision of the predictions derived from the composite *M. mobilis* genome, by obtaining a complete genomic sequence for the type strain *M. mobilis* JLW8 and comparing it to the metagenome-composite genome extracted from the metagenome (unpublished data).

27.3.2 Bacteriophage Genomes

In addition to the *M. mobilis* composite genome, bacteriophage genomes were recovered from the methylamine sample at high sequence coverage. One of these (37 kb)

Table 27.2 Presence of Major Metabolic Pathways Based on Genome–Genome Comparisons Between *M. mobilis* and *M. flagellatus*

	M. mobilis	*M. flagellatus*
Oxidation		
Methylamine dehydrogenase	Yes	Yes
Methanol dehydrogenase	No	Yes
Formaldehyde oxidation (methanopterine)	Yes	Yes
Formate dehydrogenase 2 (NAD)	Yes	Yes
Formate dehydrogenase 4 (unknown)	Yes	Yes
Carbon Assimilation		
Ribulosemonophosphate cycle	Yes	Yes
Methylcitric acid cycle	Yes	No
Electron Transfer Systems		
NADH dehydrogenase (Complex I)	Yes	Yes
Cytochrome oxidase (bb type)	Yes	Yes
Succinate dehydrogenase (Complex II)	Yes	No
NADH-ubiquinol oxidoreductase (Rnf system)	Yes	No
Ubiquinol cytochrome c reductase (bc type)	Yes	No
Cytochrome c oxidase (aa3 type)	Yes	No
Cytochrome c oxidase (o type)	Yes	No
Cytochrome c oxidase (cb type)	Yes	No
Nitric oxide reductase	Yes	No
Na/H antiporter NADH quinone dehydrogenase	No	Yes
Cytochrome oxidase (cbb type)	No	Yes
Cytochrome d ubiquinol oxidase	No	Yes
Cytochrome c oxidase (o type)	No	Yes
Cytochrome c oxidase	No	Yes
Other Features		
Urea metabolism	No	Yes
CRISPR elements	No	Yes
Prophage	No	Yes

was homologous to the genome of the *Bordetella* phage BPP1 [Liu et al., 2004] (not shown), while others (approximately 10 kb) were distantly related to the genome of a marine bacteriophage PM2, the only member of the *Corticoviridae* family [Krupovic et al., 2006], and to a prophage found in the genome of *M. flagellatus* [Chistoserdova et al., 2007] (**Fig. 27.2**). Two of the contigs of the latter type were found to contain overlapping sequences at the ends. These were trimmed and joined at the ends to produce circular phage chromosome sequences. The presence of phage chromosomes in the methylamine microcosm metagenome indicates that free-living phages were propagating during the microcosm incubation with ^{13}C-methylamine. *M. mobilis* is the most likely host for these phages, due to its dominance in the labeled microcosm community. However, the phage sequences were missing from the methane, methanol, and formaldehyde microcosms, indicating a specific association between phage and methylamine-utilizing *M. mobilis*. This was supported by the presence of a conspicuous gene cluster, also unique to the methylamine-utilizing *M. mobilis*, which encodes pilus assembly and secretion functions (*cpaABCEFtadBC*). This pilus is a possible candidate for a specific phage receptor. In addition to these, a number of other candidate phage receptors were unique to the methylamine *M. mobilis* (a biopolymer transporter, a major facilitator). The connection between methylamine metabolism and phage association is very intriguing. One hypothetical scenario could be imagined in which a specific transporter for methylamine also serves as a specific phage receptor. No phage or prophage sequences were detected in the genome of the type strain *M. mobilis* JLW8 (data not shown). However, phage transcripts were detected in the recent experiment interrogating gene expression in a methylamine-stimulated

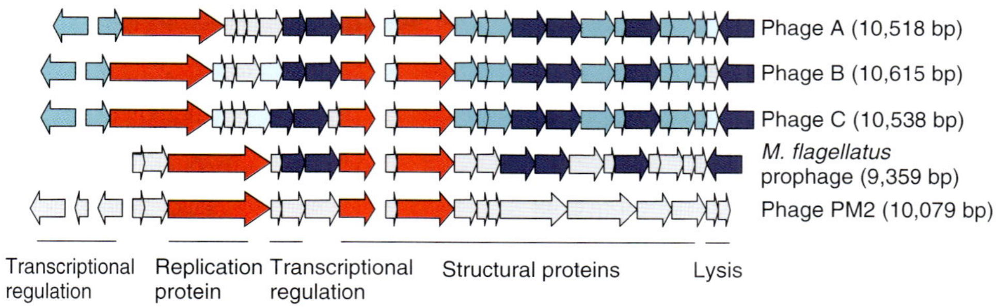

Figure 27.2 Genomic structure of novel phages from lake Washington, the *M. flagellatus* prophage and bacteriophage PM2. The circular genomes were linearized to align with the prophage. Different colors indicate different degrees of gene conservation (red, conserved in all; blue, conserved in Lake Washington phages and *M. flagellatus* prophage; turquoise, conserved in Lake Washington phages; light blue, conserved in two of the three phages; blank, unique).

microcosm using *Methylotenera*-specific microarrays [Kalyuzhnaya et al., 2010].

27.3.3 Other Genomes

We were also able to analyze other, less covered genomes by supplementing the PhyloPythia binning with protein recruitment, using related genome sequences as a reference. From comparisons with the *Methylococcus capsulatus* genome [Ward et al., 2004], we estimated that a large portion of the (composite) genome of *M. tundripaludum* was present in the methane microcosm dataset (not shown). We conducted metabolic reconstruction for this organism and mapped indels on the chromosome of *M. capsulatus* (not shown). Trends similar to the ones noted for gene conservation between *M. mobilis* and *M. flagellatus* were observed: The core parts of the genomes of *M. tundripaludum* and *M. capsulatus*, encoding central metabolism and house keeping genes, were conserved, while parts of the genomes encoding auxiliary functions were not. Notable omissions from the *M. tundripalidum* genome were gene clusters encoding the soluble methane monooxygenase, RuBisCO, and enzymes of the serine cycle. These genomic features agree with physiological analysis of the cultivated *M. tundripaludum* strain [Wartiainen et al., 2006]. In a similar fashion, a large portion of an *R. eutropha* genome was recovered from the formate microcosm metagenome. It was highly similar to the published genome of Strain H-16 [Pohlmann et al., 2006], encoding all the core functions and only missing genes for a few auxiliary pathways, such as CO dehydrogenase and polysaccharide biosynthesis. It also appeared to lack the megaplasmid found in Strain H-16 (not shown). Partial genomes were obtained for uncultivated representatives of *Burkholderiaceae, Comamonadaceae, Rhodocyclaceae*, and *Actinobacteria*, the groups that include methylotrophic representatives.

Besides the bona fide methylotrophs, our functional enrichment approach suggested that phyla not traditionally classified as methylotrophs may be involved in C_1 transformations, such as *Verrucomicrobia, Nitrospirae*, and *Planctomycetes*. The lower coverage of these strains may reflect either slower rates of metabolism or suboptimal incubation conditions. Acidophilic methane-oxidizing *Verrucomicrobia* have been described recently [Dunfield et al., 2007; Pol et al., 2007; Islam et al., 2008; Hou et al., 2008]. However, based on 16S rRNA and functional gene comparisons, *Verrucomicrobia* uncovered in this study are only distantly related to these organisms (<90% 16S rRNA identity). We analyzed the datasets containing *Verrucomicrobia* phylogenetic markers (methane, methanol and formaldehyde microcosms) for the presence of specific functional genes potentially enabling methylotrophy in these species and identified a conspicuous gene (*mtaB*) that was present in one or more copies in each dataset, predicted to encode a methanol:corrinoid methyltransferase. This enzyme has been characterized in methylotrophic archaea [Sauer and Thauer, 1997] and suggested to be involved in methanol utilization by *Clostridia* [Das et al., 2007], thus potentially implicating the *Verrucomicrobia* detected in this study in methanol utilization. While no methylotrophic *Planctomycetes* or *Nitrospirae* have been obtained at the moment in pure cultures, these organisms are often detected in environments with high rates of C1 metabolism [Lösekann et al., 2007; Mason et al., 2009].

27.4 CONCLUSIONS AND FUTURE PERSPECTIVES

The work reviewed here is a proof of principle study demonstrating that a targeted metagenomics approach can enable detailed analysis of the genomes of environmentally relevant microbes, even if the species in question comprise a minor fraction in a highly complex microbial community. A specific enrichment step, such as SIP employed in this work, is key to increasing the resolution of metagenomics, focusing the sequencing effort on specific functional types. A detailed analysis of the genome of a novel methylotroph, *M. mobilis*, comprising less than 0.5% of the total bacterial population in Lake Washington sediment, was made possible by this approach. Moreover, the study uncovered specific bacterium-phage relationships, suggesting the existence of complex population dynamics involving multiple strains of *M. mobilis* and multiple strains of novel corticoviruses. A genome of an uncultivated *M. tundripaludum* was also analyzed in detail, expanding the current genomic knowledge of methane utilizers. In addition, *Verrucomicrobia* species were identified only distantly related to the described methanotrophic isolates, suggesting that methylotrophy may be a common attribute of this phylum. Overall, this study uncovered the existence of dynamic and diverse populations responding to C1 substrates, pointing toward the existence of a complex, multi-tiered microbial food web involved in environmental C1 cycling in Lake Washington sediment and likely in other freshwater lake sediments.

This project, as most projects initiated between 2003 and 2009, involved a modest sequencing effort (Mb-scale), as dictated by the economics of Sanger-based sequencing. Recent breakthroughs in developing alternative (next-generation sequencing technologies (see Chapter 18, Vol. I)) offer a significantly higher throughput and significantly reduced cost of sequencing, setting a stage for much larger, Gb-scale metagenomics projects. The increased sequencing effort should result in better sequence coverage and thus in increased resolution of sequence analysis as well as of the derived biological knowledge. However, while scaling up the sequence production using the new technologies is straightforward, new challenges have arisen such as the diminished quality of sequence data and shorter read length, necessitating increasingly complex computation as well as special means for data handling, transfer, and storage and requiring new and advanced computer infrastructures. These are currently the bottlenecks in next-generation sequencing-based metagenomics [Hugenholtz and Tyson, 2008; Lapidus, 2009]. However, when these problems are solved and Gb-scale or even Tb-scale metagenomics become a reality, approaches employing function-specific enrichments will still remain important for connecting a specific function in the environment to specific sequence signatures. The targeted metagenomics approach described herein has the potential to be used in a wide variety of ecosystems with a wide variety of labeled substrates, as well as with other types of enrichment, in combination with next-generation sequencing technologies.

WEB RESOURCES

JGI IMG/M site: http://img.jgi.doe.gov/cgi-bin/m/main.cgi

JGI Community Sequencing Program http://www.jgi.doe.gov/CSP/index.html

Acknowledgments

The author acknowledges support by the National Science Foundation as part of the Microbial Observatories program (MCB-0604269). The sequencing was provided through the U.S. Department of Energy (DOE) Community Sequencing Program; and the work was performed, in part, under the auspices of the DOE Office of Science, Biological and Environmental Research Program, University of California, Lawrence Livermore National Laboratory, and Los Alamos National Laboratory.

REFERENCES

Auman AJ, Stolyar S, Costello AM, Lidstrom ME. 2000. Molecular characterization of methanotrophic isolates from freshwater lake sediment. *Appl. Environ. Microbiol*. **66**:5259–5266.

Chen Y, Dumont MG, Neufeld JD, Bodrossy L, Stralis-Pavese N, et al. 2008. Revealing the uncultivated majority: Combining DNA stable-isotope probing, multiple displacement amplification and metagenomic analyses of uncultivated *Methylocystis* in acidic peatlands. *Environ. Microbiol*. **10**:2609–2622.

Chistoserdova L, Lapidus A, Han C, Goodwin L, Saunders L, et al. 2007. The genome of *Methylobacillus flagellatus*, the molecular basis for obligate methylotrophy, and the polyphyletic origin of methylotrophy. *J. Bacteriol*. **189**:4020–4027.

Chistoserdova L, Kalyuzhnaya MG, Lidstrom ME. 2009. The expanding world of methylotrophic metabolism. *Annu. Rev. Microbiol*. **63**:477–499.

Chou HH, Holmes MH. 2001. DNA sequence quality trimming and vector removal. *Bioinformatics* **17**:1093–1104.

Das A, Fu ZQ, Tempel W, Liu ZJ, Chang J et al. 2007. Characterization of a corrinoid protein involved in the C1 metabolism of strict anaerobic bacterium *Moorella thermoacetica*. *Proteins: Struct. Funct. Bioinform*. **67**:167–176.

Dunfield PF, Yuryev A, Senin P, Smirnova AV, Stott MB, et al. 2007. Methane oxidation by an extremely acidophilic bacterium of the phylum *Verrucomicrobia*. *Nature* **450**:879–82.

Friedrich MW 2006. Stable-isotope probing of DNA: insights into the function of uncultivated microorganisms from isotopically labeled metagenomes. *Curr. Opin. Biotechnol*. **17**:59–66.

Ginsburg B, Chalifa I, Gun J, Dor I, Hadas O, Lev O. 1998. DMS formation by dimethylsulfoniopropionate route in freshwater. *Environ. Sci. Technol*. **32**:2130–2136.

Guenter A. 2002. The contribution of reactive carbon emissions from vegetation to the carbon balance of terrestrial ecosystems. *Chemosphere* **49**:837–844.

Hanson RS, Hanson TE. 1996. Methanotrophic bacteria. *Microbiol. Rev*. **60**:439–471.

Hou S, Makarova KS, Saw JH, Senin P, Ly BV, et al. 2008. Complete genome sequence of the extremely acidophilic methanotroph isolate V4, *Methylacidiphilum infernorum*, a representative of the bacterial phylum *Verrucomicrobia*. *Biol. Direct* **3**:26.

HUGENHOLTZ P, TYSON GW. 2008. Microbiology: metagenomics. *Nature* **455**:481–483.

ISLAM T, JENSEN S, REIGSTAD LJ, LARSEN Ø, BIRKELAND N-K. 2008. Methane oxidation at 55°C and pH 2 by a thermoacidophilic bacterium belonging to the *Verrucomicrobia* phylum. *Proc. Natl. Acad. Sci. USA* **105**:300–304.

KALYUZHNAYA MG, BOWERMAN S, LARA JC, LIDSTROM ME, CHISTOSERDOVA L. 2006. *Methylotenera mobilis* gen. nov., sp. nov, an obligately methylamine-utilizing bacterium within the family *Methylophilaceae*. *Int. J. Syst. Evol. Microbiol*. **56**:2819–2823.

KALYUZHNAYA MG, LAPIDUS A, IVANOVA N, COPELAND AC, MCHARDY AC et al. 2008. High-resolution metagenomics targets specific functional types in complex microbial communities. *Nat. Biotechnol*. **26**:1029–1034.

KALYUHZNAYA MG, MARTENS-HABBENA W, WANG T, HACKETT M, STOLYAR SM et al. 2009. *Methylophilaceae* link methanol oxidation to denitrification in freshwater lake sediment as suggested by stable isotope probing and pure culture analysis. *Environ. Microbiol. Rep*. **1**:385–392.

KALYUZHNAYA MG, BECK DAC, SUCIU D, POZHITKOV A, LIDSTROM ME, CHISTOSERDOVA L. 2010. Functioning *in situ*: Gene expression in *Methylotenera mobilis* in its native environment as assessed through transcriptomics. *ISME J*. **4**:388–398.

KRUPOVIC M, VILEN H, BAMFORD JK, KIVELÄ HM, AALTO JM, et al. 2006. Genome characterization of lipid-containing marine bacteriophage PM2 by transposon insertion mutagenesis. *J. Virol*. **80**:9270–9278.

LAPIDUS A. 2009. Genome sequence databases (overview): sequencing and assembly. In SCHAECHTER M, ed. *The Encyclopedia of Microbiology*. New York: Elsevier, pp. 196–210.

LIDSTROM ME. 2006. Aerobic methylotrophic procaryotes. In BALOWS A, TRUPER HG, DWORKIN M, HARDER W, SCHLEIFER K-H, eds. *The Prokaryotes*. New York: Springer-Verlag, pp. 613–634.

LIU M, GINGERY M, DOULATOV SR, LIU Y, HODES A et al. 2004. Genomic and genetic analysis of *Bordetella* bacteriophages encoding reverse transcriptase-mediated tropism-switching cassettes. *J. Bacteriol*. **186**:1503–1517.

LÖSEKANN T, KNITTEL K, NADALIG T, FUCHS B, NIEMANN H, BOETIUS A, AMANN R. 2007. Diversity and abundance of aerobic and anaerobic methane oxidizers at the Haakon Mosby mud volcano, Barents Sea. *Appl. Environ. Microbiol*. **73**:3348–3362.

MADSEN EL. 2006. The use of stable isotope probing techniques in bioreactor and field studies on bioremediation. *Curr. Opin. Biotechnol*. **17**:92–97.

MASON OU, DI MEO-SAVOIE CA, VAN NOSTRAND JD, ZHOU J, FISK MR, GIOVANNONI SJ. 2009. Prokaryotic diversity, distribution, and insights into their role in biogeochemical cycling in marine basalts. *ISME J*. **3**:231–242.

MCHARDY AC, MARTIN HG, TSIRIGOS A, HUGENHOLTZ P, RIGOUTSOS I. 2007. Accurate phylogenetic classification of variable-length DNA fragments. *Nat. Methods* **4**:63–72.

NERCESSIAN O, NOYES E, KALYUZHNAYA MG, LIDSTROM ME, CHISTOSERDOVA L. 2005. Bacterial populations active in metabolism of C_1 compounds in the sediment of Lake Washington, a freshwater lake. *Appl. Environ. Microbiol*. **71**:6885–6899.

POHLMANN A, FRICKE WF, REINECKE F, KUSIAN B, LIESEGANG H, et al. 2006. Genome sequence of the bioplastic-producing "Knallgas" bacterium *Ralstonia eutopha* H16. *Nat. Biotechnol*. **24**:1257–1262.

POL A, HEIJMANS K, HARHANGI HR, TEDESCO D, JETTEN MS, OP DEN CAMP HJ. 2007. Methanotrophy below pH 1 by a new *Verrucomicrobia* species. *Nature* **450**:874–878.

RADAJEWSKI S, INESON P, PAREKH NR, MURRELL JC. 2000. Stable-isotope probing as a tool in microbial ecology. *Nature* **403**:646–649.

RUSCH DB, HALPERN AL, SUTTON G, HEIDELBERG KB, WILLIAMSON S, et al. 2007. The Sorcerer II global ocean sampling expedition: Northwest Atlantic through Eastern Tropical Pacific. *PLoS Biol*. **5**:e77.

SAUER K, THAUER RK. 1997. Methanol:coenzyme M methyltransferase from *Methanosarcina barkeri*. Zinc dependence and thermodynamics of the methanol:cob(I)alamin methyltransferase reaction. *Eur. J. Biochem*. **249**:280–285.

SOMMERVILLE CC, KNIGHT IT, STRAUBE WL, COLWELL RR. 1989. Simple, rapid method for direct isolation of nucleic acids from aquatic environments. *Appl. Environ. Microbiol*. **55**:548–554.

TRINGE SG, VON MERING C, KOBAYASHI A, SALAMOV AA, CHEN K, et al. 2005. Comparative metagenomics of microbial communities. *Science* **308**:554–557.

TYSON GW, CHAPMAN J, HUGENHOLTZ P, ALLEN EE, RAM RJ, et al. 2004. Community structure and metabolism through reconstruction of microbial genomes from the environment. *Nature* **428**:37–43.

VENTER JC, REMINGTON K, HEIDELBERG JF, HALPERN AL, RUSCH D, et al. 2004. Environmental genome shotgun sequencing of the Sargasso Sea. *Science* **304**:66–74.

WARD N, LARSEN Ø, SAKWA J, BRUSETH L, KHOURI H, et al. 2004. Genomic insights into methanotrophy: the complete genome sequence of *Methylococcus capsulatus* (Bath). *PLoS Biol*. **2**:e303.

WARTIAINEN I, HESTNES AG, MCDONALD IR, SVENING MM. 2006. *Methylobacter tundripaludum* sp. nov., a methane-oxidizing bacterium from Arctic wetland soil on the Svalbard islands, Norway (78° N). *Int. J. Syst. Evol. Microbiol*. **56**:109–113.

WOYKE T, TEELING H, IVANOVA NN, HUNTEMANN M, RICHTER M, et al. 2006. Symbiosis insights through metagenomic analysis of a microbial consortium. *Nature* **443**:950–955.

Chapter 28

Gene-Targeted Metagenomics (GT Metagenomics) to Explore the Extensive Diversity of Genes of Interest in Microbial Communities

SHOKO IWAI, BENLI CHAI, EDERSON DA C. JESUS, C. RYAN PENTON, TAE KWON LEE, JAMES R. COLE, AND JAMES M. TIEDJE

28.1 INTRODUCTION

Due to extensive genetic diversity in most microbial communities, it is currently impossible to sequence any gene with sufficient coverage by shotgun metagenomic sequencing for meaningful conclusions about the gene's diversity or population characteristics. Functional screens is one means to recover more versions of targeted genes, but even this approach is not efficient for covering the extant diversity and is not feasible for genes whose functions depend on assembly of active protein complexes. In order to overcome this limitation, targeted gene sequencing approaches are needed. One approach is the use of polymerase chain reaction (PCR)-based targeting in concert with high-throughput sequencing technologies; i.e. amplicon sequencing. This approach, which we term gene-targeted metagenomics (GT metagenomics), can not only provide extensive insight into microbial gene diversity, but can also reveal functional, ecological, and evolutionary patterns of the targeted genes.

This type of approach is currently of interest for examining genes directly responsible for important ecosystem functions or ecological processes such as biogeochemical cycling, nitrogen fixation, biodegradation, pathogenesis, antibiotic resistance, and cell signaling. Considering that these ecologically important functional genes are usually present in low copy number and carried by only a fraction of the total community, an even lower relative abundance than for housekeeping genes should be expected. As such, shotgun sequencing of randomly sampled DNA is limited in its ability to obtain sufficient genetic variety of the targeted genes. While fosmids or bacterial artificial chromosome (BAC) libraries may be appropriate for exposing novel genes, they are not suitable for diversity analysis.

To illustrate the limitation of shotgun metagenomics for revealing sufficient sequence coverage of key ecologically functional genes in typically diverse microbial communities, we provide two examples. Frias-Lopez et al. [2008] found that only 1.4% out of 414,323 genomic DNA fragments were rRNA by direct pyrosequencing of a soil sample. Since functional genes are typically single copy while 16S rRNA has multiple copies—up to 15 in some species [Lee et al., 2009]—coverage of the target functional gene would be usually less than 0.1%. In addition, not all members of a community are present in the same abundance. Using rRNA amplicon library data, Quince et al. [2008] fitted parametric species-abundance curves to estimate the sampling effort necessary to observe 90% of the species in an environment. Their estimates ranged from about 35,000 16S rRNA amplicons for the upper ocean to about 30 million for a Canadian soil, with even

Handbook of Molecular Microbial Ecology, Volume I: Metagenomics and Complementary Approaches, First Edition. Edited by Frans J. de Bruijn.

higher estimates for deep-sea thermal vents. For shotgun metagenome data, if we assume a conserved gene of 1-kb size and an average genome size of 4 Mb, then to obtain some sequence from this gene from 90% of species, we would need to increase our sequencing effort by a factor of 4000—equivalent to about 120 billion reads for the Canadian soil! Fortunately, many interesting questions can be answered at sequencing depths covering less than 90% of species in a sample, but the ratio of required effort between shotgun and GT metagenomics still holds.

The potential value of more extensive sequence coverage of targeted genes in communities include (a) a more robust determination of conserved residues by a more unbiased sampling of nature and hence more insight into their functional roles; (b) a more comprehensive dataset to better establish relationships between sequence subclasses and function (e.g., for kinetic or substrate specializations; (c) a larger dataset from which to detect evolutionary adaptations—for example, to current or historical selection such as for amino acid shifts for cold- or high-temperature adaptation; and (d) a better understanding of sequence diversity in response to habitat physical, chemical, and temporal conditions. It may be some time before shotgun metagenomics provides this level of resolution in complex bacterial communities.

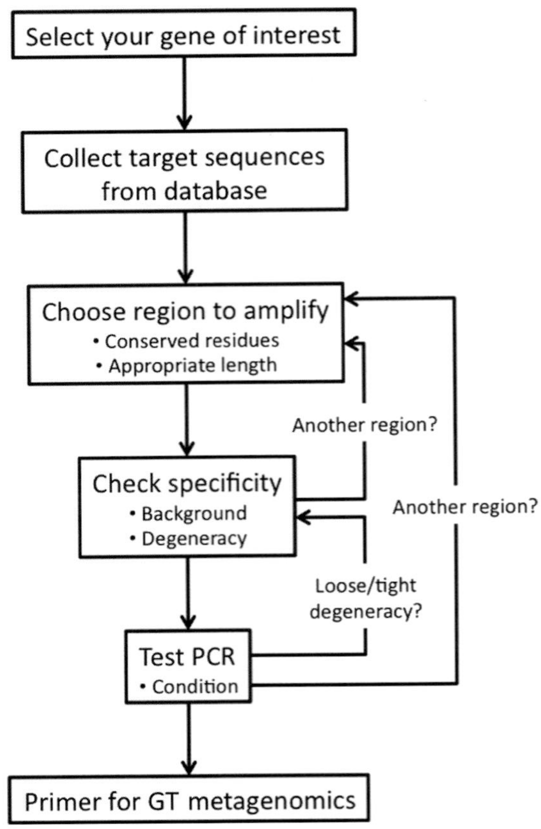

Figure 28.1 Flowchart of GT-metagenomics primer design.

28.2 GT-METAGENOMICS WORKFLOW

28.2.1 Presequencing: Primer Design

There are several considerations in identifying genes of an environmental function for study. First, if several genes are candidates for studying a specific pathway, then it is desirable to target genes that encode rate-limiting or terminal steps in product formation since they most directly control the process interest. Second, sufficient sequence information from previous studies is required for primer design. In our experience, five sequences (preferably protein but nucleotide is also acceptable) are the minimum required to build a reliable alignment and search conserved regions for primer design. If the number of sequences available is too small (<10–20), then it is difficult to design primers that will be specific enough to recover only the targeted genes, and the primers mostly likely will recover only a portion of the gene family. Third, conserved regions within suitable distances for the sequencing technology are needed. The current most common limitation to broad use of GT metagenomics is that sequence databases are still too limited for many genes of ecological interest. An overall flowchart of GT metagenomics primer design is shown in **Figure 28.1**.

28.2.1.1 Collect Previously Known Sequences of the Gene of Interest. Once the target gene is selected, sequence data of the gene should be collected for primer design. This step is critical since primer coverage is defined by the range of sequences collected. A useful tool in the initial phase of functional gene design is the Functional Gene Pipeline/Repository (FGPR). This database contains genes used as markers for antibiotic resistance, biodegradation, biogeochemical cycles, and phylogenetics that are retrieved monthly from the NCBI nonredundant database using expert submitted training sequences and a Hidden Markov Model (HMM) search program [Durbin et al., 1998]. Target genes within annotated genomes are then scanned using the NCBI gene id search function.

Metagenomics databases such as the Community Cyberinfrastructure for Advanced Marine Microbial Ecology Research and Analysis (CAMERA) and the Integrated Microbial Genomes (IMG) can be also searched. These metagenomic datasets are especially useful when the target and dataset environments are similar. Whole genome shotgun sequence similarity searches at EMBL-EBI using reliable protein sequences from organisms whose target metabolic activity has been confirmed are also useful. When available, protein family (pfam), cluster of orthologous genes (COGs), enzyme commission number (EC number), and the source (origin) are recorded for each sequence.

28.2.1.2 *Selecting Region for Primers.*

Once target gene sequences of the clade are collected, protein alignments are used to visualize trees for the determination of both outgroups and gene clusters for primer design. When there are not many sequences collected (less than ~50) due to the limitation of gene information, alignments are visually inspected for regions of protein homology that are subsequently investigated at the nucleotide level. From conserved regions of the alignment, target primer regions for GT metagenomics should be selected with the consideration of two points: the positions of conserved residues and PCR product length. If many sequences are collected, using software such as ARB [Ludwig et al., 2004; see also Chapter 46, Vol. I] (integrated software for sequence data analysis including primer design function) and CODEHOP [Rose et al., 1998] (primer design software based on protein alignments) are helpful for designing primer sets targeting the sequences.

It is preferable to select the PCR amplification regions that include known conserved residues. Although primer sets are designed to be specific to the target genes based on in silico evidence, pyrosequencing may capture nontarget genes due to its deep sequencing feature as well as undiscovered genes that have similar sequence to the primer in the environmental samples. Conserved residue positions are helpful in filtering out nontarget sequences during data processing.

For pyrosequencing, the length of the PCR product must be optimized to maximize emulsion PCR efficiency [Margulies et al., 2005]. With the 454 titanium system, an August 2009 Roche technical bulletin recommends an amplicon length between 200 and 600 bp. Longer PCR products can be attempted if the targeted conserved regions are limited. However, the amplicon length should be less than 800 bp according to the Roche technical bulletin, which agrees with our experience. We do, however, obtain a useful number of sequences from amplicons in the 400- to 500-bp range.

28.2.1.3 *Primer Specificity.*

In order to cover as many of the selected clades of target genes as possible, degenerate bases may be included in the primer. Since higher degeneracy results in decreased primer specificity, total degeneracy should be optimal for maximizing the coverage without sacrificing specificity. In the case of biphenyl dioxygenase (see below), the primer set had 48- and 16-fold degeneracy for forward and reverse primer, respectively, based on about 50 gene sequences available. This primer set provided a 60% of coverage of the known gene sequences for this function. This successfully minimized nontarget amplification but with a good coverage of target genes [Iwai et al., 2010].

The specificity of primers to background community DNA should be ascertained by performing a BLAST search to ensure that the designed primers will not hybridize and produce similar amplicon lengths. Metagenomic databases from environments similar to the samples of interest can be employed for this test by excluding the target genes from those datasets. In human microbiome studies, human genomic sequence is a significant background since a large portion of extracted DNA may be of human origin, hence minimizing cross-hybridization to and amplification of human DNA is necessary.

28.2.2 Postsequencing: Data Analysis

A significant challenge in using the GT-metagenomic approach arises in data analysis. Due to the often limited availability of sequence information prior to obtaining gene-targeted pyrosequencing results, the massive diversity generated introduces uncertainty as to whether a portion of the produced sequences is truly related to the intended target.

28.2.2.1 *Sequence Alignment.*

Sequence alignments are commonly used to carry out sequence comparisons for the estimation of diversity and protein conservation. To better estimate the diversity at a functional gene level, two different approaches are outlined that attempt to preserve protein conservation information in the alignments.

The Wise2 package [Birney et al., 2004] can be used for generating nucleotide alignments of the obtained sequences, which are guided by protein-level conservation. With the database program genewisedb, an option (-hmmer) performs sequence comparisons between nucleotide sequences and a protein HMM. The result of the comparison is output as local alignments of nucleotide sequences to HMM-derived protein consensus sequences with possible introns and frameshifting errors clearly masked. A HMM from a set of selected known target gene protein sequences (seeds) can be used to run genewisedb against the obtained sequences.

In the case of biphenyl dioxygenase [Iwai et al., 2010], the result indicated that the local alignments were too fragmented and showed insufficient coverage for all different regions of the obtained sequences. Therefore, a multiple sequence alignment could not be assembled from this genewisedb result to represent the majority of base positions across all the sequences. The lack of coverage by the local alignments from the genewisedb result was caused by the fact that the sequence diversity embodied by this dataset was not fully represented by the HMM, which was based on the seed sequences.

The other approach we used was to first translate these nucleotide sequences into peptide sequences in frames specified by the primer positions. In light of the high diversity observed from the genewisedb analysis, the peptide sequences were then pairwise-aligned using MUSCLE [Edgar, 2004] to generate dissimilarity matrices to be used for clustering. The dissimilarity values were calculated based on BLOSOM64 and were generally less than those calculated from the MUSCLE-generated multiple alignments.

28.2.2.2 Clustering and Operational Taxonomic Units.

Based on the dissimilarity matrices derived from the pairwise alignments of the translated protein reads, clustering may be performed using mothur [Schloss et al., 2009] to delineate Operational Taxonomic Units (OTUs) at different dissimilarity levels.

Mothur gives the option of using three different clustering methods: single linkage, complete linkage, and average linkage. The differences between the single and complete linkage methods is that the smallest and the largest distance between objects belonging to two groups is used to form new clusters, giving the alternative names of these methods, respectively, nearest neighbor and farthest neighbor. The average linkage is a compromise between the two, where the average of the distances for all pairs of objects belonging to two groups is used to form new clusters [Pielou, 1984]. When working with sequence data, the objects are the sequences, and individual sequences with their similarity falling within a given cutoff range are considered as belonging to the same OTU.

In contrast to the 16S rRNA gene, there are no common categories or criteria for defining OTUs of functional genes based on the sequence identity. It is debatable which cutoff is equivalent to the "species level," and the appropriate value may vary among genes. One point of reference is a cutoff of 94% nucleotide identity which is based on the finding that the average nucleotide identity (ANI) of genes shared between two bacterial genomes corresponds to the 70% DNA–DNA reassociation standard used to define a species [Konstantinidis and Tiedje, 2005; see also Chapters 2 and 11, Vol. I]. It may be, however, that certain genes are more or less conserved than the average for the genome, and hence a cutoff of 94% nt identity would not equate to the "species level."

Other authors utilize amino acid identity rather than nucleotide sequences to define OTUs. In this case, they are defined as operational protein families (OPFs). This concept was used by Lerat et al. [2003] to define homologous genes in γ-Proteobacterial genomes and by Schloss and Handelsman [2008] for the analysis of metagenomic data. Kuhn et al. [2009] recently used this concept when studying alkane monooxygenase (*alk*) genes in the environment and concluded that 82% identity in amino acid sequence was the appropriate cutoff for the OPF definition in their case. Regardless of the choice, once the OTUs or OPFs are defined, they can be used for the analysis presented below.

28.2.2.3 Diversity Estimation and Community Comparison.

After pre-processing, alignment, and clustering, sequences can be subjected to diversity estimation and bacterial community comparisons using a variety of statistical methods. The coverage of gene libraries is often calculated to evaluate how well they represent the composition of the studied communities. Collector's curves can be used to evaluate sample coverage, and rarefaction curves can be used to compare libraries with different numbers of sequences. These curves can be constructed for different richness and diversity estimators. Among the indices available in the ecological literature, the Shannon, Simpson, Chao1, and ACE indices are most used in microbial ecology studies [Hughes et al., 2001; Magurran, 2004]. Detailed information about these indices can be found in the ecological literature [Magurran, 2004; Hughes et al. 2001] and their calculations are implemented in a variety of programs, some of which were devised to work with sequence data [Singleton et al., 2001; Colwell, 2005; Kindt and Coe, 2004; Schloss et al., 2009].

Data on species composition of communities are multidimensional. Hence, multivariate techniques can be used to reduce dimensionality, thus revealing patterns underlying bacterial community structure. Among these methods are clustering, ordination, and matrix comparisons. Hierarchical clustering examines discontinuities, dividing groups of samples into subsets based on their similarity or dissimilarity, and is generally displayed in the form of a dendrogram. There are several measures of resemblance and of clustering methods, which are chosen according to the nature of the study question [Pielou, 1984; Legendre and Legendre, 1998].

Ordination methods adapt a multidimensional dataset into a projection on two- or three-dimensional space such that clustering patterns and ecological gradients may be discerned [Pielou, 1984]. Environmental information can be used directly in this process, through canonical analysis or added after the ordination is performed. This aids the explanation of ecological patterns and allows the testing of correlation between community structure and environmental variables [Legendre and Legendre, 1998]. Several methods are available, and the choice of the most appropriate method depends on the type of data and goal of the study. Among the most commonly used are: principal components analysis and correspondence analysis, along with their canonical forms; redundancy analysis; and canonical correspondence analysis. Hypotheses of correlation with environmental variables and geographic distance can also be tested by matrix comparison methods, such as the Mantel test and analysis of similarities. Details about the variety of methods for multivariate analysis of biological communities can be found in Pielou [1984], Legendre and Legendre [1998], Ter Braak and Smilauer [2002], and Kindt and Coe [2005].

Finally, new methods have been devised or adapted to compare community structure based on the use of nucleotide or amino acid sequences. Community structure can be compared by using the Cramer–von Mises statistics, which is implemented in the program ∫-libshuff [Schloss et al., 2004], as well as phylogenetic-based approaches, such as the Phylogenetic test [Martin, 2002] and Unifrac [Lozupone and Knight, 2005], which preclude the use of cutoff levels and test the significance of the difference between different communities through Monte Carlo permutation tests. However, since phylogenetic tests rely on the use of phylogenetic trees, the quality of the analysis is conditioned to the quality of the underlying phylogeny used as input. This is an especially important issue for 454 pyrosequencing data due to both the computational power required to analyze a large amount of data, and the short read lengths, which might be impediments in performing phylogenetic analysis. Additionally, the genetic diversity of two or more communities can be compared by analysis of variance-type approaches such as the analysis of molecular variance [Schloss, 2008].

28.3 EXAMPLES AS CASE STUDIES

28.3.1 Biphenyl Dioxygenase Diversity in Soils

GT metagenomics was tested on a set of dioxygenase genes important in the carbon cycle for turnover of more recalcitrant organic carbon as well as for pollutant degradation [Iwai et al., 2010]. A PCR primer set designed to target a 524-bp region that confers substrate specificity of biphenyl dioxygenases yielded 2000 and 604 sequences from 5′ and 3′ ends of the PCR products, respectively, using Genome Sequencer FLX System (454 Life Sciences) that passed the validity criteria. The sequence alignment showed three known conserved residues as well as another seven conserved residues not previously reported (**Fig. 28.2**, only 5′ sequence results are shown). This illustrates one of the useful outputs from the GT-metagenomics approach, in that it better reveals what nature has conserved that cannot be seen from either pure culture or shotgun metagenomic approaches. Ninety-five percent and 41% of the valid sequences comprised 22 and 3 novel clusters (5′ and 3′ reads, respectively) in that they did not include any previously reported sequences at 0.6 distance by Complete Linkage Clustering of the sequenced regions. The greater diversity revealed by this GT-metagenomics example provides deeper insights into genes potentially important in environmental processes to better understand their ecology, functional differences, and evolutionary origins.

28.3.2 Investigating Nitrogen Fixation in Soils

Nitrogen fixation, which is mediated by diverse phylogenetic groups of prokaryotes, is a very important ecological process, and it accounts for about 145 Tg of the nitrogen incorporated into terrestrial ecosystems per year [Gruber and Galloway, 2008]. Nitrogenase, the key enzyme in this process, contains a rather conserved subunit, coded by the *nif*H gene, which has been extensively used as a marker to study nitrogen fixing bacteria in the environment [Zehr et al., 2003]. We adapted previously published *nif*H primers [Zehr and McReynolds, 1989; Poly et al., 2001] for 454 pyrosequencing and designed a new set of consensus degenerate hybrid oligonucleotide primers [Rose et al., 1998] for that same purpose. These primers were used to assess the composition and community structure of nitrogen-fixing bacteria in the rhizosphere of switchgrass and big bluestem [McLaughlin and Kszos, 2005], two grasses with biofuel crop potential. Barcodes and 454 adaptor sequences were added to the forward primers and the samples were submitted to pyrosequencing. A total of 45,000 sequences of 150–200 bp were obtained with the Genome Sequencer FLX System (454 Life Sciences). BLASTx results showed that more than 99% of those reads matched nitrogenase reductase gene sequences in GenBank, which confirmed the 454 primer specificity. A model-based sequence alignment was generated using genewisedb [Birney et al., 2004] by comparing the individual reads to a *nif*H HMM, which was built on a set of known, full-length *nif*H protein sequences with HMMER2 [Durbin et al., 1998]. Over 90% of read positions in about 98% of the reads were alignable to the *nif*H HMM. The total number of retrieved sequences showed sufficient coverage of community diversity as indicated by the stabilization of the Shannon index accumulation curve. The resulting sequences differentiated the switchgrass and big bluestem rhizospheres. Furthermore, our results indicate that different primers revealed different specificities for populations of nitrogen fixers, as the bacterial communities also

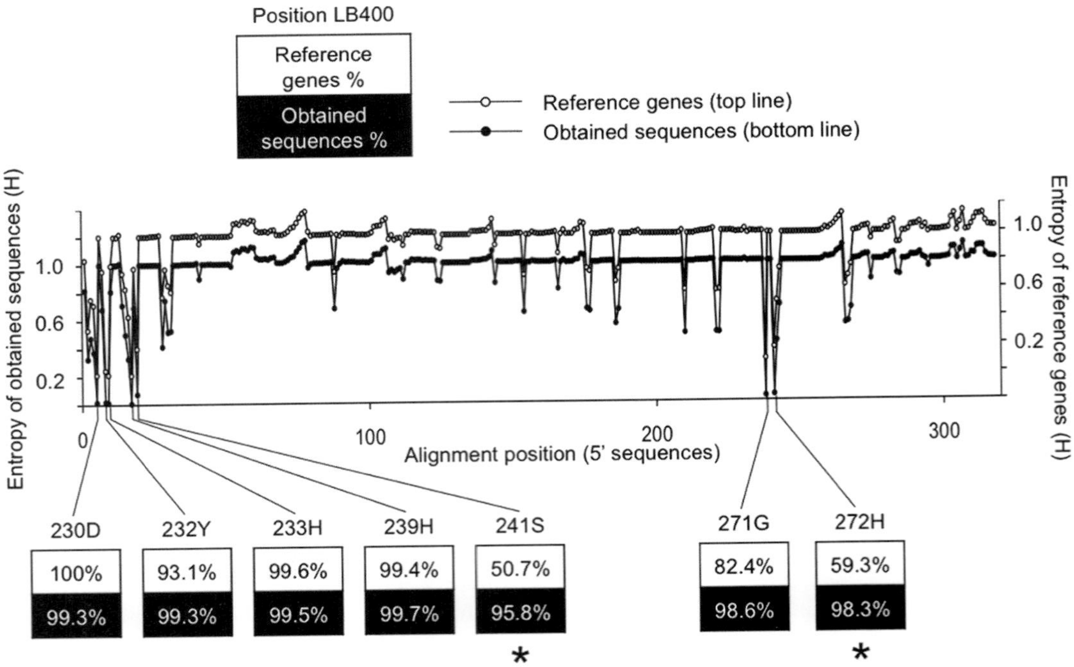

Figure 28.2 Example of Shannon entropy (H′) at each alignment position and conserved residues among obtained sequences and/or reference sequences for 5′ end sequences. (Parts redrawn from Iwai et al. [2010].) Unfilled circles (○) indicate entropy of reference sequences, and filled circles (•) indicate entropy of obtained sequences. The corresponding position numbers and the residues of *bphA1* from *Burkholderia xenovorans* LB400 are indicated. The ratio of residues conserved in either set with 495% is shown. The residues highly conserved only among obtained sequences are indicated by an asterisk. This example illustrates one of advantages of GT metagenomics, namely, letting nature—rather than the biases of isolation and choice for study—reveal the conserved amino acids in particular proteins.

clustered by primers (**Fig. 28.3**). The best BLASTx hits for representative sequences of most of the clusters group with uncultured bacteria. Some sequences were assigned to sequences belonging to species known to be present in the rhizosphere such as *Burkholderia phytofirmans, B. cepacia, Bradyrhizobium yunmingense, Azorhizobium caulinodans, Clostridium sp., Mesorhizobium sp.*, and *Paenibacillus borealis*. *nif*H is an example where the sequence database and prior primer development and testing greatly aided the development of barcoded GT metagenomics, thereby making its deployment more efficient.

28.3.3 Design and Testing of Functional Gene Primers Targeting Human Colonic Bacteria

We designed and tested three functional gene primers whose targets are involved in butyrate production and sulfite reduction, processes that might reflect or be involved in inflammatory bowel disease (IBD). As a preferred energy source for colonocytes, butyrate influences the function of a healthy human colon [Pryde et al., 2002]. Butyrate influences gene expression [Csordas, 1996], has anti-inflammatory effects and regulates cell differentiation and apoptosis in human colonic cells [Segain et al., 2000; Scheppach et al., 2001; Pajak et al., 2007; Hamer et al., 2008]. The production of butyrate from butyryl-CoA is mediated by both butyrate kinase and butyryl CoA transferase encoded by the *buk* and *but* genes, respectively. Butyrate producers are distributed throughout the human colon microbiota [Louis and Flint, 2009] and constitute 5–10% of total fecal bacteria from healthy adult human subjects. The known diversity of butyrate producers is based primarily on pure cultures, although there is indication that a large number remain uncharacterized [Lay et al., 2007]. On the other hand, dissimilatory sulfate-reducing bacteria reduce sulfate or sulfite to produce sulfide by sulfite reductase (*dsr*). This metabolite is toxic to intestinal epithelial cells and causes immune activation followed by mucosal damage [Macfarlane et al., 2009] and is also a potential inhibitor of butyrate oxidation [Roediger et al., 1993; Hulin et al., 2002].

The FGPR was used for initial reference gene screening for primer design. Whole genome shotgun sequence similarity searches at EMBL-EBI were implemented to obtain additional genetic information. In the case of *buk* and *but* genes, it was necessary to design six degenerate primers for multiplex PCR. In the case of the *dsr*

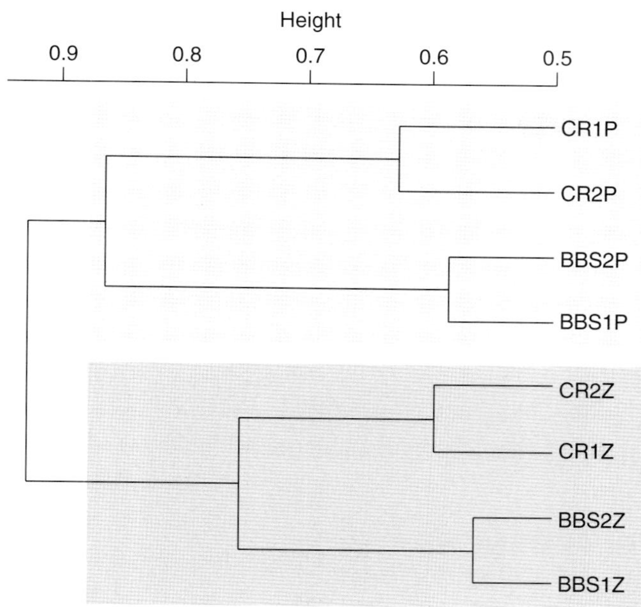

Figure 28.3 Dendrogram showing the ecological distances among big bluestem (BBS) and switchgrass cv Cave-in-Rock (CR) rhizosphere communities of N_2-fixing bacteria as evaluated by using Pol and Z primer sets [Poly et al., 2001; Zehr and McReynolds, 1989]. Operational taxonomic units were defined at the 3% distance before clustering of communities. The Jaccard index and the average linkage method were used. The letters P and Z in the labels indicate Poly and Zehr primers, respectively. Two different plants of each grass species were studied, as indicated by the numbers 1 and 2 in the labels.

gene, although the majority of the gene information was based on oceanic samples or soils, we selected three clades composed of possible human colonic bacterial species and designed three primer sets. Designed primers were tested with in silico PCR against nontarget sequences including human genome DNA, and then they were tested empirically using positive control samples for determination of the appropriate PCR conditions. Both *buk* and *but* gene targeted primers were applied on DNA extracted from IBD patients, which resulted in amplicons of the expected length. It was necessary to further refine both the template concentration and the PCR parameters after addition of the barcodes and adapters to the primers. All nine possible combinations of the *but* and *but* primers could not be multiplexed in a single reaction, possibly due to primer dimer formation as a result of primer modification. In the case of the *dsr* gene, PCR after the addition of the primer barcode and adapters resulted in no amplification in the control samples. This example is a case where the necessary primer modification for the 454 platform appeared to have a large negative influence on PCR performance.

Pyrosequencing of the *buk* and *but* amplicons produced sequences that were most similar to those of the known genes and reflected changes in community composition with disease progression. Specifically, the relative abundance of sequences most related to *Coprococcus comes* showed a 25-fold decrease with disease progression while *Bacteroides sp*. increased 20-fold. The average nucleotide identity to currently available *buk* sequences was 86%. The most abundant sequences were most closely related to *Coprococcus comes, Bacteroides sp., Bacteroides stercoris, C. perfringens, C. difficile, Coprococcus eutactus*, and *Bacteroides thetaiotaomicron*. With functional genes, primer design may be limited by the current knowledge of the functional gene of interest. This can be due either to an overall low number of sequences or a lack of sequences in a particular ecosystem of interest, such as the case with *dsr* gene in at least some intestinal samples.

28.3.4 Deep Sea Functional Gene Analysis

Another study of GT metagenomics has been undertaken on the dimethylsulfoniopropionate (DMSP) demethylase gene, *dmdA*, by Varaljay et al. [2010]. DMSP is an abundant organic sulfur compound produced by marine phytoplankton and *dmdA* encodes the first enzyme of the demethylation pathway. Thousands of *dmdA* sequences that were captured in marine metagenomic datasets, Global Ocean Survey (GOS) Sargasso Sea [Yutin et al., 2005], and 2007 GOS datasets [Rusch et al., 2007], were used for the design of PCR primer sets that targeted 10 environmental gene clades and subclades. Primer specificities were tested in silico against the GOS database after *dmdA* sequences were removed. The designed primer sets were applied to composite free-living coastal bacterioplankton DNA and sequenced using the Genome Sequencer FLX system (454 Life Sciences). An average of 6400 amplicons per primer pair grouped into more than 700 clusters with clusters defined conservatively at >90% nucleotide sequence identity (~95% protein identity). The results showed sequence richness in some *dmdA* subgroups differed significantly between size fractions, though many sequences from free-living and particle-associated bacteria were affiliated to shared clusters. This study illustrates the value of having a large shotgun metagenomic sequence dataset for more robust primer design, thereby yielding better coverage of the gene of interest in the habitat from which the sequences came.

28.4 FUTURE DESIGN

Sequencing errors now well known for pyrosequencing can greatly inflate diversity estimates (see, e.g., Reeder and Knight [2009] and Kunin [2010]; see also Chapter 19,

Vol. I). A novel method that clusters reads at the raw "flowgram" stage shows promise in removing the greater part of sequencing errors; however, the method is not yet practical because of the very high computational requirements [Quince et al., 2009]. The most common pyrosequencing errors are miscounting of homopolymer runs. For protein coding genes, these homopolymer miscounts translate to frameshift errors. Tools such as the Wise2 package can detect frameshift errors during alignment [Birney et al., 2004]. For genes too diverse to allow alignment using the Wise2 package, designing amplicons short enough to sequence through both forward and reverse primers can help with error detection. Since the correct reading frames for both forward and reverse primers will be known in advance, reads with reading frame shifts between the primers can be detected and discarded.

In the near future, Roche plans releasing improvements to its 454 technology that will increase the read length up to 1000 bp (genomeweb, September 29, 2009; http://www.genomeweb.com/sequencing/454-plans-longer-reads-other-upgrades-gs-flx-q4). To make the technology more affordable, they also plan on releasing a "GS Junior" sequencer in 2010, a lower-cost instrument that will produce approximately 100,000 reads per run, as compared to the 1 million reads possible with the GS FLX platform (genomeweb, December 1, 2009; http://www.genomeweb.com/sequencing/roche-launch-scaled-down-454-sequencer-2010-jgi-first-test-site-long-gs-flx-read). At the other end of the scale, the Illumina Genome Analyzer is capable of producing ≈300 million paired-end reads of 100–125 (or more) bp per end in a single run (Illumina product literature; see also Chapter 18, Vol I). Currently, adoption of Illumina technology for GT metagenomics appears to be limited only by lack of projects able to productively use such high-volume sequencing.

Acknowledgments

This work was supported by U.S. Department of Energy Office of Science (ER62848; ER64494), Office of Energy Efficiency and Renewable Energy (DE-AC05-76RL01830) U.S. Department of Agriculture (2008-35107-04542, 2008–04556), NIH (HMP UH2 DK83993), NIEHS (P42-ES004911, and the Korean Science Foundation's World Class University Project R33-2008-000-10076-0.

INTERNET RESOURCES

ARB: http://www.arb-home.de/

CAMERA: http://camera.calit2.net/

ClustalW: http://www.ebi.ac.uk/Tools/clustalw2/index.html

CODEHOP: http://bioinformatics.weizmann.ac.il/blocks/codehop.html

EMBL-EBI: http://www.ebi.ac.uk/Tools/fasta33/wgs.html

EstimateS: http://purl.oclc.org/estimates, Colwell, RK. 2005. EstimateS: Statistical estimation of species richness and shared species from samples. Version 7.5. User's Guide and application.

FGPR: http://fungene.cme.msu.edu

IMG: http://img.jgi.doe.gov/

NCBI: http://ncbi.nlm.nih.gov

REFERENCES

Birney E, Clamp M, Durbin R. 2004. GeneWise and genomewise. *Genome Res*. **14**:988–995.

Colwell RK. 2005. EstimateS: Statistical estimation of species richness and shared species from samples. Version 7.5.. User's Guide and application published at: http://purl.oclc.org/estimates.

Csordas A. 1996. Butyrate, aspirin and colorectal cancer. *Eur. J. Cancer Prevent*. **5**:221–231.

Durbin R, Eddy S, Krogh A, Mitchison G. 1998. *Biological Sequence Analysis: Probabilistic Models of Proteins and Nucleic Acids*. Cambridge, UK: Cambridge University Press.

Edgar RC. 2004. Muscle: A multiple sequence alignment method with reduced time and space complexity. *BMC Bioinformatics* **5**:1–19.

Frias-Lopez J, Shi Y, Tyson GW, Coleman ML, Schuster SC, Chisholm SW, DeLong EF. 2008. Microbial community gene expression in ocean surface waters. *Proc. Natl. Acad. Sci. USA* **105**:3805–3810.

Gruber N, Galloway JN. 2008. An Earth-system perspective of the global nitrogen cycle. *Nature* **451**:293–296.

Hamer HM, Jonkers D, Venema K, Vanhoutvin S, Troost FJ, Brummer RJ. 2008. Review article: The role of butyrate on colonic function. *Aliment. Pharmacol. Ther*. **27**:104–119.

Hughes JB, Hellmann JJ, Ricketts TH, Bohannan BJM. 2001. Counting the uncountable: Statistical approaches to estimate microbial diversity. *Appl. Environ. Microbiol*. **67**:4399–4406.

Hulin SJ, Singh S, Chapman MA, Allan A, Langman MJ, Eggo MC. 2002. Sulphide-induced energy deficiency in colonic cells is prevented by glucose but not by butyrate. *Aliment. Pharmacol. Ther*. **16**:325–331.

Iwai S, Chai B, Sul WJ, Cole JR, Hashsham SA Tiedje JM. 2010. Gene-targeted-metagenomics reveals extensive diversity of aromatic dioxygenase genes in the environment. *ISME J*. **4**:279–285.

Kindt R, Coe R. 2005. *Tree Diversity Analysis. A Manual and Software for Common Statistical Methods for Ecological and Biodiversity Studies*. Nairobi: World Agroforestry Centre (ICRAF).

Konstantinidis KT, Tiedje JM. 2005. Genomic insights that advance the species definition for prokaryotes. *Proc. Natl. Acad. Sci. USA* **102**:2567–2572.

Kuhn E, Bellicanta GS, Pellizari VH. 2009. New alk genes detected in Antarctic marine sediments. *Environ. Microbiol*. **11**:669–673.

Kunin V, Engelbrektson A, Ochman H, Hugenholtz, P. 2010. Wrinkles in the rare biosphere: Pyrosequencing errors lead to artifical inflation of diversity estimates. *Environ. Microbiol*. **12**:118–123.

Lay C, Doré J, Rigottier-Gois L. 2007. Separation of bacteria of the *Clostridium leptum* subgroup from the human colonic microbiota by fluorescence-activated cell sorting or group-specific PCR using 16S rRNA gene oligonucleotides. *FEMS Microbiol. Ecol*. **60**:513–520.

Lee ZM, Bussema C 3rd, Schmidt TM. 2009. rrnDB: documenting the number of rRNA and tRNA genes in bacteria and archaea. *Nucleic Acids Res*. **37**:D489–493.

Legendre P, Legendre L. Numerical Ecology. 1998. 2nd English edition ed. Amsterdam: Elsevier Scientific Publishing Company.

Lerat E, Daubin V, Moran NA. 2003. From gene trees to organismal phylogeny in prokaryotes: The case of the gamma-Proteobacteria. *PLoS Biol*. **1**:E19.

Louis P, Flint HJ. 2009. Diversity, metabolism and microbial ecology of butyrate-producing bacteria from the human large intestine. *FEMS Microbiol. Lett*. **294**:1–8.

Lozupone C, Knight R. 2005. UniFrac: a new phylogenetic method for comparing microbial communities. *Appl. Environ. Microbiol*. **71**:8228–8235.

Ludwig W, Strunk O, Westram R, Richter L, Meier H, et al. 2004. ARB: A software environment for sequence data. *Nucleic Acids Res*. **32**:1363–1371.

Macfarlane S, Steed H, Macfarlane GT. 2009. Intestinal bacteria and inflammatory bowel disease. *Crit. Rev. Clin. Lab. Sci*. **46**:25–54.

Magurran AE. 2004. *Measuring Biological Diversity*. UK: Blackwell Science Ltd.

Margulies M, Egholm M, Altman WE, Attiya S, Bader JS, et al. 2005. Genome sequencing in microfabricated high-density picolitre reactors. *Nature* **437**:376–380.

Martin AP. 2002. Phylogenetic approaches for describing and comparing the diversity of microbial communities. *Appl. Environ. Microbiol*. **68**:3673–3682.

McLaughlin SB, Kszos LA. 2005. Development of switchgrass (*Panicum virgatum*) as a bioenergy feedstock in the United States. *Biomass and Bioenergy*. **28**:515–535.

Pajak B, Orzechowski A, Gajkowska B. 2007. Molecular basis of sodium butyrate-dependent proapoptotic activity in cancer cells. *Adv. Med. Sci*. **52**:83–88.

Pielou EC. 1984. *The Interpretation of Ecological Data. A Primer on Classification and Ordination*. New York: John Wiley & Sons.

Poly F, Monrozier LJ, Bally R. 2001. Improvement in the RFLP procedure for studying the diversity of nifH genes in communities of nitrogen fixers in soil. *Res. Microbiol*. **152**:95–103.

Pryde SE, Duncan SH, Hold GL, Stewart CS, Flint HJ. 2002. The microbiology of butyrate formation in the human colon. *FEMS Microbiol. Lett*. **217**:133–139.

Quince C, Curtis TP, Sloan WT. 2008. The rational exploration of microbial diversity. *ISME J*. **2**:997–1006.

Quince C, Lanzen A, Curtis TP, Davenport RJ, Hall N, Head IM, Read LF, Sloan WT. 2009. Accurate determination of microbial diversity from 454 pyrosequencing data. *Nat. Methods* **6**:639–641.

Reeder J, Knight R. 2009. The "rare biosphere": A reality check. *Nat. Methods* **6**:636–637.

Roediger WEW, Duncan A, Kapaniris O, Millard S. 1993. Reducing sulfur-compounds of the colon impair colonocyte nutrition—Implications for ulcerative-colitis. *Gastroenterology* **104**:802–809.

Rose TM, Schultz ER, Henikoff JG, Pietrokovski S, McCallum CM, Henikoff S. 1998. Consensus-degenerate hybrid oligonucleotide primers for amplification of distantly related sequences. *Nucleic Acids Res*. **26**:1628–1635.

Rusch DB, Halpern AL, Sutton G, Heidelberg KB, Williamson S, et al. 2007. The Sorcerer II Global Ocean Sampling expedition: Northwest Atlantic through eastern tropical Pacific. *PLoS Biol*. **5**:e77.

Scheppach W, Luehrs H, Menzel T. 2001. Beneficial health effects of low digestible carbohydrate consumption. *Br. J. Nutr*. **85**:S23–S30.

Schloss, PD. 2008. Evaluating different approaches that test whether microbial communities have the same structure. *ISME J*. **2**:265–275.

Schloss PD, Handelsman J. 2008. A statistical toolbox for metagenomics: Assessing functional diversity in microbial communities. *BMC Bioinformatics* **9**:34.

Schloss PD, Larget BR, Handelsman J. 2004. Integration of microbial ecology and statistics: A test to compare gene libraries. *Appl. Environ. Microbiol*. **70**:5485–5492.

Schloss PD, Westcott SL, Ryabin T, Hall JR, Hartmann M, et al.. 2009. Introducing mothur: Open-source, platform-independent, community-supported software for describing and comparing microbial communities. *Appl. Environ. Microbiol*. **75**:7537–7541.

Segain JP, dela Bletiere DR, Boureille A, Leray V, Gervois N, et al.. 2000. Butyrate inhibits inflammatory responses through NFkappaB inhibition: implications for Crohn's disease. *Gut* **47**:397–403.

Singleton DR, Furlong MA, Rathbun SL, Whitman WB. 2001. Quantitative comparisons of 16S rRNA gene sequence libraries from environmental samples. *Appl. Environ. Microbiol*. **67**:4374–4376.

Ter Braak CJF, Smilauer P. 2002. *CANOCO Reference Manual and CanoDraw for Windows User's Guide: Software for Canonical Community Ordination (Version 4.5)*. Ithaca, NY: Microcomputer Power.

Yutin N, Suzuki MT, Beja O. 2005. Novel primers reveal wider diversity among marine aerobic anoxygenic phototrophs. *Appl. Environ. Microbiol*. **71**:8958–8962.

Varaljay VA, Howard EC, Sun S, Moran MA. 2010. Deep sequencing of a DMSP-degrading gene (*dmdA*) using PCR primer pairs designed from marine metagenomic data. *Appl. Environ. Microbiol*. **76**:609–617.

Zehr JP, McReynolds LA. 1989. Use of degenerate oligonucleotides for amplification of the nifH gene from the marine cyanobacterium *Trichodesmium thiebautii*. *Appl. Environ. Microbiol*. **55**: 2522–2526.

Zehr JP, Jenkins BD, Short SM, Steward GF. 2003. Nitrogenase gene diversity and microbial community structure: A cross-system comparison. *Environ. Microbiol*. **5**:539–554.

Chapter 29

Phylogenetic Screening of Metagenomic Libraries Using Homing Endonuclease Restriction and Marker Insertion

Torsten Thomas, Staffan Kjelleberg, and Pui Yi Yung

29.1 INTRODUCTION

One of the principal aims of metagenomics is to provide functional information on the vast microbial diversity of our planet. Molecular surveys of ribosomal RNA (rRNA) genes from the environment have been employed for taxonomic and phylogenetic classification of cultured and uncultured organisms and have uncovered in the last decades an astounding bacterial diversity (see Chapter 15, Vol. I). This is reflected by databases now housing hundreds of thousands of small- and large-subunit rRNA sequences (SSU/LSU and 16S/23S rRNA) [Tringe and Hugenholtz, 2008; see also Chapter 15, Vol. I]. However, linking the phylogenetic information to the functional nature of the organisms they represent is challenging, and therefore the majority of known bacterial diversity remains poorly characterized.

Random shotgun sequencing of environmental DNA has established itself as a core methodology in metagenomics and has the promise of describing genomes of uncultured organisms that are only known from their rRNA signature (see Chapter 20, Vol. I). Recent advances in high-throughput sequencing technology has further stimulated the field and now allows for in-depth survey of natural community within reasonable research budgets [Mardis, 2008; see also Chapter 18, Vol. I]. However, challenges in assembling genomes from heterogeneous populations and complex samples (i.e. species-rich) often results that shotgun sequencing data has to be functionally annotated on the level of individual sequence reads or assembled contigs of incomplete genomes [Kunin et al., 2008; Tringe et al., 2005]. This leads to most functional genes being physically detached (or unassembled) from genes containing taxonomic information (such as rRNA genes) and has hence limited the progress in revealing the phylotype–function relationships for many uncultured organisms.

Several computational and laboratory-based approaches have been developed to overcome this situation and provide taxonomic information for metagenomic DNA fragments [Kunin et al., 2008]. Computational methods mainly use binning of DNA fragments based on compositional features (such as oligonucletotide or k-mer frequencies) that are conserved across genomes from various taxonomic levels. This is then followed by taxonomic classification of bins. Algorithms like PhyloPythia [McHardy et al., 2007; see also Chapter 40, Vol. I], TETRA [Teeling et al., 2004], or TACOA [Diaz et al., 2009] have employed this strategy. However, their success is reliant on the availability of reference (or training) sequences that are long enough to provide sufficient compositional information to create a classifier for shorter, unknown fragments. These reference fragments should ideally also contain genes or markers suitable for taxonomic assignment, and fosmids or BAC clones containing rRNA genes would satisfy these requirements [McHardy et al., 2007; Teeling et al., 2004].

Handbook of Molecular Microbial Ecology, Volume I: Metagenomics and Complementary Approaches, First Edition. Edited by Frans J. de Bruijn.

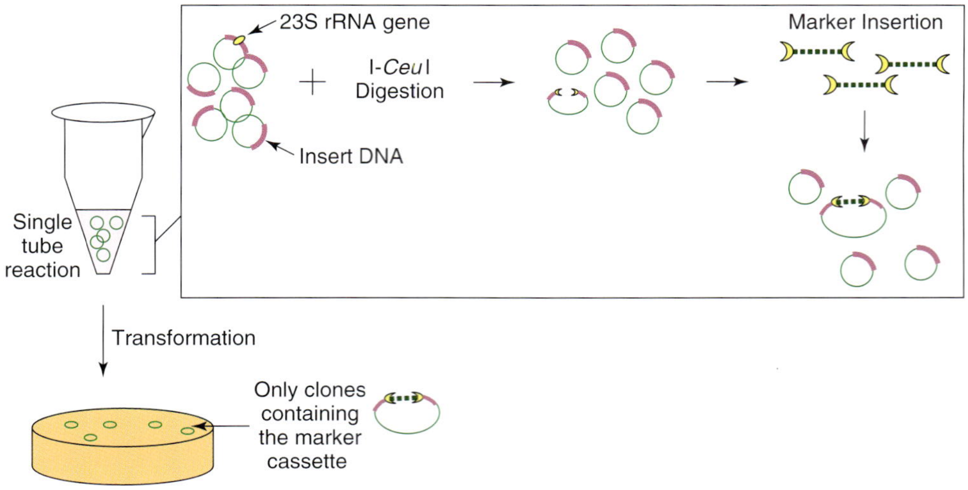

Figure 29.1 Schematic representation of the HERMI strategy.

A number of laboratory-based molecular screens have been developed to isolate fosmids or BAC clones with rRNA genes. Those approaches typically involve PCR- or fluorescent in situ hybridization-based screening of individually arrayed clones or small pools of clones (around 96 individual clones)[Moreira et al., 2006; Liles et al., 2003; Beja et al., 2000; Stein et al., 1996]. However, these methods are rather labor-intensive, potentially quite costly and often reach their practical limits with increasing clone numbers in the library.

29.2 APPROACH

We have developed an efficient phylogenetic screening strategy for metagenomic libraries using homing endonuclease restriction and marker insertion (HERMI) (**Fig. 29.1**). Its basic principle relies on the rapid selection of clones containing rRNA genes from large pools by employing a restriction digest with the homing endonuclease *I-Ceu*I (or potentially other homing endonucleases). Homing endonuclease have large, nonsymmetric recognition sites and are encoded in group I, group II, or archael introns or inteins, where they facilitate the insertion into conserved genes. *I-Ceu*I is part of an intron in the chloroplast LSU rRNA gene of the unicellular green algae *Chlamydomonas eugametos*. The recognition site of *I-Ceu*I is 26 bp long and cuts very specifically in a conserved part of the intron-free 23S rRNA gene to produce a 4-nucleotide (CTAA) 3′ overhang [Marshall and Lemieux, 1992]. The slow evolutionary rate of the recognition site, as well as the tolerance of the enzyme for minor sequence changes, means that rRNA genes from a wide range of organisms can usually be cut [Marshall and Lemieux, 1991]. This makes *I-Ceu*I an ideal enzyme to select for 23S rRNA genes. In our approach we therefore used *I-Ceu*I to digest a pool of metagenomic clones to linearize only clones that contain 23S rRNA genes. The linearized clones are subsequently isolated by inserting a selectable marker gene (i.e. antibiotic cassette), followed by retransformation and selection from the library pool (Fig. 29.1).

29.3 MATERIAL AND METHODS

For additional information on the materials and methods and their application to a metagenomic fosmid library, please refer to the original publication [Yung et al., 2009]. Here we present a detailed version of the marker production and the insertion reaction that should allow for the application of the method to any library type.

29.3.1 Construction of a Kanamycin Cassette with Flanking *I-Ceu*I Recognition Sequences

A kanamycin cassette with *I-Ceu*I recognition sequences at each end was produced through a three-step PCR strategy. The first round of PCR amplified the kanamycin resistance cassette from plasmid pACYC177 (NEB, Beverly, MA) incorporating partial *I-Ceu*I recognition sequences. A 50-μl PCR reaction contained the following: 10 ng of pACYC177 plasmid, 1× iProof HF Buffer, 200 μM of dNTPs, 0.5 μM of primers KanFCeu (5′GTCCTAAGGTAGCGAGGTTGATGAGAGCTTTG-TTG3′) and KanRCeu (5′AGGACCGTTATAGTTAAAG-TCAGCGTAATGCTCTGC3′), and 0.05 U/μl iProof DNA polymerase (Bio-Rad, Hercules, CA). Cycling conditions were: 1 min/98°C, followed by 10 s/98°C, 90 s/53°C, 1 min/72°C, 30 cycles. The second round of PCRs adds the entire *I-Ceu*I restriction site and was performed in two separate reactions; one reaction uses the primers CeuF

(5′GGGTAACTATAACGGTCCTAAGGTAGCGAG3′) and KanRCeu, and the other employs primers CeuR (5′GGGTCGCTACCTTAGGACCGTTATAGTTAAA3′) and KanFCeu. The 50-μl PCR reactions contained 3 ng of PCR product from the first round and remaining reaction mix composition as described above. Cycling conditions were: 1 min/98°C, followed by 10 s/98°C, 30 s/45°C, 90 s/72°C, 30 cycles. The two second-round PCR products are were pooled (final volume of 100 μl) and, if necessary, the 1.1-kbp product (corresponding to the kanamycin cassette plus flanking region) was purified by gel extraction. The third step involved hybridizations and fill-in to generate a product with the full *I-Ceu*I recognition site on both ends of the kanamycin cassette. For this ~50 μl of (gel-extracted) DNA from the second round, PCR was combined with 1× RedTaq PCR buffer, 10 mM of dNTPs, and 1 U of RedTaq DNA polymerase (Promega, Madison, WI) to a total volume of 100 μl, heat denatured for 2 min at 94°C and then cycled (20 times) for 20 s/94°C, 30 s/45°C, 20 s/72°C. The final product was cloned into a vector (e.g. pGEM®-T Easy Vector System, Promega, Madison, WI) and verified by sequencing and/or restriction digest. The kanamycin cassette with flanking *I-Ceu*I recognition sequences (CeuKan) was purified by gel extraction from an *I-Ceu*I restriction digest of the plasmid. A pGEM vector with the cassette (pGEM:CeuKan) ready for digestion is available from the authors.

29.3.2 Homing Endonuclease Restriction and Marker Insertion

Pooled clone libraries (e.g., fosmids or BACs) should be purified using standard techniques or commercial kits and excessive chromosomal DNA should be removed from the library preparation prior to the digest. One microgram of pooled library DNA was digested with the *I-Ceu*I homing endonuclease in a reaction containing 2 units of *I-Ceu*I enzyme (NEB), 1× NEB buffer 4, 0.5 μl of BSA (NEB), and sterile, deionized water up to a volume of 50 μl. The digest was incubated at 37°C for 3 h followed by a heat inactivation at 65°C for 20 min. The digested library was dephosphorylated using 2 units of Antarctic alkaline phosphatase (NEB) according to manufacturer's instructions. The library was then ligated with the CeuKan cassette in a 1:30 (vector:insert) molar ratio overnight at room temperature. The ligation mix was transformed into *E. coli* cells via electroporation and then plated onto LB10 agar supplemented with the appropriate antibiotic for the vector backbone and kanamycin for the CeuKan cassette insertion. To ensure a high recovery of clones containing the marker insertions, competent cells with high transformation efficiency should be used. Insertion sites of CeuKan can be determined by sequencing from the cassette with primers KanFSeq (5′GACCGTTCCGTGGCAAAG3′) or KanRSeq (5′GCTCGATGAGTTTTTCTAATC3′)

29.4 RESULTS AND DISCUSSION

29.4.1 Specificity and Selectivity of HERMI

The HERMI strategy was successful in identifying fosmids with rRNA genes from a complex metagenomic library of the microbial community associated with the marine sponge *Cymbastela concentrica* [Yung et al., 2009]. A single-tube screen of over 6000 pooled fosmid clones resulted in the isolation and identification of 12 unique fosmids containing 23S rRNA genes. None of the clones selected by the HERMI procedure had more than one marker insertion or an insertion in a site that was not the 23S rRNA gene. This indicates a high specific of the methodology and a low false-positive rate. Selectivity of the approach was evaluated by theoretical comparison of the *I-Ceu*I recognition sequence to the entire bacterial 23S rRNA Silva database [Pruesse et al., 2007; see also Chapter 45, Vol. I] applying a threshold that only one mismatch will cause a failure of recognition and/or cleavage. The analysis showed that only 1.1% (59 sequences out of 5525 sequences) of the 23S rRNA gene sequences contain mismatches that could potentially abolish the enzymes recognition or cleavage activity. Phylotypes that might not be selected for in the HERMI approach include a number of species within the genus *Chlamydophila* and *Chlamydia*, the unclassified gammaproteobacteria Candidatus *Carsonella ruddii*, the candidate division TM7 group GTL1, and a number of species in the genus *Chloroflexi* and *Roseiflexus* (all one mismatch) as well as some species within the *Planctomycetes, Coxiella burnetii* and *Francisella tularensis* subsp. (two or more mismatches). Furthermore, upon HERMI screening of the sponge metagenomic library, one clone was selected with a 23S rRNA gene that had base pair insertions between positions 4–5, 5–6, 6–7, and 7–8 (respective to the original recognition sequence). This indicates that changes in those four positions of the recognition sequence do not abolish *I-Ceu*I cleavage activity. We therefore believe that our theoretical considerations and thresholds are quite stringent and that a lower limit of selectivity of 98% of all known diversity from a wide phylogenetic spectrum is a conserved estimate.

29.4.2 Comparison to Other Methods

Handling time for the HERMI procedure is less than 2 h, and the process does not require specialized equipment and only minor amount of consumables (e.g. *I-Ceu*I

Table 29.1 Comparison of Various Methods for the Identification of Phylogenetic Marker Genes in Metagenomic Libraries

Approach	Labor	Equipment	Background Signal	Specific Groups	Pool Size
HERMI	Little	Cheap	No	Possible	Large
PCR-screening	Large	Moderate	Yes	Yes	Small
Colony hybridization	Moderate	Moderate	Yes	Yes	Small
FISH and FACS	Moderate	Expensive	Yes	Yes	Large

enzyme). Thousands of clones can be screened in a single-tube reaction, making this a high-throughput approach that compares very favorably with other strategies to screen for clones with phylogenetic markers (see **Table 29.1**).

Of the other existing strategies, PCR-based screening of metagenomic libraries clones is probably the most popular approach [Meng et al., 2009; Gilbert et al., 2008; Massana et al., 2008; Quaiser et al., 2008; Martín-Cuadrado et al., 2007]. This approach involves setting up separate PCR reactions for individual clones or small pools of clones (e.g., 96 clones) and then amplify the rRNA gene with conserved primers. Analysis of the PCR reaction is typically done by gel electrophoresis. Positive pools have to be screened again in a second round of PCRs to identify positive individual clones. The method has the drawback of unwanted cross-hybridization of the PCR primers with the host background of the library (e.g. *Escherichia coli*) causing false-positives and therefore works best for screening of groups distantly related to the library host (e.g. screen for archaeal markers in an *E. coli* host library). However, the method can be applied to select for very specific groups by designing, for example, species- or genera-specific PCR primers. Martín-Cuadrado et al. used primers that amplify the 16S rRNA gene together with the adjacent intergenic spacer (ITS) regions. This detects only bacterial fosmids containing ITSs whose size is different from that of the *E. coli* host [Martín-Cuadrado et al., 2007].

To circumvent the laborious PCR analysis, screens involving colony hybridization have been developed [Pham et al., 2008; DeLong et al., 2006]. These approaches work by arraying individual library clones onto membranes, followed by direct lysis of the cells and fixing of the DNA to the membranes. Genomic DNA may be removed by digest with nucleases specific to linear DNA to minimize cross-hybridization with the host background [Pham et al., 2008]. The membrane is then probed with a short oligonucleotide or a PCR product that targets specific groups of organisms, and positive clones are detected by a chemical light reaction or radioactivity.

A substantial increase in throughput to the previous two methods is offered by screens based on the combination of fluorescent in situ hybridization (FISH) with fluorescence-activated cell sorting (FACS) [Leveau et al., 2004]. FISH has been successfully applied to identify fosmid clones with rRNA genes in a metagenomic library [Leveau et al., 2004]. In an additional step, labeled clones could be selected by flow cytometry that can sort thousand of clones per second. The fosmid or BAC DNA of selected cells could then be retransformed into a host cell to yield a library highly enriched in rRNA genes. Technically, this process is challenging and requires access to expensive equipment.

29.4.3 Future Developments

It is clear that there are many possible variations, combinations, and optimizations for the current screening methods. For example, the selectable marker applied in the HERMI approach (i.e. kanamycin resistance marker) could be replaced with a green fluorescent protein cassette. This would then allow for the selection of clones by FACS (i.e. HERMI-FACS) rather than by growth selection.

The one disadvantage of HERMI is that it cannot target specific groups of organisms (e.g., a genus). However, Arnould et al. [2006] have recently demonstrated that a semirational design can yield *I-Cre*I mutant homing endonucleases with highly specific recognition patterns, which are distinct from the wild-type activity. As such, *I-Ceu*I or other homing endonuclease mutants could be designed that recognize and cleave other conserved motifs within the 23S rRNA gene that are specific for certain groups. Alternatively, other homing endonuclease mutants are available or could be generated that could target either (a) other common genes used for phylogenetic analysis (e.g., the 16S rRNA gene, *recA*), (b) functional genes, or (c) noncoding regions, which could then be screened for in an analogous strategy as described here. This will further enhance the utility and versatility of the HERMI approach.

Acknowledgments

This work was supported by funding from the Gordon and Betty Moore Foundation and the Australian Research Council (LP0668235).

REFERENCES

Arnould S, Chames P, Perez C, Lacroix E, Duclert A, et al. 2006. Engineering of large numbers of highly specific homing endonucleases that induce recombination on novel DNA targets. *J. Mol. Biol.* **355**:443–458.

Beja O, Suzuki MT, Koonin EV, Aravind L, Hadd A, et al. 2000. Construction and analysis of bacterial artificial chromosome libraries from a marine microbial assemblage. *Environ. Microbiol.* **2**:516–529.

DeLong E, Preston C, Mincer T, Rich V, Hallam S, et al. 2006. Community genomics among stratified microbial assemblages in the ocean's interior. *Science* **311**:496–503.

Diaz N, Krause L, Goesmann A, Niehaus K, Nattkemper T. 2009. TACOA: Taxonomic classification of environmental genomic fragments using a kernelized nearest neighbor approach. *BMC Bioinformatics* **10**:56.

Gilbert JA, Mühling M, Joint I. 2008. A rare SAR11 fosmid clone confirming genetic variability in the '*Candidatus Pelagibacter* ubique' genome. *ISME J.* **2**(7):790–793.

Kunin V, Copeland A, Lapidus A, Mavromatis K, Hugenholtz P. 2008. A bioinformatician's guide to metagenomics. *Microbiol. Mol. Biol. Rev.* **72**:557–578.

Leveau J, Gerards S, de Boer W, van Veen J. 2004. Phylogeny-function analysis of (meta)genomic libraries: screening for expression of ribosomal RNA genes by large-insert library fluorescent in situ hybridization (LIL-FISH). *Environ. Microbiol.* **6**:990–998.

Liles MR, Manske BF, Bintrium SB, Handelsman J, Goodman RM. 2003. A census of rRNA genes and linked genomic sequences within a soil metagenomic library. *Appl. Environ. Microbiol.* **69**:2684–2691.

Mardis E. 2008. Next-generation DNA sequencing methods. *Annu. Rev. Genomics Hum. Genet.* **9**:387–402.

Marshall P, Lemieux C. 1991. Cleavage pattern of the homing endomuclease encoded by the fifth intron in the chloroplast large subunit rRNA-encoding gene of *Chlamydomonas eugametos*. *Gene* **104**:241–245.

Marshall P, Lemieux C. 1992. The *I-CeuI* endonuclease recognizes a sequnce of 19 base pairs and preferentially cleaves the coding strand of the *Chlamydomonas moewusii* chloroplast large subunit rRNA gene. *Nucleic Acids Res.* **20**:6401–6407.

Martín-Cuadrado A, López-García P, Alba J, Moreira D, Monticelli L, et al. 2007. Metagenomics of the deep Mediterranean, a warm bathypelagic habitat. *PLoS ONE* **2**:e914.

Massana R, Karniol B, Pommier T, Bodaker I, Béjà O. 2008. Metagenomic retrieval of a ribosomal DNA repeat array from an uncultured marine alveolate. *Environ. Microbiol.* **10**:1335–1343.

McHardy AC, Martin HG, Tsirigos A, Hugenholtz P, Rigoutsos I. 2007. Accurate phylogenetic classification of variable-length DNA fragments. *Nat. Methods* **4**:63–72.

Meng J, Wang F, Wang F, Zheng Y, Peng X, Zhou H, Xiao X. 2009. An uncultivated crenarchaeota contains functional bacteriochlorophyll a synthase. *ISME J.* **3**:106–116.

Moreira D, Rodríguez-Valera F, L-G. P., 2006. Metagenomic analysis of mesopelagic Antarctic plankton reveals a novel deltaproteobacterial group. *Microbiology* **152**:505–517.

Pham VD, Konstantinidis KT, Palden T, DeLong EF. 2008. Phylogenetic analyses of ribosomal DNA-containing bacterioplankton genome fragments from a 4000 m vertical profile in the North Pacific Subtropical Gyre. *Environ. Microbiol.* **10**:2313–2330.

Pruesse E, Quast C, K. Knittel, B. Fuchs, W. Ludwig, J. Peplies, Glöckner. FO. 2007. SILVA: A comprehensive online resource for quality checked and aligned ribosomal RNA sequence data compatible with ARB. *Nucleic Acids Res.* **35**:7188–7196.

Quaiser A, López-García P, Zivanovic Y, Henn M, Rodriguez-Valera F, Moreira D. 2008. Comparative analysis of genome fragments of Acidobacteria from deep Mediterranean plankton. *Environ. Microbiol.* **10**:2704–2717.

Stein JL, Marsh TL, Yu KY, Shizuya H, DeLong EF. 1996. Characterization of uncultivated prokaryotes: Isolation and analysis of a 40-kilobase-pair genome fragment from a planktonic marine archaeon. *J. Bacteriol.* **178**:591–599.

Teeling H, Meyerdierks A, Bauer M, Amann R, Glockner FO. 2004. Application of tetranucleotide frequencies for the assignment of genomic fragments. *Environ. Microbiol.* **6**:938–947.

Tringe S, Hugenholtz P. 2008. A renaissance for the pioneering 16S rRNA gene. *Curr. Opin. Microbiol.* **11**:442–446.

Tringe SG, von Mering C, Kobayashi A, Salamov AA, Chen K, et al. 2005. Comparative metagenomics of microbial communities. *Science* **308**:554–557.

Yung PY, Burke C, Lewis M, Egan S, Kjelleberg S, Thomas T. 2009. Phylogenetic screening of a bacterial, metagenomic library using homing endonuclease restriction and marker insertion. *Nucleic Acids Res.* **37**:e144.

Chapter 30

ArrayOme- and tRNAcc-Facilitated Mobilome Discovery: Comparative Genomics Approaches for Identifying Rich Veins of Bacterial Novel DNA Sequences

Hong-Yu Ou and Kumar Rajakumar

30.1 INTRODUCTION

The bacterial genomics revolution kicked off in earnest with the decoding of the *Haemophilus influenzae* genome in 1995 [Fleischmann et al., 1995]. In the relatively short period since 1995, it has assumed an explosive phase of growth driven by next-generation sequencing technologies [Ansorge, 2009; MacLean et al., 2009; see also Chapter 18, Vol. I]. As of December 6, 2009, there were 957 completely and 3570 ongoing sequenced bacterial genome available that were representative of more than 604 species spanning the full taxonomic breadth of the bacterial phylum [Liolios et al., 2009]. Remarkably, these numbers are yet to reflect the true impact of high-throughput sequencing efforts because these data are only just beginning to roll of sequencing production lines in staggering quantities at numerous centers worldwide.

An initial surprise that has since become a bacterial paradigm was the discovery of marked genome diversity (genodiversity) and plasticity within many individual bacterial species, let alone genera or wider taxonomic groupings [Canchaya et al., 2004; Dobrindt and Hacker, 2001; Dorrell et al., 2005]. The notion that two organisms might belong to the same species despite sharing less than 67% of DNA, as in the case of *E. coli* strains K-12 MG1655 against O157:H7 Sakai, continues to challenge our concept of speciation [Riley and Lizotte-Waniewski, 2009; Konstantinidis et al., 2006; see also Chapters 9–13, Vol. I]. Bioinformatics-facilitated comparative genomics studies have also led at a population level to the concept of the metagenome or pan-genome specific to each bacterial species [Medini et al., 2005; Tettelin et al., 2008; see also Chapter 10, Vol. I]. The pan-genome comprises the entire set of DNA sequences that are represented invariably, frequently, sporadically, or only rarely in members of the species. As a parallel theme, the genome of each individual bacterium can be divided into a core genome, common to the entire species, and an accessory genome. This latter optional genomic repertoire, which we termed the "mobilome" (mobile genome) in 2005 [Ou et al., 2005], includes a myriad of short strain-specific sequences and longer spans of episomal and integrative plasmids, transposons, integrons, gene cassettes, prophages, and a growing list of genomic islands [Frost et al., 2005; Hacker et al., 1997; Canchaya et al., 2004]. The evolutionary and functional boundaries between integrative conjugative elements (ICEs) [Churchward, 2008], integrative plasmids, integrative nonconjugative islands, and prophages [Canchaya et al., 2004; Lima-Mendez et al., 2007] can at times be blurred, but evidence that these entities are

Handbook of Molecular Microbial Ecology, Volume I: Metagenomics and Complementary Approaches, First Edition. Edited by Frans J. de Bruijn.

horizontally acquired is irrefutable [Boyd et al., 2009]. Intriguingly, a great diversity of alien DNA appears to map to a limited number of species-specific insertion "hotspots" in many of the bacterial genomes that have been better characterized to date, with the commonest and most generic hotspots being tRNA and tmRNA sites [Ou et al., 2006; Mantri and Williams, 2004].

Despite the transformative nature of next-generation sequencing, the associated substantial equipment and analytical costs and the daunting bioinformatics and computing challenge of dealing with the resultant tsunami of digital DNA data restrict its ability to be applied widely to discover strain-specific DNA that is dispersed amongst hundreds or even thousands of members of each bacterial species [Fux et al., 2005; Field et al., 2006]. The scale of the challenge demands empowerment of a much larger community of researchers who would otherwise be constrained by the above barriers. We have developed a parallel, high-throughput, comparative genomics strategy that readily facilitates experimental and in silico discovery of genomic islands and other foreign DNA. Dubbed MobilomeFINDER [Ou et al., 2007a], our approach combines ArrayOme-facilitated *M*icroarray-*A*ssisted *m*obilome *P*rospecting (MAmP) to aid identification of bacterial strains rich in novel DNA [Ou et al., 2005], *tRNA* gene *c*ontents and *c*ontexts analysis (tRNAcc) [Ou et al., 2006] to identify tRNA-associated hotspots and to streamline design and validation of primers for tRIP (*tR*NA site *I*nterrogation for *P*athogenicity islands, prophages, and other GIs) PCR-based identification and characterization of new GIs in MAmP-selected and/or otherwise chosen test strains [Ou et al., 2005]. Novel islands can then be captured selectively by island probing [Rajakumar et al., 1997], long-range PCR, and/or the yeast-based homologous recombination system described by Wolfgang et al. [2003] and subjected to cost-effective barcode-facilitated, massively parallel full-length sequencing. In this chapter we describe the MobilomeFINDER tools ArrayOme, tRNAcc, and mGenomeSubtractor [Ou et al., 2007a], provide examples of their utility, and briefly outline selected wet-science discoveries that have been achieved through use of the simple and readily adopted tRIP PCR strategy.

30.2 MATERIALS AND METHODS

30.2.1 Databases

Experimental microarray-derived comparative genomic hybridization (CGH) data obtained using the *Lactobacillus casei* ATCC 334 microarray were recently reported by Cai et al. [2009]. CGH data for 21 *Lactobacillus casei* test strains isolated from various habitats (seven, eight, and six from cheeses, plant materials, and human sources, respectively) were taken from Gene Expression Omnibus (GEO accession no. GSE15030). The following complete and near-complete genome sequences and annotation information were downloaded from NCBI: *Lactobacillus casei* strains ATCC 334 (NC_008526) and BL23 (NC_010999); *Lactobacillus paracasei* 8700:02:00 (NZ_ABQV00000000); *Acinetobacter baumannii* strains ATCC 17978 (NC_009085), SDF (NC_010400), ACICU (NC_010611), AB0057 (NC_011586), AB307-0294 (NC_011595), and AYE (NC_010410); *A. baumannii* plasmids pAB1 (NC_009083), pAB2 (NC_009084), p1ABSDF (NC_010395), p2ABSDF (NC_010396), p3ABSDF (NC_010398), pACICU1 (NC_010605), pACICU2 (NC_010606), pAB0057 (NC_011585), p1ABAYE (NC_010401), p2ABAYE (NC_010402), p4ABAYE (NC_010403), p3ABAYE (NC_010404), pABIR (NC_010481), pMMCU2 (NC_013506), pMMA2 (NC_013277), pABVA1 (NC_012813), and pMAC (NC_006877).

30.2.2 ArrayOme, tRNAcc and mGenomeSubtractor Tools

The MobilomeFINDER web-server (http://mml.sjtu.edu.cn/MobilomeFINDER/) comprises a flexible suite of individual interactive tools that greatly facilitate high-throughput experimental and in silico discovery of bacterial mobilome in closely related bacteria (**Table 30.1**) Details on the use of these tools are available on the MobilomeFINDER web server, and the precise rationale of the strategies are described in our original papers on ArrayOme [Ou et al., 2005], tRNAcc [Ou et al., 2006], mGenomeSubtractor [Shao et al., 2010], and MobilomeFINDER [Ou et al., 2007a].

30.3 RESULTS AND DISCUSSION

30.3.1 ArrayOme- and tRNAcc-Based Analyses of the Closely Related *Lactobacillus casei* and *Lactobacillus paracasei* Genomes

Microarray-derived CGH data identifies genes common to both the microarray used and the genome of the strain under investigation. This complement of genes represents an entity we previously defined as the *m*icroarray-*v*isualised *g*enome (MVG) [Ou et al., 2005]. Briefly, ArrayOme produces an accurate estimate of MVG sizes by using a CGH dataset to generate "inferred contigs" (ICs) produced by merging adjacent genes classified as "present." The sizes of novel mobilomes to which the microarray would have been "blind" borne by individual

Table 30.1 Interactive Tools Available at the MobilomeFINDER Web Server (http://mml.sjtu.edu.cn/MobilomeFINDER/) that Facilitate the Process of Bacterial Genomic Island or Mobile Genome (Mobilome) Discovery

Tool	Description[a]
Microarray-assisted mobilome prospecting	Identify strains that are particularly rich in novel mobilome DNA; underpinned by CGH, ArrayOme and PFGE
ArrayOme (Fig. 30.1 and Table 30.2)	Estimate the cumulative size of parts of the genome that harbor genes identified as "present" by microarray-based CGH; the size of the novel mobilome is thus predicted by comparing the microarray-visualized genome size with the PFGE-measured chromosome size
Island identification[b]	Identify GIs by tRNAcc analysis of closely related bacteria (Fig. 30.1B)
IslandScreen (Table 30.2)	Identify putative islands by single-step crude analysis with combination of IdentifyIsland, LocateHotspots, and DNAnalyzer.
IdentifyIsland	Identify putative islands based on conserved flanking blocks recognized by the multiple aligner Mauve 1.2.2 [Darling et al., 2004]
TabulateIsland	Tabulate islands identified by IdentifyIsland following analysis of different subsets of genomes
LocateHotspots	Locate proposed hotspots in non-annotated chromosomal sequences using BLASTN-based searches
Primer design[b]	
ExtractFlank	Generate ClustalW-derived MSA files that contain upstream or downstream flanking regions of identified islands to serve as inputs for Primaclade-facilitated [Gadberry et al., 2005] design of conserved PCR primers
*insilico*tRIP	Interrogate identified hotspots for the presence or absence of an integrated element using a locally installed version of Electronic PCR [Schuler, 1997]; generates 0- to 500-kb virtual fragments
Island analysis[b]	
DNAnalyzer	Calculate the GC content and dinucleotide bias of identified islands, and plot the negative cumulative GC profile of genomes
mGenomeSubtractor (Fig. 30.2 and Fig. 30.3)	High-throughput mpiBLASTN-based comparison of CDS sequences against test genomes to identify strain-specific CDS based on the level of nucleotide similarity

[a] Abbreviations used: GI, genomic island; CGH, comparative genomic hybridization; PFGE, pulsed-field gel electrophoresis; MSA, multiple sequence alignment; CDS, protein coding sequence.

[b] These tools can also be used for the generic identification and preliminary characterization of putative genomic islands located at other user-specified hotspots and for the analysis of cognate flanking sequences.

strains can subsequently be estimated by comparing ArrayOme-estimated MVG sizes with PFGE-measured physical chromosome sizes to identify isolates rich in novel or highly divergent genetic material that are likely to carry unique GIs or prophages. Recently, Cai et al. [2009] reported the CGH results for 21 *Lactobacillus casei* strains isolated from various habitats using the *L. casei* strain ATCC 334 microarray that bore probes for 2661 (97%) chromosomal and 17 (85%) plasmid annotated ATCC 334 CDS. We have estimated the ATCC 334-MVG sizes of the 21 test strains using these CGH data (Table 30.2). The 2,895,264 bp chromosome *of L. casei* ATCC 334, an Emmental cheese isolate, was used as the reference template for this analysis. Based on these CGH data, the Denmark cheese isolate 7A1 was predicted to possess 2256 chromosomal and 16 plasmid-borne CDS out of the total complement of 2678 ATCC 334 CDS represented on the array. Entry of the 7A1 CGH data into ArrayOme resulted in the generation of 139 inferred contigs (ICs) and an estimated MVG size

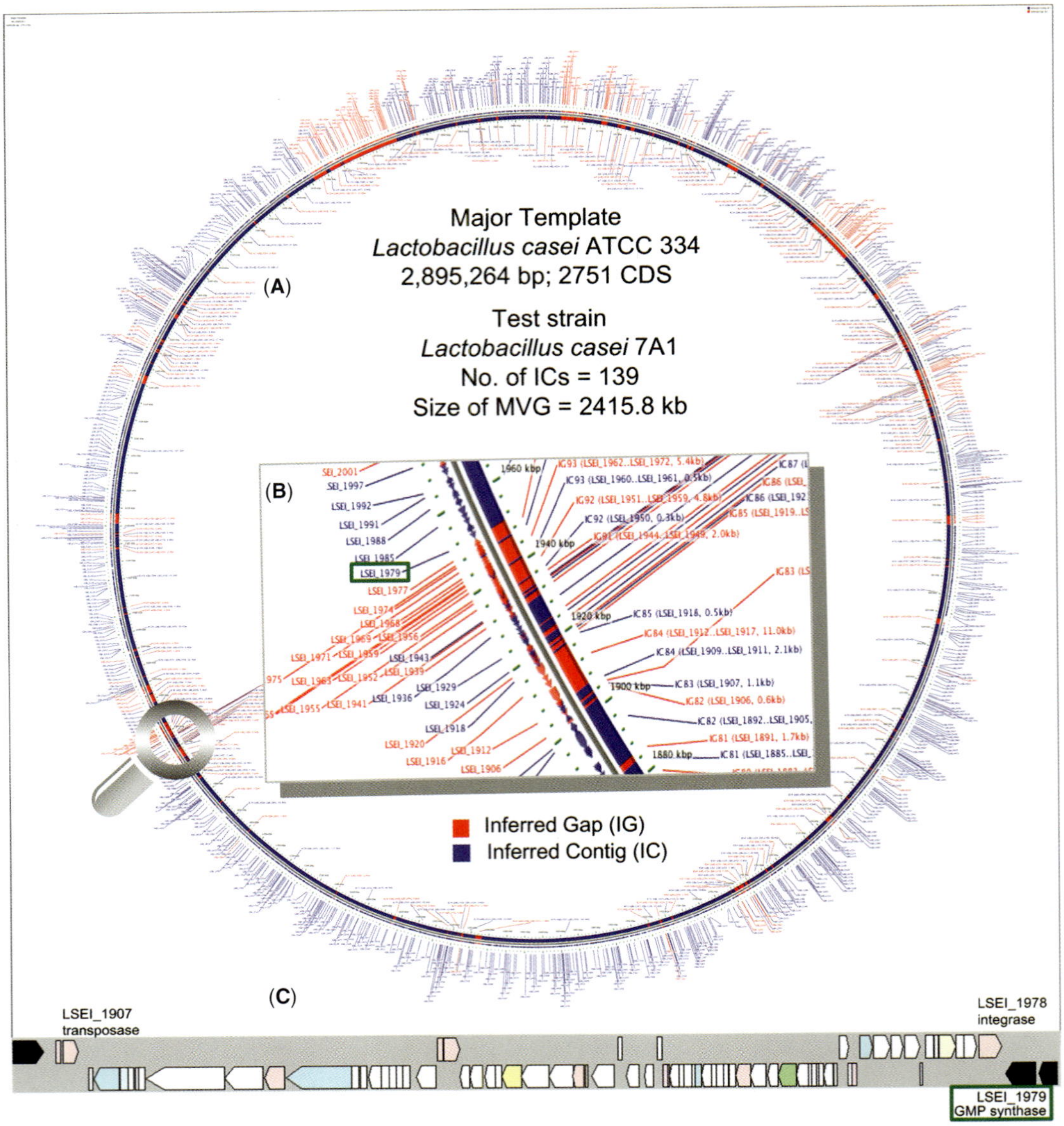

Figure 30.1 The ArrayOme-predicted microarray-visualized genome (MVG) sizes of *Lactobacillus casei* strain 7A1 based on data derived by comparative genome hybridization against the *Lactobacillus casei* ATCC 334 microarray. **(A)** Major reference template map showing inferred contigs (IGs) and inferred gaps (IGs). **(B)** A magnified view of the hypervariable region lying downstream of *LSEI_1979* showing individual ICs, IGs, and major reference template CDSs. **(C)** A much more detailed schematic of the hypervariable region (*LSEI_1906..LSEI_1979*) with individual CDS color-coded by COG assignment; putative mobility-associated genes have been labeled. The standard NCBI COG color scheme has been used.

of 2415.8 kb, including about 16 kb matching the ATCC 334 plasmid sequence. Because the physical chromosome size of 7A1 has been estimated to be about 2.9 Mb, given the determined MVG size, this strain was predicted to possess a non-ATCC 334-related genome complement of about 0.5 Mb, suggesting of the presence of about 500 additional accessory genes relative to ATCC 334. The newly updated ArrayOme web server offers intuitive visualization and exploration of output data. It was used to generate a circular map of ATCC 334 that highlighted the locations and sizes of identified ICs and inferred gaps (IGs) within the ATCC 334 major reference template (**Fig. 30.1A**); IGs comprise contiguous regions defined as "absent" following ArrayOme analysis. Hyperlinks provided within this output also allow ready access to individual NCBI Entrez Gene entries, thus permitting users to rapidly examine in depth the potential biological significance of ArrayOme-identified findings.

The individual online tRNAcc tools facilitate automated identification of genomic islands by comparative

Table 30.2 ArrayOme-Predicted Microarray-Visualized Genome (MVG) Sizes of 21 *Lactobacillus casei* Strains Based on Previously Published Comparative Genomic Hybridization Data Derived Using the *Lactobacillus casei* ATCC 334 Microarray[a]

	Strain Origin	Strain	# of CDS Classed as 'Present'[b] (Plasmid-Borne)	Number of Inferred Contigs	Size of the MVG (kb)	Cumulative Size of Inferred Gaps in the Major ATCC 334 Template (kb)
Plant	Silage	12A	2362 (0)	118	2512.3	383.3
	Silage	21/1	2406 (3)	117	2551.7	348.2
	Silage	32G	2394 (5)	119	2547.6	354.8
	Wine	A2-309	2408 (7)	128	2560.3	342.7
	Wine	A2-362	2481 (3)	113	2628.6	273.0
	Pickle	USDA-P	2422 (9)	128	2577.0	332.2
	Wine	UCD171	2344 (3)	103	2504.2	396.4
	Wine	UCD174	2376 (2)	113	2508.5	389.8
Human	Fecal	DN	2435 (5)	130	2584.2	319.1
	Fecal	L6	2433 (10)	119	2604.8	305.3
	Fecal	L9	2430 (10)	125	2588.7	321.3
	Blood	T7136	2430(12)	128	2579.7	332.9
	Blood	T71499	2396 (6)	121	2553.2	351.3
	Blood	CRF28	2372 (5)	131	2530.8	372.5
Cheese	Denmark	7A1[c]	2256 (16)	139	2415.8	503.2
	Denmark	7R1	2266 (12)	137	2420.2	494.5
	Denmark	83M4	2160 (4)	121	2327.5	574.6
	Australia	ASCC 1087	2164 (8)	125	2330.8	575.3
	USA	M36	2428 (4)	119	2578.1	324.4
	USA	UW1	2489 (6)	120	2626.5	275.9
	USA	UW4	2185 (3)	138	2330.7	568.3

[a] *L. casei* ATCC 334 microarray bore probes for 2,661 (97%) chromosomal and 17 (85%) plasmid annotated CDS. The 90 transposase-encoding CDS that were not included in the microarray design were classed as "absent" in this study.

[b] The number represents the total number of CDS (chromosome + plasmid) classed as present. The number shown in parentheses represents the number of plasmid-borne CDS classed as present.

[c] The ArrayOme-predicted MVG size of *L. casei* strain 7A1 is illustrated in Figure 30.1 as a circular map based on the major template *L. casei* ATCC 334 chromosome to highlight the locations and sizes of identified inferred contigs (ICs) and inferred gaps (IGs) identified in *L. casei* strain 7A1.

analysis of the contents and contexts of tRNA sites and other integration hotspots in closely related sequenced genomes. Briefly, IdentifyIsland exploits the multiple sequence aligner Mauve 1.2.2 [Darling et al., 2004] to investigate whether tRNA sites across multiple genomes are occupied by anomalous strain-specific DNA segments lying between the 3′ ends of tRNA genes and downstream conserved flanks. For example, when the complete genome sequences of *L. casei* strains ATCC 334 and BL23 and the near-complete genome sequence of *L. paracasei* 8700:02:00 were submitted to IslandScreen, a single-step analysis tool that integrates LocateHotspots, IdentifyIsland, and DNAnalyser, all 59 tRNA genes and the 1 tmRNA gene in ATCC 334 and their corresponding orthologues in the other two genomes were investigated to verify whether their 3′-end regions were occupied by islands. The 25 non-tRNA-associated, highly variable regions, including Region P adjacent to *LSEI_1979* shown in Figs. 30.1B and 30.1C [Cai et al., 2009], identified through ArrayOme analysis of CGH data derived from 21 *L. casei* test strains were also investigated using tRNAcc. Subsequent manual examination to exclude putative GIs of <1 kb and those assigned simultaneously to more than one closely clustered insertion site led to the identification of a final set of four tRNA and five non-tRNA hotspots among the three genomes analyzed (Table 30.3). All identified GIs had anomalous GC contents and/or significant dinucleotide biases, consistent with foreign origins. Accessory tools available on the MobilomeFINDER web site could be used to facilitate design of hotspot-flanking primers for in silico or wet-science-based interrogation of cognate loci in just-off-the-press genome data or unsequenced strains, respectively, and for further analysis of islands for features suggestive of foreign origins. Island-specific and genome-contextual features are tabulated as shown in Table 30.3 and can also be represented in schematic

Table 30.3 Putative Genomic Islands Identified by tRNAcc Analysis of 60 tRNA/tmRNA Genes and Five CGH-Defined Non-tRNA Integration Hotspots in the Genomes of the Completely Sequenced *Lactobacillus casei* Strains ATCC 334 and BL23, and the Partially Sequenced *Lactobacillus paracasei* Strain 8700:02:00[a]

Insertion Site (Homologue)	Strain	Size (kb)	DNA Coordinates (Predicted ORFs)	G+C Content (%)	$\delta^* \times 10^3$[b]	Other Features
1_Arg	ATCC 334	14.3	501,132..515,488 (LSEI_0486.. LSEI_0513)	42.6	54.8	Integrase, IS*30* family transposase, phage anti-repressor protein
	BL23	13.3	559,122..572,423 (LCABL_05480.. LCABL_05710)	42.4	49.4	Integrase, rep protein, anti-repressor
	8700:02:00	17.8	536,996..554,818	44.5	55.6	Recombinase, phage protein
28_Leu	ATCC 334	2.3	831,170..833,544 (LSEI_0857.. LSEI_0858)	48.3	68.8	Degenerate transposase, transposase
	BL23	2.3	893,693..896,067 (LCABL_09220.. LCABL_09240)	48.3	70.2	Degenerate transposase, transposase
	8700:02:00	0	—	—	—	—
32_Leu	ATCC 334	0	—	—	—	—
	BL23	34.1	928,685..962,820 (LCABL_09550.. LCABL_10020)	45.6	71.4	Integrase, phage protein
	8700:02:00	34.2	876,487..910,668	44.9	70.5	Integrase, phage protein
33_Arg	ATCC 334	1.1	946,339..947,505 (LSEI_0965)	38.5	72.4	Hypothetical protein
	BL23	35.3	1,043,259..1,078,647 (LCABL_10790.. LCABL_11280)	45.3	69.3	Integrase, phage protein, potassium/ion channel protein
	8700:02:00	2.4	990,233..992,622	38.6	99.9	Degenerate transposase
LSEI_0229[c]	ATCC 334	14.8	226,644..241,440 (LSEI_0230.. LSEI_0240)	43.3	36.8	Transposase, glycosyltransferase
(LCABL_02290)	BL23	11.8	229,044..240,804 (LCABL_02300.. LCABL_02380)	42.1	32.3	—
	8700:02:00	0.2	256,561..256,786	—	—	—
LSEI_1244[c]	ATCC 334	2.3	1,219,387..1,221,747 (LSEI_1241.. LSEI_1243)	42.8	88.3	IS*L1* transposase

(LCABL_14750)	BL23	11.9	1,385,428..1,397,364 (LCABL_14580.. LCABL_14740)	39.2	75.9	Transposase, degenerate transposase
	8700:02:00	1.2	1,260,051..1,261,286	42.4	84.7	—
LSEI_1979[c]	ATCC 334	48.5	1,902,248..1,950,810 (LSEI_1906.. LSEI_1978)	44.5	54.5	Integrase, transposase, phage Lca1
(LCABL_21540)	BL23	20.7	2,058,007..2,078,741 (LCABL_21270.. LCABL_21530)	42.6	52.5	Integrase homolog, transposase
	8700:02:00	7.8	1,935,918..1,943,769	40.0	76.2	Transposase
LSEI_2793[c]	ATCC 334	0	—	—	—	—
(LCABL_29990)	BL23	2.4	2,946,187..2,948,618 (LCABL_29970.. LCABL_29980)	42.3	103.5	Transposase
	8700:02:00	0	—	—	—	—
LSEI_1093[c]	ATCC 334	0	—	—	—	—
(LCABL_12560)	BL23	1.8	1,212,697..1,214,467 (LCABL_12570.. LCABL_12580)	50.0	114.1	IS*Sth1* transposase
	8700:02:00	0	—	—	—	—

[a]The chromosome lengths of *L. casei* ATCC 334 and BL23 are 2895.26 kb and 3079.2 kb, respectively. Eighty-nine contigs of *L. paracasei* 8700:02:00 available at the time of analysis were compared with the ATCC 334 genome using PipMaker (Schwartz et al., 2000) and ordered by coordinates of best alignment to produce a "virtual" genome with the size of 2981.34 kb.

[b]Dinucleotide bias analysis was adapted from the method proposed by Karlin [2001]. The value δ^* denotes the dinucleotide relative abundance difference between the island fragment and the reference genome. The high δ^* values of these fragments indicate likely heterologous origins. To avoid statistical fluctuations, δ^* was not determined ('—') for GI fragments less than 1000 bp. The genome-averaged δ^* value and standard deviation (SD) was calculated using a 20-kb nonoverlapping, sliding window along the entire genome sequence. The genome-averaged parameters for the three genomes are as follows: *L. casei* ATCC 334, $G + C = 46.6\%$; $\delta^* = 24.72$(SD = 11.84); *L. casei* BL23, $G + C = 46.3\%$, $\delta^* = 27.4$(SD = 12.84); *L. paracasei* 8700:02:00, $G + C = 46.2\%$, $\delta^* = 25.11$(SD = 13.41).

[c]The five non-tRNA insertion hotspots (*LSEI_0229, LSEI_1093, LSEI_1244, LSEI_1979*, and *LSEI_2793*) subjected to tRNAcc analysis were identified based on previously reported comparative genomic hybridization data derived following analysis of 21 *L. casei* strains against the *L. casei* ATCC 334 array [Cai et al., 2009].

and graphical forms using the newly incorporated visualization features found with IslandScreen.

30.3.2 The Use of mGenomeSubtractor to Highlight Genomic Mosaicism, Examine Strain Variable Regions and Define the Species Gene Pool of *Acinetobacter baumannii*

The bacterial chromosomal backbone comprises a mosaic of core genomic sequences and horizontally acquired DNA that has been termed the accessory genome or mobilome [Ou et al., 2005]. The latter flexible genomic repertoire is drawn from a gene pool dispersed amongst species-associated episomal plasmids, transposons, integrons, prophages, and a growing list of GIs. mGenomeSubtractor performs a simple and intuitive in silico "subtractive hybridization" by comparing selected closely related genomes to generate a list of strain-specific CDS which may provide clues as to the phenotype, specific environmental adaptation, and/or disease-association linked with a particular bacterium. As previously, mGenomeSubtractor uses H-value criteria that reflect the degree of similarity in terms of the length of match and degree of identity at a nucleotide level between each CDS examined and the set of comparator genomes examined [Fukiya et al., 2004]. In the example shown in Figure 30.2, the genome map of *A. baumannii* strain AYE was color-coded by mGenomeSubtractor based on the number of comparator genomes that shared each AYE CDS. Gene sequences conserved across all six fully sequenced *A. baumannii* genomes are shown in black with the legend label "5" to indicate that the corresponding AYE CDS has homologues in the other five genomes examined. *A. baumannii* is an opportunistic human pathogen which is an increasingly significant cause of hospital-associated infections, quite probably due to its frequent multi-antibiotic resistant phenotype [Towner, 2009]. In addition, there are growing data to support the notion that the *Acinetobacter* genus exhibits highly variable genomes and that interspecies, intergenus, interfamily, interorder, and even more remote DNA migration events may be relatively frequent features of this genus [Fournier et al., 2006; Vallenet et al., 2008]. These observations raise the realistic possibility of the future emergence of frankly pathogenic *A. baumannii* strains that may, like many of their predecessors, even assume endemic status [Peleg et al., 2008]. All 3607 annotated AYE CDS were analyzed against the five completely sequenced *A. baumannii* genomes. If all resulting H values for a given AYE CDS were ≤ 0.42, the CDS was classified as AYE strain-specific. Using this approach, 123 AYE CDS were found to be AYE-specific, offering significant potential for the fine-tuning of traits and/or for AYE-specific adaptations. The numbers of strain-specific CDS corresponding to each of the other five genomes and the set of 17 *A. baumannii* plasmids examined are shown in **Fig. 30.3**.

mGenomeSubtractor analysis also allowed for the ready definition of the *A. baumannii* species known gene pool based on the six completely sequenced *A. baumannii* chromosomes and 17 completely sequenced *A. baumannii* plasmids as the input. Firstly, all the 3351 annotated CDS present on the *A. baumannii* ATCC 17978 chromosome were included in the gene pool. Then each annotated CDS on the *A. baumannii* SDF chromosome was used as a query in a BLASTN similarity search against the ATCC 17978 chromosomal sequence that had been stored in the gene pool. The 553 SDF-specific CDS identified by mGenomeSubtractor (H values ≤ 0.42) based on this two chromosome comparison were added to the gene pool. A similar reiterative process was used to analyze each of the remaining four fully sequenced *A. baumannii* chromosomes, leading to the sequential inclusion of a further 1035 genes into the growing gene pool. Next, the 247 unique CDS present on 17 completely sequenced *A. baumannii* plasmids currently available in GenBank, pAB1, pAB2, p1ABSDF, p2ABSDF, p3ABSDF, pACICU1, pACICU2, pAB0057, p1ABAYE, p2ABAYE, p4ABAYE, p3ABAYE, pABIR, pMMCU2, pMMA2, pABVA1, and pMAC were also identified and placed into the gene pool. Finally, single representative unique CDSs were selected from the gene pool by removing any redundant "duplicated" genes as determined using the H-value criteria (H values >0.81). Hence, based on mGenomeSubtractor analysis of the six complete *A. baumannii* chromosomes and 17 completely sequenced *A. baumannii* plasmids, 4933 unique CDS, including 4695 chromosomal and 238 plasmid-borne CDS, were defined in the presently known *A. baumannii* gene pool, a number that was 33–89% greater than that of the unique gene complement of each of the six fully sequenced *A. baumannii* chromosomes under study.

30.4 OTHER APPLICATIONS

30.4.1 Rapid Development of a *Salmonella enterica* Serovar Paratyphi A Multiplex PCR Assay: A MobilomeFINDER-Enabled Pipeline for Future Strain-, Serovar-, or Pathotype-Specific Assays

Ou et al. [2007b] described a simple MobilomeFINDER-enabled strategy that utilized mGenomeSubtractor to identify protein coding sequences that were present in

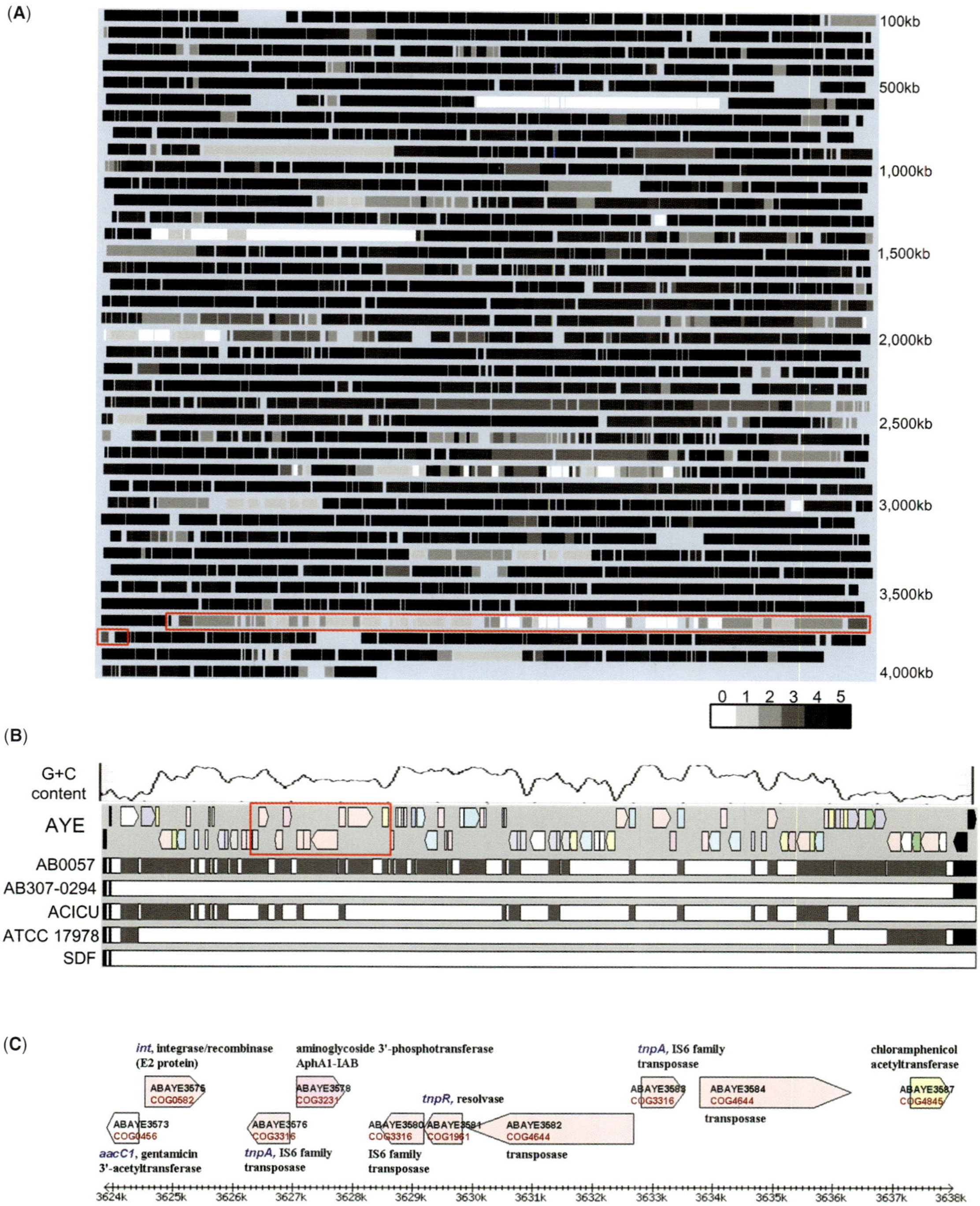

Figure 30.2 Comparative genomics based in silico "subtractive hybridization" using the mpiBLAST-based mGenomeSubtractor tool. **(A)** Genome map of *Acinetobacter baumannii* strain AYE with CDS color-coded based on the number of comparator *A. baumannii* genomes identified as harboring a nucleotide sequence-conserved homologue; CDS shown in black (5) are conserved across the five other full-sequenced *A. baumannii* comparator genomes, while at the other extreme those shown in white (0) are unique to strain AYE. AYE strain-specific CDS were identified based on an H-value cutoff of less than 0.42. The H value reflects the degree of similarity in terms of the length of match and the degree of identity at a nucleotide level between the matching reference genome sequence and the CDS examined. **(B)** An expanded view of the hypervariable region highlighted by a red rectangle in part A corresponding to the 86-kb Tn*AbaR1* resistance mega-transposons of AYE and its 2 kb upstream/downstream flanking regions. *A. baumannii* strain AYE CDS are color-coded based on their COG assignment. The five comparator genomes are shown below. Gray bars indicate the extent of AYE sequences within this region that are also present in the individual comparator genomes. The black bars at the end represent the ends of the region selected for the expanded view. A G+C profile of the selected AYE region is shown topmost. **(C)** A zoom-in view of the region of the AYE genome highlighted by the red rectangle in part B showing selected Tn*AbaR1* CDS coding for an integrase, transposases, and antibiotic resistance genes.

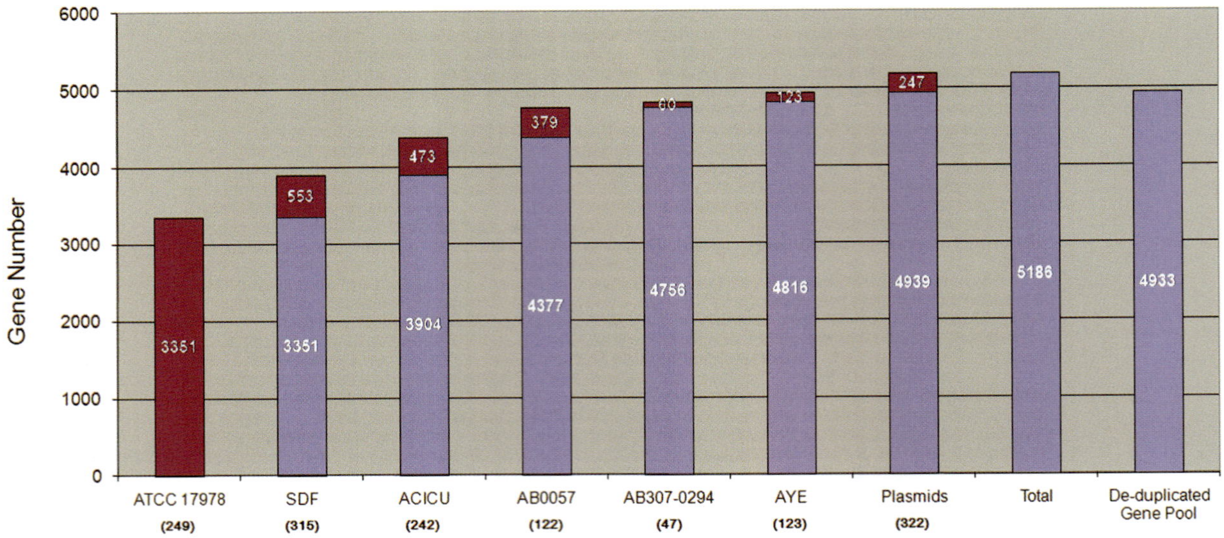

Figure 30.3 Use of mGenomeSubtractor to obtain a preliminary estimate of the known *Acinetobacter baumannii* species gene pool. As a first step, all the annotated chromosomal CDS of *A. baumannii* ATCC 17978 were added into the gene pool. Next, unique chromosomal CDS present in the five other completely sequenced *A. baumannii* genomes, SDF, ACICU, AB0057, AB307-0294, and AYE, were sequentially added to the growing gene pool following examination of each chromosome in turn in the order listed against the expanding gene pool. A similar analysis was undertaken to capture unique CDS found in 17 fully sequenced *A. baumannii* plasmids currently available in GenBank. Finally, single representative unique CDSs were selected from the gene pool by removing any redundant "duplicated" genes as determined using the H-value criteria (H values >0.81). The number of CDS specific to each of the six chromosomes and the set of 17 plasmids are shown below their labels in parentheses.

Salmonella Paratyphi A ATCC 9150 but not in any the available non-Paratyphi A *Salmonella* genome sequence information or the wider GenBank database. This in silico substractive hybridization yielded 43 ATCC 9150-unique CDS. However, 29 of these CDS were found to be variably present in 12 other Paratyphi A strains by examination of an available microarray-generated hybridization dataset. ArrayOme was used to map the remaining 14 CDS that were found to cluster into four widely separated genomic loci. Gene-specific primers were designed to target a single, empirically selected representative of each locus, and the four primer pairs were used in a multiplex PCR assay. Unexpectedly, *stkF*, one of the target genes selected, was found to exhibit a sequence polymorphism arising from the presence or absence of a duplicated 26-bp internal sequence. This finding led to the hypothesis that Stk fimbriae may be subjected to low-frequency, programmed DNA-facilitated phase variation driven by slipped strand mispairing and/or recombination events. Importantly, based on the appearance of one of two four-band pathognomonic signatures, the assay proved to be 100% sensitive and specific for Paratyphi A isolates despite no prior experimental validation, elegantly highlighting the potential of this approach as a pipeline for do-it-yourself development of strain-, serovar-, or pathotype-specific assays.

30.4.2 tRIP PCR-Based Discovery of *Klebsiella pneumoniae* Niche-Adaptation Islands

tRNAcc analysis of the first two *Klebsiella pneumoniae* genomes to be sequenced, MGH78578 & Kp342, identified five potential tRNA-associated integration hotspots in *K. pneumoniae*. Subsequent experimental tRIP PCR analysis of ~120 clinical isolates obtained from the United Kingdom and China using primers designed to anneal to conserved upstream and downstream flanks confirmed that the *pheV*, arg6, asn34 and tmRNA sites were frequently occupied by integrated DNA in these strains [Chen et al., 2009; van Aartsen, 2008; Zhang, 2011]. Chromosome walking, long-range PCR across occupied sites and/or resistance gene tagging followed by fosmid-based marker rescue have already led to the discovery of at least five novel tRNA-associated genomic islands including two islands coding for likely previously uncharacterized fimbrial structures, the KpGI-2 island that appears to confer a marked cAMP-dependent growth and/or viability phenotype that carries four novel genes plus a new variant of the *fic* gene family [Chen et al., 2009], and the small 3.7 kb KpGI-1 island coding for a putative acetyltransferase that is much more strongly associated with sputum isolates than blood, urine, or bile islands [Chen et al., 2009]. Collectively, these features suggest possible niche-adaptation roles for these islands and provide ample bioinformatics-derived substrate for targeted, hypothesis-driven experimental investigations.

30.4.3 Tn*AbaR1*-like Resistance Mega-transposons Are Widely Disseminated Among Clinical Multi-Drug-Resistant *Acinetobacter baumannii Strains*

In 2006, through full-length sequencing of two *Acinetobacter baumannii* genomes, Fournier et al. described AbaR1, a large 86-kb chromosomal island harboring more than 40 recognized resistance genes present in strain AYE [Fournier et al., 2006]. These authors also reported a much smaller, entirely unrelated island, integrated into an identical site within the *comM* gene in the second strain sequenced. Indeed, according to Fournier et al. [2006] this was an unexpected discovery. Using a targeted tRIP PCR strategy, we recently interrogated the *comM* loci of 50 multidrug-resistant *A. baumannii* isolates; 40 isolates failed to yield *comM* tRIP PCR amplicons suggestive of the presence of integrated islands within the gene. Subsequent chromosome walking into the predicted island from the 5′ and/or 3′ ends of the split *comM* gene confirmed that eight of ten genotypically distinct *A. baumannii* isolates predicted to carry *comM*-integrated elements did indeed carry such entities. However, to our surprise based on the available data, six of the eight newly discovered islands bore termini that were highly similar to those of AbaR1, a seventh possessed a divergent but closely related AbaR1-like terminus at the 5′-*comM* end, and only one appeared to be entirely novel [Shaikh et al., 2009]. These data suggested that >60% of clinical multidrug-resistant *A. baumannii* isolates harbored *comM*-borne AbaR1-like elements. Recent detailed bioinformatics analyses of the terminal segments of these elements drawn from multiple strains [Fournier et al., 2006; Post and Hall, 2009; Iacono et al., 2008; Adams et al., 2008], coupled with preliminary experimental investigations, have led us to hypothesize that these elements comprise a novel transposon family that is distantly related to Tn*7* [Rose, 2009]. Designated as Tn*AbaR1*-like transposons, typical members have accumulated multiple transposons, integrons, and resistance gene cassettes, often in a nested fashion, and, like the highly promiscuous Tn*7*, potentially exploit two distinct mobilization pathways: stable integration into a unique chromosomal site or homing onto conjugative plasmids for onward transfer to a new host [Peters and Craig, 2001; Parks and Peters, 2009]. We have hypothesized that this complex, well-orchestrated process is mediated by five novel Tn*AbaR1* transposition proteins and are currently investigating this idea.

30.4.4 Comparative Mapping of Members of the *Shigella* Resistance Locus Island Family: Clues to Improved Future Strategies

Luck et al. [2001] described the full-length characterization of the *Shigella* resistance locus island in *Shigella flexneri* 2a strain YSH6000. The SRL island was found to code for both multidrug resistance and a well-recognized virulence-associated attribute—the latter through the eight-gene *fec* locus that mediates ferric dicitrate uptake, thus offering an advantage under in vivo conditions of iron limitation. Based on a study of a collection of strains representative of the four *Shigella* species, SRL-like islands were subsequently shown to have largely, if not completely, displaced plasmids as the standard vehicle for dissemination of genes conferring resistance to the quartet of ampicillin, streptomycin, tetracycline, and chloramphenicol [Turner et al., 2003]. These findings prompted a comparative analysis of SRL-like islands in representative strains facilitated largely by an overlapping, tiling PCR approach [Turner et al., 2003; Al-Hasani et al., 2001]. We have applied a similar approach to study Tn*AbaR1*-like transposons; but faced with the much more substantial challenge of mapping large, highly divergent cargo-gene-bearing central spans of these islands, we have used mGenomeSubtractor to identify genes unique to each element, common to various subsets and those universally present amongst the full complement of Tn*AbaR1*-like elements that have been fully characterized to date. These data have been used to synthesize ~50 oligonucleotide primer pairs targeting known Tn*AbaR1*-like genes and selected other *A. baumannii* resistance genes, and they are currently being used in various semi-random permutations to facilitate a shot-gun, combinatorial, long-range PCR approach aimed at generating overlapping amplicons that span the full lengths of uncharacterized Tn*AbaR1*-like elements. Long-range PCR amplicons obtained will be subjected to additional PCR mapping and/or barcode-facilitated high-throughput sequencing to complete the comparative analysis dataset.

30.5 SUMMARY

Even with current high-throughput genomic sequencing facilities, it is not feasible to sequence hundreds of isolates to identify and decode the global gene pool accessible to a single bacterial species. Instead, we propose that the application of a strategy such as MobolimeFINDER, underpinned by ArrayOme and tRNAcc, followed by either random or targeted gene-discovery studies focused on selected strains bearing mobilome-rich regions, would make a major contribution to this effort. The highly intuitive mGenomeSubtractor tool, which offers visualization of the genomic mosaic, zoom-in options, and linkages to other databases, is ideal for the identification and exploration of the most dynamic regions of bacterial genomes. The tRIP PCR strategy that arises directly from tRNAcc analysis offers a powerful, readily accessible approach to allow for widespread engagement in the bacterial metagenome discovery game. Selection of strains representative of a greater diversity of ecological, geographic, temporal, animal–host, and/or disease-association categories could well result in the identification of dramatically more novel islands. Furthermore, initial analysis of a greater number of reference genomes by tRNAcc and mGenomeSubtractor and/or additional input from ArrayOme-analysed comparative hybridization datasets should allow for the identification of many more tRNA and non-tRNA potential hotspots, which in turn could be targeted by a broader tRIP PCR scan. Collectively, these approaches should not just help reveal the wider DNA blueprints of individual bacterial species but play a significant role in defining the functional contributions of bacterial mobilomes.

INTERNET RESOURCES

NCBI Microbial Genomes (http://www.ncbi.nlm.nih.gov/genomes/lproks.cgi)

The Genomes On Line Database (GOLD) (http://www.genomesonline.org/)

MobilomeFINDER (http://mml.sjtu.edu.cn/MobilomeFINDER/)

mGenomeSubtractor (http://bioinfo-mml.sjtu.edu.cn/mGS/)

Islander (http://kementari.bioinformatics.vt.edu/cgi-bin/islander.cgi)

xBASE (http://xbase.bham.ac.uk/)

webACT (http://www.webact.org/)

Mauve (http://gel.ahabs.wisc.edu/mauve/)

IS Finder (http://www-is.biotoul.fr/)

δρ–web (http://deltarho.amc.nl/)

PipMaker (http://pipmaker.bx.psu.edu/pipmaker/)

Primaclade (http://www.umsl.edu/services/kellogg/primaclade.html)

Acknowledgments

This work was supported by grants from National Natural Science Foundation of China to HYO (30700013, 30871345, and 30821005), as well as by the Royal Society—National Natural Science Foundation of China International Joint Program (2007/R3), British Society for Antimicrobial Chemotherapy (GA803), Action Medical Research (SP4255), and Medisearch to KR.

REFERENCES

Adams MD, Goglin K, Molyneaux N, Hujer KM, Lavender H, et al. 2008 Comparative genome sequence analysis of multidrug-resistant *Acinetobacter baumannii*. *J. Bacteriol*. **190**:8053–8064.

Al-Hasani K, Adler B, Rajakumar K, Sakellaris H. 2001. Distribution and structural variation of the she pathogenicity island in enteric bacterial pathogens. *J. Med. Microbiol*. **50**:780–786.

Ansorge WJ. 2009. Next-generation DNA sequencing techniques. *N. Biotechnol*. **25**:195–203.

Boyd EF, Almagro-Moreno S, Parent MA. 2009. Genomic islands are dynamic, ancient integrative elements in bacterial evolution. *Trends Microbiol*. **17**:47–53.

Cai H, Thompson R, Budinich MF, Broadbent JR, Steele JL. 2009. Genome sequence and comparative genome analysis of *Lactobacillus casei*: Insights into their niche-associated evolution. *Genome Biol. Evol*. **1**:239–257.

Canchaya C, Fournous G, Brussow H. 2004. The impact of prophages on bacterial chromosomes. *Mol. Microbiol*. **53**:9–18.

Chen N, Ou HY, van Aartsen JJ, Jiang X, Li M, et al. 2009. The pheV phenylalanine tRNA gene in *Klebsiella pneumoniae* clinical isolates is an integration hotspot for possible niche-adaptation Genomic Islands. *Curr. Microbiol.*, **60**:210–216.

Churchward G. 2008. Back to the future: The new ICE age. *Mol. Microbiol*. **70**:554–556.

Darling AC, Mau B, Blattner FR, Perna NT. 2004. Mauve: Multiple alignment of conserved genomic sequence with rearrangements. *Genome Res*. **14**:1394–1403.

Dobrindt U, Hacker J. 2001. Whole genome plasticity in pathogenic bacteria. *Curr. Opin. Microbiol*. **4**:550–557.

Dorrell N, Hinchliffe SJ, Wren BW. 2005. Comparative phylogenomics of pathogenic bacteria by microarray analysis. *Curr. Opin. Microbiol*. **8**:620–626.

Field D, Wilson G, van der Gast C. 2006. How do we compare hundreds of bacterial genomes? *Curr. Opin. Microbiol*. **9**:499–504.

Fleischmann RD, Adams MD, White O, Clayton RA, Kirkness EF, et al. 1995. Whole-genome random sequencing and assembly of *Haemophilus influenzae* Rd. *Science* **269**:496–512.

Fournier PE, Vallenet D, Barbe V, Audic S, Ogata H, et al. 2006. Comparative genomics of multidrug resistance in *Acinetobacter baumannii*. *PLoS Genet*. **2**:e7.

Frost LS, Leplae R, Summers AO, Toussaint A. 2005. Mobile genetic elements: The agents of open source evolution. *Nat. Rev. Microbiol*. **3**:722–732.

Fukiya S, Mizoguchi H, Tobe T, Mori H. 2004. Extensive genomic diversity in pathogenic *Escherichia coli* and *Shigella* Strains revealed by comparative genomic hybridization microarray. *J. Bacteriol*. **186**:3911–3921.

Fux CA, Shirtliff M, Stoodley P, Costerton JW. 2005. Can laboratory reference strains mirror "real-world" pathogenesis? *Trends Microbiol*. **13**:58–63.

Gadberry MD, Malcomber ST, Doust AN, Kellogg EA. 2005. Primaclade—A flexible tool to find conserved PCR primers across multiple species. *Bioinformatics* **21**:1263–1264.

Hacker J, Blum-Oehler G, Muhldorfer I, Tschape H. 1997. Pathogenicity islands of virulent bacteria: Structure, function and impact on microbial evolution. *Mol. Microbiol*. **23**:1089–1097.

Iacono M, Villa L, Fortini D, Bordoni R, Imperi F, et al. 2008. Whole-genome pyrosequencing of an epidemic multidrug-resistant *Acinetobacter baumannii* strain belonging to the European clone II group. *Antimicrob. Agents Chemother*. **52**:2616–2625.

Karlin S. 2001. Detecting anomalous gene clusters and pathogenicity islands in diverse bacterial genomes. *Trends Microbiol*. **9**:335–343.

Konstantinidis KT, Ramette A, Tiedje JM. 2006. The bacterial species definition in the genomic era. *Philos. Trans. R. Soc. Lond. B. Biol. Sci*. **361**:1929–1940.

Lima-Mendez G, Toussaint A, Leplae R. 2007. Analysis of the phage sequence space: The benefit of structured information. *Virology* **365**:241–249.

Liolios K, Chen IM, Mavromatis K, Tavernarakis N, Hugenholtz P, et al. 2009. The Genomes On Line Database (GOLD) in 2009: status of genomic and metagenomic projects and their associated metadata. *Nucleic. Acids Res.*. **38**:D346–D354.

Luck SN, Turner SA, Rajakumar K, Sakellaris H, Adler B. 2001. Ferric dicitrate transport system (Fec) of *Shigella flexneri* 2a YSH6000 is encoded on a novel pathogenicity island carrying multiple antibiotic resistance genes. *Infect. Immun*. **69**:6012–6021.

MacLean D, Jones JD, Studholme DJ. 2009. Application of 'next-generation' sequencing technologies to microbial genetics. *Nat. Rev. Microbiol*. **7**:287–296.

Mantri Y, Williams KP. 2004. Islander: A database of integrative islands in prokaryotic genomes, the associated integrases and their DNA site specificities. *Nucleic Acids Res*. **32**:D55–D58.

Medini D, Donati C, Tettelin H, Masignani V, Rappuoli R. 2005. The microbial pan-genome. *Curr. Opin. Genet. Dev*. **15**:589–594.

Ou HY, Smith R, Lucchini S, Hinton J, Chaudhuri RR, et al. 2005. ArrayOme: a program for estimating the sizes of microarray-visualized bacterial genomes. *Nucleic Acids Res*. **33**:e3.

Ou HY, Chen LL, Lonnen J, Chaudhuri RR, Thani AB, et al. 2006. A novel strategy for the identification of genomic islands by comparative analysis of the contents and contexts of tRNA sites in closely related bacteria. *Nucleic Acids Res*. **34**:e3.

Ou HY, He X, Harrison EM, Kulasekara BR, Thani AB, et al. 2007a. MobilomeFINDER: web-based tools for in silico and experimental discovery of bacterial genomic islands. *Nucleic Acids Res*. **35**:W97–W104.

Ou HY, Ju CT, Thong KL, Ahmad N, Deng Z, et al. 2007b. Translational genomics to develop a *Salmonella enterica* serovar Paratyphi A multiplex polymerase chain reaction assay. *J. Mol. Diagn*. **9**:624–630.

Parks AR, Peters JE. 2009. Tn7 elements: engendering diversity from chromosomes to episomes. *Plasmid* **61**:1–14.

Peleg AY, Seifert H, Paterson DL. 2008. *Acinetobacter baumannii:* Emergence of a successful pathogen. *Clin. Microbiol. Rev*. **21**:538–582.

Peters JE, Craig NL. 2001. Tn*7*: Smarter than we thought. *Nat. Rev. Mol. Cell. Biol*. **2**:806–814.

Post V, Hall RM. 2009. AbaR5, a large multiple-antibiotic resistance region found in *Acinetobacter baumannii*. *Antimicrob Agents Chemother*. **53**:2667–2671.

Rajakumar K, Sasakawa C, Adler B. 1997. Use of a novel approach, termed island probing, identifies the *Shigella flexneri* she pathogenicity island which encodes a homolog of the immunoglobulin A protease-like family of proteins. *Infect. Immun*. **65**:4606–4614.

Riley MA, Lizotte-Waniewski M. 2009. Population genomics and the bacterial species concept. *Methods Mol. Biol*. **532**: 367–377.

Rose A. 2009. TnAbaR1: A novel Tn*7*-related transposon in *Acinetobacter baumannii* that contributes to the accumulation and dissemination of large repertoires of resistance genes. *Biosci. Horizons* **3**:40–48.

Schuler GD. 1997. Sequence mapping by electronic PCR. *Genome Res*. **7**:541–550.

Schwartz S, Zhang Z, Frazer KA, Smit A, Riemer C, et al. 2000. PipMaker—a web server for aligning two genomic DNA sequences. *Genome Res*. **10**:577–586.

Shaikh F, Spence RP, Levi K, Ou HY, Deng Z, et al. 2009. ATPase genes of diverse multidrug-resistant *Acinetobacter baumannii* isolates frequently harbour integrated DNA. *J. Antimicrob. Chemother*. **63**:260–264.

Shao Y, He X, Harrison EM, Tai C, Ou HY, Rajakumar K, Deng Z, 2010. mGenomeSubtractor: a web-based tool for parallel in silico subtractive hybridization analysis of multiple bacterial genomes. *Nucleic. Acids. Res*. **38**:W194–200.

Tettelin H, Riley D, Cattuto C, Medini D. 2008. Comparative genomics: The bacterial pan-genome. *Curr. Opin. Microbiol*. **11**:472–477.

Towner KJ. 2009. *Acinetobacter*: An old friend, but a new enemy. *J. Hosp. Infect*. **73**:355–363.

Turner SA, Luck SN, Sakellaris H, Rajakumar K, Adler B. 2003. Molecular epidemiology of the SRL pathogenicity island. *Antimicrob. Agents Chemother*. **47**:727–734.

Vallenet D, Nordmann P, Barbe V, Poirel L, Mangenot S, et al. 2008. Comparative analysis of Acinetobacters: Three genomes for three lifestyles. *PLoS One* **3**:e1805.

van Aartsen JJ. 2008. The *Klebsiella pheV* tRNA locus: a hotspot for integration of alien genomic islands. *Biosci. Horizons* **1**: 51–60.

Wolfgang MC, Kulasekara BR, Liang X, Boyd D, Wu K, et al. 2003. Conservation of genome content and virulence determinants among clinical and environmental isolates of *Pseudomonas aeruginosa*. *Proc. Natl. Acad. Sci. USA* **100**:8484–8489.

Zhang J, van Aartsen JJ, Jiang X, Shao Y, Tai C, He X, Tan Z, Deng Z, Jia S, Rajakumar K, Ou HY, 2011. Expansion of the known *Klebsiella pneumoniae* species gene pool by characterization of novel alien DNA islands integrated into tmRNA gene sites. *J, Microbiol Methods*. **84**:283–289.

Chapter 31

Sequence-Based Characterization of Microbiomes by Serial Analysis of Ribosomal Sequence Tags (SARST)

ZHONGTANG YU AND MARK MORRISON

31.1 INTRODUCTION

The ecophysiology of microbial habitats continues to be of great interest to basic and applied microbiologists alike, and great efforts have been made to characterize the diversity, composition, and structure of microbiomes across a great range of environments (see Vol. II). Based on the evidence accumulated by far, it is recognized that microbiomes present in most environments, either natural or managed, are typically of great diversity and complex composition and structure [Curtis et al., 2002; Schloss and Handelsman, 2004]. The compositions have been demonstrated to be complex, as each microbiome can contain up to several thousand different microbial species (primarily bacteria), while the structure is complex, as different species typically exist at varying abundance over many orders of magnitude [Ashby et al., 2007]. Indeed, DNA reassociation analysis of denatured soil microbiome DNA samples revealed the presence of approximately 4000 [Torsvik et al., 1990] to 16,000 [Sandaa et al., 1999; see also Chapter 2 Vol. I] completely different genomes of standard soil bacteria per gram of wet soil, which are 200–800 times greater than those measured from isolated strains [Handelsman et al., 2002]. Cloning and sequencing of SSU rRNA (primarily 16S rRNA) gene has also revealed that soils can have several thousands or more phylogenetic species (or OTUs that share $\geq$97% sequence identity) [Fierer et al., 2007]. Recent studies using massive parallel pyrosequencing have provided further evidence that soil samples have much greater diversity than previously thought [Jones et al., 2009; Roesch et al., 2007]. Oceans, previously thought to have limited microbial diversity, were also shown to contain many thousands of phylogenetic species of microbes [Brown et al., 2009; Gilbert et al., 2009; Huber et al., 2007; Sogin et al., 2006; see also Chapters 24 and 26, Vol. II]. Managed environments, which typically have unique selective conditions, such as aeration basins [Sanapareddy et al., 2009; McLellan et al., 2010], anaerobic digesters [Krause et al., 2008; Riviere et al., 2009], the rumen of ruminant animals [Brulc et al., 2009; Kim et al., 2010; see also Chapter 23, Vol. II], and gastrointestinal tack of animals [Callaway et al., 2009; Dowd et al., 2008] and humans [Andersson et al., 2008.; Eckburg et al., 2005; see also Chapters 18–20, Vol. II], are also teeming with billions of microbes associated with several thousand species.

Full characterization of diversity is a prerequisite to understanding the function of microbiomes and to exploring and manipulating them for beneficial applications. However, it has been well-documented that cultivation-based methods only allow access to a very small portion of the microbiome in any environment [Ward et al., 1990]. Since the late 1980s, microbiologists have used nucleic acid-based molecular biology techniques in an attempt to capture the full microbial diversity [Green and Keller, 2006; Rappe and Giovannoni, 2003]. Pioneered by Pace et al. [1985], sequencing and phylogenetic analysis of 16S rRNA gene clones that were directly amplified by PCR from microbiome DNA samples became the primary method used to characterize diversity

Handbook of Molecular Microbial Ecology, Volume I: Metagenomics and Complementary Approaches, First Edition. Edited by Frans J. de Bruijn.

and composition of microbiomes. Over the past 25 years, the microbiomes in almost all types of environments have been investigated using this approach, which greatly expanded our perspectives on all environmental microbiomes [Egan et al., 2008; Hall, 2007; see also Vol. II]. However, it became increasingly evident that the datasets produced in these studies are too small to capture the full diversity [Bent and Forney, 2008; Tiedje et al., 1999]. At least hundreds of thousands of clones would need to be sequenced to characterize the full diversity of any microbiome because different species exist at very different species evenness, often over many orders of magnitude [Ashby et al., 2007; Huber et al., 2007]; and it is too costly and labor-intensive to sequence that many SSU rRNA gene sequences using Sanger technology. Thus, new methods were sought to substantially improve upon the traditional one-clone–one-sequence method with respect to throughput capacity and cost-efficiency. One strategy to reduce the cost of DNA sequencing-based analysis of microbial diversity is to sequence concatemers of SSU rRNA gene tags using the serial analysis of ribosomal sequence tags (SARST) [Kysela et al., 2005; Neufeld et al., 2004ab; Yu et al., 2006]. This approach was adapted from the serial analysis of gene expression (SAGE) [Velculescu et al., 1995], which substantially improves the analysis of gene expression in Eukaryotes [Carulli et al., 1998].

31.2 PROCEDURES OF SERIAL ANALYSIS OF RIBOSOMAL SEQUENCE TAGS (SARST)

The entire process of SARST (**Fig. 31.1**) consists of the following major steps: (i) amplification of one hypervariable region of SSU rRNA genes using two universal primers, (ii) digestion of the PCR products to cut off the primers, (iii) purification and concatenation of individual ribosomal sequence tags (RSTs), (iv) gel sizing, end repair, and cloning of the concatemers, and (v) sequencing of cloned RST concatemers and phylogenetic analysis of individual RSTs. The detailed procedures have been described elsewhere [Kysela et al., 2005; Yu et al., 2006]. Here we describe the major steps, alternatives, and cautions when warranted.

31.2.1 PCR Amplification

One hypervariable region is amplified using a pair of universal primers, each of which has an extension region that contains a type IIs restriction endonuclease recognition site, and the most 5′ nucleotide of this extension region is labeled with at least one biotin or biotin-tetraethyleneglycol (biotin-TEG) molecule. Theoretically, any one of the nine hypervariable (V) regions can be targeted in SARST, but different V regions have different sequence length and sequence divergence. Calculated from the group of representative 16S rRNA gene sequences archived in RDP Release 8.1, V1 was shown to have the highest sequence divergence [Yu and Morrison, 2004]. Conceivably, the most divergent hypervariable region should provide the highest phylogenetic resolution. Indeed, the original SARST was developed to target the V1 region [Neufeld et al., 2004a, 2004b]. The original SARST targeting V1 was modified by using new primer extensions by Yu et al. [2006] and referred to as SARST-V1. The V6 was also targeted in SARST analysis (referred to as SARST-V6) [Kysela et al., 2005]. We suggest that the same nomenclature is used when other V regions are targeted in SARST analysis. The primers used to target these two V regions are listed in **Table 31.1**.

To obtain adequate amounts of PCR products, four 50-μl PCR reactions need to be performed from each microbiome DNA sample. The quality and quantity of the PCR products should be evaluated using PAGE using an 8%T (19:1) mini gel. Subsequently, the PCR products can be purified using the QIAquick PCR Purification Kit (QIAGEN, Valencia, CA), or ethanol precipitation following one extraction with phenol/chloroform. It should be noted that hot-start PCR using either hot-start *Taq* DNA polymerase or hot-start dNTP should be used in the amplification to reduce the formation of primer dimers, which can contaminate the RSTs.

31.2.2 Digestion of PCR Products and Primer Removal

The purified PCR products are digested with the type IIs restriction endonuclease that recognizes the recognition site within the primer extension. Alternatively, the PCR products can be further purified using streptavidin-coated magnetic beads, and then the bead-bound PCR products are digested. Although the latter choice can remove contaminating DNA templates used in the PCR, digestion efficiency is typically reduced by the presence of the beads. This reduced digestion efficiency compounds the inherent low efficiency of type IIs restriction endonuclease cleavage, leading to a reduced amount of released RSTs that can be concatenated. The digestion efficiency can be assessed by running a small aliquot of the digest on a PAGE gel. If needed, a cocktail containing the fresh type IIs restriction endonuclease and the buffer can be added to further the digestion.

The released RSTs are separated from the primers using streptavidin-coated magnetic beads, such as the Dyna 280 beads (Dynal, Oslo, Norway), which immobilize the primers that have a biotin label at the 5′ end. This step is straightforward, but we noticed that dual biotin

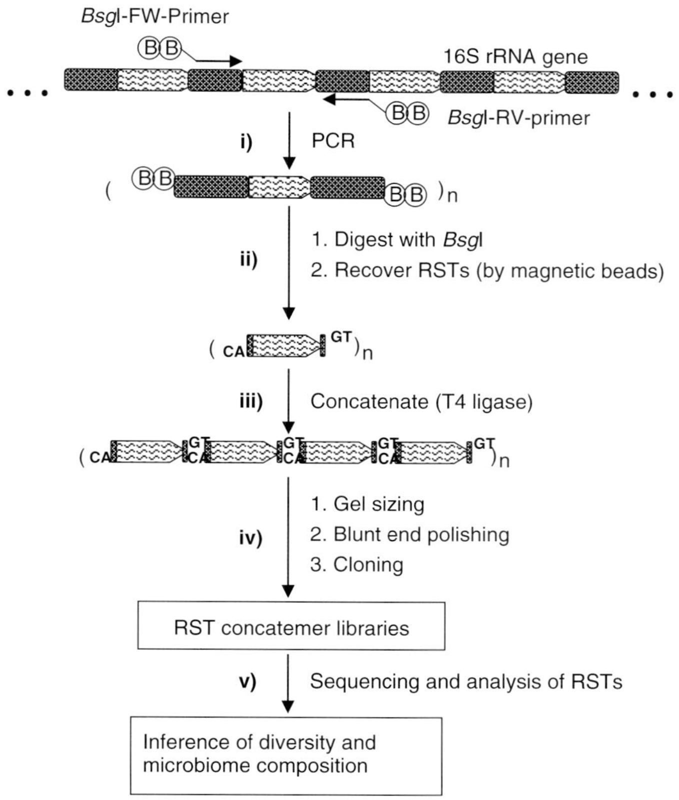

Figure 31.1 Schematic of the SARST technology. BB, dual biotin label added to the 5′ end of the primers. (Modified from Yu et al. [2006]).

labels at the 5′ end significantly increased the efficiency removing the primers [Neufeld et al., 2004b; Yu et al., 2006]. Alternatively, single biotin-TEG molecules can be conjugated to the 5′ end of each of the primers that facilitate the primer removal by streptavidin-coated beads [Kysela et al., 2005].

31.2.3 Concatenation of Individual RSTs

Each of the freed RSTs has one 2-nt 3′ overhang at either terminus, and these overhangs facilitate annealing of individual RSTs in hand-to-tail orientation in series (Fig. 31.1). Consequently, individual RSTs can be ligated head to tail to form concatemers in 5′ to 3′ orientation. Because of the short overhangs and the desire to form long (2–3 kb) concatemers, the concatenation is often performed overnight.

It should be noted that the original SARST procedures [Neufeld et al., 2004b] used two primers that contain an extension region that does not produce complementary 3′ overhangs following the digestion by the two different type IIs restriction endonucleases (*Bsg*I and *Bpm*I). Two adapters thus need to be ligated to both ends of the freed RSTs, and another round of digestion using type IIs restriction endonucleases (with *Spe*I and *Nhe*I) was needed to produce the complementary overhangs that ensure head-to-tail concatenation of individual RSTs. The new primers used in SARST-V1 and SARST-V6 were designed to place the recognition site of the type IIs restriction endonuclease in such a location that allows for compatible 3′ overhangs that enable head-to-tail concatenation of multiple RSTs. Thus, adapters are no longer required. This approach simplified the original SARST procedures. When new primers are designed to target other V regions, the same strategy should be followed. Additionally, the recognition site of the type IIs restriction endonuclease should be at such a distance from the 3′ end that digestion of the PCR products should leave at least 5 base pairs of the primers at each end of the freed RSTs. These conserved base pairs will serve as the boundary when individual RSTs are delineated from the sequenced concatemers.

31.2.4 Gel Sizing, End Repair, and Cloning of Concatemers

The concatenation of individual RSTs produces concatemers of varying lengths. If not sized, the short concatemers are cloned more efficiently than long ones. Thus, the concatemers of 0.5–2 kb are typically sized using gel (either agarose or polyacrylamide) electrophoresis. The selected concatemers can be readily recovered from the gel slice using commercial kits, such as the MiniElute Gel

Table 31.1 Primers and Linkers Used in SARST

Primers/ adapters	Sequences (5′ → 3′)	Targeted V Region	Reference
Bac64f-*Bpm*I	5′-Dual biotin-**TTTACCTGGAGC**CTWANRCATGCAAGTCG	V1	Neufeld, et al. [2004b]
Bac104r-*Bsg*I	5′-Dual biotin-**TTGCTGTGCAG**TACKCACCCGTBYGCC		
Linker A1 (*Spe*I cutting site)	5′-pCTAGTACGTGCTGGT		
Linker A2 (*Spe*I cutting site)	5′-dual biotin-AACACCAGCACGTACTAGTC		
Linker B1 (*Nhe*I cutting site)	5′-pCTAGCAACGTGCTGGT		
Linker B2 (*Nhe*I cutting site)	5′-Dual biotin-AACACCAGCACGTTGCTAGCC		
BsgI-Bact64f	5′-Dual biotin-**TTTGACCGTGCAG**CYTAAYRCATGCAAGTCG	V1	Yu, et al. [2006]
BsgI-Bact109r1	5′-Dual biotin-**TTTGACCGTGCAG**YYCACGYGTTACKCACCCGT		
SARST967F	5′-Biotin-TEG-**TTATTTTAGTGCAGTTAATA**CAACGCGAAGAACCTTACC	V6	Kysela et al. [2005]
SARST1046R	5′-Biotin-TEG-**TTATTTTAAGTGCAG**CAGCCATGCAVCACCT		

[a]Bold nt represent extension, and the underlined nt are the recognition sites of the type IIs restriction endonuclease.

Extraction Kit (QIAGEN). The purified concatemers are subjected to end repair by T4 DNA polymerase. The concatemers are then cloned by ligation into a cloning vector (e.g., pZerO-2.1 from Invitrogen and pSmartLCKan from Lucigen) that has been digested with a blunt-end restriction endonuclease. Alternatively, an adenine overhang can be added to each end of each concatemer so that the concatemers can be cloned using the TOPO TA cloning kit (Invitrogen) [Kysela et al., 2005]. Direct cloning of the blunt-ended concatemers might be preferred because Koehl et al. [2003] noted that it increased cloning efficiency of SAGE concatemers, which are essentially the same as RST concatemers in length.

31.2.5 Sequencing and Phylogenetic Analysis of Cloned RST Concatemers

The cloned concatemers are sequenced using the Sanger DNA sequencing technology. A typical Sanger sequencing read (greater than 500 bp) can determine the sequence of many RSTs, depending on the length of the V region targeted. When the V1 was targeted in SARST-V1, the longest sequenced RST concatemer contained 19 individual RSTs [Yu et al., 2006]. Individual RSTs are then delineated using the conserved base pairs that flank individual RSTs. The individual RSTs can be first grouped into OTUs and then compared to databases [Neufeld et al., 2004b; Poitelon et al., 2009; Yu et al., 2006], or compared to databases without grouping [Kysela et al., 2005]. BLASTn and SEQ MATCH are the two programs that can be used to compare RSTs to the sequences archived in GenBank (http://www.ncbi.nlm.nih.gov/) and RDP (http://rdp.cme.msu.edu/; see also Chapter 36, Vol. I), respectively. The new program ESPRIT (for details see Chapter 53, Vol. I) should also help to analyze the large number of RSTs produced by SARST. Other phylogenetic tools, such as Mothur [Schloss et al., 2009] and UniFrac [Lozupone et al., 2006], can also be used in RST analysis. Most of the RST datasets produced in previous studies can be found in either the NCBI Gene Expression Omnibus (GEO) database [Ashby et al., 2007; Neufeld et al., 2004b; Yu et al., 2006] or the Sequence Read Archive (SRA) [Huber et al., 2007].

There are some challenges in grouping RSTs into OTUs. When full-length 16S rRNA gene sequences are grouped or assigned to species, sequences sharing 97% or greater sequence identity are conventionally grouped into the same species [Ludwig et al., 1998; Stackebrandt and Goebel, 1994]. This 3% dissimilarity is, however, not distributed evenly along the 16S rRNA gene but is concentrated primarily in the nine V regions [Stackebrandt and Goebel, 1994]. Moreover, some of the V regions contribute more than others to the overall sequence divergence. As such, different cutoff values of sequence identity are probably needed to group and assign individual RSTs to RST-based OTUs.

In an attempt to determine the sequence identity cutoff values for single V regions that give rise to comparable OTU estimates as full-length 16S rRNA gene sequences, we downloaded the high-quality 16S rRNA gene sequences recovered from isolates of bacteria from a number of phyla, classes, or suborders within RDP (Release 10, Update 16, December 4, 2009). Within each of the taxa, the sequences that had been aligned by RDP were grouped into phylogenetic species and genera (based on 3% and 5% sequence divergence of full-length sequences, respectively) using the Mothur program [Schloss et al., 2009]. Then the V1, V3, and V6 regions were removed individually from the aligned full-length sequences and clustered into phylogenetic species and genera at different sequence divergence levels also using the Mothur program. The sequence identity of each V region that gave rise to the same number of phylogenetic species and genera was considered the equivalent to 97% and 95% sequence identity of full-length sequences (Table 31.2). We only estimated the equivalent for V1 and V6, both of which have been used in SARST, and V3, a much longer and phylogenetically more informative V region that can potentially be used in future studies using SARST. It is evident that different V regions have different equivalent cutoff values, and different phyla and classes also have different cutoff values for the same V region. Apparently, the cutoff values are affected by the relatedness of the sequences. Unless a truly representative sequence dataset is used, it is difficult to determine the equivalent cutoff values of individual V regions. To interpret the RST data more accurately, individual RSTs needs to be compared to rRNA gene sequence databases to identify longer sequences that can then be used to characterize the microbiomes.

Choices of V Region. By far, both V1 (65–103 nt, exclusive the flanking primers, *E. coli* numbering) [Neufeld and Mohn, 2005; Neufeld et al., 2004b; Yu et al., 2006] and V6 (987–1045 nt) [Kysela et al., 2005; Poitelon et al., 2009] have been used in SARST. The V5 region has also been targeted in analyzing agricultural soil samples by Ashby et al. [2007], who used serial analysis of rRNA genes (referred to as SARD), which is very similar to the original SAGE procedures in cloning and sequencing concatemers of very short ditags (28 bp). SARD may not provide enough phylogenetic information for reliable taxonomic assignments of the RSTs because it only generates 14 bp RSTs. As demonstrated in analyzing soil, rumen, and marine samples, V1 and V6 regions appeared to have sufficient phylogenetic information to allow for taxonomic assignments of the recovered RSTs [Neufeld and Mohn, 2005; Neufeld et al., 2004b; Poitelon et al., 2009; Yu et al., 2006]. Other V regions can be targeted in SARST. However, when choosing a V region for SARST, researchers need to consider the content of phylogenetic information that is primarily determined by the length and sequence divergence, the availability of universal primers that allow for amplification of a particular V region, and the frequency of the recognition site of the type IIs restriction endonuclease within the V region.

31.3 POTENTIAL BIAS AND LIMITATIONS OF SARST

As a PCR-based method, SARST also has the inherent PCR biases that introduce preferential amplification of some 16S rRNA genes over others [von Wintzingerode et al., 1997; see also Chapter 16, Vol. I]. In a comparison of the data obtained from the same sample in separate studies, Kysela et al. [2005] noticed that SARST-V6 and full-length clone libraries produced similar phylotype distributions. However, some phylotypes were only identified in one of the two datasets, while other phylotypes were detected at different relative frequencies. The disparities were attributed to differences in amplification between the V6 region and full-length gene [Kysela et al., 2005]. Several possible reasons can lead to differences in PCR amplification between a single V region and the full-length 16S rRNA gene, which include (a) differences in primer annealing with the templates and in requirements for template integrity [Kysela et al., 2005] and (b) differences in amplicon lengths, which affect polymerase dissociation along each template and formation of secondary structures by the denatured templates [Huber et al., 2009]. Based on a comparison between the RSTs obtained from SARST-V6 and the RSTs directly sequenced from V6 single amplicons, the concatenation step did not appear to introduce any ascertainable bias [Kysela et al., 2005].

It should also be pointed out that definitive taxonomic assignment of RSTs might be difficult, if solely based on direct comparison of RST sequences to databases, because of the highly variable nature of the RST sequences and paucity of the positions that provide high levels of evolutionary information for phylogenetic inferences [Kysela et al., 2005]. Indeed, approximately one-third of the RSTs did not return unambiguous matches from BLASTn search of GenBank [Kysela et al., 2005]. However, RSTs queries against databases (e.g., RDP, Greengenes, and GenBank) usually identify long sequences that have similar or nearly identical hypervariable sequences. The phylogenetic affiliation of these longer sequences allows for more accurate taxonomic assignment of the corresponding RSTs. Obviously, the completeness and accuracy of the databases used can have a significant impact on the taxonomic assignments of RSTs. In the same vein, habitat-specific databases can also improve taxonomic assignments of RSTs because such databases do not have sequences for microbes that do not exist in the habitat of interest.

Table 31.2 Sequence Identity Cutoff (Percent Sequence Identity) for Single Hypervariable Regions Equivalent to 97%/95% Sequence Identity of Full-Length 16S rRNA Genes

Taxa	Taxonomic Rank	Number of Sequences	V1 (65–103)[a]	V3 (358–518)[a]	V6 (987–1045)[a]
Actinobacteria except *Micrococcineae*	Phylum	15,341	96.9/86.3	99.4/97	94.6/84.2
Bacteroidetes	Phylum	1,041	93.5/90.1	96.8/95.4	94.5/89.5
Chlorobia	Phylum	66	91.9/70.9	98.0/91.6	90.8/78.1
Clostridia	Class	2,186	89.5/78.2	99.4/97.8	93.0/85.2
Cyanobacteria	Phylum	1,801	95.5/89.8	99.2/97.1	95.3/89.9
δ-Proteobacteria	Class	1,163	97.1/92.8	97.3/93.4	96.8/92.4
Fusobacteria	Phylum	143	91.2/48.0	98.5/95.4	91.0/79.4
γ-Proteobacteria	Class	21,719	93.9/81.8	97.0/93.2	98.4/94.7
Lactobaciales	Order	5,037	77.6/54.3	98.8/96.1	96.6/92.2
Micrococcineae	Suborder	2,413	94.1/87.5	99.4/97.9	97.5/90.7
Planctomyces	Phylum	98	91.2/88.0	96.4/95.0	96.8/93.4
Spirochaetes	Phylum	884	94.1/90.8	97.9/95.8	94.8/91.2
Verrucomicrobia	Phylum	96	89/76.2	96.2/93.4	94.6/86.8

[a] *Escherichia coli* numbering.

Furthermore, SARST relies on digestion of the amplified V region by the type IIs restriction endonuclease that frees the RSTs from the two flanking primers. Apparently, those RSTs that have a recognition site for the type IIs enzyme will not be recovered by SARST. Based on our previous in silico analysis, the *Bsg*I recognition sites are very rare in the V1 region, with only 8 out of the 5165 type strain 16S rRNA gene sequences (>1200 bp, RDP Release 10, Update 17) having a *Bsg*I recognition site [Yu et al., 2006]. One the other hand, 281 of these type strain sequences have a *Bsg*I recognition site within the V6 region. Therefore, SARST-V6 probably excludes more bacteria from being detected than SARST-V1. If other V regions are targeted by SARST, the frequency of the *Bsg*I (or other type IIs enzymes) recognition site within that V region needs to be determined first.

31.4 APPLICATIONS OF SARST

Although microbiomes can be profiled with cryptic community fingerprinting methods such as DGGE and T-RFLP in a high-throughput manner, these methods only detect a very small percentage of the bacteria present. Additionally, these methods rely on subsequent DNA sequencing for taxonomic assignments. Besides the massive parallel pyrosequencing that became more cost-effective recently, SARST is a method that enables the collection of taxonomically informative data for high-throughput analysis of microbial communities [Kysela et al., 2005; Neufeld et al., 2004b; Yu et al., 2006].

SARST targeting the V1 region was first used to analyze the microbial diversity of Arctic tundra soil [Neufeld et al., 2004b] and was then used in a comparative study to compare the microbial diversity of Arctic tundra soil to that of boreal forest soil [Neufeld and Mohn, 2005]. Although nearly 25% of the recovered RSTs represented unclassified bacteria, 23 bacterial phyla were detected from the Arctic tundra soil, with *Proteobacteria, Firmicutes, Actinobacteria, Bacteroidetes, Planctomycetes, Acidobacteria*, and *Verrucomicrobia* being the predominant phyla. Those two studies produced two of the largest collections of 16S rRNA gene sequences from individual environmental samples reported to date, and they revealed the "unexpectedly" higher diversity in the tundra soil than in the boreal soil samples. A later study by the same group used SARST-V1 to examine the microbial diversity in soils as affected by hexachlorocyclohexane contamination, and the recovered RSTs were successfully used in the same study to design a microarray to compare the microbial diversity in the samples [Neufeld et al., 2006]. These researchers not only demonstrated the utility of the V1-RST-based microarray, but also identified bacteria likely involved in hexachlorocyclohexane degradation in the soil.

In a study to determine a means to streamline and simplify the original SARST procedures, Yu et al. used SARST-V1 in analyzing the bacteria adherent to the feed particles recovered from the ovine rumen [Yu et al., 2006]. By sequencing more than 1000 RSTs from 192 clones of RST concatemers, 236 unique phylotypes (≥95 sequence identity) representing different bacteria broadly distributed among eight bacterial phyla were identified. This study revealed the presence of new bacteria, including bacteria belonging to phyla that had never been reported in the rumen. In addition to being the most comprehensive diversity analysis of the rumen, this study demonstrated the ability of SARST-V1 to resolve phylotypes present in specialized habitats, such as the gastrointestinal tract and the rumen, which have low microbial diversity at high taxonomic rank but have high diversity at low taxonomic levels [Backhed et al., 2005; Yu et al., 2006; see also Chapters 18–23, Vol. II].

Kysela et al. [2005] generated 526 RSTs from sequencing 96 clones of RST concatemers from a hydrothermally active sediment core sample collected from Guaymas Basin (Gulf of California). These RSTs showed taxonomic affinities with V regions of 370 bacterial and 9 archaeal rRNA genes in GenBank. This study produced a relatively small RST dataset, but the collected RSTs allowed for the identification of γ-, δ-, and ε-*Proteobacteria, Thermodesulfobacteria*, green nonsulfur bacteria, *Chlamydiae/Verrucomicrobia, Planctomyces*, Cytophaga-Flavobacter-Bacteroidetes (CFB) bacteria, OP1-, OP8-, and OP11-related bacteria, the Guaymas bacterial group, and bacteria grouped as "other." Even though only 96 clones of RST concatemers were sequenced in this preliminary study, the utility of SARST-V6 was clearly demonstrated. Also using the SARST-V6, Poitelon et al. [2009] analyzed the bacterial diversity in three finished (after ozonation and chlorination) drinking water samples that were collected from three French drinking water plants. Bacteria affiliated with nine bacterial phyla and unclassified bacteria were found in all the drinking water samples. RSTs affiliated with the order *Burkholderiales* were found to be the most predominant within the β-*Proteobacteria* class. Given the harshness of disinfections with ozonation and chlorination, it is surprising that a large diversity of bacteria were detected in the finished drinking water samples. The health implication of that finding remains to be further evaluated.

31.5 FUTURE PERSPECTIVES

SARST was developed before massive parallel pyrosequencing became available for microbiome analysis, and it significantly improved on the traditional one-clone–one-sequence approach with respect to both cost and coverage. Recently, pyrosequencing has become a powerful technique that can provide comprehensive and cost-effective diversity analysis of microbiomes. As applied until now, only single V regions (i.e., V3, V4, V6, V5, or V9) are pyrosequenced [Amaral-Zettler et al., 2009; Brown et al., 2009; Claesson et al., 2009; Huber et al., 2007; Lazarevic et al., 2009; Miller et al., 2009; Price et al., 2009] because of the limited sequence read lengths achievable by the currently pyrosequencing technologies (the GS FLX system). Because multiple RSTs can be pyrosequenced in a parallel manner, we propose referring to such analysis of RSTs as parallel analysis of ribosomal sequence tags (PARST) to distinguish it from SARST. Interested readers should consult Chapter 18 of this volume. As the GS FLX Titanium system, which can generate more than 400 bp of sequence reads, became available very recently, concatemers of multiple RSTs could be pyrosequenced. For RSTs based on V1 and V6, the throughput capacity will be increased by at least sixfold. This will be especially useful when deep coverage of multiple barcoded samples is pyrosequenced in the same run.

REFERENCES

Amaral-Zettler LA, McCliment EA, Ducklow HW, Huse SM. 2009. A method for studying protistan diversity using massively parallel sequencing of V9 hypervariable regions of small-subunit ribosomal RNA genes. *PLoS ONE* **4**:e6372.

Andersson AF, Lindberg M, Jakobsson H, Backhed F, Nyren P, Engstrand L. 2008.. Comparative analysis of human gut microbiota by barcoded pyrosequencing. *PLoS ONE* **3**:e2836.

Ashby MN, Rine J, Mongodin EF, Nelson KE, Dimster-Denk D. 2007. Serial analysis of rRNA genes and the unexpected dominance of rare members of microbial communities. *Appl. Environ. Microbiol.* **73**:4532–4542.

Backhed F, Ley RE, Sonnenburg JL, Peterson DA, Gordon JI. 2005. Host–bacterial mutualism in the human intestine. *Science* **307**:1915–1920.

Bent SJ, Forney LJ. 2008. The tragedy of the uncommon: Understanding limitations in the analysis of microbial diversity. *ISME J.* **2**:689–695.

Brown MV, Philip GK, Bunge JA, Smith MC, Bissett A, et al. 2009. Microbial community structure in the North Pacific ocean. *ISME J.* **3**:1374–1386.

Brulc JM, Antonopoulos DA, Miller ME, Wilson MK, Yannarell AC, et al. 2009. Gene-centric metagenomics of the fiber-adherent bovine rumen microbiome reveals forage specific glycoside hydrolases. *Proc. Natl. Acad. Sci. USA.* **106**:1948–1953.

Callaway TR, Dowd SE, Wolcott RD, Sun Y, McReynolds JL, et al. 2009. Evaluation of the bacterial diversity in cecal contents of laying hens fed various molting diets by using bacterial tag-encoded FLX amplicon pyrosequencing. *Poult. Sci.* **88**:298–302.

Carulli JP, Artinger M, Swain PM, Root CD, Chee L, et al. 1998. High throughput analysis of differential gene expression. *J. Cell. Biochem. Suppl.* **30–31**:286–296.

Claesson MJ, O'Sullivan O, Wang Q, Nikkila J, Marchesi JR, et al. 2009. Comparative analysis of pyrosequencing and a phylogenetic microarray for exploring microbial community structures in the human distal intestine. *PLoS ONE* **4**:e6669.

Curtis TP, Sloan WT, Scannell JW. 2002. Estimating prokaryotic diversity and its limits. *Proc. Natl. Acad. Sci. USA.* **99**:10494–10499.

Dowd SE, Callaway TR, Wolcott RD, Sun Y, McKeehan T, Hagevoort RG, Edrington TS. 2008. Evaluation of the bacterial diversity in the feces of cattle using 16S rDNA bacterial tag-encoded FLX amplicon pyrosequencing (bTEFAP). *BMC Microbiol.* **8**:125.

Eckburg PB, Bik EM, Bernstein CN, Purdom E, Dethlefsen L, Sargent M, Gill SR, Nelson KE, Relman DA. 2005. Diversity of the human intestinal microbial flora. *Science* **308**:1635–1638.

Egan S, Thomas T, Kjelleberg S. 2008. Unlocking the diversity and biotechnological potential of marine surface associated microbial communities. *Curr. Opin. Microbiol.* **11**:219–225.

Fierer N, Breitbart M, Nulton J, Salamon P, Lozupone C, et al. 2007. Metagenomic and small-subunit rRNA analyses reveal the genetic diversity of bacteria, archaea, fungi, and viruses in soil. *Appl. Environ. Microbiol*. **73**:7059–7066.

Gilbert JA, Field D, Swift P, Newbold L, Oliver A, Smyth T, Somerfield PJ, Huse S, Joint I. 2009. The seasonal structure of microbial communities in the Western English Channel. *Environ. Microbiol*. **11**:3132–3139.

Green BD, Keller M. 2006. Capturing the uncultivated majority. *Curr. Opin. Biotechnol*. **17**:236–240.

Hall N. 2007. Advanced sequencing technologies and their wider impact in microbiology. *J. Exp. Biol*. **210**:1518–1525.

Handelsman J, Liles M, Mann D, Riesenfeld C, Goodman RM. 2002. Cloninig the metagenome: Culture-independent access to the diversity and functions of the uncultured microbial world. In Wren B, Dorell N, eds. *Functional Microbial genomics, Methods in Microbiology*, Vol. 33. London: Academic Press, pp. 241–255.

Huber JA, Mark Welch DB, Morrison HG, Huse SM, Neal PR, Butterfield DA, Sogin ML. 2007. Microbial population structures in the deep marine biosphere. *Science* **318**:97–100.

Huber JA, Morrison HG, Huse SM, Neal PR, Sogin ML, Mark Welch DB. 2009. Effect of PCR amplicon size on assessments of clone library microbial diversity and community structure. *Environ. Microbiol*. **11**:1292–1302.

Jones RT, Robeson MS, Lauber CL, Hamady M, Knight R, Fierer N. 2009. A comprehensive survey of soil acidobacterial diversity using pyrosequencing and clone library analyses. *ISME J*. **3**:442–453.

Kim M, Morrison M, Yu Z. 2010. Status of the phylogenetic diversity sensus of ruminal microbiomes. *FEMS Microbiol. Ecol*. no. doi: 10.1111/j.1574-6941.2010.01029.x.

Koehl A, Friauf E, Nothwang HG. 2003. Efficient cloning of SAGE tags by blunt-end ligation of polished concatemers. *BioTechniques* **34**:692–694.

Krause L, Diaz NN, Edwards RA, Gartemann KH, Kromeke H, et al. 2008. Taxonomic composition and gene content of a methane-producing microbial community isolated from a biogas reactor. *J. Biotechnol*. **136**:91–101.

Kysela DT, Palacios C, Sogin ML. 2005. Serial analysis of V6 ribosomal sequence tags (SARST-V6): A method for efficient, high-throughput analysis of microbial community composition. *Environ. Microbiol*. **7**:356–364.

Lazarevic V, Whiteson K, Huse S, Hernandez D, Farinelli L, Osteras M, Schrenzel J, Francois P. 2009. Metagenomic study of the oral microbiota by Illumina high-throughput sequencing. *J. Microbiol. Methods* **79**:266–271.

Lozupone C, Hamady M, Knight R. 2006. UniFrac—An online tool for comparing microbial community diversity in a phylogenetic context. *BMC Bioinformatics* **7**:371.

Ludwig W, Strunk O, Klugbauer S, Klugbauer N, Weizenegger M, Neumaier J, Bachleitner M, Schleifer KH. 1998. Bacterial phylogeny based on comparative sequence analysis. *Electrophoresis* **19**:554–568.

McLellan SL, Huse SM, Mueller-Spitz SR, Andreishcheva EN, Sogin ML. 2009. Diversity and population structure of sewage-derived microorganisms in wastewater treatment plant influent. *Environ. Microbiol*. **12**:378–392.

Miller SR, Strong AL, Jones KL, Ungerer MC. 2009. Bar-coded pyrosequencing reveals shared bacterial community properties along the temperature gradients of two alkaline hot springs in Yellowstone National Park. *Appl. Environ. Microbiol*. **75**:4565–4572.

Neufeld JD, Mohn WW. 2005. Unexpectedly high bacterial diversity in Arctic Tundra relative to boreal forest soils, revealed by serial analysis of ribosomal sequence tags. *Appl. Environ. Microbiol*. **71**:5710–5718.

Neufeld JD, Yu Z, Lam W, Mohn WW. 2004a. SARST, Serial analysis of ribosomal sequence tags. In Kowalchuk GA, de Bruijn FJ, Head IM, Akkermans ADL, van Elsas JD, eds. *Molecular Microbial Ecology Manual*. Dordrecht, The Netherlands: Kluwer Academic Publishers, pp 543–568.

Neufeld JD, Yu Z, Lam W, Mohn WW. 2004b. Serial analysis of ribosomal sequence tags (SARST): A high-throughput method for profiling complex microbial communities. *Environ. Microbiol*. **6**:131–144.

Neufeld JD, Mohn WW, de Lorenzo V. 2006. Composition of microbial communities in hexachlorocyclohexane (HCH) contaminated soils from Spain revealed with a habitat-specific microarray. *Environ. Microbiol*. **8**:126–140.

Pace NR, Stahl DA, Lane DJ, Olsen GL. 1985. Analyzing natural microbial populations by rRNA sequences. *ASM News* **51**:4–12.

Poitelon JB, Joyeux M, Welte B, Duguet JP, Prestel E, Lespinet O, DuBow MS. 2009. Assessment of phylogenetic diversity of bacterial microflora in drinking water using serial analysis of ribosomal sequence tags. *Water Res*. **43**:4197–4206.

Price LB, Liu CM, Melendez JH, Frankel YM, Engelthaler D, et al. 2009. Community analysis of chronic wound bacteria using 16S rRNA gene-based pyrosequencing: impact of diabetes and antibiotics on chronic wound microbiota. *PLoS ONE* **4**:e6462.

Rappe MS, Giovannoni SJ. 2003. The uncultured microbial majority. *Annu. Rev. Microbiol*. **57**:369–394.

Riviere D, Desvignes V, Pelletier E, Chaussonnerie S, Guermazi S, et al. 2009. Towards the definition of a core of microorganisms involved in anaerobic digestion of sludge. *ISME J*. **3**:700–714.

Roesch LF, Fulthorpe RR, Riva A, Casella G, Hadwin AK, et al. 2007. Pyrosequencing enumerates and contrasts soil microbial diversity. *ISME J*. **1**:283–290.

Sanapareddy N, Hamp TJ, Gonzalez LC, Hilger HA, Fodor AA, Clinton SM. 2009. Molecular diversity of a North Carolina wastewater treatment plant as revealed by pyrosequencing. *Appl. Environ. Microbiol*. **75**:1688–1696.

Sandaa R, Torsvik VV, Enger, Daae FL, Castberg T, Hahn D. 1999. Analysis of bacterial communities in heavy metal-contaminated soils at different levels of resolution. *FEMS Microbiol. Ecol*. **30**:237–251.

Schloss PD, Handelsman J. 2004. Status of the microbial census. *Microbiol. Mol. Biol. Rev*. **68**:686–691.

Schloss PD, Westcott SL, Ryabin T, Hall JR, Hartmann M, et al. 2009. Introducing mothur: Open-source, platform-independent, community-supported software for describing and comparing microbial communities. *Appl. Environ. Microbiol*. **75**:7537–7541.

Sogin ML, Morrison HG, Huber JA, Welch DM, Huse SM, Neal PR, Arrieta JM, Herndl GJ. 2006. Microbial diversity in the deep sea and the underexplored "rare biosphere". *Proc. Natl. Acad. Sci. USA* **103**:12115–12120.

Stackebrandt E, Goebel BM. 1994. Taxonomic note: A place for DNA–DNA reassociation and 16S rRNA sequence analysis in the present species defination in Bacteriology. *Int. J.Syst. Bacteriol*. **44**:846–849.

Tiedje JM, Asuming-Brempong S, Nüsslein K, Marsh TL, Flynn SJ. 1999. Opening the black box of soil microbial diversity. *Appl. Soil. Ecol*. **13**:109–122.

Torsvik V, Goksoyr J, Daae FL. 1990. High diversity in DNA of soil bacteria. *Appl. Environ. Microbiol*. **56**:782–787.

Velculescu V, Zhang L, Vogelstein B, Kinzler K. 1995. Serial analysis of gene expression. *Science* **270**:484–487.

von Wintzingerode F, Göbel UB, Stackebrandt E. 1997. Determination of microbial diversity in environmental samples: Pitfalls of PCR-based rRNA analysis. *FEMS Microbiol. Rev*. **21**:213–229.

Ward DM, Weller R, Bateson MM. 1990. 16S-rRNA sequences reveal numerous uncultured inhabitants in a natural community. *Nature (London)* **345**:63–65.

Yu Z, Morrison M. 2004. Comparisons of different hypervariable regions of *rrs* genes for use in fingerprinting of microbial communities by PCR-denaturing gradient gel electrophoresis. *Appl. Environ. Microbiol*. **70**:4800–4806.

Yu Z, Yu M, Morrison M. 2006. Improved serial analysis of V1 ribosomal sequence tags (SARST-V1) provides a rapid, comprehensive, sequence-based characterization of bacterial diversity and community composition. *Environ. Microbiol*. **8**:603–611.

Part 4

Consortia and Databases

Chapter 32

The Metagenomics of Plant Pathogen-Suppressive Soils

Jan Dirk van Elsas, Anna Maria Kielak, and Mariana Silvia Cretoiu

32.1 INTRODUCTION

Soil is known to contain an often extreme microbial diversity per unit mass or volume [Gans et al, 2005]. By inference, the soil microbiota offers an excellent angle at novel microbial functions of ecological and industrial interest. For instance, on the basis of cultivation-based approaches, the soil microbiota has been found to harbor a wealth of antibiotic biosynthesis loci. Such functions, in particular cases, may underlie the suppressiveness of soils to plant pathogens [Steinberg et al., 2006]. Moreover, the soil microbiota is also known as a goldmine for novel biocatalysts involved in biodegradation processes, including those of human-made polluting compounds [Galvão et al., 2005; see also Chapters 43–46, Vol. II]. However, the soil microbiota as a whole has remained largely cryptic due to a phenomenon called the "Great Plate Count Anomaly" [Janssen et al., 2002; van Elsas et al., 2006], which describes the lack of direct culturability of many microorganisms in soil. Our understanding of soil functioning has thus been severely hampered, and many key traits of the soil microbiota that are involved in particular population regulatory processes (such as antibiotic production loci and particular enzymatic functions) have remained cryptic. In the light of the currently available high-throughput DNA-based technologies, the potential for examining and exploring the genetic treasures present in the soil microbiota is enormous. Thus, examination of the entire soil metagenome (here defined as the collective genomes of the microorganisms present in a soil sample) has been proposed as a means to address the issue [Rondon et al., 2000]. However, there are definite problems in this approach, being of technical as well as fundamental nature [Sjoling et al., 2006]. The fundamental caveats of soil metagenomics revolve around the relative ease to captivate the dominant soil microbiota versus the difficulty to access the so-called "rare" biosphere. Rank–abundance curves constructed for the soil microbiota have often demonstrated this rare biosphere to consist of an extremely long tail of ever rarer species. It is a fact of metagenomic life that, without a priori measures to remediate this, soil-based metagenomes are almost always biased towards the dominant community members.

A European research project denoted Metacontrol, which was executed in the early days of soil metagenomics (i.e., between 2002 and 2007), aimed to unravel the antagonistic capacities locked up in the microbiota of disease-suppressive soils. The basic idea was to find clues with respect to the involvement of such traits in the suppression of plant pathogens as well as to explore these for application purposes. The project has yielded a wealth of methodological advances and has given glimpses of the antagonistic potential of the soils studied [van Elsas et al., 2008b]. However, a full understanding of the antagonistic diversity in suppressive soils against plant pathogens is still missing, and this is largely due to the astounding diversity found in the soil microbiota at this functional level. Here we describe the major advances that have been achieved in metagenomic studies of disease-suppressive soils and address the major challenges that still lie ahead of us.

Handbook of Molecular Microbial Ecology, Volume I: Metagenomics and Complementary Approaches, First Edition. Edited by Frans J. de Bruijn.

32.2 DISEASE-SUPPRESSIVE SOILS

Disease-suppressive soils are defined by their ability to restrict the activity and/or survival of plant-pathogenic microorganisms. Some soils possess a natural ability to suppress plant pathogens [Borneman and Becker, 2007; Steinberg et al., 2007], whereas in other soils disease suppressiveness can be the result of soil management practices, such as monocropping. The key to plant disease suppression often lies in the soil microbiota—that is, the pathogen-suppressive microbiota of any kind or composition—that is present. This microbiota may be involved in competition for essential substrates the plant pathogen grows on (leading to niche exclusion), or it may be directly antagonistic to the pathogen. In the latter case, the in situ production of antibiotics that inhibit or kill the pathogen or of enzymes directly affecting the pathogen may be involved. A key example of the latter mechanism is the production of fungal pathogen-attacking chitinases or of competitive proteases, in disease-suppressive soils.

32.2.1 Antibiotic Production and Disease-Suppressive Soils

Whereas a range of antibiotics has been found to be produced by well-known culturable soil microorganisms such as pseudomonads, bacilli, streptomycetes, or particular soil fungi, another part of this antagonistic potential may be present in the as-yet-uncultured or unculturable microbiota [Steinberg et al., 2006; van Elsas et al., 2008b]. In recent work, a positive correlation was found between the level of suppression of the plant pathogen *Rhizoctonia solani* AG3 and the copy number of *prnD*, a gene that encodes an enzyme that performs a key step in the production of the antifungal compound pyrrolnitrin [Steinberg et al., 2006]. Such correlations have also been found for other antibiotic production loci like those for polyketides, which are typically produced by soil streptomycetes (unpublished results). Another piece of evidence is obtained from the implication of pseudomonads that produce particular antibiotics (e.g., phenazines, DAPG [2,4-diacetyl phloroglucinol]) in the suppression of *Geaomannomyces graminis* var. *tritici*, the causal agent of late blight in wheat. Collectively, these data strongly suggest the involvement of antibiotic production loci in the disease-suppressive power of soils. Hence, disease-suppressive soils may indeed be rich reservoirs of important antibiotic production loci that can be explored to improve plant disease suppressions as well as application in medicine.

32.2.2 Chitinolytic Activity and Disease-Suppressive Soils

Chitinases produced by soil microorganisms can also be involved in the suppression of plant disease, in this case caused by fungi that have chitineous cell walls. The level of disease suppressiveness can even be raised by adding chitin to soils [Mankau and Das, 1969; Spiegel et al., 1989]. Several studies have reported (based on the measurement of activity of nematodes and fungi) that the induction of soil suppressiveness by chitin amendment is a biotic process [Chernin et al., 1995; Kamil et al., 2007]. However, the mechanisms by which soils inhibit plant disease via chitin has not been completely elucidated. The exploration of the diversity of chitinases produced by the soil microbiota is a subject of current research, especially with respect to suppressiveness and to the possibility of manipulating this property [Downing and Thomson, 2000; Kobayashi et al., 2002]. Also, chitinolytic bacteria like *Enterobacter agglomerans, Serratia marcescens, Pseudomonas fluorescens, Stenotrophomonas maltophilia, and Bacillus subtilis* have been used as biological control agents of fungal or nematodal plant disease agents [Downing et al., 2000, Kotan et al. 2009; Kobayashi et al., 2002]. Moreover, fungi of the genera *Gliocladium* and *Trichoderma* have also been found to produce chitinolytic enzymes with protective roles for plants [di Pietro et al., 1993; Elad et al., 1982]. So far, the pathway of chitinase activity has been partially elucidated with respect to their protective role for plants [Cleveland et al., 2004].

Current insight in disease suppressiveness of soils indicates that, in most of the cases, the phenomenon is complex. That is, various mechanisms may be involved. Suppression of a particular pathogen may include (a) the production of chitinases, (b) efficient rhizosphere colonization leading to niche exclusion of the pathogen, (c) the production of one to several antibiotics, and (d) the production of different proteases.

32.3 EXPLORATION OF DISEASE-SUPPRESSIVE SOILS

A collaborative European project with acronym Metacontrol (2002–2007) had as its stated objective to examine selected phytopathogen-suppressive soils for their antagonistic potential. A range of assessments of the nature of different suppressive soils were obtained [Adesina et al., 2007; Bertrand et al., 2005; Courtois et al., 2003; Ginolhac et al., 2004; Hjort et al., 2007; Lefevre et al., 2008; Nalin et al., 2004; van Elsas et al.,

2008b]. The assumption was that the microbiota of suppressive soils would provide reservoirs of genetic loci involved in *in situ* antibiosis or antagonism. A focus was placed on genes for phytopathogen-suppressive polyketide antibiotics and chitinases. As shown in **Table 32.1**, four soils that were suppressive to varying phytopathogens were identified in the Netherlands (W, *Rhizoctonia solani* AG3 [Garbeva et al., 2004, 2006]), Sweden (U, *Plasmodiophora brassicae*), France (C, *Fusarium*), and the United Kingdom (Wy, *Fusarium*). Metagenomic libraries were constructed from these soils plus one control soil, M (Table 32.1), and screened for the occurrence of antibiotic and antagonistic functions [Adesina et al., 2007; Bertrand et al., 2005; Courtois et al., 2003; Ginolhac et al., 2004; Hjort et al., 2007; Lefevre et al., 2008; Nalin et al., 2004; van Elsas et al., 2008a,b]. In addition, a range of methodologies were developed that facilitated the preparation and exploration of the resulting libraries [Bertrand et al., 2005; Ginolhac et al., 2004; Hjort et al., 2007; Sjoling et al., 2006].

The exploration of the antagonistic potential of disease-suppressive soils by using a metagenomics-centered approach appeared straightforward at the onset of the work; however, it turned out to be utterly complex [van Elsas et al., 2008a,b]. A major issue was the prior estimation of target gene abundance, which was felt to be a strong determinant of the hit rate in the final metagenomic libraries. In the absence of a clear notion of the nature of the antagonistic compounds produced and genes involved, such an a priori assumption was very difficult to make. Other issues were of technical nature and revolved around the uncertainties and technicalities with respect to soil DNA extraction and cloning as well as the positive detection of the active compounds. In the following, we discuss the technology developed and the choices that had to be made prior to each analytical step with respect to (i) the soil DNA extraction methodology, (ii) the potential to "bias" the soil community or DNA, (iii) the suitability of the vector/host system for the objectives, (iv) the optimal screening procedure, and (v) the final analysis.

32.3.1 Soil DNA Extraction and Processing

For reliable library preparation, metagenomic soil DNA—which accurately represents the genetic make-up of the soil microbiota—is required in representative quantity [Bertrand et al., 2005, Inceoglu et al., 2010; van Elsas et al., 2008a,b]. In addition, the DNA needs to be of sufficient quality with respect to purity, integrity, and fragment length in order to be suitable for cloning into a suitable vector [Bertrand et al., 2005]. A minimal size of 40 kb will increase the chance that entire pathways—for example, those involved in the biosynthesis of polyketide antibiotics—can be cloned [Ginolhac et al., 2004, 2005; van Elsas et al., 2008a]. In several laboratories, advanced methodology that allowed us to produce pure high-molecular-weight (HMW) DNA from soil was developed [Bertrand et al., 2005; Ginolhac et al., 2004; Hjort et al., 2007; Lefevre et al., 2008; van Elsas et al., 2008a]. An efficient approach consisted of the extraction of cells from soil followed by gentle DNA extraction and purification using pulsed-field gel electrophoresis [Bertrand et al., 2005; van Elsas et al., 2008a; see also Chapters 10 and 11, Vol. II] Cushion (Percoll and/or Nycodenz) pre-separation of cells from soil was also tested as a pre-step for subsequent isolation of the HMW soil metagenomic DNA. Moreover, the microbial growth status in soil was assessed as an important determinant of the chemical quality of the extracted DNA. The quality could even be boosted by incubation with growth substrates such as glycerol [Bertrand et al., 2005; van Elsas et al., 2008a]. Typically, the approach produced adequate HMW soil DNA, often with size >60–100 kb [Bertrand et al., 2005; van Elsas et al., 2008a]. It was also found that high amounts of cells, minimally $\approx 10^{11}$, were required to yield sufficient DNA for efficient library construction [Hjort et al., 2007]. Because soils often contain on the order of 10^{8}–10^{10} cells per gram this finding sets a standard for the construction of soil metagenomic libraries. However, in spite of the improved soil DNA extractions and subsequent metagenomic library constructions, the hit rates of target genes were found to be low. Theoretically, assuming an incidence of target genes of 1% (that is, occurring once in every 100 bacterial genomes—average genome size of 4–5 Mbp), the constructed metagenome library would need to contain at least about 57,000 clones with 40-kb inserts to be able to find—with 99% probability—a single copy [Leveau, 2007]. This phenomenon has been likened to "looking for a needle in the haystack" [Kowalchuk et al., 2007] and strongly hampers the efficiency of metagenomics for bioexploration. Deliberate biasing of the habitat by applying pre-enrichment techniques has been suggested as a useful strategy that may boost hit rates [van Elsas et al., 2008b].

32.3.2 Metagenomic Libraries—Production and Screening

Clone libraries for four disease-suppressive soils (Table 32.1), each one consisting of approximately 6000 to 60,000 clones, were constructed in *Escherichia coli*

Table 32.1 Soil Metagenomic Libraries Constructed and Their Characteristics

Soil	Library Vector/ Number of Clones	Screening (Functional/Genetic)	Number of Positive Clones	Remarks/ References
W—Wildekamp grassland, suppressive to *Rhizoctonia solani* AG3	Fosmid/15,000	Functional: antagonism against *Rhizoctonia solani* AG3 and *Bacillus subtilis* Genetic: use of soil-generated PKSI[a] probe	7	Combined functional/genetic (PKS1) screening: 7 clones. Five confirmed as PKS1-positive clones. Three completely sequenced, one insert showing high similarity with *Acidobacterium sp.* sequences Van Elsas et al., 2008a,b
Wy—Wytham grassland & *Fusarium*-suppressive agricultural soil	Fosmid/100,000	Functional: antagonism against *Fusarium* sp. (agar plate based dual-culture assay)	13 (grassland) 2 (agricultural soil). Each clone distinct	Average insert size 35.6 kb. Grassland effective source of clones (high diversity). Agricultural soil low diversity and limited functional traits. End-sequencing/subcloning: mostly unidentified ORFs. Efficacy of clones lower than strains isolated from same source
C—Chateaurenard *Fusarium*-suppressive soil	Fosmid/51,000 BAC/60,000	Genetic/functional screening	22	Combined functional and genetic screening. Functional screening: *Fusarium* spore generation/hyphal production, *Aspergillus nidulans* growth, *Hebeloma cylindrosporum* hyphal generation. Genetic screening: sequencing PKS positive clones Courtois et al., 2003; Ginolhac et al., 2004
U—Uppsala *Plasmodiophora brassicae*-suppressive soil	Fosmid/8000	Functional: antagonism against *Pythium ultimum* Genetic: chitinase genes and 16S rRNA gene	— 4	Selection of *Streptomyces mutomycini, Kitosatospora, Lentzea, Oerskovia* revealed by fingerprinting. *S. mutomycini* and *S. clavifer* prevalent in library. Chitinase genes from soil, library and isolates. Cluster prevailing in soil not in library; library cluster not found in soil.
Montrond (control) soil	Fosmid/60,000	Genetic/functional screening	39	Thirty-nine novel PKS1 positive clones, most with supernatants showing antimicrobial activity, found

[a]PKS1, polyketide synthesis operon for type-I polyketides.

Source: Modified from van Elsas et al. [2008b; see also Chapter 22, Vol. I].

[van Elsas et al., 2008b]. Both large insert-size vectors, (such as bacterial artificial chromosomes (BACs), which allow cloning of inserts up to 200 kb) and fosmids (which allow insertion of 35- to 45-kb fragments) were used. BAC vectors enable the cloning of complex large operons and facilitate the analysis of a gene/operon within its original genomic context. In contrast, fosmids are able to accommodate smaller inserts and thereby only allow the cloning of smaller operons. Using a fosmid vector (such as the Epicentre pCC1FOS system) allows for the positive selection of vectors that have acquired inserts [Bertrand et al., 2005; Ginolhac et al., 2005; Nalin et al., 2004; Sjoling et al., 2006; van Elsas et al., 2008b]. Three libraries were based on fosmid vectors, the reason being the ease of obtaining appropriately sized libraries. One library, for the M soil, was constructed in a BAC vector [Courtois et al., 2003; Ginolhac et al., 2004]. The latter vector also contained a replicon that was compatible with a *Streptomyces* host, which allowed shuttling between the *E. coli* and *Streptomyces* metagenomic hosts (see also Chapter 22, Vol. I). Consequently, the probability of heterologous gene expression was enhanced for the clones obtained in this library [Courtois et al., 2003; Ginolhac et al., 2004].

Given the fact that soil metagenomic libraries are based on the random insertion of clonable DNA fragments into vectors, such libraries stochastically contain the genetic material of all genomes that were extracted from the soil microbiota and entered the DNA pool. Assuming that the prevalence of antagonistic functions across all microbial genomes in soil is low (ranging from 0.1% to 10%) and that these genes/operons may be 1–200 kb in size (over a 4- to 5-Mb average soil genome), soil-DNA-based metagenomes may contain only few clones that carry genes/operons of interest. Furthermore, there may be potential constraints to efficient gene expression in the metagenomic host strain. Hence, library screening is often a tedious task. For the metagenomes of the four disease-suppressive soils, functional as well as molecular screenings were employed in order to uncover antagonistic functions [Bertrand et al., 2005; Ginolhac et al., 2004; Nalin et al., 2004; Sjoling et al., 2006; van Elsas et al., 2008a] and, expectedly, rather low numbers of phytopathogen-suppressive clones were found [Bertrand et al.., 2005; Ginolhac et al., 2004; van Elsas et al., 2008a].

32.3.3 Functional Screening

Functional screenings of the libraries were performed using high-throughput dual-culture assays. These assays allow target phytopathogenic organisms to grow over metagenomic library clones arrayed on large Petri dishes. Scoring during and following growth was for irregularities/inhibitions in growth of the target organism [Courtois et al., 2003; Ginolhac et al., 2004; van Elsas et al., 2008a]. This experimental setup led to the detection of positive clones (up to 48 per library), amounting to <0.05% of positives for all libraries. Such low numbers can be attributed either to a rare occurrence of target genes/operons in the clones, to a lack of expression of the genes/operons in the host used or to the required molecular machineries being significantly larger than the vector inserts. The latter fact has indeed been reported for many polyketide production loci. Other factors that potentially impede the detection of function of the target genes could relate to a lack of adequate signals, such as found in expression systems that are controlled by quorum sensing (a cellular communication mechanism commonly found in bacteria). Shuttling from *E. coli* to *Streptomyces* as the metagenomic host facilitated the expression of an antibiotic (amphotericin) production locus [Courtois et al., 2003; Ginolhac et al., 2004; van Elsas et al., 2008], as indicated by high activity in resulting clones against target fungi.

In the light of the foregoing, the low numbers of functionally positive clones did not come as a surprise. Corroborating this result, a screening of forest soil libraries for antifungal traits also yielded a low hit rate; that is, one positive signal among 113,700 fosmid clones examined was found [Chung et al., 2008]. We conclude that substantial methodological improvements are required to boost the hit rates in explorations of soils for antagonistic function. Ways forward are given below.

32.3.4 Genetic (Molecular) Screening

The libraries obtained from disease-suppressive soils were also screened using molecular tools, such as hybridization and PCR-based methods [van Elsas et al., 2008b]. In this case, success in detecting novel operons, such as those involved in polyketide biosynthesis, was dependent on the application of deliberate degeneracy in the probes and primers used [Courtois et al., 2003; Ginolhac et al., 2004]. The rationale was that, using this strategy, the screening would not be restricted to exactly known genes, enabling a broader range of positive hits within the metagenomic library. The method facilitated the identification of target genes that were sufficiently similar to the mixed query sequences generated from the same soil. Using the total soil community DNA as a target, we thus amplified the KS gene of the polyketide biosynthesis operon PKS1 from soil DNA with degenerated primers. The amplicons obtained were used as probes to detect PKS1 sequences in the library [Ginolhac et al., 2004; Nalin et al., 2004; van Elsas et al., 2008a]. The approach yielded a total of seven positive clones in the W soil library, of which the majority contained genes that were likely involved in the biosynthesis of novel polyketides. This was confirmed by end-sequencing of the clones [van Elsas et al., 2008a]. In addition, the roughly 60,000 M soil clones were divided into pools, which were subsequently used as templates for PKS-based PCR screenings. This yielded over 100 positive pools (0.22% hit rate [Ginolhac et al., 2004]). The amplicons were then sequenced to check for redundancies and for known PKS sequences. In total, 39 unique PKS sequences were thus found, which all represented promising novel PKS biosynthesis operons [van Elsas et al., 2008a] The positive clones, identified using colony hybridization with relevant probes, were then tested, following shuttling into *Streptomyces*, for antagonistic activities. *Bacillus subtilis* 1A72, *Staphylococcus aureus* 21, *Enterococcus faecalis* 40, *Escherichia coli* 9, *Pseudomonas aeruginosa* 39, *Fusarium oxysporum* LNPV, *Aspergillus fumigatus* Gasp 4707, and *Neurospora crassa* HK were used. The clones exhibited 56% antimicrobial activity against at least *B. subtilis*, 13% against *S. aureus*, 4% against *E. faecalis*, and <1% partial inhibition of growth of *Neurospora crassa* mycelium [van Elsas et al., 2008b].

We conclude that the screening procedures allowed quick access to novel PKS sequences from soil. Further testing in respect of applicability is obviously required prior to any large-scale production and use in agriculture or medicine [van Elsas et al., 2008b].

32.3.5 Hit Rates of Target Genes via Pre-enrichment of Soil or Samples

In the metagenomics studies performed on disease-suppressive soils, direct (unselective) approaches were used to examine the soils for anti-phytopathogen functions [Adesina et al., 2007; Bertrand et al., 2005; Courtois et al., 2003; Ginolhac et al., 2004; Hjort et al., 2007; Lefevre et al., 2008; Nalin et al., 2004; van Elsas et al., 2008a]. In these approaches, the possibility of applying positive growth selection was dismissed, the reason being that a direct unbiased assessment of the antagonistic potential of the soil microbiota was felt to be required. Moreover, most of the selected target genes/operons were assumed not to offer an a priori growth advantage in the host used for cloning, even in the presence of the target phytopathogens [Ginolhac et al., 2004; van Elsas et al., 2008a]. However, targeted approaches were also developed to enhance the hit rates of specific targets in the microbial communities. For instance, the U soil was pretreated with chitin in order to enhance the abundance of particular chitinase producers [Hjort et al., 2007; Sjoling et al., 2006]. Starting from the premisse that the most successful antagonists in soil are those that are active in situ, attempts were also made to enrich the metabolically active bacterial cell fractions with the help of flow cytometric cell sorting [Hjort et al., 2007; van Elsas et al., 2008a]. Metabolically active cells were indeed successfully sorted from the soil [van Elsas et al., 2008a,b]. However, as a result of the limited flow rate of the cell sorter used, throughput was too limited (10^6 cells/hour) to yield sufficient biomass for library construction [van Elsas et al., 2008a], and advanced machines with higher flow rates were deemed necessary.

32.4 THE WAY AHEAD – IMPROVEMENT OF HIT RATES

What are the challenges for the further metagenomics exploration of suppressive and other soils? Let us assume that most questions revolve around enhancing the efficiency of the metagenomics-based exploitation of the soil microbiota for beneficial traits. This translates into enhancing the hit rates of target genes and may imply the application of a deliberate bias to favor target organisms and genes/operons in the starting material. In addition, strong improvements should be made in all steps in soil metagenomics—that is, in DNA extraction methodology, cloning methods, and screening methods, all aimed at increasing the throughput of metagenomics (see Chapter 22, Vol. I and Section 2, Vol. II).

32.4.1 Deliberate Bias in Sampled Communities

Deliberate manipulation of microbial communities from soil offers unique possibilities to enhance metagenomics hit rates. For instance, prior growth selection can be applied, as outlined before. Here, an intelligent selection of growth conditions will guide the bias. In this approach, fluorescence-assisted cell sorting can be applied, not only to sort the metabolically active cell fractions, but also to obtain particular interesting fractions of the community. For instance, the high-G+C% gram-positive bacteria (in which antibiotic production loci are abundantly present) can be selected following staining with specific fluorescent probes. Another promise is offered by the use of stable isotopes. Stable isotope probing (SIP) introduces ^{13}C-labeled substrates into soil communities. Members of such communities can take up the ^{13}C and incorporate it in their cellular macromolecules including DNA. The resulting "heavy" DNA is separated from ^{12}C-DNA by ultracentrifugation and sequenced, thus identifying the organisms that captured the substrate [Radajewski et al., 2003; see also Chapter 55, Vol. I]. This approach can be coupled to soil metagenomics studies [Dumont et al., 2006], resulting in the identification of a complete methane monooxygenase operon, allowing insight in this process in soil. Application in the detection of beneficials in suppressive soils will depend on unique substrates that are used by particular phytopathogens and their competitors, thuis allowing to identify the latter. As an example, by tracking the fate of ^{13}C-labeled CO_2 fixed by plants into the soil microbiota, key data on plant-responsive microorganisms—which often produce antibiotics as secondary metabolites—can be obtained [Ostle et al., 2003]. The application of SIP using other organic substrates bears great potential in future explorative metagenomic studies in which organisms with particular ecological roles are the targets.

32.4.2 Searching for Improved Metagenomic Library Hosts

Working with *E. coli* as the metagenomics host has clear advantages with respect to the ease of the laboratory work and the experience gained with it over many years. However, the use of *E. coli* is limited with regard to the screening of phenotypes from the soil metagenome, because *E. coli* is not a typical soil organism. The main restriction arises from the fact that particular promoters and associated factors required for the expression of inserted genes may be poorly recognized in this host. Moreover, essential post-translational processing and/or transport functions may be lacking in this host. Rondon et al. [1999] showed that only about 30% of *Bacillus* traits could be expressed in *E. coli*, which indicates that *E. coli* is—at best—a suboptimal host for the heterologous expression of genes from such typical soil bacteria. Bear in mind that soil microbial communities are often dominated by just five bacterial phyla: Alphaproteobacteria, Betaproteobacteria, Gammaproteobacteria, Acidobacteria, and Actinobacteria [Fierer et al., 2007]. Bacteria belonging to several of these phyla are appealing hosts for use in functional metagenomic studies of soil habitats. Thus, efforts are ongoing to develop alternative hosts, preferably within these bacterial groups [van Elsas et al., 2008b]. Recently, six novel bacterial hosts belonging to the phylum Proteobacteria (*Agrobacterium tumefaciens, Burkholderia graminis, Caulobacter vibrioides, Escherichia coli, Pseudomonas putida, and Ralstonia metallidurans*) were tested for their performance in functional metagenomic screening of a library constructed using broad-host-range cloning vectors [Craig et al., 2010]. This work, as well as ours, supports the premise that working with several hosts instead of just *E. coli* will allow us to strongly boost the metagenomics hit rates as well as expand functional and genetic diversity of target traits. However, the study of Craig et al. [2010] also revealed that not all hosts perform equally well in phenotypic screenings. It remains possible that some surrogate expression hosts are better suited for the expression of foreign genetic material than others. Furthermore, each host can be differentially sensitive to toxic compounds that are produced from inserted genes. Such host clones will disappear from the library. Obviously, these constraints are true for any host strain selected for soil metagenomics.

32.4.3 The Metagenomic Library Vector

The data discussed in this chapter and elsewhere [van Elsas et al., 2008b] affirm that critical evaluation of the host/vector system to be used in soil metagenomics is required [Sjoling et al., 2006]. For *E. coli* and for some other hosts, three types of vectors (i.e., small-, medium-, and large-insert size (see Chapter 22, Vol. I)) are available. Small-insert-size vectors, which primarily permit screening for single gene-encoded functions, are of use in shotgun sequencing approaches, allowing construction of libraries from mechanically sheared DNA. Such an approach was used for the detection of small open reading frames (ORFs) derived from uncultured prokaryotes from sediment. On the other hand, both fosmid and BAC vectors allow incorporation of larger fragments and even intact operons within their genomic context. Although this provides a better handle at gene expression from complex operons, the fact that pure HMW soil DNA is required in high amounts for efficient cloning into BAC vectors makes these less suitable for routine cloning efforts (such as required for high-throughput setups).

The identification of novel activities requires, as mentioned in the foregoing, successful transcription and translation systems. This is obviously connected with expansion of the range of bacterial hosts that allow capturing of additional expression capabilities. Preferably, complete metagenomic libraries should be transferable to different alternative hosts, which will require the development of new shuttle vectors. The recently described vector pRS44 [Aakvik et al., 2009; see also Chapter 22, Vol. I] may serve as an example of such a broad-host-range vector system. This vector can be efficiently transferred to numerous hosts by conjugation, which is spurred by the plasmid RK2 replication origin. In *E. coli*, this plasmid replicates via its plasmid F origin.

32.4.4 Improved Screening Methods

Efficient selection of clones of interest still remains a critical point in any soil metagenomics approach. The possibility of missing a target due to problems with expression of genes in the metagenomic host plus the sizes of libraries forces the development of novel genetic screening methods. For example, Demanèche et al. [2009] proposed the use of multiple probes in order to reduce the pools containing the potentially interesting clones. The simultaneous use of pooled probes increases the probability of finding clones of interest. In their study, pooled probes targeting genes associated with a range of functions (e.g., genes for antibiotic resistance, denitrification, and dehalogenation) were used for the library screening. The pooled-probes approach proved to be useful for rapid library screening.

Another method to enhance the detection frequencies of genes of interest lies in the use of "heavy" DNA from SIP experiments, as discussed in the foregoing. The use of stable isotope-labeled substrate in enrichment experiments thus increases the chances of discovery of novel enzymes from the environment [e.g., Knietsch et al., 2003; Woo et al., 2009; see also Section 7, Vol. II].

32.4.5 Direct Pyrosequencing

Currently, direct pyrosequencing technology is favored as a technology that provides a quick insight into the gene repertoire of a particular soil sample. Although this method produces only short reads (currently of about 450 bp), it compensates this limitation by its speed and simplicity and especially its tremendous output. Thus, this technology, next to being useful for studying the microbial diversity and community composition based on the 16S rRNA gene [Roesch et al., 2007; Spain et al., 2009], is also a good alternative for studying functional genes and assisting in probe design for recovery of the complete gene sequences. The most critical part in this approach is the design of a proper primer pair. On the one hand, it should cover the diversity of the gene family within the community. On the other hand, primers should not have a too-high degree of degeneracy so as not to lose specificity. Also, the amplicon cannot be too long to not reduce the emulsion PCR efficiency.

32.5 CONCLUSIONS

For the foregoing, it is thus apparent that major challenges still lie ahead of us. Granted, several interesting novel biological functions have already been uncovered in the microbiota of the suppressive soils studied, but this may be considered to represent the tip of the iceberg of the diversity that is out there. The (partial) biosynthetic machineries likely involved in the production of novel polyketide antibiotics (e.g., a leinamycin-like antibiotic, as well as other polyketides [Bertrand et al., 2005; Courtois et al., 2003; Ginolhac et al., 2004; Nalin et al., 2004; van Elsas et al., 2008a] were promising, but the work is in need of a follow-up. These discoveries were also plagued by the low hit rates of promising antibiotic biosynthesis clones, even for the disease-suppressive soils. We thus assume that, in spite of the successes, a major part of the extant antibiotic biosynthesis machineries may have been missed in the metagenomic screens for reasons explained in the foregoing. Thus, a current rule of thumb, that the search for *nonhousekeeping* functions in soil metagenomes can be compared to looking for a needle-in-a-haystack [Kowalchuk et al., 2007], was affirmed. The characteristics of the soil microbiota, exemplified by typical rank-abundance distribution of particular microbial populations, dictates this outcome and incites the application of creative tricks and tools to overcome these limitations. Such tricks and tools in all cases would need to enhance the "visibility" of target genes, allowing us to detect and get hold of these. To mention a few, soil DNA extraction could be geared towards community members or metagenome fractions that most likely contain the target genes. For instance, total plasmid DNA, the "metamobilome," can be extracted when novel biodegradation, metal, or antibiotic resistance genes (frequently found on plasmids) are targeted. Moreover, specific fractions of the chromosomal DNA pool, e.g. high G+C content DNA in the case of actinobacterial genes, may be targeted (see Chapter 23, Vol. I). Improvements in subsequent screens may also be cogitated—for example, by using a gene-centered approach [Iwai et al, 2010] or improving high-throughput formats of increased accuracy. Furthermore, the screening (gene expression) data should be increasingly linked to high-throughput sequencing.

At the level of the sample, improvements may build on the invention of new positive selection strategies for desired traits, based either on growth or on overcoming resistance. Another tool that will foster explorative metagenomics is the prescreening of habitats in respect of the incidence, abundance and expression of target genes. So-called global-scale gene mapping (GGM—analogous to the concept of environmental gene tagging) describes habitats in terms of gene abundance and/or expression [Tringe et al., 2005]. GGM can compare microbial gene pools across soils and provide a global perspective on target gene prevalence. For instance, PKS1-type polyketide biosynthetic loci are more prevalent in soil than in whale carcass, acid mine drainage, or Sargasso sea metagenomes [Tringe et al., 2005]. GGM thus allows us to predict hit rates of target genes.

These forthcoming advances will boost our capacities to finally come to grips with the astounding soil microbial diversity and harness it to our advantage. Guidance by GGM will be an important asset in the progress. The improved or finetuned soil metagenomics approaches will enable us to (1) mine soil for genes/pathways of interest to biotechnological applications, (2) decipher the identity and function of as-yet-uncultured microorganisms, and (3) obtain a characterization of soil with regard to antagonistic function, diversity, and genetic complement. The quickly increasing throughput of (pyro)sequencing technologies will also assist us in the rapid assessment of the prevalence of target genes, shedding increasing light on the soil genetic reservoir and potential for biotechnological exploration and application (see Chapter 9, 14, 15, 17, 28, Vol. II; Section 7, Vol II).

Acknowledgments

We thank all partners of the Metacontrol project for their collaboration: Kornelia Smalla (Julius Kuhn Institut, Braunschweig, Germany), Pascal Simonet and Tim Vogel (Ecole Superieure de Lyon, Lyon, France), Janet Jansson (LBL, Berkeley, USA), Mark Bailey (CEH, Oxford, UK), Sara Sjoling (Sodertorns Universitet, Stockholm, Sweden), Leo van Overbeek (Plant Research International, Wageningen, The Netherlands), and Renaud Nalin (LibraGen, Toulouse, France). This work was partially supported by the EU FP7 supported project METAEXPLORE.

REFERENCES

Aakvik T, Degnes KF, Dahlsrud R, Schmidt F, Dam R, et al. 2009. A plasmid RK2-based broad-host-range cloning vector useful for transfer of metagenomic libraries to avariety of bacterial species. *FEMS Microbiol. Lett*. **296**:149–158.

Adesina MF, Lembke A, Costa R, Speksnijder A, Smalla K. 2007. Screening of bacterial isolates from various European soils for *in vitro* antagonistic activity towards *Rhizoctonia solani* and *Fusarium oxysporum*: Site-dependent composition and diversity revealed. *Soil Biol. Biochem*. **39**:2818–2828.

Bertrand H, Polya F, Van VT, Lombarda N, Nalin R, et al. 2005. High molecular weight DNA recovery from soils prerequisite for biotechnological metagenomic library construction. *J. Microbiol. Methods* **62**:1–11.

Borneman J., Becker J. 2007. Identifying microorganisms involved in specific pathogen suppression in soil. *Annual Review of Phytopathology*. **45**:153–172.

Chernin L, Ismailov Z, Haran S, Chet I. 1995. Chitinolytic Enterobacter agglomerans antagonistic to fungal plant pathogens. *Appl. Environ. Microbiol*. **61**:1720–1726.

Chung, EJ, Lim HK, Kim JC, Choi GJ, Park EJ, et al. 2008. Forest soil metagenome gene cluster involved in antifungal activity expression in *Escherichia coli*. *Appl. Environ. Microbiol*. **74**:723–730.

Cleveland TE, Yu J, Bhatnagar D, Chen ZY, Brown RL, et al. 2004. Progress in elucidating the molecular basis of the host plant—*Aspergillus flavus* interaction, a basis for devising strategies to reduce aflatoxin contamination in crops. *J. Toxicol. Toxin Rev*. **23**(2):345–380.

Courtois S, Cappellano CM, Ball M, Francou FX, Normand P, et al. 2003. Recombinant environmental libraries provide access to microbial diversity for drug discovery from natural products. *Appl. Environ. Microbiol*. **69**:49–55.

Craig JW, Chang FY, Kim JH, Obiajulu SC, Brady SF. 2010. Expanding small-molecule functional metagenomics through parallel screening of broad-host-range cosmid environmental DNA libraries in diverse *Proteobacteria*. *Appl. Environ. Microbiol*. **76**:1633–1641.

Daniel, R. 2005. The metagenomics of soil. *Nat. Rev. Microbiol*. **3**:470–478.

Demanèche S, David MM, Navarro E, Simonet P, Vogel TM. 2009. Evaluation of functional gene enrichment in a soil metagenomic clone library. *J. Microbiol. Methods* **76**:105–107.

di Pietro A, Lorito M, Hayes CK, Broadway RM, Harman GE. 1993. Endochitinase from *Gliocladium virens*: Isolation, characterization and synergistic antifungal activity in combination with gliotoxin. *Phytopathology* **83**:308–313.

Downing KJ, Thomson JA. 2000. Introduction of the *Serratia marcescens chiA* gene into an endophitic *Pseudomonas fluorescens* for the biocontrol of phytopathogenic fungi. *Can. J. Microbiol*. **46**:363–369.

Dumont MG, Radajewski SM, Miguez CB, McDonald IR, Murrell JC. 2006. Identification of a complete methane monooxygenase operon from soil by combining stable isotope probing and metagenomic analysis. *Environ. Microbiol*. **8**:1240–1250.

Elad Y, Chet I, Henis Y. 1982. Degradation of plant pathogenic fungi by *Trichoderma harzianum*. *Can. J. Microbiol*. **28**:719–725.

Elshahed MS, Youssef N, Spain AM, Sheik CS, Najar FZ, et al. 2008. Novelty and uniqueness patterns of rare members of the soil biosphere. *Appl. Environ. Microbiol*. **74**:5422–5428.

Fierer N, Bradford MA, Jackson RB. 2007. Toward an ecological classification of soil bacteria. *Ecology* **88**:1354–1364.

Galvãao TC, Mohnb WW, de Lorenzo V. 2005. Exploring the microbial biodegradation and biotransformation gene pool. *Trends Biotechnol*. **23**:497–506.

Gans J, Wolinsky M, Dunbar J. 2005. Computational improvements reveal great bacterial diversity and high metal toxicity in soil. *Science* **309**:1387–139.

Garbeva P, Voesenek K, Elsas JD. 2004. Quantitative detection and diversity of the pyrrolnitrin biosynthetic locus in soil under different treatments. *Soil Biol. Biochem*. **36**:1453–1463.

Garbeva P, Postma J, van Veen JA, van Elsas JD. 2006. Effect of above-ground plant species on soil microbial community structure and its impact on suppression of *Rhizoctonia solani*. *Environ. Microbiol*. **8**:233–246.

Ginolhac A, Jarrin C, Gillet B, Robe P, Pujic P, et al. 2004. Phylogenetic analysis of polyketide synthase I domains from soil metagenomic libraries allows selection of promising clones. *Appl. Environ. Microbiol*. **70**:5522–5527.

Ginolhac A, Jarrin C, Robe P, Perrière G, Vogel TM, et al. 2005. Type I PKS may have evolved through horizontal gene transfer. *J. Mol. Evol*. **60**:716–725.

Hjort K, Lembke A, Speksnijder A, Smalla K, Jansson JK. 2007. Community structure of actively growing bacterial populations in plant pathogen suppressive soil. *Microb. Ecol*. **53**:399–413.

Inceoglu Ö., Hoogwout EF, Hill P, van Elsas JD. 2010. Effect of DNA extraction method on the apparent microbial diversity of soil. *Applied and Environmental Microbiology*. **76**:3378–3382.

Iwai S, Chai B, Sul WJ, Cole JR, Hashsham SA, Tiedje JM. 2010. Gene-targeted metagenomics reveals extensive diversity of aromatic dioxygenase genes in the environment. *ISME. J*. **4**:279–285.

Janssen PH, Yates PS, Grinton BE, Taylor PM, Sait M. 2002. Improved culturability of soil bacteria and isolation in pure culture of novel members of the divisions *Acidobacteria, Actinobacteria, Proteobacteria*, and *Verrucomicrobia*. *Appl. Environ. Microbiol*. **68**:2391–2396.

Kamil Z, Rizk M, Saleh M, Moustafa S. 2007. Isolation and identification of rhizosphere soil chitinolytic bacteria and their potential in antifungal biocontrol. *Global J. Mol. Sci*. **2**(2):57–66.

Knietsch A, Bowien S, Whited G, Gottschalk G, Daniel R. 2003. Identification and characterization of coenzyme B12-dependent glycerol dehydratase- and diol dehydratase-encoding genes from metagenomic DNA libraries derived from enrichment cultures. *Appl. Environ. Microbiol*. **69**:3048–3060.

Kobayashi DY, Reedy RM, Bick J, Oudemans PV. 2002. Characterization of a chitinase gene from *Stenotrophomonas maltophilia* strain 34S1 and its involvement in biological control. *Appl. Environ. Microbiol*. **68**:1047–1054.

Kotan R, Dikbas N, Bostan H. 2009. Biological control of post harvest disease caused by *Aspergillus flavus* on stored lemon fruits. *African J. Biotechnol*. **8**(2):209–214.

Kowalchuk GA, Speksnijder AG, Zhang K, Goodman RM, van Veen JA. 2007. Finding the needles in the metagenome haystack. *Microb. Ecol*. **53**:475–485.

Langer M, Gabor EM, Liebeton K, Meurer G, Niehaus F, et al. 2006. Metagenomics: An inexhaustible access to nature's diversity. *Biotechnol. J*. **1**:815–821.

Lefevre F, Robe P, Jarrin C, Ginolhac A, Zago C, et al. 2008. Drugs from hidden bugs: Their discovery via untapped resources. *Res. Microbiol*. **159**(3):153–161.

Leveau, J. H. J. 2007. The magic and menace of metagenomics: Prospects for the study of plant growth-promoting rhizobacteria. *Eur. J. Plant Pathol*. **119**:279–300.

Mankau R, Das S. 1969. The influence of chitin amendments on *Meloidogyne incognita*. *J. Nematol*. **1**:15–16.

Nalin R, Pujic P, Tuphile K, Gillet B. 2004. Method for the expression of unknown environmental DNA into adapted host cells. Patent CA2492966.

Ostle N, Whiteley AS, Bailey MJ, Sleepa D, Ineson P, et al. 2003. Active microbial RNA turnover in a grassland soil estimated using a $^{13}CO_2$ spike. *Soil Biol. Biochem*. **35**:877–885.

Radajewski S, McDonald IR, Murrell JC. 2003. Stable-isotope probing of nucleic acids: A window to the function of uncultured microorganisms. *Curr. Opin. Biotechnol*. **14**:296–302.

ROESCH LF, FULTHORPE RR, RIVA A, CASELLA G, HADWIN AK, et al. 2007. Pyrosequencing enumerates and contrasts soil microbial diversity. *ISME. J*. **1**:283–290.

RONDON MR, RAFFEL SJ, GOODMAN RM, HANDELSMAN J. 1999. Toward functional genomics in bacteria: Analysis of gene expression in *Escherichia coli* from a bacterial artificial chromosome library of *Bacillus cereus*. *Proc. Natl. Acad. Sci*. **96**:6451–6455.

RONDON MR, AUGUST PR, BETTERMANN AD, BRADY SF, GROSSMAN TH, et al. 2000. Cloning the soil metagenome: A strategy for accessing the genetic and functional diversity of uncultured microorganisms. *Appl. Environ. Microbiol*. **66**:2541–2547.

SJÖLING S, STAFFORD W, COWAN DA. 2006. Soil metagenomics: exploring and exploiting the soil. In VAN ELSAS JD, TREVORS JT, JANSSON JK, eds. *Modern Soil Microbiology II*. Boca Raton, FL: CRC Press, pp. 409–434

SPAIN AM, KRUMHOLZ LR, ELSHAHED MS. 2009. Abundance, composition, diversity and novelty of soil Proteobacteria. *ISME. J*. **8**:992–1000.

SPIEGEL Y, COHN E, CHET I. 1989. Use of chitin for controlling *Heterodera avenae* and *Tylenchulus semipenetrans*. *J. Nematol*. **21**:419–422.

STEINBERG C, EDEL-HERMANN V, ALABOUVETTE C, LEMANCEAU P. 2006. Soil suppressiveness to plant diseases. In VAN ELSAS JD, TREVORS JT, JANSSON JK, eds. *Modern Soil Microbiology II*. Boca Raton, FL: CRC Press, pp. 455–479.

SUL WJ, PARK J, QUENSEN JF 3rd, RODRIGUES JL, SELIGER L, et al. 2009. DNA-stable isotope probing integrated with metagenomics for retrieval of biphenyl dioxygenase genes from polychlorinated biphenyl-contaminated river sediment. *Appl. Environ. Microbiol*. **75**:5501–5506.

TRINGE SG, VON MERING C, KOBAYASHI A, SALAMOV AA, CHEN K, et al. 2005. Comparative metagenomics of microbial communities. *Science* **308**:554–557.

VAN ELSAS, J. D., et al. 2006. In VAN ELSAS JD, TREVORS JT, JANSSON JK, eds. *Modern Soil Microbiology II*. Boca Raton, FL: CRC Press.

VAN ELSAS, J. D. et al. 2008a. A novel protocol for the metagenomic analysis of suppressive soil. *J. Microbiol. Methods* **75**: 515–522.

VAN ELSAS JD, COSTA R, JANSSON J, SJÖLING S, BAILEY M, et al. 2008b. The metagenomics of disease-suppressive soils—Experiences from the METACONTROL project. *Trends Biotechnol*. **26**:591–601.

WHITELEY AS, GRIFFITHS RI, BAILEY MJ. 2003. Analysis of the functional diversity within water-stressed soil communities by flow cytometric analysis and CTC^+ cell sorting. *J. Microbiol. Methods* **54**:257–267.

WOO et al. 2009.

Chapter 33

Soil Metagenomic Exploration of the Rare Biosphere

Tom O. Delmont, Laure Franqueville, Samuel Jacquiod, Pascal Simonet, and Timothy M. Vogel

33.1 INTRODUCTION

Microorganisms can be considered the fundamental driving force of the biosphere and have dominated life on Earth for over 3 billion years. They have colonized all ecological niches, from caves [Pasic et al., 2010] to the stratosphere [Wainwright et al., 2003] and from deep ocean thermal vents [Huber et al., 2007] to deserts [Pointing et al., 2009] and polar snow [Larose et al., 2010; see also Vol. II]. Other forms of life are completely dependent upon these microscopic organisms. For example, humans cannot survive without the rich microbial flora inhabiting their own bodies [Turnbaugh et al., 2006]. When examining the importance of different ecosystems, soil stands out as the habitat on Earth that harbors by far the largest microbial diversity [Torsvik et al., 1990, 2002; Whitman et al., 1998; van Elsas et al., 2006; see also Chapter 9, Vol. II]. The genetic resources in a single gram of soil possess 3 million times more sequences than in the human genome; and only three grams of soil contain more bacteria than the Earth does humans.

Soil microbial communities are known to drive major geochemical cycles [Falkowski et al., 2001], to support healthy plant growth [Ortíz-Castro et al., 2009], and to degrade organic matter and pollutants [Singh et al., 2008]. However, little is known about the vulnerability of their key functions and how they respond to human-induced environmental perturbations, such as climate change and land use. Cultivation-based approaches, although limited in scope, have already shown that soil harbors diverse antibiosis-related functions [Adesina et al., 2007], pollutant degrading bacteria [Richard and Vogel, 1999], plant growth promoting bacteria [Leveau, 2007], bacteria resistant to heavy metals and antibiotics [Baker-Austin et al., 2006; Dcosta et al., 2007; Demaneche et al., 2008], and bacteria capable of surviving in extreme environments [Dib et al., 2008; Hery et al., 2003].

However, only a fraction (less than 0.5%) of the microbiota in soil has been cultivated using any of an array of techniques [Amann et al., 1995; Davis et al., 2005; Stalay and Konopka, 1995], thus limiting considerably our understanding of these microorganisms. In order to access this tremendous biodiversity, researchers have recently developed sophisticated molecular techniques that can explore and exploit a wide range of soil biodiversity. These techniques derived their pertinence from the direct analysis and exploration of microbial community DNA (metagenomics) [Courtois et al., 2003; Demaneche et al., 2008; Ginolhac et al., 2004]. In this context, soil metagenomics is the study and exploitation of the collective genome of all organisms present in a particular soil sample [Handelsman et al., 1998]. The critical technical issue today is the nonbias access to the entire soil microbial diversity.

To date, there have been only superficial, although ambitious, attempts to explore the soil metagenome. Due in part to incomplete genomic extraction techniques and incomplete sequencing, today scientists are only exploring a minute fraction of soil metagenomes. Potential errors in data interpretation due to the limited vision of these small community fractions might be driving hypotheses away from incorporating functional redundancies and the importance of the minor members of the soil community (rare biosphere). Recent advances in high-throughput cloning

Handbook of Molecular Microbial Ecology, Volume I: Metagenomics and Complementary Approaches, First Edition. Edited by Frans J. de Bruijn.

and new generation sequencing (see Chapter 18, Vol. I) might improve the prospect of completely sequencing the soil metagenome and thus obtaining a more complete picture of the soil microbial community function.

33.2 SOIL METAGENOMIC LIMITS

The soil microbial community is relatively diverse [Curtis et al., 2002, Robe et al., 2003] and appears to have the highest prokaryotic diversity of any environment [van Elsas et al., 2006; Roesch et al., 2007]. One gram of soil is reported to contain up to 10 billion microorganisms and thousands of different species [Knietch et al., 2003]. Thus, the potential access to the genomes of these microorganisms requires diverse approaches due in part to their differential location in the soil microstructure, their differences in membrane consistency, and their differences in relative numbers. In order to overcome some of these difficulties, different metagenomic associated methods have been developed. All of these methods themselves have some type of shortcoming; however, when pooled together, they will provide greater depth than any technique by itself. For example, DNA purification techniques have numerous biases that limit the quantity of the so-called metagenomic DNA [Delmont et al., 2011]. Second, the proportion of the genomes is nonuniform [Ranjard and Richaume, 2001]; thus, access to the less represented genomes is reduced statistically. In any case, these type of techniques must be applied in order to overcome cell culture limitations [Amann et al., 1995]; but because of their own limits, a complete soil metagenomic study remains an important challenge.

Classically, different approaches have been applied to estimate soil biodiversity richness. The typical method is by the amplification, cloning, and then sequencing of the ribosomal small unit RNA genes ("16S"; see Chapter 16, Vol. I). Rarefaction curves can be created based on genetic diversity and sequence proportions in order to estimate the total soil microbial diversity with (and unfortunately within) a limited sample. The 16S clone libraries constructed by Tringe et al. [2005] and Morales et al. [2009] were extrapolated to determine soil microbial diversities between 3000 and 4000 different species (based on 97% similarity of the 16S sequence for species identity) after sequencing 1700 and 5000 clones, respectively. Roesch et al. [2007] did 16S amplification and direct pyrosequencing (without the cloning step) of a part of this gene. With this approach and 55,000 16S sequences, they evaluated the diversity to be 6000 different species based on 97% similarity of the 16S sequence for species identity.

A completely different approach based on genome reassociation kinetics measures the rate of rehybridization of metagenomic DNA. The kinetic measurements are evaluated based on different hypothetical species distributions. For example, Torsvik et al. [1990]; see also Chapter 2, Vol. I estimated soil diversity to be 10,000 different species assuming that all species were equal in number. More recently, Gans et al. [2005] estimated the diversity to be 8×10^6 species in the tested soil by assuming a log-normal species number distribution.

One explanation for this wide range of predicted species numbers in soil can be the metagenomic DNA diversity represented in the 16S amplification mix used for their analyses. Most researchers amplify the 16S gene from a minor (and possibly nonrepresentative) fraction of the metagenomic DNA present in a gram of soil. Thus, the amplification mix contains only about 10 ngs (less than 1% of extracted DNA), which represent only 10^6 copies of prokaryote genomes, from about several micrograms of extracted DNA per gram of soil (quantity is dependent on the soil type and sample depth). To reduce this limitation, Roesch et al. (2007) used a total of 10 μgs of DNA (96 different 16S amplifications). However, the diversity estimate were not drastically different from the results obtained by Morales et al., [2009] with 5000 16S clones using one 16S variable region of the DNA extracted. These observations do emphasize the existence of a much more important limit. The sequence redundancy due to nonuniform species proportions leads to the dilution of the less represented ones during the first cycles of 16S amplification, thus limiting considerably the diversity obtained. This possible limitation seems more prevalent and likely than that due to the low initial amount of DNA in the PCR mix.

Another important factor that often leads to underestimated biodiversity results is the DNA extraction protocol. Most studies focus their efforts on assessing microbial diversity with only one extraction technique. The microbial diversity uncovered with different methods demonstrates the danger of such a narrow approach if the goal is to optimize access to biodiversity. The DNA extraction appears to be a crucial step in soil metagenomics because it defines in part the extent of the information available. To overcome this bias, a broad DNA recovery strategy is needed [Delmont et al., 2011; see Chapters 10 and 11, vol. II].

33.3 SOIL METAGENOMIC ADVANCES

Soil metagenomic studies have rapidly expanded since 1998 (**Fig. 33.1**) and an important number of tools and methods have been developed to recover and study prokaryotic DNA diversities [Delmont et al., 2011; Rajendhran and Gunasekaran, 2008; see Chapter 10, Vol. II]. The two major metagenomic advances were high-throughput cloning and new-generation sequencing approaches. These new technologies became precious

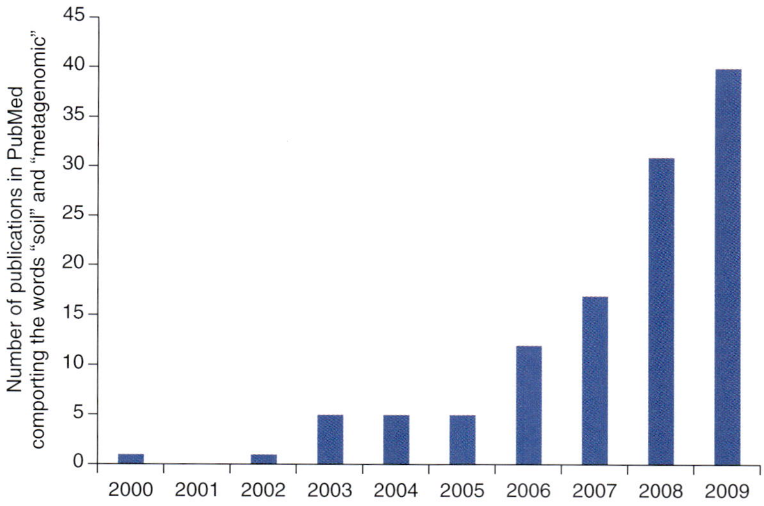

Figure 33.1 Number of publications in Pubmed with the words "soil" and "metagenomic" from 2000 to 2009.

tools for the discovery of new species, genes and functions of interest. While there is a tendency to explore the metagenomics of a soil through sequencing, the clone library approach often provides better validation of newly discovered genes.

Genomic studies stand at the vanguard of science as technological advances are providing access to the functioning of different biological systems. After the Human Genome Project (3 billion base pairs) and the Sargasso Sea marine sequencing effort (6 billion base pairs), soil exploration constitutes a new and ambitious challenge as it harbors about 10 trillion base pairs (if 10^4 species are present) to 1000 trillion base pairs of diversity (if 10^6 species are present) per gram of soil. Just as knowledge of the human genome promises to revolutionize medical science, the application of genomic technologies to microbial evolution and environmental biology promises to revolutionize microbiology. The soil microbial community represents a true goldmine for genes that encode novel biocatalysts involved in either biosynthetic or biodegradation processes, including the production of polyketide synthases [Ginolhac et al., 2004, 2005] and the degradation of human-made pollutants [Handelsman et al., 2002; Boubakri et al., 2006; Galvao et al., 2005; see also Sections 6 and 7, Vol. II]. In spite of new possibilities to study metagenomic DNA, only a minor fraction of soil metagenomes have been sequenced to date and only predominant species were discovered and studied. The majority of soil genetic richness and function still remains to be discovered.

33.4 AN INTERNATIONAL PARTNERSHIP TO OPEN THE BLACK BOX

As new sequencing technologies are increasing both in sequence length and number of sequences [Shendure and Ji, 2008], massive DNA sequencing is becoming a serious possibility. As a consequence, large national and international projects were launched to sequence environments such as the ocean and the human gut. Yet, soil nucleic explorations have been relatively limited in part due to the enormous soil heterogeneity, the vast biodiversity, and the difficulty to access this biodiversity locked up in the soil matrix. Soil microbiologists, microbial ecologists, geneticists, molecular biologists, and bioinformaticians have agreed to collaborate to provide the first complete metagenomic sequence(s) of a soil. This metagenome sequence data will constitute a "reference" to which other soils around the world could be compared. Thus, other metagenomic projects devoted to sequencing parts of different soil genomes through the world will be able to use this complete metagenome as a scaffold for annotation of "core" genes representing common soil microorganisms in other soils and as a basis for estimating differences in diversity, completeness, and richness between soils. The combination of metagenomics approaches and broad-scale sequencing will open a totally new era in soil microbiology with advances ranging from detection of climatic indicators and greenhouse gas production to drug discovery, as well as from correlating biodiversity and function to predicting the biosphere's resilience to human-induced perturbation. The soil system chosen for investigation, Park Grass, Rothamsted (UK), is a "charismatic and internationally recognized resource" [Silvertown et al., 2006]. This unique long-term ecological site includes ongoing experiments that have been running for over 150 years. The research center at Rothamsted provides a history of soil biology and chemistry as well as an archive of soil samples from detailed studies of different plot treatments (http://www.rothamsted.ac.uk/). Metadata are available concerning climate, soil use, and chemical inputs.

The success of soil metagenomics depends to a large extent on intelligent decision-making concerning sample selection, DNA extraction methods, cloning strategies, screening methods, technological advances in sequencing approaches, and data management and sharing. Recent progress in methods to capture the vast scale of genetic diversity within soil microbial communities will enable deep metagenomic sequencing [Morales et al., 2009; Delmont et al., 2011]. These advances include methods to dissect the community using DNA or cell extraction separation methods that can be optimized to detect abundant and rare members of the community. These promising new metagenomic approaches, which are currently being applied to the Rothamsted soil, rely on massive parallel high-throughput sequencing of DNA extracted from soil microorganisms. Furthermore, DNA microarrays loaded with soil-dedicated probes have been used in order to identify the diversity and relative abundance of phylogenetic and functional genes in different DNA fractions [Gebert et al., 2008; Huyghe et al., 2008; Sanguin et al., 2006; Ward et al., 2007]. Cloning of large DNA fragments from soil to create metagenomic libraries enables the examination of genetic pathways and contiguous sequences (e.g., operons). In addition, the new-generation sequencing methods, including enhanced 454 Titanium with up to 1000-bp reads, will complement the cloning-based approaches and avoid the associated cloning bias (DNA weight and cloning step). Data generated will provide insight into which soil organisms are actually involved in soil processes, especially when RNA is extracted. These approaches, coupled with novel bioinformatics methods (e.g., MG-RAST [Meyer et al., 2008; see Chapter 37, Vol. I]), provide an in-depth analysis of both generated sequence and array data. One bioinformatics limit stands out in the interpretation of this data; there are insufficient sequenced genomes or sequences that have been correctly annotated; therefore, around 50% of soil metagenome sequences cannot be used due to the lack of homologues in the databases.

33.5 THE METASOIL PROJECT

The Metasoil project, funded by the French National Research Agency, is initiating the Park Grass metagenome sequencing and cloning as well as the international consortium (Terragenome) [Vogel et al., 2009]. The first part of this project is a 100-Gbp soil metagenomic DNA sequencing (length up to 500 bp) with up to 100 titanium pyrosequencing runs. This sequencing effort corresponds to the equivalent of more than 20,000 prokaryotic genomes. Moreover, in parallel with the in-depth soil pyrosequencing, Metasoil includes the construction of a two million fosmid (40-kb inserts) library to serve the international scientific community for studying this soil metagenome and for searching for new genes of interest. The fosmid library and pyrosequencing runs will be done using various metagenomic DNA samples extracted from the same soil, but with a panel of different methods (soil, cells, and DNA separations) to access a maximum diversity [Delmont et al., 2011]. Due to DNA extraction biases, such as uneven species proportion and soil heterogeneity, a mixed DNA acquisition strategy is being evaluated in order to enhance complete soil metagenome exploration. This strategy is based on variable metagenomic DNA separation methods and aims to increase the soil DNA diversity recovery significantly. Moreover, the strategy should standardize the relative species proportion at the nucleic level as much as possible and, thus, increase access to the rare biosphere.

33.6 SOIL METAGENOMIC DNA RECOVERING STRATEGY

33.6.1 Soil Fractionation

The Park Grass untreated control plot (3d) selected for the project possesses a 249-m^2 surface (13.28 × 18.75 m). The sampling strategy consists of a randomized soil sampling in different areas of the plot [horizontal sampling] and at different depths (vertical sampling). Preliminary prokaryote diversity studies were carried out using RISA (ribosomal intergenic spacer analysis fingerprint after extracting DNA with MP Bead beating Fast Prep protocol [Griffiths et al., 2000]) and showed a stable diversity pattern horizontally. But significant differences have been discovered as a function of depth (data not shown). Further experiments were done based on different depth fraction (0–3 cm, 3–6 cm, and 18–21 cm) using a phylochip technology (**Fig. 33.2**; see Chapter 58, Vol. I).

Results show that some families are detected only in one soil fraction and that the proportion of the different families can fluctuate strongly with depth. Differences in soil depth appear to provide access to different metagenomic DNA fractions and can be helpful for increasing the diversity recovered and for limiting sequence redundancy.

33.6.2 Prokaryote Cells Extraction

After soil fractionation, we compared different methods for DNA recovery. In order to obtain metagenomic DNA, cells can be lysed within the soil sample [Orgam et al., 1987; van Elsas et al., 1997] or first removed from the matrix and then lysed [Berry et al., 2003]. This second method is commonly called indirect extraction and attempts to separate prokaryotic cells [the cell ring]

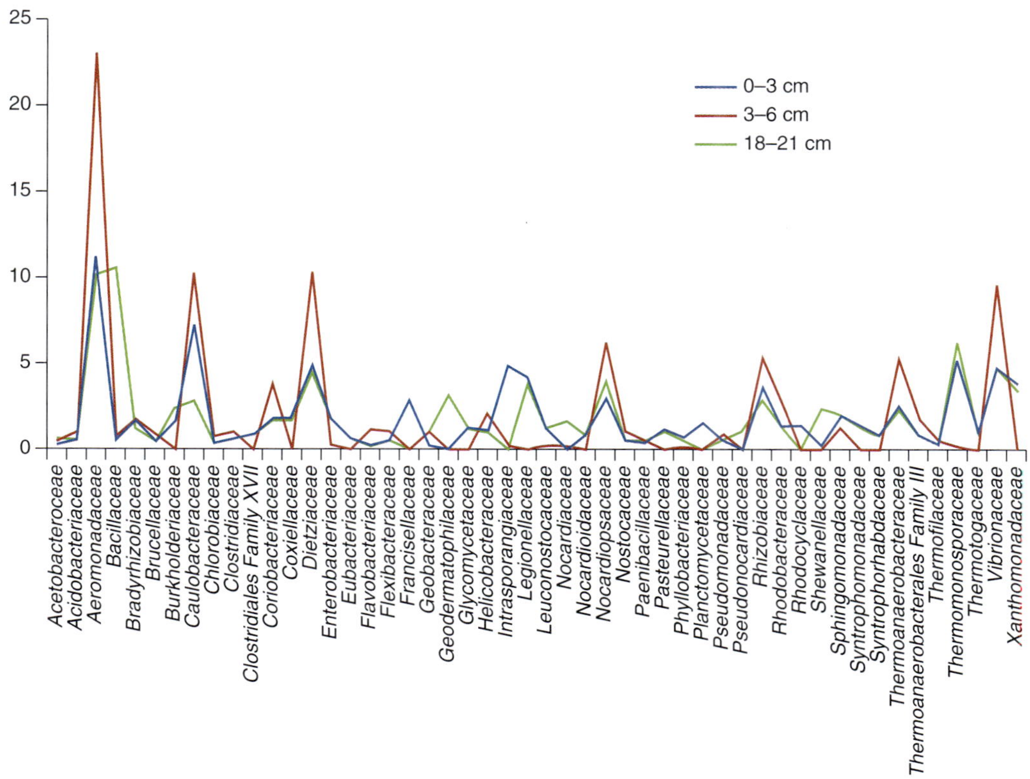

Figure 33.2 Relative proportion of families as a function of soil depth. The intensity corresponds to the microarray hybridization fluorescence signal.

from eukaryotic cells via a Nycodenz density gradient [Courtois et al., 2001; Lefevre et al., 2008]. To compare the two methods, DNA was extracted with the same protocol (MP Bead beating Fast Prep [Griffiths et al., 2000], but with direct (horizontal and vertical soil sampling) or indirect (with different centrifugation speeds during the cell ring formation) extraction approaches. The RISA profiles were then compared in a principal component analysis (PCA) (**Fig. 33.3**). Results demonstrate the differences between the direct and the indirect extraction methods in terms of metagenomic DNA. The RISA profiles are clearly different between the two methods, showing different peaks and intensities (data not shown). However, varying the centrifugation speed does not significantly change the DNA within the Nycodenz ring.

33.6.3 Prokaryote Cells Separation

Using the indirect extraction method, a prokaryote cell ring appears inside the density gradient. But interestingly, some cells stay above the ring and can be separated from the others based on their centrifugation velocity differences. The number of these cells above the Nycodenz ring was limited: Only a few nanograms of DNA could be extracted, but the relative species proportions vary strongly during the gradient (**Fig. 33.4**).

This physical cell separation increases considerably the number of detectable species, including those that would normally be undetected, due to their separation from the regular prokaryotic cell ring. Moreover, centrifugation speed can be modified because it changes relative species proportions in less dense fractions without affecting the cell ring.

33.6.4 Cells Lysis Stringency Fractionation

After separating soil and cells, extractable DNA has been further separated using cells lysis stringency. The process varies between protocols in terms of physical, chemical, and enzymatic lyses. Experiments were carried out directly from soil, or from the cell ring, in suspension or in agarose plugs. These approaches accessed different diversities with different species proportions represented in the extracted DNA (**Fig. 33.5**).

Moreover, because this step is essential in any DNA extraction protocol, it produces considerable biases in soil DNA recovering. Every DNA extraction protocol is biased to some degree; therefore, the choice of one unique approach is dependent on the researcher. An alternative would be to use several different lyses

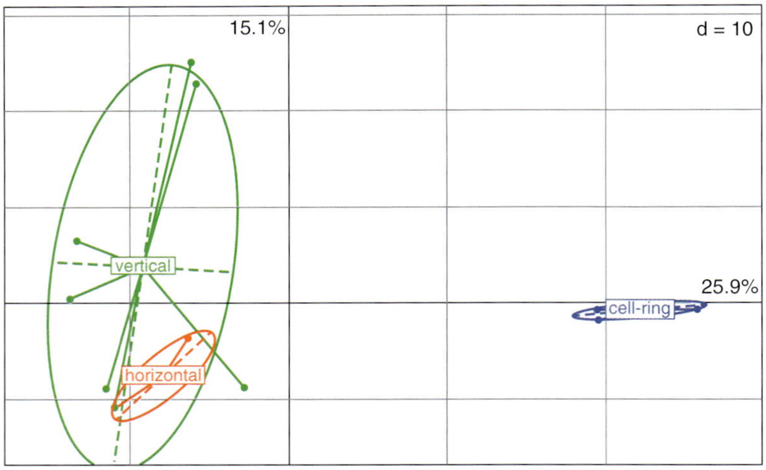

Figure 33.3 Principal component analysis [showing the first and second components] of the matrix data containing the RISA electropherograms corresponding to vertical sampling (seven fractions of three centimeters between 0 and 21 cms), horizontal sampling (fraction 3–6 cm for four distinct areas of the Park Grass 3d plot), and cell rings formed with different centrifugation speeds (1000*g*, 2000*g*, 5000*g*, and 9000*g*).

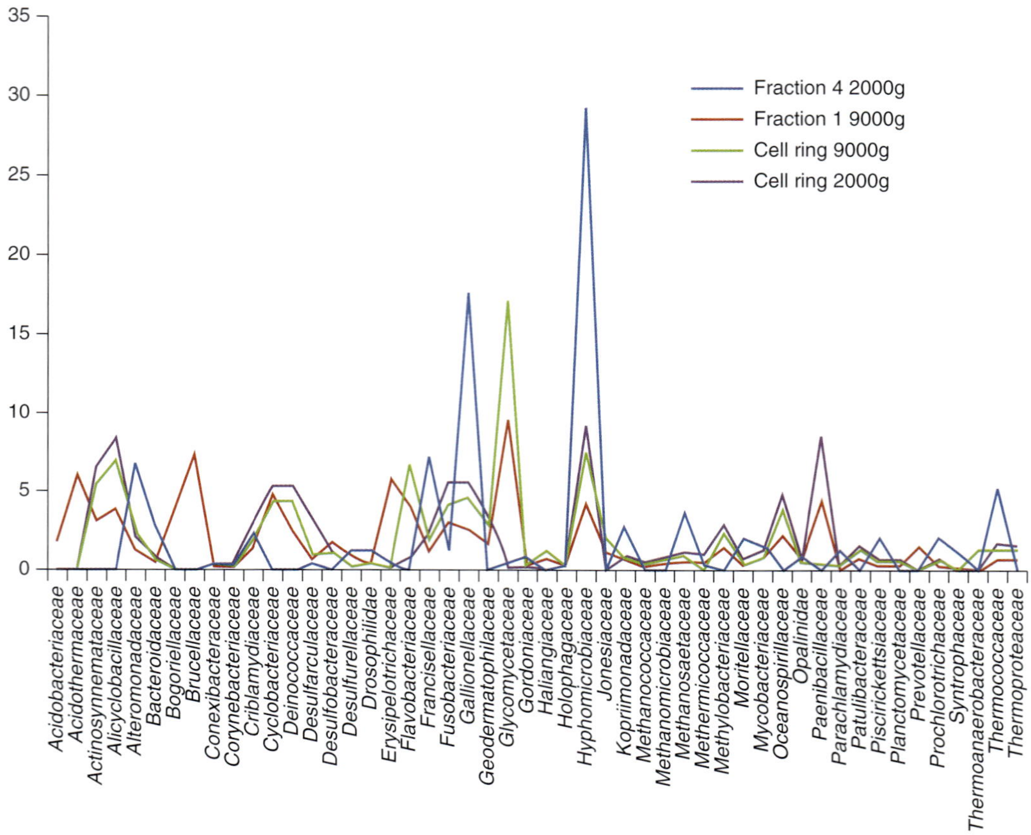

Figure 33.4 Relative proportion of families as a function of a density gradient. The intensity corresponds to the hybridization fluorescence signals.

methods before starting metagenomic studies [Delmont et al., 2011].

33.6.5 Molecular DNA Weight Separation

Finally, metagenomic DNA (agarose plug protocol) were separated as a function of DNA molecular weight in a pulsed field gel electrophoresis (PFGE). Phylochip analyses showed some prokaryote diversity differences when extracting two bands at different points of the DNA smear (50 and 250 kbp) (**Fig. 33.6**).

In particular, some families were detected only in one of these two bands. These results demonstrate that nucleic diversity separation occurred during the soil metagenomic DNA migration. The diversity differences appear to be

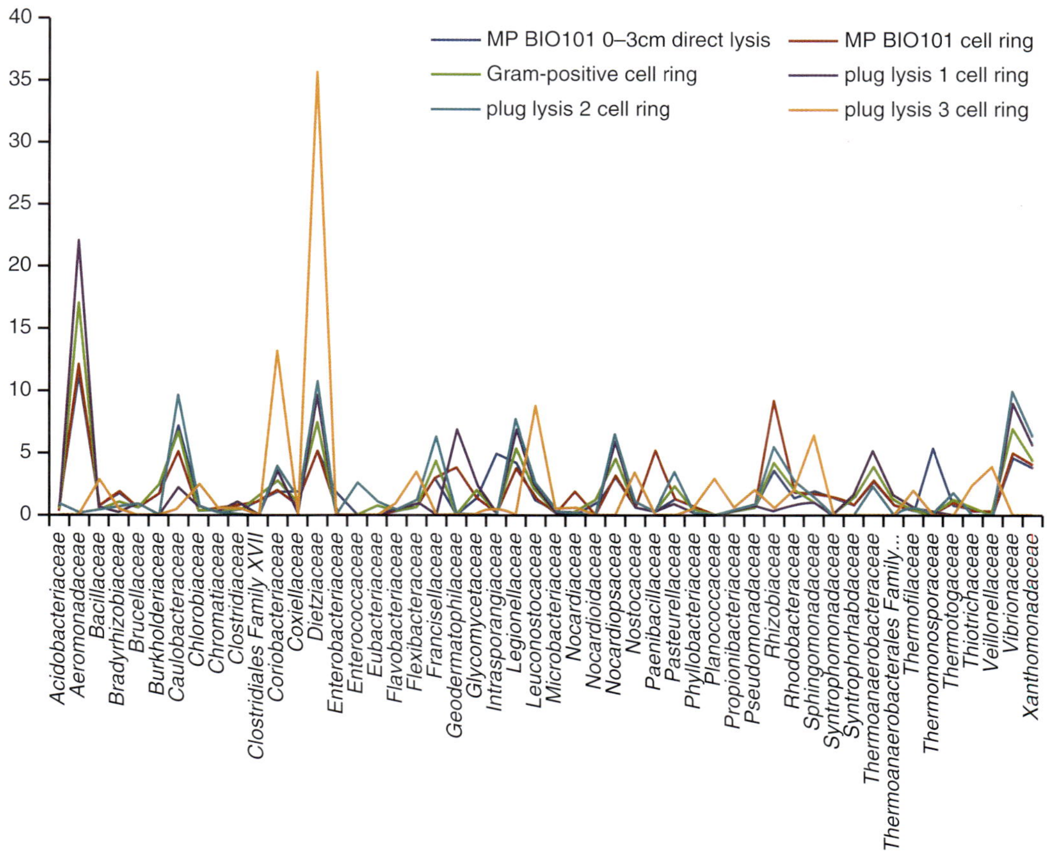

Figure 33.5 Relative proportion of families as a function of the different lyses. The intensity corresponds to the hybridization fluorescence signals.

less than for the three other DNA separation methods (soil, cells, and lysis stringency variables), but they suggest a bias involved in fosmid library constructions as DNA needs to have a molecular length between 25 and 40 kbp.

33.7 METAGENOME COMPARISONS

In spite of its ambitious objectives, Metasoil is just one of a multitude of environmental sequencing projects ongoing in laboratories from around the world [Dinsdale et al., 2008; Willner et al., 2009]. Thus, in addition to data production, there will be future requirements for analyzing large datasets. In the future, bioinformatics will be a critical tool to describe the distribution and function of the 10^{30} prokaryotes on earth. However, the specific problems, biases, and limits of these comparisons need to be well understood in order to accurately assess these data. Moreover, this understanding is crucial to choose the best normalization to compare the distribution of species and functions. A nonexhaustive list of 10 variables that might limit strong metagenomic data comparisons is suggested here:

1. The metagenome diversity can vary tremendously between environments. This diversity is known to be high in sediments and soils but limited in animals and very low in some extreme environments like acid mine biofilms.
2. The average genomes sizes change as a function of the environment.
3. The metagenome variability at a fine scale can be important in specific environments.
4. The metagenome variability over time can be high in some environments and low in stable ecosystems.
5. The metagenomic sequencing depth is unequal due to variable efforts in different projects.
6. The DNA extraction approaches used are often different, thus leading to sequencing different diversities.
7. The length of sequences varies as a function of the sequencing technology used and can influence the number of functions and species detected as well as the quality of the sequence match.
8. The phylogenetic distribution of the complete sequences used for sequence annotations is biased.

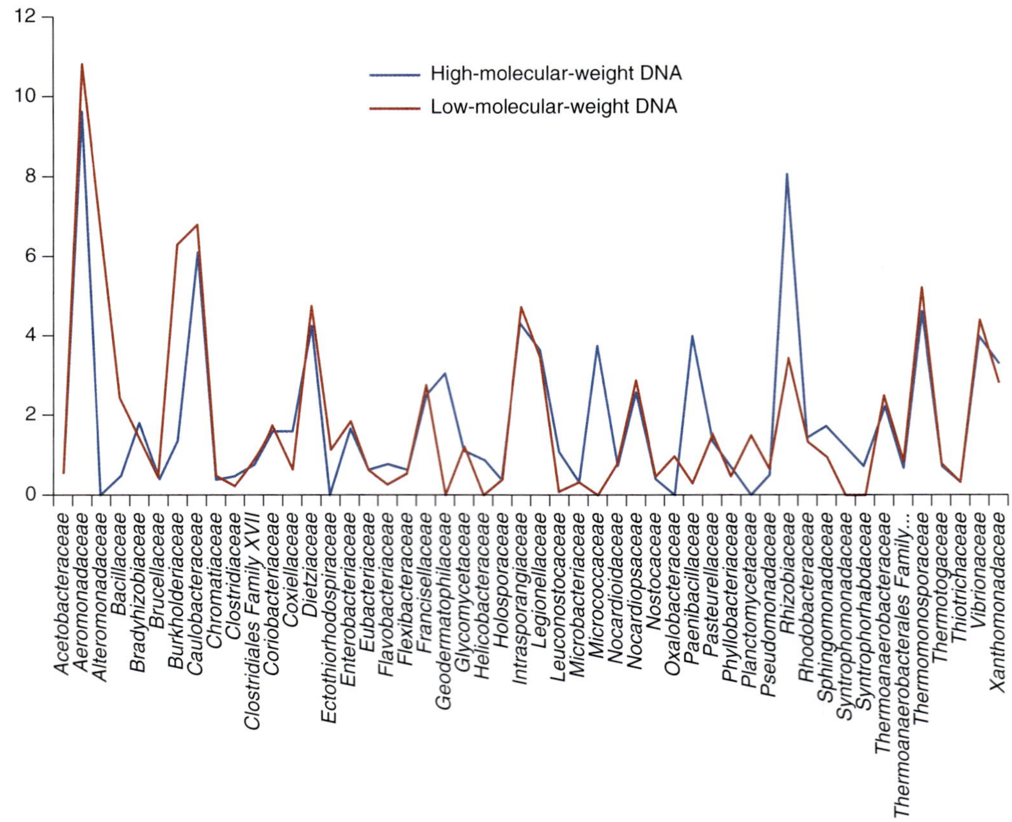

Figure 33.6 Relative proportion of families as a function of the DNA molecular weight. The intensity corresponds to the hybridization fluorescence signals.

Table 33.1 Five Characteristics Corresponding to Seven Metagenomes from Oceans, Soils, and Human Feces

Metagenome	Average of Sequences	Number of Sequences	Number of Base Pairs	Uploading Date	Functions Annotated (%)
Ocean1 (4441573.3)	1012.13	317180	321026307	Tue Oct 21 15:52:12 2008	72.18
Ocean2 (4441574.3)	1007.74	368835	371688861	Tue Oct 21 15:52:47 2008	69.71
Ocean3 (4443713.3)	238.87	217549	51966974	Wed Jun 3 16:11:04 2009	39.32
Soil1 (4445203.3)	533.8	609811	325515132	Mon Oct 26 11:57:26 2009	38.03
Soil2 (4445205.3)	538.93	552268	297632387	Mon Oct 26 12:06:43 2009	32.3
Human feces1 (4440825.3)	190.39	495865	94405318	Fri May 2 13:38:05 2008	25.64
Human feces2 (4440613.3)	335.01	302780	101434082	Thu Feb 14 10:45:23 2008	32.71

9. The databases evolved in time.
10. The percentage of annotated sequences can be highly varied as a function of the metagenomes.

Of course, these 10 variables do not all have the same influence on metagenomic comparisons, but it is difficult to know which are critical and which are not. Moreover, some of these variables are only due to the inconsistency of the metagenomes, and thus they are difficult to avoid. In order to determine if these variables are incompatible with metagenomic comparisons, seven metagenomes were analyzed (MG-RAST online server, http://metagenomics.nmpdr.org/) and the annotated function distribution was compared (E value $<10^{-5}$). As expected, the functional characteristics of the metagenomes are clearly different (**Table 33.1**).

Based on either the number of base pairs, the number of sequences, or the percentage of annotated sequences, three data normalizations can be compared based on metabolic system distributions. A PCA corresponding to the functional frequency of these metagenomes was produced as a function of these normalizations (**Fig. 33.7**). The normalization based on the total number of annotated

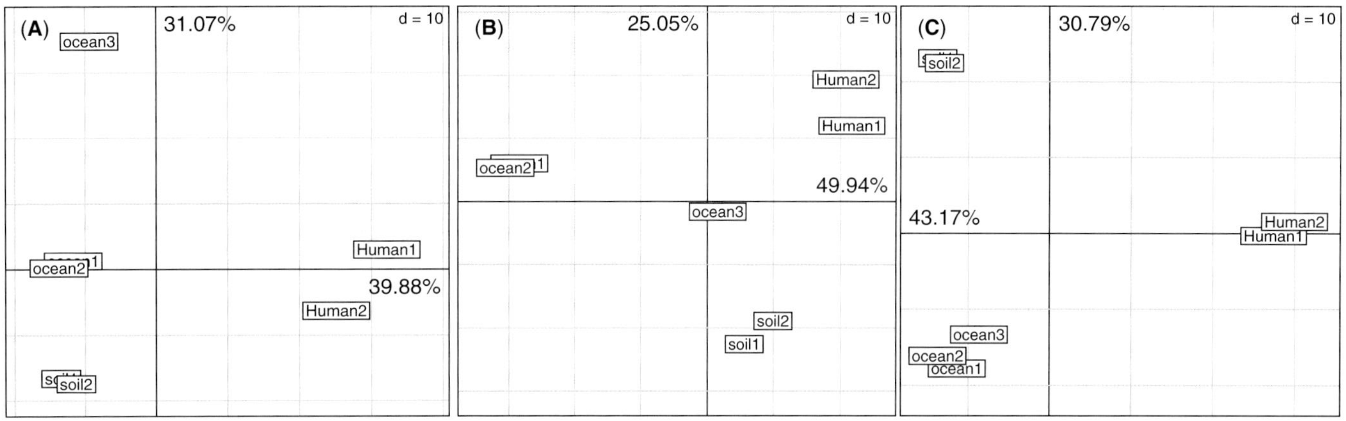

Figure 33.7 PCA corresponding to the functional distribution of seven metagenomes sequenced from oceans, soil, and human feces as a function of three normalization possibilities. (**A**) Normalization as a function of the number of base pairs sequenced; (**B**) normalization as a function of the number of sequences; (**C**) normalization as a function of the annotated sequences.

sequences appears to be the most consistent method of the three in order to reduce the impact of variation related to the different sequencing projects.

This successful comparison suggests that global comparisons of different ecosystems are possible and that the different biases are insignificant when compared to the large differences in metabolic systems between these ecosystems. This might not be the case when comparing within ecosystems.

33.8 CONCLUSION

In order to fully assess the biodiversity in one soil, deep sequencing is required. The possible sequencing depth is a function in part of the DNA extraction methods and how rare DNA can be extracted and sequenced. An initiative to sequence the Rothamsted soil was started with a French national project (Metasoil), which will provide 100 Gpb of pyrosequences and two million fosmid (40-kb inserts) clones. By pushing back the frontiers of one soil, other soil sequencing projects will be able to assess the likelihood of novel discoveries with their soils. If the rare biosphere exists in all ecosystems like in soil, then in the future these other environments need to be deeply sequenced to discover new active molecules and for global metagenomic comparisons.

For successful deep sequencing, DNA extraction methods need to be developed in the different environments. In fact, to date, only yield and purity optimization efforts are generally used by researchers [e.g., Lemarchand et al., 2005; Krasova-Wade and Neyra, 2007] to access metagenomic DNA. Other parameters like molecular weight DNA can be optimized for specific projects (e.g., fosmid library construction) [Bertrand et al., 2005]. Nevertheless, results presented in this chapter underline the incapacity of this kind of approach to access complete soil metagenomes. The international community needs to improve diversity recovery optimization strategies and particularly for deep sequencing metagenome projects, such as we have done for Metasoil. This DNA acquisition effort will decrease biases involved in metagenomic approaches and increase our understanding about the 10^{30} prokaryotes present on Earth and improve our rate of discovery of genes of interests.

In addition, it is important to enhance metatranscriptomic projects (see also Chapter 62–65, Vol. I) to study not only the genetic potential of an environment, but also the active genes. In this way, the importance of the different ecosystems (especially soil) for different services (nutrient cycling, climate change buffering, crop production and protection, drug discovery) can be accurately assessed.

Acknowledgments

Work described in this chapter that was conducted in the authors' laboratory has been supported by the French National Research Agency (ANR GMGE Metasoil), and TD was financed by the Rhone–Alpes region.

REFERENCES

Adesina MF, Lembke A, Costa R, Speksnijder A, Smalla K. 2007. Screening of bacterial isolates from various European soils for *in vitro* antagonistic activity towards Rhizoctonia solani and *Fusarium oxysporum*: Site-dependent composition and diversity revealed. *Soil Biol. Biochem*. **39**:2818–2828.

Amann RI, Ludwig W, Schleifer KH. 1995. Phylogenetic identification and *in situ* detection of individual microbial cells without cultivation. *Microbiol. Rev*. **59**:143–169.

Baker-Austin C, Wright MS., Stepanauskas R, McArthur JV. 2006. Co-selection of antibiotic and metal resistance. *Trends. Microbiol*. **14**:176–182.

Berry AE, Chiocchini C, Selby T, Sosio M, Wellington EM. 2003. Isolation of high molecular weight DNA from soil for cloning into BAC vectors. *FEMS Microbiol. Lett*. **223**:15–20.

Bertrand H, Poly F, Van VT, Lombard N, Nalin R, Vogel TM, Simonet P. 2005. High molecular weight DNA recovery from soils prerequisite for biotechnological metagenomic library construction. *J. Microbiol. Methods* **62**:1–11.

Boubakri H, Beuf M, Simonet P, Vogel TM. 2006. Development of metagenomic DNA shuffling for the construction of a xenobiotic gene. *Gene* **375**:87–94.

Courtois S, FrostegAArd A, Goransson P, Depret G, Jeannin P, Simonet P. 2001. Quantification of bacterial subgroups in soil: Comparison of DNA extracted directly from soil or from cells previously released by density gradient centrifugation. *Environ. Microbiol*. **3**:431–439.

Courtois S, Cappellano CM, Ball M, et al., 2003. Recombinant environmental libraries provide access to microbial diversity for drug discovery from natural products. *Appl. Environ. Microbiol*. **69**:49–55.

Curtis TP, Sloan WT, Scannell JW. 2002. Estimating prokaryotic diversity and its limits. *Proc. Natl. Acad. Sci. USA* **99**:10494–10499.

Davis KE, Joseph SJ, Janssen PH. 2005. Effects of growth medium, inoculum size, and incubation time on culturability and isolation of soil bacteria. *Appl. Environ. Microbiol*. **71**:826–834.

Dcosta VM, Griffiths E, Wright GD. 2007. Expanding the soil antibiotic resistome: Exploring environmental diversity. *Curr. Opin. Microbiol*. **10**:481–489.

Delmont TO, Robe P, Cecillon S, Clark IM, Constancias F, Simonet P, Hirsch PR, Vogel TM. 2011. Accessing the soil metagenome for studies of microbial diversity. *Appl. Environ. Microbiol.* **77**:1315–1324.

Demaneche S, Sanguin H, Poté J, Navarro E, Bernillon D, Mavingui P et al., 2008. Antibiotic-resistant soil bacteria in transgenic plant fields. *Proc. Natl. Acad. Sci. USA* **105**:3957–3962.

Dib J, Motok J, Zenoff VF, Ordonez O, Farias ME. 2008. Occurrence of resistance to antibiotics, UV-B, and arsenic in bacteria isolated from extreme environments in high-altitude [above 4400m] Andean wetlands. *Curr. Microbiol*. **56**:510–517.

Dinsdale, EA, Edwards RA, Hall D et al., 2008. Functional metagenomic comparison profiling of nine biomes. *Nature* **452**: 629–632.

Falkowski PG. 2001. Biogeochemical cycles. *Encyclopedia Biodivers*. **1**:437–453.

Galvao TC, Mohn WW, de Lorenzo V. 2005. Exploring the microbial biodegradation and biotransformation gene pool. *Trends Biotechnol*. **23**:497–506.

Gans J, Wolinsky M, Dunbar J. 2005. Computational improvements reveal great bacterial diversity and high metal toxicity in soil. *Science* **309**:1387–1390.

Gebert J, Stralis-Pavese N, Alawi M, Bodrossy L. 2008. Analysis of methanotrophic communities in landfill biofilters using diagnostic microarray. *Environ. Microbiol*. **10**:1175–1188.

Ginolhac A, Jarrin C, Gillet B, Robe P, et al., 2004. Phylogenetic analysis of polyketide synthase I domains from soil metagenomic libraries allows selection of promising clones. *Appl. Environ. Microbiol*. **70**:5522–5527.

Ginolhac A, et al., 2005. Type I polyketide synthases may have evolved through horizontal gene transfer. *J. Mol. Evol*. **60**:716–725.

Griffiths RI, Whitely AS, O'Donnell AG, Bailey MJ. 2000. Rapid method for co-extraction of DNA and RNA from natural environments for analysis of ribosomal DNA- and rRNA-based microbial community composition. *Appl. Environ. Microbiol*. **66**:5488–5491.

Handelsman J, Rondon MR, Brady SF, Clardy J, Goodman RM. 1998. Molecular biological access to the chemistry of unknown soil microbes: A new frontier for natural products. *Chem. Biol*. **5**: R245–R249.

Handelsman J, Wackett LP. 2002. Ecology and industrial microbiology: Microbial diversity—sustaining the Earth and industry. *Curr. Opin. Microbiol*. **5**:237–239.

Hery M, Nazaret S, Jaffre T, Normand P, Navarro E, 2003. Adaptation to nickel spiking of bacterial communities in neocaledonian soils. *Environ. Microbiol*. **5**:3–12.

Huber JA, Mark Welch DB, Morrison HG, Huse SM, Neal PR, Butterfield DA, Sogin ML. 2007. Microbial population structures in the deep marine biosphere. *Science* **318**:97–100.

Huyghe A et al. 2008. Novel microarray design strategy to study complex bacterial communities. *Appl. Environ. Microbiol*. **74**:1876–1885.

Knietch A, Waschkowitz T, Bowien S, Henne A and Daniel R. 2003. Metagenomes of complex microbial consortia derived from different soils as sources for nobel genes conferring formation of carbonyls from short-chain polyols on *Echerichia coli*. *J. Microbiol. Biotechnol*. **5**:46–56.

Krasova-Wade T, Neyra M. 2007. Optimization of DNA isolation from legume nodules. *Lett. Appl. Microbiol*. **45**:95–99.

Larose C, Berger S, Ferrari C, Navarro E, Dommergue A, Schneider D, Vogel TM. 2010. Microbial sequences retrieved from environmental samples from seasonal Arctic snow and meltwater from Svalbard, Norway. *Extremophiles* **14**:205–212.

Lefevre F, Robe P, Jarrin C, Ginolhac A, Zago C, Auriol D, et al. 2008. Drugs from hidden bugs: their discovery via untapped resources. *Res. Microbiol*. **159**:153–161.

Lemarchand K, Berthiaume F, Maynard C, Harel J, Payment P, Bayardelle P, Masson L, Brousseau R. 2005. Optimization of microbial DNA extraction and purification from raw wastewater samples for downstream pathogen detection by microarrays. *J. Microbiol. Methods* **63**:115–126.

Leveau JHJ. 2007. The magic and menace of metagenomics: Prospects for the study of plant growth-promoting rhizobacteria. *Eur. J. Plant Pathology* **11**:279–300.

Meyer F, Paarmann D, D'Souza M, Olson R, Glass EM, et al. 2008. The metagenomics RAST server—A public resource for the automatic phylogenetic and functional analysis of metagenomes *BMC Bioinformatics* **9**:386.

Morales SE, Cosart TF, Johnson JV, Holben WE. 2009. Extensive phylogenetic analysis of a soil bacterial community illustrates extreme taxon evenness and the effects of amplicon length, degree of coverage, and DNA fractionation on classification and ecological parameters. *Appl. Environn. Microbiol*. **75**:668–675.

Orgam A, Sayler GS, and Barbay T. 1987. The extraction and purification of microbial DNA from sediments. *J. Microb. Methods* **7**:57–66.

Ortíz-Castro R, Contreras-Cornejo HA, Macías-Rodríguez L, López-Bucio J. 2009. The role of microbial signals in plant growth and development. *Plant Signal Behav*. **4**:701–712.

Pasic L, Kovce B, Sket B, Herzog-Velikonja B. 2010. Diversity of microbial communities colonizing the walls of a Karstic cave in Slovenia. *FEMS Microbiol. Ecol*. **71**:50–60.

Pointing SB, Chan Y, Lacap DC, Lau MC, Jurgens JA, Farrell RL. 2009. Highly specialized microbial diversity in hyper-arid polar desert. *Proc. Natl. Acad. Sci. USA* **106**:19964–19969.

Pyrosequencing enumerates and contrasts soil microbial diversity. *ISME Journal*. **1**:283–290.

Rajendhran J, Gunasekaran P. 2008. Strategies for accessing soil metagenome for desired applications. *Biotech. Adv*. **26**:576–590.

Ranjard L, Richaume AS. 2001. Quantitative and qualitative microscale distribution of bacteria in soil. *Res. Microbiol*. **152**:707–716.

Richard JY, Vogel TM. 1999. Characterization of a soil bacterial consortium capable of degrading diesel fuel. *Intern. Biodet. Biodegrad*. **44**:93–100.

Robe P, Nalin R, Capellano C, Vogel TM, Simonet P. 2003. Extraction of DNA from soil. *Euc. J. Soil Biol*. **39**:183–190.

Roesch LL, Fulthorpe RR, Riva A, Casella G, Hadwin AKM, Kent AD, et al., 2007.

Sanguin H et al., 2006. Development and validation of a prototype 16S rRNA-based taxonomic microarray for Alphaproteobacteria. *Environ. Microbiol*. **8**:289–307.

Shendure J, Ji H. 2008. Next-generation DNA sequencing. *Nat. Biotech*. **26**:1135–1145.

Silvertown J, et al., 2006. The Park Grass Experiment 1856–2006: Its contribution to ecology. Pyrosequencing enumerates and contrasts soil microbial diversity. *ISME J*. **1**:283–290.

Singh S, Kang SH, Mulchandani A, Chen W. 2008. Bioremediation: Environmental clean-up through pathway engineering. *Curr. Opin. Biotechnol*. **19**:437–444.

Staley JT, Konopka A. 1995. Measurement of *in situ* activities of nonphotosynthetic microorganisms in aquatic and terrestrial habitats. *Annu. Rev. Microbiol*. **39**:321–346.

Torsvik V, Goksoyr J, Daae FL. 1990. High diversity in DNA of soil bacteria. *Appl. Environ. Microbiol*. **56**:782–787.

Torsvik V, Ovreas L, Thingstad TF. 2002. Prokaryotic diversity—Magnitude, dynamics, and controlling factors. *Science* **296**:1064–1066.

Tringe SG, Mering CV, Kobayashi A, Salamov AA, Chen K, Chang HW et al., 2005. Comparative metagenomics of microbial communities. *Science* **308**:554–557.

Turnbaugh PJ et al., 2006. An obesity-associated gut microbiome with increased capacity for energy harvest. *Nature* **444**:1027–1031.

van Elsas JD, Mantynen V, Wolters AC. 1997. Soil DNA extraction and assessment of the fate of *Mycobacterium cholorophenolicum* strain PC-1 in different soils by 16S ribosomal gene sequence based most probable number PCR and immunofluorescence. *Biol. Fertil. Soils* **24**:188–195.

van Elsas JD, Jansson JK, Trevors JT. 2006. *Modern Soil Microbiology II*, Boca Raton, FL: CRC Press.

Vogel TM, Simonet P, Jansson JK, Hirsch PR, Tiedje JM, van Elsas JD, et al. 2009. TerraGenome: A consortium for the sequencing of a soil metagenome. *Nat. Rev. Microbiol*. **7**:2.

Wainwright M, Wickramasinghe NC, Narlikar JV, Rajaratnam P. 2003. Microorganisms cultured from stratospheric air samples obtained at 41km. *FEMS Microbiol Lett*. **218**:161–165.

Ward BB et al. 2007. Ammonia-oxidizing bacterial community composition in estuarine and oceanic environments assessed using a functional gene microarray. *Environ. Microbiol*. **9**:2522–2538.

Whitman WB, Coleman DC, Wiebe WJ. 1998. Prokaryotes: The unseen majority. *Proc. Natl. Acad. Sci. USA* **95**:6578–6583.

Willner D, Thurber RV, Rohwer F. 2009. Metagenomic signatures of 86 microbial and viral metagenomes. *Environ. Microbiol*. **11**:1752–1766.

Chapter 34

The BIOSPAS Consortium: Soil Biology and Agricultural Production

Luis Gabriel Wall

34.1 NEW PARADIGMS IN ARGENTINE AGRICULTURE

Argentina is an agricultural country in the context of the world's economies mainly due to its geographical and cultural characteristics. The application of technology to agriculture has always been a distinctive characteristic of Argentine agricultural producers and entrepreneurs who set out to find new horizons that would combine productivity and sustainability. This shift in mentality in a context of great revenue in the international agricultural business encouraged some producers and agro-companies to allocate part of their investment toward the development of new agro-knowledge.

Of all the processes that have taken place throughout the agricultural history of Argentina, it is the incorporation of biotechnology, by means of a wide variety of genetically modified organisms and the adoption of no-till agriculture as a new soil management practice, that has entailed a real green revolution. However, these changes are not exclusive to Argentina. They also characterize a great part of South America, including countries like Brazil, Paraguay, and Uruguay [Derpsch, 2008].

No-till or conservation agriculture are different ways of referring to a series of activities that constitute a virtuous circle whose greatest virtue is that of allowing sustainable food production with an emphasis on soil preservation.

This new soil management system has marked a radical shift in paradigm for farmers, who had to abandon traditional practices of conventional farming whose main goal was weed and disease control and, supposedly, incorporation of organic matter and soil airing, practices that tend toward improving the implantation of the new crop [Derpsch, 2008; Benites, 2008].

Soil under no-till management with crop rotation is characterized by the accumulation of plant litter on its surface. Somehow no-till cultivated soil resembles soil in the woodlands where the accumulation of cuttings and litter is a distinctive characteristic of this ecosystem. Biologically speaking, the main change that occurs in soil when conventional till shifts to no-till practices is that soil trophic levels are no longer dominated by bacteria. Complexity and diversity are restored on every trophic level, and functional groups of fungi, protozoans, and meso- and macrofauna are restored [Blank, 2008].

Soil coverage has several advantages. The first to be appreciated by farmers is water management. No-till management avoids water runoff and soil humidity evaporation; it lowers surface temperature; and it creates microecological niches for the restoration of biological diversity, especially microbial, mesofaunal, and macrofaunal (worms). Soil coverage and the slow decomposition of cuttings grant gradual incorporation of nutrients and enrichment of soil organic matter, which is the basis for new biological diversity and creates a virtuous circle for biomass production. On the the level of physics, no-till farming preserves or recovers the structure of the soil on all levels. The adoption of no-till farming technology experienced an exponential growth in its area of application between 1987 and 2004. The total agricultural area under this type of management is now 59 times bigger than in 1987 [Derpsch, 2008]. Nowadays, 20,000 ha in Argentina are farmed under no-till

Handbook of Molecular Microbial Ecology, Volume I: Metagenomics and Complementary Approaches, First Edition. Edited by Frans J. de Bruijn.

systems. This represents 85% of the total agricultural production of the country (source: AAPRESID web site, www.aapresid.org.ar).

The reason for the shift in paradigm entailed by the adoption of no-till farming and the technology associated with this practice is twofold: (1) the decrease in production costs and consequent greater economic yield at a time when the international price of grain is favorable for producer countries like Argentina and (2) the awareness that, when carried out well, these practices aim toward soil conservation, which is the producers' ultimate production and profit source.

This new agriculture management rapidly highlighted the need for new working hypotheses to determine soil quality indexes, and it evidenced the lack of knowledge about how soil works under this farming system in which, through trial and error, farmers are gradually incorporating new technologies for weed, disease, and fertilizer management. Farmers define all these adopted technologies in combination with no-till management as good agricultural practices on the basis of economic yield alongside soil conservation and even enrichment as regards fertility and productivity.

But not everything has been successful in the no-till scenario. The massive adoption of technology has not always been in the hands of farmers. Agriculture is no longer exclusive to them. New economic players have joined the farming activity, but the action of those investors within a context of great agricultural development and no land-use regulations is characterized by short-term investment without care about the conservation of the resource. These activities only seek short-term investment yield on the basis of land lease and soybean monocropping.

The absence of crop rotation is a real problem in the medium- and long-term soil productivity. Monocropping acts as a selection pressure on the development of potential crop pathogens. Pathogenic fungi remain and develop in plant litter after harvest, and they are slowly incorporated into the soil until the next crop. If agriculture management is based on monocropping, then the scenario for plant disease development is perfect for that particular host, and most probably the consequences will amount to considerable economic losses in yield. For such a problem, crop rotation appears to be the best way of preventing plant diseases; at the same time, this situation is an excellent opportunity for the development of new biocontrol strategies.

34.2 CERTIFIED AGRICULTURE AND THE NEW CHALLENGES

In the words of farmers, the evolution of no-till agriculture is certified agriculture. Farmers who produce with ecological awareness aimed at sustainability of the process demand recognition of their practices. For such purpose, they have proposed a Good Agricultural Practices Certification granted on the basis of different indexes, mainly soil quality ones. At present, those indexes are physically and chemically based. Lack of biologically based soil quality indexes in no-till practices in Argentina has become patent.

34.3 PRIORITY RESEARCH AREAS IN ARGENTINA AND THE CREATION OF THE BIOSPAS CONSORTIUM

In the context previously described, the Soil Biology and Sustainable Agricultural Production—BIOSPAS (acronym of its Spanish name)—was created in 2007, and it started to work in the field in 2009. BIOSPAS was organized under the auspices of the Argentine Ministry of Science, Technology and Productive Innovation through its Strategic Area Projects. The BIOSPAS project combines the involvement of 12 research Laboratories located at different Universities or public sector Research Institutes (see **Table 34.1**). a Nongovernmental Organization coordinated by the Argentine Association of No-Till Farmers and two private companies, La Lucía S.A., and Rizobacter Argentina S.A. The BIOSPAS web site is at www.biospas.org.

The goal of BIOSPAS is to use a systemic approach to understand how soil works in this new agriculture management, which means getting a polyphasic description of biological diversity of the soil comparing three different agricultural management situations in four environment replicates. This polyphasic description includes data generated by genomics, metagenomics, and other molecular approaches (see also Chapters 5 and 28 Vol. I and Chapters 9–17, Vol. II). Thus, the treatments are as follows:

- Natural Environment: Environment with minimal human activity—that is, "sanctuaries" that have remained untouched in some farming establishments.
- Very Good Agricultural Management: Permanent no-tillage management using intensive crop rotation, fertilization, integrated pest, undergrowth, and disease control, coverage crops, and basic Certified Agriculture concepts.
- Bad Agricultural Management: No-tillage fields cultivated without crop rotation and with deficient fertilization.

The three treatments are replicated in four different environments on a west–east transect in the central part of the Argentine Pampas, where the most productive agricultural area is located. The selected locations close to the

Table 34.1 Members of the BIOSPAS Consortium and Their Roles in the Global project[a]

Lab Leader	Affiliation	Speciality within BIOSPAS
Mario Aguilar, Ph.D.	IBBM-CONICET, Facultad de Ciencias Exactas de la Universidad Nacional de La Plata	Metagenomic study of nitrogen-fixing genes and phosphorus mobilization; diversity and abundance of microorganism populations related to those metabolisms.
José Bedano, Ph.D.	Facultad de Ciencias Exactas Físico-Químicas y Naturales, Universidad Nacional de Río Cuarto	Meso- and macrofauna.
Marcelo Carmona, Agr. Eng. M. Sc.	Fitopatología de la Facultad de Agronomía, Universidad de Buenos Aires	Incidence and isolation of pathogens in the sampled environments.
Laura Domínguez, Ph.D.	IMBIV, Facultad de Ciencias Exactas Físicas y Naturales, Universidad Nacional de Córdoba	Morphological analysis of endomycorrhizae.
Daniel Ducasse, Ph.D.	Instituto de Fitopatología y Fisiología Vegetal (IFFIVE-INTA), Fitopatología, Facultad de Agronomía, Universidad Católica de Córdoba	Analysis of endomycorrhizae through molecular techniques.
Leonardo Erijman, Ph.D.	INGEBI-CONICET, Buenos Aires	Molecular analysis of soil microbial diversity and metagenomic studies of nitrogen cycle genes.
Juan Galantini, Ph.D.	CERZOS-CONICET, Bahía Blanca	Analysis of organic matter fractions.
María Iglesias, Agr. Eng. M.E.M.	Microbiología Agrícola, Facultad de Ciencias Agrarias, Universidad Nacional del Nordeste, Corrientes	Analysis of cellulolytic bacteria and free nitrogen fixers.
Ricardo Marona, Ph.D.	Departamento de Matemática, Facultad de Ciencias Exactas, Universidad Nacional de La Plata	Statistical analysis.
Héctor Morrás, Ph.D.	Instituto de Suelos, CIRN-INTA, Castelar	Physical structure of soil.
Claudio Valverde, Ph.D.	PIB, Departamento de Ciencia y Tecnología, Universidad Nacional de Quilmes	Analysis of Pseudomona bacteria group.
Luis Wall, Ph.D.	PIB, Departamento de Ciencia y Tecnología, Universidad Nacional de Quilmes	Soil biochemistry, enzymes, lipids and glomalin.
Ángeles Zorreguieta, Ph.D.	Fundación Instituto Leloir, Buenos Aires	Analysis of Burkholderia group. Biofilm formation.

[a] All of these members have equal contribution and participation in the budget. The list is organized alphabetically.

experimental fields are Bengolea (Province of Córdoba), Monte Buey (Province of Córdoba), Pergamino (Province of Buenos Aires), and Viale (Province of Entre Ríos). The soils and environments at those sites range from sandy soil with less rain in the west (Bengolea) to wetter clay soil in the east (Viale).

The soils will be sampled twice a year, in early August (winter crops in vegetative state) and in early February (summer crops in vegetative state), for four years.

34.4 NECESSARILY COMPLEMENTARY APPROACHES

Even if we know quite a lot about soil microbiology as a result of the development of modern molecular techniques, nowadays we are aware of the fact that what we ignore is much more than what we know. In other words, the black box of the soil contains inhabitants unknown to us in terms of identity, functioning, and organization within the soil system, which now is approachable using the molecular toolbox generated by molecular biologists world wide. From a more pragmatic approach, we also ignore the extent to which their existence may determine land productivity and fertility—that is to say, their food yield potential.

This is our challenge: to develop comprehensive knowledge and tools that will allow us to shed some light on the black box of the soil and to find the way to use that knowledge toward a more sensible and efficient soil management in our agricultural practices.

In order to do so, considering our limited budget and to the best of our present knowledge, we will use very different approaches to cover as much soil biology and as many biochemical functions as possible, through different analyses that are carried out on the same sites and on the same physical soil samples. The analyses are as follows.

Soil Chemical Composition, Structure, and Physical Behavior. These studies will give us an idea of the environment where the soil biology develops, which is the stage for Nature's invisible performance in soil. Soil particle–microbe interaction is very dynamic, and we should be aware of the fact that the extent to which the stage conditions microbial diversity and whether the reverse is true still remains an open question. Several soil physical parameters related to granulometry, structure, porosity, and hydric dynamics will be studied [Gil, 1998]. Soil microstructure and porosity will be analyzed in thin soil sections by microscopic analysis techniques [Stoops, 2003]. As part of our systemic approach, these soil sections will be considered tissue sections of soil taken as a whole organism. As one of the final goals of the project, we hope to be able to try different molecular tools developed in other areas of the project (see below) on these soil slides in order to identify and localize different microorganisms and/or substances and to learn about the microbiogeography at this level.

Analysis of Organic Matter Fractions. Organic matter is probably one of the most widely accepted quality indexes for agricultural soils in Argentina [Galantini, 2007]. In order to investigate detailed organic matter structure, the soil chemistry laboratory will determine the equilibrium of different organic matter fractions [Senesi and Loffredo, 1998], enriched labile, and their carbon sequestration capacity in different soil management situations, including those specific to the experimental design of the BIOSPAS project. The goal is to quantify the relation between organic matter fractions and biological properties of the soil in order to use these new indexes as diagnosis and monitoring tools.

Biochemical Analyses of Soil Samples. In order to characterize soil functioning as a whole, different biochemical substances will be quantified and characterized. Enzymatic activity will be measured using fluorometric techniques with 4MUB-derived substrates that will allow us not only to quantify enzymatic activity but also to characterize it by estimating its kinetic constants using the multi-well plate technique [García-Ruiz et al., 2008; Marx et al., 2001]. Fatty acid profile obtained from total lipids or phospholipid fraction will be analyzed to determine both taxon abundance and diversity indexes [Zelles, 1999].

Glomalin-Related Soil Proteins (GRSP). The term *glomalin* refers to a soil protein fraction related to arbuscular mycorrhizal fungi (AMF), which are important for the stabilization of soil aggregates [Rillig, 2004]. Thus, it has turned out to be used as a soil quality indicator. In the context of the BIOSPAS project, GRSP will be quantified and the biochemical characterization of this fraction will be analyzed in relation to AMF data results and the physical properties of the soils under study.

Analysis of Cellulolytic Bacteria. The no-till practice scenario is characterized by the accumulation of plant debris as crops are cultivated and harvested. This biomass is later integrated into the soil by natural decomposition as a substrate with a high C/N ratio. In order to look for new biological indicators and obtain new isolates with potential biotechnological applications, the cellulolytic bacteria group will be studied. Both classical and molecular microbiological approaches will be used to track specific genes as biological markers of these functional groups [Bloem et al, 2006]. In the case of cellulolytic bacteria, the growth on cellulose paper is characterized by the form of gums that are supposed to be a consortium of microorganisms that cannot be easily cultivated separately. In order to identify the organisms as being part of that consortium, 16S sequences will be performed on total DNA isolated from the gum after amplification and DGGE separation.

Analysis of Pseudomona Bacteria Group. This bacterial group is one of the most widely studied in soils [Cho and Tiedje, 2000; Palleroni, 2003], and this interest derives from the fact that many attractive biological activities have been found in these bacteria, either for plant growth promotion or for biocontrol purposes [Haas and Keel, 2003]. The Pseudomona group, with emphasis on the *P. fluorescens* group, will be studied in bulk soil and in rhizospheric soil using both traditional plate cultivation and molecular approaches in order to get new isolates and to estimate the size and structure of their population [Picard and Bosco, 2008]. Diversity of cultivated *P. fluorescens* will be analyzed by RFLP of particular gene characteristics for the *P. fluorescens* genus as *oprF* [Bodilis and Barray, 2006; Bodilis et al., 2006], and genes related to plant-growth-promoting activities and biocontrol will be analyzed as *gacA* [Haas et al, 2000; Costa et al, 2007; de Souza et al., 2003] and *phlD* [Mazzola et al., 2004; McSpadden Gardener et al, 2001; Raaijmakers and Weller, 2001; Ramette et al., 2001]. The same genes will be used to characterize diversity of total *Pseudomonas* population (cultivable and uncultivable) by DGGE of PCR products amplified from total DNA obtained from bulk soil or rhizospheric soil samples. The size of Pseudomonas population will be estimated by qPCR of specific genes, mentioned above, and also by most probable number applied to specific PCR [Picard and Bosco, 2008].

Analysis of Burkholderia Bacteria Group. Considering the increasing interest in this bacterial group as regards soil activity and plant interactions, a similar approach to the study of the *Pseudomona* group will be applied to the *Burkholderia* [Mahenthiralingam et al., 2005].

Development of Tools to Study Biofilm Formation in the Soil Particle. Little is known about how microbial communities are organized in the soil particles [Burmølle et al., 2007]. It is already known, although at a preliminary level, that the soil structure at the microscale is different under different agricultural managements. How are microbes organized in such soil particles and microstructures? An additional aim of the project is to develop tools to study biofilm formation and biofilm communities in the soil samples under study according to history of land use through noninvasive labeling methods [Lisle et al., 1999; Larrainzar et al., 2005].

Molecular Analysis of the Microbial Diversity Present in the Soil and Abundance of Populations of Microorganisms. Molecular analysis techniques broadly surpass many traditional cultivation limitations and allow for the procurement of more complete and detailed information on microbial community structures in natural environments [Roesch et al., 2007; Curtis and Sloan, 2004; Hong et al., 2006, see Chapters 5 Vol. I and Chapters 9-17, Vol. II. Taxon abundance patterns will be determined by quantitative PCR and also by tag-pyrosequencing of 16S gene, so as to describe microbial community structure in the different soil samples. Considering the distributions of replicates for the different soil managements under studies, we hope to be able to distinguish between a biogeographic pattern of bacterial distribution and population structures related to soil management, as a disturbance that would affect local microbial composition [Allison and Martiny, 2008]. In order to estimate the ecological strategies of bacteria in the cases of study, the rRNA/rDNA ratio will be estimated by qPCR [Klappenbach et al., 2000; Shrestha, 2007]. Using different primers, it will be possible to go from general to particular in the analysis at the domain or taxon level. The integration of these data in relation to other information obtained in the project could incorporate a conceptual basis to interpret the taxonomic information and to correlate it with soil activity, and maybe agricultural management as well.

Metagenomic Studies of Nitrogen Cycle Genes and Phosphorous Mobilization Genes. In order to look further into relevant soil functions for agricultural practices, the metagenomic approach will be applied to functional genes involved in key biogeochemical processes. Amongst them, the processes of nitrification (conversion of ammonia into nitrate) will be particularly studied considering that representatives of both domains, amonia-oxidizing bacteria (AOB) and amonia-oxidizing Archaea (AOA), are responsible for denitrification function in soils [Leininger et al., 2006]. Using specific primers for AOB [Okano et al., 2004] and for AOA [Francis et al., 2005] and also using quantitative PCR, the ratio AOB/AOA will be determined. Denitrification (reduction of nitrate into molecular nitrogen via nitrite, nitric oxide, and nitrous oxide) will be also studied with a similar approach using specific genes [Henry et al. 2006]. To study nitrogen fixation function potential in soil samples, degenerated primers will be used to amplify the *nif*H gene, which codes for one of the subunits of nitrogenase, the enzyme that reduces N_2 to ammonia. *nif*H is one of the best-documented genes in terms of available sequences related to the nitrogen fixation function, so it is a good candidate to look for functional diversity by DGGE and RFLP analyses of the amplified gene by PCR [Rösch et al., 2002; Knauth et al., 2005]. P mobilization is more complicated to study from a metagenomic approach since many different mechanisms are involved in P mobilization. Some already identified genes involved in this soil function related to phosphatase enzymes, *Phy*A, *Pho*C, and *Nap*A, or acid production from glucose, *pqq* and *gcd*, necessary to solubilize rock phosphates, will be quantified in environmental DNA samples [Richardson et al., 2001; Rodríguez et al., 2006].

Analysis of Endomycorrhizae by Means of Morphological and Molecular Techniques. The symbiotic associations of plant roots with arbuscular mycorrhizal fungi (AMF) play a vital role in the mineral nutrition of vegetables since, by means of the hyphae that colonize soil and rhizosphere, they are involved in the acquisition of essential elements such as P, K, Cu, and Zn. AMF also play an important part in the protection of plants against infectious agents, mainly root pathogens [Smith and Read, 2010]. No-till practices restore fungal populations to agricultural soils [Schalamuk et al., 2006; Pankhurst et al., 1996], and the lack of disturbance of the soil is believed to allow the restoration of fungal nets with their functions. In this scenario, another aim of the BIOSPAS project is to determine both morphologically and molecularly the species that make up the mycorrhizal fungi populations found under different crop rotation sequences. AMF populations found in the soil and in active symbiosis with roots of different crops will be quantified and compared. Spores will be separated from soil and will be quantified and identified on morphological bases. In order to confirm spores identity, molecular characterization will be performed on the analyses of 18S rDNA sequences [Schüßler et al., 2001; da Silva et al., 2006]. A molecular approach will also be performed on environmental DNA from soil and from roots. The complexity of micorrhizal population on the basis of nested amplified 18S rDNA will be analyzed by DGGE and further sequencing [da Silva et al., 2006, Franke et al., 206; Ma et al., 2005].

Qualitative–quantitative Study of Soil Meso- and Macrofauna. Agriculture significantly modifies the diversity and abundance of soil meso- and macrofauna [Edwards and Lofty, 1969]. Recent studies approach the question of how no-till practices operate in these soil functional group [Bedano and Ruf, 2007; Bedano et al., 2006]. In order to reach a more advanced step in the trophic chain in each of the different agricultural managements studied in the BIOSPAS project, meso- and macrofauna will be analyzed in each site. Litter decomposition bags will be used in the field to test for the activity of different size groups of meso- and macrofauna. A center sampler and a Berlese–Tullgren system will be used for organism extraction from soil and for further identification and quantification of groups [Southwood, 1980].

Incidence and Isolation of Pathogens within the Studied Environments. One of the problems of no-till agriculture, mainly under monocrop management, is the development of plant diseases, most of them of fungal origin. To integrate this topic into the project and within the frame of soil microbial activity, plant diseases developed in the sites of sampling will be evaluated, and putative pathogen agents will be isolated from litter, diseased plants, and soil and will be characterized in the laboratory.

34.5 CONCLUSIONS

Most probably the initial output of this project will be a huge amount of data which will help to get a comprehensive description of the biology associated with soils under different agricultural managements. A very interactive statistical analysis will be needed to learn first the structure of the data that will come out of the project in order to adjust sampling to test future hypothesis.

Drawing from these data, we hope to develop knowledge on soil biology that will allow us to identify biological indicators of soil quality and, eventually, yield potential, which will make up an essential guide for the agricultural farmer and will set the basis for the improvement of scientific and research knowledge.

These indicators will contribute to the strengthening and improvement of the good agricultural practices certification process, an initiative of utmost importance boosted by AAPRESID, the Argentine No-till Farmers Association. Also, the several bacterial isolates that will result from this project will be useful for the development of new inoculants for future agricultural use.

The metagenomic approach, as part of the global project, is a key element because it opens the doors to the unculturable microbiology living in the soil which necessarily accounts for its function and its final productivity.

Acknowledgments

LGW is Professor at the Universidad Nacional de Quilmes (Argentina) and member of CONICET (Argentina). The BIOSPAS project is granted by the ANPCyT PAE 2007-36976.

REFERENCES

Allison SD, Martiny JBH. 2008. Resistance, resilence, and redundancy in microbial communities. *Proc. Natl. Acad. Sci. USA* **105**:11512–11519.

Bedano JC, Ruf A. 2007. Soil predatory mite communities (Acari: Gamasina) in agroecosystems of central Argentina. *Appl. Soil Ecol.* **36**:22–31.

Bedano JC, Cantu MP, Doucet ME. 2006. Soil springtails (Hexapoda: Collembola), symphylans and pauropods (Arthropoda: Myriapoda) under different management systems in agroecosystems of the subhumid Pampa (Argentina). *Euro. J. Soil Biol.* **42**:107–119.

Benites J. 2008. Effect of No-Till on Conservation of the Soil and Soil Fertility. In: Goddard T, Zoebisch MA, Gan YT, Ellis W, Watson A, Sombatpanit S, (eds). 2008. *No-Till Farming Systems*. Special Publication No. 3, World Association of Soil and Water Conservation, Bangkok, p. 59–72.

Blank D. 2008. A Fresh Look at Life below the Surface. In Goddard T, Zoebisch MA, Gan YT, Ellis W, Watson A, Sombatpanit S, (eds). 2008. *No-Till Farming Systems*, Special Publication No. 3, World Association of Soil and Water Conservation, Bangkok, p. 73–82.

Bloem J, David W, Hopkins DW, Anna Benedetti A. 2006. Microbiological Methods for Assessing Soil Quality. CABI Publishing Series, CAB International, London, 336 pages.

Bodilis J, Barray S. 2006. Molecular evolution of the major outer-membrane protein gene (*opr*F) of Pseudomonas. *Microbiology* **152**:1075–1088.

Bodilis J, Hedde M, Orange N, Barray S. 2006. OprF polymorphism as a marker of ecological niche in Pseudomonas. *Environ. Microbiol.* **8**:1544–1551.

Burmølle M, Hansen LH, Sørensen SJ. 2007. Establishment and early succession of a multispecies biofilm composed of soil bacteria. *Microbial. Ecol.* **54**:352–362.

Cho J, Tiedje J. 2000. Biogeography and degree of endemicity of fluorescent Pseudomonas strains in soil. *Appl. Environ. Microbiol.* **66**:5448–5456.

Costa R, Gomes NC, Krogerrecklenfort E, Opelt K, Berg G, Smalla K. 2007. Pseudomonas community structure and antagonistic potential in the rhizosphere: Insights gained by combining phylogenetic and functional gene-based analyses. *Environ. Microbiol.* **9**:2260–2273.

Curtis TP, Sloan WT. 2004. Prokaryotic diversity and its limits: Microbial community structure in nature and implications for microbial ecology. *Curr. Opin. Microbiol.* **7**:221–226.

da Silva GA, Lumini E, Maia LC, Bonfante P, Bianciotto V. 2006. Phylogenetic analysis of Glomeromycota by partial LSU rDNA sequences. *Mycorrhiza* **16**:183–189.

de Souza JT, Mazzola M, Raaijmakers J. 2003. Conservation of the response regulator gene gacA in Pseudomonas species. *Environ. Microbiol.* **5**:1328–1340.

Derpsch R. 2008. No-Tillage and conservation Agriculture: A Progress Report. In: Goddard T, Zoebisch MA, Gan YT, Ellis W, Watson A, Sombatpanit S (eds). 2008. *No-Till Farming Systems*. Special Publication No. 3, World Association of Soil and Water Conservation, Bangkok, p. 7–42.

Edwards CA, Lofty Jr. 1969. The influence of agricultural practice on soil micro-arthropod populations. In Sheals JG. ed.), *The Soil Ecosystem*. Systematics Association Publication No. 8, London. pp. 237–247.

Francis CA, Roberts KJ, Beman JM, Santoro AE, Oakley BB. 2005. Ubiquity and diversity of ammonia-oxidizing archaea in water columns and sediments of the ocean. *Proc. Natl. Acad. Sci. USA* **102**:14683–14688.

Franke T, Beenken L, Döring M, Kocyan A, Agerer R. 2006. Arbuscular mycorrhizal fungi of the *Glomus-group* A lineage (Glomerales; Glomeromycota) detected in myco-heterotrophic plants. *Mycol. Progress* **5**:24–31.

Galantini J, ed. 2007. *Sistemas Productivos en Siembra Directa*. Rosario, Argentina: Aapresid Publicaciones, 85 pages.

García-Ruiz R, Ochoa V, Hinojosa M, Carreira J. 2008. Suitability of enzyme activities for the monitoring of soil quality improvement in organic agricultural systems. *Soil Biol. Biochem.* **40**:2137–2145.

Gil R. 1998. Effects of no-tillage on physical and chemical characteristics of soils in Argentina. JIRCAS Working Report, **13**:29–33.

Haas D, Blumer C, Keel C. 2000. Biocontrol ability of fluorescent pseudomonads genetically dissected: importance of positive feedback regulation. *Curr. Opin. Biotechnol.* **11**:290–297.

Haas D, Keel C. 2003. Regulation of antibiotic production in root-colonizing Pseudomonas spp. and relevance for biological control of plant disease. *Annu. Rev. Phytopathol.* **41**:117–153.

Henry S, Bru D, Stres B, Hallet S, Philippot L. 2006. Quantitative Detection of the nosZ Gene, encoding Nitrous Oxide Reductase, and Comparison of the Abundances of 16S rRNA, narG, nirK, and nosZ Genes in Soils. *Appl. Environ. Microbiol.* **72**:5181–5189.

Hong SH, Bunge J, Jeon SO, Epstein S, et al. 2006. Predicting microbial species richness. *Proc. Natl. Acad. Sci. USA* **103**: 117–122.

Klappenbach JA, Dunbar JM, Schmidt TM. 2000. rRNA operon copy number reflects ecological strategies of bacteria. *Appl. Environ. Microbiol.* CAB International, London. **66**:1328–1333.

Knauth S, Hurek T, Brar D, Reinhold B. 2005. Influence of different *Oryza* cultivars on expression of *nifH* gene pools in roots of rice. *Environ. Microbiol.* **7**:1725–1733.

Larrainzar E, O'Gara F, Morrisey JP. 2005. Applications of autofluorescent proteins for in situ studies in microbial ecology. *Annu. Rev. Microbiol.* **59**:257–277.

Leininger S, et al. 2006. Archaea predominate among ammonia-oxidizing prokaryotes in soils. *Nature* **442**:806–809.

Lisle JT, Stewart PS, McFeters GA. 1999. Fluorescent probes applied to physiological characterization of bacterial biofilms. *Meth-ods Enzymol.* **310**:166–78.

Ma WK, Siciliano SD, Germida JJ. 2005. A PCR-DGGE method for detecting arbuscular mycorrhizal fungi in cultivated soils. *Soil Biol. Biochem.* **37**:1589–1597.

Mahenthiralingam E, Urban TA, Goldberg JB. 2005. The multifarious, multireplicon Burkholderia cepacia complex. *Nat. Rev. Microbiol.* **3**:144–156.

Mazzola M, Funnell DL, Raaijmakers JM. 2004. Wheat cultivar-specific selection of 2,4-diacetylphloroglucinol-producing fluorescent Pseudomonas species from resident soil populations. *Microb. Ecol* **48**:338–348.

Marx MC, Wood M, Jarvis SC. 2001. A micropplate fluorimetric assay for the study of enzyme diversity in soils. *Soil Biol. Biochem.* **33**:1633–1640.

McSpadden Gardener BB, Mavrodi OV, Thomashow LS, Weller D. 2001. A Rapid Polymerase Chain Reaction-Based Assay Characterizing Rhizosphere Populations of 2,4-Diacetylphloroglucinol-Producing Bacteria. *Phytopathology* **91**:44–54.

Okano Y, Hristova KR, Leutenegger CM, Jackson LE, Ford Denison R, Gebreyesus B, Lebauer D, Scow KM. 2004. Application of real-time PCR to study effects of ammonium on population size of ammonia-oxidizing bacteria in soil. *Appl Environ Microbiol* **70**:1008–1016.

Palleroni NJ. 2003. Prokaryote taxonomy of the 20th century and the impact of studies on the genus Pseudomonas: A personal view. *Microbiology* **149**:1–7.

Pankhurst CE, Ophel-Keller K, Doube BM, Grupta VVSR. 1996. Biodiversity of soil microbial communities in agricultural systems. *Biodiversity Conserv.* **5**:197–209.

Picard C, Bosco M. 2008. Genotypic and phenotypic diversity in populations of plant-probiotic Pseudomonas spp. colonizing roots. *Naturwissenschaften* **95**:1–16.

Raaijmakers JM, Weller DM. 2001. Exploiting genotypic diversity of 2,4-diacetylphloroglucinol-producing *Pseudomonas* spp.: characterization of superior root-colonizing *P. fluorescens* strain Q8r1-96. *Appl Environ Microbiol* **67**:2545–2554.

Ramette A, Moenne-Loccoz Y, Defago G. 2001. Polymorphism of the polyketide synthase gene phID in biocontrol fluorescent pseudomonads producing 2,4-diacetylphloroglucinol and comparison of PhID with plant polyketide synthases. *Mol Plant Microbe Interact.* **14**:639–652.

Richardson AE, Hadobas PA, Hayes JE. 2001. Extracellular secretion of *Aspergillus phytase* from *Arabidopsis* roots enables plants to obtain phosphorous from phytate. *Plant J* **25**:641–649.

Rillig MC. 2004. Arbuscular mycorrhizae, glomalin, and soil aggregation. *Can. J. Soil Sci.* **84**:355–363.

Rodríguez H, Fraga R, Gonzalez T, Bashan Y. 2006. Genetics of phosphate solubilization and its potential applications for improving plant growth-promoting bacteria. *Plant and Soil* **287**:15–21.

Roesch LFW et al. 2007. Pyrosequencing enumerates and contrasts soil microbial diversity. *ISME J.* **1**:283–290.

Rösch C, Mergel A, Bothe H. 2002. Biodiversity of denitrifying and dinitrogen-fixing bacteria in an acid forest soil. *Appl. Environ. Microbiol.* **68**:3818–3829.

Schalamuk S, Velazquez S, Chidichimo H, Cabello M. 2006. Fungal spore diversity of arbuscular mycorrhizal fungi associated with spring wheat: Effects of tillage. *Mycologia* **98**:16–22.

Schüssler, A.; Schwaarzott, D. and Walker, C. 2001. A new fungal phylum, the Glomeromycota: Phylogeny and evolution. *Mycol. Res.* **105**:1413–1421.

Senesi N, Loffredo E, 1998. The chemistry of soil organic matter. In: Sparks DL. (ed.) *Soil Physical Chemistry*. Washington, DC: CRC Press, USA. p. 239–370.

Shrestha PM, Noll M, Liesack W. 2007. Phylogenetic identity, growth-response time and rrna operon copy number of soil bacteria indicate different stages of community succession. *Environ. Microbiol*. **9**:2464–2474.

Smith SE, Read DJ. 2010. *Mycorrhizal Symbiosis*, 3rd ed. London: Academic Press.

Southwood, TRE. 1980. *Ecological Methods*, 2nd ed. London: Chapman & Hall.

Stoops G. 2003. *Guidelines for Analysis and Description of Soil and Regolith Thin Sections*. WI: Soil Science Society of America, Madison, 184 pages.

Zelles L. 1999. Fatty acid patterns of phospholipids and lipopolysaccharides in the characterization of microbial communities in soil: A review. *Biol. Fertil. Soils* **29**:111–129.

Chapter 35

The Human Microbiome Project

George M. Weinstock

35.1 INTRODUCTION

The concept of a vast community of microorganisms inhabiting the human body has been around for decades. However, in studies of human–microbe interactions, most attention has been directed toward highly virulent pathogens that constitute a large burden on society, while the commensal organisms of the human body were largely ignored. Occasional exceptions include studies of dental caries—for instance, where the nonpathogenic microbial community on the tooth surface could get out of ecological balance to cause tissue damage. But for most of the modern era of research into human microbiology, the microbiome was known but unstudied.

Interest in studying the microbiome increased with the advent of metagenomic analyses of microbial communities and the ability to produce whole genome sequences of bacteria (see also Chapter 18 and 21, Vol II). By the mid-2000s, a number of small-scale projects were beginning in this area. At this time the National Institutes of Health, mainly the National Human Genome Research Institute, the National Institute of Dental and Craniofacial Research, and the National Institute of Allergy and Infectious Diseases, began to support human microbiome initiatives. Several genome centers supported by the NHGRI began large-scale research in this area. One of the first projects, the Human Gut Microbiome Initiative proposed in 2005, targeted producing genome sequences of 100 gut bacteria [Gordon et al., 2005]. This was followed in 2007 by two other proposals from NHGRI centers, one to produce another 200 bacterial genome sequences [Weinstock et al., 2007] and the other to produce large metagenomic whole genome shotgun (mWGS) datasets [Weinstock et al., 2008].

During this period the NIH was in the process of identifying new initiatives for their Roadmap program. Roadmap projects are deemed to be important to multiple institutes within the NIH, and they are funded out of the Common Fund rather than by any one institute. A series of workshops were held to define a Human Microbiome Project (HMP) that could become a Roadmap initiative (see http://nihroadmap.nih.gov/hmp/meetings.asp for summaries). This became reality in 2008 when the HMP was one of two new initiatives (out of several hundred proposals) selected for Roadmap funding. Thus was the NIH HMP was born.

35.2 DISCUSSION

As a Roadmap initiative, the NIH HMP aims to develop infrastructure, community resources, and methodology for other microbiome projects at the NIH. The concept is that each institute will develop programs to study diseases and body sites of interest to their institute, and the HMP will develop resources to empower these studies. Key elements are community resources, which are sets of DNA sequences, samples, and strains which are available without restriction and which are released, usually prior to publication, to stimulate research as rapidly as possible.

With this goal in mind, the NIH HMP comprises a series of projects addressing many areas of microbiome research. The HMP aims to produce reference genome sequences for 900 microbes, determine the structure of the microbiome in 18 body sites of healthy subjects, develop new laboratory and computational approaches, perform 15 projects investigating human disease and the microbiome, and investigate the ethical, legal, and social impact of the HMP. The funded projects are listed in **Table 35.1**.

The total investment in the HMP is over \$150 million. The reader is referred to the NIH HMP web

Handbook of Molecular Microbial Ecology, Volume I: Metagenomics and Complementary Approaches, First Edition. Edited by Frans J. de Bruijn.

Table 35.1 Funded Projects of the HMP Based on data from http://nihroadmap.nih.gov/hmp/fundedresearch.asp

Principal Investigator(s)	Institution	Project
	Genome Centers Grants	
Birren, Bruce W.	The Broad Institute	Defining the Human Microbiome
Gibbs, Richard Highlander, Sarah K. Petrosino, Joseph Frank	Baylor College of Medicine	A Microbial Genome Reference Platform for Metagenomics
Nelson, Karen E.	J. Craig Venter Institute, Inc.	Construction of A Reference Sequence Data Set For The Human Microbiome Project
Weinstock, George M.	Washington University	Sequencing the Human Microbiome
	Demonstration Projects	
Blaser, Martin J.	New York University School of Medicine	Evaluation of the Cutaneous Microbiome in Psoriasis
Buck, Gregory A. Cornelissen, Cynthia Nau Eaves, Lindon J. Strauss, Jerome Frank	Virginia Commonwealth University	The Vaginal Microbiome: Disease, Genetics, and the Environment
Bushman, Frederic D. Lewis, James D. Wu, Gary D.	University of Pennsylvania	Diet, Genetic Factors, and the Gut Microbiome in Crohn's Disease
Chang, Eugene B. Meyer, Folker Schmidt, Thomas M. Sogin, Mitchell L. Tiedje, James M. Young, Vincent B.	University of Michigan at Ann ARBOR	The Role of the Gut Microbiota in Ulcerative Colitis
Fortenberry, J. Dennis	Indiana University, Purdue Univ at Indianapolis	Urethral Microbiome of Adolescent Males
Fraser-Liggett, Claire M. Shuldiner, Alan R.	University of Maryland, Baltimore	The Thrifty Microbiome: The Role of the Gut Microbiota in Obesity in the Amish
Fraser-Liggett, Claire M.	University of Maryland, Baltimore	Metagenomic Analysis of the Structure and Function of the Human Gut Microbiota in Crohn's Disease
Li, Ellen	Washington University	Effect of Crohn's Disease Risk Alleles on Enteric Microbiota
Li, Huiying	University of California, Los Angeles	Metagenomic Study of the Human Skin Microbiome Associated with Acne
Nelson, Karen E. Pei, Zhiheng	New York University School of Medicine	Foregut Microbiome in Development of Esophageal Adenocarcinoma
Ravel, Jacques	University of Maryland, Baltimore	The Microbial Ecology of Bacterial Vaginosis: A Fine Scale Resolution Metagenomic
Segre, Julia	U.S. National Human Genome Research Institute	Skin Microbiome in Disease States: Atopic Dermatitis and Immunodeficiency
Storch, Gregory A.	Washington University	The human virome in children and its relationship to febrile illness
Tarr, Phillip I.	Washington University	The Neonatal Microbiome and Necrotizing Enterocolitis
Versalovic, James	Baylor College of Medicine	The Human Microbiome in Pediatric Abdominal Pain and Intestinal Inflammation
	Data Analysis and Coordination Center	
White, Owen R.	University of Maryland Baltimore	A Quality Assurance, Coordination, and Analysis Center for HMP

	Development of New Technologies	
Bradbury, Andrew Han, Cliff	Los Alamos NAT SECTY-Los ALAMOS NAT LAB	Species-by-Species Dissection of Microbiomes using Phage Display and Flow Sorting
Podar, Mircea	UT Battelle, LLC-Oak Ridge National Lab	Targeted Genomic Characterization of Uncultured Bacteria from the Human Microbiot
Singh, Anup K.	Sandia Corp-Sandia National Laboratories	FISH 'N' Chips: A Microfluidic Processor for Isolating and Analyzing Microbes
Doktycz, Mitchel	UT Battelle, LLC-Oak Ridge National Lab	Functional Sorting of Microbial Cells from Complex Microbiota
Worthen G. Scott	Children's Hospital of Philadelphia	Multi-Dimensional Separation of Bacteria
Lo, Yu-Hwa Zhang, Kun	University of California, San Diego	An Integrated Lab-on-Chip System for Genome Sequencing of Single Microbial Cells
Marziali, Andre	Boreal Genomics Inc.	SCODA DNA Extraction to Normalize Species Representation
Relman, David A.	Stanford University	Optimization of a Microfluidic Device for Single Bacterial Cell Genomics
Schmidt, Thomas Mitchell Young, Vincent B.	Michigan State University	Cultivation and Characterization of Microaerobes from the Human Microbiome
Chang, Eugene B.	University of Chicago	Technologies for the Discovery of Novel Human Colonic Mucosal-Associated Microbes
	Development of New Computational Technologies	
Haft, Daniel H.	J. Craig Venter Institute, Inc.	Algorithmically Tuned Protein Families, Rule_Base and Characterized Proteins
Knight, Robin	University of Colorado at Boulder	New Tools for Understanding the Composition and Dynamics of Microbial Communities
Pop, Mihai	University of Maryland, College Park Campus	Assembly and Analysis Software for Exploring the Human Microbiome
Ye, Yuzhen	Indiana University, Bloomington	Fragment Assembly and Metabolic/Species Diversity Analysis for Human Microbiome
	Studies of Ethical, Legal, and Social Implications	
Cho, Mildred K. Sankar, Pamela L.	Stanford University	Toward a Framework for Policy Analysis of Microbiome Research
Foster, Morris W.	University of Oklahoma, Norman	Indigenous Communities and Human Microbiome Research
Hoffmann, Diane E.	University of Maryland, Baltimore	Federal Regulation of Probiotics: An Analysis of Existing Regulatory Framework
McGuire, Amy Lynn	Baylor College of Medicine	Ethical, Legal, and Social Dimensions of Human Microbiome Research
Rhodes, Rosamond	Mount Sinai School of Medicine of NYU	Human Microbiome Research and the Social Fabric
Farrell, Ruth Sharp, Richard R.	Cleveland Clinic Lerner College of Medicine of CWRU	Patient Perceptions of Bioengineered Probiotics and Clinical Metagenomics

site at http://nihroadmap.nih.gov/hmp/and the web site of the Data Analysis and Coordination Center (DACC, at the University of Maryland) for the HMP at http://www.hmpdacc.org/where relevant sources of information can be found.

35.2.1 A Catalog of Reference Genomes

Whole genome reference sequences are important for interpreting mWGS reads: Their species source and the implications for other genes that may be present as part of that genome. Before the HMP was selected as a Roadmap project, there was already considerable effort going into producing 300 reference genome sequencing by the NHGRI centers, the Human Genome Sequencing Center at Baylor College of Medicine (BCM), the Broad Institute (BI), and the Genome Center at Washington University (WU) [Gordon et al. 2005; Weinstock et al. 2007]. This was increased to 900 genomes with funding from the HMP, and the J. Craig Venter Institute (JCVI) joined the group of centers performing this work.

Selection and acquisition of strains for sequencing has involved a series of working groups led by members of different NIH institutes to ensure that members of the research community have input into the process and that interests of different NIH components are represented. Currently, there are 945 strains that have been recommended, and over 360 have been sequenced (unpublished). Information on the selection and current status of reference strain sequencing is available on the DACC web site at http://www.hmpdacc.org/reference_genomes.php. A report of the first 178 strains to be sequenced along with analysis has been published [Nelson et al., 2010].

Associated with the development of this project, the four genome centers involved have worked to define standards to ensure uniformity and quality in the genomes produced. A quantifiable definition of a high-quality draft genome, the initial target of the project, has been implemented. The provisional HMP draft genome standards, with their rationales, are as follows [Nelson et al. 2010]:

1. >90% of the genome included in contigs (≥500 bp) so as to ensure completeness for identification of source species for metagenomic sequences. Genome size is estimated as the sum of contigs for fragment assemblies without read pairs, or sum of scaffold spans for assemblies with read pairs. Only contigs submitted to NCBI (≥500 bp and containing at least two reads) were used in genome size estimation.
2. >90% of bases at greater than 5× read coverage to give assurance of high base quality in the consensus sequence (this allows base quality to be assessed independent of sequencing platform as use of quality values may be inconsistent between platforms).
3. >5-kb contig N50 length to ensure long enough contiguous sequences so the most genes are intact.
4. >20-kb scaffold N50 length to ensure long enough scaffolds to capture large operons.
5. Average contig length >5 kb to provide uniformity throughout the assembly; that is, assembly is not a few large contigs and many small ones.
6. >90% of "core genes" present in the gene list, to ensure completeness. The core genes comprise single copy genes conserved among all sequenced genomes in the superkingdom Bacteria. A similar set of core genes for Archaea was derived.

In addition to the high-quality draft genome definition, a series of levels of upgrading a HQD genome have been defined. At least 15% of the genomes will be upgraded from high-quality draft status. These HMP upgrade definitions are similar to those recently proposed by a multicenter group [Chain et al., 2009], but include more detail to specify HMP goals. The grades—Improved High-Quality Draft, Annotation-Directed Finishing, Noncontiguous Finished, and Finished (see below)—are provisional until more genomes can be assessed and, like the HQD metrics, are independent of sequencing platform and assembly software.

Improved High-Quality Draft. A sequence grade characterized by automated or manual work involving manipulation of existing shotgun data or addition of automated directed reads. With minimal work per genome, this standard may be applied to a wide subset of the HMP reference genome collection. Unclosed areas require no annotation. HMP genomes with this designation will exhibit a minimum 50-kb contig N50 and are free of N base calls.

Annotation-Directed Finishing. Finishing work is targeted to clearly defined areas identified by an automated annotation pipeline. A coordinate key is included with the submission describing boundaries of finished versus draft sequence. Such annotation includes information regarding improved areas not meeting finished standards. Assemblies subjected to Annotation grade will exhibit a 50-kb minimum contig N50 and will carry a representational full-length or attempted full-length 16S rRNA copy. These genomes will be subject to a second automated annotation after improvement is complete to confirm improvement in quality of gene content.

Noncontiguous Finished. This intermediate finished sequence level reflects the comparative grade finished sequence previously applied to BACs. Noncontiguous finished sequence will be subject to manual closure approaches for all sequence problems. Minimal effort, however, will be expended on areas of low complexity. Full annotation of any areas not meeting finished standard is required. HMP noncontiguous finished assemblies are limited to a maximum of 3 scaffolds/Mb, must cover 97% of the captured genome, require identification and processing of bacterial plasmids, and contain one finished 16S rRNA gene. Base quality is expected at finished quality unless otherwise noted, including (a) removal of low confidence data at contig ends and (b) resolution of ambiguous bases and potentially misassembled regions. It is expected that most finished HMP genomes will fall into this category.

Finished. Finished reflects the traditional understanding of finished sequence, where the genome is completely deciphered along with extra-chromosomal elements such as plasmids. Consensus quality is upgraded to 10^{-5} maximum error rate. The assembly is expected to be free of misassembly and has been subjected to a QC review after completion. Any exceptions to completely finished sequence are noted with the submission.

Additional efforts to also provide consistent standards and definitions for annotation of these genomes are nearly mature. In these standards, the philosophy has been that the genome centers should be able to use any methods they wish (e.g., sequencing platform, assembly software) as long as the final product achieves the standards set by the consortium.

35.2.2 The Normal Human Microbiome

A second major activity of the HMP is to characterize the human microbiome in healthy individuals. The term "healthy" is hard to define, but a series of exclusionary criteria are evaluated before a subject enters into the study. These basically ensure subjects that have no obvious disease (e.g., no rashes, no significant tooth loss, no recent antibiotic use, etc.). In order to keep the study somewhat focused, the subjects are limited to young adults. These subjects are being recruited at two clinical sampling locales, Baylor College of Medicine and Washington University. The goal is to sample 300 subjects and over 280 have been recruited as of this writing.

The subjects are sampled at 15/18 body sites (male/female), listed in **Table 35.2**. The goal is for equally matched gender sampling, with 200 subjects sampled at two visits and 100 subjects sampled at three visits. This will amount to 11,550 specimens collected at 700 sampling visits. At present, nearly all 300 subjects have been sampled at least once, and about half of the total number of specimens have been collected.

Table 35.2 Body Sites Being Sampled in the HMP

Oral Sites	Skin Sites
Saliva	Retroauricular crease, left and right ear
Tongue dorsum	Antecubital fossa (inner elbow), left and right
Hard palate	Nasal sites
Buccal mucosa	Anterior right and left nares (pooled)
Keratinized (attached) gingiva	Gastrointestinal site
Palatine tonsils	Stool
Throat	Vaginal sites
Supragingival plaque	Posterior fornix
Subgingival plaque	Midpoint
	Vaginal introitus

Two approaches are planned for sequencing the samples. All samples are to be subjected to 16S rRNA gene sequencing on the Roche-454 platform to a depth of 5000 reads per sample (on average). The 454 sequencing will be directed at the V3–V4–V5 regions of the 16S gene, although a number of samples will also be sequenced at the V1–V2–V3 regions. Sequencing has been completed on the samples obtained to date (about half of the total), and these data are being released to the public domain at the NCBI.

The second dataset being produced is mWGS on the Illumina platform. Currently, 572 specimens have been selected, focusing on six body sites, although other body sites will be sampled as well. Each of these will be sequenced to a depth of 10 Gb, and it is expected that this dataset will be released by the fall of 2010. A Data Analysis Working Group comprising about 80 analysts from the four genome centers, the DACC, and the research community is initiating analysis of this and the 16S dataset.

35.2.3 The Human Microbiome and Disease

A series of 15 Demonstration Projects, each designed to investigate the role of the microbiome in disease, were also initiated as part of the HMP (Table 35.1). After one year, these projects are to be evaluated and a subset will be selected for further funding for another three years. These projects represent a broad set of studies, described on the HMP web site.

35.2.4 Other Projects

As mentioned, there are a number of technology development projects, computational projects, and ethical, legal, and social implications projects also funded as part of the HMP (Table 35.1). These smaller projects add to the infrastructure being developed and address key issues, such as how to produce a whole genome sequence from an uncultured strain.

35.2.5 Conclusion

The NIH HMP is the most ambitious and aggressive of many human microbiome projects worldwide. These projects have joined to form an International Human Microbiome Consortium, representing hundreds of millions of dollars of research investment in understanding the human microbiome. These efforts are only a couple of years old at this point, and large datasets are only now appearing for analysis and guidance of future projects. The rapid development of this research field is unmatched since the acceleration of the Human Genome Project in 1999 with a concomitant large investment by both NIH and industry. This marks the beginning of a very exciting era of human microbiome research.

Acknowledgments

I would like to thank the extremely talented team at the Genome Center at Washington University for much of the progress in applying NGS to metagenomic problems. In particular, it is a pleasure to acknowledge Dr. Erica Sodergren, leader of the HMP at the Genome Center. Many of the ideas presented here have also been part of the NIH HMP consortium effort, and my thanks go to colleagues at the participating institutions: the Human Genome Sequencing Center at Baylor College of Medicine, the Broad Institute, The J. Craig Venter Institute, and the Data Analysis and Coordination Center at the University of Maryland. I also express my thanks to the NIH for their generous funding.

REFERENCES

Chain PS, Grafham DV, Fulton RS, Fitzgerald MG, Hostetler J, Muzny D, Ali J, Birren B, Bruce DC, Buhay C et al. 2009. Genomics. Genome project standards in a new era of sequencing. *Science* **326**(5950):236–237.

Gordon JI, Ley RE, Wilson RK, Mardis ER, Xu J. 2005. Extending Our View of Self: The Human Gut Microbiome Initiative (HGMI). http://wwwgenomegov/Pages/Research/Sequencing/SeqProposals/HGMISeqpdf

Nelson KE, Weinstock GM, Highlander SK, Worley KC, Creasy HH, Wortman JR, Rusch DB, Mitreva M, Sodergren E, Chinwalla AT, et al. 2010. A catalog of reference genomes from the human microbiome. *Science* **328**(5981):994–999.

Weinstock GM, Gibbs RA, Wilson RK, Gordon JI, Clifton SW, Birren B, Nusbaum C. 2007. Pilot Project to Expand the Number of Sequences of Culturable Microbes from the Human Body. http://wwwgenomegov/Pages/Research/Sequencing/SeqProposals/HMPP_Proposalpdf.

Weinstock GM, Gibbs RA, Wilson RK, Gordon JI, Clifton SW, Birren B, Ward D, Nusbaum C. 2008. Pilot Experiments for Metagenomic Sample Sequencing. http://wwwgenomegov/Pages/Research/Sequencing/SeqProposals/MetagenomicPilotFinalWhitePaperSAPApprovedpdf.

Chapter 36

The Ribosomal Database Project: Sequences and Software for High-Throughput rRNA Analysis

JAMES R. COLE, QIONG WANG, BENLI CHAI, AND JAMES M. TIEDJE

36.1 INTRODUCTION

Our view of the evolutionary relationships among life forms on Earth has been revolutionized by the comparative analysis of ribosomal RNA sequences [Woese et al., 1990]. Life is now viewed as belonging to one of three primary lines of evolutionary descent: the Archaea, Bacteria, or Eucarya. This shift in paradigm has not only challenged our understanding of life's origin [Woese, 1998], but has also provided an intellectual framework for studying extant life—particularly the vast diversity of microorganisms. Ribosomal RNA diversity analysis had become such a standard and relied-upon methodology that by 2008, 77% of all INSDC (GenBank, DDBJ, EMBL) bacterial DNA sequence submissions described an rRNA sequence, and only 2% of these entries had a Latin name attached (valid or otherwise [Christen, 2008])! Cultivated organisms represent only a fraction of observed rRNA diversity (**Fig. 36.1**), and currently available genome sequences cover an even smaller slice of this cultivated fraction [Wu et al., 2009]. Because of availability of a large and diverse collection of rRNA gene sequences, the use of rRNA as the bedrock of phylogenetic and diversity analysis in bacteria and archaea is unlikely to change in the near future (see also Chapter 15, Vol. I).

The RDP arose out of research conducted by two University of Illinois at Urbana-Champaign (UIUC) faculty members: Carl R. Woese and Gary J. Olsen. Woese recognized that, due to rRNA's conserved nature and universal presence, ribosomal RNA could be used to elicit phylogenetic relationships between organisms. They foresaw that making a collection of rRNA sequences available would be useful to the research community and stimulate research in this area. Argonne National Laboratory first hosted the RDP ftp and public sites and on January 5, 1992, 636 small-subunit rRNA sequences, many of which were generated in Woese's laboratory, were made available to the public in the first release of the RDP. In 1995, a collaboration developed between researchers at Michigan State University (MSU) and UIUC aimed at developing infrastructure for RDP data. On July 31, 1998, the RDP officially moved to the MSU Center for Microbial Ecology, where it is still housed.

RDP is a value added database available to the research community through its web site (http://rdp.cme.msu.edu). The RDP organizes raw sequences into alignments, annotates rRNA sequence data, provides a phylogenetic overview of life, and offers a suite of services and tools to assist in the handling and analysis of the data. These RDP products are widely used in molecular phylogeny and evolutionary biology, microbial ecology, bacterial identification, characterizing microbial populations, and understanding the diversity of life.

In the last decade, the RDP has evolved dramatically. In 2002, switching from manual alignment to an automated alignment strategy and using a new Naïve Bayesian Classifier to rapidly place the sequences to bacterial taxonomy made it possible to update RDP monthly. Many new tools have been added; many of the classic RDP tools are still available, but have been

Handbook of Molecular Microbial Ecology, Volume I: Metagenomics and Complementary Approaches, First Edition. Edited by Frans J. de Bruijn.

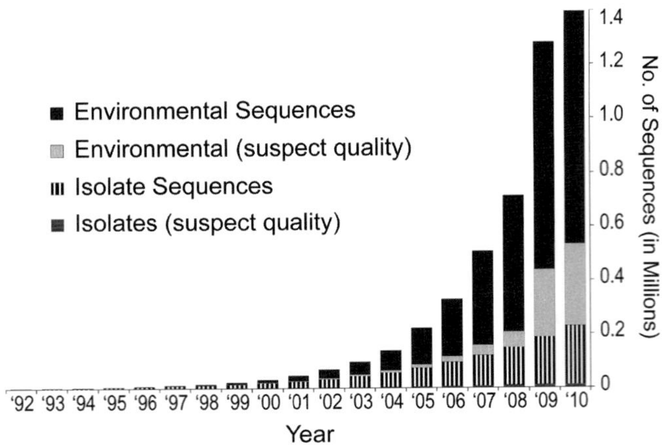

Figure 36.1 Growth of RDP public sequences. Sequences were labeled as environmental or isolate based on GenBank annotation. Sequences that failed anomaly checks with the Pintail program [Ashelford et al., 2005] are marked as suspect quality. Numbers are based on RDP Release 10.19 (March 2010).

completely reimplemented to handle the exponential increase in public rRNA sequence data. RDP introduced a user account system to help researchers integrate the analysis of their private data in the context of the public data. In 2007, RDP developed a suite of new tools to meet the challenges of ultra-high-throughput next-generation sequencing technology.

36.1.1 RDP Sequence Data

The RDP maintains aligned and annotated collections of bacterial and archaeal small-subunit (16S) rRNA gene sequences. These collections are updated monthly from the INSDC sequence repositories by a combination of search strategies. New 16S gene sequences or sequences that have been modified since the last search are added to the RDP collection, while sequences that have been removed from INSDC are removed from the public RDP collection. At the current time, the RDP collections are limited to sequences derived from the main INSDC collection and not from the trace archive or SRA Single-Read Archive, as the quality of these sequences is often low and their length often too short to serve in a reference collection.

In addition to sequence data, the RDP extracts annotation information from the INSDC sequence records, including INSDC accession and version, modification date, organism information including Latin name, where available, source-related features, and references. The Genome Project ID is kept if the sequence is derived from a genome or metagenome project. Bacterial and archaeal names in the INSDC records are updated using "Bacterial Nomenclature Up-to-Date," a compilation of valid bacterial names and nomenclatural updates maintained by Deutsche Sammlung von Mikroorganismen und Zellkulturen (DSMZ). To determine whether a sequence originates from an uncultured environmental sample, the GenBank taxonomic lineage for each sequence is examined for an "environmental" or similar term indicating that the sequences were not derived from pure culture.

Quality Testing. Not all rRNA sequences submitted to the INSDC are of high quality. All RDP public sequences are tested for anomalies using the Pintail chimera detection program [Ashelford et al., 2005]. Chimeric sequences are derived from multiple template sequences during PCR amplification of mixed community DNA containing rRNA genes from multiple organisms. Pintail incorporates a model of the relative rate of evolutionary change at different positions in the 16S gene. It compares the number of changes between a query sequence and a related high-quality control sequence to detect regions with anomalous amounts of change, as would be expected for regions of a chimeric sequence where the regions were derived from different parents. Unlike most chimera-detection programs, Pintail can also detect other types of anomalies, such as errors in sequence assembly. The RDP tests each public sequence at least twice with high-quality control sequences from different publications and considers sequences as "suspect" if they fail Pintail in both tests. The sequences are not removed from the RDP collection, but simply annotated to alert researchers to potential quality issues. In addition, several RDP tools allow the user to choose whether to include or exclude suspect sequences from analysis.

Sequence Alignments. The secondary structure of rRNA is more highly conserved than the rRNA primary sequence. This conserved secondary-structure (and higher-order) information can be used in rRNA alignments to make sure that homologous positions are aligned together. Up until 2002, the RDP manually incorporated secondary-structure information into its alignments.

The RDP now uses the Infernal Stochastic Context-Free Grammar (SCFG)-based aligner [Nawrocki et al., 2009]. This aligner incorporates secondary-structure information in a probabilistic framework, akin to how Hidden Markov Model (HMM)-based aligners incorporate conserved sequence information. Model-based aligners like HMM and SCFG have some significant advantages for alignments composed of large numbers of sequences collected over time. A new sequence can be aligned to a model and added to an existing alignment without having to realign or modify the already aligned sequences. In addition, the time required to align a set of sequences is directly proportional to the number of sequences. With a model-based aligner such as Infernal, only residues that are present in a large fraction of the molecules are "modeled"; other residues are treated as inserts. For the models used by RDP, positions are "modeled" if they are present in a minimum of 95% of rRNA sequences [Cannone et al., 2002]. The Infernal aligner was trained by RDP using a small hand-curated set of high-quality full-length rRNA sequences derived mainly from genome sequencing projects. For the average bacterial sequence, 92.5% of nucleotides are in modeled positions. Hypervariable regions with variable length and unconserved secondary structure are not aligned. These regions, while potentially useful for comparing very closely related sequences, are not generally useful for phylogenetic analysis, because most phylogenetic methods only consider base substitution events and not the insertion and deletion events that predominate in these hypervariable regions. The concept of modeled alignment positions is similar in concept to the *alignment mask*, normally used in phylogenetic analysis to restrict comparisons to sequence regions where homology between residues can be asserted with reasonable confidence (see, e.g., Lane [1991]).

36.1.2 RDP Analysis Tools

In addition to its collection of sequences, RDP provides a complete set of tools for analysis and comparison of rRNA sequences. Many of these tools can be applied to one unknown to many thousands of unknown sequences at once. Most tools can be applied either to directly uploaded files of sequences or to public or private sequences selected into the RDP Sequence Cart (SeqCart). Most RDP tools work in a web session, and many of the tools remember their states. For example, while browsing RDP Classifier results, you can switch over to another tool such as Sequence Match and then back to the RDP Classifier. Your previous results will be displayed.

Hierarchy Browser. The main entry point for browsing the RDP datasets is the RDP Hierarchy Browser. Users can browse the dataset by taxonomy, publication, or sequenced genome. In the default Hierarchy View mode, sequences are arranged in an expandable taxonomic hierarchical tree. The hierarchy nodes correspond to the formal taxonomic ranks from Domain to Genus. The bottom-level Genus nodes can be opened to display lists of individual sequences. *Hot tip:* From an open Genus node, a Fasta or GenBank representation of a single sequence can be displayed in a new window by mousing over its sequence id. Several other RDP tools display results in a similar browsable taxonomic hierarchy.

Browsing can be restricted to specific subsets of the RDP sequences by applying one or more of four different filters: type or non-type, uncultured or isolate, length ≥ 1200 or <1200 bases, and good or suspect quality. The type/non-type filter refers to sequences that RDP has determined are derived from bacterial or archaeal species type strains, while uncultured and suspect quality are described above under RDP Sequence Data. These filters can be applied to other RDP tools such as Sequence Match and Probe Match.

Hierarchy View incorporates a powerful search tool that allows both simple and complex searches. Search terms can be applied to specific fields of the sequence annotation. Advanced search features include exact and partial matching, range queries, fuzzy searching, and proximity searches. The syntax for complex searches can be found in the help page.

From the Hierarchy View, users can switch to a Publication View that groups sequences by publication (actual publications and submission "publications" listed in INSDC records). Clicking the "View" link will display the sequences published in that reference in Hierarchy View, where individual sequences from the publication or sequences from the publication corresponding to specific taxonomic groups can be selected into SeqCart for download or further analysis. Another convenient feature, Genome Browser, allows users to focus on rRNA sequences derived from genome projects. Information about the genomes and reference links are also provided. All the Browsers allow selection of sequences individually or by taxa. The selections in SeqCart will be maintained even when switching between different taxonomies.

RDP SeqCart. Many RDP tools act on a selection of sequences held in the SeqCart. Both public and private *my*RDP sequences (described below) can be added to the cart. From the Hierarchy Browser, you can add either single sequences or all sequences assigned to a taxon to SeqCart. Other tools, such as Sequence Match, allow selection of analysis results into the cart. Sequences remain in your cart until the end of your session or until you reset your cart. Your SeqCart can be saved between sessions by downloading a list of identifiers for the sequences in your cart. Uploading this file reloads the sequences into your cart. You can create your own list of INSDC identifiers to upload into your SeqCart. The corresponding RDP sequences will be selected into your cart. For INSDC identifiers that refer to records with more than one rRNA operon (e.g., a complete genome sequence), all corresponding RDP sequences will be added to your cart. *Hot tip:* If using the same standards for many analyses, download a cart with just the standards, then upload at the beginning of each session.

RDP Classifier. The RDP Naïve Bayesian Classifier provides rapid and accurate assignments of 16S rRNA sequences to the bacterial taxonomy down to genus level [Wang et al., 2007]. The Classifier is trained on known type strain 16S sequences and a small number of other sequences representing regions of bacterial diversity with few named organisms. The hierarchy model and the training sequences have been updated several times to reflect the changes in the bacterial and archaeal taxonomies. The current training set is version 6, and it will continue to be updated as our understanding of bacterial and archaeal taxonomy evolves.

For each query sequence, a subset of 8-base words is randomly chosen from the set of all overlapping words. The joint probability of observing the words in this subset is calculated for each genus. The sequence is assigned to the genus giving the highest probability based on the naïve Bayesian assumption. Taxa above genus are assigned based on the genus assignment. This sampling is repeated for 100 bootstrap trials. The number of times a taxon is selected is used as an estimate of confidence in the assignment to that taxon.

If a sequence fails to be assigned to the same taxon enough times out of the 100 bootstrap trials, the query is not assigned at this rank. Modifying this bootstrap threshold changes the percentage of sequences that can be assigned at lower taxonomic ranks at the expense of classification accuracy. The default threshold is 80% for near-full-length sequences. For partial sequences of length shorter than 250 bps (minimum 50 bps), a bootstrap cutoff of 50% was shown to be sufficient to accurately classify sequences at the genus level and provides higher percentage assignments at genus level than at the 80% cutoff [Claesson et al., 2009].

The Classifier automatically checks the orientation of each sequence and returns only the results for the correct orientation. The web interface offers two different views once the sequences have been classified. The hierarchy view displays all taxonomic nodes with assigned sequences in hierarchical order, similar to the Hierarchy View in the Hierarchy Browser. The top taxon under the hierarchy view is the current taxon root. The user can browse the results by clicking any of the taxa. If you click the "show details" link next to the root taxon, a detailed classification result of the sequences that are assigned to that taxon will be displayed. The symbol "−" after the sequence name indicates that the query is in reverse orientation. Users can change the confidence threshold at any time. Changing the bootstrap confidence threshold may change the number of sequences assigned to a taxon. Sequences that don't pass the threshold are displayed under an artificial "unclassified" taxon. Both the hierarchy assignment and assignment detail can be downloaded for local use.

The Classifier works well on both full-length and short partial rRNA gene sequences and is fast. This makes it a popular choice for processing reads from high-throughput sequencing. In addition to the web interface and SOAP interface (see below), the RDP Classifier is freely available in a stand-alone version. The stand-alone program allows for an unlimited number of query sequences, offers more output format choices, and can be easily integrated with other tools. The stand-alone Classifier can be trained on a different taxonomy if provided. The accuracy of the Classifier trained on NCBI's bacterial taxonomy, based on leave-one-out testing, is close to that seen with the RDP's bacterial taxonomy.

RDP Library Compare. RDP Library Compare reports taxa that are statistically differentially represented between two environmental libraries. It uses the RDP Classifier for rapid classification of sequences from the two libraries into the bacterial and archaeal taxonomies. For each taxon, the likelihood that the frequency of membership is the same in the two libraries is estimated using a statistical test first developed for comparing transcript levels in "digital Northern" analysis [Audic and Claverie, 1997] when the frequencies are small. For higher frequencies, the standard two-population proportions test assuming approximately a standard normal distribution is used.

Results are presented in several interactive views, including a tabular view with comparison results displayed in a table sorted by significance value, a hierarchy view displaying taxa with sequences assigned to them in the hierarchical order, and a detail view with classification results for individual queries. The hierarchy view provides a graphical summary of the relative representation among subtaxa of any chosen taxon. A taxon that differs significantly with a p value <0.01 between two libraries is flagged with a red "*". The user can drill down to any lower rank to view the assignments below a taxon by clicking the link on a taxon name. On the tabular view page, click the button "download compare results as text" to download the results to your local machine for additional analysis. The results can be imported into a spreadsheet program to make pie charts using the percentage of assignments of all the taxa at a given rank, or used with EstimateS (http://viceroy.eeb.uconn.edu/estimates) with minor format changes.

Sequence Match. Searching for related sequences is often one of the first analyses carried out on a newly generated sequence dataset. Sequence Match finds the sequences in the RDP public dataset sharing the most 7-base words with query sequences. Sequence Match can be run either on sequences in the SeqCart, on sequences directly uploaded on this tool page, or through the RDP web service interface (see below).

The current Sequence Match is a complete Java reimplementation of the original Similarity Rank algorithm developed at the RDP by Nils Larsen [Larsen et al., 1993]. Compared to Blast, Sequence Match is faster and can be more accurate [Cole et al., 2005]. The comparison metric calculated by Sequence Match is called S_ab score after the association coefficient (S_{AB}) used by Carl Woese and George Fox as a measure of the shared rRNA digestion products in their landmark paper delineating the three primary kingdoms of life [Woese and Fox, 1977]. The S_ab score is the number of (unique) 7-base oligomers shared between the query sequence and a given RDP sequence divided by the lower number of unique oligos in either of the two sequences.

In addition to providing nearest neighbors to unknowns, Sequence Match can be used as a k nearest-neighbor classifier. Sequence match results are initially displayed in a Summary View showing such classification results for the unknowns presented in an expandable taxonomic hierarchy, with each sequence identified to the lowest taxonomic rank shared by all k nearest neighbors. The value of k can be adjusted on the results page within the range of one to 20 nearest neighbors. In addition, the sequences from which the nearest neighbors are chosen can be restricted using any of the filter options described under Hierarchy Browser. Each taxon is annotated with the number of queries classified under that taxon. From the summary page, the user can navigate to a Detail View for each individual query sequence. In this view, the nearest neighbors for the query are displayed, again in a taxonomic hierarchy. From this view, nearest-neighbor sequences can be selected into SeqCart for further analysis. You can select all matches, only those classified under a specific taxon, or individual neighbors. From the Summary View, you can download a text file with a summary of the nearest neighbors or navigate to a printer-friendly Results View that displays multiple hierarchies, each similar to Detail View, for all queries under a selected node. From both Results View and Detail View, a new window can be opened to display any neighbor in Fasta or GenBank format.

RDP Probe Match. RDP's Probe Match is not meant as a probe design tool; there are several good programs available for primer design (e.g., Primer3 [Rozen and Skaletsky, 2000], CODEHOP [Rose et al., 1998; see also Chapter 15, Vol. II]). Instead, Probe Match provides a quick way to check primer and probe coverage and specificity. This tool uses a fast bit-vector algorithm for approximate string matching [Myers, 1999] to allow fast searches of the entire RDP database. Probe Match accepts probe sequences up to 64 bases in length. To match oligonucleotides with mixed positions, Probe Match search strings can include ambiguity codons in addition to the four standard bases. You can choose to search for matches on either the forward or reverse strand. Since many RDP sequences are partial, an option allows the users to restrict the search to only sequences containing the expected probe binding site region.

Results from Probe Match are presented in a browsable taxonomic hierarchy. Results can be filtered using the same four filters used in Hierarchy Browser and Sequence Match. An option on the results page allows you to see only sequences exactly matching your probe sequence or to see sequences with any number of differences (mismatches plus inserts and deletes), up to 10% of the probe length. Changing to the List View displays all matches under the root taxon and includes an alignment between the probe and RDP sequence highlighting differences in red. A download button will download a list of all matches under the current root taxon. This list can be uploaded into SeqCart to select the matching sequences for further analysis.

RDP Tree Builder. The Tree Builder allows users to produce publication-quality phylogenetic trees with bootstrap branch confidence estimates. Tree Builder combines the Weighbor Weighted Neighbor Joining program [Bruno et al., 2000] with a custom Java applet for tree viewing and manipulation. The Weighbor algorithm is an advanced algorithm that corrects for phylogenetic artifacts such as long-branch-length attraction and produces a more accurate estimate of the underlying phylogeny than the commonly used Neighbor Joining method.

To generate a tree with Tree Builder, first load your SeqCart with between four and 200 RDP public or private *my*RDP sequences. These sequences need to all share positions for comparison, so including short partial sequences without overlap will return an error message. *Hot tip:* Since species type strains link taxonomy and phylogeny, type strain sequences chosen using the "type" filter in Sequence Match can make good landmarks to include when building a tree. Tree Builder builds a rooted tree and requires an outgroup sequence. The Hierarchy Browser includes one bacterial sequence in the archaeal alignment and one archaeal sequence in the bacterial alignment for use as outgroups, but a less-distant outgroup may give a more accurate root. To estimate confidence in the branching order, tree construction is repeated with 100 bootstrap resamplings of the sequence data. The tree is displayed in a custom Java applet with functions to flip branches and switch leaf labels between sequence id, organism name, and sequence description. Other viewing options allow zooming and highlighting of nodes. The modified tree can be downloaded in either PostScript or Newick/NHX format. The latter can be imported to third-party programs such as MEGA (Molecular Evolutionary Genetics Analysis) [Kumar et al., 2008] for viewing and further manipulation.

RDP Taxomatic. Taxomatic allows users to create interactive heatmaps based on the pairwise distances between sequences to visually explore the relationships between large numbers of aligned sequences. Users can interact with the heatmap by panning and zooming to take a broad survey of their data or to see detailed similarities between select sequences. Taxomatic also allows users to view taxonomic information in their analysis by optionally overlaying taxonomic boundaries on the image. Taxonomic overlays help users compare the existing taxonomy for their sequences with actual pairwise similarity. Taxa that are phylogenetically incoherent or misplaced in the hierarchy stand out visually in the representation, as do individual misplaced sequences. The overlays make it easy to identify these outliers.

Taxomatic takes a variety of flexible input data sources and taxonomies. The heatmaps can be generated from RDP and *my*RDP sequences, a user's own aligned sequences, or even a distance matrix. When using aligned sequence data, a distance matrix is generated using uncorrected pairwise distance. Taxonomic overlay information can be automatically generated from RDP's taxonomy for RDP or *my*RDP sequences. Uploaded sequences can be classified by the RDP Classifier to produce overlay data or a user-specified taxonomy can be uploaded in XML format. Taxomatic allows users to download a full image of their heatmap at any zoom level, the underlying distance matrix used to generate the heatmap, and the taxonomic overlay data in XML format.

RDP Assignment Generator. This educational tool is intended for professors or instructors who want to teach rRNA analysis methods to large classes. It provides a lesson plan along with individualized material that can be easily distributed to students. The instructor enters information in a simple form that asks questions about the number of students, the number of sequences to be assigned to each student, and other information about the class. The tool modifies existing RDP public sequences on-the-fly to generate a unique set of sequences for each student. The modifications made to the sequences follow evolutionary principles and preserve the rRNA secondary structure and highly conserved positions. The instructor has options to choose from three sets of sequences, *Bacteria*, *Bacteria* and *Archaea*, or medically important *Bacteria*. The instructor assigns one set of sequences to each student with the instructions and evaluates the student response based on an answer key provided by the tool. The students have the opportunity to perform sequence analysis using *my*RDP, Classifier, Sequence Match, and Tree Builder with the assigned material.

RDP Web Service Interfaces. Web services are standard Application Programming Interfaces (API) that allow programmers to run RDP tools over the web and incorporate the results in their own workflows. These interfaces are especially intended for programmers who wish to automate or easily parse the results of RDP tools. The RDP has SOAP web service interfaces for many RDP tools and provides documentation, along with the sample code, on the RDP web site. The list of tools with SOAP interfaces includes RDP Classifier, Sequence Match, Probe Match, and *my*RDP. With a few exceptions, these services return the same information presented on the tools' web pages.

36.1.3 The *my*RDP Personal Account System

Although most RDP tools are available without using *my*RDP, a *my*RDP account can help simplify rRNA sequence analysis by removing the need to re-upload sequences for each analysis step. Also, many RDP tools provide additional features for *my*RDP sequences that are not available for sequences uploaded directly to the individual tools (**Fig. 36.2**). The *my*RDP account system was designed for sequence groups of up to a few thousand Sanger sequences. For next-generation sequences the RDP Pyrosequencing Pipeline is more appropriate.

Sequence Groups. The first step in analyzing sequences with *my*RDP is to upload a new Sequence Group. A group can consist of one to a few thousand sequences. A group normally consists of sequences from an experiment that a researcher might want to analyze together. *Hot tip:* You'll be able to choose RDP public sequences to include with your analysis, so there's no need to upload public sequences in your sequence group. When you upload your sequences, *my*RDP automatically schedules your sequences for alignment using the Infernal aligner and for taxonomic classification using the RDP Classifier. It may take from a few minutes to several hours for newly uploaded sequences to be ready, depending on the server load. When first uploaded, an Overview page will show the newly uploaded sequence group with the total number of sequences uploaded and their alignment status listed as "pending." Over time, the number pending alignment will decrease and the numbers aligned (or failed) will increase. Sometimes some aberrant sequences

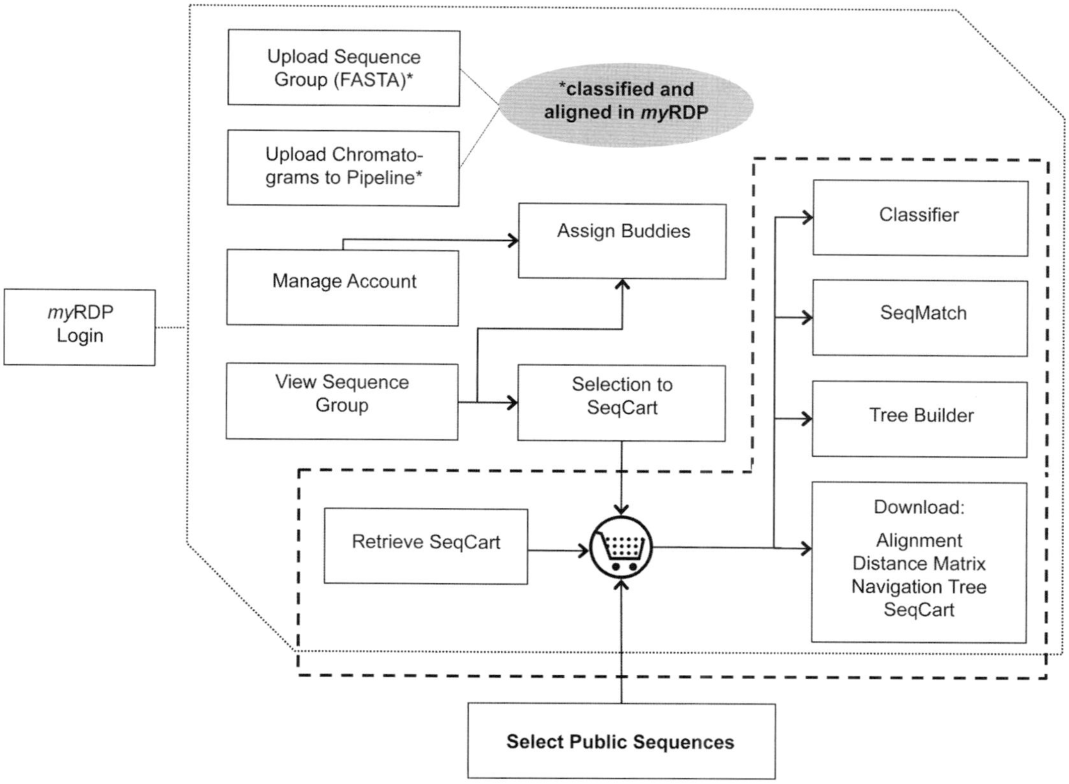

Figure 36.2 *my*RDP. Features offered by the *my*RDP user account system. Dashed line indicates tools available both inside and outside *my*RDP that have extra features when used in *my*RDP.

can't be aligned with the Infernal aligner and are marked as failed. These sequences may not be rRNA or may have gross sequencing artifacts, such as rearrangements. (It's important to note that many aberrant sequences can be "aligned" with Infernal; however, inspection may reveal a very poor quality alignment with large stretches of bases treated as inserts.)

When the initial processing is finished, the sequences can be analyzed further using other RDP tools. Entire sequence groups can be selected into the SeqCart from the Overview page, or individual sequences from a group can be selected from the Group List page. Viewing the classification of large sequence groups in the RDP Classifier takes only a few seconds because they were preclassified as part of the initial processing steps. *Hot tip:* The Group View page has a shortcut button to the RDP Classifier. With this button, you can display a group in the RDP Classifier without modifying your current SeqCart selection. In addition to the standard S_ab scoring metric, Sequence Match calculates the pairwise similarity score between *my*RDP sequences and their nearest matches since both *my*RDP and the target sequences are prealigned. The RDP Tree Builder program only works with *my*RDP and public RDP sequences, as does the Distance Matrix generator on the RDP Download page.

Research Buddies. Users can grant access to their data to other *my*RDP users (Research Buddies). When you grant access to some of your data to a Research Buddy, the data will appear on their *my*RDP Overview page, exactly like their own data, but with your account name listed under submitter id. The data can be included in their analysis just like their own *my*RDP data. A buddy *cannot* modify your sequence data, but he/she can update the project name and add notes to a particular sequence group. You can choose to hide or view any buddy data.

There are two ways to grant access to your data to your buddies: (1) Grant Super Buddy privileges to a *my*RDP user, which grants access to all your data (like having a "Master key" to all the locks in a building). You can do that by selecting "Account Info" from the top menu and entering the buddy's *my*RDP user id. (2) Grant Research Buddy privileges to an individual Sequence Group. You do this on the Group View page. You can remove a buddy for the Sequence Group by entering their account name a second time. Adding or removing research buddies is easy; but remember, you can only grant access to your own data, not your buddy's data.

my*RDP Pipeline.* This tool provides an automated pipeline to process entire libraries of Sanger single read survey sequences from raw sequence chromatograms to Sequence Groups ready for analysis. The Pipeline incorporates Phred [Ewing et al., 1998; Ewing and Green, 1998] for base calling and incorporates Lucy [Chou and Holmes, 2001] for read trimming. The input parameters for these programs were optimized for single-read sequence data to provide a balance between post-trimming lengths and sequence quality.

Prior to uploading chromatograms to the *my*RDP Pipeline, a user creates a new Library Run to group sequence reads derived from the same clone library and generated from the same sequencing run. Each Library Run is automatically assigned a unique identifier. Information about the clone library and the sequencing parameters are entered when the Library Run is created. Each Library Run can contain sequence reads from one or more 96-well sequencing plates.

When a new Library Run is created, users can provide mappings of read ids to wells. The well mapping allows the pipeline to locate each sequence read by well position on the 96-well sequencing plate, but it imposes certain naming requirements on the chromatogram files. If the well mapping is provided, an interactive visual representation of the plates will be available after processing the chromatograms on the Plate Detail page. This can provide clues to help troubleshoot sequencing problems. The chromatograms corresponding to a Library Run can all be uploaded at one time or in batches as they become available. After upload, the chromatograms are processed to trimmed reads by *my*RDP Pipeline and automatically submitted to *my*RDP for further processing. After processing, both trimmed and raw reads, along with quality scores and additional processing information, will become available through *my*RDP Pipeline, while the trimmed, aligned, and classified sequences will become available as a Sequence Group through the *my*RDP Overview page.

36.1.4 RDP Pyro Pipeline

Measuring microbial diversity in environmental samples was, until recently, constrained by the limited sequencing depth achievable with traditional Sanger sequencing. Newer technologies now allow ultra-high-throughput sequencing at lower cost (see Chapter 18, Vol. I). A single run of 454 Titanium generates over 400 thousand reads with read length of 400–500 bases (Roche 454 company literature), while Illumina sequencing technology can produce over 100 million sequence reads per run with read length of 35–125 bases (Illumina company literature). Although new technologies are reducing the cost of sequencing, many tools built to analyze sequences from the Sanger era are no longer suitable or even computationally feasible for the large numbers of short reads generated by these new technologies. New bioinformatics tools are needed to handle these types of data.

The RDP Pyrosequencing Pipeline (RDP Pyro Pipeline) was developed to meet these challenges [Cole et al., 2009]. It provides a collection of tools to facilitate the data processing and simplify the computationally intensive analysis of such large sequencing libraries. Analysis tools include initial data processing, fast alignment, complete-linkage clustering, and ecological metrics (**Fig. 36.3**).

Initial Process. The RDP Pyro Pipeline is designed to work with the user's choice of amplicon region and sequencing technologies, so initial processing steps are designed to be flexible. Because an entire "run" of these high-throughput sequencing technologies would produce more coverage depth than needed for most environmental samples, barcodes (keys; also called multiplex identifiers) are usually included at the end of the raw sequence reads to allow mixtures of multiple samples to be sequenced together. The Initial Process takes the raw reads, sorts them by barcode, optionally trims off primer sequences, and filters out low-quality reads based on several parameters. If the gene is bacterial or archaeal 16S rRNA, the orientation of the reads is checked and reverse complemented if needed. From the Initial Process page, users upload the read files, primers, and key file. The user can control the quality of the output sequences by manipulating the maximum forward primer edit distance (number of errors), maximum reverse primer edit distance, maximum number of ambiguity characters allowed in a read, minimum read length, and minimum average exponential quality score (Q score). The user should use a stringent Q score filter with caution because it may introduce taxonomic bias. We recommend testing for bias with a mock community of knowns for any new amplicon. An output directory with the trimmed sequence file and quality file will be created for each barcode. A summary stats file and chart are created for each barcode (sample) output by the initial processing tool. The stats chart is a histogram of sequence lengths after filtering and trimming (generated from the data in the stats file). The reason why an individual sequence failed the Initial Process is also recorded in another file. These additional files are useful to spot errors and adjust the quality filtering. If the reads don't align well—for example, reads targeting the short V6 region—keeping primers may help the alignment.

Taxonomy-Based Analysis. The quality filtered reads can be analyzed in two different paths. For taxonomy-based analysis, as mentioned in the RDP general tool section, the RDP Classifier is a fast and reliable classification tool for short 16S rRNA sequence reads.

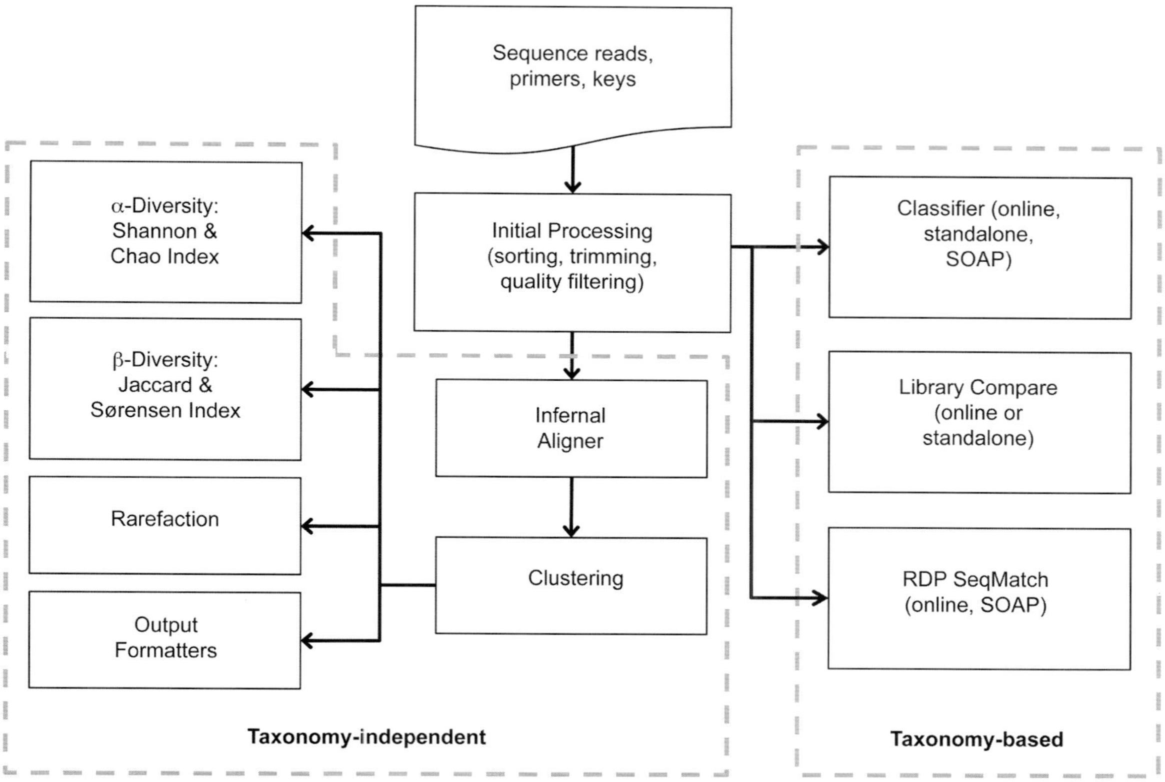

Figure 36.3 Pyro Pipeline. Analysis tools available in RDP's Pyrosequencing Pipeline.

Comparisons between samples can be made using RDP Library Compare. Sequence Match can be used to find the nearest neighbor to each read in the RDP database. Choosing the isolate dataset for Sequence Match can help determine how many reads are close to known isolates. For large sets of sequences, we recommend using the stand-alone RDP Classifier and Sequence Match Web Service Interface (see above).

Taxonomy-Independent Analysis. Sequences are aligned using the Infernal aligner. Users can upload one or multiple sample files returned from Initial Process, and they have a choice of generating one alignment file for each uploaded sample file or generating a combined alignment file for all the samples. Sequences will be reverse-complemented if necessary before submitting to the aligner. When the aligner job is finished, an email will be sent to the user with a link to download a zip file containing the aligned Fasta files and a summary stats file and chart for each alignment file produced by the aligner. The summary file and the chart can be useful to spot poor alignment of certain sequences. *Hot tip:* If you are not sure about merging the files when submitting the alignment job, you can always obtain one alignment file for each sequence file and use the Alignment Merger tool to merge multiple alignment files into one file.

The Pyro clustering tool is a fast implementation of the in-memory complete-linkage clustering algorithm. Users can upload one or multiple aligned files in either Fasta or Stockholm format [Nawrocki et al., 2009]. The user can specify the maximum cluster distance and an increment between cluster distances and can create a separate cluster file for each uploaded file or a single cluster file for all uploaded files. A summary stats file and a chart are created for each cluster file produced. The clustering tool can handle up to 150,000 unique sequences. In our hands, this has proven to be equivalent to about 1.5 million reads from human gut samples, or an entire run of 454 reads from soil samples.

After clustering analysis, for alpha diversity estimates, the Shannon and Chao1 [Shannon, 1948; Chao, 1984] estimators can be computed from a cluster file containing a single sample. Another choice produces rarefaction analyses. To measure the beta diversity among the samples, Jaccard and Sørensen [Chao et al., 2005] similarity matrices can be calculated from either a cluster file of multiple samples or a distance matrix file. A set of Jaccard and Sørensen UPGMA dendrograms and heatmaps are generated with the R statistical computing package, one for each distance within a specified range. For the convenience of users, RDP Pyro Pipeline offers several tools to generate or convert data to formats suitable for common ecological and statistical packages such as Mothur [Schloss et al., 2009], SPADE (Species Prediction and Diversity Estimation; http://chao.stat.nthu.edu.tw/softwareCE.html), EstimateS [Colwell, 2009], and the R statistical package (http://www.r-project.org/).

Care must be taken to understand the limitations of current ultra-high-throughput data. Several recent papers have highlighted the potential for small numbers of errors to inflate alpha diversity estimates (e.g., Reeder and Knight [2009]; see also Chapter 19, Vol. I). A tool to correct sequencing errors in 454 data is available (PyroNoise; [Quince et al., 2009]); however, the current implementation is very CPU intensive and not practical for any but small datasets.

Experimental Choices. The choice of primers and amplified region can make a profound difference in the ease of analysis of high-throughput data. As mentioned previously, the Infernal aligner uses conserved secondary structure information to make a high-quality alignment, but, as a model-based aligner, it does not produce alignments for regions missing in many sequences or regions with highly variable secondary structure. For short partial sequences, the choice of region can greatly affect the alignment information obtained with Infernal. For example, for a V4 variable region amplicon, 90% of the sequences ranged between 206 and 208 bases in length across the RDP public sequence data, with 205 of these positions in the Infernal alignment used by RDP. However, for the shorter V6 region used in early 454 studies [Sogin et al., 2006], the 90% sequence range in the RDP database was between 56 and 65 bases in length, and only 47 were modeled in RDP's Infernal alignment. One way of handling such data is to not use a taxonomy-independent method to measure diversity, but to use a taxonomy-based method. This was the choice taken by the designers of the Global Alignment for Sequence Taxonomy (GAST) comparison system designed originally for analysis of the short V6 region [Huse et al., 2008]. Another tactic is to produce pairwise alignments between each possible pair of reads in a sample [Sun et al., 2009]. However, the same residues may not be considered homologous across pairwise alignments. It also greatly increases the analysis time and limits practical analysis to relatively small sets of sequences.

Amplicon length is another important variable under experimental control. Counterintuitively, longer amplicons are not necessarily better. Both Illumina and 454 pyrosequencing have limits on the template length; but before these limits, longer amplicons may be subject to taxonomic bias at the PCR step [Huber et al., 2009]. Sequence reads tend to have a higher likelihood of errors as the distance increases, and the error rate near the maximum length of a given technology may be unacceptably high. One solution is to truncate all sequences to the same length [Kunin et al., 2010]; however, because of hypervariable length regions, this may mean that the reads, although the same length, cover different portions of the 16S rRNA gene. A better approach may be to truncate the reads at a conserved landmark position on the gene, perhaps after alignment. However, it would seem that the best approach might be, wherever feasible, to use an amplicon that can be completely sequenced comfortably with a given technology. When the reads extend through the distal primer, it has the benefit of guaranteeing that all reads cover the same portion of the gene. In addition, since the distal primer sequence is known, it can serve as an important additional quality-control measure. This is used by the Pyro Pipeline to help filter out low-quality reads.

36.2 CONCLUSIONS

Current sequencing technologies have greatly increased throughput over those available just a few years ago. The 454 Pyrosequencing technology used in the first deep sequencing experiments by Mitch Sogin and colleagues [Sogin et al., 2006] has been subject to improvements in both number of reads and read length. Current 454 read lengths are at over 400 bases and the company plans on improving this to 1 kb in the near future (Roche product literature). Up to the time of this writing, this technology has been the main technology used in published ultra-high-throughput 16S rRNA gene sequencing experiments. Illumina technology, although available and in production for several years, is just now starting to be applied to in-depth amplicon sequencing (see, e.g., Lazarevic et al. [2009]). This technology brings in a new set of challenges along with a much higher number of reads per sequencing run (**Fig. 36.4**). Although the read lengths are still shorter

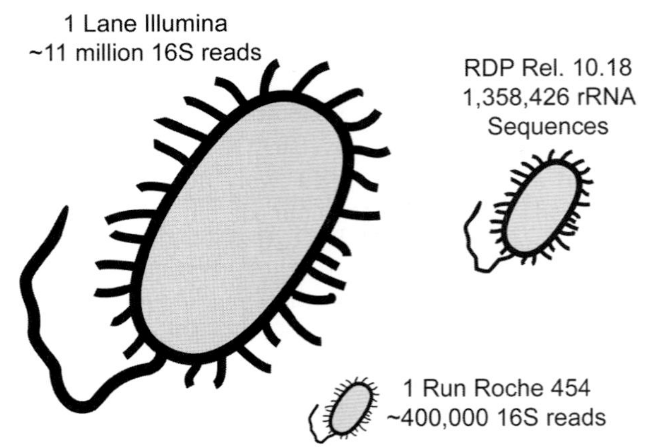

Figure 36.4 Sequence throughput. Comparison of the number of public sequences in RDP and the capacity of two high-throughput sequencing technologies.

than with pyrosequencing, Illumina read length is now at over 100 bases, comparable with the first-generation 454 G20 read length (Illumina product literature). With this increase in throughput, it becomes even more important to choose analysis methods that scale efficiently. While some of the tools discussed here were never intended to be used on datasets of this size, many of RDP's tools are highly efficient and are up to the immediate challenge. However, third-generation technologies on the horizon promise even higher throughput or longer read lengths or both. Meeting the demands of these new technologies may require completely different paradigms for sequence analysis (see Chapter 18, Vol. I). The RDP will continue to evolve to meet these challenges.

Acknowledgments

Work described in this chapter has been supported by the Office of Science (BER), U.S. Department of Energy, Grant DE-FG02-99ER62848 and Bioenergy Center Grant DE-FC02-07ER64494; National Research Initiative 2008-35107-04542 from the USDA National Institute of Food and Agriculture; and National Institute of Environmental Health Sciences Grant 5P42 ES004911-18/19. We also thank recent contributors to RDP's development, including Terry Marsh, George Garrity, Erick Cardenas, Ryan Farris, Jordan Fish, Donna McGarrell, Abubacker Kulam Syed Mohideen, and Woo Jun Sul.

REFERENCES

Ashelford KE, Chuzhanova NA, Fry JC, Jones AJ, Weightman AJ. 2005. At least 1 in 20 16S rRNA sequence records currently held in public repositories is estimated to contain substantial anomalies. *Appl. Environ. Microbiol*. **71**:7724–7736.

Audic S, Claverie JM. 1997. The significance of digital gene expression profiles. *Genome Res*. **7**:986–995.

Bruno WJ, Socci ND, Halpern AL. 2000. Weighted neighbor joining: a likelihood-based approach to distance-based phylogeny reconstruction. *Mol. Biol. Evol*. **17**:189–197.

Cannone JJ, Subramanian S, Schnare MN, Collett JR, D'Souza LM, et al. 2002. The comparative RNA web (CRW) site: An online database of comparative sequence and structure information for ribosomal, intron, and other RNAs. *BMC Bioinform*. **3**:2; erratum in **3**:15.

Chao A. 1984. Non-parametric estimation of the number of classes in a population. *Scand. J. Stat*. **11**:265–270.

Chao A, Chazdon RL, Colwell RK, Shen T-J. 2005. A new statistical approach for assessing similarity of species composition with incidence and abundance data. *Ecol. Lett*. **8**:148–159.

Chou HH, Holmes MH. 2001. DNA sequence quality trimming and vector removal. *Bioinformatics* **17**:1093–1104.

Christen R. 2008. Global Sequencing: A Review of Current Molecular Data and New Methods Available to Assess Microbial Diversity. *Microbes Environ*. **23**:253–268.

Claesson MJ, O'Sullivan O, Wang Q, Nikkilä J, Marchesi JR, et al. 2009. Comparative analysis of pyrosequencing and a phylogenetic microarray for exploring microbial community structures in the human distal intestine. *PLoS One* **4**:e6669.

Cole JR, Chai B, Farris RJ, Wang Q, Kulam SA, et al. 2005. The Ribosomal Database Project (RDP-II): Sequences and tools for high-throughput rRNA analysis. *Nucleic Acids Res*. **33**:D294–D296.

Cole JR, Wang Q, Cardenas E, Fish J, Chai B, et al. 2009. The Ribosomal Database Project: Improved alignments and new tools for rRNA analysis. *Nucleic Acids Res*. **37**:D141–D145.

Colwell RK. 2009. EstimateS: Statistical estimation of species richness and shared species from samples. Version 8.2. User's Guide and application. Available at http://purl.oclc.org/estimates.

Ewing B, Green P. 1998. Base-calling of automated sequencer traces using phred. II. Error probabilities. *Genome Res*. **8**:186–194.

Ewing B, Hillier L, Wendl MC, Green P. 1998. Base-calling of automated sequencer traces using phred. I. Accuracy assessment. *Genome Res*. **8**:175–185.

Huber JA, Morrison HG, Huse SM, Neal PR, Sogin ML, Mark Welch DB. 2009. Effect of PCR amplicon size on assessments of clone library microbial diversity and community structure. *Environ. Microbiol*. **11**:1292–1302.

Huse SM, Dethlefsen L, Huber JA, Mark Welch D, Relman DA, Sogin ML. 2008. Exploring microbial diversity and taxonomy using SSU rRNA hypervariable tag sequencing. *PLoS Genet*. **4**:e1000255.

Kumar S, Nei M, Dudley J, Tamura K. 2008. MEGA: A biologist-centric software for evolutionary analysis of DNA and protein sequences. *Brief Bioinform*. **9**:299–306.

Kunin V, Engelbrektson A, Ochman H, Hugenholtz P. 2010. Wrinkles in the rare biosphere: pyrosequencing errors can lead to artificial inflation of diversity estimates. *Environ. Microbiol*. **12**:118–123.

Lane DJ. 1991. 16S/23S rRNA sequencing. In Stackebrandt E, Goodfellow M, eds. *Nucleic Acid Techniques in Bacterial Systematics*. New York: John Wiley & Sons, pp. 115–175.

Larsen N, Olsen GJ, Maidak BL, McCaughey MJ, Overbeek R, et al. 1993. The ribosomal database project. *Nucleic Acids Res*. **21**:3021–3023.

Lazarevic V, Whiteson K, Huse S, Hernandez D, Farinelli L, et al. 2009. Metagenomic study of the oral microbiota by Illumina high-throughput sequencing. *J. Microbiol. Methods* **79**:266–271.

Myers G. 1999. A fast bit-vector algorithm for approximate string matching based on dynamic programming. *J. ACM* **46**:395–415.

Nawrocki EP, Kolbe DL, Eddy SR. 2009. Infernal 1.0: Inference of RNA alignments. *Bioinformatics* **25**:1335–1337.

Quince C, Lanzén A, Curtis TP, Davenport RJ, Hall N, et al. 2009. Accurate determination of microbial diversity from 454 pyrosequencing data. *Nat. Methods* **6**:639–641.

Reeder J, Knight R. 2009. The "rare biosphere": A reality check. *Nat. Methods*. **6**:636–637.

Rose TM, Schultz ER, Henikoff JG, Pietrokovski S, McCallum CM, Henikoff S. 1998. Consensus-degenerate hybrid oligonucleotide primers for amplification of distantly related sequences. *Nucleic Acids Res*. **26**:1628–1635.

Rozen S, Skaletsky H. 2000. Primer3 on the WWW for general users and for biologist programmers. In Krawetz S, Misener S, eds. *Bioinformatics Methods and Protocols* in the series *Methods in Molecular Biology*. Totowa, NJ: Humana Press, pp. 365–386.

Schloss PD, Westcott SL, Ryabin T, Hall JR, Hartmann M, et al. 2009. Introducing mothur: Open-source, platform-independent, community-supported software for describing and comparing microbial communities. *Appl. Environ. Microbiol*. **75**:7537–7541.

Shannon CE. 1948. A mathematical theory of communication. *Bell System Technical J*. **27**:379–423, 623–656.

Sogin ML, Morrison HG, Huber JA, Mark Welch D, Huse SM, et al. 2006. Microbial diversity in the deep sea and the underexplored "rare biosphere". *Proc. Natl. Acad. Sci. USA* **103**:12115–12120.

Sun Y, Cai Y, Liu L, Yu F, Farrell ML, et al. 2009. ESPRIT: Estimating species richness using large collections of 16S rRNA pyrosequences. *Nucleic Acids Res*. **37**:e76.

Wang Q, Garrity GM, Tiedje JM, Cole JR. 2007. Naïve Bayesian Classifier for rapid assignment of rRNA sequences into the new bacterial taxonomy. *Appl. Environ. Microbiol*. **73**:5261–5267.

Woese C. 1998. The universal ancestor. *Proc. Natl. Acad. Sci. USA* **95**:6854–6859.

Woese CR, Fox GE. 1977. Phylogenetic structure of the prokaryotic domain: The primary kingdoms. *Proc. Natl. Acad. Sci. USA* **74**:5088–5090.

Woese CR, Kandler O, Wheelis ML. 1990. Towards a natural system of organisms: Proposal for the domains Archaea, Bacteria, and Eucarya. *Proc. Natl. Acad. Sci. USA* **87**:4576–4579.

Wu D, Hugenholtz P, Mavromatis K, Pukall R, Dalin E, et al. 2009. A phylogeny-driven genomic encyclopaedia of Bacteria and Archaea. *Nature* **462**:1056–1060.

Chapter 37

The Metagenomics RAST Server: A Public Resource for the Automatic Phylogenetic and Functional Analysis of Metagenomes

ELIZABETH M. GLASS AND FOLKER MEYER

37.1 INTRODUCTION

Studying uncultivable microorganisms has been a major obstacle to understanding natural microbial populations within the context of their environment. Metagenomics is presently entering a new phase of expansion as the implementation of the high-throughput provided by second-generation sequencing approaches become more widespread and applied to an increasing number of environments. Sanger sequencing of metagenomic samples traditionally relied on large insert libraries propagated in an *E. coli* host. Next-generation sequencing technology, however, allows for access to genes that may have been otherwise unstable or toxic in the cloning vector host cell without cloning bias in the sequencing profile. Moreover, Sanger-based sequencing underrepresents the DNA sequence space present in most natural environments. In order to circumvent the need for cloning, second-generation sequencing technology enables a new approach to comparative metagenomics [Ronaghi et al., 1996, 1998; Margulies et al., 2005; see also Chapter 18, Vol. I]. Now sequence abundance can be used to contextualize datasets for driving pattern recognition and uncovering unique properties within natural microbial communities.

Regardless of the sequencing approach used to generate the data, the first steps in analysis of any metagenome involve comparative analysis against various taxonomic and protein databases. These comparisons are computationally costly but provide the basic data types for many subsequent analyses, including phylogenetic comparisons, functional annotations, binning of sequences, phylogenomic profiling, and metabolic reconstructions and modeling.

MG-RAST is available over the web to researchers, and access is not limited to specific groups or data types. Over 4000 metagenomes have been processed through the MG-RAST pipeline so far.

37.2 METHODS

The MG-RAST [Meyer et al., 2008] server is an open source system based on the SEED [Overbeek et al., 2005] framework for comparative genomics. The system was designed with a modular framework to accommodate new methods that may be developed. This allows the rapid addition of new analysis steps or comparative data at any stage of the analysis. Users provide metadata related to their sample as well as upload their raw sequence data in fasta format; the sequences are then normalized and processed and summaries automatically generated. The server provides several methods to access different data types, including phylogenetic and metabolic reconstructions, and the ability to compare the metabolism and annotations of one or more metagenomes and genomes. In addition, the server offers browsing of data and a comprehensive search

Handbook of Molecular Microbial Ecology, Volume I: Metagenomics and Complementary Approaches, First Edition. Edited by Frans J. de Bruijn.

capability. Access to the data is password protected, and all data generated by the automated pipeline are available for download and analysis in a variety of common formats.

37.2.1 User Registration and Management

The user registration serves two functions: to limit access to each dataset to the user and their colleagues and to secure a valid email address in case correspondence is required (e.g., contact needed due to a data-processing problem). Once logged in, users can view their own metagenomes, those to which others have granted them rights to view, and the default set of publicly available metagenomes. In addition, data owners can release their metagenomes to the public at any point, allowing all users of the system to view their data. Registration is not required to browse public metagenomes.

37.2.2 Data Types

The pipeline accepts fasta formatted sequence files from 454, Solexa and Sanger. The pipeline will also accept assembled sequences in fasta format. Sequence data may be compressed by one of several common computer programs to speed upload. Users may choose to upload raw unassembled reads or assembled contigs.

It is becoming increasingly apparent that the full potential of comparative metagenome analysis can be achieved only in the context of the metadata (information describing the sample). The Genomics Standards Consortium has proposed a minimal set of data, called the Minimum Information about a (Meta)Genome Sequence (MIGS/MIMS) [Kottman et al., 2008] that should be collected with every metagenome sequence. Although this is an evolving standard, the MG-RAST server is MIGS/MIMS-compliant. Metadata are requested from the user at the time of sequence submission to MG-RAST (**Fig. 37.1**). These data are stored with the user's data and are made available to them. If the user chooses to publicly release the sequence data, the metadata are also included.

37.2.3 Implementation

The pipeline is implemented in Perl and Python by using a number of open source components, such as BLAST [Altschul et al., 1997], SQLite, AWE (Argonne Work Engine), and Postgres relational database. The system also uses data from the publicly available SEED subsystems, and it uses protein databases from SEED, KEGG [Kanehisa et al., 2004], UniProt [UniProt Consortium, 2010], IMG [Markowitz et al., 2008], NCBI [Sayers et al., 2010], and FIGfam protein families [Meyer et al., 2009].

The analysis steps are implemented to provide a flexible, extensible processing pipeline. Relational database technology is used to efficiently provide a mapping of sequences in a metagenome to both organisms and metabolic functions and at the same time allow the user to change the parameters for the underlying sequence matches. The user interface enables the download of the analysis results.

After uploading the data, a normalization step is executed, generating unique internal IDs and removing duplicate sequences. In the next step, the sequences are screened for potential protein encoding genes (PEGs) via a blastx search against the MG-RAST nonredundant database composed of various public databases mentioned previously. An expect value cutoff of 0.001 is used to pick up potentially coding elements. (This was chosen empirically to increase the number of potentially coding elements while not being overwhelming for data analysis.) In parallel with the blastx searches, the sequence data are compared to other databases by using the appropriate algorithms and significance selection criteria. These databases include: several ribosomal databases, including GREENGENES [DeSantis et al., 2006], RDP-II [Cole et al., 2007; see also Chapter 36, Vol. I], and Silva [Pruesse et al., 2007; see also Chapter 52, Vol. I]; specialized databases such as chloroplast and mitochondrial databases; and the ACLAME database of mobile elements [Leplae et al., 2010]. The search criteria are specific for each database. For example, using blastn against the rDNA databases enables users to screen for ribosomal RNA genes, but much more stringent selection criteria are used to identify candidate RNA genes than for identifying protein-encoding genes (by default, the similarity must exceed 50 bp in length and have an expect value less than 1×10^{-5}). Lastly, these matches to external databases are used to compute the derived data. A phylogenomic reconstruction of the sample is computed by using both the phylogenetic information contained in the nonredundant database and the similarities to the ribosomal RNA database. Functional classifications of the PEGs are computed by projecting against FIGfams and subsystems based on these similarity searches. These annotations become the raw input to both (a) an automatically generated initial metabolic reconstruction of the sample and (b) a subsequent metabolic model for the sample by providing suggestions for metabolic fluxes and flows, reactions, and enzymes.

The user interface for MG-RAST was designed to provide easy navigation and use of comparative tools. There are multiple views for browsing and analysis of the data, as well as a means to download all result tables and

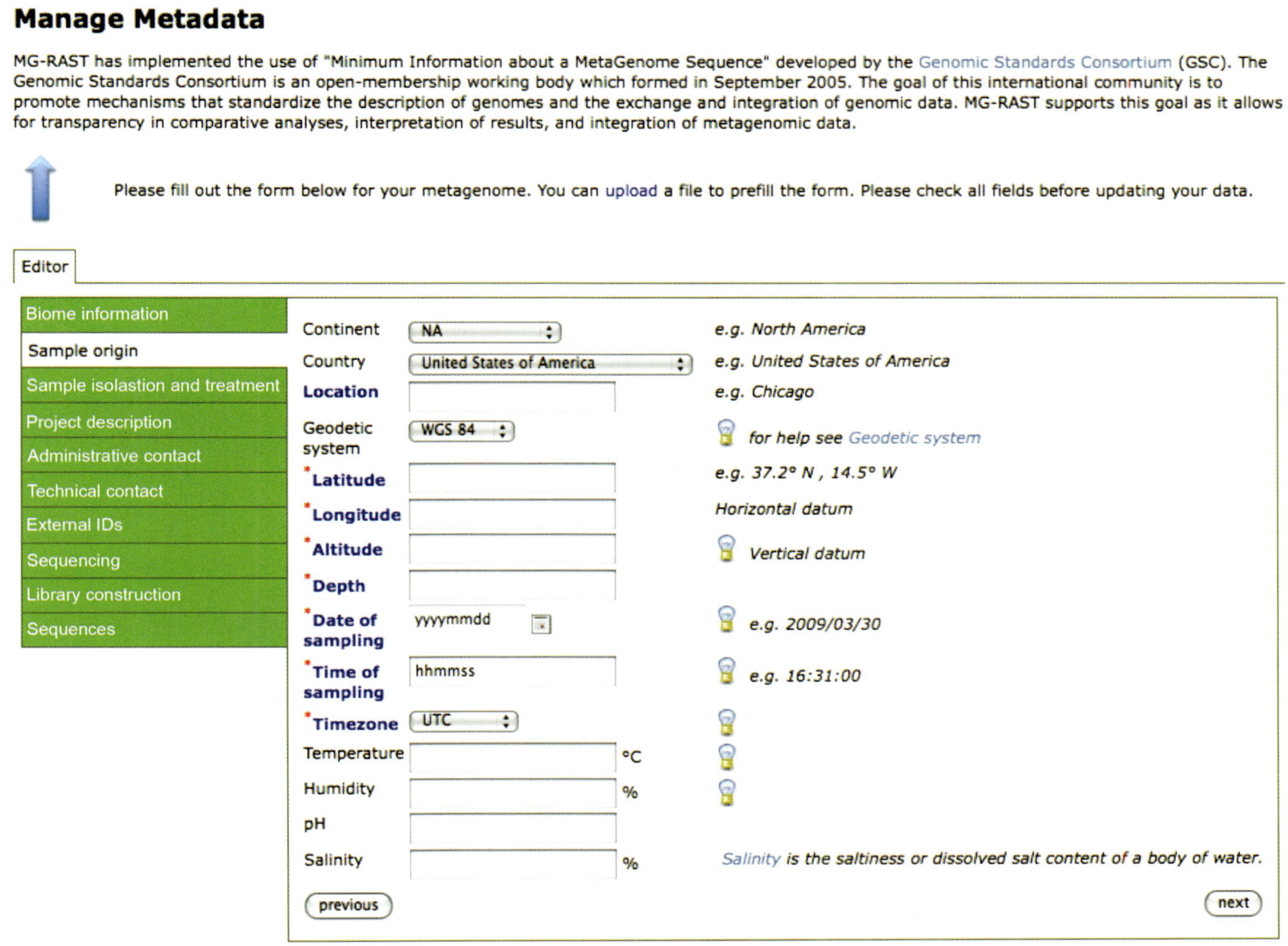

Figure 37.1 MIGS/MIMS-compliant metadata editor. Upon upload, users are requested to include pertinent data related to their sample. Required fields are in bold text.

the sequences for every subset displayed. Users are also enabled to modify the displayed results by modifying search parameters used to compute the functional, metabolic, and phylogenetic reconstruction. This allows more stringent match criteria (e.g., expectation value, score, overall percent identity, length of match, and number of mismatches); and, by restricting the matches, the derived data are dynamically changed. The default parameters have been chosen by empirical testing and represent a tradeoff between accuracy and specificity.

37.2.4 Annotations

Users can view and search their annotated genome through a couple of avenues. Metagenome Overview provides a summary of the sequence and annotation statistics against the various databases. More details are presented in the Sequence Profiles, which display the metabolic and phylogenetic distributions in a given sample. Views are in the form of charts and tables, and data are downloadable for each profile. Like all analyses in MG-RAST, the user can modify inclusion parameters and export results. Each metabolic or phylogenetic/phylogenomic profile can also be viewed singularly or compared with other metagenomes using a circular tree comparison tool. Each branch is colored to represent what is similar and what is different between samples (**Fig. 37.2**). Each node is expandable and displays details of the distributions. If only viewing one metagenome, the branches are in blue and the depth of the color denotes the number of fragments assigned to the corresponding node.

37.2.5 Comparative Metagenomics

In addition to viewing the analysis results of a given metagenome, the system supports comparative analysis, considering that it is the core driver for discovery-based biology. Therefore, comparative metagenomics tools

Figure 37.2 An example of a comparative view in MG-RAST. A tree representing metabolic profiles from two samples is compared. Each node can be expanded to get detailed information about the distribution for each sample.

are central to the utility of the MG-RAST platform. Several tools have been developed and integrated into the MG-RAST framework, allowing users to compare a metagenome to either (1) other metagenomes, (2) individual genomes, or (3) both metagenomes and genomes.

37.2.6 Comparative Heat Maps

Metabolic. PEGs identified to have functions belonging to a subsytem(s) are mapped to that subsystem. A "subsystem" is defined as a set of abstract "functional roles" that together implement a specific biological process or structural complex [Overbeek et al., 2005]. When these functional roles are linked to specific genes across metagenomes, a populated subsystem emerges. The utility of this organization is extended by subsystem connections that allow linkage of genes between subsystems.

Each subsystem present in a sample is scored by counting the number of sequences that are similar to a protein in each subsystem. This score is divided by the total number of sequences from the sample that are similar to any protein in a subsystem, to give a fraction of sequences in subsystems that are in a given subsystem. This approach allows comparisons between samples that have different numbers of sequences. Since the fractions tend to be small, the scores can be factored for display purposes. Moreover, the display can be limited or expanded to include various levels in the subsystem hierarchy (**Fig. 37.3**), specific areas of metabolism, or other subsystem groups, as chosen by the user.

Phylogenetic. The taxonomic heat map works in an analogous fashion but highlights the different taxonomic profiles in each sample, as determined by the phylogenetic or phylogenomic approaches selected by the end user (e.g., 16S comparisons, phylogenomics from BLAST

Subsystem Hierarchy 1 (all)	ID 4441096.3	ID 4441095.3	ID 4443729.3
Amino Acids and Derivatives	0.0745	0.0685	0.0966
Carbohydrates	0.0710	0.1051	0.1242
Cell Division and Cell Cycle	0.0332	0.0211	0.0164
Cell Wall and Capsule	0.0304	0.0512	0.0369
Clustering-based subsystems	0.1755	0.1741	0.1452
Cofactors, Vitamins, Prosthetic Groups, Pigments	0.0768	0.0632	0.0682
DNA Metabolism	0.0955	0.0644	0.0416
Dormancy and Sporulation	0	0.0003	0
Fatty Acids and Lipids	0.0097	0.0129	0.0165
Macromolecular Synthesis	0	0.0003	0.0015
Membrane Transport	0.0203	0.0149	0.0272
Metabolism of Aromatic Compounds	0.0039	0.0114	0.0152
Miscellaneous	0.0069	0.0088	0.0035
Motility and Chemotaxis	0.0131	0.0140	0.0163
Nitrogen Metabolism	0.0078	0.0094	0.0071
Nucleosides and Nucleotides	0.0404	0.0252	0.0333
Phosphorus Metabolism	0.0083	0.0117	0.0184
Photosynthesis	0.0007	0.0003	0.0028
Potassium Metabolism	0.0048	0.0059	0.0040
Protein Metabolism	0.0976	0.0708	0.0963
RNA Metabolism	0.0602	0.0577	0.0456
Regulation and Cell signaling	0.0058	0.0108	0.0140
Respiration	0.0574	0.0307	0.0436
Secondary Metabolism	0.0009	0.0003	0.0009
Stress Response	0.0125	0.0401	0.0208
Sulfur Metabolism	0.0106	0.0117	0.0148
Unclassified	0.0286	0.0304	0.0494
Virulence	0.0533	0.0849	0.0396

Figure 37.3 An example of a subsystem heat map that compares three public metagenomes from similar biomes. Shown is the top most level of the subsystem hierarchy. This view can be expanded to show subcategories and subsystems. Phylogenetic and phylogenomic heat maps also have the option to collapse the taxonomic nodes.

results). Again, samples may be grouped in a nonquantitative fashion to rapidly highlight particular phylogenetic groups that predominate in different samples.

37.2.7 Recruitment Plot

The recruitment plot tool is set up to provide a selected sequenced microbial genome as a scaffold to map metagenome-derived sequences to (**Fig. 37.4**). As in the heat map, sequences that have been annotated from a metagenome are used as the queries. The initial view provides a ranked list of microbial genomes that contain the highest number of matched sequences from the metagenome. This gives an indication of the relative representations in terms of genomic content found within the metagenome.

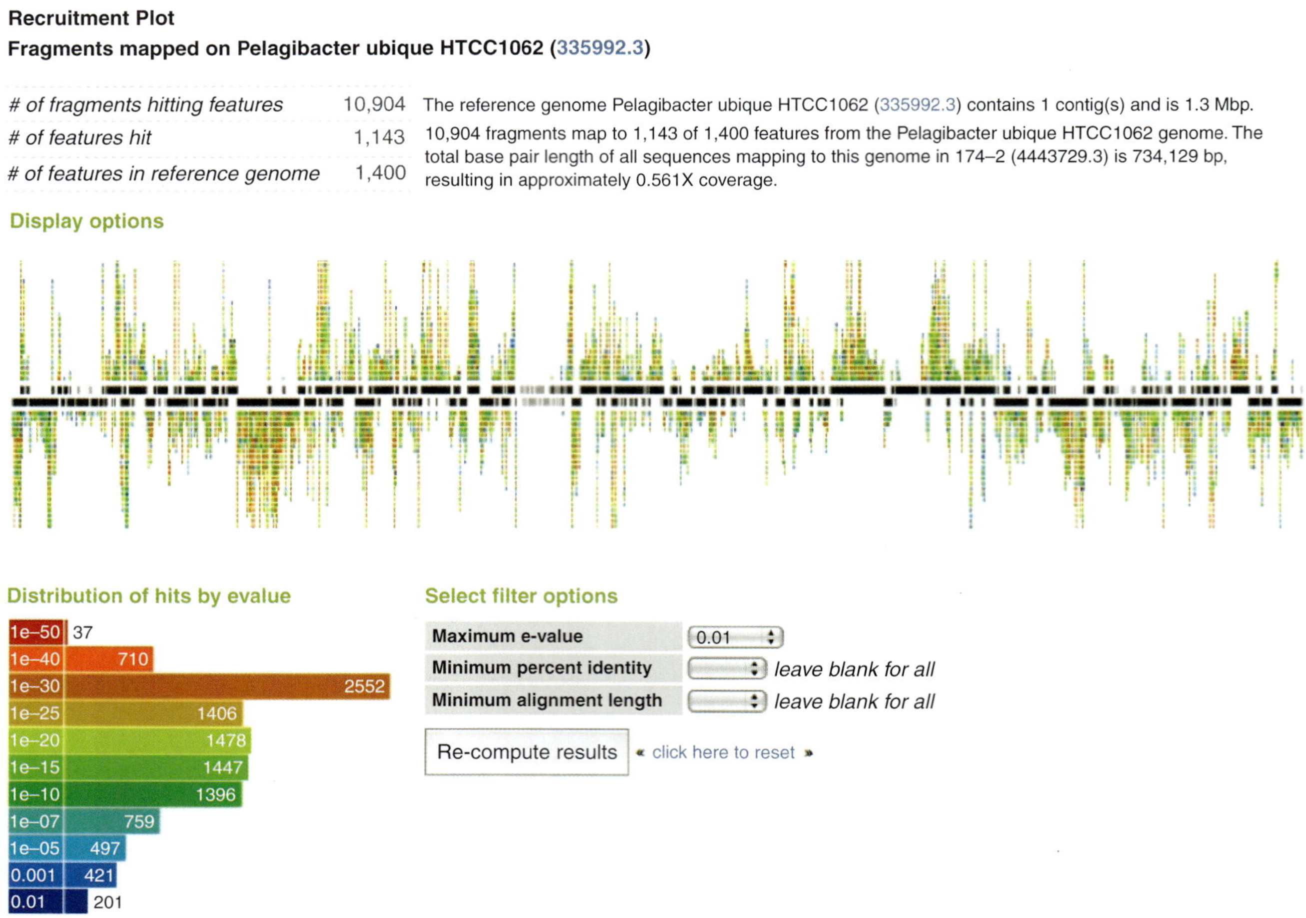

Figure 37.4 A recruitment plot showing the distribution of fragments against the genome of Pelagibacter ubique HTCC10062. The metagenome had the greatest number of fragments hit this organism. Expect values are color-coded in terms of range.

37.2.8 Metabolic Reconstructions and Models

Metagenomics also has the potential to provide insights into the critical biochemical mechanisms in each environment. Models in the MG-RAST are based on the initial metabolic reconstructions constructed. The functional roles from the reconstruction are then mapped to reactions in the SEED biochemistry database, and this mapping is then used to assemble a reaction list for the model. Models are based on a steady state and undergo flux balance analysis.

37.3 RESULTS AND DISCUSSION

Improved sampling of diverse environments, combined with advances in the development and application of next-generation sequencing technologies, is set to vastly accelerate the pace at which new metagenomes are generated. For example, as of January 2010, 4255 user-generated metagenomes had been submitted to the MG-RAST annotation server, more than 90% of which were produced by using high-throughput sequencing methodologies.

The process of building MG-RAST is the result of several years of planning and engineering. The system provides integration of metagenome data, microbial genomics, and manually curated annotations. The metagenomics analysis pipeline was designed to allow for interactive analysis, and the system as a whole has been built by using an extensible format allowing the integration of new datasets and algorithms without a need for recomputation of existing results.

The MG-RAST server handles both assembled and unassembled data. Each approach has advantages that should be considered when comparing metagenomes. For example, a case where sequences should be assembled is when comparisons between samples is being calculated, as the assembly process loses the frequency information critical for determining differences between samples. In contrast, assembled sequences tend to be longer and therefore more likely to accurately identify gene function or phylogenetic source from binning [McHardy et al., 2007; see also Chapters 40 and 42, Vol. II].

The analytical methods integrated into MG-RAST provide core annotations and analysis tools to compare and contrast sets of metagenomes [Edwards et al., 2006; Fierer et al., 2007; Mou et al., 2008]. The approach underlying the subsystems-based functional analysis of metagenomes has been validated with 90 different samples from nine major biomes. The analysis demonstrated that the biomes could clearly be separated by their functional composition [Dinsdale et al., 2008]. All of the metagenomes present in that study are included in the publicly available datasets visible on the MG-RAST server.

Although the service contains core functionality for the annotation and analysis of metagenomes, many of the techniques traditionally used for genome analysis either do not work with metagenomes or show a significant performance degradation [Krause et al., 2006]. Therefore, new analytical methods are needed to fully understand metagenomics data. The most obvious problem is with the large number of unknown sequences in any sample. Others and we are developing new binning, clustering, and coding region prediction tools to handle these unknown sequences, and effective tools will be incorporated into the pipeline when available. Another problem is that the rapid pace with which sequence data are being generated outpaces increases in computational speed, and therefore improvements in common search algorithms are required to ensure that sequence space can be accurately and efficiently searched. A third problem, common to all annotation platforms, is that metabolic reconstructions and analyses are dependent on the underlying quality of the data and that comprehensive subsystem coverage, especially of metabolism specific to diverse environments, is required to truly comprehend those data sets.

INTERNET RESOURCES

MG-RAST (http://metagenomics.anl.gov)

Acknowledgments

This work was supported in part by the U.S. Department of Energy, under Contract DE-AC02-06CH11357.

REFERENCES

Altschul SF, Madden TL, Schaffer AA, Zhang J, Zhang Z, et al. 1997. Gapped BLAST and PSI-BLAST: A new generation of protein database search programs. *Nucleic Acids Res*. **25**:3389–3402.

Cole JR, Chai B, Farris RJ, Wang Q, Kulam-Syed-Mohideen AS, et al. 2007. The ribosomal database project (RDP-II): Introducing myRDP space and quality controlled public data. *Nucleic Acids Res*. **35**(Database issue):D169–D172.

DeSantis TZ, Hugenholtz P, Larsen N, Rojas M, Brodie EL, et al. 2006. Greengenes, a chimera-checked 16S rRNA gene database and workbench compatible with ARB. *Appl. Environ. Microbiol*. **72**(7):5069–72.

Dinsdale EA, Edwards RA, Hall D, Angly F, Breitbart M, et al. 2008. Functional metagenomic profiling of nine biomes. *Nature* **452**(7187):629–632.

Edwards RA, Rodriguez-Brito B, Wegley L, Haynes M, Breitbart M, et al. 2006. Using pyrosequencing to shed light on deep mine microbial ecology. *BMC Genomics*. **7**:57.

Fierer N, Breitbart M, Nulton J, Salamon P, Lozupone C, et al. 2007. Metagenomic and small-subunit rRNA analyses reveal the genetic diversity of bacteria archaea, fungi, and viruses in soil. *Appl. Environ. Microbiol*. **73**(21):7059–7066.

Kanehisa M, Goto S, Kawashima S, Okuno Y, Hattori M. 2004. The KEGG resource for deciphering the genome. *Nucleic Acids Res*. **1**:32(Database issue):D277–D280.

Kottmann R, Gray T, Murphy S, Kagan L, Kravitz S, et al. 2008. A standard MIGS/MIMS compliant XML Schema: Toward the development of the Genomic Contextual Data Markup Language (GCDML) OMICS. **12**(2):115–121.

Krause L, Diaz NN, Bartels D, Edwards RA, Puhler A, et al. 2006. Finding novel genes in bacterial communities isolated from the environment. *Bioinformatics* **22**(14):e281–e289.

Leplae R, Lima-Mendez G, Toussaint A. 2010. ACLAME: A CLAssification of Mobile genetic Elements, update 2010. *Nucleic Acids Res*. **38**(Database issue):D57–D61.

Margulies M, Egholm M, Altman WE, Attiya S, Bader JS, et al. 2005. Genome sequencing in microfabricated high-density picolitre reactors. *Nature* **437**:376–380.

Markowitz VM, Szeto E, Palaniappan K, Grechkin Y, Chu K, et al. 2008. The integrated microbial genomes (IMG) system in 2007: Data content and analysis tool extensions. *Nucleic Acids Res*. **36**(Database issue):D528–D533.

McHardy AC, Martin HG, Tsirigos A, Hugenholtz P, Rigoutsos I. 2007. Accurate phylogenetic classification of variable-length DNA fragments. *Nat. Methods* **4**(1):63–72.

Meyer F, Overbeek R, Rodriguez A. 2009. FIGfams: Yet another set of protein families. *Nucleic Acids Res*. **37**(20):6643–6654.

Meyer F, Paarmann D, D'Souza M, Olson R, Glass EM, et al. 2008. The metagenomics RAST server—a public resource for the automatic phylogenetic and functional analysis of metagenomes. *BMC Bioinformatics*. **9**:336.

Mou XSS, Edwards RA, Hodson RE, Moran MA. 2008. Bacterial carbon processing by generalist species in the coastal ocean. *Nature* **451**(7179):708–711.

Overbeek R, Begley T, Butler RM, Choudhuri JV, Chuang HY, et al. 2005. The subsystems approach to genome annotation and its use in the project to annotate 1000 genomes. *Nucleic Acids Res*. **33**(17):5691–5702.

Pruesse E, Quast C, Knittel K, Fuchs BM, Ludwig W, et al. 2007. SILVA: A comprehensive online resource for quality checked and aligned ribosomal RNA sequence data compatible with ARB. *Nucleic Acids Res*. **35**(21):7188–7196.

Ronaghi M, Karamohamed S, Pettersson B, Uhlen M, Nyren P. 1996. Real-time DNA sequencing using detection of pyrophosphate release. *Anal. Biochem*. **242**:84–89.

Ronaghi M, Uhlen M, Nyren P. 1998. A sequencing method based on real-time pyrophosphate. *Science* **281**:363, 365.

Sayers EW, Barrett T, Benson DA, Bolton E, Bryant SH, et al. 2010. Database resources of the National Center for Biotechnology Information. *Nucleic Acids Res*. **38**(Database issue): D5–D16.

UniProt Consortium. 2010. The Universal Protein Resource (UniProt) in 2010. *Nucleic Acids Res*. **38**(Database issue):D142–D148.

Chapter 38

The EBI Metagenomics Archive, Integration and Analysis Resource

C. Hunter, G. Cochrane, R. Apweiler, and S. Hunter

38.1 INTRODUCTION

The term *metagenomics* was coined in 1998 by Handelsman et al. describing the creation of a bacterial artificial chromosome (BAC) library from noncultured DNA that had been extracted directly from a soil sample. Since then the term has been broadly used to discuss aspects of modern genetic techniques employed in the study of uncultured environmental DNA samples. Metagenomics is considered to encompass the genomic, transcriptomic, and proteomic studies of environmental communities, which are typically microbial in nature—that is, prokaryotes, viruses, Archaea, and small eukaryotic organisms [see Chapter 18, Vol. I].

Traditionally, microbial genome sequencing relied upon the laboratory culture of clonal microbes, but it is now known that only a very small percentage of microbes are amenable to these techniques. An advantage of metagenomics is that it enables the study of microbes that are either difficult or impossible to culture. It also allows the study of low abundance microbes in their natural environment, unaffected by the conditions found in artificial culture.

Metagenomic techniques can also be adapted to the study of multicellular organisms, such as plants and insects. The Windshield splatter analysis [Kosakovsky Pond et al., 2009] used the same techniques of environmental sample DNA extraction to identify eukaryotic organisms (plant pollen and insects) that had flown or been blown into the path of a moving vehicle. This was one of the first studies of higher eukaryotic species composition in a random-community environmental DNA sample.

The generation of sequence data from environmental samples is only a small part of the field. The varied analyses of these data also fall under the umbrella term of metagenomics. We therefore consider metagenomics to encapsulate the study of any community in terms of their genes, genomes, transcriptomes, member taxa, functions, and pathways.

While the study of random communities has been possible for some years by methods similar to those described by Handelsman et al. [1998] and by locus-specific cloning (e.g., 16S rRNA gene amplification), these types of studies usually have the aim of minimizing the amount of sequencing required. The advent of next-generation sequencing (NGS) technologies (see also Chapter 18, Vol. I) has provided the means by which researchers can acquire large volumes of sequence data from a single sample at a reasonable cost. Coupled with the current trends in environmental awareness, the result has been a rapid growth in the field of metagenomics.

38.2 THE MINIMUM INFORMATION STANDARD

A particular facet of metagenomic data is the desire/requirement for extensive sampling information. This information referred to as "metadata" includes everything from geographical sampling location, temperature, depth/altitude, and host–organism to how the sample was collected and stored and even the method

Handbook of Molecular Microbial Ecology, Volume I: Metagenomics and Complementary Approaches, First Edition. Edited by Frans J. de Bruijn.

of DNA extraction. In the past, this information has held less interest to researchers and data archiving resources alike. Recently, the research community has unified under the umbrella of the Genome Standards Consortium (GSC) in order to outline a set of minimal information standards (MIGS/MIMS/MIMARKS), to which all genomic researchers are strongly encouraged to adhere [Field et al., 2009]. These standards describe the metadata being collected at the point of sampling and encourage its dissemination. The European Nucleotide Archive (ENA) is committed to (a) asserting minimal submissions requirements for raw data and sampling metadata and (b) supporting compliance where possible with community standards.

The recent advances in sequencing technologies have brought about dramatic reductions in cost per base of sequence to researchers, this brings with it the question "Why archive the data when it's just as cheap to regenerate them?" We believe that the value of the data is far greater than the monetary cost of the sequencing; the sample itself (which cannot necessarily be stored and accessed for re-sequencing) has high value that can be preserved in the digital domain. This is particularly true with metagenomic data, where information about the diversity of a particular site is being monitored over a period of time. Even when this was not the initial goal of a study, if the data and accompanying metadata are archived, then researchers can do a comparison of previously collected data from the same environment.

The analysis of metagenomic data presents new challenges beyond those traditionally encountered for isolated genomes. In this chapter we describe a typical metagenomic study highlighting the challenges encountered at each stage. Finally we present an overview of a selection of available analysis tools and the features being incorporated into the EBI metagenomics portal.

38.3 OUTLINE OF A TYPICAL METAGENOMIC STUDY

The sample diversity and analysis complexities present problems for the implementation of generic pipeline analysis solutions. Here we outline a typical metagenomic study from sampling through to data analysis. There are many ways to analyze the data and metadata, and it should be noted that the outline below is not definitive (see **Fig. 38.1**).

38.3.1 Sample and Metadata Collection

The utility of any metagenomic study will be limited by the quantity and quality of the metadata that goes with it. It is good practice to record all available metadata at the time of sampling, including information that may not be of direct relevance to the study in progress. Recording additional information may allow a greater degree of relevance to be assigned to cross-sample comparisons.

38.3.2 Sample Diversity and Sequence Coverage

In order to estimate the depth of sequencing required, one must have a means to estimate the degree of organism diversity within a sample, which can be used in conjunction with estimates of genome size(s). If the sample is diverse in taxa, more sequencing will be required to achieve any given taxonomic coverage. However, if there is one or a few highly predominant species, then one may wish to acquire a level of sequencing that will allow assembly of large parts or the entirety of the genome(s). An initial survey of community diversity is usually conducted either by a PCR-based approach using 16S rRNA (or other selective markers) or by microarray analysis [Chen et al., 2009; Avarre et al., 2007; see also Chapters 57–61, Vol. I]. The community survey results are used to estimate the effective genome size (EGS), which can vary dramatically depending not only on the type of sample [Hugenholtz et al., 1998] but also on the method of calculation [Sun et al., 2009].

38.3.3 16S rRNA Analysis

The inference of phylogenetic relationships in bacteria has long been carried out by comparison of 16S ribosomal RNA gene sequences [Woese, 1987]. The 16S rRNA gene is widely accepted as the ultimate molecular measure of relatedness because it is omnipresent and functionally constant, with a degree of sequence variability [see Chapter 14, Vol. I].

There are drawbacks to this technique, namely that any technique to select only 16S rRNA genes for sequencing may introduce bias. Many prokaryotic organisms contain multiple copies of the 16S rRNA gene, and there is evidence that these can vary within a single organism's genome by as much as 1.26% [Coenye and Vandamme, 2003]. This intragenic variability has led to controversy over the degree of similarity required to call sequences as representative of the same or different species. Typically, a value of 3% dissimilarity is used to determine species while $<5\%$ is used as the cutoff for genus-level identifications [Sun et al., 2009].

The fact that species identification by this method is controversial has led to the adoption of a term to represent species in metagenomic studies: operational taxonomic unit (OTU). These OTUs are taxonomically independent groups of sequences that are classified on the basis of sequence variation.

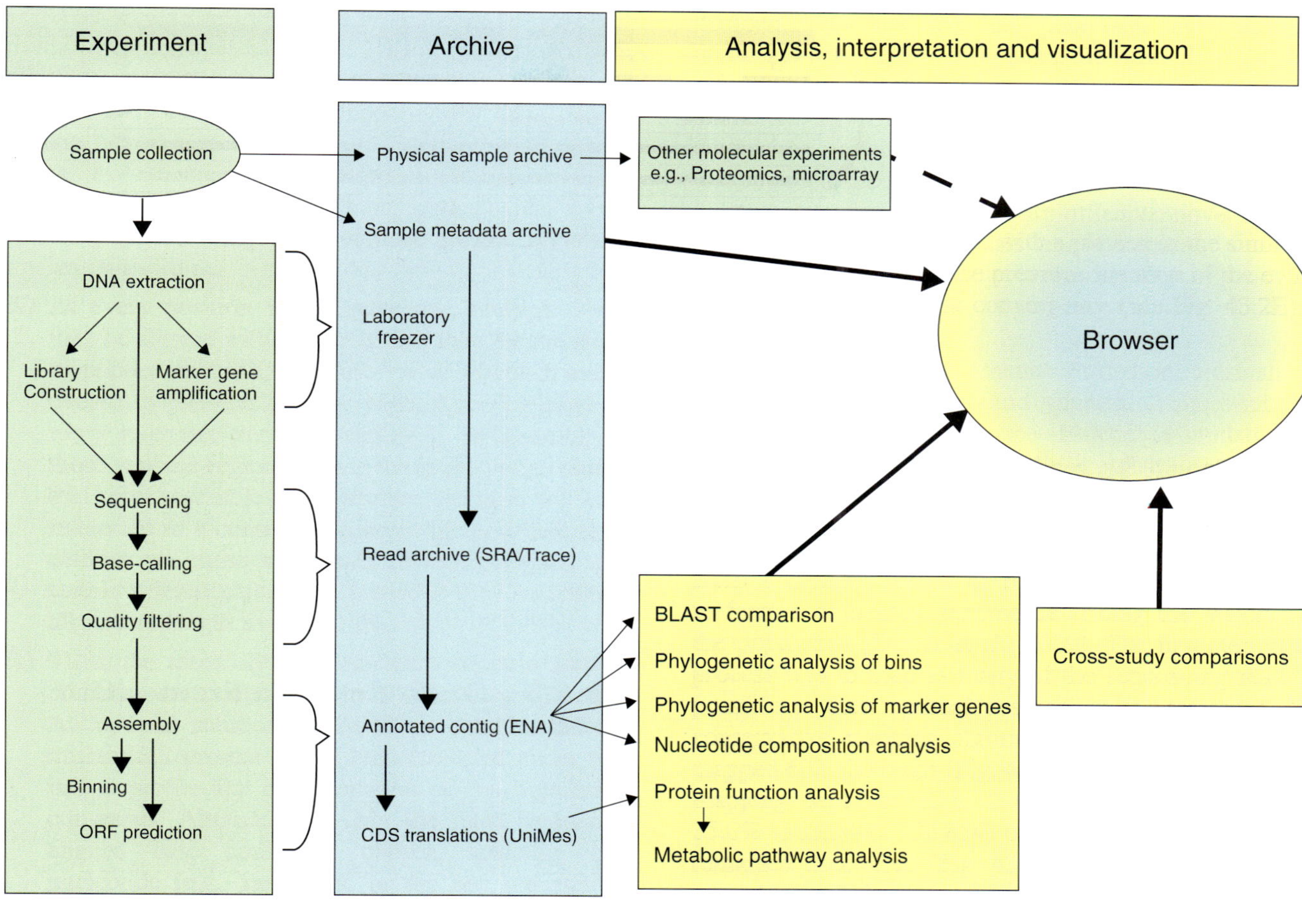

Figure 38.1 Outline of a typical metagenomics study. Green indicates the experimental parts of the study, where samples are collected, prepared, cloned, and sequenced. Blue indicates the items that can be archived during the study. The variable end analysis is shown in yellow.

38.3.4 Sequencing the Sample

Once the scale of sequencing required has been determined, the environmental sample is either subcloned to create a library prior to sequencing or simply sequenced directly from the prepared DNA sample. Depending on volumes of sequence required and the platform used, this can be carried out as multiplexed pools of samples to share costs between studies if required [Parameswaran et al., 2007].

The technology of choice at the time of writing is the 454-GS20 Titanium from Roche, primarily because the read lengths that can be achieved are among the longest of all NGS at present. However, it should be noted that other platforms, including but not limited to Illumina and SOLiD, are gaining credence in the field [see Chapter 18, Vol. I] The major advantage that most NGS have over the traditional Sanger-based approaches is that samples can be sequenced directly without the need for cloning and the inherent bias that this can introduce. This does, however, have the disadvantage of not producing a clone library to refer to when genes of particular interest are discovered and deeper analysis is required.

38.3.5 Preanalysis and Assembly

Once the sequencing stage has been accomplished, the reads are base-called, screened to remove vector contaminants (if any) and barcodes (if appropriate), and quality checked to remove uncalled and poor-quality bases/reads. These steps are often thought of as unimportant and are sometimes not carried out, but can have a significant impact on the utility/validity of the downstream results and should be given appropriate consideration by scientists.

Next, it can be useful to attempt an assembly on reads called and cleaned in this way, whereby overlapping sequences are joined to create longer sequences called contigs. Assembly of metagenomic reads is not trivial and often results in a significantly higher final assembly error rate than would be accepted with isolated genomes [Kunin et al., 2008]. The problems include both over-assembly (i.e., assembly of reads originating from different taxonomic groups creating chimaeric contigs) and under-assembly (i.e., single reads remain unassembled, also known as "singletons"). Alternative approaches to assembly for traditional genome sequencing are reference-based methods that make use of genome sequences from similar organisms. These methods are not necessarily suited to metagenomic sequence assembly due to the presence of multiple uncultured (and therefore previously unsequenced) organisms.

There are numerous free assemblers available, the relative merits of each are beyond the scope of this chapter. Here we outline two assembly tools that have been used in at least one published metagenomics study.

Tyson et al. [2004] used the JAZZ [Aparicio et al., 2002] assembler for their study of community structure in an acid mine drainage community. This is a low taxonomic complexity study completed using Sanger sequencing. At the time of publication, JAZZ was a well-established assembly tool used for traditional genome assemblies and would remain a reasonable choice for such a dataset.

A much larger, more diverse dataset was assembled by the Global Ocean Sampling (GOS) researchers [Rusch et al., 2007] using the Celera Assembler. The taxonomic diversity represented in this dataset meant that the assembly was only partial, with more than half of the reads remaining as singletons after assembly and only 9% being placed into scaffolds >10 kb. The Celera assembler was originally conceived as a traditional genome assembly tool and was adjusted to deal with this low coverage dataset. The authors used stringent cutoffs to reduce over-assembly of the sequences because this would lead to misleading analysis results.

More recently, there has been a shift in opinion toward the use of De Bruijn graph-based assembly methods, such as Velvet [Zerbino and Birney, 2008] and AllPaths [Butler et al., 2008], for large datasets. This is partly due to the higher fidelity shown by these methods when used on the very short sequence lengths seen in the early days of NGS but also due to the reduced computing time. A further advantage of the graph-based approaches is that they can handle any read length data, so a mixture of both traditional sequencing and NGS is possible. Ultimately, the choice of assembler will depend on the type of sequence data and the estimated coverage available in the dataset.

38.3.6 Bioinformatic Data Analysis

The analysis carried out for metagenomic studies will vary depending on the type of study and the assembled contigs therein. For a gene-specific locus survey study (e.g., 16S rRNA gene sequences), the diversity can be studied by comparison to locus-specific databases (e.g., the 16S rRNA database Silva [Pruesse et al., 2007; see also Chapter 45, Vol I]) or by direct clustering within the dataset for phylogenetic analysis. For whole metagenome studies with a high sequence volume to diversity ratio, where long contigs have been generated, the analysis will encompass the following: phylogeny of marker genes and database comparisons at the whole genome nucleotide level; open reading frame (ORF) prediction and translations with subsequent comparison to existing protein databases and domain databases by similarity search; potential function inference from protein and domain analysis; and ab initio structure prediction to determine putative identity and/or function of the encoded proteins. Once functions are determined, they can be mapped to metabolic pathways to indicate those pathways potentially present in the metagenome. Other analysis may include: noncoding gene predictions or comparison to other experiments carried out on the same sample—for example, transcriptomic sequencing or microarray studies [see Chapters 62–65 and 57–61, Vol. I].

38.3.7 Open Reading Frame (ORF) Prediction

Traditional isolated genome sequences are assembled into large contigs that are more amenable to gene predictions based on algorithms that are trained on complete genes. These algorithms may take into account various transcription signals as well as start and stop codons. Metagenomic data are often made up of much smaller contigs that do not encompass complete genes, restricting the utility of such algorithms. Instead, the usual approach for these small fragments is simple ORF prediction.

Prediction of ORFs in metagenomic data is complicated by the short sequence fragments and assembly errors that are inevitable from low coverage sequencing. The gene fragmentation, seen in these short contigs, results in poor calling rates for many gene prediction tools, overcome by allowing partial ORFs that continue beyond the end of a sequence; that is, there is no requirement for the ORF to begin with a start codon and end with a stop codon. This relaxed stringency results in multiple ORFs being predicted for every sequence. To reduce the number of ORFs, one would usually impose a size filter; a minimum ORF length of 30 amino acids has been used by some researchers [Gilbert et al., 2008], but no empirical optimum has yet been defined.

38.3.8 Sample Diversity Analysis

Sequences can be binned into groups of data that are believed to be derived from the same or closely related organisms (OTU) within the original sample to give an indication of organism diversity within the sample. This is commonly carried out by amino acid level comparisons of the ORFs to archives of known proteins from classified organisms, an example of which is the method employed by CARMA [Gerlach et al., 2009]. Another common approach is to use a nucleotide similarity method; however, this is only successful if appropriate reference organism genomes are available. The lack of reference genomes that are representative of all branches in the taxonomic tree is a limiting factor with the latter method, is something being addressed by multiple groups including GEBA [Wu et al., 2009], and is likely to eventually become standard practice. Alternative methods of sequence binning are based on the frequency of oligonucleotide sequence words (2–8 bases in length), although currently this method is only viable on longer sequence fragments (>1 kb). The resultant bins of sequences are unidentified at the taxonomic level, but can referred to as OTU1, OTU2, and so on. There has been a degree of success with a combinatorial approach comparing oligonucleotide frequency of known organism genome sequences to that of the unknown OTUs (Phylopythia [McHardy et al., 2007; see also Chapter 40, Vol. I]). However, there is still a need for being able to accurately define taxonomic groups from much shorter sequences.

38.3.9 Functional Elucidation

Nucleotide sequence comparison (e.g., BLAST) can be carried out on the raw reads or assembled contig sequences. The comparison is made to existing nucleotide databases e.g. the INSDC databases at the European Nucleotide Archive (ENA), DNA DataBank of Japan (DDBJ), or GenBank in the USA, where submitted sequences are from known organisms and close matches can be assumed to be from the same genus or even species. This has two major drawbacks:

- The databases are biased toward organisms that are classically more accessible to sequencing.
- It does not allow for horizontal gene transfer, which is known to be commonplace in many microorganisms.

Additionally, there is a computation consideration to nucleotide sequence comparisons, where large datasets demand extensive computation.

38.3.10 Data Archiving and Retrieval

Data and metadata archives not only allow a dissemination of information that aids the furthering of fundamental scientific knowledge, but also facilitate the search for novel species and help those researchers attempting to catalogue life on Earth, through an accumulation of overlapping sequences from the same and/or similar environments and/or samples [Cochrane et al., 2007].

The efficient integration of archived data and metadata can be used to illuminate habitat distributions of known organisms, showing researchers where particular species appear in high abundance.

38.3.10.1 Nucleotide Sequence Archiving. Raw sequences and their assembled, binned, and annotated contigs are the levels of data that are traditionally accepted and accessioned by the INSDC databases [Leinonen et al., 2010]. The ENA at the EBI has long been a leading resource for archiving nucleotide sequence data and is perfectly positioned (with its partners in the INSDC) to continue this standard into the world of metagenomics. A three-tiered view of ENA can be presented where sequencing information is classed as "reads," "assembly," or "annotation" [Cochrane et al., 2009]. Associated with any particular sequence is information (metadata) relating to the provenance and treatment of the biological sample(s) sequenced. An overarching concept of "project" exists, where information dispersed across the three ENA tiers can be collected to facilitate both deposition and access.

The current flatfile convention of the INSDC nucleotide archives is not well-suited to the association of extensive sample metadata to the sequences. The Sequence Read Archive (EBI-SRA), however, has an extensible sample object that can easily facilitate the storage of all sample metadata. The sample object is accessioned and can be referenced by any number of archived sequences, thereby allowing downstream applications using the ENA sequences to acquire the metadata.

The association of metadata with sequences will become increasingly important as researchers realize the potential of cross-sample analysis with preexisting data. Currently, this is difficult due to the lack of consistent and complete metadata associated with sequences. The consistency of metadata is already being improved by the use of community driven standards such as those discussed previously.

The services currently offered by most other public repositories for metagenomic data allow researchers to archive preassembled data, with some offering to archive FASTA format read data. Even fewer have provisions in place to accept more raw forms of the underlying NGS data (which may prove beneficial for future use as analysis tools are developed). No service currently offers submission or retrieval tools for such data. The ENA (and other INSDC partners) are committed to providing archival resources for raw forms of NGS data that may include richer data types, such as per-base quality information.

Table 38.1 Metagenomic Resources Summary

Resource Name	GOLD	CAMERA	MEGAN	MG-RAST	CARMA	MegX[a]	IMG/M[b]
Access (web/local/both)	Web	Web	Local	Web	Both	Both	Web
Archiving of metadata (with public access)	Yes	Yes	No	Yes	No	Yes	Yes
Archiving of nucleotide data (with public access)	No	Yes	No	Yes	No	No	Yes
Archiving of protein translations (with public access)	No	Yes	No	Yes	No	No	Yes
Format of sequences accepted	None	Any	Fasta	Fasta	Fasta	None	N/A[b]
Includes sequence filtering steps	No	No	No	No	No	No	Yes
Includes assembly step	No	No	No	No	No	No	Yes
ORF predictions	No	Yes	No	No	Yes	No	Yes
Binning by oligonucleotide frequency	No	No	No	No	No	Yes	No
Binning by sequence comparison to reference genomes	No	Yes	No	No	No	No	Yes
BLASTx[c]	No	Yes	Yes	Yes	Yes	Yes	Yes
Phylogenetic analysis based on BLASTx results	No	No	Yes	Yes	Yes	Some	Yes
Protein function predictions by results of BLASTx	No	Yes	Yes	Yes	Yes	Yes	Yes
Metabolic pathway usage from protein function predictions	No	No	Yes	Yes	Yes	No	Yes
Cross-sample comparisons	No	Yes	Yes	Yes	No	No	Yes
Sequence data search tools	No	Yes	No	No	No	No	Yes
Metadata search tools	Some	Yes	No	No	No	Some	Yes
Protein data search tools	No	Yes	No	Yes	No	No	Yes

[a] MegX links to multiple separate programs for each task.
[b] IMG/M only takes sequences from JGI and its collaborators.
[c] BLASTx is carried out against a variety of different protein databases. See text for references.

A number of repositories exist that offer utility beyond data archiving in the form of access to analysis tools. Here we list a number of these integrated archival and analysis services freely available to researchers; GOLD [Liolios et al., 2010], CAMERA [Seshadri et al., 2007], MEGAN [Huson et al., 2007; see Chapter 39, Vol. I], MG-RAST [Meyer et al., 2008; see Chapter 37, Vol. I], CARMA [Gerlach et al., 2009], MegX [Kottmann et al., 2009], IMG/M [Markowitz et al., 2008] and GALAXY [Kosakovsky Pond et al., 2009]. A summary of the user discernable services provided by each of the most commonly used resources is given in **Table 38.1**.

38.3.10.2 Protein Archiving. All nucleotide sequences submitted to the ENA that are annotated with a CDS (complete or partial) have their resulting amino acid sequence automatically included into the UniProt Archive (UniParc). UniParc maintains a unique set of the vast majority of publicly available protein sequences in the world (at the time of writing, over 20 million unique sequences). It is arguably the most comprehensive protein sequence resource available and contains cross-references to the myriad of accession numbers and identifiers which have been used to refer to these sequences in the literature and in other protein databases.

Protein sequences from UniParc feed into the analysis pipelines of the InterPro database. InterPro (http://www.ebi.ac.uk/interpro/) contains predictive models (also known as "signatures") that classify proteins into functional families and predict the presence of protein domains and sites [Hunter et al., 2009]. Where possible, molecular function, biochemical process, and component terms from the gene ontology (GO) are associated with InterPro's entries. This means that when a signature matches to a protein, its associated entry's GO terms are also implicitly associated and can therefore be used to describe that protein's function in a structured manner. Consequently, InterPro can be used to automatically summarize the potential function of a large set of sequences and is well-suited to genomic and metagenomic sequence characterization.

38.3.10.3 The EBI Metagenome Archive, Integration, and Analysis Resource. Some of the currently available metagenomics services have arisen from the need of researchers to archive and present their data and metadata in ways previously not possible within the existing INSDC archives. The EBI is addressing this issue and has recently begun development of its own metagenomics resource portal. Progress of these services will be announced on the EBI web site http://www.ebi.ac.uk/metagenomics.

The initialization of a metagenomic portal at the EBI is a natural progression of the ongoing development of bioinformatics services at the institute. The portal will bring together the archiving of metadata and sequence data in the nucleotide archives (at the level of ENA) and a selection of generic protein function and domain family analyses that are currently performed as standard on traditional annotated genomic sequences [Quevillon et al., 2005].

Close collaborations with many data providers around the world position the EBI well to implement the shift in metadata collection and archiving practices required in the field of metagenomics. A primary aim of the portal will be to provide researchers with simple tools to prepare metadata and sequence data for submission to the ENA. The metadata will be fully indexed and searchable using a dedicated browser to display datasets. Where nucleotide sequences are submitted with ORF predictions, these will be parsed by InterProScan to provide annotation of protein identity and function, and where appropriate the translations will be stored in UniMES, the subsection of UniParc dedicated to metagenomics data. For datasets submitted without ORF predictions, we will provide the option of assembly and ORF prediction, which will facilitate the flow of data into the same pipeline. The EBI metagenomics database will hold references to the relevant nucleotide, protein, sample, and project identifiers in the existing databases (initially at EBI with the intention to extend to external databases as well) and, where appropriate, store additional assembly and ORF prediction information.

Once the initial portal has been established and dataflow is apparent, the EBI intend to extend the services offered to afford users the ability to analyze their own data in a variety of different ways including, but not limited to, sequence assembly, ORF predictions, diversity estimates based on phylogeny in amplicon-based studies, and some statistical-based diversity estimates of metagenome survey sequences (mGSS).

38.4 CONCLUSION

The last decade has seen an explosion in the exploration of the genetic world surrounding us. With the additional, rapid progress in sequencing technologies expected in the short term, this expansion looks set to continue for the foreseeable future. A strong community-level interest in standardization of metadata collection and dissemination has impelled the EBI to step up and provide the services to allow researchers to make use of this information and, in doing so, contribute to our understanding of the natural world.

GLOSSARY

EBI	European Bioinformatics Institute
ENA	European Nucleotide Archive
INSDC	International Nucleotide Sequence Database Consortium
(p)ORF	(Partial) open reading frame
NGS	Next-generation sequencer (sequencing)
OTU	Operational taxonomic unit
SRA	Sequence read archive
mGSS	metaGenome survey sequence
CDS	Protein-coding region of sequence

REFERENCES

Aparicio S, Chapman J, Stupka E, Putnam N, Chia JM, et al. 2002. Whole-genome shotgun assembly and analysis of the genome of Fugu rubripes. *Science* **297** (5585):1301–1310.

Avarre JC, de Lajudie P, Bena G. 2007. Hybridization of genomic DNA to microarrays: A challenge for the analysis of environmental samples. *J. Microbiol. Methods* **69**(2):242–248.

Butler J, MacCallum I, Kleber M, Shlyakhter IA, Belmonte MK, Lander ES, Nusbaum C, Jaffe DB. 2008. ALLPATHS: De novo assembly of whole-genome shotgun microreads. *Genome Res.* **18**:810–820.

Chen Q, Yin H, Luo H, Xie M, Qiu G, Liu X. 2009. Micro-array based whole-genome hybridization for detection of microorganisms in acid mine drainage and bioleaching systems. *Hydrometallurgy* **95**(1–2):96–103.

Cochrane G, Akhtar R, Bonfield J, Bower L, Demiralp F, et al. 2009. Petabyte-scale innovations at the European Nucleotide Archive. *Nucleic Acids Res*. **37**:D19–D25.

Cochrane G, Martin MJ, Apweiler R. 2007. Public data resources as a foundation for a worldwide metagenomics data infrastructure. In Marco D, ed. *Metagenomics: Theory Methods and Applications*. Norfolk, UK: Caister Academic Press, pp. 79–105.

Coenye T, Vandamme P. 2003. Intragenomic heterogeneity between multiple 16S ribosomal RNA operons in sequenced bacterial genomes. *FEMS Microbiol. Lett*. **228**(1):45–49.

Field D, Sterk P, Kyrpides N, Glöckner FO, Hirschman L, Garrity G, Wooley J, Gilna P. 2009. Meeting Reports from the Genomic Standards Consortium (GSC) Workshops 6 and 7. Stand. *Genomics Sci*. **1**:68–71.

Gerlach W, Jünemann S, Tille F, Goesmann A, Stoye J. 2009. WebCARMA: A web application for the functional and taxonomic classification of unassembled metagenomic reads. *BMC Bioinformatics* **10**:430.

Gilbert JA, Field D, Huang Y, Edwards R, Li W, Gilna P, Joint I. 2008. Detection of large numbers of novel sequences in the metatranscriptomes of complex marine microbial communities. *PLoS ONE* **3**(8):e3042.

Handelsman J, Rondon MR, Brady SF, Clardy J, Goodman RM. 1998. Molecular biological access to the chemistry of unknown soil microbes: A new frontier for natural products. *Chem. Biol*. **5**:245–249.

Hugenholtz P, Goebel BM, Pace NR. 1998. Impact of culture-independent studies on the emerging phylogenetic view of bacterial diversity. *J. Bacteriol*. **180**(18):4765–4774.

Hunter S, Apweiler R, Attwood TK, Bairoch A, Bateman A, et al. 2009. InterPro: the integrative protein signature database. *Nucleic Acids Res*. **37**(Database issue):D211–D215.

Huson DH, Auch AF, Qi J, Schuster SC. 2007. MEGAN analysis of metagenomic data. *Genome Res*. **17**:377–386.

Kosakovsky Pond S, Wadhawan S, Chiaromonte F, Ananda G, Chung WY, Taylor J, Nekrutenko A, Galaxy Team. 2009. Windshield splatter analysis with the Galaxy metagenomic pipeline. *Genome Res*. **19**(11):2144–2153.

Kottmann R, Kostadinov I, Duhaime MB, Buttigieg PL, Yilmaz P, Hankeln W, Waldmann J, Glockner FO. 2009. Megx.net: integrated database resource for marine ecological genomics. *Nucleic Acids Res*. **38**(Database issue):D391–D395.

Kunin V, Copeland A, Lapidus A, Mavromatis K, Hugenholtz P. 2008. A bioinformatician's guide to metagenomics. *Microbiol Mol. Biol. Rev*. **72**(4):557–578.

Leinonen R, Akhtar R, Birney E, Bonfield J, Bower L, et al. 2010. Improvements to services at the European Nucleotide Archive. *Nucleic Acids Res*. **38**(Database issue):D39–D45.

Liolios K, Chen IMA, Mavromatis K, Tavernarakis N, Hugenholtz P, Markowitz VM, Kyrpides NC. 2010. The Genomes On Line Database (GOLD) in 2009: Status of genomic and metagenomic projects and their associated metadata. *Nucleic Acids Res*. **38**(Database issue):D346–D354.

Markowitz VM, Ivanova N, Szeto E, Palaniappan K, Chu K, et al. 2008. IMG/M: A data management and analysis system for metagenomes. *Nucleic Acids Res*. **36**:D534–D538.

McHardy AC, Garcia-Martin H, Tsirigos A, Hugenholtz P, Rigoutsos I. 2007. Accurate phylogenetic classification of variable-length DNA fragments. *Nat. Methods* **4**:63–72.

Meyer F, Paarmann D, D'Souza M, Olson R, Glass EM, et al. 2008. The Metagenomics RAST server—A public resource for the automatic phylogenetic and functional analysis of metagenomes. *BMC Bioinform*. **9**:386.

Parameswaran P, Jalili R, Tao L, Shokralla S, Gharizadeh B, Ronaghi M, Fire AZ. 2007. A pyrosequencing-tailored nucleotide barcode design unveils opportunities for large-scale sample multiplexing. *Nucleic Acids Res*. **35**(19):e130.

Pruesse E, Quast C, Knittel K, Fuchs B, Ludwig W, Peplies J, Glöckner FO. 2007. SILVA: A comprehensive online resource for quality checked and aligned ribosomal RNA sequence data compatible with ARB. *Nucleic Acids Res*. **35**(21):7188–7196.

Quevillon E, Silventoinen V, Pillai S, Harte N, Mulder N, Apweiler R, Lopez R. 2005. InterProScan: Protein domains identifier. *Nucleic Acids Res*. **33**(Web Server issue):W116–W120.

Rusch DB, Halpern AL, Sutton G, Heidelberg KB, Williamson S, et al. 2007. The Sorcerer II Global Ocean Sampling Expedition: Northwest Atlantic through Eastern Tropical Pacific. *PLoS Biol*. **13**:5(3):e77.

Seshadri R, Kravitz SA, Smarr L, Gilna P, Frazier M. 2007. CAMERA: A community resource for metagenomics. *PLoS Biol*. **5**(3):e75.

Sun Y, Cai Y, Liu L, Yu F, Farrell ML, McKendree W, Farmerie W. 2009. ESPRIT: Estimating species richness using large collections of 16S rRNA pyrosequences. *Nucleic Acid Res*. **37**(10):e76.

Tyson GW, Chapman J, Hugenholtz P, Allen EE, Ram RJ, et al. 2004. Community structure and metabolism through reconstruction of microbial genomes from the environment. *Nature* **428**:25–26.

The UniProt Consortium. 2010. The Universal Protein Resource (UniProt) in 2010. *Nucleic Acids Res*. **38**(Database issue):D142–D148.

Woese CR. 1987. Bacterial evolution. *Microbiol. Rev*. **51**:221–271.

Wu D, Hugenholtz P, Mavromatis K, Pukall R, Dalin E, et al. 2009. A phylogeny-driven genomic encyclopaedia of Bacteria and Archaea. *Nature* **462**(7276):1056–1060.

Zerbino DR, Birney E. 2008. Velvet: Algorithms for de novo short read assembly using de Bruijn graphs. *Genome Res*. **18**:821–829.

Part 5

Computer-Assisted Analysis

Chapter 39

Comparative Metagenome Analysis Using MEGAN

Daniel H. Huson and Suparna Mitra

39.1 INTRODUCTION

Metagenomics, also known as environmental genomics or community genomics, is the study of microbial organisms using sequencing applied directly to environmental samples rather than to isolated cultures. Technological advances in next-generation sequencing methods are fueling a rapid increase in the number and scope of metagenome projects. As of December 2009, nearly 100 metagenome projects have been completed, 100 are underway, and many more are planned [Bernal et al., 2001]. Typical applications include the study of the marine biosphere [Venter et al., 2004; Gilbert et al., 2008; see also Chapters 24–31, Vol. II], soil metagenomics [Schloss and Handelsman, 2006; van Elsas et al., 2008; see also Chapters 9–17, Vol. II], ancient DNA studies [Poinar et al., 2006] and medical applications [Gill et al., 2006; Turnbaugh et al., 2006; see also Chapters 18–21, Vol. II].

The initial aim of the computational analysis of a metagenomic dataset is to answer the following three questions:

1. *Who is out there?* Determine the taxonomical content of a dataset by estimating which taxa are present in which relative proportions. Also, determine the presence or absence of key species of interest.
2. *What are they doing?* Determine the functional content of a dataset by estimating what types of genes are present in which relative proportions. Also, determine which metabolic pathways of interest are supported.
3. *How do different datasets compare?* Compare the taxonomical and functional content of multiple datasets, and attempt to correlate significant differences of their genomic content to environmental differences.

There is a need for new computational tools to answer these questions. In 2007, we published and released the first stand-alone analysis tool for metagenomic data, called MEGAN (MEta Genome ANalyzer) [Huson et al., 2007]. Initially, our aim was to provide a tool for studying the taxonomical content of a single dataset. We have subsequently extended the program to support comparative analyses of multiple datasets (MEGAN 2) and also to allow a functional analysis of multiple datasets based on the Gene Ontology (GO) [Ashburner et al., 2000].

The goal of this chapter is to provide an introduction to comparative metagenome analysis using MEGAN. We will use a number of published mouse gut datasets as a running example. After first describing how to get started using the program, we will discuss how to perform a taxonomical analysis of a dataset, based on the NCBI taxonomy. We will then discuss how to perform a functional analysis using COGs [Tatusov et al., 1997] and the GO [Ashburner et al., 2000]. This will be followed by a discussion of how to compare the taxonomical and functional content of multiple metagenome datasets. In the final section, we briefly describe some new features that are currently under development. All the methods and the ideas described in this chapter are applicable to both metagenome and metatranscriptome datasets.

Handbook of Molecular Microbial Ecology, Volume I: Metagenomics and Complementary Approaches, First Edition. Edited by Frans J. de Bruijn.

39.2 GETTING STARTED

Throughout this chapter, we will use six published mouse gut datasets, obtained using random shotgun sequencing as a running example. In more detail, these include (a) an "obese mouse" dataset (ob1: 677,384 reads) and a "lean mouse" dataset (lean1: 1,046,611 reads) from a study reported in Turnbaugh et al. [2006] and (b) four datasets from a study of diet-induced obesity in mice (Western1: 9,072 reads; Western3: 10,997 reads; FAT-R1: 10,681 reads and CARB-R1: 10,773 reads) [Turnbaugh et al., 2008].

Given a file of DNA sequence reads obtained by sequencing an environmental sample using environmental shotgun sequencing [Staden, 1979; Mardis, 2008; Shendure and Ji, 2008; see also Chapter 20, Vol. I], the first computational step is to compare the reads against one or more reference databases using a tool such as BLAST. This is usually the most demanding step of any computational analysis of metagenomic data. For example, one gigabase of sequence requires around 10,000 CPU hours for a BLASTX comparison against the NCBI-NR database. Typical analyses involve comparing the sequence against the NCBI-NR database [Benson et al., 2005] using BLASTX [Altschul et al., 1990], against the NCBI-NT database [Benson et al., 2005] using BLASTN [Altschul et al., 1990], or against one or more whole-genome sequences using BLASTZ [Schwartz et al., 2003]. However, MEGAN is not tied to any particular comparison method or database.

MEGAN needs to parse the header lines associated with entries in the reference database to extract taxonomical and functional information. Thus the program requires that input is provided either in the BLAST 'standard format' (plain text, -m 0) or in the "XML format" (-m 7), but not in the "tab delimited format" (-m 8). The ideal setup for performing metagenome analyses using MEGAN is a powerful desktop workstation with at least 8 GB of main memory and a large, fast local disk.

All the methods described in this chapter are implemented in the latest version of MEGAN.

The program is written in Java and requires JRE version 1.5 or newer. Installers for all major operating systems are available from www-ab.informatik.uni-tuebingen.de/software/megan.

Upon launch, MEGAN first loads its own version of the NCBI taxonomy and then displays the first three levels of the taxonomy. Once this step is completed, the user can start a new analysis by importing a BLAST file using the "Import from BLAST" option. MEGAN will parse the BLAST file (and the reads file, if present) and will then perform an initial taxonomical and functional analysis of the data. All reads, matches, and results of the analysis are saved in an "RMA" file (RMA stands for Read-Match-Archive). This is a compressed binary format especially designed for metagenomic data. The size of an RMA file is typically about 20–30% the size of the original reads and BLAST file. The initial analysis of a dataset by MEGAN can take a number of hours and may require a lot of computer memory, depending on the size of the dataset. However, once the initial analysis has been completed, opening and working with multiple RMA files is very fast and memory-efficient.

39.3 TAXONOMICAL ANALYSIS

MEGAN uses the NCBI taxonomy to bin all reads of a given metagenome dataset. The NCBI taxonomy provides names and IDs for over 550,000 taxa, including approximately 270,000 eukaryota, 25,000 bacteria, and 52,000 viruses. The species are hierarchically classified at the levels of superkingdom, kingdom, phylum, class, order, family, genus, and species (and some unofficial clades in between such as groups or subspecies).

To perform a taxonomical analysis, MEGAN places each read of a given dataset onto one of the taxa (or "nodes") of the NCBI taxonomy, based on the BLAST matches provided for the read. If a read has significant matches to sequences in more than one species, MEGAN will assign the read to the "lowest common ancestor" (LCA) node of the species in the taxonomy, thus using a simple LCA algorithm [Huson et al., 2007]. This is an attractive approach because it is conservative and will tend to make an unspecific assignment to a higher-level taxon (such as class or phylum, for example) rather than a false-positive assignment to an incorrect species. In essence, the LCA algorithm places reads by "gene content" rather than sequence similarity, thus avoiding some of the pitfalls caused by horizontally transferred genes, integrated viruses, or other misleading sequences.

We will now describe how to perform a taxonomical analysis with MEGAN using the obese mouse data from the mouse gut datasets mentioned above [Turnbaugh et al., 2006]. The first step is to compare the 677,384 reads against the NCBI-NR database using BLASTX. The result of this step is a file of size 10.57 GB consisting of 311,755,132 BLAST alignments of reads to sequences in the database. The second step is then to process the BLAST file and reads using MEGAN to obtain an RMA file *Obese.rma* of size 3.22 GB, in which all reads have been taxonomically placed using the LCA algorithm.

MEGAN can then be used to explore the dataset interactively. In **Figure 39.1**, we show how reads have been assigned to the NCBI taxonomy, down to the taxonomical rank of class. In this figure, each node is labeled by the number of reads assigned to it and the size of the

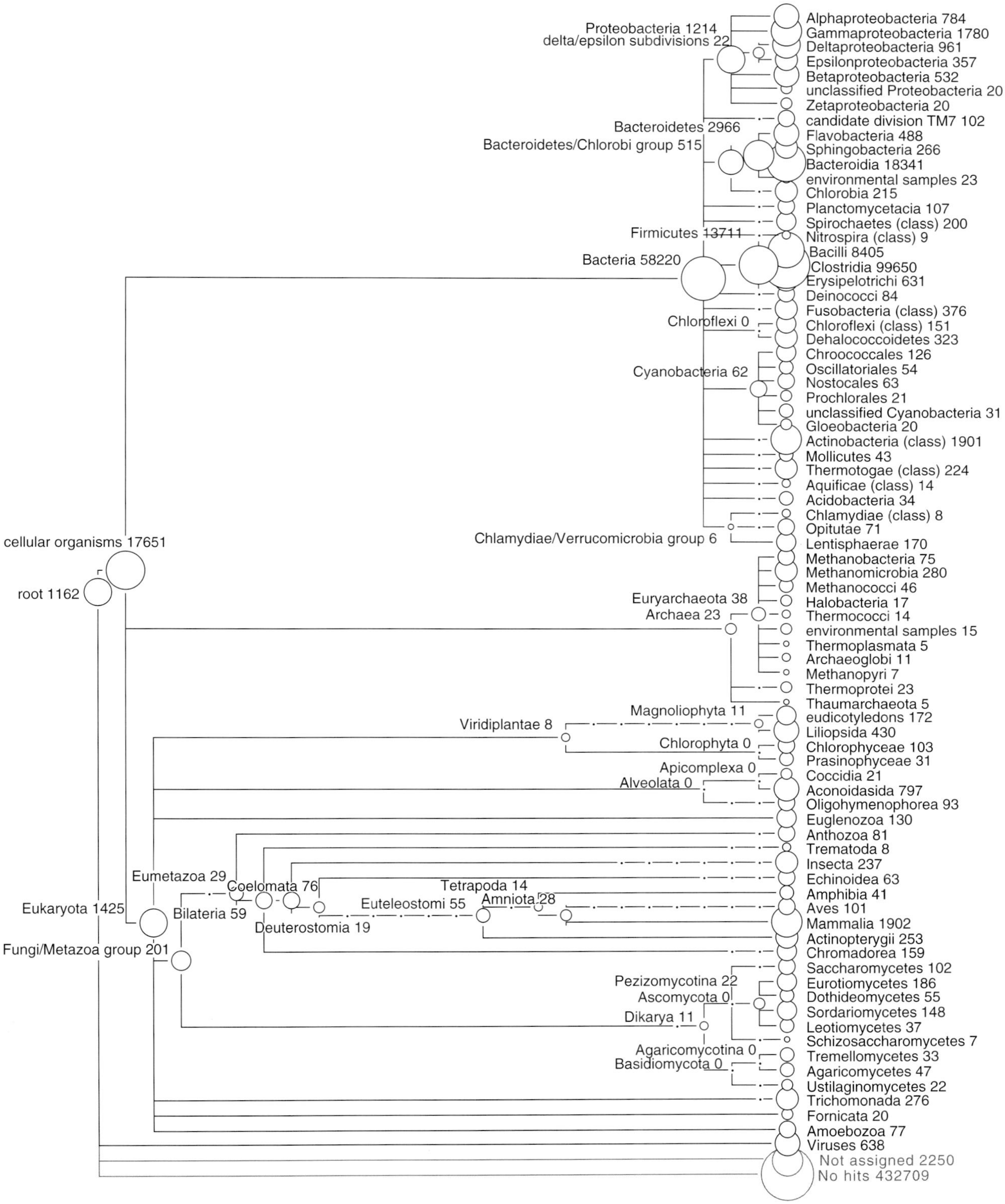

Figure 39.1 Taxonomical analysis of 677,384 reads of the obese mouse gut dataset (TurnbaughLeyMahowaldMagriniMardis et al., 2006) by MEGAN. Each circle represents a node of the NCBI taxonomy and is labeled by the number of reads assigned to the node. The sizes of the circles are logarithmically scaled to represent the numbers visually.

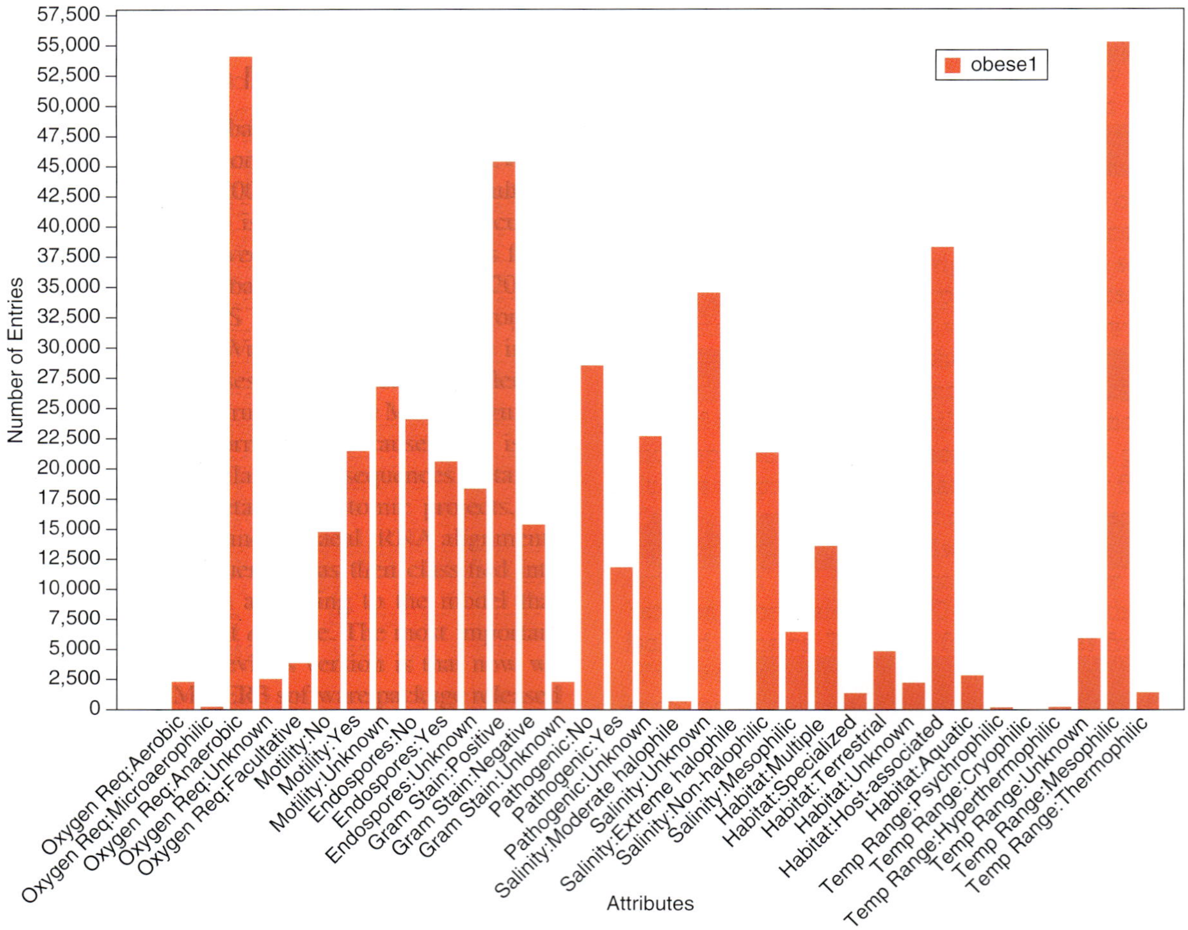

Figure 39.2 A summary of microbial attribute data for the obese mouse gut dataset.

nodes is scaled logarithmically to represent the number of reads. The program provides the exact numbers of reads assigned to any given node, and the number of hits on any nodes in the subtree rooted at the node. Moreover, with different chart tools one can visualize the distribution of the assigned reads. The program also allows one to interactively inspect the assignment of reads to a specific node, to zoom in to the individual BLAST hits that support the assignment of a read to a node, and to export all reads (and their matches, if desired) that were assigned to a specific part of the NCBI taxonomy.

MEGAN provides a number of additional tools for exploring the content of a dataset. For example, MEGAN can provide an estimation of the microbial content based on the "Prokaryotic Attributes Table" of NCBI. These attributes describe the physiology and habitat types of the sample (see **Fig. 39.2**). As another example, MEGAN contains a fast algorithm for computing a rarefaction curve showing the number of nodes that would be present in the analysis if based on only 10%, 20%, . . . , 90% of the reads.

39.4 FUNCTIONAL ANALYSIS

Initially, MEGAN was designed to provide only a taxonomical analysis of a dataset. Recent versions of the program provide two different types of functional analysis. The first approach is based on NCBI's Clusters of Orthologous Groups (COG) classification [Tatusov et al., 1997; 2003], which was developed to cluster annotated genes into functionally related groups. As an example, in **Figure 39.3** we display a chart of abundances for several COG categories for the obese mouse gut dataset as computed by MEGAN. This chart corresponds closely to a similar analysis reported in Turnbaugh et al. [2006].

A COG analysis is easy for MEGAN to perform when given the result of a BLAST comparison against the NCBI-NR database because the representatives of the COG families are all present in that database. However, while COGs are still used in publications, the COG classification is no longer curated and has thus become outdated.

To provide a more up-to-date and sophisticated functional analysis, MEGAN also uses the Gene Ontology

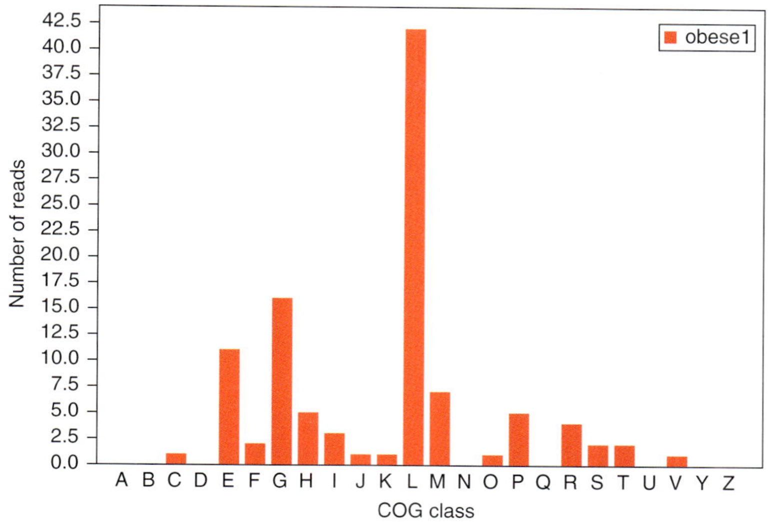

COG Codes	Description
	Information storage and processing
J	Translation, ribosomal structure and biogenesis
K	Transcription
L	DNA replication, recombination and repair
	Cellular processes
D	Cell division and chromosome partitioning
O	Posttranslational modification, protein turnover, chaperones
M	Cell envelope biogenesis, outer membrane
N	Cell motility and secretion
P	Inorganic ion transport and metabolism
T	Signal transduction mechanisms
	Metabolism
C	Energy production and conversion
G	Carbohydrate transport and metabolism
E	Amino acid transport and metabolism
F	Nucleotide transport and metabolism
H	Coenzyme metabolism
I	Lipid metabolism
Q	Secondary metabolites biosynthesis, transport and catabolism
	Poorly characterized
R	General function prediction only
S	Function unknown

Figure 39.3 Abundances of several COG categories for the obese mouse gut dataset.

(GO) [Ashburner et al., 2000] as a classification structure for binning environmental sequences. GO provides three sets of structured vocabularies (ontologies). *Molecular function* describes what a gene product does at the molecular level, *biological process* describes which processes it is involved in, and *cellular component* refers to the location in the cell where the gene product is usually found. GO is regularly updated and widely used in many biological databases, gene expression, and annotation studies and has been referred to as "the most successful example of systematic description of biology" [Rhee et al., 2008].

The GO currently contains around 28,000 terms. As the large number of terms and relationships can be unwieldy for some applications, the Gene Ontology Consortium also provides reduced, summarizing ontologies called "GO slims." At present, four slims are provided on the GO web site: the *generic GO slim, GOA and whole proteome analysis, plan GO slim*, and *yeast GO slim*. Additionally, a prokaryotic subset of all GO terms is also provided. MEGAN has the ability to summarize results based on GO slims.

MEGAN uses an LCA-like approach to assign each read to at most one node in each of the three GO onotologies. Since GO identifiers are not reported directly in a BLAST result file, MEGAN employs ref-seq identifiers [Pruitt et al., 2009] and a lookup-table to assign GO terms to read matches. Along with the assignment, the specificity information of a GO term, which maps GO identifiers to a specificity score, is also provided. The specificity is based on the number of annotated genes for a single GO term and its descendants compared to the total number of annotated genes [Alterovitz et al., 2007; Gaudet et al., 2009].

In **Figure 39.4**, we show part of a MEGAN analysis focused on the *metabolic process* ontology and *generic GO slim*, for the obese mouse dataset. As in the case of a taxonomical analysis, in a functional analysis, the user can search for specific nodes of interest, view all reads (and their matches) assigned to a given node or save all reads (and their matches, if desired) assigned to a given node.

39.5 COMPARING DATASETS

In this section, we will describe three different methods for comparing metagenome datasets. First we will describe how to compare multiple datasets visually. Second, we will describe a method of detecting taxa for which the number of assigned reads differs in a statistically significant way in the comparison of exactly two datasets. Third, we will briefly mention a new method that we are developing for analyzing and comparing multiple metagenomic datasets.

To illustrate the comparison methods, we will also use the other mouse gut metagenomes of [Turnbaugh et al., 2006, 2008], after first processing them to obtain RMA files as described above for the obese mouse sample.

39.5.1 Visual Comparison

To perform a visual comparison of multiple datasets using MEGAN, first the RMA files of interest must all be opened in MEGAN. Then the "Compare" menu item can be used to generate a new document that contains a comparison of all datasets, using either absolute counts or normalized values. The latter choice is of interest when

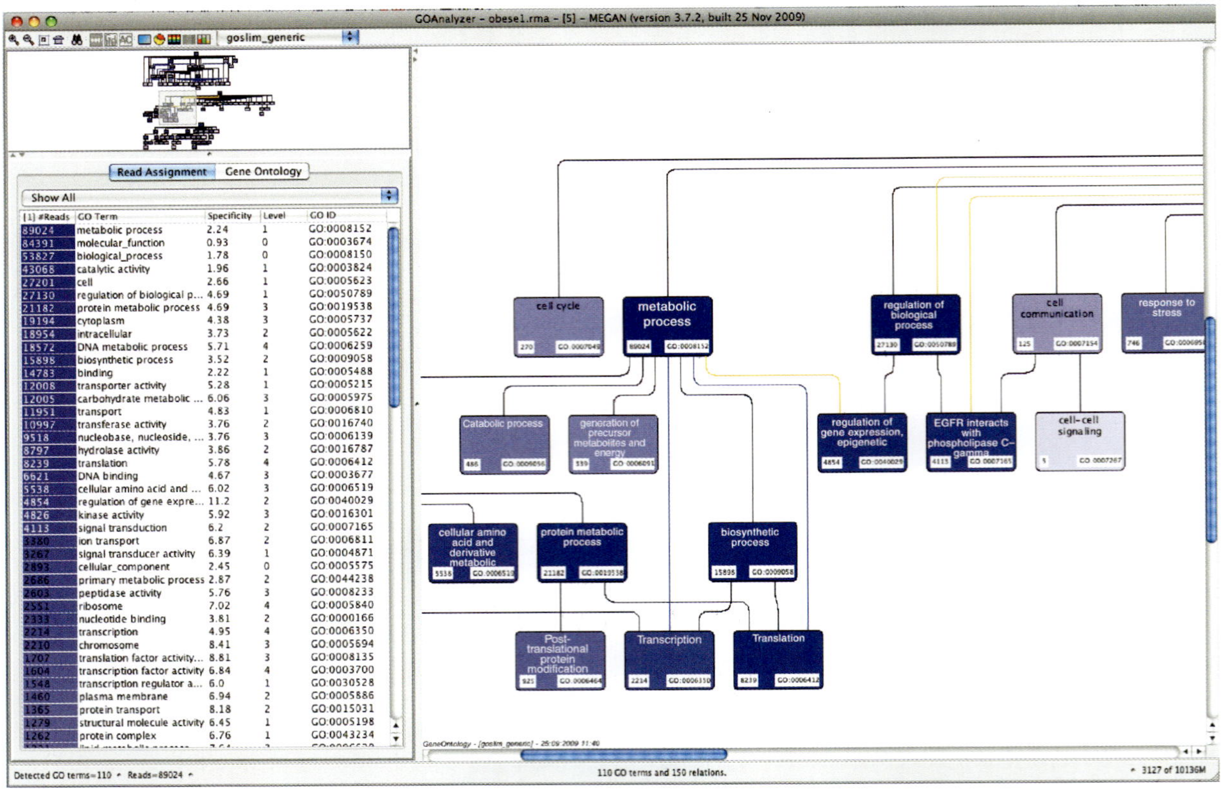

Figure 39.4 Example of a functional analysis, showing a part of the metabolic process ontology of the obese mouse gut dataset. The GOAnalyzer window is divided into three parts: At the top left, an overview panel shows the whole GO graph; on the left, all the GO terms are listed; and on the right, the three GO ontologies are displayed in separate subgraphs.

the compared datasets are very different in size. (Note that normalization can be problematic when applied to datasets of very different sizes as the numbers for rare taxa may be skewed.) The program provides a number of different options to determine how the comparison should be drawn on the nodes of the taxonomy or GO graph (whether as pie charts, heat maps, etc.).

To demonstrate the visual comparison, we opened all six previously mentioned mouse gut datasets and then produced a comparison document for them. Because the obese and the lean mouse gut datasets are much larger than the other four mouse gut samples, we used randomly selected subsamples of 10,000 reads from the obese and lean mouse datasets, and then we compared all datasets with normalization.

39.5.2 Comparison of Taxonomical Content

The default comparison view is displayed as a hierarchical tree, where each node shows the taxonomic assignments obtained for the datasets under consideration. In **Figure 39.5**, we display the comparison of six mouse gut datasets, collapsed at the *class* level of the NCBI taxonomy.

39.5.3 Charts for Comparison

MEGAN provides a range of chart tools to visualize the distribution of assigned reads. In **Figure 39.6**, we show the taxon chart for the Bacteroitedes and Firmicutes divisions for all six datasets.

39.5.4 Comparison of Functional Content

MEGAN provides a comparison view of the functional content of multiple datasets based on the GO classification. Once the user has conducted a taxonomical comparison, pressing the "GOAnalyzer" button will open the functional comparison view. **Figure 39.7** shows a functional comparison of the six mouse gut datasets. The numbers of the assigned reads listed in the table are normalized to 100,000 reads per dataset. A heat map gradient helps the user to differentiate the datasets easily.

39.5.5 Finding Significant Differences in the Pairwise Comparison of Taxonomical Content

To get an impression of how significantly two datasets differ, we have developed a *directed homogeneity* test, which

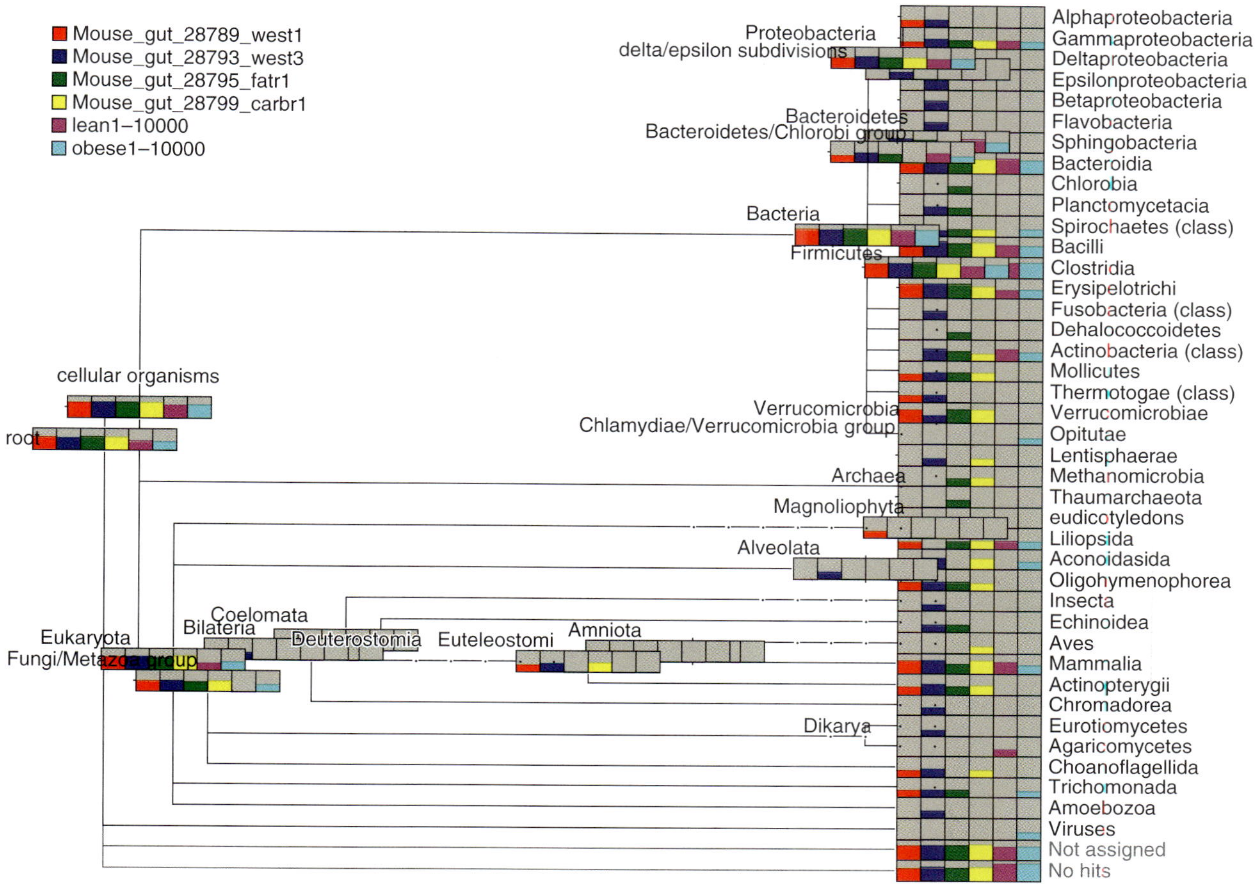

Figure 39.5 Comparative visualization of six mouse gut datasets with the NCBI taxonomy collapsed at the class level of the NCBI taxonomy.

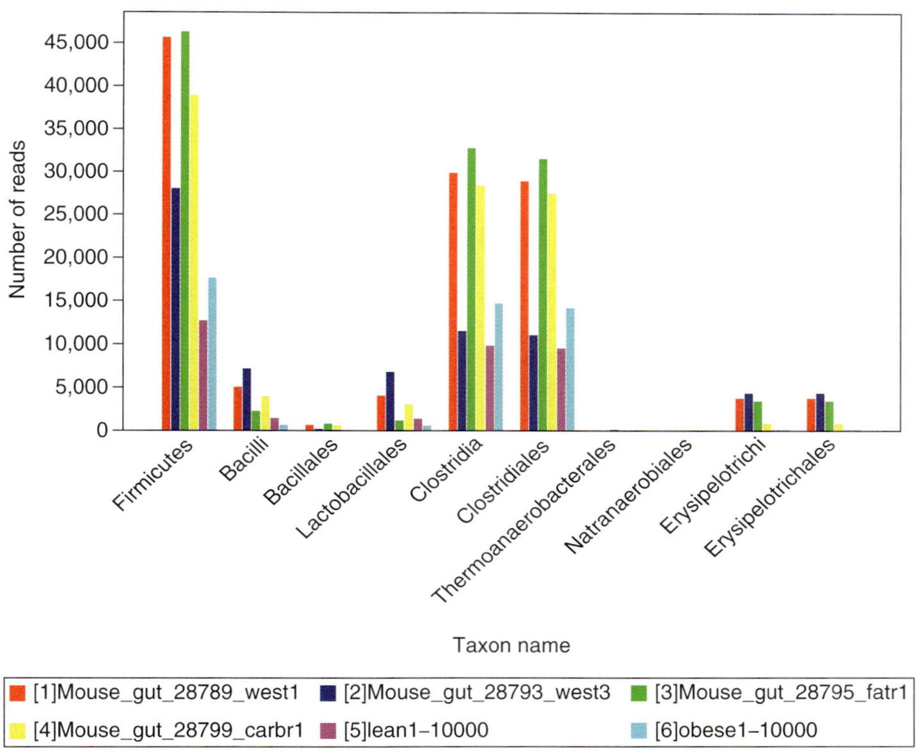

Figure 39.6 Taxon chart for the Firmicutes division for six mouse gut datasets generated by MEGAN.

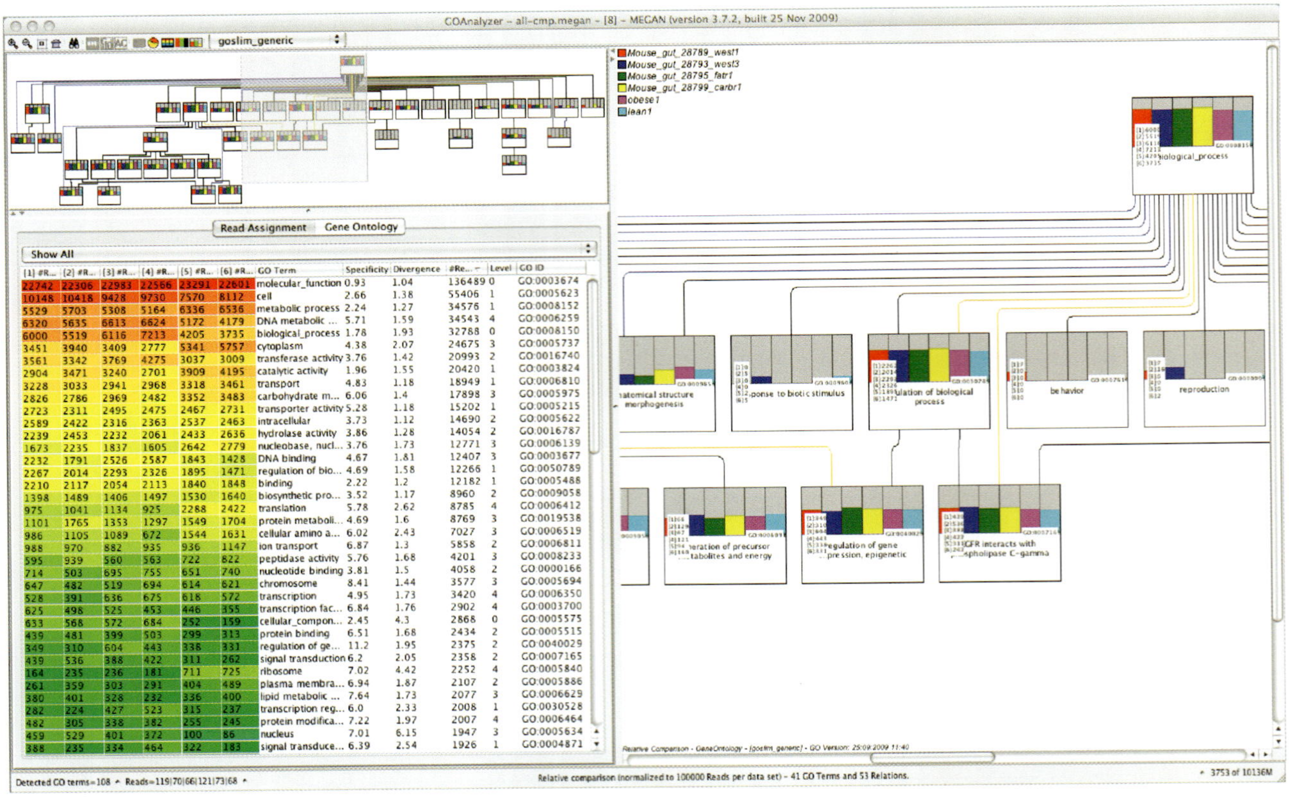

Figure 39.7 Comparative visualization of six mouse gut datasets for a part of the metabolic process ontology.

is based on simple statistics. The test provides answers to the two following questions:

1. Is there any significant difference in the proportions of occurrences of a particular node in two datasets?
2. Is there a significant difference in the distribution of reads among the children of a particular node in two datasets?

The answers are given by the *up* and the *down* test of the directed homogeneity test in MEGAN. The program uses the two tests to highlight all nodes for which either test asserts a statistically significant difference. To be precise, if the p-value of the up test is below a critical level (e.g., 0.01), then the part of the node that faces the parent will be highlighted, whereas a significant p-value for the down test will result in the part of the node that faces the children being highlighted (see Mitra et al. [2009] for details of the test). Note that such a difference may arise from a number of causes, such as different sampling or sequencing technologies, and thus will not necessarily reflect a biologically interesting difference.

In **Figure 39.8**, we show the result of a pairwise comparison between the obese and lean mouse gut datasets [Turnbaugh et al., 2006]. From the black highlighting and the up p-values, we can easily see that the significant differences between the two datasets mainly lie in the Bacteroidetes/Chlorobi group and Firmicutes classes. From the down p-values, we can say that the difference in read numbers for the Bacteroidetes node is mostly due to the difference in read numbers for the Favobacteria class and that the difference for the Firmicutes is mostly due to the difference for the Bacilli class. These differences are similar to the differences reported in Turnbaugh et al. [2006].

39.5.6 Similarity Analysis of Multiple Datasets

Besides the visual comparison of different nodes within datasets, it would also be very helpful to have a method for obtaining an overview of the similarity and the dissimilarity between multiple datasets. To address this, we are developing a new approach that combines the use of taxonomical analysis, ecological indices, and nonhierarchical clustering to provide a network representation of the relationships between multiple metagenome datasets.

39.6 SUMMARY AND OUTLOOK

MEGAN is a widely used tool for analyzing and comparing metagenomic datasets. Its main advantages are that is

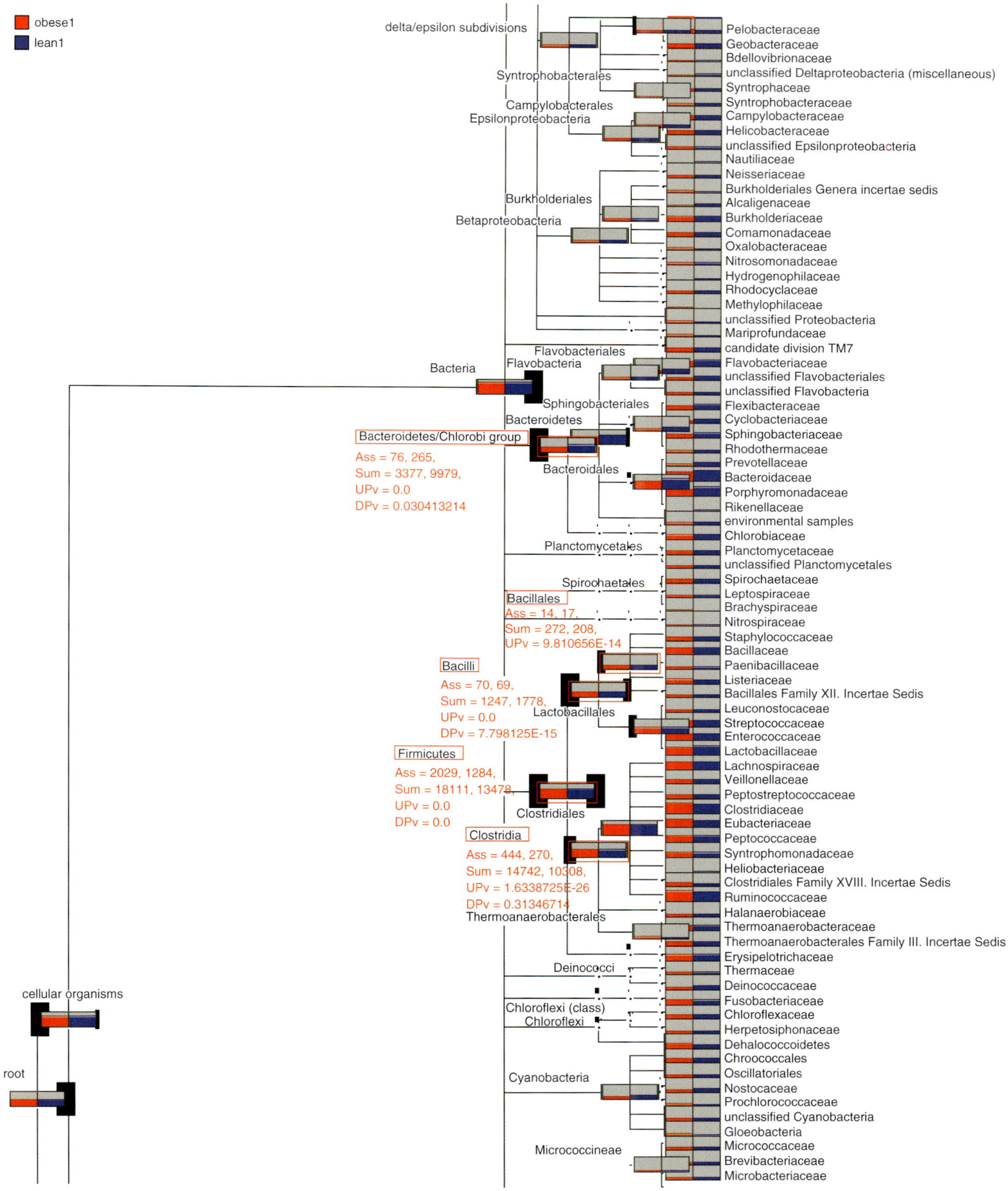

Figure 39.8 Pairwise comparison of the lean and obese mouse datasets. Black highlighting on the left side of a node implies that the up-test of the directed homogeneity test indicates a significant difference, whereas black highlighting on the right side indicates a significant difference detected by the down test. By selecting nodes, one can query the exact up and down p-values (UPv and DPv) together with the counts of assigned and summarized reads (as shown in red).

easy to deploy and to use, requiring only a BLAST file to perform both a taxonomical and functional analysis, and allows one to interactively explore, analyze, and compare a number of large datasets simultaneously. As MEGAN runs on a local computer, data do not have to be uploaded to a third-party server.

We are currently developing MEGAN-DB, a new version of MEGAN that will serve as a rich-client front end to a database of precomputed RMA files of published metagenome datasets. With MEGAN-DB, users will be able to analyze their own data locally on a desktop and will be able to compare their data with published data available over the internet without having to download that data (or upload their own data).

We are currently developing a third option for functional analysis in MEGAN based on subsystem technology [Overbeek et al., 2005; Meyer et al., 2008; see also Chapter 37, Vol. I], and this will allow comparative analyses using multiple functional classification systems (COG, GO, and subsystem technology).

INTERNET RESOURCES

MEGAN software (http://www-ab.informatik.uni-tuebingen.de/software/megan)

MEGAN DB (http://www.megan-db.org)

BLAST (http://blast.ncbi.nlm.nih.gov/Blast.cgi)

Refseq (http://www.ncbi.nlm.nih.gov/RefSeq)

COG database (http://www.ncbi.nlm.nih.gov/COG/)

GO slims (http://www.geneontology.org/)

GO slims (http://www.geneontology.org/GO.slims.shtml)

Acknowledgment

The authors would like to thank Daniel Richter and Stephan Schuster for helpful discussions.

REFERENCES

Alterovitz G, Xiang M, Mohan M Ramoni MF. 2007. GO PaD: The gene ontology partition database. *Nucleic Acids Res*. **35**:D322–D327.

Altschul SF, Gish W, Miller W, Myers EW Lipman DJ. 1990. Basic Local Alignment Search Tool. *J. Mol. Biol*. **215**:403–410.

Ashburner M, Ball CA, Blake JA, Botstein D, Butler H, et al. 2000. Gene Ontology: Tool for the unification of biology. *Nat. Genet*, **25**:25–29.

Benson DA, Karsch-Mizrachi I, Lipman DJ, Ostell J Wheeler DL. 2005. GenBank. *Nucleic Acids Res*. **33**:D34–D38.

Bernal A, Ear U Kyrpides N 2001. Genomes OnLine Database (GOLD): A monitor of genome projects world-wide. *Nucleic Acids Res*. **29**:126–127.

Gaudet P, Chisholm R, Berardini T, Dimmer E, Engel SR, et al. 2009. The Gene Ontology's Reference Genome Project: A Unified Framework for Functional Annotation across Species. *PLoS Comput. Biol*. **5**.

Gilbert JA, Field D, Huang Y, Edwards R, Li W, et al. 2008. Detection of large numbers of novel sequences in the metatranscriptomes of complex marine microbial communities. **3**:e3042.

Gill SR, Pop M, DeBoy RT, Eckburg PB, Turnbaugh PJ, et al. 2006. Metagenomic analysis of the human distal gut microbiome. *Science* **312**:1355–1359.

Huson DH, Auch AF, Qi J Schuster SC 2007. MEGAN analysis of metagenomic data. *Genome Res*. **17**:377–386.

Mardis ER 2008. Next-generation DNA sequencing methods. *Annu. Rev. Genomics Hum. Genet*. **9**:387–402.

Meyer F, Paarmann D, D'Souza M, Olson R, Glass EM, et al. 2008. The metagenomics RAST server—A public resource for the automatic phylogenetic and functional analysis of metagenomes. **9**:386.

Mitra S, Klar B Huson DH. 2009. Visual and statistical comparison of metagenomes. *Bioinformatics* **25**:1849–1855.

Overbeek R, Begley T, Butler RM, Choudhuri JV, Chuang HY, et al. 2005. The subsystems approach to genome annotation and its use in the project to annotate 1000 genomes. *Nucleic Acids Res*. **33**:5691–5702.

Poinar HN, Schwarz C, Qi J, Shapiro B, MacPhee RDE, et al. 2006. Metagenomics to paleogenomics: Large-scale sequencing of mammoth DNA. *Science* **311**:392–394.

Pruitt KD, Tatusova T, Klimke W Maglott DR 2009. NCBI reference sequences: Current status, policy and new initiatives. *Nucleic Acids Res*. **37**:D32–D36.

Rhee SY, Wood V, Dolinski K Draghici S. 2008. Use and misuse of the gene ontology annotations. *Nat. Rev. Genet*. **9**:509–515.

Schloss PD Handelsman J 2006. Toward a census of bacteria in soil. *PLoS Comput. Biol*. **2**:e92.

Schwartz S, Kent WJ, Smit A, Zhang Z, Baertsch R, et al. 2003. Human–mouse alignments with BLASTZ. *Genome Res*. **13**:103–107.

Shendure J, Ji HL. 2008. Next-generation DNA sequencing. *Nat. Biotechnol*. **26**:1135–1145.

Staden R. 1979. A strategy of DNA sequencing employing computer programs. *Nucleic Acids Res*. **6**:2601–2610.

Tatusov RL, Koonin EV, Lipman DJ. 1997. A genomic perspective on protein families. *Science* **278**:631–637.

Tatusov RL, Fedorova ND, Jackson JD, Jacobs AR, Kiryutin B, et al. 2003. The COG database: An updated version includes eukaryotes. *BMC Bioinform*. **4**.

Turnbaugh PJ, Ley RE, Mahowald MA, Magrini V, Mardis ER, et al. 2006. An obesity-associated gut microbiome with increased capacity for energy harvest. *Nature* **444**:1027–1031.

Turnbaugh PJ, Baeckhed F, Fulton L Gordon JI. 2008. Diet-induced obesity is linked to marked but reversible alterations in the mouse distal gut microbiome. *Cell Host Microbe* **3**:213–223.

van Elsas JD, Costal R, Jansson J, Sjoling S, Bailey M, et al. 2008. The metagenomics of disease-suppressive soils—Experiences from the METACONTROL project. *Trends Biotechnol*. **26**:591–601.

Venter JC, Remington K, Heidelberg JF, Halpern AL, Rusch D, et al. 2004. Environmental genome shotgun sequencing of the Sargasso Sea. *Science* **304**:66–74.

Chapter 40

Phylogenetic Binning of Metagenome Sequence Samples

Alice Carolyn McHardy and Kaustubh Patil

40.1 INTRODUCTION

Metagenomics, the culture-independent community level sequencing of microbial organisms, is rapidly changing our knowledge of the microbial world. Traditional genome sequencing requires pure cultures of the organism of interest and is thus limited to the study of those organisms that can be cultivated as a clonal culture in the laboratory with standard techniques. This, however, is the case only for very few members of the microbial world [Hugenholtz, 2002]. Metagenomics overcomes this restriction by direct DNA sequencing of a mixture of organisms found in an environment of interest. This milestone development has given researchers the capability to study the world of uncultivable microorganisms by genome sequencing; furthermore, it provides system-level insights into the composition, structure, and functioning of microbial communities from various environments [Gill et al., 2006; Martin et al., 2006; Poinar et al., 2006; Turnbaugh et al., 2006; Tyson et al., 2004; Venter et al., 2004; Warnecke et al., 2007; Woyke et al., 2006].

A metagenome sequence sample initially consists of a large set of reads obtained by sequencing a mixture of populations of different organisms. Using assembly algorithms, the reads are assembled based on overlaps into longer continuous pieces of sequence, or contigs. The complexity of the analyzed community in terms of the number of populations contained therein, their relative abundances, and the size and architecture of the individual genomes have a strong effect on the average length of the resulting contigs. Higher complexity and low abundance translates into less extensive sampling of genomic regions of the individual populations, along with shorter fragment lengths after assembly. Because the obtained fragments originate from the various populations of a community, a key step that is subsequently required is assignment of the sample sequences to the respective organisms or higher-level phylogenetic clades from which they were derived, a procedure known as "binning." Various methods have been developed for this task.

40.2 METHODS FOR THE PHYLOGENETIC BINNING OF METAGENOME SEQUENCE SAMPLES

Methods for the phylogenetic characterization of metagenome sequence samples can be classified as either homology-based or sequence-composition-based approaches. We describe each approach briefly in the following subsections.

40.2.1 Sequence Homology-Based Approaches

Due to their highly conserved nature and ubiquitous presence, phylogenetic assignment based on small-subunit ribosomal RNA (16S rRNA) genes is considered the most accurate for phylogenetic assignment [see Chapter 16, Vol. I]. Fragments with 16S rRNA genes can be identified in a metagenome sample, and their placement in a reference phylogeny can be evaluated. 16S rRNA phylotyping is performed extensively, and large reference libraries exist [DeSantis et al., 2006; see also Chapter 36, Vol. I]. However, such marker genes do not allow an extensive

Handbook of Molecular Microbial Ecology, Volume I: Metagenomics and Complementary Approaches, First Edition. Edited by Frans J. de Bruijn.

characterization of most metagenome sequence samples, in particular so for samples of higher organismal complexity. For instance, sequence fragments carrying rRNA genes correspond to only 0.071%, 0.06%, and 0.17% of the total for samples from agricultural soil, the Sargasso Sea, and enhanced biological phosphorus-removing sludge [McHardy and Rigoutsos, 2007]. Extended lists of markers of 25–30 highly conserved genes have been defined [Ciccarelli et al., 2006] and can be used for a more extensive characterization. Because most other markers have not been sequenced as comprehensively as rRNA, however, the reference system for such genes does not always allow unambiguous phylotyping, and this approach does not greatly extend sample coverage. The inclusion of further, less conserved genes in the analysis is possible, but will result in a loss of specificity due to phenomena such as gene duplications, gene loss, horizontal gene transfer, and, in general, differences of individual gene trees to the species tree. A similar issue arises for clade-specific markers [Graham et al., 2000], of which currently only few are known and which are not defined for novel, unstudied clades.

However, given that genome sequences of related organisms to a sample population are available as reference, a taxonomic analysis of the results from homology searches using BLAST, FASTA, and so on, can be very effective. Generally, assignments may be made with a resolution up to the lowest taxonomic rank shared by query sequence and queried databases. For instance, in Huson et al. [2007] it is shown that homology-based assignments are accurate for very short fragments if the reference genome is contained in the queried database. For the Sargasso Sea sample, homology searches allowed the assignment of fragments from *Shewanella* and *Burkholderia*, for which sequenced genomes were available. Similarly, individual reads can be used in combination with mate-pair information for genome reconstruction if closely related reference genomes are available [Rusch et al., 2007]. Alternatively, the results of protein domain searches can be taxonomically analyzed and used for phylogenetic sequence assignment [Krause et al., 2008]. However, the content of public databases in terms of sequenced genomes is strongly biased toward a few clades containing most cultivable organisms (**Fig. 40.1**) and, within those, toward organisms of biomedical relevance. Thus, the required information is frequently not available and homology-based assignment can become inaccurate, as demonstrated in a large-scale benchmark study [Mavromatis, et al., 2007]. For the taxonomic characterization of viral metagenome samples reference genomes are required, as they lack universal markers or well-defined phylogenetic compositional signatures [Breitbart and Rohwer, 2005; see also Chapters 2–8, Vol. II]. Furthermore, for some groups of viruses, conserved clade-specific genes can be used to characterize a sample.

40.2.2 Sequence Composition-Based Approaches

Another approach assigns fragments based on sequence composition, which refers to the frequencies of the constituent substrings (oligomers) with lengths from two to nine base pairs [Abe et al., 2002, 2003; Deschavanne et al., 1999; McHardy et al., 2007; Nakashima et al., 1998; Sandberg et al., 2001]. Various studies have concluded that different species have distinct compositional signatures [Deschavanne et al., 1999; Karlin and Burge, 1995; Karlin et al., 1997; Mrazek, 2009] and that sequence composition can be used to distinguish fragments from different species [Abe et al., 2003; Nakashima et al., 1998; Sandberg et al., 2001; Teeling et al., 2004; McHardy et al., 2007]. Normalization of oligomer frequencies can be achieved in various ways—for example, by dividing by sequence length or by the expected frequency of the embedded substrings [Karlin and Burge, 1995; Karlin et al., 1997; Mrazek, 2009; Pride et al., 2003; Teeling et al., 2004]. The accuracy and coverage of assignments are affected by factors such as the organismal complexity of the analyzed community, fragment lengths, and the method used for assignment.

Compositional signatures represent sequence fragments as fixed-length vectors of oligomer usage, which allows to directly apply state-of-the-art data mining and machine learning methods. From a technical perspective, such methods can be grouped into unsupervised (clustering) and supervised methods. Unsupervised methods seek to group similar items of dataset into distinct clusters and can be applied in the absence of reference information (i.e., training data). For the phylogenetic binning of metagenome sequence samples such approaches partition the fragments of the sample into a user-specified number of clusters (bins), based on the compositional signatures of individual fragments [Abe et al., 2005; Chatterji et al., 2008; Teeling et al., 2004; Tyson et al., 2004]. Assignments of the bins to phylogenetic clades can subsequently be performed based on the presence of fragments with phylogenetically informative marker genes or fragments of known organisms of a clade within a bin. As an example, a clustering analysis based on fragment sequence composition enabled the phylogenetic characterization of a metagenome of four marine sea worm symbionts [Woyke et al., 2006; see also Chapter 32, Vol. II].

A crucial decision that influences the outcome of a clustering analysis is the choice of the number of clusters which are to be formed, which is not trivial. This number may be chosen based on prior knowledge about the sample or by careful evaluation of the possible parameter settings.

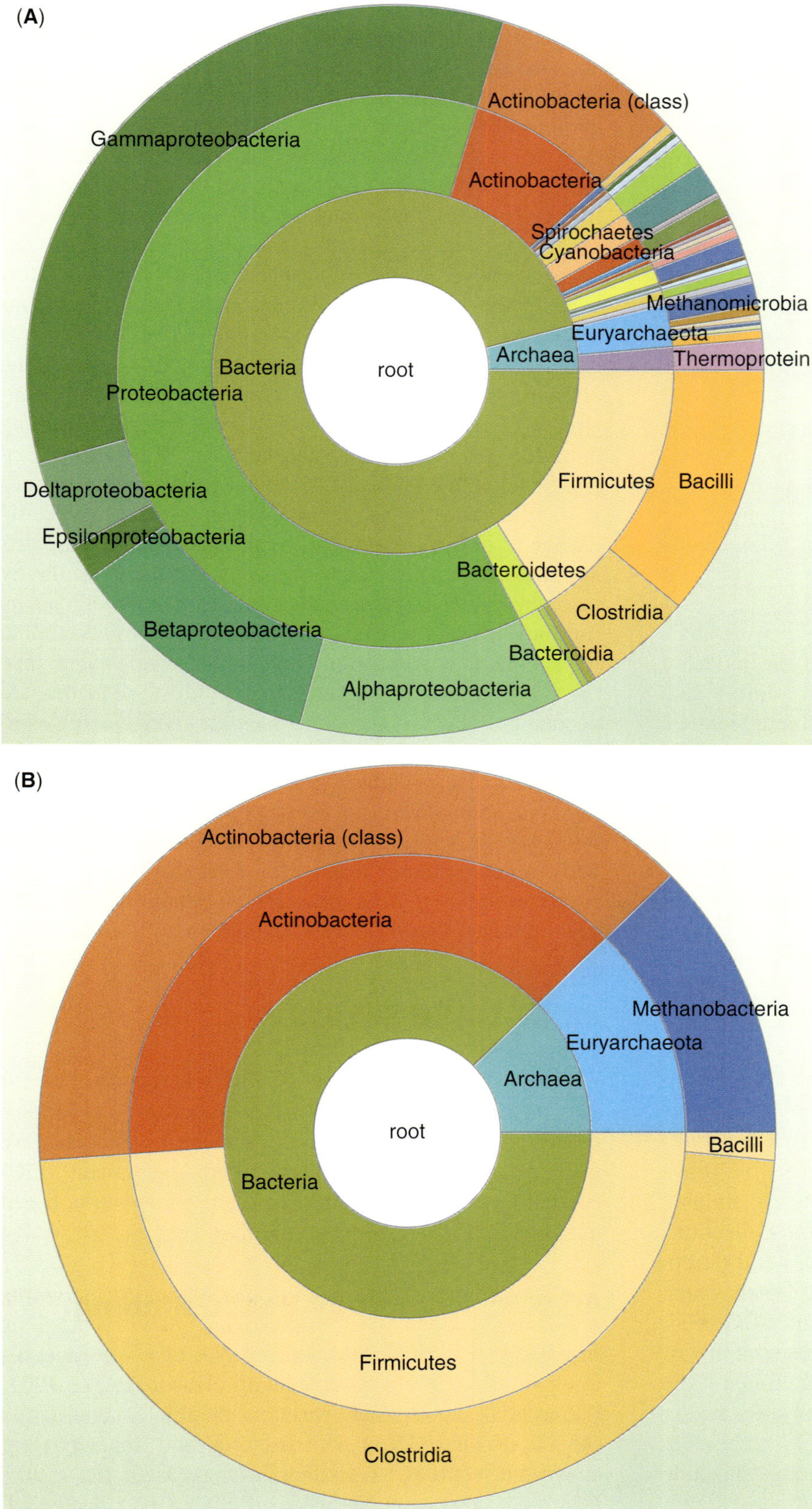

Figure 40.1 Comparison of the taxonomic composition of public database content to a microbial community analyzed by metagenome sequencing. (**A**) Taxonomic composition of finished genomes present in Genbank, state of May 2009: The large bias toward Gammaproteobacteria is caused predominantly by 164 genome sequences of *Escherichia coli* strains. (**B**) Taxonomic composition of microbial populations in the human gut estimated by 16S rRNA phylotyping [Gill et al., 2006].

Moreover, the data contain noise, caused by sequencing errors, chimeric contigs [Mavromatis et al., 2007], and other signals, such as leading/lagging strand bias, gene expression rates, and so on, which influences fragment sequence composition besides the phylogenetic origin. Generally, misassignments become more likely with increasing organismal complexity of a sample. However, for samples of high complexity, the characterized portion of the sample can be increased by assigning fragments to higher-level clades.

Higher-level clades are defined by more subtle signals in sequence composition than clades at the species level. Because supervised methods generally feature a higher accuracy than unsupervised approaches, they represent an attractive option for analyzing complex datasets. In supervised classification the compositional signature of fragments of known phylogenetic origin, such as fragments from sequenced genomes or marker-gene carrying fragments within a metagenome, is used as a reference to model a number of classes, which are the phylogenetic clades represented in a metagenome sample.

Once a model is trained, a sample can then be characterized by assigning fragments to one of the modeled classes. Some supervised classification algorithms also allow us to determine the importance of the input features, the oligomers, and input items (the sequence fragments) for discrimination between different phylogenetic classes, which reduces the effect of noisy/irrelevant features (and items) on the classification accuracy.

For supervised classification the quantity and quality of the available training data (i.e., taxonomically labeled reference data from a given clade) also play a critical role. No complete reference genomes are required for composition-based assignment of fragments, because the data need only be representative for the sequence composition of a given clade. For higher-level clades comprising more distantly related organisms, the compositional signals are less pronounced than at the species level, likely due to the evolutionary processes that have shaped genomic sequence composition since divergence from the ancestral species. However, using suitable techniques, these signals can be detected from collections of genomic sequences for phylogenetic clades at different ranks [Abe et al., 2002; Chapus et al., 2005; Deschavanne et al., 1999; McHardy et al., 2007; Pride et al., 2003], and their value for phylogenetic characterization of metagenome sequence samples has been quantified [Abe et al., 2005; Mavromatis et al., 2007; McHardy et al., 2007].

In our work we use a support vector machine (SVM) to model the fragment compositional signatures of phylogenetic clades for the characterization of metagenome sequence samples [Mavromatis et al., 2007; McHardy et al., 2007; Warnecke et al., 2007]. SVMs exhibit generally high classification accuracy among the available methods for supervised classification. They are able to learn accurate classifiers from high-dimensional input spaces such as the space of fragment sequence composition and from relatively small samples of training data. In our previous work we found that fragments with length of more than 1 kb can be accurately assigned to clades from the genus to the domain level, using models trained in the absence of any data from the same organism that we were seeking to classify. Thus, even in the absence of reference data from a particular organism or population, metagenome sequence fragments can be assigned to higher-level taxonomic ranks. The assignment accuracy can be further enhanced by constructing models that include sample-specific training data for the sample populations. For populations that are taxonomically distant from the sequenced genomes in public repositories, the inclusion of sample-specific reference data is particularly important to increase the assignment accuracy and allow assignment at lower taxonomic levels. Here conserved marker genes can be used to identify contigs or scaffolds as training data for the sample populations and higher-level clades. For the SVM-based models, around 100 kb of sequence are needed per clade to construct accurate models. However, depending on sequencing depth and on the composition of a particular community, 100 kb of training sequence may not be available for every population of interest. In this case, an option is to identify and sequence fosmids with larger genomic inserts for relevant sample clades based on phylogenetic markers to attain the required training data [Warnecke et al., 2007].

40.3 CASE STUDIES

Here we illustrate the use of different approaches for the phylogenetic binning of metagenome sequences based on several metagenome sequencing studies. We give examples for homology-based assignment, GC-content-based binning, and sequence composition-based unsupervised and supervised binning.

40.3.1 Homology-Based Binning

In a sequence sample of ancient DNA of a wooly mammoth obtained by pyrosequencing [Huson et al., 2007; Poinar et al., 2006], there was considerable contamination (45.5% of total DNA) with bacterial and nonelephantid eukaryotic sources. A taxonomic analysis of the results from homology searches with MEGAN (see Chapter 39, Vol. I) placed reads at the domain level to be of nonvertebrate eukaryotic (4.15%), bacterial (5.76%), and archaeal (0.24%) origin.

40.3.2 GC Content and Read Coverage-Based Binning

A metagenome of biofilm from acid mine drainage samples was binned based on GC content (see Chapter 23, Vol. I) and sequencing coverage [Tyson et al., 2004]. The biofilm community comprises six archaeal and bacterial populations, two of which predominate. Sample fragments with lengths of more than 2 kb could be consistently assigned to bins based on GC content and read coverage, in accordance with the distribution of markers on the respective fragments, for all but the least abundant population.

40.3.3 Composition-Based Unsupervised Binning

A metagenome of four co-occurring symbionts of the marine worm *Olavius algarvensis* was binned with an unsupervised approach using compositional signatures [Woyke et al., 2006]. The obtained bin assignments for scaffolds larger than 5 kb were in agreement with the distribution of phylogenetic marker genes and a PCA-based visualization of the fragment distribution in a space of GC content, read depth, and oligomer usage.

40.3.4 Composition-Based Supervised Binning

A metagenome of a microbial community from the hindgut of a wood-feeding higher termite [Warnecke et al., 2007] was binned using a phylogenetic model with sample-specific clades created with a supervised technique based on compositional signatures. A 16S rRNA survey of the community complexity identified 216 distinct phylotypes from 12 phyla which covered an estimated 80% of diversity samples. Two phyla comprising various phylotypes of the genus *Treponema* and the phylum Fibrobacteres predominated. Due to the high complexity of the sample, the longest sequenced fragment obtained was approximately 15 kb in length. The composition-based models for phylogenetic fragment assignment were trained on available genome sequences from public repositories and, furthermore, with fully sequenced fosmids for the predominant sample phyla identified using phylogenetic marker genes. Subsequent phylogenetic characterization of the metagenome sequence fragments allowed assignment of 36% of the fragments larger than 1 kb to the predominant phyla (3% to *Treponema*, 33% to *Fibrobacteres*), in agreement with the prevalence of phylogenetic marker genes. Overall 99% of the fragments larger than 1 kb was assigned to phyla at ranks of class to domain.

40.4 RECENT DEVELOPMENTS AND CHALLENGES

Computational methods are essential for the metagenomic analysis of microbial communities and will continue to play an important role in the field. For the phylogenetic assignment of sequence fragments to the respective sample clades, various methods have been developed [Abe et al., 2002; Brady and Salzberg, 2009; Chatterji et al., 2008; Diaz et al., 2009; Krause et al., 2008; McHardy et al., 2007; Sandberg et al., 2001; Teeling et al., 2004]. Most recently, methods have been designed for the characterization of short fragments of less than 1 kb in length [Brady and Salzberg, 2009; Chatterji et al., 2008; Krause et al., 2008]. The ability to accurately classify short fragments considerably increases the characterized portion of metagenome samples, especially for samples of complex communities generated by next generation sequencing [Edwards et al., 2006; Quail et al., 2008]. These recent developments bring us closer toward this goal, even though the obtained accuracy and coverage are not comparable to those for longer fragments yet. Short fragments need to be assigned based on less information, and are in particular affected by local variations in genomic sequence composition, which makes their assignment so challenging.

An important issue is the choice of the evaluation setup for phylogenetic binning methods. If evaluations are based on simulated datasets generated within the scope of a study, this makes a comparison to subsequently developed approaches a difficult task. Benchmark datasets of metagenome sequence samples with varying complexity exist [Mavromatis et al., 2007]. A more extensive use of such benchmarks could facilitate comparison of different new techniques. A further issue is that benchmarks are typically constructed from a random sample of genomic sequences available in public databases such as GenBank. This introduces additional, not immediately obvious hurdles to generating a suitable evaluation setup: Great care has to be taken to exclude all (to-be-tested) benchmark data from any training or reference data that are used.

Secondly, sequenced genomes are skewed toward a few phyla that comprise most cultivable organisms (Fig. 40.1). However, the taxonomic sample composition, organismal complexity, population structure, available training data/reference genomes for sample-specific clades, and amount of noise introduced by the different sequencing technologies may differ greatly for any real sample from such a simulated benchmark. Therefore, the accuracy on simulated benchmarks may not be representative for a methods performance on real samples. It is thus recommended to also use real metagenome sequence samples to assess and compare performance. Classical performance measures such as accuracy, sensitivity, and specificity can only be applied to a small subset of the sample fragments, namely those that can be characterized with a reliable gold standard such as conserved marker genes. However, indirect performance measures, such as the assignment consistency of contigs originating from a common scaffold, can be applied [McHardy et al., 2007]. The increased use of such measures and evaluation on real data could be greatly beneficial for assessing and comparing performance of different techniques for the phylogenetic characterization of metagenome datasets.

Acknowledgment

This work was partly supported by the Max Planck Society.

REFERENCES

Abe T, Kanaya S, Kinouchi M, Ichiba Y, Kozuki T, et al. 2002. A novel bioinformatic strategy for unveiling hidden genome signatures of eukaryotes: self-organizing map of oligonucleotide frequency. *Genome Inform. Ser. Workshop Genome Inform*. **13**:12–20.

Abe T, Kanaya S, Kinouchi M, Ichiba Y, Kozuki T, et al. 2003. Informatics for unveiling hidden genome signatures. *Genome Res*. **13**:693–702.

Abe T, Sugawara H, Kinouchi M, Kanaya S, Ikemura T. 2005. Novel phylogenetic studies of genomic sequence fragments derived from uncultured microbe mixtures in environmental and clinical samples. *DNA Res*. **12**:281–290.

Brady A, Salzberg SL. 2009. Phymm and PhymmBL: Metagenomic phylogenetic classification with interpolated Markov models. *Nat. Methods* **6**:673–676.

Breitbart M, Rohwer F. 2005. Here a virus, there a virus, everywhere the same virus? *Trends Microbiol*. **13**:278–284.

Chapus C, Dufraigne C, Edwards S, Giron A, Fertil B, et al. 2005. Exploration of phylogenetic data using a global sequence analysis method. BMC *Evol. Biol*. **5**:63.

Chatterji S, Yamazaki I, Bai Z, Eisen J. 2008. CompostBin: A DNA Composition-Based Algorithm for Binning Environmental Shotgun Reads. *Res. Comput. Mol. Biol*. 17–28.

Ciccarelli FD, Doerks T, von Mering C, Creevey CJ, Snel B, et al. 2006. Toward automatic reconstruction of a highly resolved tree of life. *Science* **311**:1283–1287.

DeSantis TZ, Hugenholtz P, Larsen N, Rojas M, Brodie EL, et al. 2006. Greengenes, a chimera-checked 16S rRNA gene database and workbench compatible with ARB. *Appl. Environ. Microbiol*. **72**:5069–5072.

Deschavanne PJ, Giron A, Vilain J, Fagot G, Fertil B. 1999. Genomic signature: Characterization and classification of species assessed by chaos game representation of sequences. *Mol. Biol. Evol*. **16**:1391–1399.

Diaz NN, Krause L, Goesmann A, Niehaus K, Nattkemper TW. 2009. TACOA: Taxonomic classification of environmental genomic fragments using a kernelized nearest neighbor approach. *BMC Bioinform*. **10**:56.

Edwards RA, Rodriguez-Brito B, Wegley L, Haynes M, Breitbart M, et al. 2006. Using pyrosequencing to shed light on deep mine microbial ecology. *BMC Genomics* **7**:57.

Gill SR, Pop M, Deboy RT, Eckburg PB, Turnbaugh PJ, et al. 2006. Metagenomic analysis of the human distal gut microbiome. *Science* **312**:1355–1359.

Graham DE, Overbeek R, Olsen GJ, Woese CR. 2000. An archaeal genomic signature. *Proc. Natl. Acad. Sci. USA* **97**:3304–3308.

Hugenholtz P. 2002. Exploring prokaryotic diversity in the genomic era. *Genome Biol*. **3**:00–00.

Huson DH, Auch AF, Qi J, Schuster SC. 2007. MEGAN analysis of metagenomic data. *Genome Res*. **17**:377–386.

Karlin S, Burge C. 1995. Dinucleotide relative abundance extremes: A genomic signature. *Trends Genet*. **11**:283–290.

Karlin S, Mrazek J, Campbell AM. 1997. Compositional biases of bacterial genomes and evolutionary implications. *J. Bacteriol*. **179**:3899–3913.

Krause L, Diaz NN, Goesmann A, Kelley S, Nattkemper TW, et al. 2008. Phylogenetic classification of short environmental DNA fragments. *Nucleic Acids Res*. **36**:2230–2239.

Martin HG, Ivanova N, Kunin V, Warnecke F, Barry KW, et al. 2006. Metagenomic analysis of two enhanced biological phosphorus removal (EBPR) sludge communities. *Nat. Biotechnol*. **00**:000–000.

Mavromatis K, Ivanova N, Barry K, Shapiro H, Goltsman E, et al. 2007. Use of simulated data sets to evaluate the fidelity of metagenomic processing methods. *Nat. Methods* **4**:495–500.

McHardy AC, Rigoutsos I. 2007. What's in the mix: phylogenetic classification of metagenome sequence samples. *Curr. Opin. Microbiol*. **10**:499–503.

McHardy AC, Garcia-Martin H, Tsirigos A, Hugenholtz P, and I. R., 2007. Accurate Phylogenetic Classification of Variable-length DNA fragments. *Nat. Methods* **4**:63–72.

Mrazek J. 2009. Phylogenetic signals in DNA composition: Limitations and prospects. *Mol. Biol. Evol*. **26**:1163–1169.

Nakashima H, Ota M, Nishikawa K, Ooi T. 1998. Genes from nine genomes are separated into their organisms in the dinucleotide composition space. *DNA Res*. **5**:251–259.

Poinar HN, Schwarz C, Qi J, Shapiro B, Macphee RD, et al. 2006. Metagenomics to paleogenomics: Large-scale sequencing of mammoth DNA. *Science* **311**:392–394.

Pride DT, Meinersmann RJ, Wassenaar TM, Blaser MJ. 2003. Evolutionary implications of microbial genome tetranucleotide frequency biases. *Genome Res*. **13**:145–158.

Quail MA, Kozarewa I, Smith F, Scally A, Stephens PJ, et al. 2008. A large genome center's improvements to the Illumina sequencing system. *Nat. Methods* **5**:1005–1010.

Rusch DB, Halpern AL, Sutton G, Heidelberg KB, Williamson S, et al. 2007. The Sorcerer II Global Ocean Sampling Expedition: Northwest Atlantic through Eastern Tropical Pacific. *PLoS Biol*. **5**:e77.

Sandberg R, Winberg G, Branden CI, Kaske A, Ernberg I, et al. 2001. Capturing whole-genome characteristics in short sequences using a naive Bayesian classifier. *Genome Res*. **11**:1404–1409.

Teeling H, Meyerdierks A, Bauer M, Amann R, Glockner FO. 2004. Application of tetranucleotide frequencies for the assignment of genomic fragments. *Environ. Microbiol*. **6**:938–947.

Turnbaugh PJ, Ley RE, Mahowald MA, Magrini V, Mardis ER, et al. 2006. An obesity-associated gut microbiome with increased capacity for energy harvest. *Nature* **444**:1027–1031.

Tyson GW, Chapman J, Hugenholtz P, Allen EE, Ram RJ, et al. 2004. Community structure and metabolism through reconstruction of microbial genomes from the environment. *Nature* **428**:37–43.

Venter JC, Remington K, Heidelberg JF, Halpern AL, Rusch D, et al. 2004. Environmental genome shotgun sequencing of the Sargasso Sea. *Science* **304**:66–74.

Warnecke F, Luginbuhl P, Ivanova N, Ghassemian M, Richardson TH, et al. 2007. Metagenomic and functional analysis of hindgut microbiota of a wood-feeding higher termite. *Nature* **450**: 560–565.

Woyke T, Teeling H, Ivanova NN, Huntemann M, Richter M, et al. 2006. Symbiosis insights through metagenomic analysis of a microbial consortium. *Nature* **443**:950–955.

Chapter 41

Gene Prediction in Metagenomic Fragments with Orphelia: A Large-Scale Machine Learning Approach

Katharina H. Hoff, Maike Tech, Thomas Lingner, Rolf Daniel, Burkhard Morgenstern, and Peter Meinicke

41.1 INTRODUCTION

Metagenomic sequencing projects generate large datasets. A Global Ocean Sampling dataset was, for instance, assembled from 7.7 million reads [Yooseph et al., 2007]. Usually, the central goal of all projects are to find out which species are present in each metagenome, which pathways are encoded in their genomes, which are possible metabolites produced or consumed by the organisms, which genes are characteristic for living in the studied environment, and how the species interact. Computational analysis of metagenomes helps to answer these questions. A central point is the identification of protein coding genes (PCGs).

In many metagenomic studies, PCGs were identified through similarity of reads to genes in public databases using BLAST [Altschul et al., 1990]. However, this approach is limited to the discovery of already known genes and cannot detect novel ones. Through clustering of open reading frames (ORFs) within a metagenomic sample, sequence similarity can be used to identify conserved novel genes [Krause et al., 2006; Yooseph et al., 2007]. Considering the size of most metagenomes, computational cost of these methods is a limiting factor for groups that are not equipped with suitable computational facilities.

For the detection of novel genes at acceptable computational cost, statistical models can be used. For ab initio prediction of genes in complete microbial genomes, numerous methods have been developed and successfully applied (e.g., GLIMMER [Delcher et al., 1999] or GeneMark.hmm [Lukashin and Borodovsky, 1998]). Common to these methods is an initial training phase on data from the target genome or from the genome of a closely related species. Such tools are, in principle, applicable to metagenomic data, if the input sequence is of sufficient length. In contrast to genome sequencing projects of single species, where sufficient sequencing depth for read assembly into long contigs is provided, the sequencing coverage of a single species in a metagenome depends on its dominance in a community, the number of species in a community, and the applied sequencing depth. Usually, assembly of reads into long contigs is only possible for a fraction of reads, although binning reads into taxonomic categories (e.g., see Chapter 42, Vol. I) can improve assembly quality. For most metagenomes, a considerable proportion remains as single reads. Therefore, specialized methods that are able to predict PCGs in short and anonymous sequencing reads are required.

This chapter describes the metagenomic gene prediction tool Orphelia [Hoff et al., 2008, 2009]. After outlining the algorithm, we analyze prediction accuracy and compare it with the other available methods

Handbook of Molecular Microbial Ecology, Volume I: Metagenomics and Complementary Approaches, First Edition. Edited by Frans J. de Bruijn.

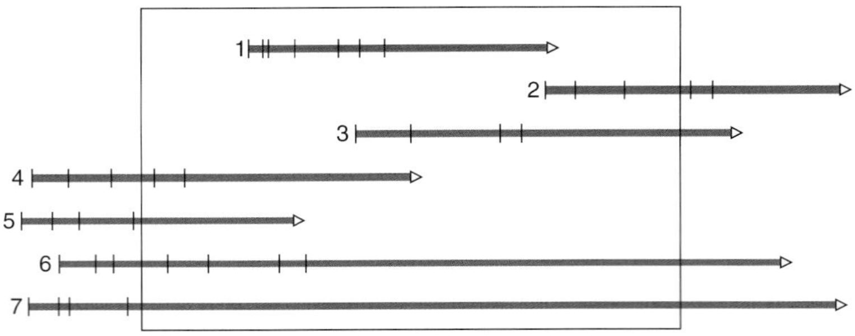

Figure 41.1 Illustration of possible locations of open reading frames (ORFs) in a DNA fragment (shown only for the forward strand). ORFs are symbolized by gray bars, ▷ are stop codons, and | are positions of translation initation candidates. ORFs that share a common stop codon are grouped and referred to as ORF sets. The box denotes a fragment range. Everything outside the box will be invisible to a gene prediction program. (This figure was originally published in Hoff et al. [2008].)

for metagenomic gene prediction, namely MetaGene [Noguchi et al., 2006; see Chapter 57, Vol. I], MetaGeneAnnotator [Noguchi et al., 2008; see also Chapter 50, Vol. I], and GeneMark with heuristic models [Besemer and Borodovsky, 1999].

41.2 MATERIALS AND METHODS

In this chapter, we concisely describe the state-of-the-art Orphelia gene finding method [Hoff et al., 2009], which was developed from the *original method* described in Hoff et al. [2008]. For a full description of materials and methods, please refer to the original publications [Hoff et al., 2008, 2009]. An abbreviated version is shown below.

41.2.1 Open Reading Frame Concept

Prokaryotic PCGs generally begin with a start codon (ATG, CTG, GTG, or TTG) that is followed by a variable number of consecutive codons and is terminated by a stop codon (TAG, TGA, or TAA). This arrangement is commonly known as open reading frame (ORF) and not to be confused with "gene" because only a fraction of ORFs is protein coding. Very often, several possible start codons in one reading frame are associated with the same stop codon. We term this arrangement an ORF set. Due to limited sequencing read length, PCGs are often only partially contained in a fragment and PCG end(s) exceed fragment end(s). Therefore, the ORF definition needs to be extended to incomplete ORFs, which are ORFs that exceed one or both fragment ends. This ORF definition is illustrated in **Figure 41.1**. Orphelia considers only ORFs that are equal to or longer than 60 nucleotides (nt) in length in both training and prediction.

Orphelia is designed for discriminating protein coding from noncoding ORFs. At first, all ORFs ($\geq$60 nt) in an input sequence are identified. Subsequently, they are scored with a model that is based on machine learning techniques. After scoring, a final and high-scoring gene combination is selected by a greedy method with a constraint that allows at most 60 nt overlap between predicted genes.

41.2.2 Gene Prediction Method

ORF scoring is achieved through a two-step machine learning approach: at first, monocodon usage, dicodon usage, and translation initiation site (TIS) features are extracted using linear discriminants. In a second step, an artificial neural network combines the extracted features with ORF length, TIS coverage, and fragment GC content and produces a posterior probability of an ORF to encode a PCG.

Linear discriminants and neural networks were trained on data derived from 131 annotated genomes that represented "one species per genus" in GenBank in 2006 (see Hoff et al. [2008], Additional file 1).

41.2.2.1 Codon and Dicodon Usage. Probably the most important feature for discrimination of coding from noncoding ORFs is derived from monocodon and dicodon frequencies. These frequencies represent the occurrences of successive, nonoverlapping tri- and half-overlapping hexanucleotides. We use two linear discriminants for computing features from monocodon and dicodon frequencies.

For training these discriminants, we randomly sampled 50% of protein coding ORFs (1.9×10^5 examples) and 50% of the longest noncoding ORFs in an ORF set (2.8×10^6 examples), respectively. Mono- and dicodon frequencies were encoded in vectors of 4^3 and 4^6 frequencies, respectively.

For the ith example, a monocodon frequency vector is denoted as $\vec{x}_M^i \in \mathbb{R}^{64}$, which is the ith column of data matrix X_M, which contains all training vectors. For removal of length information, training vectors are normalized to unit Euclidean norm. A corresponding label $y_M^i \in \{-1, 1\}$, which is the ith element of label vector $\vec{y}_M$, indicates whether an example represents a coding ($y = 1$) or noncoding ($y = -1$) region. A linear support vector machine (SVM) implementation with soft margin and quadratic loss function as described in Chappelle [2007] is used for training of a discriminant weight vector $\vec{w}_M$ and a corresponding bias parameter b_m. Dicodon frequency discriminant $\vec{w}_D$ and bias b_d are obtained in the same way. Thus, we have two discriminant scores for each ORF that serve as the first two input features of the neural network:

$$x_1 = \vec{w}_M \cdot \vec{x}_M + b_m, x_2 = \vec{w}_D + \vec{x}_D + b_d \qquad (41.1)$$

An additional regularization parameter λ required for SVM training is adjusted by minimizing the area under "precision recall curve" (PRC). The PRC shows for each possible score threshold the relation of sensitivity (proportion of examples that are classified as genes in the amount of existing true examples, x axis) and specificity (proportion of examples that are classified as genes that are actually true, y axis). For validation of parameter λ, the examples for monocodon and dicodon data were randomly split into 50% for training and 50% for validation. The maximum area under PRC on the validation data over a wide range of λ values was used to adjust the regularization parameter.

41.2.2.2 Translation Initiaton Site. A linear discriminant for scoring TIS is derived from a 60-nt window centered around a potential start codon. Sequence windows of annotated start codons of 50% of the genes in the training genomes served as positive examples (1.9×10^5), while other possible start codons of the same ORF sets were used as negative examples (5.6×10^6). The set of all training examples was randomly split into 50% for discriminant training and 50% for validation of the regularization parameter. Trinucleotide occurrences in the TIS window are encoded to yield a binary indicator vector. In each of its 3712 dimensions (64 trinucleotides $\times$ 58 positions), the vector indicates whether a particular trinucleotide is present or absent. The TIS discriminant was trained using a regularized least-squares approach (see Hoff et al. [2008]). A suitable regularization parameter λ was determined by validation as described above. The TIS score achieved with this linear discriminant is converted to a posterior TIS probability (see Hoff et al. [2008]) that is used as input x_3 for the neural network.

In addition to the linear discriminant, we used the "TIS coverage" as input feature x_4 for the neural network. TIS coverage is the fraction of the TIS region which is actually contained in the sequence fragment and accounts for incomplete TIS regions and completely missing start codons that imply a zero coverage (this feature was added in Hoff et al. [2009]).

41.2.2.3 Length Feature. Sequence length is another important feature for the discrimination between coding and noncoding ORFs. In the case of incomplete ORFs, the observable "incomplete length" is merely a lower bound for the unobservable "complete length." Therefore, lengths of complete and incomplete ORFs are treated separately as two features, x_5 and x_6. For each ORF, only that feature that corresponds to the type of ORF (complete/incomplete) has a nonzero value. The length feature itself is the observed sequence length divided by the maximal length $l_{\max}$, which was here set to 700 nt (original neural network in [Hoff et al., 2008] and the network "Net700" from [Hoff et al., 2009]) and 300 nt (Net 300 from [Hoff et al., 2009]) in order to keep the range of possible input values constant.

41.2.2.4 GC Content. The last feature $x_7 \in [0, 1]$ is the GC content estimated from the entire input fragment in which the ORF occurs.

41.2.2.5 Neural Network. We use standard multilayer perceptrons with one layer of k hidden nodes and with a single logistic output [Bishop, 1995]. In case of binary classification with labels $\gamma = 1$ (true) or $\gamma = -1$ (false), the output can be seen as an approximation of the posterior probability of the true class. Here, the true class represents coding ORFs, and therefore the network output can be interpreted as a "gene probability."

For an input feature vector $\vec{x}$, the k hidden layer activation values z_i are

$$z_i = \tanh\,(\vec{w}_1^i \cdot \vec{x} + b_1^i) \qquad (41.2)$$

where $\vec{w}_1^i$ denotes the $i = 1, \ldots, k$ input weight vector and b_1^i denotes the corresponding bias parameter. Putting z_i into a vector $\vec{z}$, the output of the neural network based on vector $\vec{w}_O$ and bias b_O is

$$g(\vec{z}) = \frac{1}{1 + \exp(-\vec{w}_O \cdot \vec{z} - b_O)} \qquad (41.3)$$

Corresponding to training set $\vec{x}_1, \ldots, \vec{x}_N$ and a network with weight and bias parameters collected in vector $\vec{\theta}$, we can write the network output as $f(\vec{x}_1; \vec{\theta}), \ldots, f(\vec{x}_N; \vec{\theta})$.

Training objective is the minimization of the following regularized error:

$$E(\vec{\theta}) = \sum_{i=1}^{N} (f(\vec{x}_i, \vec{\theta}) - \gamma_i)^2 + \vec{\theta}^T A \vec{\theta} \qquad (41.4)$$

with diagonal matrix $A = \text{diag}(\alpha_1 \ldots, \alpha_1, \alpha_2, \ldots, \alpha_2, \alpha_3, \ldots, \alpha_3, \alpha_4)$ containing the regularization parameters. Four hyperparameters $\alpha_1, \alpha_2, \alpha_3, \alpha_4 > 0$ are used for separate scaling of $\vec{w}_1^i, b_1^i, \vec{w}_O$, and b_O. The evidence framework [MacKay, 1992], which is based on a Gaussian approximation of the posterior distribution of network weights, is used for hyperparameter adaptation. This hyperparameter adaptation can be incorporated into the network training procedure and does not require additional validation data. A scaled conjugate gradient scheme (implemented in Nabney [2001]) is used for minimization of (41.6) with respect to weight and bias parameters. Weight and bias parameters were initially set randomly. Hyperparameters were initially set to $\alpha_1 = \alpha_2 = \alpha_3 = \alpha_4 = 0.01$. Training was performed for 50 iterations with each iteration comprising 50 gradient steps and two successive hyperparameter adaptation steps.

We trained two separate, length-specific neural networks, one for a target length of ~300 nt and one for a target length of ~700 nt. The neural networks were trained with extracted features from ORFs in 300-nt and 700-nt fragments that were randomly excised to a 1-fold coverage from each training genome. An n-fold coverage is here defined as the amount of sampled DNA that is in total length (nt) n times longer than the original genome sequence. Annotated genes in these fragments served as positive examples for coding regions (2.6×10^6), and one candidate out of each non-coding ORF set was randomly selected for the negative examples (4.5×10^6). The datasets were randomly split into 50% for neural network training and 50% for validation of the network size, which was finally set to $k = 25$ (see Hoff et al. [2008], section entitled "Network Validation").

41.2.3 Final Candidate Selection

After scoring, ORFs that obtained a score higher than 0.5, corresponding to a posterior probability of 50%, are kept in a list of "likely genes." However, most of these ORFs overlap by too many nucleotides—for example, because they belong to the same ORF set and differ in their start codon position only. A list of final, high-scoring genes G is therefore selected by overlap and score from the list of all candidate ORFs C using an iterative greedy strategy:

While C is nonempty, do the following:

- Determine the highest scoring ORF i_{max} in in list C with score $P_i > 0.5$.
- Remove i_{max} from C and add it to G.
- Remove all candidates from C that overlap with ORF i_{max} by more than o_{max} nt.

Here, we set o_{max} to 60 nt, which corresponds to the minimal ORF length but it can be defined by the user in the command line tool and web-server application of Orphelia.

41.2.4 Accuracy Evaluation

Gene prediction accuracy is estimated by comparing predicted genes to an existing annotation. In metagenomics, we have to use datasets that were simulated from complete genomes for this purpose because reliably and exhaustively annotated metagenomes are not available yet. In this chapter, we present accuracy results on four simulated datasets (none of the species used for accuracy evaluation data simulation was contained in the training dataset of Orphelia) containing:

(a) Fixed-length, error-free 700-nt fragments that were randomly excised to a 5-fold coverage from the genomes of *A. fulgidus, M. jannaschii, N. pharaonis, B. aphidicola, B. pseudomallei, B. subtilis, C. jeikeium, C. tepidum, E. coli, H. pylori, P. aeruginosa, P. marinus*, and a *Wolbachia* endosymbiont (for details, see Hoff et al. [2008]). One species in this dataset was used in the training dataset of MetaGene/MetaGeneAnnotator, and several species were used for building the heuristic GeneMark models.

(b) Fixed-length, error-free fragments that were randomly sampled to a 1-fold coverage from genomes of the same species as in dataset a with lengths of 100–2000 nt in 100-nt intervals.

(c) Fixed-length, error-free 300-nt and 700-nt fragments that were randomly excised to a 1-fold coverage from the genomes in dataset a except for the species *P. aeruginosa*. Here, this species was omitted because it is contained in the training set of MetaGene and MetaGeneAnnotator (for details, see Hoff et al. [2009]). For a length evaluation, similar fragments were also sampled for lengths of 100–500 nt in 25-nt intervals.

(d) With MetaSim [Richter et al., 2008; see Chapter 48, Vol. I] simulated Sanger and pyrosequencing (also termed "454") reads with mean lengths of 450 nt (pyrosequencing) and 700 nt (Sanger sequencing) and several sequencing error rates that were obtained from literature (for details see Hoff [2009a]). For Sanger reads, we simulated error rates of 0%, 0.0015%, 0.015%, 0.15%, and 1.5%. For pyrosequencing reads, we simulated error rates of 0%, 0.22%, 0.49%, and 2.8%.

The most important measures for assessing gene prediction accuracy are sensitivity and specificity (see Hoff et al. [2008] for further measures and the according results):

$$\text{sensitivity} = \frac{\text{TP}}{\text{TP} + \text{FN}} \tag{41.5}$$

$$\text{specificity} = \frac{\text{TP}}{\text{TP} + \text{FP}} \tag{41.6}$$

They can be combined to one measure by the harmonic mean:

$$\text{harmonic mean} = \frac{2 * \text{sensitivity} * \text{specificity}}{\text{sensitivity} + \text{specificity}} \tag{41.7}$$

Depending on what exactly we measure, the counts for true positives (TP), false negatives (FN), and false positives (FP) can be of different nature.

- For calculating gene prediction sensitivity and specificity, we regard every predicted gene that overlaps by at most 60 nt in the same reading frame on the same strand with a gene in the annotation as a TP. Consequently, we count all predicted genes that do not match the TP criterion as FP, and all annotated genes that were not predicted as FN.
- In the presence of sequencing errors, the above measure is unsuitable to determine gene prediction accuracy because frame shifts might alter the reading frame of a predicted gene in comparison to the annotated gene in an error-free read. A sequence alignment between predicted and annotated sequence with a close to 100% sequence identity requirement over the entire alignment length is able to compensate minor errors like a rare frame shift and can therefore be used to reliably identify correctly predicted genes. We used BLAT [Kent, 2002] alignments that were generated with standard parameters. Predicted genes that have an alignment of at least 20 amino acids (aa) length to an annotated gene in a read were counted as TP, annotated aa sequences that did not fall in this category are called FN, and predicted aa sequences that are not part of a TP are called FP. This way, we calculate an amino acid sequence prediction sensitivity and specificity (see Hoff [2009a] for details; in the original publication the measures are termed "gene prediction accuracy," but in this chapter they are termed "amino acid sequence prediction accuracy" to avoid confusion).

41.3 RESULTS

41.3.1 Gene Prediction Tool

Orphelia is available as command line tool and through a web-server application at http://orphelia.gobics.de/. Currently, two neural networks are available and can be specified by the user, Net300 that was trained on 300-nt DNA fragments and Net700 that was trained on 700-nt DNA fragments. The maximal overlap is by default set to 60 nt but can also be modified. Applicability of the different neural networks is described in Section 41.3.3.

41.3.2 Gene Prediction Accuracy in Reads from Different Species

We evaluated the accuracy of Orphelia with a neural network trained for 700 nt (original Orphelia version) on randomly excised 700-nt fragments from three archaeal and 10 bacterial genomes by comparing predictions to the GenBank annotation of protein coding genes (dataset a). Results are shown in **Table 41.1**. The neural network has a sensitivity ranging from 82% to 92% and a specificity ranging from 85% to 97%. No major difference in accuracy could be observed between archaeal and bacterial species, but, in general, variation across different species was large.

Compared to MetaGene, the neural network generally has a higher specificity (on average 4.6% higher) while MetaGene has a higher sensitivity in fragments from most species (on average 3.8% higher). Overall accuracy measured by harmonic mean is similar for both tools.

41.3.3 Accuracy on Different Read Lengths

Predictions of the neural network that was trained for 700 nt (with the original TIS model) were evaluated on randomly sampled DNA fragments from 13 prokaryotic species with lengths ranging from 100 to 2000 nt (dataset b). Mean gene prediction sensitivity and specificity for all fragment lengths are shown in **Figure 41.2**. In 700-nt fragments, we observed an average sensitivity of 89% and an average specificity of 93%. A slight increase of sensitivity and specificity with growing fragment size can be observed, probably because ORFs in longer fragments are frequently longer and therefore carry more distinct codon usage patterns and TIS signals than ORFs in shorter fragments. On fragments shorter than 200 nt, we observe a sharp drop in accuracy.

In order to improve gene prediction accuracy in short DNA fragments, we trained a neural network for 300 nt (Net300) [Hoff et al., 2009] and tested both a network trained with 700 nt fragments (Net700) and a network trained on 300-nt fragments (Net300) on fragments that were randomly sampled from 12 test species (dataset c) with lengths ranging from 200 to 500 nt. Gene prediction sensitivity and specificity are shown in **Figures 41.3** and **41.4**, respectively. With Net700, Orphelia achieves an average sensitivity of 88% and a specificity of 93% on 700 nt fragments. On 300-nt fragments, Orphelia with Net300 shows a sensitivity of 82% and a specificty of 92%. On fragments with 300 nt long or shorter, Net300 is superior to Net700; but on fragments longer 300 nt, Net700 shows much better accuracy.

41.3.4 Accuracy in Reads with Sequencing Errors

In order to test the accuracy of Orphelia in "realistic" sequencing reads, we simulated Sanger and pyrosequencing reads with a wide range of sequencing errors using

Table 41.1 Mean and Standard Deviation for Gene Prediction Performance of Our Method (Neural Net) and MetaGene[a]

	Sensitivity		Specificity		Harmonic Mean	
Species	Neural Net	MetaGene	Neural Net	MetaGene	Neural Net	MetaGene
Archaeoglobus fulgidus	87.2 ± 0.21	93.7 ± 0.15	93.4 ± 0.16	92.7 ± 0.16	90.2 ± 0.17	93.2 ± 0.14
Methanococcus jannaschii	91.7 ± 0.17	95.8 ± 0.14	96.2 ± 0.13	92.7 ± 0.19	93.9 ± 0.10	94.3 ± 0.15
Natronomonas pharaonis	87.9 ± 0.22	95.1 ± 0.09	93.9 ± 0.10	92.7 ± 0.17	90.8 ± 0.16	93.9 ± 0.12
Buchnera aphidicola	90.6 ± 0.37	96.7 ± 0.24	95.3 ± 0.31	91.1 ± 0.29	92.9 ± 0.28	93.8 ± 0.21
Burkholderia pseudomallei	87.9 ± 0.11	94.1 ± 0.11	90.1 ± 0.09	85.1 ± 0.13	89.0 ± 0.08	89.4 ± 0.10
Bacillus subtilis	91.4 ± 0.16	89.8 ± 0.14	95.3 ± 0.09	89.3 ± 0.19	93.3 ± 0.10	89.5 ± 0.14
Corynebacterium jeikeium	89.7 ± 0.24	91.9 ± 0.12	93.8 ± 0.19	89.2 ± 0.21	91.7 ± 0.19	90.5 ± 0.13
Chlorobium tepidum	82.1 ± 0.25	85.7 ± 0.27	91.2 ± 0.17	88.4 ± 0.26	86.4 ± 0.19	87.0 ± 0.22
Escherichia coli	91.7 ± 0.16	93.3 ± 0.07	95.3 ± 0.09	90.9 ± 0.10	93.5 ± 0.12	92.1 ± 0.07
Helicobacter pylori	92.1 ± 0.11	90.2 ± 0.14	96.6 ± 0.15	89.6 ± 0.23	94.3 ± 0.11	89.9 ± 0.15
Pseudomonas aeruginosa	90.4 ± 0.14	96.2 ± 0.07	92.5 ± 0.11	91.4 ± 0.09	91.4 ± 0.12	93.7 ± 0.07
Prochlorococcus marinus	87.2 ± 0.21	93.7 ± 0.25	95.9 ± 0.14	90.8 ± 0.20	91.4 ± 0.15	92.2 ± 0.19
Wolbachia endosymbiont	87.2 ± 0.27	90.6 ± 0.42	85.2 ± 0.44	71.2 ± 0.54	86.2 ± 0.29	79.7 ± 0.45

[a]Performance was measured on 700-bp fragments that were randomly excised from each test genome to 5-fold coverage (10 replications per species, dataset a). The harmonic mean is a measure that combines sensitivity and specificity.

Source: Hoff et al. [2008].

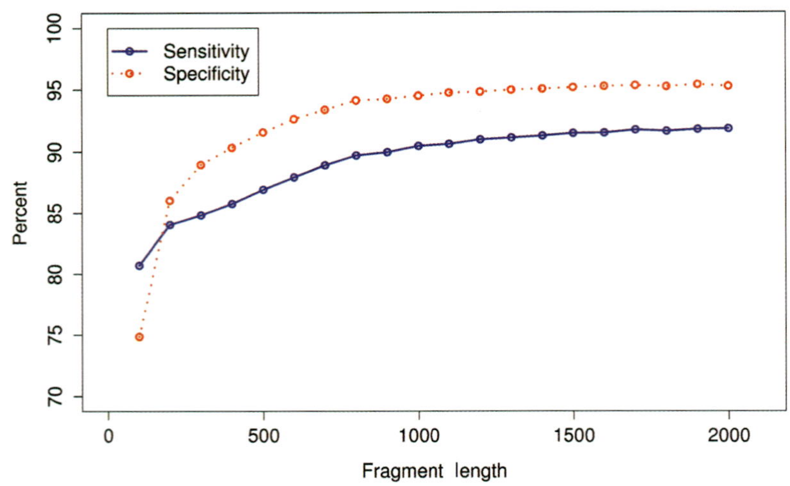

Figure 41.2 Average gene prediction accuracy of the neural network in fragments of the lengths 100–2000 bp (dataset b). Accuracy values from 13 test species were averaged by arithmetic mean. (This figure was originally published in Hoff et al. [2008].)

MetaSim (dataset d). Orphelia with Net300 was applied to reads shorter or equal in length to 300 nt, and Net700 was applied to all reads longer than 300 nt. Amino acid sequence prediction accuracy (harmonic mean) in both read types is shown in **Figures 41.5** and **41.6**, respectively. On error-free Sanger reads, Orphelia has a harmonic mean of 93%. At an error rate of 0.15%, this values drops to ~91%, and with an error rate of 1.5%, the harmonic mean lies at ~77%. On this dataset, the accuracy of Orphelia is almost equal to the accuracy of heuristic GeneMark (except for reads with an error rate of 1.5%, where GeneMark has a higher harmonic mean than Orphelia). MetaGene and MetaGeneAnnotator generally have a slightly higher harmonic mean than Orphelia. In 450-nt pyrosequencing reads, we observe similar results: Generally, the amino acid sequence prediction is slightly lower than on Sanger reads. At low error rates, Orphelia and GeneMark have an almost identical harmonic mean, but for Orphelia we observe a bigger drop in accuracy for reads with error rates of 0.49% and 2.8%. MetaGene and MetaGeneAnnotator have a higher accuracy than Orphelia on reads across all error rates.

41.4 DISCUSSION

Orphelia is a metagenomic gene prediction tool with several unique features:

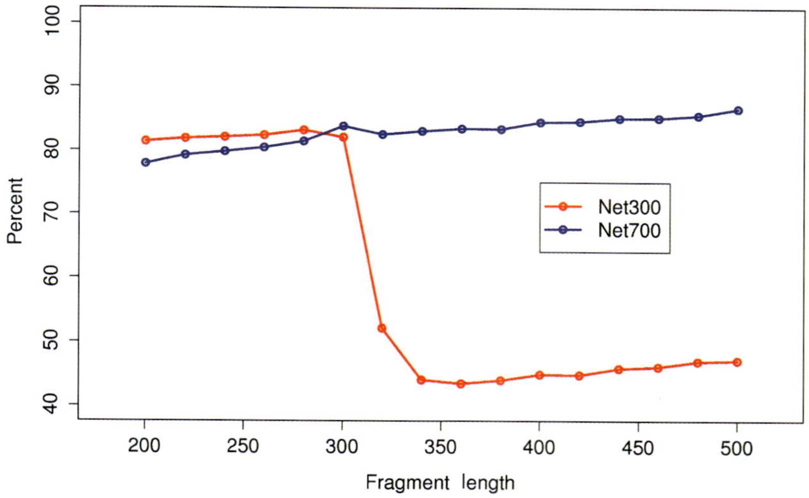

Figure 41.3 Average gene prediction sensitivity of neural networks trained with 300-nt (Net300) and 700-nt (Net700) DNA fragments in fragments with variable lengths (dataset c). Sensitivity values of 12 test species were averaged by arithmetic mean. (This figure was originally published in Hoff et al. [2009].)

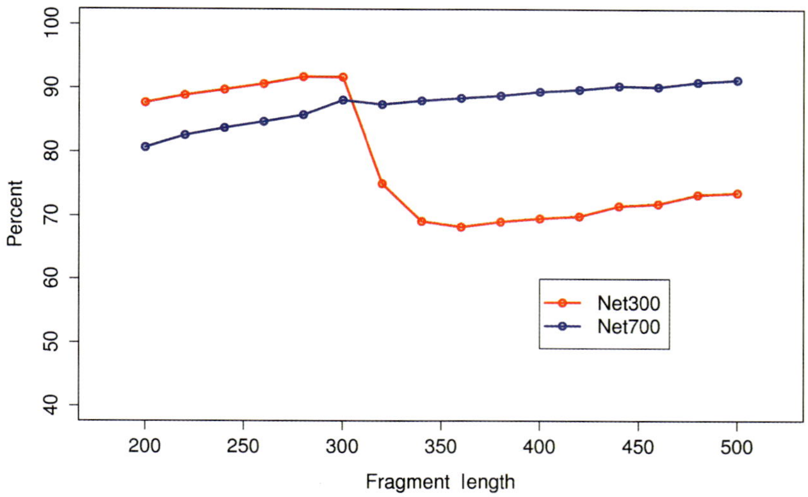

Figure 41.4 Average gene prediction specificity of neural networks trained with 300-nt (Net300) and 700-nt (Net700) DNA fragments in fragments with variable lengths (dataset c). Specificity values of 12 test species were averaged by arithmetic mean. (This figure was originally published in Supplementary Materials of Hoff et al. [2009].)

- Although mono- and dicodon usage of genes in a sequence strongly correlates with the GC content of a sequence [Besemer and Borodovsky, 1999; Noguchi et al., 2006], Orphelia achieves competitive gene prediction accuracy without taking advantage of this correlation. Orphelia utilizes the same models (linear discriminants) for scoring codon usage of ORFs in input sequence with all GC contents. However, accuracy might be increased if this correlation was taken into account.
- In contrast to MetaGene and MetaGeneAnnotator, Orphelia shows that a good gene prediction accuracy can be achieved for *Bacteria* and *Archaea* without utilizing domain-specific scoring models for the two domains.
- Due to its generalized codon usage model, Orphelia tends to predict 'typical' genes with a high specificity while it overlooks some "atypical" genes that more sensitive methods still detect (for details, see Tables 2 and 3 in Hoff [2009a]).

However, Orphelia has a few shortcomings. Firstly, the training dataset for linear discriminants and artificial neural networks utilized in this tool comprises an 'outdated' set of species that represented 'one species per genus' over all prokaryotic genomes present in GenBank in 2006. For MetaGene, the set of training species has been extended significantly (together with other updates) to form the tool MetaGeneAnnotator. In Figures 41.5 and 41.6, MetaGeneAnnotator shows a higher amino acid sequence prediction accuracy than MetaGene, and we assume that Orphelia's accuracy could also be improved by extending the training set.

Another question is whether the "one species per genus" criterion is in general suitable for the selection of training species. It is known that taxonomy does not reflect phylogeny properly. In some cases, species

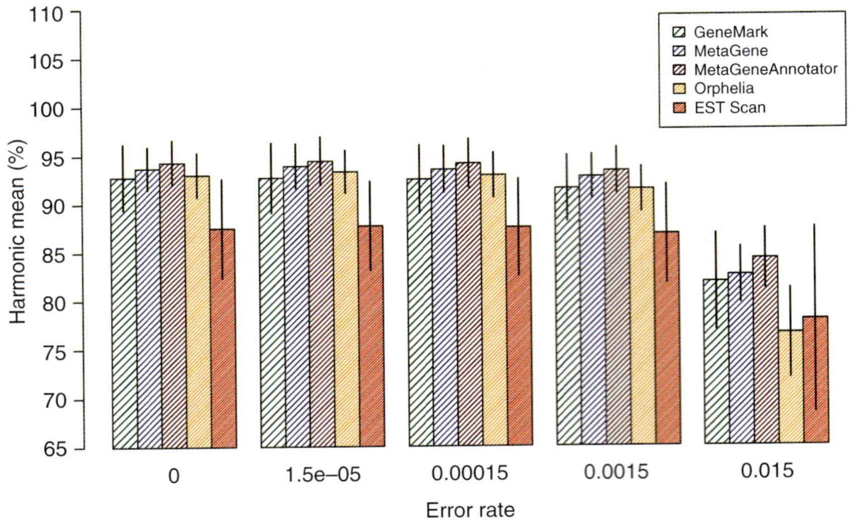

Figure 41.5 Average amino acid sequence prediction accuracy on simulated Sanger reads. (This figure was originally published in Supplementary Materials of Hoff [2009a].)

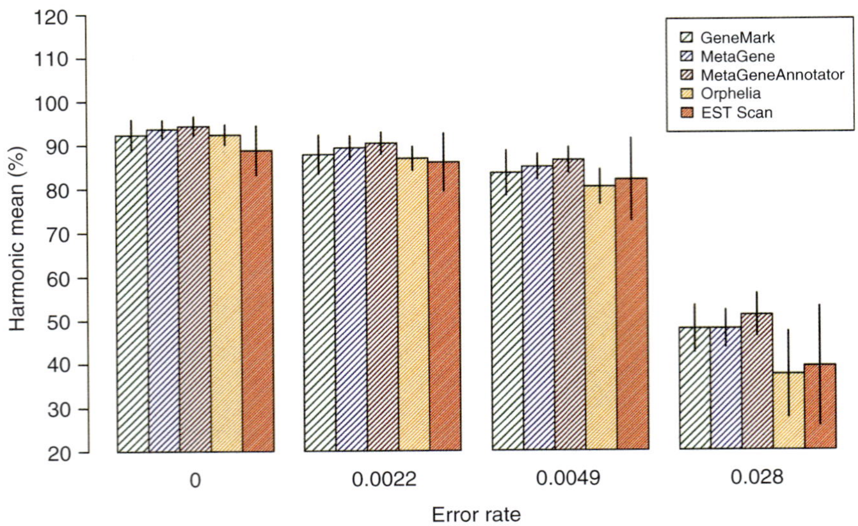

Figure 41.6 Average amino acid sequence prediction accuracy on simulated pyrosequencing reads. (This figure was originally published in Supplementary Materials of Hoff [2009a].)

of different genera show highly similar codon usage patterns. In addition, the collection of species whose genomes were sequenced and deposited in public databases (e.g., GenBank) is biased because most of those species are culturable under laboratory conditions. In contrast, many species in metagenomes are (a) yet unknown and (b) probably not culturable. This bias could be reduced by selecting training genomes according to other criteria—for example, GC content, oligonucleotide frequencies, or mono-/dicodon frequencies in PCGs [Hoff, 2009b].

In general, one might consider integrating further signals of the translation and transcription machinery into metagenomic gene prediction tools. However, most signals of the transcription machinery—for instance, transcription start or termination signals—are not necessarily located in close proximity to the protein coding region. With the limited read length of current sequencing techniques, these features might therefore not be contained in most gene-carrying metagenomic sequencing reads.

From data presented in this chapter and in Hoff et al. [2009], it remains unclear which metagenomic gene prediction method has overall the highest accuracy. On some datasets (e.g., dataset a), it looks like Orphelia and MetaGene are overall equal in accuracy where MetaGene has a higher sensitivity and Orphelia a higher specificity. On dataset d), MetaGene shows higher accuracy than Orphelia. On the same dataset, it looks like MetaGeneAnnotator is the method with overall highest accuracy but on dataset c, heuristic GeneMark has an equally high accuracy as MetaGeneAnnotator (see Table 1 in Hoff et al. [2009]).

However, when it comes to data with a high expected sequencing error rate, Orphelia is not a suitable tool for gene prediction. Other tools perform better on such data, but they also lose accuracy with increasing sequencing error rates. Up to now, no model-based metagenomic gene prediction method that can really cope with sequencing errors is available [Hoff, 2009a]. Only sequence-homology-based methods that explicitly consider frame shifts (e.g., Krause et al. [2006]) may be able to compensate sequencing errors to some extent because they tolerate small differences between related PCGs.

41.5 SUMMARY

Orphelia is a metagenomic gene prediction tool that utilized linear discriminants and an artificial neural network. Separate neural networks were trained for input sequences shorter 300 nt (Net300) and input sequences longer than 300 nt (Net700). Orphelia is available as a command line tool and through a web-server application. Orphelia's gene/amino acid sequence prediction specificity is higher than that of other metagenomic gene prediction tools. On the other hand, other tools are more sensitive. Orphelia performs competitively with the other methods on short reads up to a sequencing error rate of ~0.5%.

INTERNET RESOURCES

Orphelia: http://orphelia.gobics.de

GenBank: http://www.ncbi.nlm.nih.gov/Genbank

Acknowledgments

Work described in this chapter was conducted in the Department for Bioinformatics in the Institute for Microbiology and Genetics at University of Göttingen and has been supported by a Georg Christoph Lichtenberg stipend to KJH, a BMBF project MediGrid (01AK803G) to TL, and a fellowship within the Postdoc-program of the German Academic Exchange Service (DAAD) to TL. We thank Dr. Mario Stanke for proofreading of this manuscript.

REFERENCES

Altschul SF, Gish W, Miller W, Myers EW, Lipan DJ. 1990. Basic local alignment search tool. *J. Mol. Biol*. **215**:403–410.

Besemer J, Borodovsky M. 1999. Heuristic approach to deriving models for gene finding. *Nucleic Acids Res*. **27**(19): 3911–3920.

Bishop CM. 1995. *Neural Networks for Pattern Recognition*. Oxford: Clarendon Press.

Chapelle O. 2007. Training a support vector machine in the primal. *Neural Comput*. **19**(5):1155–1178.

Delcher AL, Harmon D, Kasif S, White O, Salzberg SL. 1999. Improved microbial gene identification with GLIMMER. *Nucleic Acids Res*. **27**(23):4636–4641.

Hoff KJ. 2009a. The effect of sequencing errors on metagenomic gene prediction. *BMC Genomics* **10**:520.

Hoff KJ. 2009b. Gene prediction in metagenomic sequencing reads. Ph.D. thesis, Georg August Universität Göttingen, Göttingen, Germany. Electronic dissertation at Georg August Universität Göttingen: http://webdoc.sub.gwdg.de/diss/2009/hoff/hoff.pdf

Hoff KJ, Tech M, Lingner T, Daniel R, Morgenstern B, Meinicke P. 2008. Gene prediction in metagenomic fragments: A large scale machine learning approach. *BMC Bioinform*. **9**:217.

Hoff KJ, Lingner T, Meinicke P, Tech M. 2009. Orphelia: Predicting genes in metagenomic sequencing reads. *Nucleic Acids Res*. **37**:W101–W105.

Kent WJ. 2002. BLAT—The BLAST-like alignment tool. *Genome Res*. **12**(4):656–664.

Krause L, Diaz NN, Bartels D, Edwards RA, Pühler A, Rohwer F, Meyer F, Stoye J. 2006. Finding novel genes in bacterial communities isolated from the environment. *Bioinformatics* **22**(14):e281–e289.

Lukashin A, Borodovsky M. 1998. GeneMark.hmm: New solutions for gene finding. *Nucleic Acids Res*. **26**(4): 1107–1115.

MacKay DJC. 1992. A practical Baysian framework for backpropagation networks. *Neural Comput*. **4**(3):448–472.

Nabney IT. 2001. *Netlab: Algorithms for Pattern Recognition*. New York,Springer-Verlag.

Noguchi H, Park J, Takagi T. 2006. MetaGene: Prokaryotic gene finding from environmental shotgun sequences. *Nucleic Acids Res*. **34**(19):5623–5630.

Noguchi H, Taniguchi T, Itoh T. 2008. MetageneAnnotator: detecting species-specific patterns of ribosomal binding site for precise gene prediction in anonymous prokaryotic and phage genomes. *DNA Res*. **15**(6):387–396.

Richter DC, Ott F, Auch AF, Schmid R, Huson DH. 2008. MetaSim—A sequencing simulator for genomics and metagenomics. *PloS ONE* **3**(10):e3373.

Yooseph S, Sutton G, Rusch DB, Halpern AL, Williamson SJ, Remington K, Eisen JA, Heidelberg KB, Manning G, Li W, Jaroszewski L, Cieplak P, Miller CS, Li H, Mashiyama ST, Joachimiak MP, van Belle C, Chandonia J-M, Soergel DA, Zhai Y, Natarajan K, Lee S, Raphael BJ, Bafna V, Friedman R, Brenner SE, Godzik A, Eisenberg D, Dixon JE, Taylor SS, Strausberg L, Frazier M, Venter JC. 2007. The SorcererII global ocean sampling expedition: Expanding the universe of protein families. *PloS Biol*. **5**(3):e16.

Yooseph S, Li W, Sutton G. 2008. Gene identification and protein classification in microbial metagenomic sequence data via incremental clustering. *BMC Bioinform*. **9**:182.

Chapter 42

Binning Metagenomic Sequences Using Seeded GSOM

Ching-Hung Tseng, Chon-Kit Kenneth Chan, Arthur L. Hsu, Saman K. Halgamuge, and Sen-Lin Tang

42.1 INTRODUCTION

As metagenomes are generally composed of sequences from a community of organisms, how to categorize these sequences based on biological features will radically affect the accuracy and sensitivity of the downstream analyses. Thus the sequence binning process is a bottleneck step in the early process of the pipeline of metagenomic characteristics analysis. Several binning approaches have been proposed by using different strategies and features: BLAST using sequence similarity search; *k*-mer [Sandberg et al., 2001], Self-Organizing Map (SOM) [Abe et al., 2003], and TETRA [Teeling et al., 2004b] using oligonucleotide frequency; and PhyloPythia [McHardy et al., 2007; see also Chapter 40, Vol. I], which is based on Support Vector Machine (SVM), using oligonucleotide frequency and pattern similarity. These methods are either supervised or unsupervised but have their own limitations. Supervised learning methods, such as BLAST, *k*-mer, SOM, and PhyloPythia, require training datasets, such as completed genomes or labeled long contigs; the unsupervised learning method, TETRA, is intractable for huge datasets because of the required computation on its all-versus-all pairwise comparison matrix. Although these methods have demonstrated their great feasibility on binning task, they were incapable to resolve clusters without identifiable labels. So, we proposed a semisupervised learning method combining a seeding strategy with Growing Self-Organizing Map (GSOM) [Alahakoon et al., 2000], called Seeded GSOM and abbreviated as S-GSOM, for sequence binning. In this study, we have compared S-GSOM with other semisupervised learning methods on the task of sequence clustering. Furthermore, in benchmark datasets, we have shown a higher binning fidelity result with S-GSOM in comparison to other binning methods.

42.2 MATERIALS AND METHODS

The following section has been concentrated and rearranged in order to introduce our S-GSOM. Readers can refer to the original publication [Chan et al., 2008a] for full details.

42.2.1 Datasets of Clustering Performance Comparison

Seven sets of randomly sampled metagenomic fragments were prepared as described in Section 42.2.3: Three sets are composed of 10 species, three sets are composed of 20 species, and one is composed of 40 species. For convenience, we named these datasets "XSp-SetY," where "X" is the number of species and "Y" represents the serial number for the unique set. The lists of bacterial species involved in each dataset can be found in the supplementary material of the original paper.

42.2.2 Datasets of Binning Fidelity Comparison

We benchmarked S-GSOM against other binning methods using simulated datasets, which were generated by the Mavromatis et al. [2007] group, of communities with

Handbook of Molecular Microbial Ecology, Volume I: Metagenomics and Complementary Approaches, First Edition. Edited by Frans J. de Bruijn.

different complexities. The simLC, low-complexity dataset simulates the environment of single dominant species with a few low-abundance ones. The simMC, medium-complexity dataset contains more than one dominant species flanked by low-abundance ones. On the other hand, the simHC, high-complexity dataset does not have any dominant species.

42.2.3 Seeds and Metagenomic Fragments Preparation

In either randomly selected bacterial genomes or published metagenomic benchmark datasets, the seeds were first identified as the continuous stretch of DNA in the flanking regions of 16S rRNA with a length between 8 and 13 kb. Flanking regions of 16S rRNA, which are often similar within the closely related species in terms of nucleotide frequency, can provide enough separating resolution at the species level, and in reality they also are easy to obtain by sequencing with universal primers. To avoid possible interferences imposed by highly similar sequence contents and nucleotide frequencies, tRNA, rRNA, and seed sequences are excluded from the bacterial genomes or benchmark datasets, from which we generate metagenomic fragments. The length restriction of 8–13 kb is used to provide a standardized rule for either seeds or metagenomic fragments [Mavromatis et al., 2007], but with the outlook for Single Molecule Sequencing techniques on the horizon [Clarke et al., 2009], these are definitely achievable lengths.

42.2.4 Self-Organizing Map (SOM) and Growing SOM (GSOM) Methods

The Self-Organizing Map (SOM) [Kohonen, 1990] is an unsupervised clustering algorithm that can visualize unlabeled high-dimensional feature vectors into groups on a lattice grid map that has a fixed map size and shape throughout the training process. In the map, every lattice point represents a node whose weight vector has the same dimension as the input vectors. The SOM algorithm separates training into three phases: initializing, ordering, and tuning. In the initialization phase, the initial weight vectors of each node can be either generated from random values or sampled evenly from the plane spanned by the first two principal vectors [Kohonen, 1999]. The number of nodes needs to be determined by the user. In the ordering and tuning phases, each input identifies a winning node, the node of smallest Euclidean distance to the input, in the given map. Then the weight vectors of the winning node and its neighboring nodes are updated by

$$w(t+1) = w(t) + \alpha \times h \times [x(k) - w(t)]$$

where w is the weight vector of the node, x is the input vector ($w, x \in R^D$, where D is the dimension), k is the index of the current input vector, α is the learning rate, and h is the neighborhood kernel function.

The Growing SOM (GSOM) [Alahakoon et al., 2000; Hsu and Halgamuge, 2003] is an extension of SOM that overcomes SOM's weakness of a static map structure; that is, GSOM initiates its training with the minimum single lattice grid to facilitate the dynamic growth of the map in the training process. GSOM is also used for clustering and employs the same weight adaptation and neighborhood kernel function as SOM. The size of a perfectly trained GSOM map is controlled by a global parameter of growth called growth threshold (GT), which is defined as

$$GT = -D \times \ln(SF)$$

where D is the dimensionality of data and SF is the user-defined spread factor that takes value (0, 1], with 0 representing minimum and 1 representing maximum growth.

There are four training phases of GSOM: initializing, growing, and two smoothing phases. In the initializing phase, weight vectors of nodes in the minimum lattice grid are initialized by random values, and the GT is calculated according to data dimension and user-required SF. During the growing phase, every node keeps an accumulated error counter and the counter of the winning node (E_{winner}) is updated by

$$E_{\text{winner}}(t+1) = E_{\text{winner}}(t) + \|x(k) - w_{\text{winner}}(t)\|$$

When E_{winner} exceeds GT, the winning node that is at the boundary of the current map will grow new nodes to its neighboring vacant slots and initialize weight vectors of the new nodes by interpolating or extrapolating weight vectors of existing neighboring nodes of the winning node. However, if the winning node is not at the boundary, the accumulated error (E_{winner}) is evenly distributed outwards to its neighbors. The two smoothing phases are for fine-tuning the weights of nodes by employing a smaller learning rate.

42.2.5 Seeded Growing Self-Organizing Map (S-GSOM) Method

In binning, metagenomic fragments that belong to closely related species are most likely to have homologous sequences present in between clusters, and this fact usually makes the identification of clustering boundaries much more difficult. Therefore, a modified strategy is needed to identify clusters so that GSOM can be improved as a practical approach for binning.

The Seeded GSOM (S-GSOM), which allows identifying clusters automatically in the feature map using sparse labeled samples as seeds, is our proposed modification of GSOM. There are three core steps in S-GSOM.

Firstly, the very small amounts of labeled seeds (labeled feature vectors) are combined with unlabeled samples (unlabeled feature vectors). Secondly, the combined input vectors are fed into GSOM training, in which the seeds are treated the same as the unlabeled data. Finally, after the normal phases of GSOM training, S-GSOM identifies clusters based on the location of seeds in the final map and the specified amount of clustered nodes (**Fig. 42.1a**).

In the last step of S-GSOM training, the cluster identification phase, the nodes that have seeds are identified and assigned as clustered nodes. Following that, the S-GSOM is going to assign the unclustered nodes, one by one, to clusters iteratively until the specified clustering percentage (CP), the percentage of the number of clustered nodes to the total number of nodes in the map, is reached. In each iteration, a set of unclustered nodes that are adjacent to the clustered nodes is identified. The node within the set of the shortest Euclidean distance to its adjacent clustered node will be assigned to the same cluster with the clustered node. However, nodes that do not contain any sample are most likely representing a cluster boundary. So a penalty factor greater than one is multiplied to the actual distance when calculating the distance between empty nodes and clustered nodes. This will not only force the S-GSOM to avoid clustering empty nodes, but also helps the algorithm completing current cluster before jumping into other clusters. (Fig. 42.1b). According to the empirical observation that the clustering results are not very sensitive to the penalty factor between value of 2 and 5, the penalty factor value of 2 was used in all our experiments.

Before the assigning process is initiated, the nodes containing seeds must be assigned to a specific taxon for taxonomy identification. When all seeds are coming from the same taxon or there is only a single seed, it is trivial for S-GSOM to assign the seeded node to the specific taxonomy of its containing seeds. If the seeds belong to multiple taxa, the seeded node is assigned to the taxon to which the majority seeds belong. However, when seeds are composed of multiple taxa and have equal amounts (e.g., 2 seeds are in the same node) and one belongs to taxon A but the other belongs to taxon B, all seeds are discarded.

To illustrate the role of S-GSOM in binning, **Figure 42.2** depicts the schematic diagram that explains how S-GSOM fits into the whole binning process. There is another, similar algorithm in Computer Vision called Seeded Region Growing [Adams and Bischof, 1994]; please refer to the original publication [Chan et al., 2008a] for the comparison between S-GSOM and SRG.

42.2.6 Training feature Used Throughout This Research

We applied tetranucleotide frequency, which has a better species separation [Abe et al., 2003] and is highly similar between intragenomic fragments but different between intergenomic fragments [Teeling et al., 2004a], as our training feature. The tetranucleotide frequencies were computed using a four-base sliding window and normalized by the full length of the corresponding fragments, which are of different lengths. Because each position in four-base sliding window has four possible nucleotides, the feature vector contains 256 dimensions.

42.3 RESULTS

42.3.1 Clustering Percentage (CP) Determination

Because sequence fragments of closely related species occur mostly at the cluster boundary [Abe et al., 2003; Chan et al., 2008b] and are very likely to disrupt the cluster identification and binning accuracy, an appropriate CP value is necessary for S-GSOM to avoid assigning too many sequence fragments, thereby avoiding the incorrect assignment of highly ambiguous fragments. It was noted in our binning experiments that the clustering performance of S-GSOM declined when the CP was higher than 55% (**Fig. 42.3**). However, S-GSOM, at CP = 55%, assigned more than 80% of sequence fragments in most cases, thus CP = 55% was used throughout the experiments in this research. One can view the CP value, analogous to the p-value in PhyloPythia, as a confidence threshold and opt to use a higher CP value to assign more contigs to bins with lower confidence, where a high CP value is equivalent to a low p-value in PhyloPythia.

42.3.2 Comparison of Semisupervised Algorithms for Binning

To test the feasibility of semisupervised binning methods, other four notable semisupervised clustering algorithms—COP K-means, Constrained K-means, Seeded K-means, and Transductive Support Vector Machine (TSVM)—were used alongside S-GSOM. In the above methods, different runs of random initiation of the COP K-means and S-GSOM can lead to diverse results, which is not an issue for Constrained K-means and Seeded K-means because they use the labeled sample for initiation. So, the best results of COP K-means in 100 runs of random initiations were reported, and to ensure repeatability, all the vectors in S-GSOM's initialization were fixed with the mid-value 0.5 in all dimensions.

Two indices were used to measure clustering performance: adjusted Rand index (ARI) [Hubert and Arabie, 1985] and weighted F-measure (WF) [Van Rijsbergen, 1979]. A higher index indicates a better clustering accuracy.

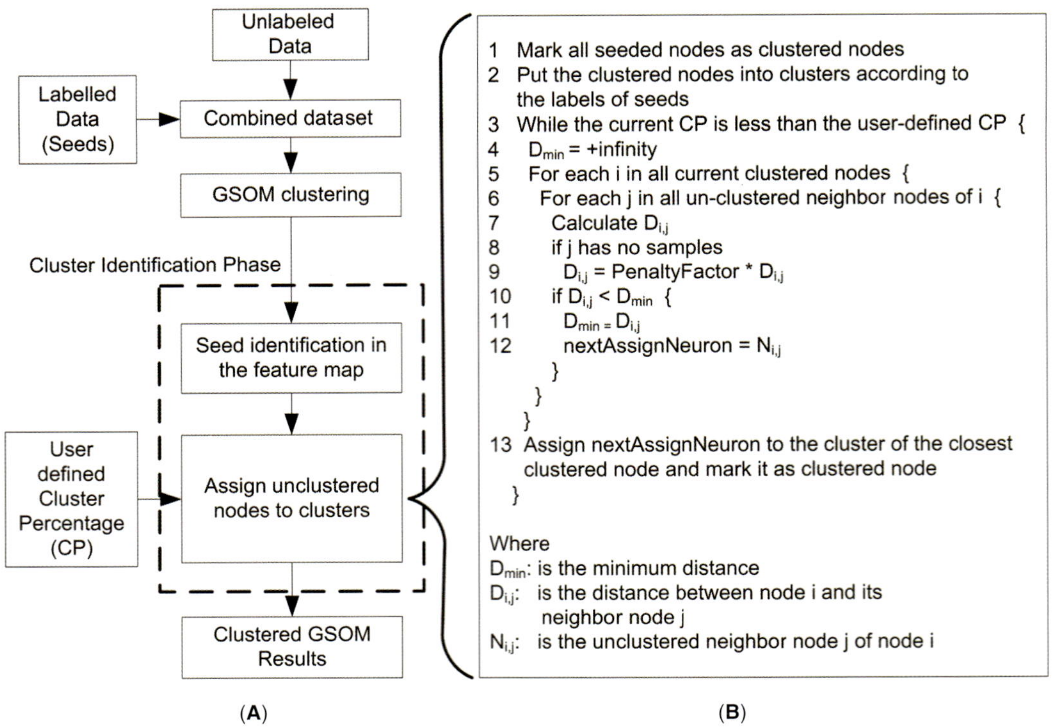

Figure 42.1 The S-GSOM algorithm. **(a)** Schematic diagram of the clustering process of S-GSOM. **(b)** The pseudo-code for node assigning process in S-GSOM. (Reprinted from Chan et al. [2008a], with permission of BioMed Central.)

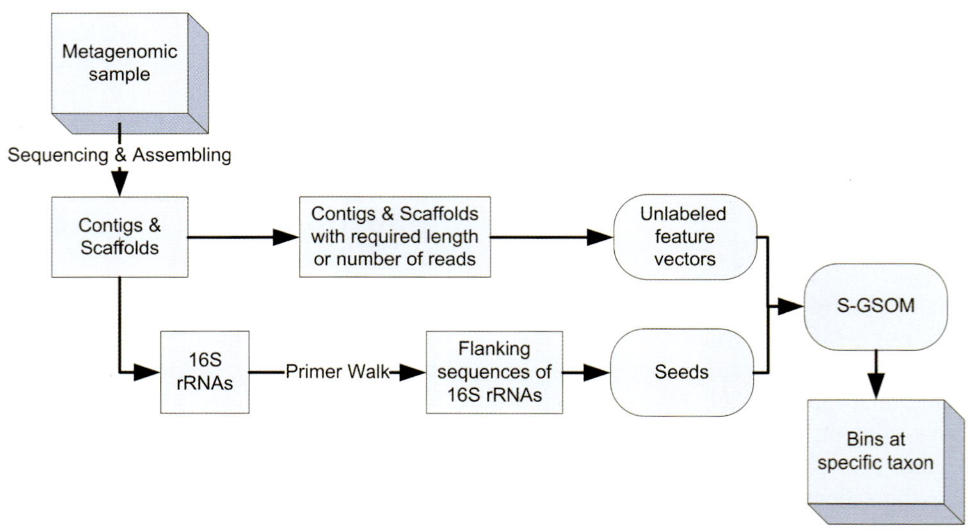

Figure 42.2 An overview of binning process using S-GSOM. (Reprinted from Chan et al. [2008a], with permission of BioMed Central.)

S-GSOM manifested consistently superior performance on both measures, ARI and WF, with the exception of Constrained K-means on the ARI measure for the 10Sp_Set3 dataset (**Table 42.1**). We suspect that the considerable worse performance of TSVM is a result of insufficient labeled data. The superior performance of S-GSOM, which accurately assigned 75% to 90% of all fragments at CP = 55%, clearly demonstrates that the adjustable CP value effectively helps S-GSOM to achieve better clustering by not assigning those ambiguous fragments. The S-GSOM visualization of binning sequences of 10Sp_Set1, 20Sp_Set1 and 40Sp are provided in **Figure 42.4**.

We have considered the 20-species datasets as an example to analyze the resolution of binning with S-GSOM. At CP = 55%, an average 82% of sequence

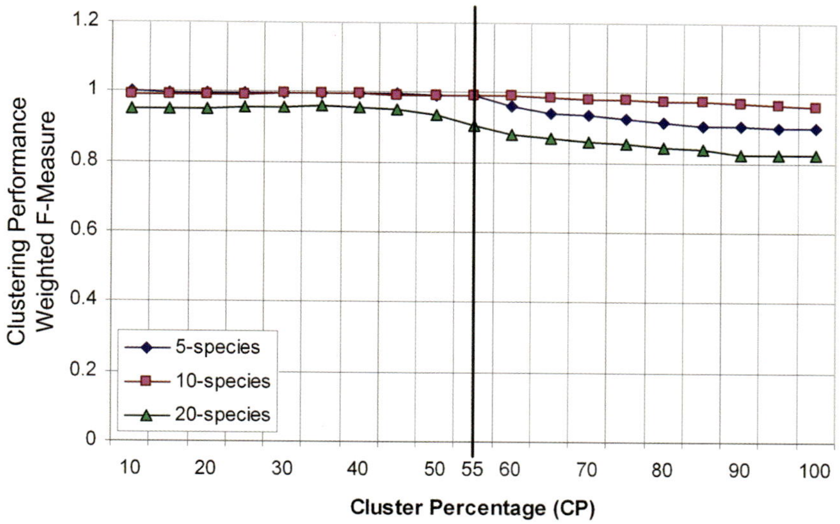

Figure 42.3 Identification of an appropriate clustering percentage (CP). Five datasets for each of 5, 10, and 20 species are randomly sampled. The average of S-GSOM's clustering performance for the datasets are plotted against clustering percentage (CP). A trend of decreasing in clustering performance with increasing CP can be noted. A compromised value of CP=55% is marked where both the number of assigned nodes and clustering performance are high. (Reprinted from Chan et al. [2008a], with permission of BioMed Central.)

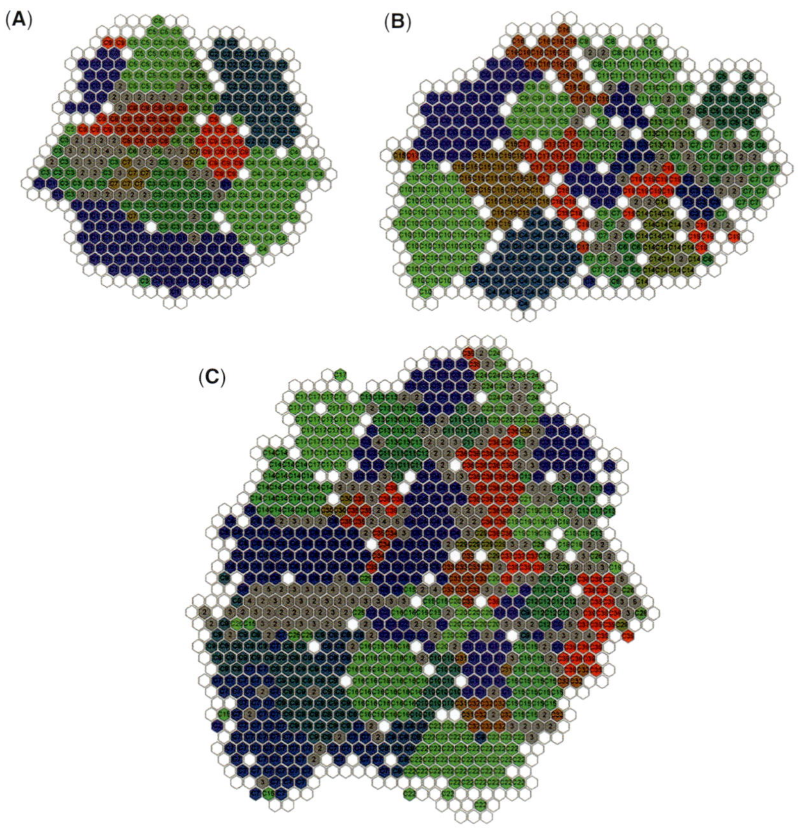

Figure 42.4 Resulted growing self-organizing maps of randomly sampled species. The figure illustrates the GSOM results of clustering sequence fragments according to species: **(a)** 10Sp_Set1, **(b)** 20Sp_Set1, and **(c)** 40Sp. Each hexagon represents a single node. If it only contains a single species, it is displayed in a color that uniquely identifies the species. A node without a letter means that there is no sample located in it. The gray node represents two or more species in the node, and the number of species is displayed on the node. (Reprinted from Chan et al. [2008a], with permission of BioMed Central.)

fragments were assigned with 92% accuracy to their original seeds (species). The distribution of sequence fragments according to species is shown in Figure 42.4b. Nodes that contain seeds from more than one species are colored grey and numbered with the number of species it represents. A significantly higher number of grey nodes around 'C6' and 'C7', representing *Haemophilus influenzae* 86-028NP and *Haemophilus somnus* 129PT respectively, demonstrates that fragments with similar tetranucleotide frequency, resulting from closely related species, tend to be clustered without a clear boundary. This highlights the importance of obtaining seeds in non-boundary regions.

In addition to the distinguished clustering performance, S-GSOM possesses a prominent advantage brought by the seeding method to cluster sequence

Table 42.1 Clustering Performance of Semisupervised Algorithms[a]

	COP K		Constrained K		Seeded K		TSVM		S-GSOM-55	
	ARI	WF	ARI	WF	ARI	WF	ARI	WF	ARI	WF
10Sp_Set1	0.84	0.94	0.84	0.94	0.84	0.93	0.25	0.59	**0.85**	**0.95**
10Sp_Set2	0.89	0.96	0.79	0.90	0.78	0.90	0.41	0.69	**0.93**	**0.97**
10Sp_Set3	0.58	0.83	**0.85**	**0.93**	0.84	**0.93**	0.27	0.62	0.83	**0.93**
20Sp_Set1	0.91	0.90	0.77	0.82	0.76	0.82	0.45	0.65	**0.97**	**0.96**
20Sp_Set2	0.76	0.82	0.70	0.79	0.67	0.79	0.43	0.62	**0.83**	**0.89**
20Sp_Set3	0.81	0.89	0.75	0.86	0.75	0.86	0.46	0.67	**0.97**	**0.98**
40Sp	0.58	0.76	0.71	0.85	0.68	0.84	0.24	0.56	**0.83**	**0.91**

[a]Performance is measured by the adjusted Rand index (ARI) and weighted F-measure (WF). Results for COP K-means are the best results in 100 runs with different initial k cluster centers. The highest values of ARI and WF among different algorithms are shown in bold.

Source: Reprinted from Chan et al. [2008a], with permission of BioMed Central.

fragments of unseeded species, an unknown species. To demonstrate this advantage, an iso-CP (constant CP) contour is delineated in **Figure 42.5a**, generated with sequence fragments from 5 species (for clarity of presentation) in which there are only four seeds, represented by unique colors. By applying different CP values, a group of nodes were rapidly assigned to cluster at CP = 77%. This situation is most likely when a species is relatively abundant, but does not have a seed. Figure 42.5b shows the allocation of nodes to seeds at CP = 55%. However, a protrusion of species "1" into the unassigned region, which belongs to species "5," is an incorrect assignment that sometimes happens to nodes without a correct seed, even at a low CP, 55% (Fig. 42.5b).

42.3.3 Binning Fidelity Comparison

In this section, we tested the performance of S-GSOM in binning against three binning methods: BLAST, k-mer, and PhyloPythia, reported on the datasets, which have been assembled by commonly used assemblers, Archne [Batzoglou et al., 2002], Phrap [Green, 1996], and JAZZ [Aparicio et al., 2002], of different complexities [Mavromatis et al., 2007].

Two considerations were taken into account when comparing the reported binning results with S-GSOM. Firstly, Jazz produced a very small number of binned contigs compared to the other two assemblers [Mavromatis et al., 2007], so fragments assembled by JAZZ are excluded. Secondly, because the simHC, a community without any dominant species, has insufficient contigs of the same species for composition-based analysis [Mavromatis et al., 2007; Teeling et al., 2004a], we also excluded the simHC dataset from our analysis.

For the purpose of fair comparison, all methods need to be compared at the same taxonomic level of binning. Binning at a very high level (e.g., kingdom), clearly has no significance; therefore the results are compared at the order level here, and results for comparing at other taxonomic levels are included in the supplementary materials of original publication [Chan et al., 2008a].

At the order level, the results for simLC and simMC are shown in two separated tables, one for binning contigs greater or equal to 8-kb length and another one for binning contigs with at least 10 reads. To evaluate the performance, rather than using simple averages of all bins [Mavromatis et al., 2007], we used weighted average that gives higher weighting to larger bins to better reflect the amount of correctly binned contigs.

In both low- and medium-complexity datasets, S-GSOM performed reasonably for binning contigs longer than 8 kb, where it is more accurate than all settings of k-mer and BLAST methods, but was outperformed by PhyloPythia in both confidence settings (CP = 75% versus p-value = 0.5 and CP = 55% vs. p-value = 0.85) regardless of the assembler used (**Tables 42.2 and 42.3**). Nevertheless, S-GSOM still outperformed PhyloPythia for the simMC, particularly in terms of sensitivity (i.e., having a higher true positive rate) at the family level (refer to the supplementary materials of original publication).

At the order level, while PhyloPythia performed best for all binning tests on contigs larger than or equal to 8 kb, our S-GSOM was the best-performing method when used to bin contigs that contains at least 10 reads (**Tables 42.4 and 42.5**).

42.4 DISCUSSION AND CONCLUSIONS

S-GSOM enables the clustering of sequence fragments with phylogenetical meanings that are given by using flanking regions of highly conserved genes as seeds. The application of these seeds, which are parts of a completed genome and easier to obtain through molecular

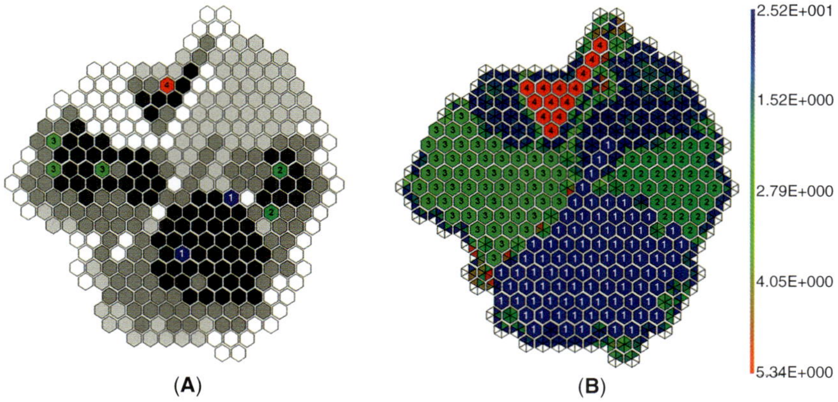

Figure 42.5 Illustration of exploring an unseeded cluster. **(a)** The 5-species S-GSOM map. The seeded nodes are shown with unique colors and labels. Nodes in charcoal color represent nodes that will be assigned when CP = 27% , dark gray nodes at CP = 55% , light gray at CP = 77%, and white at CP = 100% . **(b)** Internode distance map with nodes assigned at CP = 55%. (Reprint from Chan et al. [2008a], with permission of BioMed Central.)

Table 42.2 Binning Summary for Low-Complexity Datasets for Contigs Larger than 8 kb[a]

Assembler	Method	Bins	Binned Contigs	Total # Contigs[a]	% of Bin Contigs	# of Pred Not In Act	wSp	wSn
Arachne	kmer (7 mer)	0	0	202	0	85	—	0.000
Arachne	kmer (8 mer)	0	0	202	0	149	—	0.000
Arachne	BLAST distr 1	0	0	202	0	0	—	0.000
Arachne	BLAST distr 2	0	0	202	0	0	—	0.000
Arachne	S-GSOM (CP = 55%)	1	141	202	69.8	0	1.000	0.698
Arachne	gen PhyloPythia (p:0.85)	1	168	202	83.17	0	1.000	0.832
Arachne	ssp PhyloPythia (*p*: 0.85)	1	186	202	92.08	0	1.000	0.921
Arachne	S-GSOM (CP = 75%)	1	180	202	89.11	0	1.000	0.891
Arachne	**gen PhyloPythia (*p*: 0.5)**	**1**	**201**	**202**	**99.5**	**0**	**1.000**	**0.995**
Arachne	**ssp PhyloPythia (*p*: 0.5)**	**1**	**201**	**202**	**99.5**	**0**	**1.000**	**0.995**
Phrap	kmer (7 mer)	0	0	229	0	129	—	0.000
Phrap	kmer (8 mer)	0	0	229	0	154	—	0.000
Phrap	BLAST distr 1	0	0	229	0	0	—	0.000
Phrap	BLAST distr 2	0	0	229	0	0	—	0.000
Phrap	S-GSOM (CP = 55%)	1	157	229	68.56	0	1.000	0.686
Phrap	gen PhyloPythia (*p*: 0.85)	1	185	229	80.79	0	1.000	0.808
Phrap	ssp PhyloPythia (*p*: 0.85)	1	205	229	89.52	0	1.000	0.895
Phrap	S-GSOM (CP = 75%)	1	204	229	89.08	0	1.000	0.891
Phrap	**gen PhyloPythia (*p*: 0.5)**	**1**	**227**	**229**	**99.13**	**0**	**1.000**	**0.991**
Phrap	**ssp PhyloPythia (*p*: 0.5)**	**1**	**227**	**229**	**99.13**	**0**	**1.000**	**0.991**

[a]Total # Contigs: Total number of contigs in the dataset; % of Bin Contigs: The percentage of contigs binned; # of Pred Not In Act: The number of contigs predicted as a taxon that is not present in the dataset, which are treated as the unbinned contigs; wSp: Weighted specificity; wSn: Weighted sensitivity. The methods with the best performance, highest wSp and Wsn, using contigs assembled by different assemblers are shown in bold.

Source: Reprinted from Chan et al. [2008a], with permission of BioMed Central.

biology techniques, makes S-GSOM more feasible when dealing with samples containing many unknown species. Furthermore, due to the visualization property, S-GSOM allows the user to visually identify clusters of unseeded species. However, it is necessary to have a large number of sequence fragments in the unseeded clusters, at least as many as in the seeded clusters; that is, the species are relatively abundant. Otherwise, the unseeded cluster may be wrongly assigned to an unrelated species at low CP value, or be considered as part of the boundary of neighboring clusters and thus become hardly detectable. On the other hand, if the unseeded clusters have far more samples than the seeded clusters, a low CP should be applied to reduce incorrect assignments.

In addition to the feasible application of S-GSOM, it is also an efficient algorithm in terms of the training time. S-GSOM utilizes the CP value to control the number of nodes going to be assigned in the cluster identification

Table 42.3 Binning Summary for Medium Complexity Datasets for Contigs Larger than 8 kb[a]

Assembler	Method	Bins	Binned Contigs	Total # Contigs[a]	% of Bin Contigs	# of Pred Not In Act	wSp	wSn
Arachne	kmer (7 mer)	0	0	301	0	47	—	0.000
Arachne	kmer (8 mer)	0	0	301	0	191	—	0.000
Arachne	BLAST distr 1	0	0	301	0	0	—	0.000
Arachne	BLAST distr 2	0	0	301	0	0	—	0.000
Arachne	S-GSOM (CP = 55%)	2	220	301	73.09	0	1.000	0.731
Arachne	gen PhyloPythia (*p*: 0.85)	2	242	301	80.4	0	1.000	0.804
Arachne	ssp PhyloPythia (*p*: 0.85)	2	242	301	80.4	0	1.000	0.804
Arachne	S-GSOM (CP = 75%)	2	279	301	92.69	0	1.000	0.927
Arachne	**gen PhyloPythia (*p*: 0.5)**	**2**	**301**	**301**	**100**	**0**	**1.000**	**1.000**
Arachne	**ssp PhyloPythia (*p* :0.5)**	**2**	**301**	**301**	**100**	**0**	**1.000**	**1.000**
Phrap	kmer (7 mer)	0	0	401	0	84	—	0.000
Phrap	kmer (8 mer)	0	0	401	0	271	—	0.000
Phrap	BLAST distr 1	0	0	401	0	0	—	0.000
Phrap	BLAST distr 2	0	0	401	0	0	—	0.000
Phrap	S-GSOM (CP = 55%)	2	318	401	79.3	0	1.000	0.793
Phrap	gen PhyloPythia (*p*: 0.85)	2	301	401	75.06	0	1.000	0.751
Phrap	ssp PhyloPythia (p:0.85)	2	295	401	73.57	0	1.000	0.736
Phrap	S-GSOM (CP = 75%)	2	367	401	91.52	0	1.000	0.915
Phrap	**gen PhyloPythia (*p*: 0.5)**	**2**	**399**	**401**	**99.5**	**1**	**1.000**	**0.995**
Phrap	**ssp PhyloPythia (*p*: 0.5)**	**2**	**399**	**401**	**99.5**	**1**	**1.000**	**0.995**

[a]See footnote *a* in Table 42.2 for explanation of abbreviations in column headings. The methods with the best performance, highest wSp and Wsn, using contigs assembled by different assemblers are shown in bold.

Source: Reprinted from Chan et al. [2008a], with permission of BioMed Central.

phase (Fig. 42.1a), so users can easily adjust the CP value, which is indirectly related to confident assignation, to retrieve different clustering results without waiting for GSOM's growing. The nature of self-organizing indeed forms S-GSOM an automated process that can be improved when new seeds are available.

It is very likely that the 16S rRNA fragments of some species were not sampled though seeds are easier to obtain compared to completed genomes. In such circumstances, we can still obtain the sequence fragments in the possible bins, which have been identified by using the iso-CP contour map and then comparing the sequences with existing databases by BLAST searching. If any conserved marker gene is detected, such as elongation factors and cytochrome oxidase, then we may assess the clusters of these sequences by phylogenetic analysis.

Even though these composition-based binning methods have shown good results, currently they are hindered by the requirement of long sequence length. This limitation of length is partially due to the occurrence of chimeric sequences from cloning procedures of experiments and from the incorrect assembly of sequences. The former source of chimeric sequences can be reduced by advanced cloning-free sequencing—for example, Roche 454 genome sequencer FLX. However, the latter source of chimeric sequence is derived from the incompatible design of current assembler, which assembles all reads into one single genome and does not satisfy the requirement of metagenomic samples of poor sequencing coverage and containing multiple genomes. Therefore, if the number of chimeric sequences is reduced, the required sequence length in S-GSOM can also be reduced. To help the reduction of chimeric sequences, we suggest including the compositional information in the assembling level.

INTERNET RESOURCES

FAMeS database, http://fames.jgi-psf.org

Phrap Assembler, http://www.phrap.org

Acknowledgments

We thank K. Mavromatis of the DOE Joint Genome Institute for evaluating the results for S-GSOM on their simulated metagenome data and for discussing and clarifying their results. We also thank the Australian Research Council for funding the research on S-GSOM with multiple Discovery grants and the grant of Taiwanese National Science Council (NSC 97-2621-B-001-004-MY2).

Table 42.4 Binning Summary for Low-Complexity Datasets for Contigs with at Least 10 reads[a]

Assembler	Method	Bins	Binned Contigs	Total # Contigs[a]	% of Bin Contigs	# of Pred Not In Act	wSp	wSn
Arachne	kmer (7 mer)	0	0	367	0	168	—	0.000
Arachne	kmer (8 mer)	0	0	367	0	312	—	0.000
Arachne	BLAST distr 1	0	0	367	0	0	—	0.000
Arachne	BLAST distr 2	0	0	367	0	0	—	0.000
Arachne	S-GSOM (CP = 55%)	3	295	367	80.38	0	1.000	0.798
Arachne	gen PhyloPythia (p: 0.85)	2	214	367	58.31	0	1.000	0.583
Arachne	ssp PhyloPythia (p: 0.85)	2	236	367	64.31	0	1.000	0.638
Arachne	**S-GSOM (CP = 75%)**	**3**	**343**	**367**	**93.46**	**0**	**0.950**	**0.926**
Arachne	gen PhyloPythia (p: 0.5)	2	292	367	79.56	0	1.000	0.796
Arachne	ssp PhyloPythia (p: 0.5)	2	296	367	80.65	0	1.000	0.798
Phrap	kmer (7 mer)	2	3	482	0.62	159	1.000	0.000
Phrap	kmer (8 mer)	3	17	482	3.53	281	1.000	0.000
Phrap	BLAST distr 1	0	0	482	0	0	—	0.000
Phrap	BLAST distr 2	0	0	482	0	1	—	0.000
Phrap	S-GSOM (CP = 55%)	8	381	482	79.05	9	1.000	0.728
Phrap	gen PhyloPythia (p: 0.85)	3	236	482	48.96	0	1.000	0.488
Phrap	ssp PhyloPythia (p: 0.85)	3	272	482	56.43	0	1.000	0.560
Phrap	**S-GSOM (CP = 75%)**	**8**	**443**	**482**	**91.91**	**9**	**1.000**	**0.840**
Phrap	gen PhyloPythia (p: 0.5)	4	368	482	76.35	1	1.000	0.759
Phrap	ssp PhyloPythia (p: 0.5)	5	387	482	80.29	1	1.000	0.797

[a] See footnote *a* in Table 42.2 for explanation of abbreviations in column headings.

Source: Reprinted from Chan et al. [2008a], with permission of BioMed Central.

Table 42.5 Binning Summary for Medium-Complexity Datasets for Contigs with at Least 10 Reads[a]

Assembler	Method	Bins	Binned Contigs	Total # Contigs[a]	% of Bin Contigs	# of Pred Not In Act	wSp	wSn
Arachne	kmer (7 mer)	1	2	1372	0.15	133	1.000	0.000
Arachne	kmer (8 mer)	0	0	1372	0	1241	—	0.000
Arachne	BLAST distr 1	0	0	1372	0	0	—	0.000
Arachne	BLAST distr 2	0	0	1372	0	1	—	0.000
Arachne	S-GSOM (CP = 55%)	5	1061	1372	77.33	0	0.998	0.768
Arachne	gen PhyloPythia (p: 0.85)	3	562	1372	40.96	0	1.000	0.409
Arachne	ssp PhyloPythia (p: 0.85)	3	657	1372	47.89	0	1.000	0.478
Arachne	**S-GSOM (CP = 75%)**	**5**	**1253**	**1372**	**91.33**	**0**	**0.983**	**0.897**
Arachne	gen PhyloPythia (p: 0.5)	4	1036	1372	75.51	6	1.000	0.753
Arachne	ssp PhyloPythia (p: 0.5)	4	1102	1372	80.32	4	1.000	0.802
Phrap	kmer (7 mer)	1	1	1980	0.05	163	1.000	0.000
Phrap	kmer (8 mer)	2	391	1980	19.75	1457	1.000	0.000
Phrap	BLAST distr 1	0	0	1980	0	2	—	0.000
Phrap	BLAST distr 2	0	0	1980	0	3	—	0.000
Phrap	S-GSOM (CP = 55%)	8	1409	1980	71.16	9	0.995	0.686
Phrap	gen PhyloPythia (p: 0.85)	3	799	1980	40.35	1	1.000	0.404
Phrap	ssp PhyloPythia (p: 0.85)	3	844	1980	42.63	1	1.000	0.426
Phrap	**S-GSOM (CP = 75%)**	**8**	**1708**	**1980**	**86.26**	**9**	**0.991**	**0.816**
Phrap	gen PhyloPythia (p: 0.5)	5	1484	1980	74.95	6	1.000	0.745
Phrap	ssp PhyloPythia (p: 0.5)	5	1524	1980	76.97	4	1.000	0.767

[a] See footnote *a* in Table 42.2 for explanation of abbreviations in column headings.

Source: Reprinted from Chan et al. [2008a], with permission of BioMed Central.

REFERENCES

ABE T, KANAYA S, KINOUCHI M, ICHIBA Y, KOZUKI T, et al. 2003. Informatics for unveiling hidden genome signatures. *Genome Res*. **13**:693–702.

ADAMS R, BISCHOF L. 1994. Seeded region growing. *IEEE Trans. Pattern Anal. Mach. Intell*. **16**:641–647.

ALAHAKOON D, HALGAMUGE SK, SRINIVASAN B. 2000. Dynamic self-organizing maps with controlled growth for knowledge discovery. *IEEE Trans. Neural Networks* **11**:601–614.

APARICIO S, CHAPMAN J, STUPKA E, PUTNAM N, CHIA JM, et al. 2002. Whole-genome shotgun assembly and analysis of the genome of Fugu rubripes. *Science* **297**:1301–1310.

BATZOGLOU S, JAFFE DB, STANLEY K, BUTLER J, GNERRE S, et al. 2002. ARACHNE: A whole-genome shotgun assembler. *Genome Res*. **12**:177–189.

CHAN CK, HSU AL, HALGAMUGE SK, TANG, SL. 2008a. Binning sequences using very sparse labels within a metagenome. *BMC Bioinform*. **9**:215.

CHAN CKK, HSU AL, TANG SL, HALGAMUGE SK. 2008b. Using growing self-organising maps to improve the binning process in environmental whole-genome shotgun sequencing. *J. Biomed. Biotechnol*. (http://www.hindawi.com/journals/jbb/2008/513701.html)

CLARKE J, WU, HC, JAYASINGHE L, PATEL A, REID S, et al. 2009. Continuous base identification for single-molecule nanopore DNA sequencing. *Nat. Nanotechnol*. **4**:265–270.

GREEN P. 1996. Documentation for PHRAP. http://bozeman.mbt.washington.edu/

HSU AL, HALGAMUGE SK. 2003. Enhancement of topology preservation and hierarchical dynamic self-organising maps for data visualisation. *Int. J. Approx. Reasoning* **32**:259–279.

HUBERT L, ARABIE P. 1985. Comparing partitions. *J. Classif*. **2**:193–218.

KOHONEN T. 1990. The self-organizing map. *Proc. IEEE* **78**:1464–1480.

KOHONEN T. (1999). Analysis of processes and large data sets by a self-organizing method. *Intelli. Proc. Manufact. Mater*. **1**:27–36.

MAVROMATIS K, IVANOVA N, BARRY K, SHAPIRO H, GOLTSMAN E, et al. 2007. Use of simulated data sets to evaluate the fidelity of metagenomic processing methods. *Nat. Methods* **4**:495–500.

MCHARDY AC, MARTIN HG, TSIRIGOS A, HUGENHOLTZ P, RIGOUTSOS I. 2007. Accurate phylogenetic classification of variable-length DNA fragments. *Nat. Methods* **4**:63–72.

SANDBERG R, WINBERG G, BRANDEN C.I, KASKE A, ERNBERG, I, et al. 2001. Capturing whole-genome characteristics in short sequences using a naive Bayesian classifier. *Genome Res*. **11**: 1404–1409.

TEELING H, MEYERDIERKS A, BAUER M, AMANN R, GLOCKNER FO. 2004a. Application of tetranucleotide frequencies for the assignment of genomic fragments. *Environ. Microbiol*. **6**:938–947.

TEELING H, WALDMANN J, LOMBARDOT T, BAUER, M, GLOCKNER, FO. 2004b. TETRA: A web-service and a stand-alone program for the analysis and comparison of tetranucleotide usage patterns in DNA sequences. *BMC Bioinform*. **5**:163.

VAN RIJSBERGEN CJ. 1979. *Information Retrieval*. London: Butterworths.

Chapter 43

Iterative Read Mapping and Assembly Allows the Use of a More Distant Reference in Metagenome Assembly

Bas E. Dutilh, Martijn A. Huynen, Jolein Gloerich, and Marc Strous

43.1 INTRODUCTION

Sequencing a microbial genome requires isolation and clonal amplification of the strain, which is impossible for most microorganisms because the large majority of all microbes cannot be cultured in the laboratory [Rappe and Giovannoni, 2003]. A promising solution to this problem is to perform selective enrichment in continuous culture, where the conditions favorable for the species' growth can be approximated more closely. For example, in enrichment culture, interdependency between species (e.g., for the exchange of cofactors) is not problematic. A culture that is inoculated with a natural sample can yield a population that is highly enriched for a single species within a few months (e.g., Ettwig et al. [2008]). Sequencing of such a metapopulation could yield a near-complete genome if the resulting population is enriched and if the sequencing coverage is high (degree of enrichment times average depth times coverage $\gtrsim$20).

Currently, high sequencing coverage is achieved most cost effectively with massively parallel sequencing methods that produce short reads (SOLiD sequencing [http://solid.appliedbiosystems.com] and Illumina/Solexa sequencing [http://www.illumina.com/pages.ilmn?ID=203; Bentley, 2006; see also Chapter 20, Vol. I]). Such reads are usually processed with mapping algorithms such as Eland [Bentley et al., 2008] or Maq [Li et al., 2008] if a reference genome from a closely related species is available. Truly de novo assembly directly from short reads (e.g., Velvet [Zerbino and Birney, 2008]) remains difficult, although innovative techniques that use, for example, conservation at the gene level [Salzberg et al., 2008] are promising for species with high coding densities like Bacteria and Archaea.

The first short-read mapping algorithms that were developed were highly conservative: They permitted no more than one or two mismatches per read, and did not allow for the presence of gaps in the alignment. This means that any read derived from a region with a lower conservation than 30/32 94% identity will not be mapped, restricting the use of a reference genome to highly similar species. These algorithms are suitable for resequencing and SNP finding, where the bulk of the sequenced reads is almost identical to the reference. More permissive mapping algorithms are being developed, but to obtain a high enough assembly depth a closely related reference genome remains indispensable. Mapping reads to a reference has two limitations. Firstly, it depends on the availability of a closely related complete genome that can function as a reference (>94% identity). Secondly, an enriched microbial community culture often contains multiple related strains with similar fitness (quasispecies), and the sequence diversity between such strains can be quite high [Wilmes et al., 2009; Venter et al., 2004]. Such a polymorphic population can be expected to confound conservative mapping and/or assembly programs leading

Handbook of Molecular Microbial Ecology, Volume I: Metagenomics and Complementary Approaches, First Edition. Edited by Frans J. de Bruijn.

to unnecessary fragmentation of the assembly, as well as a large fraction of the reads not being used (low assembly depth).

In a recent paper, we have shown that by iteratively mapping the sequencing reads and assembling them into a consensus genome, it is possible to decipher the consensus genome of the parallel populations of a quasispecies present in a sampled community [Dutilh et al., 2009]. The population was enriched in a continuous culture, under conditions that favored our particular species of interest: Candidatus *Methylomirabilis oxyfera*, a denitrifying, methanotrophic bacterium from the NC10 phylum. After 7 months, this species made up about 70% of the bacterial community, based on fluorescence in situ hybridization (FISH) counts (Figure 4 in Ettwig et al. [2009]). DNA isolated from this enriched community was sequenced with short-read Illumina/Solexa sequencing of 32-nt reads. From a batch culture enriched under similar circumstances, but inoculated with a sample from a different location, we obtained the complete genome of a related organism by combined next-generation sequencing and traditional Sanger sequencing [Ettwig et al., 2010]. Using this genome as a scaffold, we first mapped the short reads to their best possible position on this reference. Then, for each reference position we asked which nucleotide was the most highly represented in the population of strains. Because the resulting assembly was already a better approximation of the sequences in the strain population than the external reference, we iterated the mapping and assembly procedure to increase the coverage. For several iterations, we observed an increase in the performance of the assembly to explain proteomic peptides obtained by tandem mass spectrometry, (see Chapters 71 and 74, Vol. II) and we terminated the iteration process when this independent measure reached a maximum. The final consensus assembly captures the majority vote of the genomes in the multi-strain population.

43.2 METHODS

43.2.1 Data

The short reads were obtained by performing one single-end Illumina sequencing run of a metapopulation, that was dominated by a single taxon. Sequencing yielded 6,667,153 32-nt reads. A complete reference genome (length 2,752,854 nt) was obtained from another continuous culture by combined Illumina sequencing and 454 pyrosequencing, and the remaining gaps were closed by Sanger sequencing. For validation of this approach, we used a single metaproteomic sequencing run carried out on a nanoLC-MS/MS system. Using Mascot version 2.2 (Matrix Science Inc., USA), we mapped the peptides to a single database containing all translated ORFs from all assemblies, and the number of peptides mapped to each iteration was parsed from the results. For a full description of the material and the data, please refer to Ettwig et al. [2010].

43.2.2 Mapping Reads to Their Optimal Position

The first step is to map the sequenced reads to the reference genome. In principle, this can be done with any read mapping program. What is important is that the mapping program is permissive enough to map even quite dissimilar reads to their optimal position on the distant reference. The rationale is that most of the sequenced reads from the enriched community will be derived from the dominant quasispecies, and it should therefore be possible to map them back to the reference genome. As an example, we present here the results obtained with a permissive mapping algorithm (the local sequence similarity search algorithm BlastN v2.2.20 [Altschul et al., 1990]) and a conservative one (the global read mapping program Maq 0.7.1 [Li et al., 2008]). For the search parameters and the details of the mapping procedures, please refer to [Dutilh et al., 2009].

43.2.3 Assembly

Next, we assembled the mapped reads to form a consensus genome. Again, this can be done with any assembly program, but it is important to consider the choices the program will have to make. For the details of the assembly method as we implemented it, please refer to Dutilh et al. [2009].

(i) We already mapped the reads to their optimal position on the reference (above), but what if a read maps equally well to more than one position on the reference? Where this happened, our approach took these positions into account equally at all the top-scoring positions, but this choice may introduce conflict with other mapped reads at some sites. Other options are to remove the (minority of) multi-mapped reads from the dataset, to distribute them randomly, or to take them into account only at sites where they do not introduce conflict.

(ii) Do we take into account mapped reads whose alignment score to their best-scoring reference position is low? Although most of the sequenced reads should come from the dominant quasispecies, there will always be reads derived from other species and from unique genomic regions not present in the reference plus sequencing errors. It seems wise to include some threshold on the mapping quality. In our approach, we discarded reads that were mapped

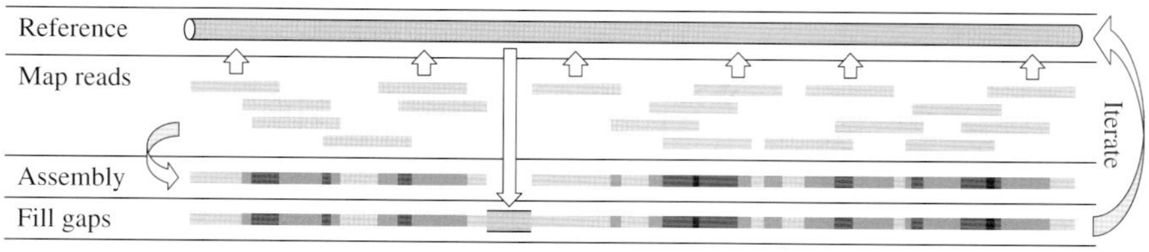

Figure 43.1 Overview of the procedure for iterated mapping of sequencing reads to a reference.

to the reference genome with an aligned region of less than 20 nt.

(iii) At every position in the reference, multiple reads may be aligned with a score higher than the threshold mentioned under (ii). A conservative approach would choose to consider only the highest scoring reads at every reference position, leading to an assembly of those genome regions among the quasispecies strains that most resemble the reference. Because we were interested in a consensus genome of the whole population, we chose to include all reads aligned at a certain reference position, provided they passed the alignment length threshold (ii).

(iv) Then, the assembly program has to choose how to combine the nucleotides in all the mapped reads. An optimal solution would be to combine all the nucleotides present in the reads selected under (iii) into a profile-like representation of the population consensus. Alternatively, only the most frequently aligned base could be chosen, selecting one nucleotide (e.g., randomly, or the one represented in the reference) if there are several with equal abundance. As an intermediate approach, we chose to select the nucleotide with the highest occurrence in the community, replacing draws by their IUPAC multi-nucleotide code [Cornish-Bowden, 1985].

(v) Because we have no information about the reference positions without any mapped reads (zero depth), these should either be marked as "unknown" (N) or left out.

43.2.4 Iteration

After assembling a genome from the mapped reads, we iterated the whole procedure. Reference Positions with zero depth in the assembly (the Ns) were replaced with the corresponding reference nucleotides, as this is still our best bet for these positions, and all the reads were re-queried against this new reference (as above). We carried out 10 iterations with each read mapping algorithm. This procedure is illustrated in **Figure 43.1**. The question is of course when to stop iterating, and one option is to continue until the genome converges (note that this convergence may be cyclic). We chose to iterate until we reached an optimum in the number of proteomic peptides mapped to the assembly's translated ORFs. This independent measure of assembly quality coincided with a plateau in the similarity of the mapped reads to the previous iteration of the evolving assembly, indicating convergence (see Fig. 43.2E).

43.3 RESULTS

43.3.1 Initial Mapping of Reads to the Reference Genome

We combined the short 32-nt Illumina sequencing reads from a metapopulation of related strains (quasispecies) to form a consensus genome describing the majority of the population [Dutilh et al., 2009]. The first step in the process was to map as many of the sequencing reads as possible to their optimal position on the reference. The conservative mapping algorithm Maq [Li et al., 2008] mapped 602,120 reads (Fig. 43.2A), leading to an average mapping depth of 10.8 in the assembled regions, but 35.0% (963,544/2,752,854) of the reference genome still remained uncovered (Fig. 43.2B). The large gaps in the assembly and the many unmapped reads that remained with this conservative mapping algorithm already showed that the reference is distant enough from the community to require a more relaxed sequence similarity search.

We used BlastN as an example of a permissive read mapping algorithm. We used very relaxed search parameters, allowing even quite distant reads to be mapped to their optimal position in the reference. However, this approach did require that we employ a filter for spurious short hits, so we selected only those reads that were aligned to the reference over at least 20 nt. In this preliminary search, BlastN mapped 1,598,549 reads (Fig. 43.2A), leading to an average depth in the assembled regions of 18.5, while 14% (387,421/2,752,854) of the reference nucleotides remained uncovered by reads (Fig. 43.2B).

43.3.2 Depth and Coverage Increase by Iteration

Any available mapping algorithm will suffice to map highly identical reads to a reference. The challenge is to also map the more divergent reads and obtain a higher coverage of the polymorphic community on the divergent reference. The initial coverage of the BlastN-based assemblies are already higher than the conservative Maq

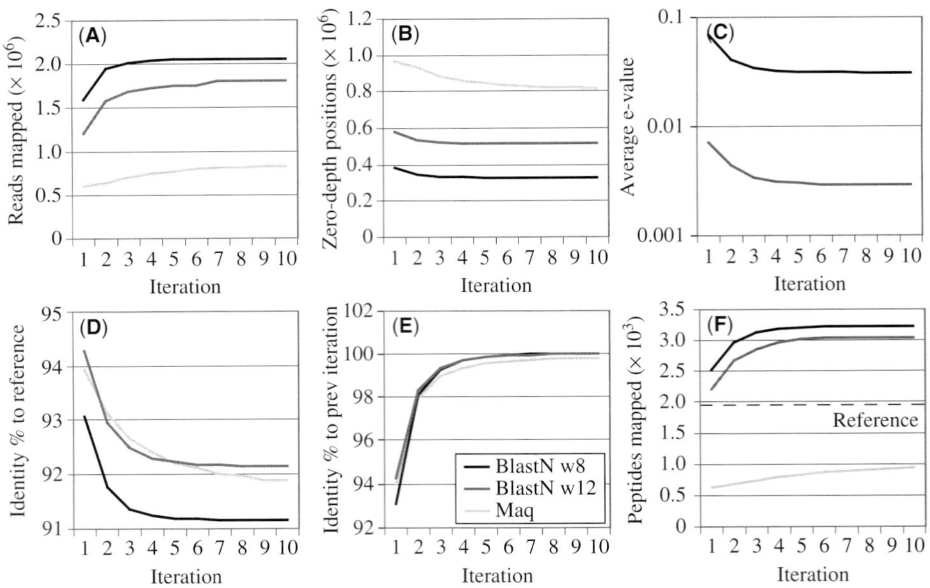

Figure 43.2 Statistics of the evolving assembly for the first ten iterations of mapping and assembly using Blast (word lengths 8 and 12) or Maq. The legend displayed in panel E is valid for all panels. (**A**) Number of 32-nt reads mapped, the total assembly depth follows these curves closely. (**B**) Number of reference positions without any aligned reads. (**C**) Average of the top-scoring *e*-values over all mapped reads (only for the BlastN mappings). (**D**) Identity of the assembly to the reference genome. (**E**) Identity of the assembly to the assembly from the previous iteration. (**F**) Number of proteomic peptides mapped to the assembly's ORFs. The dashed line indicates the number of peptides mapped to the ORFs in the reference.

assembly (Fig. 43.2A), but the average assembly depth is still quite low. However, since this first assembly is composed of the metagenomic reads themselves, we expected that using it iteratively, as a new mapping scaffold, would yield a higher coverage. Indeed, the number of mapped reads and therewith the assembly depth clearly increased after a second round of querying and assembling the reads to the consensus genome (Fig. 43.2A). Additional iterations gradually increased the number of reads that could be mapped for both algorithms, while at the same time increasing the fraction of reference bases that were covered (Fig. 43.2B). These results show that more reads can be mapped as the reference is adjusted to the reads, implying that the assembly becomes more similar to the consensus genome of the community.

43.3.3 Quality Check

As we have seen above, the evolving sequence changes with the iterations to accommodate more of the reads, therewith increasing both the coverage and depth of the assembly. A valid concern is whether the assembly evolves in the right direction: Does it indeed approach the consensus genome of the sequenced population, or could it be that it changes while in fact drifting away from the sequenced population? To address this question, we performed a number of quality checks.

Firstly, we assessed in each of the iterations how well the average read mapped to its optimal position by assessing the average *e*-value of the mapped reads (this was only possible in the BlastN mappings). Figure 43.2C shows a marked decrease in *e*-value, indicating that the reads better map to the evolving assemblies than to the original reference. This shows that the assembly approaches the consensus sequence of the population from which the reads were derived.

Secondly, we assessed the similarity of each assembly with the original reference genome (Fig. 43.2D) and with the assembly from the previous iteration (Fig. 43.2E). While the percent identity of the assemblies to the reference genome decreased, the identity to the assembly from the previous iteration increased and plateaued at almost 100% identity, indicating convergence of the sequence.

Thirdly, we attempted to confound our approach by adding an equal number of 32-nt Illumina reads from a different source (a genome sequencing run of *Aspergillus fumigatus*) to the query data, to see how many of these reads, which should not map to our selected reference genome, would be incorporated. Figure 43.3 shows that using the strict mapping tool Maq, only a very small fraction of $\sim 10^{-3}\%$ of the mapped reads originated from *A. fumigatus*. As could be expected, the more permissive mapping with BlastN also incorporated more noise into the assembly: ~6% of the reads originated from *A.*

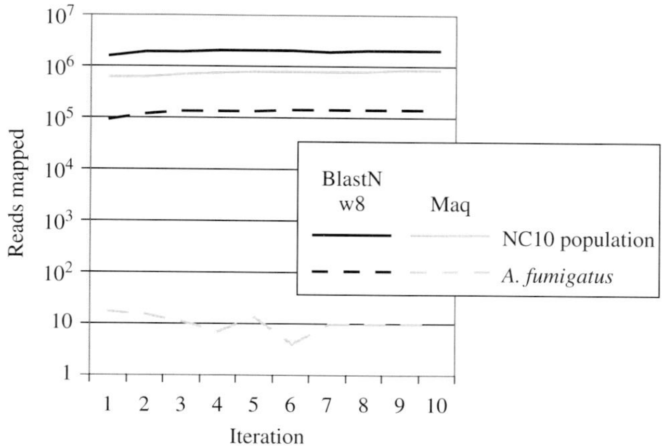

Figure 43.3 Incorporation of contaminating reads. NC10 population reads incorporated in the assembly versus an equal number of contaminating reads derived from Illumina sequencing of an *Aspergillus fumigatus* genome for BlastN- and Maq-based mapping.

fumigatus. It may be expected that some of these reads are simple repeat regions, while others may map to highly conserved regions in genes. In general, this experiment shows that the iterative mapping and assembly procedure draws the assembled consensus toward the signal in the reads that is consistent with the reference genome, rather than toward a read average.

Finally, we carried out a protein sequencing run of the population's metaproteome on a LC-MS/MS system, and we mapped the resulting peptides to the predicted ORFs in each of the iterations (Fig. 43.2F). The results of this mapping show an increase in the number of peptides that can be explained by the assembly's ORFs. The great advantage of these independent validation data is that they give us an objective criterion when to stop the iterations. The number of peptides mapped to the translated ORFs increased until iteration 7, where it reached a plateau of 3,229 mapped peptides, dropping again very slightly to 3,228 in iteration 9. Thus, we considered iteration 7 the optimal assembly.

43.3.4 Quasispecies Diversity

In nature, microbes occur as a quasispecies [Eigen and Schuster, 1979], which can be visualized in multidimensional sequence space as a cloud of related genotypes centered around a consensus that can be considered the "wild type." To assess the diversity in the population we sequenced from our bioreactor, we looked in detail at the relation between single-nucleotide polymorphisms (SNPs) and sequencing errors in the assembled reads. Polymorphisms are sites in the assembled consensus genome where reads supporting alternative nucleotides other than the majority consensus are aligned. It may be expected that frequently occurring polymorphisms represent the true sequence of one or more of the strains in the population, whereas rare polymorphisms (i.e., supported by only few reads) are likely sequencing errors. Figure 43.4A displays the frequency distribution of the number of reads that support alternative nucleotides in the optimal assembly. If we assume that every polymorphism supported by at least two reads is reliable, then 171,509 SNPs can be identified, including multiple SNPs at the same genomic position (Fig. 43.4B).

43.3.5 Consensus Genome

Our iterative mapping and assembly approach allowed us to increase the average assembly depth by 30.9%, decrease the number of uncovered reference bases by 14.8%, and enabled us to explain 28.3% more of the peptides obtained by metaproteomic sequencing (these numbers are based on the seventh iteration of the BlastN-based approach with word length 8). Observing this clear improvement of the assembly quality, we decided to take a look in detail at how the consensus sequence changed with the iterations. Figure 43.5 shows a small part of the genome, illustrating some of the changes that occur as the iterations progress. For example, position 2,314,927 in the alignment (indicated with a black arrowhead) contains a cytosine in the reference. In the first round of mapping and assembly, this site mainly attracts reads that agree with the consensus, but in iteration 2 the reads supporting cytosine are matched in number by reads aligning a thymine, and a Y is called (i.e., cytosine or thymine [Cornish-Bowden, 1985]). This trend is subsequently confirmed in iteration 3, and the consensus nucleotide present in the population of reads is settled as a thymine. At the same time, as we can see in the bottom panel of Figure 43.5, the number of reads on which the call is based (the mapping coverage at position 2,314,927) greatly increases as soon as they agree on voting for a thymine. Another example is the uncovered region (stretch of Ns) in the first and second iteration assemblies that is filled in the subsequent iterations. It should be noted that we map the complete set of 6,667,153 reads against the reference or previous assembly in every iteration, and there is no source of new reads. It is possible that reads are remapped to a different region (e.g., to the region of Ns in Fig. 43.5) if (i) the new region has altered and gained similarity with reads that were not mapped before or that were mapped to another part of the reference or (ii) the region where these reads were mapped before has altered and lost similarity with the reads so that they now map to this new position instead. However, as we see that the reads generally gain similarity with the evolving genome (Figs. 43.2C and 43.2E), scenario (i) seems to occur most frequently.

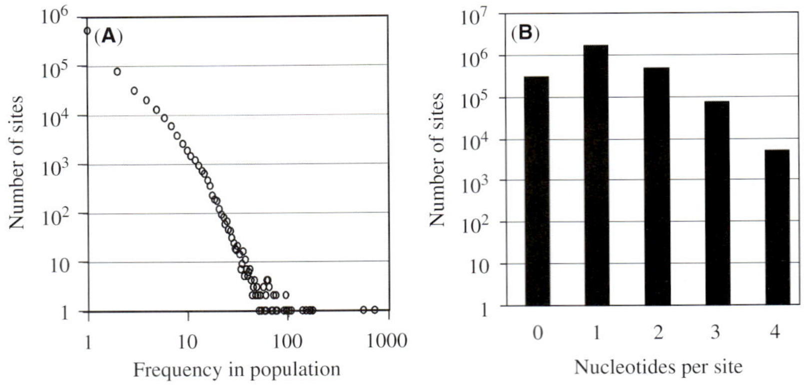

Figure 43.4 Diversity in the population. (**A**) Number of sites in the assembled consensus genome where polymorphisms occur, plotted against their frequency among the population of reads. Rare polymorphisms may represent sequencing errors, while frequent polymorphisms are likely real variations in the community. (**B**) Histogram of the number of different nucleotides called per site.

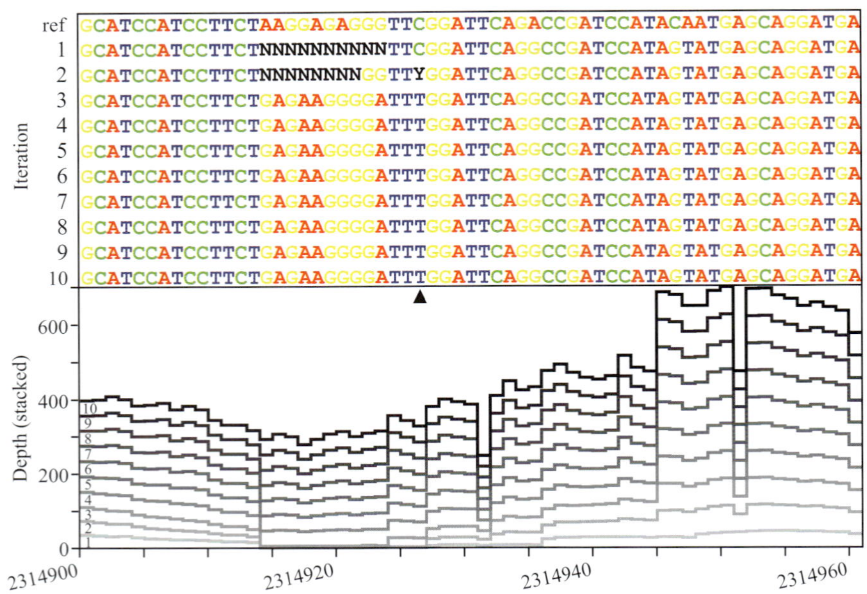

Figure 43.5 Example of how the assembly changes. A region of the assembled sequence showing some of the changes that occur with the iterations. Gaps in the assembly are filled, and single nucleotides are settled. The depth per position in every iteration is shown in the bottom panel.

In general, we observe that the assembly slowly drifts away from the reference genome, as measured by the percentage identity of the mapped regions (i.e., regions with nonzero coverage) to the original reference (Fig. 43.2D). At the same time, the assembly becomes more coherent, as measured by the percentage identity of the mapped regions to the assembly from the previous iteration (Fig. 43.2E). Moreover, a larger fraction of the reads is mapped with a lower *e*-value (Figs. 43.2A and 43.2C). This indicates that the consensus genome of the population of strains is gradually approached.

43.4 CONCLUSION

We have shown how a consensus genome can be composed by mapping metagenomic sequencing reads from a community of strains to a reference [Dutilh et al., 2009]. The consensus genome better represents the community after several iterations of the mapping and assembly procedure, and this increase is independent of the read mapping algorithm. A strict mapping and assembly program such as Maq initially mapped only 602,120 reads (9.0%), but this number increased to 835,328 reads (12.5%) in iteration 10. A more permissive mapping algorithm like BlastN mapped 1,598,549 reads (24.0%) and 2,051,404 reads (30.8%) in the first and tenth iteration, respectively. Note that there is no (artificial) evolution in this method, and no optimality criteria are used. The higher coverage solely results from the fact that the assembly better accommodates the reads. Thus, we profit from the best of both worlds: We use a reference to scaffold the reads, yet the iterated assembly allows the sequence to drift away from the scaffold and approach the consensus genome of the population. Iterative read mapping and assembly has previously been applied in the reconstruction of a bacterial genome from environmental sequence data [Pelletier et al., 2008], but the sequencing reads in that experiment had a much longer mean size of 633 nt, and the idea was not systematically analyzed. We show that our approach can be used with very short 32-nt reads, and the results can only be expected to improve with longer read length.

The sequence we create can be interpreted as the consensus genome of the metapopulation of strains. As always when mapping short sequencing reads, the structure of the genome is scaffolded onto the reference and therefore does not necessarily reflect the genome structure of any particular strain in the sequenced community. This approach is suited to construct the consensus genome of the most abundant lineages in the sample. Moreover, the DNA sequence at any site within the genome is not even necessarily an existing sequence, but rather the consensus of the most abundant sequences. However, we note that, generally, this is also the case for the genome sequencing projects of species that cannot be amplified clonally for sequencing, like animals. For example, the first human genome was composed of the combined DNA of several individuals [Lander et al., 2001]. Therefore, we expect that the consensus genome we obtain using our iterated assembly method can still provide meaningful information about the encoded proteins and other genomic features. Indeed, we verified the approach by mapping peptides derived from metaproteomic sequencing to the translated ORFs of each iteration, and we observed an increase in the number of peptides that could be mapped. In our case, the optimal assembly is the sequence from iteration 7 with BlastN read mapping, which mapped 2,050,700 reads (30.8%). For the in-depth analysis of this sequence, please refer to Ettwig et al. [2010].

INTERNET RESOURCES

Illumina/Solexa sequencing (http://www.illumina.com/pages.ilmn?ID=203)

SOLiD sequencing (http://solid.appliedbiosystems.com)

Acknowledgments

This research was funded in part by Dutch Science Foundation (NWO) Horizon Project 050-71-058.

REFERENCES

Altschul SF, Gish W, Miller W, Myers EW, Lipman DJ. 1990. Basic local alignment search tool. *J. Mol. Biol*. **215**: 403–410.

Bentley DR. 2006. Whole-genome re-sequencing. *Curr. Opin. Genet. Dev*. **16**:545–552.

Bentley DR, Balasubramanian S, Swerdlow HP, Smith GP, Milton J, et al. 2008. Accurate whole human genome sequencing using reversible terminator chemistry. *Nature* **456**:53–59.

Cornish-Bowden A. 1985. Nomenclature for incompletely specified bases in nucleic acid sequences: Recommendations 1984. *Nucleic Acids Res*. **13**:3021–3030.

Dutilh BE, Huynen MA, Strous M. 2009. Increasing the coverage of a metapopulation consensus genome by iterative read mapping and assembly. *Bioinformatics* **25**:2878–2881.

Eigen M, Schuster P. *The Hypercycle: A Principle of Natural Self-Organization*. Berlin: Springer, 1979.

Ettwig KF, Shima S, van de Pas-Schoonen KT, Kahnt J, Medema MH, et al. 2008. Denitrifying bacteria anaerobically oxidize methane in the absence of Archaea. *Environ. Microbiol*. **10**: 3164–3173.

Ettwig KF, van Alen T, van de Pas-Schoonen KT, Jetten MS, Strous M. 2009. Enrichment and molecular detection of denitrifying methanotrophic bacteria of the NC10 phylum. *Appl. Environ. Microbiol*. **75**:3656–3662.

Ettwig KF, Butler MK, Le Paslier D, Pelletier E, Mangenot S, et al. 2010. Nitrite-driven anaerobic methane oxidation by oxygenic bacteria. *Nature* **464**:543–548.

Lander ES, Linton LM, Birren B, Nusbaum C, Zody MC, et al. 2001. Initial sequencing and analysis of the human genome. *Nature* **409**:860–921.

Li H, Ruan J, Durbin R. 2008. Mapping short DNA sequencing reads and calling variants using mapping quality scores. *Genome Res*. **18**:1851–1858.

Mardis ER. 2008. Next-generation DNA sequencing methods. *Annu. Rev. Genomics Hum. Genet*. **9**:387–402.

Pelletier E, Kreimeyer A, Bocs S, Rouy Z, Gyapay G, et al. 2008. "*Candidatus Cloacamonas acidaminovorans*": Genome sequence reconstruction provides a first glimpse of a new bacterial division. *J. Bacteriol*. **190**:2572–2579.

Rappe MS, Giovannoni SJ. 2003. The uncultured microbial majority. *Annu. Rev. Microbiol*. **57**:369–394.

Salzberg SL, Sommer DD, Puiu D, Lee VT. 2008. Gene-boosted assembly of a novel bacterial genome from very short reads. *PLoS Comput. Biol*. **4**:e1000186.

Venter JC, Remington K, Heidelberg JF, Halpern AL, Rusch D, et al. 2004. Environmental genome shotgun sequencing of the Sargasso Sea. *Science* **304**:66–74.

Wilmes P, Simmons S, Denef V, Banfield J. 2009. The dynamic genetic repertoire of microbial communities. *FEMS Microbiol. Rev*. **33**:109–132.

Zerbino DR, Birney E. 2008. Velvet: Algorithms for de novo short read assembly using de Bruijn graphs. *Genome Res*. **18**:821–829.

Zhang Z, Schwartz S, Wagner L, Miller W. 2000. A greedy algorithm for aligning DNA sequences. *J. Comput. Biol*. **7**: 203–214.

Chapter 44

Ribosomal RNA Identification in Metagenomic and Metatranscriptomic Datasets

YING HUANG, WEIZHONG LI, PATRICIA W. FINN, AND DAVID L. PERKINS

44.1 INTRODUCTION

The emerging field of metagenomics and metatranscriptomics can give us a more comprehensive and complete picture of the microbial communities. Many projects have been reported with metagenomic approaches to study microbes and microbial communities that live in many different environmental conditions [Tringe and Rubin, 2005; see also the chapters in Vol. II]. The sequence data generated by these projects impose several new challenges on analysis algorithms [Raes et al., 2007].

Prokaryotic ribosomes can be separated into two subunits. The large subunit contains the 5S and 23S rRNAs, and the small subunit contains the 16S rRNAs. Because rRNA genes are highly conserved and the least variable genes, they have been widely used as markers for phylogenetic analysis and microbial diversity estimation (see Chapter 15, Vol. I). Also, due to their conservation, they serve as the focal points for misassembly of reads from different genomes [DeLong, 2005; Kunin et al., 2008]; removal of rRNA sequences may help the assembly analysis of metagenomic datasets. For metatranscriptomic projects, a large fraction of total RNAs extracted from bacteria and archaea are rRNA genes. To increase the signal of mRNAs, several protocols have been used to remove rRNAs before sequencing [Frias-Lopez et al., 2008; Gilbert et al., 2008; Poretsky et al, 2009; Shi et al., 2009; see also Chapters 62–64, Vol. I]. The percentage of rRNA genes in final sequencing results could be used to estimate the efficiency of these protocols, and these rRNAs were often identified and removed from the raw reads before further analysis. Therefore, identification and separate analysis of rRNA genes remains an important step for both metagenomic and metatranscriptomic projects.

Researchers have proposed several methods to predict noncoding RNA genes [Meyer, 2007], but Freyhult et al. [2007] indicated that the results of most commonly used methods are less than encouraging. Recently, Lagesen et al. [2007] also proposed RNAmmer, a program based on Hidden Markov Models for annotation of rRNA genes in complete genomic sequences. Most of these methods are focused on full genomic sequences; however, for a typical metagenomic study, most of the reads remain unassembled due to insufficient coverage. The full lengths of most of 16S and 23S rRNAs are much longer than the length of raw reads generated by different sequencing platforms (35–120 bp for Illumina. 100–450 bp for 454 pyrosequencing, and 700 bp for Sanger sequencing). Therefore, most rRNA genes in metagenomic sequencing reads are incomplete and will be overlooked by methods that focus on full-length rRNAs. To overcome this limitation, we used HMMs to discover incomplete rRNA gene fragments and achieved improved results [Huang et al., 2009]. In this chapter, we present an updated version of our algorithm based on HMMER3. We then demonstrate its advantage based on simulated sequence reads and also analysis of real datasets.

Handbook of Molecular Microbial Ecology, Volume I: Metagenomics and Complementary Approaches, First Edition. Edited by Frans J. de Bruijn.

44.2 METHODS

44.2.1 Algorithm Descriptions

Figure 44.1 illustrates the basic steps of our algorithms, and the readers can get more detail from our previous publication [Huang et al., 2009]. A high-quality multiple sequence alignment (MSA) is required to build accurate HMMs; therefore we retrieved MSAs of 5S rRNAs from the 5S ribosomal RNA database [Szymanski et al., 2002], and MSAs of 16S and 23S rRNAs from the European ribosomal RNA database [Wuyts et al., 2004]. The input sequences in these databases should meet strict length requirement. Also, the construction of the MSAs is guided by secondary structure information. Because there is no taxonomy information available for sequences obtained from metagenomic and metatranscriptomic projects, we built HMMs from bacterial and archaeal rRNA alignments separately. Each query sequence was then classified into bacteria or archaea domain according to the model that reported the most significant *e*-value. The most important change compared to the previous version is that now we use a beta version of HMMER3 software package released in 2009 to build and search models. Therefore we called the updated version Meta_RNA(H3) and previous version Meta_RNA(H2). Compared to HMMER 2.3.2 [Eddy, 1998], HMMER3 [Rivas and Eddy, 2008] has several advantages, such as consideration of alignment uncertainty and introduction of log-odds scores for whole sequences. It is also reported that it could be as fast as BLAST for protein sequence comparison.

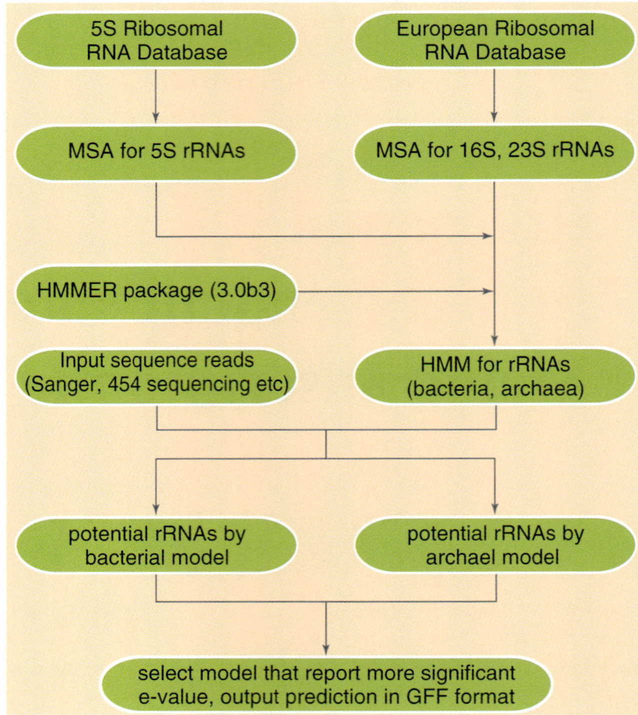

Figure 44.1 Overview of the procedure for HMM construction and sequence annotation. A diagram to demonstrate how we obtain MSAs from rRNA databases, use HMM3.0b3 to build Hidden Markov Models and annotate reads from metagenomic and metatranscriptomic projects.

44.2.2 Evaluation of Algorithms

We downloaded all fully sequenced archael and bacterial genomes from NCBI on September 30, 2008. Next we generated artificial error-free DNA fragments of lengths 100–800 bp (in intervals of 100 bp) from the GenBank files of full genomes. To mimic the conditions of analysis of metagenomic sequences, we did not include all species that appear in training sets in our analysis. The annotation information of rRNA genes was based on feature fields of GenBank files. There are different terms used to annotate rRNA genes; for example, "23S large subunit ribosomal RNA," "23SrRNA," "23S_rRNA," and "23S-rRNA" all point to 23S rRNA genes. Therefore, a manual check was performed to ensure accuracy. We defined a sequence fragments as a positive sample if it had an overlap (>40 nt) with a known rRNA gene in the same strand. A prediction was considered to be a true-positive if it overlaps with a known rRNA. We then calculated sensitivity and specificity according to equations (44.1) and (44.2) to measure the performance of different algorithms.

$$\text{Sensitivity} = \text{true positives}/(\text{true positives} + \text{false negatives}) \tag{44.1}$$

$$\text{Specificity} = \text{true positives}/(\text{true positives} + \text{false positives}) \tag{44.2}$$

Sequence comparison based on BLASTN [Altschul et al., 1997] is a popular choice for screening rRNA genes. The MG-RAST server [Meyer et al., 2008; see also Chapter 37, Vol. I] provides an automated rRNA annotation service using BLASTN to compare query sequences against several rDNA databases. This strategy is also used by Frias-Lopez et al. [2008] and Poretsky et al. [2009]. Prediction results based on BLAST comparison may be problematic due to database inconsistency [Lagesen et al., 2007]. For comparison, our benchmark datasets were also analyzed by BLASTN against the 5S Ribosomal Database and the SILVA database [Pruesse et al., 2007; see also Chapter 45, Vol. I] to identify rRNA genes (with *e*-value of 10^{-5} or less, which is suggested by the MG-RAST).

44.3 RESULTS AND DISCUSSION

44.3.1 Algorithm Performance

Tables 44.1 and **44.2** show the prediction sensitivities and specificities for all fragment lengths. The result of Meta_RNA(H2) can be found in our previous publication [Huang et al., 2009]. Our algorithm can achieve very high sensitivities and specificities (>90%) for almost all configurations. In comparison to BLASTN, the Meta_RNA(H3) achieved better performance for both sensitivities and specificities. For 5S RNA prediction, the average sensitivity improvement is 6.3% and the average specificity improvement is 0.8%. For 16S RNA prediction, the average sensitivity improvement is 1.4% and the average specificity improvement is 7.4%.

44.3.2 Computation Speed

Figure 44.2 illustrates how average running time per read changes with read length. For Meta_RNA(H2), the running time increases almost linearly with the read length, while for the BLASTN comparison and Meta_RNA(H3) the running times don't have significant changes when the read length is longer than 600 nt. Meta_RNA(H3) is much faster than Meta_RNA(H2) and BLASTN at all configurations. With a single 2.33G Xeon® CPU, the average running time per 800-bp read for Meta_RNA(H3), Meta_RNA(H2) and BLASTN is 22 ms, 744 ms, and 145 ms, respectively. For the real application on the Sargasso Sea metagenomic project (Venter et al., 2004), the total size of 811,372 input sequences is over 800 Mbp. The search speed was 20 s/Mbp for Meta_RNA(H3), compared to 1088 s/Mbp for Meta_RNA(H2). In summary, the Meta_RNA(H3) can be 35–70 times faster than its previous version and 10–50 times faster than BLASTN.

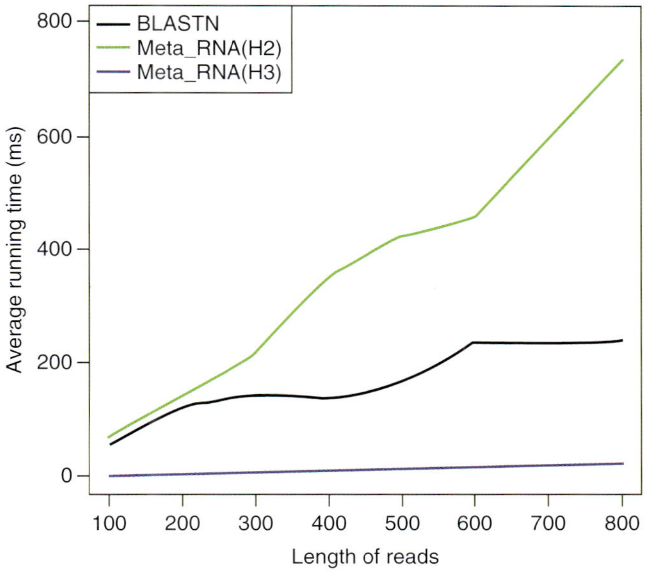

Figure 44.2 Computation speed of BLASTN, Meta_RNA(H2) and Meta_RNA(H3) in benchmark datasets. The speed is measured as running time in terms of milliseconds per sequence. We respectively estimate the computation speed (*Y* axis) for sequences with different length (*X* axis). Therefore, we can check the effect of sequence length on the running time.

44.3.3 Application on Real Datasets

Application of Meta_RNA(H3) on the Sargasso Sea dataset generated 607 5S, 1,217 16S, and 2,360 23S rRNA genes or fragments of genes. These numbers are slightly different from previous predictions made by Meta_RNA(H2). A detailed check of the difference in the predicted results indicates that the updated version may provide more reliable results. For example, contig IBEA_CTG_UEABT56TR was predicted by Meta_RNA(H2) to have a 5S rRNA fragment in the interval [943,975], while there is no prediction by Meta_RNA(H3). Considering the length of this contig (1055 nt) and prediction e-value (0.0015), this prediction may be a false positive. In addition, Meta_RNA(H3) could also generate more continuous predictions than Meta_RNA(H2). There are four 16S rRNA fragments predicted on contig IBEA_CTG_UBAAF70TR produced by Meta_RNA(H2), and all these fragments are combined into a single large fragment in our new analysis.

For the metatranscriptomic dataset from Gilbert et al. [2008; see also Chapter 27, Vol. II], our algorithm identified 2326 rRNA gene fragments in 992,224 DNA sequences and 405 rRNA gene fragments in 506,353 mRNA sequences. The percentage of rRNA genes is 0.24% (DNA) and 0.08% (mRNA), respectively, which is very close to the 0.25% (DNA) and 0.08% (mRNA) estimation based on BLASTN [Gilbert et al, 2008]. This result shows the efficacy of Gilbert et al.'s protocol for metatranscriptome analysis and also indicates the reliability of our algorithm. More importantly, Meta_RNA(H3) used only 954s (DNA) and 387s (cDNA) to analyze this huge dataset with almost one million sequences.

44.4 SUMMARY

In this chapter, we present an updated version of our rRNA gene identification algorithm, Meta_RNA(H3). Testing on benchmark datasets demonstrates its advantage on running speed and prediction performance compared to commonly used BLASTN methods. Our algorithm is implemented as an open source package in Python language. It can accept a FASTA sequence file as the input and generate widely

Table 44.1 Prediction Sensitivities for Different Fragment Lengths

Prediction Method	Meta_RNA(H3)[a]			BLASTN		
Length of Reads	5S	16S	23S	5S	16S	23S
100	86.4	98.6	95.0	79.4	89.9	94.8
200	91.6	97.9	98.2	85.7	96.7	97.8
300	93.5	99.3	98.7	88.3	99.0	98.2
400	95.0	98.3	98.8	89.1	97.5	98.5
500	96.3	99.2	98.5	89.2	99.2	98.4
600	96.3	98.8	98.7	89.5	98.4	98.5
700	96.7	99.5	98.8	90.3	99.5	98.7
800	97.2	99.2	99.4	90.8	99.2	99.1

[a] Here Meta_RNA(H3) represents the updated version of our algorithm. Sensitivities are represented in percentage (%).

Table 44.2 Prediction Specificities for Different Fragment Lengths

Prediction Method	Meta_RNA(H3)[a]			BLASTN		
Length of Reads	5S	16S	23S	5S	16S	23S
100	94.2	94.0	94.9	92.8	91.5	94.8
200	94.5	93.4	94.6	93.0	88.1	94.6
300	95.4	94.3	94.8	94.9	86.9	94.8
400	95.2	95.9	94.9	94.2	88.6	94.9
500	95.7	94.0	94.1	95.0	84.4	94.1
600	94.6	93.2	94.7	94.1	86.5	94.6
700	96.0	93.8	95.4	95.6	85.5	95.6
800	94.4	94.5	95.0	94.1	82.3	94.9

[a] Here Meta_RNA(H3) represents updated version of our algorithm. Specificities are represented in percentage (%).

used GFF format text as the output. The package is available at http://tools.camera.calit2.net/camera/meta_rna, and has been incorporated into RAMMCAP metagenomic annotation pipeline [Li, 2009] in CAMERA project [Seshadri et al., 2007]. A limitation of our study is that current analysis was performed on error-free DNA fragments; therefore, estimated performances may be slightly overoptimistic. How the sequencing error rate will affect prediction performance is an important question for practical application of our algorithms. For a more comprehensive comparison, we plan to incorporate additional sequencing error models into future studies. Now we are also implementing our prediction method through a web-server application to facilitate the biologists who are not familiar with common line running of Meta_RNA program.

INTERNET RESOURCES

5S ribosomal RNA database: http://www.man.poznan.pl/5SData/

CAMERA: http://camera.calit2.net

European ribosomal RNA database: http://bioinformatics.psb.ugent.be/webtools/rRNA/

HMMER: http://hmmer.janelia.org

Meta_RNA : http://tools.camera.calit2.net/camera/meta_rna

RAMMCAP metagenomic annotation pipeline: http://tools.camera.calit2.net/camera/rammcap

SILVA: http://www.arb-silva.de/

Acknowledgments

Ying Huang, Finn W. Patricia, and David L. Perkins were supported by NIH grant R01AI075317. Weizhong Li was supported by NIH grant R01RR025030 and Gordon and Betty Moore Foundation.

REFERENCES

Altschul SF, Madden TL, Schäffer AA, Zhang J, Zhang Z, et al. 1997. Gapped BLAST and PSI-BLAST: A new generation of protein database search programs. *Nucleic Acids Res*. **25**: 3389–3402.

DeLong EF. 2005. Microbial community genomics in the ocean. *Nat. Rev. Microbiol*. **3**:459–469.

Eddy SR. 1998. Profile hidden Markov models. *Bioinformatics* **14**:755–763.

Freyhult EK, Bollback JP, Gardner PP. 2007. Exploring genomic dark matter: A critical assessment of the performance of homology search methods on noncoding RNA. *Genome Res*. **17**:117–125.

Frias-Lopez J, Shi Y, Tyson GW, Coleman ML, Schuster SC, et al. 2008. Microbial community gene expression in ocean surface waters. *Proc. Natl. Acad. Sci. USA* **105**:3805–3810.

Gilbert JA, Field D, Huang Y, Edwards R, Li W, et al. 2008. Detection of large numbers of novel sequences in the metatranscriptomes of complex marine microbial communities. *PLoS ONE* **3**:e3042.

Huang Y, Gilna P, Li W. 2009. Identification of ribosomal RNA genes in metagenomic fragments. *Bioinformatics* **25**:1338–1340.

Kunin V, Copeland A, Lapidus A, Mavromatis K, Hugenholtz P. 2008. A bioinformatician's guide to metagenomics. *Microbiol. Mol. Biol. Rev*. **72**:557–578.

Lagesen K, Hallin P, Rødland EA, Staerfeldt H-H, Rognes T, et al. 2007. RNAmmer: Consistent and rapid annotation of ribosomal RNA genes. *Nucleic Acids Res*. **35**:3100–3108.

Li W. 2009. Analysis and comparison of very large metagenomes with fast clustering and functional annotation. *BMC Bioinform*. **10**:359.

Meyer F, Paarmann D, D'Souza M, Olson R, Glass EM, et al. 2008. The metagenomics RAST server—A public resource for the automatic phylogenetic and functional analysis of metagenomes. *BMC Bioinformatics* **9**:386.

Meyer IM. 2007. A practical guide to the art of RNA gene prediction. *Brief. Bioinform*. **8**:396–414.

Poretsky RS, Hewson I, Sun S, Allen AE, Zehr JP, et al. 2009. Comparative day/night metatranscriptomic analysis of microbial communities in the North Pacific subtropical gyre. *Environ. Microbiol*. **11**:1358–1375.

Pruesse E, Quast C, Knittel K, Fuchs BM, Ludwig W, et al. 2007. SILVA: A comprehensive online resource for quality checked and aligned ribosomal RNA sequence data compatible with ARB. *Nucleic Acids Res*. **35**:7188–7196.

Raes J, Foerstner KU, Bork P. 2007. Get the most out of your metagenome: computational analysis of environmental sequence data. *Curr. Opin. Microbiol*. **10**:490–498.

Rivas E, Eddy Sr. 2008. Probabilistic phylogenetic inference with insertions and deletions. *PLoS Comput. Biol*. **4**:e1000172.

Seshadri R, Kravitz SA, Smarr L, Gilna P, Frazier M. 2007. CAMERA: A community resource for metagenomics. *PLoS Biol*. **5**: e75.

Shi Y, Tyson GW, DeLong EF. 2009. Metatranscriptomics reveals unique microbial small RNAs in the ocean's water column. *Nature* **459**:266–269.

Szymanski M, Barciszewska MZ, Erdmann VA, Barciszewski J. 2002. 5S Ribosomal RNA Database. *Nucleic Acids Res*. **30**:176–178.

Tringe SG, Rubin EM. 2005. Metagenomics: DNA sequencing of environmental samples. *Nat. Rev. Genet*. **6**:805–814.

Venter JC, Remington K, Heidelberg JF, Halpern AL, Rusch D, et al. 2004. Environmental genome shotgun sequencing of the Sargasso Sea. *Science* **304**:66–74.

Wuyts J, Perrière G, Peer, YVD. 2004. The European ribosomal RNA database. *Nucleic Acids Res*. **32**:D101–D103.

Chapter 45

SILVA: Comprehensive Databases for Quality Checked and Aligned Ribosomal RNA Sequence Data Compatible with ARB

ELMAR PRÜSSE, CHRISTIAN QUAST, PELIN YILMAZ, WOLFGANG LUDWIG, JÖRG PEPLIES, AND FRANK OLIVER GLÖCKNER

45.1 INTRODUCTION

Initiated by the pioneering studies of Fox and Woese [Fox et al., 1977] 30 years ago and later on pursued by Pace, Olsen, Giovannoni, and Ward [Pace et al., 1985; Olsen et al., 1986; Giovannoni et al., 1988; Ward et al., 1990], the ribosomal RNA (rRNA) molecule has been established as the "gold standard" for the investigation of the phylogeny and ecology of microorganisms [Amann et al., 1995; Pace, 2009; see Chapter 15, Vol. I]. Today, more than 1,900,000 publicly available small and large subunit (SSU and LSU) rRNA sequences demand appropriate software tools and specialized quality controlled databases. In anticipation of this impending deluge of rRNA data, the development of the ARB software suite and the curation of its associated databases began more than 15 years ago [Ludwig et al., 2004; see Chapter 46, Vol. I]. ARB offers a graphical user interface and a wide variety of interacting software tools built around a common database. It is estimated that ARB is currently employed by several thousand users worldwide, coming from both academia and industry. Since 2007, the corresponding SILVA database project provides structured, integrative knowledge datasets for SSU and LSU rRNAs fully compatible with ARB [Pruesse et al., 2007; see Chapter 46, Vol. I)]. Besides the SILVA project, there are currently two projects offering access to a set of curated rRNA sequences and alignments: the Ribosomal Database Project II at Michigan State University in East Lansing, MI [Cole et al., 2009] (See Chapter 36, Vol. I) and the greengenes project maintained by the Lawrence Berkeley National Laboratory in Berkeley, CA [DeSantis et al., 2006]. All projects offer at least one 16S rRNA dataset, but vary in the amount of sequences, quality checks, alignments, taxonomies, and update procedures. However, the SILVA project is the only platform that actively incorporates homologous SSU as well as LSU sequences from all three domains of life, the *Bacteria, Archaea* (16S/23S), and *Eukarya* (18S/28S). To compensate for the limited phylogenetic resolution of the SSU rRNA [Peplies et al., 2004; Ludwig et al., 2005], the twofold larger LSU rRNA should now also be included in the rRNA approach [Amann et al., 1995; see also Chapter 3, Vol. I]. Especially for *Eukaryotes*, the highly variable regions in the LSU rRNA are already commonly used for species discrimination [Wuyts et al., 2001].

The recent introduction of accelerated and less expensive sequencing technologies, such as pyrosequencing [Margulies et al., 2006] and their application in microbial ecology [Tringe and Hugenholtz, 2008; Reeder and Knight, 2009; see also Chapter 18, Vol. I], further substantiates the need for comprehensive quality

Handbook of Molecular Microbial Ecology, Volume I: Metagenomics and Complementary Approaches, First Edition. Edited by Frans J. de Bruijn.

controlled datasets for comparisons. The SILVA web site was officially launched in January 2007, and this chapter is an updated version of the corresponding publication by Pruesse et al. [2007].

45.2 MATERIALS AND METHODS

45.2.1 Sequence Data Retrieval and rRNA Extraction

The SILVA release cycle and numbering corresponds to that of the EBI-EMBL database, a member of the International Nucleotide Sequence Database Collaboration. A complex combination of keywords including all permutations of 16/18S, 23/28S, SSU, LSU, ribosomal, and RNA is used to retrieve a comprehensive subset of all available SSU and LSU rRNA sequences. Additionally, the complete EBI-EMBL database is searched for rRNAs using Hidden Markov Models provided by RNAmmer [Lagesen et al., 2007]. The internal reference database providing the seed alignment for the automatic alignment of the SSU sequences includes a representative set of 56,354 aligned rRNA sequences from *Bacteria, Archaea*, and *Eukarya* with 50,000 alignment positions. The database providing the LSU reference alignment contains 2868 sequences with 150,000 alignment positions. Both datasets were iteratively cross-checked by expert curators during database buildup.

45.2.2 Quality Checks

Every imported SSU and LSU sequence has to pass a multistage quality inspection. Sequences are rejected if they are shorter than 300 unaligned nucleotides, if they are composed of more than 2% of ambiguous bases, if homopolymers longer than four bases comprise more than 2% of the sequence, or if they have more than 5% identity to vector sequences. The identity is checked by querying a database of commonly used vector sequences, based on the EMVEC and UniVec databases using the blastn tool [Korf et al., 2003]. All thresholds to reject sequences were defined based on statistical analysis of the retrieved SSU and LSU sequences. Each sequence in the SILVA databases carries the percentages of ambiguities, homopolymers, and vector contamination. A summary "sequence quality" score is calculated according to the following formula (with Sq = sequence quality, $A = \%\text{ambiguities}, H = \%\text{homopolymers}$ and $V = \%\text{vector identity}$):

$$\text{Sq} = 1 - \left(\frac{\frac{A}{A_{\max}} + \frac{H}{H_{\max}} + \frac{V}{V_{\max}}}{3} \right) \times 100$$

This score represents the mean of the three individual parameters, such that 100 is the best possible value.

45.2.3 Aligner

To guarantee the specificity of the SILVA databases and a high-quality alignment of the rRNAs, the fast and accurate sequence aligner SINA (SILVA incremental aligner) was developed. In the first step the aligner uses the suffix tree index of ARB [Ludwig et al., 2004; see also Chapter 46, Vol. I] to find up to 40 closely related sequences within the reference alignment. These reference sequences are then transferred into a partial-order graph as used in Lee et al. [2002], but preserving the positional identity from the reference alignment. The graph concept allows "jumping" between the different references to find an optimal alignment for different sequence regions. To further improve the alignment quality, a variability statistic is applied to give more weight to conserved positions. Results of each step of the aligner are reported to the database and shown in the corresponding fields of the exported ARB file. The "alignment quality" score is a measure of the similarity with the reference sequences that are taken into account for the alignment process. High values (>90) indicate that very similar sequences have been found within the seed alignment, resulting in a high likeliness for the alignment to be accurate. Due to the size of the seed alignment, low values are rather rare and suggest manual inspection of the particular sequences. The "base pair" score is calculated from the number of bases involved in helix binding according to the secondary structure model of Gutell et al. [1994]. To fit our unified scoring scheme, the alignment quality and the base pair score were normalized to values between 0 and 100, such that 100 represents the maximum score. After alignment, the constraint on the sequence length is tightened to at least 300 aligned bases within the rRNA gene boundaries.

45.2.4 Anomaly Check

To check for sequence anomalies, a customized version of the Pintail software [Ashelford et al. 2005] is used. The software was initially adapted for batch processing by the RDP II team (see Chapter 36, Vol. I). Pintail checks whether a pair of sequences is mutually anomalous by computing a distance profile and comparing it to a predicted distance profile. The result is "yes," "likely," or "no," depending on the amount of measured deviation from expectation. From this operation, the SILVA Pintail score is constructed by running each sequence against the 10 most similar sequences within a cleaned reference set. Sequences that have passed all tests with "no" (not anomalous) get a score of "100%," whereas all tests returning "likely" would yield a 50% score. Only SSU sequences are checked for anomalies because the Pintail software does not contain profiles for sequences other than 16S rRNA.

45.2.5 Taxonomy and Type Strain Information

Every sequence in the SILVA databases carries the EBI-EMBL taxonomy assignment. Where available, the greengenes and RDP taxonomies are added for comparison. The EMBL taxonomy is retrieved simultaneously with the sequences, whereas the other taxonomies are assigned to the sequences based on accession numbers. For LSU rRNA sequences, no additional up-to-date datasets are available. A substantial revision of the classification of all sequences in the Ref datasets was first published with SILVA release 100. Based on the guide trees, all phylogenetic assignments are manually curated, taking into account taxonomic information provided by *Bergey's Taxonomic Outline of the Prokaryotes* [Garrity et al., 2004], the taxonomic outlines for Volumes 3, 4, and 5 of *Bergey's Manual*, and the List of Prokaryotic names with Standing in Nomenclature [Euzeby, 1997]. Furthermore, extensive effort is spent to represent prominent uncultured and not-validly published environmental clades, groups, and taxa, respectively. The majority of these clades and groups are annotated in the guide tree for the SSU Ref dataset based on literature surveys and personal communications. Taxonomic groups consisting only of sequences from uncultured organisms are named after the clone sequence submitted earliest. Due to this exhaustive manual approach, SILVA currently contains the most up-to-date and detailed bacterial and archaeal taxonomic classification.

Type strain information for *Bacteria* and *Archaea* is added to the field "strain" and indicated by "[T]." Mapping is based on the "All-Species Living Tree" project [Yarza et al., 2008], the Straininfo.net database [Dawyndt et al., 2005], and RDP II [Cole et al., 2009; see also Chapter 36, Vol. I].

45.2.6 Nomenclature and rDNAs from Genome Projects

With every release, all organism names are synchronized with the "Nomenclature up-to-date" web site of the "Deutsche Sammlung für Mikroorganismen und Zellkulturen" (DSMZ, http://www.dsmz.de/download/bactnom/names.txt) and the "All-Species Living Tree" project [Yarza et al., 2008]. All rRNA sequences marked by EBI-EMBL as genome projects are labeled by "[G]" in the "strain" field. Manually curated information about the isolation environment (habitat) of the rRNAs of genome sequences is added based on the EnvO-Lite annotations in the megx.net database [Kottmann et al., 2009].

45.2.7 SSU and LSU rRNA databases for ARB

Two types of pre-compiled databases for both SSU and LSU rRNA sequences are available in ARB format: the high-quality Ref databases and the comprehensive Parc databases. Each Ref database is based on a subset of its Parc database comprising only full-length or nearly full-length 16/18S and 23/28S rRNA sequences. A SSU sequence is considered "full length" if it contains at least 1200 aligned bases within the gene boundaries. This constrained is loosened to 900 bases for sequences belonging to the domain *Archaea* because applying a strict cutoff at 1200 bases would result in the loss of the majority of these sequences. LSU sequences are considered full length if they are at least 1900 bases long. For quality assurance, sequences that could not be unambiguously aligned (alignment quality score <50 for SSU or <30 for LSU) are removed from the Ref databases. Both Ref databases are supplemented with a fully classified guide tree. The trees are incrementally built using the ARB parsimony tool with filters to remove highly variable positions.

The rRNA Parc databases are a collection of all quality checked and automatically aligned rRNA sequences longer than 300 bases of the aligned rRNA gene. All sequences in the SILVA databases are associated with a rich set of sequence and process parameters, including information taken directly from the EBI-EMBL sequence record, as well as information from the initial quality checks of the alignment process. Using the search and query features of the web site or of ARB, one can quickly locate problematic sequences and generate individual high- or low-quality sequence subsets.

45.2.8 Availability/Website

The SILVA databases are available via a web-based interface. Downloads of the complete Parc and Ref datasets in the ARB file format are available in the download section. Subsets of aligned sequences from the Parc dataset can be retrieved from the taxonomic browser and with the advanced search functions. After selecting a database and the desired taxonomy in the browser, the user can navigate through the taxonomy by clicking on the respective nodes. A cart system is used to easily select and download subsets of single sequences, complete groups, or even whole phyla.

The advanced search functionality offered on the SILVA web site allows the user to easily compile custom subsets of sequences. In addition to simple searches (e.g. for accession numbers, organism names, taxonomic entities, or publication DOI/PubMed IDs), complex queries over several database fields using constraints such as sequence length or quality values are possible. All sequences or subsets can be added to the cart for subsequent export in the ARB and multi-FASTA file formats.

The colored bars on the search page and in the short and detailed sequence views of the browser give a fast overview of the different quality aspects assigned to each sequence. A wealth of additional information about the current status of the databases, as well as FAQs, are available in the background section of the web site. Furthermore, the SILVA web site hosts a set of projects like "The All-Species Living Tree" project [Yarza et al., 2008] and the "Standard Operating Procedure for Phylogenetic Inference (SOPPI)" [Peplies et al., 2008] and is part of the international Genomic Standards Consortium [Field et al., 2008] currently developing the Minimum Information about an Marker Gene Sequence (MIMARKS) checklist and standard.

45.3 RESULTS AND DISCUSSION

45.3.1 Data Retrieval and Processing

Cross checks with RDP II and greengenes indicate a sensitivity of the SILVA rDNA sequence retrieval procedure of >99%. A comparison of the length distribution immediately after importing the SSU sequences with the length distribution of sequences after the specific alignment for SILVA releases 89–100 shows that partial sequences between 300 and 800 bases were more frequently rejected than longer ones (>900 bases) (**Fig. 45.1**). The short "problematic" sequences may be generated in diversity studies based on single-strand sequencing. The high number of rejected sequences with less than 300 bases is an indicator for the increased number of projects employing tag sequencing based on next-generation sequencing technologies (see Chapter 18, Vol. I).

As expected, the peaks of the SSU sequence length distribution follow the prominent primer sets used to sequence specific conserved regions on the 16S/18S rRNA gene [Marchesi et al., 1998] (Fig 45.1). The large number of sequences with 300 and 600 bases is typical for diversity studies that use single reads or fingerprinting techniques. It is interesting to note that up to SILVA release 94, the 500 base peaks clearly dominated over the full-length sequences. Recent releases show a trend toward the submission of higher-quality, nearly full-length rRNA sequences.

It has to be emphasized that the primary intention of the SILVA project is to provide reliable rRNA datasets with an informative set of processing and quality values assigned to each sequence. Such quality values enable users to easily evaluate sequences in order to create subsets of sequences for specific applications, or to identify sequences that need further attention with respect to sequence and/or alignment quality or anomalies. The alternative taxonomies and type strain information, as well as the latest nomenclature will facilitate the daily workflow of diversity analysis using classical clone-based and high-throughput sequencing approaches. Additionally, SILVA provides two LSU databases to support the increasing use of molecular markers with a higher resolution than the SSU rRNA [Ludwig and Schleifer 2005]. A taxonomic breakdown of the LSU Parc database contents shows that 91% of the sequences are of eukaryotic origin. A closer look indicates that the LSU rRNA is becoming more and more attractive for the molecular identification of, for example, Fungi.

45.3.2 Alignment

The current SILVA alignment is based on 50,000 and 150,000 alignment positions for the small and large subunit rRNA, respectively. The reasons for the large amount of alignment positions are (1) large insertions often present in *Eukarya* and (2) sequencing errors, such as additional artificial bases often found in homopolymeric sequence stretches. Such errors are common and require placement to be filtered before phylogenetic tree reconstruction, without corrupting the rest of the alignment.

To further improve the quality of the SSU and LSU seed databases, a manual curation process is performed and, over time, additional curated sequences are added to underrepresented sections of the seed. The SSU seed currently includes over 1000 unpublished sequences that primarily cover the domain *Archaea*. The SILVA team highly appreciates the return of manually inspected and corrected alignments of sequence subsets for inclusion in the SILVA seed. This will allow to further increase the quality of future alignments.

45.4 CONCLUSIONS

The SILVA system provides comprehensive, quality-controlled, richly annotated and aligned reference rRNA datasets to support the molecular assessment of biodiversity, as well as investigations of the evolution of organisms. Applications of the datasets range from basic research in microbiology and molecular ecology to the detection of contaminants and pathogens in biotechnology and medicine. Molecular taxonomy and diagnostics have already revolutionized our view on microbial diversity on Earth [Hong et al., 2006; Pedros-Alio, 2006; Tringe and Hugenholtz, 2008], and the added value of molecular techniques for the determination of eukaryotic diversity has recently been documented by Tautz et al. [2002]. The SILVA databases combined with the ARB software

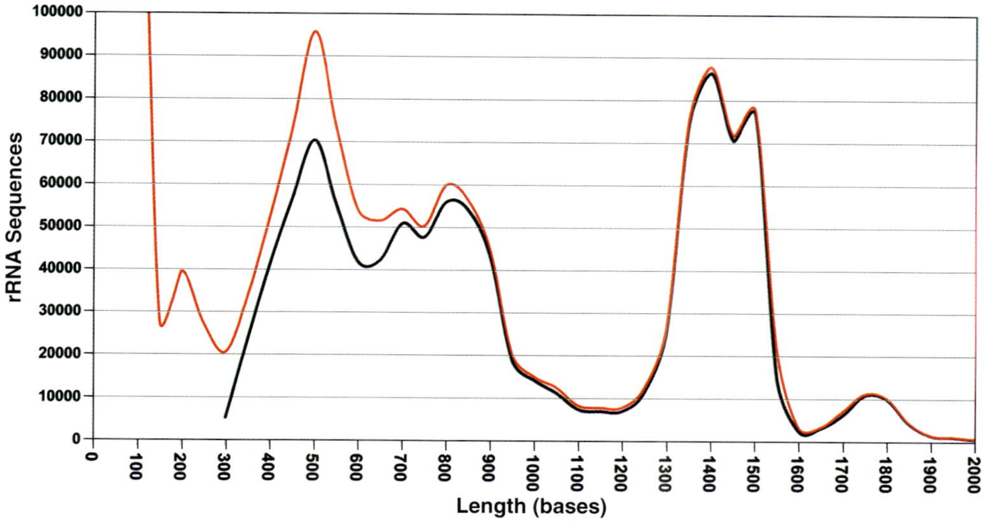

Figure 45.1 Sequence length distribution of rRNA genes in the SILVA 100 SSU database. The red line represents the sequence distribution directly after importing, and the black line represents it after quality checks and alignment. The huge amount of sequences up to 200 bases reflects the impact of tag sequencing approaches.

suite (see Chapter 46, Vol. I) provide a stable and easy to use workbench for researchers worldwide to perform in-depth sequence analysis and phylogenetic reconstructions. They are designed as specialist databases to assist in the daily effort to keep pace with the increasing amount of data flooding the general-purpose primary databases.

INTERNET RESOURCES

The SILVA project (www.arb-silva.de)

The Ribosomal Database Project II (http://rdp.cme.msu.edu/)

The greengenes project (http://greengenes.lbl.gov/)

The International Nucleotide Sequence Database Collaboration (http://www.insdc.org).

The EMVEC database (http://www.ebi.ac.uk/blastall/vectors.html)

The UniVec database (http://www.ncbi.nlm.nih.gov/VecScreen/VecScreen.html)

Documentation of the SILVA database fields in ARB (http://www.arb-silva.de/documentation/faqs/).

Bergey's Manual (http://www.bergeys.org/outlines.html)

List of Prokaryotic names with Standing in Nomenclature (http://www.bacterio.net/)

The Megx.net database (www.megx.net)

The Minimum Information about an MARKER gene Sequence (MIMARKS) checklist and standard (http://gensc.org/gcwiki/index.php/MIMARKS).

Acknowledgments

We would like to thank Ralf Westram for expert assistance with the ARB software suite, and we are grateful to all colleagues and students who helped with the manual curation of the databases. We would also like to thank James Cole, George Garrity, and the RDP II team for help with Pintail and fruitful discussions. We are grateful for funding from the Max Planck Society.

REFERENCES

AMANN RI, LUDWIG W, SCHLEIFER KH. 1995. Phylogenetic identification and in situ detection of individual microbial cells without cultivation. *Microbiol. Rev*. **59**:143–169.

ASHELFORD KE, CHUZHANOVA NA, FRY JC, JONES AJ, WEIGHTMAN AJ. 2005. At least 1 in 20 16S rRNA sequence records currently held in public repositories is estimated to contain substantial anomalies. *Appl. Environ. Microbiol*. **71**:7724–7736.

COLE JR, et al. 2009. The Ribosomal Database Project: improved alignments and new tools for rRNA analysis. *Nucleic Acid Res*. **37**:D141–D145.

DAWYNDT P, VANCANNEYT M, DE MEYER H, SWINGS J. 2005. Knowledge accumulation and resolution of data inconsistencies during the integration of microbial information sources. *IEEE Trans. Knowledge Data Eng*. **17**:1111–1126.

DESANTIS TZ, et al. 2006. Greengenes, a chimera-checked 16S rRNA gene database and workbench compatible with ARB. *Appl. Environ. Microbiol*. **72**:5069–5072.

EUZEBY JP. 1997. List of bacterial names with standing in nomenclature: A folder available on the Internet. *Int. J. Syst. Bacteriol*. **47**:590–592.

FIELD D, et al. 2008. The minimum information about a genome sequence (MIGS) specification. *Nat. Biotechnol*. **26**:541–547.

FOX GE, PECHMAN KR, WOESE CR. 1977. Comparative cataloging of 16S ribosomal ribonucleic acid: Molecular approach to procaryotic systematics. *International Journal of Bacteriol*. **27**:44–57.

Garrity GM, Bell JA, Lilburn TG. 2004. Bergey's *Taxonomic Outline of the Prokaryotes*. 5th ed. New York: Springer Verlag.

Giovannoni SJ, DeLong EF, Olsen GJ, Pace NR. 1988. Phylogenetic groupspecific oligodeoxynucleotide probes for identification of single microbial cells. *J. Bacteriol*. **170**:720–726.

Gutell RR, Larsen N, Woese CR. 1994. Lessons from an evolving rRNA: 16S and 23S rRNA structures from a comparative perspective. *Microbiol. Rev*. **58**:10–26.

Hong SH, Bunge J, Jeon SO, Epstein SS. 2006. Predicting microbial species richness. *Proc. Natl. Acad. Sci. USA* **103**:117–122.

Korf I, Yandell M, Bedell J. 2003. *BLAST*. Beijing, Cambridge, Farnham, Köln, Paris, Sebastopol, Taipei, Tokyo: O'Reilly & Associates.

Kottmann R et al. (2009) Megx.net: Integrated database resource for marine ecological genomics. *Nucleic Acids Res*. online.

Lagesen K, Hallin P, Rodland EA, Staerfeldt HH, Rognes T, Ussery DW (2007) RNAmmer: Consistent and rapid annotation of ribosomal RNA genes. *Nucleic Acid Res*. **35**:3100–3108.

Lee C, Grasso C, Sharlow MF. 2002. Multiple sequence alignment using partial order graphs. *Bioinformatics* **18**:452–464.

Ludwig W, Schleifer KH. 2005. Molecular phylogeny of bacteria based on comparative sequence analysis of conserved genes. In Sapp J, ed. *Microbial Phylogeny and Evolution, Concepts and Controversies*. New York: Oxford University Press, pp. 70–98.

Ludwig W, et al. 2004. ARB: A software environment for sequence data. *Nucleic Acid Res*. **32**:1363–1371.

Marchesi JR, Sato T, Weightman AJ, Martin TA, Fry JC, Hiom SJ, Wade WG. 1998. Design and evaluation of useful bacterium-specific PCR primers that amplify genes coding for bacterial 16S rRNA. *Appl. Environ. Microbiol*. **64**:795–799.

Margulies M et al. 2006. Genome sequencing in microfabricated high-density picolitre reactors. *Nature* **441**:120–120.

Olsen GJ, Lane DJ, Giovannoni SJ, Pace NR, Stahl DA. 1986. Microbial ecology and evolution: A ribosomal RNA approach. *Annu. Rev. Microbiol*. **40**:337–365.

Pace NR. 2009. Mapping the tree of life: Progress and prospects. *Microbiol. Mol. Biol. Rev*. **73**:565–576.

Pace NR, Stahl DA, Olsen GJ, Lane DJ. 1985. Analyzing natural microbial populations by rRNA sequences. *ASM News* **51**:4–12.

Pedros-Alio C. 2006. Marine microbial diversity: Can it be determined? *Trends Microbiol*. **14**:257–263.

Peplies J, Glöckner FO, Amann R, Ludwig W. 2004. Comparative sequence analysis and oligonucleotide probe design based on 23S rRNA genes of Alphaproteobacteria from North Sea bacterioplankton. *Syst. Appl. Microbiol*. **27**:573–580.

Peplies J, Kottmann R, Ludwig W, Glöckner FO. 2008. A standard operating procedure for phylogenetic inference (SOPPI) using (rRNA) marker genes. *Syst. Appl. Microbiol*. **31**:251–257.

Pruesse E, Quast C, Knittel K, Fuchs BM, Ludwig WG, Peplies J, Glöckner FO. 2007. SILVA: a comprehensive online resource for quality checked and aligned ribosomal RNA sequence data compatible with ARB. *Nucleic Acid Res*. **35**:7188–7196.

Reeder J, Knight R. 2009. The "rare biosphere": A reality check. *Nat. Methods* **6**:636–637.

Tautz D, Arctander P, Minelli A, Thomas RH, Vogler AP. 2002. DNA points the way ahead of taxonomy—In assessing new approaches, it's time for DNA's unique contribution to take a central role. *Nature* **418**:479–479.

Tringe SG, Hugenholtz P. 2008. A renaissance for the pioneering 16S rRNA gene. *Curr. Opin. Microbiol*. **11**:442–446.

Ward DM, Weller R, Bateson MM. 1990. 16S rRNA sequences reveal numerous uncultured microorganisms in a natural community. *Nature* **345**:63–65.

Wuyts J, De Rijk P, Van de Peer Y, Winkelmans T, De Wachter R. 2001. The European Large Subunit Ribosomal RNA Database. *Nucleic Acid Res*. **29**:175–177.

Yarza P et al. 2008. The All-Species Living Tree project: A 16S rRNA-based phylogenetic tree of all sequenced type strains. *Syst. Appl. Microbiol*. **31**:241–250.

Chapter 46

ARB: A Software Environment for Sequence Data

RALF WESTRAM, KAI BADER, ELMAR PRÜSSE, YADHU KUMAR, HARALD MEIER, FRANK OLIVER GLÖCKNER, AND WOLFGANG LUDWIG

46.1 INTRODUCTION

Comparative sequence analysis of evolutionary conserved marker molecules nowadays is the standard procedure for assigning organisms to phylogenetic groups and/or taxonomic units. The current prokaryotic taxonomic framework is mainly based on rRNA-based phylogenetic conclusions [Ludwig and Klenk, 2001; Ludwig et al., 2009]. This approach provides the basis for identification or new description in pure culture investigations or culture-independent studies of complex environmental samples [Amann et al., 1995]. Furthermore, comparative analysis of appropriate markers allows assigning contigs to taxa in metagenomics studies.

Powerful interoperating bioinformatics tools are prerequisites for sound utilization of the data flood for identification and phylogenetic inference in the genomics era. Such tools were missing or only available as stand-alone programs when the ARB project was initiated about 16 years ago [Ludwig et al., 2004]. Given this situation, two major goals were formulated in the early days of the ARB project and are maintained to the present: (1) the maintenance of a structured integrative secondary database combining processed primary structures and any type of additional data assigned to the individual sequence entries and (2) a comprehensive selection of software tools directly interacting with one another as well as the central database which are controlled via a common graphical interface. Initially, the ARB package was designed for handling and analyzing rRNA data. Later, it was extended by developing and/or including software tools for managing protein sequences as well as contigs and genomes.

Currently, the ARB project is maintained by members of the institutions with which the authors of this chapter are affiliated. The ARB package [Ludwig et al., 2004; Ludwig, 2005; Kumar et al., 2005; Kumar et al., 2006] as well as expert-curated rRNA databases [Pruesse et al., 2007] are freely available via http://www.arb-home.de and http://www.arb-silva.de.

46.2 THE ARB SOFTWARE PACKAGE

The ARB software package provides a set of cooperating tools for database maintenance and managing as well as data handling and analysis. These tools directly interact with a central database of processed sequence and various types of sequence associated meta data. A common graphical user interface allows data access, modification, and analysis. The database structure as well as the mode and parameters of interaction of the software tools are customizable by the user to a large extent.

46.2.1 The ARB Main Window

After database selection and ARB program start, the ARB main window provides the turnip for accessing the various software tools and facilities of the ARB package via the respective menus and buttons (**Fig. 46.1**). Furthermore, a user-selected tree is shown in radial or (two different) dendrogram formats. Primary data and metadata can be visualized at the terminal nodes. Compression of the view is possible by depicting user defined (phylogenetic)

Handbook of Molecular Microbial Ecology, Volume I: Metagenomics and Complementary Approaches, First Edition. Edited by Frans J. de Bruijn.

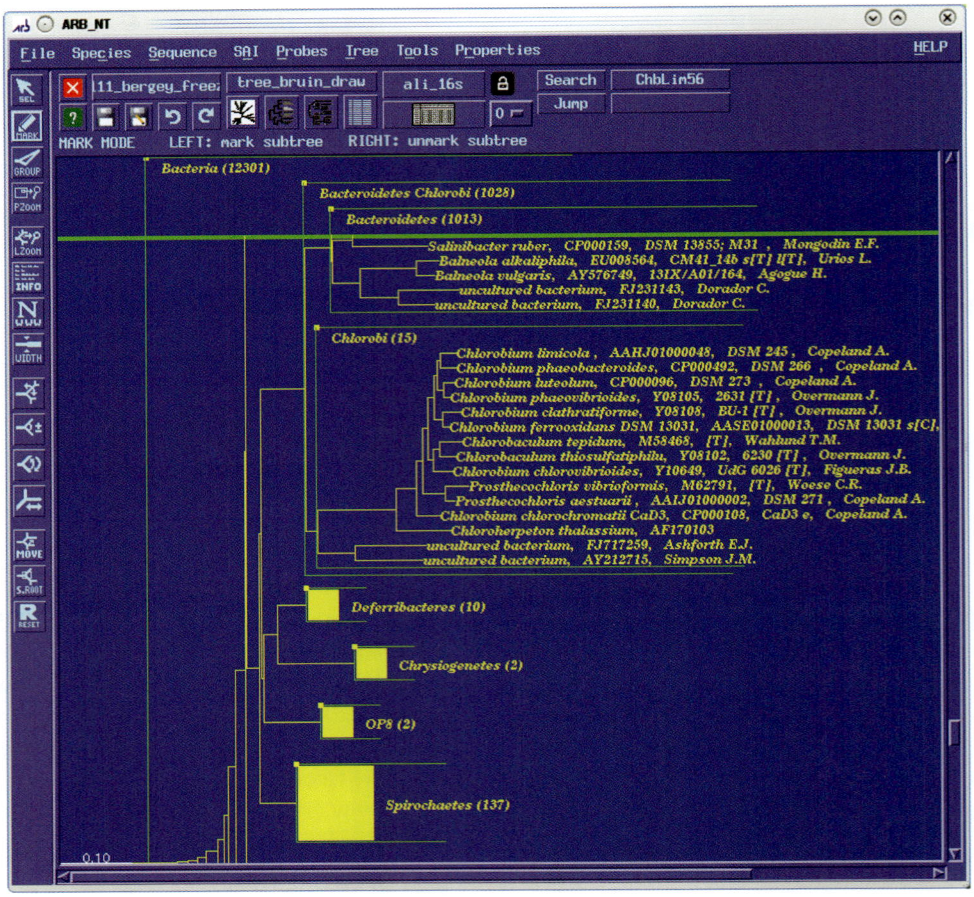

Figure 46.1 The ARB main window. Buttons in top and left panels provide access to the various ARB tools. Phylogenetic groups are indicated by brackets, and condensed groups are represented by rectangles along with numbers of terminal nodes hidden. NDS (node display setup)-controlled database field entries at terminal nodes indicate the names, accession numbers, strain designations of the respective organisms (master entries), and first authors of the respective bibliography.

groups as triangles or rectangles in radial trees or dendrograms, respectively. Alternatively, these data can be shown by simple listing. Datasets for further analyses can be selected by mouse button directed "marking" of the respective internal or terminal nodes. Opening a slave window for tree comparisons is also possible. The respective trees can be exported to xfig—a simple open source graphics program (http://www.xfig.org)—for further modification and/or transformation into various formats.

46.2.2 The Central Database

The central component of the ARB package is a special hierarchical and highly compressed database. During operation, it is loaded in the main memory ensuring rapid access by the peripheral software tools. The sequences representing organisms, genes, or gene products are stored in individual database fields. Different sequences (genes, contigs, nucleic acid, and protein sequences) of the same organism can be stored in individual containers (alignments) assigned to the same master entry (organism). A unique identifier (short_name) is automatically generated and assigned to each master entry under the control of a "name server." Following the ARB concept of an integrative database, any type of additional data can be assigned to the individual master entry and stored within default or user defined database fields. Besides a set of default database fields, additional ones can be created, deleted, and renamed by the user. The metadata can either be intrinsic parts of the database or linked to it via local networks or the internet. In the latter case the path to the respective file or the URL of an external database—optionally including commands and search strings—have to be defined using the ARB WWW (world wide web) tool. The default hierarchy of the database entries is according to the phylogeny of the organisms derived from the respective sequence data. However, it can also be changed according to other criteria defined by database field entries. This hierarchy is used by special algorithms for highly effective data compression. Different protection levels (0–6) can be assigned to the individual database fields. Database as well as security management is facilitated by this tool. Data import and export is possible in various common flat file formats. Default or user-defined parsing filters control the storage or extraction of data and features into and from defined ARB database fields, respectively. A versatile merge tool allows data merging and exchanging between different ARB databases. A similar tool can be used for exporting of data subsets in the ARB format.

46.2.3 Data Access and Visualization

Multiple alternative ways provide data access, selection, visualization, modification, and analysis using the ARB package. As mentioned above (see Section 46.2.1), the tree or list shown in the ARB main window can be used for browsing the data. Phylogenetic trees generated by intrinsic ARB tree reconstruction tools or imported from external sources are stored in the database and can be visualized in different formats within the ARB main window. Any (combination of) database field entries can be visualized at the terminal nodes of the tree currently shown (Fig. 46.1). Selection and order of data entries, the results of data analysis, or extraction to be visualized are defined by the NDS (node display settings) tool. Irrespective of the visualization mode used, the ARB SRT (search and replacement tool), ACI (ARB command interpreter), and RGE (regular expressions) tools can be used for extraction of combinations of (sub)strings as well as for analysis of database field entries, respectively.

A powerful search tool allows simple (strings and combination of strings) and complex (default or user-defined algorithms) searches in one or more (up to three) of the database fields. The matching master entries are shown in a hit list along with restricted information on the respective hits. Selecting from this list provides access to the information in all or user-defined selections of database fields.

The "info" window displayed starting the standard tool for data visualization lists the database fields along with the respective stored information for one master entry. Database field selection and order in this list can be customized by the user. Furthermore, editing of the filed entries is possible using this tool. Multiple windows can be opened allowing simultaneous data access for different master entries. Besides this standard procedure, raw and processed data visualization is possible via "user masks." The layout of the visualization windows (i.e., selection, size, and positioning) of database field entries can be customized by the user. Furthermore, simple algorithms for modifying and analyzing of database field entries (SRT, ACI, RGE) can be included when designing "user masks."

46.2.4 Sequence Editors

A powerful editor provides versatile user access to primary structure (nucleotide or amino acid sequences) visualization, arrangement, and modification (**Fig. 46.2**). The set of sequences to be displayed can be interactively defined as well as stored in user-defined "configurations." The arrangement of the primary structures depends on the tree displayed in the ARB main window or is taken from a "configuration" selected while starting the editor. The original data as well as virtually transformed (e.g., purine-pyrimidine, in silico translated amino acid sequences, or simplified amino acid presentation) data are displayed in user-defined color codes. Keyboard customization is possible for data entry and modification. Two different editing modes can be selected. The "Align" mode allows inserting/removing alignment gaps and moving sequence characters or stretches, while character changes are only possible after switching to the "Edit" mode. The rights to overcome protection of the individual sequence entries can be given for the two modes independently. This helps to prevent unwanted character changes when manually modifying the sequence data or the alignment. A set of hot keys in combination with (alignment, sequence, reference, or helix specific) cursor positioning facilities support easy navigation. Block operations are available for modifying the respective primary structure or alignment regions. Sets of search strings can be defined and optionally stored. Perfect or partial matches can be visualized within the displayed sequences by user-defined background colors (Fig. 46.1). Virtual compression—removal of alignment gaps common to all or a certain fraction of the displayed sequences—is possible. This makes data inspection and editing more convenient in case of large insertions occurring in only part of the sequences. Groups of sequences can be interactively defined or are automatically shown if defined according to the tree selected while starting the editor. Consensus sequences are determined for each defined group of sequences according to default or user defined criteria and optionally visualized along with or instead of the individual sequences. This consensus can be edited, and changes made concern any sequence in the group. A special feature of the editor is the simultaneous secondary structure check if rRNA (gene) data are visualized. Symbols indicating the presence or absence as well as the character of base pairings are shown below the individual nucleotide symbols and immediately refreshed during sequence editing. A (three-domain) consensus secondary structure mask established according to commonly accepted secondary structure models [Cannone et al., 2002] functions as a guide for this tool.

Figure 46.2 The ARB primary structure editor. Buttons in top and left panels provide access to the various editor-associated ARB tools. Subwindows in the upper part indicate cursor positioning, error messages, and search strings. SAI (sequence-associated information) lines show the *E. coli* reference sequence as well as secondary structure mask and helix numbering. Condensed groups as shown in Figure 46.1 are represented by the respective consensus. The "Probe" search string is highlighted in the respective primary structures. Positional base pairing (~,−,+,=) or consensus secondary structure violation (#) is indicated below the base symbols.

The ARB (nucleotide) secondary structure editor fits any sequence selected by cursor positioning in the primary structure editor into the common consensus model (**Fig. 46.3**). The layout of the structure—that is, color coding of base paired, nonpaired, and loop positions as well as the arrangement, shape, and size of helices and loops—can be customized according to the user's preferences. Any of the search strings or SAIs ("sequence associated information"; see above) activated in the primary structure editor can be visualized by background colors in the secondary structure model [Kumar et al., 2005]. The structure can be exported to xfig—a simple open source graphics program (http://ww.xfig.org)—for further modification and/or transformation into various formats.

Three-dimensional (3-D) presentation of the respective sequence optionally with search string and SAI visualization is also possible [Kumar et al., 2006]. Color coding can be customized as described for the secondary structure editor. The 3-D structure is based on x-ray structure data for the rRNA molecules of *Escherichia coli* [Ban et al., 2000; Tung et al., 2002].

The primary structure editor contains a "protein viewer" component allowing in silico translation and virtual presentation of database inherited nucleic acid sequences in selected or all frames. Two- and three-letter as well as user-defined color code presentation is possible. This tool helps when performing primary structure quality checking and optimizing the respective alignment. For further analyses of the in silico translated amino acid, sequences have to be stored in a separate protein sequence alignment (database field; see Section 46.2.1). The respective nucleic and amino acid alignments can be synchronized (see Section 46.2.8).

46.2.5 Profiles, Masks, and Filters

Conservation or base composition profiles, higher-order structure masks, and filters including or excluding particular alignment positions are important tools for sequence data analyses, especially for phylogenetic inference [Ludwig and Klenk, 2001; Peplies et al., 2008]. The ARB package provides tools for determining such profiles based upon the full database or user-defined subsets. These profiles, masks, and filters are stored in the central database as so-called SAIs (sequence associated information) and can be visualized and modified by the primary structure editor. The filter selection tool not only allows us to choose sets of particular filters but also allows to perform a fine tuning

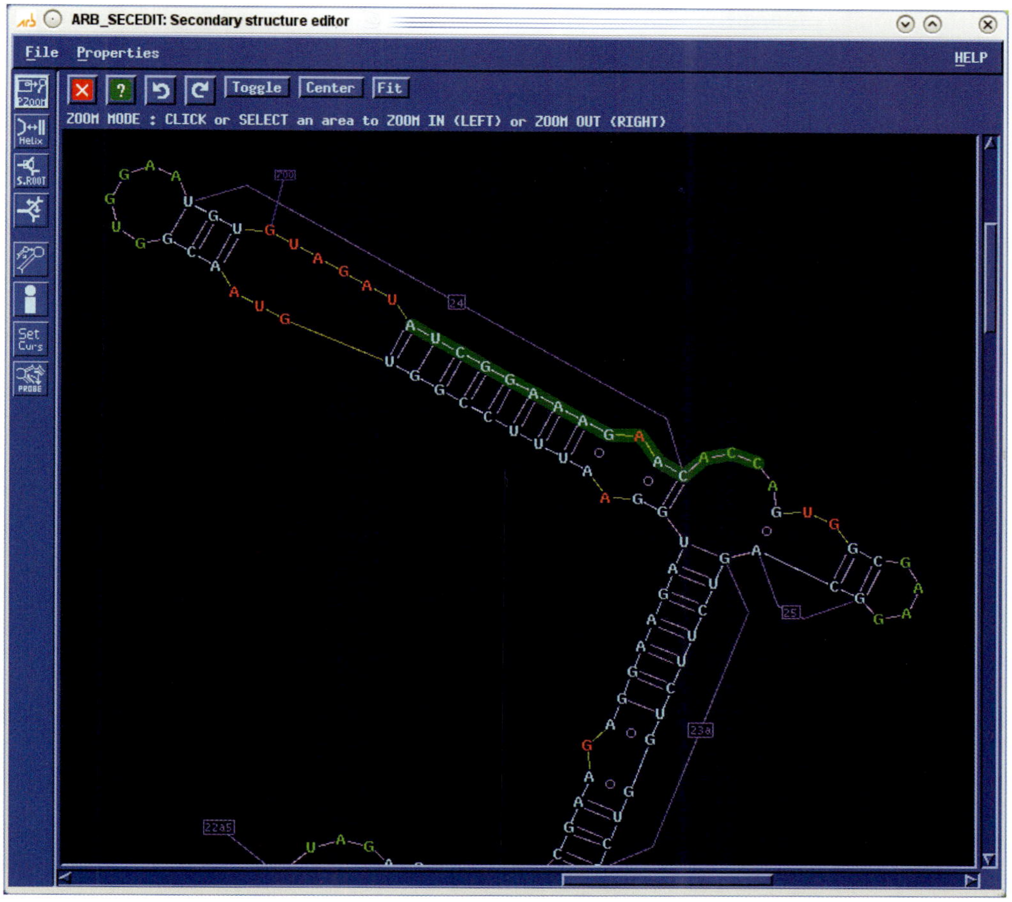

Figure 46.3 The ARB secondary structure editor. Buttons in top and the left panels provide access to the editor associated layout tools. The "Probe" search string is highlighted (see Fig. 46.1).

with respect to the inclusion or exclusion of alignment positions in case of multiple character filters. Besides SAIs derived from the primary structures, any other information that can be assigned to sequence/alignment positions or regions can be stored and used as SAIs. Examples are rRNA–protein interaction sites or "in situ" accessibility maps for FISH (fluorescence in situ hybridization) [Amann et al., 1995; Kumar et al., 2005] probes.

46.2.6 Phylogenetic Treeing

Software tools for nucleotide and amino acid sequence-based tree reconstruction according to the three most commonly used approaches (i.e., distance matrix, maximum likelihood, and maximum parsimony-based procedures) are incorporated in the package. They cooperate as intrinsic tools with the respective ARB components and database elements such as alignment and filters.

The central treeing tool of the package—ARB parsimony—is a special development for the handling of several thousand sequences (more than 400.000 in the current small subunit (SSU Ref) rRNA SILVA database [Pruesse et al., 2007). New sequences are successively added to an existing tree according to the parsimony criterion. A special software component superimposes branch lengths to the parsimony generated tree topology. These branch lengths reflect the significance of the individual "tetra-furcations" by expressing the difference of the most and the two less parsimonious solutions when performing NNI (nearest-neighbor interchange of adjacent branches or sub trees). These relative distances are normalized according to a distance matrix deduced from primary structure comparison. Thus branch lengths in ARB-parsimony-generated trees in the first instance visualize the significance of topologies, while in the second instance they reflect a degree of estimated sequence divergence. A special feature of ARB parsimony allows adding sequences to an existing tree without permitting any changes in the initial tree. This enables the user to include partial, low-quality or preliminary aligned sequences without perturbing the topology of an optimized tree based upon optimally aligned full-

and high-quality data. Another peculiarity of this treeing software concerns the tree optimization by performing cycles of NNI (nearest-neighbor interchange) and KL [Kernigham and Lin, 1970] topology modifications. These optimizations not only can be performed for the complete tree but also can be confined to user-selected subtrees. Thus tree optimization is possible by applying the appropriate filters for the respective phylogenetic levels and groups.

The ARB-neighbor tool for generating distance matrix trees is an accelerated and improved version of the respective component of Felsenstein's [Felsenstein, 1989] PHYLIP package.

Selected stand-alone tools of the former package can be used in the ARB environment in combination with all respective ARB features.

The various facilities of the currently most powerful maximum likelihood program RAxML [Stamatakis, 2006] can also be operated from the ARB user interface applying parameters and filters generated by the respective ARB features. Besides RaxML, also TREE-PUZZLE [Schmidt et al., 2002] and PhyML [Guindon and Gascuel, 2003] versions can be used for ARB controlled tree reconstruction.

A "concatenation" tool allows merging alignments of different genes or gene products for multiple marker-based phylogenetic studies. The full spectrum of filter and parameter setting is available for analyzing or controlling the influences of the individual markers in the concatenated set.

46.2.7 The Positional Tree Server

Once established, the ARB PT server (positional tree) allows rapid and exact searching for sequence identity or peculiarity. Thus, it represents the central tool for fast searching of closest relatives for automated sequence alignment or to define diagnostic sequence stretches for primer and probe design. Establishing a prefix tree server of any oligonucleotide sequence up to 100-mers occurring in the underlying database and assignment of the individual oligonucleotides to the sequences or organisms containing them is the basis for these procedures. PT-server-based analyses do not rely upon aligned sequences. The PT server is not provided with the ARB program package or ARB database. It has to be established for the respective database locally. The PT server is used for rapid finding of the most similar reference sequences indicating the closest relative of the query organism. This also helps finding appropriate templates for adding new sequences to existing alignments (see Section 46.2.8). The PT server is also used for finding (taxon- or group-specific) diagnostic sequence stretches for probe and primer design and evaluation (see Section 46.2.9).

46.2.8 Sequence Alignment and Quality Checks

For de novo-generating nucleic or amino acid sequence alignments, ClustalW [Thompson et al., 1994] was added to the peripheral tools of the ARB package. However, in the context of database maintenance, new sequence entries have to be integrated in an already existing database of aligned sequences. For this purpose the ARB fast aligner was developed. This aligner uses a (set of) selected aligned reference sequences as template(s) for rapid integration of a (set of) unaligned sequence(s). Individual entries—that is, sequences or consensus defined by the user or automatically determined by PT-server-based search for most similar reference sequences—are used as template.

In case of protein coding nucleic acid sequences, the alignment usually is optimized on the amino acid level (given that the phylogenetic information is stored there) [Peplies et al., 2008]. The underlying nucleic acid alignment can then be adapted to the amino acid alignment by a back-translation based tool taking into consideration all known codon usages.

Once a reasonable data set of high quality and optimally aligned primary structures is reasonably structured (grouped) according to the results of careful phylogenetic analyses, further sequence and alignment quality checking is possible using the respective ARB tools. A component of the primary structure editor takes into account SAIs (sequence associated information; see Section 46.2.4) expressing positional variability as well as phylogenetic tree topologies for estimating reasonability of a certain monomer (nucleotide or amino acid) at a certain alignment position. The degree of "(miss)-fit" is optionally indicated by user defined background colors in the editor window. Another tool determines a quality score for the individual sequences by estimating degrees of deviation from group specific primary and secondary structure consensus, conservation profiles, sequence sizes, and completeness.

46.2.9 Probe Design and Evaluation

Taxon- or gene-specific probes or primers certainly play a central role in many molecular biological research and analysis projects—for example, the identification and detection of organisms in complex environmental samples or expression studies within the scope of genome projects. The ARB "Probe Design" and "Probe Match" tools are searching the PT server to identify short (10–100 monomers) diagnostic sequence stretches that are evaluated against the background of all sequences in the database the PT server has been built from. In principle, no alignment of the sequence data is needed for specific probe design. However, in the case of taxon-specific probes, alignment and phylogenetic analyses are necessary for defining groups of phylogenetically (taxonomically) related organisms as the targets of specific probes. The design of taxon-specific oligonucleotide probes with ARB is performed in three steps. First, the (group of) target organism(s), gene(s), or sequence(s) has to be defined ("marked," see Section 46.2.1). Second, potential target sites are searched by the "Probe Design" tool with the aid of a PT server. The results are shown in a ranked list of proposed targets, probes, and additional information. The ranking is according to in silico-predicted probe quality. Third, the proposed oligonucleotide probes are evaluated against the whole database by using the program "Probe Match." Local alignments are determined between the probe target sequence(s) and the most similar reference sequences (optionally from 0 to 5 mismatches) in the respective database. Furthermore, these sequence strings can automatically be visualized in the primary and secondary structure editors (see Section 46.2.4). A special advancement is the ARB multiprobe software component. It determines sets of up to five probes optimally identifying the target group. Color-coded visualization of target master entries (see Section 46.2.2) and matching probe combinations is possible in the ARB main window.

46.2.10 Further Useful ARB Tools

A large fraction of sequences in the currently available rRNA sequence databases [Pruesse et al., 2007] comprises clusters of highly similar to identical primary structures most often retrieved by culture independent environmental studies. Commonly, such "sequence clouds" are represented as OTUs (operational taxonomic units) in further data analyses. Such OTUs are defined either manually or by applying respective software tools [Schloss et al., 2009]. Using the ARB package OTUs can be defined and automatically grouped in the selected tree by a newly developed component. The OTU definition according to user provided parameters is deduced from the topology of a selected tree and reassessed using distance methods. A (best) representative is proposed by the software.

ARB can also function as a simple genome viewer allowing comparison of annotated contigs or genomes. Data access is possible by "search" and "info" tools, alternatively via genome maps similarly as described in Section 46.2.2. Extraction of (sets of) genes into ARB gene databases can also be managed by this ARB facility.

46.2.11 Availability and Training

The ARB software has been designed for Linux operating systems. Tested versions for SuSE and Ubuntu Linux distributions are available at http://www.arb-home.de and http://www.arb-silva.de. The binaries, source code, and some documentation are provided in the download area of these web pages. The latter URL also provides access to the current release of the SILVA LSU and SSU rRNA databases. Furthermore, there is a mailing list of the world wide ARB users community. Subscription is needed for those interested in joining (subscribe@arb-home.de). Basic and advanced ARB training courses are offered by the Ribocon company in Bremen (Germany, http://www.ribocon.com). Mac users interested in ARB should contact http://www.haloarchaea.com/resources/arb/.

46.3 CONCLUDING REMARKS

The ARB software package provides a powerful and comprehensive set of directly cooperating software tools for managing and analyzing integrative databases of sequences. It is in use worldwide. The ARB software and database maintaining teams try to keep it up to date and compatible with the ongoing hardware developments. Given more than 16 years of ARB development by different computer scientists and a large number of students of computer science, a huge and heterogeneous source code would have to be cleaned and at least partially redesigned. However, it is difficult to get funding or sponsoring of software redesign.

INTERNET RESOURCES

ARB software (http://www.arb-home.de)

ARB databases (http://www.arb-silva.de)

Acknowledgments

ARB software and database maintenance is partially supported by the Deutsche Forschungsgemeinschaft and the Bayerische Forschungstiftung.

REFERENCES

AMANN R, LUDWIG W, SCHLEIFER KH. 1995. Phylogenetic identification and in situ detection of individual microbial cells without cultivation. *Microbiol. Rev.* **59**:143–169.

BAN N, NISSEN P, HANSEN J, MOORE PB, STEITZ TA. 2000. The complete atomic structure of the large ribosomal subunit at 2.4 A resolution. *Science* **289**:905–920.

Cannone JJ, Subramanian S, Schnare MN, Collett JR, D'Souza LM, et al. 2002. The comparative RNA Web (CRW) site: An online database of comparative sequence and structure information for ribosomal, intron, and other RNAs. *BMC Bioinform*. **3**:2.

Felsenstein J. 1989. PHYLIP—Phylogeny inference package (version 3.2). *Cladistics* **5**:164–166.

Guindon S, Gascuel O. 2003. A simple, fast, and accurate algorithm to estimate large phylogenies by maximum likelihood. *Syst. Biol*. **52**:696–704.

Kernigham BW, Lin S. 1970. An efficient heuristic procedure for partitioning graphs. *Bell Syst. Tech. J*. **49**:291–307.

Kumar Y, Westram R, Behrens S, Fuchs B, Gloeckner FO, et al, 2005. Graphical representation of ribosomal RNA probe accessibility data using ARB software package. *BMC Bioinform*. **6**:61.

Kumar Y, Westram R, Kipfer P, Meier H, Ludwig W. 2006. Evaluation of sequence alignments and oligonucleotide probes with respect to three-dimensional structure of ribosomal RNA using ARB software package. *Bioinformatics* **7**:240–251.

Ludwig W, Klenk HP 2001. Overview: A phylogenetic backbone and taxonomic framework for prokaryotic systematics. In Garrity, G. M. ed. *Bergey's Manual of Systematic Bacteriology*, 2nd ed. Vol. **1**. New York: Springer, pp. 49–65.

Ludwig W, Strunk O, Westram R, Richter L, Meier H, et al. 2004. ARB: A software environment for sequence data. *Nucleic Acids Res*. **32**:1363–1371.

Ludwig W. 2005. Bioinformatics and web resources for the microbial ecologist. In Osborn AM, Smith CJ. eds. *Molecular Microbial Ecology*. Abingdon: Taylor and Francis, pp. 345–371.

Ludwig W, Schleifer KH, Whitman WB. 2009. Revised road map to the phylum *Firmicutes*. In Whitman WB, ed. *Bergey's Manual of Systematic Bacteriology*, 2nd ed, Vol. **3**. New York: Springer, pp. 1–13.

Peplies J, Kottmann R, Ludwig W, Glöckner FO. 2008. A standard operating procedure for phylogenetic inference (SOPPI) using (rRNA) marker genes. *Syst. Appl. Microbiol*. **31**:251–257.

Pruesse E, Quast C, Knittel K, Fuchs B M, Ludwig W, et al. 2007. SILVA: A comprehensive online resource for quality checked and aligned ribosomal RNA sequence data compatible with ARB. *Nucleic Acids Res*. **35**:7188–7196.

Schloss PD, Westcott S L, Ryabin T, Hall JR, Hartmann M, et al. 2009. Introducing mothur: Open source, platform-independent, community-supported software for describing and comparing microbial communities. *Appl. Environ. Microbiol*. **75**:7537–7541.

Schmidt HA, K. Strimmer M, Vingron, von Haeseler A. 2002. TREE-PUZZLE: Maximum likelihood phylogenetic analysis using quartets and parallel computing. *Bioinformatics* **18**:502–504.

Stamatakis A. 2006. RAxML-VI-HPC: Maximum likelihood-based phylogenetic analyses with thousands of taxa and mixed models. *Bioinformatics* **22**:2688–2690.

Thompson JD, Higgins DG, Gibson DJ (1994) CLUSTAL W: improving the sensitivity of progressive multiple sequence alignment. *Comput. Appl. Biosci*. **8**:189–191.

Tung C S, Joseph S, Sanbonmatsu KY. 2002. All-atom homology model of the *Escherichia coli* 30S ribosomal subunit. *Nat. Struct. Biol*. **9**:750–755.

Chapter 47

The Phyloware Project: A Software Framework for Phylogenomic Virtue

Daniel N. Frank and Charles E. Robertson

47.1 INTRODUCTION

Don't waste clean thoughts on dirty sequences.

—To paraphrase Efraim Racker

47.1.1 Problem Statement

The recent explosions in culture-independent studies of environmental DNA sequences (i.e., metagenomics) and especially environmental ribosomal RNA (rDNA) sequences [Peplies et al., 2008], combined with rapidly advancing automated DNA sequencing capabilities [Mardis, 2008; see also Chapter 18, Vol. I], have altered profoundly the volume of sequence data which must be processed for each study. When culture-independent techniques were originated, sequencing effort and expense dominated any considerations as to how much sequencing "was enough" for a culture-independent dataset. Consequently, most early publications were based on datasets consisting of less than 100 sequences. As sequencing technology became more commonplace, propelled mainly by the human genome project, diminished sequencing costs propelled an expansion of metagenomic dataset size. As recently as two years ago, a 1000-sequence dataset was considered substantial. However, the latest generation of commercial high-throughput sequencing systems has inflated dataset size by yet another three or four orders of magnitude. Concurrent with sequencing technology advances, the general understanding of how to accurately analyze culture-independent gene sequences has resulted in an increase in the number of software analyses that must be performed in order to validate, classify, and (hopefully) uniquely identify each sequence. The actual computing volume required for each culture-independent dataset likely has increased by four orders of magnitude within the last 15 years, forcing a transition from human-intensive manual methodologies to predominately computer-based approaches of metagenomic analysis. These developments have prompted the creation of numerous software applications designed to make sense of a sometimes overwhelming avalanche of sequence data.

Many sequence analysis software packages have been developed, but are available as disparate pieces of software, which often use different data formats and sometimes are implemented solely as web services. Use of these tools often requires some facility with the UNIX/Linux operating system and/or specialized scripting languages to either manipulate files in batch or link the output of one analysis step to the input of the succeeding analysis step. These issues have caused some recent systems to be created along the lines of an automobile assembly line with each sequence sequentially processed by well-integrated analysis modules. (This type of system often is termed a sequence-processing pipeline, and each analysis module is called a pipe step.) Ironically, the availability of software systems capable of high-volume, mostly automatic sequence analyses has created an unexpected problem, namely, a tendency of users to unconditionally accept results of these highly automated analyses verbatim. Thus, the complexity of the metagenomics software environment may mean that users unfamiliar with either the biological or computational framework under which a tool operates may default to a black box mentality and unconditionally accept the results of a program's outputs. Alternatively, the user may simply forego performing some

Handbook of Molecular Microbial Ecology, Volume I: Metagenomics and Complementary Approaches, First Edition. Edited by Frans J. de Bruijn.

analytic steps (e.g., chimera checking), despite their theoretical impact on data quality. In our opinion, the trend toward building turnkey software pipelines that import raw sequence data and export statistical analyses complete with publication-ready figures can promote an aversion to critical thinking on the part of the user community.

The very rapid increase in sequence accumulation has put additional strains on public sequence databases. Our practical experience indicates that these databases increasingly are populated with low-quality, poorly annotated sequences. As a database accumulates such sequences, its overall quality is diminished. Consequently, subsequent analyses of newly generated sequences are compromised. The net result may be an error-catastrophe in which the information content of the database degrades over time to the point of being nonfunctional. This is of particular concern because the rapidly expanding scale of high-throughput sequencing has significantly decreased the amount of labor allocated to the analysis of any particular sequence. In other words, sequences often are dumped into public databases with little quality assurance. In analogy to the open-source software community principle that "given enough eyeballs, all bugs are shallow" [Raymond, 2001], the metagenomics community is rapidly approaching a state in which the bugs (sequences of poor quality) outnumber the eyeballs.

47.1.2 Project Goals

The overarching goal of the Phyloware project is to develop a set of best practices for performing metagenomic research and to implement these rules in robust, user-friendly software that follows a few basic principles. To improve final database quality, our practical experience has demonstrated that sequences should be removed from the pipeline as early in the process as possible. This tactic has the added benefit of reducing the overall amount of computation required to process a dataset. We envision that such software can be used as a routine quality assurance front-end to public sequence databases and thus provide consumers with some confidence of sequence quality. Project design and software development are guided by an open-source philosophy in order to better reflect the needs of the user community and thereby increase the rate of adoption of the technology. A crucial design goal of the Phyloware project is to produce software that is easy to administer and use, even by computer novices, without the support of dedicated system administration or programming staff. Presenting a unified user experience simplifies sequence analysis projects and reduces the effort for users to teach themselves computer-based sequence analysis. This is achieved through development of user-friendly glue code that links existing software tools (the XplorSeq model, developed below) into a seamless pathway. Additional analytic tools can be developed and incorporated as needs arise. To the end user, these tools will be made available as a single software package (ideally including a GUI) in which data analyses are relatively automated. To promote critical analyses of datasets (to avoid the "black box belief" propensity), Phyloware encourages users to participate in all phases of data analysis, by requiring that they actively choose inputs, outputs, and specify program parameters, while making the process seamless and relatively painless by hiding esoteric implementations details. In sum, the aim of Phyloware is to enable biological scientists to focus on their chosen science, rather than the underlying computer science of their software tools.

This chapter begins to outline a formal approach to quality assurance in metagenomics projects that currently is underdeveloped. We present these ideas in hopes to stimulate discussion and welcome comment on the issues raised, as well as those that have been neglected.

47.2 BEST PRACTICES FOR METAGENOMIC SEQUENCE ANALYSIS

From a practical perspective, we can begin to codify a set of best practices for metagenomic sequence analysis (see box). Although these principles can be easily stated, details of their implementation have not been established. A difficulty is that the phylogenomics practices that have developed are guided by folk tradition that is passed from person to person, rather than evidence-based rules. The sequential order in which these principles are listed roughly reflects the typical temporal order in which projects proceed. Considered individually, each structure poses a distinct arena that involves its own problems of optimization. Additionally, the sequential order of project dataflow means that solutions in one problem area will propagate onward. Because all steps are coupled, the optimization problem becomes immense. How, for instance, should sequences be trimmed in order to optimize good alignment and good tree-building? Would trimming with higher-quality scores (shorter sequence lengths) be better than the use of lower-quality scores (lower quality, longer sequence lengths)? Does the extent of trimming impact alignment? What is the best way to score the quality of alignments?

BEST PRACTICES FOR METAGENOMICS

1. Sacrifice quantity for quality.
2. Carefully trim poor quality bases from new sequences.
3. Remove chimeric sequences.
4. Align new sequences to case-appropriate alignments.
5. Assign taxonomy and function wisely.
6. Build trees with case-appropriate datasets.
7. Doubt any single phylogenetic tree.

Because the "best practices" of the field are still in flux, an equally important design goal is to use Phyloware software as a test bed for establishing evidence-based rules for tuning code to particular problem arenas (e.g., 16S rRNA based phylogenetics). Questions relating to optimization and coupling of analytic steps can then be explored in a uniform environment. Consequently, our software architecture is made as modular as possible in order to readily incorporate and compare the results of competing tools (e.g., alignment software).

47.3 IMPLEMENTATION

47.3.1 Overview

The Phyloware package is implemented through four components (**Fig. 47.1**): (1) third-party software, (2) Phyloware-specific C and C++ libraries encoding common sequence manipulation functions, (3) platform-independent programs written in C and/or C++ using the phyloware libraries, and (4) XplorSeq, a graphical-user interface (GUI)-based program that links the underlying components (currently available only for the Macintosh OS X operating system). Components 2 and 3 comprise the XplorSeq Toolkit (XSTK).

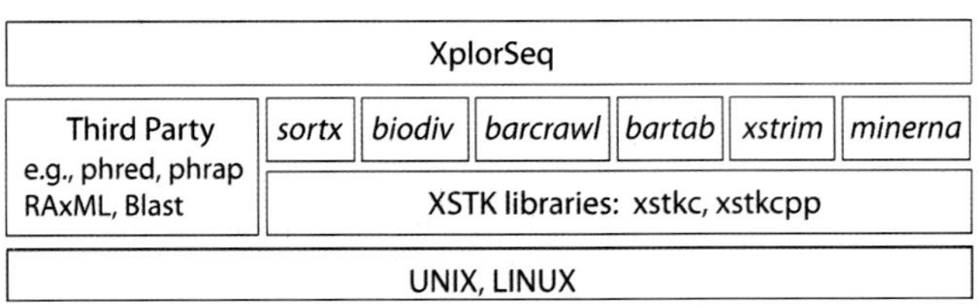

Figure 47.1 Architecture of Phyloware package. Four components makeup the Phyloware package, built on the UNIX/LINUX operating system: (1) Third-party software; (2) Phyloware-specific C and C++ libraries; (3) platform-independent programs written in C and/or C++ using the phyloware libraries; and (4) the XplorSeq graphical-user interface program that links the underlying components.

The software is freely available for noncommercial use at http://www.phyloware.com, however, full implementation of XplorSeq requires separate acquisition and installation of phred and phrap (www.phrap.org). The following sections describe (in greater detail) XplorSeq and software built using the XplorSeq toolkit.

47.3.2 XplorSeq

The GUI front end of the Phyloware package is XplorSeq [Frank, 2008], which was written for the Macintosh OS X operating system (**Fig. 47.2**). XplorSeq integrates the use of several popular third-party UNIX-based DNA sequence analysis applications with XplorSeq-specific code in order to track, annotate, and analyze sequence information in a manner conducive to high-throughput metagenomics. XplorSeq was developed and has been used primarily for analysis of rDNA clone libraries, but should be applicable to any sequencing project.

Although several commercial and noncommercial software packages implement some of the same basic functionalities as XplorSeq, XplorSeq implements several domain-specific software tools (e.g., for state-of-the-art phylogenetic tree inference, OTU clustering, biodiversity estimates) that are not available in general-purpose DNA analysis packages. **Table 47.1** lists many of the features available in XplorSeq v2.

Many published studies, from a variety of laboratories engaged in metagenomics, have used XplorSeq, and thereby established its stability, ease-of-use, and capabilities [Baumgartner et al., 2006, 2009; Dalby et al., 2006; Feazel et al., 2008, 2009; Frank, 2008, 2009; Frank and Pace, 2008; Frank et al., 2003, 2007, 2009a,b; Harris et al., 2007; Isenbarger et al., 2008; Lee et al., 2007; Ley et al., 2005, 2006, 2008; McManus and Kelley, 2005; Papineau et al., 2005; Perkins et al., 2009; Peterson et al., 2008; Rawls et al., 2006; Robertson et al., 2009; Sahl et al., 2008; Salmassi et al., 2006; Spear et al., 2005a,b, 2006, 2007; Turnbaugh et al., 2006, 2008; Walker and Pace, 2007; Walker et al., 2005]. XplorSeq is written in Objective-C using the Cocoa application framework (Apple Inc.), which implements a Model-View-Controller (MVC) design pattern. Releases are compiled for the OS X operating system (current versions require OS 10.4.x, OS 10.5.x, or 10.6.x) as universal binaries, which run natively on Macintosh computers with Intel or PowerPC microprocessors.

Following the original publication of XplorSeq [Frank, 2008], several new features have been added and other capabilities substantially enhanced, culminating in the release of XplorSeq v2 in 2009. In general, though, the look-and-feel of XplorSeq has remained

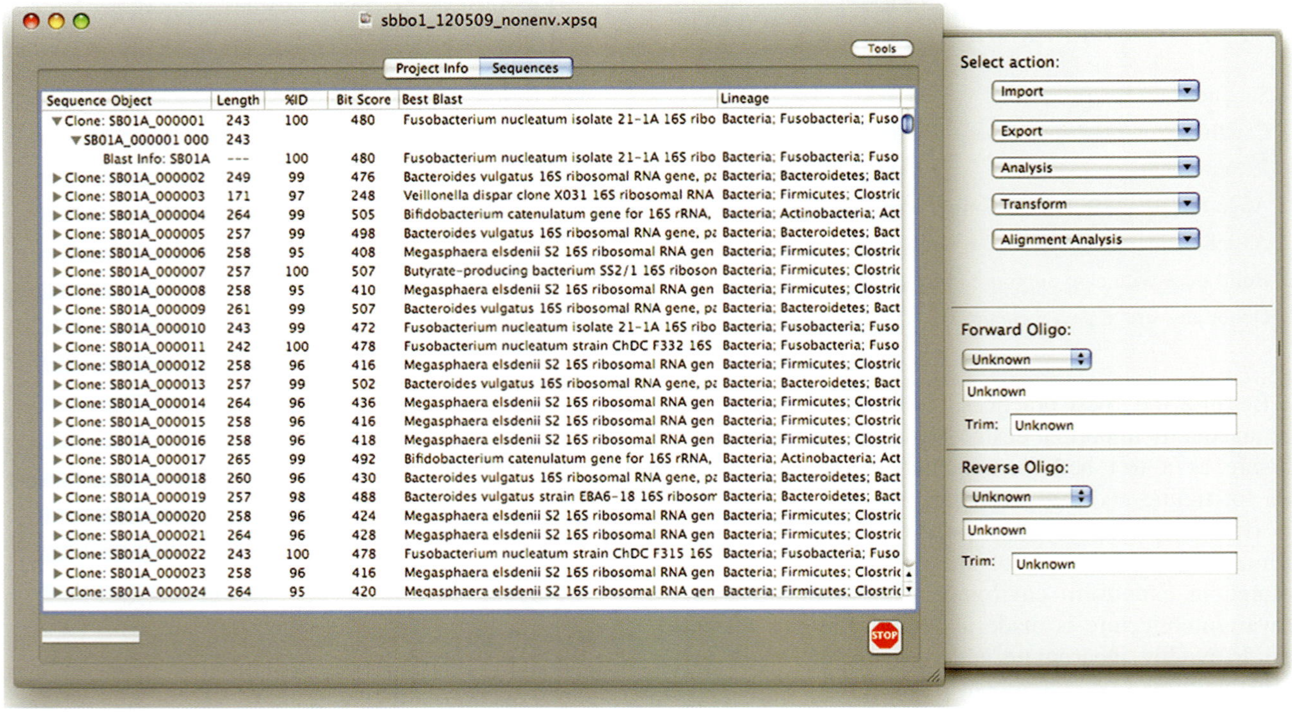

Figure 47.2 Main XplorSeq window. Screen shots of XplorSeq main window. Import, export, data analysis, and data transformation options are presented in menus within the tool drawer adjacent to the main window.

unchanged. We focus here on the recent evolution of XplorSeq.

47.3.3 Parallel Computing

Many of the data analysis steps employed by XplorSeq (e.g., base-calling, contig assembly, blast) are embarrassingly parallel (i.e., free of dependencies) and thus can be invoked straightforwardly in separate threads or processes. To exploit the increasing availability of multicore central processing units, XplorSeq has been refactored significantly to leverage concurrent programming primitives (e.g., NSOperation and NSOperationQueue) available through the Cocoa framework in Mac OS 10.5 onward (as well as Grand Central Dispatch functionality available in OS 10.6). Consequently, XplorSeq can exploit multicore architectures and scale-up analysis throughput as hardware evolves, without the need for any software reconfiguration by end users. For instance, basecalling (phred), contig assembly (phrap), vector and quality score-based trimming, and blast automatically use the full complement of cores available on a particular system. Because the Cocoa framework is not entirely thread-safe, trade-offs were required between providing rapid GUI updates and avoiding thread collisions on multicore systems. Consequently, screen updates are less frequent when multiple processors are in use.

47.3.4 Batch Processing

BATCH XplorSeq FUNCTIONS

1. Creation of XplorSeq files from fasta.
2. Sequence export.
3. BLAST analysis.
4. SINA alignment.
5. Post-processing of SINA alignment output.
6. Design of barcoded primers (*barcrawl*).

Although XplorSeq is architected primarily for single-document manipulation, certain methods lend themselves to batch processing of multiple files with little or no user intervention. Functions called on multiple files are accessed through the "Options" menu of the XplorSeq application menu bar (i.e., the menu bar on the top of a Macintosh screen). In these operations, documents are automatically opened, edited, saved, and closed as needed. GUI outputs are minimized in order not to clutter the user's workspace and to improve thread safety.

Table 47.1 Summary of XplorSeq Functionality

Function	Description[a]
Import	
Chromatogram ...	Import DNA chromatograms (.esd,.scf,.abi, etc.): phred
PHD ...	Import DNA sequences in Phd format
Contig ...	Import DNA sequences and quality scores in FastA format
Blast ...	Parse Blast records
FastA ...	Import DNA sequences in FastA format
XplorSeq Library ...	Import XplorSeq document
Phylogenetic Lineage ...	Import phylogenetic lineage information from entrez
Metadata ...	Import metadata in key-value format
Export	
Sequences ...	Export DNA sequences in variety of formats
FastA + Qual ...	Export DNA sequences and quality scores
Blast Info ...	Export summary of Blast records
Cluster Table ...	Enumerate OTUs belonging to groups of sequences
OTU Diversity ...	Calculate OTU richness for set of sequences
Quality Scores ...	Export summary of quality scores
Blast Accession #'s ...	Export accession numbers of top Blast hits
Sequin Script ...	Export data in format for Genbank submission (sequin)
Blast Database ...	Create a Blast database (formatdb)
XML File ...	Export data in XML format
Metadata ...	Summarize and export metadata
Selection Tree ...	List selected sequences in Newick format
Analyze	
Basecall → Blast...	Pipe data from chromatogram through Blast analysis
Contig → Blast...	Pipe data from contig assembly to Blast analysis
Basecall ...	Perform base calling (phred or ttuner)
Contig ...	Perform contig assembly (phrap or $TIGR_A$ssembler)
Blast NCBI ...	Blast query of Genbank (blastall)
Blast Local ...	Blast query of local blast database (blastcl3)
Get Entrez Lineage Info ...	Download entrez phylogenetic lineage information (idfetch)
Align ...	Perform multiple sequence alignment (clustal)
Biodiversity (biodiv) ...	Calculates biodiversity indices with random resampling (biodiv)
XplorSeq Doc Difference ...	Generate differences between two XplorSeq documents
SINA align ...	Align sequences with the Silva SINA tool (sina & si2fa)
NAST align ...	Align sequences with the NAST alignment tool
Transform	
Edit Sequence Names ...	Alter names of sequences
Edit Lineage Names ...	Edit phylogenetic lineage information
Edit Metadata ...	Edit metadata associated with sequence
Edit Metadata Keys ...	Edit all metadata keys in document
Group ...	Group sequences and contigs
UnGroup ...	Ungroup sequences and contigs
Clean ...	Delete blast information, contigs
Sort ...	Sort records in document
Set Oligos ...	Associate primer sequences with sequence objects
Trim ...	Trim sequences based on quality score and primer
UnTrim ...	Remove trimming information
Rev.-Complement	Reverse complement sequence

(*continued*)

Table 47.1 Summary of XplorSeq Functionality (*Continued*)

Function	Description[a]
DNA → RNA	Convert DNA sequence to RNA sequence
RNA → DNA	Convert RNA sequence to DNA sequence
UPPERCASE	Convert sequence to uppercase
lowercase	Convert sequence to lowercase
Alignment Analysis	
OTU clustering ...	Cluster operational taxonomic units (sortx)
Phylip distance matrix ...	Calculate distance matrix (dnadist)
Phylip NJ/UPGMA Tree ...	Calculate neighbor joining or UPGMA trees (neighbor)
Phylip parsimony ...	Calculate parsimony tree (dnapars)
Phylip seqboot ...	Generate bootstrap replicates of alignment (seqboot)
Phylip consense ...	Generate consensus of multiple trees (consense)
Clearcut NJ Tree ...	Fast neighbor joining trees (clearcut)
Fasttree ...	Fast maximum likelihood tree (fasttree)
RAxML ...	Generate maximum likelihood tree (raxmlHPC)

[a]Command line executables are listed in parentheses.

47.3.5 Third-Party Software Integration

XplorSeq v2 provides integrated access to several third-party software packages in addition to those available in earlier releases. Both the NAST [DeSantis et al., 2006a,b] and SILVA/SINA [Pruesse et al., 2007] automated rRNA alignment web services can be accessed through the XplorSeq "Analyze" menu. Output from SINA is automatically transformed into fasta format and can be filtered based on alignment quality scores. *Fasttree* [Price et al., 2009] computes robust minimum evolution trees and can be applied to large (e.g., >100,000 taxa) sequence alignments.

47.3.6 XplorSeq Toolkit

Several of the software components underlying XplorSeq are implemented as platform-independent, command-line utilities that can be compiled and used independently of XplorSeq. These programs along with a pair of libraries comprise the XplorSeq Toolkit (XSTK). This software is written primarily in C and C++ and should compile on any computer system for which a GNU compiler is available. The following sections present synopses of the tools in the current XSTK release.

47.3.7 Barcoded Primer design: Barcrawl

An effective means of increasing metagenomic project throughput is by sample multiplexing during the sequencing phase. Samples can be distinguished by tagging input DNAs (genomic DNA or PCR amplicons) with short, sample-specific barcode sequences. A popular strategy is to include barcode sequences in the PCR primers used for primary target gene amplification. *Barcrawl* facilitates the design of barcoded primers [Frank, 2009]. *Barcrawl* constructs a set of barcodes, each separated in sequence space from all other barcodes by a minimum number of base substitutions (default value of three events). Barcodes are evaluated within the context of other sequences contained in the forward and reverse PCR primers in order to cull potentially problematic sequences (e.g., due to inhibitory homopolymers, potential hairpins, or heteroduplex formation between primers). Finally, *barcrawl* can sort the set of barcodes by the number of 454 GS-FLX nucleotide flows required to pyrosequence each barcode sequence. Use of more efficiently sequenced barcodes (i.e., those with minimal flow patterns) should help maximize read lengths within the template regions of amplicons.

barcrawl: BARCODED PRIMER DESIGN SOFTWARE

Input file(s): None

Output file(s):

- List of acceptable barcodes and program settings
- Log file summarizing outputs from multiple barcrawl runs

47.3.8 Deconvolution of Multiplexed Sequencing Output: bartab

To facilitate manipulation of barcoded sequence datasets following multiplexed sequencing, we developed the program *bartab* as a general-purpose tool for polishing and annotating sequence datasets [Frank, 2009]. Inputs to *bartab* are sequence and quality score files (fasta format) along with a tab-delimited file that maps barcode sequences to library metadata. Command line options can be set to rename sequences and split them into multiple files, based on barcode sequences. Metadata specified in the input barcode file are appended to fasta definition lines in the form of *key* = *value* pairs. XplorSeq recognizes this data format and will parse metadata appropriately during fasta file import so that it is available for use by other XplorSeq functions. If sequences are written to multiple, barcode-specific files, the batch fasta import function of XplorSeq can be used to create documents from each output sequence files. Finally, *bartab* also can dereplicate sequence files to create files only of the unique, representative sequences in a dataset. In this case, an additional metadata tag is created that records the number of sequences represented by a particular sequence in the dereplicated dataset. Upon import of a dereplicated file into XplorSeq, this metadata is tracked so that downstream analyses (e.g., *biodiv*) reflect the makeup of the full dataset.

bartab: UTILITY TO DECONVOLUTE AND ANNOTATE BARCODED SEQUENCES

Input file(s):

- Sequences (fasta format)
- Quality scores (fasta format)
- Mapping of barcodes to library/sample names and metadata

Output file(s):

- Accepted sequences (.fa file; fasta format)
- Quality scores for accepted sequences (.qual file; fasta format)
- Rejected sequences (.rej.fa file; fasta format)
- Quality scores of rejected sequences (.rej.qual file; fasta format)
- Log file listing program settings and statistics for run
- If file dereplication was invoked:
    - Representative sequences (.rep.fa; fasta format)
    - Mapping of representative sequence names to other identical sequences (.dec file)
- If file splitting was invoked:
    - Directory of sequences and quality scores, sorted by term set by "-spl" option

47.3.9 Sequence Polishing: xstrim

The output of any automated sequencing platform should be post-processed to separate good from bad reads. The tool *xstrim* provides a number of filtering functions that can be used to cull short sequences—that is, those with high levels of ambiguous basecalls (e.g., multiple "N" basecalls) and/or low-quality scores. Multiple algorithms for trimming poor-quality bases from sequences are provided in order to tailor *xstrim* functionality to the performances of different sequencing platforms. In addition to producing files of acceptable and rejected sequences, *xstrim* compiles basic statistics for the sequencer run under analysis, including distributions of sequence lengths, trimmed sequence lengths, average quality score per read, and number of low-quality bases per sequence. Thus, *xstrim* can be used to monitor and evaluate the qualities of datasets obtained through high-throughput sequencing.

xstrim: TRIMS DNA SEQUENCES BASED ON QUALITY SCORES

Input file(s):

- Sequences (fasta format)
- Quality scores (fasta format)

Output file(s):

- Accepted sequences (.fa file; fasta format)
- Quality scores for accepted sequences (.qual file; fasta format)
- Rejected sequences (.rej.fa file; fasta format)
- Quality scores of rejected sequences (.rej.qual file; fasta format)
- Log file listing program settings and statistics for run
- Summary statistics for sequence dataset (.sum file)

47.3.10 Sequence Clustering: sortx and x2x

Clustering of related sequences into operational taxonomic units (OTUs) is implemented through the program *sortx*. The input for *sortx* consists of a fasta formatted multiple-sequence alignment file. *Sortx* uses a fast radial clustering algorithm to bin aligned sequences based on uncorrected pairwise sequence distances. Clusters can be assembled based on complete-, average-, or single-linkage rules (although single-linkage is available, it is not generally recommended due to its propensity to chain together unrelated clusters). Following cluster formation, *sortx* selects a representative sequence for each OTU, which maximizes both pairwise similarity to other OTU members *and* sequence length (simply choosing the sequence with minimum pairwise distance could select for short, but well-conserved sequences, which would not necessarily be representative of the cluster). The user also can select a range of pairwise sequence distance thresholds by which to assemble OTUs in order to create multiple datasets at different phylogenetic depths.

The outputs of *sortx* consist of the following files: (1) files ("filename.grp") that map sequences to OTUs (one file is produced for each percent identity threshold); (2) a file ("filename.sum") that tabulates the number of OTUs formed at each percent identity threshold); (3) a fasta-formatted file ("filename.rep") of sequences representative of each OTU (fasta deflines are annotated to specify the name of the OTU from which the representative sequence was extracted); and (4) a directory that stores pairwise distances for sequences in the input dataset. The file that lists the OTU to which each sequence was assigned ("filename.grp") can be imported directly into XplorSeq, through the "Import Metadata..." menu option, thereby annotating the sequences with respect to OTUs. Other XplorSeq functions (e.g., *biodiv*) can then use this metadata to analyze sequences organized by OTU. Alternatively, multiple sortx. grp files can be compiled into a single tab-separated-value table via the program *x2x* and used as input into XplorSeq and/or *biodiv*.

sortx: FAST CLUSTERING OF ALIGNED SEQUENCES

Input file(s):

- Multiple sequence alignment (fasta format)

Output file(s):

- Mapping of sequence names to OTUs (.grp file)
- Representative sequences for OTUs (.rep file; fasta format)
- Summary of OTUs formed for each percent identity threshold (.sum file)
- Pairwise distances for each sequence

47.3.11 Biodiversity Measures: biodiv and biodivprep

Estimates of basic biodiversity statistics (species richness, diversity, evenness) are available through the program *biodiv*. Inputs for biodiv are provided by two tab-delimited files that (1) map sequences to OTUs (.grp file) and (2) map sequences to treatment groups (.exp file), for comparisons between subsets of the data. The utility *biodivprep* greatly facilitates preparation and formatting of these input files, although the output from a spreadsheet may also suffice. *Biodiv* performs random resampling of OTUs (bootstraps without replacement) and calculates collector's curves and associated biodiversity indices (S_{obs}, S_{chao1}, C_{ACE}, Good's coverage, Shannon diversity, Shannon evenness, Simpson diversity) as a function of sampling effort [Magurran, 2003]. *Biodiv* also reports rarefied biodiversity indices, based on bootstrap resampling, with 95% confidence intervals for each type of environment [Magurran, 2003]. Results can be imported directly into spreadsheet applications for further data manipulation, graphing, and so on.

biodiv: BOOTSTRAP ESTIMATION OF BIODIVERSITY STATISTICS

Input file(s):

- Mapping of sequence names to OTUs (.grp file)
- Mapping of sequences to treatment groups (.exp file)

Output file(s):

- Collector's curve data for each biodiversity index (e.g., .sobs,.chao,.goods,.shannon files)
- Summary of rarefaction data (.sum file)

47.3.12 Parse rRNA Sequences from GenBank files: minerna

This program extracts rRNA sequences (SSU, LSU, 5S, and 5.8S) from GenBank files. The user can supply a query string to select particular sequences for output. For instance, the string "(env | Env | uncult | Uncult) & (16S | SSU)" selects and outputs all 16S/SSU sequences annotated with terms such as "environmental" or "uncultured." Filters for minimum and maximum acceptable sequence lengths can be set on the command line. Another command line option ("-t") directs *minerna* to write sequence metadata that is available in GenBank-formatted (but not fasta-formatted) records to the fasta definition lines of outputted sequences in XML format. Accession numbers, taxonomic lineages, and submission dates are included in this tagged format if specified. We use fasta-formatted *minerna* output files along with the NCBI *formatdb* utility to create domain-specific BLAST databases.

minerna: EXTRACT rRNA SEQUENCES FROM GenBank FILE

Input file(s):

- Sequences (GenBank format)

Output file(s):

- rRNA sequences (GenBank or fasta format)

47.3.13 Miscellaneous Utilities

Finally, XSTK provides a number of utility programs for reformatting and otherwise manipulating sequence data:

1. *xsub*. Select sub-alignment from multiple sequence alignment, based on input of 5′ and 3′ sequences or positions.
2. *xsel*. Select sub-sequences from unaligned sequences, based on input of 5′ and 3′ sequences or positions.
3. *xsdstcmp*. Calculate pairwise sequence distances between two multiple sequence alignments. For each sequence in the query alignment, report the most similar sequence in the target alignment.
4. *xsgap*. Remove gaps from aligned sequences.
5. *xsrep*. Dereplicate sequence file, based on unique sequences (default) or taxon names. Can annotate output sequences with number of representatives.
6. *si2fa*. Translate SINA output to fasta format. Accept/reject sequences based on quality scores.
7. *fa2fq*. Translate from fasta to fastq format.
8. *fq2fa*. Translate from fastq to fasta format.
9. *fa2nw*. Construct newick-formatted selection tree from taxon names in fasta file. Can be imported into ARB to select taxa in fasta file.
10. *nm2nw*. Construct newick formatted selection tree from list of taxon names. Can be imported into ARB to select taxa.

47.4 CONCLUSIONS

The Phyloware package provides an integrated suite of programs that are designed to facilitate DNA sequence analysis, particularly as it relates to phylogenetic analysis of metagenomic sequences. Although the Phyloware package was developed to expedite batch analysis of ribosomal RNA (rRNA) gene libraries, it should prove useful in any sequencing project. Our goal is to help users focus on their science by alleviating the burden of mastering multiple software packages and computer platforms. Future development efforts will be directed toward increased software parallelization in order to keep pace with the evolution of multicore commodity computers. Because XplorSeq, the GUI component of the package, is coded in a highly modular form, any UNIX-based DNA sequence analysis tool that can be ported to Mac OSX can be incorporated readily into XplorSeq. Suggestions for the addition of other third-party software tools to XplorSeq or novel features to the Phyloware package are most welcome.

Acknowledgments

The authors wish to thank Professor Norman R. Pace for encouragement and support. This work was generously supported by the Butcher Foundation of Colorado and the NIH Human Microbiome Project grant 1UH2DK083994-01.

REFERENCES

ALTSCHUL SF, GISH W, MILLER W, MYERS EW, LIPMAN DJ. 1990. Basic local alignment search tool. *J. Mol. Biol*. **215**:403–410.

ALTSCHUL SF, MADDEN TL, SCHAFFER AA, ZHANG J, ZHANG Z, et al. 1997. Gapped BLAST and PSI-BLAST: A new generation of protein database search programs. *Nucleic Acids Res*. **25**:3389–3402.

BAUMGARTNER LK, REID RP, DUPRAZ C, DECHO AW, BUCKLEY DH, et al. 2006. Sulfate reducing bacteria in microbial mats: Changing paradigms, new discoveries. *Sediment. Geol*. **185**:131–145.

BAUMGARTNER LK, SPEAR JR, BUCKLEY DH, PACE NR, REID RP, et al. 2009. Microbial diversity in modern marine stromatolites, Highborne Cay, Bahamas. *Environ. Microbiol*. **11**:2710–2719.

DALBY AB, FRANK DN, ST AMAND AL, BENDELE AM, PACE NR. 2006. Culture-independent analysis of indomethacin-induced alterations in the rat gastrointestinal microbiota. *Appl. Environ. Microbiol*. **72**:6707–6715.

DESANTIS TZ, HUGENHOLTZ P, LARSEN N, ROJAS M, BRODIE EL, et al. 2006a. Greengenes, a chimera-checked 16S rRNA gene database and workbench compatible with ARB. *Appl. Environ. Microbiol*. **72**:5069–5072.

DESANTIS TZ, Jr., HUGENHOLTZ P, KELLER K, BRODIE EL, LARSEN N, et al. 2006b. NAST: A multiple sequence alignment server for comparative analysis of 16S rRNA genes. *Nucleic Acids Res*. **34**:W394–W399.

FEAZEL LM, SPEAR JR, BERGER AB, HARRIS JK, FRANK DN, et al. 2008. Eucaryotic diversity in a hypersaline microbial mat. *Appl. Environ. Microbiol*. **74**:329–332.

FEAZEL LM, BAUMGARTNER LK, PETERSON KL, FRANK DN, HARRIS JK, PACE NR. 2009. Opportunistic pathogens enriched in showerhead biofilms. *Proc. Natl. Acad. Sci. USA* **106**:16393–16399.

FRANK DN. 2008. XplorSeq: A software environment for integrated management and phylogenetic analysis of metagenomic sequence data. *BMC Bioinform*. **9**:420.

FRANK DN. 2009. BARCRAWL and BARTAB: Software tools for the design and implementation of barcoded primers for highly multiplexed DNA sequencing. *BMC Bioinform*. **10**:362.

FRANK DN, PACE NR. 2008. Gastrointestinal microbiology enters the metagenomics era. *Curr. Opin. Gastroenterol*. **24**:4–10.

FRANK DN, SPIEGELMAN GB, DAVIS W, WAGNER E, LYONS E, PACE NR. 2003. Culture-independent molecular analysis of microbial constituents of the healthy human outer ear. *J. Clin. Microbiol*. **41**:295–303.

Frank DN, St Amand AL, Feldman RA, Boedeker EC, Harpaz N, Pace NR. 2007. Molecular-phylogenetic characterization of microbial community imbalances in human inflammatory bowel diseases. *Proc. Natl. Acad. Sci. USA* **104**:13780–13785.

Frank DN, Wilson SS, St. Amand AL, Pace NR. 2009a. Culture-independent microbiological analysis of Foley urinary catheter biofilms. *PLoS One* **4**:e7811.

Frank DN, Wysocki A, Specht-Glick DD, Rooney A, Feldman RA, et al. 2009b. Microbial diversity in chronic open wounds. *Wound Repair Regen*. **17**:163–172.

Harris JK, De Groote MA, Sagel SD, Zemanick ET, Kapsner R, et al. 2007. Molecular identification of bacteria in bronchoalveolar lavage fluid from children with cystic fibrosis. *Proc. Natl. Acad. Sci. USA* **104**:20529–20533.

Isenbarger TA, Finney M, Rios-Velazquez C, Handelsman J, Ruvkun G. 2008. Miniprimer PCR, a new lens for viewing the microbial world. *Appl. Environ. Microbiol*. **74**:840–849.

Lee L, Tin S, Kelley ST. 2007. Culture-independent analysis of bacterial diversity in a child-care facility. *BMC Microbiol*. **7**:27.

Ley RE, Backhed F, Turnbaugh P, Lozupone CA, Knight RD, Gordon JI. 2005. Obesity alters gut microbial ecology. *Proc. Natl. Acad. Sci. USA* **102**:11070–11075.

Ley RE, Harris JK, Wilcox J, Spear JR, Miller SR, et al. 2006. Unexpected diversity and complexity of the Guerrero Negro hypersaline microbial mat. *Appl. Environ. Microbiol*. **72**:3685–3695.

Ley RE, Hamady M, Lozupone C, Turnbaugh PJ, Ramey RR, et al. 2008. Evolution of mammals and their gut microbes. *Science* **320**:1647–1651.

Magurran AE. 2003. *Measuring Biological Diversity*. Malden, MA: Blackwell.

Mardis ER. 2008. Next-generation DNA sequencing methods. *Annu. Rev. Genomics Hum. Genet*. **9**:387–402.

McManus CJ, Kelley ST. 2005. Molecular survey of aeroplane bacterial contamination. *J. Appl. Microbiol*. **99**:502–508.

Papineau D, Walker JJ, Mojzsis SJ, Pace NR. 2005. Composition and structure of microbial communities from stromatolites of Hamelin Pool in Shark Bay, Western Australia. *Appl. Environ. Microbiol*. **71**:4822–4832.

Peplies J, Kottmann R, Ludwig W, Glockner FO. 2008. A standard operating procedure for phylogenetic inference (SOPPI) using (rRNA) marker genes. *Syst. Appl. Microbiol*. **31**:251–257.

Perkins SD, Mayfield J, Fraser V, Angenent LT. 2009. Potentially pathogenic bacteria in shower water and air of a stem cell transplant unit. *Appl. Environ. Microbiol*. **75**:5363–5372.

Peterson DA, Frank DN, Pace NR, Gordon JI. 2008. Metagenomic approaches for defining the pathogenesis of inflammatory bowel diseases. *Cell Host Microbe* **3**:417–427.

Price MN, Dehal PS, Arkin AP. 2009. FastTree: Computing large minimum evolution trees with profiles instead of a distance matrix. *Mol. Biol. Evol*. **26**:1641–1650.

Pruesse E, Quast C, Knittel K, Fuchs B, Ludwig W, et al. 2007. SILVA: A comprehensive online resource for quality checked and aligned ribosomal RNA sequence data compatible with ARB. *Nucleic Acids Res*. **35**:7188–7196.

Rawls JF, Mahowald MA, Ley RE, Gordon JI. 2006. Reciprocal gut microbiota transplants from zebrafish and mice to germ-free recipients reveal host habitat selection. *Cell* **127**:423–433.

Raymond ES. 2001. *The Cathedral & the Bazaar: Musings on Linux and Open Source by an Accidental Revolutionary*. Sebastopol, CA: O'Reilly Media, Inc.

Robertson CE, Spear JR, Harris JK, Pace NR. 2009. Diversity and stratification of archaea in a hypersaline microbial mat. *Appl. Environ. Microbiol*. **75**:1801–1810.

Sahl JW, Schmidt R, Swanner ED, Mandernack KW, Templeton AS, et al. 2008. Subsurface microbial diversity in deep-granitic-fracture water in Colorado. *Appl. Environ. Microbiol*. **74**:143–152.

Salmassi TM, Walker JJ, Newman DK, Leadbetter JR, Pace NR, Hering JG. 2006. Community and cultivation analysis of arsenite oxidizing biofilms at Hot Creek. *Environ. Microbiol*. **8**:50–59.

Spear JR, Barton HA, Robertson CE, Francis CA, Pace NR. 2007. Microbial community biofabrics in a geothermal mine adit. *Appl. Environ. Microbiol*. **73**:6172–6180.

Spear JR, Walker JJ, McCollom TM, Pace NR. 2005a. Hydrogen and bioenergetics in the Yellowstone geothermal ecosystem. *Proc. Natl. Acad. Sci USA* **102**:2555–2560.

Spear JR, Walker JJ, Pace NR. 2005b. Hydrogen and primary productivity: Inference of biogeochemistry from phylogeny in a geothermal ecosystem. In Inskeep WP, McDermott TR, eds. *Geothermal Biology and Geochemistry in Yellowstone National Park*. Bozeman, MT: Thermal Biology Institute, Montana State University, pp. 113–128.

Spear JR, Walker JJ, Pace NR. 2006. Microbial ecology and energetics in yellowstone hotsprings. *Yellowstone Sci*. **14**:17–24.

Turnbaugh PJ, Ley RE, Mahowald MA, Magrini V, Mardis ER, Gordon JI. 2006. An obesity-associated gut microbiome with increased capacity for energy harvest. *Nature* **444**:1027–1031.

Turnbaugh PJ, Backhed F, Fulton L, Gordon JI. 2008. Diet-induced obesity is linked to marked but reversible alterations in the mouse distal gut microbiome. *Cell Host Microbe* **3**:213–223.

Walker JJ, Pace NR. 2007. Phylogenetic composition of Rocky Mountain endolithic microbial ecosystems. *Appl. Environ. Microbiol*. **73**:3497–3504.

Walker JJ, Spear JR, Pace NR. 2005. Geobiology of a microbial endolithic community in the Yellowstone geothermal environment. *Nature* **434**:1011–1014.

Wheeler DL, Barrett T, Benson DA, Bryant SH, Canese K, et al. 2007. Database resources of the National Center for Biotechnology Information. *Nucleic Acids Res*. **35**:D5–D12.

Chapter 48

MetaSim: A Sequencing Simulator for Genomics and Metagenomics

Daniel C. Richter, Felix Ott, Alexander F. Auch, Ramona Schmid, and Daniel H. Huson

48.1 INTRODUCTION

The recent developments of next-generation sequencing (NGS) technologies opened the flood gates to an extensive amount of sequence data (see Chapter 18, Vol. I). Prior to any biological interpretation of the studied DNA and its features, the set of short sequencing reads has to be processed, aligned, assembled, or classified depending on the type of biological analysis. For example, for a typical whole-genome sequencing project of a single organism, the reads have to be filtered by quality and then assembled to obtain the final genome sequence. Once larger assembled fragments (contigs) are obtained, analyses like gene prediction or motif finding become reasonable. In the case of re-sequencing projects, the reads have to be mapped to a closely related reference genome by efficient short read mapping software [Trapnell and Salzberg, 2009]. A different scenario of processing reads is the analysis of environmental DNA obtained from a community sequencing project. In contrast to single genome studies, the new discipline of metagenomics focuses on the analysis of the microbial diversity found in various habitats, such as ocean [Rusch et al., 2007], soil [Tringe et al., 2005], mines [Tyson et al., 2004; Edwards et al., 2006], or the human microbiome [Turnbaugh et al., 2007; see also Vol. II].

Although high-throughput sequencing technologies promise faster and relatively inexpensive generation of reads, Sanger sequencing still has been used in environmental genome projects [Turnbaugh et al., 2006] to avoid the drawbacks of shorter read lengths. Due to the high complexity of ecological systems, the assembly of sequencing reads is very challenging; that is, the assembly of reads into contigs belonging to only one species fails or is misleading. Commonly, genome assembly approaches are only suitable for environmental sequences under special conditions—for example, in low-complexity populations [Tringe et al., 2005; Pachter, 2007]. It has been shown that it is very difficult to assemble reads from highly diverse ecologic systems [Mavromatis et al., 2007].

Furthermore, the fast and cost-effective generation of sequencing data enables researchers to perform a series of measurements to compare, for instance, the taxonomical composition of samples derived from the same location within several time points under varying environmental conditions [Gilbert et al., 2008]. Such comparative studies again depend on powerful, statistical techniques and analysis tools that are able to deal with the highly variable data [Mavromatis et al., 2007; Mitra et al., 2009; see also Chapter 39, Vol. I]. This overview of different types of read processing presents only an incomplete list of all common analysis strategies. For other examples see Chapters 39–53, Vol. I. But it should give an idea of how the progress of next-generation sequencing technologies is spurring the field of bioinformatic software development.

Regarding metagenomic studies, the data size generated (measured in base pairs) occasionally exceeds common single genome sequencing projects. However, it is striking that the number of specialized software and algorithms for processing environmental sequences is surprisingly low. As a consequence, many studies use the classic methods, software, or web services that originally were not intended for metagenomic data.

Handbook of Molecular Microbial Ecology, Volume I: Metagenomics and Complementary Approaches, First Edition. Edited by Frans J. de Bruijn.

To sum it up, there is a great demand for improved and specialized software solutions in both research fields of genomics and metagenomics that keep up with the rapid developments and improvements of NGS technologies. The vast collection of current assembly and mapping software for genomics and the upcoming tools and methods for metagenomics need to be compared and benchmarked to evaluate their performance and applicability. Standardized test scenarios using simulated and verifiable data are useful for developers to analyze their programs and for users to select the software that optimally fits their needs. These considerations motivated us to develop MetaSim, a DNA sequencing simulator software for the generation of synthetic reads based on given genome sequences. A prior study [Mavromatis et al., 2007] provided three datasets with varying complexity by selecting original sequence reads from 113 isolated genomes. The authors anticipated that the community uses these precomputed datasets as standard test cases for software testing.

In contrast to this "static" approach, our software MetaSim allows researchers to create their own test data by choosing from a set of source genomes and error models derived from several sequencing technologies (Sanger [Sanger et al., 1977], Roche's 454 [Margulies et al., 2005], and Illumina [Bentley, 2006]; see also Chapter 18, Vol. I). These error models (or error probabilities per base position) are used to modify the original sequence at certain base positions to reflect real sequencing error patterns of each technology. Furthermore, MetaSim offers the possibility to load individual error models based on empirical data. MetaSim takes as input a set of known DNA sequences and an abundance profile. The profile determines the source DNA sequences and their relative abundances for the simulation of read sequences. The abundance values can be used to reflect the variable species composition of a metagenome. This general approach allows us to use MetaSim flexibly as read simulator for single genomes or for metagenomes.

48.2 METHODS

For a full outline of this section, please refer to the original publication [Richter et al., 2008].

The main steps of MetaSim's simulation processing pipeline are the selection of source genome sequences, the configuration of an abundance profile, the sampling of sequence fragments, and the subsequent generation of synthetic reads according to a chosen error model.

48.2.1 Configuration of Species Profiles

At the beginning, whole genome sequences available from public databases can be stored locally as source sequences in an internal database (based on http://hsqldb.org). The user specifies the relative abundance of each genome sequence in a text-based profile file. An interesting feature is the possibility to set abundance values not only to species (leaf) nodes (single organisms in the database) but also to inner nodes in the taxonomy (e.g., at genus level). The abundance value of an inner node is split and applied to its descendant species which are available from the database. Alternatively, MetaSim provides a visual component that allows us to set the abundance values for genome sequences directly in a taxonomic tree. Therefore, an "induced" tree viewer (TaxEditor) of the NCBI taxonomy [Wheeler et al., 2008] is integrated that displays the genomes in the database as nodes in a rooted tree according to their taxonomical relationships (**Fig. 48.1**).

48.2.2 Population Sampler

Typical environmental samples contain a vast variety of microbial species. Most of these organisms (or bacterial strains) are usually still unknown because they could not be cultured and isolated in the laboratory before. As a consequence, one can hardly estimate their occurrence and, in general, the genetic diversity of a sample. When it comes to simulating metagenomes, one has to keep that in mind. Thus, to mimic the complexity of real-world datasets, MetaSim provides a population sampler that generates evolved (mutated) offsprings of single-source genome sequences. The calculation is based on a mathematical model of DNA evolution and a given evolutionary tree that determines how the offsprings descend from the source genome. By default, we use the Yule–Harding model to generate phylogenetic trees [Yule, 1925; Harding, 1971], but the user may load individual trees as well. For the model of DNA evolution, the widely known Jukes–Cantor model [Jukes and Cantor, 1969] has been implemented. It defines the probability of a change for each base pair, with an adjustable transition rate α (0.001 by default) and time t based on the edge weights of the tree. After applying the population sampler to a genome sequence, the desired number of evolved genomes are added to the internal database.

48.2.3 Read Sampling

MetaSim uses different statistical models to simulate the frequency of simulated reads, the distribution of the read lengths, and the probability of occurring mate pairs. First, larger fragments called clones are extracted from the set of source genomes with normally or uniformly distributed lengths. These clones are the basis for either the read or mate-pair sampling. If only a single genome sequence is included in a profile file, the clones are sampled randomly from this genome sequence. In contrast, a metagenome

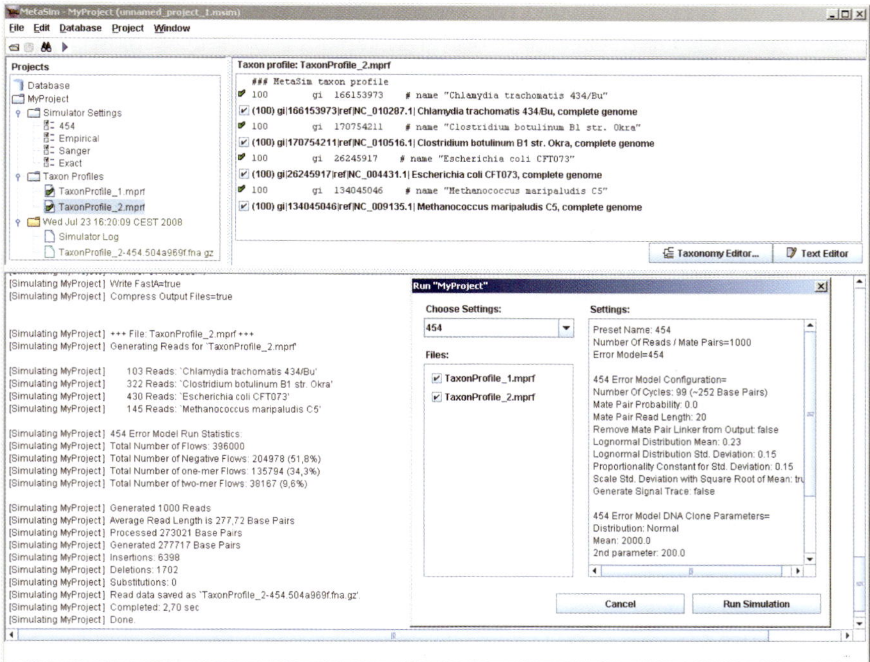

Figure 48.1 A clipping of the taxonomy editor view is shown. Three taxa are assigned an abundance value (number in parentheses). These settings can be either determined in a text-based abundance profile file or directly in the taxonomy editor by right-clicking on a node. (Taken from Richter et al. [2008].)

consists of many genomes with different lengths and assigned abundances, and therefore the clones have to be sampled from many sequences. So, each genome sequence s is assigned a weight $w_s = l_s \times c_s \times a_s$, where l_s is the length, c_s is the copy number, and a_s is the specified relative abundance of the genome sequence s as determined in the profile file. The copy number can be used, for instance, to model the abundance of plasmids versus the organism genomes. For each length of the clone length distribution, the weights of all sequences are summed up to obtain the summarized weight, w_{sum}, that is needed to compute a sequence probability $p_s = w_s/w_{\text{sum}}$. Considering the overall lengths distribution, a frequency value for each sequence is then obtained. After the clone sampling, the ends of the clones are the basis for the subsequent sampling of the reads or mate pairs, respectively. Again, read lengths can be either uniformly or normally distributed. Finally, read sequences are processed and modified by applying the selected error model.

48.2.4 Simulation of Sanger Sequencing

A widely used approach to sequencing large DNA molecules is the Sanger technology, using a shotgun approach that involves cloning small pieces (or inserts) of DNA and then determining their sequence using fluorescent dideoxynucleotides for termination and capillary electrophoresis. MetaSim uses a similar approach as reported in Myers [1999]. The empirical observation is that the base quality decreases toward the end of the read. The error rates for insertion, deletions, and substitutions are fixed values, whereas the general error rate per base position at the beginning of a read is lower than the rate at the read end. The read length is distributed either normally or uniformly. Optionally, mate pairs can be simulated with a certain probability.

48.2.5 Simulation of Sequencing by Synthesis

The main concepts of the 454 sequencing system are reported in [Margulies et al., 2005; see also Chapter 18, Vol. I]. Due to chemical and technical issues, the signal is subject to fluctuation that may lead to sequencing errors. MetaSim's pyrosequencing read simulator is based on data published in 2005. To that time, an average error rate of 3% was reported [Margulies et al., 2005; see also Chapter 19, Vol. I]. Our strategy to simulate 454 reads is to model the process of light emission and the detection of the observed base sequence in the base-calling procedure (for details see Richter et al. [2008]). Given a source read sequence, for each simulated flow of single nucleotides, all its homopolymers are extracted. In a second step, the homopolymer lengths are converted into virtual light emissions using a normal distribution. During the base-calling, the algorithm then calculates which probable length is to set for the observed homopolymer length. The optional generation of 454 mate pairs follows the protocol reported in Korbel et al. [2007]: two read sequences are generated, a fixed linker sequence is concatenated connecting both reads. Finally, the error simulator produces a synthetic mate pair.

48.2.6 Simulation of Reads Using Empirical Error Models

The characterization of empirical error rates of DNA sequencing technologies relies on the divergence between the observed and the expected nucleotide at specific base positions. In other words, the error rate probability is not directly determined by the specifications of the sequencer (e.g., light detection) but rather by the analysis of alignment experiments (empirical data). Such an experiment to detect possible base indels or substitution may be, for instance, a re-sequencing project: Reads are mapped to an already completed reference genome to detect sequence variations. To enable the import of such statistics, a new file format was created that enables the definition of individual error probabilities for deletions, insertions, or substitutions per base position adopting the general approach of the program GenFrac [Engle and Burks, 1994]. An empirical error model is based on several mappings, each consisting of three parameters: type of error, base at the position where the error occurs, and base preceding the position where the error occurs. In this way, 48 independent mappings can be individually determined. MetaSim currently provides error models for 36 bp, 62 bp, and 80 bp (paired-end) reads. By adding or removing error mappings for specific base positions, users may create error models for other read lengths.

48.3 RESULTS AND DISCUSSION

MetaSim is a Java program and installers for different operating systems are available. Besides the interactive graphical user interface **(Fig. 48.2)**, MetaSim can be controlled via command-line for automatic simulation runs. To perform a simulation, the user initially creates an abundance profile and then chooses one of the four preconfigured simulator settings (Sanger, 454, Illumina, exact reads). The simulator settings are all adjustable (e.g., number of reads, length of clones, probability of mate pairs, etc.). For each successful run, a new result folder is added to the current simulation project containing a log file and a preview of the final (optionally compressed) multiFASTA file. A typical simulation run that generates 100 Mbp of sequence (e.g., 400,000 454 reads of length 250 bp) takes less than 80 seconds on a single-processor computer.

In the original publication [Richter et al., 2008], the results of a simulation study using MetaSim are presented (see Chapter 48, Vol. I). The software was used to benchmark the performance of the MEGAN software [Huson et al., 2007; see also Chapter 39, Vol. I] that classifies environmental read sequences taxonomically.

48.4 SUMMARY

The MetaSim software fills the gap of missing simulation software for applications in genomics (regarding genome assembly and resequencing projects) and metagenomics. New technical improvements and innovative, pioneering studies in both research fields still spur software developers to implement new tools and algorithms. The testing of applications using simulated, and therefore verifiable datasets facilitate the comparison of different software tools. The overall success of a software tool can be measured by its adaptability and applicability to new or changed conditions and specifications. Besides the built-in error models for the Sanger and Roche's 454 sequencing technologies, MetaSim provides functionality for simulations based on empirical data. A comprehensive set of error mappings allows for the individual design of error models independent of the sequencer technology

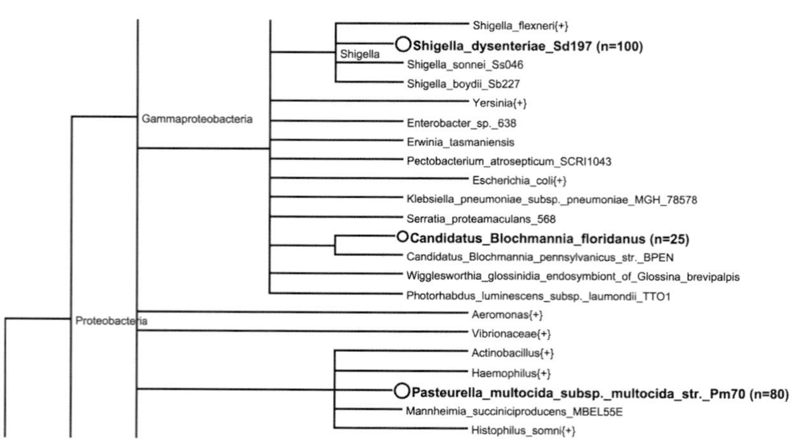

Figure 48.2 The graphical user interface of MetaSim is divided into three panels: a project tree on the left containing all simulation settings and axon profiles, an overview and edit panel on the right, and a message panel at the bottom. Additionally, a configuration window is shown. (Taken from Richter et al. [2008].)

or read lengths. MetaSim is able to integrate and provide simulation settings for any upcoming sequencing technology that will be available in the foreseeable future.

INTERNET RESOURCES

MetaSim Download: http://www-ab.informatik.uni-tuebingen.de/software/metasim

NCBI Taxonomy: http://www.ncbi.nlm.nih.gov/Taxonomy

Acknowledgments

We would like to thank Korbinian Schneeberger and Stephan Ossowski for providing the empirical error models for the Illumina sequencing.

REFERENCES

Bentley DR. 2006. Whole-genome re-sequencing. *Curr. Opin. Genet. Dev*. 16:545–552.

Edwards RA, Rodriguez-Brito B, Wegley L, Haynes M, Breitbart M, et al. 2006. Using pyrosequencing to shed light on deep mine microbial ecology. *BMC Genomics* 7:57.

Engle ML, Burks C. 1994. GenFrag 2.1: New features for more robust fragment assembly benchmarks. *Comput. Appl. Biosci*. 10: 567–568.

Gilbert JA, Field D, Huang Y, Edwards R, Li W, et al. 2008. Detection of large numbers of novel sequences in the metatranscriptomes of complex marine microbial communities. *PLoS ONE* 3(8): e3042.

Harding EF. 1971. The probabilities of rooted tree-shapes generated by random bifurcation. *Adv. Appl. Prob*. 3:44–77.

Huson DH, Auch AF, Qi J, Schuster SC. 2007. MEGAN analysis of metagenomic data. *Genome Res*. 17:377–386.

Jukes TH, Cantor CR (1969) Evolution of Protein Molecules. In Munro HN, ed. *Mammalian Protein Metabolism*. New York: Academic Press, pp. 21–132.

Korbel JO, Urban AE, Aourtit JP, et al. 2007. Paired-end mapping reveals extensive structural variation in the human genome. *Science* 318(5849):420–426.

Margulies M, Egholm M, Altman WE, Attiya S, Bader JS, et al. 2005. Genome sequencing in microfabricated high-density picolitre reactors. *Nature* 437:376–380.

Mavromatis K, Ivanova N, Barry K, et al. 2007. Use of simulated data sets to evaluate the fidelity of metagenomic processing methods. *Nat. Methods* 4(6):495–500.

Myers G. 1999. A dataset generator for whole genome shotgun sequencing. *Proc. Int. Conf. Intell. Syst. Mol. Biol*. 202–210.

Mitra S, Klar B, Huson DH, 2009. Visual and statistical comparison of metagenomes. *Bioinformatics* 25(15):1849–1855.

Pachter L. 2007. Interpreting the unculturable majority. *Nat. Methods* 4:479–480.

Richter DC, Ott F, Auch AF, Schmid R, Huson DH. 2008. MetaSim—A sequencing simulator for genomics and metagenomics. *PLoS ONE* 3(10):e3373.

Rusch DB, Halpern AL, Sutton G, et al. 2007. The Sorcerer II Global Ocean Sampling Expedition: Northwest Atlantic through eastern tropical Pacific. *PLoS Biol*. 5(3):e77.

Sanger F, Nicklen S, Coulson AR. 1977. DNA sequencing with chainterminating inhibitors. *Proc. Natl. Acad. Sci. USA* 74(12):5463–5467.

Trapnell C, Salzberg SL. 2009. How to map billions of short reads onto genomes. *Nat. Biotechnol*. 27(5):455–457.

Tringe SG, von Mering C, Kobayashi A, et al. 2005. Comparative metagenomics of microbial communities. *Science* 308:554–557.

Turnbaugh PJ, Ley RE, Mahowald MA, Magrini V, Mardis ER, et al. 2006. An obesity-associated gut microbiome with increased capacity for energy harvest. *Nature* 444(7122):1027–1031.

Turnbaugh PJ, Ley RE, Hamady M, Fraser-Liggett CM, Knight R, et al. 2007. The human microbiome project. *Nature* 449(7164):804–810.

Tyson GW, Chapman J, Hugenholtz P, Allen EE, Ram RJ, et al. 2004. Community structure and metabolism through reconstruction of microbial genomes from the environment. *Nature* 428(6978): 37–43.

Wheeler DL, Barrett T, Benson DA, et al. 2008. Database resources of the National Center for Biotechnology Information. *Nucleic Acids Res*. 36(Database issue):D13–D21.

Yule GU. 1925. A mathematical theory of evolution, based on the conclusions of Dr. J.C. Willis. *Philos. Trans. R. Soc. Lond. Ser B Biol. Sci*. 213:21–87.

Chapter 49

ClustScan: An Integrated Program Package for the Detection and Semiautomatic Annotation of Secondary Metabolite Clusters in Genomic and Metagenomic DNA Datasets

JOHN CULLUM, ANTONIO STARCEVIC, JANKO DIMINIC, JURICA ZUCKO, PAUL F. LONG, AND DASLAV HRANUELI

49.1 INTRODUCTION

Secondary metabolites are a very important source of chemical diversity for commercial exploitation by the pharmaceutical, food, and agrochemical industries; they are also fascinating for evolutionary and ecological studies [Jenke-Kodama and Dittmann, 2009]. Secondary metabolites are produced by most classes of organisms, but are particularly prevalent in microorganisms [Demain, 2009]. An important general principle that has been borne out by numerous publications is that the genes for the biosynthesis of a particular secondary metabolite almost invariably occur contiguously in the DNA of the producer organism; that is, they form a cluster (e.g., reviews from Hopwood and Chater [1980] and Nett et al. [2009]). This greatly simplifies the detection and analysis of the genes, and it means that clusters can be dealt with as units of evolution.

Traditionally, secondary metabolites have been discovered by screening to detect biological activity in a test system [Demain and Sanchez, 2009]. Once an interesting activity is found, the compound is purified in large enough quantities for chemical identification. This can be a very labor-intensive and expensive process, and with the compounds isolated often turning out to be known compounds or closely related to known compounds. There are also several other reasons for believing that this approach will fail to detect many interesting or novel metabolites. The physiology of secondary metabolite production is complex, and the media used may not be suitable for producing a particular compound in detectable amounts. An even more serious problem is that it is necessary to use an organism that can be cultured in the laboratory, and it is estimated that less than 1% of bacteria in most environmental samples are cultivable [Dunlap et al., 2006]. This has led to a lot of interest in detecting secondary metabolite clusters in DNA sequences, including genome sequences. The rapidly falling price of genome sequencing is leading to an explosive growth in the number of genome sequences of bacteria (at least 912 complete genomes and 2123 in progress) and fungi (at least 10 complete genomes and 201 in progress) [NCBI Genome, December 9, 2009

Handbook of Molecular Microbial Ecology, Volume I: Metagenomics and Complementary Approaches, First Edition. Edited by Frans J. de Bruijn.

Table 49.1 Internet Resources Used

Internet Resources	URLs
National Center for Biotechnology Information, ENTREZ Genome Project	http://www.ncbi.nlm.nih.gov/
Perl Foundation	http://www.perl.org/
Extensible Markup Language (XML) 1.0 (Fifth Edition)	http://www.w3.org/TR/xml/
Sun Microsystems, Inc.	http://java.sun.com/
Jupitermedia Corporation	http://javascript.internet.com/
Jmol: an open-source Java viewer for chemical structures in 3D	http://www.jmol.org/
Daylight Chemical Information Systems, Inc.	http://www.daylight.com/dayhtml/doc/theory/theory.smarts.html

(see **Table 49.1**)]. There are also many metagenomic datasets being produced, which also offer the chance of examining the secondary metabolite potential of noncultivable organisms from diverse ecological niches (see Li [2009] and references therein; see Chapter 44, Vol. I). It is relatively easy to recognize secondary metabolite clusters using bioinformatics methods such as similarity searches (e.g., BLAST [Altschul et al., 1990]). However, it is a much more challenging problem to make predictions about the chemical structure of the product.

The program package *ClustScan* ("*Clust*er *Scan*ner") was developed to help prediction of chemical structures of secondary metabolites [Starcevic et al., 2008], particularly of modular biosynthetic clusters: polyketide synthases (PKS) and nonribosomal peptide synthetases (NRPS). Modular PKS clusters are responsible for the synthesis of many commercially important products, including the antibiotic erythromycin, the antiparasitic avermectin, and the immunosuppressant rapamycin [Hranueli et al., 2005]. The biosynthesis consists of a series of addition steps catalyzed by individual enzymic domains that can be grouped into modules on a polypeptide; this is illustrated by the well-known biosynthesis pathway for erythromycin [Khosla et al., 2007] (**Fig. 49.1**). The sequences of the individual modules can be recognized in the DNA and, in principle, the nature of the added unit and, thus, the chemistry of the product should be predictable. The catalytic domains that determine the chemistry of the addition reaction include (Fig. 49.1) an acyl transferase domain (AT) that is responsible for substrate selection. The substrate is an acyl-CoA ester, and AT domains are usually specific for a single substrate. The most common substrates are malonyl-CoA and methylmalonyl-CoA, but there are at least eight known substrates [Chan et al., 2009]. The AT domain transfers the acyl substrate for the next synthesis step to the acyl carrier protein (ACP) domain. The growing polyketide chain, which was synthesized by the previous step, is present on the ACP domain of the previous module. The ketosynthase domain (KS) transfers this chain from the previous ACP domain to the new substrate molecule with elimination of CO_2. The KS, AT, and ACP domains are the minimal unit for an extender module and result in a keto group in the final product (e.g., erythromycin module 3, Fig. 49.1); this keto group is derived from the acyl substrate introduced by the previous module. However, there are often additional reduction domains present in a module that reduce the keto group from the preceding module. If only a ketoreductase domain (KR) is present, the keto group is reduced to a hydroxyl group (e.g., erythromycin module 2, Fig. 49.1). If a dehydratase domain (DH) is present, the hydroxyl group is further reduced to a double bond. If there is also an enoyl reductase domain (ER), there is full reduction (e.g., erythromycin module 4, Fig. 49.1). Thus, there are a large number of different module types using different acyl-CoA substrates, as well as having different degrees of reduction. In addition, there may be different possible stereochemical outcomes. An important domain for the control of stereochemistry is the ketoreductase domain (see Starcevic et al. [2007] and references therein). It controls not only the stereochemistry of the hydroxyl group (if present), but also that of the α-carbon atom. Further stereochemical choices occur if a double bond is present (*cis*–*trans*), or if full reduction of the double bond occurs by the enoylreduction domain [Kwan et al., 2008]. The chemical diversity that can be generated by the large number of choices helps account for the significance of modular PKS for the production of commercially important compounds. The diversity in the extender modules is supplemented by the large number of starter molecules (at least a dozen naturally occurring substrates with many more possible through precursor directed biosynthesis [Moore and Hertweck, 2002] that can be used. These are incorporated by special starter modules (often called "loading domains"). At the end of synthesis the polyketide chain is released from the synthase complex; this is often achieved by a thioesterase domain (TE). The chain usually undergoes chemical rearrangements to produce ring structures (Fig. 49.1).

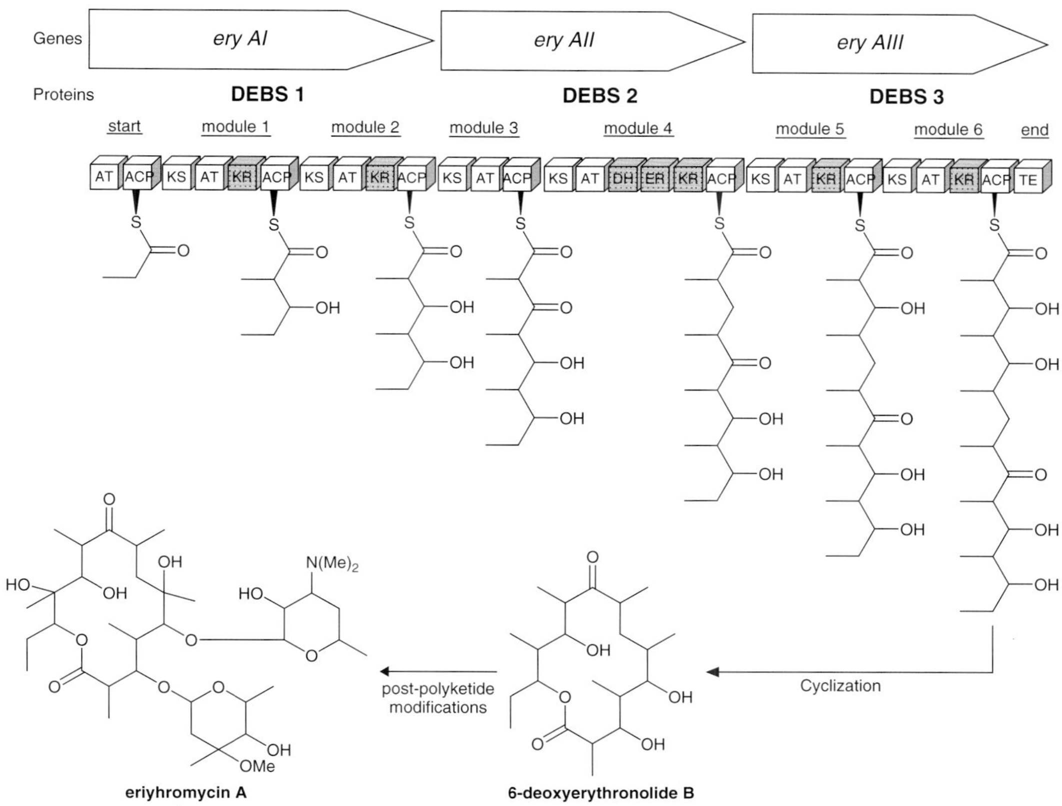

Figure 49.1 Biosynthesis of erythromycin. The starter module and the six extender modules are distributed over three polypeptides DEBS1-DEBS3 encoded by the three genes *eryAI-III*. In this case, each extender module uses methylmalonyl-CoA as a substrate resulting in incorporation of a C3 unit. The different modules differ in reduction domains ranging from module 3 with no reduction domains resulting in a keto group to module 4, where the presence of all three reduction domains results in complete reduction. Release of the polyketide chain by the TE domain results in cyclization. There are subsequent chemical modifications including hydroxylation and glycosylation to produce the final bioactive product.

Modular NRPS clusters have a similar architecture [Koglin and Walsh, 2009], but incorporate amino acids instead of acyl groups. The minimal extender module contains an adenylation domain (A), which is responsible for substrate selection, a peptidyl carrier protein domain (PCP), to which the new substrate is coupled, and a condensation domain (C), which transfers the growing chain from the PCP domain of the previous module to the newly incorporated substrate. Some modules have additional domains that modify the incorporated amino acid (e.g., epimerase or methylation domains). PKS and NRPS modules not only have similar architectures, but can also interact so that mixed clusters containing both sorts of modules are possible [Du et al., 2001].

The aim of *ClustScan* was to predict the chemistry of products from DNA sequence. The PKS literature includes much work on the substrate specificity of AT domains and on determinants for activity and stereochemistry of KR domains (see Starcevic et al. [2008] and references therein), yet predicting biochemical outcomes with any degree of certainty still remains elusive. There is also another problem. The modules have a collinear arrangement, spanning several polypeptides. The ends of these polypeptides ("dockers") interact specifically to ensure that the correct biosynthetic order is maintained between polypeptides [Weissman, 2006]. Predicting docker interaction is not yet possible, so the biosynthetic order is not clear from the DNA sequence. There are also many exceptions to the simple general picture given above, particularly within myxobacteria [Weissman and Müller, 2009]. For instance, modules are known that are either skipped or used more than once in successive biosynthetic steps, and extender modules are known that use external (*trans*) AT proteins that are not part of the module [Nguyen et al., 2008]. These uncertainties mean that a fully automatic prediction program is unrealistic and *ClustScan* is a semiautomatic program that gives the user the ability to examine the basis for assignment of functions and to override the default predictions of the program if necessary.

49.2 MATERIALS AND METHODS

This is an abbreviated version of the materials and methods used. For a full outline of the materials and methods, please refer to the original publication [Starcevic et al., 2008]. The server part of *ClustScan* uses Perl (Table 49.1). The HMMER package [Eddy, 1998] is used for identifying domains and GeneMark-PS [Besemer and Borodovsky, 2005], and Glimmer [Delcher et al., 2007] are used to identify protein-coding genes. Each user has a password-protected workspace. A special XML (Table 49.1) format was developed to coordinate information about DNA sequence, genes, modules, domains, and biosynthetic order and transfers this information between server and client. The predicted chemical structure is generated by parsing the XML description. The client is written in Java and JavaScript (Table 49.1) with versions for PC, Mac and Linux operating systems. The Jmol window (see Table 49.1) or ChemAxon [Csizmadia, 2000] can be used to display chemical structures. All Internet resources used are shown in Table 49.1.

49.3 RESULTS

49.3.1 Initial Steps of Analysis

ClustScan has a client–server architecture. The major analysis is carried out on the server, and the user can view and manipulate the results using the client Java graphical interface, which has versions for PC, Mac, and Linux operating systems. The data and results of analyses are held in a workspace, which can be saved and loaded for further work. A workspace can be used to analyze a single cluster or a larger sequence such as a genome sequence or metagenomic data. The starting point for any analysis is the uploading of the DNA sequence to be analyzed, which can be in different standard formats (FASTA, GenBank, EMBL). Once the sequence has been uploaded to the workspace, it can be analyzed. The most important analysis tool uses Hidden Markov Model (HMM) profiles to locate domains. The program automatically translates the DNA sequence in all six reading frames and carries out the analysis using the HMMER program package [Eddy, 1998]. It is possible to use any profiles for this analysis such as profiles from the Pfam database [Bateman et al., 2002] or profiles constructed by the user. However, for analysis of modular clusters, it is usual to use a standard set of custom profiles implemented in *ClustScan* to detect PKS and NRPS domains. This is illustrated with a 203 kb contig (AACY020563593) from the J. Craig Venter Institute Global Ocean Sampling (GOS) Expedition [Rusch et al., 2007]. After uploading the DNA sequence and carrying out the HMMER analysis, the results can be viewed in the annotation editor window (**Fig. 49.2**), which shows the location of domains in a graphical form for all six reading frames. In this case, there is a clustering of PKS and NRPS domains over a region of about 50 kb, which allows easy recognition of a potential modular biosynthetic cluster. The zoom and movement functions allow efficient navigation of long sequences.

The HMMER program package produces a bit score and a statistic (the expected value) that shows how many hits would be expected by chance. It also gives the alignment of the analyzed sequence with the profile. The profiles of the domains are built from well-known clusters and, especially for metagenomic data, it is possible that clusters are present that deviate significantly from the profile. *ClustScan* sets a default cutoff that is very relaxed so that there will be many false positives. In Fig. 49.2, it can be seen that in addition to the cluster of domains between coordinates 150000 and 200000 there are also isolated domains. These are false positives from the point of view of finding modular clusters, but may well be genes encoding biochemical functions related to those of modular biosynthesis domains. The user can decide whether to accept or reject domains. One criterion is whether a group of domains is present that is typical of a module; thus, if a KS-AT-ACP sequence is seen, it would indicate that the hits are valid. A further help is obtained by clicking on the potential domain to open the details window (**Fig. 49.3A** and 49.3B). This shows the score and E-value as well as the alignment from the HMMER analysis.

In some cases a domain may be split. This can arise in eukaryotic sequences due to the presence of introns or in bacterial sequences due to sequencing errors giving an apparent frameshift. *ClustScan* uses local HMM profiles so that splitting of domains is efficiently detected. An example (**Fig. 49.4**) is shown from genomic DNA of the eukaryotic slime mold *Dictyostelium discoideum* [Zucko et al., 2007], where such analyses helped define the intron positions in PKS genes. This sort of analysis is particularly useful for metagenomic datasets, which may include exotic eukaryotic species with unusual intron splice sequences.

The second major analysis that is available is gene detection. *ClustScan* uses the Markov model-based programs GeneMark-PS [Besemer and Borodovsky, 2005] and Glimmer [Delcher et al., 2007] to find genes in prokaryotic DNA. These programs use models derived from training data. It is possible to use preexisting models, but this presupposes some information on the taxonomic position of the organism to select the correct model. Glimmer can also construct a model on the basis of the input DNA sequence; however, this requires a fairly long sequence to give good results. The location of

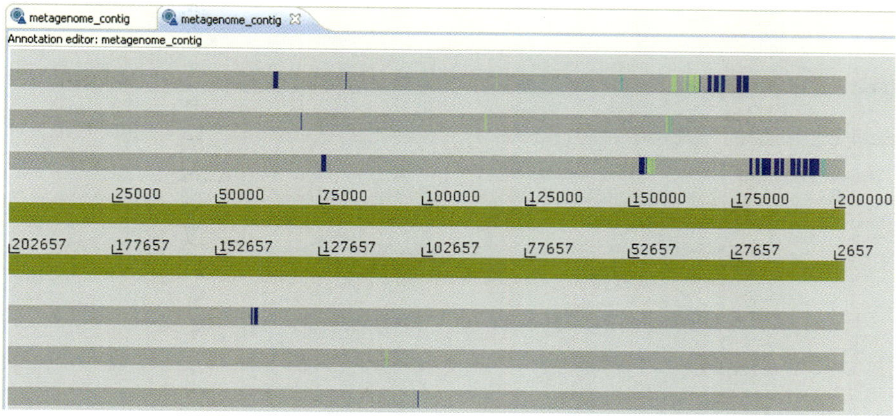

Figure 49.2 Searching for clusters in a 203 kb contig of marine metagenomic origin (AACY020563593). A HMMER analysis was run with PKS domains (green) and NRPS domains (blue). The annotation editor window allows an overview of the hits in all six reading frames.

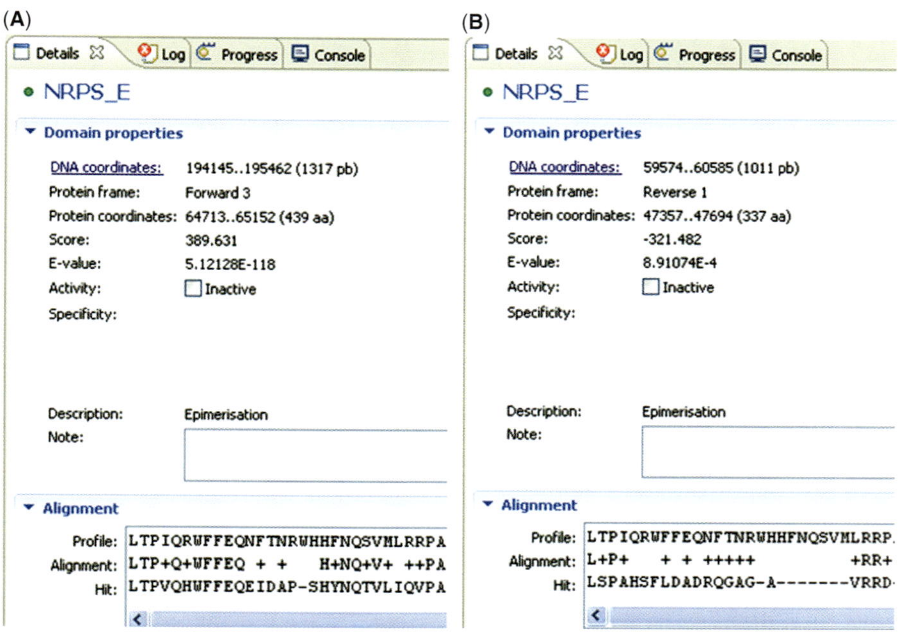

Figure 49.3 Using the details window to examine the quality of hits. (**A**) An epimerase domain from the suspected cluster with a good hit. (**B**) An isolated epimerase domain with a low score that is unlikely to be an NRPS synthethase domain.

genes is shown in the workspace window as well as in the annotation editor window, where it can be compared with the location of domains.

49.3.2 Definition of Clusters

The location of clusters can be seen in the annotation editor window as a collection of appropriate domains. The user can use this information to define a cluster as a group of contiguous genes. Once the cluster has been defined, the cluster editor window (**Fig. 49.5**) becomes available, which shows the cluster in a cartoon form. The user can then perform a semiautomatic annotation of the cluster. A set of domains is selected and used to define a module. User control at this point is important, because modules can deviate from the typical KS-AT-reduction domains-ACP structure. In the case of the metagenomic cluster (Fig. 49.5) a PKS module is shown that appears to lack an AT domain. This suggests that it may be a *trans*-AT module, where the AT function is provided by a separate polypeptide. The cluster editor can be used to define the biosynthetic order of the genes. If a starter module has been detected, the polypeptide containing it will be first; and if a TE domain is detected in a polypeptide, it will be the last. The default is that polypeptides are used in the same order as in the DNA, which corresponds to most known clusters. However, the user can change the order if desired.

In addition to finding domains, *ClustScan* also makes predictions about the functions of domains. In the case of AT domains, the program uses sets of fingerprints to specific amino acids of the alignment to predict the substrate specificity. This is shown in the details window (see Fig. 49.3) and the user can override the prediction

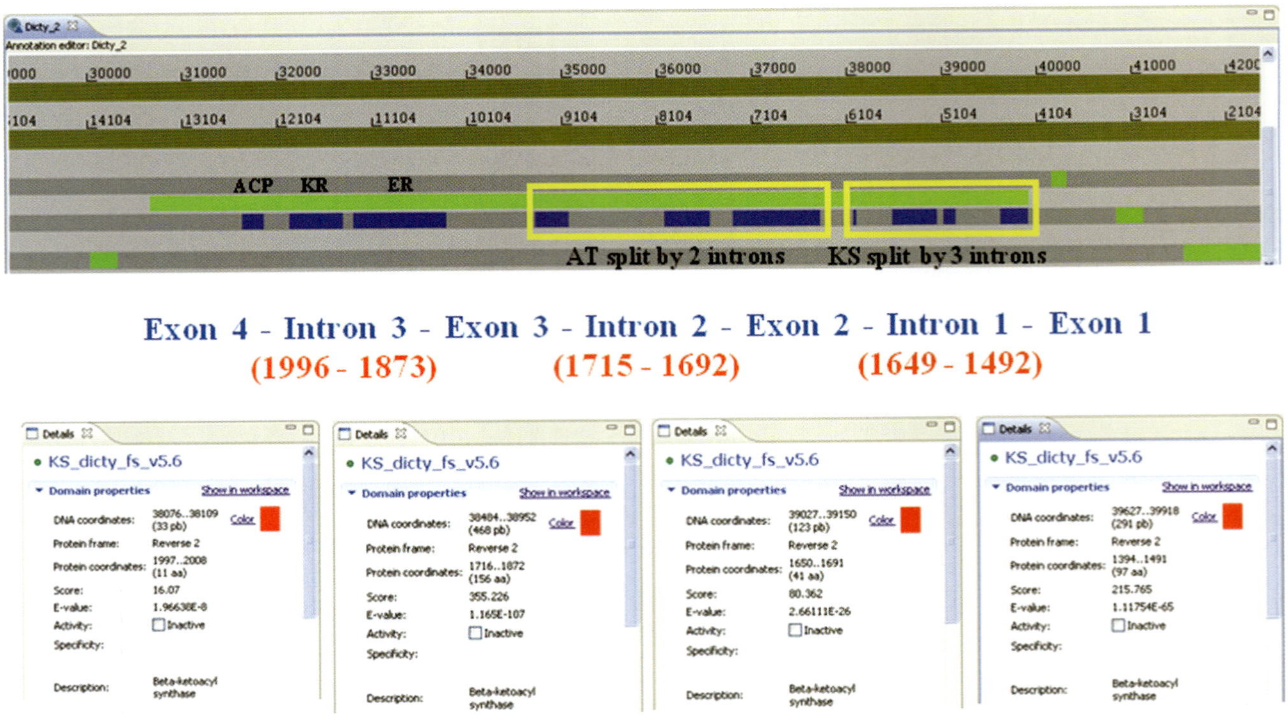

Figure 49.4 Use of *ClustScan* to recognize introns. Analysis of part of the genome sequence of the slime mold *Dictyostelium discoideum* is shown. As local profiles are used, domains that are split by introns give several hits to the profile. Use of the details windows allows the user to check that the hits are to successive parts of the profile and to identify the ends of the introns to within a few base pairs.

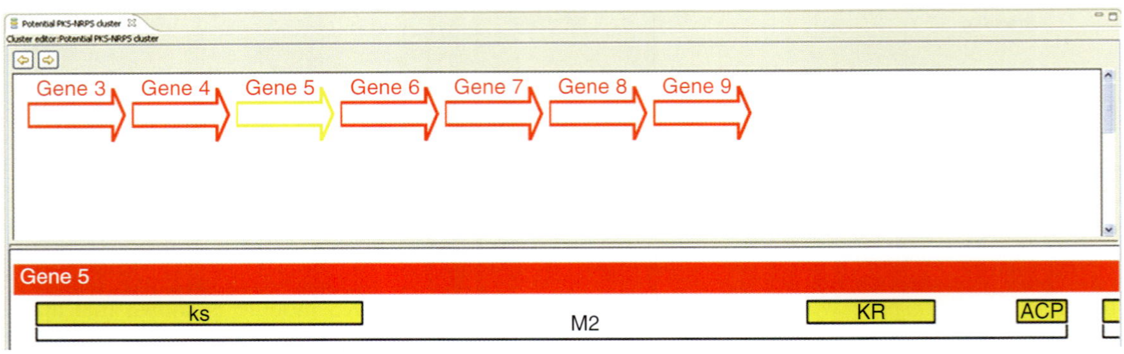

Figure 49.5 The cluster editor window during analysis of the metagenomic sequences shown in Figure 49.2. Modules can be defined by selecting sets of domains. The top part of the window shows an overview of the genes present, and the bottom part shows the domain and module structures of a selected gene. The PKS module shown lacks an AT domain, which would be expected to be present between the KS and KR domains.

if desired. For the KR domains, the program uses fingerprints to classify them as either active or inactive and can predict the stereochemistry of the α-carbon atom and hydroxyl group if relevant. The program also predicts whether the DH and ER domains are active. It is quite common in known clusters for inactive reduction domains to be present.

A graphical overview of the annotated cluster is given in the annotation editor window, and there is also an overview in a collapsible tree form in the workspace window (**Fig. 49.6**). In many metagenomic datasets, sequencing errors are common and complicate the analysis. The cluster sequence shown in Fig. 49.6 probably has sequencing errors, which would explain some anomalies; however, it is also possible that the cluster is nonfunctional because of mutation. It is a mixed PKS-NRPS cluster with an NRPS loading module in gene 1. This is followed by a PKS domain (KS), but there is no ACP domain. However, gene 2 has no KS domain, but an ACP domain. This suggests that a sequencing error

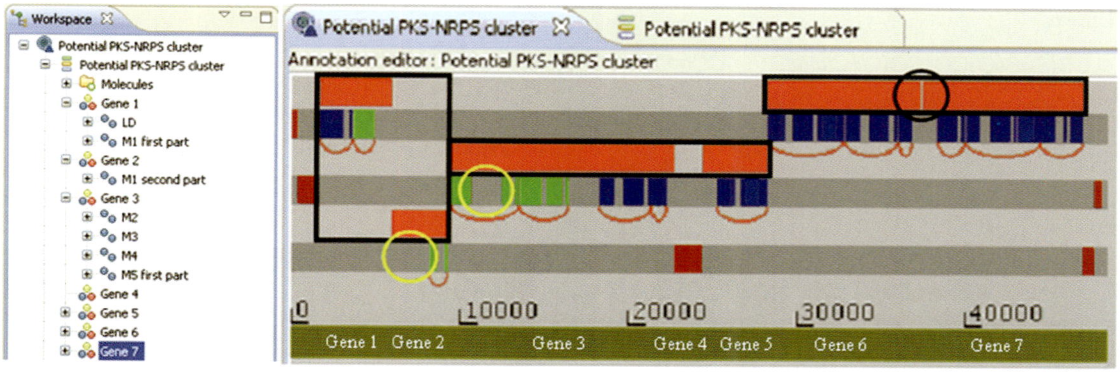

Figure 49.6 The annotated cluster. The cluster can be viewed both as collapsible trees in the workspace window or in a graphical form in the annotation editor window. The domains (PKS in green and NRPS in blue) and predicted genes (in red) are shown, and the modules deduced during the annotation process are shown in the annotation editor by lines connecting the single domains. The two regions, where expected AT domains were not detected, are ringed in yellow and the region containing the unexpected stop codon is ringed in black. The three probable genes (after correction for sequencing mistakes) are boxed in black.

has led to an apparent frameshift and that there is a single gene with the NRPS loading module and a PKS extender module. Gene 3 has two intact PKS extender modules (one lacking an AT domain) and one intact NRPS extender module; however, there is part of a second NRPS module, which would be completed by domains in gene 5. This suggests that there is a single gene corresponding to genes 3–5 with two PKS and two NRPS modules. Similarly, there is an NRPS module distributed between genes 6 and 7, which suggest that a sequencing error has introduced a stop codon; gene 7 has a terminal NRPS thioesterase domain. Thus, there is probably a mixed cluster with an NRPS loading module, three PKS extender modules, and seven NRPS extender modules distributed over three polypeptides [Starcevic et al., 2008].

49.3.3 Prediction of Product Chemistry

Once the modules of a cluster have been defined, the functions of the domains predicted, and the order of biosynthesis decided, the chemistry of the product can be predicted. *ClustScan* uses the standard chemical notation of isomeric SMILES [Weininger, 1988] to describe the product chemistry; this works well because the SMILES of the extender unit of each biosynthetic step can be simply concatenated. *ClustScan* also implements the extension of SMARTS (see Table 49.1) that allows incorporation of generic units for steps that cannot be precisely predicted. The SMILES description can be viewed in a window (**Fig. 49.7**) and exported. It is also possible to view the molecule using a Jmol window (see Table 49.1) or ChemAxon [Csizmadia, 2000] (Figs. 49.7A and 49.7B). SMILES are a standard format, so the predicted molecules can be imported into any standard chemistry program and subjected to any desired analyses so as to identify potentially interesting molecules. *ClustScan* can be used to examine the strength of prediction for critical parts of the molecule.

Most natural polyketides are not linear chains as produced during the synthesis reactions, but undergo cyclization to ring structures. *ClustScan* also has a ring prediction function that will take the linear SMILES description and produce a cyclic structure (Figs. 49.7A and 49.7B). This gives reasonable results with macrolides, but is less effective for other classes of polyketides.

49.4 DISCUSSION

49.4.1 *ClustScan* as a Productivity Tool for Metagenomics

The major driving force for metagenomics is the wish to access biodiversity that will not be revealed if only the minority of culturable organisms is examined. In the context of secondary metabolite biosynthesis clusters, it is expected that novel clusters will be revealed that differ in architecture and sequence from well-known clusters. Most classes of biochemical reactions in secondary metabolite biosynthesis are also present in primary metabolism. Because of the search for novelty, it is necessary to use nonstringent search parameters, which will result in a lot of false positives. A major advantage of *ClustScan* is that any search using a set of profiles typical for a particular class of secondary metabolite will distinguish between isolated hits and a cluster of domains typical for such a secondary metabolite. This is illustrated in Fig. 49.2 where a search with a set of PKS and NRPS domains is shown. In this case, there is a mixed PKS-NRPS cluster.

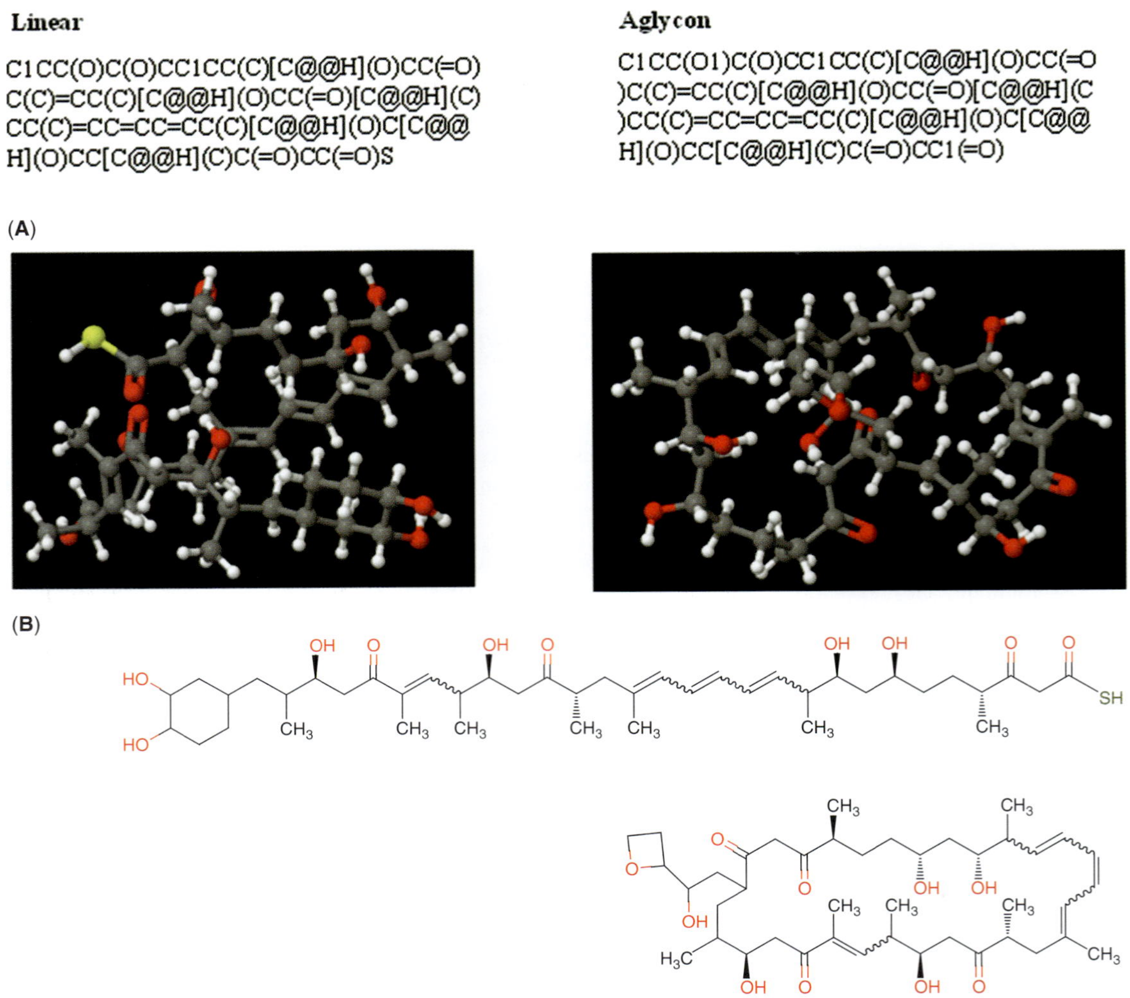

Figure 49.7 The predicted chemical structure of polyketides (in this case rapamycin) is shown as isomeric SMILES. There is also a function for predicting the ring structure. The SMILES can be used by standard programs such as Jmol (**A**) and ChemAxon (**B**) to display the chemical structures.

Once clusters are found, *ClustScan* is an effective way of annotating them in order to obtain maximal knowledge about the likely chemical structures of products. This is, at present, most highly developed for modular PKS systems. Manual annotation of such a cluster is a labor-intensive activity that requires a lot of expert knowledge and is error-prone. We estimate that a researcher skilled in the art could annotate a PKS cluster in 1–4 hours, depending on the detail of detail required and the extent of problems caused by sequencing mistakes or the presence of unusual cluster architectures, a process that could take several weeks manually. The generation of chemical structures allows rapid screening of DNA sequences and evaluation of clusters for potential novel compounds.

Metagenomic datasets will often contain sequencing errors, which complicate analysis; in many cases such errors will involve apparent insertion or deletion of a nucleotide, resulting in an apparent frameshift. The use of local HMM profiles with in silico-translated DNA in *ClustScan* makes it fairly robust. This is shown for a metagenomic cluster (Fig. 49.6), where there appears to be at least three apparent frameshifts and an artificial stop codon in a 50 kb sequence. It is also easy to recognize non-typical modules such as the two PKS extender modules lacking AT domains [Starcevic et al., 2008].

49.4.2 Future Developments of *ClustScan*

The present version of *ClustScan* recognizes domains of modular NRPS clusters, but does not make further prediction of chemistry. A major determinant of chemistry is the adenylation domain (A), which is responsible for substrate selection. There has been considerable work on prediction of substrate [Rausch et al., 2005], and the open structure of *ClustScan* will make it easy to incorporate this knowledge in an analogous way to that used for PKS prediction. The prediction of NRPS substrates is considerably more difficult than for PKS substrates because there are about 400 known ones [Caboche et al., 2008] in contrast to 8 [Chan et al., 2009]. However, there are some common substrates that can be recognized reasonably efficiently. Modules in NRPS clusters may also contain domains (e.g., epimerization and methylation domains) that modify the substrate, and it should be possible to treat such domains with methods similar to those used for reduction domains in PKS clusters. Most modular polyketides have chemical modifications that occur after synthesis of the polyketide backbone (e.g., addition of sugars, hydroxylation, methylation). The presence of such "tailoring" enzymes can be detected by the use of suitable profiles, but the prediction of the precise chemical modification (e.g., nature of a sugar unit and position of incorporation in the polyketide skeleton) is not possible with present knowledge. Similarly, the presence of other types of secondary metabolite clusters can be recognized by using appropriate collections of profiles (e.g., aromatic polyketide synthases, amino glycoside synthases, terpene synthases). However, in nonmodular systems there is little prospect in the near future of predicting the chemistry of products from the DNA sequence.

49.4.3 Limitations of Metagenomic Data

Secondary metabolite biosynthesis clusters are often large (20–200 kb), with many interesting clusters being larger than 50 kb in size. Much present metagenomics data consists of fairly small contigs (<10 kb). Although this allows detection of parts of clusters, much of the most interesting information requires an overview of the total cluster architecture. It is likely that improving technology will yield a larger number of large contigs in the next few years. It is also possible that effective methods for enriching genes for particular classes of secondary metabolites will be developed. The evolution of secondary metabolite clusters is a fascinating topic [Jenke-Kodama and Dittmann, 2009], which would be greatly enriched by the possibility of extending analysis to nonculturable species. Such clusters can be viewed as mobile elements that are indulging in chemical manipulation of their immediate environment. A better understanding of their evolution is also likely to assist in searching for useful new compounds.

49.5 CONCLUSIONS

Secondary metabolites are chemically very diverse, and the opportunity to access uncultivable strains via metagenomics holds the promise of significantly expanding our knowledge of chemical biodiversity in this field. Genes for secondary metabolite biosynthesis occur in clusters, and technical progress in improving lengths of contigs is starting to make it possible to find and analyze clusters in metagenomic data. The program package *ClustScan* ("*Clust*er *Scan*ner") greatly simplifies the task of finding such clusters amongst the mass of sequence data. The use of HMM profiles with in silico-translated DNA allows the recognition of significant clusters of enzymic domains in a robust way that is resistant to problems caused by sequencing errors or the existence of unusual intron sequences. A rapid annotation of modular biosynthetic clusters is possible (needing hours instead of weeks) with maximal information about the chemical structures of putative products being generated. In the absence of such a program, the labor needed to detect and analyze secondary metabolite clusters is prohibitive and the resulting knowledge will rarely be extracted from the sequence data.

REFERENCES

Altschul SF, Gish W, Miller W, Myers EW, Lipman DJ. 1990. Basic local alignment search tool. *J. Mol. Biol*. **215**:403–410.

Bateman A, Birney E, Cerruti L, Durbin R, Etwiller L, et al. 2002. The Pfam protein families database. *Nucleic Acids Res*. **30**:276–280.

Besemer J, Borodovsky M. 2005. GeneMark: Web software for gene finding in prokaryotes, eukaryotes and viruses. *Nucleic Acids Res*. **33**(Web Server issue): W451–W454.

Caboche S, Pupin M, Leclère V, Fontaine A, Jacques P, et al. 2008. NORINE: A database of nonribosomal peptides. *Nucleic Acids Res*. **36**(Database issue): D326–D331.

Chan YA, Podevels AM, Kevany BM, Thomas MG. 2009. Biosynthesis of polyketide synthase extender units. *Nat. Prod. Rep*. **26**:90–114.

Csizmadia F. 2000. JChem: Java applets and modules supporting chemical database handling from web browsers. *J. Chem. Inf. Comput. Sci*. **40**:323–324.

Delcher AL, Bratke KA, Powers EC, Salzberg SL. 2007. Identifying bacterial genes and endosymbiont DNA with Glimmer. *Bioinformatics* **23**:673–679.

Demain AL. 2009. Antibiotics: Natural products essential to human health. *Med. Res. Rev*. **29**:821–842.

Demain AL, Sanchez S. 2009. Microbial drug discovery: 80 years of progress. *J. Antibiot. (Tokyo)* **62**:5–16.

Du L, Sánchez C, Shen B. 2001. Hybrid peptide–polyketide natural products: Biosynthesis and prospects toward engineering novel molecules. *Metab. Eng*. **3**:78–95.

Dunlap WC, Jaspars M, Hranueli D, Battershill CN, Perić-Concha N, et al. 2006. New methods for medicinal chemistry-universal gene cloning and expression systems for production of marine bioactive metabolites. *Curr. Med. Chem*. **13**:697–710.

Eddy SR. 1998. Profile hidden Markov models. *Bioinformatics* **14**:755–763.

Hopwood DA, Chater KF. 1980. Fresh approaches to antibiotic production. *Philos. Trans. R. Soc. Lond. B Biol. Sci*. **290**:313–328.

Hranueli D, Cullum J, Basrak B, Goldstein P, Long PF. 2005. Plasticity of the *Streptomyces* genome—Evolution and engineering of new antibiotics. *Curr. Med. Chem*. **12**:1697–1704.

Jenke-Kodama H, Dittmann E. 2009. Evolution of metabolic diversity: Insights from microbial polyketide synthases. *Phytochemistry* **70**:1858–1866.

Khosla C, Tang Y, Chen AY, Schnarr NA, Cane DE. 2007. Structure and mechanism of the 6-deoxyerythronolide B synthase. *Annu. Rev. Biochem*. **76**:195–221.

Koglin A, Walsh CT. 2009. Structural insights into nonribosomal peptide enzymatic assembly lines. *Nat. Prod. Rep*. **26**:987–1000.

Kwan DH, Sun Y, Schulz F, Hong H, Popovic B, et al. 2008. Prediction and manipulation of the stereochemistry of enoylreduction in modular polyketide synthases. *Chem. Biol*. **15**: 1231–1240.

Li W. 2009. Analysis and comparison of very large metagenomes with fast clustering and functional annotation. *BMC Bioinform*. **10**: 359.

Moore BS, Hertweck C. 2002. Biosynthesis and attachment of novel bacterial polyketide synthase starter units. *Nat. Prod. Rep*. **19**:70–99.

Nett M, Ikeda H, Moore BS. 2009. Genomic basis for natural product biosynthetic diversity in the actinomycetes. *Nat. Prod. Rep*. **26**:1362–1384.

Nguyen T, Ishida K, Jenke-Kodama H, Dittmann E, Gurgui C, et al. 2008. Exploiting the mosaic structure of trans-acyltransferase polyketide synthases for natural product discovery and pathway dissection. *Nat. Biotechnol*. **26**:225–233.

Rausch C, Weber T, Kohlbacher O, Wohlleben W, Huson DH. 2005. Specificity prediction of adenylation domains in nonribosomal peptide synthetases (NRPS) using transductive support vector machines (TSVMs). *Nucleic Acids Res*. **33**:5799–5808.

Rusch DB, Halpern AL, Sutton G, Heidelberg B, Williamson S, et al. 2007. The Sorcerer II Global Ocean Sampling expedition: Northwest Atlantic through Eastern Tropical Pacific. *PLoS Biol*. **5**:e77.

Starcevic A, Cullum J, Jaspars M, Hranueli D, Long PF. 2007. Predicting the nature and timing of epimerisation on a modular polyketide synthase. *ChemBioChem* **8**:28–31.

Starcevic A, Zucko J, Simunkovic J, Long PF, Cullum J, et al. 2008. *ClustScan*: An integrated program package for the semi-automatic annotation of modular biosynthetic gene clusters and in silico prediction of novel chemical structures. *Nucleic Acids Res*. **36**:6882–6892.

Weininger D. 1988. SMILES, a chemical language and information system. 1. Introduction to methodology and encoding rules. *J. Chem. Inf. Comput. Sci*. **28**:31–36.

Weissman KJ. 2006. Single amino acid substitutions alter the efficiency of docking in modular polyketide biosynthesis. *ChemBioChem* **7**:1334–1342.

Weissman KJ, Müller R. 2009. A brief tour of myxobacterial secondary metabolism. *Bioorg. Med. Chem*. **17**:2121–2136.

Zucko J, Skunca N, Curk T, Zupan B, Long PF, et al. 2007. Polyketide synthase genes and the natural products potential of *Dictyostelium discoideum*. *Bioinformatics* **23**:2543–2549.

Chapter 50

MetaGene: Prediction of Prokaryotic and Phage Genes in Metagenomic Sequences

HIDEKI NOGUCHI

50.1 INTRODUCTION

One of the most important advantages of the metagenomic approach is that genomic information of all organisms in an environment can be made available [Handlesman, 2004, Streit et al., 2004]. Microorganisms form complex communities in natural environments, and exhaustive investigation of genes from all community members including uncultivated microorganisms is fundamental to elucidating the ecological systems of the environments. Whole-genome shotgun sequencing of environmental-pooled DNA enables us to obtain genomic sequences from various organisms without isolation and cultivation of individual species. This approach is an efficient solution for environmental genomics, but it also includes basic problems: The obtained genomic sequences are unidentified sequences derived from diverse organisms, and many sequence reads remain as unassembled one-pass sequences especially when a great variety of microorganisms are in the environment [Chen and Pachter, 2005]. These limitations make gene finding difficult, because the conventional gene finding tools utilize species-specific statistics calculated from known genes of the target microorganism.

Ab initio gene finding is a powerful yet simple process for detecting genes from genomic DNA sequences and is widely applied to the microbial genome annotations. Unlike the homology search, the ab initio method can predict novel genes having no sequence similarities to any known genes by using statistical properties of protein-coding sequences (CDSs) [Gribskov et al., 1984, Staden, 1984]. On the other hand, the statistical properties, such as codon usages, differ from species to species, and one's own statistics should be prepared prior to the gene finding. This means that it is difficult for the ab initio method to predict genes from a mixture of anonymous genomic sequences whose statistics of genes are unknown. If a target (anonymous) genomic sequence is long enough, the statistics of the species' genes can be calculated from long open reading frames (ORFs), most of which are thought to be genes [Hayes and Borodovsky, 1998, Delcher et al., 1999]. However, metagenomic sequences are very short fragments, and each fragment provides little information about genes of the original species. These difficulties restrict the use of the ab initio methods for metagenomic gene finding, and the only realistic alternative is the homology searches of the known genes or the metagenome itself. To overcome the difficulties, we developed MetaGene [Noguchi et al., 2006], which is an ab initio gene finding tool specialized for metagenome analyses.

MetaGene (and its enhanced version, MetaGeneAnnotator (MGA) [Noguchi et al., 2008]) utilizes relationships between the GC contents of the sequences and the statistics of genes, and it estimates the approximated values of the statistics for the anonymous sequences. By using the estimated statistics as the parameters of the gene finding, MetaGene accurately detects genes on anonymous genome fragments without information about the original species of the fragments.

Handbook of Molecular Microbial Ecology, Volume I: Metagenomics and Complementary Approaches, First Edition. Edited by Frans J. de Bruijn.

50.2 MATERIALS AND METHODS

For more details on the materials and methods, please refer to the original publications [Noguchi et al., 2006, 2008]. A general outline is given in the following section.

50.2.1 Dataset

The complete genomic sequences and the annotations for 116 bacteria, 15 archaea, and 244 phages were obtained from GenBank to construct the statistical models. In this dataset, an arbitrary species was selected from members of the same genus (prokaryotes), or close relatives having less than 0.02 of Euclidean distances between the mono-codon usages were removed (phages). The complete sequences were randomly split into fixed-length fragments having 1 × genomic coverage and used as the artificial metagenomic sequences.

50.2.2 Basic Architecture of MetaGene

MetaGene predicts genes by the following steps: (i) extracting all possible ORFs as candidates, (ii) scoring each candidate ORF based on the statistical futures of CDSs, and (iii) searching for an optimal (high-scored) combination of the ORFs. In the second step, an ORF score is calculated by using the dicodon frequencies, the length distribution of ORFs, and the distance distribution from the annotated start codon to the leftmost start codon. In the current version of MetaGene (MGA), a score of RBS is also calculated and added to the ORF score. Although these statistics vary by species, they can be estimated by the GC contents of the sequences. MetaGene has the logistic regression models of the di-codon frequencies for bacteria and archaea, and MGA also has the models for prophage to estimate the statistics. In addition to the estimated parameters by GC%, MGA also uses self-trained parameters if an input sequence had an enough size to perform self-training. In the third step, an optimal combination of ORFs is calculated according to directions of adjacent ORFs and their distances.

50.2.3 Estimation of Codon Frequencies by GC%

It was known that use frequencies of some amino acids in a microorganism are related to the GC% of its genome, and this observation was used in GeneMark.hmm [Besemer and Borodovsky, 1999], which is an unsupervised gene finding tool, to estimate codon frequencies. As a more natural way of the codon frequency estimation, codon (and dicodon) frequencies were directly approximated by the genome GC% using logistic regression analysis in MetaGene. Since the codon frequency–GC% relationships were slightly different among bacteria, archaea, and phage, these were separately interpolated. By using a sigmoid function, the frequency of a codon (or dicodon) i for a sequence having GC% $= x$ can be represented as

$$f_i(x) = \frac{f_i^{\max} - f_i^{\min}}{1 + e^{-ax+b}} + f_i^{\min}$$

where $f_i^{\max}$ and $f_i^{\min}$ are the maximal and minimal values of the observed frequencies for the codon i in the training data. The parameters of the sigmoid function, a and b, were determined by the Marquardt–Levenberg algorithm. Similarly, frequencies of triplets in the genome sequences were interpolated by GC% as the background frequencies.

50.2.4 Length and Distance Distributions of Genes

We found that the length distributions of genes, the ratios of directions of neighboring genes, and the (direction-dependent) distance distributions of intergenic regions were basically common to all species. This is consistent with the fact that the numbers of genes are proportional to the sizes of the genomes. On the other hand, the length distribution of the random genes differed according to the GC% of the genomes. This is because the length distributions of the random genes are determined by the (GC%-dependent) frequencies of the start and stop codon-like triplets in the genomes. Similarly, the distance distributions between the real start codons and their leftmost start codons are related to the GC% of the genomes. As a result, almost all parameters used in MetaGene are known or can be estimated by the GC% of the input genomic sequences except parameters of the RBSs.

50.2.5 Construction of the RBS Model

A ribosomal binding site (RBS), which is also known as Shine–Dalgarno sequence [Shine and Dalgarno, 1974], is a sequence located upstream of the start codon, and is recognized by 16S rRNA for the translation. The motifs of the RBSs and their positions relative to the start codons differ from species to species and are not related to the GC% of the genomes unlike the other parameters of genes. However, we found that all RBS motifs found in the training genomes including both bacterial and archaeal species were complementary to narrow area of a 3′ tail of 16S rRNA. So we defined nine hexamers derived from the sequence, G(A/T)(A/T)AGGAGGT(G/A)ATC, as the potential RBS motifs (starting from the left, motif 1, . . . , 9) for all species. An exact match or one-base mismatch sequence of the motifs was searched against an upstream region of a start codon (ranged from −3 to −19), and a two-dimensional (9 motifs × 16 positions) frequency distribution of RBSs (named RBS map) was calculated. For the gene finding from metagenomic sequences, in which there are not enough RBSs to calculate the reliable RBS maps, the average map derived from all prokaryotes were used as the general RBS model to calculate anonymous species' RBS scores. When a significant number of genes was predicted in an input sequence, the RBS map for the sequence was estimated by the upstream subsequences of the predicted genes, and the gene finding was iteratively performed by using the reconstructed RBS map.

50.3 RESULTS AND DISCUSSION

50.3.1 Regression Analysis of Codon Frequencies

Codon frequencies of the bacterial and archaeal genomes were plotted against the genome GC%. Most codon (and dicodon) frequencies were correlated to the GC%, and their gradients strongly depended on the third nucleotides of the codons. As a result, frequencies of some synonymous codons showed inverse correlations (**Fig. 50.1A,C,D**), although they code the same amino acid. In some codon frequencies, bacteria and archaea showed significant differences, and these differences were also inversely correlated in the synonymous codons. On the other hand, the background frequencies were not necessarily different between bacteria and archaea (Fig. 50.1B). Since scores of genes are calculated by the ratios of codons and their background frequencies, in the case of "ATA" the codon has a positive score in archaea but has a neutral or negative score in bacteria. Therefore, bacterial and archaeal codon frequencies were independently interpolated by the sigmoid functions. In addition, regression models for phage codons were also prepared and used for predicting genes in MetaGene.

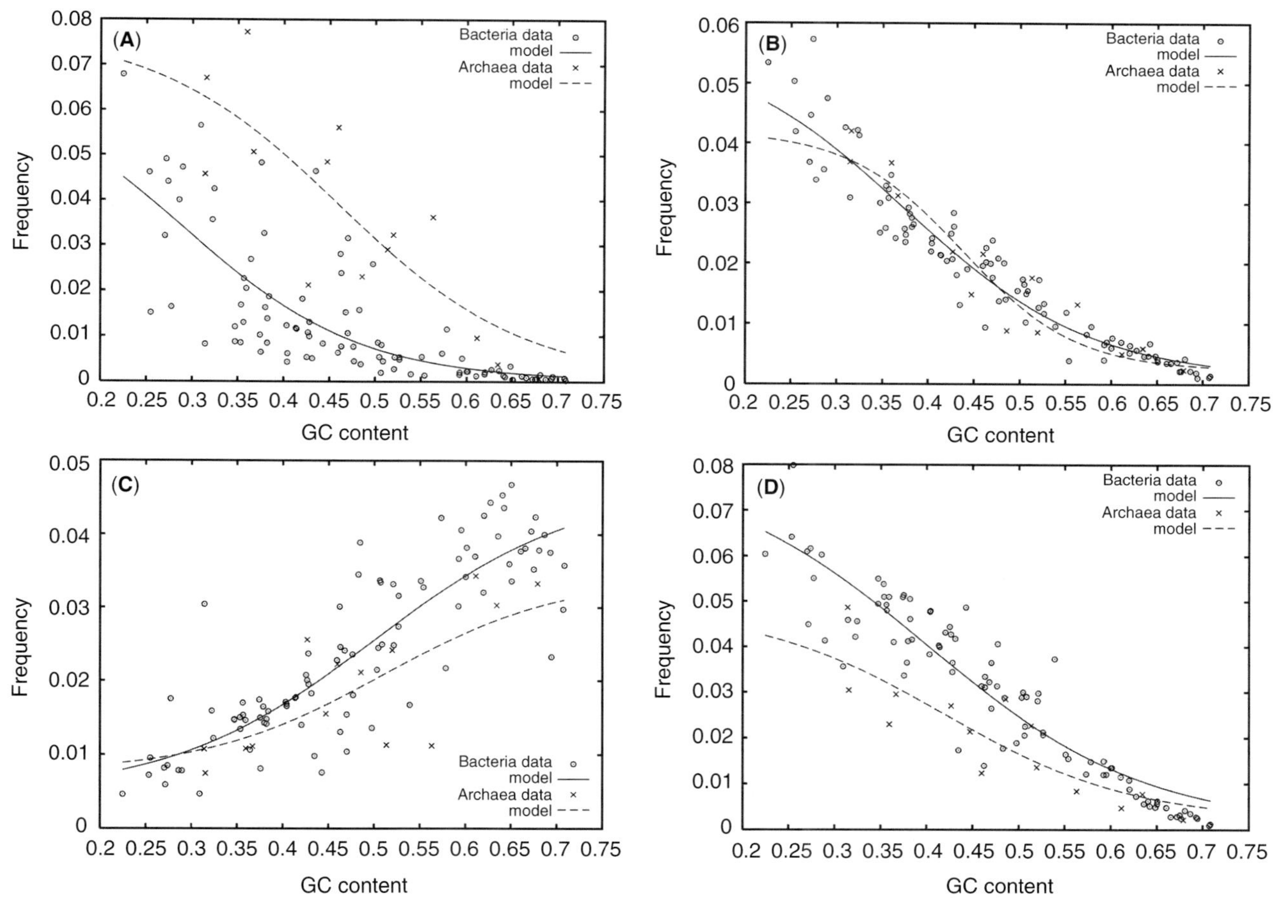

Figure 50.1 Relationships between codon frequencies and GC% in the bacterial and archaeal genomes. Frequencies of three mono-codons, which code isoleucine, are plotted (circle: bacteria, cross: archaea). The results of regression analysis are also indicated by solid lines (bacteria) and dashed lines (archaea). (**A**) Frequencies of ATA in CDSs. (**B**) Frequencies of ATA in the genomes (background frequencies). (**C**) Frequencies of ATC in CDSs. (**D**) Frequencies of ATT in CDSs.

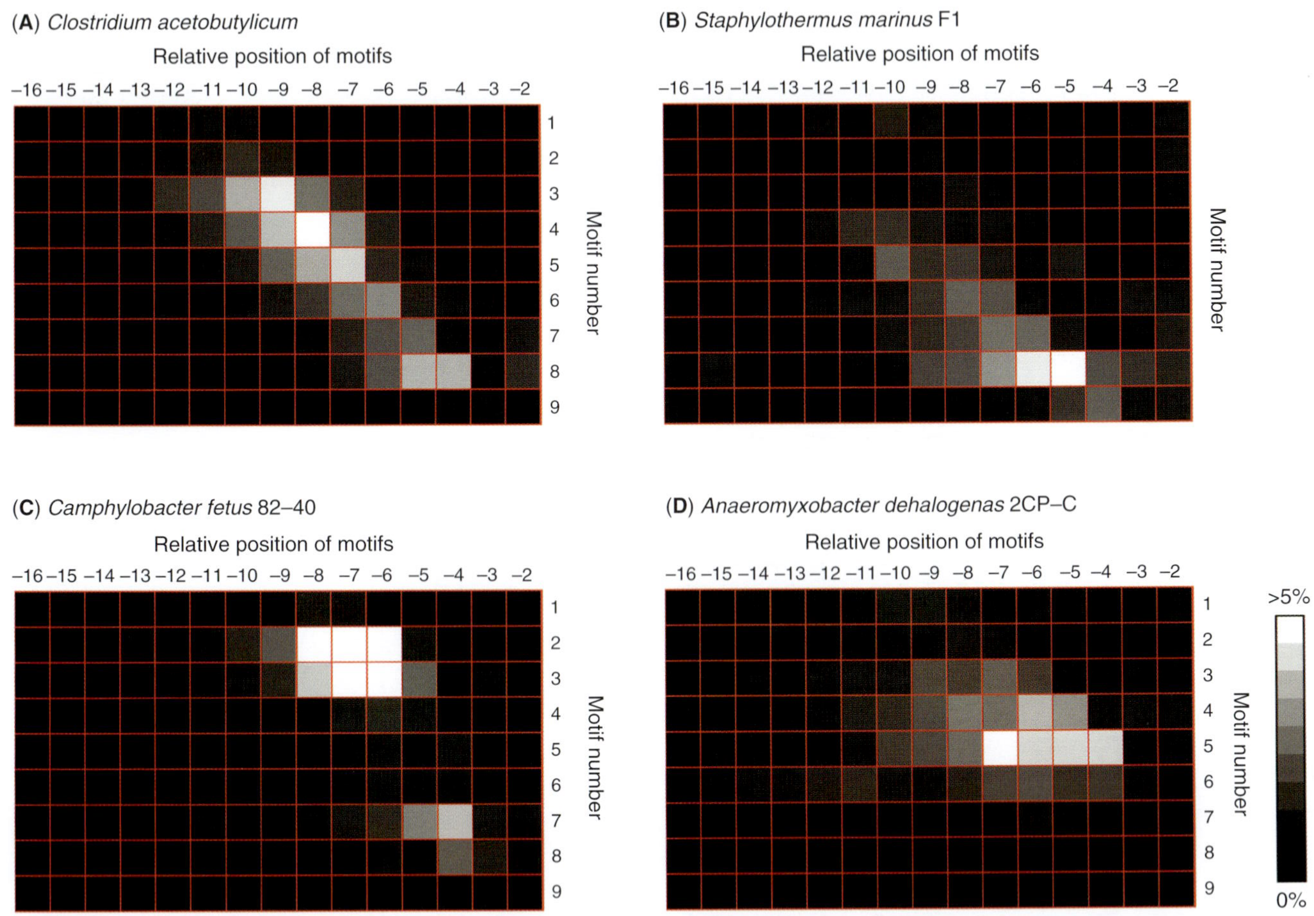

Figure 50.2 The RBS maps of four microorganisms. Appearance frequencies of the motifs are plotted against the motif number (vertical axis) and the position relative to the start codon (horizontal axis). **(A)** *Clostridium acetobutylicum*, **(B)** *Staphylothermus marinus* F1, **(C)** *Campylobacter fetus* 82-40 [Moolhuijzen et al., 2009], **(D)** *Anaeromyxobacter dehalogenas* 2CP-C [Sanford et al., 2002].

50.3.2 Species-Specific Patterns of RBS

In the various species' genomes, the upstream regions of the annotated start codons were examined for the presence of the nine RBS motifs we defined. Various patterns of the RBS maps, which represent the preference of the motifs and their positions in the species, were observed **(Fig. 50.2)**. Broadly speaking, the bacterial species prefer the motifs of the 3′ side of 16S rRNA (e.g., motif 2–4) and the archaeal species prefer those of the 5′ side (e.g., motif 6–8). The appearance positions of the motifs were inversely correlated with the motif number (= interacting positions of 16S rRNA), and so positions of the main body of 16S rRNA (or ribosome) were almost constant with the used motifs. Some species lacked any of the nine motifs. In such species, no significant motif was found even when Gibbs sampling algorithm was used to extract unknown motifs. In the MetaGene prediction, the RBS scores were multiplied by the existence probability of RBSs. Therefore, if no motif was found in upstream regions of the predicted genes, the RBS score is set to zero and the scores of genes are determined by the other parameters.

50.3.3 Prediction Performance on the Complete Genomes

To evaluate validity of the parameter estimation by GC% and the RBS map, prediction performances of MetaGene were tested on the complete genomic sequences **(Table 50.1)**. The performances of GeneMarkS [Besemer et al., 2001] and Glimmer3 [Delcher et al., 2007], which are unsupervised gene finding tools designed to predict genes on the long genomic sequences, were also tested for comparison.

On average, sensitivities to the annotated genes were almost identical among the gene finders, while specificity was higher in MetaGene compared with the others. This result suggests that the parameter estimation in MetaGene worked correctly and is suitable for gene finding even in the long genomic sequences. In addition, the proportion of exactly predicted genes in MetaGene was significantly higher than those in the other tools. In particular, the start codons of *Clostridium acetobutylicum* [Nölling et al., 2001], in which various sequences are used as RBSs (Fig. 50.2A), were precisely predicted in MetaGene. Although our RBS model uses just nine motifs as RBSs, the model accommodated various types of RBSs. The RBS model of MetaGene was also contributed to the improvement of the prediction specificity, and the comprehensive performance of MetaGene outperformed that of the others even in gene finding from the complete genomes.

50.3.4 Prediction Performance on the Short Genomic Fragments

Artificial genomic fragments were made from the complete genomes and were used for the performance test of MetaGene **(Fig. 50.3)**. Since the parameters (frequency information) for the gene finding were not calculated

Table 50.1 Prediction Performances on the Complete Genomes

			MetaGeneAnnotator		GeneMarkS		Glimmer3	
Species	GC%	RBS%[a]	Sn[b] (exact[c]) (%)	Sp[d] (%)	Sn (exact) (%)	Sp (%)	Sn (exact) (%)	Sp (%)
M. jannaschii	31.4	87.6	99.3 (62.7)	95.0	99.3 (62.9)	91.5	99.1 (61.7)	93.2
S. marinus	35.7	85.4	99.4 (87.8)	94.5	99.6 (87.2)	92.5	99.8 (87.6)	90.8
A. fulgidus	48.6	61.7	97.8 (72.6)	93.9	97.9 (72.9)	92.0	97.2 (70.3)	91.7
N. pharaonis	63.4	39.6	98.0 (85.7)	98.3	98.7 (86.3)	97.6	98.5 (83.5)	96.4
C. acetobutylicum	30.9	93.7	98.3 (92.1)	96.1	98.5 (74.1)	92.8	98.0 (90.9)	94.5
F. nodosum	35.0	90.2	99.6 (91.2)	94.8	99.8 (90.6)	92.8	99.7 (91.1)	94.0
L. lactis	35.3	81.1	98.5 (88.0)	95.1	98.9 (88.4)	92.7	98.2 (86.2)	93.2
D. radiodurans	67.0	47.9	97.8 (63.5)	93.6	96.3 (56.7)	93.1	96.5 (58.3)	92.1
A. caulinodans	67.3	64.8	99.2 (66.2)	95.4	98.8 (61.5)	95.8	98.6 (63.6)	93.6
Average			98.7 (78.9)	95.2	98.6 (75.6)	93.4	98.4 (77.0)	93.3

[a]Proportion of genes having representative RBSs.

[b]Sensitivity.

[c]Sensitivity of the exact prediction.

[d]Specificity.

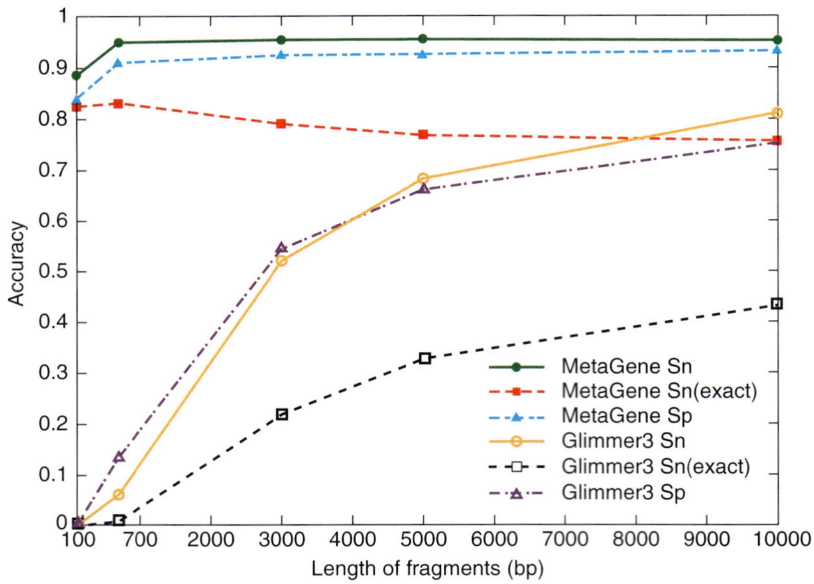

Figure 50.3 Prediction accuracies of MetaGene and Glimmer3. Sensitivities (filled and unfilled circles), sensitivities by exact predictions (rectangles), and specificities (triangles) are plotted against lengths of the fragments. In the Glimmer3 prediction, a script "g3-iterated.csh" is used.

directly from each sequence in MetaGene, the prediction performance was not so influenced by lengths of the fragments. Most genes on the short genomic fragments were small partial genes (e.g., 92% of genes on 700-bp fragments were partial), and there were slight reductions both in the sensitivity and specificity. However, the accuracies were sufficiently high even in the 100-bp fragments (89% sensitivity and 84% specificity). In such short fragments, 5′ parts (including start codons) of most genes were truncated, and this is one of the reasons why the sensitivities of the exact predictions were increased. Even allowing for the fact, the accuracies were remarkably high, and it indicates that our general RBS model was useful for scoring of anonymous RBSs. Glimmer3 and GeneMarkS require a long input sequence to calculate the parameters of the gene finding, and so their performances on the short fragments were very low. These results suggest that the parameter estimation models implemented in MetaGene are adequate for gene finding from very short genomic sequences, and MetaGene can be applied to a wide variety of metagenomic sequences containing bacterial, archaeal, and phage genes.

Table 50.2 Prediction Performances on the *Mycoplasma* Genomes

	MGA for *Mycoplasma*		Original MGA	
Species	Sn (exact) (%)	Sp (%)	Sn (exact) (%)	Sp (%)
M. agalactiae	96.6 (86.7)	91.6	54.7 (18.3)	29.1
M. arthritidis	97.9 (89.5)	96.0	56.3 (19.5)	29.8
M. capricolum	99.3 (93.3)	95.2	50.4 (19.3)	27.0
M. conjunctivae	95.5 (80.6)	97.1	53.8 (17.5)	30.9
M. gallisepticum	98.9 (34.8)	89.3	61.2 (8.7)	30.5
M. genitalium	99.2 (88.4)	89.9	65.3 (26.1)	35.1
M. hyopneumoniae	96.1 (81.3)	97.1	53.3 (16.9)	30.8
M. mobile	99.7 (88.9)	94.0	56.4 (21.0)	29.9
M. mycoides	98.1 (86.4)	85.1	50.6 (19.5)	26.7
M. penetrans	99.0 (81.0)	96.8	52.7 (15.3)	25.2
M. pneumonia	95.9 (79.2)	89.0	67.9 (26.1)	35.1
M. pulmonis	95.3 (78.1)	96.5	47.2 (14.8)	26.6
M. synoviae	98.9 (52.4)	91.8	48.3 (10.6)	27.0
Average	97.7 (78.5)	93.0	55.2 (18.0)	29.5

50.3.5 Dealing with *Mycoplasma* Genes

MetaGene presupposes all (anonymous) species use the standard genetic code for encoding amino acids, although alternative codes are rarely utilized in some species. In *Mycoplasma*, the codon UGA is translated as tryptophan rather than as a stop signal, and MetaGene misses the prediction of their genes because of the apparent stop codons. On the other hand, use frequencies of the other codons conform to the codon frequency-GC% models constructed by the other bacterial species. Consequently, MetaGene was modified to just ignore the stop signal of UGA (the log-odds scores for UGA-related dicodons were set to zero), and its prediction performance was tested on some *Mycoplasma* species' complete genomes. As the result, both sensitivities and specificities to all *Mycoplasma* species were remarkably improved **(Table 50.2)**. (Sensitivity to the start codons in *Mycoplasma gallisepticum* R (NC_004829) was abnormally low because of the inaccurate CDS annotation, in which leftmost start codons were used as the start codons of almost all genes [Papazisi et al., 2003].) In the original MetaGene, most genes were split at the internal TGA, while the modified version detected such genes correctly. At this point, MetaGene can't automatically detect *Mycoplasma* genes in metagenomic mixture sequences, but can be applied to the annotation process when the target genome is known to belong to the *Mycoplasma* genus.

INTERNET RESOURCES

MetaGene webserver (http://metagene.cb.k.u-tokyo.ac.jp/)

Acknowledgments

The original MetaGene was developed with Park Jungho at Toshihisa Takagi Laboratory (The University of Tokyo) and was supported by Grant-in-Aid for Scientific Research on Priority Areas "Systems Genomics" from the Ministry of Education, Culture, Sports, Science and Technology of Japan. MetaGeneAnnotator was developed and tested with Takehiko Itoh and Takeaki Taniguchi at Mitsubishi Research Institute, Inc.

REFERENCES

Besemer J, Borodovsky M. 1999. Heuristic approach to deriving models for gene finding. *Nucleic Acids Res*. **27**:3911–3920.

Besemer J, Lomsadze A, Borodovsky M. 2001. GeneMarkS: A self-training method for prediction of gene starts in microbial genomes. *Nucleic Acids Res*. **29**:2607–2618.

Chen K, Pachter L. 2005. Bioinformatics for whole-genome shotgun sequencing of microbial communities. *PloS Comput. Biol*. **1**:106–112.

Delcher AL, Harmon D, Kasif S, White O, Salzberg SL. 1999. Improved microbial gene identification with GLIMMER. *Nucleic Acids Res*. **27**:4636–4641.

Delcher AL, Bratke KA, Powers EC, Salzberg SL. 2007. Identifying bacterial genes and endosymbiont DNA with Glimmer. *Bioinformatics* **23**:673–679.

Glibskov M, Devereux J, Burgess RR. 1984. The codon preference plot: Graphic analysis of protein coding sequences and prediction of gene expression. *Nucleic Acids Res*. **12**:539–549.

Handelsman J. 2004. Metagenomics: Application of genomics to uncultured microorganisms. *Microbiol. Mol. Biol. Rev*. **68**:669–685.

Hayes WS, Borodovsky M. 1998. How to interpret an anonymous bacterial genome: Machine learning approach to gene identification. *Genome Res*. **8**:1154–1171.

Moolhuijzen PM, Lew-Tabor AE, Wlodek BM, Agüero FG, Comerci DJ, et al. 2009. Genomic analysis of *Campylobacter fetus* subspecies: Identification of candidate virulence determinants and diagnostic assay target. *BMC Microbiol*. **9**:86.

Noguchi H, Park J, Takagi T. 2006. MetaGene: Prokaryotic gene finding from environmental genome shotgun sequences. *Nucleic Acids Res*. **34**:5623–5630.

Noguchi H, Taniguchi T, Itoh T. 2008. MetaGeneAnnotator: Detecting species-specific patterns of ribosomal binding site for precise gene prediction in anonymous prokaryotic and phage genomes. *DNA Res*. **15**:387–396.

Nölling J, Breton G, Omelchenko MV, Makarova KS, Zeng Q, et al. 2001. Genome sequence and comparative analysis of the solvent-producing bacterium *Clostridium acetobutylicum*. *J. Bacteriol*. **183**:4823–4838.

Papazisi L, Gorton TS, Kutish G, Markham PF, Browning GF, et al. 2003. The complete genome sequence of the avian pathogen *Mycoplasma gallisepticum* starin R(low). *Microbiology* **149**:2307–2316.

Shine J, Dalgarno L. 1974. The 3′-terminal sequence of *Escherichia coli* 16S ribosomal RNA: Complementary to nonsense trplets and ribosome binding sites. *Proc. Natl. Acad. Sci. USA* **71**:1342–1346.

Sanford RA, Cole JR, Tiedje JM. 2002. Characterization and description of *Anaeromyxobacter dehalogenas* gen. nov., sp. nov., an aryl-halorespiring facultative anaerobic myxobacterium. *Appl. Environ. Microbiol*. **68**:893–900.

Staden R. 1984. Measurements of the effects of that coding for a protein has on a DNA sequence and their use for finding genes. *Nucleic Acids Res*. **12**:551–567.

Streit WR, Schmitz RA. 2004. Metagenomics—the key to the uncultured microbes. *Curr. Opin. Microbiol*. **7**:492–498.

Chapter 51

Primers4clades: A Web Server to Design Lineage-Specific PCR Primers for Gene-Targeted Metagenomics

BERNARDO SACHMAN-RUIZ, BRUNO CONTRERAS-MOREIRA, ENRIQUE ZOZAYA, CRISTINA MARTÍNEZ-GARZA, AND PABLO VINUESA

51.1 INTRODUCTION

This chapter presents a practical overview and case study of the usage and capabilities of the primers4clades version 1.0 web server [Contreras-Moreira et al., 2009], a software tool useful to design degenerate PCR primers for gene-targeted analysis of metagenomic DNA (see also Chapter 28, Vol. I) and multilocus sequence analysis. Although direct shotgun sequencing of community DNA is now feasible [Venter et al., 2004], polymerase chain reaction (PCR) remains the most widely used technology to gain molecular markers for studies in molecular ecology and systematics. With the ongoing accumulation of fully sequenced genomes in public sequence databases, the focus of microbial ecology studies is increasingly shifting to the analysis of protein coding genes and sequences (CDSs) to understand ecological, metabolic, and evolutionary processes in nature [Falkowski et al., 2008; Frias-Lopez et al., 2008; Hunt et al., 2008]. This trend is reflected in the huge interest of studying the diversity and expression patterns of "functional genes" in the environment, such as those encoding for antibiotic resistance and virulence [Castiglioni et al., 2008; Manning et al., 2008], photosynthesis [Yutin and Beja, 2005], or enzymes involved in the nitrogen cycle [Smith et al., 2007; Zehr et al., 2003], to mention a few. Furthermore, multilocus sequence analysis (MLSA) and typing (MLST) of protein-coding genes are the new standards in molecular systematics [Edwards, 2009; Gevers et al., 2005; Vinuesa, 2010] and molecular epidemiology [Maiden, 2006; Smith et al., 2009]. However, it still remains a major challenge to design optimal PCR primers to specifically amplify CDSs from target lineages directly from environmental DNA samples or from novel organisms. Primers4clades is a publicly available web server that uses phylogenetic trees to aid in the design of lineage-specific degenerate PCR primers for the above-mentioned purposes, which takes into account both protein and the corresponding codon multiple sequence alignments [Contreras-Moreira et al., 2009].

The major advantages of using a lineage-specific gene-targeted approach to metagenomics lie in the high sampling coverage and long sequence reads that can be achieved by sequencing a relatively low number of clones using classic Sanger sequencing technology. Furthermore, amplification biases derived from sequence composition heterogeneity [Polz and Cavanaugh, 1998] are reduced due to the relatively homogeneous composition of the sequences targeted with lineage-specific PCR primers. Their lower degeneracy also contributes to a reduction in amplification biases.

Here we present an empirical validation study that demonstrates the highly specific amplification of *Mycobacterium* spp. *rpoB* sequences from complex metagenomic DNA samples gained from two tropical Mexican forest soils: (i) the humid evergreen forest reserve of Los Tuxtlas, Veracruz, and (ii) a conserved patch of seasonally dry deciduous forest in the Biosphere

Handbook of Molecular Microbial Ecology, Volume I: Metagenomics and Complementary Approaches, First Edition. Edited by Frans J. de Bruijn.

Reserve of Sierra de Huautla, Morelos. Rarefaction curves for a random subsample of 50 clones of each library are compared with rarefaction curves for a 16S rRNA-based library of 98 clones obtained from Amazonian pasture and forest soils using universal primers [Borneman and Triplett, 1997], which illustrates the usefulness of the lineage-specific metagenomics analysis approach to improve sampling coverage and statistical power.

51.2 METHODS

51.2.1 Implementation, Input Data Processing, and Overview of the Two Run Modes Using the Server's *Actinobacteriales* Demonstration Dataset

Primers4clades (primers for clades) is an easy-to-use web server developed for researchers interested in the design of PCR primers for cross-species amplification of novel sequences from metagenomic DNA or from uncharacterized organisms belonging to user-specified phylogenetic clades or taxa. Degenerate PCR primers are derived from conserved motifs in protein multiple sequence alignments using the CODEHOP primer design strategy described below, but considering also the underlying codon alignments to obtain what we call "corrected CODEHOP" primer formulations. This is one of the diverse components of the extended CODEHOP algorithm that the server implements [Contreras-Moreira et al., 2009].

Primers4clades was mainly written in Perl and uses several BioPerl modules [Stajich et al., 2002] along with the open source software cited below to perform the different computations involved in the extended CODEHOP algorithm. The input for the server is a set of homologous protein-coding genes in FASTA format and in frame +1, which may be aligned or not, with or without introns (**Fig. 51.1**).

We will demonstrate the usage of the software by choosing the server's *Actinobacteriales* dataset, one of the three provided as tutorial material, which can be selected by clicking on the "*demo*" button. Although no registration is required, we recommend that the user provide his email address, which will allow access for one week to the analyses stored on the server and also avoids potential loss of results due to eventual browser timeout events due to often long computation times (see FAQ section in the online tutorial for the details). Note that the taxon names (Latin binomials) should be placed within brackets, which is how the header from FASTA-formatted sequences are downloaded from GenBank, as shown in Figure 51.1.

This is required for the server to parse the taxon names in order to select the corresponding codon usage tables, as explained below. The server excises introns if their coordinates are indicated in the FASTA header (see the server's documentation and the fungal alpha-tubulins demo dataset), collapses redundant sequences to haplotypes, translates the CDSs with user-selected translation tables, and aligns them using Muscle [Edgar, 2004]. The alignment step is skipped, if the server detects that the uploaded DNA sequences are aligned. The protein alignment is projected on the underlying DNA sequences to compute the corresponding codon alignment, along with maximum likelihood (ML) distance matrices from the protein (WAG+G) or the codon (HKY85+G) alignments, using Tree-Puzzle [Schmidt et al., 2002]. If the server is run in the "advanced" and interactive "*cluster sequences*" mode, the above-mentioned distance matrices are used to compute and display a neighbor-joining (NJ) tree with "neighbor" from the PHYLIP package [Felsenstein, 2004], based either on the protein or codon alignments. The user can then select a clade from the displayed NJ tree to target the primer design toward its members (**Fig. 51.2**). Note that the first sequence in the alignment will be used to root the tree. In the default, noninteractive "*get primers*" run mode, all the uploaded sequences will be considered to compute the primer formulations.

51.2.2 The Standard and Extended CODEHOP Primer Design Strategies

Primers4clades implements an extended and fully automated CODEHOP (Consensus Degenerate Hybrid Oligonucleotide Primer) design strategy, which is based on both DNA and protein multiple sequence alignments of protein-coding sequences [Contreras-Moreira et al., 2009]. The standard CODEHOP design strategy was first published by Rose et al. [1998, 2003]. It represents a novel PCR primer design method devised to amplify distantly related members of viral protein families. The primers are derived from conserved amino acid BLOCKS within a protein multiple sequence alignment that don't contain gaps. Each hybrid primer consists of (a) a short (11–12 nucleotides long) 3′ degenerate core region that binds to the codons encoding 3–4 highly conserved amino acids and (b) a longer 5′ consensus clamp region that contains the most abundant codon for each amino acid as predicted on the basis of a user-selected codon usage table (CUT). The basic structure of a CODEHOP can be seen from the link # 1 provided in the Internet Resources section. Amplification initiates by annealing and extension of primers in the pool with the highest similarity in the 3′ core region, which confers the specificity of the amplification, and their stabilization by the consensus clamp, which only partially matches the target template. Because all primers are identical at their 5′ consensus clamp region, they will all anneal at high

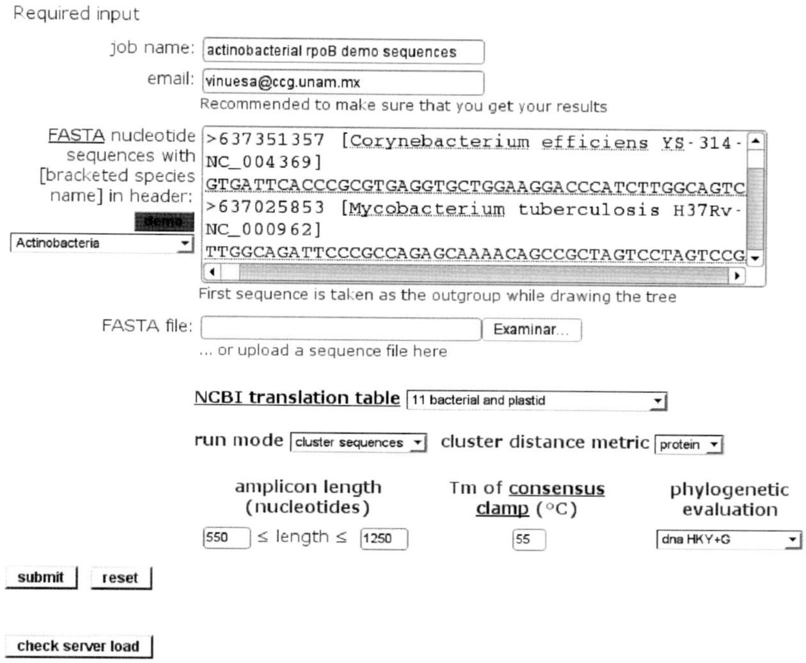

Figure 51.1 Data input and analysis customizations page of the primers4clades server. Three demo datasets are provided for rapid testing of the server's interface and capabilities. The tutorial presented in this chapter will focus on the actinobacterial dataset and the advanced, interactive "*cluster sequences*" run mode. The selected "*cluster distance metric*" option indicates to the server that the reference NJ tree should be computed from the protein sequence alignment generated from the uploaded coding sequences using NCBI's translation table number 11. The user can also change the preferred size range for the amplicons, T_m of the consensus clamp, and whether the phylogenetic information content of the resulting amplicon sets should be evaluated at the DNA or protein levels under a variety of substitution models.

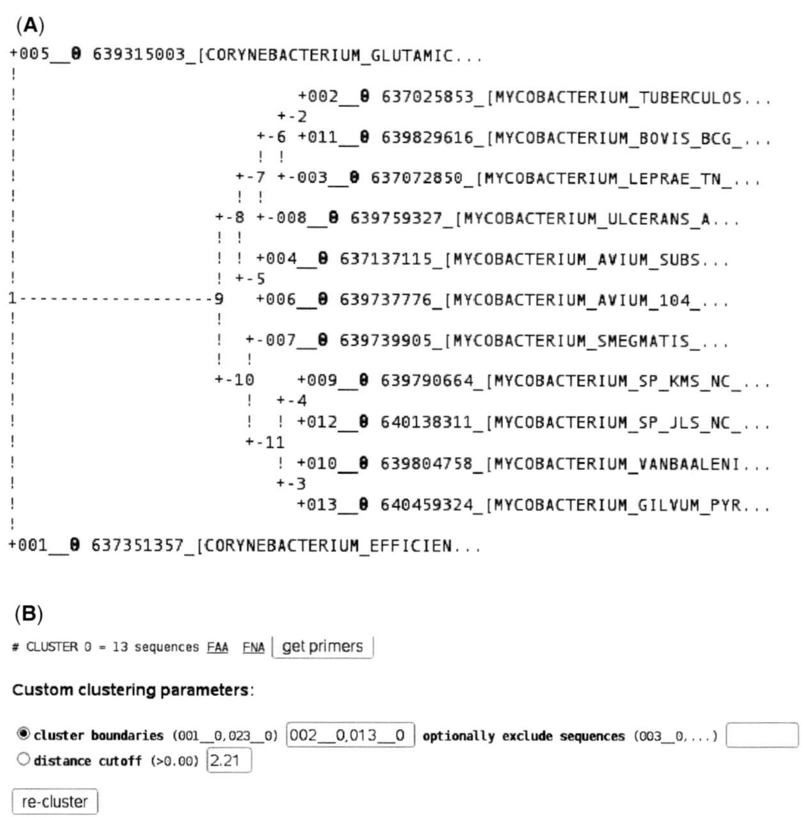

Figure 51.2 Reference NJ tree displayed on the first output page generated by primers4clades run in the interactive "*cluster sequences*" mode using the actinobacterial demonstration dataset. (**A**) Labeled NJ tree computed from the protein alignment of the input dataset. (**B**) Cluster selection panel based on the labels displayed on the NJ tree. Hitting the *re-cluster button* parses the alignment to use only the selected sequences. Hitting the *get primers button* starts the primer design and evaluation steps.

stringency during subsequent rounds of amplification. This increases the efficiency of the PCR amplification, making it in principle more efficient than PCR with standard consensus or degenerate oligonucleotides and allowing the amplification of divergent sequences of protein families. If no oligonucleotide formulations are returned by the server, or these are very long, it may be necessary to decrease the T_m of the consensus clamp (set to 55°C), as explained below. Primers4clades automatically uses the CUTs for all species identified in the input dataset for which a CUT is available at the Codon Usage Database [Nakamura et al., 2000]. It also computes an alignment-specific CUT on the fly. Taking into account several nonredundant CUTs notably increases the amount and quality of the primers found for a given dataset, as shown by our genome-scale benchmark analysis results reported in the original primers4clades publication [Contreras-Moreira et al., 2009]. This represents a notable performance improvement of our extended CODEHOP algorithm with respect to that implemented in the original CODEHOP and recent iCODEHOP servers (links #2 and #3 of Internet Resources).

51.2.3 Customization of the Primers4clades Server Behavior

As mentioned above, the server can be run either in the basic "*get primers*" or in the advanced "*cluster sequences*" modes. However, further customization of the server's behavior may be required for optimal primer design in either mode. For demonstration purposes, we will use the advanced and interactive "*cluster sequences*" run mode. Clicking on the "*customize settings*" button opens a window with self-explanatory options (Fig. 51.1). First of all, the input DNA protein-coding sequence can be translated by any of the current NCBI translation tables, with the "bacterial plastid" translation table 11 being the correct choice for our actinobacterial demonstration data. Next, the user can select whether to work at the DNA or protein levels to estimate a first reference NJ tree using the "*cluster metric*" option. As shown in Figure 51.1, we selected the default protein choice. Next, the user can choose a desired size range for the amplicons, as well as the T_m for the consensus clamp. Lastly, we choose the HKY85 + G DNA substitution model for the evaluation of the phylogenetic information content of the different amplicon sets, as explained below. Please read the servers FAQ section of the online documentation if you need some background information on selecting DNA substitution models.

A unique feature of the primers4clades server is its ability to compute a robust phylogenetic information content parameter for each theoretical amplicon set, which is based on a recently developed Shimodaira–Hasegawa (SH)-like test for the significance of branches in a maximum likelihood tree, originally implemented in PhyML v2.4.5 [Anisimova and Gascuel, 2006]. In brief, the test assesses whether the branch being studied provides a significant likelihood gain, in comparison with the null hypothesis that involves collapsing that branch, but leaving the rest of the tree topology identical. The resulting SH-like branch support values therefore indicate the probability that the corresponding split is significant. The phylogenetic information content of each amplicon set is calculated as the mean and median SH-like branch support values for the corresponding maximum likelihood tree inferred under the user-specified substitution model or matrix. As a rule of thumb, the phylogenetic information content of the targeted sequence region increases as branch support values approach to 1. This strategy of calculating the phylogenetic information content of different sequence alignments was recently proposed by Vinuesa et al. [2008].

Once all customization choices have been properly set, we are ready to hit the submit button to let the server start its work. After a few seconds, we get a new page displaying the NJ phylogeny (see Fig. 51.2A) based on the translation products of the actinobacterial *rpoB* sequences with some summary information of the alignment and distance matrix written on top of it (not shown). Notice that the two *Corynebacterium* sequences form an outgroup clade to the target *Mycobacterium* spp. ingroup clade on which we will now instruct the server to focus the search for PCR primers. To do so, we provide the cluster boundary identification numbers of the target clade in the corresponding box, as shown in Figure 51.2B. This can be simply done by copying and pasting the numbers from the displayed tree that demarcate the target clade, separating them with a comma without spaces. In our example, this is 002_0,013_0. After hitting the re-cluster button a new page will be presented, displaying the same tree but only with the target sequences selected. We could still at this point exclude further sequences or define a different subcluster for primer design, but in our example we are now ready to hit the "*get-primers button*."

51.2.4 Results Returned to the User by the Primers4clades Server

A very useful feature of primers4clades is that it returns a nonredundant set of primer pair formulations, ranked according to their thermodynamic properties (the theoretically best primers are displayed first). The server checks that the resulting amplicon sets for the primer pairs do not overlap more than 80%, ensuring a high coverage of the target locus, but filtering excessive redundancy, as shown on the amplicon distribution maps generated by the server after some seconds or minutes (**Fig. 51.3**), depending on

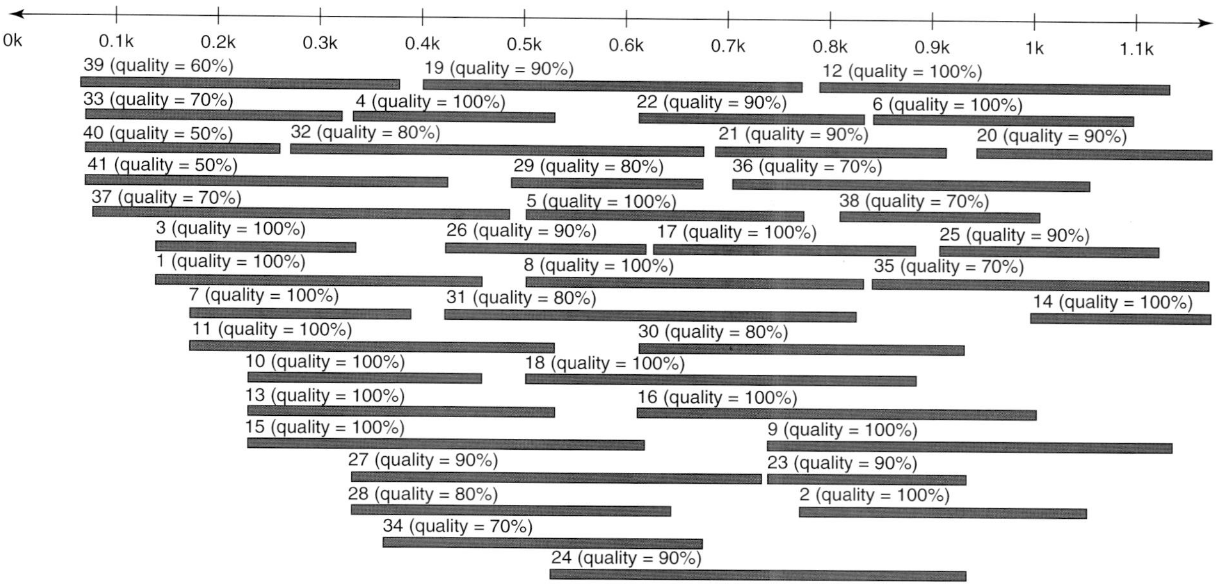

Figure 51.3 Amplicon distribution map for the actinobacterial demonstration dataset generated using the parameters shown in Figure 51.1 for the selected strains as shown in Fig. 51.2. The positions of the different amplicons are mapped on the first sequence of the input alignment at the protein level. A quality value is assigned to each amplicon, with 100% being best. This value is computed based on thermodynamic properties of the primer pair (see text for the details). The amplicon distribution map and all primer pair formulations along with their thermodynamic properties can be downloaded from the server.

the size of the alignment and number of primer pairs found. During the process of primer pair calculation, the server displays information about the redundancy among the codon usage tables (CUTs) for each taxon and the number of primer pairs found for each nonredundant CUT (see the section on "Technical Information" of the server's online documentation page for a detailed explanation of how it computes the nonredundant CUTs).

The amplicon distribution map can be downloaded from the server as well as a TAB-delimited text file containing all the primer pair formulations, amplicon size, and a comprehensive set of their thermodynamic properties such as degeneracy, T_m, hairpin-formation potential, and cross-hybridization potential computed with subroutines borrowed from the Amplicon software [Jarman, 2004]. The quality % value shown on top of each bar representing an amplicon indicates how "good" the corresponding primer pairs are, based on the above-mentioned thermodynamic properties. A 100% score indicates that none of the property values is worse than a predetermined cutoff value for each parameter that we have empirically chosen based on the evaluation of dozens of well-working primer pairs. Note also that the bars representing amplicons are color-coded according to their corresponding primer pair quality scores.

In addition to the standard CODEHOP, the primers4clades server computes three additional primer formulations: the corrected CODEHOP, relaxed corrected CODEHOP, and fully degenerate oligonucleotides. The former two are based on the original CODEHOP formulation, but adjusting the degeneracy of the latter by taking into account the sequences of the underlying codon alignment, as shown in **Figure 51.4**. The relaxed corrected CODEHOP extends the 3′ core region of the corrected CODEHOP toward the 5′ end until it reaches a degeneracy level of 24. If the corrected CODEHOP already has a degeneracy equal or greater than 24, then the latter two formulations will be identical. The fully degenerated oligonucleotide formulation is computed to reflect the full nucleotide sequence variation of the targeted binding site. Table 1 of the server's online documentation page summarizes the recommended uses of the four oligonucleotide formulations computed by the server. We recommend the use of the corrected CODEHOP formulation for most purposes.

After displaying the amplicon distribution map, the server enters the second computing intensive phase. Maximum likelihood phylogenies (in our case study using the HKY85+G model) will be estimated from each of the theoretical amplicon sets displayed on the amplicon distribution map. It is from these phylogenies that the phylogenetic information content of the amplicons is calculated based on Shimodaira–Hasegawa-like branch support values [Anisimova and Gascuel, 2006; Guindon et al., 2009], as we have described recently [Vinuesa, 2010; Vinuesa et al., 2008]. The best scoring amplicons (based on the quality parameter explained above) are evaluated first.

```
GCATCGATTTCGtgytcytcdat 5'->3' C 505 775 (aligned residues)
gcatcgatctcatgctcctcgat >001 637351357 [CORYNEBACTERIUM EFFICIENS YS-314-NC_004369]
gcatcgatctcatgctcctcgat >002 637025853 [MYCOBACTERIUM TUBERCULOSIS H37RV-NC_000962]
gcgtcgatctcatgctcctcaat >003 637072850 [MYCOBACTERIUM LEPRAE TN-NC_002677]
gcgtcgatctcgtgctcctcgat >004 637137115 [MYCOBACTERIUM AVIUM SUBSP. PARATUBERCULOSIS K-10-NC_002944]
gcatcgatctcgtgctcctcgat >005 639315003 [CORYNEBACTERIUM GLUTAMICUM ATCC 13032-NC_006958]
gcgtcgatctcgtgctcctcgat >006 639737776 [MYCOBACTERIUM AVIUM 104-NC_008595]
gcatcgatctcgtgctcctcgat >007 639739905 [MYCOBACTERIUM SMEGMATIS STR. MC2 155-NC_008596]
gcgtcgatctcgtgttcctcgat >008 639759327 [MYCOBACTERIUM ULCERANS AGY99-NC_008611]
gcatcgatctcgtgctcttcgat >009 639790664 [MYCOBACTERIUM SP. KMS-NC_008705]
gcatcgatctcgtgctcttcgat >010 639804758 [MYCOBACTERIUM VANBAALENII PYR-1-NC_008726]
gcatcgatctcatgctcctcgat >011 639829616 [MYCOBACTERIUM BOVIS BCG STR. PASTEUR 1173P2-NC_008769]
gcatcgatctcgtgctcttcgat >012 640138311 [MYCOBACTERIUM SP. JLS-NC_009077]
gcatcgatctcgtgctcttcgat >013 640459324 [MYCOBACTERIUM GILVUM PYR-GCK-NC_009338]
..?.....!..?........?..
GCATCGATCTCGtgytcytcrat codeh_corr bea4cfef_intronless_cluster_0_amp3_C775
GCAtcgatctcrtgytcytcrat relax_corr bea4cfef_intronless_cluster_0_amp3_C775
gcrtcgatctcrtgytcytcrat degen_corr bea4cfef_intronless_cluster_0_amp3_C775

# primer pair quality = 100%
# expected PCR product length (nt) = 813
# fwd: minTm = 63.8 maxTm = 68.4
# rev: minTm = 60.0 maxTm = 65.1
#--
# phylogenetic amplicon evaluation:
# subs.model = HKYG
# alpha = 0.28
# mean aLRT = 0.80 median aLRT = 0.95
# amplicon FASTA file
# aLRT likelihood tree (newick)
# end_of_amplicon
```

Figure 51.4 Example primer formulation output panel (only the reverse primer of the first pair returned by the server is shown) for the actinobacterial example dataset. The server returns the standard CODEHOP (bold) and three codon-based primer formulations, aligned with the underlying codons. An '!' sign denotes positions corrected by the system based on the codon alignment. Also shown are the summary outputs for key primer pair and corresponding amplicon characteristics, as well as the associated phylogenetic information content of the theoretical amplicon sets (see the text for the details).

After the server is done with the ML search for the first amplicon set, the corresponding primer pairs are shown on screen, aligned with the underlying codons (Fig. 51.4). A header line indicates the codon usage table used to calculate the corresponding primer pair, in this case a data-specific one computed from the input alignment (not shown on Fig. 51.4). The first line after the header shows the sequence of the CODEHOP formulation, as calculated by the original CODEHOP program, showing also the coordinates of the forward (fw) and reverse (rev) oligos on the original (full) protein alignment. The lowercase letters on the 3′ end of the oligo correspond to the degenerated portion of the CODEHOP, while the uppercase residues toward the 5′ end correspond to the consensus clamp. The lines below the CODEHOP formulation are the corresponding target sites at the DNA level. This codon alignment is used to calculate the corrected CODEHOP formulation. The ′!′ and ′?′ characters just above the corrected CODEHOP formulation line indicate the discrepancies between the CODEHOP formulation and the codon alignment that are considered to formulate the corrected CODEHOP sequence, the ′!′ indicating nondegenerate sites, and the ′?′ highlighting degenerate sites (Fig. 51.4). The ′.′ symbol indicates matches between the original CODEHOP formulation and the codon alignment. The relaxed corrected CODEHOP formulation and the fully degenerated oligonucleotide formulations are also displayed (Fig. 51.4).

A primer pair quality summary report is displayed on screen after each primer pair, based on computations performed on the corrected CODEHOP formulations. The first line provides a summary score of the "primer quality" in thermodynamic terms for each primer pair (range 100% to 0%, best to worst). The next three lines indicate the expected amplicon length in bps and the computed T_m ranges for the fw- and rev-corrected codehop formulations (Fig. 51.4). If the quality score is $<100\%$, this means that at least one oligo has a thermodynamic parameter value worse than the cutoff values used by the server (these values are presented in Table 2 of the server's online documentation page).

Lastly, we get a summary of the phylogenetic evaluation of the corresponding amplicon (Fig. 51.4), which indicates the substitution matrix/model used for the evaluation, and the value of the corresponding shape parameter (alpha). The next line is very useful for users interested in using the new molecular markers for phylogenetics, since it provides the mean and median Shimodaira–Hasegawa-like bipartition support values computed by the PhyML maximum likelihood algorithm. The closer to 1, the higher the resolution level of the tree inferred from the theoretical amplicon set and therefore the higher its phylogenetic information content. A FASTA-formatted file containing the aligned amplicon sets can be downloaded (in this case at the protein level). The ML trees for each amplicon can be either directly visualized or downloaded.

51.2.5 Purification of Soil DNA, rpoB Amplification, Clone Library Construction, and Analysis

Ten soil samples of approximately 100 g were randomly sampled within a 10-m × 10-m area in both the tropical humid forest Reserve of Los Tuxtlas, Estación Biológica de la UNAM, Veracruz, Mexico (18°34′23.24″N, 95°9′40.10″O), and a conserved patch of seasonally dry deciduous tropical forest near "El Limón" in the Biosphere Reserve of Sierra de Huautla, Morelos, Mexico (18°32′33.20″N, 98°56′11.19″O). The 10 soil samples were from each site were pooled, mixed, and sieved (2 mm) to obtain two composite samples. A 2-g aliquot from each of the composite samples was used for DNA extraction using the MoBio PowerMax Soil DNA Extraction Kit. DNA integrity was checked on an agarose gel, and the lack of strong PCR inhibitors was confirmed by amplification of 16S rRNA genes using the universal fD1 and rD1 primers [Weisburg et al., 1991]. The *rpoB* amplicons were generated using a hot-start protocol and *Taq*-platinum (Invitrogen), with annealing at 60°C, and cloned using the PCR Topo cloning system 2.1 (Invitrogen). Plasmids were purified using a 96-well plate format and commercially sequenced at LANGEBIO, Mexico, using the Sanger method at 2 × coverage with the M13f and M13rev primers that bind at vector positions flanking its cloning site. The two sequence reads for each plasmid were automatically assembled and validated by using a custom Perl script that runs Phred/Phrap [Ewing and Green, 1998; Ewing et al., 1998] followed by a BlastX [Altschul et al., 1997] search of each contig against a locally formatted *rpoB* sequence database of *Actinobacteria* assembled from fully sequenced genomes of these bacteria. Only sequences aligning globally with entries from the database were further processed by subjecting them to a sequence-to-profile alignment using Clustalw 2.1 [Larkin et al., 2007]. The reference profile consists of a codon alignment of 25 nonredundant *rpoB* sequences of reference *Mycobacterium* strains, trimmed to the segment corresponding to the expected amplicon. The resulting multiple sequence alignment was further corrected manually.

Phylogenetic analyses were performed with PhyML v. 3.0 [Guindon et al., 2009], as previously described [Vinuesa, 2010; Vinuesa et al., 2008]. All community ecology analyses of the environmental *rpoB*-amplicon libraries were performed with Mothur version 1.7 [Schloss et al., 2009].

51.3 RESULTS

51.3.1 Design and Evaluation of the Specificity of *Mycobacterium* spp. PCR Primers Targeting the RNA Polymerase Beta Subunit (*rpoB*) Gene Using Soil Metagenomic DNA

Using the primers4clades server and selected reference sequences (a few more than those provided at the server's actinobacterial demonstration sequence set), we designed the primer pair shown in **Table 51.1**. These primers were used to generate PCR amplicons from metagenomic DNA purified from the two contrasting tropical forest soils in Mexico (humid evergreen forest at Los Tuxtlas, Veracruz; deciduous seasonally dry forest from Sierra de Huautla, Morelos), as described in Materials and Methods.

Specific amplification products could be obtained for both soils, as shown in Figure 10 of the server's online documentation page. We sequenced 50 clones from each of the two amplicon libraries to evaluate

Table 51.1 Basic Thermodynamic Parameters and Sequence Formulation of the Corrected Codehop Primers Used in This Study to Amplify *Mycobacterium*-Specific Sequences from Two Contrasting Soil Samples

Name	L^a	Sequence	T_m	Corrected Degeneracy[b]	Full Degeneracy[c]	h_pot[d]	c_pot[e]
rpoB2_N326	26	GAGAACCTGTTCTTCaaggagaagcg	63.3	0	0	0.52	0.50
rpoB2_C757	25	CGTCCTCGTAGTTGtgrccytccca	66.6	4	4	0.00	0.50

[a] Oligonucleotide length.
[b] Degeneracy of the corrected codehop oligonucleotide (see technical details in the server's tutorial page).
[c] Full degeneracy of the target alignment site (see technical details in the server's tutorial page).
[d] Hairpin formation potential.
[e] Cross-hybridization potential.

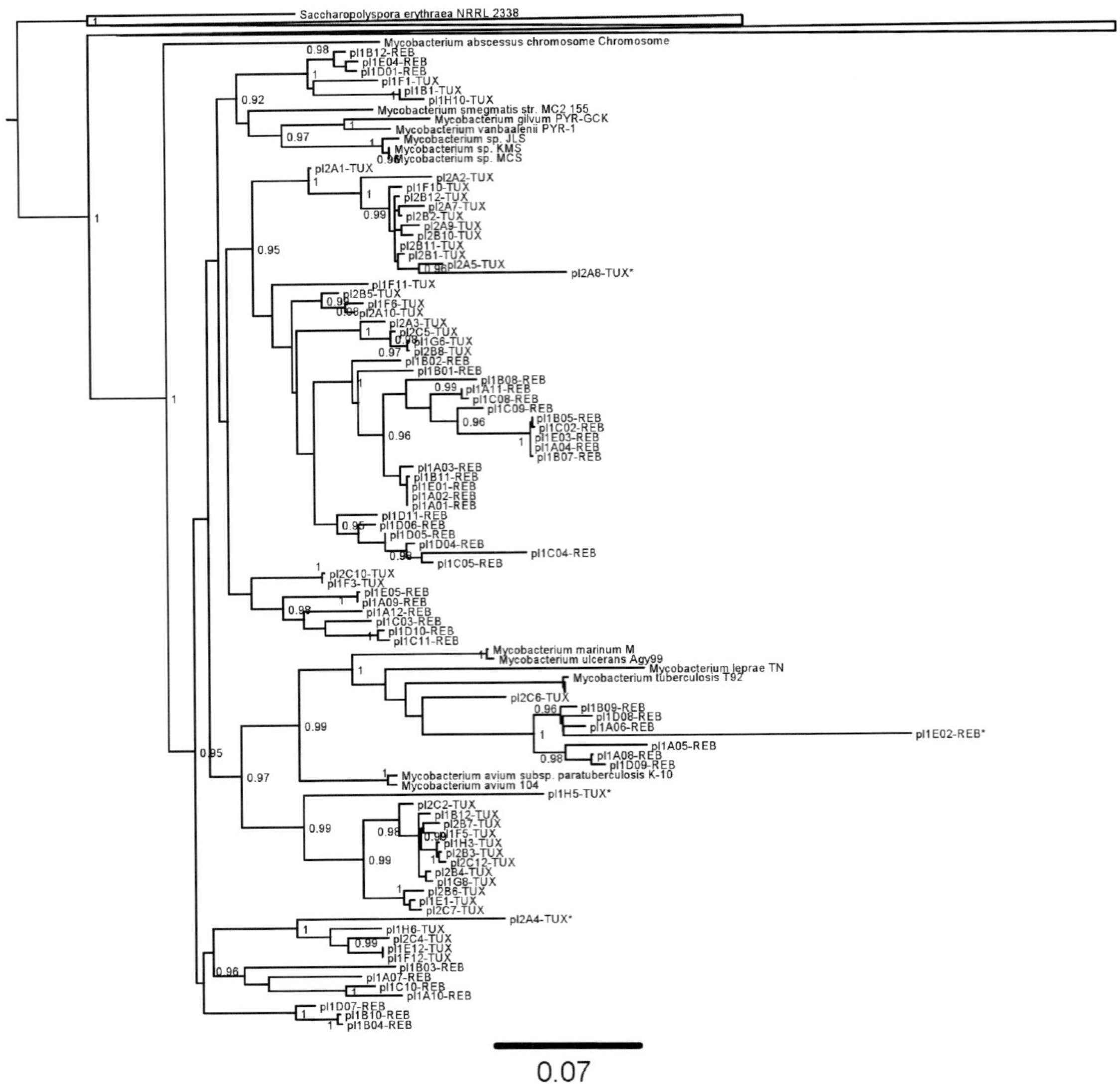

Figure 51.5 Maximum likelihood phylogeny under the GTR+G+I substitution model for 32 reference sequences [*Corynebacterium* (7), *Mycobacterium* (17), *Nocardia* (1), *Rhodococcus* (3), *Saccharopolyspora* (1), *Salinispora* (2), and *Tsukamurella* (1)] and 100 soil clones (50 from the Reserve of Los Tuxtlas, labeled as TUX; 50 from the Sierra de Huautla Reserve, labeled as REB). For the sake of clarity, the non-*Mycobacterium* sequence clades are collapsed. Eighty-nine percent of the soil clones clustered within the *Mycobacterium* spp. clade, demonstrating the high selectivity of the *rpoB* primers shown in Table 51.1 to target this group of organisms in complex metagenomic DNA samples.

the specificity of the oligonucleotides in a phylogenetic context. **Figure 51.5** shows a maximum likelihood phylogeny (GTR+G) of the 100 environmental sequences along with 15 reference outgroup strains in the genera *Corynebacterium* (7), *Nocardia* (1), *Rhodococcus* (3), *Saccharopolyspora* (1), *Salinispora* (2), and *Tsukamurella* (1), which are shown as collapsed clades in Figure 51.5, and 17 *Mycobacterium* spp. ingroup reference strains. From this phylogeny we could unambiguously define the perfectly supported *Mycobacterium* spp. clade, which clustered 89 of the environmental clones. The remaining 11 environmental sequences clustered with the closely related actinobacterial genera used as outgroups.

In summary, 89% of the environmental *rpoB* amplicons could be unambiguously assigned to the genus *Mycobacterium*, indicating that the primers have a high specificity for the target genus. Only ~5% of the environmental *Mycobacterium* sequences had suspiciously long branches, which might reflect sequence artifacts such as chimeras, heteroduplexes, or *Taq* errors.

51.3.2 Statistical Evaluation of the Sampling Effort Achieved by the Lineage-Targeted Approach

A recent review on the use of *rpoB* sequences in *Mycobacterium* molecular systematics suggests that 97.7–98.2% similarity cutoff values for *rpoB* sequences could be used to delimit species within the genus [Adekambi et al., 2009]. However, in this study we use a slightly more conservative cutoff value (3% divergence) to define operational taxonomic units (OTUs; approximately equivalent to species) for our *Mycobacterium* spp. diversity study.

Figures 51.6A and 51.6B show the rarefaction curves [Hughes et al., 2001] generated for the 50 *Mycobacterium* spp. and closely related actinobacterial sequences obtained from both *rpoB* soil clone libraries at different cutoff values. Notice that for both soils the rarefaction curves are close to reach an asymptote at the 3% divergence OTU definition, indicating that the sequencing effort has been reasonably good. As a reference point, Figure 51.6C shows the rarefaction curves at the same cutoff values for a classical 16S rRNA gene clone library made from a Brazilian tropical soil using universal primers [Borneman and Triplett, 1997]. These curves indicate that the 98 clones sequenced in that study are far from being representative of the diversity at the species level, which is classically defined by a cutoff level of 97–98% of sequence similarity (2–3% divergence). Figures 51.6D and 51.6E show collector curves for the two libraries using the nonparametric Chao1 richness estimator [Hughes et al., 2001], which indicate that the *Mycobacterium* community at seasonally dry deciduous forest of El Limón is more diverse (19 predicted species, with a 95% credibility interval 14–46) than that found in the humid tropical forest of Los Tuxtlas (12 predicted species, with a 95% credibility interval 12–32). Interestingly, not a single OTU is shared among the communities at this cutoff level, indicating a strongly differentiated species composition of both communities, which together bring about 31 OTUs. The Chao1 collector curves shown in Figures 51.6D and 51.6E indicate that the sampling effort has not been sufficient to evaluate the diversity at the two lower cutoff values since they have not stabilized [Hughes et al., 2001], but those for the 0.03 and 0.05 cutoff values seem to have stabilized. Therefore, inferences at these cutoff levels are robust.

51.4 DISCUSSION

Primers4clades is currently the only publicly available server that integrates alternative primer-design strategies with phylogenetic trees to (a) interactively target the search for oligonucleotide formulations to specific sequence clusters and (b) evaluate the phylogenetic information content of the new molecular markers. Together, these features are very useful to make an informed choice among alternative, nonredundant primer pair formulations, taking into account both the thermodynamic properties of the primers and the phylogenetic information content of the expected amplicon sets. These attributes make of primers4clades a novel and powerful tool for the targeted design and informed selection of PCR primers for lineage-targeted metagenomic studies of microbial communities, as demonstrated herein for environmental mycobacteria.

The genus *Mycobacterium* currently has 147 validly described species (see link #4 in Internet resources), most of which are poorly characterized ubiquitous environmental mycobacteria (EM) [Bland et al., 2005; Jacobs et al., 2009]. These are known as "nontuberculous" or "atypical" mycobacteria among clinicians [van Ingen et al., 2009]. The most "famous" mycobacterial species are *M. tuberculosis* and *M. leprae*, obligate parasites and the causative agents of tuberculosis (TB) and leprosy, respectively. *M. avium, M. abscessus, M. chelonae, M. fortuitum, M. intracellulare, M. kansasii, M. mucogenicum, M. phocaicum, M. scrophulaceum, M. ulcerans*, and *M. vanbalenii* are among the most common nontuberculous or atypical mycobacterial species causing opportunistic infections in immunocompromised humans and have a broad environmental distribution [Falkinham, 2009; Primm et al., 2004]. Understanding the diversity, abundance, and distribution of EM is of prime interest from a biomedical standpoint because immune responses following exposure to these bacteria may influence susceptibility to tuberculosis and leprosy and may severely interfere with the proliferation and protective effect of the attenuated BCG vaccine used against TB [Black et al., 2001; Weir et al., 2006]. Their functional role in ecosystems remains largely unknown despite their environmental ubiquity.

The results presented in this study demonstrate that a good sampling effort can be achieved for the genus *Mycobacterium* with the primers described in Table 51.1 by sequencing as few as 50 clones per site using the lineage-targeted approach to gene-based metagenomic analyses of microbial communities. Further sequencing and statistical analysis of these and additional libraries (>200 clones each) generated with the mentioned primers from soil and water samples [Sachman-Ruiz et al., 2009] demonstrated that sequencing around 200 clones provides very strong statistical power, allowing for the first time to make fine-grained inferences about

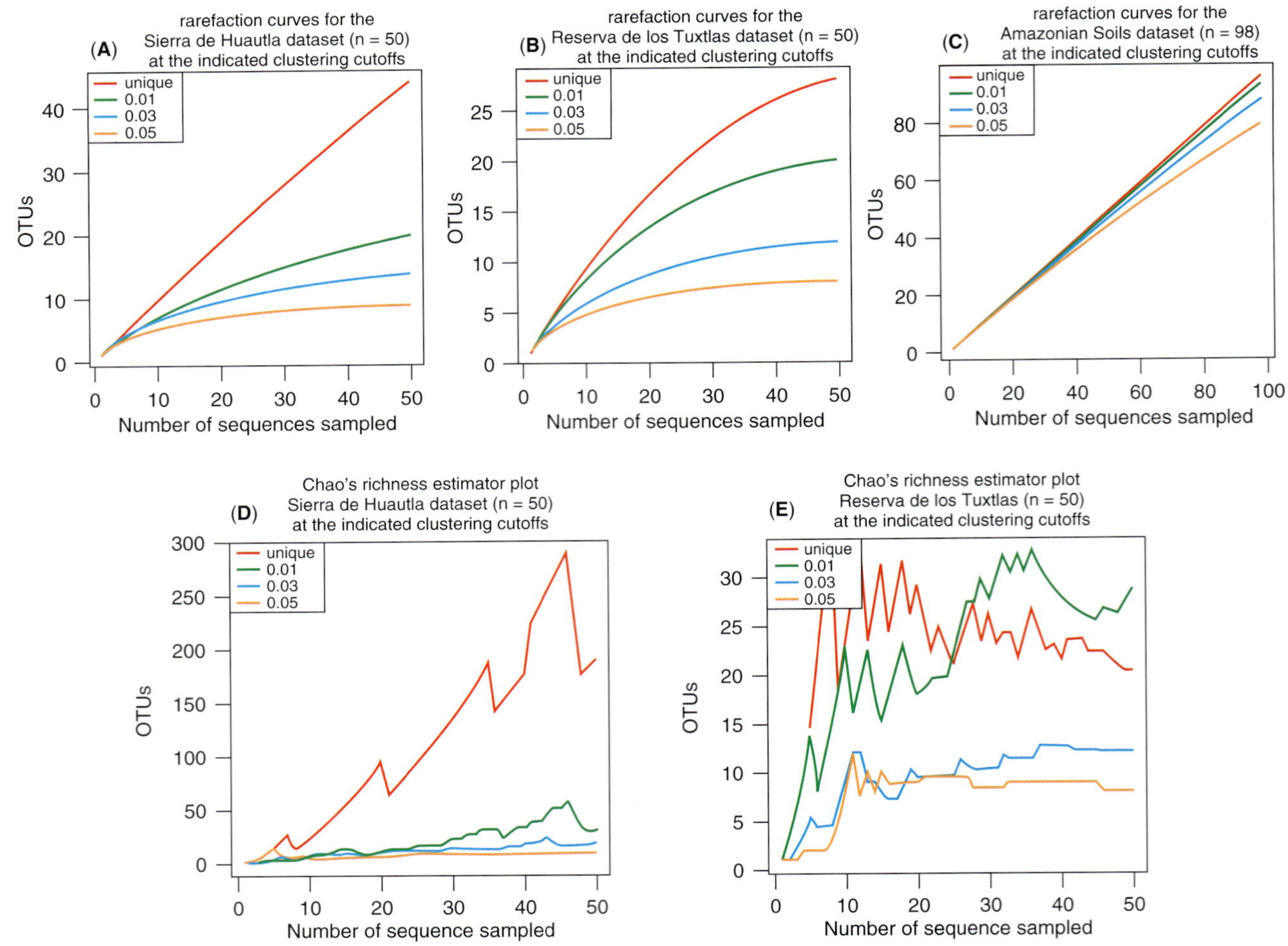

Figure 51.6 Rarefaction (**A–C**) and collector curves for the nonparametric Chao1 richness estimator (**C, D**) at four clustering levels (0.0–0.05 *P* distance) for OTU definition. Panel C shows for comparative purposes the rarefaction curves for 98 SSU rRNA Amazonian soil clones generated with universal primers from the classical study by Borneman and Triplett [1997].

the richness and structure of mycobacteria communities in diverse natural environments. Furthermore, the latter studies have revealed the magnitude of the bias in diversity estimation introduced by classical *Mycobacterium* isolation procedures and have also revealed that we have a strongly distorted notion of the diversity of this species-rich bacterial genus of great environmental and clinical relevance.

These results demonstrate the utility and strong statistical power of the lineage-targeted approach to microbial community ecology and show that primers4clades [Contreras-Moreira et al., 2009] (link #5 in the Internet Resources) is a useful tool to develop primers for such gene-targeted metagenomic analyses of microbial diversity. The development of the tool is now coupled to its recent implementation in a phylogenomics analysis pipeline to construct an interactive primer database for phylogenetic clades at different taxonomic and phylogenetic depths. The graphical interface, analysis options, and parameter evaluation procedures will be improved, extended, and refined in future versions, which will also include faster ML algorithms.

INTERNET RESOURCES

Link 1: CODEHOP structure (http://www.ncbi.nlm.nih.gov/pmc/articles/PMC168931/figure/gkg524f1/)

Link 2: CODEHOP server (http://blocks.fhcrc.org/codehop.html)

Link 3: iCODEHOP server (https://icodehop.cphi.washington.edu/i-codehop-context/iCODEHOP/view/PrimerAnalysis)

Link 4: *Mycobacterium* taxonomy (http://www.bacterio.cict.fr/m/mycobacterium.html)

Link 5: Primers4clades web server (http://maya.ccg.unam.mx/primers4clades)

Acknowledgments

Romualdo Zayas-Laguna and Víctor del Moral from the Information Technology Administration UNIT at CCG-UNAM are acknowledged for their technical support. This work was supported by DGAPA/PAPIIT-UNAM [grant number IN201806-2], CONACyT-Mexico [grant number P1-60071], and by Consejo Superior de Investigaciones Científicas [grant number 200720I038]. UNAM's Ph.D. Program in Biomedical Sciences and CONACyT-Mexico are acknowledged for the financial support offered to BSR for his Ph.D. studies.

REFERENCES

ADEKAMBI T, DRANCOURT M, RAOULT D. 2009. The *rpoB* gene as a tool for clinical microbiologists. *Trends Microbiol*. **17**(1):37–45.

ALTSCHUL SF, MADDEN TL, SCHÄFFER AA, ZHANG J, ZHANG Z, et al. 1997. Gapped BLAST and PSI-BLAST: A new generation of protein database search programs. *Nucleic Acids Res*. **25**:3389–3402.

ANISIMOVA M, GASCUEL O. 2006. Approximate likelihood-ratio test for branches: A fast, accurate, and powerful alternative. *Syst. Biol*. **55**(4):539–352.

BLACK GF, DOCKRELL HM, CRAMPIN AC, FLOYD S, WEIR RE, et al. 2001. Patterns and implications of naturally acquired immune responses to environmental and tuberculous mycobacterial antigens in northern Malawi. *J. Infect. Dis*. **184**(3):322–329.

BLAND CS, IRELAND JM, LOZANO E, ALVAREZ ME, PRIMM TP. 2005. Mycobacterial ecology of the Rio Grande. *Appl. Environ. Microbiol*. **71**(10):5719–5727.

BORNEMAN J, TRIPLETT EW. 1997. Molecular microbial diversity in soils from eastern Amazonia: Evidence for unusual microorganisms and microbial population shifts associated with deforestation. *Appl. Environ. Microbiol*. **63**(7):2647–2653.

CASTIGLIONI S, POMATI F, MILLER K, BURNS BP, ZUCCATO E, et al. 2008. Novel homologs of the multiple resistance regulator *marA* in antibiotic-contaminated environments. *Water Res*. **42**(16):4271–4280.

CONTRERAS-MOREIRA B, SACHMAN-RUIZ B, FIGUEROA-PALACIOS I, VINUESA P. 2009. Primers4clades: A web server that uses phylogenetic trees to design lineage-specific PCR primers for metagenomic and diversity studies. *Nucleic Acids Res*. **37**(Web Server issue):W95–W100.

EDGAR RC. 2004. MUSCLE: Multiple sequence alignment with high accuracy and high throughput. *Nucleic Acids Res*. **32**(5):1792–1797.

EDWARDS SV. 2009. Is a new and general theory of molecular systematics emerging? *Evolution* **63**(1):1–19.

EWING B, GREEN P. 1998. Base-calling of automated sequencer traces using phred. II. Error probabilities. *Genome Res*. **8**(3):186–194.

EWING B, HILLIER L, WENDL MC, GREEN P. 1998. Base-calling of automated sequencer traces using phred. I. Accuracy assessment. *Genome Res*. **8**(3):175–185.

FALKINHAM JO, 3rd. 2009. Surrounded by mycobacteria: nontuberculous mycobacteria in the human environment. *J. Appl. Microbiol*. **107**(2):356–367.

FALKOWSKI PG, FENCHEL T, DELONG EF. 2008. The microbial engines that drive Earth's biogeochemical cycles. *Science* **320**(5879):1034–1039.

FELSENSTEIN J. 2004. PHYLIP (Phylogeny Inference Package), 3.6 ed. Distributed by the author. Seattle: Department of Genetics, University of Washington.

FRIAS-LOPEZ J, SHI Y, TYSON GW, COLEMAN ML, SCHUSTER SC, et al. 2008. Microbial community gene expression in ocean surface waters. *Proc. Natl. Acad. Sci. USA* **105**(10):3805–3810.

GEVERS D, COHAN FM, LAWRENCE JG, SPRATT BG, COENYE T, et al. 2005. Opinion: Reevaluating prokaryotic species. *Nat. Rev. Microbiol*. **3**(9):733–739.

GUINDON S, DELSUC F, DUFAYARD JF, GASCUEL O. 2009. Estimating maximum likelihood phylogenies with PhyML. *Methods Mol. Biol*. **537**:113–137.

HUGHES JB, HELLMANN JJ, RICKETTS TH, BOHANNAN BJ. 2001. Counting the uncountable: Statistical approaches to estimating microbial diversity. *Appl. Environ. Microbiol*. **67**(10):4399–4406.

HUNT DE, DAVID LA, GEVERS D, PREHEIM SP, ALM EJ, et al. 2008. Resource partitioning and sympatric differentiation among closely related bacterioplankton. *Science* **320**(5879):1081–1085.

JACOBS J, RHODES M, STURGIS B, WOOD B. 2009. Influence of environmental gradients on the abundance and distribution of *Mycobacterium* spp. in a coastal lagoon estuary. *Appl. Environ. Microbiol*. **75**(23):7378–7384.

JARMAN SN. 2004. Amplicon: Software for designing PCR primers on aligned DNA sequences. *Bioinformatics* **20**(10):1644–1645.

LARKIN MA, BLACKSHIELDS G, BROWN NP, CHENNA R, MCGETTIGAN PA, et al. 2007. Clustal W and Clustal X version 2.0. *Bioinformatics* **23**(21):2947–2948.

MAIDEN MC. 2006. Multilocus sequence typing of bacteria. *Annu. Rev. Microbiol*. **60**:561–588.

MANNING SD, MOTIWALA AS, SPRINGMAN AC, QI W, LACHER DW, et al. 2008. Variation in virulence among clades of *Escherichia coli* O157:H7 associated with disease outbreaks. *Proc. Natl. Acad. Sci. USA* **105**(12):4868–4873.

NAKAMURA Y, GOJOBORI T, IKEMURA T. 2000. Codon usage tabulated from international DNA sequence databases: Status for the year 2000. *Nucleic Acids Res*. **28**(1):292.

POLZ MF, CAVANAUGH CM. 1998. Bias in template-to-product ratios in multitemplate PCR. *Appl. Environ. Microbiol*. **64**(10):3724–3730.

PRIMM TP, LUCERO CA, FALKINHAM JO, 3rd. 2004. Health impacts of environmental mycobacteria. *Clin. Microbiol. Rev*. **17**(1):98–106.

ROSE TM, SCHULTZ ER, HENIKOFF JG, PIETROKOVSKI S, MCCALLUM CM, et al. 1998. Consensus-degenerate hybrid oligonucleotide primers for amplification of distantly related sequences. *Nucleic Acids Res*. **26**(7):1628–1635.

ROSE TM, HENIKOFF JG, HENIKOFF S. 2003. CODEHOP (COnsensus-DEgenerate Hybrid Oligonucleotide Primer) PCR primer design. *Nucleic Acids Res*. **31**(13):3763–3766.

SACHMAN-RUIZ B, CASTILLO-RODAL AI, LÓPEZ-VIDAL Y, MARTÍNEZ-ROMERO E, VINUESA P. 2009. Diversity of environmental mycobacteria in Mexican rivers assessed by cultivation and metagenomics approaches. Poster abstract U-007. 109th General Meeting of the American Society for Microbiology. May 17–21 2009. Philadelphia.

SCHLOSS PD, WESTCOTT SL, RYABIN T, HALL JR, HARTMANN M, et al. 2009. Introducing mothur: open-source, platform-independent, community-supported software for describing and comparing microbial communities. *Appl. Environ. Microbiol*. **75**(23):7537–7541.

SCHMIDT HA, STRIMMER K, VINGRON M, VON HAESELER A. 2002. TREE-PUZZLE: maximum likelihood phylogenetic analysis using quartets and parallel computing. *Bioinformatics* **18**(3):502–504.

SMITH CJ, NEDWELL DB, DONG LF, OSBORN AM. 2007. Diversity and abundance of nitrate reductase genes (*narG* and *napA*), nitrite reductase genes (*nirS* and *nrfA*), and their transcripts in estuarine sediments. *Appl. Environ. Microbiol*. **73**(11):3612–3622.

SMITH GJ, VIJAYKRISHNA D, BAHL J, LYCETT SJ, WOROBEY M, et al. 2009. Origins and evolutionary genomics of the 2009 swine-origin H1N1 influenza A epidemic. *Nature* **459**(7250):1122–1125.

Stajich JE, Block D, Boulez K, Brenner SE, Chervitz SA, et al. 2002. The Bioperl toolkit: Perl modules for the life sciences. *Genome Res*. **12**(10):1611–1618.

van Ingen J, Boeree MJ, Dekhuijzen PN, van Soolingen D. 2009. Environmental sources of rapid growing nontuberculous mycobacteria causing disease in humans. *Clin. Microbiol. Infect*. **15**(10):888–893.

Venter JC, Remington K, Heidelberg JF, Halpern AL, Rusch D, et al. 2004. Environmental genome shotgun sequencing of the Sargasso Sea. *Science* **304**(5667):66–74.

Vinuesa P. 2010. Multilocus sequence analysis and bacterial species phylogeny estimation. In Oren A, Papke RT, eds. *Molecular Phylogeny of Microorganisms*. Norwich, UK: Caister Academic Press, pp. 41–64.

Vinuesa P, Rojas-Jimenez K, Contreras-Moreira B, Mahna SK, Prasad BN, et al. 2008. Multilocus sequence analysis for assessment of the biogeography and evolutionary genetics of four *Bradyrhizobium* species that nodulate soybeans on the Asiatic continent. *Appl. Environ. Microbiol*. **74**(22):6987–6996.

Weir RE, Black GF, Nazareth B, Floyd S, Stenson S, et al. 2006. The influence of previous exposure to environmental mycobacteria on the interferon-gamma response to bacille Calmette–Guerin vaccination in southern England and northern Malawi. *Clin. Exp. Immunol*. **146**(3):390–399.

Weisburg WG, Barns SM, Pelletie DA, Lane DJ. 1991. 16S ribosomal DNA amplification for phylogenetic study. *J. Bacteriol*. **173**:697–703.

Yutin N, Beja O. 2005. Putative novel photosynthetic reaction centre organizations in marine aerobic anoxygenic photosynthetic bacteria: Insights from metagenomics and environmental genomics. *Environ. Microbiol*. **7**(12):2027–2033.

Zehr JP, Jenkins BD, Short SM, Steward GF. 2003. Nitrogenase gene diversity and microbial community structure: A cross-system comparison. *Environ. Microbiol*. **5**(7):539–554.

Chapter 52

A Parsimony Approach to Biological Pathway Reconstruction/Inference for Metagenomes

YUZHEN YE AND THOMAS G. DOAK

52.1 INTRODUCTION

Microbial whole genome sequencing has become a routine practice in recent years, because of the rapid advances of DNA sequencing technologies [Morozova and Marra, 2008; see Chapter 18, Vol. I]. One of the first analyses that biologists attempt, once they obtain a complete genome sequence, is to *reconstruct* the biological pathways encoded by the organism, which is usually accomplished in silico by mapping the protein coding genes onto reference pathway collections, such as KEGG [Kanehisa and Goto, 2000] or SEED [Overbeek et al., 2005], based on their homology to reference genes with previously characterized functions. For example, KAAS, the pathway annotation system based on the KEGG database [Moriya et al., 2007], first annotates K numbers [each K number represents an ortholog group of genes and is directly linked to an object (a biochemical step) in the KEGG pathway map] and then reconstructs pathways based on the assigned K numbers. Similarly, the RAST server (and MG-RAST) first annotates FIG families and then maps the identified FIG families onto the SEED subsystems [Aziz et al., 2008; Meyer et al., 2008; see also Chapter 37, Vol. I]. These automatic methods are promising for the analysis of most genomes, although they may leave "holes" in the reconstructed pathways, due to either *missing* genes (i.e., the genes are nonhomologous to reference genes of the same specific functions, and thus cannot be identified by a homology-based method, or were simply not annotated as ORFs by annotation pipelines) [Osterman and Overbeek, 2003] or alternative and novel pathways (i.e., the target organism adopts variant pathways, which are different from the reference pathway, to accommodate a specific niche or a lifestyle) [Ye et al., 2005; see also Chapter 52, Vol. I]. After all, many bacterial genomes have fewer than 60% of their genes assigned to a proposed function [Friedberg et al., 2006; Sivashankari and Shanmughavel, 2006].

We note that *pathway reconstruction* is essential for understanding the biological functions that a newly sequenced genome encodes. For instance, in a recently published report, the coupling of N_2 fixation to cellulolysis was revealed within protist cells in the termite gut, based solely on the in silico pathway reconstruction of the complete genome sequence of an endosymbiont [Hongoh et al., 2008; see also Chapter 22, Vol. II].

Moreover, pathway reconstruction based on new high-throughput techniques must provide conclusions from explicitly incomplete information, which poses fresh challenges. For example, in a typical *proteomics* experiment, the proteins represent a particular biological sample collected under a specific physiological condition or from a specific tissue (e.g., from yeast cells after the heat shock), which are in high enough abundance to be identified by tandem mass spectrometry [Koller et al., 2002; Gilchrist et al., 2006; see also Chapters 71–74, Vol. I]. Based on these data, one may ask, What biological pathways were activated (or suppressed) under the physiological condition? A similar but more complicated case is pathway analysis of *metagenomic* data, to characterize the aggregate metabolic processes of

Handbook of Molecular Microbial Ecology, Volume I: Metagenomics and Complementary Approaches, First Edition. Edited by Frans J. de Bruijn.

microbial communities in a given environment [Galperin, 2004]. Metagenomic profiling data can be viewed as a sampling of the genomic sequences from many kinds of microbes living in a specific environment. Again, the incompleteness of the data makes it difficult to reconstruct the entire pathways encoded by a metagenome. Nevertheless, it is becoming routine to "reconstruct" pathways for proteomic [Yates et al., 2009] and metagenomic data [Dinsdale et al., 2008; Turnbaugh et al., 2009; by best similarity matches (often derived from BLAST searches); a pathway is inferred to be absent or present in a dataset if highly confident similarity hits identify one or more of the protein functions associated with the pathway in other organisms.

In addition to the problems that arise from incomplete data, existing methods of pathway reconstruction or inference may overestimate the number of pathways because of redundancy in the protein pathway, at four levels: (1) Different pathways may share the same biological functions. The partition of pathways (as the entire cellular network is partitioned into several hundreds of biological pathway entities in KEGG database) is extremely important for understanding of biological processes, even though there is only a single large biological network within any cell, and all pathways are to some extent connected [Okuda et al., 2008]. It is not surprising that many pathways defined in the pathway databases are overlapping. (2) Some proteins carry out multiple biological functions [Rosin et al., 2005]—for example, through different protein domains, active sites, or substrate specificities. (3) Neither organisms nor communities are closed boxes, and the products or intermediates of pathways may be exogenously supplied. An interesting case involves endosymbiotic and infectious bacteria, which take advantage of host functions and in some cases are still in the process of losing their own commensurate functions [Ochman and Davalos, 2006]. (4) Homology-based protein searching may map one protein to multiple homologous proteins with different biological functions (i.e., paralogous proteins). In summary, it cannot be safely concluded that a pathway is present, even if one or more proteins are mapped to it.

Even for single complete genomes, pathway reconstruction does not always give a clear picture, and human curation and experimental verification are often needed [Francke et al., 2005; Oberhardt et al., 2008]. We illustrate this by a rather extreme example found in the pathway analysis of the human genome. The KEGG pathway annotation of the human genome includes the reductive carboxylate cycle, with proteins annotated to six steps in this pathway (http://www.genome.jp/kegg-bin/show_organism?menu_type=pathway_maps&org=hsa) (as of July 2, 2009). The Calvin cycle is the most common method of carbon fixation, while the reductive carboxylate cycle is an alternative carbon fixation pathway, currently found only in certain autotrophic microorganisms. In fact, the reductive carboxylate cycle is essentially the reverse of the Krebs cycle (citric acid or tricarboxylic acid cycle), the final common pathway in aerobic metabolism for the oxidation of carbohydrates, fatty acids, and amino acids, so they share reactions and functional roles. For this reason, the proteins responsible for the normal function of the Krebs cycle can be mistakenly taken as evidence that the reductive carboxylate cycle is also encoded by the human genome.

To address these problems, we [Ye and Doak, 2009] developed a pathway reconstruction/inference method in which we do not attempt to reconstruct entire pathways from a given set of protein sequences (e.g., identified in a proteomics experiment, or encoded by the sequences sampled in a metagenomic project), but instead attempt to determine the minimal set of biological pathways that must exist in the biological system to explain the protein sequences sampled from it. In this context, we note that *pathway inference* is a more suitable terminology than *pathway reconstruction*. However, considering that pathway inference has been used in a different context to infer networks or pathways from gene expression data [Ourfali et al., 2007], and pathway reconstruction is commonly used in the field, we use both pathway inference and pathway reconstruction in this chapter. Our parsimony approach to the pathway reconstruction/inference problem, called MinPath (*Min*imal set of *Path*ways), can be roughly described as follows: Given a set of reference pathways and a set of proteins (and their predicted functions) that can be mapped to one or more pathways, we attempt to find the minimum number of pathways that can explain all proteins (functions) (see **Fig. 52.1**). Although this problem is NP-hard in general, we provide an integer programming (IP) framework to solve it.

We focus on analyzing complete genomes in this study because there is a relatively good understanding of the pathways that actually exist in organisms with completely sequenced genomes (as compared to the emerging metagenomes), making this analysis a good test of our method. Besides, the pathway annotations for these genomes are still far from perfection, as shown in the example of a carbon fixation pathway in the human genome (as well as chickens, mosquitoes, etc.). We also applied MinPath to the analyses of several metagenomic datasets, to demonstrate the potential applications of MinPath in metagenome annotation.

52.2 MATERIALS AND METHODS

First we will briefly describe the naive mapping approach that is commonly used in current automatic biological

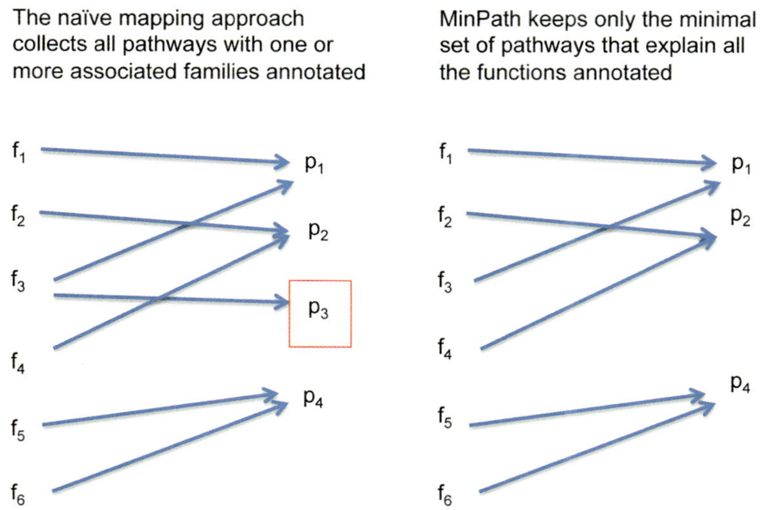

Figure 52.1 Schematic illustration of MinPath, which is to deduce the minimal set of pathways. Assume six families (or orthologous groups, $f_1, \ldots, f_6$) are identified from a given sample of genes (e.g., the genes could be from a genome, or sampled from a metagenome). The naïve mapping approach (shown on the left) will lead to a reconstruction with four pathways annotated (p_1, p_2, p_3, and p_4). Due to the overlapping nature of the biological pathways (see text for more details), pathway p_3 shares function f_3 with pathway p_2. We claim that only three pathways, p_1, p_2, and p_3, are sufficient to explain the existence of the six families annotated in the dataset, and a conservative reconstruction of pathways should have only three pathways (shown on the right). As we show in Ye and Doak [2009], such a conservative estimation of pathways provides a more reliable estimation of the functional diversity of a sample.

pathway reconstruction services (e.g., the KAAS and RAST servers), as well as for pathway reconstruction for metagenomic sequences. Then we present a novel minimal pathway reconstruction approach, based on a simple yet efficient algorithm, for solving this problem that we published in Ye and Doak [2009].

52.2.1 The Naive Mapping Approach to Pathway Reconstruction

Pathway reconstruction has become routine in functional annotation of genomes and metagenomes, in which KEGG pathways (or other biological pathways such as SEED subsystems) are reconstructed based on homology. KEGG and SEED databases collect pathways (or subsystems) curated by experts, with each pathway/subsystem consisting of a series of functional roles (enzymes, transporters, etc.). Pathway reconstruction consists of two key steps: (1) predicting the functions (represented by protein families) of proteins encoded by the DNA sequences, which is often achieved by similarity searching of the predicted proteins against reference proteins from previously characterized genomes; and (2) predicting the presence or absence of pathways in the query dataset, based on the identified functions associated with the pathways. Conventional pathway reconstruction usually adopts a simple criterion in this second step (herein referred to as the naïve mapping approach); that is, a pathway is considered to be present if one or more functions in the pathway are identified in the first step. We will show in this chapter that this approach may lead to the identification of spurious pathways and an overestimation of functional ability, which motivated us to develop a novel approach to pathway reconstruction based on the parsimony principle presented below.

52.2.2 Minimal Pathway Reconstruction Problem

We define the *minimal pathway reconstruction problem* as the following: Given a list of functions annotated for a set of genes (which can be an incomplete set, as we encounter in metagenomic analysis, or a nearly complete set, as in complete genome analysis), find the *minimal* set of pathways that include all given functions (see Fig. 52.1). Note that this formulation is different from the conventional formulation of the pathway reconstruction problem, which attempts either to reconstruct the complete pathways encoded by a given genomic dataset (in a sense, the pathway holes should be minimized) or to identify the set of pathways that have at least one associated function annotated (i.e., the naïve mapping approach).

52.2.3 Integer Programming Algorithm

We use integer programming to solve the minimal pathway reconstruction problem. Linear programming (LP) is an algorithm for finding the maximum or minimum of a linear function of variables (objective function) that are subject to linear constraints [Bertsimas and Tsitsiklis, 1997]. Simplex and interior point methods are widely used for solving LP problems. The related problem of integer programming (IP) requires some or all of the variables to take integer (whole number) values. Some of the most powerful algorithms for finding exact solutions of combinatorial optimization problems [Cook et al., 1998] are based on IP. LP and IP have been applied to many fields in the biological sciences, such as the maximum contact map overlap problem for protein structure comparison [Caprara et al., 2004], optimal protein threading [Xu et al., 2004], probe design for microarray experiments [Klau et al., 2004], and pathway variant problem [Ye et al., 2005].

Here we transform the minimal pathway reconstruction problem to an integer programming problem: Denote the number of functions (protein families) that are annotated in a dataset as n. Let the total number of putative pathways that have at least one component function annotated be p. Denote the mapping of protein functions to the pathways as M, where $M_{ij} = 1$ if function i is involved in pathway j, otherwise 0 (note one function may map to multiple pathways or subsystems). Denote if a pathway j is selected in the final list or not as P_j, with $P_j = 1$ if selected, $P_j = 0$ otherwise. The set of pathways with $P_i = 1$ composes the minimal set of pathways that can explain all the functions that are annotated for a dataset.

The objective function for integer programming is

$$\min \sum_{j=1}^{p} P_j$$

$$\text{s.t.} \quad \sum_{j=1}^{p} M_{ij} P_j \geq 1 \qquad \forall i \in [1, n]$$

That is, our goal is to find the minimum number of pathways that can explain all the functions carried by at least one protein from a dataset.

52.2.4 Protein Function and Function Annotation

We use the KO and FIG protein families defined in the KEGG database and the SEED subsystems, respectively, for this study. Many of the mappings of KO families to KEGG pathways were carried out manually in the KEGG database. These families are the basic units for pathway reconstruction (or subsystem reconstruction in SEED), in which a pathway (or a subsystem) is composed of a list of functional roles.

52.2.5 Implementation Details

We use the GLPK package (GNU Linear Programming Kit; http://www.gnu.org/software/glpk/glpk.html) for solving the integer-programming problem; all the other functions are implemented in Python.

The input for MinPath is a list of protein families (e.g., KO and FIG families) annotated in a given dataset of genes (from a genome, or a metagenome), and the output is the list of pathways reconstructed/inferred for the dataset.

Note that in some cases, two pathways may share most of their functional roles (for example, the biosynthesis and degradation pathway of the same biological molecule, such as the lysine biosynthesis and degradation pathways). MinPath will keep one of these pathways, because that is sufficient to explain the functional roles identified. We added a post-processing step here to add those pathways that have more than 50% of their functional roles identified back to the pathway pool, even when these functional roles appear in another pathway that is already predicted by MinPath.

52.2.6 Benchmarking Experiments

We revisited the pathway reconstruction for the 854 genomes in the KEGG database (as of December 2008) that have at least 20 KEGG pathways annotated for each of these genomes. For these genomes, the function (or protein families) annotations were downloaded from the KEGG database (ftp://ftp.genome.jp/pub/kegg/release/current/).

We also applied MinPath to reanalyze the pathways for nine biome metagenomic datasets [Dinsdale et al., 2008. The FIG family annotations for the metagenomic sequences were downloaded from the MG-RAST server (http://metagenomics.theseed.org/); (see Chapter 37, Vol. I). We conducted the KO family annotations of the sequences based on the best blast hits with E-value cutoff of 1e-5, a typical E-value cutoff used for KEGG pathway reconstruction in metagenomes [Turnbaugh et al. 2009].

52.3 RESULTS

We first revisited the pathway reconstruction of individual genomes using MinPath. The results show that MinPath gives a conservative but reliable estimation of the pathways of a genome, and therefore the functional ability/diversity encoded by a genome. In addition, MinPath will flag suspicious pathways that were manually annotated for individual genomes in KEGG. Then we applied MinPath to a set of metagenomic datasets, and the results indicate that the current estimation of functional diversity/ability of studied microbial communities might be greatly overestimated.

52.3.1 Pathway Reconstruction for Genomes

52.3.1.1 Overview of the Performance of MinPath. The functional annotations of gene sets for individual genomes were extracted from the KEGG database, and they were used as input for MinPath to reconstruct the pathways encoded by each genome. A total of 854 genomes were studied, and the overall performance of the MinPath, compared with the curated KEGG pathways and the pathway reconstruction based on the naïve mapping approach (see METHODS for details), is shown in **Figure 52.2**. The comparison shows that MinPath gives an estimation of functional diversity (measured by the number of pathways constructed) that is closer to the curated KEGG database, as compared to simple pathway construction based on the appearance of families. MinPath gives a more conservative estimation of the pathways than even KEGG in most genomes (with fewer annotated biological pathways), but we argue that even some of the pathways collected in KEGG should be removed (such as the ascorbate and aldarate metabolism pathway in human, as we discuss below).

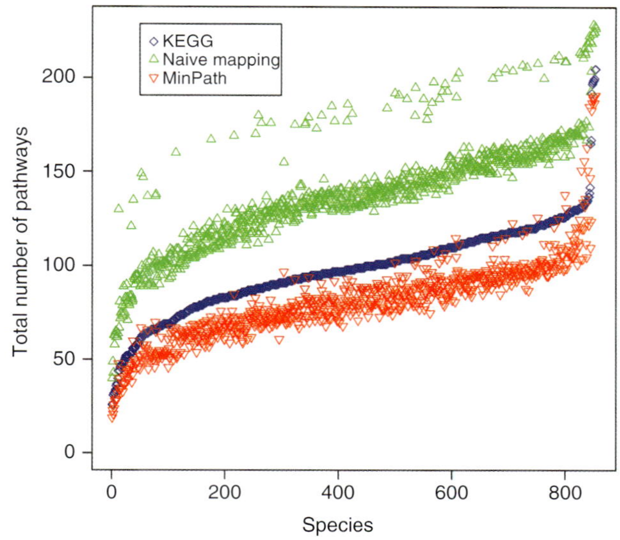

Figure 52.2 Comparison of the number of pathways reconstructed for various genomes by different methods. The coloring schema is as following: MinPath (red triangles), naive mapping approach (green triangles), and the pathway annotation maintained in KEGG database after human evaluation (blue triangles).

52.3.1.2 The Human Genome. For the human genome, there are 205 predicted KEGG pathways (as of December 2008), while the naive mapping approach identifies 227 pathways. MinPath identified only 191 pathways; these pathways are necessary and sufficient to explain all the annotated human proteins in the KEGG database.

Many of the pathways that are identified by the naïve mapping approach are spurious and are not curated in the KEGG database (e.g., the penicillin and cephalosporin biosynthesis pathway, the two-component systems, and the type II secretion system), indicating that MinPath will remove pathways that are mistakenly annotated using the naïve mapping approach. More examples are listed in the supplementary web site.

Some of the pathways that are curated in the KEGG database are marked by MinPath as spurious (see Table 1 in Ye and Doak [2009]). For example, the ascorbate and aldarate metabolism pathway (see Fig. 3 in Ye and Doak [2009]) is annotated in KEGG as a biological pathway in humans, but not by MinPath. In humans there are only three functions (out of 24) annotated for this pathway, and these three functions are not unique to the pathway: EC 1.2.1.3 (aldehyde dehydrogenase 2 family) is involved in 15 other pathways, EC 1.1.1.22 (UDP-glucose dehydrogenase) is involved in three other pathways, and myo-inositol oxygenase (EC:1.13.99.1) is involved in both this pathway and the inositol phosphate metabolism pathway. Based on the sparseness of the genes assigned to this pathway and their ubiquitous nature, and the fact that humans require vitamin C in the diet, we believe that the ascorbate and aldarate metabolism pathway should be removed from the pathways assigned to the human genome.

52.3.1.3 Overlap Between MinPath-Derived Pathways and Pathways Collected in KEGG. We further note that while MinPath always inferred fewer pathways than the naive method, the MinPath pathways are not simply a subset of the assigned pathways listed on KEGG (for KEGG pathways). Rather, there are pathways that are uniquely assigned by one or the other method. For example, the overlap between MinPath-derived pathways and KEGG pathways for human genome is 86%. This clearly complicates the task of determining the actual pathways encoded in a genome or metagenome. Also, alternative solutions may exist for pathway reconstruction using the parsimony approach; that is, there could be another set of pathways with the same total number of pathways to explain all the predicted functions.

52.3.1.4 Pathway Reconstruction for Metagenomes. We used MinPath to reanalyze the biological pathways of several metagenomes [Dinsdale et al., 2008], which were previously analyzed by a naïve mapping approach. The results are summarized in **Table 52.1**. We used both the KEGG and SEED databases in this experiment. For KEGG pathways, we did local BLAST searches, using the criteria as shown in Turnbaugh et al., [2009] for KO family identification. For SEED

Table 52.1 Comparison of Biological Pathway Reconstruction Based on MinPath and the Naive Mapping Approach for Selected Metagenomes[a]

Environmental Samples	Naïve Mapping (KEGG)[c]	MinPath (KEGG)	Naïve Mapping (SEED)[d]	MinPath (SEED)
Coral-Mic (7)[b]	188/232[e]	109/171	497/629	144/318
Coral-Vir (6)	174/211	105/140	594/667	238/377
Marine-Mic (8)	221/236	146/174	695/730	442/518
Marine-Vir (10)	213/236	154/175	680/733	457/536
Freshwater-Mic (4)	196/220	137/165	678/739	423/550
Freshwater-Vir (4)	113/154	57/90	392/559	89/219
Hyper-saline-Mic (9)	196/221	146/170	724/763	497/590
Hyper-saline-Vir (12)	164/181	105/137	613/697	292/442

[a] Metagenomes sampled from different environments [Dinsdale et al., 2008] (-Mic, and -Vir are for microbial and viral metagenomes, respectively, as shown in the table).

[b] Microbial metagenomes sampled from coral, with the total number of sequencing datasets shown in brackets.

[c] Based on the KEGG pathways (the KEGG database used in this study was downloaded in December 2008, which has 345 pathways).

[d] Based on the SEED subsystems (we used FIGfams release 6, which has more subsystems than reported in Dinsdale et al. [2008], and the total number of subsystems included is 898).

[e] The two numbers represent the total number of pathways (or subsystems) found in at least two of the datasets (e.g., two out of seven for Coral-Mic) and in at least one of the datasets for each environmental location, respectively.

subsystems, the FIG annotations were downloaded from the MG-RAST server (http://metagenomics.theseed.org/).

For all the datasets we tested, MinPath reduced the total number of annotated pathways (or subsystems) significantly (as shown in Table 52.1). For example, for the metagenome sampled from a coral microbial community (Coral-Mic), there are in total 232 KEGG biological pathways annotated in at least one of the seven sequencing datasets. Based on MinPath, however, only 160 KEGG biological pathways are sufficient to explain all the predicted functions. These results indicate that the naïve mapping of the biological pathways may overestimate the biological pathways (so the functional diversity) of those microbial communities, and we need to be cautious when interpreting the results from such an analysis [Dinsdale et al., 2008; Turnbaugh et al., 2009].

We also examined the details of pathway reconstruction for a single sequence dataset from the coral biome (4440319.3.dna.fa). The naïve mapping approach identified 224 KEGG pathways, whereas MinPath identified only 143 KEGG pathways. The pathways eliminated by MinPath include the inositol metabolism pathway, the androgen and estrogen metabolism pathway, and the caffeine metabolism pathway (see **Fig. 52.3**; See more examples at the supplementary web site). Obviously, comparisons of microbial communities or other biomes will be more telling if spurious pathways are eliminated, and our results suggest that as many as 40% of the 224 pathways could be wrong.

52.3.1.5 Viral Metagenomes Encode More Pathways than Bacterial Metagenomes?. As reported by Dinsdale et al., [2008] (Table 52.1), the coral viral community seemed have more subsystems annotated than the coral bacterial community, whereas in other environments, viral and bacterial subsystems are often roughly equivalent. This is provocative, given that scientists have recently observed that some viruses do carry important metabolic genes, such as genes involved in photosynthesis [Sharon et al., 2007, 2009]. However, pathway reconstruction based on the KEGG biological pathway database revealed that there are clearly more (KEGG) biological pathways annotated for bacterial functions in all nine communities studied, including in the coral community (see Table 52.1). The discrepancy between the published results (based on SEED) and our reanalysis of the relative functional diversity (based on KEGG) can be partially explained by the fact that the SEED subsystem database includes many viral subsystems (and prophage subsystems), whereas KEGG does not contain any specific/explicit pathways for viruses (KEGG also has no annotations for viral genomes). The viral (or prophage-related) subsystems collected in SEED subsystems include staphylococcal phi-Mu50B-like, and Listeria_phi-A118-like_prophages; the staphylococcal phi-Mu50B-like prophages subsystem collects proteins related to these phages, including tail fiber proteins [SA bacteriophages 11, Mu50B] (family id: FIG104219). SEED collected annotations for viruses (more than 1000) outnumbered the included bacterial genomes. Our analysis suggests that the choice of databases can strongly

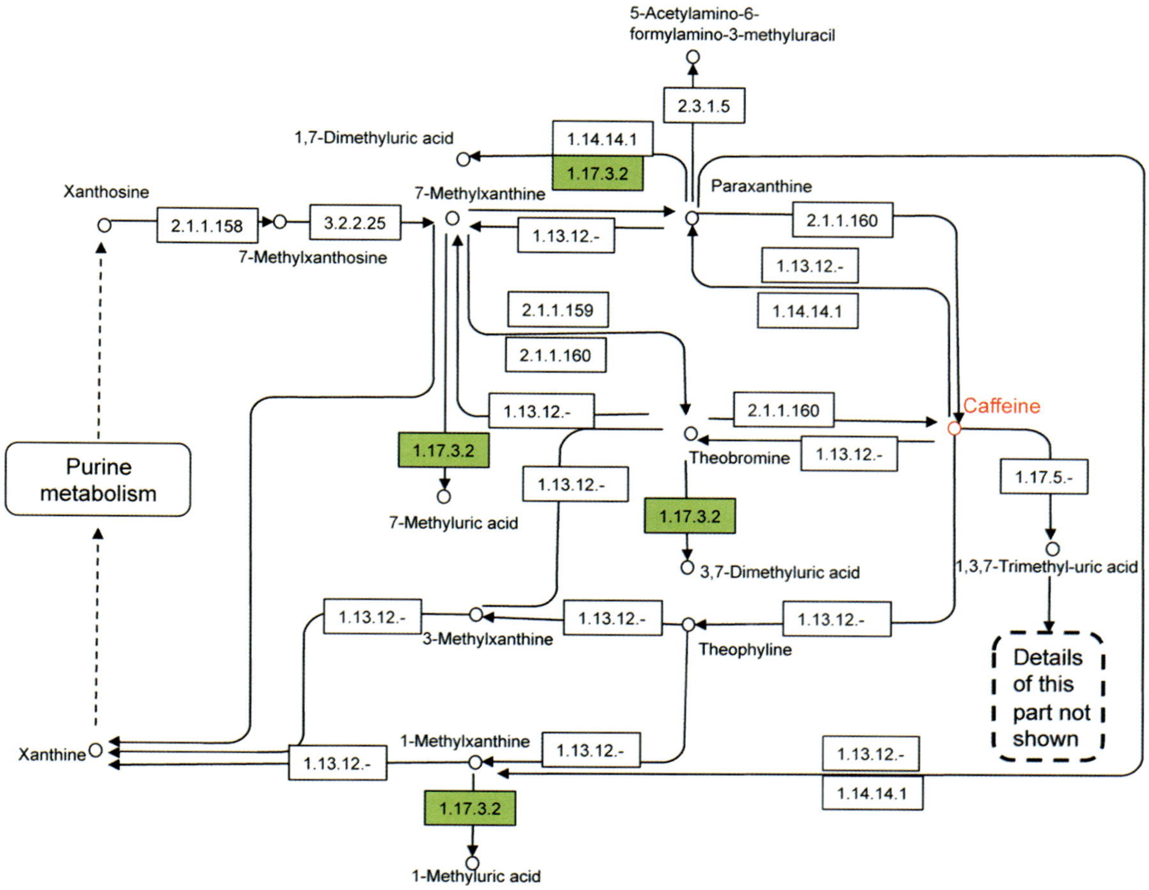

Figure 52.3 The caffeine metabolism pathway eliminated by MinPath. The diagram was prepared based on the corresponding KEGG pathway (ko00232). As shown in the diagram, the enzyme (EC 1.17.3.2) that is annotated in a coral metagenome (4440319.3.dna.fa) is involved in multiple reactions (highlighted in green). This enzyme is not unique to this pathway—it is involved in other pathways as well, including the purine metabolism pathway.

skew the outcome of the analysis, independent of the biological samples.

52.4 SUMMARY

We developed the MinPath approach [Ye and Doak, 2009] to provide more conservative and more reliable estimations of biological pathways from a sequence dataset, and we applied this approach to the biological pathway reconstruction problem for genomes as well as metagenomes. Our results show that without curated post-processing of the reconstructed pathways, the naïve mapping strategy will overestimate the biological pathways that are encoded by a genome or metagenome, jeopardizing any conclusions drawn from the constructed biological pathways (such as the metabolic diversity/capacity of an environmental microbial or viral community, as measured by the Shannon index) [Dinsdale et al., 2008; Turnbaugh et al., 2009] or other downstream analyses based on constructed pathways [Gianoulis et al., 2009]. It was noted in Turnbaugh et al., [2009] that most of the microbial communities in that study were approaching saturation for known pathways. More conservative estimates of pathways for each environment may allow actual functional differences between the samples to be detected.

Note that MinPath is not designed to directly improve the still imperfect definition of pathways and/or functions in databases such as KEGG or SEED. For example, as a result of how some pathways are grouped in the KEGG database, peptidoglycan biosynthesis is listed for the human genome by KEGG annotation, and MinPath does not eliminate this pathway from the list of annotated pathways from the human genome. In this sense, efforts are still needed to improve the elucidation and annotation of extent biochemical pathways. But given a database of reference pathways, we feel that MinPath provides a sensible method for inferring the pathways represented in biological sequence samples.

COMPUTER RESOURCES

KEGG (http://www.genome.jp/kegg/)

SEED (http://theseed.uchicago.edu/)

MG-RAST (http://metagenomics.theseed.org/)

GLPK (http://www.gnu.org/software/glpk/glpk.html)

MinPath server (http://omics.informatics.indiana.edu/MinPath/)

Supplementary material and MinPath source codes are available for download at the MinPath web site.

Acknowledgments

This work was supported by NIH grant 1R01HG004908-01 and NSF CAREER award DBI-084568. The authors would like to thank Dr. Haixu Tang for inspiring discussions, as well as Drs. Alex Rodriguez and Ross Overbeek for their help with using the Figfam database.

REFERENCES

AZIZ RK, BARTELS D, BEST AA, DEJONGH M, DISZ T, et al. 2008. The RAST Server: rapid annotations using subsystems technology. *BMC Genomics* **9**:75.

BERTSIMAS D, TSITSIKLIS JN. 1997. *Introduction to Linear Optimization*. Nashua, Athena Scientific.

CAPRARA A, CARR R, ISTRAIL S, LANCIA G, WALENZ B. 2004. 1001 optimal PDB structure alignments: integer programming methods for finding the maximum contact map overlap. *J. Comput. Biol*. **11**(1):27–52.

COOK WJ, CUNNINGHAM WH, PULLEYBLANK WR, SCHRIJVER A. 1998. *Combinatorial Optimization*. New York: John Wiley & Sons.

DINSDALE EA, EDWARDS RA, HALL D, ANGLY F, BREITBART M, et al. 2008. Functional metagenomic profiling of nine biomes. *Nature* **452**(7187):629–632.

FRANCKE C, SIEZEN RJ, TEUSINK B. 2005. Reconstructing the metabolic network of a bacterium from its genome. *Trends Microbiol*. **13**(11):550–558.

FRIEDBERG I, JAMBON M, GODZIK A. 2006. New avenues in protein function prediction. *Protein Sci*. **15**(6):1527–1529.

GALPERIN M. 2004. Metagenomics: From acid mine to shining sea. *Environ. Microbiol*. **6**: 543–545.

GIANOULIS TA, RAES J, PATEL PV, BJORNSON R, KORBEL JO, et al. 2009. Quantifying environmental adaptation of metabolic pathways in metagenomics. *Proc. Natl. Acad. Sci. USA* **106**(5):1374–1379.

GILCHRIST A, AU CE, HIDING J, BELL AW, FERNANDEZ-RODRIGUEZ J, et al. 2006. Quantitative proteomics analysis of the secretory pathway. *Cell* **127**(6):1265–1281.

HONGOH Y, SHARMA VK, PRAKASH T, NODA S, TOH H, et al. 2008. Genome of an endosymbiont coupling N_2 fixation to cellulolysis within protist cells in termite gut. *Science* **322**(5904):1108–1109.

KANEHISA M, GOTO S. 2000. KEGG: Kyoto encyclopedia of genes and genomes. *Nucleic Acids Res*. **28**(1):27–30.

KLAU GW, RAHMANN S, SCHLIEP A, VINGRON M, REINERT K. 2004. Optimal robust nonunique probe selection using Integer Linear Programming. *Bioinformatics* **20**(Suppl 1):i186–i193.

KOLLER A, WASHBURN MP, LANGE BM, ANDON NL, DECIU C, et al. 2002. Proteomic survey of metabolic pathways in rice. *Proc. Natl. Acad. Sci. USA* **99**(18):11969–11974.

MEYER F, PAARMANN D, D'SOUZA M, OLSON R, GLASS EM, et al. 2008. The metagenomics RAST server—A public resource for the automatic phylogenetic and functional analysis of metagenomes. *BMC Bioinform*. **9**:386.

MORIYA Y, ITOH M, OKUDA S, YOSHIZAWA AC, KANEHISA M. 2007. KAAS: An automatic genome annotation and pathway reconstruction server. *Nucleic Acids Res*. **35**(Web server issue):W182–W185.

MOROZOVA O, MARRA MA. 2008. Applications of next-generation sequencing technologies in functional genomics. *Genomics* **92**(5):255–264.

OBERHARDT MA, PUCHALKA J, FRYER KE, MARTINS DOS SANTOS VA, PAPIN JA. 2008. Genome-scale metabolic network analysis of the opportunistic pathogen *Pseudomonas aeruginosa* PAO1. *J. Bacteriol*. **190**(8):2790–2803.

OCHMAN H, DAVALOS LM. 2006. The nature and dynamics of bacterial genomes. *Science* **311**(5768):1730–1733.

OKUDA S, YAMADA T, HAMAJIMA M, ITOH M, KATAYAMA T, et al. 2008. KEGG Atlas mapping for global analysis of metabolic pathways. *Nucleic Acids Res*. **36**(Web server issue):W423–W426.

OSTERMAN A, OVERBEEK R 2003. Missing genes in metabolic pathways: A comparative genomics approach. *Curr. Opin. Chem. Biol*. **7**(2):238–251.

OURFALI O, SHLOMI T, IDEKER T, RUPPIN E, SHARAN R 2007. SPINE: A framework for signaling-regulatory pathway inference from cause–effect experiments. *Bioinformatics* **23**(13):i359–i366.

OVERBEEK R, BEGLEY T, BUTLER RM, CHOUDHURI JV, CHUANG HY, et al. 2005. The subsystems approach to genome annotation and its use in the project to annotate 1000 genomes. *Nucleic Acids Res*. **33**(17):5691–5702.

ROSIN FM, WATANABE N, LAM E 2005. Moonlighting vacuolar protease: Multiple jobs for a busy protein. *Trends Plant Sci*. **10**(11):516–518.

SHARON I, TZAHOR S, WILLIAMSON S, SHMOISH M, MAN-AHARONOVICH D, et al. 2007. Viral photosynthetic reaction center genes and transcripts in the marine environment. *ISME J*. **1**(6):492–501.

SHARON I, ALPEROVITCH A, ROHWER F, HAYNES M, GLASER F, et al. 2009. Photosystem I gene cassettes are present in marine virus genomes. *Nature* **461**(7261):258–262.

SIVASHANKARI S, SHANMUGHAVEL P. 2006. Functional annotation of hypothetical proteins—A review. *Bioinformation* **1**(8):335–338.

TURNBAUGH PJ, HAMADY M, YATSUNENKO T, CANTAREL BL, DUNCAN A, et al. 2009. A core gut microbiome in obese and lean twins. *Nature* **457**(7228):480–484.

XU J, LI M, KIM D, XU Y. 2004. RAPTOR: optimal protein threading by linear programming. *J. Bioinform. Comput. Biol*. **1**(1):95–117.

YATES J, RUSE C, NAKORCHEVSKY A. 2009. Proteomics by mass spectrometry: Approaches, advances, and applications. *Annu. Rev. Biomed. Eng*. **11**:49–79.

YE Y, DOAK TG. 2009. A parsimony approach to biological pathway reconstruction/inference for genomes and metagenomes. *PLoS Comput. Biol*. **5**(8):e1000465.

YE Y, OSTERMAN A, OVERBEEK R, GODZIK A. 2005. Automatic detection of subsystem/pathway variants in genome analysis. *Bioinformatics* **21**(Suppl 1):i478–i486.

Chapter 53

ESPRIT: Estimating Species Richness Using Large Collections of 16S rRNA Data

YIJUN SUN, YUNPENG CAI, LI LIU, FAHONG YU, AND WILLIAM FARMERIE

53.1 INTRODUCTION

The latest development of massively parallel pyrosequencing technology allows researchers to study genetic materials recovered directly from environmental samples, bypassing the needs for isolation and lab cultivation of individual species, and thus opens a new window to probe the hidden world of microbial communities [Eisen, 2007, Rothberg and Leamon, 2008; see also Chapters 15 and 18, Vol. I]. This technique has been successfully used in several 16S rRNA-based metagenomics analyses of various environments. For example, Sogin et al. [2006] provided one of the first global in-depth descriptions of microbial diversities and their relative abundance in the ocean, and Keijser et al. [2008] were among the first to study oral microbial populations. It has been shown that microbial communities are much more diverse than previously reported. These estimation results, however, were computed through extrapolation. In order to obtain more accurate estimates, surveys that are several orders of magnitude larger than those reported in the literature may be required to uncover sequences from minor components [Sogin et al. 2006, Keijser et al., 2008]. However, analyzing large collections of 16S ribosomal sequences poses a serious computational challenge for existing algorithms.

Existing algorithms can be generally categorized into taxonomy-dependent or taxonomy-independent analyses. In the former methods, query sequences are first compared against a reference database and then assigned to the organism of the best-matched reference sequences. However, since the genomes of the vast majority of microbes have not been sequenced yet, these methods are inherently limited by the completeness of a database. In this chapter, we focus on taxonomy independent analysis where sequences are classified into operational taxonomic units (OTUs) of specified sequence variations, based on which various ecological metrics are estimated. Typically, sequences with <3% dissimilarity are assigned to the same species, while those with <5% dissimilarity are assigned to the same genus, although these distinctions are controversial [Borneman and Triplett, 1997, Sait et al., 2002, Schloss and Handelsman, 2005].

One outstanding challenge of taxonomy-independent analysis is the alignment of sequences of a sample for the calculation of sequence variations. One commonly used method in the literature is multiple sequence alignment (MSA) (see, for example, Sogin et al. [2006], Keijser et al. [2008], and Roesch et al. [2007]). While significant improvement has been made in the last decade to reduce the computational complexity of MSA (e.g., MUSCLE [Edgar, 2004a] and MAFFT [Katoh et al., 2005]), it is still computationally intractable to align hundreds of thousands of sequences. Moreover, the use of MSA for aligning hypervariable regions of 16S rRNA gene has not been well-justified in the literature. We experimentally found through a benchmark study that the use of MSA leads to an inflated estimate of genetic distances and microbial diversities (see Section 53.4 for details). Another major

Handbook of Molecular Microbial Ecology, Volume I: Metagenomics and Complementary Approaches, First Edition. Edited by Frans J. de Bruijn.

issue of taxonomy independent analysis is the high computational complexity and memory requirement associated with assigning sequences into OTUs. DOTUR [Schloss and Handelsman, 2005] is a commonly used algorithm for this purpose. It conveniently integrates the tasks of performing clustering and statistical interference of species richness into one program, which frees researchers from manually manipulating a large distance matrix. However, DOTUR does not scale well to handle extremely large datasets. The recently released MOTHUR significantly improved the computational performance of DOTUR. As with DOTUR, MOTHUR loads a distance matrix into memory before proceeding to perform clustering. Hence, it does not fundamentally address the computational issue associated with processing massive pyrosequencing data. Given a full run of 454 data, a full distance matrix can be as large as 1000 GB. Even if we remove duplicated sequences and sequence pairs that have a pairwise distance larger than a cutoff value (say 0.1), the resulting distance matrix in a sparse format can be 20 GB in size, which is too big to be directly loaded into the memory in most computers.

We proposed a new algorithm, referred to as ESPRIT, which addresses several computational limitations with prior work. ESPRIT uses the Needleman–Wunsch algorithm [Needleman and Wunsch, 1970] to optimally align each pair of sequences. Through a benchmark study, we demonstrated that global pairwise sequence alignment can provide a much more accurate estimate of microbial richness than MSA. A more important reason, however, is that pairwise alignment allows for parallel computing, while most MSA algorithms can only be used in a single computer. We also developed a new cluster algorithm, referred to as Hcluster, within the ESPRIT framework to handle large-scale clustering problems. Unlike a brute-force method, Hcluster groups sequences into OTUs on-the-fly, while keeping track of linkage information, to overcome memory limitations. By assigning a computational task to hundreds of nodes, ESPRIT is not computationally constrained by the number of sequences to be analyzed, but by the capacity of a computer cluster. Hence, ESPRIT is suitable for applications such as global ocean survey of microbial populations that may require the analysis of data collected from thousands of locations.

We conducted large-scale experiments to demonstrate the effectiveness of the proposed algorithm. We first performed a simulation study by using a dataset generated by sequencing PCR amplicon libraries of 43 known 16S rRNA genes. To our knowledge, this is the first benchmark study reported in the literature to assess the estimation accuracy of an algorithm for taxonomy-independent analysis. Our experimental results justified the use of global pairwise alignment for the purpose of estimating microbial community compositions, and showed that the commonly used MUSCLE+DOTUR pipeline may overestimate biodiversity. We then applied ESPRIT to eight seawater samples. The results are very consistent with that of the benchmark study. We finally used ESPRIT to analyze a recently collected air sample consisting of about 350K short reads. It was stated in Gans et al. [2005] that microbial communities may be too diversified to be practically tested by amplification and sequencing of 16S rRNA genes. Our results suggest that at least for air samples, it is computationally practical to use the current sequencing technology to conduct 16S rRNA-based biodiversity surveys.

53.2 METHOD

The ESPRIT algorithm consists of four modules: (i) removes low-quality reads using various criteria, (ii) computes pairwise distances of reads, (iii) groups reads into OTUs at different dissimilarity levels, and (iv) performs statistical inference to estimate species richness. We developed two versions of ESPRIT, one for personal computers (PC) and one for computer clusters (CC). The PC version can process several tens of thousands sequences within a few minutes, while the CC version is able to analyze several hundreds of thousands of reads within one day. The code and user guide are freely available at http://www.biotech.ufl.edu/people/sun/esprit.html. Each module is detailed in the following sections.

53.2.1 Removing Low-Quality Reads

Pyrosequence data contain a certain percentage of sequencing errors, which can lead to an overestimation of microbial diversity [Huse et al., 2007; see also Chapter 19, Vol. I]. To minimize the effects of random sequencing errors, we remove (a) the reads that contain ambiguous nucleotides (N) and (b) those with more than one mismatch with the PCR primer at the beginning of a read. Also, we eliminate the sequences with atypical lengths. The default setting of the algorithm is to retain the reads with a length within one standard deviation from the mean length. In order to reduce computational complexity, by following the strategy used in Huber et al. [2007], if two sequences are identical or one sequence is a subset of the other, only the longer sequence is retained, and the number of occurrence of each retained sequence is recorded and used in the statistical inference. A similar trimming procedure was also used in Huse et al. [2007]. ESPRIT allows users to bypass this module to use a sample filtered by a customized trimming procedure.

53.2.2 Computing Pairwise Distances

We use the Needleman–Wunsch algorithm to optimally align each pair of sequences in a sample, and we use the quickdist algorithm [Sogin et al., 2006] to compute pairwise distances. More specifically, each pairwise distance equals mismatches, including indels, divided by a sequence length. To avoid overestimating distances between sequences from rapidly diverging variable regions, end gaps are ignored and gaps of any length are treated as a single evolutionary event or mismatch.

The naive application of the Needleman–Wunsch algorithm to an environmental sample, however, is computationally very expensive. For example, given one full run of 454 data with >400K reads, it is estimated that it would take about 3 years to align all 80 billion pairs of sequences and would need about 1000 GB to store the resulting distance matrix in the PHYLIP format. However, we noticed that for the purpose of estimating biodiversity, only the sequence pairs with distances <0.10 are of most interest, which only account for a small fraction of all possible pairs (about 1–5%, **Fig. 53.1**). This means possibly 20- to 100-folds of speedup. Moreover, by removing unwanted pairs, distance information can be stored in a sparse matrix that requires much less memory. In order to rapidly identify the sequence pairs of interest, we compute the k-mer distance of each pair of sequences. Below we give a brief description of how it works. A k-mer, also known as k-tuple, is a sub-sequence consisting of k possible nucleotides. By specifying the value of k, a complete alphabet of k-mer is constructed. The number of occurrences of each k-mer in alphabet Ω, also called genomics profile [Karlin and Burge, 1995, Karlin et al., 1997], is computed for each sequence. Then, the k-mer distance of a pair of sequences is derived as

$$d = 1 - \sum_{i=1}^{\Omega} \min(f_1(i), f_2(i))/(\min(L_1, L_2) - k + 1)$$

where $f_1(i)$ and $f_2(i)$ are the numbers of the occurrence of the ith k-mer of two sequences, respectively, and L_1 and L_2 are the lengths of two sequences, respectively. It has been shown that k-mer distances are highly correlated with genetic distances [Edgar, 2004b]. The concept of k-mer counting has been used for various applications, including sequence alignment [Edgar, 2004a], phylogenetic analysis [Mavromatis et al., 2007, Sun et al., 2007], and detection of horizontal gene transfer [Dalevi et al., 2006]. We used this technique to remove unwanted sequence pairs, and found that by using a k-mer distance of 0.4, most of the sequence pairs of interest can be rapidly identified **(Fig. 53.2a)**. The default setting in ESPRIT is 0.5. Figure 53.2b shows that removing sequence pairs with k-mer distances larger than the default parameter has a negligible impact on the estimation accuracy for the distance levels <0.1. It should be noted that though the default k-mer threshold works well for all of the 10 datasets tested in this chapter, it is possible that the parameter may vary for different applications. Hence, we provide an auxiliary program in the software package that allows users to determine the parameter by comparing k-mer distances against genetic distances using a small subset of their samples.

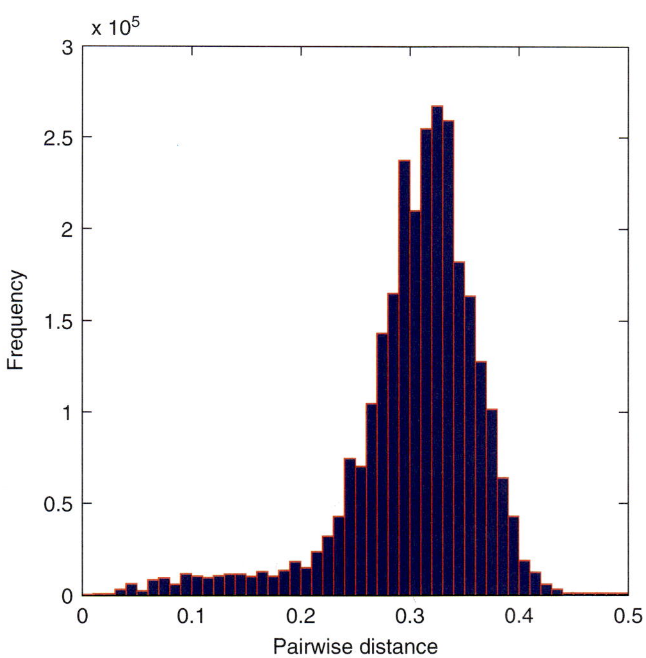

Figure 53.1 Distribution of pairwise distances in a seawater sample. Sequence pairs with distances less than 0.10 only account for a very small fraction of all possible pairs (2.25% for this example). The experiment was performed on the 53R seawater sample.

In addition to a PC version, we also developed a CC version for calculating both k-mer and genetic distances based on globally aligned sequences. Given one full run of 454 sequences, by using 100 computer nodes, it takes only 3–5 h to finish this module, depending on the availability of computer nodes.

53.2.3 Assigning Sequences to OTUs

After a distance matrix is computed, hierarchical clustering [Duda et al., 2001] is performed to group sequences into OTUs of defined sequence variations. Although ESPRIT uses a sparse distance matrix, given one million reads, the size of the matrix can still be quite large. We developed a new clustering algorithm, referred to as Hcluster, within the ESPRIT framework to handle large-scale clustering operations. Unlike conventional

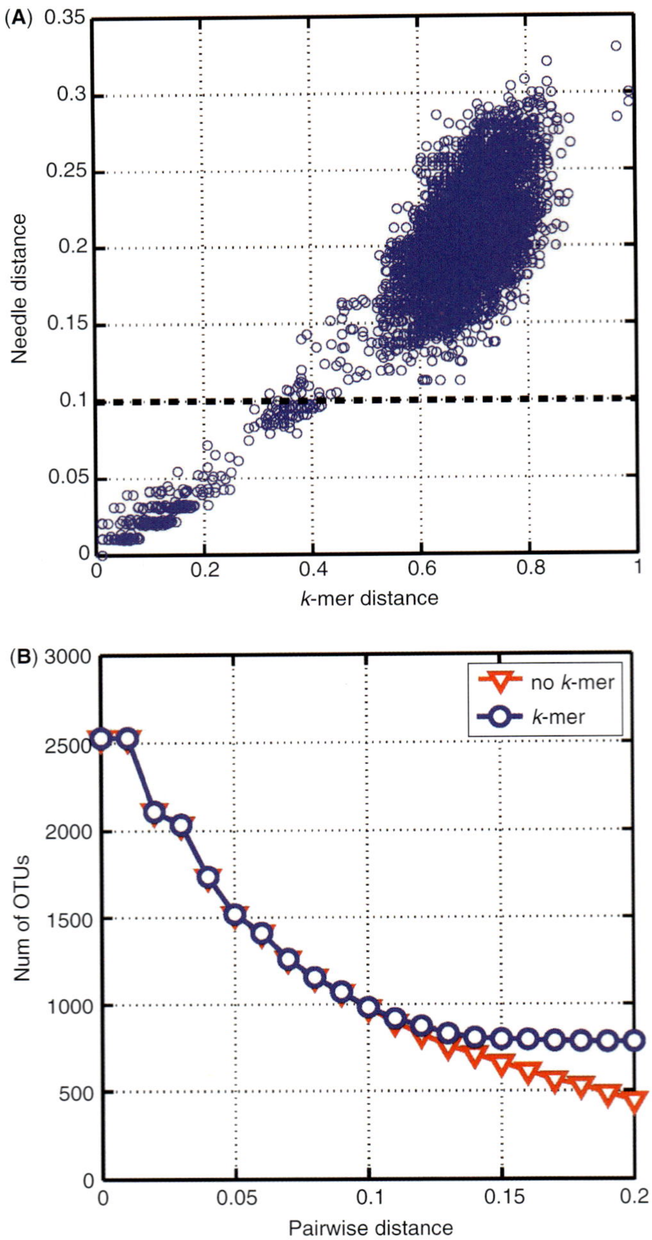

Figure 53.2 (**a**) k-mer distances are highly correlated with genetic distances. (**b**) Removing sequence pairs with k-mer distances larger than the default threshold has a negligible impact on the estimation accuracy for the distance levels of interest. The experiment was performed on the 53R seawater sample.

methods that load a distance matrix directly into memory, Hcluster groups sequences into OTUs *on-the-fly* (i.e., processing one distance at a time, also called *online learning* in machine learning) while keeping track of linkage information. At each step, we classify the clusters being analyzed into *active* or *inactive* clusters. Active clusters are those with known distance information to other clusters, but the information is insufficient to decide whether to merge the clusters together. Inactive clusters are those without information or those already merged with other clusters. We only need to maintain the linkage information for active clusters, which is updated when new distance information is processed. The memory required to store linkage information grows only linearly.

53.2.4 Statistical Inference of Species Richness

ESPRIT supports three richness estimators: rarefaction analysis [Hurlbert, 1971], Chao1 [Chao, 1984], and ACE [Chao and Lee 1992; Chao et al., 1993]. Rarefaction allows the calculation of the species richness for a given number of sampled individuals and constructs known as rarefaction curves. The curve is the number of observed OTUs as a function of the number of sampled sequences. In case of a steep slope, it means that a large fraction of the species diversity is not sampled yet, and more exhaustive sampling will yield a significant number of additional species. Chao1 and ACE are two abundance-based coverage estimators that predict the species richness based on the number of rarely occurrent OTUs. The cluster information generated by ESPRIT allows users to compute other ecological metrics, to derive a consensus sequence of each cluster, and to align the sequences of rarely occurrent OTUs against a database, which may lead to the identification of novel organisms.

53.3 RESULTS

We conducted a large-scale experiment to demonstrate the effectiveness of the newly proposed ESPRIT algorithm. When the PC version of ESPRIT was used, the experiment was performed on a server with eight E5345 Xeon CPUs and 16 GB memory operated on Linux 5.2 system. When the CC version was used, the experiment was performed on a computer cluster administrated by the High-Performance Computing Center at the University of Florida. The detailed computer configuration can be found at http://wiki.hpc.ufl.edu/index.php/Operating_Environment.

53.3.1 Benchmark Study

The purpose of the simulation study is twofold: (i) to investigate which strategy, pairwise or MSA, is more suitable for taxonomy-independent analysis and (ii) to assess the estimation accuracy of the ESPRIT algorithm. Although MUSCLE+DOTUR has been used in several metagenomics studies, its performance has never been benchmarked in the literature. This is in part due to the fact that under current technologies, one may never know precisely the ground-truth information of the compositions of a microbial community. The benchmark information, however, is critical to evaluate the performance of an algorithm performed on real-world data and to make a meaningful comparison of taxonomic distributions of different environments.

The simulation data consisted of ~340K short sequences, generated by pyrosequencing two PCR amplicon libraries of 43 known 16S rRNA gene fragments using the Roche GS20 system. They were originally used in Huse et al. [2007] (see also Chapter 19, Vol. I) to study the sequencing error rate of the system. We applied both ESPRIT and MUSCLE+DOTUR to the 43 reference gene sequences to estimate the numbers of OTUs defined at various distance levels, also known as lineage-through-time curve in the literature [Schloss and Handelsman, 2005]. Parameters -maxiters 1, -diags 1, and -sv were used in MUSCLE. The two so-obtained curves were used as the ground truth to benchmark the performance of the two algorithms. In order to study how the two algorithms perform in the presence of sequencing errors, we generated two datasets by mapping each read to the 43 reference sequences and retaining the reads that have less than 3% or 5% mismatches with the closest reference sequence. To eliminate statistical variations, each algorithm was run 10 times for each dataset. In each run, 10K reads were randomly sampled from each dataset. The lineage-through-time curves, averaged over 10 runs, are plotted in **Figure 53.3**. Since the ground-truth curves generated by ESPRIT and MUSCLE+DOTUR are very similar, for ease of presentation, only the curve of ESPRIT is plotted. From the figure, we observe that while both algorithms always lead to an overestimate of the number of OTUs due to the presence of sequencing errors, pairwise alignment can provide a more accurate estimate of microbial richness than multiple alignment. For example, for the first dataset, at the 0.05 distance level, the numbers of OTUs estimated by ESPRIT and MUSCLE+DOTUR are 44 (95% CI: 43–45) and 92 (95% CI: 82–102), respectively, while the ground truth is 42 (Fig. 53.3a).

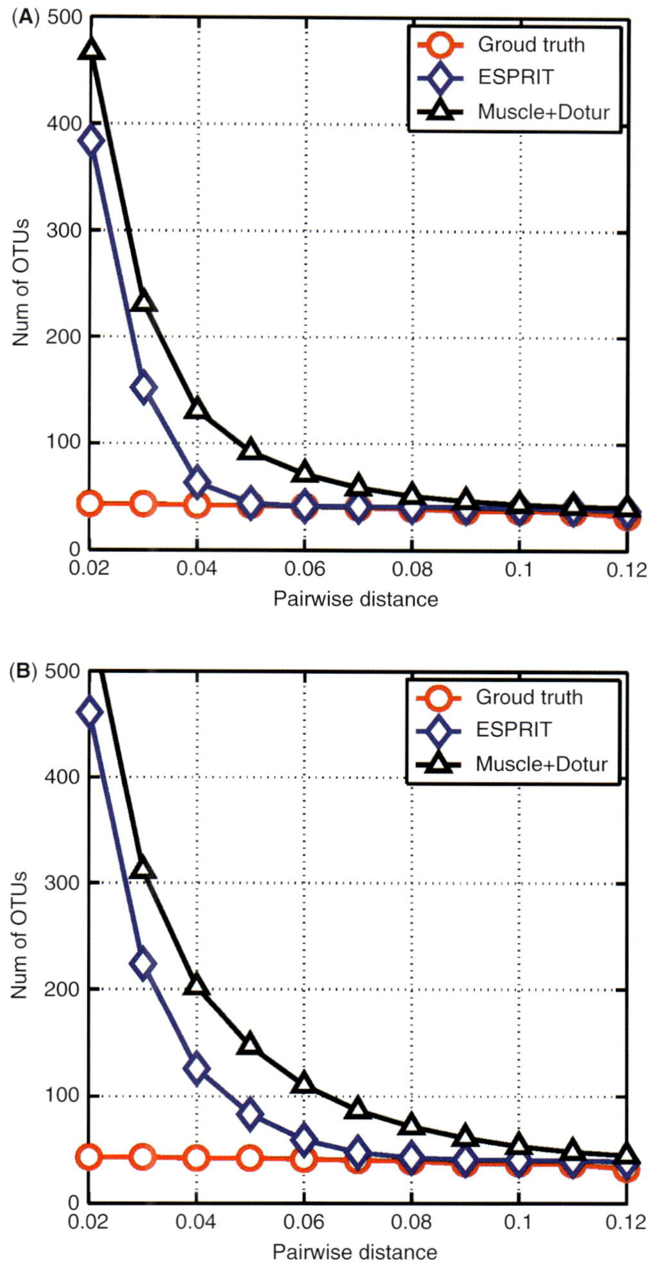

Figure 53.3 Lineage-through-time curves generated by using ESPRIT and MUSCLE + DOTUR algorithms performed on simulation data with each read containing up to (*a*) 3% and (**b**) 5% sequencing errors.

53.3.2 Experiments on Eight Seawater Samples

We applied ESPRIT to reanalyze eight seawater samples downloaded from Sogin et al. [2006]. The DNA materials within environmental samples were collected from eight different locations in the Atlantic and Pacific Oceans, respectively, as part of efforts to develop a global description of microbial diversities in the ocean. A total of 118, 000 PCR amplicons were sequenced that covered the V6 hyperviable region of rRNAs. The 454 reads have undergone a systematic trimming process. Thus, we bypassed the trimming procedure. The number of sequences for each sample ranges from 5,000 to 17,666 (see **Table 53.1**). The interested reader may refer to Sogin et al. [2006] for a detailed description of the data and sample preparations. We re-ran MUSCLE + DOTUR by using the parameters provided by Sogin et al. [2006]. The default parameters of ESPRIT were used. The lineage-through-time curves, ACE and Chao1 estimates at three

Table 53.1 Running Time of the PC Version of ESPRIT Performed on Eight Seawater Samples

	Datasets							
	112R	115R	137	138	53R	55R	FS312	FS396
Number of Reads	9,282	11,005	13,097	14,374	5,000	13,902	4,835	17,666
CPU time	2 min 54 s	4 min 13 s	4 min 28 s	6 min 19 s	59 s	7 min 31 s	49 s	6 min 34 s

different distance levels are depicted in **Figures 53.4** and **53.5**. We observe that ESPRIT gives a much lower estimate of species richness than the competing method, which is consistent with the result of the simulation study described above. For example, at the distance of 0.05, the number of observed OTUs and the ACE and Chao1 estimates obtained by ESPRIT are only about 50–70% of those obtained by MUSCLE + DOTUR. This becomes more evident in **Figure 53.6**, where the results of the rarefaction analysis of FS396 are presented. Though the existing pipeline overestimates the diversity of microbes, the conclusion of Sogin et al. [2006] still holds that the microbial diversity of seawater samples is much larger than previously reported, and the steep slopes of the rarefaction curves, even at relatively large genetic distances (e.g., 0.1), suggest that a large fraction of species have not been sampled yet. We notice that the disparity between the results of the two methods is much larger than that in the simulation study (Fig. 53.3–53.5). One possible reason is that the sequences in the seawater samples are more diverse than those in the simulation data and the inflated estimates by multiple sequence alignment become more prominent. A more detailed discussion is given in Section 53.4.

53.3.3 Experiments on an Air Sample

We applied ESPRIT to an air sample recently collected from Iraq. The small subunit rRNA gene fragments that cover the V6 hyperviable region of rRNAs were amplified, cloned, and pyrosequenced using 454 Life Sciences systems. A total of 348,952 sequences of an average of 275 bp in length were obtained. This is one of the largest 16S rRNA-based biodiversity surveys conducted using air samples. The number of sequences is one order of magnitude larger than those of the seawater samples we considered in the previous section. The CC version of ESPRIT with the default parameters was used. After the trimming procedure, 40,928 (~13%) reads were removed as low-quality reads. **Figure 53.7** presents the results of the rarefaction analysis of the OTUs defined at four different sequence variations, ranging from 0.01 to 0.1. We observe that, in contrast to the results of the seawater samples, the rarefaction curves saturate even at a relatively small genetic distance of 0.03, indicating that additional sampling may not lead to significantly increased estimates of total species diversity. The results of ACE and Chao1 estimators are reported in **Table 53.2**. At the distance levels of 0.05 and 0.1, the numbers of observed OTUs are already very close to the ACE and Chao1 estimates. These experimental results suggest that, at least for air samples, the current sequencing technology is sufficient to conduct 16S rRNA-based biodiversity surveys. The experiment was performed on a small CC consisting of 100 nodes, and it took about 10 h to finish the entire analysis.

53.4 DISCUSSION

53.4.1 Pairwise or Multiple Sequence Alignment

In contrast to many existing metagenomics studies, we used pairwise sequence alignment, instead of multiple sequence alignment, to align 16S rRNA sequences. This is a critical issue that merits more explanations. MSA is used to infer homology segments of input sequences. An underlying assumption is that input sequences should share some similarities, which may not be valid for 16S rRNA-based studies that target on hypervariable regions of rRNA gene (e.g., V6 and V3 regions). It has been reported in the literature that microbial communities are much more diverse than expected [Sogin et al., 2006, Keijser et al., 2008, Gans et al., 2005]. To further demonstrate this, we performed a simulation study where we calculated the pairwise distances of the sequences of a seawater sample and plotted a histogram of the distances in Figure 53.1. We report that only about 2% sequence pairs have a genetic distance smaller than 0.1. In other words, for every two randomly selected sequences, there is a 98% probability that the two sequences are from two distantly related OTUs. The sequences were aligned by using the Needleman–Wunsch algorithm. We will shortly see that if MSA is used, the histogram will be further skewed toward the right. We then performed another simulation study to demonstrate how the presence of a large proportion of highly diverse sequences affects the alignment of sequences with a small genetic distance. The following describes in detail how the experiment was conducted. First we randomly selected two sequences

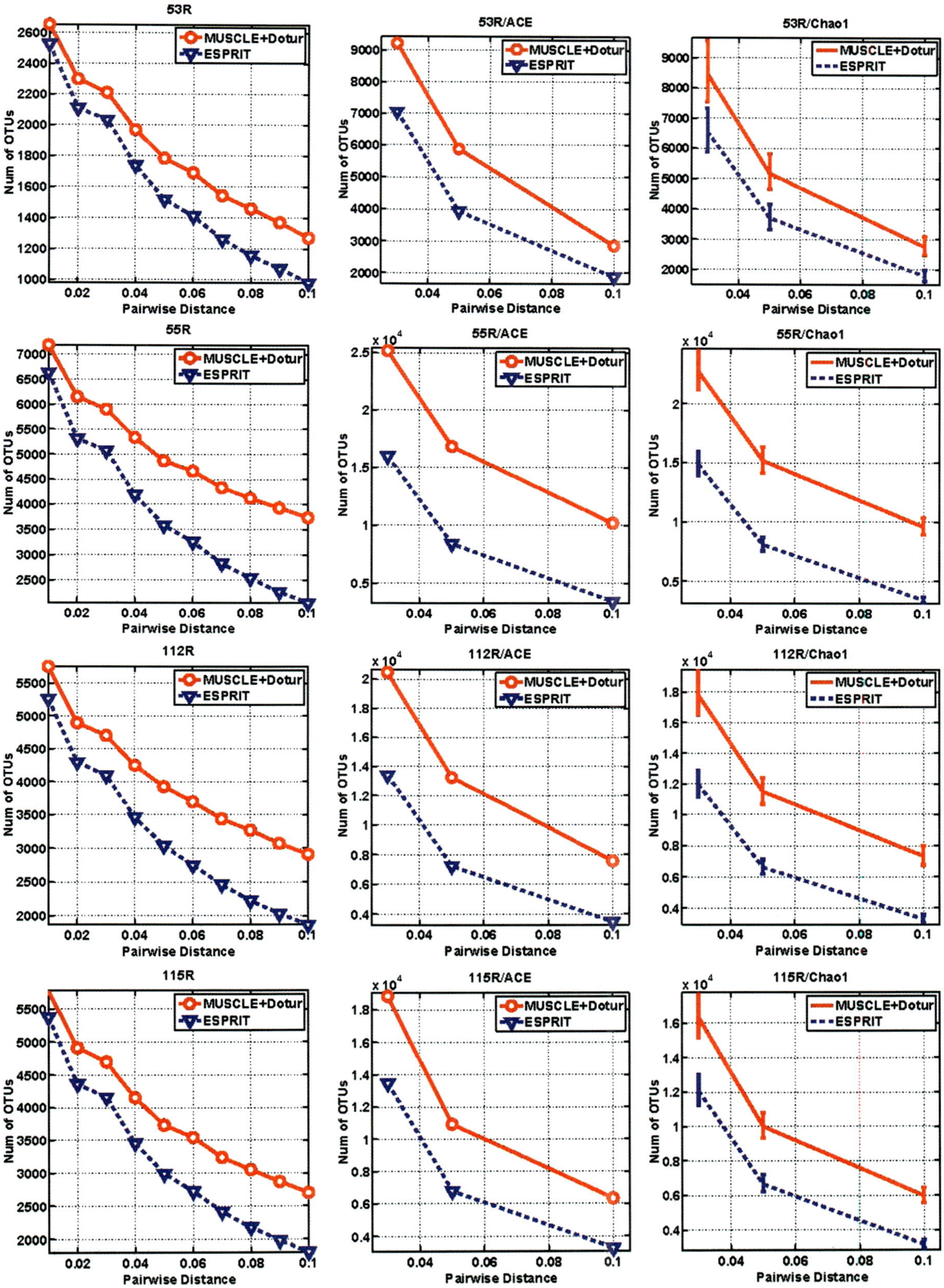

Figure 53.4 Lineage-through-time curves, ACE and Chao1 estimates generated by using ESPRIT and MUSCLE + DOTUR algorithms performed on 53R, 55R, 112R, and 115R datasets.

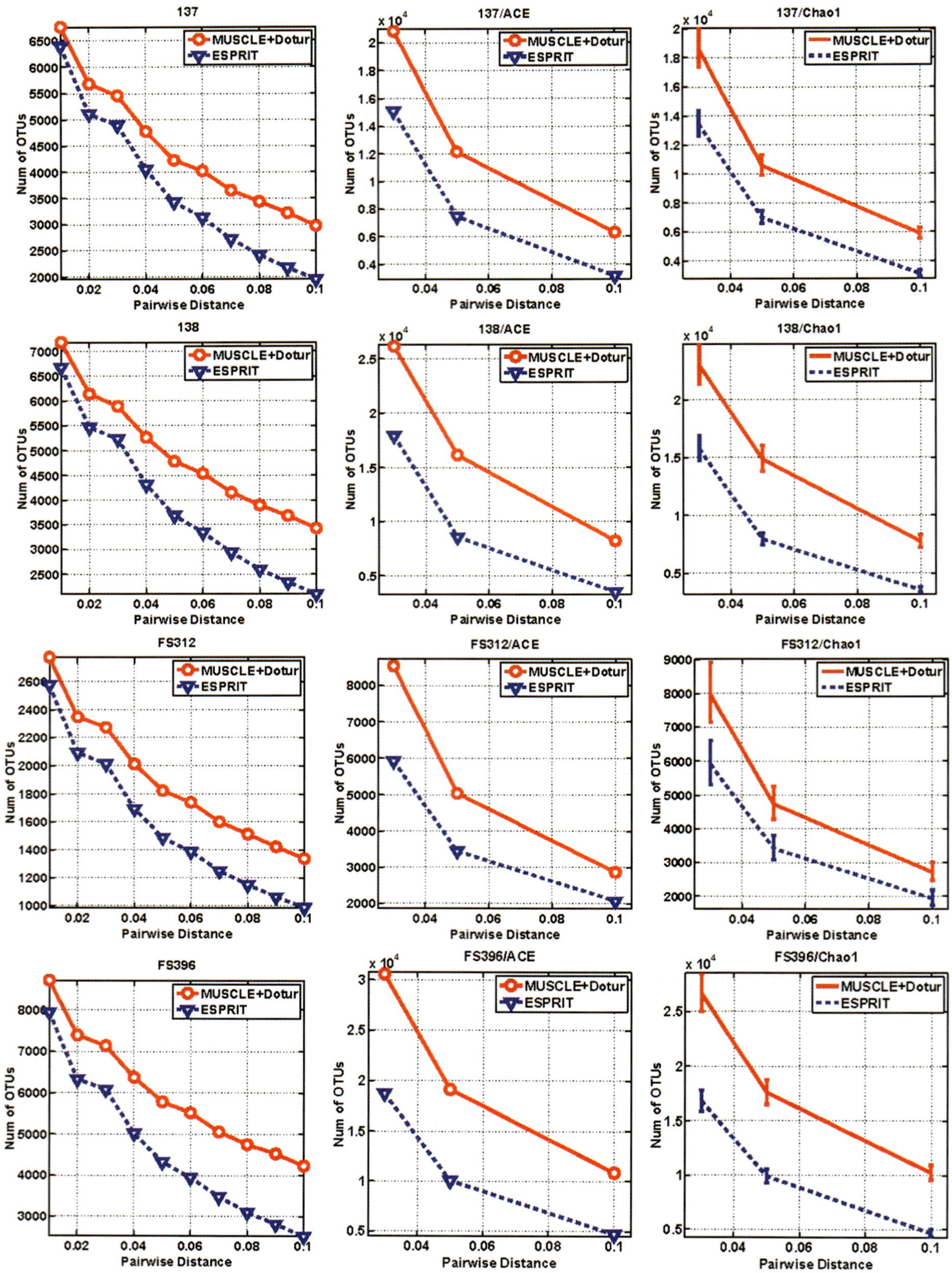

Figure 53.5 Lineage-through-time curves, ACE and Chao1 estimates generated by using ESPRIT and MUSCLE + DOTUR algorithms performed on 137, 138, FS312, and FS396 datasets.

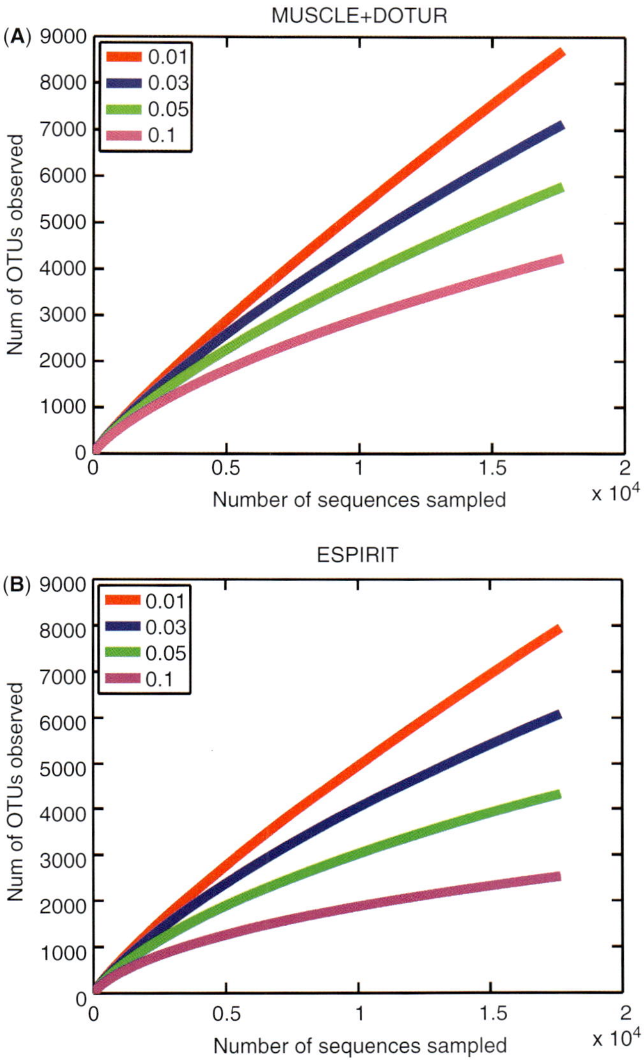

Figure 53.6 Rarefaction curves generated by using (**a**) MUSCLE + DOTUR and (**b**) ESPRIT performed on the FS396 data.

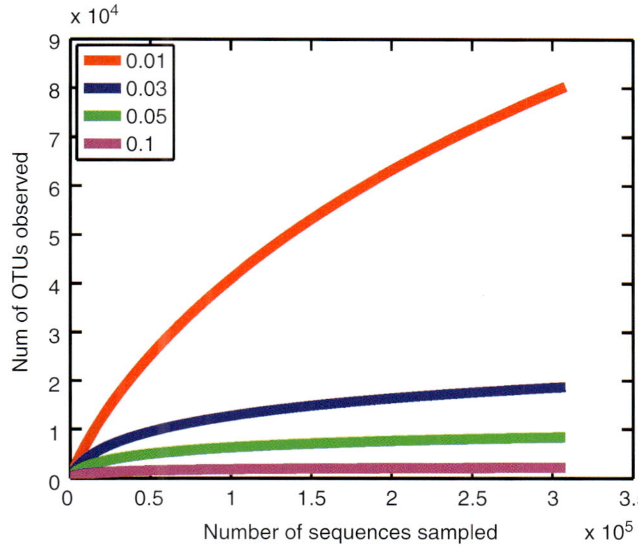

Figure 53.7 Rarefaction analysis of an air sample. Rarefaction curves are shown for OTUs with sequence variations that do not exceed 1%, 3%, 5%, or 10%.

Table 53.2 The Number of Observed OTUs, the ACE and Chao1 Estimates of an Air Sample at Four Different Distance Levels[a]

	Pairwise Distance			
	0.01	0.03	0.05	0.1
OTUs	80,238	18,686	8,344	2,109
ACE	147,266	23,894	9,664	2,262
Chao1	138,376	23,921	9,748	2,293
Upper	139,911	24,302	9,932	2,362
Lower	136,881	23,566	9,585	2,242

[a] The 95% confidence intervals of the Chao estimates are also provided.

with a distance of 0.06 calculated based on pairwise alignment, and then we performed a multiple sequence alignment of the two sequences along with 100 sequences randomly selected from the data (using MUSCLE with the default parameter) and recorded the pairwise distance of the first two sequences. In order to eliminate statistical variations, the experiment was repeated 100 times. The so-obtained distances are plotted in **Figure 53.8**, from which we can see that the pairwise distances computed based on MSA are much larger than 0.06. The maximum distance is 0.22 and the average is 0.09. This can be explained by the fact that MSA is aimed to minimize the sum of pairwise scores, and in order to align the 100 highly diverse sequences, the alignment quality of the first two sequences has to be sacrificed, leading to an inflated estimate of genetic distance. Note that we include only 100 randomly selected sequences in this experiment. With more unrelated sequences and/or if unrelated sequences are more diverse, the distance between the first two sequences would be even larger. This leads to an interesting observation: whether two sequences are from the same OTU (e.g., species or genus) should be physically determined by their sequence composition. Now with the use of MSA, the distance and thus whether the two sequences belong to the same OTU are also determined by sequencing depth (i.e., the number of sequences sampled) and the environment from which they are extracted (i.e., how diverse other sequences are), which is a highly undesirable property. Another interesting thing we should point out is that the harder an MSA algorithm tries to minimize the sum of pairwise scores by aligning highly diverse sequences, the larger the distance between the two first sequence

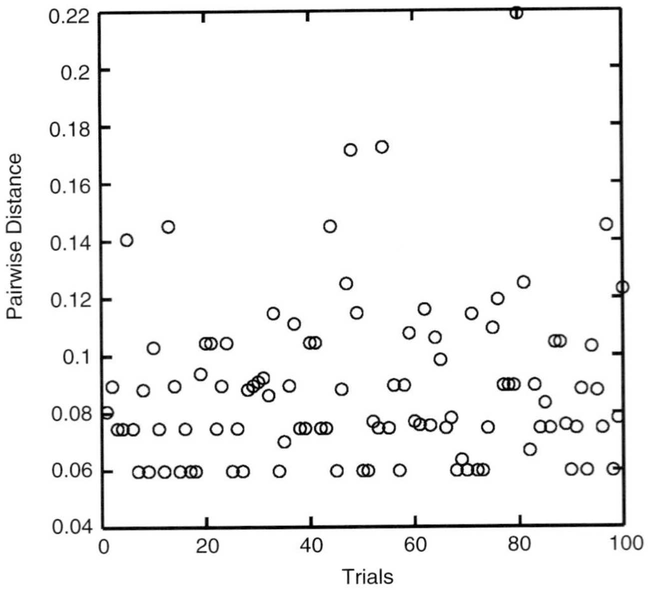

Figure 53.8 Pairwise distances computed based on multiple sequence alignment are much larger than those obtained by using the Needleman–Wunsch algorithm, which is 0.06 in this example. The maximum distance is 0.22 and the average is 0.09.

can be. For example, in the benchmark study described above, we observed that MUSCLE with the default parameter achieves better alignments, but yields an even worse result than that obtained by using the simple parameter [Sun et al., 2009]. From the above analysis, we conclude that MSA should not be used for the purposes of estimating microbial diversity, even if the optimal MSA could be performed.

53.4.2 Efficiency of ESPRIT

Equipped with next-generation sequencing technology, researchers can now obtain millions of sequences for diverse applications such as large-scale human epidemiological studies and global ocean surveys. Parallel computing that distributes computation to hundreds or even thousands of nodes seems to be a viable approach. We have demonstrated that the newly proposed ESPRIT algorithm can process several hundreds of thousands of sequences within 10 h by using a relatively small CC. The two key components of ESPRIT are (a) the use of pairwise instead of multiple sequence alignment, which allows for parallel computing, and (b) the development of Hcluster that performs hierarchical clustering using online learning to address the memory issue. We have demonstrated that MSA is not only computationally expensive, but also tends to overestimate microbial diversity. In order to process many millions of reads, researchers may need to collaborate with a supercomputing center (e.g., IBM Roadrunner, which has about 20,000 nodes). Our ESPRIT algorithm can easily be modified to work on any cluster. At Roadrunner's full capacity, we estimate that it would take 1 s, 10 h, and 40 h to process 1, 50, and 100 million reads, respectively. Of course, this is an overoptimistic estimation, since it focuses only on computational complexity and assumes that the resources of the entire supercomputing center are available for this singular task. ESPRIT, however, does provide a promising direction for analyzing large collection of 16S rRNA data. We will continue to optimize the performance of ESPRIT and release the software to the community. We believe that ESPRIT will evolve to be a powerful tool for metagenomics study.

INTERNET RESOURCES

The code and user guide are freely available at http://www.biotech.ufl.edu/people/sun/esprit.html.

REFERENCES

Borneman J, Triplett EW. 1997. Molecular microbial diversity in soils from eastern Amazonia: Evidence for unusual microorganisms and microbial population shifts associated with deforestation. *Appl. Environ. Microbiol*. **63**:2647–2653.

Chao A. 1984. Non-parametric estimation of the number of classes in a population. *Scand. J. Stat*. **11**:265–270.

Chao A, Lee SM. 1992. Estimating the number of classes via sample coverage. *J. Am. Stat. Assoc*. **87**:210–217.

Chao A, Ma MC, Yang MCK. 1993. Stopping rules and estimation for recapture debugging with unequal failure rates. *Biometrika* **80**:193–201.

Dalevi D, Dubhashi D, Hermansson M. 2006. Bayesian classifiers for detecting HGT using fixed and variable order markov models of genomic signatures. *Bioinformatics* **22**:517–522.

Duda RO, Hart PE, Stork DG. 2001. *Pattern Classification*, 2nd ed. New York: John Wiley & Sons.

Edgar RC. 2004a. MUSCLE: Multiple sequence alignment with high accuracy and high throughput. *Nucleic Acids Res*. **32**:1792–1797.

Edgar RC. 2004b. Local homology recognition and distance measures in linear time using compressed amino acid alphabets. *Nucleic Acids Res*. **32**:380–385.

Eisen JA. 2007. Environmental shotgun sequencing: Its potential and challenges for studying the hidden world of microbes. *PLoS Biol*. **5**:e82.

Gans J, Woilinsky M, Dunbar J. 2005. Computational improvements reveal great bacterial diversity and high metal toxicity in soil. *Science* **309**:1387–1390.

Huber JA, Welch DBM, Morrison HG, Huse SM, Neal PR, et al. 2007. Microbial population structures in the deep marine biosphere. *Science* **318**:97–100.

Hurlbert SH. 1971. The non-concept of species diversity: A critique and alternative parameters. *Ecology* **52**:577–586.

Huse SM, Huber JA, Morrison HG, Sogin ML, Welch DM. 2007. Accuracy and quality of massively parallel DNA pyrosequencing. *Genome Biol*. **8**:R143.

Karlin S, Burge C. 1995. Dinucleotide relative abundance extremes: A genomic signature. *Trends Genet*. **11**:283–290.

KARLIN S, MRAZEK J, CAMPBELL A. 1997. Compositional biases of bacterial genomes and evolutionary implications. *J. Bacteriol.* **179**:3899–3913.

KATOH K, KUMA K, TOH H, MIYATA T. 2005. MAFFT version 5: Improvement in accuracy of multiple sequence alignment. *Nucleic Acids Res.* **33**:511–518.

KEIJSER B, ZAURA E, HUSE SM, VAN DER VOSSEN J, SCHUREN F, et al. 2008. Pyrosequencing analysis of the oral microflora of healthy adults. *J. Dent. Res.* **87**:1016–1020.

MAVROMATIS K, IVANOVA N, BARRY K, SHAPIRO H, GOLTSMAN E, et al. 2007. Use of simulated data sets to evaluate the fidelity of metagenomic processing methods. *Nat. Methods* **4**: 495–500.

NEEDLEMAN SB, WUNSCH CD. 1970. A general method applicable to the search for similarities in the amino acid sequence of two proteins. *J. Mol. Biol.* **48**:443–453.

ROESCH LFW, FULTHORPE RR, RIVA A, CASELLA G, et al. 2007. Pyrosequencing enumerates and contrasts soil microbial diversity. *ISME J.* **1**:283–290.

ROTHBERG JM, LEAMON JH. 2008. The development and impact of 454 sequencing. *Nat. Biotechnol.* **26**:1117–1124.

SAIT M, HUGENHOLTZ P, JANSSEN PH. 2002. Cultivation of globally distributed soil bacteria from phylogenetic lineages previously only detected in cultivation-independent surveys. *Environ. Microbiol.* **4**:654–666.

SCHLOSS PD, HANDELSMAN J. 2005. Introducing DOTUR, a computer program for defining operational taxonomic units and estimating species richness. *Appl. Environ. Microbiol.* **71**:1501–1506.

SOGIN ML, MORRISON HG, HUBER JA, WELCH DM, HUSE SM, et al. 2006. Microbial diversity in the deep sea and the underexplored "rare biosphere". *Proc. Natl. Acad. Sci. USA* **103**:12115–12120.

SUN Y, YU F, LIU L, FARMERIE W. 2007. Estimating microbial population densities based on genomic signatures. *Proc. Int. Conf. Bioinform. Comput. Biol.* **1**:163–168.

SUN Y, CAI Y, LIU L, YU F, FARRELL M, et al. 2009. ESPRIT: Estimating species richness using large collections of 16S rRNA pyrosequences. *Nucleic Acids Res.* **37**:e76.

Part 6

Complementary Approaches

Chapter 54

Metagenomic Approaches in Systems Biology

MARÍA-EUGENIA GUAZZARONI AND MANUEL FERRER

54.1 INTRODUCTION

Cell differentiation and maintenance are intrinsically complex processes requiring the coordinated dynamic expression and action of hundreds of genes and proteins in precise response to external transcriptional and metabolic signalling cues in connection to the production and transformation of thousands of metabolites that, all together, capitalize the cell cycle [Richard and Mobashery, 2003]. Numerous recent reports have used both experimental and computational techniques to dissect this complexity, an objective of Systems Microbiology [Raes and Bork, 2008; Vieites et al., 2009]. These reports suggest that the control of cell functioning has both deterministic and stochastic elements: Complex underlying regulatory networks define stable molecular states toward which individual cells are drawn over time, whereas stochastic fluctuations in gene and protein expression and macromolecule production levels drive transitions between coexisting attractors, ensuring robustness at the population level [Feist et al., 2009].

During the last few years, we have seen enormous strides in our abilities to sequence single genomes, and the information that has poured out of these sequences is quite astonishing: With more than 150 complete genome sequences now available and many laboratories rushing into microarray analysis (see Chapters 57, 58, 60, 61, Vol. I), proteomic (see Chapters 66–70, Vol. I) and metabolomic (see Chapters 71–74, Vol. I) initiatives (**Fig. 54.1**), still it is paramount to understand the underlying interspecies metabolic and regulatory networking studied as integrated units in complex communities. Thus, genome-wide modeling of communities has remained largely untouched, namely because the shortcoming of traditional microbiology applied to System Biology derives from the fundamental maxim that the greater part of biodiversity is represented by a largely untapped reservoir of uncultured microbial genomes [Hugenholtz, 2002].

Metagenomics is the application of modern genomics techniques to the study of communities of microbial organisms directly in their natural environments, bypassing the need for isolation and lab cultivation of individual species [DeLong, 2002; Handelsman, 2004; Riesenfeld et al., 2004; Edwards and Rohwer, 2005]. The field has its roots in the culture-independent retrieval of 16*S* rRNA genes, pioneered by Pace and colleagues two decades ago [Olsen et al., 1986; see Chapter 15, Vol. I]. Since then, metagenomics has revolutionized (systems) microbiology by shifting focus away from single microbial isolates toward the estimated 99% of microbial species that cannot currently be cultivated [Hugenholtz, 2002; Rappe and Giovannoni, 2003].

Over the past few years, the major challenge associated with metagenomics shifted from generating to analyzing sequences: As compared to conventional sequencing projects, metagenomics generates massive amounts of genome information [Raes et al., 2007; Sterk et al., 2010]. Bottlenecks become data exchange, analysis, and visualization, which require paired work of systems biologists and microbiologists. This is because the new-generation sequencing technologies have made the data production even more high-throughput, several orders of magnitude less expensive, and, most importantly, cloning- and culturing-independent [Valdivia-Granda, 2008;

Handbook of Molecular Microbial Ecology, Volume I: Metagenomics and Complementary Approaches, First Edition. Edited by Frans J. de Bruijn.

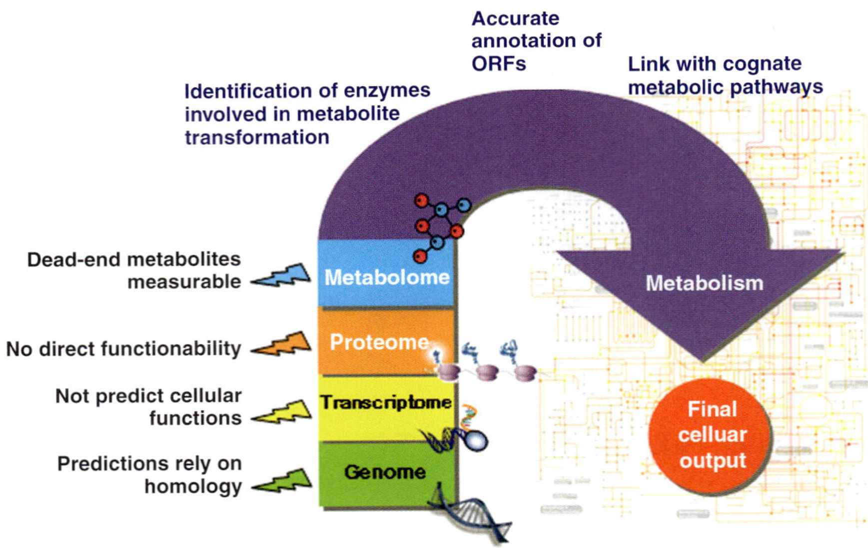

Figure 54.1 "-Omics" approaches used in Systems Biology of living organisms. Despite major advances in understanding triggers of cell metabolism, there is a dearth of information on the global metabolic events that take place in reprogrammed cells. Genomic, proteomic, and metabolomic methods are partially available, but those still have some limitations for metabolic understanding—that is, difficulties in identifying and localizing metabolites and proteins and low ability to predict microbial functions (except for only up- or down-regulation of genes or proteins levels).

Fox et al., 2009; see also Chapter 27, Vol. I]. Accordingly, new advances in sequencing technologies bring random shotgun sequencing of ecosystems within reach of smaller labs, but the complexity of metagenomics data can be overwhelming [Kunin et al., 2008]. Estimates of the average DNA content per cell (converted to megabases [Mb] for comparison) range from 1.5 to 8.0 Mb for soil and from 1.5 to 9.5 Mb for aquatic environments. Accordingly, the DNA from a single organism's genome requires 1×10^7 bytes of storage, increasing to 1×10^{10} bytes when the genome is annotated, whereas, by contrast, 1×10^7 bytes of metagenome from a single sample and its associated metadata might require 1×10^{12} bytes or nine terabytes of disk space [Valdivia-Granda, 2008]. Furthermore, because of its direct link with the functional repertoire, microbial genome size is an important ecologic parameter that is believed to be closely coupled to the functional complexity: The storage and management of transcriptome, proteome, and metabolic data (directly linked to functions) of a metagenome might require from 20,000 to 10^8-fold more disk space as compared to single organisms [Vieites et al., 2009] (**Fig. 54.2**).

Apart from the amount of metagenomic sequence data, one should take into consideration that metagenomics just tells the blueprint of what's there and what could possibly be functioning, but not every gene is expressed: Very often, certain portions [of microbes' genomes] will be expressed under certain conditions to respond to the environment [Kalyuznaya et al., 2008; Giuanoulis et al., 2009]. Therefore, from a Systems

Size	KT2440	PAO1	Human	ENV.SAMPLE and HUMAN MICROBIOME
Species	1	1	1	> 20,000
Genome	6.18 Mbp	6.3 Mbp	3.4 Bbp	1–100 Gbp
Transcriptome & Proteome	5437	5570	25000	~10^{12}
Reactome	1482 reactions representing 1216 compounds and 838 enzymes	1576 reactions representing 1339 compounds and 876 enzymes	2250 reactions representing 1445 compounds and 2289 enzymes	~10^{12} reactions representing 10^5compounds and 10^9enzymes

Figure 54.2 Complexity of bacterial metagenomes. This figure illustrates how the genetic and metabolic complexity increases when increasing genome size and diversity.

Biology perspective, a major and innovative thrust of metagenome-scale reconstruction and modeling of the metagenome's metabolic space and interspecies interactions is the development of computational workflow for data handling and model generation, using genomic, thermodynamic, and biochemical information, to generate genome-scale constraint-based models of the metabolism of the sequenced metagenomes, as previously has been done for individual bacteria [Raes and Bork, 2008]. Extensive experimental validation by a number of methods, including "-omics," enables iterative refining of the models. Such models define the organism's global metabolic space and network structural properties, and they enable identification of critical and structural nodes and flux distribution potential across and within cells [Giuanoulis et al., 2009].

We are just starting to exploit metagenomics to understand the microbial world as a whole, its role in climate and biochemical processes and potential for bioenergy sources, and its implications in human health, disease susceptibilities pathophysiology of infectious and immune-mediated diseases and a broad range of biomedical, industrial, and environmental applications [Desai et al., 2009; Petrosino et al., 2009; Qi et al., 2009]. However, in order to simultaneously compare metagenomic samples at different levels of resolution and discover new life forms as well as unravel their diversity and their genetic and metabolic shifts, new computational biology tools are required for both storage and analysis using comparative methods [Chen and Pachter, 2005; Kunin et al., 2008; Ansorge, 2009; Huson et al., 2009; Li et al., 2009; Sterk et al., 2010; see also Sections 6 and 7, Vol. II].

Following on from this, the chapter thus aims at filling this knowledge gap by providing a conceptual and technical basis for the translation of metagenome information into a model framework accounting for the complex interactions in complete ecosystems and allowing the modeling of multispecies metabolic networks in a Systems Biology perspective. In consideration of these issues, here some computational biology and cyber-infrastructure challenges that need to be addressed in order to translate metagenomic information into biomedical, agricultural, environmental, and industrial applications will be highlighted. In addition, issues of metagenomic-, raw-, and meta- data exchange formats, algorithms for sequence assembly, comparative analysis, pathway reconstruction, phylogenomics, molecular barcoding, and diversity estimation will be presented. Finally, some emerging technologies that bioinformaticians and metagenomic researchers should consider during the implementation of new research programs will be summarized.

54.1.1 Systems Biology: Moving from Single Genomes to Metagenomes

Cell functioning is a complex process, in which function declines while damage accumulates [Partridge and Gems, 2002]. Functioning has, nonetheless, proved to be responsive to simple genetic and environmental interventions [Kalyuznaya et al., 2008]. One of the numerous discoveries have been that mutations in single genes or the external incorporation of nutrients can substantially modify (increase or decrease) both lifespan and cell metabolism. Many of the genes involved encode components of evolutionarily conserved nutrient-sensing pathways, known for their role in control of growth and metabolism. The discovery of the role of conserved nutrient-sensing pathways in cell functioning has opened the way to using simpler organisms to understand cell biology [see examples of ongoing single organisms Systems Biology projects in http://www.sysmo.net/index.php?index=52] and has also provided powerful tools to address the many challenges posed by experimental analysis of mechanisms of the cell cycle. These pathways are highly pleiotropic in their effects, and can influence growth and development, metabolism, and resistance to various stresses, as well as lifespan. The pathways may act at multiple points in the life course to influence development and survival, and their protective effects may also involve changes at the molecular, cellular, tissue (in case of multicellular organisms) and systemic levels [Hartwell and Weinnert, 1989; Manning and Cantley, 2007].

The understanding of such complexity requires a diversity of experimental approaches directed at integrating events at different levels of organization (**Fig. 54.3**) and on different time scales [VerBerkmoes et al., 2009]. Ultimately, a systems approach is required to understand how the signaling pathways alter biochemistry and organismal function so as to maintain function and protect against damage, a general objective of Systems Microbiology [Kalyuzhnaya et al., 2008]. This is an important issue because we know that cell growth [the increase in cell mass through macromolecular synthesis] requires the synthesis of cellular components in precise, stoichiometric quantities and must be subject to tight coordinate control as response to environmental cues [Jorgensen and Tyers, 2004].

We have entered an era of "megasequencing projects" that now include funded projects like the Genomic Encyclopedia of Bacteria and Archaea (GEBA) project [http://www.jgi.doe.gov/programs/GEBA/] and the Human Microbiome Project [http://nihroadmap.nih.gov:hmp//; see also Chapter 35, Vol. I], with many more equally visionary projects on the horizon. While a genome represents the full genetic (DNA) complement of

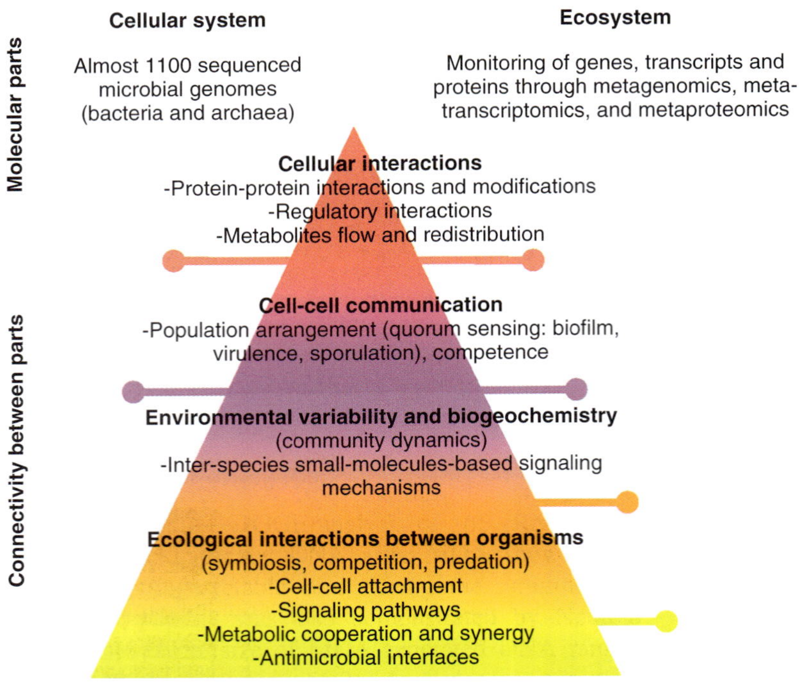

Figure 54.3 Molecular components of complex biological systems. As shown here, the monitoring of component networking in single cells is less understood than interaction at a global scale (the last two pyramid's levels).

a single organism, metagenomes represent the DNA of an entire community of organisms. There are now thousands of genomes and metagenomes available for study (see the Genomes Online Database [Liolios et al., 2008]). Interest in improved sampling of diverse environments (e.g., ocean, soil, sediment, and a range of hosts), combined with advances in the development and application of ultra-high-throughput sequencing methods, is vastly accelerating the pace at which new metagenomes are generated [Ansorge, 2009; Fox et al., 2009]. For example, in 2007, the Global Ocean Survey published scientific analyses of 41 metagenomes [Rush et al., 2007]; and as of October 2008, the submission of user-generated metagenomes to the public MG-RAST annotation server surpassed 1300 [Meyer et al., 2008; see also Chapter 37, Vol. I] (**Fig. 54.4**).

The objective of a metagenomic approach is to propose a model explaining the interaction between the different members of a simplified microbial community [Tyson et al., 2004; Angly et al, 2005, 2006; Desnues et al., 2008; Woyke et al., 2006; Bench et al., 2007; Warnecke and Hugenholtz, 2007; Bertin et al., 2008; Biddle et al., 2008; Kalyuzhnaya et al., 2008]—in particular, to address the following specific questions: (1) What are the metabolic functional redundant and unique features of the selected members of the simplified microbial community? (2) What are the synergistic interactions existing between bacterial community members resulting from complementary metabolic capacities? (3) Can we predict and/or explain based on in silico metabolic reconstructions what will be the impact of presence/absence of a particular community member on ecosystem metabolites and community functioning?

The analysis of such components require a number of actions interweaving model-driven experimentation and experiment-driven model development, including: (i) systematic data management [Chen and Pachter, 2005], (ii) a server infrastructure that allows easy local installations by outsourcing all computations to grid computing service centers [Meyer et al., 2008; Desai et al., 2009], (iii) software components for high-throughput analysis of large metagenome data sets [Mavromatis et al., 2007; Valdivia-Granda, 2008; DeLong, 2009], and (iv) advanced comparative metagenomics [Kalyuzhnaya et al., 2008] and interactive visualization [Letunic et al., 2008; Gianoulis et al., 2009]. In an iterative process of model-building and experimentation, these methods are normally validated stepwise, through the sequence and analysis of a natural and relatively well-characterized short member consortium of environmental relevance [Parro et al., 2007; Kalyuzhnaya et al., 2008; Moran, 2009; VerBerkmoes et al., 2009]. Subsequently, the computational methods already available are then tested for a higher complexity system through the (deep) sequencing, analysis, and constraint-based reconstruction and modeling of metagenomic samples from a series of studies with a simplified microbial composition measuring by the various x-omics and by monitoring parameters related to energy metabolism and weight management, among others.

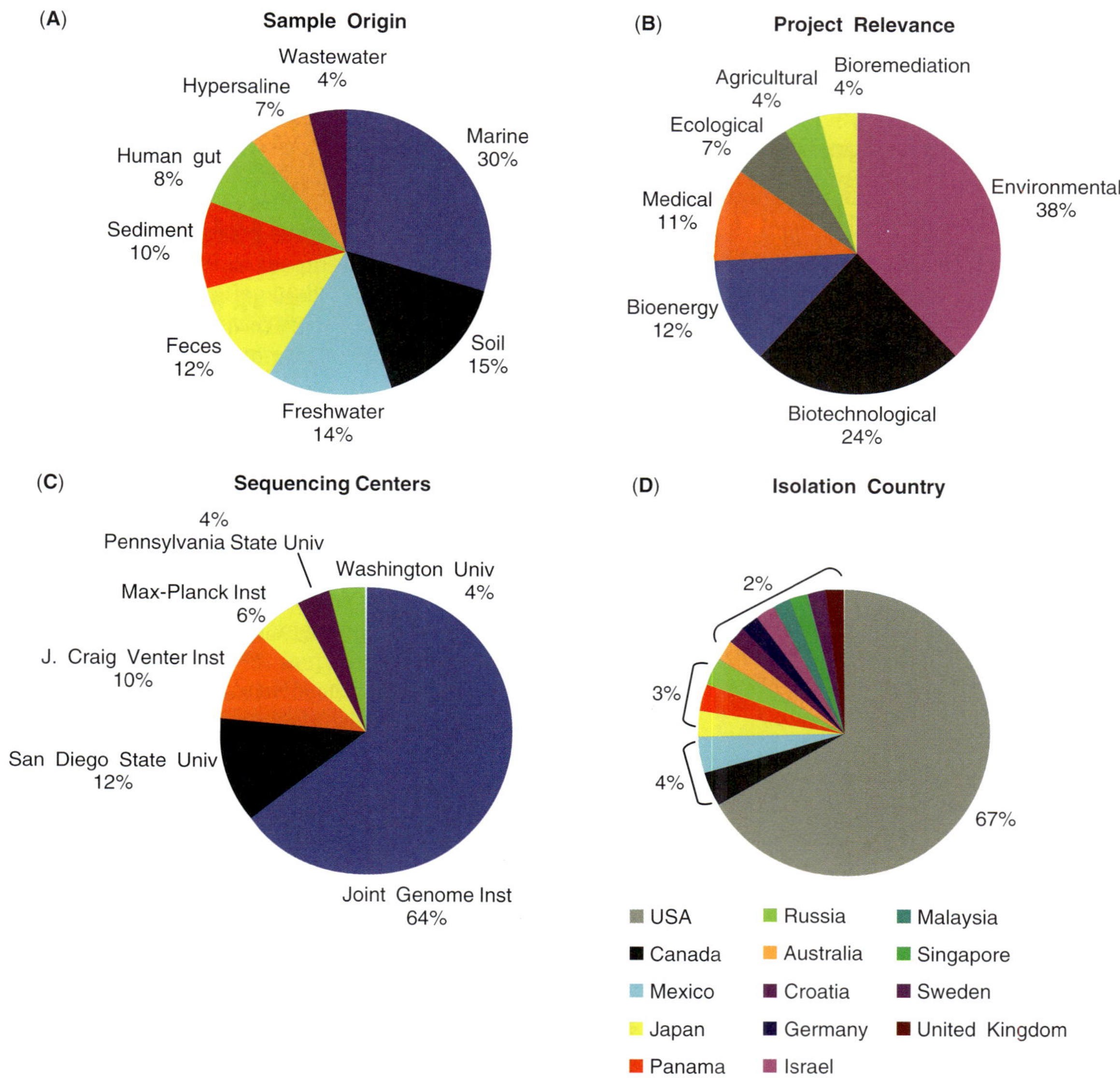

Figure 54.4 Summary of some features of metagenomics projects related to microbial characterization listed in the Genomes online database (GOLD). Currently (database updated on January 24, 2010) the GOLD database enumerates 207 projects, of which 63 are complete and 144 are incomplete. The sequencing statuses of the later (which are detailed in the database) are: 14, draft; 38, in progress; and 11, awaiting DNA. (**A**) Sample origin distribution. (**B**) Area of relevance of the listed projects. (**C**) Top sequencing centres involved in metagenomics projects. (**D**) Diagram showing percentages of countries' contribution regarding to sample isolation.

54.1.2 Computational Approaches in Systems Biology

Development of automated and high-throughput sequencing technologies, as well as advancement in bioinformatics, has allowed sequencing of entire communities in short periods of time at reasonable cost, generating giga- and tera-bases of new sequence data [Margulies et al. 2005; Huson et al., 2007]. Thus, an annotated collection of all publicly available nucleotide and amino acid sequences exists: More than 190 billion bases are released to databases, and this number doubles approximately every 18 months. On one hand, this opens up an unprecedented opportunity to dig into the gold mine of "new" sequences; on the other, such large datasets raise several processing problems and drive current bioinformatics tools to their limits. The current analyses of metagenomics data largely rely on the computational tools originally designed for microbial genomics [Valdivia-Granda, 2008]. These data can be then complemented with additional data applying high-throughput approaches (see examples in **Table 54.1**).

However, a primary bottleneck is in the lack of scalable algorithms and open source software for large-scale

Table 54.1 Summary of the Main High-Throughput Approaches Used in the Study of Microbial Community Functioning and Major Inherent Limitations[a]

Goal of Studying	Approach and Macromolecular Studied Part (MSP)	Applied Techniques	Shortcomings
Assortment of specific groups of organisms in an ecosystem	Single gene analysis such as small subunit (SSU) rRNA gene MSP: marker gen	• Amplyfing and sequencing of a single gene of all organisms—for 16S rRNA gene (for archaea and bacteria) and 18S rRNA gene (for eukaryotes). • Direct pyrosequencing	• Some particular phylogenetic groups have cloning limitations and may be under-represented. • SSU data in isolation are limited, and the physiological role of individual species cannot be detailed.
Diversity and function of genes in a particular ecosystem	Metagenomics MSP: genes	• Shotgun cloning and high-throughput sequencing • Pyrosequencing • Multiple Displacement Amplification (MDA) and sequencing	• Cloning biases, some specific genes may be underrepresented in the metagenomic library. • In silico function predictions rely on homology, many wrong annotations in the databases.
Genes that are expressed in a specific ecosystem	Metatranscriptomics MSP: mRNA	• cDNA sequencing through shotgun cloning and sequencing (Differential Display-PCR) • cDNA pyrosequencing • Microarray hybridizations • cDNA-AFLP • Suppressive subtractive hybridization • CARD-FISH	• Inherent instability of RNA molecules • Low quantities of transcripts recovered from samples. • Presence of environmental contaminants that affect the quality of RNA and DNA hybridization. • Does not reflect the protein levels and does not predict the cellular functions.
Direct signal of activities of genes in specific environments	Metaproteomics MSP: proteins	• Protein separation by 2D-gel electrophoresis SDS-PAGE followed by protein identification by Mass spectrometry (MS). MALDI-ToF-MS, MALDI-ToF/ToF-MS, or ESI-MS/MS are preferentially used for protein identification of totally sequenced organisms.	• Low resolution and no direct functionality. • Difficulty in identifying peptides through database searches. • Ambiguity in interpreting data in the absence of any corresponding metagenomic information
Metabolic network between species of a ecosystem	Meta-metabolomics MSP: small-molecule metabolites	• NMR spectroscopy, HPLC/GC-MS, and capillary electrophoresis-MS	• Only accumulating/dead-end metabolites measurable • Complex ecosystems cannot be analyzed. At the moment, there is not any analytical methodology for deconvoluting all the metabolites present in a complex mixture.

[a] Shortcomings of each methodology show that just a multifaceted approach will lead to a global and comprehensive picture of the complexity and functioning of an ecosystem.

data processing. This is of special interest, because we know that species distribution impacts sequence sampling; and as such, one sacrifices the depth of sampling obtained by pyrosequencing for accuracy in describing the relatively more numerous (yet still rarely sampled) members of the community [Raes and Bork, 2008].

The computational task of analyzing this massive volume of complex data is quickly outpacing the capacity for existing software and hardware capabilities. Current stand-alone sequence analysis implementations generally take a single dedicated machine days or weeks to complete the analysis of a single microbial genome against the nonredundant protein database (a growing curated collection of proteins from all sequenced organisms). The grand challenge in this "genomes-to-life" sequence analysis effort is much larger, putting it beyond the capacity of most biologists to complete in a reasonable time see Section 4, Vol. I. The burgeoning number of sequenced genomes presents computational challenges that are just being met with existing methods and infrastructure [Pop, 2009].

The challenge of assembling metagenomic sequences arises mainly from the high species complexity of the community [Vieites et al., 2009]. To solve this, the majority of Systems Biology studies have been centred in low community members, such as the simple acid mine drainage biofilm community comprising <10 species [Tyson et al., 2004]. Key insights into the biofilm community function and interaction could be inferred from a metabolic reconstruction of the genomic data, such as the partitioning of community essential functions, such as nitrogen fixation [Tyson et al., 2004]. Other ecosystems currently under investigation include termite hindgut communities for bioenergy production (see Chapter 22, Vol. II), activated sludges, soils (see Chapters 9–17, Vol. II), and terephthalate-degrading communities [Vieites et al., 2009]. Bioremediation is also an issue of interest from a Systems Biology perspective since it involves the exposure of a whole mixture of chemical structures to an intricate multispecies metabolic network present in a polluted scenario [Desai et al., 2009; see also Chapters 43–46, Vol. II)]. The complexity involved in such events is growingly amenable to the conceptual frame and the tools of systems biology. The availability of genes, genomes, and metagenomes of biodegradative microorganisms makes it possible to model and even predict the fate of chemicals through the global metabolic network that results from connecting all known biochemical transactions. Microbial communities thus embody a landscape of pan-enzymes that is shaped by the freely diffusible metabolic pool (epimetabolome) [de Lorenzo, 2008].

The main focus of computational biology is to decipher and understand the biology of metagenomes. Genomes, as the molecule of life, contain the necessary information for the development and evolution of living organisms. Despite the fact that this information is encoded in a four-letter code, it presents different levels of complexity (Fig. 54.3).

The comparison of metagenomic and derived sequences allows us to uncover the general rules and patterns that control and make life possible. Ongoing research activity uses computational and experimental approaches to reach these goals by finding answers to specific molecular, evolutionary, and applied questions and by using bioinformatics tools, a discipline that is the crossroads of different scientific disciplines—in particular, biology, mathematics, and computer sciences [Kunin et al., 2008]. **Tables 54.2** and **54.3** summarized a list of bioinformatics tools and web sources available for metagenomics applications. As shown, a wide range of tools are available not only for assembly and binning of metasequences [Pop, 2009] but also for (a) the automatic annotation and analysis of metagenome sequence fragments and (b) functional and phylogenetic classification [Dick et al., 2009; Gerlach et al., 2009; Monzoorul et al., 2009; Vieites et al., 2009]. This analysis is facilitated by the creation of a number of repository and publicly available databases that archive most microbial metagenome projects.

54.1.3 Computer Programs for Metagenomic Data Analysis

The remarkable advance of metagenomics presents significant new challenges in data analysis. Several computational approaches have been applied for the analysis of large collections of metagenomic datasets from anonymous species in different environments. Thus, several high-throughput pipelines have been constructed to provide high-performance computing to all researchers interested in using metagenomics [Meyer et al., 2008; see also Chapter 37, Vol I]. The pipelines produce automated functional assignments of sequences in the metagenome by comparing both protein and nucleotide databases. Phylogenetic and functional summaries of the metagenomes are then generated, and tools for comparative metagenomics are incorporated into the standard views to infer the properties of a habitat through the analysis of metagenome hints (see examples in Venter et al. [2004], Edwards et al. [2006], and Biddle et al. [2007]). Furthermore, metagenomic sequence information is complemented with related "-omics" data covering multiple environmental variables (many-to-many) to achieve the metabolic footprint of heterogeneous communities [Gianoulis et al., 2009].

A number of methods and resources have been developed that provides automatic annotation and analysis of millions of fragmented metagenome sequences (see Tables 54.1 and 54.2). These services have removed one of the primary bottlenecks in metagenome sequence analysis: the availability of high-performance computing for annotating the data [Li, 2009]. The open-source IMG/M [Markowitz et al., 2008], CAMERA [Seshadri et al., 2007], MG-RAST [Aziz et al., 2008; Meyer et al., 2008], RAMMCAP [Li, 2009], WebCARMA [Gerlach et al., 2009] and Sort-ITEMS [Monzoorul Haque et al., 2009] tools are example of freely available services that provides fast automatic annotation and analysis of metagenomes. Additionally, a number of tools (i.e., TETRA) have been specially designed for the phylogenetic analysis of metagenomes based on the

Table 54.2 Bioinformatics Tools and Web Resources Available for Metagenomics Applications

Software Name	Function	URL	Reference
	Sequencing/Assembly		
AMOS (A Modular, Open-Source whole genome assembler)	Manipulation of input and output files related to whole-genome shotgun assembly.	http://sourceforge.net/projects/amos/files/	Phillippy et al. [2008]
Elvira (Executive for Large-scale VIRal Assembly)	High-throughput assembly of amplicon reads for virus-sized genomes.	http://elvira.sourceforge.net/	The Institute for Genomic Research (TIGR)
Lucy	A tool for cleaning data produced by automated Sanger DNA sequencers prior to sequence assembly and other downstream uses.	http://lucy.sourceforge.net/	Chou and Holmes [2001]
Velvet	Sequence assembler for very short reads.	http://www.ebi.ac.uk/~zerbino/velvet/	Zerbino and Birney [2008]
Celera Assembler	Can reconstruct long sequences of genomic DNA given the fragmentary data produced by whole-genome shotgun sequencing.	http://sourceforge.net/apps/mediawiki/wgs-assembler/index.php?title = Main-Page	Myers et al. [2000]
TIGR (The Institute for Genomic Research) Assembler	An approach to assembling large, random shotgun sequencing projects.	ftp://ftp.jcvi.org/pub/software/assembler/	Sutton et al. [1995]
Phrap (Phragment assembly program)	Program for assembling of shotgun DNA sequence data which allows the use of the entire read and not just the trimmed high-quality part.	http://www.phrap.org/phredphrapconsed.html#block-phrap	Gordon et al. [2001]
ABySS (Assembly by Short Sequences)	This tool is a de novo, parallel, paired-end sequence assembler that is designed for short reads.	http://www.bcgsc.ca/platform/bioinfo/software/abyss	Simpson et al. [2009]
Newbler	Software package designed specifically for assembling de novo DNA sequence data generated by the 454 GS series of pyrosequencing platforms.	http://www.454.com/	454 Life Sciences (Roche Diagnostics)
MetaORFA (Metagenomic ORFome Assembly)	ORFome assembly approach first applies gene finding in the unassembled reads, and then assembles only those annotated (partial) ORFs into peptides.	Tool to be released soon: http://omics.informatics.indiana.edu/hmp/projects.php	Ye and Tang [2009]
	Annotation		
MG-RAST (Metagenome Rapid Annotation using Subsystem Technology)	Fully automated service for annotating metagenome sequence fragments, their phylogenetic classification, metabolic reconstructions and comparison tools.	http://metagenomics.nmpdr.org	Meyer et al. [2008]
Manatee	Web-based tool used to perform manual functional annotation.	http://manatee.sourceforge.net/	The Institute for Genome Sciences
PASA (Program to Assemble Spliced Alignments)	A eukaryotic genome annotation tool that exploits spliced alignments of expressed transcript sequences to automatically model gene structure.	http://pasa.sourceforge.net/	Haas et al. [2003]
CAMERA (Community Cyberinfrastructure for Advanced Marine Microbial Ecology Research and Analysis)	CAMERA is developing the cyberinfrastructure necessary to support the data, tools, and resources that will be needed to enable the scientific community to use the rapidly growing treasure of metagenomic information.	http://camera.calit2.net/	Seshadri et al. [2007]

Table 54.2 Bioinformatics Tools and Web Resources Available for Metagenomics Applications *(Continued)*

Software Name	Function	URL	Reference
RAMMCAP (Rapid Analysis of Multiple Metagenomes with a Clustering and Annotation Pipeline)	This software identifies raw read clusters and protein clusters that may include novel gene families, and compares metagenomes using clusters or functional annotations calculated by RAMMCAP.	http://tools.camera.calit2.net/camera/rammcap/	Li [2009]
	Analysis		
MEGAN (MEtaGenome Analyzer)	Computer program that allows laptop analysis of large metagenomic datasets. In a preprocessing step, the set of DNA reads (or contigs) is compared against databases of known sequences using BLAST or another comparison tool.	http://www-ab.informatik.uni-tuebingen.de/software/megan/	Huson et al. [2007]
IGM/M (Integrated Microbial Genomes/Metagenomes)	Provides tools for analyzing the functional capability of microbial communities based on their metagenome sequence, in the context of reference isolate genomes included from the IGM system.	http://img.jgi.doe.gov/cgi-bin/m/main.cgi	Markowitz et al. [2008]
CAMERA (Cyberinfrastructure for Advanced Marine Microbial Ecology Research and Analysis)	Cyberinfrastructure for Metagenomics, data repository and tools for metagenomics research.	http://camera.calit2.net/	Seshadri et al. [2007]
PhyloPythia	Binning tool for doing a phylogenetic classification of variable-length DNA fragments.	http://cbcsrv.watson.ibm.com/phylopythia.html	McHardy et al. [2007]
TETRA	Is a web-service and a stand-alone program for the analysis and comparison of tetranucleotide usage patterns in DNA sequences.	http://www.megx.net/tetra/	Teeling et al. [2004]
WebCarma	Web application for the taxonomic and functional classification of unassembled ultrashort reads from metagenomic communities.	http://webcarma.cebitec.unibielefeld.de	Gerlach et al. [2009]
SOrt-ITEMS (Sequence orthology based approach for improved taxonomic estimation of metagenomic sequences)	The method uses alignment parameters besides the bit score to first identify an appropriate taxonomic level where the read can be assigned.	http://metagenomics.atc.tcs.com/binning/SOrt-ITEMS	Monzoorul Haque et al. [2009]
MetaBioME (Comprehensive Metagenomic BioMining Engine)	Web resource to find novel homologs for known Commercially Useful Enzymes (CUEs) in metagenomic datasets and completed bacterial genomes.	http://metasystems.riken.jp/metabiome/	Sharma et al. [2009]

genomic composition of oligonucleotides of different lengths (motifs), named genomic signatures [Teeling et al., 2004; McHardy et al., 2007; Krause et al., 2008].

The challenge of assembling metagenomic sequences arises mainly from the short reads and the high species complexity of the community. Alternatively, individual (short) reads may be searched directly against databases of known genes (or proteins) to identify homologous sequences. The latter approach may have low sensitivity and specificity in identifying homologous sequences, which may further bias the subsequent diversity analysis. To solve this problem, novel approach to metagenomic data analysis have been designed to improve gene predictions. One is the so-called metagenomic ORFome Assembly (MetaORFA) program [Ye and Tang, 2009]. This computational framework consists of three steps. First, each read from a metagenomic project is first annotated with putative open reading frames (ORFs) that likely encode proteins. Next, the predicted ORFs are assembled into a collection of peptides using a EULER assembly method. Finally, the assembled peptides (i.e., ORFome) are used for database searching of homologues and subsequent diversity analysis.

Recently, a new computer program that allows computational analysis of large metagenomic datasets, named MEGAN, has been designed [Huson et al., 2007;

Table 54.3 Relevant Databases Related to Metagenomics Applications

Database	Description	URL
GOLD (Genomes Online Database)	Includes all prokaryotic and eukaryotic genome projects, with links to their respective webpages and data, when available. It also includes metagenomic projects.	http://genomesonline.org/index2.htm
GenBank	GenBank® is the NIH (National Institutes of Health) genetic sequence database, an annotated collection of all publicly available DNA sequences.	http://www.ncbi.nlm.nih.gov/Genbank/index.html
EMBL - Nucleotide Sequence Database (also known as EMBL-Bank)	EMBL constitutes Europe's primary nucleotide sequence resource. Main sources for DNA and RNA sequences are direct submissions from individual researchers, genome sequencing projects and patent applications.	http://www.ebi.ac.uk/embl/
ENA (European Nucleotide Archive)	ENA (which belongs to EMBL Bank) accepts data generated by next-generation sequencing methodologies such as 454, Illumina, and ABI SOLiD. This is a new service that will provide utility for those generating and using data for such applications as genome assembly, resequencing for polymorphism analysis, gene expression, epigenomics, and others.	http://www.ebi.ac.uk/embl/Documentation/ENA-Reads.html
COGs (Clusters of Orthologous Groups of proteins)	Phylogenetic classification of proteins by comparing protein sequences encoded in complete genomes. Each COG consists of individual proteins or groups of paralogs from at least three lineages and thus corresponds to an ancient conserved domain. Proteins are grouped in Functional categories.	http://www.ncbi.nlm.nih.gov/COG/
GenomeNet Database Resources	Japanese network of database and computational services for genome research and related research areas in biomedical sciences, operated by the Kyoto University Bioinformatics Center. It includes KEGG and DBGET databases.	http://www.genome.jp/
JCVI (J. Craig Venter Institute)	Compile many research areas related to Metagenomics and include The Comprehensive Microbial Resource (CMR), which is a free tool that allows researchers to access all of the publicly available bacterial genome sequences completed to date.	http://cmr.jcvi.org/tigr-scripts/CMR/CmrHomePage.cgi
UniMES (UniProt Metagenomic and Environmental Sequences)	UniMEs is a repository of the UniProtKB database (Protein knowledgebase) which is specifically developed for metagenomic and environmental data.	http://www.uniprot.org/
DDBJ (DNA Data Bank of Japan)	DDBJ is the sole nucleotide sequence data bank in Asia, which is officially certified to collect nucleotide sequences from researchers and to issue the internationally recognized accession number to data submitters.	http://www.ddbj.nig.ac.jp/Welcome-e.html
RDP (Ribosomal Database Project)	The RDP provides ribosome-related data and services to the scientific community, including online data analysis and aligned and annotated Bacterial and Archaeal small-subunit 16S rRNA sequences.	http://rdp.cme.msu.edu/
JGI (Joint Genome Institute)	The JGI provides integrated high-throughput sequencing and computational analysis.	http://www.jgi.doe.gov/
DAC (Danish Archaea Centre)	The DAC has a long-term basic research program focused on understanding the basic molecular and cellular biology of Archaea and in characterizing their archaea-specific and many proto-eukaryotic cellular mechanisms and processes.	http://dac.molbio.ku.dk/
Greengenes	This web application provides access to the current and comprehensive 16S rRNA gene sequence alignment for browsing, blasting, probing, and downloading. The data and tools presented by greengenes can assist the researcher in choosing phylogenetically specific probes, interpreting microarray results, and aligning/annotating novel sequences.	http://greengenes.lbl.gov/cgi-bin/nph-index.cgi

see also Chapter 39, Vol. I]. In a preprocessing step, the set of DNA sequences is compared against databases of known sequences using BLAST or another comparison tool [Altschul et al., 1997]. MEGAN is then used to compute and explore the taxonomical content of the dataset, employing the NCBI taxonomy to summarize and order the results. A simple lowest common ancestor algorithm assigns reads to taxa such that the taxonomical level of the assigned taxon reflects the level of conservation of the sequence. The software allows large datasets to be dissected without the need for assembly or the targeting of specific phylogenetic markers, thus providing graphical and statistical output for comparing different datasets.

Many novel computational tools have also been developed to unravel ecosystem properties starting from fragmented sequences—that is, not fully completed metagenomes. In addition, the so-called "comparative metagenomics" approaches have allowed the discovery of specific genomic and community adaptations to environmental factors (see review in Vieites et al. [2009]). However, many of the parameters extracted from these data to describe the environment at hand (e.g., genomic features, functional complement, phylogenetic composition) are interdependent and influenced by technical aspects of sample preparation and data treatment, leading to various pitfalls during analysis. To avoid this and complement existing initiatives in data standards, a minimal standard for metagenomics data analysis ("MINIMESS") to be able to take full advantage of the power of comparative metagenomics in understanding microbial life on Earth has been proposed [Raes et al., 2007].

The amount of metagenomic sequence data is growing fast while computational methods for metagenome analysis are still in their infancy. In contrast to genomic sequences of single species, which can usually be assembled and analyzed by many available methods, a large proportion of metagenome data remains as unassembled anonymous sequencing read. One of the aims of all metagenomic sequencing projects is the identification of novel genes that can be then integrated within a Systems Biology pipeline. Short length (for example, Sanger sequencing will yield, on average, 700-bp fragments) and unknown phylogenetic origin of most fragments require approaches to gene prediction that are different from the currently available methods for genomes of single species. In particular, the large size of metagenomic samples requires fast and accurate methods with small numbers of false-positive predictions. In this context, a recent study has introduced a novel gene prediction algorithm for metagenomic fragments based on a two-stage machine learning approach (see Chapter 41, Vol. I). Linear discriminants for mono- and dicodon usage and translation initiation sites to extract features from DNA sequences are used. In the second stage, an artificial neural network combines these features with open reading frame length and fragment GC content to compute the probability that this open reading frame encodes a protein. This probability is used for the classification and scoring of gene candidates. This method is specially designed for the analysis of fragmented genomic DNA. Additionally, this method is able to predict translation initiation sites accurately and distinguishes complete from incomplete genes with high reliability.

As shown above, the large-scale machine learning methods are well-suited for gene prediction in metagenomic DNA fragments. Therefore, the metagenomic analysis of complex biological samples through Systems Biology has traditionally been sequence-centric; that is, functional assignments have depended on both (a) the metagenome sequence of the organism(s) in question and (b) bioinformatic analysis based on homologies to known genes [Vieites et al., 2009]. New advances in Systems Biology require the formation of consortia of experimental and computational biologists that collaborate directly in order to experimentally determine high-priority genes of unknown function or specificity. Central to this effort is the creation of a community web-based database (portal) that allow computational and experimental scientists to communicate easily, assist experimentalists in identifying those high-priority genes for which there are the highest-quality computational predictions for their molecular function, and provide feedback to the computational biologist, since it remains true that the insights and experience of the dedicated biochemist can be essential in guiding the development of algorithmic sophistication [Roberts, 2004].

Following the above observations, by far the most serious problem is the quality and degree of completeness of the annotation of metagenomes [Gomez-Alvarez et al., 2009; Hoff, 2009]. Most troublesome are the large numbers of open reading frames that have been identified by computer programs, but remain labeled as a "conserved hypothetical protein" when they occur in more than one genome or simply a "hypothetical protein" when they appear unique to the metagenome in question [Roberts, 2004]. As an example, molecular diversity in a wastewater treatment plant was large, and a considerable amount of contigs were related to hypothetical proteins (21%) or could not be assigned (25%) [García-Martin et al., 2006; Sanapareddy et al., 2009; Wu et al., 2009]. Moreover, many-to-many mutations in the amino acid sequences of protein encoded by metagenes may provide a new biochemistry not known from the sequence, and, more importantly, multiple mutations in the same gene may constitute a problem both for traditional genome sequencing and for the understanding of their biological implications. The metagenomic scenario may therefore be extremely complex if one wants to extract biochemical information from sequence and incorporate them in Systems Biology studies. Unless we rectify this situation, it is likely to undermine many of the other "-omic" efforts currently underway (Fig. 54.4). In this context, databases such as MetaBioME [Sharma et al., 2009] are specially designed to facilitate homology-based computational identification of novel enzymes homologous from metagenomic and bacterial genomic dataset.

54.1.4 Function-Directed Metaomics

Systems microbiology treats microbial communities as a whole, integrating fundamental biological knowledge with metagenomics and other experimental data, to create an integrated picture of how a microbial community operates. Systems biology is not just genome sequence data. Thus, flux analysis, integrated with quantitative information on cell, mRNA, and protein concentrations, contributes to a better understanding on how microbes deal with perturbations, at the species level [Daran-Lapujade et al., 2007], and at the community level [Roling, 2007].

Gene-centric multivariate analysis that is not dependent on the placement of genes in the context of an individual genome can still reveal environmental characteristics and biological processes. metabolic capabilities of microbial communities are habitat-specific. Gianoulis et al. [2009] introduced the concept of metabolic footprints: the relationship between metabolism and quantitative environmental parameters by identifying environmentally dependent or co-varying metabolic pathways.

Metaproteomics informs on actual system functioning [Vieites et al., 2009]. For example, the identification of highly expressed proteins from an activated sludge system for enhanced biological phosphorus removal (EBPR) [Wilmes et al, 2008b] was aided by metagenomic analysis of other EBPR systems [Garcia-Martin et al., 2006] to suggest an essential role for genetic diversity in process stability [Wilmes et al., 2008a].

Likely, not all parts of the microbial network contribute equally to ecosystem functioning: To interpret a puzzle of the Eiffel tower, the position of the sun and clouds in the sky need not to be known. The development of ecological control analysis (ECA) provided a first step into quantifying the control of individual functional groups on flux through a microbial network [Roling et al., 2007]. ECA is a sensitivity analysis combined with strong theorems. Counterintuitively, calculations suggested that halorespirers themselves do not control the rate of anaerobic degradation of chlorinated compounds. Fermenting organisms that provide halorespirers with acetate and hydrogen are dominant in flux control.

Information on which functional groups are flux-controlling, or otherwise of interest, can be used to specifically target these groups. Stable isotope probing (SIP, see also Chapter 55, Vol I) identifies microorganisms that consume an isotope-enriched substrate, by incorporation of isotopes into DNA or RNA and subsequent separation from unlabeled nucleic acids by density centrifugation [Radajewsky et al., 2000]. Combining pyrosequencing with SIP allowed for high-resolution metagenomics to target a minor functional type present in complex microbial communities: A nearly complete genome of a novel methylotroph revealed its major metabolic pathways and energy-generating systems [Kalyuznaya et al., 2008].

Metabolic flux analysis (^{13}C-MFA) which quantifies intracellular fluxes by iteratively fitting the measured distribution of amino acid isotopomers in cell protein to a stoichiometric model of central carbon metabolism [Sauer, 2006] provides adequate quantitative predictions of functional characteristics on the basis of a metabolic network derived from the metagenome only and without requiring kinetic data [Edwards and Palsson, 2000]. Therewith, MFA, combined with sequence data, has guided the metabolic engineering of environmental important species [Izallalen et al., 2008]. Solution space can be further constrained by using transcriptomics and proteomics data to delineate the active metabolic network and by excluding reactions that are thermodynamically unfeasible [Henry et al., 2007].

The above information revealed that the constraint-based and stoichiometric analyses using constraints such as mass balances, flux limitations, biomass composition, mRNA and protein concentrations, together with the integration of metagenome sequences, required the design of new computational approaches. These new applications may integrate all data from a Systems Biology perspective especially designed to quantitative predict functional characteristics on the basis of genetic and metabolic network for their application at the community level.

54.2 ADDITIONAL INTERNET RESOURCES (NOT INCLUDED IN TABLES 54.2 AND 54.3)

Webserver of the Committe on Metagenomics (http://www.nap.edu/catalog/11902.html)

Systems Biology projects webserver (http://www.sysmo.net/index.php?index=52)

Genomic Encyclopedia of Bacteria and Archaea (GEBA) project webserver (http://www.jgi.doe.gov/programs/GEBA/)

Human Microbiome Project webserver (http://nihroad map.nih.gov/hmp/) NCBI blast (http://blast.ncbi.nlm.nih.gov/)

Acknowledgments

The authors are supported by CSD2007-00005, GEN2006-27750-C-4-E, BFU2008-04398-E/BMC, KBBE-226977 (MAMBA), KBBE-245226 (MAGICPAH), KBBE-266473 (ULIXES) and FEDER funds. M-E.G. thanks the CSIC for a JAE fellowship.

REFERENCES

Altschul SF, Madden TL, Schaffer AA, Zhang J, Zhang Z, Miller W, Lipman DJ. 1997. Gapped BLAST and PSI-BLAST: A new generation of protein database search programs. *Nucleic Acids Res*. **25**:3389–3402.

Angly F, Rodriguez-Brito B, Bangor D, McNairnie P, Breitbart M, et al. 2005. PHACCS, an online tool for estimating the structure and diversity of uncultured viral communities using metagenomic information. *BMC Bioinform*. **6**:41.

Angly FE, Felts B, Breitbart M, Salamon P, Edwards RA, et al. 2006. The marine viromes of four oceanic regions. *PLoS Biol*. **4**:e368.

Ansorge WJ. 2009. Next-generation DNA sequencing techniques. *Nat. Biotechnol*. **25**:195–203.

Aziz RK, Bartels D, Best AA, DeJongh M, Disz T, et al. 2008. The RAST server: Rapid annotations using subsystems technology. *BMC Genomics* **9**:75.

Bench SR, Hanson TE, Williamson KE, Ghosh D, Radosovich M, Wang K, Wommack KE. 2007. Metagenomic characterization of Chesapeake Bay virioplankton. *Appl. Environ. Microbiol*. **73**:7629–7641.

Bently DR, Dunham I. 1995. Mapping human chromosomes. *Curr. Opin. Genet. Dev*. **5**:328–334.

Bertin PN, Medigue C, Normand P. 2008. Advances in environmental genomics: Towards an integrated view of micro-organisms and ecosystems. *Microbiology* **154**:347–359.

Biddle JF, Fitz-Gibbon S, Schuster SC, Brenchley JE, House CH. 2008. Metagenomic signatures of the Peru Margin subseafloor biosphere show a genetically distinct environment. *Proc. Natl. Acad. Sci. USA* **105**:10583–10588.

Bidle KD, Lee S, Marchant DR, Falkowski PG. 2007. Fossil genes and microbes in the oldest ice on earth. *Proc. Natl. Acad. Sci. USA* **104**:13455–13460.

Chen K, Pachter L. 2005. Bioinformatics for whole-genome shotgun sequencing of microbial communities. *PLoS Comput. Biol*. **1**:106–112.

Chou HH, Holmes MH. 2001. DNA sequence quality trimming and vector removal. *Bioinformatics* **17**:1093–1104.

Daran-Lapujade P, Rossell S, van Gulik WM, Luttik MA, de Groot MJ, et al. 2007. The fluxes through glycolytic enzymes in *Saccharomyces cerevisiae* are predominantly regulated at posttranscriptional levels. *Proc. Natl. Acad. Sci. USA* **104**:15753–15758.

de Lorenzo V. 2008. Systems biology approaches to bioremediation. *Curr. Opin. Biotechnol*. **19**:579–589.

DeLong EF. 2002. Microbial population genomics and ecology. *Curr. Opin. Microbiol*. **5**:520–524.

DeLong EF. 2009. The microbial ocean from genomes to biomes. *Nature* **459**:200–206.

Desai C, Pathak H, Madamwar D. 2009. Advances in molecular and "-omics" technologies to gauge microbial communities and bioremediation at xenobiotic/anthropogen contaminated sites. *Bioresour. Technol*. **101**:1558–1569.

Desnues C, Rodriguez-Brito B, Rayhawk S, Kelley S, Tran T, et al. 2008. Biodiversity and biogeography of phages in modern stromatolites and thrombolites. *Nature* **452**:340–343.

Dick GJ, Andersson AF, Baker BJ, Simmons SL, Thomas BC, Yelton AP, Banfield JF. 2009. Community-wide analysis of microbial genome sequence signatures. *Genome Biol*. **10**:R85.

Edwards JS, Palsson BO. 2000. Robustness analysis of the *Escherichia coli* metabolic network. *Biotechnol. Prog*. **16**:927–939.

Edwards RA, Rohwer F. 2005. Viral metagenomics. *Nat. Rev. Microbiol*. **3**:504–510.

Edwards RA, Rodriguez-Brito B, Wegley L, Haynes M, Breitbart M, et al. 2006. Using pyrosequencing to shed light on deep mine microbial ecology. *BMC Genomics* **7**:57.

Feist AM, Herrgard MJ, Thiele I, Reed JL, Palsson BO. 2009. Reconstruction of biochemical networks in microorganisms. *Nat. Rev. Microbiol*. **7**:129–143.

Fox S, Filichkin S, Mockler TC. 2009. Applications of ultra-high-throughput sequencing. *Methods Mol. Biol*. **553**:79–108.

Garcia Martin H, Ivanova N, Kunin V, Warnecke F, Barry KW, et al. 2006. Metagenomic analysis of two enhanced biological phosphorus removal (EBPR) sludge communities. *Nat. Biotechnol*. **24**:1263–1269.

Gerlach W, Junemann S, Tille F, Goesmann A, Stoye J. 2009. WebCARMA: A web application for the functional and taxonomic classification of unassembled metagenomic reads. *BMC Bioinform*. **10**:430.

Gianoulis TA, Raes J, Patel PV, Bjornson R, Korbel JO, et al. 2009. Quantifying environmental adaptation of metabolic pathways in metagenomics. *Proc. Natl. Acad. Sci. USA* **106**:1374–1379.

Gomez-Alvarez V, Teal TK, Schmidt TM. 2009. Systematic artifacts in metagenomes from complex microbial communities. *ISME J*. **3**:1314–1317.

Gordon D, Desmarais C, Green P. 2001. Automated finishing with autofinish. *Genome Res*. **11**:614–625.

Haas BJ, Delcher AL, Mount SM, Wortman JR, Smith RK, Jr., et al. 2003. Improving the *Arabidopsis* genome annotation using maximal transcript alignment assemblies. *Nucleic Acids Res*. **31**:5654–5666.

Handelsman J. 2004. Metagenomics: application of genomics to uncultured microorganisms. *Microbiol. Mol. Biol. Rev*. **68**:669–685.

Hartwell LH, Weinert TA. 1989. Checkpoints: Controls that ensure the order of cell cycle events. *Science* **246**:629–634.

Henry CS, Broadbelt LJ, Hatzimanikatis V. 2007. Thermodynamics-based metabolic flux analysis. *Biophys. J*. **92**:1792–1805.

Hoff KJ. 2009. The effect of sequencing errors on metagenomic gene prediction. *BMC Genomics* **10**:520.

Hugenholtz P. 2002. Exploring prokaryotic diversity in the genomic era. *Genome Biol*. **3**:0003.

Huson DH, Auch AF, Qi J, Schuster SC. 2007. MEGAN analysis of metagenomic data. *Genome Res*. **17**:377–386.

Huson DH, Richter DC, Mitra S, Auch AF, Schuster SC. 2009. Methods for comparative metagenomics. *BMC Bioinform*. **10**(Suppl 1):S12.

Izallalen M, Mahadevan R, Burgard A, Postier B, Didonato R, Jr., Sun J, Schilling CH, Lovley DR. 2008. *Geobacter sulfurreducens* strain engineered for increased rates of respiration. *Metab. Eng*. **10**:267–275.

Jorgensen P, Tyers M. 2004. How cells coordinate growth and division. *Curr. Biol*. **14**:R1014–R1027.

Kalyuzhnaya MG, Lapidus A, Ivanova N, Copeland AC, McHardy AC, et al. 2008. High-resolution metagenomics targets specific functional types in complex microbial communities. *Nat. Biotechnol*. **26**:1029–1034.

Krause L, Diaz NN, Goesmann A, Kelley S, Nattkemper TW, Rohwer F, Edwards RA, Stoye J. 2008. Phylogenetic classification of short environmental DNA fragments. *Nucleic Acids Res*. **36**:2230–2239.

Kunin V, Copeland A, Lapidus A, Mavromatis K, Hugenholtz P. 2008. A bioinformatician's guide to metagenomics. *Microbiol. Mol. Biol. Rev*. **72**:557–578.

Letunic I, Yamada T, Kanehisa M, Bork P. 2008. iPath: Interactive exploration of biochemical pathways and networks. *Trends Biochem. Sci*. **33**:101–103.

Li W. 2009. Analysis and comparison of very large metagenomes with fast clustering and functional annotation. *BMC Bioinform*. **10**:359.

Liolios K, Mavromatis K, Tavernarakis N, Kyrpides NC. 2008. The Genomes On Line Database (GOLD) in 2007: Status of genomic and metagenomic projects and their associated metadata. *Nucleic Acids Res*. **36**:D475–D479.

Manning BD, Cantley LC. 2007. AKT/PKB signaling: Navigating downstream. *Cell* **129**:1261–1274.

Margulies M, Egholm M, Altman WE, Attiya S, Bader JS, et al. 2005. Genome sequencing in microfabricated high-density picolitre reactors. *Nature* **437**:376–380.

Markowitz VM, Ivanova NN, Szeto E, Palaniappan K, Chu K, et al. 2008. IMG/M: A data management and analysis system for metagenomes. *Nucleic Acids Res*. **36**:D534–D538.

Mavromatis K, Ivanova N, Barry K, Shapiro H, Goltsman E, et al. 2007. Use of simulated data sets to evaluate the fidelity of metagenomic processing methods. *Nat. Methods* **4**:495–500.

McHardy AC, Martin HG, Tsirigos A, Hugenholtz P, Rigoutsos I. 2007. Accurate phylogenetic classification of variable-length DNA fragments. *Nat. Methods* **4**:63–72.

Meyer F, Paarmann D, D'Souza M, Olson R, Glass EM, et al. 2008. The metagenomics RAST server—A public resource for the automatic phylogenetic and functional analysis of metagenomes. *BMC Bioinform*. **9**:386.

Monzoorul Haque M, Ghosh TS, Komanduri D, Mande SS. 2009. SOrt-ITEMS: Sequence orthology based approach for improved taxonomic estimation of metagenomic sequences. *Bioinformatics* **25**:1722–1730.

Moran AM. 2009. Metatranscriptomics: Eavesdropping on complex microbial communities. *Microbe* **4**:329–335.

Myers EW, Sutton GG, Delcher AL, Dew IM, Fasulo DP, et al. 2000. A whole-genome assembly of *Drosophila*. *Science* **287**:2196–2204.

Olsen GJ, Lane DJ, Giovannoni SJ, Pace NR, Stahl DA. 1986. Microbial ecology and evolution: a ribosomal RNA approach. *Annu. Rev. Microbiol*. **40**:337–365.

Parro V, Moreno-Paz M, Gonzalez-Toril E. 2007. Analysis of environmental transcriptomes by DNA microarrays. *Environ. Microbiol*. **9**:453–464.

Partridge L, Gems D. 2002. A lethal side-effect. *Nature* **418**:921.

Petrosino JF, Highlander S, Luna RA, Gibbs RA, Versalovic J. 2009. Metagenomic pyrosequencing and microbial identification. *Clin. Chem*. **55**:856–866.

Phillippy AM, Schatz MC, Pop M. 2008. Genome assembly forensics: Finding the elusive mis-assembly. *Genome Biol*. **9**:R55.

Pop M. 2009. Genome assembly reborn: recent computational challenges. *Brief Bioinform*. **10**:354–366.

Qi W, Nong G, Preston JF, Ben-Ami F, Ebert D. 2009. Comparative metagenomics of *Daphnia* symbionts. *BMC Genomics* **10**:172.

Radajewski S, Ineson P, Parekh NR, Murrell JC. 2000. Stable-isotope probing as a tool in microbial ecology. *Nature* **403**:646–649.

Raes J, Foerstner KU, Bork P. 2007. Get the most out of your metagenome: Computational analysis of environmental sequence data. *Curr. Opin. Microbiol*. **10**:490–498.

Raes J, Bork P. 2008. Molecular eco-systems biology: Towards an understanding of community function. *Nat. Rev. Microbiol*. **6**:693–699.

Rappe MS, Giovannoni SJ. 2003. The uncultured microbial majority. *Annu. Rev. Microbiol*. **57**:369–394.

Richard JP, Mobashery S. 2003. Complex biological processes and their central chemical events. *Curr. Opin. Chem. Biol*. **7**:525–527.

Riesenfeld CS, Schloss PD, Handelsman J. 2004. Metagenomics: Genomic analysis of microbial communities. *Annu. Rev. Genet*. **38**:525–552.

Roberts RJ. 2004. Identifying protein function—a call for community action. *PLoS Biol*. **2**:E42.

Roling WF. 2007. Do microbial numbers count? Quantifying the regulation of biogeochemical fluxes by population size and cellular activity. *FEMS Microbiol. Ecol*. **62**:202–210.

Roling WF, van Breukelen BM, Bruggeman FJ, Westerhoff HV. 2007. Ecological control analysis: Being(s) in control of mass flux and metabolite concentrations in anaerobic degradation processes. *Environ. Microbiol*. **9**:500–511.

Rusch DB, Halpern AL, Sutton G, Heidelberg KB, Williamson S, et al. 2007. The Sorcerer II Global Ocean Sampling expedition: Northwest Atlantic through eastern tropical Pacific. *PLoS Biol*. **5**:e77.

Sanapareddy N, Hamp TJ, Gonzalez LC, Hilger HA, Fodor AA, Clinton SM. 2009. Molecular diversity of a North Carolina wastewater treatment plant as revealed by pyrosequencing. *Appl. Environ. Microbiol*. **75**:1688–1696.

Sauer U. 2006. Metabolic networks in motion: 13C-based flux analysis. *Mol. Syst. Biol*. **2**:62.

Seshadri R, Kravitz SA, Smarr L, Gilna P, Frazier M. 2007. CAMERA: A community resource for metagenomics. *PLoS Biol*. **5**:e75.

Sharma VK, Kumar N, Prakash T, Taylor TD. 2009. MetaBioME: A database to explore commercially useful enzymes in metagenomic datasets. *Nucleic Acids Res*. **38**:D468–D472.

Simpson JT, Wong K, Jackman SD, Schein JE, Jones SJ, Birol I. 2009. ABySS: A parallel assembler for short read sequence data. *Genome Res*. **19**:1117–1123.

Sterk P, Hirschman L, Field D, Wooley J. 2010. Genomic standards consortium workshop: Metagenomics, metadata and metaanalysis (m3). *Pac. Symp. Biocomput*. 481–484.

Sutton GG, White O, Adams MD, Kerlavage AR. 1995. Tigr Assembler: A new tool for assembling large shotgun sequencing projects. *Genome Sci. Technol*. **1**:9–19.

Teeling H, Waldmann J, Lombardot T, Bauer M, Glockner FO. 2004. TETRA: A web-service and a stand-alone program for the analysis and comparison of tetranucleotide usage patterns in DNA sequences. *BMC Bioinform*. **5**:163.

Tyson GW, Chapman J, Hugenholtz P, Allen EE, Ram RJ, et al. 2004. Community structure and metabolism through reconstruction of microbial genomes from the environment. *Nature* **428**:37–43.

Valdivia-Granda W. 2008. The next meta-challenge for Bioinformatics. *Bioinformation* **2**:358–362.

Venter JC, Remington K, Heidelberg JF, Halpern AL, Rusch D, et al. 2004. Environmental genome shotgun sequencing of the Sargasso Sea. *Science* **304**:66–74.

VerBerkmoes NC, Denef VJ, Hettich RL, Banfield JF. 2009. Systems biology: Functional analysis of natural microbial consortia using community proteomics. *Nat. Rev. Microbiol*. **7**:196–205.

Vieites JM, Guazzaroni ME, Beloqui A, Golyshin PN, Ferrer M. 2009. Metagenomics approaches in systems microbiology. *FEMS Microbiol. Rev*. **33**:236–255.

Warnecke F, Hugenholtz P. 2007. Building on basic metagenomics with complementary technologies. *Genome Biol*. **8**:231.

Wilmes P, Andersson AF, Lefsrud MG, Wexler M, Shah M, et al. 2008a. Community proteogenomics highlights microbial strain-variant protein expression within activated sludge performing enhanced biological phosphorus removal. *ISME J*. **2**: 853–864.

Wilmes P, Wexler M, Bond PL. 2008b. Metaproteomics provides functional insight into activated sludge wastewater treatment. *PLoS One* **3**:e1778.

Woyke T, Teeling H, Ivanova NN, Huntemann M, Richter M, et al. 2006. Symbiosis insights through metagenomic analysis of a microbial consortium. *Nature* **443**:950–955.

Wu D, Hugenholtz P, Mavromatis K, Pukall R, Dalin E, et al. 2009. A phylogeny-driven genomic encyclopaedia of Bacteria and Archaea. *Nature* **462**:1056–1060.

Ye Y, Tang H. 2009. An ORFome assembly approach to metagenomics sequences analysis. *J. Bioinform. Comput. Biol*. **7**:455–471.

Zerbino DR, Birney E. 2008. Velvet: Algorithms for de novo short read assembly using de Bruijn graphs. *Genome Res*. **18**:821–829.

Chapter 55

Towards "Focused" Metagenomics: A Case Study Combining DNA Stable-Isotope Probing, Multiple Displacement Amplification, and Metagenomics

YIN CHEN, MARC G. DUMONT, JOSHUA D. NEUFELD, AND J. COLIN MURRELL

55.1 INTRODUCTION

The fact that the majority of microorganisms in the environment are difficult to cultivate has led to the development of cultivation-independent methods for analyzing microbial diversity and distribution. Metagenomics is such a method to study microorganisms without the prerequisite of cultivation [Handelsman et al., 1998]. Genetic information from environmental samples is retrieved by either (a) large-scale sequencing of DNA extracted directly from that environment, (b) PCR- or hybridization-based screens of large-insert libraries, or (c) functional screening of relevant bioactive compounds from clone libraries made with the DNA [Handelsman, 2005; Lasken, 2007; Schmeisser et al., 2007; and references therein]. Since their development, metagenomic approaches have been widely used and have contributed to some major advances in environmental microbiology in the last few years. For example, large-scale sequencing projects of environmental DNA have allowed the almost complete reconstruction of potential metabolic pathways of microorganisms that have never been cultivated [Tyson et al., 2004; Venter et al., 2004].

Although a powerful approach, the major concern of traditional sequence-based metagenomic studies is that the functions of the associated microorganisms are still uncertain due to the lack of direct link between the DNA sequences retrieved and the functions that they may have encoded in the uncultivated microbes themselves. A number of methods have recently been developed in order to decipher the functions of uncultivated microorganisms in the environment (see review by Wagner [2009] and references therein). These include the technique of DNA stable-isotope probing (DNA-SIP) [Radajewski et al., 2000]. Labeling of uncultivated microorganisms with stable isotopes (e.g., ^{13}C, ^{15}N), followed by the physical separation of their DNA from that which is not labeled, establishes a direct link between their identity and function, by indicating which substrates they use for growth [Dumont and Murrell, 2005]. There is considerable potential in combining DNA-SIP with metagenomics to act as a "filter" in isolating DNA from functionally relevant microorganisms for the construction of metagenomic libraries [e.g., Schloss and Handelsman, 2003; Friedrich, 2006; Neufeld et al., 2007a]. This so-called "focused" metagenomics approach has been developed very recently, and a few studies have demonstrated its

Handbook of Molecular Microbial Ecology, Volume I: Metagenomics and Complementary Approaches, First Edition. Edited by Frans J. de Bruijn.

usefulness and feasibility [Dumont et al., 2006; Schwarz et al., 2006; Chen et al., 2008b; Neufeld et al., 2008; Kalyuzhnaya et al., 2008].

The focused metagenomics approach relies on the incorporation of sufficient ^{13}C-isotope into the DNA of target microbes. In order to obtain large quantities of ^{13}C-labeled 'heavy' DNA (>500 ng) for constructing a metagenome library, the environmental samples may need to be exposed to ^{13}C-isotopes for a relatively long time, which might result in potential cross-feeding of the ^{13}C to other microbes. However, DNA-SIP has been successfully carried out using near in situ concentrations of ^{13}C-substrate [Neufeld et al., 2008]; in this case, only a few nanograms of heavy DNA were recovered from the cesium chloride (CsCl) gradient. A method called multiple displacement amplification (MDA) [Blanco et al., 1989; Dean et al., 2002] was used to amplify the ^{13}C-labeled DNA in order to generate sufficient quantities for metagenomic library construction. In this study, a methanol dehydrogenase gene cluster from a novel uncultivated *Methylophaga* sp. was retrieved from seawater, implying *Methylophaga*-like bacteria as significant players in the metabolism of methanol in marine environment. The authors also reported that the MDA approach introduced chimeric DNA sequences into the subsequent libraries and that the formation of these chimeric sequences needed to be eliminated or at least minimized [Neufeld et al., 2008].

Peatlands harbor diverse groups of methanotrophs that consume methane before it is released into the atmosphere. Using culture-independent molecular methods, studies have shown that a group of peat-associated *Alphaproteobacteria*, closely related to *Methylocystis*, are particularly abundant in peatlands [Dedysh et al., 2001–2006; McDonald et al., 1996, 1999; McDonald and Murrell 1997a, 1997b; Morris et al., 2002; Jaatinen et al., 2005]. It is evident that they are directly involved in methane oxidation in situ since they express particulate methane monooxygenase (pMMO) [Chen et al., 2008a], the key enzyme catalyzing the first step in the methane oxidation pathway in aerobic methanotrophic bacteria [Murrell et al., 2000]. Suitable cultivation conditions for these acidophilic *Methylocystis* present in peatlands have not yet been established, and little is known about these environmentally important microorganisms. Here we describe a previously published case study where focused metagenomics (i.e., combining DNA-SIP with metagenomics) has been used to study a group of uncultivated *Methylocystis* spp. growing on methane in an acidic peatland [Chen et al., 2008b]. A detailed protocol for minimizing chimera formation during cloning of MDA generated DNA was also established.

55.2 MATERIALS AND METHODS

Detailed materials and methods have been published in the original publication [Chen et al., 2008b]. The following is the abbreviated outline of the materials and methods used.

Five grams of soil were taken from Moor House (England) and incubated in a 125-ml crimp-top serum vial with 2% (v/v) of $^{13}CH_4$. The incubation was stopped when 140 μmol g^{-1} (wet weight soil) $^{13}CH_4$ was consumed. High-molecular-mass DNA was extracted from SIP incubations using the method of Dumont et al. [2006]. Two micrograms of the DNA was then added to CsCl solutions and subjected to isopycnic ultracentrifugation to separate ^{13}C-DNA from ^{12}C-DNA, as described previously [Neufeld et al., 2007b]. The densities of all fractions were determined using a digital refractometer (Reichert AR200, Reichert). DNA was precipitated from each fraction and suspended in 50 μl of TE buffer.

To determine the bacterial diversity associated with heavy DNA, PCR amplifications were performed using the primer set 341f_GC and 907r [Muyzer et al., 1993] and denaturing gradient gel electrophoresis (DGGE) was performed as described previously [Chen et al., 2007; see also Chapter 5, Vol. I] to compare the fingerprint profiles between heavy and "light" DNA fractions. Bands from DGGE gels (indicated in **Fig. 55.1**) were excised, reamplified, purified, and sequenced.

To generate sufficient quantities of DNA for the construction of a metagenomic library, the ^{13}C-DNA from the heavy fraction (i.e., fraction 7, containing a few nanograms of DNA) was subjected to MDA, using the GenomiPhi V2 DNA Amplification Kit (GE Heathcare) as described previously [Neufeld et al., 2008]. To test the potential bias in amplification of DNA using the Phi29 enzyme, serial dilutions of total DNA (DNA extracted from Moor House SIP incubations), with 10, 1, 0.1, and 0.01 ng of template, was used in MDA reactions. The MDA product was diluted tenfold, and 1 μl was used as template for PCR/DGGE analysis as described earlier to determine the potential bias of the MDA process.

To minimize potential chimera formation introduced by MDA during cloning of MDA DNA into *Escherichia coli*, an enzymatic treatment process was applied to the MDA-generated DNA [Zhang et al., 2006]. Briefly, five reactions (20 μl each) of MDA-generated "heavy" DNA (fraction 7) were combined and purified using a Microcon YM-30 column (Millipore), then incubated with 10 U of Phi29 DNA polymerase (Fermentas) at 30°C for 2 h without adding random hexamer primers to de-branch single- and double-stranded DNA. The reaction was stopped by heating at 65°C for 10 min, and the DNA was then purified and incubated with 200 U of S1 nuclease

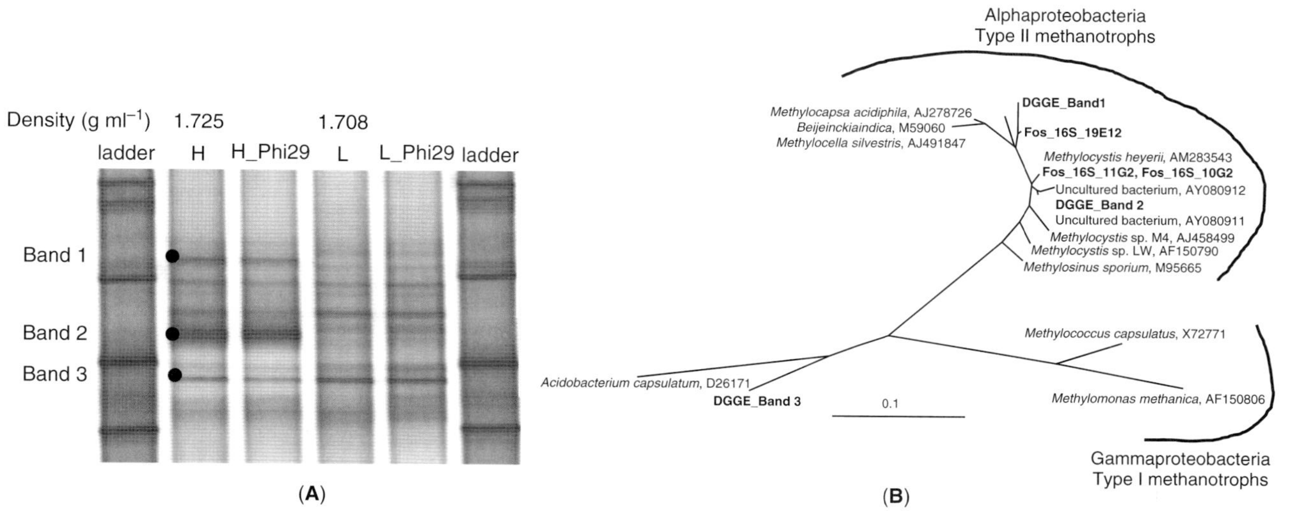

Figure 55.1 (**A**) DGGE fingerprint analysis of 16S rRNA genes from ^{13}C-DNA (fraction 7) and ^{12}C-DNA (fraction 11) before (lane H and lane L) and after MDA (lane H_Phi29 and L_Phi29), showing minimum bias during MDA. The sequences from band 1, band 2, and band 3 were closely related to *Methylocella* (97%, AJ555244)/*Methylocapsa* (96%, AJ278726), uncultivated *Methylocystis* (98%, AM162432), and an *Acidobacterium* clone (93%, EF076217), respectively. (**B**) Unrooted neighbor-joining phylogenetic analysis of 16S rRNA gene sequences (511 nucleotides) amplified from the DGGE gel and the fosmids 10G2, 11G2, and 19E12. The scale bar represents 1 substitution per 10 nucleotides.

(Fermentas) at 37°C for 30 min to digest single-stranded DNA. The reaction was stopped by adding 50 μl of 0.5 M EDTA and heating at 70°C for 10 min. The treated DNA was extracted with phenol:chloroform:isoamyl alcohol (25:24:1 v/v), again with chloroform:isoamyl alcohol (24:1 v/v), and then precipitated with ethanol. The pellet was air-dried and dissolved in 100 μl of nuclease-free water, to which 20 U of DNA polymerase I was added. The reaction was incubated at 25°C for 1 h and then stopped by heating at 75°C for 10 min. The enzyme-treated MDA-generated DNA was then end-repaired using the End-Repair enzyme mix from CopyControl Fosmid Library Production Kit (Epicentre).

The end-repaired, MDA-generated DNA was then subjected to pulsed field gel electrophoresis (PFGE) (**Fig. 55.2**) as described previously [Neufeld et al., 2008]. Fosmid library construction was carried out according to the instruction manual (Epicentre) with 6 μl (~50 ng) of purified size-fractionated DNA. The resulting 2300 clone fosmid library was screened as described previously [Dumont et al., 2006; Neufeld et al., 2008]. From this screen by colony PCR using specific primers for *mxaF*, a single *mxaF*-containing fosmid insert was subcloned, sequenced, and assembled. The open reading frames (ORFs) were identified using the ORF finder program (http://www.ncbi.nlm.nih.gov/projects/gorf/) and compared with the GenBank database. DNA sequence alignment and phylogenetic analyses were performed as previously described [Chen et al., 2007].

55.3 RESULTS AND DISCUSSION

Our previous study suggested that *Methylocystis*-related species were actively involved in methane oxidation in peatlands [Chen et al., 2008a]. Using samples taken from Moor House peat (England), a primary site for our previous investigations [McDonald et al., 1996; Chen et al., 2008a], DNA-SIP was applied to label these microorganisms with ^{13}C-methane. The 16S rRNA gene DGGE fingerprint of heavy ^{13}C-DNA (fraction 7) and light ^{12}C- DNA (fraction 11) revealed that two distinct bands (band 1 and band 2) appeared after $^{13}CH_4$ labeling (Fig. 55.1A). The sequence of band 2 was closely related (97–98% identity) to 16S rRNA gene sequences from a group of uncultivated *Methylocystis* retrieved from many acidic environments [Dedysh et al., 2006; Radajewski et al., 2002] (Fig. 55.1B). These uncultivated peat-affiliated *Methylocystis* spp. are of global importance and are present in a number peatlands worldwide [McDonald et al., 1996, 1999; McDonald and Murrell, 1997a, 1997b; Dedysh et al., 2001–2006; Morris et al., 2002; Radajewski et al., 2002; Jaatinen et al., 2005; Raghoebarsing et al., 2005].

In order to gain access to the genome information of these uncultivated *Methylocystis* spp., microgram quantities of ^{13}C-DNA were required to construct a metagenomic library using the ^{13}C-DNA. However, the minimum time required to achieve successful labeling of DNA with ^{13}C had to be used in order to minimize potential cross-feeding of ^{13}C. As a result, only low nanogram quantities of ^{13}C-labeled heavy (fraction 7) were obtained. To generate sufficient DNA for constructing a metagenomic library, MDA was performed [Blanco et al., 1989; Dean et al., 2002].

Before MDA of heavy ^{13}C-DNA (fraction 7), potential bias associated with Phi29 DNA polymerase was assessed using a serial dilution of the total DNA as template. DGGE fingerprints of total DNA generated from MDA with 0.1 to 10 ng of template were virtually identical (data not shown). However, bias toward DNA from certain organisms was observed when using 0.01 ng of template. This is similar to findings from other studies [Kalyuzhnaya et al., 2006; Zhang et al., 2006; Neufeld et al., 2008]. Therefore, 1 μl (~1–5 ng) of the heavy DNA (i.e., DNA from fraction 7) was used as template for MDA. DGGE analysis of 16S rRNA genes from bacteria in heavy DNA after MDA showed that there was no detectable bias during the amplification (Fig. 55.1A). This was also demonstrated by Neufeld et al. [2008]. The size of MDA-generated DNA ranged from 5 to 50 kb (Fig. 55.2). In order to minimize potential chimera formation during cloning of the MDA-generated DNA into *Escherichia coli*, the amplified DNA was subjected to several enzyme treatments to digest single-stranded DNA and for de-branching as recommended by Zhang et al. [2006]. Enzyme treatments reduced the yield of MDA-generated DNA, but did not appear to reduce the fragment size of the amplified ^{13}C-DNA (Fig. 55.2). After enzyme treatments, the MDA-generated DNA from fraction 7 (heavy DNA) was size-selected (30–50 kb) and 50 ng was used to construct a fosmid library (2300 clones, average insert size ~30 kb).

Screening of the 2300 clones for *mxaF, mmoX, pmoA* (encoding a subunit of methanol dehydrogenase, soluble methane monooxygenase, and particulate methane monooxygenase, respectively) and 16S rRNA genes for type II methanotrophs using primer set type IIF and type IIR [Chen et al., 2007] yielded four positive clones. Three clones contained 16S rRNA genes (fosmid 10G2, 11G2, and 19E12) related to type II methanotrophs and one contained *mxaF* (fosmid 16A2). The 16S rRNA fragments of fosmid 10G2 and 11G2 were identical and closely related to the 16S rRNA gene sequence from band 2 (Fig. 55.1B) previously identified as derived from *Methylocystis*. The deduced sequence of MxaF from fosmid 16A2 was closely related (96–98%) to MxaF sequences retrieved from several different peatland soils. The *mxaF*-containing fosmid (16A2) was sequenced, and this allowed the assembly and identification of a 14-kb contig containing a cluster of genes (*mxaFJGIRSAC*) known to be involved in bacterial methanol utilization (**Table 55.1**, **Fig. 55.3**). A similar arrangement of these methanol oxidation genes has also been observed in other methylotrophs [e.g., Amaratunga et al., 1997; Ward et al., 2004; Chistoserdova et al., 2007; see also Chapter 27, Vol. I].

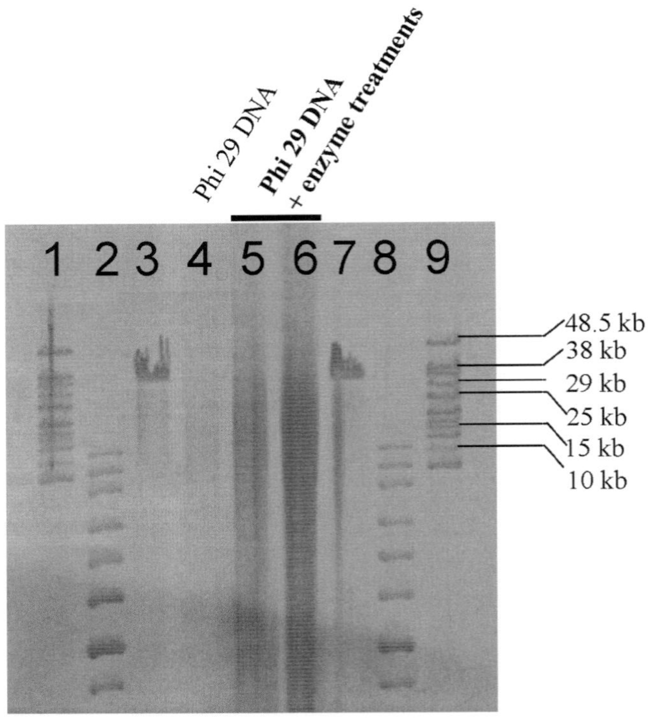

Figure 55.2 PFGE analysis of enzyme-treated MDA-generated DNA from fraction 7. Lane 1 and 9: High-molecular-weight ladder (Invitrogen); Lane 2 and 8; 1 kb ladder (Invitrogen); Lanes 3 and 7: Lamda DNA (40 kb). Lane 4: MDA-generated DNA from fraction 7. lanes 5 and 6: enzyme-treated, MDA-generated DNA.

55.4 CONCLUDING REMARKS

We have demonstrated that it is feasible to construct a fosmid library using MDA-generated heavy DNA from DNA-SIP [Chen et al., 2008b]. This methodological approach identified a novel gene cluster involved in the utilization of the ^{13}C-substrate added to the SIP incubations. The successful application of MDA to generate sufficient DNA for constructing a metagenomic library has exciting implications for future applications of DNA-SIP-enabled metagenomics. A much lower substrate concentration can be used during SIP incubations, and incubation times can be substantially reduced. Both improvements will help minimize potential cross-feeding of stable isotopes through a microbial population. Further research is required to investigate the mechanism of chimera formation during MDA and to eliminate chimera formation.

The demonstration that relatively large inserts (up to 50 kb) can be obtained in clone libraries constructed

Table 55.1 Putative Genes Corresponding to Each ORF on Fosmid 16A2

Name	Protein Name or Role of Closest BLASTX Match	Source	% Identity[b]
Orf1	Putative transposase	*Methylobacterium dichloromethanicum*	76
Orf2	Hypothetical protein, EF-hand domain protein	*Bradyrhizobium japonicum*	59
Orf3	Diheme cytochrome c	*Rhodobacter sphaeroides*	37
Orf4	Hypothetical protein	*Methylobacterium dichloromethanicum*	56
mxaF	Methanol dehydrogenase, large subunit	*Methylobacterium nodulans*	82
		Uncultivated bacterium (AY080930)[a]	99
mxaJ	Methanol oxidation protein, MxaJ	*Methylobacterium extorquens*	57
mxaG	Cytochrome c, class I	*Methylobacterium extorquens*	54
mxaI	Methanol dehydrogenase, small subunit	*Methylococcus capsulatus*	66
mxaR	ATPase	*Methylobacterium extorquens*	65
mxaS	Conserved hypothetical protein, MxaS	*Methylobacterium extorquens*	54
mxaA	Methanol oxidation protein, MxaA	*Methylobacterium extorquens*	40
mxaC	Methanol utilization protein, MxaC	*Methylobacterium extorquens*	60

[a]BLASTX searched using the nucleotide region between the primers 1003F/1555R [Neufeld et al., 2007b]. Gene abbreviations and close BLASTX relatives were obtained from Genbank.

[b]Amino acid sequence identity.

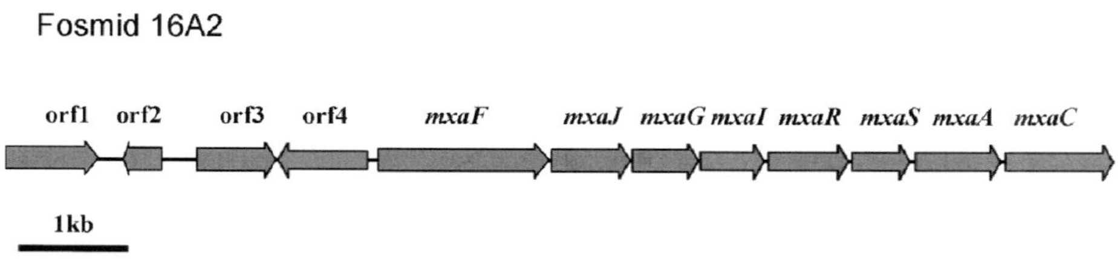

Figure 55.3 Gene organization of *mxaF*-containing fosmid 16A2 contig generated from Phi29-amplified SIP heavy DNA (fraction 7). Putative gene abbreviations are shown above each ORF.

using MDA-generated heavy DNA may also be of interest for identifying bioproduct-related genes for industry. For example, the method might be useful for gene mining for novel bioactive compounds since many useful secondary metabolites are synthesized by metabolic pathways encoded by large gene clusters [Piel et al., 2004]. A recent study has shown that using heavy DNA from SIP experiments can improve the frequency of isolation of target genes during gene mining [Schwarz et al., 2006], and another study found that a novel aromatic ring hydroxylating dioxygenase has been retrieved from a cosmid library constructed using heavy DNA from SIP [Sul et al., 2009].

The combination of DNA-SIP with metagenomics provides a novel tool for studying the functions of environmental microorganisms and is of considerable utility for the discovery of novel enzymes and bioactive compounds. Environmental genomics is a rapidly expanding field of research. The combination of DNA-SIP and metagenomics enables an additional window opened into the diversity and function of uncultivated microorganisms from complex microbial communities.

REFERENCES

Amaratunga K, Goodwin PM, O'Connor CD, Anthony C. 1997. The methanol oxidation genes *mxaFJGIR(S)ACKLD* in *Methylobacterium extorquens*. *FEMS Microbiol. Lett*. **146**:31–38.

Blanco L, Bernad A, Lazaro JM, Martin G, Garmendia C, Salas M. 1989. Highly efficient DNA synthesis by the phage phi 29 DNA polymerase. Symmetrical mode of DNA replication. *J. Biol. Chem*. **264**:8935–8940.

Chen Y, Dumont MG, Cébron A, Murrell JC. 2007. Identification of active methanotrophs in a landfill cover soil though detection of expression of 16S rRNA and functional genes. *Environ. Microbiol*. **9**:2855–2869.

Chen Y, Dumont MG, McNamara NP, Chamberlain PM, Bodrossy L, Stralis-Pavese N, Murrell JC. 2008a. Diversity of the active methanotrophic community in acidic peatlands as assessed by mRNA and SIP-PLFA analyses. *Environ. Microbiol*. **10**:446–459.

Chen Y, Dumont MG, Neufeld JD, Bodrossy L, Stralis-Pavese N, McNamara NP, Ostle N, Briones MJI, Murrell JC. 2008b. Revealing the uncultivated majority: Combining stable isotope probing, multiple displacement amplification and metagenomic analysis of uncultivated *Methylocystis* in acidic peatlands. *Environ. Microbiol*. **10**:2609–2622.

Chistoserdova L, Lapidus A, Han C, Goodwin L, Saunders L, Brettin T, et al. 2007. Genome of *Methylobacillus flagellatus*, molecular basis for obligate methylotrophy, and polyphyletic origin of methylotrophy. *J. Bacteriol*. **189**:4020–4027.

DEAN FB, HOSONO S, FANG L, WU X, FARUQI AF, BRAY-WARD P, et al. 2002. Comprehensive human genome amplification using multiple displacement amplification. *Proc. Natl. Acad. Sci. USA* **99**:5261–5266.

DEDYSH SN, DERAKSHANI M, LIESACK W. 2001. Detection and enumeration of methanotrophs in acidic *Sphagnum* peat by 16S rRNA fluorescence in situ hybridization, including the use of newly developed oligonucleotide probes for *Methylocella palustris*. *Appl. Environ. Microbiol*. **67**:4850–4857.

DEDYSH SN, DUNFIELD PF, DERAKSHANI M, STUBNER S, HEYER J, LIESACK W. 2003. Differential detection of type II methanotrophic bacteria in acidic peatlands using newly developed 16S rRNA-targeted fluorescent oligonucleotide probes. *FEMS Microbiol. Ecol*. **43**:299–308.

DEDYSH SN, PANKRATOV TA, BELOVA SE, KULICHEVSKAYA IS, LIESACK W. 2006. Phylogenetic analysis and in situ identification of *Bacteria* community composition in an acidic *Sphagnum* peat bog. *Appl. Environ. Microbiol*. **72**:2110–2117.

DUMONT MG, RADAJEWSKI SM, MIGUEZ CB, MCDONALD IR, MURRELL JC. 2006. Identification of a complete methane monooxygenase operon from soil by combining stable isotope probing and metagenomics analysis. *Environ. Microbiol*. **8**:1240–1250.

DUMONT MG, MURRELL JC. 2005. Stable isotope probing—Linking microbial identity to function. *Nat. Rev. Microbiol*. **3**:499–504.

FRIEDRICH MW. 2006. Stable-isotope probing of DNA: Insights into the function of uncultivated microorganisms from isotopically labeled metagenomes. *Curr. Opin. Biotechnol*. **17**:59–66.

HANDELSMAN J. 2005. Metagenomics: Application of genomics to uncultured microorganisms. *Microbiol. Mol. Biol. Rev*. **68**:669–685.

HANDELSMAN J, RONDON MR, BRADY SF, CLARDY J, GOODMAN RM. 1998. Molecular biological access to the chemistry of unknown soil microbes: A new frontier for natural products. *Chem. Biol*. **5**:245–249.

JAATINEN K, TUITTILA ES, LAINE J, YRJALA K, FRITZE H. 2005. Methane-oxidizing bacteria in a Finnish raised mire complex: Effects of site fertility and drainage. *Microb. Ecol*. **50**:429–439.

KALYUZHNAYA MG, ZABINSKY R, BOWERMAN S, BAKER DR, LIDSTROM ME, CHISTOSERDOVA L. 2006. Fluorescence in situ hybridization flow cytometry cell sorting based method for separation and enrichment of type I and type II methanotroph populations. *Appl. Environ. Microbiol*. **72**:4293–4301.

KALYUZHNAYA M G, LAPIDUS A, IVANOVA N, COPELAND AC, et al. 2008. High-resolution metagenomics targets specific functional types in complex microbial communities. *Nat. Biotechnol*. **26**:1029–1034.

LASKEN RS. 2007. Single-cell genomic sequencing using multiple displacement amplification. *Curr. Opin. Microbiol*. **10**:510–516.

MCDONALD IR, HALL GH, PICKUP RW, MURRELL JC. 1996. Methane oxidation potential and preliminary analysis of methanotrophs in blanket bog peat using molecular ecology techniques. *FEMS Microbiol. Ecol*. **21**:197–211.

MCDONALD IR, UPTON M, HALL G, PICKUP RW, EDWARDS C, SAUNDERS JR, et al. 1999. Molecular ecological analysis of methanogens and methanotrophs in blanket bog peat. *Microb. Ecol*. **38**:225–233.

MCDONALD IR, MURRELL JC. 1997a. The particulate methane monooxygenase gene *pmoA and* its use as a functional gene probe for methanotrophs. *FEMS Microbiol. Lett*. **156**:205–210.

MCDONALD IR, MURRELL JC. 1997b. The methanol dehydrogenase structural gene *mxaF and* its use as a functional gene probe for methanotrophs and methylotrophs. *Appl. Environ. Microbiol*. **63**:3218–3224.

MORRIS SA, RADAJEWSKI S, WILLISON TW, MURRELL JC. 2002. Identification of the functionally active methanotroph population in a peat soil microcosm by stable isotope probing. *Appl. Environ. Microbiol*. **68**:1446–1453.

MURRELL JC, GILBERT B, MCDONALD IR. 2000. Molecular biology and regulation of methane monooxygenase. *Arch. Microbiol*. **173**:325–332.

MUYZER G, DE WAAL EC, UITTERLINDEN AG. 1993. Profiling of complex microbial populations by denaturing gradient gel electrophoresis analysis of polymerase chain reaction-amplified genes coding for 16S rRNA. *Appl. Environ. Microbiol*. **59**: 695–700.

NEUFELD JD, DUMONT MG, VOHRA J, MURRELL JC. 2007a. Methodological considerations for the use of stable isotope probing in microbial ecology. *Microb. Ecol*. **53**:435–442.

NEUFELD JD, VOHRA J, DUMONT MG, LUEDERS T, MANEFIELD M, FRIEDRICH MW, MURRELL JC. 2007b. DNA stable-isotope probing. *Nat. Protoc* **2**:860–866.

NEUFELD JD, CHEN Y, DUMONT MG, MURRELL JC. 2008. Marine methylotrophs revealed by stable-isotope probing, whole genome amplification and metagenomics. *Environ. Microbiol*. **10**: 1526–1535.

PIEL J, HUI D, FUSETANI N, MATSUNAGA S. 2004. Targeting modular polyketide synthases with iteratively acting acyltransferases from metagenomes of uncultured bacterial consortia. *Environ. Microbiol*. **6**:921–927.

RADAJEWSKI S, INESON P, PAREKH NR, MURRELL JC. 2000. Stable-isotope probing as a tool in microbial ecology. *Nature* **403**:646–649.

RADAJEWSKI S, WEBSTER G, REAY DS, MORRIS SA, INESON P, NEDWELL DB, et al. 2002. Identification of active methylotroph populations in an acidic forest soil by stable-isotope probing. *Microbiology* **148**:2331–2342.

RAGHOEBARSING AA, SMOLDERS AJP, SCHMID MC, RIJPSTRA WIC, WOLTERS-ARTS M, DERKSEN J, et al. 2005. Methanotrophic symbionts provide carbon for photosynthesis in peat bogs. *Nature* **436**:1153–1156.

SCHLOSS PD, HANDELSMAN J. 2003. Biotechnological prospects from metagenomics. *Curr. Opin. Biotechnol*. **14**:303–310.

SCHMEISSER C, STEELE H, STREIT WR. 2007. Metagenomics, biotechnology with non-culturable microbes. *Appl. Environ. Microbiol*. **75**:955–962.

SCHWARZ S, WASCHKOWITZ T, DANIEL R. 2006. Enhancement of gene detection frequencies by combining DNA-based stable-isotope probing with the construction of metagenomic DNA libraries. *World J. Microb. Biot*. **22**:363–367.

SUL WJ, PARK J, QUENSEN JF 3rd, RODRIGUES JL, SELIGER L, TSOI TV, ZYLSTRA GJ, TIEDJE JM. 2009. DNA-stable isotope probing integrated with metagenomics for retrieval of biphenyl dioxygenase genes from polychlorinated biphenyl-contaminated river sediment. *Appl. Environ. Microbiol*. **75**: 5501–5506.

TYSON GW, CHAPMAN J, HUGENHOLTZ P, ALLEN EE, RAM RJ, RICHARDSON PM, et al. 2004. Community structure and metabolism through reconstruction of microbial genomes from the environment. *Nature* **428**:37–43.

VENTER JC, REMINGTON K, HEIDELBERG JF, HALPERN AL, RUSCH D, EISEN JA, et al. 2004. Environmental genome shotgun sequencing of the Sargasso Sea. *Science* **304**:66–74.

WARD N, LARSEN O, SAKWA J, BRUSETH L, KHOURI H, DURKIN AS, et al. 2004. Genomic insights into methanotrophy: The complete genome sequence of *Methylococcus capsulatus* (Bath). *PLoS Biol*. **2**: e303.

WAGNER M. 2009. Single-cell ecophysiology of microbes as revealed by Raman microspectroscopy or secondary ion mass spectrometry imaging. *Annu. Rev. Microbiol*. **63**:411–429.

ZHANG K, MARTINY AC, REPPAS NB, BARRY KW, MALEK J, CHISHOLM SW, CHURCH GM. 2006. Sequencing genomes from single cells by polymerase cloning. *Nat. Biotechnol*. **24**: 680–686.

Chapter 56

Suppressive Subtractive Hybridization Reveals Extensive Horizontal Transfer in the Rumen Metagenome

ELIZABETH A. GALBRAITH, DIONYSIOS A. ANTONOPOULOS, KAREN E. NELSON, AND BRYAN A. WHITE

56.1 INTRODUCTION

Molecular-based approaches have traditionally targeted the small subunit ribosomal RNA (SSU rRNA) encoding gene to study microbial diversity in environmental systems; and although these methods provide a wealth of information about microbial community structure, they are limited to inferring function by using the SSU rRNA sequence as a proxy for the organisms present and their known physiologies [Torsvik et al., 1990; Pace, 1997; Theron and Cloete, 2000; see also Chapter 15, Vol. I]. Metagenomic-enabled approaches take advantage of the plethora of complete genome sequences [Doolittle, 1997; Fraser et al., 2000] and the functional genes contained therein to infer community function in a more direct manner. Metagenomics addresses the collective community structure and functional gene composition of a microbial community without relying on cultivation of the microorganisms present. Owing to the high cost of first-generation Sanger-based DNA sequencing and the relatively large sequence search space of environmental samples, metagenomic libraries relied on large-insert vector systems such as bacterial artificial chromosomes (BAC) and fosmid vectors to capture sufficient sequence length for gene identification and functional inference [Béjà et al., 2000; Rondon et al., 2000; see also Chapter 22, Vol. I]. A shotgun sequencing strategy for metagenomics of environmental samples is now more widely available by the combination of second-generation DNA sequencing technologies and high-throughput data analysis platforms (see Chapters 18 and 20, Vol. I). However, a key challenge remains in comparative studies—namely, identifying the differential components between samples. Profound changes in community structure may belie subtle changes in functional gene content, and therefore a great deal of sequencing effort is focused on sequencing the same abundant sequences between samples (e.g., genes involved in core metabolism or information processing). Suppressive subtractive hybridization (SSH) is one approach that can direct sequencing effort toward differential metagenomic components.

The core concept of subtractive hybridization was first documented by Bautz and Reilly [1966] when they demonstrated that a bacteriophage T4 deletion mutant could be used to isolate mRNA from a specific deleted gene found in the wild-type counterpart without knowledge of the gene content (i.e., without benefit of a reference genome sequence). Approximately 20 years later this idea was systematically applied to isolating differences in genomic DNA from eukaryotes, using a modified approach to that of Bautz and Reilly—"genomic subtraction" [Lamar and Palmer, 1984; Kunkel et al., 1985; Nussbaum et al., 1987]. However, the widest application found for subtractive hybridization at the time was in the normalization of cDNA libraries [Duguid and Dinauer, 1990; Cecchini et al., 1993]. Since cDNA libraries are based on extracted mRNA from a specific eukaryotic tissue, certain mRNA species may be

Handbook of Molecular Microbial Ecology, Volume I: Metagenomics and Complementary Approaches, First Edition. Edited by Frans J. de Bruijn.

underrepresented in one tissue type versus another; subtractive hybridization would allow the recovery of these low-expression mRNA species, thereby providing a more complete repertoire of the potentially expressed mRNA species [Diatchenko et al., 1996; Hubank and Schatz, 1999; Ji et al., 2002].

The work up to this time was successful in identifying differential genomic regions, yet a nontrivial theoretical description of the subtraction process was lacking, considering that several dynamic processes were employed and all were solution-based. Computer-based modeling was soon developed [Ermolaeva and Wagner, 1995; Ermolaeva et al., 1996; Ermolaeva and Sverdlov, 1996] for predicting idealized parameters for the technique that allowed a further refinement that came to be called suppressive subtractive hybridization (SSH; other descriptions in the literature also refer to it as "suppression subtractive hybridization") and introduced the use of specially designed adaptors [Diatchenko et al., 1996]. The formation of "panhandle"-like structures during PCR amplification by these adaptors would efficiently decrease the detection of background fragments and allow the identification of larger DNA fragments, thereby increasing the available "informational" length for analysis [Akopyants et al., 1998]. The subtractive hybridization approach has been used extensively to describe strain-to-strain differences for many microbial species [Emmerth et al., 1999; Ferreira et al., 1999; Sawada et al., 1999; Vinnemeier and Hagemann, 1999; Lai et al., 2000; Agron et al., 2001; Herd and Kocks, 2001; Ahmed et al., 2002; Pettigrew et al., 2002; Pradel et al., 2002; Smoot et al., 2002; Parsons et al., 2003]. As genome sequencing became more routine with the streamlining of methodologies and decrease in cost, the technique has been instrumental in screening specific strains relative to a fully sequenced reference genome [Frohme et al., 2001; Nesbø et al., 2002; Nesbø and Doolittle, 2003; Antonopoulos et al., 2004]. More recently, we demonstrated the application of subtractive hybridization to comparing the metagenomes of microbial communities from the rumen [Galbraith et al., 2004]. This particular study utilized the same rumen samples for which our group had amassed phylogenetic data from microbial fingerprints (based on DGGE; see Chapter 5, Vol. I) and SSU rRNA-based clone libraries [Kocherginskaya et al., 2001]. Moreover, rumen-derived samples were chosen because of a considerable knowledge base regarding their microbial diversity, and information regarding their function was available based on both culture-based and molecular-based approaches [Hungate, 1966; Raskin et al., 1994; Hespell et al., 1997; White et al., 1999].

In this study, SSH was employed to isolate DNA fragments present in one rumen sample, but absent from another. The subtracted fragments were cloned and then sequenced. Based on simple quality control assessments via probing of the SSH libraries constructed, the technique appears capable of enriching the DNA fragments unique to one complex microbial community relative to another. Additionally, a surprising level of Archaeal sequences were identified that tracked with the structure of the overall community based on rRNA inference, lending support that the SSH approach is representative of the community dynamics.

56.2 MATERIALS AND METHODS

For a full outline of the materials and methods, please refer to the original publication [Galbraith et al., 2004] as well as the chapter entitled "Environmental Genomics" in *Methods in Molecular Biology* [Galbraith et al., 2008]. What follows is an abbreviated version.

56.2.1 Sample Collection and Total DNA Extraction

Rumen contents were obtained from cannulated steers maintained at the University of Illinois at Urbana-Champaign Research Farm [Kocherginskaya et al., 2001]. Steers fed a hay-based diet (#218 and #256) were given medium-quality grass-legume hay *ad libitum* while the animal fed a grain-based diet (J277) was fed 52% corn, 20% hay, 5% corn steep liquor, and a 3% mixture of vitamins and mineral salts. The additional 20% of the grain-based diet consisted of various corn derivatives including modified corn fiber, distillers dried grains, corn starch, or corn gluten feed. Samples were collected and processed as previously reported [Peter et al., 2000; Kocherginskaya et al., 2001]. In summary, fresh rumen contents were strained through two layers of cheesecloth to remove large particulate matter prior to extraction of total metagenomic DNA [Tsai and Olson, 1991] (amended by the addition of proteinase K [50 mg ml^{-1}] to samples following the freeze–thaw cycles).

56.2.2 Suppressive Subtractive Hybridization

The summarized protocol is based on the manufacturer's protocol for CLONTECH's PCR-Select™ Bacterial Genome Subtraction Kit. SSH was performed using total metagenomic DNA isolated from rumen fluid samples of three steers (two fed a hay diet, #256 and #218, and one fed a grain-based diet, J277), alternately designated as tester or driver. These were then digested individually with *Alu* I. The TaKaRa *ExTaq* PCR kit (TaKaRa) and a GeneAmp PCR System 2400 thermal cycler (Perkin Elmer) were used for both primary and secondary PCR amplifications in the procedure. Primary PCR cycling conditions were 94°C for 5 min followed by 30 cycles of 94°C for 10 s, 66°C for 30 s, and 72°C for 1.5 min, and one final incubation of 72°C for 7 min. The secondary round of PCR amplification followed a similar program, but with a raised annealing temperature of 68°C and 12 amplification cycles. The secondary PCR acts as an additional enrichment of SSH products. Secondary PCR products were then cloned with the pCR4-TOPO (Invitrogen) vector and transformed into *E. coli* TOP10 cells (Invitrogen) using electroporation.

56.2.3 Dot Blot Hybridization

In the Southern dot blot procedure, cloned SSH products were spotted onto a nylon membrane using standard procedures [Ausubel et al., 2001] and hybridized with tester metagenomic DNA acting as a complex probe and later to a driver metagenome-based probe (see **Fig. 56.1**). PCR amplified inserts from the SSH clones were purified using the Qiaquick PCR Purification Kit (Qiagen), diluted, denatured, and spotted directly onto Hybond-N + nylon membranes (Amersham) using a 96-well manifold (Gibco BRL). Membranes were then UV cross-linked using a Stratalinker (Stratagene).

Prehybridization of the membranes was performed in a hybridization oven with 7 ml of warm (68°C) Perfect Hyb Plus (Sigma) for 15 minutes. Probes were prepared with 100 ng of purified, *Alu* I-digested, driver or tester DNA and labeled with [^{32}P-α]-dCTP using the Random Primers DNA Labeling System (Invitrogen). Probe

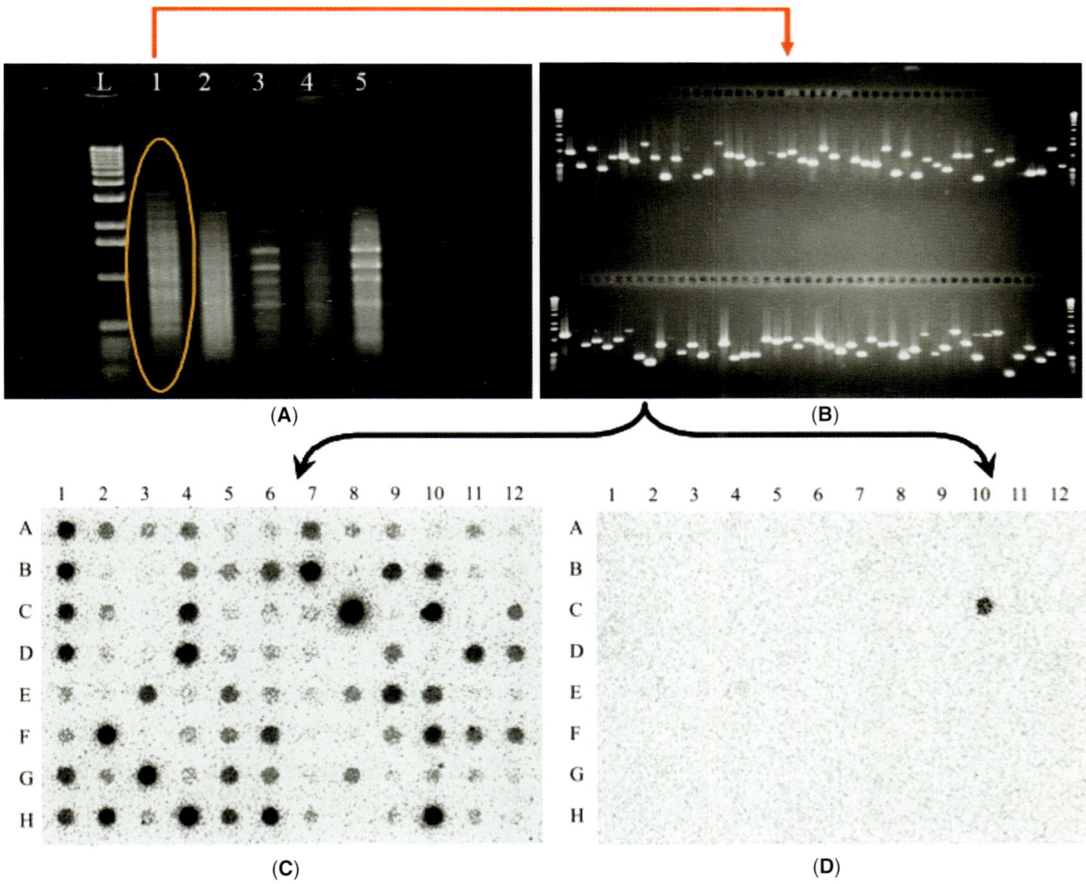

Figure 56.1 Flowchart of metagenomic identity quality control work of randomly selected clones from the subtracted library of rumen metagenome sample #256. (**A**) Results and quality control checks of PCR-based subtractive hybridization utilizing hay-fed rumen metagenome #256 as the tester and hay-fed rumen metagenome #218 as the driver. Lane 1 represents the PCR-amplified pool of final subtraction products used for subsequent cloning and sequencing. Lane 2 represents unsubtracted DNA from the metagenome of rumen sample #256. Lane 3 represents control subtraction DNA with defined ϕX174/*Hae*III digestion fragments recovered from whole *E. coli* genomic DNA (control provided in CLONTECH PCR-Select™ Bacterial Genome Subtraction Kit). Lane 4 represents unsubtracted *E. coli* genomic DNA (also provided in the kit). Lane 5 represents ϕX174/*Hae*III digestion fragments provided as a PCR control (also provided in the kit). A 1-kb DNA standard ladder (L) (Invitrogen) is also shown. (**B**) PCR amplicons of 96 randomly selected clones from the subtracted library of rumen metagenome #256. (**C, D**) Southern dot blot hybridizations of PCR amplicons from 96 randomly selected clones hybridized with ^{32}P–labeled genomic DNA from the tester metagenome (**C**) or the driver metagenome (**D**).

hybridization was then performed for 16 h, at 68°C in a rotating hybridization oven. Standard protocols for washing [Ausubel et al., 2001] were followed: twice under low stringency conditions (50 ml of 2xSSC + 0.1% SDS, 10 min, room temperature) and twice under moderately stringent conditions (50 ml of 0.2xSSC + 0.1% SDS, 15 min, 42°C). Membranes were then exposed to phosphorimaging plates for 20 min, and the signal was visualized using the BAS-1800 II system (Fujifilm).

56.2.4 Sequencing and Data Analysis

Sanger sequencing of cloned SSH products was performed at the W.M. Keck Center for Comparative and Functional Genomics at the University of Illinois at Urbana-Champaign (UIUC). Sequence editing was performed by the Bioinformatics Unit at the W.M. Keck Center, UIUC. Data were uploaded to the MetaGenome Rapid Annotation using Subsystem Technology (MG-RAST) server for additional analysis [Meyer et al., 2008; see also Chapter 37, Vol. I].

56.3 RESULTS

56.3.1 Analysis of Domain Composition of SSH Clone Libraries

Suppressive subtractive hybridization was performed using total genomic DNA isolated from rumen fluid samples of three steers (two fed a hay diet, #256 and #218, and one fed a grain-based diet, J277) and was alternately designated as tester or driver. Clone libraries of enriched unique fragments from tester rumen samples were constructed and used for Southern blots to test the efficacy of subtraction by hybridizing labeled metagenomic DNA from the tester sample to the blot followed by DNA from the driver sample. Tester-specific SSH fragments were found on average in 95 of 96 randomly selected clones. Without a priori information about the sequence content of the metagenomes, the SSH technique displays a high level of efficiency as measured by dot blot hybridizations.

Protein-encoding genes detected in subtractive hybridization fragments specific to sample #256 following hybridization with sample #218 (hay versus hay diet) exhibited significant similarity to Bacterial proteins (32.1%), whereas 18.5% of them had significant similarity to proteins from Archaea [Galbraith et al., 2004]. For the subtractions between rumen samples #256 and J277 (hay versus grain), 35.7% of subtraction fragments showed similarity to Bacterial proteins when #256 was used as the tester (hay-specific versus grain; see **Fig. 56.2**), and 43.8% of subtraction fragments showed similarity to Bacterial proteins when J277 was used as the tester (grain-specific versus hay; see **Fig. 56.3**). In contrast, 8.6% of subtraction fragments showed similarity to Archaeal proteins when #256 was used as the tester (hay-specific versus grain) and 1.6% of subtraction fragments showed similarity to Archaeal proteins when J277 was used as the tester (grain-specific versus hay).

56.3.2 Analysis of Hay-Specific SSH Clone Libraries

Clone library data were assembled to create a hay-specific dataset (186 sequences) and a grain-specific dataset (78 sequences). These were then further analyzed using the MG-RAST sequence analysis platform [Glass et al., 2010; Meyer et al., 2008; see also Chapter 37, Vol. I]. Over half of the observed bacterial protein similarity matches from the hay-specific assembled data (31/56 or 55.4%) were from members of the Firmicutes. This result is consistent with DNA sequence data collected from previous analysis of the microbial community SSU rDNA in our rumen samples [Kocherginskaya et al., 2001]. Proteins from one particular member of the Firmicutes, the highly cellulolytic *Clostridium thermocellum*, provided the closest similarity matches from seven of the subtractive hybridization assembly fragments. Three sequences contained protein-encoding genes that were classified within the DNA metabolism subsystem, representing subsystems for restriction-modification systems, DNA repair, and DNA replication. Another observed trend from the hay-specific dataset, is a relatively large number of subtraction fragments with similarities to transposase proteins.

Individual clones resembling genes for putative Bacterial proteins involved in a variety of cell functions, including (a) energy production and conversion, (b) cell envelope biogenesis, (c) translation, and (d) ribosomal structure and biogenesis, were also found. One DNA fragment contained a protein-encoding gene with similarity to a surfactin synthetase (subunit I) from *Bacillus subtilis*. This protein functions as part of a multienzyme complex to catalyze the synthesis of a powerful biosurfactant with antimicrobial activities [Vollenbroich et al., 1994; Symmank et al., 2002].

From the hay-specific dataset, 19 of the assembled sequences (10.2%) contained matches with protein-encoding genes from Archaeal genomes. The protein-encoding genes in this dataset encompassed a variety of cell functions, including (a) amino acid transport and metabolism, (b) transcription, (c) translation, and (d) ribosomal structure and biogenesis. One in particular showed a high level of sequence similarity to an anaerobic ribonucleotide triphosphate reductase from *Methanothermobacter thermautotrophicus*. The Archaeal enzyme, involved in the reduction of nucleotide diphosphates

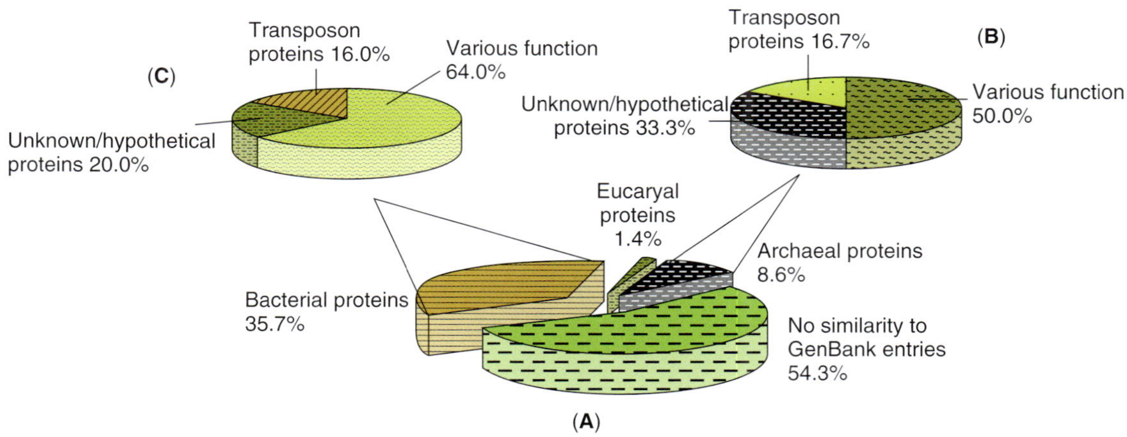

Figure 56.2 Distribution of hay-specific protein similarity matches from the subtractive hybridization library of rumen metagenome #256 after subtraction of rumen metagenome J277. The center pie chart (**A**) organizes SSH fragments according to the domain of their highest sequence similarity matches. Classifications of subtraction fragments by function are depicted for fragments with highest sequence similarity to Archaeal proteins (**B**) and to Bacterial proteins (**C**).

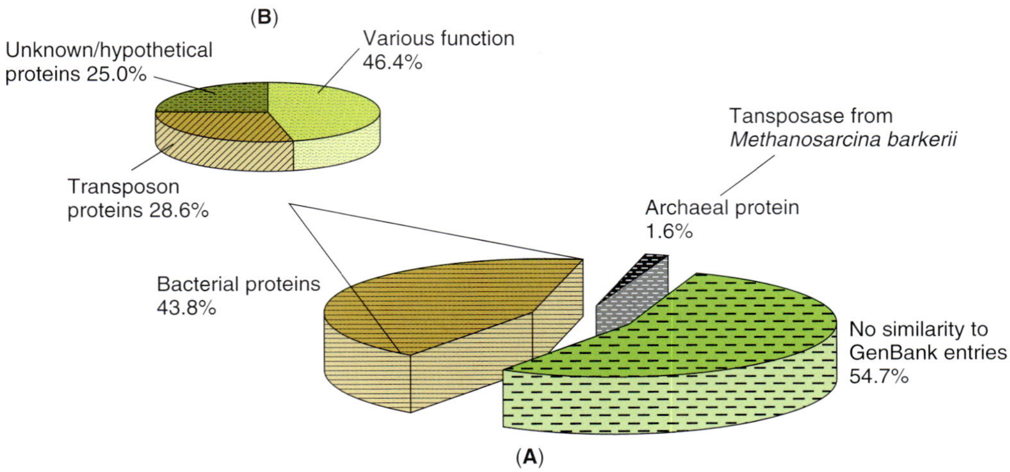

Figure 56.3 Distribution of grain-specific protein similarity matches from the subtractive hybridization library of rumen metagenome J277 after subtraction of rumen metagenome #256. The center pie chart (**A**) organizes SSH fragments according to the domain of their highest sequence similarity matches. Classifications of subtraction fragments by function are depicted for fragments with highest sequence similarity to Bacterial proteins (**B**).

to nucleotide triphosphates, has distinguishing characteristics that set it apart from its Bacterial counterparts such as susceptibilities to *S*-adenosylmethionine, oxygen, and azide, and an independence from coenzyme B12, manganese, and iron [Sprengel and Follmann, 1981; Sze et al., 1992]. Moreover, it has been suggested that the Archaeal ribonucleotide reductase may have a different activation mechanism than Bacterial ribonucleotide reductases [Sze et al., 1992].

A functional trend regarding transport proteins was also observed in the hay-specific dataset. Subtraction fragments resembled transport proteins including ABC-transport system proteins, a polysaccharide transport protein, and a Na^+-dependent phosphate transport protein. In addition, a putative glycosyltransferase protein had closest similarity to one subtraction fragment. This may reflect interesting differences in the surface proteins of bacteria from a hay-fed animal versus a grain-fed animal. Furthermore, two subtraction fragments had highest similarity to proteins regulating sporulation in *Clostridium sp*. One fragment resembled a DNA segregation ATPase that is involved in chromosome segregation during sporulation and is necessary for sporulation to occur [Sharp and Pogliano, 2003]. In contrast, another fragment resembled a sporulation initiation inhibitor involved in the negative regulation of sporulation [Ireton et al., 1994].

56.3.3 Analysis of Grain-Specific SSH Clone Libraries

Of the annotated 78 assembled sequences specific to the grain-fed rumen sample (J277), 20 (69.0% of the annotated protein encoding genes) had closest similarity to Bacterial proteins, while only three assembled DNA fragments (10.3%) resembled proteins from the Archaea. When viewing the non-assembled clone library data a high frequency of transposase matches was strikingly apparent from the Bacterial similarity matches (Fig. 56.3). Twenty-nine individual cloned sequences could be assembled into eight unique, contiguous sequences that each resembled a transposase. Mainly, these transposases were from *Clostridium* and *Lactobacillus sp*.

An additional fragment contained a protein encoding gene that had a high similarity to rubrerythrin, from *Desulfitobacterium hafniense*. This protein is a nonheme iron protein typically found in anaerobic sulfate-reducing bacteria, that has the ability to catalyze the reduction of hydrogen peroxide to water and may be part of an alternative system to the superoxide dismutase-catalase oxidative stress defense system [Sztukowska et al., 2002].

56.3.4 Analysis of rRNA from Rumen Metagenomes

Whole RNA was extracted from rumen samples #218 and #256 (hay-fed animals), blotted and probed with universal (S-*-Universal-1392-a-A15), Bacterial-specific (S-D-Bact-0338-a-A-18), and Archaeal-specific (S-D-Arch-0915-a-A-20) probes to characterize the community composition and their relative activity at the domain level. Following signal normalization with organisms representing the Bacteria (*Escherichia coli* B13711) and Archaea (*M. acetivorans* C2A), approximate percentages of domain-specific 16S rRNA were calculated for each rumen sample (**Table 56.1**). Bacterial 16S rRNA comprised 55.5% and 57.4% of the total 16S rRNA for samples #218 and #256, respectively, while the Archaeal contributions to the total 16S rRNA were approximately 25.3% (#218) and 24.4% (#256). Although a direct correlation between transcript-level and gene-level signal is not possible, this assay did provide support for the robust Archaeal signal detected in the SSH clone library.

Table 56.1 Bacterial and Archaeal Contributions to the Total 16S rRNA Isolated from Rumen Samples of Hay Fed Steers #218 and #256

Rumen Sample	Bacterial 16S rRNA	Archaeal 16S rRNA
#218	55.5%	25.3%
#256	57.4%	24.4%

56.4 DISCUSSION

The application of first-generation DNA sequencing technology in a high-throughput manner to environmental metagenomes has provided a wealth of information about the structure and function of microbial communities from a variety of natural environments [Béjà et al., 2000; Rondon et al., 2000; DeLong, 2002]. Shotgun metagenomics has now entered a new phase of system description with the incorporation of second-generation DNA sequencing platforms. Although the depth of sequencing necessary for circumnavigating complex environmental microbial remains to be seen, this problem emphasizes the need for approaches to partition the DNA sequence space. The rapid and cost-effective methods of SSH or representational difference analysis [Lisitsyn et al., 1993] have yet to be applied extensively to the metagenomic analysis of environmental microbial communities.

The present experiment served as a pilot study for using SSH to identify unique DNA sequences found in the metagenome of one rumen sample relative to another, thus demonstrating the utility of this technique for broader-scale explorations of natural environmental microbial communities. Although our study employed relatively low-throughput first-generation Sanger-based DNA sequencing technology and would have required hundreds of thousands of clones in order to gain a comprehensive appraisal of the genetic diversity existing between complex metagenomes, second-generation DNA sequencing platforms achieve this scale of sampling routinely. Rather, the intent of our initial study was not to exhaustively survey all of the genetic difference between these metagenomes, but to determine whether the sensitivity of SSH was sufficient to enrich for DNA sequences unique to one sample and absent from another in the setting of a complex microbial community. We were able to demonstrate that SSH is an effective method for capturing fragments representing a range of cellular functions.

More surprising, however, was the distribution of cloned SSH products with significant database similarities between different domains of life. Shifts in the Bacterial community structure between samples were expected, given the great diversity of Bacterial species present within the rumen. In contrast, the high proportion of matches to sequences from Archaeal genomes in the hay-specific dataset was unexpected (see Fig. 56.2), because estimates of rumen Archaeal populations range from only 0.5% to 3% of the total microbial population [Lin et al., 1997; Ziemer et al., 2000]. Traditionally, *Methanobrevibacter ruminantium*, *Methanobacterium formicicum*, *Methanomicrobium mobile*, and *Methanosarcina barkeri* are the most commonly found ruminal Archaeal organisms [Hespell et al., 1997]. The majority of the cloned

SSH products matching Archaeal sequences were with those from *Methanothermobacter thermautotrophicus*, a genus not found in the rumen. Only two of the cloned SSH products matched *Methanosarcina barkeri*, which is found in the rumen.

Quantitative dot blots of rRNA from these rumen samples were performed to verify the high proportion of Archaeal sequences recovered from our subtraction study. Corresponding high percentages of Archaeal 16S rRNA were found in samples #218 (25.3%) and #256 (24.4%), which was in contrast to previous estimates in the literature [Lin et al., 1997; Ziemer et al., 2000]. This observed fraction of Archaeal signal provides evidence that the corresponding high proportion of Archaeal sequence similarities in the SSH data is likely not an artifact of the technique. The signal contributed by Bacteria was roughly half of the total 16S rRNA signal in both rumen samples (#218 [55.5%] and sample #256 [57.4%]). Considering that similar ratios of domain-specific 16S rRNA were detected in both of the samples suggests that differences in rumen community structure revealed by SSH are not a result of unbalanced domain-level differences.

This metagenomic approach has exposed an unexpectedly large difference in Archaeal community structure between the rumen microbial populations of two steers fed identical diets and housed together. 16S rRNA dot blot hybridizations revealed similar proportions of Bacteria and Archaea in both rumen samples and suggest that the differences uncovered by SSH are due to varying community structural composition. Variations in the rumen microbial populations of steers fed different diets were also found, including differences in (a) genes coding for surface or membrane proteins and (b) proteins related to sporulation.

Suppressive subtractive hybridization represents an effective methodology for focusing sequencing efforts on unique fractions of genes present in complex microbial communities. In the context of rumen microbial communities, the extent and nature of Archaeal population disparities between animals fed different diets would be of considerable interest. Coupling SSH-enabled approaches with the sampling depth afforded by second-generation DNA sequencing technologies may offer one way of exploring the dynamics of these populations in greater detail.

INTERNET RESOURCES

MG-RAST metagenomics analysis server (http://metagenomics.anl.gov)

Acknowledgments

This study was supported by USDA NRI grant no. 970-326 and by the Agricultural Experimental Station of the University of Illinois. The authors gratefully acknowledge Professor Adrian Egan of the University of Melbourne for spawning this idea at the RRI-INRA meeting in Aberdeen, Scotland. The authors also thank the staff of the W.M. Keck Center for Comparative and Functional Genomics at the University of Illinois at Urbana-Champaign.

REFERENCES

Agron PG, Walker RL, Kinde H, Sawyer SJ, Hayes DC, Wollard J, et al. 2001. Identification by subtractive hybridization of sequences specific for *Salmonella enterica* serovar enteritidis. *Appl. Environ. Microbiol*. **67**:4984–4991.

Ahmed IH, Manning G, Wassenaar TM, Cawthraw S, Newell DG. 2002. Identification of genetic differences between two *Campylobacter jejuni* strains with different colonization potentials. *Microbiology* **148**:1203–1212.

Akopyants NS, Fradkov A, Diatchenko L, Hill JE, Siebert PD, et al. 1998. PCR-based subtractive hybridization and differences in gene content among strains of *Helicobacter pylori*. *Proc Natl Acad Sci USA* **95**:13108–13113.

Antonopoulos DA, Nelson KE, Morrison M, White BA. 2004. Strain specific genomic regions of *Ruminococcus flavefaciens* FD-1 as revealed by combinatorial random-phase genome sequencing and suppressive subtractive hybridization. *Environ. Microbiol*. **6**:335–346.

Ausubel FM, Brent R, Kingston RE, Moore DD, Seidman JG, et al. 2001. *Current Protocols in Molecular Biology*. New York: John Wiley & Sons.

Bautz EK, Reilly E. 1966. Gene-specific messenger RNA: Isolation by the deletion method. *Science* **151**:328–330.

Béjà O, Suzuki MT, Koonin EV, Aravind L, Hadd A, et al. 2000. Construction and analysis of bacterial artificial chromosome libraries from a marine microbial assemblage. *Environ. Microbiol*. **2**:516–529.

Cecchini E, Dominy PJ, Geri C, Kaiser K, Sentry J, Milner JJ. 1993. Identification of genes up-regulated in dedifferentiating *Nicotania glauca* pith tissue, using an improved method for constructing a subtractive cDNA library. *Nucleic Acids Res*. **21**:5742–5747.

DeLong EF. 2002. Microbial population genomics and ecology. *Curr Opin Microbiol*. **5**:520–524.

Diatchenko L, Lau YF, Campbell AP, Chenchik A, Moqadam F, et al. 1996. Suppression subtractive hybridization: A method for generating differentially regulated or tissue-specific cDNA probes and libraries. *Proc.Natl. Acad. Sci*. **93**:6025–6030.

Doolittle RF. 1997. Microbial genomes opened up. *Nature* **392**:339–342.

Duguid JR, Dinauer MC. 1990. Library subtraction of in vitro cDNA libraries to identify differentially expressed genes in scrapie infection. *Nucleic Acids Res*. **18**:2789–2792.

Emmerth M, Goebel W, Miller SI, Hueck CJ. 1999. Genomic subtraction identifies *Salmonella typhimurium* prophages, F-related plasmid sequences, and a novel fimbrial operon, stf, which are absent in *Salmonella typhi*. *J. Bacteriol*. **181**:5652–5661.

Ermolaeva OD, Sverdlov ED. 1996. Subtractive hybridization, a technique for extraction of DNA sequences distinguishing two closely related genomes: Critical analysis. *Genet. Anal*. **13**:49–58.

Ermolaeva, OD, Wagner, MC. 1995. SUBTRACT: A computer program for modeling the process of subtractive hybridization. *CABIOS* **11**:457–462.

Ermolaeva OD, Lukyanov SA, Sverdlov ED. 1996. The mathematical model of subtractive hybridization and its practical application. *Proc. Int. Conf. Intell. Syst. Mol. Biol*. **4**:52–58.

Ferreira H, Rodrigues Neto J, Goncalves ER, Rosato YB. 1999. A simplified subtractive hybridization protocol used to isolate DNA sequences specific to *Xylella fastidiosa*. *Microbiol*. **145**:967–1975.

Fraser, CM, Eisen JA, Salzberg SL. 2000. Microbial genome sequencing. *Nature* **406**:799–803.

Frohme M, Camargo AA, Czink C, Matsukuma AY, Simpson AJ, et al. 2001. Directed gap closure in large-scale sequencing projects. *Genome Res*. **11**:901–903.

Galbraith EA, Antonopoulos DA, White BA. 2004. Suppressive subtractive hybridization as a tool for identifying genetic diversity in an environmental metagenome: the rumen as a model. *Environ. Microbiol*. **6**:928–937.

Galbraith EA, Antonopoulos DA, White BA. 2008. Suppressive subtractive hybridization as a tool for the identification of genetic diversity in an environmental metagenome. In Martin, C. ed. *Methods in Molecular Biology. Environmental Genomics*. Totowa, NJ: Humana Press, pp. 295–333.

Glass EM, Wilkening J, Wilke A, Antonopoulos D, Meyer F. 2010. Using the metagenomics RAST server (MG-RAST) for analyzing shotgun metagenomes. *CSH Protocol* **2010**:pdb prot5368.

Herd M, Kocks C. 2001. Gene fragments distinguishing an epidemic-associated strain from a virulent prototype strain of *Listeria monocytogenes* belong to a distinct functional subset of genes and partially cross-hybridize with other *Listeria* species. *Infect. Immun*. **69**:3972–3979.

Hespell RB, Akin DE, Dehority BA. 1997. Bacteria, fungi, and protozoa of the rumen. In Mackie, RI, White, BA, Isaacson, RE eds. *Gastrointestinal Microbiology*. New York: Chapman and Hall, pp. 59–141.

Hubank M, Schatz DG. 1999. cDNA representational difference analysis: A sensitive and flexible method for identification of differentially expressed genes. *Methods Enzymol*. **303**:325–349.

Hungate RE. 1966. *The Rumen and Its Microbes*. New York and London: Academic Press.

Ireton K, Gunther NW, Grossman AD. 1994. *spo0J* is required for normal chromosome segregation as well as the initiation of sporulation in *Bacillus subtilis*. *J. Bacteriol*. **176**:5320–5329.

Ji W, Wright MB, Cai L, Flament A, Lindpaintner K. 2002. Efficacy of SSH PCR in isolating differentially expressed genes. *BMC Genomics* **3**:12.

Kocherginskaya SA, Aminov RI, White BA. 2001. Analysis of the rumen bacterial diversity under two different diet conditions using denaturing gradient gel electrophoresis, random sequencing, and statistical ecology approaches. *Anaerobe* **7**:119–134.

Kunkel LM, Monaco AP, Middlesworth W, Ochs HD, Latt SA. 1985. Specific cloning of DNA fragments absent from the DNA of a male patient with and X chromosome deletion. *Proc. Natl. Acad. Sci. USA* **82**:4778–4782.

Lai YC, Yang SL, Peng HL, Chang HY. 2000. Identification of genes present specifically in a virulent strain of *Klebsiella pneumoniae*. *Infect. Immun*. **68**:7149–7151.

Lamar EE, Palmer E. 1984. Y-encoded, species-specific DNA in mice: Evidence that the Y chromosome exists in two polymorphic forms in inbred strains. *Cell* **37**:171–177.

Lin C, Raskin L, Stahl DA. 1997. Microbial community structure in gastrointestinal tracts of domestic animals: comparative analyses using rRNA-targeted oligonucleotide probes. *FEMS Microbiol. Ecol*. **22**:281–294.

Lisitsyn N, Lisitsyn N, Wigler M. 1993. Cloning the differences between two complex genomes. *Science* **259**:946–951.

Meyer F, Paarmann D, D'Souza M, Olson R, Glass EM, et al. 2008. The metagenomics RAST server—A public resource for the automatic phylogenetic and functional analysis of metagenomes. *BMC Bioinformatics* **9**:386.

Nesbø CL, Doolittle WF. 2003. Targeting clusters of transferred genes in *Thermotoga maritima*. *Environ. Microbiol*. **5**:1144–1154.

Nesbø CL, Nelson KE, Doolittle WF. 2002. Suppressive subtractive hybridization detects extensive genomic diversity in *Thermotoga maritima*. *J. Bacteriol*. **184**:4475–4488.

Nussbaum RL, Lesko JG, Lewis RA, Ledbetter SA, Ledbetter DH. 1987. Isolation of anonymous DNA sequences from within a submicroscopic X chromosomal deletion in a patient with choroideremia, deafness, and mental retardation. *Proc. Natl. Acad. Sci. USA* **84**:6521–6525.

Pace NR. 1997. A molecular view of microbial diversity and the biosphere. *Science* **276**:734–740.

Parsons YN, Banasko R, Detsika MG, Duangsonk K, Rainbow L, Hart CA, et al. 2003. Suppression-subtractive hybridisation reveals variations in gene distribution amongst the *Burkholderia cepacia* complex, including the presence in some strains of a genomic island containing putative polysaccharide production genes. *Arch. Microbiol*. **179**:214–223.

Peter CM, Faulkner DB, Merchen NR, Parrett DF, Nash TG, Dahlquist JM. 2000. The effects of corn milling coproducts on growth performance and diet digestibility by beef cattle. *J. Anim. Sci*. **78**:1–6.

Pettigrew MM, Foxman B, Marrs CF, Gilsdorf Jr. 2002. Identification of the lipooligosaccharide biosynthesis gene *lic2B* as a putative virulence factor in strains of nontypeable *Haemophilus influenzae* that cause otitis media. *Infect. Immun*. **70**:3551–3556.

Pradel N, Leroy-Setrin S, Joly B, Livrelli V. 2002. Genomic subtraction to identify and characterize sequences of Shiga toxin-producing *Escherichia coli* O91:H21. *Appl. Environ. Microbiol*. **68**:2316–2325.

Raskin L, Stromley JM, Rittmann BE, Stahl DA. 1994. Group-specific 16S rRNA hybridization probes to describe natural communities of methanogens. *Appl. Environ. Microbiol*. **60**:1232–1240.

Rondon MR, August PR, Betterman AD, Brady SF, Grossman TH, et al. 2000. Cloning the soil metagenome: A strategy for accessing the genetic and functional diversity of uncultured microorganisms. *Appl. Environ. Microbiol*. **66**:2541–2547.

Sawada K, Kokeguchi S, Hongyo H, Sawada S, Miyamoto M, et al. 1999. Identification by subtractive hybridization of a novel insertion sequence specific for virulent strains of *Porphyromonas gingivalis*. *Infect. Immun*. **67**:5621–5625.

Sharp MD, Pogliano K. 2003. The membrane domain of SpoIIIE is required for membrane fusion during *Bacillus subtilis* sporulation. *J. Bacteriol*. **185**:2005–2008.

Smoot LM, Franke DD, McGillivary G, Actis LA. 2002. Genomic analysis of the F3031 Brazilian purpuric fever clone of *Haemophilus influenzae* biogroup aegyptius by PCR-based subtractive hybridization. *Infect. Immun*. **70**:2694–2699.

Sprengel G, Follmann H. 1981. Evidence for the reductive pathway of deoxyribonucleotide synthesis in an archaebacterium. *FEBS Lett*. **132**:207–209.

Symmank H, Franke P, Saenger W, Bernhard F. 2002. Modification of biologically active peptides: Production of a novel lipohexapeptide after engineering of *Bacillus subtilis* surfactin synthetase. *Protein Eng*. **15**:913–921.

Sze IS, McFarlan SC, Spormann A, Hogenkamp HP, Follmann H. 1992. A possible new class of ribonucleotide reductase from *Methanobacterium thermoautotrophicum*. *Biochem. Biophys. Res. Commun*. **184**:1101–1107.

Sztukowska M, Bugno M, Potempa J, Travis J, Kurtz DM, Jr. 2002. Role of rubrerythrin in the oxidative stress response of *Porphyromonas gingivalis*. *Mol. Microbiol*. **44**:479–488.

Theron J, Cloete TE. 2000. Molecular techniques for determining microbial diversity and community structure in natural environments. *Crit. Rev. Microbiol*. **26**:37–57.

Torsvik V, Salte K, Sorheim R, Goksoyr J. 1990. Comparison of phenotypic diversity and DNA heterogeneity in a population of soil bacteria. *Appl. Environ. Microbiol*. **56**: 776–781.

Tsai Y, Olson BH. 1991. Rapid method for direct extraction of DNA from soil and sediments. *Appl. Environ. Microbiol*. **57**: 1070–1074.

Vinnemeier J, Hagemann M. 1999. Identification of salt-regulated genes in the genome of the cyanobacterium *Synechocystis* sp. strain PCC 6803 by subtractive RNA hybridization. *Arch. Microbiol*. **172**:377–386.

Vollenbroich D, Mehta N, Zuber P, Vater J, Kamp RM. 1994. Analysis of surfactin synthetase subunits *in srfA* mutants of *Bacillus subtilis* OKB105. *J. Bacteriol*. **176**:395–400.

White BA, Cann IKO, Kocherginskaya SA, Aminov RI, Thill LA, et al. 1999. Molecular analysis of Archaea, Bacteria, and Eucarya communities in the rumen. Asian-Australasian *J. Anim. Sci*. **12**:129–138.

Ziemer CJ, Sharp R, Stern MD, Cotta MA, Whitehead TR, Stahl DA. 2000. Comparison of microbial populations in model and natural rumens using 16S ribosomal RNA-targeted probes. *Environ. Microbiol*. **2**:632–643.

Part 6A

Microarrays

Chapter 57

GeoChip: A High-Throughput Metagenomics Technology for Dissecting Microbial Community Functional Structure

Joy D. van Nostrand, Zhili He, and Jizhong Zhou

57.1 INTRODUCTION

A key goal in the study of microbial ecology is to understand how microorganisms impact their environment and to understand how microbial communities change in structure and function spatially and temporally. However, efforts to study these communities are hampered by the vast diversity of these communities (estimated 2000 and 50,000 species per gram of soil [Torsvik et al., 1990; see also Chapter 2, Vol. I; Schloss and Handelsman, 2006; Hong et al., 2006; Roesch et al., 2007]) and the very small number of microorganisms that have been cultivated (~1%) [Whitman et al., 1998]. Focusing only on culturable organisms presents a very limited view of microbial communities. As such, in order to overcome these issues, culture-independent approaches are needed to examine the vast majority of environmental microorganisms and provide a more complete picture of these communities. However, many of the culture independent approaches available (e.g., 16S rRNA gene cloning, terminal-restriction fragment length polymorphism (T-RFLP), quantitative PCR) have a PCR amplification step, which introduces well-known biases [Warnecke et al., 1997; Lueders and Friedrich, 2003; Suzuki and Giovannoni, 1996; see Chapter 16, Vol. I]. In addition, if many genes are examined, these methods can be time-consuming and expensive. Microarrays provide a simple, rapid, and high-throughput method for examining thousands of genes at one time and can be used to overcome the limitations of many culture independent approaches.

The first microarray was developed to examine gene expression in the plant *Arabidopsis thaliana* [Schena et al. 1995], and similar arrays have since been developed to study gene expression in numerous microorganisms. Microarrays were later used on microbial communities [Guschin et al., 1997], greatly expanding the potential uses of this technology. Also, microarrays are ideal for comparing environmental microbial communities because all samples are interrogated against a common set of genes [Zhou et al., 2008]. Currently, several different arrays have been developed for studying environmental microbial communities [Zhou and Thompson, 2002; Zhou, 2003; Gentry et al., 2006]. Phylogenetic oligonucleotide arrays (POA) are composed of probes for 16S rRNA or other phylogenetically informative genes (e.g., 23S rRNA gene [Lee et al., 2006; see Chapter 3 Vol. I]) and can be used to examine the composition of microbial communities or the phylogenetic relatedness of different microbial strains [Small et al., 2001; Loy et al., 2002; Wilson et al., 2002]. PhyloChip is the most comprehensive POA available and contains approximately 300,000 perfect-match and mismatch 16S rRNA gene probes for 842 subfamilies [Brodie et al., 2006; see also Chapter 58, Vol. I]. Additional information on the PhyloChip can be found in Chapter 58, 60 and 61, Vol. I. Whole-genome open reading frame (ORF) arrays (WGA) contain probes

Handbook of Molecular Microbial Ecology, Volume I: Metagenomics and Complementary Approaches, First Edition. Edited by Frans J. de Bruijn.

for all ORFs in one or more genomes [Wilson et al., 1999] and can be used to examine the relatedness of microbial strains [Murray et al., 2001]. A virulence factor WGA was used to examine over 100 *Pseudomonas syringae* strains to determine which genes were related to host specificity and several genes statistically associated with specific hosts were identified [Sarker et al., 2006]. Metagenomic arrays (MGA) have probes obtained from environmental DNA clone libraries [Sebat et al., 2003] and can be used for high-throughput screening of clone libraries. Community genome arrays (CGA) use whole genomic DNA from individual strains as probes and can be used to evaluate DNA–DNA relatedness of microbial strains or to detect microorganisms in the environment [Wu et al. 2004, 2008b; Zhang et al. 2004]. Functional gene arrays (FGAs) are made up of probes for functional genes involved in specific processes [Wu et al., 2001; Gentry et al., 2006; He et al., 2007, et al., 2010a]. With FGAs, thousands of functional genes from multiple functional gene groups can be examined simultaneously [Wu et al., 2001; Zhou and Thompson, 2002; Gentry et al., 2006; Wu et al., 2006; He et al., 2007; Wagner et al., 2007; Zhou et al., 2008; Wang et al., 2009; He et al., 2010a], unlike conventional molecular techniques such as PCR-based cloning, in situ hybridization, and quantitative PCR, which severely limit the number of genes that can be examined at one time. The remainder of this chapter will focus on FGAs and the Phylo Chip will be reviewed in Chapter 58, Vol. I.

57.2 DEVELOPMENT OF FUNCTIONAL GENE ARRAYS

The first FGA was described by Wu et al. [2001]. The prototype array used 104 PCR-amplicon probes: 89 probes of *nirS, nirK, amoA*, and *pmoA* from pure culture isolates and clone libraries created from marine sediments as probes, 10 positive control probes for 16S rRNA genes, and 5 negative control probes for yeast genes [Wu et al., 2001]. The array was both specific and sensitive based on hybridization conditions. For example, gene-specific hybridizations occurred at 45°C for both pure cultures and environmental samples, while at 65°C the array was strain-specific. Sensitivity was 1 ng for pure culture DNA and 10–25 ng for community DNA. When tested with microbial communities from marine sediment and soil communities, the array was able to detect *nirS, nirK*, and *amoA* or *pmoA* genes from all samples with higher abundances detected in soil compared to marine sediment. While these results proved that this array could be used to analyze environmental microbial communities, the use of PCR-based probes puts severe limits on this array. First, conserved PCR primers are required; however, conserved PCR primers cannot be designed for many functional genes of interest. Consequently, those PCR amplification-based approaches cannot be used to detect and quantify many functional genes of interest. Even though fairly conserved primers can be designed for some functional genes of interest, it is sometimes difficult to amplify these genes from many environmental samples and hence such genes fail to be detected. Second, the overall comprehensiveness of the array is limited because of the impracticality of creating amplicons for all known bacterial strains and creating clone libraries from diverse communities. Oligonucleotide probes overcome these limitations since only gene sequences are needed rather than the actual DNA and they can be easily modified to improve specificity [Denef et al., 2003; Zhou, 2003; Gentry et al., 2006]. For these reasons, most FGAs are now constructed using oligonucleotide probes [Rhee et al., 2004; Tiquia et al., 2004; He et al., 2007].

Several FGAs targeting specific functional groups, genes, environments, or populations have been described since this first report:

57.2.1 Functional Group-Specific Arrays

Two oligonucleotide (70-mer) arrays were designed to monitor N cycling (denitrification, N-fixation, ammonia oxidation, nitrite reduction; *amoA, nifH, nirS*, and *nirK* genes) [Taroncher-Oldenburg et al., 2003]. Highest specificity (84–90% sequence similarity) was observed for hybridizations at 65°C, with a corresponding sensitivity of 5 pg ($\sim 10^7$ DNA copies). These arrays were evaluated using sediment samples from upper and lower areas of the Choptank River. The results indicated differences in denitrifying communities based on location, consistent with differences observed in the overall community.

A *nifH* oligonucleotide (15- to 25-mers) array was designed to examine N-fixing communities [Zhang et al., 2007a]. An initial array comprised of 56 probes, with melting temperatures (T_m) of 65–69°C, was used to evaluate the probe design strategy. This array was tested using 5 bacterial strains and 26 clones from an environmental sample, and the probes were found to be specific for their corresponding targets when hybridized at 25°C and 50% formamide. The array was then expanded to 194 probes, with a coverage of over 900 *nifH* sequences, and used to examine microbial communities (DNA and RNA) from roots of *Oryza longistaminata*, a wild rice species. Gene variants from β-Proteobacteria, the Omega subcluster, the *anf* subcluster, and gram-positive bacteria were detected, and RNA hybridization results indicated that the gram-positive bacteria were active within this system.

Bodrossy et al. [2003] developed a methanotroph array comprised of oligonucleotide probes (17- to 26-mers) for 59 *pmoA/amoA* genes. The array was tested using amplified RNA from landfill samples that were hybridized at 55°C, and the results showed a positive hybridization for 50 probes. Under these hybridization conditions, the array was able to detect populations comprising ~5% of the original population. Stralis-Pavese et al. [2004] later expanded this array to 68 probes and used it to examine the effects of plant cover on methanotrophic communities from simulated landfills, and their results indicated that the quantity of all methanotrophs increased when a CH_4 and CO_2 mixture (3:2; biogas) was added. In addition, the results from DNA and RNA hybridizations showed that while *Methylocystis* and *Methylocaldum* spp. were the dominant methanotrophs based on DNA [Stralis-Pavese et al., 2004], their activity was extremely low or nonexistent, respectively, based on mRNA results [Bodrossy et al., 2006].

While most FGAs are printed onto glass slides, nylon membranes have also been used. Steward et al. [2004] designed a macroarray using PCR amplicons (318–333 bp) of *nifH* genes to study nitrogenase diversity in the environment. The array was specific to sequences with >78% similarity at 60°C and 40% formamide with a corresponding sensitivity of <0.13 ng ml^{-1}. Samples from Chesapeake Bay sediments were used to test the array and hybridizations were found to be very reproducible. This macroarray was later expanded and then used to examine Chesapeake Bay diazotroph communities [Jenkins et al., 2004]. The array indicated that this system had a high diversity of *nifH* containing microorganisms and that there were changes in this diversity across the area of the Bay examined.

57.2.2 Pathogen Detection

Several FGAs have been developed for use in pathogen detection since microarrays would be an ideal tool for use in medical diagnosis and water and food safety testing. Arrays would provide a high-throughput method of testing and could be used to screen hundreds of pathogens at one time. Microarrays could also provide rapid information on antibiotic resistance, even for extremely slow-growing strains which would be of great benefit for early treatment of infection. Several prototype pathogen detection arrays have been designed. Most of these pathogen-specific arrays have focused on detecting pathogen-specific virulence factors.

Kostić et al. [2005] developed an array comprised of (a) 70-mer oligonucleotide probes for virulence factors specific to *Salmonella* (*invA* and *sopB*) and (b) a positive control probe for *Enterobacteriaceae* 16S rRNA genes. This array was tested using whole-genome amplified DNA from *Escherichia coli* DH5α as background DNA and *Salmonella enterica* serovar Senftenberg as the target DNA. Hybridizations at 42°C and 40% formamide resulted in positive hybridizations for *Salmonella* probes when the target sequences represented as little as 1% of the original community. Cleven et al. [2006] designed an array using PCR amplicon probes (200–800 bp) for 120 virulence genes, antibiotic resistance genes, and metabolic or structural genes from *Staphylococcus aureus*, *Escherichia coli*, and *Pseudomonas aeruginosa*. Forty-five clinical and reference strains were used to test the array, and it was found to be specific for all tested strains. Array hybridization results for *S. aureus* were confirmed using traditional culture-based methods, indicating that this array has the potential for use as a diagnostic tool. A PCR-amplicon (~550 bp)-based array for detecting tetracycline resistance in isolated cultures was developed by Call et al. [2003]. The array was tested against 48 isolates, and the results indicated that 38 of the isolates were positive for at least one *tet* gene, which agreed with previous findings [DePaola and Roberts, 1995; Miranda et al., 2003].

Another application for pathogen arrays is for water quality testing. A virulence and marker gene (VMG) array was designed for this purpose [Miller et al., 2008]. The VMG array was comprised of two sets of probes: 791 targeted probes for 35 VMGs from 12 pathogens and 2034 nontargeted probes for 67 VMGs from 17 pathogens. DNA from tap water, wastewater treatment plant effluent, and river water samples were amplified using multiplex PCR of all genes covered on the array and then hybridized to the array. Genes comprising as little as 0.01% of the original population were detected using this amplification method.

A much larger array was recently reported by Palka-Santini et al. [2009]. This array screened for 10 pathogens (*Staphylococcus* spp., *Streptococcus* spp., *Enterococcus* spp., *Proteus* spp., *Klebsiella* spp., *Stenotrophomonas* spp., *Enterobacter* spp., *Acinetobacter* spp., *E. coli*, *Pseudomonas aeruginosa*, and *Candida albicans*) and used tens of genes from each organism including probes for virulence factors, antimicrobial resistance genes, non-bacterial genes as negative controls and bacterial 16S rRNA genes as positive controls ($n = 930$). To increase the amount of target sequences in the sample, a multiplex PCR amplification was performed using primers specific for all genes represented on the array (800 primer pairs). Using this amplification method, as little as 1 ng of DNA was detectable. The array and amplification method were tested with an actual clinical specimen, an infected wound swab, and multiple *Enterococcus faecium* and *Staphylococcus epidermidis* genes were detected. Culturing of the swab confirmed the presence of these pathogens.

57.2.3 Specific Environments

An FGA has been developed to examine rhizobial and plant nodule isolates for known and novel *nodC* sequences [Bontemps et al., 2005]. The arrays contained probes (41- and 50-mers) for 130 *nodC* variants and were designed based on two conserved regions of the *nodC* gene. To test the array, 14 rhizobial strains were hybridized to the array; of these, 10 were positive for *nodC*. A partial *nodC* sequence was amplified and sequenced from nine of these isolates and found to be previously unknown sequences.

An array specific for acid mine drainage (AMD) sites has also been developed. This array contains 1071 50-mer oligonucleotide probes for 16S rRNA and functional genes related to AMD and bioleaching systems [Yin et al., 2007]. Samples from AMD (pH 2.0), a bioleaching system (pH 2.5), and a higher pH comparison site (pH 5.0) were hybridized to the array and more 16S genes and higher functional diversities were observed in the low pH samples, as would be expected. Further examples of the use of Phylogenetic microarrays can be found in Chapters 58 and 61, Vol. I, including microarrays to screen bacterial contamination in spacecrafts at NASA (see Chapter 60, Vol. I).

57.2.4 GeoChips

Since many of these FGAs were designed to examine specific environmental niches or to address specific questions, they contained only a limited number of probes covering a small number of genes and were of limited value in other studies, so have only been used in one or two studies. To provide a more comprehensive view of microbial communities from a wide range of environmental resources, different versions of GeoChips have been developed. The probe sets for the GeoChip 1.0 were reported in 2004 [Rhee et al., 2004; Tiquia et al., 2004]. This FGA contained 2006 50-mer oligonucleotide probes covering genes for nitrification, denitrification, nitrogen fixation, methane oxidation, sulfate reduction [Tiquia et al., 2004], organic contaminant degradation and metal resistance [Rhee et al., 2004]. This FGA was the first to provide such a functionally diverse probe set. The GeoChip 1.0 has been used in several studies that have shown the benefit of having a more comprehensive FGA for environmental microbiology studies [Wu et al., 2006, 2008a; Tiquia et al., 2006; Zhang et al., 2007b; Waldron et al., 2009].

While GeoChip 1.0 provided a greater diversity of gene coverage than other FGAs, it still lacked a truly comprehensive probe set. The GeoChip 2.0 was developed to provide both a comprehensive probe set and the specificity necessary to distinguish highly homologous gene variants [He et al. 2007]. GeoChip 2.0 is a high-density array, containing 24,243 50-mer oligonucleotide probes, covering ~10,000 functional genes from 150 gene families involved in the geochemical cycling of C, N, and P cycling, sulfate reduction, metal reduction and resistance, and organic contaminant degradation. Recently, a more comprehensive GeoChip, called GeoChip 3.0, has been developed [He et al., 2010a]. GeoChip 3.0 contains about 28,000 probes and covers approximately 57,000 gene sequences from 292 gene families, greatly increasing the functional gene coverage. Functional gene groups involved in antibiotic resistance and energy processes have been added, and additional functional genes have been added and updated from GeoChip 2.0. In addition, the phylogenetic marker, *gyrB* has been included so that both functional and phylogenetic analyses can be performed together. Since they provide such comprehensive coverage, GeoChips are discussed in more detail in the following sections. In addition to increasing gene coverage, new features have been added, including a common oligo reference standard (CORS) for normalization and data comparison across experiments [Liang et al, 2010]. A computational pipeline was also developed for GeoChip probe design and data analysis [He et al., 2010a].

57.2.4.1 GeoChip Design. The first step in designing a probe set for GeoChips, or other FGAs, is to determine what functional genes would be included on the array. The selected genes should encode for proteins that are key to the function of interest. Next, keywords are used to automatically search public sequence databases (e.g., GenBank) and the resulting sequences are downloaded. The selected sequences are then compared to seed sequences using HMMER alignment (http://hmmer.wustl.edu/) to confirm that the included sequences are valid. Seed sequences are those sequences that have been experimentally determined to encode the specific function of interest. Probes (50-mer oligonucleotide) are then designed using a new version of CommOligo software [Li et al., 2005] based on experimentally predetermined criteria of sequence homology, continuous stretches and free energy [He et al., 2005, 2007]. The designed probes are then checked against the GenBank database to ensure specificity, and the best probes are then synthesized and used to construct the array. An overview of this process is shown in **Figure 57.1**.

57.2.4.2 Sample Preparation and Hybridization. Nucleic acid extraction and purification are the most important steps for hybridization success (see Chapters 10–11, Vol. II). For GeoChip analysis, environmental DNA is generally extracted using a well-established freeze-grind method combined with detergent lysis [Zhou et al, 1996; Hurt et al, 2001;

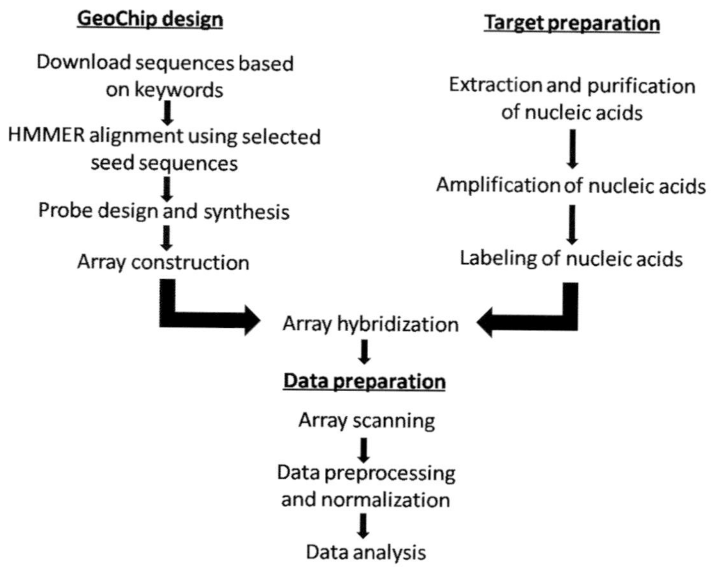

Figure 57.1 Overview of GeoChip design, sample preparation, and data analysis. See text for complete description of all steps.

see Chapter 10–11, Vol. II] since this method produces large fragments of DNA, which is important for subsequent amplification steps. DNA purification can be carried out using a column or gel purification method. Ideally, the purified DNA should have $A_{260} : A_{280} > 1.8$ and $A_{260} : A_{230} > 1.7$; the $A_{260} : A_{230}$ ratio is most important for hybridization success [Ning et al., 2009]. While more challenging, environmental RNA can also be used for GeoChip hybridizations. A few methods have been described for extracting environmental RNA [Hurt et al., 2001; Burgmann et al, 2003; Poretsky et al, 2009; see also Chapters 62–63, Vol. I, Chapter 10, Vol. II].

Since large amounts of DNA (2–5 μg) or RNA (10–20 μg) are needed for GeoChip hybridization, an amplification step is often necessary. For GeoChip analysis, whole community genome or RNA amplification (WCGA/WCRA) is generally used for samples with low amounts of nucleic acids. Relatively small quantities of DNA (1–100 ng) or RNA (50–100 ng) can be used to obtain a representative and even amplification with minimal bias [Wu et al., 2006; Gao et al., 2007]. Some studies have used multiplex PCR to increase nucleic acid concentration and increase sensitivity [Miller et al., 2008; Palka-Santini et al., 2009]. While this strategy worked for these particular studies, this would be both time- and cost-prohibitive with GeoChip because of the large number of genes covered on the array in addition to the inability to design primers for most genes, as mentioned previously. WCGA can increase sensitivity to 10 fg of DNA (one to two bacterial cells) [Wu et al., 2006].

The amplified DNA or RNA is then labeled with a fluorescent dye (e.g., Cy3, Cy5) and hybridized to the array at 42–50°C and 50% formamide [He et al., 2007; Mason et al., 2009; Liang et al., 2009b; Van Nostrand et al, 2009; Waldron et al, 2009]. The method of labeling can affect signal intensity in that higher fluorescent signals can occur when hybridization occurs near an end-labeled nucleotide; this effect is ameliorated when evenly labeled targets are used [Zhang et al., 2007a]. Changing the hybridization temperature or formamide concentration changes the hybridization stringency, which in turn affects probe specificity, although changing the formamide concentration may affect specificity in a more reliable manner than changing temperature [Zhang et al., 2007a].

57.2.4.3 Post-Hybridization Analysis. After hybridization, the array is imaged and then digitally analyzed using microarray analysis software to quantify the signal intensity (pixel density) of each spot and the background. Spot quality is evaluated at this time using predetermined criteria and flagged for later removal. The raw data are uploaded to the GeoChip data analysis pipeline (http://ieg.ou.edu/). The overall hybridization quality is assessed based on (a) the uniformity of control spot hybridization signals across the array and (b) the evenness of background levels. Poor spots and outliers (positive spots with (signal minus mean signal intensity of all replicate spots) greater than three times the replicate spots' signal standard deviation [He and Zhou, 2008]) are removed and the data are normalized. The remaining spots are evaluated to determine whether a positive signal is present. The signal-to-noise ratio [SNR; SNR = (signal mean − background mean)/background standard deviation] is generally used, although other measures can also be used, such as signal-to-both-standard-deviations ratio [SSDR; SSDR = (signal mean − background mean)/(signal standard deviation background standard deviation)] [He and Zhou, 2008].

57.2.4.4 Data Analysis. One of the main challenges for data analysis is the vast amounts of data generated with the GeoChip. Several statistical methods are recommended for GeoChip studies, including multivariate statistical methods such as principal component analysis (PCA) or detrended correspondence analysis (DCA) [He et al, 2008], which reduce the number of explanatory variables and emphasize the variability between samples. Cluster analysis (CA), which groups samples based on gene pattern similarity, and neural network analysis (NNA), which is used to visualize relationships between genes or gene groups, can also be used [He et al, 2008]. Canonical correspondence analysis (CCA) [ter Braak, 1986] and variance partitioning analysis (VPA) [Økland and Eilertsen, 1994; Ramette and Tiedje, 2007] can be used to correlate environmental variables with the functional community structure.

57.2.5 GeoChip Examples

Numerous studies have been performed using GeoChips to examine microbial communities from a variety of environments (**Table 57.1**). It has been used to (a) study bioremediation systems involved in U(VI) reduction [Wu et al., 2006; Waldron et al., 2009; He et al., 2007; Van Nostrand et al., 2009; Xu et al., 2010] or oil degradation [Liang et al., 2009a; Rodríguez-Martínez et al., 2006], oil [Liang et al.], pesticide/pollutant [Taş et al., 2009; Liebich et al., 2009], or metal [Xiong et al., 2010] contaminated environments, marine [Wu et al., 2008a; Tiquia et al., 2006], lake [Parnell et al., 2010], and deep sea environments [Mason et al., 2009], different land-use strategies [Zhang et al., 2007b; Reeve et al., 2010], and sediment systems [Yergeau et al., 2007], in addition to coral mucus-associated communities [Kimes et al., 2010], (b) examine microbial activity [Leigh et al., 2007; Gao et al., 2007], or (c) probe microorganisms for metal resistance genes [Van Nostrand et al., 2007]. A few recent studies have used GeoChip to examine the impacts of global change on microbial communities [He et al., 2010b], to answer overarching ecological questions related to taxa-area relationships [Zhou et al., 2008], and to examine extreme environments such as deep sea hydrothermal vents [Wang et al., 2009].

Understanding the responses of biological communities to elevated atmospheric CO_2 is a central issue in ecology and for society. The responses of below-ground microbial communities to elevated CO_2 are critical to determine whether and how much the fertilization effects will lead to carbon loss or sequestration in terrestrial ecosystems, but they are poorly understood and controversial. The response of soil microbial communities to elevated CO_2 in a grassland system were examined using GeoChip 3.0 [He et al., 2010b]. The results indicated that elevated CO_2 significantly altered the below-ground microbial community functional structure after a 10-year field exposure of grass species to elevated CO_2. In addition, the community changes detected by GeoChip were significantly correlated to soil carbon and nitrogen content and to plant productivity. Finally, genes involved in nitrogen fixation and carbon fixation significantly increased under elevated CO_2, while genes involved in degrading recalcitrant soil carbon remained unchanged. These results have important implications for the feedback responses of ecosystems to atmospheric CO_2 and hence to global climate change modeling needed for reliable prediction of future atmospheric CO_2.

Understanding the spatial patterns of organisms and the underlying mechanisms shaping biotic communities is a central goal in community ecology. One of the most well-documented spatial patterns in plant and animal communities is the positive-power law relationship between species (or taxa) richness and area. Such taxa-area relationships (TARs) are one of the principal generalizations in ecology and are fundamental to our understanding of the distribution of global biodiversity. However, TARs remain elusive in microbial communities, especially in soil habitats, due to inadequate sampling methodologies. GeoChip 2.0 was used to determine spatial scaling differences across various functional and phylogenetic groups at a whole-community level [Zhou et al., 2008]. The results indicated that the forest soil microbial community exhibited a relatively flat gene-area relationship, but varied considerably across different functional and phylogenetic groups. The results revealed that the turnover in space of microorganisms might be, in general, lower than that of plants and animals.

Deep-sea hydrothermal vents are one of the most unique and fascinating ecosystems on Earth. Although phylogenetic diversity of vent communities has been extensively examined, their physiological diversity is poorly understood. GeoChip 2.0 was used to examine microbial communities from deep-sea hydrothermal vent chimneys including the inner and outer parts of a newly grown five day's premature chimney and a mature chimney [Wang et al., 2009]. The results indicated that microbial functional diversity is much lower in the inner part than the outer part or the mature chimney. Also the array results revealed that deep sea vent microbial communities appear to be capable of carbon fixation, anaerobic methane oxidation, methanogenesis, and nitrogen fixation, which had not previously been observed in these environments. Overall, the array analyses indicated that the hydrothermal microbial communities are metabolically and physiologically highly diverse, and the communities

Table 57.1 GeoChip-Based Studies of Microbial Communities

GeoChip version	Study description	Reference
GeoChip 1.0	Used samples from contaminated and uncontaminated sites at the Oak Ridge Field Research Center (OR-FRC) to test the use of whole community genome amplification	Wu et al. [2006]
GeoChip 1.0	Examined the functional structure of microbial communities from contaminated sites within the OR-FRC	Waldron et al. [2009]
GeoChip 1.0	Examined microbial communities from soils exposed to different land usage strategies	Zhang et al. [2007b]
GeoChip 1.0	Examined communities from Gulf of Mexico sediments at different depths	Wu et al. [2008a]
GeoChip 1.0	Examined microbial communities from Puget Sound sediments at different depths and nutritional levels	Tiquia et al. [2006]
GeoChip 2.0	Examined communities of a pilot-scale system for biostimulation of U(VI) reduction at the OR-FRC	He et al. [2007]
GeoChip 2.0	Examined communities of a pilot-scale system for biostimulation of U(VI) reduction at the OR-FRC during exposure to dissolved oxygen and starvation	Van Nostrand et al. [2009]
GeoChip 2.0	Examined functional gene composition of the sediment communities of a pilot-scale system for biostimulation of U(VI) reduction at the OR-FRC	Xu et al. [2010]
GeoChip 2.0	Examined communities of a pilot-scale system for biostimulation of U(VI) reduction at the OR-FRC during the initial active U(VI) reduction and maintenance phases	Van Nostrand et al. [2011]
GeoChip 2.0	Examined communities in a diesel fuel bioremediation system in Puerto Rico	Rodríguez-Martínez et al. [2006]
GeoChip 2.0	Examined microbial communities in a laboratory-scale crude oil bioremediation system using ozone to degrade recalcitrant oil contaminants	Liang et al. [2009a]
GeoChip 2.0	Examined microbial communities from contaminated oil fields in China	Liang et al. [2009b]
GeoChip 2.0	Examined Antarctic sediments to characterize the microbial communities in this environment	Yergeau et al. [2007]
GeoChip 2.0	Functional gene data was used to examine gene-area relationships of microbial communities from forest soils	Zhou et al. [2008]
GeoChip 2.0	Examined communities from deep sea hydrothermal vents using samples collected from the inner and outer portions of a five-day-old chimney and a mature chimney	Wang et al. [2009]

(continued)

Table 57.1 GeoChip-Based Studies of Microbial Communities *(Continued)*

GeoChip version	Study description	Reference
GeoChip 2.0	Examined deep sea basalt communities to determine their functional abilities	Mason et al. [2009]
GeoChip 2.0	Examined metal resistance genotypes of four Ni-resistant actinomycetes	Van Nostrand et al. [2007]
GeoChip 2.0	Used stable isotope probing (biphenyl) to detect active PCB-degrading microbial populations from a hydrocarbon-contaminated aquifer	Leigh et al. [2007]
GeoChip 2.0	Used whole community RNA amplification to examine the activity of microbial communities in a denitrifying fluidized bed reactor at the OR-FRC	Gao et al. [2007]
GeoChip 2.0	Examined microbial communities associated with coral mucus from diseased and healthy corals in order better understand the role the microbial community plays in disease	Kimes et al. [2010]
GeoChip 2.0	Examined pollutant and pesticide impacted locations of the Ebro River in Spain to determine the impact to the microbial communities	Taş et al. [2009]
GeoChip 2.0	Examined samples from a hypersaline lake to determine levels of gene transfer within this system	Parnell et al. [2010]
GeoChip 2.0	Examined microbial communities from a grassland and a *Eucalyptus* plantation, formerly a grassland	Berthong et al. [2009]
GeoChip 2.0	Examined microbial communities associated with strawberry fields farmed using either organic or commercial methods	Reeve et al. [2010]
GeoChip 3.0	Examined microbial communities from the BioCON (Biodiversity, CO_2, and Nitrogen deposition) experimental site at the Cedar Creek Ecosystem Science Reserve in Minnesota, USA to determine how increased CO_2 impacts microbial functions	He et al. [2010b]
GeoChip 3.0	Examined microbial communities associated with microbial electrolysis cells to determine functional characteristics associated with high H_2 production	Wang et al. [2010]
GeoChip 3.0	Examined rhizosphere microbial communities associated with As contaminated sites to examine the relationship between the plant, rhizosphere microorganisms and As	Xiong et al. [2010]

appear to be undergoing rapid dynamic succession and adaptation in response to the steep temperature and chemical gradients across the chimney see Chapters 37, 41, Vol. II.

GeoChip has been used extensively for monitoring the progress of contaminant bioremediation and assessing bioremediation efficiency. Unlike other methods, the GeoChip can quickly identify key functional genes and microbial populations in response to changes in the environment, such as heavy metal or oxygen concentrations and pH. Studies on an ethanol-fed in situ U(VI) bioremediation system indicated that sulfate-, nitrate/nitrite- and metal-reducing communities were responsible for the rapid U(VI) reduction observed in the system [Van Nostrand et al., 2009; Van Nostrand et al., 2011]. The ethanol, provided as an electron donor and carbon source, was an important driver in determining community structure [Van Nostrand et al., 2009]. While subsequent system permutations (introduction of oxygen or nitrate, cessation of ethanol) resulted in changes to the community structure, the overall functional diversity remained relatively constant. Thus, all results indicated that GeoChip is a powerful tool for tracking bioremediation processes.

Great advances have been made over the past decade in FGA development and microarray technology. Studies with the GeoChip have shown its great potential in the study of microbial ecology and for linking microbial community function with ecosystem processes. The GeoChip has been shown to provide sensitive and specific information on the function, structure, and dynamics of microbial communities from a wide range of environments. However, new technologies and techniques are needed to increase array sensitivity and specificity in order to examine the true diversity of environmental communities and monitor low abundance species.

Acknowledgments

The effort for preparing this chapter was supported by ENIGMA, through the Office of Science, Office of Biological and Environmental Research, of the U. S. Department of Energy under Contract No. DE-AC02-05CH11231 the Virtual Institute for Microbial Stress and Survival (http://VIMSS.lbl.gov) supported by the U.S. Department of Energy, Office of Science, Office of Biological and Environmental Research, Genomics Program:GTL through contractDE-AC02-05CH11231 between Lawrence Berkeley National Laboratory and the U.S. Department of Energy, Environmental Remediation Science Program (ERSP), Office of Biological and Environmental Research, Office of Science, and Oklahoma Applied Research Support (OARS), Oklahoma Center for the Advancement of Science and Technology (OCAST), the State of Oklahoma through the Project AR062-034.

REFERENCES

Berthong ST, Schadt CW, Piñeiro G, Jackson RB. 2009. Afforestation alters the composition of functional genes in soil and biogeochemical processes in South American grasslands. *Appl. Environ. Microbiol.* **75**:6240–6248.

Bodrossy L, Stralis-Pavese N, Murrell JC, Radajewski S, Weilharter A, Sessitsch A. 2003. Development and validation of a diagnostic microbial microarray for methanotrophs. *Environ. Microbiol.* **5**:566–582.

Bodrossy L, Stralis-Pavese N, Konrad-Köszler M, Weilharter A, Reichenauer TG, et al. 2006. mRNA-based parallel detection of active methanotroph populations by use of a diagnostic microarray. *Appl. Environ. Microbiol.* **72**:1672–1676.

Bontemps C, Goldier G, Gris-Liebe C, Carere S, Talini L, Boivin-Masson C. 2005. Microarray-based detection and typing of the rhizobium nodulation gene nodC: Potential of DNA arrays to diagnose biological functions of interest. *Appl. Environ. Microbiol.* **71**:8042–8048.

Brodie EL, DeSantis TZ, Joyner DC, Baek SM, Larsen JT, et al. 2006. Application of a high-density oligonucleotide microarray approach to study bacterial population dynamics during uranium reduction and reoxidation. *Appl. Environ. Microbiol.* **72**:6288–6298.

Burgmann H, Widmer F, Sigler WV, Zeyer J. 2003. mRNA extraction and reverse transcription-PCR protocol for detection of nifH gene expression by *Azotobacter vinelandii* in soil. *Appl. Environ. Microbiol.* **69**:1928–1935.

Call DR, Bakko MK, Krug MJ, Roberts MC. 2003. Identifying antimicrobial resistance genes with DNA microarrays. *Antimicrob. Agents Chemother.* **47**:3290–3295.

Cleven BEE, Palka-Santini M, Gielen J, Meembor S, Krönke M, Krut O. 2006. Identification and characterization of bacterial pathogens causing bloodstream infections by DNA microarray. *J. Clin. Microbiol.* **44**:2389–2397.

Denef VJ, Park J, Rodrigues JLM, Tsoi TV, Hashsham SA, Tiedje JM. 2003. Validation of a more sensitive method for using spotted oligonucleotide DNA microarrays for functional genomics studies on bacterial communities. *Environ. Microbiol.* **5**:933–943.

DePaola A, Roberts MC. 1995. Class D and E tetracycline resistance determinants in gram-negative bacteria from catfish ponds. *Mol. Cell. Probes* **9**:311–313.

Fitter AH, Gilligan CA, Hollingworth K, Kleczkowski A, Twyman RM, et al. 2005. Biodiversity and ecosystem function in soil. *Funct. Ecol.* **19**:369–377.

Gao H, Yang ZK, Gentry TJ, Wu L, Schadt CW, Zhou J. 2007. Microarray-based analysis of microbial community RNAs by whole-community RNA amplification. *Appl. Environ. Microbiol.* **73**:563–571.

Gentry TJ, Wickham GS, Schadt CW, He Z, Zhou J. 2006. Microarray application in microbial ecology research. *Microbial Ecol.* **52**:159–175.

Guschin DY, Mobarry BK, Proudnikov D, Stahl DA, Rittmann BE, Mirzabekov AD. 1997. Oligonucleotide microchips as genosensors for determinative and environmental studies in microbiology. *Appl. Environ. Microbiol.* **63**:2397–2402.

He Z, Zhou J. 2008. Empirical evaluation of a new method for calculating signal to noise ratio (SNR) for microarray data analysis. *Appl. Environ. Microbiol.* **74**:2957–2966.

He Z, Wu LY, Li XY, Fields MW, Zhou JZ. 2005. Empirical establishment of oligonucleotide probe design criteria. *Appl. Environ. Microbiol.* **71**:3753–3760.

He Z, Gentry TJ, Schadt CW, Wu L, Liebich J, et al. 2007. GeoChip: A comprehensive microarray for investigating biogeochemical, ecological and environmental processes. *ISME J.* **1**:67–77.

He Z, Van Nostrand JD, Wu L, Zhou, J. 2008. Development and application of functional gene arrays for microbial community analysis. *Trans. Nonferrous Met. Soc. China* **18**:1319–1327.

He Z, Deng Y, Van Nostrand JD, Tu Q, Xu M, et al. 2010a. GeoChip 3.0 as a high throughput tool for analyzing microbial community structure, composition, and functional activity. *ISME J*. **4**:1167–1179.

He Z, Xu M, Deng Y, Kang S, Kellogg L, et al. 2010b. Metagenomic analysis reveals a marked divergence in the functional structure of belowground microbial communities at elevated CO_2. *Ecol. Lett*., **13**:564–575.

Hong SH, Bunge J, Jeon SO, Epstein SS. 2006. Predicting microbial species richness. *Proc. Natl. Acad. Sci. USA* **103**:117–122.

Hurt RA., Qiu X, Wu L, Roh Y, Palumbo AV, et al. 2001. Simultaneous recovery of RNA and DNA from soils and sediments. *Appl. Environ. Microbiol*. **67**:4495–4503.

Jenkins BD, Steward GF, Short SM, Ward BB, Zehr JP. 2004. Fingerprinting diazotroph communties in the Chesapeake Bay. *Appl. Environ. Microbiol*. **70**:1767–1776.

Kimes NE, Van Nostrand JD, Weil E, Zhou J, Morris PJ. 2010. The microbial functional structure of *Montastraea faveolata*, an important Caribbean reef-building coral, differs between healthy and yellow-band diseased (YBD) colonies. *Environ. Microbiol*. **12**:541–556.

Kostić T, Weilharter A, Sessitsch A, Bodrossy L. 2005. High-sensitivity, polymerase chain reaction-free detection of microorganisms and their functional genes using 70-mer oligonucleotide diagnostic microarray. *Anal. Biochem*. **346**:333–335.

Lee DY, Shannon K, Beaudette LA. 2006. Detection of bacterial pathogens in municipal wastewater using an oligonucleotide microarray and real-time quantitative PCR. *J. Microbiol. Methods* **65**:453–467.

Leigh MB, Pellizari VH, Uhlík O, Sutka R, Rodrigues J, et al. 2007. Biphenyl-utilizing bacteria and their functional genes in a pine root zone contaminated with polychlorinated biphenyls (PCBs). *ISME J*. **1**:134–148.

Li X, He Z, Zhou J. 2005. Selection of optimal oligonucleotide probes for microarrays using multiple criteria, global alignment and parameter estimation. *Nucleic Acids Res*. **33**:6114–6123.

Liang Y, Li G, Van Nostrand JD, He Z, Wu L, et al. 2009a. Microarray-based analysis of microbial functional diversity along an oil contamination gradient in oil field. *FEMS Microbiol*. **70**:168–177.

Liang Y, Wang J, Van Nostrand JD, Zhou J, Zhang X, Li G. 2009b. Microarray-based functional gene analysis of soil microbial communities in ozonation and biodegradation of crude oil. *Chemosphere* **75**:193–199.

Liang Y, He Z, Wu L, Deng Y, Li G, Zhou J. 2010. Development of a common oligo reference standard (CORS) for microarray data normalization and comparison across different microbial communities. *Appl. Environ. Microbiol*. **76**:1088–1094.

Liebich J, Wachtmeister T, Zhou J, Burauel P. 2009. Degradation of diffuse pesticide contaminants: Screening for microbial potential using a functional gene microarray. *Vadose Zone J*. **8**:703–710.

Loy A, Lehner A, Lee N, Adamczyk J, Meier H, et al. 2002. Oligonucleotide microarray for 16S rRNA gene-based detection of all recognized lineages of sulfate-reducing prokaryotes in the environment. *Appl. Environ. Microbiol*. **68**:5064–5081.

Lueders T, Friedrich MW. 2003. Evaluation of PCR amplification bias by terminal restriction fragment length polymorphism Analysis of small-subunit rRNA and *mcrA* genes by using defined template mixtures of methanogenic pure cultures and soil DNA extracts. *Appl. Envrion. Microbiol*. **69**:320–326.

Mason OU, DiMeo-Savoie CA, Van Nostrand JD, Zhou J, Fisk MR, Giovannoni SJ. 2009. Prokaryotic diversity, distribution, and preliminary insights into their role in biogeochemical cycling in marine basalts. *ISME J*. **3**:231–242.

Miller SM, Tourlousse DM, Stedtfeld RD, Baushke SW, Herzog AB, et al. 2008. In situ-synthesized virulence and marker gene biochip for detection of bacterial pathogens in water. *Appl. Environ. Microbiol*. **74**:2200–2209.

Miranda CD, Kehrenberg C, Ulep C, Schwarz S, Roberts MC. 2003. Diversity of tetracycline resistance genes in bacteria isolated from Chilean salmon farms. *Antimicrob. Agents Chemother*. **47**:883–888.

Murray AE, Lies D, Li G, Nealson K, Zhou J, Tiedje JM. 2001. DNA–DNA hybridization to microarrays reveals gene-specific differences between closely related microbial genomes. *Proc. Natl. Acad. Sci. USA* **98**:9853–9858.

Ning J, Liebich J, Kästner M, Zhou J, Schäffer A, Burauel P. 2009. Different influences of DNA purity indices and quantity on PCR-based DGGE and functional gene microarray in soil microbial community study. *Appl. Microbiol. Biotechnol*. **82**:983–993.

Økland RH, Eilertsen O. 1994. Canonical correspondence analysis with variation partitioning: Some comments and an application. *J. Veg. Sci*. **5**:117–126.

Palka-Santini M, Cleven BE, Eichinger L, Krönke M, Krut O. 2009. Large scale multiplex PCR improves pathogen detection by DNA microarrays. *BMC Microbiol*. **9**:1.

Parnell JJ, Rompato G, Latta IV LC, Pfrender ME, Van Nostrand JD, et al. 2010. Functional biogeography as evidence of gene transfer in hypersaline microbial communities. *mBio*. PLoS One. 5:e12919.

Poretsky RS, Gifford S, Rinta-Kanto J, Vila-Costa M, Moran MA. 2009. Analyzing gene expression from marine microbial communities using environmental transcriptomics. *J Visualized Exp.* **24**.http://www.jove.com/index/Details.stp?ID=1086, doi: 10.3791/1086.

Ramette A, Tiedje JM. 2007. Multiscale responses of microbial life in spatial distance and environmental heterogeneity in a patchy ecosystem. *Proc. Natl. Acad. Sci. USA* **104**:2761–2766.

Reeve J, Schadt CW, Carpenter-Boggs L, Kang S, Zhou J, Reganold JP. 2010. Effects of soil type and farm management on soil ecological functional genes and microbial activities. *ISME J*. **4**:1099–1107.

Rhee SK, Liu X, Wu L, Chong SC, Wan X, Zhou J. 2004. Detection of genes involved in biodegradation and biotransformation in microbial communities by using 50-mer oligonucleotide microarrays. *Appl. Environ. Microbiol*. **70**:4303–4317.

Rodríguez-Martínez EM, Pérez EX, Schadt CW, Zhou J, Massol-Deyá AA. 2006. Microbial diversity and bioremediation of a hydrocarbon-contaminated aquifer (Vega Baja, Puerto Rico). *Int. J. Environ. Res. Public Health* **3**:292–300.

Roesch LFW, Fulthorpe RR, Riva A, Casella G, Hadwin AKM, et al. 2007. Pyrosequencing enumerates and contrasts soil microbial diversity. *ISME J*. **1**:283–290.

Sarkar SF, Gordon JS, Martin GB, Guttman DS. 2006. Comparative genomics of host-specific virulence in *Pseudomonas syringae*. *Genetics* **147**:1041–1056.

Schena M, Shalon D, Davis RW, Brown PO. 1995. Quantitative monitoring of gene expression patterns with a complementary DNA microarray. *Science* **270**:467–470.

Schloss PD, Handelsman J. 2006. Toward a census of bacteria in soil. *PLOS Comput. Biol*. **2**:786–793.

Sebat JL, Colwell FS, Crawford RL. 2003. Metagenomic profiling: Microarray analysis of an environmental genomic library. *Appl. Environ. Microbiol*. **69**:4927–4934.

Small J, Call DR, Brockman FJ, Straub TM, Chandler DP. 2001. Direct detection of 16S rRNA in soil extracts by using oligonucleotide microarrays. *Appl. Environ. Microbiol*. **67**:4708–4716.

Steward GF, Jenkins BD, Ward BB, Zehr JP. 2004. Development and testing of a DNA macroarray to assess nitrogenase (nifH) gene diversity. *Appl. Environ. Microbiol.* **70**:1455–1465.

Stralis-Pavese N, Sessitsch A, Weilharter A, Reichenauer T, Riesing J, et al. 2004. Optimization of diagnostic microarray for application in analysing landfill methanotroph communities under different plant covers. *Environ. Microbiol.* **6**:347–363.

Suzuki MT, Giovannoni SJ. 1996. Bias caused by template annealing in the amplification of mixtures of 16S rRNA genes by PCR. *Appl. Envrion. Microbiol.* **62**:625–630.

Taroncher-Oldenburg G, Griner EM, Francis CA, Ward BB. 2003. Oligonucleotide microarray for the study of functional gene diversity in the nitrogen cycle in the environment. *Appl. Environ. Microbiol.* **69**:1159–1171.

Taş N, van Eekert MHA, Schraa G, Zhou J, de Vos WM, Smidt H. 2009. Tracking functional guilds: Dehalococcoides spp. in European river basins contaminated with hexachlorobenzene. *Appl. Environ. Microbiol.* **75**:4696–4704.

ter Braak CJF. 1986. Canonical correspondence analysis: A new eigenvector technique for multivariate direct gradient analysis. *Ecology* **67**:1167–1179.

Tiquia SM., Gurczynski S, Zholi A, Devol A. 2006. Diversity of biogeochemical cycling genes from Puget Sound sediments using DNA microarrays. *Environ. Technol.* **27**:1377–1389.

Tiquia SM, Wu L, Chong SC, Passovets S, Xu D, et al. 2004. Evaluation of 50-mer oligonucleotide arrays for detecting microbial populations in environmental samples. *Biotechniques* **36**:1–8.

Torsvik V, Goksoyr J, Daae FL. 1990. High diversity in DNA of soil bacteria. *Appl. Environ. Microbiol.* **56**:782–787.

Van Nostrand JD, Khijniak TV, Gentry TJ, Novak MT, Sowder AG, et al. 2007. Isolation and characterization of four gram-positive nickel-tolerant microorganisms from contaminated sediments. *Microb. Ecol.* **53**:670–682.

Van Nostrand JD, Wu WM, Wu L, Deng Y, Carley J, et al. 2009. GeoChip-based analysis of functional microbial communities during the reoxidation of a bioreduced uranium-contaminated aquifer. *Environ. Microbiol.* **11**:2611–26.

Van Nostrand JD, Wu L, Wu WM, Gentry TJ, Huang Z, et al. 2011. Dynamics of microbial community composition and function during the in situ bioremediation of a uranium-contaminated aquifer. *Appl Environ Microbiol.* In press.

Wagner M, Smidt H, Loy A, Hou Z. 2007. Unraveling microbial communities with DNA-microarrays: Challenges and future directions. *Microb. Ecol.* **53**:498–506.

Waldron PJ, Wu L, Van Nostrand JD, Schadt C, Watson D, et al. 2009. Functional gene array-based analysis of microbial community structure in groundwaters with a gradient of contaminant levels. *Environ. Sci. Technol.* **43**:3529–3534.

Wang A, Liu W, Cheng S, Logan BE, Yu H, et al. 2010. GeoChip-based functional gene analysis of anodophilic communities in microbial electrolysis cells under different operational modes. *Environ. Sci. Technol.* **44**:7729–7735.

Wang F, Zhou H, Meng J, Peng X, Jiang L, et al. 2009. GeoChip-based analysis of metabolic diversity of microbial communities at the Juan de Fuca Ridge hydrothermal vent. *Proc. Natl. Acad. Sci. USA* **106**:4840–4845.

Warnecke PM, Stirzaker C, Melki JR, Millar DS, Paul CL, Clark SJ. 1997. Detection and measurement of PCR bias in quantitative methylation analysis of bisulphite-treated DNA. *Nucleic Acids Res.* **25**:4422–4426.

Whitman WB, Coleman DC, Wiebe WJ. 1998. Prokaryotes: The unseen majority. *Proc. Natl. Acad. Sci. USA* **95**:6578–6583.

Wilson M, DeRisi J, Kristensen HH, Imboden P, Rane S, et al. 1999. Exploring drug-induced alterations in gene expression in Mycobacterium tuberculosis by microarray hybridization. *Proc. Natl. Acad. Sci. USA* **96**:12833–12838.

Wilson WJ, Strout CL, DeSantis TZ, Stilwell JL, Carrano AV, Andersen GL. 2002. Sequence-specific identification of 18 pathogenic microorganisms using microarray technology. *Mol. Cell. Probes* **16**:119–127.

Wu L, Thompson DK, Li G, Hurt RA, Tiedje JM, Zhou J. 2001. Development and evaluation of functional gene arrays for detection of selected genes in the environment. *Appl. Environ. Microbiol.* **67**:5780–5790.

Wu L, Thompson DK, Liu X, Fields MW, Bagwell CE, et al. 2004. Development and evaluation of microarray-based whole genome hybridization for detection of microorganisms within the context of environmental applications. *Environ. Sci. Technol.* **38**:6775–6782.

Wu L, Liu X, Schadt CW, Zhou, J. 2006. Microarray-based analysis of submicrogram quantities of microbial community DNAs by using whole-community genome amplification. *Appl. Environ. Microbiol.* **72**:4931–4941.

Wu L, Kellogg L, Devol AH, Tiedje JM, Zhou J. 2008a. Microarray-based characterization of microbial community functional structure and heterogeneity in marine sediments from the Gulf of Mexico. *Appl. Environ. Microbiol.* **74**:4516–4529.

Wu L, Liu X, Fields MW, Thompson DK, Bagwell CE, et al. 2008b. Microarray-based whole-genome hybridization as a tool for determining prokaryotic species relatedness. *ISME J.* **6**:642–655.

Xiong J, Wu L, Tu S, Van Nostrand JD, He Z, et al. 2010. Microbial communities and functional genes associated with soil arsenic contamination and rhizosphere of the arsenic hyper-accumulating plant *Pteris vittata* L. *Appl. Environ. Microbiol.* doi:10.1128/AEM.00500-10

Xu M, Wu W, Wu L, He Z, Van Nostrand JD, et al. 2010. Responses of microbial community functional structures to pilot-scale uranium in situ bioremediation. *ISME J.* **4**:1060–1070.

Yergeau E, Kang S, He Z, Zhou J, Kowalchuk GA. 2007. Functional microarray analysis of nitrogen and carbon cycling genes across an Antarctic latitude transect. *ISME J.* **1**:1–17.

Yin H, Cao L, Qiu G, Wang D, Kellogg L, et al. 2007. Development and evaluation of 50-mer oligonucleotide arrays for detecting microbial populations in acid mine drainages and bioleaching systems. *J. Microbiol. Methods* **70**:165–178.

Zhang L, Srinivasan U, Marrs CF, Ghosh D, Gilsdorf JR, Foxman B. 2004. Library on a slide for bacterial comparative genomics. *BMC Microbiol.* **4**:12–18.

Zhang L, Hurek T, Reihold-Hurek B. 2007a. A nifH-based oligonucleotide microarray for functional diagnostics of nitrogen-fixing microorganisms. *Microb. Ecol.* **53**:456–70.

Zhang Y, Zhang X, Liu X, Xiao Y, Qu L, et al. 2007b. Microarray-based analysis of changes in diversity of microbial genes involved in organic carbon decomposition following land use/cover changes. *FEMS Lett.* **266**:144–151.

Zhou J. 2003. Microarrays for bacterial detection and microbial community analysis. *Curr. Opin. Microbiol.* **6**:288–294.

Zhou J, Bruns MA, Tiedje JM. 1996. DNA recovery from soils of diverse composition. *Appl. Environ. Microbiol.* **62**:316–322.

Zhou J, Thompson DK. 2002. Challenges in applying microarrays to environmental studies. *Curr. Opin. Biotechnol.* **13**:204–207.

Zhou J, Kang S, Schadt CW, Garten CT Jr. 2008. Spatial scaling of functional gene diversity across various microbial taxa. *Proc. Natl. Acad. Sci. USA* **105**:7768–7773.

Chapter 58

Phylogenetic Microarrays (PhyloChips) For Analysis of Complex Microbial Communities

Eoin L. Brodie

58.1 INTRODUCTION

Microbial communities in general are complex and dynamic entities, with potentially thousands of interacting members active intermittently [Quince et al., 2008; Hibbing et al., 2010]. Given the critical role that microbial communities play in our health and the health of our planet, it is surprising how little we know of their complexity. Such knowledge will be critical if we are to accurately determine the primary factors influencing microbial community structure and function and how future perturbations, such as changing climate or host diet, may alter their composition and activities. This information has historically been limited by our relatively narrow focus on only those organisms cultivable under standard laboratory conditions, whereas the majority of diversity in most ecosystems remains uncultured [Hugenholtz et al., 1998]. The latter decades of the twentieth century ushered in the molecular revolution, where nucleic acid-based tools became the primary approach to study microbial diversity without the need for isolation. This revealed a hidden planet, teeming with seemingly unlimited diversity and created problems of its own. How do we survey such complexity? At this point it became clear that standard approaches based on cloning a biomarker gene, such as 16S rRNA, into a host and serially sampling those clones for sequence diversity were in many cases a futile effort, given the number of clone samples required to adequately represent the true diversity of a community [Curtis and Sloan, 2005; Gans et al., 2005; Quince et al., 2008].

Recently, a number of approaches have been developed to overcome some of the limitations of serial cloning-sequencing. Massively parallel sequencing using next-generation technologies such as pyrosequencing (Roche-454) or reversible terminator chemistry (Illumina-Solexa) are rapidly becoming important tools in the analysis of microbial community composition and function and are discussed further in see Chapters 18, 19 and 20 Vol. I. Even with the availability of such deep-sequencing technologies, DNA microarrays remain an attractive and cost-effective alternative approach to biomarker sampling by serial cloning-Sanger sequencing. Microarray hybridization of target sequences, such as 16S rRNA PCR amplicons, to an array of probes allows a far greater number of molecules to be assayed simultaneously compared to the thousands feasible with Sanger sequencing [DeSantis et al., 2007]. One distinct advantage of a microarray approach is that the detection of lower abundance organisms is less impacted by dominance within a microbial community compared to typical sequencing approaches as each hybridization assay is theoretically an independent test [DeSantis et al., 2007]. There are many formats of phylogenetic microarrays that vary substantially in format (e.g., spotted probes or in situ synthesized probes), probe density (tens to hundreds of thousands), and target range (species to domains). For excellent reviews of published microarray applications

Handbook of Molecular Microbial Ecology, Volume I: Metagenomics and Complementary Approaches, First Edition. Edited by Frans J. de Bruijn.

in microbial ecology see Bodrossy and Sessitsch [2004], Gentry et al. [2006], Andersen et al. [2010], and Chapters 58, 61, Vol. I in this volume.

For our research at Lawrence Berkeley National Laboratory, which includes soil microbial communities, bioaerosols, biofuels, bioremediation processes, plant pathogens, and the human microbiome, we designed a microarray (16S rRNA PhyloChip) that would permit analysis of the response of thousands of microbial (bacterial and archaeal) taxa across many samples in a high-throughput manner [Wilson et al., 2002a,b; Brodie et al., 2006; DeSantis et al., 2007]. In this chapter we will discuss briefly the design of such an array and highlight key applications in environmental and clinical microbial ecology.

58.2 METHODS

58.2.1 Development of a High-Density Phylogenetic Oligonucleotide Microarray

The full details of the PhyloChip G2 array design and validation are available elsewhere [Brodie et al., 2006, 2007; DeSantis et al., 2007]. In this chapter we briefly describe the platform we chose and primary steps in array design. The Affymetrix GeneChip platform employs short (25-mer) DNA oligonucleotides probes that hybridize to biotinylated DNA or RNA targets for gene or gene expression assays [Fodor et al., 1993; Lockhart et al., 1996]. This platform uses multiple probe pairs (termed a "probeset") to assay the response of a single gene. Key to this approach is the perfect-match (PM) mismatch (MM) probe-pair concept. Here, one probe of each pair matches a fragment of the gene exactly (PM) while the other contains a single mismatching nucleotide in the central (13th nucleotide) position (MM) with the difference in fluorescence signals (PM-MM) used to account for nonspecific hybridization. In theory, if the hybridization signal emanates from a specific target interaction, then the MM probe (compared with the PM probe) should have much reduced fluorescence signal due to the mismatching base. If a nonspecific hybridization events occurs, the PM probe will also have a mismatching base and thus the difference between the PM and MM probes will be less—a threshold is determined based on this difference relative to the noise. This probe-pair/probeset approach requires a high density of oligonucleotides such as that available on Affymetrix GeneChips (500,000 at the time of PhyloChip G2 array design and currently up to 6.5 million).

Design of the G2 PhyloChip followed these major steps: (1) Creation of a database of all available 16S rRNA gene sequences in GenBank. (2) Quality filtering of sequences based upon sequence length, quality (ambiguous bases), and chimeric properties using Bellerophon [Huber et al., 2004]. (3) Clustering of sequences into OTUs or taxa based on 17-mers in common to result in clusters with approximately 97% sequence identity, (4) Classification of sequences using Bergey's Taxonomic Outline [Garrity, 2001] and the Hugenholtz taxonomy [DeSantis et al., 2006]. (5) Selection of probe sets containing 11 or more 25-mer probe pairs to distinguish a given taxon from all others through the combined probe-set hybridization response. Here, each sequence in an OTU/taxon was split into all possible overlapping 25-mers and frequency of 25-mer occurrence across an alignment of all 16S rRNA gene sequences was compared. Then potential probes were filtered based on capacity to cross-hybridize to other taxa (if the central 17 bases in a probe match sequence, in more than one taxon, cross-hybridization is expected) and paired with a mismatch probe (MM) as described above, avoiding mismatch probes that contained a central 17-mer complementary to any taxon. (6) Synthesis by photolithography of an array of 712 rows and columns, each containing a 25-mer probe with a total of 506,944 oligonucleotides of which 297,851 were either PM or MM probes targeting 16S rRNA genes. The remaining probes are used for image orientation, normalization, or some selective non-16S target detection.

58.2.2 Sample Preparation

A typical PhyloChip experiment begins with the extraction of genomic DNA from a sample. There are many approaches for this step, and typically a single extraction procedure is carried out for all samples in a study to ensure consistency. Examples of procedures are given in Chapters 10 and 11 of Vol. II. An effective procedure based upon a combination of physical (bead beating) and chemical (surfactant, organic solvent) lysis used by our group has been described in detail elsewhere [Ivanov et al., 2009]. Once purified, 16S rRNA genes are amplified from the genomic DNA using broad-coverage primers. Our typical amplification protocol involves the use of nondegenerate primers: 27F (5′-AGAGTTTGATCCTGGCTCAG-3′) with 1492R (5′-GGTTACCTTGTTACGACTT-3′) for bacteria and 4Fa (5′ TCCGGTTGATCCTGCCRG 3′) with 1492R for archaea. A gradient of annealing temperatures (48–58°C) bracketing the estimated annealing temperature of the primers is used to allow mismatching templates to amplify. In total, 8–12 PCR reactions are combined for each sample, which also permits the random effects of PCR amplification to be diluted [Ivanov et al., 2009] and allows the number of PCR cycles to be reduced. Hong et al. [2009] have recently demonstrated that a combination of extraction methods and primer pairs may be desirable when estimating richness using cloning-sequencing approaches, and the same concepts should apply to microarray detection.

Following PCR amplicon concentration and purification, products are quantified using gel electrophoresis; aliquots (typically 100–2000 ng) are combined with spike-in controls, fragmented by DNase I digestion, biotin-labeled using terminal transferase, and hybridized to G2 PhyloChips at 48°C and 60 rpm for 16 h. After hybridization, washing and staining (streptavidin-phycoerythrin) occurs according to standard Affymetrix protocols.

58.2.3 PhyloChip Data Analysis

Each PhyloChip is scanned and recorded as a pixel image, and initial data acquisition and intensity determination is performed using standard Affymetrix software (GeneChip microarray analysis suite, version 5.1). Background subtraction, data normalization, and probe pair scoring are performed as reported previously [Brodie et al., 2006, 2007; DeSantis et al., 2007]. Briefly, for a probe pair to be scored positive, the PM intensity must be 1.3 times the MM intensity and the absolute difference between the two intensities must be 130 times the squared noise value. On the G2 PhyloChip, for a probe set (taxon) to be considered present, typically 90% or more of the probe pairs in a set must be considered positive. Intensities are summarized for each taxon/probe set using a trimmed average (highest and lowest values removed before averaging) of the intensities of the PM probes minus their corresponding MM probes.

PhyloChip intensity data are normalized to the internal spike mix using either a scaling procedure [Brodie et al., 2006; DeSantis et al., 2007] or a maximum-likelihood procedure [Ivanov et al., 2009], thus adjusting for variations in fragmentation, labeling, hybridization, washing, staining, and scanning. Once normalized for technical variation, the PhyloChip intensity data may be treated like expression array data or taxonomic community data, although including phylogenetic information, such as with the *Unifrac* distance metric [Lozupone and Knight, 2005], may provide greater resolution in distinguishing shifts in community composition. Some third-party software has recently been developed to allow independent analysis of PhyloChip data (http://www.phylotrac.org/) and provides an output compatible with a rapid version of Unifrac (FastUnifrac [Hamady et al., 2009]) allowing intersample phylogenetic distances to be calculated with ease. Other software for microarray and ecological analyses, such as the *siggenes* and *vegan* packages respectively within the R programming environment (http://www.r-project.org/), are particularly useful. Environmental variables can be related to qualitative (presence/absence) or quantitative (taxon intensities) PhyloChip data using R packages like *cca* (for canonical correlation analysis) or *mvpart* (for multivariate regression trees) [De'ath, 2002], or functions such as "envfit" or "adonis" within the *vegan* package [Oksanen et al., 2008] and the tree-wide relationships between microbial composition and treatment groups or other metadata may be visualized using online tools such as the Interactive Tree of Life [Letunic and Bork, 2007].

To date, the G2 PhyloChip has been used effectively for analysis of microbial communities in many distinct ecosystems including soil/sediment microbial communities [Wan et al., 2005; Brodie et al., 2006; Tokunaga et al., 2008; Cruz-Martinez et al., 2009; DeAngelis et al., 2009; Yergeau et al., 2009], heavy metal contaminated aquifers [Faybishenko et al., 2008], urban aerosols [Brodie et al., 2007], deep subsurface fracture water and biofilms [Lin et al., 2006; MacLean et al., 2007; Chivian et al., 2008], microbial fuel cells [Wrighton et al., 2008], spacecraft assembly clean rooms [La Duc et al., 2009], human respiratory tracts [Flanagan et al., 2007], mouse [Ivanov et al., 2009] and human intestines [Cox et al., 2010], birds eggs [Shawkey et al., 2009], coral [Sunagawa et al., 2009], and diseased plants [Sagaram et al., 2009]. In the following section we will highlight some of these prior applications.

58.2.4 Application of a High-Density Oligonucleotide Microarray Approach to Study Bacterial Population Dynamics during Uranium Reduction and Reoxidation

A product of cold war era weapons production, groundwater contamination by uranium is a significant problem within the Department of Energy (DOE) complex [McCullough et al., 2004]. A promising strategy for containment of uranium within the boundaries of such sites is through the process of reductive immobilization whereby soluble uranium U(VI) is reduced to the less-soluble uranium U(IV) [Finneran et al., 2002; Holmes et al., 2002; Senko et al., 2002; Anderson et al., 2003; Suzuki et al., 2003; Istok et al., 2004; Tokunaga et al., 2005; Wu et al., 2006]. This process may be accelerated through the addition of electron donor, typically organic carbon, which is oxidized by indigenous sediment or groundwater microorganisms which in turn reduce U(VI), decreasing its mobility. It had generally been expected that, under low redox conditions, reduced U(IV) would remain immobile; however, we previously demonstrated that reoxidation of U(IV) could occur under continuous reducing conditions [Wan et al., 2005]. One unresolved question was whether or not alterations in microbial community composition, such as decreased relative abundance of metal-reducing bacteria, was a factor in U(IV) reoxidation. To test this hypothesis, we analyzed bacterial community composition using 16S rRNA PhyloChips to follow changes during this remediation process. Full details of this work are described in Brodie et al. [2006].

Column experiments exhibiting an initial U(VI) reducing phase but switching to a U(IV) reoxidizing after approximately 100 days were destructively sampled and DNA was extracted from the sediment [Wan et al., 2005]. 16S rRNA genes were amplified from sediment sampled prior to stimulation ("Area 2"), during net U reduction ("Red"), and during net U reoxidation ("Ox"). Amplicons from the pre-stimulated sediment were analyzed by cloning and sequencing over 740 clones and by hybridization to PhyloChip microarrays. By comparing PhyloChip data with standard 16S rRNA gene clone libraries, we determined that most classifiable clone groups were indeed detected by the PhyloChip; however, the PhyloChip detected far greater diversity, probably due to undersampling of the community by cloning highlighted by nonasymptotic rarefaction curves and Chao1 richness estimates (**Fig. 58.1**). This demonstrated that the PhyloChip approach could provide a more complete view of microbial population dynamics and was observed for other types of environments including urban aerosols and groundwater [DeSantis et al., 2007]. During the transition from U(VI) reducing conditions to U(IV) oxidizing conditions, microbial biomass declined, whereas microbial activity (measured as dehydrogenase) increased substantially (**Fig. 58.2**). Using hierarchical clustering to compare the response of the 100 most dynamic bacterial subfamilies (defined as groups of ~94% sequence divergence) detected by the PhyloChip, we detected five primary response groups (**Fig. 58.3**). Response group 1 was stimulated during the reduction phase but decreased during the reoxidation phase and consisted primarily of known nitrate reducers within the alpha- and beta-proteobacteria, in addition to some Actinobacteria, Firmicutes, Deinococcus-Thermus, and Planctomycetes taxa. Response group 2 was potentially the most interesting and relevant to U mobility because it contained known metal-reducing bacteria, including *Geothrix fermentans, Geobacter*, and *Anaeromyxobacter* species. These taxa were stimulated during U(VI) reduction but retained their increased relative abundance during the period when U(IV) reoxidation was observed. Response group 3 comprised three taxa with greater relative abundances during U(IV) reoxidation; two of these were Acidobacteria and the other was from the Desulfovibrionaceae family. This was interesting because non-U-reducing sulfate reducers have been shown to

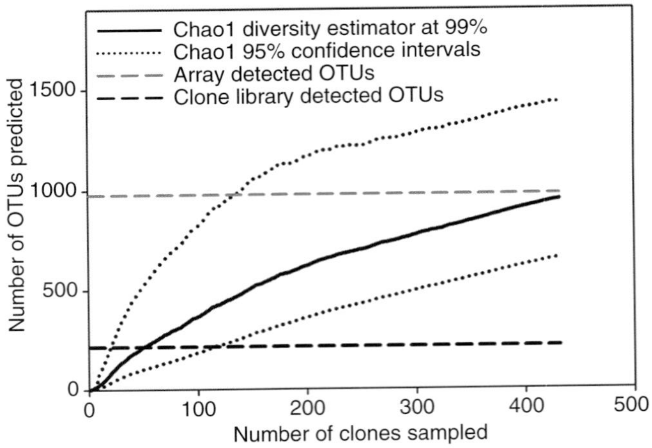

Figure 58.1 Chao1 richness estimates of Area 2 sediment bacterial communities based on clone library sampling. Richness estimates were made using a 99% sequence homology group classification equivalent to array-defined OTUs. The dashed gray line denotes the mean number of OTUs detected by array in Area 2 samples. The dashed black line denotes the number of OTUs detected by clone sequence analysis in Area 2 sediments. (Reproduced from Brodie et al. [2006]. Copyright © American Society for Microbiology.)

outcompete U-reducing *Geobacter* species in field applications [Anderson et al., 2003]; however, *Geobacter* species did not decline in relative abundance in our study. Therefore the reoxidation of U(IV) occurred under anaerobic conditions, despite a highly active community and a continued enrichment of metal-reducing bacteria. Work is continuing to determine whether this reoxidation phenomenon is observed in other uranium contaminated systems and if it is biologically mediated or primarily under geochemical control.

58.2.5 Urban Aerosols Harbor Diverse and Dynamic Bacterial Populations

Atmospheric aerosols can have significant implications for human health, agricultural productivity, and ecosystem stability yet until recently our knowledge of their microbial components was limited to those organisms that could be cultured on standard laboratory media [Lighthart, 1973; Lighthart and Frisch, 1976; Marthi and Lighthart, 1990; Lighthart and Shaffer, 1995; Shaffer and Lighthart, 1997; Kellogg and Griffin, 2006]. Despite the lower phylogenetic coverage of these approaches, these data had shown that the microbial composition of aerosols was highly dynamic and can be impacted by temperature and wind speed [Lindemann and Upper, 1985] and by global weather patterns such as El Niño events [Shinn et al., 2003]. Significantly, the microbial components arising from events such as African dust storms have been associated with human (meningitis)

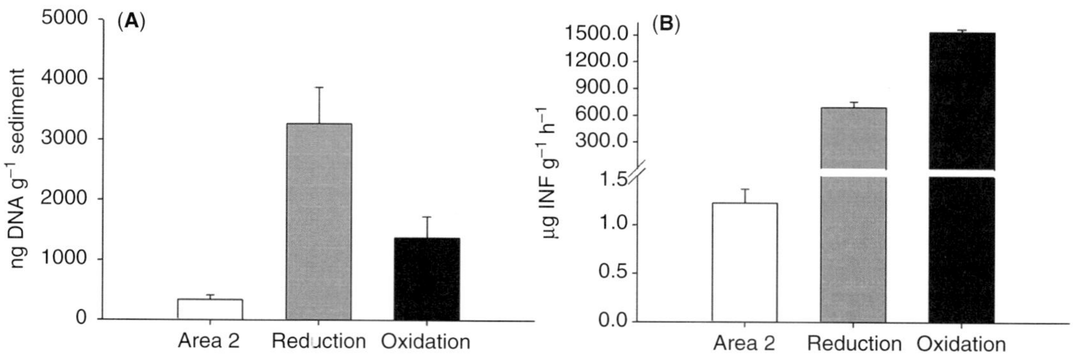

Figure 58.2 Plots show (**A**) DNA concentration (**B**) and microbial activity (dehydrogenase) in soil samples taken from the original Area 2 sediment prior to column packing and carbon stimulation and during the net U(VI) reduction and U(IV) reoxidation phases. Bars show means ± standard errors (n = 3). (Reproduced from Brodie et al. [2006]. Copyright © American Society for Microbiology.)

and animal (coral-bleaching) disease [Garrison et al., 2003]. In 2001, in the wake of Anthrax letter attacks, the Department of Homeland Security BioWatch program was created to monitor aerosols for future biological attacks. As part of routine monitoring, filters were obtained from U.S. Environmental Protection Agency air quality monitoring stations. In 2003, some of the deployed detectors in Houston, Texas gave positive results on consecutive days for the causative agent of tularemia ("rabbit fever"), *Francisella tularensis* (http://www.houstontx.gov/health/NewsReleases/bacteria%20detection.htm). *F. tularensis* has been considered a potential biological weapon due to its high infectivity (less than 10 organisms can cause disease) and potentially high mortality (30–60% if untreated) [Dennis et al., 2001]. In the case of the Houston positive detections, no intentional release of *F. tularensis* was suspected. Rather, naturally occurring strains of this bacterium present in water and soils of the region [Barns et al., 2005] were thought to be the cause. The extent of such natural diversity, the variability in biological composition of urban aerosols, and their ability to trigger biological weapons detectors was relatively undefined. Therefore, in order to evaluate the natural bacterial background in urban aerosols and how it might differ by location and climate, we began by surveying the bacterial composition of aerosols in two Texas cities, San Antonio and Austin, for 17 weeks from May 2003. Full details of this study are described by Brodie et al., [2007].

Using PhyloChip microarrays, we detected at least 1800 diverse bacterial taxa, a similar level of richness found in some soils and again detected greater diversity by PhyloChip than by standard cloning methods, much of which was confirmed by target specific PCR and sequencing (**Fig. 58.4**). Due to the turbulent and well-mixed nature of the atmosphere, in addition to the numerous sources contributing microbial components, dynamic microbial populations were expected in urban aerosols. We detected microbial signatures related to both freshwater and marine systems, plant surfaces, soil, and potentially sewage/wastewater treatment systems. Plant chloroplasts were also consistently detected, presumably from pollen. Bacterial families with pathogenic members including environmental relatives of select agents of bioterrorism significance such as *Bacillus anthracis* and *Francisella* spp. were also consistently detected. In order to relate the bacterial composition of aerosols to climatic conditions over the 17-week period, we used multivariate regression tree analysis [De'ath, 2002; David and Paul, 2004] and demonstrated that time and climate variables (temperature, wind speed, and particulate matter <2.5 µm) could explain more of the variance in aerosol bacterial composition than city location (**Fig. 58.5**). While our study evaluated differences between two cities over many weeks, a more recent study at a single site in Boulder, Colorado determined that bacterial composition may also vary significantly on a daily basis [Fierer et al., 2008]. Overall, we demonstrated that urban aerosols contain a diverse consortium of bacteria that likely represent the union of numerous and distinct point sources. Furthermore, the composition of such aerosols is highly variable and appears to be regulated by climatic conditions.

58.2.6 Induction of Intestinal Th17 Cells by Segmented Filamentous Bacteria

The mammalian gastrointestinal tract is colonized by hundreds to thousands of microbial taxa representing the gut microbiome [Ley et al., 2008; see also Chapters 18, 20 and 21, Vol. I]; most of these organisms are commensals and exist in mutualistic relationships with their host, assisting in the extraction of nutrients and energy from the host's diet [Ley et al., 2005; Turnbaugh et al., 2006]. In humans, the gut microbiome evolves from infancy through

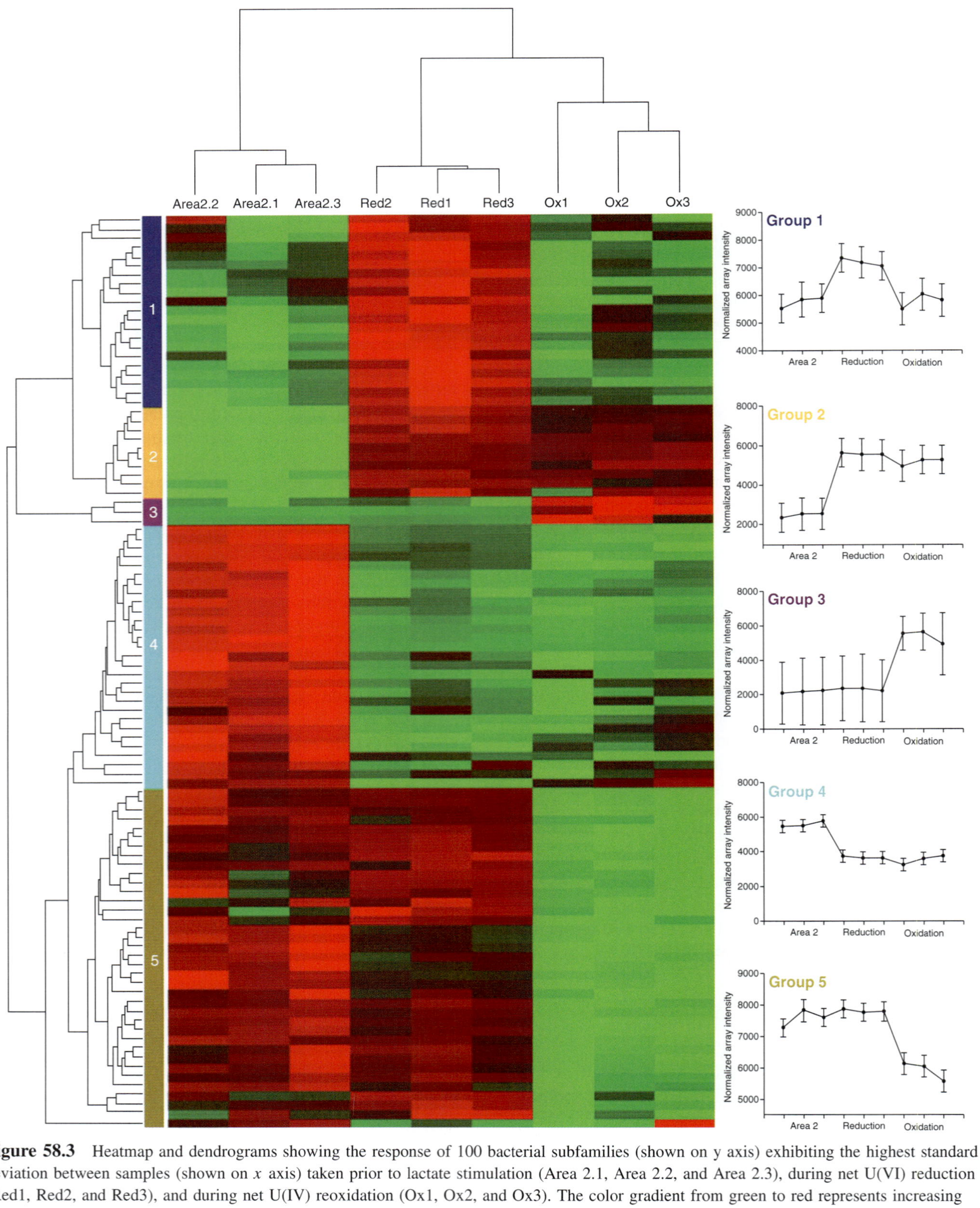

Figure 58.3 Heatmap and dendrograms showing the response of 100 bacterial subfamilies (shown on y axis) exhibiting the highest standard deviation between samples (shown on *x* axis) taken prior to lactate stimulation (Area 2.1, Area 2.2, and Area 2.3), during net U(VI) reduction (Red1, Red2, and Red3), and during net U(IV) reoxidation (Ox1, Ox2, and Ox3). The color gradient from green to red represents increasing array hybridization intensity. Five main response groups were detected, and the average intensity (HybScore) of the cluster group response is presented in line plots to the right of the heatmap. Error bars represent standard errors in HybScore differences between all subfamilies in a cluster. (Reproduced from Brodie et al. [2006]. Copyright © American Society for Microbiology.)

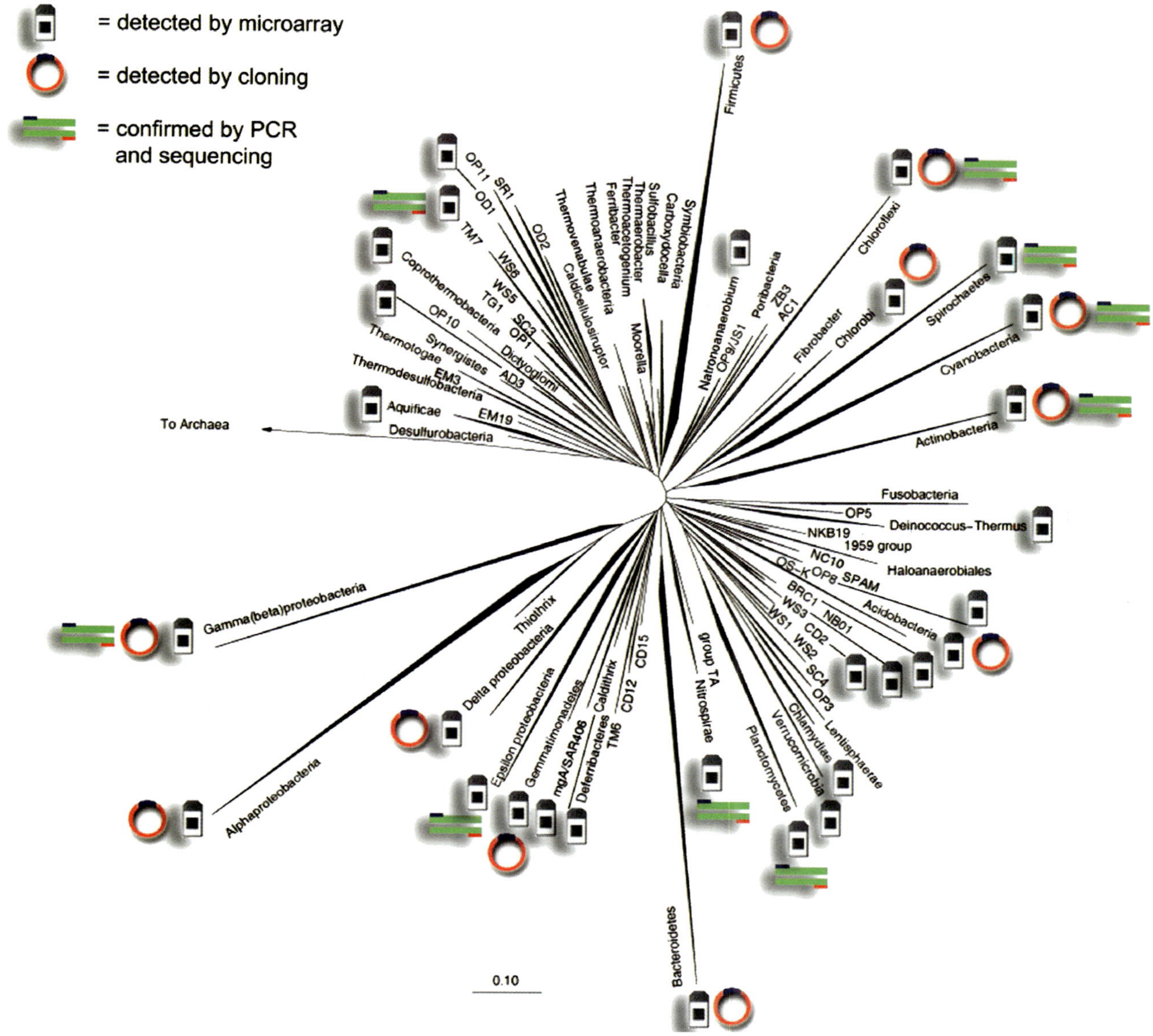

Figure 58.4 Representative phylogenetic tree showing all bacterial phyla (and individual classes in the case of proteobacteria) represented on the G2 PhyloChip. Branches are labeled to display 16S rRNA gene sequences detected in an urban aerosol by microarray or cloning or both. Phyla detected by microarray only that subsequently were confirmed by targeted PCR and sequencing are also marked. The Archaea are used as an outgroup. (Scale bar: 0.1 changes per nucleotide.). (Reproduced from Brodie et al. [2007]. Copyright © National Academy of Sciences, USA.)

adulthood [Kurokawa et al., 2007; Palmer et al., 2007; Cox et al., 2010] along with a concomitant development of the host immune system [Hattori and Taylor, 2009]. While intestinal microbiota as a whole are known to influence the development and balance of the host immune system [Rakoff-Nahoum et al., 2004], our understanding of how specific members of the commensal microbiota impact the immune system is less well understood.

Th17 cells are critical components of the immune system, providing protection against bacterial and fungal infections at mucosal surfaces [Aujla et al., 2007]; however, they are also key mediators of autoimmune disease due to their significant inflammatory potential [Bettelli et al., 2007]. In a previous study, Ivanov et al. [2008] identified that a specific component of the intestinal microbiota in B6 mice from a specific supplier (Taconic Farms) was capable of inducing Th17 cell differentiation in germ-free B6 mice and B6 mice from another supplier (Jackson Lab) which lacked Th17 cell production. To determine the exact components of the microbiome responsible for Th17 cell induction, we used PhyloChip microarrays followed by confirmatory

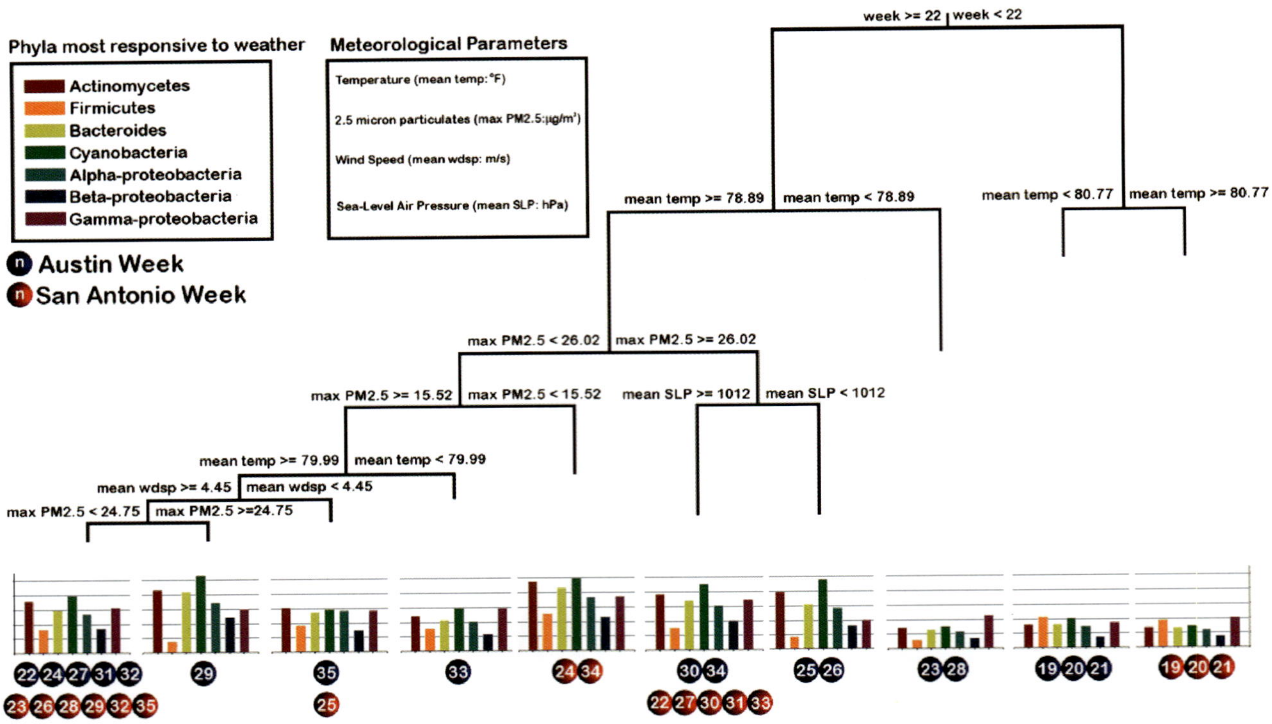

Figure 58.5 Multivariate regression tree analysis of the interaction between aerosol bacterial dynamics (PhyloChip fluorescence intensity) and environmental parameters. The model explains 89.1% of variance in bacterial relative abundance. Bars plotted under each cluster represent mean of normalized array intensities of phylogenetically-related bacteria shown to be significantly correlated with environmental/temporal parameters. (Reproduced from Brodie et al. [2007]. Copyright © National Academy of Sciences, USA.)

experiments with identified candidate microorganisms. This work is described in detail by Ivanov et al. [2009].

DNA was extracted from dissected small intensities of B6 mice from both Taconic and Jackson suppliers (four replicates each); 16S rRNA genes were amplified, fragmented, biotinylated, and hybridized to G2 PhyloChips. Bacterial taxa that were significantly different in relative abundance (based on normalized PhyloChip fluorescence intensity) were identified by t-test and fold changes in copy number estimated (**Fig. 58.6**). Due to the depth of coverage and phylogenetic breadth of the PhyloChip, we could assess differences in the small-intestine microbial communities at multiple phylogenetic levels. The sensitivity of the PhyloChip (detection limit ~0.01% of community) permitted detection of lower-abundance organisms in dominated populations [DeSantis et al., 2007] such as these gut communities. Overall, we detected 766 bacterial taxa in at least three out of four replicates from either strain of mice, and the relative abundance of 479 of these taxa was significantly different ($p \leq 0.05$) between the mice from the two suppliers; 107 and 372 taxa had greater relative abundance in Taconic and Jackson mice, respectively. When comparing the magnitude of the differences in relative abundance, most differences were subtle, and only 52 taxa were more than 5-fold different (17 greater in Taconic and 35 greater in Jackson), while only two taxa were >25-fold more abundant. These two taxa were members of the Lactobacillaceae and Clostridiaceae families and were of significantly greater relative abundance in the Taconic mice that had exhibited Th17 cell differentiation. They were *Lactobacillus murinus* ASF361 (~94-fold greater) and a segmented filamentous species of the candidate genus *Arthromitus* (~40-fold greater) (Fig. 58.6). Significantly, overrepresentation of close phylogenetic relatives was not observed.

Lactobacillus murinus ASF361 is a component of the Altered Schaedler's Flora (ASF) inoculum which is used to colonize Taconic B6 mice but not Jackson B6 mice. Previously, Ivanov et al. [2008] investigated whether this inoculum (containing *L. murinus*) could induce Th17 cell production: It did not, and therefore we focused on the segmented filamentous bacterium (SFB—*Candidatus Arthromitus*) as a likely inducer of Th17 cell differentiation. Q-PCR and electron microscopy confirmed the relative difference in SFB abundance between Taconic and Jackson mice and also between Jackson mice controls and Jackson mice co-housed with Taconic mice (**Fig. 58.7**). This demonstrated that SFB could be horizontally transferred and that Th17 cells were induced in the Jackson B6 mice within 10 days.

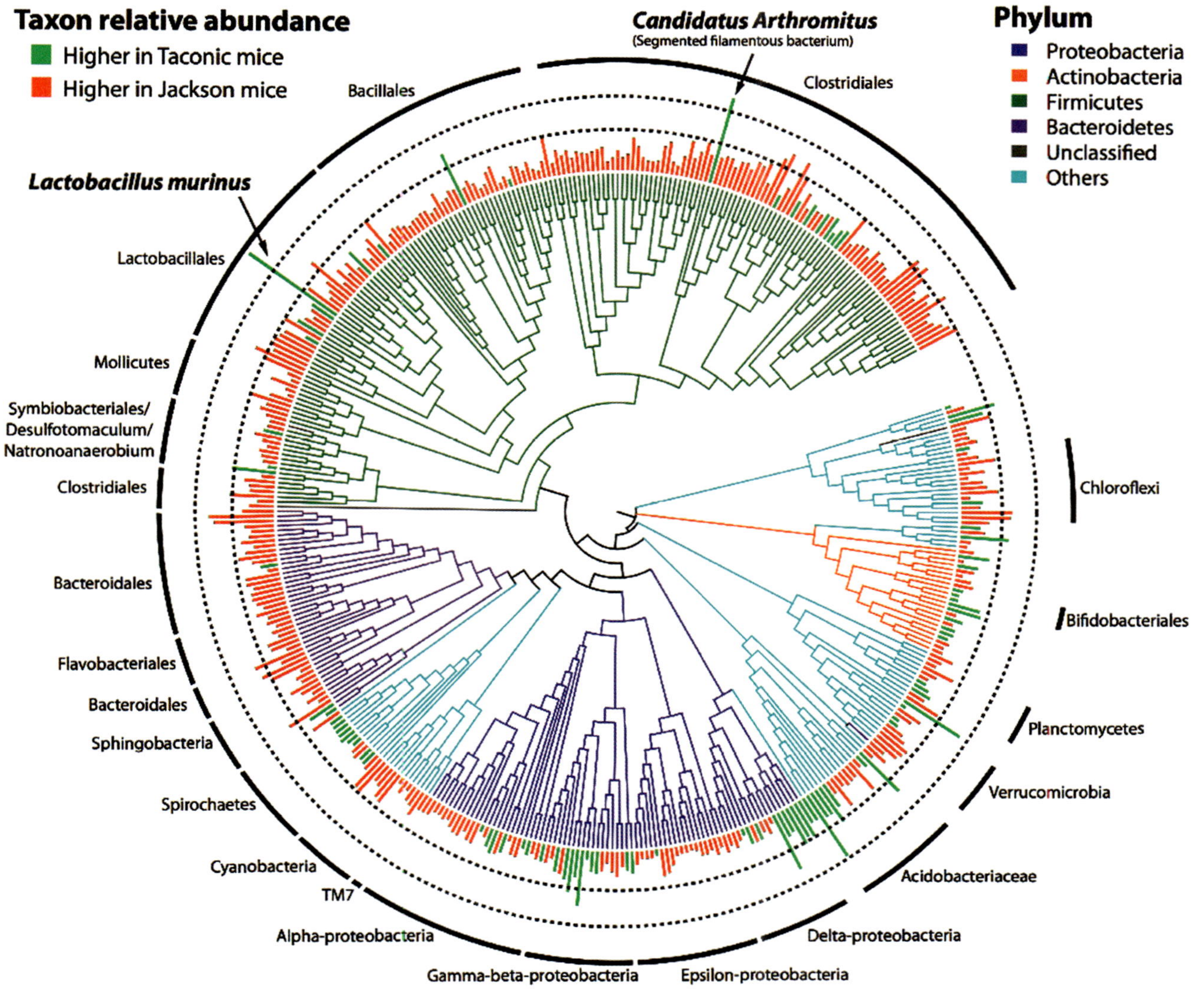

Figure 58.6 Phylogenetic tree based on 16S rRNA gene sequences of bacterial taxa detected in the terminal ileum showing significantly different relative abundances (PhyloChip fluorescence intensity) between the suppliers, Taconic and Jackson. Branches of the tree are color-coded according to phylum, while green and red bars display taxa with significantly greater relative abundance in Taconic and Jackson mice, respectively. The inner and outer dotted rings represent intensities corresponding to 5-fold and 25-fold differences in 16S copy number. The two taxa with the greatest difference between Taconic and Jackson mice, *Lactobacillus murinus* (~94-fold difference) and Candidatus *Arthromitus* (~40-fold difference) are noted by arrows. (Reproduced and modified from Ivanov et al. [2009]. Copyright © with permission from Elsevier Inc.)

Although a strong relationship with SFB and Th17 cell differentiation was established, to confirm their role, inoculation of germ-free mice with a monoculture of SFBs was required. However, SFBs have yet to be cultured; therefore, to obtain pure SFB material, fecal material from mice mono-colonized with SFB was used as the inoculum [Umesaki et al., 1995]. Flow cytometry demonstrated that previously germ-free mice and Jackson B6 mice inoculated with SFBs accumulated Th17 cells, thereby confirming SFBs role in inducing Th17 cell differentiation. Further work determined that SFB colonization resulted in elevated expression and production of serum amyloid A (SAA), a protein typically induced during infection, tissue damage, or inflammatory disease. In turn, SAA was shown to directly induce Th17 cell differentiation. The protective effects of SFB were also highlighted by the finding that inoculation with SFB resulted in reduced growth of the mouse intestinal pathogen *Citrobacter rodentium*, reduced infiltration of this pathogen into the colonic wall, and decreased colonic inflammation. This stimulation of intestinal T cells by SFB was also reported by Gaboriau-Routhiau et al. [2009].

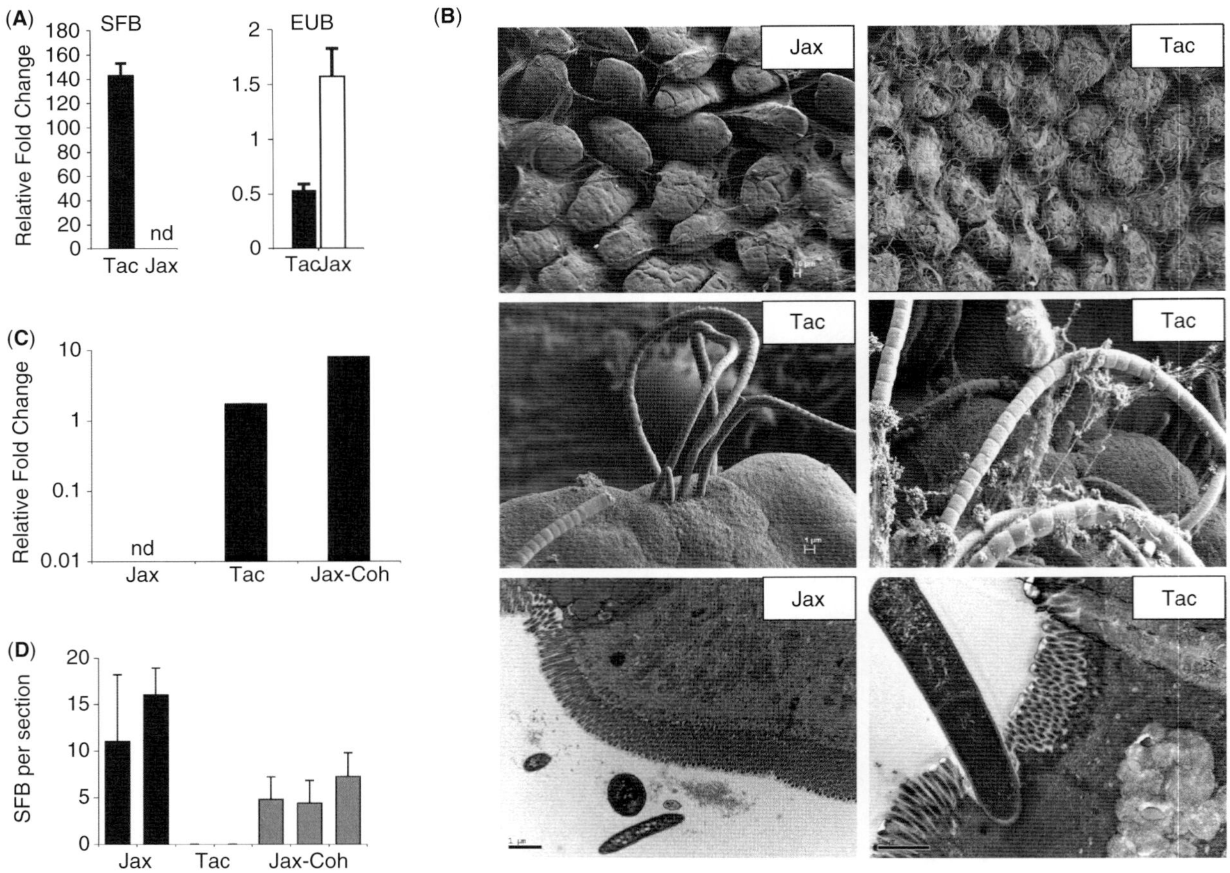

Figure 58.7 Segmented filamentous bacteria in the intestinal tract of Th17 cell-sufficient and Th17 cell-deficient mice. (**A**) Quantitative PCR (qPCR) analysis of segmented filamentous bacteria (SFB) and total bacterial (EUB) 16S rRNA genes in mouse feces from Taconic (Tac) and Jackson (Jax) B6 mice. Error bars represent the standard deviation (SD) across 4 replicate mice. (**B**) Scanning (SEM) and transmission (TEM) electron microscopy of terminal ileum of 8-week-old Jackson (Jax) and Taconic (Tac) C57BL/6 mice. (**C**) Credit qPCR analysis for SFB presence in Jackson B6 mice after 14 days of co-housing with Taconic B6 mice (Jax-Coh). Error bars represent the SD across 3–4 replicate mice. (**D**) SFB colonization of terminal ileum of Jackson B6 mice after 14 days of co-housing with Taconic B6 mice (Jax-Coh). Each bar represents a separate mouse, and error bars represent the SD across 4–5 imaged sections. (Reproduced from Ivanov et al. [2009]. Copyright © with permission from Elsevier Inc.)

While SFB are the first confirmed component of the commensal microbiota that induces a particular helper T-cell population in the lamina propria, further work in this area is necessary to determine whether other organisms found to be of differential abundance are also associated with immune system modulation.

58.2.7 The Path Forward

Phylogenetic microarrays have been successfully used to provide highly sensitive and rapid analyses of complex microbial communities at low cost relative to standard Sanger-sequence-based approaches. However, the use of microarrays in biological science as a whole is at an interesting juncture; speculation as to whether next-generation sequencing approaches will replace microarrays as quantitative instruments is common, as are suggestions that microarrays may evolve into primarily preparative tools (e.g., sequence capture for high-throughout sequencing). Currently, microarrays based on photolithography, or those using digital masks, are an attractive approach for microbial profiling due to their combined properties of low cost, high density, high sensitivity, standardized work flow, and low technical variability. Future developments in high-density arrays with significantly more features or those on flexible platforms such as NimbleGen should expand the utility of microarray applications in microbial ecology.

Acknowledgments

Work described in this chapter that was performed in the author's laboratory was supported by the Department of Energy's Environmental Remediation Sciences Program, the Department of Homeland Security, and the National Institutes of Health. This work was performed under the auspices of the U.S. Department of Energy by the University of California, Lawrence Berkeley National Laboratory, under Contract DE-AC02-05CH11231.

REFERENCES

Andersen GL, He Z, DeSantis TZ, Brodie EL, Zhou J. 2010. The use of microarrays in microbial ecology. In Liu W-T, Jansson JK, eds. *Environmental Molecular Microbiology*. Norwich, UK: Horizon Scientific Press, pp. 87–109.

Anderson RT, Vrionis HA, Ortiz-Bernad I, Resch CT, Long PE et al. 2003. Stimulating the in situ activity of *Geobacter* species to remove uranium from the groundwater of a uranium-contaminated aquifer. *Appl. Environ. Microbiol*. **69**:5884–5891

Aujla SJ, Dubin PJ, Kolls JK. 2007. Th17 cells and mucosal host defense. *Semin. Immunol*. **19**:377–382.

Barns SM, Grow CC, Okinaka RT, Keim P, Kuske CR. 2005. Detection of diverse new *Francisella*-like bacteria in environmental samples. *Appl. Environ. Microbiol*. **71**:5494–5500.

Bettelli E, Oukka M, Kuchroo VK. 2007. Th-17 cells in the circle of immunity and autoimmunity. *Nat. Immunol*. **8**:345–350.

Bodrossy L, Sessitsch A. 2004. Oligonucleotide microarrays in microbial diagnostics. *Curr. Opin. Microbiol*. **7**:245–254.

Brodie EL, Desantis TZ, Joyner DC, Baek SM, Larsen JT et al. 2006. Application of a high-density oligonucleotide microarray approach to study bacterial population dynamics during uranium reduction and reoxidation. *Appl. Environ. Microbiol*. **72**:6288–6298.

Brodie EL, DeSantis TZ, Parker JP, Zubietta IX, Piceno YM, Andersen GL. 2007. Urban aerosols harbor diverse and dynamic bacterial populations. *Proc. Natl. Acad. Sci. USA* **104**:299–304.

Chivian D, Brodie EL, Alm EJ, Culley DE, Dehal PS, et al. 2008. Environmental genomics reveals a single-species ecosystem deep within earth. *Science* **322**:275–278.

Cox MJ, Huang YJ, Fujimura KE, Liu JT, McKean M, et al. 2010. *Lactobacillus casei* abundance is associated with profound shifts in the infant gut microbiome. *PloS ONE* **5**:e8745.

Cruz-Martinez K, Suttle KB, Brodie EL, Power ME, Andersen GL, Banfield JF. 2009. Despite strong seasonal responses, soil microbial consortia are more resilient to long-term changes in rainfall than overlying grassland. *ISME J*. **3**:738–744.

Curtis TP, Sloan WT. 2005. Exploring microbial diversity—A vast below. *Science* **309**:1331–1333.

David RL, Paul LS. 2004. Multivariate regression trees for analysis of abundance data. *Biometrics* **60**:543–549.

De'ath G. 2002. Multivariate regression trees: A new technique for modeling species environment relationships. *Ecology* **83**:1105–1117.

DeAngelis KM, Brodie EL, DeSantis TZ, Andersen GL, Lindow SE, Firestone MK. 2009. Selective progressive response of soil microbial community to wild oat roots. *ISME J*. **3**:168–178.

Dennis DT, Inglesby TV, Henderson DA, Bartlett JG, Ascher MS et al. 2001. Tularemia as a biological weapon: Medical and public health management. *J. Am. Med. Assoc*. **285**:2763–2773.

DeSantis TZ, Hugenholtz P, Larsen N, Rojas M, Brodie EL et al. 2006. Greengenes, a chimera-checked 16S rRNA gene database and workbench compatible with ARB. *Appl. Environ. Microbiol*. **72**:5069–5072.

DeSantis TZ, Brodie EL, Moberg JP, Zubieta IX, Piceno YM, Andersen GL. 2007. High-density universal 16S rRNA microarray analysis reveals broader diversity than typical clone library when sampling the environment. *Microb. Ecol*. **53**:371–383.

Faybishenko B, Hazen TC, Long PE, Brodie EL, Conrad ME, et al. 2008. In situ long-term reductive bioimmobilization of Cr(vi) in groundwater using hydrogen release compound. *Environ. Sci. Technol*. **42**:8478–8485.

Fierer N, Liu Z, Rodriguez-Hernandez M, Knight R, Henn M, Hernandez MT. 2008. Short-term temporal variability in airborne bacterial and fungal populations. *Appl. Environ. Microbiol*. **74**:200–207.

Finneran KT, Housewright ME, Lovley DR. 2002. Multiple influences of nitrate on uranium solubility during bioremediation of uranium-contaminated subsurface sediments. *Environ. Microbiol*. **4**:510–516.

Flanagan JL, Brodie EL, Weng L, Lynch SV, Garcia O et al. 2007. Loss of bacterial diversity during antibiotic treatment of intubated patients colonized with *Pseudomonas aeruginosa*. *J. Clin. Microbiol*. **45**:1954–1962.

Fodor SPA, Rava RP, Huang XC, Pease AC, Holmes CP, Adams CL. 1993. Multiplexed biochemical assays with biological chips. *Nature* **364**:555–556.

Gaboriau-Routhiau V, Rakotobe S, LÈcuyer E, Mulder I, Lan A, et al. 2009. The key role of segmented filamentous bacteria in the coordinated maturation of gut helper t cell responses. *Immunity* **31**:677–689.

Gans J, Wolinsky M, Dunbar J. 2005. Computational improvements reveal great bacterial diversity and high metal toxicity in soil. *Science* **309**:1387–1390.

Garrison VH, Shinn EA, Foreman WT, Griffin DW, Holmes CW, et al. 2003. African and asian dust: From desert soils to coral reefs. *BioScience* **53**:469–480.

Garrity GM. 2001. *Bergey's Manual of Systematic Bacteriology*. New York: Springer-Verlag.

Gentry T, Wickham G, Schadt C, He Z, Zhou J. 2006. Microarray applications in microbial ecology research. *Microb. Ecol*. **52**:159–175.

Hamady M, Lozupone C, Knight R. 2009. Fast Unifrac: Facilitating high-throughput phylogenetic analyses of microbial communities including analysis of pyrosequencing and PhyloChip data. *ISME J*. **4**:17–27.

Hattori M, Taylor TD. 2009. The human intestinal microbiome: A new frontier of human biology. *DNA Res*. **16**:1–12.

Hibbing ME, Fuqua C, Parsek MR, Peterson SB. 2010. Bacterial competition: Surviving and thriving in the microbial jungle. *Nat. Rev. Microbiol*. **8**:15–25.

Holmes DE, Finneran KT, O'Neil RA, Lovley DR. 2002. Enrichment of members of the family Geobacteraceae associated with stimulation of dissimilatory metal reduction in uranium-contaminated aquifer sediments. *Appl. Environ. Microbiol*. **68**:2300–2306.

Hong S, Bunge J, Leslin C, Jeon S, Epstein SS. 2009. Polymerase chain reaction primers miss half of rRNA microbial diversity. *ISME J*. **3**:1365–1373.

Huber T, Faulkner G, Hugenholtz P. 2004. Bellerophon: A program to detect chimeric sequences in multiple sequence alignments. *Bioinformatics* **20**:2317–2319.

Hugenholtz P, Goebel BM, Pace NR. 1998. Impact of culture-independent studies on the emerging phylogenetic view of bacterial diversity. *J. Bacteriol*. **180**:4765–4774.

Istok JD, Senko JM, Krumholz LR, Watson D, Bogle MA, et al. 2004. In situ bioreduction of technetium and uranium in a nitrate-contaminated aquifer. *Environ. Sci. Technol*. **38**:468–475.

Ivanov II, Frutos RdL, Manel N, Yoshinaga K, Rifkin DB et al. 2008. Specific microbiota direct the differentiation of Il-17-producing T-helper cells in the mucosa of the small intestine. *Cell Host Microbe* **4**:337–349.

Ivanov, II, Atarashi K, Manel N, Brodie EL, Shima T, et al. 2009. Induction of intestinal Th17 cells by segmented filamentous bacteria. *Cell* **139**:485–498.

Kellogg CA, Griffin DW. 2006. Aerobiology and the global transport of desert dust. *Trends Ecol. Evol*. **21**:638–644.

Kurokawa K, Itoh T, Kuwahara T, Oshima K, Toh H, et al. 2007. Comparative metagenomics revealed commonly enriched gene sets in human gut microbiomes. *DNA Res*. **14**:169–181.

La Duc MT, Osman S, Vaishampayan P, Piceno Y, Andersen G, Spry JA, Venkateswaran K. 2009. Comprehensive census of bacteria in clean rooms by using DNA microarray and cloning methods. *Appl. Environ. Microbiol*. **75**:6559–6567.

Letunic I, Bork P. 2007. Interactive tree of life (itol): An online tool for phylogenetic tree display and annotation. *Bioinformatics* **23**:127–128.

Ley RE, Backhed F, Turnbaugh P, Lozupone CA, Knight RD, Gordon JI. 2005. Obesity alters gut microbial ecology. *Proc. Natl. Acad. Sci. USA* **102**:11070–11075.

Ley RE, Hamady M, Lozupone C, Turnbaugh PJ, Ramey RR, et al. 2008. Evolution of mammals and their gut microbes. *Science* **320**:1647–1651.

Lighthart B. 1973. Survival of airborne bacteria in a high urban concentration of carbon monoxide. *Appl. Microbiol*. **25**:86–91.

Lighthart B, Frisch AS. 1976. Estimation of viable airborne microbes downwind from a point source. *Appl. Environ. Microbiol*. **31**:700–704.

Lighthart B, Shaffer BT. 1995. Airborne bacteria in the atmospheric surface layer: Temporal distribution above a grass seed field. *Appl. Environ. Microbiol*. **61**:1492–1496.

Lin LH, Wang PL, Rumble D, Lippmann-Pipke J, Boice E, et al. 2006. Long-term sustainability of a high-energy, low-diversity crustal biome. *Science* **314**:479–482.

Lindemann J, Upper CD. 1985. Aerial dispersal of epiphytic bacteria over bean plants. *Appl. Environ. Microbiol*. **50**:1229–1232.

Lockhart DJ, Dong H, Byrne MC, Follettie MT, Gallo MV, et al. 1996. Expression monitoring by hybridization to high-density oligonucleotide arrays. *Nat. Biotech*. **14**:1675–1680.

Lozupone C, Knight R. 2005. Unifrac: A new phylogenetic method for comparing microbial communities. *Appl. Environ. Microbiol*. **71**:8228–8235.

MacLean LCW, Pray TJ, Onstott TC, Brodie EL, Hazen TC, Southam G. 2007. Mineralogical, chemical and biological characterization of an anaerobic biofilm collected from a borehole in a deep gold mine in South Africa. *Geomicrobiol. J*. **24**:491–504.

Marthi B, Lighthart B. 1990. Effects of betaine on enumeration of airborne bacteria. *Appl. Environ. Microbiol*. **56**:1286–1289.

McCullough J, Hazen TC, Benson SM, Metting FB, Palmisano AC. 2004. *Bioremediation of Metals and Radionuclides: What It Is and How It Works*. Berkeley, CA: Lawrence Berkeley National Laboratory.

Oksanen J, Kindt R, Legendre P, O'Hara B, Simpson GL, et al. 2008. Vegan: Community ecology package. R package version 1.13–2. http://vegan.r-forge.r-project.org.

Palmer C, Bik EM, DiGiulio DB, Relman DA, Brown PO. 2007. Development of the human infant intestinal microbiota. *PLoS Biol*. **5**:e177.

Quince C, Curtis TP, Sloan WT. 2008. The rational exploration of microbial diversity. *ISME J*. **2**:997–1006.

Rakoff-Nahoum S, Paglino J, Eslami-Varzaneh F, Edberg S, Medzhitov R. 2004. Recognition of commensal microflora by toll-like receptors is required for intestinal homeostasis. *Cell* **118**:229–241.

Sagaram US, DeAngelis KM, Trivedi P, Andersen GL, Lu SE, Wang N. 2009. Bacterial diversity analysis of huanglongbing pathogen-infected citrus, using PhyloChip arrays and 16S rRNA gene clone library sequencing. *Appl. Environ. Microb*. **75**: 1566–1574.

Senko JM, Istok JD, Suflita JM, Krumholz LR. 2002. In-situ evidence for uranium immobilization and remobilization. *Environ. Sci. Technol*. **36**:1491–1496.

Shaffer BT, Lighthart B. 1997. Survey of culturable airborne bacteria at four diverse locations in Oregon: Urban, rural, forest, and coastal. *Microb. Ecol*. **34**:167–177.

Shawkey MD, Firestone MK, Brodie EL, Beissinger SR. 2009. Avian incubation inhibits growth and diversification of bacterial assemblages on eggs. *PloS ONE* **4**:e4522.

Shinn EA, Griffin DW, Seba DB. 2003. Atmospheric transport of mold spores in clouds of desert dust. *Arch. Environ. Health* **58**:498–504.

Sunagawa S, DeSantis TZ, Piceno YM, Brodie EL, DeSalvo MK, et al. 2009. Bacterial diversity and white plague disease-associated community changes in the Caribbean coral *Montastraea faveolata*. *ISME J*. **3**:512–521.

Suzuki Y, Kelly SD, Kemner KM, Banfield JF. 2003. Microbial populations stimulated for hexavalent uranium reduction in uranium mine sediment. *Appl. Environ. Microbiol*. **69**:1337–1346.

Tokunaga TK, Wan J, Pena J, Brodie EL, Firestone MK, et al. 2005. Uranium reduction in sediments under diffusion-limited transport of organic carbon. *Environ. Sci. Technol*. **39**:7077–7083.

Tokunaga TK, Wan JM, Kim YM, Daly RA, Brodie EL, et al. 2008. Influences of organic carbon supply rate on uranium bioreduction in initially oxidizing, contaminated sediment. *Environ. Sci. Technol*. **42**:8901–8907.

Turnbaugh PJ, Ley RE, Mahowald MA, Magrini V, Mardis ER, Gordon JI. 2006. An obesity-associated gut microbiome with increased capacity for energy harvest. *Nature* **444**: 1027–1031.

Umesaki Y, Okada Y, Matsumoto S, Imaoka A, Setoyama H. 1995. Segmented filamentous bacteria are indigenous intestinal bacteria that activate intraepithelial lymphocytes and induce MHC class II molecules and fucosyl asialo GM1 glycolipids on the small intestinal epithelial cells in the ex-germ-free mouse. *Microbiol. Immunol*. **39**:555–562.

Wan JM, Tokunaga TK, Brodie E, Wang ZM, Zheng ZP, et al. 2005. Reoxidation of bioreduced uranium under reducing conditions. *Environ. Sci. Technol*. **39**:6162–6169.

Wilson KH, Wilson WJ, Radosevich JL, DeSantis TZ, Viswanathan VS, Kuczmarski TA, Andersen GL. 2002a. High-density microarray of small-subunit ribosomal DNA probes. *Appl. Environ. Microb*. **68**:2535–2541.

Wilson WJ, Strout CL, DeSantis TZ, Stilwell JL, Carrano AV, Andersen GL. 2002b. Sequence-specific identification of 18 pathogenic microorganisms using microarray technology. *Mol. Cell. Probe* **16**:119–127.

Wrighton KC, Agbo P, Warnecke F, Weber KA, Brodie EL, et al. 2008. A novel ecological role of the Firmicutes identified in thermophilic microbial fuel cells. *ISME J*. **2**:1146–1156.

Wu WM, Carley J, Gentry T, Ginder-Vogel MA, Fienen M, et al. 2006. Pilot-scale in situ bioremedation of uranium in a highly contaminated aquifer. 2. Reduction of U(VI) and geochemical control of U(VI) bioavailability. *Environ. Sci. Technol*. **40**:3986–3995.

Yergeau E, Schoondermark-Stolk SA, Brodie EL, Dejean S, DeSantis TZ, et al. 2009. Environmental microarray analyses of Antarctic soil microbial communities. *ISME J*. **3**:340–351.

Chapter 59

Phenomics and Phenotype Microarrays: Applications Complementing Metagenomics

BARRY R. BOCHNER

59.1 "WHO IS OUT THERE?" AND "WHAT ARE THEY DOING?"

Interest in metagenomics among microbial ecologists grew from a desire to know "who is out there." There was a general recognition that surveying microbes by existing culture methods was recovering probably less than 1% of the viable population [Rappe and Giovannoni, 2003], and there was justifiable unease that this was resulting in a severe bias. Metagenomics was seen as a potential technology to give an unbiased and complete census of complex microbial populations. This potential has started to become a reality as the increasing speed and decreasing cost of DNA sequencing along with advances in bioinformatics and systems biology modeling continue to improve the efficiency of microbial genome sequencing and annotation [Wu et al., 2009; see also Chapters 18–20, Vol. 1].

Recent years have produced an explosive growth in our knowledge of genes and genomes. But now, what is lagging severely is our knowledge of the biology and physiology that corresponds to this enormous and rapidly expanding genetic inventory. In this review I will present data and my opinions and suggestions about why and how phenomics has a very important role to play in advancing microbial ecology studies and in improving metagenomic analysis. I think everyone would agree that metagenomics is doing a very good job of answering the first question, "Who are out there?", but is doing a much less complete and satisfactory job of answering the second question, "What are they doing?".

59.2 METAGENOMICS NEEDS PHENOMICS

The field of metagenomics faces the same challenges and problems as the field of genome annotation: (1) there are a lot of genes of unknown function, (2) there are a lot of genes where the extrapolated function is partially or entirely incorrect, (3) we are unable to predict gene regulation, so just because a pathway is present in the genome, one cannot conclude if and when it is active. Biological assays are fundamentally needed to provide experimental data to prove or disprove the predictions of metagenomics.

The phenomic technology, phenotype microarrays (PMs, see **Fig. 59.1**), can be used to address all of these issues with culturable cells. This has been discussed in a recent review [Bochner, 2009] so details do not need to be reiterated here. To summarize very briefly, PMs allow a scientist to test nearly 2000 phenotypes of a cell line in a single experiment. This provides an overview of cellular carbon, nitrogen, phosphorus, and sulfur metabolism, sensitivity to various salts and ions and total osmolarity, sensitivity to pH, and sensitivity to 240 chemicals with capabilities for inducing different types of stresses due to their widely varying modes of action and propensity for disrupting different cellular pathways.

Handbook of Molecular Microbial Ecology, Volume I: Metagenomics and Complementary Approaches, First Edition. Edited by Frans J. de Bruijn.

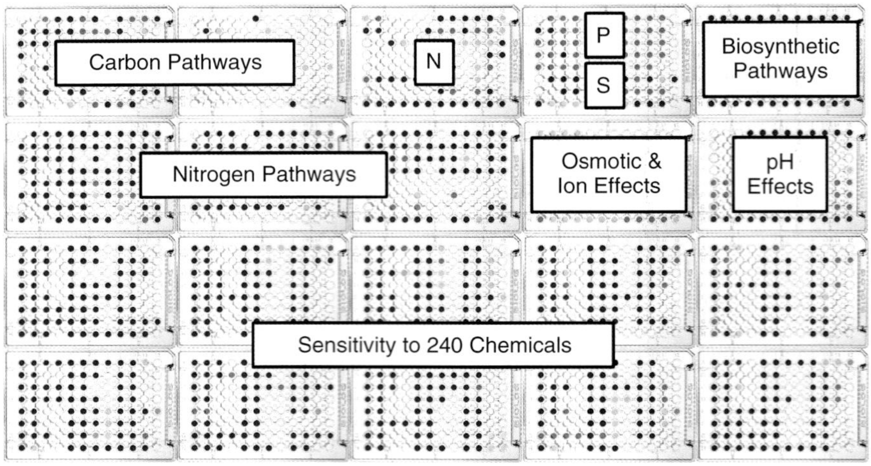

Figure 59.1 Phenotype Microarrays are sets of phenotypic assays performed in 96-well microplates. The microplate wells contain chemicals dried on the bottom to create unique culture conditions after rehydration. Assays are initiated by inoculating all wells with cell suspensions. After incubation, some of the wells turn various shades of purple due to reduction of a tetrazolium dye as the cells produce energy (NADH). The variable level of purple color indicates that the cells are metabolically active and producing energy in some wells but not others. Other colors such as yellow and orange are the colors of the inhibitory chemicals in the wells. Microplates in the PM set are organized into functional groups as labeled in the figure. Assays of C, N, P, and S metabolism provide information about which metabolic pathways are present and active in the cells. Assays of ion, pH, and chemical sensitivities provide information on stress and repair pathways that are present and active in cells.

59.3 APPLICATION #1: GETTING CELLS TO GROW IN CULTURE

Ultimately, it is essential to culture microbial cells so that we can study their properties in pure culture and in detail. Gradual progress is being made in this, and PM technology can also help in these endeavors. An overview article describes using PM technology to perform high-throughput growth curves from PM experiments [Jacobsen et al., 2007].

When an organism is difficult to culture, the traditional approach was to keep adding more nutrients. This approach has enabled cultivation of most human pathogenic bacteria, but has failed to enable cultivation of microbes from other important environments such as soil and aquatic environments of low or high salinity. We now realize that many environmental bacteria are oligotrophic, so a "less is more" approach is likely to be more successful.

PMs take an oligotrophic approach to defining the nutritional needs of a cell by employing a minimal basal medium. As far as I am aware, all cells need micromolar or higher levels of the following six elements (in addition to H and O): C, N, P, S, K, Mg. Therefore, the basal medium in the PMs is based on a formulation containing a carbon source plus low levels of NH_3, PO_4, SO_4, K, and Mg [Bochner et al., 2001]. Because it is easiest to determine, the first thing we do to characterize a novel microbe is to test it for carbon metabolism in PM1 and PM2. This tells us the range of carbon sources utilized as well as which are most favored.

Next, using a good carbon source selected from the PM1 and PM2 results, we use PM3, PM6, PM7, and PM8 to determine the range and preference for nitrogen source. In our experience, we frequently find microorganisms that cannot use NH_3. They most commonly need glutamate or glutamine or arginine instead, or in addition. We also find that microbial cells frequently prefer peptides to individual amino acids. This may be because peptides would be found more often than single amino acids in real-world environments and/or because peptides are easier to transport, having less charge. Furthermore, broad specificity peptide transporters can potentially transport all amino acids and eliminate the need for 20 or more individual amino acid transporters.

After determining the required carbon and nitrogen source, we then attempt to determine the phosphorus and sulfur requirements. In our experience, most microorganisms will use PO_4 for phosphorus, although some can require nucleotides. We frequently encounter microorganisms that cannot use SO_4 for sulfur. It is very common to find species that need one or more forms of reduced sulfur—for example, thiosulfate or cysteine, or methionine. Requirement for alternative sulfur compounds was recently found to be the key factor for culturing the SAR 11 bacterium from ocean environments. This bacterium was shown to have an obligatory need for methionine or 3-dimethylsulfoniopropionate [Tripp et al., 2008]. It is not surprising that many microorganisms need reduced sulfur. The main sulfur needs of cells are for cysteine and methionine. A great deal of energy is required to transport the SO_4 anion across the membrane, and then to sequentially reduce sulfate to sulfite, sulfide, and ultimately to cysteine.

In most attempts to culture cells, pH buffering is also helpful. The choice of buffer will of course depend on the pH of the environment being sampled. Some useful alternatives are inorganic buffers such as bicarbonate or phosphate, and some organic buffers such as triethanolamine, HEPES, and tricarballylic acid.

Often the use of oligotrophic culture media and a strong measure of patience are needed to culture microbes. Hattori and colleagues pioneered this approach in their lifelong work on soil bacteria [Hattori and Hattori, 1976; Hattori et al., 1997; Mitsui et al., 1997] showing that the use of oligotrophic media, low temperatures, and extended cultivation led to a large increase in viable cell counts recovered from soil. In a more recent published study [Joseph et al., 2003], many previously uncultured soil bacteria were successfully cultured on regular oligotrophic media, but in some cases the agar media were incubated up to 3 months. The introduction of better oligotrophic media such as R2A agar [Reasoner and Geldreich, 1985] likewise has led to important improvements in culture and recovery of microbes from rivers and lakes. Now 1/10 diluted R2A media are used to help culture bacteria even from saline aquatic environments [Cho and Giovannoni, 2004].

The reasons why media with higher levels of nutrients are toxic have not been studied and defined well enough, but some general observations can be made. Microbiology laboratories typically prepare organic supplements to culture media by autoclaving, and this excessive heat can transform many organic molecules into toxic chemicals. Most laboratories culture bacteria on media with agar. Agar can contain small amounts of fatty acids or inorganic salts that can be toxic to some microbes. Use of ultrapure agar grades, agarose, or other gelling substances can also be tried. Inorganic salts and anions must also be selected with care. It is clear that chloride ions are relatively toxic to many soil bacteria and to most yeast and fungi. With these microorganisms, sulfates and phosphates are the preferred anions to use. Yet even a chemical as seemingly innocuous and necessary as PO_4 can be toxic. When we studied the preferred nutrients of the halophilic bacterium *Halobacterium salinarum*, we found that it preferred nutrients with glycerol: glycerol as carbon source and glycerol-phosphate as phosphorus source; and that inorganic phosphate was toxic [Bochner, 2009]. Media containing nitrate can also be toxic. If the pH of nitrate-containing media drops even slightly below pH 7, nitrate is spontaneously converted to nitrite, nitrous acid, and nitric oxide [Yoon et al., 2006], all of which can be toxic to microbes lacking nitroreductase defensive enzymes. In my experience, the best approach is to keep levels of nutrients as low as possible and avoid heating culture media. Cells have efficient transporters and usually need essential nutrients at levels ranging from about 25 μM for sulfur to 2 mM for carbon.

Another important factor limiting cultivation is iron. Kim Lewis and colleagues have done a lot of work recently to cultivate previously nonculturable antibiotic producing *Streptomyces* isolated from soil. They report that the key to their successful cultivation is the supplementation with one or more siderophores [Lewis, 2009]. Expansion of our knowledge of how to supply iron supplements properly to microbes is certainly another area for productive focus.

Gas environments must also be considered. Many bacteria require low or no oxygen, or elevated carbon dioxide. One reason why bacteria may appear to "not be culturable on agar" is that on agar media colonies arise from a single cell, and a cell taken from its native environment and deposited onto an agar surface can get a substantial oxidation shock. Much better success may be obtainable by plating cells onto agar in a controlled low-oxygen chamber. Other gases found in nature such as hydrogen, nitrogen, nitric oxide, sulfur dioxide, hydrogen sulfide, and ethylene should perhaps also be considered for their effects on growth.

Although it is quite often the case that "less is more," sometimes microbes do require a supplement. Many bacteria lack one or more biosynthetic pathways and require supplementation. **Figure 59.2** shows an example of a strain of *Yersinia pseudotuberulosis* that displayed a diagnosable defect when tested with PMs. It showed a stimulation on PM4 by phospho-L-arginine and on PM5 by L-arginine, L-ornithine, and L-citrulline, indicating that it was an arginine auxotroph. On the PM1 carbon source panel, this strain showed a very weak result. However, when it was supplemented with L-arginine, it gave the expected pattern of diverse carbon source utilization.

A key advantage to using PMs to determine the culture characteristics of a microbial cell is that the redox color reaction can provide essential information about cellular metabolism under experimental conditions where the cell is not growing. If a cell needs more than one special condition, one can never determine what is needed by doing growth-based assays because one change in the medium is not sufficient to enable growth. However, using redox dyes, subtle and slight stimulation of energy production by nongrowing cells can often provide the needed clues.

A beautiful illustration of this point is the recent breakthrough leading to successful culture of *Coxiella*

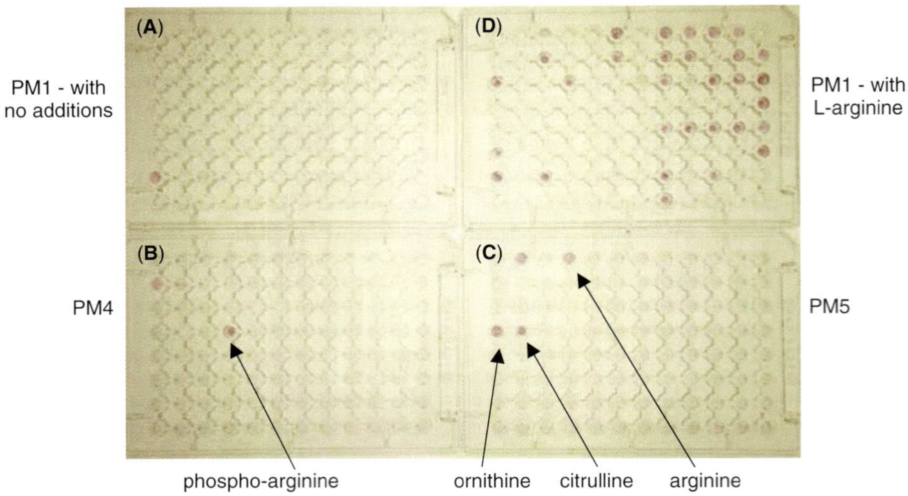

Figure 59.2 (**a**) A strain of *Yersinia pseudotuberulosis* exhibited a very weak and obviously incomplete metabolic reaction pattern in the PM1 carbon source metabolism assay panel.(**b**) The strain showed a stimulation in the PM4 phosphorus source panel by phospho-L-arginine. (**c**) The strain also showed a stimulation in the PM5 biosynthetic pathway assay panel by L-arginine, L-ornithine, and L-citrulline. The results of parts b and c suggested that the strain required supplementation with L-arginine. (**d**) When supplemented with L-arginine, the strain gave the pattern of diverse carbon source utilization in the PM1 panel expected of *Yersinia pseudotuberulosis*.

burnetii [Omsland et al., 2009]. *Coxiella* is an important pathogen to cattle that can also infect humans. It has been very difficult to study because it could only be cultured inside of animal cells. PM technology allowed the researchers to gradually optimize the medium in several sequential steps by optimizing energy production and redox dye reduction. *Coxiella* cells were recovered from lysed animal cells and tested promptly in various PMs. First they optimized the dye (Dye A) and pH (4.5), then the carbon source (succinate), then the inorganic salts mixture, and finally the level of oxygen (2.5%) and cysteine. With all of these components optimized, they can now culture *Coxiella* in a test tube.

59.4 APPLICATION #2: IMPROVING GENE AND GENOME ANNOTATION

One can examine gene function using PM technology by comparing the biological properties of cells that differ by a single gene, as exemplified by a study of two component regulatory genes [Zhou et al., 2003]. This mutant cell approach also allows testing of genes where function has already been predicted, as exemplified by a study of *Pseudomonas* transporter genes [Johnson et al., 2008]. Gene and pathway regulation can be partially addressed, for example, by looking at phenotypic changes in strains with altered regulatory genes [Zhou et al., 2003, Jones et al., 2007] or through direct metabolic assay to examine the effect of the principle carbon source on nitrogen metabolism [Bochner, 2009] or the effect of any other global parameter of interest such as temperature, pH, osmolarity, a toxic chemical, or even effects of light [Freidl et al., 2007, 2008].

59.5 APPLICATION #3: IMPROVING METABOLIC MODELING

Numerous attempts have been made to model the metabolic properties and growth rates of various culturable bacteria, based on either (a) assembly of pathways from genome annotation or (b) more complex rate-based models. All such models and extrapolations must be challenged and verified or disproven by real cellular assays, and PM technology has become a popular approach to doing this efficiently. Several published examples can now be cited [AbuOun et al., 2009; Covert et al., 2004; Feist et al., 2007; Jones et al., 2007; Mols et al., 2007; Oberhardt et al., 2008; Oh et al., 2007].

The best way to illustrate the problems with genome sequence-based predictions is with an example. The important human pathogen, *Helicobacter pylori*, was one of the very first bacteria to have its genome sequenced. This organism was difficult to culture, and its metabolism had not been well-studied. The groups that performed the sequencing proposed a metabolic model based on the pathways that they found [Doig et al., 1999; Marais et al., 1999]. In **Figure 59.3** I show a simplified version of their prediction (Fig. 59.3a) along with a listing of the predicted carbon sources that would be utilized (Fig. 59.3b). Then I show the actual carbon sources metabolized as determined by PM assay (Fig. 59.3c). Essentially, their predictions were accurate but rather incomplete. More importantly, however, there was no way to tell from their predictions the orientation of *H. pylori*'s metabolism—in other words, which were the most and least preferred carbon sources. Most bacteria that can metabolize glucose will use it as the most preferred carbon source. However, in the case of *H. pylori*, PM analysis showed the preferred carbon sources to be some carboxylic and amino acids, especially (lactic acid, α-hydroxybutryric acid, α-keto-glutaric acid, L-serine, and L-glutamine, see Fig. 59.3c). A major deficit in current genome-based metabolic modeling is the inability to predict metabolic regulation. Furthermore, PM assay produced another very interesting and unpredicted finding. The hierarchy of carbon sources utilized by *H. pylori* was strongly affected by the concentration of serum albumin in the culture environment [Lei and Bochner, 2008].

In addition to orientation of metabolism by regulatory hierarchies, there are also other important factors that can have strong effects on metabolism and physiology. One such factor is temperature. In a previous review we have shown examples and discussed the important biological significance of temperature regulation of hexose phosphate metabolism in *Yersinia pseudotuberculosis* and *Listeria monocytogenes* [Bochner, 2009]. Another example, recently published, shows major shifts in carbon metabolism by a strain of *Campylobacter jejuni* comparing 42°C to 37°C. Metabolism of TCA cycle chemicals is much stronger at 42°C, whereas metabolism of α-hydroxybutyric acid is much stronger at 37°C [Line et al., 2010]. This probably also has biological importance because *C. jejuni* lives in the cecum of poultry at 42°C, where short-chain carboxylic acid by-products are plentiful.

Examples of the utility of PM technology in providing gene function information are not limited to carbon metabolism. Successes have also been demonstrated in nitrogen metabolism with regulatory genes of *Pseudomonas syringae* [Jones, et al., 2007], genes of unknown function of *Escherichia coli* [Loh et al., 2006], and nutrient transporter genes of *Pseudomonas aeruginosa* [Johnson et al., 2008]. Other technologies such as transcriptome [Guell et al., 2009] and growth curve analysis [Yus et al., 2009] can also provide important information as shown in the recent comprehensive tour de force analysis of the metabolism of *Mycoplasma pneumoniae*.

59.6 APPLICATION #4: ANALYZING SECONDARY METABOLISM AND CHEMICAL RESISTANCE

Understanding the ability of microbes to compete and survive in different environments requires not only understanding their capabilities to utilize nutritional chemicals in that environment, but also their capabilities to survive and produce toxic chemicals. PM technology has been used in creative ways also in these areas of study. Viti and colleagues [Decorosi et al., 2009; Tremaroli et al., 2009; Viti et al., 2007, 2009] have used PM technology to examine resistance of soil bacteria to inorganic chemicals such as chromate and tellurite. In a very different application, two groups have recently used PM technology to determine culture conditions that induce the synthesis of antibiotics and toxins. To determine antibiotic production, microscale LC-mass spectrometry was used to profile the secondary metabolites in the microwell culture supernatants [Singh, 2009]. To determine culture conditions inducing synthesis of a tricothecene toxin produced by the wheat pathogen *Fusarium graminarum*, fluorescence was measured in the microwells after the *Fusarium* was engineered so that the first enzyme in the toxin synthesis pathway was fused to the green fluorescent protein. This has led to the breakthrough discovery that polyamines and low pH are required for induction of toxin synthesis [Gardiner et al., 2009a, 2009b]. A fourth example of a novel use of PM technology was the detection of a protective effect of nitric oxide on antibiotic susceptibility of *Bacillus subtilis* [Gusarov et al., 2009].

59.7 APPLICATION #5: RECONSIDERATION OF COMMUNITY LEVEL PHYSIOLOGICAL PROFILES USING UPDATED PM TECHNOLOGY

A final topic is the use of phenotypic assay plates to directly assay and compare microbial communities from different natural environments. This approach was pioneered in 1991 by Garland and Mills using Biolog GN MicroPlates [Garland and Mills, 1991] and later EcoPlates [Insam and Rangger, 1997], and it became popular and widely used (see the bibliography section of the Biolog web site, www.biolog.com). Subsequently, this approach was criticized for reflecting primarily the metabolism of the fast growing gram-negative bacteria in these communities [Haack et al., 1995]. The bias is engendered by the use of tetrazolium violet (Dye A) as the redox dye in these assays. This dye is toxic to most gram-positive and slow-growing bacteria, especially at low cell densities and high oxygen concentrations. Recently, however, Biolog has developed new redox dyes that work well with a very wide range of both gram-positive and gram-negative bacteria (Biolog Dyes D, G, and H). These redox dyes exhibit little or no toxicity and can be used with PM plates to examine the Community Level Physiological Profiles not only for carbon sources (PM1, PM2) but also for

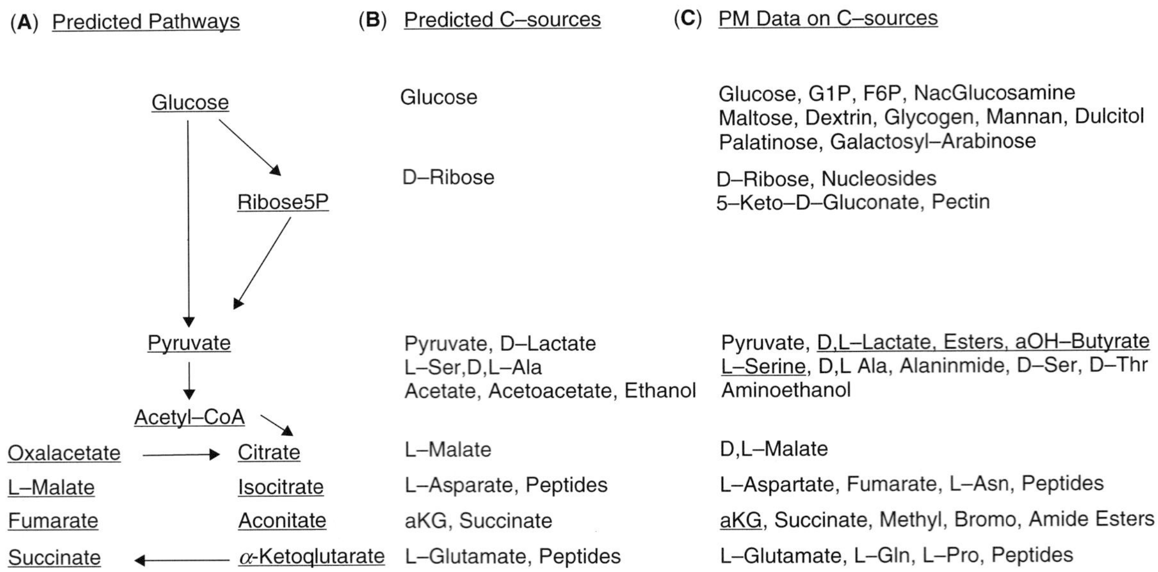

Figure 59.3 The major catabolic pathways of *Helicobacter pylori* were extrapolated, based upon annotation of genes found from genomic sequencing. (**a**) Evidence was found for three pathways: glycolysis pathway, pentose phosphate pathway, and citric acid cycle. (**b**) From these pathways, most likely carbon sources metabolized can be predicted, but their order of relative preference cannot be predicted. (**c**) Phenotype microarray analysis of *H. pylori* indicates more carbon sources metabolized than were predicted. More importantly, phenotype microarray analysis shows that the most preferred carbon sources for *H. pylori* are carboxylic and amino acids, especially those that are underlined: lactic acid, α-hydroxy-butryric acid, α-keto-glutaric acid, L-serine, and L-glutamine.

nitrogen sources (PM3, PM6, PM7, PM8) and phosphorus and sulfur sources (PM4). Furthermore, the OmniLog instrument is now available to continuously record kinetic data from these PM assays, automating and facilitating the determination of key parameters such as the Average Well Color Development [Garland et al., 2001].

PM technology and its precursor phenotypic assays have also been used by a number of research groups to look at evolution of phenotypic traits within experimental microbial communities [Cooper and Lenski, 2000; MacLean and Bell, 2002, 2003; Venail at al., 2008].

59.8 CONCLUDING REMARKS

At one level, PM technology helps scientists examine the metabolic and physiologic properties of microbial cells. Because of its unique capabilities, it can be used to determine optimal conditions for culturing novel microbes. At another level, PM plates provide an in vitro simulation of diverse environments by providing nearly 2000 different culture conditions with culture environments based on either nutritional stimulation or stress and toxic conditions. This technology and other phenomic approaches provide numerous ways to assist microbiologists in cataloging and understanding the microbial diversity of our planet.

INTERNET RESOURCES

Biolog web site, http://www.biolog.com

Acknowledgments

I gratefully acknowledge and thank my colleagues that have participated in the development of PM technology, especially Amalia Franco-Buff, Xiang-He Lei, Vanessa Gomez, Lawrence Wiater, Jeffrey Carlson, Michael Ziman, Peter Gadzinski, Eric Olender, Grace Chou, Eugenia Panomitros, and Luhong He. I also thank NIH for supporting the work to develop phenotype microarray technology through SBIR grants from NIGMS (GM62107) and NIAID (AI57232).

REFERENCES

AbuOun M, Suthers PF, Jones GI, Carter BR, Saunders MP, Maranas CD, et al. 2009. Genome scale reconstruction of a *Salmonella* metabolic model. *J. Biol. Chem*. **284**:29480–29488.

Bochner BR. 2009. Global phenotypic characterization of bacteria. *FEMS Microbiol. Rev*. **33**:191–205.

Bochner BR, Gadzinski P, Panomitros E. 2001. Phenotype microarrays for high-throughput phenotypic testing and assay of gene function. *Genome Res*. **11**:1246–1255.

Cho J-C, Giovannoni SJ. 2004. Cultivation and growth characteristics of a diverse group of oligotrophic marine *Gammaproteobacteria*. *Appl. Environ. Microbiol*. **70**:432–440.

Cooper VS, Lenski RE. 2000. The population genetics of ecological specialization in evolving *Escherichia coli* populations. *Nature* **407**:736–739.

Covert MW, Knight EM, Reed JL, Herrgard MJ, Palsson BO. 2004. Integrating high-throughput and computational data elucidates bacterial networks. *Nature* **429**:92–96.

Decorosi F, Tatti E, Mini A, Giovannetti L, Viti C. 2009. Characterization of two genes involved in chromate resistance in a Cr(VI)-hyper-resistant bacterium. *Extremophiles* **13**:917–923.

Doig P, de Jonge BL, Alm RA, Brown ED, Uria-Nickelsen M, Noonan B, et al. 1999. *Helicobacter pylori* physiology predicted from genomic comparison of two strains. *Microbiol. Mol. Biol. Rev*. **63**:675–707.

Feist, AM, Henry CS, Reed JL, Krummenacker M, Joyce AR, Karp PD, et al. 2007. A genome-scale metabolic reconstruction for *Escherichia coli* K-12 MG1655 that accounts for 1260 ORFs and thermodynamic information. *Mol. Syst. Biol*. **3**:12_–138.

Friedl MA, Kubicek CP, Druzhinina IS. 2007. C-source dependence and photostimulation of conidiation in *Hypocrea atroviridis*. *Appl. Environ. Microbiol*. **74**:245–250.

Friedl MA, Schmoll M, Kubicek CP, Druzhinina IS. 2008. Photostimulation of *Hypocrea atroviridis* growth occurs due to a cross-talk of carbon metabolism, blue light receptors and response to oxidative stress. *Microbiology* **154**:1229–1241.

Gardiner DM, Kazan K, Manners JM. 2009a. Nutrient profiling reveals potent inducers of trichothecene biosynthesis in *Fusarium graminearum*. *Fungal Genet. Biol*. **46**:604–613.

Gardiner DM, Osborne S, Kazan K, Manners JM. 2009b. Low pH regulates the production of deoxynivalenol by *Fusarium graminearum*. *Microbiology* **155**:3149–3156.

Garland JL, Mills AL. 1991. Classification and characterization of heterotrophic microbial communities on the basis of patterns of community-level sole-carbon-source utilization. *Appl. Environ. Microbiol*. **57**:2351–2359.

Garland JL, Mills AL, Young JS. 2001. Relative effectiveness of kinetic analysis vs single point readings for classifying environmental samples based on community-level physiological profiles (CLPP). *Soil Biol. Biochem*. **33**:1059–1066.

Guell M, van Noort V, Yus E, Chen W-H, Leigh-Bell J, et al. 2009. Transcriptome complexity in a genome-reduced bacterium. *Nature* **326**:1268–1271.

Gusarov I, Shatalin K, Starodubtseva M, Nudler E. 2009. Endogenous nitric oxide protects bacteria against a wide spectrum of antibiotics. *Science* **325**:1380–1384.

Haack SK, Garchow H, Klug MJ, Forney LJ. 1995. Analysis of factors affecting the accuracy, reproducibility, and interpretation of microbial community carbon source utilization patterns. *Appl. Environ. Microbiol*. **61**:1458–1468.

Hattori T, Hattori R. 1976. The physical environment in soil microbiology: An attempt to extend principles of microbiology to soil microorganisms. *CRC Crit. Rev. Microbiol*. **4**:423–461.

Hattori T, Mitsui H, Haga H, Wakao N, Shikano S, Gorlach K, et al. 1997. Advances in soil microbial ecology and the biodiversity. *Antonie Van Leeuwenhoek* **72**:21–28.

Insam H, Rangger A. 1997. *Microbial Communities: Functional Versus Structural approaches*. Heidelberg: Springer-Verlag.

Jacobsen JS, Joyner DC, Borglin SE, Hazen TC, Arkin AP, Bethel ES. 2007. Visualization of growth curve data from phenotype microarray experiments. In *11th International Conference on Information Visualization (IV'07)*, Zurich, Switzerland, July 4–6. Washington, DC: IEEE Computer Society.

Johnson DA, Tetu SG, Phillippy K, Chen J, Ren Q, Paulsen, IT. 2008. High throughput phenotypic characterization of *Pseudomonas aeruginosa* membrane transport genes. *PLoS Genet*. **4**:1–11.

Jones J, Studholme DJ, Knight CG, Preston GM. 2007. Integrated bioinformatic and phenotypic analysis of RpoN-dependent traits in the plant growth-promoting bacterium *Pseudomonas fluorescens* SBW25. *Environ. Microbiol*. **9**:3046–3064.

Joseph SJ, Hugenholtz P, Sangwan P, Osborne CA, Janssen PH. 2003. Laboratory cultivation of widespread and previously uncultured soil bacteria. *Appl. Environ. Microbiol*. **69**:7210–7215.

Lei X-H, Bochner B. 2008. Phenotype microarray analysis of the metabolism of *Helicobacter pylori*. Abstract PBMG 28, Society for General Microbiology, Spring Meeting, Edinburgh.

Lewis K, Epstein SS. 2009. Persisters, biofilms, and the problem of culturability. In *Uncultivated Microorganisms*, Vol. 10. Berlin/Heidelberg: Springer, pp. 181–194.

Line JE, Hiett KL, Guard-Bouldin J, Seal BS. 2010. Differential carbon source utilization by *Campylobacter jejuni* 11168 in response to growth temperature variation. *J. Microbiol. Methods* **80**:198–202.

Loh KD, Gyaneshwar P, Papadimitriou EM, Fong R, Kim K-S, Parales R, et al. 2006. A previously undescribed pathway for pyrimidine catabolism. *Proc. Natl. Acad. Sci. USA* **103**:5114–5119.

MacLean RC, Bell G. 2002. Experimental adaptive radiation in *Pseudomonas*. *Am. Naturalist* **160**:569–581.

MacLean RC, Bell G. 2003. Divergent evolution during an adaptive radiation. *Proc. Biol. Sci*. **270**:1645–1650.

Marais A, Mendz GL, Hazell SL, Megraud F. 1999. Metabolism and genetics of *Helicobacter pylori*: The genome era. *Microbiol. Mol. Biol. Rev*. **63**:642–674.

Mitsui H, Gorlach K, Lee H, Hattori R, Hattori J. 1997. Incubation time and media requirements of culturable bacteria from different phylogenetic groups. *J. Microbiol. Methods* **30**:103–110.

Mols M, de Been M, Zwietering MH, Moezelaar R, Abee T. 2007. Metabolic capacity of *Bacillus cereus* strains ATCC 14579 and ATCC 10987 interlinked with comparative genomics. *Environ. Microbiol*. **9**:2933–2944.

Oberhardt MA, Puchalka J, Fryer KE, Martins dos Santos VA, Papin JA. 2008. Genome-scale metabolic network analysis of the opportunistic pathogen *Pseudomonas aeruginosa* PAO1. *J. Bacteriol*. **190**:2790–2803.

Oh YK, Palsson BO, Park SM, Schilling CH, Mahadevan R. 2007. Genome-scale reconstruction of metabolic network in *Bacillus subtilis* based on high-throughput phenotyping and gene essentiality data. *J. Biol. Chem*. **282**:28791–28799.

Omsland A, Cockrell DC, Howe D, Fischer ER, Virtaneva K, Sturdevant DE, et al. 2009. Host cell-free growth of the Q fever bacterium *Coxiella burnetii*. *Proc. Natl. Acad. Sci. USA* **106**:4430–4434.

Rappe MS, Giovannoni SJ. 2003. The uncultured microbial majority. *Annu. Rev. Microbiol*. **57**:369–394.

Reasoner DJ, Geldreich EE. 1985. A new medium for the enumeration and subculture of bacteria from potable water. *Appl. Environ. Microbiol*. **49**:1–7.

Singh MP. 2009. Application of Biolog FF MicroPlate for substrate utilization and metabolite profiling of closely related fungi. *J. Microbiol. Methods* **77**:102–108.

Tremaroli V, Workentine ML, Weljie AM, Vogel HJ, Ceri H, Viti C, et al. 2009. Metabolomic investigation of the bacterial response to a metal challenge. *Appl. Environ. Microbiol*. **75**:719–728.

Tripp HJ, Kitner JB, Schwalbach MS, Dacey JW, Wilhelm LJ, Giovannoni SJ. 2008. SAR11 marine bacteria require exogenous reduced sulfur for growth. *Nature* **452**:741–744.

Venail PA, MacLean RC, Bouvier T, Brockhurst MA, Hochberg ME, Mouquet N. 2008. Diversity and productivity peak at intermediate dispersal rate in evolving metacommunities. *Nature* **452**:210–214.

Viti C, Decorosi F, Tatti E, Giovannetti L. 2007. Characterization of chromate-resistant and -reducing bacteria by traditional means and by a high-throughput phenomic technique for bioremediation purposes. *Biotechnol. Prog.* **23**:553–559.

Viti C, Decorosi F, Mini A, Tatti E, Giovannetti L. 2009. Involvement of the *oscA* gene in the sulphur starvation response and in Cr(VI) resistance in *Pseudomonas corrugata* 28. *Microbiology* **155**:95–105.

Wu D, Hugenholtz P, Mavromatis K, Pukall R, Dalin E, Ivanova NN, et al. 2009. A phylogeny-driven genomic encyclopedia of bacteria and archaea. *Nature* **462**:1056–1060.

Yoon SS, Coakley R, Lau GW, Lymar SV, Gaston B, Karabulut AC, et al. 2006. Anaerobic killing of mucoid *Pseudomonas aeruginosa* by acidified nitrite dervatives under cystic fibrosis airway conditions. *J. Clin. Invest.* **116**:436–446.

Yus E, Maier T, Michalodimitrakis K, van Noort V, Yamada T, Chen W-H, et al. 2009. Impact of genome reduction on bacterial metabolism and its regulation. *Nature* **326**:1263–1268.

Zhou L, Lei X-H, Bochner BR, Wanner BL. 2003. Phenotype microarray analysis of *Escherichia coli* K-12 mutants with deletions of all two-component systems. *J. Bacteriol.* **185**:4956–4972.

Chapter 60

Microbial Persistence in Low-Biomass, Extreme Environments: The Great Unknown

Parag Vaishampayan, James N. Benardini, Myron T. La Duc, and Kasthuri Venkateswaran

60.1 EXTREME ENVIRONMENTS

Life has been shown to flourish in the midst of varying physical and chemical extremes, such as microgravity, pressure, radiation, temperature, acidity and alkalinity, salinity, radiation, desiccation, and oxygen tension—factors that, until recently, were thought to preclude life. The Atacama Desert is one of the oldest, driest, hottest deserts on this planet, while Antarctic dry valleys are the coldest, driest places on Earth. In both environments, despite environmental extremes, microbial life forms endure. Available carbon and nutrients are scarce, and extreme starvation conditions result in cell doubling times ranging from hundreds to thousands of years [Phelps et al., 1994, Lin et al., 2006]. As a result, resident microorganisms either (a) tailor their genomes to that of streamlined specialists capable of tolerating the challenging conditions or (b) exist in a dormant state of suspended animation, anxiously awaiting the return of favorable conditions.

60.1.1 Extreme Environments with Low Microbial Diversity

Approximately one nonillion (10^{30}) microorganisms (excluding virions) are in intimate association with the Earth, of which only 2000 strains have been genetically decoded thus far. A population this large must surely have a profound influence on its environment. Yet the scientific community has barely begun to scratch the surface of understanding with regards to the genetic diversity and metabolic capacities of the microbial world. This is primarily a consequence of our inability to culture the vast majority of microorganisms known to exist. Furthermore, the eubacterial and archaeal domains have remained quite enigmatic and have not been explored in great depth until now, due in large part to limitations in gene-sequencing and computational analysis.

A single-species ecosystem was reported from a deep South African gold mine, which represents an extreme example of a low-biodiversity environment [Chivian et al., 2008]. *Candidatus Desulforudis audaxviator* was shown to be capable of developing an independent lifestyle well-suited to long-term isolation from the photosphere deep within Earth's crust. The presence of this organism deep within the Earth offers an example of a natural ecosystem with all the processes necessary for life encoded within a single genome.

60.1.2 Extreme Environments with Low Biomass

Some habitats of particular interest to microbiologists, such as the biosphere–lithosphere interface, highly polluted environments, and others that are physicochemically extreme, seldom support significant microbial assemblages (e.g., the deep hypersaline anoxic basins of the Mediterranean Sea [van der Wielen et al.,

Handbook of Molecular Microbial Ecology, Volume I: Metagenomics and Complementary Approaches, First Edition. Edited by Frans J. de Bruijn.

2005] and heavy metal- and nitrate-contaminated soils [Abulencia et al., 2006]). Though the genetic diversity of such environments is most often very low with regard to novel metabolic physiologies, these environments remain particularly interesting and attract a great deal of research-oriented attention. Due to the intrinsically low microbial community size and overall biomass of these ecosystems, these habitats are not as easily studied as other environments by metagenomic approaches [Ferrer et al., 2009]. One possible solution to the low-biomass conundrum is to amplify the available biomass of these environments by enrichment (e.g., Ferrer et al. [2005]), but this inevitably results in loss of biodiversity and significant biases in downstream libraries.

It is generally held that the more extreme an environment is and the more isolated the habitat, the higher the level of microbial endemism and the lower the microbial diversity and biomass [Fierer and Jackson, 2006]. However, a recent study showed that Antarctic Dry Valley cryptoendolithic communities (a) are rich in prokaryotic diversity, limited in eukaryotic diversity, and house significant levels of biomass, (b) exist in surprisingly complex assemblages, and (c) harbor a considerable number of previously unreported microbial taxa [Pointing et al., 2009]. In this environment, biomass determination based on ATP, lipid, and DNA quantification suggested the presence of 10^6 to 10^8 cells/g soil, which was orders of magnitude higher than that determined by microscopic and culture-dependent estimates. The presence of these high-biomass, complex cryptoendolithic communities contradict the aforementioned assumption.

It is widely acknowledged among environmental molecular microbiologists that genetic biosignatures identified from an environment represent not only active, functional members of the community, but also dormant microbes and exogenous DNA from dead cells. The latter is generally not considered a significant factor due to adsorption, degradation, and reutilization of exogenous nucleic acids. However, the cold, arid nature of Antarctic desert soils slows the rate of turnover and degradation of naked DNA and works to preserve these signature biomolecules.

If we were to contemplate the upper limits of extreme dryness, then interstellar space would represent the most extreme "desert" imaginable. The theory of lithopanspermia posits that evolutionary adaptations to chronic desiccation and radiation would be absolutely requisite for microbial life forms to survive interplanetary transfer. This hypothesis is being tested experimentally in space simulation facilities in the United States and Germany, as well as concurrently aboard manned (ISS) and unmanned space-flight endeavors. Indeed, NASA's Long Duration Exposure Facility and the European Space Agency's BioPan space experiments have shown that microbes can survive direct exposure to the unaltered conditions of space [Demets et al., 2005]. Spores of a radiation-resistant lineage of *Bacillus pumilus* (strain SAFR-032), which were isolated from a spacecraft assembly environment, were flown (February 2008 to September 2009) to the International Space Station (ISS) and exposed to a variety of space conditions using the *European Technology Exposure Platform and Experiment Facility* (EuTEF). The exposure conditions were (i) space vacuum, (ii) solar extraterrestrial UV radiation including vacuum-UV, (iii) simulated Martian UV radiation regime, and (iv) galactic cosmic radiation. During ground simulation, desiccated spores exhibited only a 2-log reduction in viability following full Martian UV (200–400 nm) exposure ($87h, 30W\ m^{-2}$). After 18 months of exposure in the EuTEF facility under dark space conditions, SAFR-032 spores showed 10–40% survivability, whereas the survival rate (85–100%) increased when these spores were kept aboard the ISS under dark simulated Mars atmospheric conditions. These spores were not immune to space UV conditions (210- to 300-nm range), but they could survive when the spores were shadowed by the geometry of the space vehicle or when they were inadvertently hidden in the pits and faults of the spacecraft materials.

60.1.3 Is a Low-Biomass Cleanroom an Extreme Environment?

NASA's planetary protection efforts work toward (a) protecting solar system bodies from contamination by terrestrial biological material (forward contamination), thus preserving opportunities for future scientific investigation, and (b) protecting the Earth from harmful contamination by materials returned from outer space (back contamination) [COSPAR, 2002]. These approaches apply directly to the control and eradication of microorganisms present on the surfaces of spacecraft intended to land, orbit, fly-by, or be in the vicinity of extraterrestrial bodies. Consequently, current planetary protection policies require that spacecraft be assembled and readied for launch in controlled cleanroom environments. To achieve these conditions and maintain compliance with Good Manufacturing Practice regulations, robotic spacecraft components are assembled in ultraclean facilities. Much like facilities in the medical, pharmaceutical, and semiconductor sectors, NASA spacecraft assembly cleanrooms are kept extremely clean and are maintained to the highest of industry standards [http://www.engineeringtoolbox.com/clean-rooms-iso-d_933.html]. A better understanding of the distribution and frequency at which high-risk contaminant microbes are encountered on spacecraft surfaces would significantly aid in assessing the threat of forward contamination [Space Science Board & National Research Council, 1992].

60.1.4 Microbial Diversity of Spacecraft Assembly Cleanrooms

Cleanroom facilities with controlled conditions for airflow, humidity, temperature, and air particulate concentrations, coupled with periodic cleaning using chemical detergents [NASA-KSC, 1999], result in a nutrient-limited and harsh environment not conducive for microbial growth [La Duc et al., 2007]. Several investigations, both culture-based and culture-independent, have demonstrated that a variety of bacterial taxa are repeatedly isolated under cleanroom conditions [La Duc et al., 2007; Moissl et al., 2007, 2008; Vaishampayan et al., 2010]. However, although the stringent conditions within these facilities are effective in reducing the overall microbial load [Venkateswaran et al., 2001, La Duc et al., 2007], they also select for "hardy" microorganisms capable of tolerating prolonged periods of desiccation, extremes of temperature, and exposure to UV light or hydrogen peroxide [Puleo et al., 1978; La Duc et al., 2003; Kempf et al., 2005; Newcombe et al., 2005; La Duc et al., 2007]. A number of microbes have been isolated that not only survive the nutrient-limiting conditions of cleanroom environments but also can withstand even more inhospitable environmental stresses and have been reported to survive under simulated and actual space conditions [Newcombe et al., 2005; Osman et al., 2008]. Due to their potential capabilities to survive and proliferate in Martian and other extreme conditions, screening of anaerobes and archaea on spacecraft surfaces and instruments associated with future life-detection missions is crucial. It has been documented that a wide range of anaerobic [Moissl et al., 2007] and archaeal populations [Moissl et al., 2008] persist in the cleanrooms, even after imposition of rigorous maintenance programs, and continue to pose a challenge to planetary protection implementation activities. However, despite a growing understanding of the diverse microbial populations present in these cleanrooms, predicting the true risk of any such microbes compromising the findings of extraterrestrial life-detection efforts remains a significant challenge [Rummel, 2001].

A comprehensive census of cleanroom surface-associated bacterial populations was derived using the cloning and sequencing of 16S rRNA genes and DNA microarray (PhyloChip) analyses during various assembly phases of the spacecraft [La Duc et al., 2009]. Clone library-derived analyses in agreement with PhyloChip (see Chapter 58, Vol. I) results detected a larger bacterial diversity prior to the arrival of spacecraft hardware in these cleanroom facilities. PhyloChip results unveiled the presence of 9- to 70-fold more bacterial taxa than cloning approaches. In our earlier studies, molecular bacterial community composition of geographically distinct spacecraft-associated cleanrooms exhibited significantly different bacterial populations, revealing that only a small subset of microorganisms were common to all locations [Moissl et al., 2007].

60.1.5 Microbial Community Structures of Several NASA Missions

60.1.5.1 Mars Odyssey Orbiter Mission. Microbial characterization of the Mars Odyssey spacecraft and its Spacecraft Assembly and Encapsulation Facility II (SAEF-II) was carried out by both culture-based and molecular methods [La Duc et al., 2003]. The most dominant cultivable microbes were species of *Bacillus*, with comamonads, microbacteria, and actinomycetales also represented. Several spore-forming isolates were resistant to γ-radiation, UV, H_2O_2, and desiccation, and one *Acinetobacter radioresistens* isolate and several *Aureobasidium*, isolated directly from the spacecraft, survived various conditions. Sequences arising in clone libraries were fairly consistent between the spacecraft and facility; predominant genera included *Variovorax, Ralstonia*, and *Aquaspirillum*.

60.1.5.2 Genesis Mission. The *Genesis* discovery mission was the first sample-return mission to be performed beyond the lunar orbit. Its objective was the collection of solar wind samples that could be compared with known compositions of the planets to help elucidate the origins of the solar system. Because the goal of the *Genesis* mission was particulate collection, meticulous cleaning and assembling of spacecraft components took place in very-low-particulate Class 10 cleanrooms. The microbial burden of samples collected from these cleanrooms was assessed via traditional culture-based and molecular biomarker-targeted methods. The Cleaning Laboratory and Assembly Laboratory had extremely low bioburdens, with most (but not all) samples below the detection limits of the assays employed. Spikes in microbial incidence did occur on plenums, on light fixtures, and in the subfloors. Ultimately, the data suggested that the Genesis Cleaning Laboratory class 10 cleanroom cleaning procedures, coupled with air filtration, maintained a very low bioburden, thus decreasing the likelihood of spacecraft contamination. Nevertheless, the detection of viable bacteria in these cleanrooms, despite stringent cleaning and air-handling protocols, exemplifies the potential for microbial survival.

60.1.5.3 Mars Reconnaissance Orbiter Mission. Launched in August 2005, the main goals of the *Mars Reconnaissance Orbiter* (MRO) were to (a) provide data that could determine the extent to which water persisted on the surface of Mars and (b) help locate potential landing sites for future lander missions. To carry out these objectives, the orbiter underwent several months of aerobraking, using the Martian atmosphere to slow the craft and ease it into a circular orbit low enough for scientific observations. This process, as well as the resulting operational altitude in Mars, required compliance with more stringent NASA planetary protection cleanliness requirements than typical Mars orbiters. Prior to launch, samples from the spacecraft and two of its assembly facilities were made available for cultivation and molecular assays of bioburden. Traditional cultivation revealed that spore-forming bacteria were common to both the spacecraft and its assembly facilities; these bacteria included *B. pumilus, B. licheniformis*, and *B. megterium*, which have been repeatedly isolated in previous studies of spacecraft assembly facilities. Although spore formers were present in all sample sets, they constituted only 16% of the total cultivable isolates within the MRO spacecraft-assembly facilities ($\sim 10^2$ CFU m^2); however, the 16S rRNA copy numbers suggested a higher degree of bioburden ($\sim 10^7$ CFU m^2). The DNA-based biodiversity study of the MRO and its assembly facilities revealed a broad spectrum of microbes that varied depending on the geographic location of the spacecraft. The vast majority of clones sequenced from the MRO at the Lockheed Martin Assembly Facility (LMA-MTF, 89%) and from the LMA-MTF itself (99%) were microbes associated with human skin. The majority of clones from the MRO spacecraft at KSC were predominantly α- and β-proteobacteria (40% and 51%, respectively), while clones from the KSC Payload Hazard Spacecraft Facility had an even distribution of clones among the bacterial classifications. The absence of gram-positive bacteria, particularly of the genus *Bacillus*, from the MRO spacecraft, while present at LMA or KSC, further illustrated the low concentration of *Bacillus* with respect to other bacterial genera.

60.1.5.4 Mars Exploration Rovers Mission. In an effort to minimize the probability of forward contamination of pristine extraterrestrial environments, NASA requires that all U.S. robotic spacecraft undergo assembly, testing, and launch operations (ATLO) in controlled cleanroom environments. When several locations and phases of the spacecraft assembly were analyzed, surprisingly, the greatest estimates of airborne bioburden were observed prior to the commencement of MER ATLO activities and gradually declined from the initiation of ATLO on through to launch. Increased cleaning and maintenance initiated immediately prior to the start of ATLO activity resulted in a decline in both airborne bioburden and microbial diversity. Proteobacterial sequences were common in 16S rDNA clone libraries. Conspicuously absent were members of the Firmicutes phylum, which includes the genus *Bacillus*. In previous studies, species of this genus were repeatedly isolated from the surfaces of spacecraft and cleanroom assembly facilities.

60.1.5.5 Phoenix Mission. The bacterial diversity and comparative community structure of the Phoenix spacecraft assembly environment, Kennedy Space Center Payload Hazardous Servicing Facility (PHSF), was characterized throughout the spacecraft-assembly process using cultivation-based techniques, 16S rRNA gene cloning (see Chapter 15, Vol. I) , and DNA-microarray (PhyloChip) technologies [Ghosh et al., 2010, Vaishampayan, et al., 2010; see also Chapter 58, Vol. I]. Extremo-tolerant bacteria that could potentially survive conditions experienced en route to Mars or on the planet's surface were isolated with a series of cultivation-based assays that promoted the growth of a variety of organisms, including spore-formers, mesophilic heterotrophs, anaerobes, thermophiles, psychrophiles, alkaliphiles, and bacteria resistant to UVC radiation and hydrogen peroxide exposure. Analysis of hundreds of isolates from the facility demonstrated that there was also a shift in predominant cultivable bacterial populations accompanied by a reduction in diversity during and after assembly. It is suggested that this shift was a result of increased cleaning when Phoenix was present in the assembly facility and that certain species, such as *Acinetobacter johnsonii* and *Brevundimonas diminuta*, may be better adapted to environmental conditions found during and after assembly. In addition, problematic bacteria resistant to multiple extreme conditions, such as *Bacillus pumilus*, were able to survive these periods of increased cleaning.

Although cleanroom facilities are extremely inhospitable environments for microbial survival and proliferation, 16S rRNA gene sequences were recovered from all major bacterial phyla, with a high percentage of sequences originating from yet-to-be-identified organisms. Bacterial diversity retrieved from samples before Phoenix was housed was found to be statistically different from the samples collected during and after assembly.

Due to stringent cleaning and decontamination protocols during spacecraft assembly, the bacterial diversity was dramatically reduced. Comparative community analysis based on Phylochip results revealed overall trends similar to those seen in clone libraries, but the high-density phylogenetic microarray detected larger diversity in all sampling events. The bacterial community of the sample during Phoenix assembly was less diverse and was dominated by human-associated populations such as *Streptococcus* sp., indicating a possible consequence of increased human activity during spacecraft assembly. The decrease in community complexity in the samples when Phoenix was assembled compared to before and the subsequent recurrence of these organisms after assembly speaks to the effectiveness of NASA cleaning protocols. However, the persistence of a subset of bacterial signatures throughout all spacecraft-assembly phases underscores the need for continued refinement of sterilization technologies and the implementation of safeguards that monitor and inventory microbial contaminants.

60.1.5.6 Closed Habitat System. The Regenerative Enclosed Life Support Module Simulator (REMS) was designed to simulate the conditions aboard the International Space Station (ISS). This unique terrestrial, encapsulated environment for humans and their associated organisms allowed investigations into the microbial communities within an enclosed habitat system, primarily with respect to diversity, phylogeny, and the possible impact on human health. Traditional culture methods exhibited that only gram-positive and α-proteobacteria grew under tested culture conditions, with a predominant occurrence of *Methylobacterium radiotolerans* and *Sphingomonas yanoikuyae*. Direct DNA extraction and 16S rDNA sequencing methodology revealed a broader diversity of microbes present in the REMS air (51 species). Unlike culture-dependent analysis, both gram-positive and proteobacteria were equally represented, while members of a few proteobaterial groups dominated (*Rhodopseudomonas, Sphingomonas, Acidovorax, Ralstonia, Acinetobacter, Pseudomonas, and Psychrobacter*). Although the presence of several opportunistic pathogens warrants further investigation, the results demonstrated that routine maintenance such as controlling the humidity, the crew's daily cleaning, and air filtration was effective in reducing the microbial burden in the REMS.

60.1.5.7 International Space Station Water System. Molecular analyses were carried out on several pre- and post-flight International Space Station (ISS)-associated potable water samples at various stages of purification, storage, and transport, to ascertain their associated microbial diversities and overall microbial burdens [La Duc et al., 2004b]. Following DNA extraction, PCR amplification, and molecular cloning procedures, rDNA sequences closely related to pathogenic species of *Acidovorax, Afipia, Brevundimonas, Propionibacterium, Serratia*, and others were recovered in varying abundance. Retrieval of sequences arising from the iodine (biocide)-reducing *Delftia acidovorans* in post-flight waters is a concern. Regardless of innate biases in sample collection and analysis, such circumstantial evidence for the presence of viable, intact pathogenic cells should not be taken lightly. Implementation of new cultivation approaches and/or viability-based assays are needed to confirm such an occurrence; most importantly, such studies will enable the development of suitable antibacterial systems.

60.1.5.8 International Space Station Internal Active Thermal Coolant System. Since the water processed for use in the ISS is very low in biomass, the conventional PCR system did not yield appropriate products to further analyze via cloning or DNA microarray analyses. Hence, these samples were initially subjected to whole-genome amplification (WGA) to increase the DNA content, and then the target molecules such as ribosomal RNA or other housekeeping genes were amplified. In such an attempt, molecular microbial community analysis of the fluids associated with the Internal Active Thermal Coolant System (IATCS) intended for use in the ISS was carried out using state-of-the art molecular methods. Only 17% of the samples collected from Ground Support Equipment (GSE) and potential flight hardware of the IATCS yielded PCR amplification products for regular cloning. In contrast, the WGA amplification, followed by 16S rRNA gene analyses, amplified all samples. The WGA-cloning and sequencing of 16S RNA gene showed the presence of bacteria representing members of the Firmicutes β- and γ-Proteobacteria in both GSE and flight hardware libraries. However, α-proteobacterial sequences and yet-to-be-classified phyla were retrieved only from the flight hardware and GSE samples, respectively. These results illustrated that certain microbial populations were present in IATCS fluids and were probable metal reducers, biofilm formers, and opportunistic pathogens. It is possible that these microbes might have been unnoticed when the metagenomics approaches such as the WGA-cloning method were not attempted.

Subsequent to this study, the efficiency of heat-based (GenomiPhi, GE Healthcare) and alkaline-based (repli-g, Qiagen) DNA-denaturation whole-genome-amplification (WGA) kits in the amplification of target gene 16S rRNA was tested. DNA extracted from known bacterial consortium (10 different kinds of bacterial species) [La Duc et al., 2008] was serially diluted from 40 ng to 0.004 ng, and PCR was performed from the naked DNA and WGA-amplified chromosomal DNA. As expected, the 1.5-kb 16S rRNA gene product was repeatedly amplified from as low as 0.004 ng from WGA-amplified DNA, while the traditional PCR methods yielded products only up to 0.04 ng.

Over the past decade, efforts aimed at elucidating the microbial diversity of spacecraft and associated cleanrooms have progressed collinearly with the advent and availability of state-of-the-art molecular techniques. Cloning and sequencing techniques significantly bolstered the level at which microbial diversity could be assayed, resulting in up to a 14-fold increase in diversity data as compared to cultivation-based methods (**Table 60.1**).

While 16S rRNA cloning was at the time a significant improvement, and even considered to be the golden standard for elucidating microbial diversity, DNA microarray techniques are now empowering our understanding of microbial diversity at a seemingly unimaginable rate. Phylochip DNA microarray technologies can yield up to 70-fold more phylogenetic data than standard cloning regimes, (Table 60.1). The field of molecular biology is evolving at an astronomical rate, and it requires no major stretch of the imagination to posit that metagenomics will provide a quantum leap forward in our ability to detect and estimate the widest possible spectrum of microbial diversity (phylogenetic and functional) present in ultraclean spacecraft-associated environments.

60.2 CHALLENGES TO STUDY LOW-BIOMASS ENVIRONMENTS

Natural environments such as soil and sea contain vast numbers of diverse microorganisms, with more than 99% of the population not yet cultured in the laboratory. A current estimate is that the Earth's oceans are teeming with some 3.6×10^{29} microbial cells [Whitman, et al., 1998]. This massive microbial biodiversity increases dramatically if viruses are considered, since approximately 10^7 particles of virus size per milliliter of seawater have been found, most of which are bacteriophages [Breitbart and Rohwer, 2005, Angly et al., 2006]. In the case of simple microbial communities comprising only a few microbial species, the metagenomics approach allows their genome reconstruction and annotation [Tyson et al., 2004, Strous et al., 2006], which in turn provides further understanding of community composition, functionalities, and new insight into functional interactions of communities. However, in the case of complex microbial communities, it is very difficult to achieve complete or even near-complete genome reconstruction of a significant portion of community members.

In the wake of new-generation sequencing facilities, efforts such as the Genomic Encyclopedia of Bacteria and Archaea (GEBA), Sorcerer II Global Ocean Sampling expedition [Rusch et al., 2007], and the human microbiome project have paved the way into microbial "dark matter." In spite of such efforts, physicochemical extreme environments such as ice [Simon et al., 2009], highly polluted environments [Abulencia et al., 2006], or deep hypersaline anoxic basins [van der Wielen et al., 2005] represent a widely unexplored ecological niches with a vast potential for retrieving novel biocatalysts suitable for industrial use [Abulencia et al., 2006].

Despite advances in the specificity and sensitivity of molecular biological technologies, the efficient recovery of DNA from low-biomass samples remains extremely challenging. Optimal methods to purify biomolecules from such environments should (1) achieve the greatest total yield and (2) reflect the true microbial diversity of the sample. To overcome this limitation, multiple displacement amplification (MDA) of environmental DNA using the ϕ29 polymerase can be applied (see Chapter 77, Vol. I). In this way, high-throughput metagenomic approaches from small quantities of DNA as starting material are feasible. The whole-genome amplification technique leads to the formation of chimeric artifacts and short amplification products, and it introduces selective amplification of certain community members due to template inaccessibility or low priming efficiency [Abulencia et al., 2006]. Despite these drawbacks, this approach has been successfully employed in several metagenomic studies of different environments, including contaminated sediments [Abulencia et al., 2006], the Soudan mine [Edwards et al., 2006], scleratinian corals [Yokouchi et al., 2006], the marine viral metagenomes of four oceanic regions [Angly et al., 2006], and glacier ice [Simon et al., 2009]. Recent progress has revealed that the capture of genetic resources of complex microbial communities in metagenome libraries allows the discovery of a rich new enzymatic diversity that had not previously been imagined. This new diversity, the surface of which has thus far only been scratched, constitutes potential for a wealth of new and improved applications in industry, medicine, agriculture, and so on, and promises to facilitate in a significant manner our transition to a sustainable society by aiding the transition to renewable sources of energy [Ferrer et al., 2009].

Indeed, recent successful examples of MDA-driven whole-genome amplification from single cells obtained from environmental probes [Spits et al., 2006, Stepanauskas and Sieracki, 2007] suggest that WGA could not only provide access to the biodiversity of biomass-poor environments, but could also extend options for automation and high-throughput practices in library construction and screening. Recently, a fluorescence-activated cell sorting and multiple displacement amplification approach was employed to obtain genomic DNA from individual, uncultured cells of two marine flavobacteria from the Gulf of Maine [Woyke et al., 2009]. This single-cell sequencing approach can be used to obtain high-quality genome assemblies

Table 60.1 Bacterial Diversity of Clean Rooms During Spacecraft Assembly Phases of Different NASA Missions

Spacecraft Assembly Phase/ Mission	Year	Type of Samples	Number of OTUs Detected by Cultivation Techniques	Number of OTUs Detected by Cloning	Fold Increase in OTU (Cloning/ Cultivation Techniques)	Number of OTUs Detected by PhyloChip	Fold Increase in OTU (Phylochip/ Cloning)
International Space Station							
Closed habitat system-Water	2002	Drinking	6	37	6.2	NA	—
Closed habitat system-Water	2002	Condensate	2	14	7.0	NA	—
IATCS system[a]	2007	Fluids	7	20	2.9	NA	—
Mock-up habitation	2004	Air	16	77	4.8	NA	—
Mars Odyssey Orbiter Mission							
Spacecraft	2003	Surface	14	7	0.5	NA	—
Cleanroom floors	2003	Surface	6	12	2.0	NA	—
Genesis Mission	2003	Surface	21	25	1.2	NA	—
Mars Reconnaissance Orbiter Mission	2005	Surface	34	53		NA	—
Mars Exploration Rovers Mission	2008	Air	6	23	3.8	NA	—
Mars Sampling Return R&D Program							
Jet Propulsion Laboratory	2007	Surface	6	38	6.3	NA	—
Johnson Space Center	2007	Surface	9	78	8.7	NA	—
Kennedy Space Center	2007	Surface	17	77	4.5		
Phoenix Mission							
LMA-MTF-During	2007	Surface	NA	33	—	1222	37.0
KSC-PHSF-During	2007	Surface	25	26	1.0	1079	0.0
KSC-PHSF-Before	2007	Surface	43	132	3.1	1788	0.1
KSC-PHSF-After	2007	Surface	7	100	14.3	1120	2.0
MSL Mission							
JPL-SAF-Before	2007	Surface	NA	91	—	1492	16.4
JPL-SAF-During	2008	Surface	NA	7	—	491	70.1
JPL-SAF-During[a]	2009	Surface	NA	NA	—	226	—

LMA-MTF, Lockheed Martin Aeronautics—Multiple Testing Facility.

KSC-PHSF, Kennedy Space Center Payload Hazardous Servicing Facility.

JPL-SAF, Jet Propulsion Laboratory-Spacecraft assembly facility.

IATCS, International Space Station internal active thermal coolant system.

NA, not attempted.

[a] Metagenomics approach was attempted which yielded 16S rRNA gene amplification in 100% of the samples compared to <10% in conventional cloning technique.

from numerically trivial microorganisms from low-biomass extreme environments.

60.3 THE FUTURE OF LOW-BIOMASS CLEANROOM ENVIRONMENTAL METAGENOMICS

The molecular bacterial diversity surveys carried out as part of the aforementioned NASA missions improve our understanding of fluctuations in microbial community structure, while traditional culture-based approaches provide valuable information pertaining to the survival capabilities of bacteria in these extreme cleanroom environments. However, phylogenetic approaches do not reveal the complete picture, and emerging molecular approaches must be evaluated and utilized to explore the functional capabilities of both cultivable and viable-but-not-cultivable microbes. It is likely that a wide spectrum of microbial diversity has yet to be detected due to biases and limitations in current techniques. State-of-the-art metagenomics approaches promise to provide a more in-depth understanding of the phylogenetic and functional diversity of microbial genes present in low-biomass cleanrooms and other extreme environments.

There is a need to refine and apply emerging molecular strategies that will enable a comprehensive, in-depth cataloguing of microbial DNA sequences retrieved from spacecraft and associated cleanroom facilities. Future approaches should address all of the aspects of sample collection, processing, and analysis samples arising from these low-biomass environments. Because routine cleaning activities significantly reduce the microbial burden in these facilities to very low levels, the enumeration and phylogenetic analyses of such contaminant populations remain extremely challenging tasks. Accurate identification of the microbial constituents in environmental samples will rely heavily on sound, reliable sampling methodologies. Through optimizations and comparative evaluations of methods designed to collect, process, analyze, and archive biomolecules sampled from low-biomass environments, it will be possible to increase downstream sensitivity of laboratory assays and bolster the current understanding of microbial population dynamics in cleanrooms. Improvements in sample collection, including recovery and yield from sampling device, will significantly enable more sophisticated methods for analyzing microbial communities. Currently, in accordance with regulations outlined in international treaties, NASA employs a cultivation-based assay to gauge spacecraft cleanliness [NASA, 2005; NRC, 2006]. While this single assay provides a valid estimation of microbial abundance, it offers only a limited assessment of the phylogenetic diversity and physiological breadth of the microbial population present [La Duc et al., 2003, 2004a]. There is a need to adapt novel technologies, such as "rRNA gene variable region tag sequencing" and DNA microarrays, to comprehensively elucidate the microbial diversity profiles (bacterial, archaeal, and eukaryotic) associated with these extremely low-biomass environments. Conventional rRNA gene cloning and sequencing techniques are also required in parallel to generate confirmatory data. In addition to overall profiles of microbial populations, it is necessary to determine which microbial taxa pose particular threats to planetary protection concerns and astrobiology objectives. Metagenomic approaches will identify and screen for functional gene sequences conferring resistance to desiccation, UV and gamma radiation, oxidative stress, and exposure to other such conditions relevant to interplanetary transfer and/or extraterrestrial environments. Massively parallel sequencing approaches will compare genomes of resistant strains to their closest known nonresistant relatives in an attempt to identify genetic origins linked to resistance traits. Any such metagenomic approach will also allow examination of the adaptation of bacteria to extreme UV radiation in order to determine the genetic mechanisms of resistance and their prevalence. With likely resistance factors identified, hybridization arrays and sequencing approaches will detect the relative proportions of resistant strains to nonresistant strains in spacecraft assembly environments. Such knowledge will further NASA's ability to gauge the probability of spacecraft transfer of organisms with sequences conferring attributes relevant to survival in extraterrestrial environments. Furthermore, a metagenomically derived database will facilitate the development of technologies capable of identifying, capturing, and/or eliminating microbial targets. Metagenomic approaches present an opportunity to meet specific and vital NASA astrobiology objectives while simultaneously producing discoveries of tremendous relevance and introducing forward-reaching bioinformatics tools to the broader scientific community.

Acknowledgments

Part of the research described in this publication was carried out at the Jet Propulsion Laboratory, California Institute of Technology, under a contract with the National Aeronautics and Space Administration. This research was funded by NASA Research Announcement (NRA) ROSS 2006 awarded to Kasthuri Venkateswaran. We are thankful to J. Spry for valuable advice and encouragement. We thank S. Nelson for critically reviewing the manuscript.

REFERENCES

Abulencia CB, Wyborski DL, Garcia JA, et al. 2006. Environmental whole-genome amplification to access microbial populations in contaminated sediments. *Appl. Environ. Microbiol*. **72**:3291–3301.

Angly FE, Felts B, Breitbart M, et al. 2006. The marine viromes of four oceanic regions. *PLoS Biol*. **4**:e368.

Breitbart M, Rohwer F. 2005. Here a virus, there a virus, everywhere the same virus ? *Trends Microbiol*. **13**:278–284.

Chivian D, Brodie EL, Alm EJ, et al. 2008. Environmental genomics reveals a single-species ecosystem deep within Earth. *Science* **322**:275–278.

COSPAR. 2002. Planetary Protection Policy, October 2002, as amended, March 2005. http://www.cosparhq.org/scistr/PPPolicy.htm. COSPAR, Houston, TX.

Demets R, Schulte W, Baglioni P. 2005. The past, present and future of Biopan. *Adv. Space Res*. **36**:6.

Edwards RA, Rodriguez-Brito B, Wegley L, et al. 2006. Using pyrosequencing to shed light on deep mine microbial ecology. *BMC Genomics* **7**:57.

Ferrer M, Golyshina OV, Chernikova TN, et al. 2005. Microbial enzymes mined from the Urania deep-sea hypersaline anoxic basin. *Chem. Biol*. **12**:895–904.

Ferrer M, Beloqui A, Timmis KN, Golyshin PN. 2009. Metagenomics for mining new genetic resources of microbial communities. *J. Mol. Microbiol. Biotechnol*. **16**:109–123.

Fierer N, Jackson RB. 2006. The diversity and biogeography of soil bacterial communities. *Proc. Natl. Acad. Sci. USA* **103**:626–631.

Ghosh S, Osman S, Vaishampayan PA, Venkateswaran K. 2010. Recurrent isolation of extremo-tolerant bacteria from the clean room where Phoenix Spacecraft components are assembled. *Astrobiology* **10**(3):325–335.

Kempf MJ, Chen F, Kern R, Venkateswaran K. 2005. Recurrent isolation of hydrogen peroxide-resistant spores of *Bacillus pumilus* from a spacecraft assembly facility. *Astrobiology* **5**:391–405.

La Duc MT, Nicholson W, Kern R, Venkateswaran K. 2003. Microbial characterization of the Mars Odyssey spacecraft and its encapsulation facility. *Environ. Microbiol*. **5**:977–985.

La Duc MT, Kern R, Venkateswaran K. 2004a. Microbial monitoring of spacecraft and associated environments. *Microb. Ecol*. **47**:150–158.

La Duc MT, Sumner R, Pierson D, Venkat P, Venkateswaran K. 2004b. Evidence of pathogenic microbes in the International Space Station drinking water: Reason for concern? *Habitation (Elmsford)* **10**:39–48.

La Duc MT, Dekas AE, Osman S, Moissl C, Newcombe D, Venkateswaran K. 2007. Isolation and characterization of bacteria capable of tolerating the extreme conditions of clean-room environments. *Appl. Environ. Microbiol*. **73**:2600–2611.

La Duc M, Osman S, Venkateswaran K. 2008. Comparative analysis of methods for the purification of DNA from low-biomass samples based on total yield and conserved microbial diversity. *J. Rapid Methods Autom. Microbiol*. **17**:9.

La Duc MT, Osman S, Vaishampayan P, Piceno Y, Andersen G, Spry JA, Venkateswaran K. 2009. Comprehensive census of bacteria in clean rooms by using DNA microarray and cloning methods. *Appl. Environ. Microbiol*. **75**:6559–6567.

Lin LH, Wang PL, Rumble D, et al. 2006. Long-term sustainability of a high-energy, low-diversity crustal biome. *Science* **314**: 479–482.

Moissl C, La Duc MT, Osman S, Dekas AE, Venkateswaran K. 2007. Molecular bacterial community analysis of clean rooms where spacecraft are assembled. *FEMS Microbiol. Ecol*. **61**: 509–521.

Moissl C, Bruckner JC, Venkateswaran K 2008. Archaeal diversity analysis of spacecraft assembly clean rooms. *ISME J*. **2**:115–119.

NASA-KSC. 1999. Launch Site Requirement Planning Group. *Facilities Handbook of Payload Hazardous Servicing Facility (PHSF)*, Vol. K-STSM-14.1.15 rev D. Cape Canaveral, FL: NASA.

NASA. 2005. Planetary Protection Provisions for Robotic Extraterrestrial Missions. NPR 8020.12C, April 2005. Washington, D.C.: National Aeronautics and Space Administration.

Newcombe DA, Schuerger AC, Benardini JN, Dickinson D, Tanner R, Venkateswaran K. 2005. Survival of spacecraft-associated microorganisms under simulated martian UV irradiation. *Appl. Environ. Microbiol*. **71**:8147–8156.

NRC. 2006. Preventing the forward contamination of Mars. In *Committee on Preventing the Forward Contamination of Mars*. National Research Council. Washington, D.C.: National Academies Press.

Osman S, Peeters Z, LaDuc MT, Mancinelli R, Ehrenfreund P, Venkateswaran K. 2008. Effect of shadowing on the survival of bacteria to conditions simulating Martian atmosphere and UV-radiation. *Appl. Environ. Microbiol*. **74**:959–970.

Phelps TJ, Murphy EM, Pfiffner SM, White DC. 1994. Comparison between geochemical and biological estimates of subsurface microbial activities. Microb. *Ecol*. **V28**:335–349.

Pointing SB, Chan Y, Lacap DC, Lau MC, Jurgens JA, Farrell RL. 2009. Highly specialized microbial diversity in hyper-arid polar desert. *Proc. Natl. Acad. Sci. USA* **106**:19964–19969.

Puleo JR, Bergstrom SL, Peeler JT, Oxborrow GS. 1978. Thermal resistance of naturally occurring airborne bacterial spores. *Appl. Environ. Microbiol*. **36**:473–479.

Rummel JD. 2001. Planetary exploration in the time of astrobiology: Protecting against biological contamination. *Proc. Natl. Acad. Sci. USA* **98**:2128–2131.

Rusch DB, Halpern AL, Sutton G, et al. 2007. The Sorcerer II Global Ocean Sampling expedition: Northwest Atlantic through eastern tropical Pacific. *PLoS Biol*. **5**:e77.

Simon C, Herath J, Rockstroh S, Daniel R 2009. Rapid identification of genes encoding DNA polymerases by function-based screening of metagenomic libraries derived from glacial ice. *Appl. Environ. Microbiol*. **75**:2964–2968.

Space Science Board & National Research Council. 1992. Biological contamination of Mars: Issues and recommendations. In *Task Group on Planetary Protection*. Washington, D.C.: National Academy of Sciences.

Spits C, Le Caignec C, De Rycke M, Van Haute L, Van Steirteghem A, Liebaers I, Sermon K. 2006. Whole-genome multiple displacement amplification from single cells. *Nat. Protoc*. **1**:1965–1970.

Stepanauskas R, Sieracki ME. 2007. Matching phylogeny and metabolism in the uncultured marine bacteria, one cell at a time. *Proc. Natl. Acad. Sci. USA* **104**:9052–9057.

Strous M, Pelletier E, Mangenot S, et al. 2006. Deciphering the evolution and metabolism of an anammox bacterium from a community genome. *Nature* **440**:790–794.

Tyson GW, Chapman J, Hugenholtz P, et al. 2004. Community structure and metabolism through reconstruction of microbial genomes from the environment. *Nature* **428**:37–43.

Vaishampayan P, Osman S, Andersen G, Venkateswaran K. 2010. High-density 16S microarray and clone library based microbial community composition of the Phoenix Spacecraft assembly clean room. *Astrobiology* **10**:499–508.

van der Wielen PW, Bolhuis H, Borin S, et al. 2005. The enigma of prokaryotic life in deep hypersaline anoxic basins. *Science* **307**:121–123.

Venkateswaran K, Satomi M, Chung S, Kern R, Koukol R, Basic C, White DC. 2001. Molecular microbial diversity of a spacecraft assembly facility. *Syst. Appl. Microbiol*. **24**:311–320.

Whitman WB, Coleman DC, Wiebe WJ. 1998. Prokaryotes: The unseen majority. *Proc. Natl. Acad. Sci. USA* **95**:6578–6583.

Woyke T, Xie G, Copeland A, et al. 2009. Assembling the marine metagenome, one cell at a time. *PLoS ONE* **4**:e5299.

Yokouchi H, Fukuoka Y, Mukoyama D, Calugay R, Takeyama H, Matsunaga T. 2006. Whole-metagenome amplification of a microbial community associated with scleractinian coral by multiple displacement amplification using phi29 polymerase. *Environ. Microbiol.* **8**:1155–1163.

Chapter 61

Application of Phylogenetic Oligonucleotide Microarrays in Microbial Analysis

PANKAJ TRIVEDI AND NIAN WANG

61.1 INTRODUCTION

The prokaryotic species evolved more than 3.8 billion years ago and they occupy almost every niche on the planet Earth [Wintzingerode et al., 1997]. Microbes perform a variety of processes such as nutrient cycling, organic matter formation and decomposition, soil structure formation, growth promotion, and pathogenicity, and thus microbial diversity is critical for ecosystem function and dynamics [Kennedy, 1999]. Recent molecular biological studies have shown that the assemblage of organisms present in an environmental sample can contain thousands of distinct biotypes of bacteria and that many of these organisms represent undescribed microbial lineages [Torsvik and Øvreås, 2002; see also Chapter 2, Vol. I]. For example, 1 gram of soil may harbor up to 10 billion bacteria of possibly 4000–7000 different species and a biomass density of 300–30,000 kg ha^{-1} [Rosello-Mora and Amann, 2001]. It has been postulated that the total number of bacteria are on the order of 10^{30}, but still we have only a vague idea of the extent of microbial diversity, with estimates differing by factors of 1000 or more [Pace, 1997; Rappe and Giovannoni, 2003]. This complexity is further compounded by the high spatial and temporal variability of species represented in many natural communities [Dubey et al., 2006]. The characterization of natural microbial communities is clearly a daunting task.

Traditional exploration of microbial diversity has relied on serial dilution assays and colony formation by individual cells on agar based growth medium. Such studies provided useful information but suffered bias, resulting from the media and cultivation condition used and the inability to grow and isolate ~99% of the natural microbial communities [Prosser, 2002]. Broad-scale methods, such as (a) phospholipid fatty acid analysis (PLFA) and (b) DNA hybridization and reassociation kinetics, have provided limited culture independent analysis of microbial communities [Soderberg et al., 2004]. However, these methods cannot provide a higher resolution of microbial diversity where it is necessary to identify key microbial species at the community level or elucidate their role in the ecosystem [Torsvik and Øvreås, 2002; see also Chapter 2, Vol. I]. These limitations have been overcome to some extent by using rRNA gene analysis for microbial diversity. Polymerase chain reaction (PCR) amplification of rRNA genes from the environmental DNA samples, combined with fingerprinting techniques, such as denaturing gradient gel electrophoresis (DGGE), terminal restriction fragment analysis (T-RFLP), amplified rDNA restriction analysis (ARDRA), cloning, and sequencing, have yielded unprecedented advances in our appreciation for the breadth and depth of microbial diversity [Singh et al., 2004; Torsvik and Øvreås, 2002; see also Chapters 15, Vol. I]. Fluorescent *in situ* hybridization (FISH) has also been used to evaluate the distribution and function of bacterial populations in situ [Dubey et al., 2006]. These have served to quantify the striking abundance and genetic diversity of all microbes in all of life's recognized reaches [Leadbetter, 2003]. While these

Handbook of Molecular Microbial Ecology, Volume I: Metagenomics and Complementary Approaches, First Edition. Edited by Frans J. de Bruijn.

techniques are being used with considerable success, there are various limitations and biases that need to be addressed (reviewed in Dubey et al. [2006], Lynch et al. [2004], Nannipieri et al. [2002], and Singh et al. [2004]).

Since it was first proposed as a universal phylogenetic marker of life on Earth, sequence analysis of the small subunit ribosomal RNA (usually 16S rRNA) gene has become the gold standard for assessing the microbial diversity within environmental samples [Pace, 1997; see also Chapter 15, Vol. I]. This molecular nucleic acid-based approach provides the means to study microbes in their natural surroundings, typically by cloning and sequencing a few hundred 16S rRNA genes from the ~1 trillion generated from a PCR amplification [Curtis and Solan, 2004; Lunn et al., 2004]. Although species richness may be roughly predicted by extrapolation of these very few observations, reproducible discovery of species composition may require >10,000 sequencing reactions per sample [Narang and Dunbar, 2004]. As it is currently practiced, cloning and sequencing the 16S rRNA gene pool of environmental microbial communities is an essential but inefficient process, where the most abundant phylotypes or species mask less abundant but potentially significant members [Dunbar et al., 2002; Hong et al., 2009]. The use of advanced sequencing techniques such as 454 pyrosequencing, Solexa, and SOLiD has opened new perspectives in microbial ecology, and readers are referred to one recent excellent publication on that issue [Demaneche et al., 2009; Chapter 18, Vol. I]. By pyrosequencing, adapters are added to each DNA fragment obtained from the environmental sample, and subsequently these fragments are added to a bead: one fragment per bead. The 454 pyrosequencing technique allows a large amount of DNA to be sequenced in less time and at low cost. Currently there are some limitations associated with the use of 454 pyrosequencing, Solexa, and SOLiD in metagenomics, such as shorter sequence reads as opposed to ~800 nucleotides for Sanger sequencing and high cost per run (though they are cheaper and faster per base) [Sorensen et al., 2009]. This is particularly problematic when analyzing environmental samples. Thus, additional high-throughput and hybridization based quantitative methods for analyzing the bacterial structure and functions in ecosystem are warranted [Bae and Park, 2006; Gentry et al., 2006].

Owing to their extremely high diversity and their as-yet uncultivated status, microbial detection and quantification remain challenging, especially on a large scale and in a parallel and high-throughput fashion [He et al., 2007]. In order to get access to this microbial "black box," the development of powerful tools such as microarrays is necessary. Microarrays can be broadly defined as tools for massively paralleled ligand binding assays where features (e.g., oligonucleotides) are placed on a solid support (e.g., a glass slide) at high density for recognizing a complex mixture of target molecules [Ekins and Chu, 1999]. Microarray technology in principle allows the estimation of target abundance and the detection of biological interactions at molecular or cellular levels [Hoheisel, 2006]. Microarray technology is regarded as one of the major methodological advances that have propelled the field of molecular biology into the postgenomic era [see Chapters 57, 58, 59 and 60, Vol. I]. Microarray technology was originally designed for large–scale DNA sequencing by hybridization, clinical diagnostics (e.g., detection of single-nucleotide polymorphism), and genetic analysis [Chee et al., 1996; Pease et al., 1994; Richmond et al., 1999; Schena et al., 1995; Wang et al., 2003b; Yershov et al., 1996], but this technology likewise offers tremendous potential for microbial community analysis, pathogen detection, and process monitoring in both basic and applied environmental sciences [Li and Liu, 2003; Zhou, 2003; Chapter 57, Vol. I; Loy and Bodrossy, 2006; Sessitsch et al., 2006; Wagner et al., 2007].

61.2 TYPE OF MICROARRAYS APPLIED IN MICROBIAL ECOLOGY

DNA microarray technology has enormous potential in the analysis of microbial community structure, function, and dynamics on the one hand and as a specific, sensitive, quantitative, parallel, and high-throughput tool in microbial ecology [Lucchini et al., 2001]. The microarrays applied to microbial ecology research can be broadly divided into five categories based on the targeted genes [Loy and Bodrossy, 2006; Sessitsch et al., 2006; Wagner et al., 2007].

1. **Phylogenetic oligonucleotide microarrays (POAs)** use a conserved marker such as the 16S rRNA gene as probe template and are used to compare the phylogenetic relationship of microbial communities in different environments.
2. **Functional gene arrays (FGAs)** are designed for key functional genes that code for proteins involved in various biogeochemical processes and may also provide information on the microbial populations controlling these processes. The most comprehensive FGA available is the GeoChip, which targets ~10,000 genes involved in the nutrient cycling, metal reduction and resistance, and organic contaminant degradation [He et al., 2007; see also Chapter 57, Vol. I].
3. **Community genome arrays (CGAs)** contain whole genomic DNA of cultured organisms and can describe a community based on its relationship to these cultivated organisms.

4. **Metagenomic arrays (MGAs)** contain probes produced directly from environmental DNA and can be applied with no previous knowledge of the sequences present in the samples.
5. **Whole-genome open reading frame (ORF) arrays (WGAs)** contain probes for all of the ORFs in one or multiple genomes. These have primarily been used for functional genomic analysis of individual organisms, but can also be useful for comparative genomic analysis or to investigate the interactions of multiple organisms at the transcription level.

In this review, we will discuss the specific application of POA to microbial ecology research along with the challenges of applying this technology to environmental samples and some of the latest work on addressing various issues related to the use of this technology.

61.3 DEVELOPMENT AND ANALYTICAL PERFORMANCE OF POA

61.3.1 Choice of Marker Genes and Probe Length

One of the main parameters affecting the resolution of POA is the degree of conservation of the marker gene. Pioneering work by Woese and colleagues [Woese, 1987] described bacterial rRNA genes as uniquely suited for molecular phylogenetic analysis due to features such as universality, activity in cellular functions, and extremely conserved structure and nucleotide sequence. Out of the three types of rRNA genes (5S, 16S, and 23S) present in bacteria, the 16S rRNA gene has become a standard in taxonomic classification because it is more readily sequenced [Spiegelman et al., 2005]. Because no lateral gene transfer seems to occur between 16S rRNA genes [Olsen et al., 1986] and because their structure contains both highly conserved and variable regions with different evolution rates, the relationship between 16S rRNA genes reflects evolutionary relationships between organisms. Most of the POA to date have contained short oligonucleotide probes complementary to specific regions of 16S rRNA genes [Brodie et al., 2007; Guschin et al., 1997; Ludwig and Schleifer, 1999; Loy et al., 2002; 2005; Sanguin et al., 2006a]. Also, the existence of large and regularly updated sequence [Cole et al., 2003; DeSantis et al., 2006a, 2006b; Ludwig et al., 2004) and probe databases [DeSantis et al., 2006a, 2006b; Loy et al., 2003, 2008] make this molecule ideal for designing probes for POA. Another significant advantage of using 16S rRNA gene sequences for microarray analysis is that their extremely high probe capacity enables researchers to create highly redundant (multiple distinct probes that target the same organisms or groups) and hierarchically nested (probes targeting organisms at multiple phylogenetic levels) probe sets [Kelly, 2009]. There are various limitations of using the 16S rRNA gene as a marker in POA (discussed later) which include the limited resolution at species level or above. Alternative probe targets with species level resolution include large-subunit ribosomal RNA (LSU rRNA; see also Chapter 3, Vol. I); the SSU-LSU rRNA intergenic spacer region (see Chapter 4, Vol. I) and various housekeeping genes [Sessitsch et al., 2006]. However, individual sequence database for these alternative markers (if available) contain considerably fewer identities than the SSU rRNA databases, constraining the development and evaluation of encompassing probe sets for microarrays [Hashsham et al., 2004; Sessitsch et al., 2006].

The length of probe is also one of the critical categories in setting up a microarray assay. Oligonucleotide probes used to target the marker gene can be long (typically 50- to 100-mers) or short (typically 15- to 30-mer) [Huges et al., 2001; Wang et al., 2003a]. Long oligonucleotide probes offer an advantage in terms of higher target-binding capacities and their use in combination with universal amplification strategies or sometimes without amplification. In most of the POA the researchers prefer to use short oligonucleotide probes because they allow the discrimination of single nucleotide differences under optimal conditions and provide a higher threshold of differentiation [Loy and Bodrossy, 2006; Taroncher-Oldenburg et al., 2003]. However, there is no universal answer to the question of which probe-target combination is best, because this will depend on the intended application.

61.3.2 Probe Design

Probe design for microarray experiments is not a trivial computational task [Militon et al., 2007]. To obtain an efficient probe selection, previously described parameters have to be considered. Militon et al. [2007] have stated that probe designing for bacterial identification is basically the same problem as probe selection for classical gene expression experiments. The only difference is the specificity test. In gene expression experiments, a probe identifying a given gene must be specific among all the other gene sequences of the studied organism. When designing oligonucleotide probes from the 16S rRNA gene, each probe must be specific among all the sequences that may be present in the samples during the hybridization step. If the mixture composition is totally unknown, the specificity can only be checked against all known 16S rRNA gene sequences. These sequences can be obtained from various major primary databases (GenBank/EMBL/DDBJ) or curated secondary databases [Cole et al., 2005; DeSantis et al., 2006a, 2006b; Ludwig et al., 2004; Wuyts et al., 2004; see also Chapter 36, Vol.

I and Chapter 46, Vol. II]. However, because the majority of the microorganisms are still unidentified, it can be said that most classical oligonucleotide design software uses incomplete datasets to design species-specific probes. Thus, in reality a small fraction of known microbes can be studied with the use of these probes. However, a few design tools try to decrease this bias by allowing the selection of probes targeting higher bacterial taxa. ARB probe design tools [Ludwig et al., 2004; see also Chapter 46, Vol. I]; PRIMROSE software [Ashelford et al., 2002] can generate these kinds of taxon specific primers. ARB is used in most of the biodiversity studies that use phylogenetic microarrays [Franke-Whittle et al., 2005; Loy et al. 2005; Sanguin et al. 2006a]. Schliep and Rahmann [2006] used a statistical group-testing approach with nonunique probes to detect targets related by a phylogenetic tree. This method can detect unknown targets, but it has been validated only on simulations of hybridization experiments. Militon et al. [2007] designed a new probe design algorithm called "PhylArray" that is able to select microarray probes targeting the 16S rRNA gene at any phylogenetic level. Application of combined PhyArray/GoArrays strategy has been shown to optimize the hybridization performances of short probes and can draw attention to even previously unknown bacteria. Another approach entitled "primers-4-clade" has been designed for lineage-specific primers (see Chapter 51, Vol. I).

61.3.3 Validation

Thorough, rigorous validation, involving an evaluation and refinement of in silico predictions as well as the adjustment of hybridization conditions, is a key in the development of POA. However, the current in silico approaches for predicting the hybridization behavior of microarray probes are limited in their accuracy. Several software packages have been developed to assist in validation process. ARB, PRIMROSE, OligoCheck, CalcOligo, Mfold, and HyTher can be used to detect the specificity, sensitivity, and/or homogeneity of the designed probes. CalcOligo, Mfold, and HyTher can be used for predicting the hybridization behavior. At best, these tools will lead to a prefiltered set of candidate probes whose true experimental performance will only be uncovered after extensive empirical testing [Sessitsch et al., 2006; Loy and Bodrossy, 2006]. In practice, a suitable set of test targets should contain at least one perfectly matched target sequence (but ideally three) for each probe on the microarray. After testing the probe set by individual hybridizations with each test target, 'bad' probes showing low sensitivity and/or specificity are removed or replaced. Subsequently, a concentration series of targets perfectly matching those probes that have displayed the highest and lowest duplex yield should be hybridized to the microarray, giving an impression of the range of sensitivities achievable for the individual probe [Loy et al., 2003; Stralis-Pavese et al., 2004]. Proper validation of arrays covering a broad phylogenetic diversity of microbes presents a serious challenge as the sequences released in databases are frequently not readily available as clones, genomic DNA, or microbial isolates [Franke-Whittle et al., 2005; Sanguin et al., 2006a].

Gary Andersen's Laboratory at Lawrence Berkeley National Laboratory (LBNL) designed a second generation of the PhyloChip (referred to as Affimetrix PhyloChip in subsequent sections) intended for widespread use [Brodie et al., 2007; DeSantis et al., 2007]. Manufactured by Affymetrix, the high-density oligonucleotide microarray containing 500,000 probes is a reliable tool for the comprehensive identification, detection, and quantification of all known prokaryotic 16S rRNA gene sequences (Bacterial and Archea). The PhyloChip targets the variation in the 16S rRNA genes of over 30,000 database sequences totaling almost 9000 distinct taxonomic groups. These groups are assayed by a set of 11 or more pairs of perfectly matching (PM) and mismatching probes (MM), making the PhyloChip (Next generation PhyloChip 3.0, with greater number of probes and higher sensitivity has recently been released) highly sensitive and specific. The validation of such high-density microarray is clearly unrealistic, but its high probe density offers ample compensation to this problem. Also, this chip is validated by comparing the detected bacterial groups by clone library sequencing and analysis [DeSantis et al., 2007].

61.3.4 Data Analysis

In spite of the growing number of applications for POA, they represent only a small proportion of all DNA-microarray related work. Most publications describe expression studies (e.g., Csako [2006], Lockhart and Winzeler [2000], Rensink and Buell [2005], and Stoughton [2005]). Consequently, the majority of protocols are optimized for applications related to expression analysis. However, the application of POA for species identification in environmental samples presents technical challenges that are not encountered in gene expression studies of laboratory samples [Call et al., 2005; Peplies et al., 2003]. There are numerous commercial and noncommercial programs for the analysis of expression studies (e.g., Dondrup et al. [2003] and Vaquerizas et al. [2005]), but few programs exist for phylochip analysis. One example is the Unix-based program ChipChecker [Loy et al., 2002], which is dedicated to data interpretation from phylochips. It calibrates signal to noise ratios to a set threshold determined by the user and finds positive signals with respect to that threshold based on the fact that a positive signal can only be located where there is a fully complementary probe to its target. However, in a hierarchical probe set, a signal is only considered truly positive if all probes in the hierarchy are positive. Therefore, the analysis of hierarchically organized phylochips requires an additional step in comparison to the functions provided by ChipChecker. The positive signals must be tested for their robustness in relation to the hierarchy on the phylochip. In summary, a program for the analysis of hierarchically organized phylochips has to provide (a) an algorithm for the calculation of a signal-to-noise value and (b) a tool that allows setting positive signals in relation to the hierarchy inherent in the design of the probe set. Metfies et al. [2008] developed a program called PhylochipAnalyzer, that implements the calculation of signal-to-noise ratios and the evaluation of phylochip data with respect to probe hierarchy. PhyloTrac is an application for the visualization and analysis of Affymetrix PhyloChip microarrays. This freely available software enables deep analysis of environmental samples through integrated visualizations, on-the-fly clustering, and dynamic filtering. Researchers can load the raw probe signal intensity values recorded from the Affimetrix PhyloChip into PhyloTrac and can then can explore and analyze the phylogenetic diversity. PhyloTrac is capable of displaying data from multiple Phylochip and traditional 16S sequencing projects in a variety of styles, including heatmap, time series/parallel coordinates, probe intensity display, difference plot, phylogenetic tree, and textual spreadsheets. Incorporation of Fast UniFrac in PhyloTrac offers orders-of-magnitude improvement over the original version [Hamady et al., 2010]. New 3D visualization of principal coordinates analysis results, with the option to view multiple coordinate axes simultaneously, provides a powerful way to quickly identify patterns that relate vast numbers of microbial communities.

61.4 CHALLENGES FOR MICROARRAY TECHNOLOGY

61.4.1 Specificity

The application of POA to the assessment of microbial diversity in the environment poses a number of technical challenges. The specificity of the probes to discriminate between target and nontarget is a major challenge. Short oligonucleotide probes used in POA generally offer greater specificity to distinguish single-nucleotide polymorphism and discern spliced variants [Hughes et al., 2001]. However, it has been noted that these probes often have poor hybridization properties. Poor signal intensities may pose difficulties in the detection of certain microbial communities which are present in very low numbers in samples. In membrane-based hybridizations, this problem is overcome by optimizing the hybridization and wash conditions (buffer composition, salt concentration, and temperature) for each probe [Stahl et al., 1988]. However, the challenge of high-density microarrays is that a large number of probes, which can vary in duplex stability due to differences in length and base composition, are hybridized and washed simultaneously under the same condition, making it impossible to optimize the conditions for all of the probes on the array [Kelly, 2009].

To discriminate between target and nontarget sequences, various approaches have been applied. In the nonequilibrium dissociation approach, the dissociation processes of both perfect PM) and MM duplexes, which are independent of time, are performed by increasing the temperature from low to high at a relatively fast rate (e.g., $0.5^{\circ}C\ min^{-1}$ to $1^{\circ}C\ min^{-1}$) [Liu et al., 2001; Urakawa et al., 2002]. This approach is performed at the washing stage, and it allows entire dissociation curves of every probe-target duplex of interest in a single experiment. Another approach to optimize specificity is the inclusion of tetramethylammonium chloride [Maskos and Southern, 1993] or betaine [Rees et al., 1993]. These compounds equalize the melting points of oligonucleotides with different base compositions by stabilizing AT base pairs [Peplies et al., 2003] so that a single wash temperature can be used for all probes [Loy et al., 2002]. Another approach is to design probe sets with identical melting temperature so that the microarray can be hybridized and washed under conditions that will minimize nonspecific hybridizations. Although effective [Bodrossy et al., 2003; Zhang et al., 2007], this technique places significant constraints on probe design. We should keep in mind that there is no single universal condition to ensure absolute specificity of all the probes during simultaneous hybridization. However, microarrays offer the possibility to compensate for lack of specificity on the single probe level by inclusion of a multiplicity of redundant probes [Loy and Bodrossy, 2006]. To increase the specificity, the Affymetrix PhyloChip was designed to have a minimum of 11 different, short oligonucleotide probes for each taxonomic grouping, allowing the failure of one or more probes. Specificity was ensured by the inclusion of MM-control probes on the microarray [Brodie et al., 2007; see also Chapter 58, Vol. I]. Comparison of signal intensities from PM and MM probes allows cross-hybridization to be identified and its extent is further estimated.

61.4.2 Sensitivity

The sensitivity of POA (and other environmental microarrays) is usually defined as the lowest relative abundance of the target group detectable within the analyzed community [Bodrossy and Sessitsch, 2004]. The sensitivity is normally limited by the relative abundance of the microbial population within the targeted community, with reported detection limits being 1–5%. Most of the

POA are glass-based and have a lower hybridization sensitivity as compared to membrane-based hybridization. This is because the probe binding capacity of glass is much lower than that on porous membranes. Increasing binding capacity is one way to enhance microarray hybridization sensitivity. Some enzyme-based labeling methods have also been reported to increase sensitivity [Loy and Bodrossy, 2006; Sessitsch et al., 2006]. Further research is warranted on the development and use of new slide chemistries including ultrathin three-dimensional platforms, which will enable increased binding capacities with high density arrays and increase sensitivity [Guschin et al., 1997; Urakawa et al., 2003; Ji et al., 2004]. The most common approach used to increase sensitivity and detect less dominant populations is to PCR amplify the community DNA. However, this potentially introduces other well-documented biases and limitations [Crosby and Criddle, 2003; Reysenbach et al., 1992; see also Chapters 16–17; Chapter 58, Vol. I]. Sensitivity can also be increased by using stringent detection and quantification criteria. In the case of the Affymetrix PhyloChip, positive probe prime pairs should meet two criteria: (i) Intensity of fluorescence from the PM probe should be greater than 1.3 times the intensity from MM control; and (ii) the difference in intensity, PM minus MM, should be at least 500 times greater than the squared noise value ($>500\ N^2$). These two empirically chosen criteria provide stringency while maintaining sensitivity to the amplicons known to be present from cloning [Brodie et al., 2007].

61.4.3 Quantification

It should be possible to extract quantitative information from microarray hybridization if the amount of hybridization to a probe (fluorescent intensity) correlates with the amount of target present in the sample. There have been some concerns regarding the quantitative ability of the environmental microarrays given the potential variability in steps including DNA extraction, labeling, hybridization, and analysis. However, recent research indicates that FGAs and CGAs can be quantitative within a range of concentrations [Rhee et al., 2004; Wu et al., 2004]. Determination of the quantitative potential of POA and WGAs is difficult because they are based on the perfect mismatch probes. Ward et al. [2007] have demonstrated that different probes can yield dramatically different hybridization signals even when hybridized to equal amounts of their respective targets. These situations make it extremely challenging to extract quantitative information on relative target abundance from multiple probes within a microarray. Brodie et al. [2007] tested the ability of the Affimetrix PhyloChip to track 16S rRNA amplicon dynamics quantitatively by using a Latin Square-type study containing mixtures of amplified 16S rRNA gene from a variety of microbes applied to the microchip, in rotating concentrations. A strong linear relationship, spanning five orders of magnitude between Affimetrix PhyloChip intensities and quantities of bacterial 16S rRNA gene signatures applied, was observed. Real-time quantitative PCR demonstrated that changes in Affimetrix PhyloChip intensities were representative of the dynamics of selected microorganisms. This underlines the value of microarray not only as a tool for qualitative microbial ecology but also to provide quantitative information about the abundance of specific organisms.

61.4.4 Cost

Another major constraint lies in the expense of microarray chips and the equipment used (printing and imaging). Apparently a single chip and the reagents needed costs about $1000, and they can also be used only once. Although this may seem expensive, the unparalleled resolution and detection power as compared to any other technology can compensate the associated cost of using microarrays. In addition to the identification of millions of microbes present in a sample, basic genotype and semi-quantitative information such as relative abundance can be obtained. In the long run, full-scale automation of microarray technology and the increase in demand and supply will further reduce the cost as observed for various other genome-based technologies. Also a previous thought process on the nature of the biological question, treatments and comparisons that minimize the unwanted variations or noise, and a sound experimental design will maximize the information return while minimizing the cost [Churchill, 2002; Yang and Speed, 2002].

61.5 APPLICATION OF POAs

As outlined above, POAs are based on the use of conserved phylogenetic markers and are used in order to detect specific bacteria such as pathogens [Franke-Whittle et al., 2005] or to study diversity and structure of the microbial community in a given environment [Günther et al., 2005; Loy et al. 2002, 2004; Sagaram et al., 2009). In addition of addressing phylogenetic relatedness of microbes in a sample, POAs also yield substantial functional information on a particular ecosystem [Loy et al., 2002].

Günther et al. [2005] developed a microarray to detect members of the genus *Kitasatospora* and to differentiate them from the related genus *Streptomyces*. Because the 16S/23S rRNA genes do not resolve a close phylogenetic relationship between these two groups of actinomycetes, the ITS region was used to design probes. The microarray contained 29 specific oligonucleotides and was applied to detect *Kitasatospora* in spiked and forest soil. The study shows that a microarray could be used for an accelerated screening of soils, for sorting already known species, and for further processing interesting samples [Günther et al., 2005]. To develop a molecular tool that would allow screening for the presence or absence of different microorganisms within compost, a microarray using variable regions of the 16S rRNA gene was designed and validated [Franke-Whittle et al., 2005]. The developed microarray contained 12 probes with differing levels of specificity targeting actinomycetes and other organisms with significance as degraders and 35 probes specific to pathogens. The application of the microarray to compost samples indicated the presence of *Streptococcus, Acinetobacter lwoffii*, and *Clostridium tetani* in various compost samples. The microarray was suggested to offer potential for process monitoring and the detection of pathogens and beneficial bacteria. Schönmann et al. [2009] developed an oligonucleotide microarray consisting of 131 hierarchically nested 16S rRNA gene-targeted oligonucleotide probes for cultivation-independent and highly parallel analysis of members of the genus *Burkholderia*. The evaluated microarray was applied to investigate shifts in the *Burkholderia* community structure in acidic forest soil upon addition of cadmium, a condition that selected for *Burkholderia* species. The microarray results were in agreement with those obtained from phylogenetic analysis of *Burkholderia* 16S rRNA gene sequences recovered from the same cadmium-contaminated soil, demonstrating the value of the *Burkholderia* phylochip for determinative and environmental studies.

One of the most illustrative ecological applications of POAs comes from the study of Loy et al. [2002]. They developed an oligonucleotide microarray consisting of 132 oligonucleotide probes (18-mers) targeting the 16S rRNA gene, having hierarchical and parallel (identical) specificity for the detection of all known lineages of sulfate-reducing prokaryotes (SRP-PhyloChip). The chip was subsequently evaluated with 41 suitable pure cultures of SRPs. The applicability of SRP-PhyloChip for diversity screening of SRPs in environmental and clinical samples was tested by using samples from periodontal tooth pockets and from the chemocline of a hypersaline cyanobacterial mat from Solar Lake (Sinai, Egypt). SRP-PhyloChip indicated the occurrence of *Desulfomicrobium* spp. in the tooth pockets and the presence of *Desulfonema*- and *Desulfomonile*-like SRPs (together with other SRPs) in the chemocline of the mat. The SRP-PhyloChip results were confirmed by several DNA microarray-independent techniques, including specific PCR amplification, cloning, and sequencing of SRP 16S rRNA genes and the genes encoding the dissimilatory (bi)sulfite reductase (*dsrAB*). The development of the SRP-PhyloChip was made possible by the availability of a large sulfate-reducing bacteria-specific rRNA probe database known as "ProbeBase." This is particularly noteworthy because it shows the importance and power of a comprehensive probe database in POA development and application. The same group developed another 16S rRNA gene-targeted oligonucleotide microarray (RHC-PhyloChip) consisting of 79 probes for the simultaneous identification of members of the betaproteobacterial order "*Rhodocyclales*" in environmental samples [Loy et al., 2004]. The 16S rRNA gene sequences from all cultured and as yet uncultured members of the "*Rhodocyclales*" were used for probe designing. The implementation of a newly designed "*Rhodocyclales*"-selective PCR amplification system prior to microarray hybridization greatly enhanced the sensitivity of the RHC-PhyloChip and thus enabled the detection of "*Rhodocyclales*" populations with relative abundances of less than 1% of all bacteria (as determined by FISH) in the activated sludge. The presence of as yet uncultured bacteria in the industrial activated sludge, as indicated by the RHC-PhyloChip analysis, was confirmed by retrieval of their 16S rRNA gene sequences and subsequent phylogenetic analysis, demonstrating the suitability of the RHC-PhyloChip as a novel monitoring tool for environmental microbiology [Loy et al., 2004].

The usability of the DNA microarray format for the specific detection of bacteria based on their 16S rRNA genes was systematically evaluated with a model system composed of six environmental strains and 20 oligonucleotide probes [Peplies et al., 2004]. The overall aim of this study was to investigate the influence of parameters such as secondary structure and steric hindrance on the specificity of hybridization. Because the oligonucleotide probes were originally designed for FISH protocol, they also established the transfer ability of this procedure to the DNA chip system but with the added flexibility of multiple probes and allowing the validation of both methods by each other. The addition of helper oligonucleotides, intended to attach adjacent to a probe recognition location and to open the secondary structure of the target molecule, improved the sensitivity. With adequate hybridization conditions, false-positive signals could be almost completely prevented, resulting in clear data interpretation. Among 199 potential nonspecific hybridization events, only one false-positive signal was observed, whereas false-negative results were more common (17 of 41). The results showed that, compared to standard hybridization formats such as FISH, a large number of oligonucleotide probes with different characteristics can be applied in parallel in a highly specific way without extensive experimental effort.

A new approach to study functional activities by using a phylochip was presented by Adamczyk et al. [2003]. A small microarray consisting of 16S rRNA gene-based oligonucleotide probes targeting ammonia-oxidizing bacteria was used to identify cells that consume ^{14}C-labeled substrates and that were hybridized with the microarray and subsequently scanned for fluorescence as well as for radioactivity. The suitability of the approach was demonstrated for monitoring community composition and CO_2 fixation activity of ammonia-oxidizers in two nitrifying activated sludge samples [Adamczyk et al., 2003].

A prototype of a taxonomic 16S rRNA gene-based microarray was developed for high-throughput analysis of the microbial community by providing snapshots of the microbial diversity under different environmental conditions [Sanguin et al., 2006a]. The prototype microarray was composed of 122 probes that target bacteria at various taxonomic levels from phyla to species (mostly *Alphaproteobacteria*). The array was validated by using a range of bacterial strains as hybridization targets as well as by analyzing *Agrobacterium* diversity in the rhizosphere of maize and by comparing microarray results with those obtained by clone libraries. The prototype was further modified by including 41 new probes that targeted essentially the *Proteobacteria* [Sanguin et al., 2006b]. The hierarchically nested probes were found to be reliable, but the level of taxonomic identification was variable, depending on the probe set specificity. A comparison of the maize rhizosphere and bulk soil hybridization results showed a significant rhizosphere effect, with a higher predominance of *Agrobacterium* spp. in the rhizosphere, as well as a lower prevalence of *Acidobacteria, Bacteroidetes, Verrucomicrobia*, and *Planctomycetes*. Because Proteobacteria play an important role in plant–microbe interactions, such a taxonomic microarray offered extensive possibilities for systematic exploration and understanding the ecology of plant-associated microorganisms.

Recently, the HITChip, an oligonucleotide microarray for the phylogenetic profiling of human intestinal tract communities, was developed [Rajilic-Stojanovic et al., 2009]. The 4800 probes on this 16S rRNA gene tailing microarray consist of three 18- to 30-nt-long overlapping nucleotides targeting the V1 and V6 region sequences from 1140 phylotypes, respectively. Using HITChip, Rajilic-Stojanovic et al. [2009] confirmed a previous finding that the adult fecal microbiota is highly individual specific and relatively stable over time. With the aid of this technology, it was also shown that a multispecies probiotic cocktail alleviated symptoms of irritable bowel syndrome. Claesson et al. [2009] validated HITChip and reported that chip hybridizations and resulting community profiles correlate well with pyrosequensing-based composition.

Affimetrix PhyloChip has been applied to a wide variety of environments (**Table 61.1**; also see Section 61.6). In each case, it was demonstrated and validated to reveal a much broader diversity than typical 16S rRNA gene clone library. Because of its capability, low cost, and high-throughput processing, the technology is revolutionizing the field of microbial ecology and metagenomics by providing a broader picture of microbial diversity in a rapid and low-cost way.

61.6 CASE STUDY: *CANDIDATUS* LIBERIBACTER ASIATICUS AS THE PATHOGEN RESPONSIBLE FOR HUANGLONGBING (HLB) DISEASE IN FLORIDA

For a full outline, please refer to the original publication [Sagaram et al., 2009]. What follows is an abbreviated version.

We have reported the use of two 16S rRNA gene-based culture independent methods, clone library sequencing (the current "gold standard" in microbial ecology) and a high-density oligonucleotide microarray (Affymetrix PhyloChip), to evaluate the microbial community composition of the leaf midribs of citrus HLB (citrus greening) symptomatic and asymptomatic citrus. HLB is a highly destructive, fast-spreading disease of citrus and is linked to a fastidious, gram-negative, phloem-limited bacterium, *Candidatus* Liberibacter spp. Out of three species of HLB pathogens, only *Candidatus*. Liberibacter asiaticus (Las) is reported in Florida. Koch's postulates have not been completed to prove Las to be the sole causal agent due to the inability to culture the pathogen. Also unknown is the relationship of Las to the native microbial communities present in citrus.

As mentioned previously, clone libraries involve the sequencing of a few hundred PCR amplified 16S rRNA gene. Although the method provides great specificity, it may profile only the dominant phylotypes in a complex environmental mixture. Therefore, less abundant species that may contribute to disease pathogenesis will remain undetected. For this reason, the PhyloChip approach was applied in parallel with 16S rRNA gene clone library sequencing to determine changes in the bacterial community composition in HLB symptomatic and asymptomatic leaf mibribs of citrus collected from two different groves.

The clone library analysis using PCR products amplified by bacteria specific primers (799f-1492r) showed the presence of 20 orders of bacteria belonging to eight recognized phyla (**Fig. 61.1**). Rarefaction curves showed that none of the four clone libraries reached its plateau, and additional cloning reactions are necessary to have a glimpse of the total bacterial diversity of citrus (**Fig. 61.2**). The presence of Las was detected in all the symptomatic samples and was verified by Q-PCR-based analysis using Las-specific primer probe combination (results not present in original text). Clones representing Las were also observed in a few of the asymptomatic samples (**Fig. 61.3**). This suggests that a certain number

Table 61.1 Highlights from Selected Studies that Applied Affymetrix PhyloChip Analyses to Microbial Ecology Research

Sample	Major Objective(s)	Principle Finding(s)	Reference
Soil	To determine if the remobilization of soluble uranium U (VI) was associated with alterations with microbial populations.	The observed reoxidation of uranium under reducing conditions occurred despite elevated microbial activity and consistent presence of metal reducing bacteria.	Brodie et al. [2006]; see also Chapter 58, Vol. I
Alkaline saline groundwater	The microbial diversity and metabolic activity of a 3- to 4-km-deep fracture in the 2.7-billion-year-old Ventersdorp Supergroup metabasalt, in which fracture water ages of tens of millions of years was accessed to determine the long-term sustainability of this deep terrestrial environment.	Geochemical, microbiological, and molecular analyses of alkaline saline groundwater at 2.8-km depth in Archaean metabasalt revealed a microbial biome dominated by a single phylotype affiliated with thermophilic sulfate reducers belonging to Firmicutes. These sulfate reducers were sustained by geologically produced sulfate and hydrogen at concentrations sufficient to maintain activities for millions of years with no apparent reliance on photosynthetically derived substrates.	Lin et al. [2006]
Urban aerosols	To access the bacterial composition of environment aerosols and how it changes overtime and with location.	The richness of microbes in urban aerosol was equivalent to soil bacterial community. Some bacterial families with pathogenic members including environmental relatives of select agents of bioterrorism significance were consistently observed.	Brodie et al. [2007]
Endotracheal aspirates	To determine changes in the bacterial community composition of the lungs of intubated patients during antibiotic treatment *Pseudomonas aeruginosa*.	Loss of bacteria diversity under antibiotic selection is highly associated with the development of pneumonia in ventilated patients colonized with *P. aeruginosa*.	Flanagan et al. [2007]
Urban aerosols, subsurface soils, subsurface waters	To determine the breadth and accuracy of microarray in accessing microbial diversity as compared to clone library analysis.	Although the microarray is unreliable in identifying novel bacterial taxa, it reveals greater diversity as compared to cloning. Furthermore, the microarray allowed samples to be rapidly evaluated with replication.	DeSantis et al. [2007]
Deep sediment cores	To determine the quantity, diversity, and distribution of microbial communities in the context of abiotic and geochemical properties in gas-rich marine sediments.	The PhyloChip detected the presence of methanogens, sulfate reducers, sulfur oxidizers, and other metal reducers, as more prevalent taxa.	Briggs et al. [2008]
Fracture water	To determine the microbial community composition of low biodiversity fracture water collected at 2.8-km depth in a South African gold mine.	*Candidatus Desulforudis audaxviator* was found to comprise >99.9% of the microorganisms inhabiting the fluid phase of this particular fracture. The bacterium is capable of an independent lifestyle well-suited to long-term isolation from the photosphere deep within the Earth's crust, and it offers the first example of a natural ecosystem that appears to have its biological component entirely encoded within a single genome.	Chivian et al. [2008]
Microbial fuel cells	To identify organisms enriched only in current-producing reactors. First application of the PhyloChip to anode biofilm communities.	Research illustrates the importance of using a variety of molecular and culture-based methods to reliably characterize bacterial communities. Consequently, a previously unidentified functional role for gram-positive bacteria in MFC current generation was observed.	Wrighton et al. [2008]

(continued)

Table 61.1 Highlights from Selected Studies that Applied Affymetrix PhyloChip Analyses to Microbial Ecology Research *(Continued)*

Sample	Major Objective(s)	Principle Finding(s)	Reference
Rhizosphere	To determine the changes in microbial community composition in response to environmental changes accompanying root movement through soils.	Changes in relatively small subset of soil microbial community are sufficient to produce substantial changes in functions observed earlier in progressively more mature rhizosphere zones.	DeAngelis et al. [2009]
Leaf midrib	Detection of microbial community structure associated with Huanglongbing symptomatic and asymptomatic citrus leaves.	Data implicate "*Candidatus* Liberibacter asiaticus" as the pathogen responsible for greening disease in Florida.	Sagaram et al. [2009]
Soil	To determine suitability of PhyloChip to monitor Antarctic soil communities and to assess the feasibility of linking GeoChip and PhyloChip analysis.	Integration of PhyloChip data with other complementary methods provide an unprecedented understanding of the microbial community structure and function.	Yergeau et al. [2009]
Mining-impacted soils	To characterize microbial diversity in mining-impacted soils collected from two abandoned uranium mine sites	Highly diverse microbial populations were present in these uranium mine sites.	Rastogi et al. [2010]
Cleanroom surface	To perform a census of cleanroom surface-associated bacterial populations.	A larger bacterial diversity was present prior to the arrival of spacecraft hardware in cleanroom facilities. The robust nature and high sensitivity of DNA microarray technologies should prove beneficial to a wide range of scientific, electronic, homeland security, medical, pharmaceutical, and any other ventures with vested interest in monitoring and controlling contamination in exceptionally clean environments.	La Duc et al. [2009]; see also Chapter 60, Vol. I
Avian Egg	To determine the effect on avian incubation on the risk reduction of trans-shell infection.	Incubation inhibits all of the relatively few bacteria that grow on eggshells, and does not appear to promote growth of any bacteria.	Shawkey et al. [2009]
Coral reef	To determine bacterial diversity and white plague disease-associated community changes in the Caribbean coral *Montastraea faveolata*.	Although an ecological succession of bacteria during disease progression after causation by a primary agent represents a possible explanation for changes in diversity, the possibility that a disease of yet-to-be-determined etiology may have affected *M. faveolata* colonies and resulted in an increase in opportunistic pathogens.	Sunagawa et al. [2009]

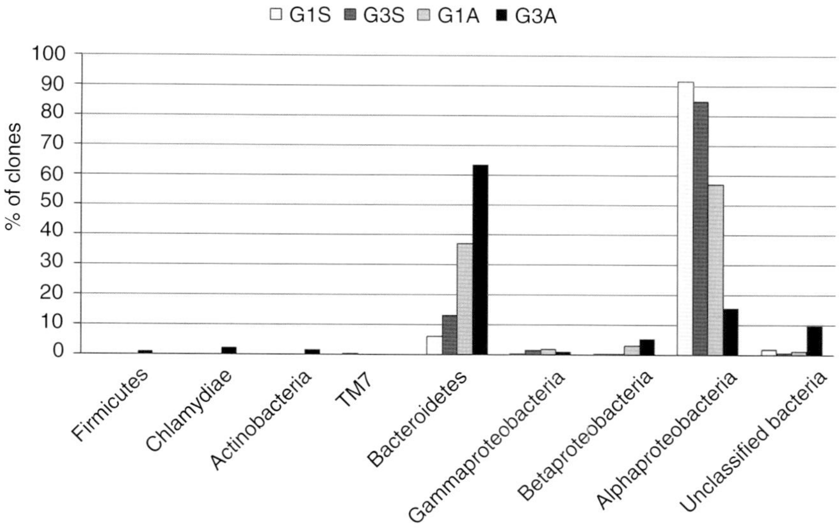

Figure 61.1 16S rRNA gene clone library compositions at the phylum level (class level for Proteobacteria) using the RDP Classifier tool with 80% confidence level. The *y* axis represents the abundance (%) of each taxon within a given library. The numbers of clones in the libraries and dataset were as follows: Grove 1 Asymptomatic (G1A), 190; Grove 1 Symptomatic (G1S), 407; Grove 2 (G3A) Asymptomatic, 188; and Grove 2 (G3S) symptomatic, 501. Totally there were 776 clones matching chloroplast or mitochondria products which were not included in this analysis.

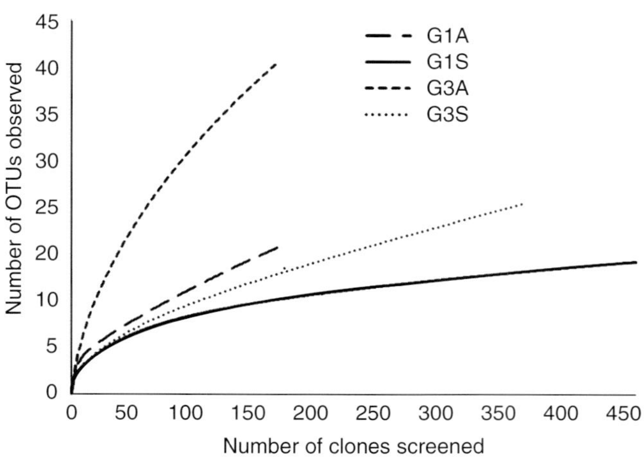

Figure 61.2 Rarefaction curves of bacterial 16S rRNA gene clone libraries of HLB symptomatic and asymptomatic citrus leaves. G1A, Grove 1 Asymptomatic; G1S, Grove 1 Symptomatic; G3A, Grove 2 Asymptomatic; G3S, Grove 2 (G3S) symptomatic.

of Las bacteria are required for symptom development. An important point to be noticed is that in many samples (both asymptomatic and symptomatic) the majority of phylotypes represented in the clone library were Las. This shows the limitation of clone library while analyzing the overall bacterial diversity as it reports mostly the dominant bacteria present in the samples.

A comparison between clone library sequencing and PhyloChip analysis of the microbial community showed that PhyloChip analysis detected a broader richness of taxa than cloning (**Table 61.2**). This resolution was significantly greater as compared to clone library analysis. The only striking similarity was observed in the abundance of Alphaprotobacteria, especially the OTU showing similarity to Las. The taxon otu_{7603}, representing Las, was detected at a very low level in asymptomatic plants, and it was 200 times more abundant in symptomatic plants. This, along with clone library results, could establish Las as the causal agent of HLB in Florida. In addition to Las, PhyloChip also revealed eight other taxa that were observed to be more abundant in symptomatic samples (**Fig. 61.4**). These variations were not observed in the clone libraries.

Overall, both methods showed that Las is the dominant organism in the symptomatic leaves compared to asymptomatic leaves, implicating the organisms as the causal agent of HLB disease. The phylochip was clearly superior to clone library sequencing for assessing the complete spectrum of bacteria present in the community including low-abundant species. Unlike sequencing, the PhyloChip method does not require cloning and is therefore rapid enough that it could be used to provide information about bacterial adundance and diversity in complex environmental settings.

61.7 CONCLUSIONS

Microarrays are powerful genomic tools, designed to illuminate differences in the expression of genes within cells. Despite being a relatively new technology, the scientific community has quickly adopted its use in a variety of fields including microbial ecology studies. Various types of microarrays have a potential to revolutionize the field of microbial ecology via high-throughput analysis of microbial community structure, function, and population dynamics. There are significant challenges to the use of microarray technology which include optimization of specificity and sensitivity and quantification of targets. Because the microarray technology is still in the early phase of development in the field of microbial ecology, intensive studies with careful

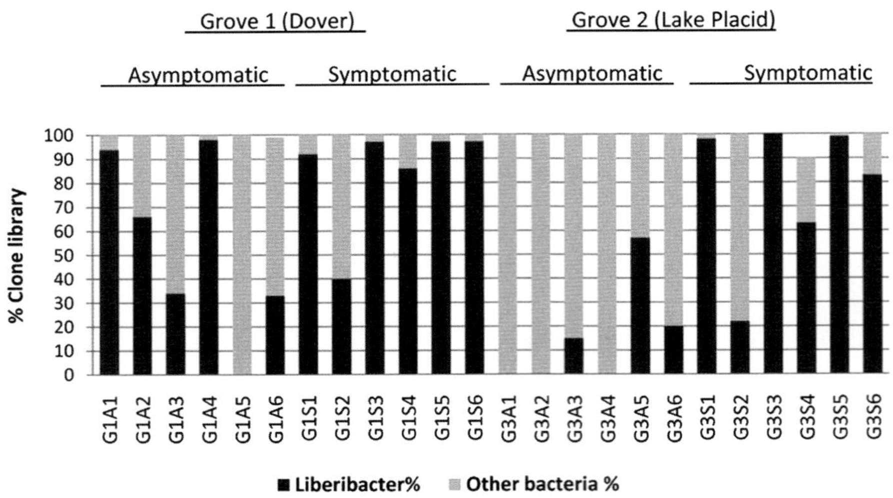

Figure 61.3 Prevalence of Liberibacter in clone libraries from asymptomatic and symptomatic trees in each of the two groves sampled.

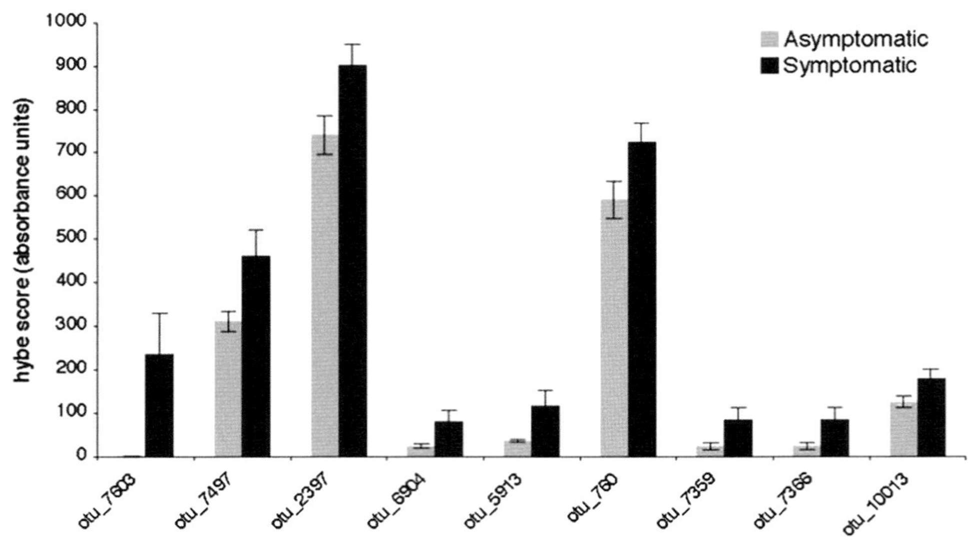

Key:

otu id	Phylum	Class	Order	Nearest neighbor
otu_7603	Proteobacteria	Alphaproteobacteria	Rhizobiales	*Liberibacter asiaticus*
otu_7497	Proteobacteria	Alphaproteobacteria	Rhizobiales	*Phyllobacter loti*
otu_2397	Chloroflexi	Dehalococcoidetes	Unclassified	*Dehalococcoides ethenogenes*
otu_6904	Proteobacteria	Alphaproteobacteria	Caulobacterales	*Brevundimonas vesicularis*
otu_5913	Bacteroidetes	Sphingobacteria	Sphingobacteriales	Sphingobacterium clone
otu_760	Verrucomicrobia	Verrucomicrobiae	Verrucomicrobiales	Verrucomicrobia clone
otu_7359	Proteobacteria	Alphaproteobacteria	Caulobacterales	*Caulobacter bacteroides*
otu_7366	Proteobacteria	Alphaproteobacteria	Caulobacterales	*Brevundimonas subvibroides*
otu_10013	Proteobacteria	Deltaproteobacteria	Syntrophobacterales	*Geobacter metallireducens*

Figure 61.4 Mean hybridization scores (hybe score) for 9 of 117 taxa detected in the leaf midrib microbial community. These nine taxa were significantly different ($P < 0.05$) for the symptomatic and asymptomatic leaves in each grove. The error bar indicates standard errors.

selection of probes, along with vigorous and systematic optimization of hybridization conditions, should continue to modify and improve the POA technique. Research is needed for the development of additional genetic markers of community diversity to enhance the phylogenetic and functional resolution of microbial communities. Large databases, comparable in size to that of the 16S rRNA gene databases, are critically needed for high-resolution phylogenetic markers. It is also critical to develop highly sensitive and specific direct nucleic acid isolation and amplification methods. Breakthroughs in various aspects of the technology from

Table 61.2 Phyla Detected in Different Samples by High-Density PhyloChip Analysis or by Cloning and Sequencing

	PhyloChip				Cloning and Sequencing			
	Asymptomatic		Symptomatic		Asymptomatic		Symptomatic	
Phylum/Class	G1A	G3A	G1S	G3S	G1A	G3A	G1S	G3S
Proteobacteria	Y	Y	Y	Y	Y	Y	Y	Y
Alphaproteobacteria	Y	Y	Y	Y	Y	Y	Y	Y
Betaproteobacteria		Y	Y	Y	Y	Y	Y	Y
Deltaproteobacteria	Y			Y			Y	Y
Episilonproteobacteria				Y				
Gammaproteobacteria	Y		Y	Y	Y	Y		Y
Acidobacteria	Y	Y	Y	Y				
Actinobacteria	Y	Y	Y	Y		Y	Y	Y
AD3	Y		Y	Y				
Bacteroidetes	Y		Y	Y	Y	Y	Y	Y
BRC1	Y	Y	Y	Y				
Chlamydiae	Y	Y	Y	Y		Y		
Chlorobi	Y	Y	Y	Y				
Chloroflexi		Y	Y	Y				
Dictyoglomi					Y	Y	Y	
Firmicutes				Y		Y		
Gemmatimonadetes				Y				
NC10				Y				
Planctomycetes	Y	Y	Y	Y				
TM7	Y			Y		Y		
Unclassified Bacteria	Y		Y	Y	Y	Y	Y	Y
Verrucomicrobia	Y	Y	Y	Y	Y			

Y denotes positive; blank denotes negative.

fabrication to commercialization will help in large-scale application of POA in the exploration of a wide variety of ecological niches. Advancements in software and robotics will help POA to become more inexpensive, robust, and reliable. Further modifications in the form of an integrated platform like lab-on-a-chip (a system that combines multiple manipulations including sample mixing, labeling, and separation) can have a profound effect on the efficiency of POA. Integration of POA analysis with that of other microarrays such as FGA and MGA will greatly contribute to the elucidation of structure and function of microbial communities. Further coupling of these datasets with physical and chemical processes will certainly contribute to a better understanding of ecosystem dynamics. Although the overall microbial diversity is yet to be fully known and the number of species analyzed is likely to be far lower than the complexity encountered in a typical environmental sample, this review illustrates the future use of microarray technology to monitor microbial communities and their interactions under different environments. The outcome of such studies will provide insights for understanding not only the dynamics of microbial communities but also the effects of environmental parameters on microbial adaptation.

SOFTWARE INDICATED IN THE TEXT AND SOURCE

ARB: http://www.arb-home.de/

PRIMROSE: http://www.bioinformatics-toolkit.org/Primrose/

PhylArray: http://fc.isima.fr/~rimour/phylarray/

GoArray: http://www.isima.fr/bioinfo/goarray/

OligoCheck: http://www.bioinformatics-toolkit.org/Dandelion/

Mfold: http://mfold.bioinfo.rpi.edu/

HyTher: http://ozone2.chem.wayne.edu Hyther/hyther m1main.html

CalcOligo: http://www.calcoligo.org/

ChipChecker: http://chip.chemie.uni-karlsruhe.de/chipcheck/

PhylochipAnalyzer: http://www.awi.de/en/go/phylochipanalyzer

PhyloTrac: http://www.phylotrac.org

Fast UniFrac: http://bmf2.colorado.edu/fastunifrac/index.psp

ProbeBase: http://www.microbial-ecology.de/probebase

Acknowledgments

This work has been supported by Florida Citrus Production Research Advisory Council (FCPRAC).

REFERENCES

ADAMCZYK J, HESSELSOE M, IVERSEN N, HORN M, LEHNER A, et al. 2003. The isotope array, a new tool that employs substrate-mediated labeling of rRNA for determination of microbial community structure and function. *Appl. Environ. Microbiol.* **69**:6875–6887.

ASHELFORD KE, WEIGHTMAN AJ, FRY JC. 2002. PRIMROSE: A computer program for generating and estimating the phylogenetic range of 16S rRNA oligonucleotide probes and primers in conjunction with the RDP-II database. *Nucleic Acids Res.* **30**:3481–3489.

BAE JW, PARK YH. 2006. Homogeneous versus heterogeneous probes for microbial ecological microarrays. *Trends Biotechnol.* **24**:318–323.

BODROSSY L, SESSITSCH A. 2004. Oligonucleotide microarrays in microbial diagnostics. *Curr. Opin. Microbiol.* **7**:245–254.

BODROSSY L, STRALIS-PAVESE N, MURRELL JC, RADAJEWSKI S, WEILHARTER A, SESSITSCH A. 2003. Development and validation of a diagnostic microbial microarray for methanotrophs. *Environ. Microbiol.* **5**:566–582.

BRIGGS B, COLWELL F, CARINI P, TORRES M. 2008. Distribution of the dominant microbial communities in marine sediments containing high concentrations of gas hydrates. In *Proceedings of the 6th International Conference on Gas Hydrates* (*ICGH 2008*), Vancouver, British Columbia, Canada, July 6–10, 2008.

BRODIE EL, DESANTIS TZ, JOYNER DC, BAEK SM, LARSEN JT, et al. 2006. Application of a high-density oligonucleotide microarray approach to study bacterial population dynamics during uranium reduction and reoxidation. *Appl. Environ. Microbiol.* **72**:6288–6298.

BRODIE EL, DESANTIS TZ, PARKER JP, ZUBIETTA IX, PICENO YM, ANDERSEN GL. 2007. Urban aerosols harbor diverse and dynamic bacterial populations. *Proc. Natl. Acad. Sci. USA* **104**:299–304.

CALL DR. 2005. Challenges and opportunities for pathogen detection using DNA microarrays. *Crit. Rev. Microbiol.* **31**:91–99.

CHEE M, YANG R, HUBBELL E, BERNO A, HUANG XC, STERN D, WINKLER J, LOCKHART DJ, MORRIS MS, FODOR SPA. 1996. Accessing genetic information with high-density DNA arrays. *Science* **274**:610–614.

CHIVIAN C, BRODIE EL, ALM EJ, CULLEY DE, DEHAL PS, et al. 2008. Environmental genomics reveals a single-species ecosystem deep within earth. *Science* **322**:275–278.

CHURCHILL GA. 2002. Fundamentals of experimental design for cDNA microarrays. *Nature Genet.* **32**(Suppl): 490–495.

CLAESSON MJ, O'SULLIVAN O, WANG Q, NIKKILÄ J, MARCHESI JR, et al. 2009 Comparative analysis of pyrosequencing and a phylogenetic microarray for exploring microbial community structures in the human distal intestine. *PLoS ONE* **4**(8):6669.

COLE JR, CHAI B, MARSH TL, FARRIS RJ, WANG Q, KULAM SA, CHANDRA S, MCGARRELL DM, SCHMIDT TM, GARRITY GM., et al. 2003. The Ribosomal Database Project (RDP-II): Previewing a new autoaligner that allows regular updates and the new prokaryotic taxonomy. *Nucleic Acids Res.* **31**:442–443.

COLE R, CHAI B, FARRIS RJ, WANG Q, KULAM SA, MCGARRELL DM, GARRITY GM, TIEDJE JM. 2005. The Ribosomal Database Project (RDP-II): Sequences and tools for high-throughput rRNA analysis. *Nucleic Acids Res.* **33**(Suppl 1): D294–D296.

CROSBY LD, CRIDDLE CS. 2003. Understanding systematic error in microbial community analysis techniques as a result of ribosomal RNA (rrn) operon copy number. *BioTechniques* **34**:790–803.

CSAKO G. 2006. Present and future of rapid and/or high-throughut methods for nucleic acid testing. *Clin. Chim. Acta* **363**:6–31.

CURTIS TP, SOLAN WT. 2004. Prokaryotic diversity and its limits: Microbial community structure in nature and implications for microbial ecology. *Curr. Opin. Microbiol.* **7**:221–226.

DEANGELIS KM, BRODIE EL, DESANTIS TZ, ANDERSEN GL, LINDOW SE, FIRESTONE MK. 2009 Selective progressive response of soil microbial community to wild oat roots. *ISME J.* **3**:168–178.

DEMANECHE S, DAVID MM, NAVARRO E, SIMONET P, VOGEL TM. 2009. Evaluation and functional gene enhrichment in a soil metagenomic clone library. *J. Microbiol. Methods* **76**:105–107.

DESANTIS TZ, HUGENHOLTZ P, KELLER K, BRODIE EL, LARSEN N, PICENO YM et al. 2006a. NAST: A multiple sequence alignment server for comparative analysis of 16S rRNA genes. *Nucleic Acids Res.* **34**: D394–D399.

DESANTIS, TZ, HUGENHOLTZ P, LARSEN N, ROJAS M, BRODIE EL, KELLER K et al. 2006b. Greengenes, a chimera-checked 16S rRNA gene database and workbench compatible with ARB. *Appl. Environ. Microbiol.* **72**:5069–5072.

DESANTIS TZ, BRODIE EL, MOBERG JP, ZUBIETA IX, PICENO YM, ANDERSEN GL. 2007. High-density universal 16S rRNA microarray analysis reveals broader diversity than typical clone library when sampling the environment. *Microb. Ecol.* **53**:371–383.

DONDRUP M, GOESMANN A, BARTELS D, KALINOWSKI J, KRAUSE L, et al. 2003. EMMA: A platform for consistent storage and efficient analysis of microarray data. *J. Biotechnol.* **106**:135–146.

DUBEY SK, TRIPATHI AK, UPADHYAY SN. 2006. Exploration of soil bacterial communities for their potential as bioresource. *Bioresour. Technol.* **97**:2217–2224.

DUNBAR J, BARNS SM, TICKNOR LO, KUSKE CR. 2002. Empirical and theoretical bacterial diversity in four Arizona soils. *Appl. Environ. Microbiol.* **6**:3035–3045.

EKINS R, CHU FW. 1999. Microarrays: Their origins and applications. *Trends Biotechnol.* **17**:217–218.

FLANAGAN JL, BRODIE EL, WENG L, LYNCH SV, GARCIA O, et al. 2007. Loss of bacterial diversity during antibiotic treatment of intubated patients colonized with *Pseudomonas aeruginosa*. *J. Clin. Microbiol.* **45**:1954–1962.

FRANKE-WHITTLE I, KLAMMER S, INSAM H. 2005. Design and application of an oligonucleotide microarray for the investigation of compost microbial communities. *J. Microbiol. Methods* **62**:37–56.

GENTRY TJ, WICKHAM GS, SCHADT CW, HE Z, ZHOU J. 2006. Microarray applications in microbial ecology research. *Microb. Ecol.* **52**:159–175.

GÜNTHER S, GROTH I, GRABLEY S, MUNDER T. 2005. Design and evaluation of an oligonucleotide microarrays for the detection of different species of the genus *Kitasatospora*. *J. Microbiol. Methods* **65**:226–236.

GUSCHIN DY, MOBARRY BK, PROUDNIKOV D, STAHL DA, RITTMANN BE, MIRZABEKOV AD. 1997. Oligonucleotide microchips as genosensors for determinative and environmental studies in microbiology. *Appl. Environ. Microbiol.* **63**:2397–2402.

HAMADY M, LOZUPONE C, KNIGHT R. 2010. Fast UniFrac: facilitating high-throughput phylogenetic analyses of microbial communities including analysis of pyrosequencing and PhyloChip data. *ISME J*. **4**:17–27.

HASHSHAM SA, WICKS LM, ROUILLARD JM, GULARI E, TIEDJE JM. 2004. Potential of DNA microarrays for developing parallel detection tools (PDTs) for microorganisms relevant to biodefense and related research needs. *Biosensors Bioelectron*. **20**:668–683.

HE Z, GENTRY TJ, SCHADT CW, WU L, LIEBICH J, et al. 2007. GeoChip: A comprehensive microarray for investigating biogeochemical, ecological and environmental processes. *ISME J*. **1**:67–77.

HOHEISEL JD. 2006. Microarray technology: beyond transcript profiling and genotype analysis. *Nat. Rev. Genet*. **7**:200–210.

HONG SH, BUNGE J, LESLIN C, JEON S, EPSTEIN SS. 2009. Polymerase chain reaction primers miss half of rRNA microbial diversity. *ISME J*. **3**:1365–1373.

HUGHES T, MAO M, JONES A, BURCHARD J, MARTON M, et al. 2001. Expression profiling using microarrays fabricated by an ink-jet oligonucleotide synthesizer. *Nat. Biotechnol*. **19**:342–347.

JI W, ZHOU W, GREGG K, YU N, DAVIS S, DAVIS S. 2004. A method for cross-species gene expression analysis with high-density oligonucleotide arrays. *Nucleic Acids Res*. **32**:e93.

KELLY JJ. 2009. Application of DNA microarrays to microbial ecology research: history, challenges, and recent developments. *Environ. Res. J*. **3**:357–384.

KENNEDY AC. 1999. Bacterial diversity in agroecosystems. *Agric. Ecosyst. Environ*. **74**:65–76.

LA DUC MT, OSMAN S, VAISHAMPAYAN P, PICENO Y, ANDERSEN G, SPRY JA, VENKATESWARAN K. 2009. Comprehensive bacterial census of cleanrooms using DNA microarray and cloning methods. *Appl. Environ. Microbiol*. **75**:2559–2567.

LEADBETTER Jr. 2003. Cultivation of recalcitrant microbes: Cells are alive, well and revealing their secrets in the 21st century laboratory. *Curr. Opin. Microbiol*. **6**:274–281.

LI ESY, LIU WT. 2003. DNA microarray technology in microbial ecology studies-principle, applications and current limitations. *Microbes Environ*. **18**:175–187.

LIN LH, WANG PL, RUMBLE D, LIPPMANN-PIPKE J, BOICE E, et. al. 2006. Long-term sustainability of a high-energy, low-diversity crustal biome. *Science* **314**:479.

LIU WT, MIRZABEKOV AD, STAHL DA. 2001. Optimization of an oligonucleotide microchip for microbial identification studies: A non-equilibrium dissociation approach. *Environ. Microbiol*. **3**:619–629.

LOCKHART DJ, WINZELER EA. 2000. Genomics, gene expression and DNA arrays. *Nature* **405**:827–836.

LOY A, BODROSSY L. 2006. Highly parallel microbial diagnostics using oligonucleotide microarrays. *Clin. Chim. Acta* **363**:106–119.

LOY A, LEHNER A, LEE N, ADAMCZYK J, MEIER H, ERNST J, SCHLEIFER K-H, WAGNER M. 2002. Oligonucleotide microarray for 16S rRNA gene-based detection of all recognized lineages of sulphate-reducing prokaryotes in the environment. *Appl. Environ. Microbiol*. **68**:5064–5081.

LOY A, HORN M, WAGNER M. 2003. probeBase: An online resource for rRNA-targeted oligonucleotide probes. *Nucleic Acids Res*. **31**:514–516.

LOY A, KÜSEL K, LEHNER A, DRAKE HL, WAGNER M. 2004. Microarray and functional gene analyses of sulphate-reducing prokaryotes in low-sulfate, acidic fens reveal co-ocurrence of recognized genera and novel lineages. *Appl. Environ. Microbiol*. **70**:6998–7009.

LOY A, SCHULZ C, LÜCKER S, SCHÖPFER-WENDELS A, STOECKER K, BARANYI C, LEHNER A, WAGNER M. 2005. 16S rRNA gene-based oligonucleotide microarray for environmental monitoring of the betaproteobacterial order "*Rhodocyclales*". *Appl. Environ. Microbiol*. **71**:1373–1386.

LOY A, ARNOLD R, TISCHLER P, RATTEI T, WAGNER M, HORN M. 2008. ProbeCheck—A central resource for evaluating oligonucleotide probe coverage and specificity. *Environ. Microbiol*. **10**:2894–2896.

LUCCHINI S, THOMPSON A, HINTON JCD. 2001. Microarrays for microbiologists. *Microbiology* **147**:1403–1414.

LUDWIG W, SCHLEIFER K-H. 1999. Phylogeny of bacteria beyond the 16S rRNA standard. *ASM News* **65**:752–757.

LUDWIG W, STRUNK O, WESTRAM R, RICHTER L, MEIER H, YADHUKUMAR H. et al. 2004. ARB: A software environment for sequence data. *Nucleic Acids Res*. **32**:1363–1371.

LUNN M, SLOAN WT, CURTIS TP. 2004. Estimating bacterial diversity from clone libraries with flat rank abundance distributions. *Environ. Microbiol*. **6**:1081–1085.

LYNCH JM, BENEDETTI A, INSAM H, NUTI MP, SMALLA K, TORSVIK V, NANNIPIERI P. 2004. Microbial diversity in soil: Ecological theories, the contribution of molecular techniques and the impact of transgenic plants and transgenic microorganisms. *Biol. Fertil. Soils* **40**:363–385.

MASKOS, U, SOUTHERN EM. 1993. A study of oligonucleotide reassociation with large arrays of oligonucleotides synthesised on a glass support. *Nucleic Acids Res*. **21**:4663–4669.

METFIES K, BORSUTZKI P, GESCHER C, MEDLIN LK, FRICKENHAUS S. 2008. PhylochipAnalyzer—A program for analysing hierarchical probe-sets. *Mol. Ecol. Notes* **8**:99–102.

MILITON C, RIMOUR S, MISSAOUI M, BIDERRE C, BARRA V, et al. 2007. PhylArray: Phylogenetic probe design algorithm for microarray. *Bioinformatics* **23**:2550–2557.

NANNIPIERI P, ASCHER J, CECCHERINI MT, LANDI L. 2002. Microbial diversity and soil functions. *Eur. J. Soil Sci*. **25**:655–670.

NARANG R, DUNBAR J. 2004 Modeling bacterial species abundance from small community surveys. *Microbial Ecol*. **97**:396–406.

OLSEN J, LANE EJ, GIOVANNONI SJ, PACE NR, STAHL DA. 1986. Microbial ecology and evolution: A ribosomal RNA approach. *Annu. Rev. Microbiol*. **40**:331–355.

PACE NR. 1997. A molecular view of microbial and the biosphere. *Science* **276**:734–740.

PEASE AC, SOLAS D, SULLIVAN EJ, CRONIN MT, HOLMES CP, FODOR SP. 1994. Light-generated oligonucleotide arrays for rapid DNA sequence analysis. *Proc. Natl. Acad. Sci. USA* **91**:5022–5026.

PEPLIES J, GLÖCKNER FO, AMANN R. 2003. Optimization strategies for DNA microarray-based detection of bacteria with 16S rRNA-targeting oligonucleotide probes. *Appl. Environ. Microbiol*. **69**:1397–407.

PEPLIES J, LAU SC, PERNTHALER J, AMANN R, GLOCKNER FO. 2004. Application and validation of DNA microarrays for the 16S rRNA-based analysis of marine bacterioplankton. *Environ. Microbiol*. **6**:638–645.

PROSSER JI. 2002. Molecular and functional diversity in soil microorganisms. *Plant Soil*. **244**:9–17.

RAJILIC-STOJANOVIC M, HEILIG HG, MOLENAAR D, KAJANDER K, SURAKKA A, SMIDT H, DE VOS WM. 2009. Development and application of the human intestinal tract chip, a phylogenetic microarray: analysis of universally conserved phylotypes in the abundant microbiota of young and elderly adults. *Environ Microbiol*. **11**:1736–1751.

RAPPE MS, GIOVANNONI SJ. 2003. The uncultured microbial majority. *Annu. Rev. Microbiol*. **57**:369–394.

RASTOGI G, OSMAN S, VAISHAMPAYAN PA, ANDERSEN GL, STETLER LD, SANI RK. 2010. Microbial diversity in uranium mining-impacted soils as revealed by high-density 16S microarray and clone library. *Microbial Ecol*. **59**:94–108

REES WA, YAGER TD, KORTE J, VON HIPPEL PH. 1993. Betaine can eliminate the base pair composition dependence of DNA melting. *Biochemistry* **32**:137–144.

RENSINK WA, BUELL CR. 2005. Microarray expression profiling resources for plant genomics. *Trends Plant Sci*. **10**:603–609.

Reysenbach AL, Giver LJ, Wickham GS, Pace NR. 1992. Differential amplification of rRNA genes by polymerase chain reaction. *Appl. Environ. Microbiol.* **58**:3417–3418.

Rhee S-K, Liu X, Wu L, Chong SC, Wan X, Zhou J. 2004. Detection of genes involved in biodegradation and biotransformation in microbial communities by using 50-mer oligonucleotide microarrays. *Appl. Environ. Microbiol.* **70**:4303–4317.

Richmond CS, Glasner JD, Mau R, Jin H, Blattner FR. 1999. Genome-wide expression profiling in *Escherichia coli* K-12. *Nucleic Acids Res.* **27**:3821–3835

Rosello-Mora R, Amann R. 2001. The species concept for prokaryotes. *FEMS Microbiol. Rev.* **25**:39–67.

Sagaram US, DeAngelis KM, Trivedi P, Andersen GL, Lu SE, Wang N. 2009. Bacterial diversity analysis of Huanglongbing pathogen-infected citrus using PhyloChips and 16S rDNA clone library sequencing. *Appl. Environ. Microbiol.* **75**:1566–1574.

Sanguin H, Herrera A, Oger-Desfeux C, Dechesne A, Simonet P, et al. 2006a. Development and validation of a prototype 16S rRNA-based taxonomic microarray for *Alphaproteobacteria*. *Environ. Microbiol.* **8**:289–307.

Sanguin H, Remenant B, Dechesne A, Thioulouse J, Vogel TM, et al. 2006b. Potential of a 16S rRNA-based taxonomic microarray for analyzing the rhizosphere effects of maize on *Agrobacterium* spp. and bacterial communities. *Appl. Environ. Microbiol.* **72**:4302–4312.

Schena M, Shaldon D, Davis RW, Brown PO. 1995. Quantitative monitoring of gene expression patterns with a complementary DNA microarray. *Science* **270**:467–470.

Schliep A, Rahmann S. 2006. Decoding non-unique oligonucleotide hybridization experiments of targets related by a phylogenetic tree. *Bioinformatics* **22**: e424–e430.

Schönmann S, Loy A, Wimmersberger C, Sobek J, Aquino C, Vandamme P, Frey B, Rehrauer H, Eberl L. 2009. 16S rRNA gene-based phylogenetic microarray for simultaneous identification of members of the genus *Burkholderia*. *Environ. Microbiol.* **11**:779–800.

Sessitsch A, Hackl E, Wenzl P, Kilian A, Kostic T, Stralis-Pavese N, Sandjong BT, Bodrossy L. 2006. Diagnostic microbial microarrays in soil ecology. *New Phytol.* **171**:719–735.

Shawkey MD, Firestone MK, Brodie EL, Beissinger SR. 2009. Avian incubation inhibits growth and diversification of bacterial assemblages on eggs. *PLoS ONE* **4**:e4522.

Singh BK, Millard P, Whiteley AS, Murrell JC. 2004. Unravelling rhizosphere-microbial interactions: Opportunities and limitations. *Trends Microbiol.* **12**:386–393.

Soderberg KH, Probanza A, Jumpponen A, Baath E. 2004. The microbial community in the rhizosphere determined by community level physiological profile (CLPP) and direct-soil and cfu-PLFA techniques. *Appl. Soil Ecol.* **25**:135–145.

Sorensen J, Nicolaisen MH, Ron E, Simonet P. 2009. Molecular tools in rhizosphere microbiology-from single-cell to whole-community analysis. *Plant Soil* **321**:483–512.

Spiegelman D, Whissell G, Greer CW. 2005. A survey of the methods for the characterization of microbial consortia and communities. *Can. J. Microbiol.* **51**:355–386.

Stahl DA, Flesher B, Mansfield HR, Montgomery L. 1988. Use of phylogenetically based hybridization probes for studies of ruminal microbial ecology. *Appl. Environ. Microbiol.* **54**:1079–1084.

Stoughton RB. 2005. Applications of DNA microarrays in biology. *Annu. Rev. Biochem.* **74**:53–82.

Stralis-Pavese N, Sessitsch A, Weilharter A, et al. 2004. Optimisation of diagnostic microarray for application in analysing landfill methanotroph communities under different plant covers. *Environ. Microbiol.* **6**:347–363.

Sunagawa S, DeSantis TZ, Piceno YM, Brodie EL, DeSalvo MK, et al. 2009. Bacterial diversity and white plague disease-associated community changes in the Caribbean coral *Montastraea faveolata*. *ISME J.* **3**:512–521.

Taroncher-Oldenburg G, Griner EM, Francis CA, Ward BB. 2003. Oligonucleotide microarray for the study of functional gene diversity in the nitrogen cycle in the environment. *Appl. Environ. Microbiol.* **69**:1159–1171.

Torsvik V, Øvreås L. 2002. Microbial diversity and functions in soil: From genes to ecosystem. *Curr. Opin. Microbiol.* **5**:240–245.

Urakawa H, Noble PA, El Fantroussi S, Kelly JJ, Stahl DA. 2002. Single-base-pair discrimination of terminal mismatches by using oligonucleotide microarrays and neural network analyses. *Appl. Environ. Microbiol.* **68**:235–244.

Urakawa H, El Fantroussi S, Noble PA, Kelly JJ, Stahl DA. 2003. Optimization of single-base-pair mismatch discrimination in oligonucleotide microarrays. *Appl. Environ. Microbiol.* **69**:2848–2856.

Vaquerizas JM, Conde L, Yankilevich P, Cabezon A, Minguez P, Diaz-Ulriarte R, Al- Shhrour F, Herrero J, Dopazo J. 2005. GEPAS, an experiment-oriented pipeline for the analysis of microarray gene expression data. *Nucleic Acids Res.* **33**(Web Server issue): W616–W620.

Wagner M, Smidt H, Loy A, Zhou J. 2007. Unravelling microbial communities with DNA-microarrays: challenges and future directions. *Microb. Ecol.* **53**:498–506.

Wang HY, Malek RL, Kwitek AE, Greene AS, Luu TV, et al. 2003a. Assessing unmodified 70-mer oligonucleotide probe performance on glass-slide microarrays. *Genome Biol.* **4**:R5.

Wang X, Hessner MJ, Wu Y, Pati N, Ghosh S. 2003b. Quantitative quality control in microarray experiments and the application in data filtering, normalization and false positive rate prediction. *Bioinformatics* **19**:1341–1347.

Ward BB, Eveillard D, Kirshtein JD, Nelson JD, Voytek MA, Jackson GA. 2007. Ammonia-oxidizing bacterial community composition in estuarine and oceanic environments assessed using a functional gene microarray. *Environ. Microbiol.* **9**:2522–2538.

Wintzingerode F, Göbel UB, Stackebrandt E. 1997. Determination of microbial diversity in environmental samples: Pitfalls of PCR-based rRNA analysis. *FEMS Microbiol. Rev.* **21**:213–229.

Woese CR. 1987. Bacteria evolution. *Microbiol. Rev.* **51**:221–271.

Wrighton KC, Agbo P, Warnecke F, Weber KA, Brodie EL, et al. 2008. A novel ecological role of the *Firmicutes* identified in thermophilic microbial fuel cells. *ISME J.* **2**:1146–1156.

Wu L, Thompson DK, Liu X, Fields MW, Bagwell CE, Tiedje JM, Zhou J. 2004. Development and evaluation of microarray-based whole genome hybridization for detection of microorganisms within the context of environmental applications. *Environ. Sci. Technol.* **38**:6775–6782.

Wuyts J, Perriere G, Van De Peer Y. 2004. The European ribosomal RNA database. *Nucleic Acids Res.* **32**: D101–D103.

Yang YH, Speed T. 2002. Design issues for cDNA microarray experiments. *Nat. Rev. Genet.* **3**:579–588.

Yergeau E, Schoondermark-Stolk SA, Brodie EL, Déjean S, DeSantis TZ, et al. 2009. Functional microarray analysis of nitrogen and carbon cycling genes across an Antarctic latitudinal transect. *ISME J.* **1**:163–169.

Yershov G, Barsky V, Belgovskiy A, Kirillov E, Kreindlin E, et al. 1996. DNA analysis and diagnostics on oligonucleotide microchips. *Proc. Natl. Acad. Sci. USA* **93**:4913–4918.

Zhang L, Hurek T, Reinhold-Hurek B. 2007. A *nif*H-based oligonucleotide microarray for functional diagnostics of nitrogen-fixing microorganisms. *Microb. Ecol.* **53**:456–470.

Zhou J. 2003. Microarrays for bacterial detection and microbial community analysis. *Curr. Opin. Microbiol.* **6**:288–294.

Part 6B

Metatranscriptomics

Chapter 62

Isolation of mRNA From Environmental Microbial Communities for Metatranscriptomic Analyses

PEER M. SCHENK

62.1 INTRODUCTION

Microbial life is rapidly becoming a center of interest as humanity realizes the importance of the fundamental and global processes that are driven by microbial communities in the environment. One way to increase our understanding of these processes is the establishment of microbial activity profiles from environmental communities by analyzing the actively transcribed genes. For a recent review on the use of this metatranscriptomics approach for biodiscovery, see Warnecke and Hess [2009]. This chapter presents an updated version of a simple and cost-effective method for mRNA extraction from environmental microbial communities and is based on the report by McGrath et al. [2008].

Isolation of total RNA from microbial communities has been described in many publications [e.g., Hurt et al., 2001; Zoetendal et al., 2006; see also Chapter 10, Vol. II]. Microbial mRNA has been analyzed using different experimental strategies, such as expression profiling using individual genes [Burgmann et al., 2003], differential display [Fleming et al., 1998; Brzostowicz et al., 2003], subtracted libraries [Poretsky et al., 2005; see also Chapter 63 and 64, Vol. I], and fluorescent in situ hybridization [Pernthaler and Amann, 2004]. However, the functions of genes from bacteria and archaea have been difficult to study due to technical difficulties with the isolation of good-quality mRNA. Total RNA comprises primarily ribosomal RNA, with approximately 1–5% mRNA [Neidhardt and Umbarger, 1996]. Furthermore, transcripts from bacteria or archaea typically do not possess poly(A) tails, and because of simultaneous transcription and translation, mRNA is usually fragmented and unstable [Nakazato et al., 1975]. Hence the isolation of mRNA from total RNA from unknown species has been difficult, and cDNA libraries are often dominated by rRNA clones [Botero et al., 2005]. Several methods are available that allow removal of 16S and 23S ribosomal RNA to some extent (e.g., MICROB*Express*™, Ambion; mRNA-ONLY, Epicentre), but these can be limited in their species range and ability to remove all forms of rRNA [Poretsky et al., 2005]. Recently, mRNA has been enriched by combining both methods, which included pretreatment with terminator exonuclease (to remove 5′ monophosphate-capped RNA; mRNA-ONLY protocol; Epicentre) followed by subtractive hybridization using rRNA capture probes attached to magnetic beads (MICROB*Express*™, Ambion) and subsequent precipitation [Hewson et al., 2009; Poretsky et al., 2009; see also Chapter 63, Vol. I]. rRNA capture probes alone had varying success for different microbial communities but were used to synthesize cDNA data with less than 20% [Poretsky et al., 2005] and even 0.08% rRNA sequence output [Gilbert et al., 2008; see also Chapter 27, Vol. I]. Another method uses the *Escherichia coli* poly(A) polymerase, which preferentially polyadenylates mRNA followed by cDNA synthesis using poly-dT

Handbook of Molecular Microbial Ecology, Volume I: Metagenomics and Complementary Approaches, First Edition. Edited by Frans J. de Bruijn.

primers (MessageAmp II-Bacteria kit, Ambion) which yielded mRNA enrichment of 47% of total sequence reads [Frias-Lopez et al., 2008]. Recently, a protocol for subtractive hybridization of bacterial 16S and 23S rRNA that uses sample-specific probes has been developed for mRNA enrichment (Stewart et al., 2010). Currently, there is no method without any bias during mRNA isolation or that warrants complete removal of rRNA. In any case, an analytical RNA gel should be run to check the removal efficiency and integrity of the remaining RNA.

Unfortunately, the protocols described above typically yield small amounts of mRNA and often require an additional RNA amplification step. Several methods have been used for microbial transcript amplification, including random hexamers with T7 RNA polymerase [Frias-Lopez et al., 2008] and multiple strand displacement amplification (MDA) [Gilbert et al., 2008]. A cDNA amplification kit based on random primers and MDA ('Quanti-Tect', Qiagen) is also available. A promising approach to amplify small amounts of cDNA, potentially without amplification bias, is emulsion PCR where each template is amplified in a single droplet without competing for reagents [Blow et al., 2008]. Several metatranscriptomics projects are underway at the Joint Genome Institute using this approach [Warnecke and Hess, 2009].

Here, we describe a simple and robust method for isolating high-quality mRNA from diverse microbial communities from terrestrial, commensal, and aquatic sources, and we discuss the construction of cDNA libraries without RNA amplification. Sequence analysis of a fraction of the clones revealed a high degree of diversity and novelty [McGrath et al., 2008]. This technique has already assisted research in the field of metatranscriptomics, and our results emphasize the need for large-scale transcriptional analyses of microbial communities in different environments.

62.2 MATERIALS AND METHODS

For full details of materials and methods, please refer to the original paper by McGrath et al. [2008]. What follows is an abbreviated version.

62.2.1 Total RNA Isolation

Different soil types were tested for mRNA isolation from terrestrial samples. These included sugarcane field soil with different fertilizer application histories, organic composts, and topsoil from garden beds. Immediate processing of samples was carried out on site to minimize RNA degradation and changes to its composition, and a field centrifuge was powered from a car battery for this purpose. Briefly, 20 g of soil was placed in a 50-ml container with 20 ml of sterile distilled water and shaken vigorously for approximately 20–30 s until the soil sample was completely suspended. Larger particles (e.g., sand and stones) were allowed to settle for 10 s, and the supernatant was transferred into single 2-ml microcentrifuge tubes and centrifuged for 2 min (14,000 × g) to pellet the microbial contents. Pellets were snap-frozen in liquid nitrogen or dry ice and stored in the laboratory at −80°C. The PowerSoil™ RNA extraction kit (MoBio, USA) was used for total RNA isolation from frozen pellets according to the manufacturer's instructions with the modification that the first two buffers were added to the frozen pellet simultaneously. Aquatic samples (10 Falcon tubes with 50 ml each) were collected from activated flocculant at a waste water treatment plant and from a eutrophic freshwater lake. Tubes were centrifuged for 2 min (14,000 × g), and combined unfrozen pellets were used immediately for total RNA extraction using the SV Total RNA Isolation System (Promega). Human fecal and oral samples from teeth, inner cheek, and tongue, as well as bovine rumen material, were collected to obtain RNA from mammalian-associated (commensal) microbial communities. Samples of 200–400 mg were snap-frozen in 2 ml microcentrifuge tubes and stored at −80°C before RNA isolation (SV Total RNA Isolation System, Promega).

62.2.2 mRNA Isolation

mRNA isolation was carried out using RNase-free equipment (RNase away™, Invitrogen) and with solutions made from DEPC-treated water. RNA samples were subjected to preparative 1.5% agarose gel electrophoresis with ethidium bromide at 100 V for 45 min in TAE buffer. UV illumination was used to check RNA quantity and integrity and to excise the regions between the 16S and 5S and between the 23S and 16S rRNA bands to isolate the mRNA. RNA was extracted from the excised agarose gel fragments using the Wizard SV Gel and PCR Clean-Up System (Promega). Alternatively, the SV Total RNA Isolation System (Promega) was used with the modification that five volumes of the first buffer (SV RNA Lysis buffer) were added to the gel fragment that was incubated at 50°C until melting occurred before addition of the second buffer (SV RNA Dilution buffer).

62.2.3 cDNA Synthesis, Cloning, and Sequence Analysis

Approximately 200–500 ng of isolated mRNA was vacuum-concentrated to 15 μl, and first-strand cDNA was produced by addition of 0.3 μg of random hexamers and SuperScript III Reverse Transcriptase (Invitrogen) following the manufacturer's instructions. Double-stranded cDNA was produced by addition of 13 μl of 10X *E. coli* ligase buffer, 1 μl of 10 U/μl *E. coli* ligase, 4 μl of 10 U/μl Klenow fragment, 1 μl of 1 U/μl RNase H, 3 μl of 10 mM dNTP's, and 106 μl of water to a final volume of 150 μl, followed by incubation at 16°C for 2 h. Blunt ends were produced by adding 1.5 μl T4 DNA polymerase (7.9 U/μl) to the reaction and incubation for an additional 5 min. Double-stranded cDNA was purified (Wizard SV Gel and PCR Clean-Up System, Promega) and then vacuum-concentrated to 5 μl. cDNA libraries (1000–4000 colonies each) were constructed using the pCR-Blunt system (Invitrogen). *E. coli* cultures obtained from >13,000 white colonies were grown in 2 ml of LB broth, and cDNA from >100 clones was sequenced (NCBI accession numbers ES544490-ES544589). Sequence analyses included blastn and blastx homology alignments to Genbank entries and have been summarized by McGrath et al. [2008].

62.3 RESULTS

62.3.1 Optimization of RNA Extraction from Different Environmental Microbial Communities

Many different methods were tested to obtain high-quality RNA from microbial communities from a diverse range of environments. Finally, three protocols were developed for three general microbial environments: terrestrial, aquatic, and mammalian-associated (commensal). For terrestrial soil samples it was important to bring the majority of microbes into suspension to remove larger fragments (e.g., sand, small stones) before snap-freezing of a concentrated microbial pellet. This required on-site pre-extraction processing, including a centrifugation step, that was kept to minimum time (maximal 3 min) to avoid major changes in RNA composition or degradation (see Section 62.2). To the contrary, mammalian-associated (commensal) samples (bovine rumen, human oral, and human intestinal) could be used for RNA isolation with minimal pre-processing when samples were snap-frozen with a large surface area (e.g., as a smear on the inside of a tube) and then rapidly thawed while resuspending the sample in the presence of RNA Lysis buffer. The large surface area was found to be important to avoid RNA degradation, because it allows immediate contact with RNA Lysis buffer that contains RNase-inhibiting chemicals. Aquatic samples (e.g., pond water and waste water) contain a much lower concentration of microorganisms than the commensal samples and required a rapid on-site centrifugation step for concentration. A minimum volume of 500 ml of sample was required to obtain sufficient amounts for RNA extraction (see Section 62.2). Once centrifuged and combined in pellets, RNA could be extracted using the same method as used for commensal samples. High-quality RNA was indicated by minimal rRNA degradation, distinct 23S, 16S, and 5S rRNA bands, and the absence of fragmentation (smearing) after gel electrophoresis (**Fig. 62.1**).

Yields for RNA obtained from soil samples were compared to RNA yields obtained from cultured *E. coli* cells. A total of 2.3 μg of RNA per gram of soil (115 ng mRNA/g soil) was isolated when using fresh samples, compared to 60 μg of total RNA per milliliter of late log-phase *E. coli* culture. To test whether chemical factors of soil contributed to RNA yield, 25 μl of *E. coli* culture was added to 2 g of fresh or autoclaved soil before RNA extraction. Autoclaved soil did not seem to have a major negative effect on RNA yield from *E. coli* culture (46 μg of total RNA per milliliter of culture), while the addition of *E. coli* culture to fresh soil had an additive effect on yield [McGrath et al., 2008].

62.3.2 mRNA Isolation from Total RNA

Microbial RNA contains primarily ribosomal RNA and only approximately 1–5% mRNA (Neidhardt et al., 1996). We tested the subtractive hybridization method using the rRNA capture probes attached to magnetic

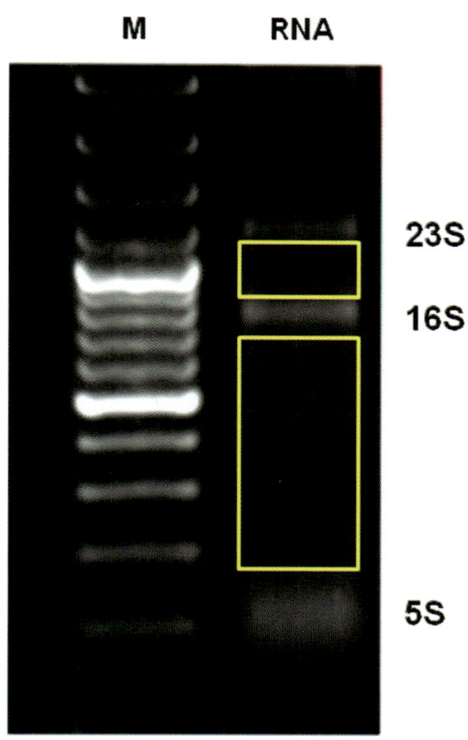

Figure 62.1 Agarose gel image showing separation of total RNA isolated from human feces. Highlighted regions between rRNA bands (23S, 16S, and 5S), containing mRNA, were used for extraction and cDNA synthesis. The left lane (M) includes GeneRuler™ 100-bp DNA ladder (Fermentas).

beads (MICROB*Express*™, Ambion) on soil and human fecal samples to remove rRNA from total RNA. But quality control after gel electrophoresis revealed that rRNA bands were still present, suggesting that this protocol was unsuitable for the removal of all rRNA from our samples (data not shown). However, using agarose gel electrophoresis for removal of rRNA by size fractionation was successful for all samples from environmental microbial communities when RNA was not degraded and displayed distinct 23S, 16S, and 5S rRNA bands. The mRNA between rRNA bands was efficiently removed by excision of the agarose fragments (highlighted in Fig. 62.1 for fecal RNA) and purified, yielding an mRNA mixture of various sizes. Using this technique, the isolated, enriched mRNA fraction contributed to approximately 5% of the total RNA from the microbial communities that were tested in this study. Although not visible by gel electrophoresis, it can be expected that the enriched mRNA may still contain traces of rRNA that migrate outside of the three major rRNA bands.

62.3.3 Sequence Analysis of mRNA Transcripts from Environmental Microbial Communities

mRNA transcripts from terrestrial, aquatic, and commensal communities that were isolated from total RNA by electrophoretic size fractionation were used to produce cDNA libraries in *E. coli* plasmids. A total of 13,056 clones were cultured from single colonies, and a random sample of 100 cDNA clones was analyzed by sequencing. These clones contained cDNA from each of the three major groups of environmental microbial communities (59 from terrestrial soil, 19 from aquatic environments, and 22 from commensal samples). All cDNA sequences were submitted to Genbank (accession numbers ES544490-ES544589) and compared to Genbank entries using the BLAST algorithms for proteins (blastx) and nucleotides (blastn). Full details of the analysis of the BLAST results are published as supplementary material by McGrath et al. [2008]. The lowest E-value returned during blastn and blastx analyses was used to assign each microbial cDNA sequence to a group of organisms. Most (72) of the sequence reads matched to Genbank entries for bacteria, compared to animals (14), fungi (3), protista (4), and plants (6). When separating data by different environments, matches to bacteria were produced for the majority for all microbial communities, but none of the cDNA sequences had a close match to an existing sequence entry from archaea (**Table 62.1**). As expected, transcript matches to fungal genes were highest represented in soil, while human- and bovine-associated microbial samples harbored a high proportion of sequences with homology to animals (mainly to the host species). Despite size separation from rRNA bands by gel electrophoresis, 11% of the sequences analyzed still showed homology to rRNA, although most of these matched to eukaryotic organisms, and hence may not have been precisely removed during the mRNA isolation step by gel electrophoresis. Other rRNA clones may originate from partial rRNA products co-migrating with mRNA or incomplete separation by gel electrophoresis. Nearly all of the cDNA sequences were between 700 and 800 bp or less than 450 bp in size [McGrath et al., 2008]. These sizes correspond roughly to the two regions of the excised gel fragments used to separate mRNA from total RNA. Conversely, this means that mRNA transcripts that are of the same size as rRNA molecules and co-migrate cannot be recovered using this technique. This may not result in a strong underrepresentation of bacterial and archaeal transcripts because the majority of prokaryotic mRNA does not occur in full length due to simultaneous transcription and translation/degradation.

Table 62.1 Distribution of 100 cDNA Sequences from Different Environmental Communities into Taxonomic Groups Based on the Highest BLAST Homology Match to Genbank (NCBI) Entries

	Bacteria	Protista	Archaea	Fungi	Plantae	Animalia
Soil	40	4	0	3	4	8
Aquatic	17	0	0	0	1	1
Commensal	15	1	0	0	1	5

The study by McGrath et al. [2008] and our subsequent sequence analyses of additional cDNA clones revealed that mostly new transcribed sequences were recovered from environmental microbial communities. Only 7% of the sequences had perfect matches to NCBI Genbank entries, while 71% had E-values higher than 10^{-50}, and 19% had an E-value over 0.1, signifying the ability of this technique to obtain new mRNA sequence reads directly from environmental microbial communities. Large-scale next generation sequencing of cDNA clones should be carried out to more accurately reveal the diversity of expressed genes in the environmental microbial communities described here.

62.4 DISCUSSION

Metagenomics is rapidly increasing our information on microbial genes and genomes, and broadens our understanding of functional significance and genetic variability of environmental microbial communities (see Chapters 14–31, Vol. I; also see throughout Vol. II). However, there is currently limited data available regarding the regulation and activity of microbial genes in the environment.

Functional and regulatory information on genes has been obtained for individual organisms over the past decade mostly by transcriptional profiling using microarray or mutational analyses. Using this approach for a whole population with thousands and potentially millions of different species, as encountered in many environmental microbial communities, can prove difficult because most of these organisms cannot be cultured and live in very closely linked associations in their natural environment.

The successful extraction of high-quality mRNA from different environmental samples is an important step to increase our knowledge of the complex processes of microbial ecology. Metatranscriptomics or environmental transcriptomics obtain information on functional gene expression within microbial communities without bias toward known sequences, and they present a new strategy for identifying and characterizing community-specific variants of key functional genes [Poretsky et al., 2005; Frias-Lopez et al., 2008; Warnecke and Hess, 2009]. Rohwer [2007] predicted that "massive sequencing of RNA populations will become routine and replace the current array technologies." Indeed, recent studies on metatranscriptomics using high-throughput sequencing [e.g., Frias-Lopez et al., 2008, Gilbert et al., 2008, Hewson et al., 2009, Poretsky et al., 2009, Shi et al., 2009; Chapters 63–65, Vol. I; Chapters 27 and 54, Vol. II] elegantly demonstrate the feasibility of this approach. To contribute to this rapidly growing area, we have developed a cost-effective and simple protocol to isolate mRNA from environmental samples that contain many currently unculturable organisms. This technique was suitable for extracting viable mRNA from a range of natural microbial communities from diverse environments, including terrestrial (garden topsoil, sugarcane field soils, organic compost soil), commensal (cow rumen, human oral, human fecal) and aquatic (activated flocculant of communal wastewater; eutrophic freshwater lake) samples. Other methods (e.g., using CTAB or LiCl or other commercial kits for nucleic acid extraction) or variations from our technique (e.g., extraction from fresh or frozen soil without pre-processing step) resulted in low-quality RNA, indicated by a smear of rRNA when viewed on a gel. Occasionally, RNA was partly degraded for unknown reason, but a repeat of the extraction was successful for the communities tested in this study. The mRNA isolation protocol described in this chapter can be carried out using standard laboratory equipment, and it yields less rRNA sequences in cDNA libraries than other methods [Poretsky et al., 2005]. Limitations of this method include a possible bias toward mRNA transcripts that do not degrade easily and do not co-migrate with rRNA bands during gel electrophoresis and toward microorganisms that are more resilient and easier to pellet and lyse. In addition, smaller fragments are easier to convert into cDNA and to clone into libraries. In its current form, this method may also not be suitable to isolate microbial small RNAs [Shi et al., 2009]. Our recent results showed that the method of mRNA isolation by size fractionation using gel electrophoresis can also be combined with other approaches (e.g., using rRNA capture probes; MICROB*Express*™, Ambion), but each step reduces the amount of mRNA available for cDNA synthesis, and amplification steps may subsequently be required. However, it is likely that the next-generation sequencing platforms that are currently under development will be able to handle even very small amounts of cDNA.

Only a small number of clones were analyzed in this study, and the number of reads matching to kingdoms (Table 62.1) cannot be representative for the communities analyzed. For example, none of the sequence reads in our study showed a good homology match to an existing archaea sequence entry. This could also be due to a low abundance of transcriptionally active archaea in the microbial samples, or a shortage of sequence entries for nonextremophile archaea in the public databases. Indeed, most BLAST searches did not find a close match to any existing Genbank sequence and our homology annotations are clearly biased toward the organisms that provided most sequence entries.

Using the metatranscriptomics approach to discover functional genes and their regulation from environmental microbial communities holds the promise to provide a wide range of biotechnological and medical applications. The use of this and other techniques to isolate mRNA transcripts from natural microbial communities will unquestionably help researchers to identify actively expressed genes with significant functions and their regulation (see Chapter 18, Vol I). For example, apart from next-generation sequencing, anonymous or sequenced cDNA clones from microbial communities can be used to construct a custom microarray to profile the expression of thousands of microbial genes from environmental microbial communities in parallel (see also Chapters 57, 58, 60 and 61, Vol. I). Using these "environmental functional microarrays" that contained all 13,056 cDNA clones, for transcription profiling of sugarcane soil with different cultivation histories, we identified environmental marker genes and obtained reproducible expression patterns for independent biological replicates [McGrath et al., 2010]. Significantly, it has been pointed out that functional genes from highly diverse environmental microbial communities may encode many new biologically active peptides and enzymes [Blake, 2004; Warnecke and Hess, 2009]. Large-scale expression libraries constructed from mRNA transcripts expressed in their natural environment may provide highly diverse, natural compound collections suited for biodiscovery projects.

Acknowledgments

I wish to acknowledge the co-authors of the original study by McGrath et al. [2008] that this chapter is based upon. The described methods were developed for research funded by the Australian Greenhouse Office, Sugar Research and Development Corporation, and Australian Research Council (DP1094749).

REFERENCES

Blake D. 2004. Biodiscovery—From reef to outback. *Nature* **429**:15–17.

Blow MJ, Zhang T, Woyke T, Speller CF, Krivoshapkin A, et al. 2008. Identification of ancient remains through genomic sequencing. *Genome Res*. **18**:1347–1353.

Botero LM, D'Imperio S, Burr M, McDermott TR, Young M, et al. 2005. Poly(A) polymerase modification and reverse transcriptase PCR amplification of environmental RNA. *Appl. Environ. Microbiol*. **71**:1267–1275.

Brzostowicz PC, Walters DM, Thomas SM, Nagarajan V, Rouviere PE. 2003. mRNA differential display in a microbial enrichment culture: Simultaneous identification of three cyclohexanone monooxygenases from three species. *Appl. Environ. Microbiol*. **69**: 334–342.

Burgmann H, Widmer F, Sigler WV, Zeyer J. 2003. mRNA extraction and reverse transcription-PCR protocol for detection of nifH gene expression by *Azotobacter vinelandii* in soil. *Appl. Environ. Microbiol*. **69**:1928–1935.

Fleming JT, Yao WH, Sayler GS. 1998. Optimization of differential display of prokaryotic mRNA: Application to pure culture and soil microcosms. *Appl. Environ. Microbiol*. **64**:3698–3706.

Frias-Lopez J, Shi Y, Tyson GW, Coleman ML, Schuster SC, et al. 2008. From the cover: Microbial community gene expression in ocean surface waters. *Proc. Natl. Acad. Sci. USA* **105**: 3805–3810.

Gilbert JA, Field D, Huang Y, Edwards R, Li W, et al. 2008. Detection of large numbers of novel sequences in the metatranscriptomes of complex marine microbial communities. *PLoS ONE* **3**:e3042.

Hewson I, Poretsky RS, Dyhrman ST, Zielinski B, White AE, et al. 2009. Microbial community gene expression within colonies of the diazotroph, *Trichodesmium*, from the Southwest Pacific Ocean. *ISME J*. **3**:1286–1300.

Hurt RA, Qiu X, Wu L, Roh Y, Palumbo AV, et al. 2001. Simultaneous recovery of RNA and DNA from soils and sediments. *Appl. Environ. Microbiol*. **67**:4495–4503.

McGrath KC, Thomas-Hall SR, Cheng CT, Leo L, Alexa A, et al. 2008. Isolation and analysis of mRNA from environmental microbial communities. *J. Microbiol. Methods* **75**:172–176.

McGrath KC, Mondav R, Sintrajaya R, Slattery B, Schmidt S, Schenk PM, 2010. Development of an environmental functional gene microarray for soil microbial communities. *Appl. Environ. Microbiol*. **76**:7161–7170.

Nakazato H, Venkatesan S, Edmonds M. 1975. Polyadenylic acid sequences in *E. coli* messenger RNA. *Nature* **256**:144–146.

Neidhardt FC, Umbarger HE. 1996. In Neidhardt, FC, ed. *Escherichia coli and Salmonella*: *Cellular and Molecular Biology*. Vol. **1**. Washington, D.C.: ASM Press, pp. 13–16.

Pernthaler A, Amann R. 2004. Simultaneous fluorescence in situ hybridization of mRNA and rRNA in environmental bacteria. *Appl. Environ. Microbiol*. **70**:5426–5433.

Poretsky RS, Bano N, Buchan A, LeCleir G, Kleikemper J, et al. 2005. Analysis of microbial gene transcripts in environmental samples. *Appl. Environ. Microbiol*. **71**:4121–4126.

Poretsky R, Hewson I, Sun S, Allen AE, Moran MA, et al. 2009. Comparative day/night metatranscriptomic analysis of microbial communities in the North Pacific subtropical gyre. *Environ. Microbiol*. **11**:1358–1375.

Rohwer F. 2007. Real-time microbial ecology. *Environ. Microbiol*. **9**:10.

Shi Y, Tyson GW, DeLong EF. 2009. Metatranscriptomics reveals unique microbial small RNAs in the ocean's water column. *Nature* **459**:266–269.

Stewart FJ, Ottesen EA, DeLong EF, 2010. Development and quantitative analyses of a universal rRNA-subtraction protocol for microbial metatranscriptomics. *ISME J*. **4**:896–907.

Warnecke F, Hess M. 2009. A perspective: Metatranscriptomics as a tool for the discovery of novel biocatalysts. *J. Biotechnol*. **142**:91–95.

Zoetendal EG, Booijink CC, Klaassens ES, Heilig HG, Kleerebezem M, et al. 2006. Isolation of RNA from bacterial samples of the human gastrointestinal tract. *Nat. Protoc*. **1**:954–959.

Chapter 63

Comparative Day/Night Metatranscriptomic Analysis of Microbial Communities in the North Pacific Subtropical Gyre

RACHEL S. PORETSKY AND MARY ANN MORAN

63.1 INTRODUCTION

Our estimates based on microbial gene expression patterns and abundance indicate that at any given time, aquatic microorganisms are expressing $\sim10^{30}$ transcripts globally. These expressed genes directly mediate many ecologically relevant processes; studying controls and patterns of gene expression is therefore one of the best ways to explore how these processes are regulated and how they respond to changing environmental conditions. Technological and computational advancements have made metatranscriptomics an increasingly popular method for accessing in situ gene expression in natural microbial communities. This approach is analogous to metagenomics in that it retrieves and sequences environmental mRNAs from a microbial assemblage in a way that is less biased than traditional expression methods such as microarrays and qPCR [Bürgmann et al., 2003; Wawrik et al., 2002; Zhou, 2003] that rely on specific primers, probes, and the prerequisite knowledge of which genes might be expressed. Environmental transcriptomics protocols are technically difficult (see Chapter 62, Vol. I), however, since prokaryotic mRNAs generally lack the poly(A) tails that are used for specific capture of eukaryotic mRNAs [Bailly et al., 2007; Shendure, 2008] and degrade more rapidly (half-lives of minutes; [Rauhut and Klug, 1999]). The high rRNA content of prokaryotic cells (>80%; Ingraham et al. [1983]) is another challenge, because rRNAs are abundant in total RNA extracts and can overwhelm mRNA signals.

The first analysis of environmental transcriptomes by creating clone libraries using random primers to reverse-transcribe and amplify environmental mRNAs [Poretsky et al., 2005] had biases resulting from the selection of the random primers used to initiate cDNA synthesis, but nevertheless laid the conceptual groundwork for the utility of transcriptomics. Recently developed high-throughput pyrosequencing technologies allow direct sequencing (without cloning) [Margulies et al., 2005] and was first coupled with transcript analysis in a grassland [Leininger et al., 2006]. This approach, along with techniques that linearly amplify mRNA without random primers, is the basis for environmental metatranscriptomics protocols used in the study summarized here as well as in other recent efforts [Frias-Lopez et al., 2008; Gilbert et al., 2008; Hewson et al., 2009; Poretsky et al., 2009a; Urich et al., 2008; (see Chapter 27, Vol. II]).

This chapter summarizes [Poretsky et al., 2009b] a comparative environmental transcriptomics approach used to elucidate day/night differences in gene expression in surface waters of the North Pacific subtropical gyre [Karl and Lukas, 1996]. The analysis provided information on the dominant metabolic processes within the bacterioplankton assemblages and revealed changes

Handbook of Molecular Microbial Ecology, Volume I: Metagenomics and Complementary Approaches, First Edition. Edited by Frans J. de Bruijn.

in expression patterns of biogeochemically relevant processes. Characterizing expression patterns of microbial genes and identifying what factors induce their expression is a critical step understanding diverse ecosystems.

63.2 METHODS

For a full description of the methods used, see the original publication [Poretsky et al., 2009b] and additional technical descriptions [Poretsky et al., 2008, 2009a]. What follows is an abbreviated outline of the methods.

63.2.1 Sample Collection

Samples were collected at the Hawaiian Ocean Time-series (HOT) Station ALOHA, defined by the 6-nautical-mile radius circle centered at 22°45′N, 158°W in November 2005 (HOT-175). Seawater was collected from a depth of 25 m using Niskin bottles on a conductivity–temperature–depth (CTD) rosette sampler and processed as quickly as possible. A night sample was collected at 03:00 on November 11, 2005, and a daytime sample was collected at 13:00 on November 13, 2005. Seawater (80 L for the night sample and 40 L for the day sample) was prefiltered through a 5-μm, 142-mm polycarbonate filter (GE Osmonics, Minnetonka, MN) followed by a 0.2-μm, 142-mm Durapore (Millipore) filter using positive air pressure. The 0.2-μm filters were placed in a 15-ml tube containing 2 ml of Buffer RLT (containing β-mercaptoethanol) from the RNeasy kit (Qiagen, Valencia, CA) and flash-frozen in liquid nitrogen for RNA extraction. For DNA extraction, an additional 20 L of seawater was simultaneously filtered using the protocol outlined above at both time points. The 0.2-μm filters were placed in Whirlpack bags and flash frozen.

63.2.2 RNA and DNA Preparation

DNA was extracted using a phenol:chloroform-based protocol [Fuhrman et al., 1988]. RNA was extracted using a modified version of the RNeasy kit (Qiagen) that results in high RNA yields from material on polycarbonate filters [Poretsky et al., 2008]. Following extraction, RNA was treated with DNase using the TURBO DNA-*free* kit (Ambion, Austin, TX). Two methods were employed for rRNA removal: Total RNA was first treated enzymatically with the mRNA-ONLY Prokaryotic mRNA Isolation Kit (Epicentre Biotechnologies, Madison, WI) that uses a 5′-phosphate-dependent exonuclease to degrade rRNAs. The MICROB*Express* kit (Ambion) subtractive hybridization with capture oligonucleotides hybridized to magnetic beads was subsequently used as an additional mRNA enrichment step. In order to obtain the microgram quantities of mRNA required for pyrosequencing, approximately 500 ng of RNA was linearly amplified using the MessageAmp II-Bacteria Kit (Ambion) according to the manufacturer's instructions. Finally, the amplified, antisense RNA (aRNA) was converted to double-stranded cDNA with random hexamers using the Universal RiboClone cDNA Synthesis System (Promega, Madison, WI). The cDNA was purified with the Wizard DNA Clean-up System (Promega). The quality and quantity of the total RNA, mRNA, aRNA, and cDNA was assessed by measurement on the NanoDrop-1000 Spectrophotometer (NanoDrop Technologies, Wilmington, DE) and the Experion Automated Electrophoresis System (Bio-Rad, Hercules, CA).

63.2.3 16S rRNA Gene Libraries

PCR amplification of ribosomal DNA was carried out using primers 27F and 1522R [Johnson, 1994]. PCR products were cleaned using the QIAquick PCR Purification Kit (Qiagen). Multiple PCR reactions were pooled and cloned into pCR2.1 vector using the TOPO TA cloning kit (Invitrogen, Carlsbad, CA). Clones from each sample (192) were sequenced at the University of Georgia Sequencing Facility on an ABI 3100 (Applied Biosystems, Foster City, CA).

63.2.4 cDNA Sequencing

cDNAs from each sample (night and day) were sequenced using the GS 20 sequencing system by 454 Life Sciences (Branford, CT) [Margulies et al., 2005], resulting in 10,682,120 bp from 106,907 reads for the night sample and 13,255,704 bp from 133,515 reads for the day sample. The average sequence length was 99 bp. The sequences can be found in the NCBI Short Read Archive with the Genome Project ID 33463.

63.2.5 cDNA Sequence Annotation

For rRNA sequence identification, the sequences were clustered at an identity threshold of 98% based on a local alignment (number of identical residues divided by length of alignment) using the program Cd-hit [Li and Godzik, 2006]. Ribosomal RNA sequences were identified by BLASTN queries of the reference sequence of each cluster against the noncurated, GenBank nucleotide database (nt) [Benson et al., 2007] using cutoff criteria of E-value$\leq 10^{-3}$, nucleic acid length $\geq$69, and percent identity $\geq$40% previously established with in silico tests for rRNA sequence predictions of short pyrosequences

[Frias-Lopez et al., 2008; Mou et al., 2008]. A sequence was conservatively identified as rRNA-derived and was removed from the analysis pipeline if any of the top three BLASTN hits were to an rRNA gene.

The criteria for protein predictions generated using BLASTX against the NCBI curated, nonredundant reference sequence database (RefSeq) [Pruitt et al., 2005] were established with in silico tests to determine suitable cutoff limits for reliable functional prediction (E-value$<$0.01, identity $>$40%, and overlapping length $>$23 aa to the corresponding best hit [Mou et al., 2008]). Sequences with hits to RefSeq were assigned functional protein or pathway predictions based on the COG database [Tatusov et al., 2000] or KEGG database [Kanehisa and Goto, 2000]. The cutoff criteria for functional protein prediction based on orthologous groups using BLASTX analysis against the COG database (E-value$<$0.1, identity $>$40%, and overlapping length $>$23 aa to the corresponding best hit) were established using the same in silico approach [Mou et al., 2008]. Taxonomic binning of the sequences was carried out using MEGAN with the default settings for all parameters [Huson et al., 2007; see also Chapter 39, Vol. I]. The taxonomic affiliations of the putative mRNA sequences were predicted using (a) MEGAN to the family level and (b) the top BLAST hit for any higher-resolution taxonomic assignments. All non-rRNA sequences that had no RefSeq hits were BLASTX-queried against the nr database as well as against CAMERA unassembled ORFs predicted from the Global Ocean Survey (GOS) reads [Seshadri et al., 2007].

63.2.6 Eukaryotic Sequence Annotation

Eukaryotic transcripts were binned by MEGAN. Sequences were queried (BLASTX) against a curated database of protein sequences derived from all available complete eukaryotic organelle and nuclear genomes (currently, 46 eukaryotic genomes). Transcripts that matched a reference protein sequence with $>$60% identity and an E-value $< e^{-10}$ were retained, and the reference protein for the cluster was used for functional annotation. Functional annotation was performed using Java-based Blast2go [Conesa et al., 2005] that annotates genes based on similarity searches with statistical analysis and highlighted visualization on directed acyclic graphs.

63.2.7 Predicted Highly Expressed Genes

Predicted highly expressed (PHX) genes were determined for cultured representatives of three prokaryotic taxa that were well represented in the transcript libraries (*Prochlorococcus*, Roseobacter, and SAR11) using an algorithm developed by Karlin and Mrázek [2000]. The algorithm is based on comparisons to codon usage patterns in genes expected to be frequently transcribed in a prokaryotic genome (ribosomal proteins, chaperone proteins, etc.). Environmental transcript sequences that had best BLAST hits to one of the PHX genes were similarly designated as PHX.

63.2.8 Statistical Analysis

A statistical program designed for comparing gene frequency in metagenomic datasets [Rodriguez-Brito et al., 2006] was used to compare the night and day mRNA sequences categorized based on COGs, KEGGs, and proteins. The program was run with 20,000 repeated samplings with a sample size of 10,000 for COGs, 9000 for KEGGs, and 25,000 for proteins. The significance level (p) was set at $<$0.05.

63.2.9 qPCR Verifications

To confirm that the composition of the pyrosequence library was representative of the initial mRNAs, transcripts of five genes that were top hits to multiple sequences in both transcript pools were quantified in the total RNA pool using quantitative PCR (qPCR) primer sets designed for these genes. Reverse transcription reactions were carried out on 200 ng of RNA using the Omniscript RT kit (Qiagen). The day:night ratio of each gene transcript in the RNA pools was determined by (a) qPCR amplification of a serial dilution of cDNAs in triplicate and (b) calculation of the difference in cycle threshold values (ΔC_T) between the two samples. Quantitative amplification was done using the iCycler iQ real-time PCR detection system (Bio-Rad).

63.3 RESULTS AND DISCUSSION

63.3.1 cDNA Sequence Annotation

A major challenge for any genomic-based effort is our limited ability to accurately assign functions to sequences, because only a small percentage of genes in the environment have been correctly annotated and existing databases are mostly comprised of abundant genes from a small subset of the world's microbes. In the case of the hundreds of thousands of reads generated in the study described here [Poretsky et al., 2009b], annotation was accomplished by passing the sequences through an pipeline using empirically derived criteria for gene predictions from ~100-bp pyrosequences [Mou et al., 2008]. Multiple sequence databases were queried in order to identify as many sequences as possible. The first step in the pipeline was the removal of 88,916 (31,402 night and 57,514 day; **Table 63.1**) predicted rRNA sequences based on sequence similarity to the nt database using BlastN. Accurate identification of rRNAs is crucial because of numerous misidentified sequences in the RefSeq protein database (that is, rRNA sequences that are incorrectly annotated as putative proteins). Relatively low rRNA sequence contamination (29% in the night library and 43% in the day) compared to the >80% rRNA content of prokaryotic cells [Ingraham et al., 1983], indicated that the steps for excluding rRNAs through selective degradation and subtractive hybridization were largely successful.

Following rRNA removal, 151,504 possible protein-encoding sequences remained (75,946 night and 75,558 day; Table 63.1). Putative functions were established by BlastX against RefSeq (the NCBI curated, nonredundant reference sequence database); approximately one-third of these possible protein-encoding sequences (24,515 night and 24,133 day; Table 63.1) had hits in RefSeq that met the established annotation criteria. This is nearly twice the fraction of reads identified in metagenomic efforts with similar pyrosequencing read lengths [Frias-Lopez et al., 2008; Mou et al., 2008], as might be expected for sequences biased toward coding regions of genomes. These sequences were further classified based on similarities to the Clusters of Orthologous Groups (COG) categories and Kyoto Encyclopedia of Genes and Genomes (KEGG) pathways, with 24,474 and 35,927 sequences, respectively, meeting the annotation criteria for these databases. Some of the sequences without hits in RefSeq were similar to proteins in the Global Ocean Sampling (GOS) database, indicating that similar sequences have been found in other marine bacterioplankton communities, but functional annotation is not currently possible.

Any remaining sequences were checked for similarity to the nonredundant (nr) database, which is not curated and includes more sequences than RefSeq. This rescued a few more sequences, although typically these were unannotated hits to sequences from "uncultured bacteria." At the end of the annotation pipeline, half of the possible protein-encoding sequences in each library (~38,000; Table 63.1) had no significant hits to previously sequenced genes. An in silico test (detailed in the original paper) showed that short sequences from uncultured marine bacterial taxa that are known to occur in this environment had no hits in RefSeq. Many unannotated sequences in the HOT libraries are therefore likely to be transcripts from unknown taxa, but also include some transcripts from poorly conserved regions of known genes. In support of this, a preliminary analysis of a marine environmental transcriptome consisting of longer reads (~200 bp; 454 GS FLX sequencing platform; Poretsky et al. [2010]) resulted in twice the frequency of annotated sequences as the HOT metatranscriptome. A final analysis estimated that ~4% of the 76,327 unidentified sequences were from non-protein-coding, primarily intergenic regions. Overall, although only 32% of the 151,000 possible transcript sequences could confidently be assigned to a known function, this study was able to show that there was a reasonable amount of coverage and that these sequences provided a fairly unbiased sample of the mRNA pool in the microbial community (see below). As more metagenome and metatranscriptome sequences are generated and as gene annotations improve, sequence data become more biologically relevant. Ultimately, these large databases of environmental transcript sequences are useful resources that are available to probe with questions as we learn more about microbial processes in the environment.

63.3.2 Metatranscriptome Quality Analysis

Several estimates of transcriptome coverage include a taxon-abundance model based on 16S rRNA clone library data and the Chao1 index of diversity among annotations. These estimates, detailed in the original paper, showed that the community transcriptomes provided reasonable

Table 63.1 Annotation Pipeline Results for Night and Day Transcriptomes

	Night	% Night	Day	% Day
Total Reads	106,907	100	133,515	100
Ribosomal RNA Reads	31,402	29	57,514	43
Possible protein-encoding sequences	75,946	71	75,558	57
RefSeq Identified	24,515	23	24,133	18
GOS (non-RefSeq) Identified	13,222	12	13,144	10
nr Identified	92	0	71	0
Unidentified in RefSeq, GOS, and nr	38,117	36	38,210	29
KEGG-assigned Reads	19,273	18	16,654	12
COG-assigned Reads	12,487	12	11,987	9

coverage of mRNAs from the dominant organisms. Although increased sequencing depth would have been required to fully capture some specialized processes carried out by rarer members of the HOT community, frequently transcribed genes from abundant taxa were well represented. In support of this, transcript mapping to several abundant reference genomes showed sequences with homology to approximately half the genes, at coverage depths ranging from 1 to nearly 500 hits per gene (**Fig. 63.1A**). The reference genes with high coverage appeared to mediate processes expected to be dominant in the HOT bacterioplankton community (e.g., the photosynthesis genes *psaA* and *psaB*, the light-harvesting complex and RuBisCo, ammonium transporters, and transcription-related genes) or processes hypothesized to be dominant (Fig. 63.1A). In addition to evidence of good coverage, the composition of these transcriptomes was consistent with models of prokaryotic gene expression. This was verified with operon-based patterns of transcription; that is transcripts were more often found with their adjacent genes than would be expected by chance (Fig. 63.1B). Furthermore, a bias toward transcripts from predicted highly expressed (PHX) genes, identified based on patterns in codon usage [Karlin and Mrázek, 2000], was observed (Fig. 63.1C). A larger proportion of PHX transcripts were typically found in the day metatranscriptome, suggesting that highly expressed genes more frequently mediate daytime-biased processes.

Finally, relative representation of transcripts was corroborated by RT qPCR-based expression analyses. Quantification of five genes showed a strong positive correlation between night and day ratios in the original RNA pool and the pyrosequence datasets ($r = 0.94$), indicating that the sequenced metatranscriptome was representative of the unamplified mRNA pool despite linear amplification, cDNA preparation, and pyrosequencing steps.

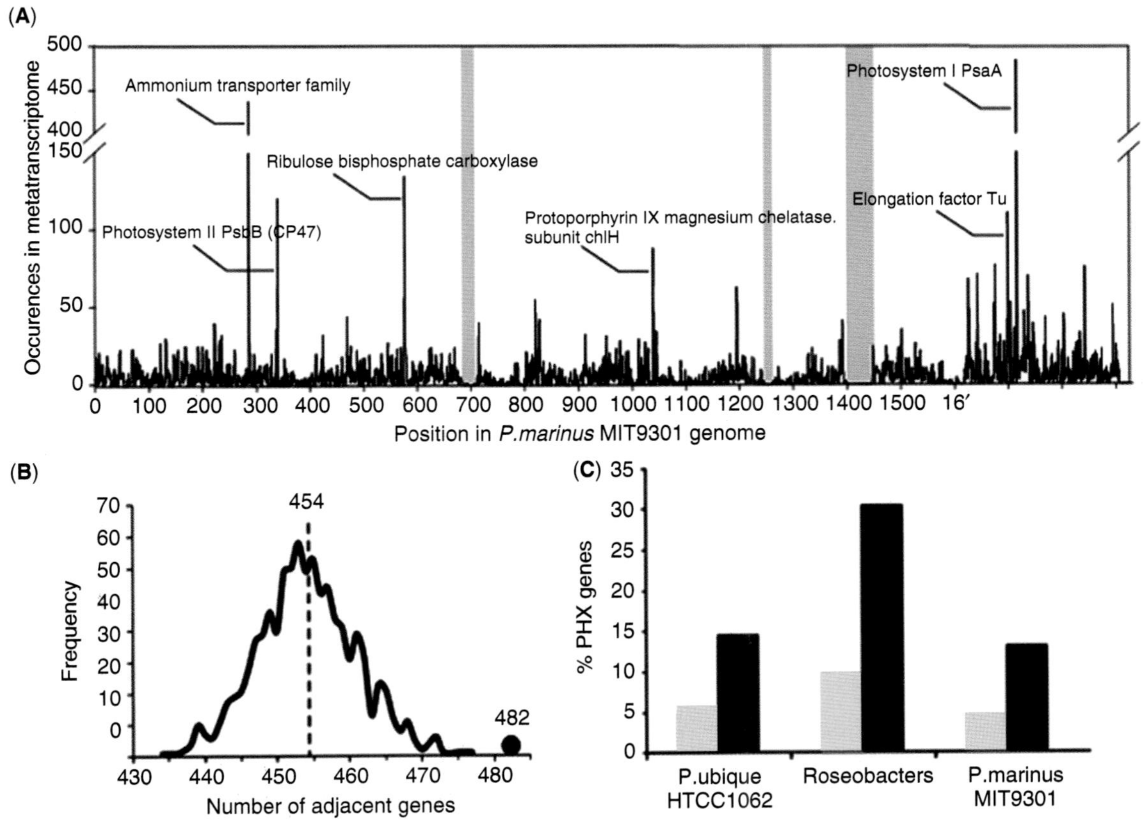

Figure 63.1 Evidence for prokaryotic gene expression patterns in the community transcriptome based on reference genome bins. (**A**) Mapping of transcripts to a *P. marinus* reference genome where shaded areas represent possible hypervariable regions with few mapped transcripts. (**B**) Operon-based expression was evaluated by comparing the number of adjacent transcripts (filled circle) to the number of adjacent genes found in 1000 random samples of the same size from the *P. ubique* HTCC1002 reference genome (black line). (**C**) Preferential representation of transcripts from genes predicted to be highly expressed was evaluated by comparing the percent of PHX genes in the reference genome (gray bar) to the percent in the transcript pool (black bar). Differences between transcript pools and reference genomes were significant for both operon and PHX analyses (Wilcoxon signed-rank test; $p < 0.05$). (Adapted with permission from the original publication [Poretsky et al., 2009b].)

63.3.3 Community Composition and Taxonomic Origin of Transcripts

In addition to functional assignments, the putative taxonomic origin of sequences can be informative. The predicted taxonomic affiliations of the putative mRNA sequences described here were predicted using MEGAN, a program that assigns likely taxonomic origin to sequences based on the NCBI taxonomy of closest Blast hits [Huson et al., 2007]. Despite the annotation limitations discussed above, in silico analyses (described in the original publication) established that taxonomic assignments are almost always correct at the phylum or subphylum level and the assignments represent the best current sequence matches for some of the more abundant environmental transcripts.

The transcriptionally active populations can be further analyzed with respect to overall community composition to determine whether taxa are equally transcriptionally active on a per-cell basis. This is accomplished via comparisons to the rRNAs in the total RNA pool (provided that no removal steps are taken) (see Urich et al. [2008]; see also Chapter 64, Vol. I), via a metagenomic dataset obtained in parallel (see Frias-Lopez et al. [2008]), or by cell counts and independent PCR-based 16S rRNA libraries (as in Poretsky et al., [2009b]). In the Poretsky et al. study [2009b], the taxonomic affilitations of the HOT transcripts were consistent with 16S rRNA clone libraries and abundance measurements; that is, most transcripts had highest identity to genes from taxa known to be present in the community. The abundant taxa included *Prochlorococcus*, the most abundant Cyanobacteria at Station ALOHA (>95% of photosynthetic picoplankton cells [Campbell and Vaulot, 1993]), which occurred as approximately 2×10^5 cell ml^{-1} (~30% of the total microbial community present at the time of sampling) and ~20% of the 16S rRNA sequences. Heterotrophic bacteria (including phototrophs) were numerically dominant with $\sim 5 \times 10^5$ cell ml^{-1} (~65% of the microbial community) and ~80% of 16S rRNA sequences (**Fig. 63.2**). Among these heterotrophic 16S rRNA sequences, Proteobacteria were most abundant (41%; Fig. 63.2) and were dominated by α-proteobacteria (22%), β-proteobacteria (8%), and γ-proteobacteria (8%). Bacteroidetes (8%) and Firmicutes (12%, biased toward the day sample) were also well represented.

Cyanobacteria also dominated the transcript libraries (55% of sequences; Fig. 63.2) with the highest per-cell transcriptional activity [about twofold higher representation than in the 16S rRNA amplicons or the cell count data (Fig. 63.2)]. This may reflect an advantage of autotrophy for maintaining cellular activity levels in the oligotrophic ocean. These Cyanobacterial transcripts were primarily *Prochlorococcus*-like sequences most similar to *P. marinus* AS9601, *P. marinus* MIT 9301, and *P. marinus* MIT 9312 (**Table 63.2**). Conversely, the heterotrophic groups generally had similar contributions to the transcript pool and amplicon pool, suggesting comparable levels of transcriptional activity on a per-gene basis (Fig. 63.2). Proteobacteria contributed the second largest number of transcript sequences (28%), most of which were attributed to α-proteobacteria (19%) and γ-proteobacteria (4%). The α-proteobacteria were the most transcriptionally active among the heterotrophic groups and contained sequences with similarity to the SAR11 group members *P. ubique* HTCC1002 and *P. ubique* HTCC1062 (~10% of prokaryotic transcripts; Table 63.2). Roseobacter-like sequences were also represented and were primarily assigned to *Dinoroseobacter shibae* DFL 12, *Jannaschia* sp. CCS1, *Silicibacter pomeroyi* DSS-3, *Roseobacter denitrificans* Och 114, and *Silicibacter* sp. TM1040 (Table 63.2). Approximately 2% of the total transcripts were of eukaryotic origin. The majority of eukaryotic transcripts were most closely affiliated with sequences from green-lineage organisms (Viridiplantae) such as the picoeukaryotic prasinophytes *Ostreococcus* spp. [Derelle et al., 2006] and *Micromonas* spp. A large number of transcripts also appeared to be most closely related to genes in Chromalveoltae (Stramenopile or Alevolate) genomes. These groups are major components of the picoeukaryotic phytoplankton [McDonald et al., 2007] and are small enough to pass the 5-μm prefilter used in this study. Most groups of autotrophic and heterotrophic microorganisms contributed equally to the day and night transcriptome (Fig. 63.2).

Table 63.2 Number of Sequences from the Community Transcriptome with Highest Homology to the Listed Reference Genomes, as Determined by top BLASTX Hit to RefSeq

	Night	Day
Prochlorococcus marinus str. MIT 9301	6309	6292
Prochlorococcus marinus str. AS9601	3214	2849
Pelagibacter ubique HTCC1002	2541	1851
Prochlorococcus marinus str. MIT 9312	1430	1264
Pelagibacter ubique HTCC1062	1308	944
Dinoroseobacter shibae DFL 12	48	34
Jannaschia sp. CCS1	41	27
Silicibacter pomeroyi DSS-3	39	30
Roseobacter denitrificans Och 114	30	28
Silicibacter sp. TM1040	19	26

63.3.4 Metatranscriptomic Comparison of Day and Night Samples

One of the promises of metatranscriptomics is the ability to discern metabolic changes in the community driven by different environmental conditions. In the paper reviewed here, comparative metatranscriptomics was used to explore microbial processes that are differentially expressed over a day/night cycle. To this end, transcript abundance was analyzed as relative abundance within the collective community transcriptome rather than per-gene expression levels (as in Frias-Lopez et al. [2008]). Overall, transcript pools were dominated by genes necessary for the maintenance of basic cellular machinery for growth and metabolism (e.g., transcription, translation, and oxidative phosphorylation; **Fig. 63.3**), DNA replication and repair, protein folding, and export. Although these processes were identified both in the day and night, there was also substantial evidence of dominant gene expression in the presence and absence of solar irradiation. Among the 167 KEGG metabolic pathways represented in the annotated sequences, four pathways were significantly more abundant among the night transcripts and six were better represented in the day. Similarly, among the 1577 COGs represented, statistical comparisons identified 12 that were better represented at night and 13 that were better represented in the day. Some KEGG pathways had significant diel differences in frequency for individual taxonomic bins such as histidine biosynthesis by *P. marinus* and *P. ubique* at night, vitamin B6 metabolism by *P. ubique* at night, and transfer of methyl groups for C1 metabolism by *P. ubique* and Roseobacter in the day.

As expected, many transcripts involved in light-mediated processes, such as photosynthesis and proteorhodopsin activity, were among those overrepresented in the community transcriptome in the day by both prokaryotes and photosynthetic picoeukaryotes. Transcripts involved in protection or repair of light-induced DNA and protein damage (e.g., catalase, chaperones, photolyases, superoxide dismutase, and various DNA repair proteins) were also common in the day sample. Evidence of daytime C1 utilization by some heterotrophs suggests a source of C1 compounds or methyl groups in this ecosystem (**Table 63.3**). Compounds such as methanol and formaldehyde [Carpenter et al., 2004; Giovannoni et al., 2008; Heikes et al., 2002], methane [Ward et al., 1987], and methylhalides [Schaefer et al., 2002; Woodall et al., 2001] may be available to heterotrophic bacterioplankton in surface seawater. Recovery of nearly four times as much mRNA per volume of seawater in the day (~30 ng L^{-1}) compared to night (~8 ng L^{-1}) is consistent with high relative abundance of RNA polymerase transcripts in the day and likely reflects increased gene expression when solar radiation is available.

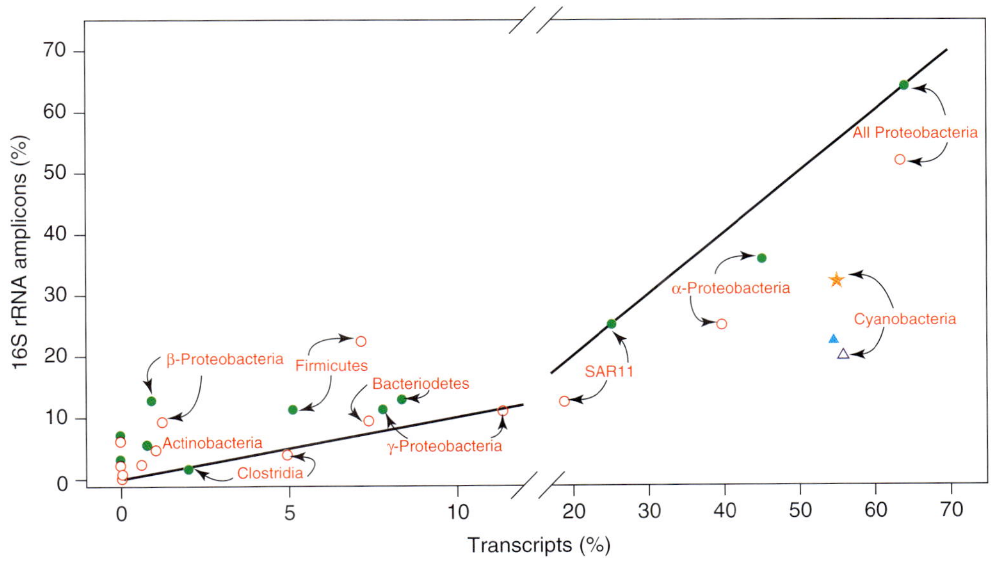

Figure 63.2 Contribution of taxa to the 16S rRNA amplicon pool compared to the transcript pool in the day (unfilled symbols) and night (filled symbols) communities. Cyanobacterial counts (triangles) are displayed as a percentage of total sequences, while the heterotrophic bacterial counts (circles) are displayed as a percentage of heterotrophic sequences only. Cyanobacterial transcript contribution abundance as determined by flow cytometry is indicated (star). α = α-proteobacteria; β = β-proteobacteria; γ = γ-proteobacteria; AP = all Proteobacteria; F = Firmicutes, Cyano = Cyanobacteria; A = Actinobacteria; C = Clostridia; S = SAR11; B = Bacteroidetes/Chlorobi. The line shows a **1:1** relationship.

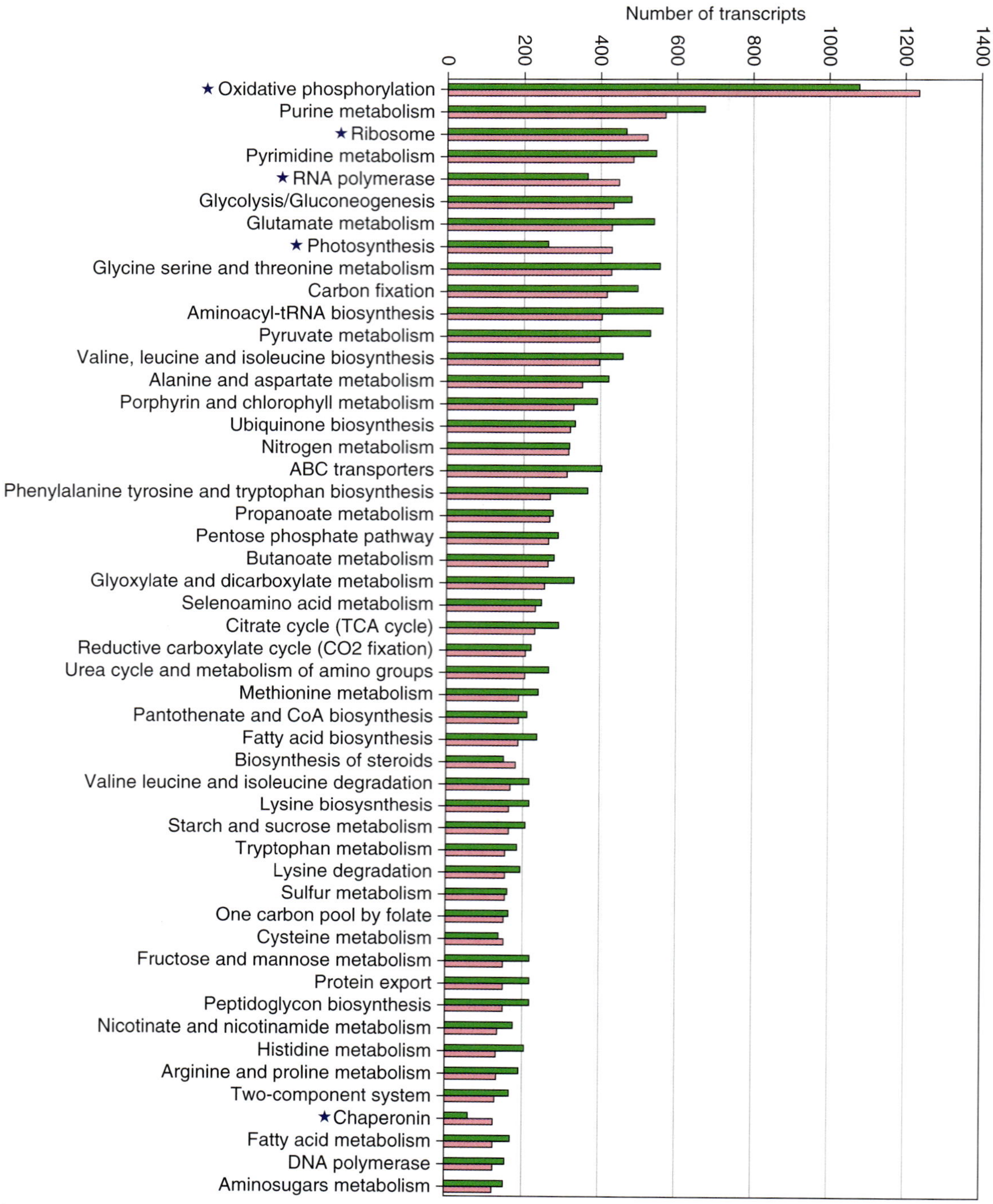

Figure 63.3 The 50 most abundant KEGG pathways in the day (green) and night (pink) transcriptomes. The pathways marked with stars were significantly overexpressed in one of the pools as determined by comparisons with $p<0.05$ [Rodriguez-Brito et al., 2006]. (Reprinted with permission from the original publication [Poretsky et al., 2009b].)

Table 63.3 Selected Biogeochemically Relevant Genes in the HOT Metatranscriptome

			Night[a]	Day[a]
Nitrogen	Nitrogenase (N fixation)	*nifH, nifU, nifS, nifB*	+	+
	Ammonium transport	*amt*	+	+*
	Ammonia monooxygenase	*amoA*		
	Assimilatory nitrate reductase	*narB*	+	
	Hydroxylamine oxidoreductase	*hao*		
	Nitrate permease	*napA*	+	
	Nitrite reductase	*nirA*	+	
	Dissimilatory nitrite reductase	*nirK, nirS*		
	Nitric oxide reductase	*norQ*	+	
	Nitrate transporter	*narK*	+	
	Urease	*ureC, ureE, ureF*	+	+
Methylotrophy	Serine-glyoxylate aminotransferase		+	+
	Formate dehydrogenase	*fdh, fdsD*	+	+
	Methylene tetrahydrofolate reductase	*metF*	+	+
	Methane monooxygenase	*mmo*		
	Methanol dehydrogenase	*mxa*		+
	Methenyltetrahydromethanopterin cyclohydrolase	*mch*	+	+
	Crotonyl-CoA reductase		+	+
	Formaldehyde-activating enzyme	*fae*		+
Polyamine degradation	Deoxyhypusine synthase	*dys2*	+*	+
	Spermidine/putrescine transport system permease	*potC*	+*	+
	Acetylpolyamine aminohydrolase	*aphA*		
Sulfur cycle	Sulfur oxidation	*soxB, soxC, soxA, soxZ, soxF*	+	+
	Dimethylsulfoniopropionate (DMSP) demethylase	*dmdA*		
Glycine betaine	Dimethylglycine dehydrogenase	*dmgdh*	+	+
	Glycine cleavage system (amnomethyltransferase)	*gcvT*	+*	+
Aromatic compounds	Aromatic ring hydroxylase	*chlP*	+	+*
	protocatechuate 3,4-dioxygenase	*pcaH*		
	Benzoyl-CoA oxygenase	*boxA*		+
Carbon monoxide	Carbon monoxide dehydrogenase	*cosS, coxM, coxL*	+	+
Phototrophy and C fixation	Photosystem I	multiple	+	+*
	Photosystem II	multiple	+	+*
	Rubisco	*rbcL, rbcS*	+	+*
	Photosynthetic reaction center, M subunit	*pufM*		+
	Proteorhodopsin		+	+*
Phosphate assimilation	Phosphonate uptake	*phnD, phnC*	+	+
	Alkaline phosphatase	*phoA*	+	+
	Phosphate uptake	*pstA, pstS*	+	+
Amino acid metabolism	Glutamate synthase	*gltB*	+	+
	Glutathione reductase	*gor*	+*	+
	Histidine kinase	*baeS*	+*	+
	Threonine synthase	*thrC*	+*	+
Trace metal uptake	Selenium		+*	+
	Iron	*tonB*	+	+
	Arsenite		+	
	Arsenate reductase	*arsC*	+	+

[a] A plus sign indicates occurrence in the night or Day Sample. An asterisk indicates significantly higher transcript frequency in one

Night-biased synthesis of vitamin B6, essential for a variety of amino acid conversions including transaminations, decarboxylations, and dehydrations, in conjunction with evidence for other nighttime activities such as the γ-glutamyl pathway for amino acid uptake, the overrepresentation of amino acid transport and metabolism genes, and the histidine synthesis pathway (Table 63.3), indicate that amino acid acquisition in general may be a relatively more important metabolic activity in the night. *P. marinus* has recently been shown to exhibit diel patterns of amino acid uptake, with acquisition occurring predominantly at dusk [Mary et al., 2008]. The transcriptome data agree with this and further suggest that heterotrophic taxa also devote a greater percentage of their transcriptome to transporting and synthesizing amino acids at night. Nighttime accumulation of amino acids might be a mechanism for nitrogen storage by many organisms, particularly for *P. marinus*, which undergoes cell division at night. Histidine, the amino acid with the most consistent signal for synthesis at night by both autotrophs and heterotrophs, is one of the most nitrogen-rich amino acids.

63.3.5 Summary

Metatranscriptomics enables an assessment of activity, response, and regulation in complex microbial communities. This can be used both for descriptive studies to establish baseline gene expression information and for experimental comparisons of gene expression resulting from differing conditions. In the HOT system described here, comparative metatranscriptomics captured microbial processes expected to be differentially expressed over a diel cycle, such as photosynthesis, oxidative phosphorylation, and synthesis of light-driven cellular machinery such as proteorhodopsins and photosynthetic pigments. Other, less anticipated processes that emerged from the comparative diel analyses included uptake and utilization of multiple nitrogen (e.g., polyamines), carbon (C1), sulfur, phosphorus, and aromatic compounds, many of which are relevant to major biogeochemical cycles. Overall, bacterial community investment in this oligotrophic ocean system was skewed toward energy acquisition and metabolism (specifically photosynthesis and C1 compound utilization) during the day, while biosynthesis (specifically of membranes, amino acids, and vitamins) received relatively greater investments at night. This comparative analysis of microbial community transcripts provided an inventory of ongoing metabolic processes, offered insights into their temporal patterns, and supplied a new type of data for predictive modeling of environmental controls on ecosystem properties.

INTERNET RESOURCES

NCBI Sequence Read Archive: http://www.ncbi.nlm.nih.gov/Traces/sra/

The Hawaii Ocean Time-Series database: http://hahana.soest.hawaii.edu/hot/hot-dogs/

COG database: http://www.ncbi.nlm.nih.gov/COG/

KEGG database: http://www.genome.jp/kegg/

CAMERA: http://camera.calit2.net/index.php

Acknowledgments

Sampling at HOT was conducted with the support of the Captain and crew of the R/V Kilo Moana and Dr. David Karl. Jennifer Oliver assisted with sample processing. Funding was provided by The Gordon and Betty Moore Foundation, National Science Foundation grants MCB-0702125 (Mary Ann Moran) and OCE-0425363 (Jonathan P. Zehr), and the NSF C-MORE Center for Microbial Oceanography.

REFERENCES

Bailly J, Fraissinet-Tachet L, Verner MC, Debaud JC, Lemaire M, Wesolowski-Louvel M, Marmeisse R. 2007. Soil eukaryotic functional diversity, a metatranscriptomic approach. *ISME J*. **1**:632–642.

Benson DA, Karsch-Mizrachi I, Lipman DJ, Ostell J, Wheeler DL. 2007. GenBank. *Nucleic Acids Res*. **35**:D21–D25.

Bürgmann H, Widmer F, Sigler WV, Zeyer J. 2003. mRNA extraction and reverse transcription-PCR protocol for detection of *nifH* gene expression by *Azotobacter vinelandii* in soil. *Appl. Environ. Microbiol*. **69**:1928–1935.

Campbell L, Vaulot D. 1993. Photosynthetic picoplankton community structure in the subtropical North Pacific Ocean near Hawaii (Station ALOHA). *Deep Sea Res*. (*I Oceanogr Res. Pap*.) **40**: 2043–2060.

Carpenter LJ, Lewis AC, Hopkins JR, Read KA, Longley ID, Gallagher MW. 2004. Uptake of methanol to the North Atlantic Ocean surface. *Global Biogeochem. Cycles* **18**:GB4027.

Conesa A, Gotz S, Garcia-Gomez JM, Terol J, Talon M, Robles M. 2005. Blast2GO: A universal tool for annotation, visualization and analysis in functional genomics research. *Bioinformatics* **21**:3674–3676.

Derelle E, Ferraz C, Rombauts S, Rouze P, Worden AZ, et al. 2006. Genome analysis of the smallest free-living eukaryote *Ostreococcus tauri* unveils many unique features. *Proc. Natl. Acad. Sci. USA* **103**:11647–11652.

Frias-Lopez J, Shi Y, Tyson GW, Coleman ML, Schuster SC, Chisholm SW, DeLong EF. 2008. Microbial community gene expression in ocean surface waters. *Proc. Natl. Acad. Sci. USA* **105**:3805–3810.

Fuhrman JA, Comeau DE, Hagstrom A, Chan AM. 1988. Extraction from natural planktonic microorganisms of DNA suitable for molecular biological studies. *Appl. Environ. Microbiol*. **54**: 1426–1429.

Gilbert JA, Field D, Huang Y, Edwards R, Li W, Gilna P, Joint I. 2008. Detection of large numbers of novel sequences in the metatranscriptomes of complex marine microbial communities. *PLoS ONE* **3**:e3042.

Giovannoni SJ, Hayakawa DH, Tripp HJ, Stingl U, Givan SA, et al. 2008. The small genome of an abundant coastal ocean methylotroph. *Environ. Microbiol*. **10**:1771–1782.

Heikes BG, Chang WN, Pilson MEQ, Swift E, Singh HB, et al. 2002. Atmospheric methanol budget and ocean implication. *Global Biogeochem. Cycles* **16**:80.1–80.13.

Hewson I, Poretsky RS, Beinart RA, White AE, Shi T, et al. 2009. *In situ* transcriptomic analysis of the globally important keystone N2-fixing taxon *Crocosphaera watsonii*. *ISME J*. **3**:618–631.

Huson DH, Auch AF, Qi J, Schuster SC. 2007. MEGAN analysis of metagenomic data. *Genome Res*. **17**:377–386.

Ingraham JL, Maaløe O, Neidhardt FC. 1983. *Growth of the Bacterial Cell*. Sunderland, MA: Sinauer Associates.

Johnson JL, 1994, Similarity analysis of rRNAs, In Gerhardt P, Murray RGE, Wood WA, Krieg NR, eds. *Methods for General and Molecular Bacteriology*. Washington, D.C.: American Society for Microbiology, pp. 683–700.

Kanehisa M, Goto S. 2000. KEGG: Kyoto Encyclopedia of Genes and Genomes. *Nucleic Acids Res*. **28**:27–30.

Karl DM, Lukas R. 1996. The Hawaii Ocean Time-series (HOT) program: Background, rationale and field implementation. *Deep Sea Res. (II Top. Stud. Oceanogr.)* **43**:129–156.

Karlin S, Mrázek J. 2000. Predicted highly expressed genes of diverse prokaryotic genomes. *J. Bacteriol*. **182**:5238–5250.

Leininger S, Urich T, Schloter M, Schwark L, Qi J, Nicol GW, Prosser JI, Schuster SC, Schleper C. 2006. Archaea predominate among ammonia-oxidizing prokaryotes in soils. *Nature* **442**:806–809.

Li W, Godzik A. 2006. Cd-hit: a fast program for clustering and comparing large sets of protein or nucleotide sequences. *Bioinformatics* **22**:1658–1659.

Margulies M, Egholm M, Altman WE, Attiya S, Bader JS, et al. 2005. Genome sequencing in microfabricated high-density picolitre reactors. *Nature* **437**:376–380.

Mary I, Garczarek L, Tarran GA, Kolowrat C, Terry MJ, et al. 2008. Diel rhythmicity in amino acid uptake by *Prochlorococcus*. *Environ. Microbiol*. **10**:2124–2131.

McDonald SM, Sarno D, Scanlan DJ, Zingone A. 2007. Genetic diversity of eukaryotic ultraphytoplankton in the Gulf of Naples during an annual cycle. *Aquat. Microb. Ecol*. **50**:75–89.

Mou X, Sun S, Edwards RA, Hodson RE, Moran MA. 2008. Bacterial carbon processing by generalist species in the coastal ocean. *Nature* **451**:708–711.

Poretsky RS, Bano N, Buchan A, LeCleir G, Kleikemper J, Pickering M, Pate WM, Moran MA, Hollibaugh JT. 2005. Analysis of microbial gene transcripts in environmental samples. *Appl. Environ. Microbiol*. **71**:4121–4126.

Poretsky RS, Bano N, Buchan A, Moran MA, Hollibaugh JT. 2008. Environmental transcriptomics: A method to access expressed genes in complex microbial communities, In Kowalchuk GA, de Bruijn FJ, Head IM, Akkermans ADL, van Elsas JD, eds. *Molecular Microbial Ecology Manual*. Netherlands: Springer, pp. 1892–1904.

Poretsky RS, Gifford S, Rinta-Kanto J, Vila-Costa M, Moran MA. 2009a. Analyzing gene expression from marine microbial communities using environmental transcriptomics. *J. Vis. Exp*. pii: 1086. doi: 10.3791/1086

Poretsky RS, Hewson I, Sun SL, Allen AE, Zehr JP, Moran MA. 2009b. Comparative day/night metatranscriptomic analysis of microbial communities in the North Pacific subtropical gyre. *Environ. Microbiol*. **11**:1358–1375.

Poretsky RS, Sun S, Mou X, Moran MA, 2010. Transporter genes expressed by coastal bacterioplankton in response to dissolved organic carbon. *Environ. Microbiol*. **12**:616–627.

Pruitt KD, Tatusova T, Maglott DR. 2005. NCBI Reference Sequence (RefSeq): A curated nonredundant sequence database of genomes, transcripts and proteins. *Nucleic Acids Res*. **33**:D501–D504.

Rauhut R, Klug G. 1999. mRNA degradation in bacteria. *FEMS Microbiol. Rev*. **23**:353–370.

Rodriguez-Brito B, Rohwer F, Edwards R. 2006. An application of statistics to comparative metagenomics. *BMC Bioinform*. **7**:162.

Schaefer JK, Goodwin KD, McDonald IR, Murrell JC, Oremland RS. 2002. *Leisingera methylohatidivorans* gen. nov., sp nov., a marine methylotroph that grows on methyl bromide. *Int. J. Syst. Evol. Microbiol*. **52**:851–859.

Seshadri R, Kravitz SA, Smarr L, Gilna P, Frazier M. 2007. CAMERA: A community resource for metagenomics. *PLoS Biol*. **5**:394–397.

Shendure J. 2008. The beginning of the end for microarrays? *Nat. Methods* **5**:585–587.

Tatusov RL, Galperin MY, Natale DA, Koonin EV. 2000. The COG database: A tool for genome-scale analysis of protein functions and evolution. *Nucleic Acids Res*. **28**:33–36.

Urich T, Lanzén A, Qi J, Huson DH, Schleper C, Schuster SC. 2008. Simultaneous assessment of soil microbial community structure and function through analysis of the metatranscriptome. *PLoS ONE* **3**:e2527.

Ward BB, Kilpatrick KA, Novelli PC, Scranton MI. 1987. Methane oxidation and methane fluxes in the ocean surface-layer and deep anoxic waters. *Nature* **327**:226–229.

Wawrik B, Paul JH, Tabita FR. 2002. Real-time PCR quantification of rbcL (ribulose-1,5-bisphosphate carboxylase/oxygenase) mRNA in diatoms and pelagophytes. *Appl. Environ. Microbiol*. **68**: 3771–3779.

Woodall CA, Warner KL, Oremland RS, Murrell JC, McDonald IR. 2001. Identification of methyl halide-utilizing genes in the methyl bromide-utilizing bacterial strain IMB-1 suggests a high degree of conservation of methyl halide-specific genes in gram-negative bacteria. *Appl. Environ. Microbiol*. **67**:1959–1963.

Zhou JH. 2003. Microarrays for bacterial detection and microbial community analysis. *Curr. Opin. Microbiol*. **6**: 288–294.

Chapter 64

The "Double-RNA" Approach to Simultaneously Assess the Structure and Function of a Soil Microbial Community

TIM URICH AND CHRISTA SCHLEPER

64.1 INTRODUCTION

Soils cover almost all of the terrestrial area on Earth and have an indispensable ecological function in the global cycles of carbon, nitrogen, and sulfur. Due to their physicochemical complexity with many microniches, they teem with biodiversity, both phylogenetically and functionally. A single gram of soil has been estimated to contain thousands to millions of different bacterial, archaeal, and eukaryotic species [Torsvik et al., 2002; Gans et al., 2005; see Chapter 2, Vol. I] interwoven in extremely complex food webs. Communities of soil microbes carry out a multitude of small-scale processes that underlie many environmentally important functions. However, the explicit functional and ecological roles of individual taxa remain uncertain because most microbes withstand laboratory cultivation [Fierer et al., 2007b; Prosser, 2007]. Therefore the most basic questions in microbial ecology—"Who is out there?" and "What are they doing?"—are still often unanswered for many environments and for many microbial taxa. Ideally, especially the second question requires simultaneously information about the identity of taxa within a community and about functional processes performed.

With metagenomic technologies, new dimensions in the characterization of complex microbial communities have been reached. Large-scale shotgun sequencing approaches allow the discovery of many novel genes found in the environments independent of cultivation efforts [Venter et al., 2004; Tringe et al., 2005; Martin-Cuadrado et al., 2007; Rusch et al., 2007]. In almost all of the metagenomic studies, a separate accompanying molecular typing method—usually based on PCR-amplified 16S rRNA genes—is needed to characterize the gene discoveries in the context of the microbial phylogenetic diversity of the sample [Venter et al., 2004; Tringe et al., 2005]. These PCR-based typing methods—though very powerful, in particular when combined with the novel pyro-sequencing technology [Sogin et al., 2006; Huber et al., 2007]—have some well-known drawbacks: (1) Bias is introduced by primers and/or exponential amplification, (see Chapters 16–17, Vol. I) (2) simultaneous quantitative assessment of all three domains of life is impossible, and (3) persistence of free DNA can bias the measurement of community responses to environmental changes.

In analogy to postgenomic studies of cultivated organisms, a logical next step in the metagenomic area included metatranscriptomic technology. Since the pioneering study of Poretsky and colleagues in 2005 [Poretsky et al., 2005; see Chapter 63, Vol. I], several studies have been performed, initially with small clone libraries of a few hundred sequences [Poretsky et al., 2005; Bailly et al., 2007], but soon combined with the upcoming next-generation sequencing technologies [Leininger et al., 2006]. Since 2008 the first studies on marine environments were published combining in vitro

Handbook of Molecular Microbial Ecology, Volume I: Metagenomics and Complementary Approaches, First Edition. Edited by Frans J. de Bruijn.

enrichment strategies for bacterial and archaeal mRNA with pyrosequencing [Frias-Lopez et al., 2008; Gilbert et al., 2008; Poretsky et al., 2009; see Chapter 63, Vol. I and Chapter 27, Vol. II]. All involved in vitro amplification steps to maintain small sample size and fast preparation.

In 2008 we introduced the "double-RNA approach" for in-depth characterization of a soil microbial community by studying mRNA and rRNA molecules simultaneously from the same sample [Urich et al., 2008]. Alternative to the previous studies, which focused exclusively on mRNA as a functional marker, we explored the possibility to analyze the total RNA pool of a community, because it is naturally enriched not only in functionally but also taxonomically relevant molecules—that is, mRNA and rRNA, respectively. This offered the opportunity to obtain information on the community structure and function from a single experiment. Since highly parallel GS-20 pyrosequencing was used for this approach, which produced many (but short) reads (100 bp), we have set up an appropriate bioinformatic analysis pipeline that allowed us to reliably extract functional and taxonomic information from this dataset.

64.2 MATERIALS AND METHODS

For a full outline of the materials and methods, please refer to the original publication [Urich et al., 2008]. An abbreviated version follows below.

64.2.1 Site Description and Soil Sample Processing

The soil samples were obtained from a sandy lawn in the environmentally protected area "Am Rotböll" (Germany, Hessen, 49°55′34″N 8°37′21.6″E [Leininger et al., 2006). The top soil was sampled at a depth of 0–10 cm in January 2006 after a snow thawing period (soil temperature 5.5°C). Three replicates were withdrawn and sieved (2-mm mesh size), and equal amounts of the triplicates were subsequently mixed. Six grams of soil were then processed for RNA extraction.

64.2.2 cDNA Synthesis and Pyrosequencing

Nucleic acid extractions (RNA and DNA) were performed using a modification of the method of Griffiths et al. [2000] and as briefly described in Leininger et al. [2006]. Total nucleic acids were subjected to DNase treatment, and the remaining total RNA was used as template for random hexamer-primed reverse transcription performed independently at 37°C and 50°C, respectively. The subsequently generated double-stranded (ds) cDNA ranged from approximately 100 to 1500 bp in length (not shown). Approximately 1 μg of ds cDNA was generated. Sequencing was performed on a Roche GS20 sequencer (Roche Applied Sciences/454 Life Sciences, Branford, CT).

64.2.3 Bioinformatic Analysis

Two rRNA reference sequence databases were constructed for community structure analysis using MEGAN [Huson et al., 2007; see Chapter 39, Vol. II]. The sequences therein were linked to taxa in the NCBI Taxonomy (as of June 2007), extended with 28 archaeal candidate divisions [Schleper et al., 2005] (see supplementary tables ST1 and ST2). The SSUrdb includes sequences from the RDP-II release 9.39 [Cole et al., 2007; see Chapter 36, Vol. I] (bacteria), the SILVA SSURef database release 89 [Pruesse et al., 2007] (eukaryotes and archaea), and sequences from the dataset described in [Schleper et al., 2005] (archaea). The LSUrdb includes sequences from the Silva LSURef database release 90 [Pruesse et al., 2007]. Sequences from the RDP-II were selected based on affiliation to well-defined taxa in the NCBI taxonomy (order level or more specific). Only from taxa with comparably few reference sequences or a less well resolved taxonomy were less-defined sequences included. Similar selections were made for sequences from SILVA, using the ARB software package [Ludwig et al., 2004; see Chapter 45, Vol. I]. Sequences where the alignment did not reflect the taxonomic (NCBI) affiliation or sequences with comparably low taxonomic resolution were omitted.

All RNA tags were compared to the reference databases using BLASTN [Altschul et al., 1997]. The MEGAN program see Chapter 39, Vol. I (v1.18) subsequently placed each RNA-tag on a node in the NCBI taxonomy according to the BLAST comparison. This was performed using the "LCA algorithm" that assigns each read to the lowest common ancestor out of the most similar taxa identified by BLAST (absolute cutoff: BLAST bitscore 86, relative cutoff: 10% of the top hit).

All unassigned RNA-tags were aligned against the All nonredundant Genbank CDS database (release of June 25, 2007) using BLASTX and analyzed with MEGAN. RNA-tags with a BLAST bitscore higher than 40 were considered as putative mRNA-tags. These were submitted to the MG-RAST server (http://metagenomics.nmpdr.org; see also Chapter 37, Vol. I) along with two soil metagenome datasets (see supplementary table ST7). MG-RAST produces a functional annotation based on subsystems (groups of functional roles) as defined in the SEED database [Overbeek et al., 2005]. Taxonomic binning of mRNA-tags was performed using MEGAN.

64.2.4 Data Deposition

The SSUrdb and LSUrdb are available for download from http://www.bioinfo.no/services/community-profiling. The sequences reported in this paper have been submitted to GenBank under accession number SRA001014.

64.3 RESULTS AND DISCUSSION

64.3.1 The Analysis Pipeline

The soil studied here was from a conservation area and has been the subject of a variety of studies. Many molecular data are therefore available which facilitate result interpretation [Quaiser et al., 2002, 2003; Ochsenreiter et al., 2003; Treusch et al., 2004, 2005; Leininger et al., 2006]. We generated cDNA from the transcriptome of this soil community by a nontargeted approach applying random-hexamer primed reverse transcription on total RNA. The cDNA was subjected directly to pyrosequencing without any prior PCR or cloning steps. Several potentially biasing steps were therefore avoided. One experiment resulted in 258,411 RNA-derived sequence tags of ~98 bp (**Table 64.1**).

We set up a two-step bioinformatic analysis pipeline to identify rRNA- and mRNA-derived tags ("ribo-tags" and "mRNA-tags," respectively) for efficiently and reliably obtaining taxonomic and functional information (**Fig. 64.1**). In the first step, all RNA-tags were compared against a small subunit (SSU) and a large subunit (LSU) rRNA reference database (SSUrdb and LSUrdb, respectively) that we compiled from publicly available sources (SILVA [Pruesse et al., 2007], RDP-II [Cole et al., 2007; see Chapter 36, Vol. I], Genbank nucleotide). These two databases contain sequences from all three domains of life (SSUrdb: 137,160 sequences; LSUrdb: 6247 sequences). This approach enables the first simultaneous taxonomic analysis of communities based on the two most commonly used taxonomic marker molecules. The output file was analyzed with the program MEGAN [Huson et al., 2007].

We have extensively verified the taxonomic resolution power of short (100 bp), random-primed rRNA sequences as well as our analysis pipeline [Urich et al., 2008]. For most bacterial and archaeal taxa tested, assignments until the taxonomic rank "order" was possible. Since this study, this resolution has improved further because 454 read lengths increased to 250 bp and even 450 bp and the rRNA reference databases contain more sequences. A current study with 454 Titanium rRNA-tags (450 bp) and an updated rRNA reference database shows a taxonomic resolution of rRNA-tags until the family and genus level for many taxa [Lanzén and Urich, unpublished].

Table 64.1 Statistics of Soil Community RNA-Tag Analysis

	No. of RNA-Tags	% of RNA-Tags
Total RNA-tags	258,411	100
Total ribo-tags	193,227	74.8
SSU ribo-tags	99,061	38.3
LSU ribo-tags	94,166	36.4
mRNA-tags	21,133	8.2
Unassigned RNA-tags	43,889	17.0

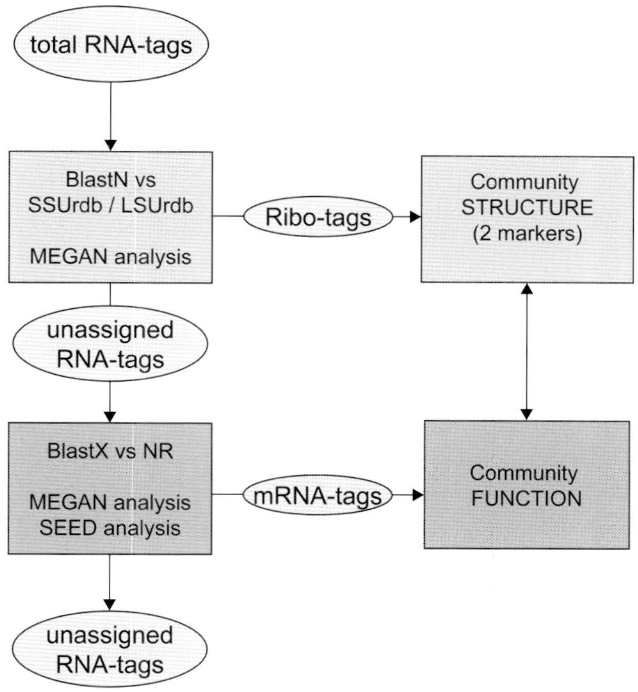

Figure 64.1 Overview about the double RNA analysis pipeline. Refer to text for details. (Reproduced from Urich et al. [2008] with permission from the authors.)

64.3.2 Community Profiling Based on SSU and LSU rRNA-Tags

The dataset contained 193,219 RNA-tags, which had significant BLAST hits against the rRNA reference databases (Table 64.1 and Fig. 64.2a). This dataset, generated by random reverse transcription, is as large as, or even larger than, recently published soil-derived 16S rRNA PCR pyrosequencing datasets (e.g., Roesch et al. [2007]). The high abundance of rRNA in the total RNA pool made it possible to avoid PCR technologies combined with specific primers for target enrichment, as is usually necessary with DNA. This prevented limitations to a single domain due to primer restrictions, as well as amplification biases. In particular, lineages that are currently not picked up by "universal" primers—like, for example, the Poribacteria or the Nanoarchaeota [Huber et al., 2002]—could potentially be identified via this approach.

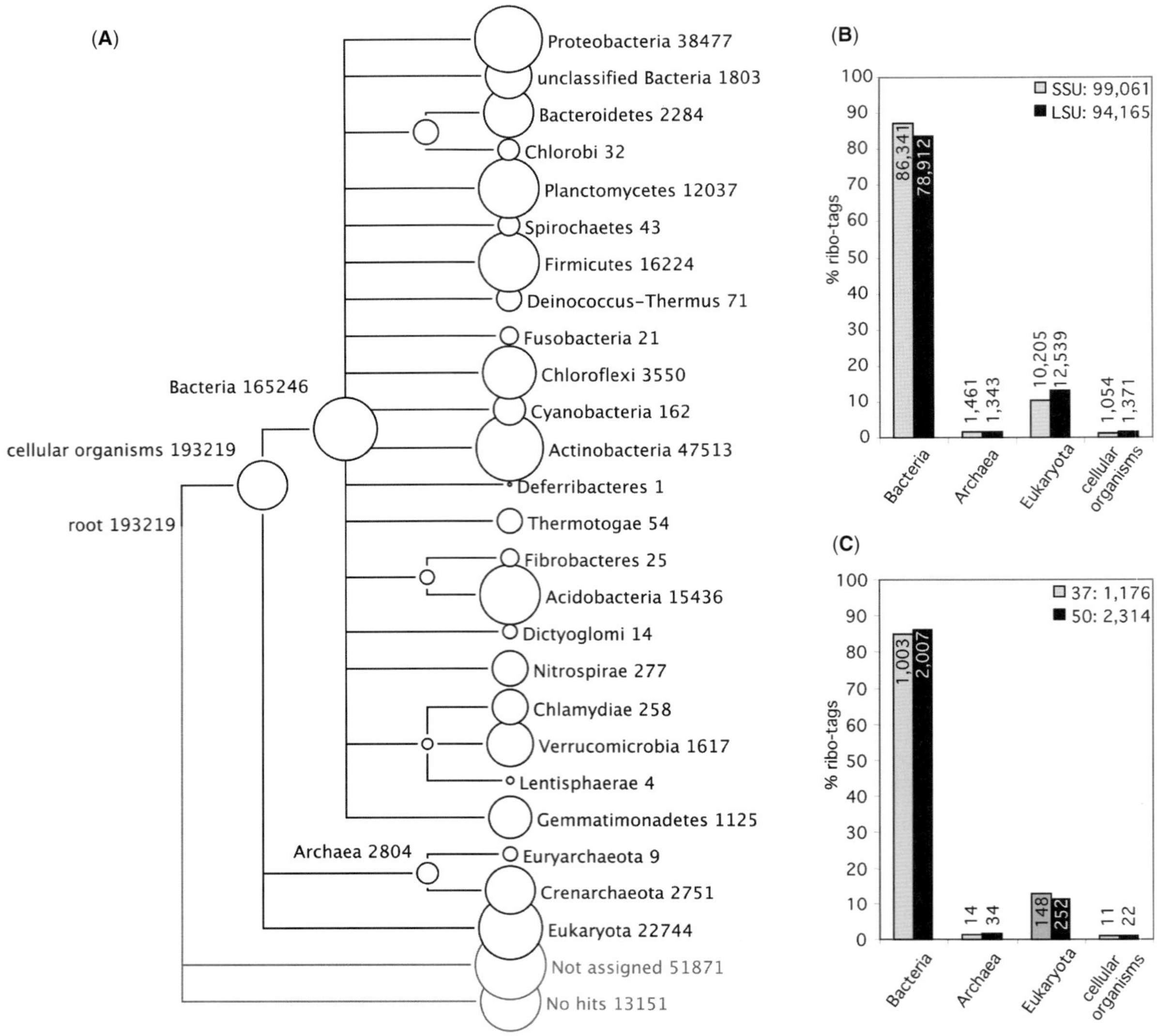

Figure 64.2 In-depth taxonomic community profiling based on SSU and LSU rRNA. (**a**) MEGAN tree with the taxonomic affiliation of ribo-tags identified by BLASTN of all RNA-tags against our SSUrdb and LSUrdb according to the NCBI taxonomy. The numbers and sizes of the circles at the tree nodes refer to the ribo-tags affiliated to the respective taxon. (**b**) Relative ribo-tag distribution of the different cellular domains of life in the SSUrdb and LSUrdb. Both reference databases report similar fractions, with bacteria-derived ribo-tags being most prominent. (**c**) Comparison of relative ribo-tag distribution from two independent cDNA syntheses at 37°C and 50°C. Values represent the mean of SSU and LSU ribo-tags. (Reproduced from Urich et al. [2008] with permission from the authors.)

One hallmark of our approach is the capturing of rRNA from all three domains of life, resulting a previously impossible holistic community profiling. The SSU rRNA and LSU rRNA databases reported very similar relative community proportions on the domain level, with 10.3% and 13.3% of ribo-tags stemming from Eukarya, 87.2% and 83.8% from Bacteria and 1.5% and 1.4% from Archaea, respectively (Fig. 64.2b), which shows the reliability of our analytical approach. We further confirmed the experimental reproducibility by performing two independent cDNA syntheses from the same RNA pool (Fig. 64.2c). In addition, abundances measured previously for some bacterial and archaeal groups in the same environment using various quantitative real-time PCR or metagenomic methodologies were in agreement with this study [Quaiser et al., 2002, 2003; Ochsenreiter et al., 2003; Treusch et al., 2004; Leininger et al., 2006]. With 165,246 ribo-tags (85.5% of total ribo-tags), bacteria were by far the most abundant domain in the sample; they were also extraordinarily diverse. Nineteen out of 24 validly described bacterial phyla were identified, in addition to 20 candidate divisions [Urich et al., 2008]. The dominant bacterial phyla such as Actinobacteria (47,513 ribo-tags) and Proteobacteria (38,477 ribo-tags) are typically found in high numbers in soils. Other prominent phyla were the Acidobacteria, Planctomycetes, Firmicutes, and Chloroflexi. Even rare phyla, such as Chlorobi and Dictyoglomi (~0.02% of all bacterial ribo-tags) were reliably detected by SSU and LSU ribo-tags. The predominant archaeal taxon (>98% of SSU ribo-tags) was the Crenarchaeota groupI.1b, which was consistent with earlier studies in the same habitat [Urich et al., 2008]. More than 20,000 ribo-tags were identified as of eukaryotic origin, with more than 80% of the tags having a taxonomic resolution at least to the kingdom level. The numerically dominant kingdoms were the fungi, but also ribo-tags of plants (viridiplantae) and animals (metazoa) were abundant [Urich et al., 2008], with ~50%, ~20%, and ~10% of the ribo-tags being consistently assigned by both reference databases. In addition, a considerable number of ribo-tags were assigned to the different Protist kingdoms [Urich et al., 2008].

Using the SSU ribo-tags allowed a holistic display of the heterotrophic soil community, including the different microbial groups and additionally lower and higher eukaryotes. Therefore several functional groups and trophic levels were covered simultaneously (**Fig. 64.3**). Solely using SSU ribo-tags allowed informative taxonomic assignment at higher resolution, which is currently not possible with LSU ribo-tags—especially for many eukaryotic groups due to the much smaller LSUrdb. As mentioned previously, bacterial ribosomes were by far dominating, followed by fungi and archaea. Within the fungi, the phylum Ascomycota accounted for two-thirds of the ribo-tags, followed by Glomeromycota and Basidiomycota. Ascomycota are typically saprotrophs, degrading dead plant material. The different Protist kingdoms are mostly grazers of prokaryotes, having therefore an important role in the soil food web by regulating bacterial and archaeal populations. In this soil, slime molds (Mycetozoa) were the most abundant group, followed by Cercozoa, Plasmodiophorida, and Alveolata. Other higher-level trophic groups belonged to the Metazoa. This included members of the microfauna (Nematoda, Rotifera) and mesofauna (Enchytraeidae, Tardigrada, Collembola, and Acari), occupying a variety of trophic positions in the soil food web [Schaefer, 1990].

The community profiles presented here were derived from the counting of ribosomes, but not their genes, as it is the case with DNA-based approaches. Unlike ribosomal genes, ribosomal RNA is often used as an indicator for activity [Weller and Ward, 1989]. The ribosome content can vary considerably in an organism, likely reflecting its physiological state. However, cells that have recently been active might keep their ribosomes, although having changed to a dormant state. Therefore, ribosomal content might not always be a direct indicator for activity. Currently, few studies are available; *E. coli* cells were estimated to contain between 6800 and 72,000 ribosomes, depending on the growth phase [Bremer and Dennis, 1996]. In addition, the ribosome content probably also differs between taxa. Unicellular eukaryotes likely have

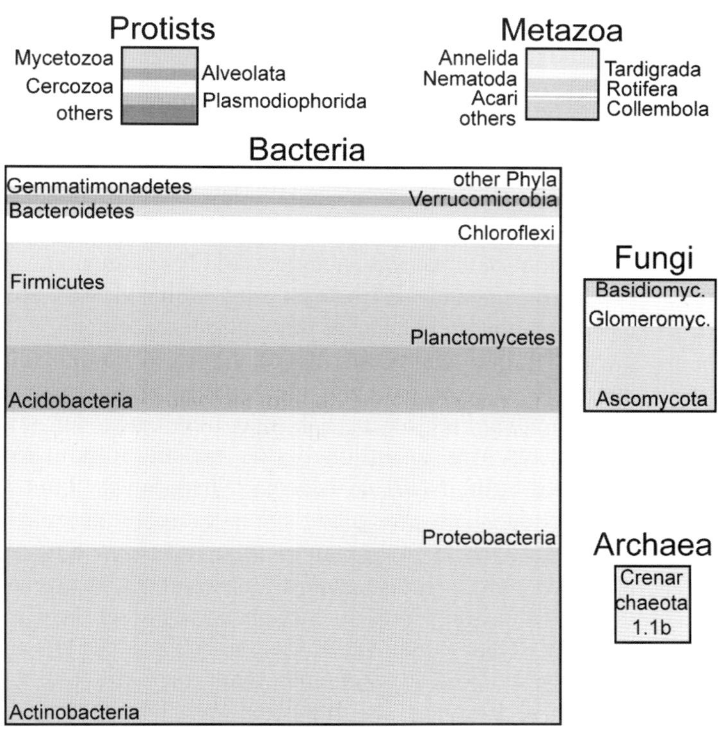

Figure 64.3 A holistic view onto the soil (microbial) community from an SSU rRNA perspective. The postion of the taxa tentatively refers to their trophic level in the soil foodweb. The area of the boxes is proportional to the number of SSU ribo-tags for each group. Bacterial, archaeal, and fungal groups with >1% and Protist groups with >5% abundance are shown; low-abundant groups are shown as "others," if present in sufficient numbers. For metazoa, "others" refers to ribo-tags of low-abundant groups and ribo-tags with insufficient taxonomic resolution power.

more ribosomes per cell than prokaryotes, due to their size (although no quantitative studies exist). Even higher is the ribosomal content of multicellular organisms like fungi, metazoans, or plants. It is clear that more research is needed to understand the relationship of the numbers of ribosomes and (a) the activity of microbes and (b) the biomass in different taxa. We consider the amount of ribo-tags as determined in this study to be a measure of a taxon's cellular biomass within a community.

Another biomarker, phospholipid fatty acid (PLFA), is often used to estimate the relative proportion of fungal and bacterial biomass in soils [Joergensen and Wichern, 2008]. Different factors (ranging from 14 to 40) are applied to calculate biomass carbon from the experimentally determined PLFAs for fungi and bacteria, to account for the higher concentration of PLFAs in the smaller bacterial cells. A conversion factor likely needs to be taken into account to relate the numbers of ribosomes to the biomass of prokaryotes and eukaryotes as well. Compared to PLFAs, ribo-tags have a higher taxonomic resolution power.

64.3.3 Global Functional Analysis of Putative mRNA-Tags

65,192 RNA-tags, which did not give a significant hit against the rRNA reference databases, were aligned against the Genbank nonredundant protein database. Homologues to 21,133 RNA-tags were found, showing that a considerable amount of assignable mRNA had been reversely transcribed (Table 64.1). We subjected those presumable mRNA-tags (2.1 Mbp) to a global functional analysis using the SEED database [Overbeek et al., 2005] and compared the metatranscriptomic data to metagenomic data from (1) the same soil habitat (4.3 Mbp [Treusch et al., 2004]) and (2) a different farm soil (145 Mbp [Tringe et al., 2005]). Overall, the DNA-based (metagenomic) functional repertoire in both soils was surprisingly similar, as judged from the relative distribution of the functional subsystems (**Fig. 64.4a**). This hints for a generally similar "pool of functions" present in both soil communities, thus indicating that functional investigations of soil communities based on DNA might always give similar global patterns. In contrast to this, categories involved in RNA and protein metabolism (transcription, translation, protein folding, and degradation) were significantly overrepresented in the metatranscriptome compared to both metagenomes (2.7- to 4.0-fold; Rodriguez-Brito et al., 2006), as one would expect to see for actively growing organisms. Further differences between the transcriptome and the metagenomes were related to carbohydrate metabolism: while transcripts of proteins involved in the aerobic degradation of mono-, di-, and oligosaccharides and amino-sugars seemed to be less frequent than suggested by the metagenomes [Urich et al., 2008], transcripts for fermentation, degradation of sugar alcohols, and CO_2 fixation were equally represented.

64.3.4 A Critical Assessment of Taxonomic Binning

Metagenomic analyses frequently assign protein-encoding genes to taxonomic groups by comparing them against the content of sequenced genomes to derive a taxonomic community profile [Martin-Cuadrado et al., 2007; Rusch et al., 2007]. The simultaneously obtained rRNA and mRNA data provided us with the unique opportunity to validate these procedures. For this we followed the simple assumption that the number of ribosomes and mRNAs are proportional for a given taxon. When we compared the ribo-tag profile with the community profile derived from taxonomic binning of the mRNA-tags using MEGAN, we observed a considerable shift for the five dominant bacterial phyla [Urich et al., 2008]. These differences correlated strongly with the number of sequenced genomes of the different phyla (Fig. 64.4b). This suggests that taxonomic binning solely based on protein encoding genes currently generates an artificial bias against groups with few sequenced genomes and correspondingly overrepresents phyla with many sequenced genomes. This problem, which is inherent to all metagenomic studies, will likely be overcome as more genome sequences of less represented phyla become available.

64.3.5 Probing the Metabolism of a Low-Abundant Uncultured Group

The metabolism of groupI.1b Crenarchaeota from soil was largely unknown due to the lack of cultured representatives. Through taxonomic binning, 237 mRNA-tags were identified as of archaeal origin. This accounted for 1.1% of all mRNA-tags, which was similar to the proportion of archaeal ribo-tags (1.5%). Remarkably, two-thirds of the mRNA-tags were affiliated with lineages within the Euryarchaeota and only one-third with Crenarchaeota, which was in strong contrast to the SSU and LSU ribo-tag-derived community profile [Urich et al., 2008, Fig. S3 in SI]. The root cause for this discrepancy was again likely the genetic information deposited in the NCBI nr database. There is very limited information about the genomic repertoire of groupI.1b Crenarchaeota, as no genome sequence is published, and only few genomic fragments have been deposited in the databases, obtained through metagenomic studies [Quaiser et al., 2002; Treusch et al., 2004, 2005]. This probably led to the taxonomic binning of mRNA-tags to other archaeal groups. Based on the ribo-tag analysis

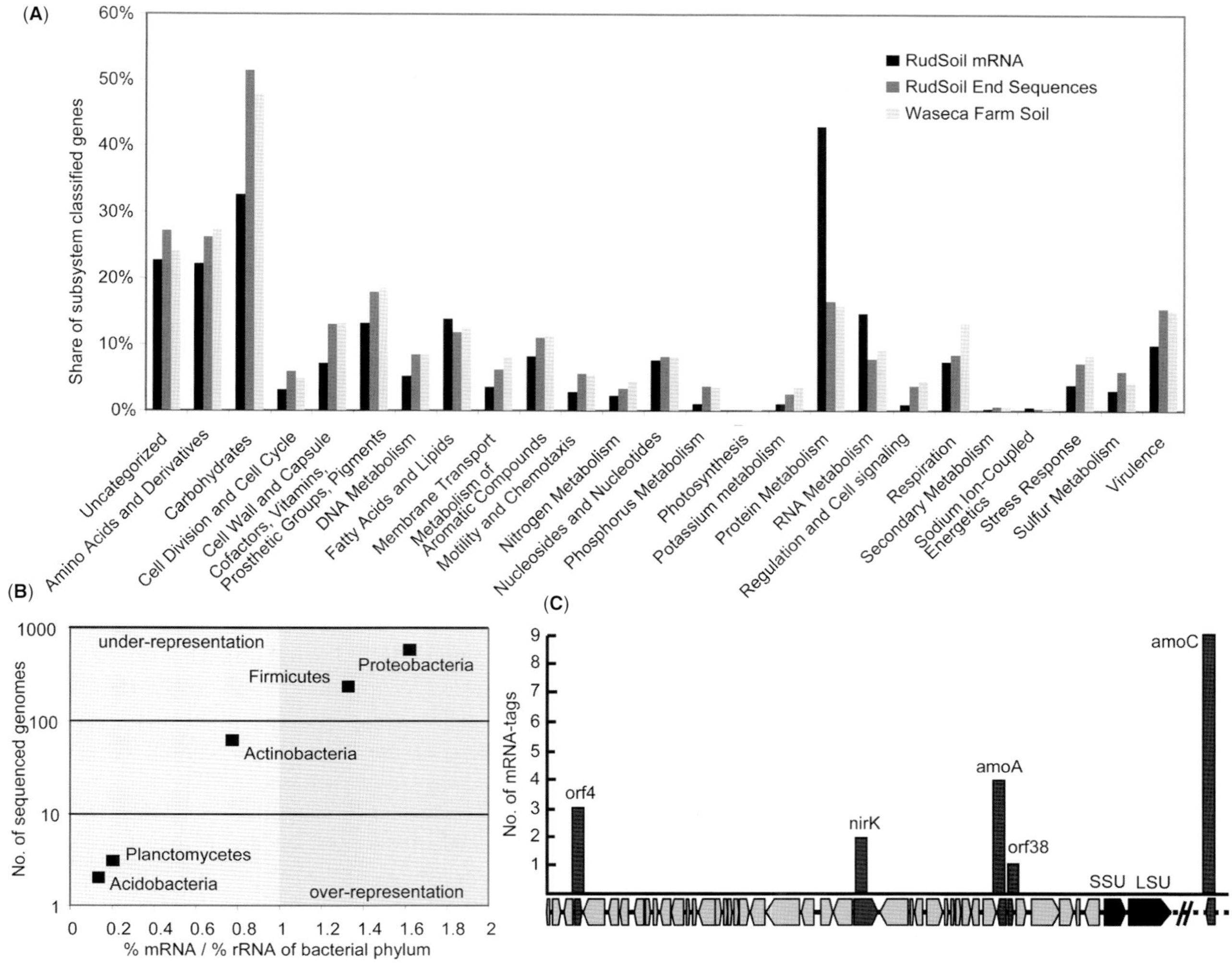

Figure 64.4 Functional analysis of mRNA-tags. (**a**) global functional analysis of mRNA-tags, fosmid-derived end sequences from DNA of the same community [Treusch et al., 2005] and shotgun-cloned DNA from a farm soil community [Tringe et al., 2005]. All three datasets were subjected to automated analysis using the MG-RAST annotation procedure (http://metagenomics.theseed.org). Percentages are expressed as the number of mRNA-tags assigned to a subsystem category, divided by the total number of mRNA-tags assigned to subsystems. (**b**) Abundance-dependent plot of presumably archaeal mRNA-tags onto the crenarchaeal fosmid clone 54d9 isolated from the same soil habitat (accession number: AJ627422). The *x* axis represents the annotated open reading frames (orfs). Note that an *amoC* gene was not found on the fosmid and is therefore indicated as loosely affiliated. (**c**) Logarithmic bivariance plot of the number of publicly available genomes (as of September 2007) of the five numerically dominant bacterial phyla, as judged from the ribo-tags, versus the mRNA-tag based over- and underrepresentation, compared to the mean of SSU and LSU ribo-tag fraction. A ratio of one means that mRNA- and ribo-tags report the same fraction for the respective phylum. (Reproduced from Urich et al. [2008] with permission from the authors.)

from the same experiment, where the crenarchaeal groupI.1b consistently accounted for more than 98% of the LSU and SSU ribo-tags [Urich et al., 2008, Table S6 in SI], we assumed that most of the archaeal mRNA-tags were indeed derived from this group and not from euryarchaeal or other crenarchaeal lineages. Although most of the identified homologues encoded hypothetical proteins, approximately 80 homologues were functionally annotated, which allowed a first glimpse into the in situ activity of this yet uncultured group. Besides transcripts of typical archaeal housekeeping gene products, those involved in ammonia oxidation were predominant (Fig. 64.4c); 13 mRNA-tags were derived from transcripts of the key metabolic enzyme ammonia monooxygenase (*amoA and amoC*). Furthermore, mRNA-tags of a homologue of copper-containing nitrite reductase gene (*nirk*) [Treusch et al., 2005] indicate an important function for this enzyme. The NirK homologue could be involved in the process of ammonia oxidation under either aerobic or anaerobic conditions, as postulated for ammonia oxidation bacteria [Beaumont et al., 2004]. These findings again hint for ammonia oxidation being the main energy metabolism in soil Crenarchaeota [Konneke et al., 2005]. In addition, 10 mRNA-tags could be related to the potential carbon metabolism. One mRNA-tag was derived from a homologue of methyl-malonyl-CoA mutase (MCM) and two from 4-hydroxybutyryl-CoA dehydratase (4-HBDH) homologues. These two gene products are, together with acetyl-CoA/propionyl-CoA carboxylase, diagnostic for a CO_2 fixation pathway recently characterized in hyperthermophilic Crenarchaeota and suggested for marine crenarchaeota [Berg et al., 2007]. This indicates that a similar pathway of CO_2 fixation might act in the soil crenarchaeota. Taken together, the metatranscriptomic data provide evidence for a chemolithoautotrophic lifestyle of this yet poorly characterized group.

Different from PCR-dependent pyrosequencing approaches [Sogin et al., 2006; Huber et al., 2007; Roesch et al., 2007] that are confined to specific regions in the SSU rRNA molecule, our method allows the in silico reassembly of a full-length "composite community" rRNA molecule for certain, abundant taxa. Such a 1502-bp SSU rRNA gene for groupI.1b of Crenarchaeota, assembled from 1105 ribo-tags, was 3.1% and 5.4%, respectively, different from the SSU rRNA sequences of two fosmid clones, 29i4 and 54d9, that had been isolated earlier from this habitat [Quaiser et al., 2002; Treusch et al., 2005]. Full-length SSU or LSU rRNA can subsequently be used for proper phylogenetic classification, rather than taxonomic assignment based on a reference database.

In silico reassembly was also possible with mRNA-tags. A nearly full-length "composite" archaeal *amoC* transcript was assembled [Urich et al., 2008, supplementary figure SF4]. The deduced amino acid sequence covered 146 out of 189 positions (77%) of the archaeal sponge symbiont *Cenarchaeum symbiosum* AmoC homologue (belonging to the archaeal sister groupI.1a [Hallam et al., 2006; see also Chapter 25, Vol. II] and had 88% identity to *C. symbiosum* AmoC, similar to the 84% identity for the AmoA protein sequences of both groups [Treusch et al., 2005; Leininger et al., 2006].

64.3.6 The Nitrifier Community

Nontargeted in-depth taxonomic and functional profiling using metatranscriptomics enables the analysis of microbial communities from various functional perspectives. We demonstrate this for the process of nitrification, the conversion of ammonia to nitrate via nitrite. The quantification of ribo-tags of the different groups of bacteria and archaea harboring ammonia- and nitrite-oxidation capabilities showed that the groupI.1b Crenarchaeota were, from a numerical standpoint, the major ammonia-oxidizing group in this soil (**Fig. 64.5**). The subsequent

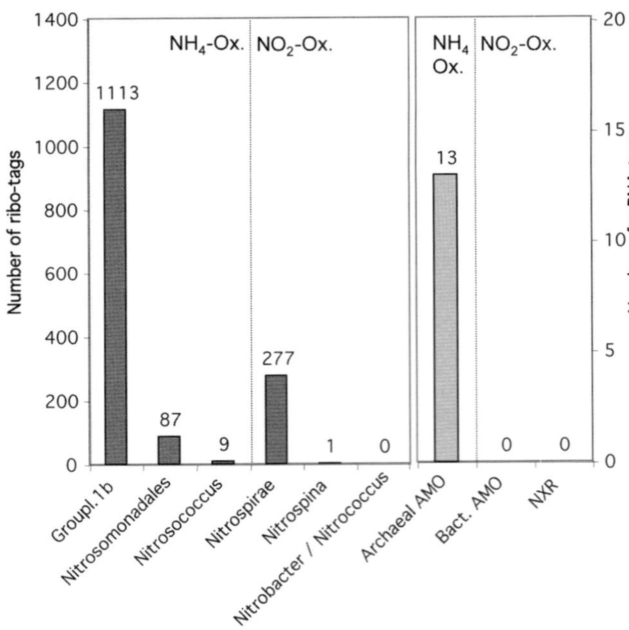

Figure 64.5 The nitrifying community, as seen from the ribo-tag and mRNA-tag perspective.

nitrite-oxidation step appeared to be mainly performed by members of the Nitrospirae phylum. Quantification of ammonia monooxygenase (AMO) mRNA-tags from archaea and bacteria supported the 16S rRNA data, because only transcripts of the archaeal variant were found (Fig. 64.5). The ribo-tag and mRNA-tag ratios between archaea and ammonia-oxidizing bacteria were very similar to archaeal and bacterial amoA transcript ratios determined from the same cDNA preparation using quantitative real-time PCR (12 vs. 16 [Leininger et al., 2006]).

64.3.7 Conclusions

In conclusion, we have presented the "double RNA approach" as a rapid experimental and analytical method that uses rRNA and mRNA to characterize microbial community structure and in situ function in depth and simultaneously. This methodology will help to (1) identify microbial groups in complex communities, (2) relate taxonomic groups to their ecological function (as demonstrated for the soil crenarchaeota), and (3) efficiently monitor structural and functional community shifts caused by environmental changes. Furthermore, this approach enables for the first time the simultaneous quantitative assessment of the abundance of members of all three domains of life.

We are currently using the approach presented here in parallel with mRNA enrichment procedures [Poretsky et al., 2005; Frias-Lopez et al., 2008] to ensure an efficient global analysis of the activity of naturally occurring assemblages, with one approach covering all trophic levels and domains as well as reasonable numbers of mRNA-tags in one sequencing step (total RNA) and the other one adding in depth functional information from the enriched mRNA fraction.

Acknowledgments

Work described in this chapter was financially supported by an EMBO long-term fellowship to TU and by funding of the University of Bergen, Norway to CS. We wish to acknowledge as well the financial support of the University of Vienna, Austria.

REFERENCES

Altschul SF, Madden TL, Schaffer AA, Zhang J, Zhang Z, Miller W, Lipman DJ. 1997. Gapped BLAST and PSI-BLAST: A new generation of protein database search programs. *Nucleic Acids Res*. **25**:3389–3402.

Bailly J, Fraissinet-Tachet L, Verner MC, Debaud JC, Lemaire M, Wesolowski-Louvel M, Marmeisse R. 2007. Soil eukaryotic functional diversity, a metatranscriptomic approach. *ISME J*. **1**:632–642.

Beaumont HJ, Lens SI, Reijnders WN, Westerhoff HV, van Spanning RJ. 2004. Expression of nitrite reductase in *Nitrosomonas europaea* involves NsrR, a novel nitrite-sensitive transcription repressor. *Mol Microbiol*. **54**:148–158.

Berg IA, Kockelkorn D, Buckel W, Fuchs G. 2007. A 3-hydroxypropionate/4-hydroxybutyrate autotrophic carbon dioxide assimilation pathway in Archaea. *Science* **318**:1782–1786.

Bremer H, Dennis PP. 1996. Modulation of chemical composition and other parameters of the cell by growth rate. In Neidhardt FC, Curtiss RI, Ingraham JL, Lin ECC, Low KB, et al., eds. Escherichia coli and Salmonella: Cellular and Molecular Biology. Washington, D.C.: ASM Press, pp. 1553–1569.

Cheung F, Haas BJ, Goldberg SM, May GD, Xiao Y, Town CD. 2006. Sequencing Medicago truncatula expressed sequenced tags using 454 Life Sciences technology. *BMC Genomics* **7**:272.

Cole JR, Chai B, Farris RJ, et al. 2007. The ribosomal database project (RDP-II): Introducing myRDP space and quality controlled public data. *Nucleic Acids Res*. **35**:D169–D172.

Ewing B, Hillier L, Wendl MC, Green P. 1998. Base-calling of automated sequencer traces using phred. I. Accuracy assessment. *Genome Res*. **8**:175–185.

Fierer N, Bradford MA, Jackson RB. 2007a. Toward an ecological classification of soil bacteria. *Ecology* **88**:1354–1364.

Fierer N, Breitbart M, Nulton J, et al. 2007b. Metagenomic and small-subunit rRNA analyses reveal the genetic diversity of bacteria, archaea, fungi, and viruses in soil. *Appl. Environ. Microbiol*. **73**:7059–7066.

Frias-Lopez J, Shi Y, Tyson GW, Coleman ML, Schuster SC, Chisholm SW, Delong EF. 2008. Microbial community gene expression in ocean surface waters. *Proc. Natl. Acad. Sci. USA* **105**:3805–3810.

Gans J, Wolinsky M, Dunbar J. 2005. Computational improvements reveal great bacterial diversity and high metal toxicity in soil. *Science* **309**:1387–1390.

Gilbert JA, Field D, Huang Y, Edwards R, Li W, Gilna P, Joint I. 2008. Detection of large numbers of novel sequences in the metatranscriptomes of complex marine microbial communities. *PLoS One* **3**:e3042.

Griffiths RI, Whiteley AS, O'Donnell AG, Bailey MJ. 2000. Rapid method for coextraction of DNA and RNA from natural environments for analysis of ribosomal DNA- and rRNA-based microbial community composition. *Appl. Environ. Microbiol*. **66**:5488–5491.

Hallam SJ, Mincer TJ, Schleper C, Preston CM, Roberts K, Richardson PM, DeLong EF. 2006. Pathways of carbon assimilation and ammonia oxidation suggested by environmental genomic analyses of marine Crenarchaeota. *PLoS Biol*. **4**:e95.

Huang X, Madan A. 1999. CAP3: A DNA sequence assembly program. *Genome Res*. **9**:868–877.

Huber H, Hohn MJ, Rachel R, Fuchs T, Wimmer VC, Stetter KO. 2002. A new phylum of Archaea represented by a nanosized hyperthermophilic symbiont. *Nature* **417**:63–67.

Huber JA, Welch DB, Morrison HG, Huse SM, Neal PR, Butterfield DA, Sogin ML. 2007. Microbial population structures in the deep marine biosphere. *Science* **318**:97–100.

Huson DH, Auch AF, Qi J, Schuster SC. 2007. MEGAN analysis of metagenomic data. *Genome Res*. **17**:377–386.

Jeanmougin F, Thompson JD, Gouy M, Higgins DG, Gibson TJ. 1998. Multiple sequence alignment with Clustal X. *Trends Biochem Sci*. **23**:403–405.

Joergensen RG, Wichern F. 2008. Quantitative assessment of the fungal contribution to microbial tissue in soil. *Soil. Biol. Biochem*. **40**:2977–2991.

Konneke M, Bernhard AE, de la Torre JR, Walker CB, Waterbury JB, Stahl DA. 2005. Isolation of an autotrophic ammonia-oxidizing marine archaeon. *Nature* **437**:543–546.

Leininger S, Urich T, Schloter M, et al.. 2006. Archaea predominate among ammonia-oxidizing prokaryotes in soils. *Nature* **442**:806–809.

Ludwig W, Strunk O, Westram R, et al. 2004. ARB: A software environment for sequence data. *Nucleic Acids Res*. **32**:1363–1371.

Martin-Cuadrado AB, Lopez-Garcia P, Alba JC, et al. 2007. Metagenomics of the deep mediterranean, a warm bathypelagic habitat. *PLoS ONE* **2**:e914.

Ochsenreiter T, Selezi D, Quaiser A, Bonch-Osmolovskaya L, Schleper C. 2003. Diversity and abundance of Crenarchaeota in terrestrial habitats studied by 16S RNA surveys and real time PCR. *Environ. Microbiol*. **5**:787–797.

Overbeek R, Begley T, Butler RM, et al.. 2005. The subsystems approach to genome annotation and its use in the project to annotate 1000 genomes. *Nucleic Acids Res*. **33**:5691–5702.

Poretsky RS, Bano N, Buchan A, et al. 2005. Analysis of microbial gene transcripts in environmental samples. *Appl. Environ. Microbiol*. **71**:4121–4126.

Poretsky RS, Hewson I, Sun S, Allen AE, Zehr JP, Moran MA. 2009. Comparative day/night metatranscriptomic analysis of microbial communities in the North Pacific subtropical gyre. *Environ. Microbiol*. **11**:1358–1375.

Prosser JI. 2007. Microorganisms cycling soil nutrients and their diversity. In van Elsas JD, Jansson JD, Trevors JT, eds. *Modern Soil Microbiology*, 2nd ed. Boca Raton, FL: CRC Press, pp. 237–262.

Pruesse E, Quast C, Knittel K, Fuchs BM, Ludwig W, Peplies J, Glockner FO. 2007. SILVA: A comprehensive online resource for quality checked and aligned ribosomal RNA sequence data compatible with ARB. *Nucleic Acids Res*. **35**:7188–7196.

Quaiser A, Ochsenreiter T, Klenk HP, et al.. 2002. First insight into the genome of an uncultivated crenarchaeote from soil. *Environ. Microbiol*. **4**:603–611.

QUAISER A, OCHSENREITER T, LANZ C, SCHUSTER SC, TREUSCH AH, ECK J, SCHLEPER C. 2003. Acidobacteria form a coherent but highly diverse group within the bacterial domain: Evidence from environmental genomics. *Mol. Microbiol.* **50**: 563–575.

RODRIGUEZ-BRITO B, ROHWER F, EDWARDS RA. 2006. An application of statistics to comparative metagenomics. *BMC Bioinform.* **7**:162.

ROESCH LFW, FULTHORPE RR, RIVA A, et al.. 2007. Pyrosequencing enumerates and contrasts soil microbial diversity. *ISME J.* **1**:283–290.

RUSCH DB, HALPERN AL, SUTTON G, et al. 2007. The Sorcerer II Global Ocean Sampling expedition: northwest Atlantic through eastern tropical Pacific. *PLoS Biol.* **5**:e77.

SCHAEFER M. 1990. The soil fauna of a beech forest on limestone: trophic structure and energy budget. *Oecologia* **82**:128–136.

SCHLEPER C, JURGENS G, JONUSCHEIT M. 2005. Genomic studies of uncultivated archaea. *Nat. Rev. Microbiol.* **3**:479–488.

SOGIN ML, MORRISON HG, HUBER JA, et al.. 2006. Microbial diversity in the deep sea and the underexplored "rare biosphere". *Proc. Natl. Acad. Sci. USA* **103**:12115–12120.

TORSVIK V, OVREAS L, THINGSTAD TF. 2002. Prokaryotic diversity—Magnitude, dynamics, and controlling factors. *Science* **296**:1064–1066.

TREUSCH AH, KLETZIN A, RADDATZ G, et al. 2004. Characterization of large-insert DNA libraries from soil for environmental genomic studies of Archaea. *Environ. Microbiol.* **6**:970–980.

TREUSCH AH, LEININGER S, KLETZIN A, SCHUSTER SC, KLENK HP, SCHLEPER C. 2005. Novel genes for nitrite reductase and Amo-related proteins indicate a role of uncultivated mesophilic crenarchaeota in nitrogen cycling. *Environ. Microbiol.* **7**:1985–1995.

TRINGE SG, VON MERING C, KOBAYASHI A, et al.. 2005. Comparative metagenomics of microbial communities. *Science* **308**:554–557.

URICH T, LANZÉN A, QI J, HUSON DH, SCHLEPER C, et al. 2008. Simultaneous assessment of soil microbial community structure and function through analysis of the meta-transcriptome. *PLoS ONE* **3**:e2527.

VENTER JC, REMINGTON K, HEIDELBERG JF, et al. 2004. Environmental genome shotgun sequencing of the Sargasso Sea. *Science* **304**:66–74.

WELLER R, WARD DM. 1989. Selective recovery of 16S rRNA sequences from natural microbial communities in the form of cDNA. *Appl. Environ. Microbiol.* **55**:1818–1822.

Chapter 65

Soil Eukaryotic Diversity: A Metatranscriptomic Approach

ROLAND MARMEISSE, JULIE BAILLY, CORALIE DAMON, FRÉDÉRIC LEHEMBRE, MARC LEMAIRE, MICHELINE WÉSOLOWSKI-LOUVEL, AND LAURENCE FRAISSINET-TACHET

65.1 INTRODUCTION

A detailed description of the data reported in this chapter was presented in Bailly et al. [2007]. The global analysis of mRNA (metatranscriptomics) directly extracted from environmental samples is receiving increasing interest because it provides information on the genes being transcribed and therefore on the activities being performed globally by a microbial community [Urich et al., 2008; see Chapter 64, Vol. I, Frias-Lopez et al., 2008, Gilbert et al., 2008, (see Chapter 27, Vol. II); Poretsky et al., 2009 (see also Chapter 63, Vol. I); Shrestha et al., 2009] or, alternatively, by specific taxonomic groups [Grant et al., 2006; Todaka et al., 2007]. Ideally, to get a global picture of a complex microbial community, metatranscriptomic analysis should be carried out in parallel with analyses specifically targeting the ribosomal genes and/or the metagenome. By combining these different approaches, it should be possible to simultaneously address some of the following classical questions in ecology: Who is there?, Who is active?, What is the functional potential or breadth of the community? and Who is doing what and when?

As for metagenomic analyses, most published works on metatranscriptomics were performed using as starting material total environmental RNA or preparation enriched in mRNA either by in vitro polyadenylation of the mRNA [Frias-Lopez et al., 2008] or by the specific elimination of the rRNA from the preparations [Poretsky et al., 2009; Shrestha et al., 2009; see also Chapters 62–64, Vol. I]. As a consequence, these studies did not target a specific taxonomic group. Furthermore, most of these studies concentrated on the systematic pyrosequencing of the cDNAs to get information on the genes being transcribed in the community [Urich et al., 2008; Frias-Lopez et al., 2008, Poretsky et al., 2009; Shrestha et al., 2009; see also Chapters 63–64, Vol. I, and Chapter 27, Vol. II].

Metatranscriptomics can, however, be limited to one taxonomic group, the eukaryotes, which produce 5′-polyadenylated mRNA, a unique feature that allows their specific isolation by capture on a solid matrix, on which poly-dT oligonucleotides have been grafted [Grant et al., 2006]. Metatranscriptomic analysis of eukaryotic communities can be justified on both scientific and practical grounds. Targeting the eukaryotes specifically should allow an in-depth analysis of major ecological processes carried out essentially by these organisms. This is the case, for instance, for the primary degradation of plant litter in soils or, of photosynthetic carbon fixation in some aquatic environments. Furthermore, analysis of eukaryotic cDNAs rather than of their environmental genomic DNA makes greater sense in both an ecological context and in terms of environmental biotechnology. The frequent occurrence of noncoding introns in eukaryotic coding sequences renders their identification more difficult and strongly limits their use for the characterisation and production of novel biocatalysts and other peptides of interest in heterologous hosts (cell factories).

For these different reasons, we evaluated the feasibility and potential of the metatranscriptomic approach to study the eukaryotic community of a temperate forest soil.

Handbook of Molecular Microbial Ecology, Volume I: Metagenomics and Complementary Approaches, First Edition. Edited by Frans J. de Bruijn.

cDNAs made from poly-A mRNA extracted from soil were cloned in a shuttle bacterial (*E. coli*)-yeast (*Saccharomyces cerevisiae*) plasmid vector. Random sequencing of partial 18S rDNA genes amplified from this soil and of cDNAs allowed identification of both the major groups of active eukaryotic organisms and of the main functional gene categories they expressed. Functional complementation of yeast mutants allowed identification of full-length environmental eukaryotic genes, thus opening the door to the exploitation of the biotech potential of eukaryotic genes from the environment, potentially present in the genomes of new and/or presently uncultivable taxa.

65.2 METHODS

In autumn 2004, soil samples were collected in a monospecific *Pinus pinaster* forest stand located in Southwest France. Samples of this sandy, organic-matter-poor (<0.5%) soil were, within 12 h after collection, sieved and pooled to get a composite soil sample that was frozen at -70°C to limit changes in gene expression and RNA degradation.

After two consecutive grinding steps, the first one at low temperature (-70°C or below) in an agate cup mill used to grind rock and the second by bead beating at room temperature in the extraction buffer, total soil RNA and DNA were co-extracted. The protocol included the use of an SDS, guanidinium isothiocyanate, ß-mercaptoethanol-based buffer, and acid phenol–chloroform extractions followed by successive ethanol and LiCl precipitations. Quality of the DNase I-digested RNA preparation was assessed by capillary electrophoresis using a RNA 6000 nano-lab on chips (Agilent technologies).

Eukaryotic poly-A mRNA was isolated by capture on magnetic beads coated with poly-dT (Dynal), reverse transcribed using poly-dT as the primer and converted into double-stranded cDNAs using the SMART cDNA synthesis strategy (Clontech). The procedure included a PCR amplification of the cDNAs which are bordered by *Sfi*IA and *Sfi*IB restriction sites allowing their directional cloning in plasmid vectors containing these sites. Cloning was carried out using the pYESfi bacterial/yeast cloning and expression plasmid. This derivative of pYES2 contains the colE1 and 2μ origins of replication and the *bla* and URA3 genes for autonomous replication and selection in *E. coli* and *S. cerevisiae*, respectively. cDNAs were cloned downstream of the strong, glucose-repressed, galactose-induced *S. cerevisiae* GAL1 promoter. Ligation mixes were transformed into competent *E. coli* cells to obtain the primary cDNA library, which was subsequently used to transform *S. cerevisiae* cells. Cloned cDNAs were sequenced from their 5′-end using the Sanger chemistry.

The 5′-end of 18S rDNA was amplified from soil DNA and reverse-transcribed soil rRNA using primers described in Diez et al. [2001]. PCR products were cloned and sequenced. The putative taxonomic origin of the cDNA sequences was determined by Blast searches against the GenBank protein (Blastx) and EST (tBlastx) databases. For functional annotation, cDNA sequences were assigned to eukaryotic clusters of orthologous groups (KOG, http://www.ncbi.nlm.nih.gov/COG/grace/kognitor.html). The 18S ribosomal sequences were screened for the presence of chimeras and assigned to major eukaryotic phyla based on Blastn searches and phylogenetic reconstructions.

Search for full-length functional environmental cDNA sequences was performed by complementation of yeast mutations. Two mutant phenotypes were complemented. The first one was the inability to synthesize histidine resulting from the deletion of the *HIS3* gene encoding an imidazole-glycerol phosphate dehydratase in strain BY4741. The second mutant phenotype was the inability to synthesize methionine resulting from the deletion of the *TRX1* and *TRX2* genes, encoding the two yeast cytoplasmic thioredoxins in strain EMY63 [Muller, 1991]. Yeast transformation and rescue of the pYESfi2 autonomously replicating plasmid in *E. coli* was carried out using standard procedures [Rose et al., 1990].

65.3 RESULTS AND DISCUSSION

More than 100 μg of RNA were extracted from 300 g of soil. Working on the largest possible amount of soil is necessary not only to get the largest possible amount of RNA but also to be representative of the diversity of soil eukaryotic microorganisms present in a forest soil. This is particularly true in the case of the micro- and mesofauna, which cannot be eliminated from a raw soil sample. For these organisms, only a few individuals of a particular species are often present per gram of soil. The electrophoretic profile of the soil RNA was revealed by three discrete bands that co-migrated with a bacterial 16S rRNA, a fungal 18S rRNA and the bacterial and eukaryotic large rRNA subunits. The relative intensities of these different bands suggested that bacterial RNA dominated the sample.

The taxonomic diversity of the eukaryotes present in the soil sample and therefore contributing to the extracted RNA was estimated by the cloning and sequencing of partial 18S rDNA sequences amplified from either soil DNA or reverse-transcribed soil RNA. Sequences from all major eukaryotic phyla [Adl et al., 2005; Lopez-Garcia and Moreira, 2008] were identified. Sequences from the unicellular protists belonging to the Rhizaria (Cercozoas),

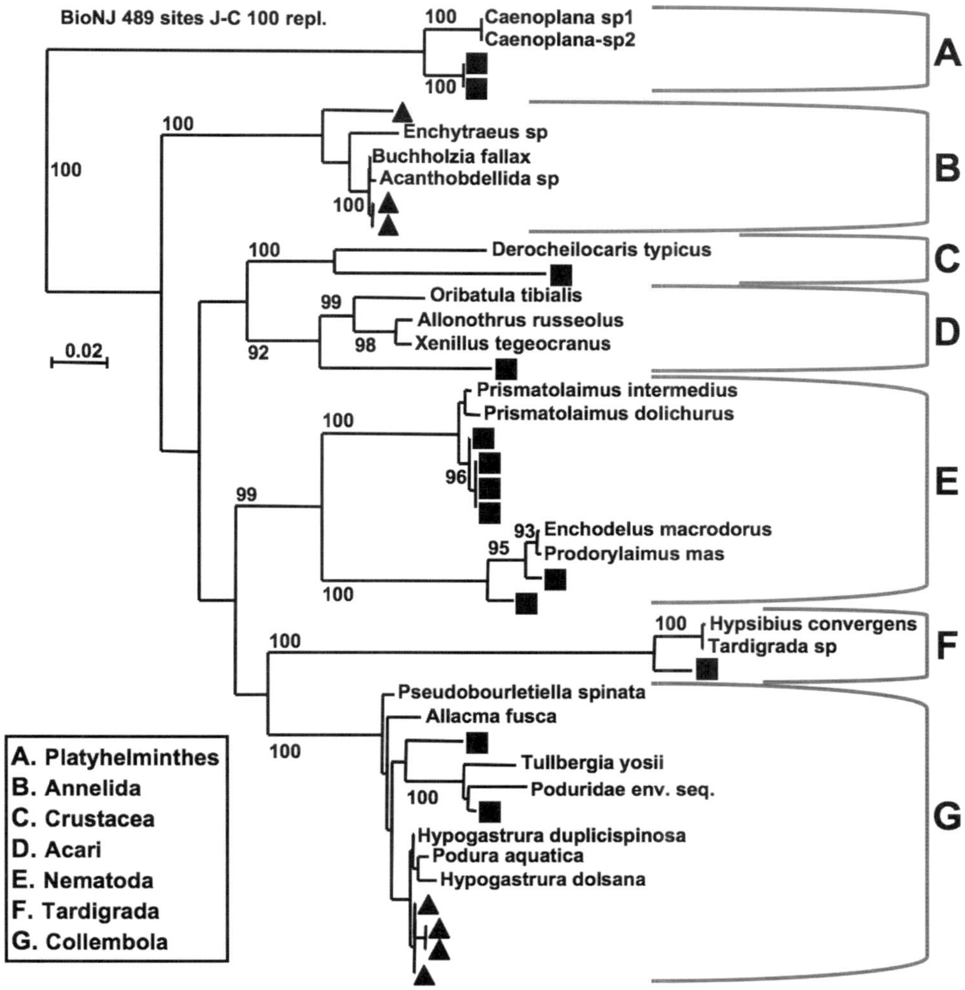

Figure 65.1 Neighbor joining phylogenetic tree of the partial forest soil 18S rDNA animal sequences amplified from either soil DNA (black squares) or reverse-transcribed soil RNA (black triangles). Reference sequences from representatives of different animal taxonomic groups were included in the analysis. Bootstrap values (1000 replicates) above 90% are given.

Amoebozoa, Alveolata (Ciliates and Apicomplexa), and Excavata (Euglenozoas) represented 30% of the total number of sequences. The largest group of Eukarya was the Opisthokonts, with 40% of sequences from fungi (essentially Basidio and Ascomycotina) and 27% from metazoas. As shown in **Figure 65.1**, several groups of the soil mesofauna were represented in the sequence dataset. If we except elimination by sieving of objects larger than 2 mm, it is indeed not possible to sort the different soil taxonomic groups prior to RNA extraction from soil samples. Therefore, metatranscriptomics takes into account gene expression by very different eukaryotic organisms, from unicellular protists to multicellular animals.

The global soil eukaryotic species richness at the scale of a forest stand is not known, especially since unicellular eukaryotes in soils are understudied. Clustering of the 118 18S rDNA sequences defined 84 different phylotypes (or MOTUs, molecular operational taxonomic units). Rarefaction curves clearly indicated that we identified only a fraction of the phylotypes present in the soil sample. Calculation of the nonparametric estimators S_{Chao1} and S_{ACE} [Kemp and Aller, 2004)] suggested that the total species richness in the studied soil sample could be in the range of 180–300.

Starting from 100 μg of total soil RNA, less than 1 μg of polyA mRNA was recovered by capture on poly-dT coated magnetic beads. The mRNA were converted into double-stranded cDNAs that were size fractionated to eliminate molecules smaller than 400 bp prior to cloning in the yeast expression vector pYESfi. An environmental cDNA library containing more than 8×10^6 independent clones was obtained.

Inserts from 119 randomly selected plasmids were sequenced and annotated with respect to their potential taxonomic origins and molecular functions. Thirty-two percent of the sequences were not homologous to any sequence present in GenBank and were not analyzed further (**Fig. 65.2**, "new hypothetical proteins"). As for the 18S rDNA, the cDNA dataset was dominated by sequences from fungi (30%) and metazoas (6%), while sequences from unicellular organisms were almost completely absent (only two sequences; i.e., less than 2%). This result highlights one of the main challenges that metatranscriptomics will have to face in the future. At present, accurate taxonomic affiliation of environmental coding sequences relies on the comprehensiveness of public sequence databases in term of coverage of the phylogenetic diversity of eukaryotes. This is, however, not the case at present. Indeed, many major eukaryotic phylogenetic groups are poorly studied while they constitute a significant fraction of the diversity both in terrestrial (soils) and aquatic ecosystems. Noteworthy, the rhizaria, essentially represented by the cercozoas in the soil ecosystem, have not yet been included in genome sequencing programmes. In other groups such as the Alveolata or the Excavata, mostly obligate intracellular parasites (e.g., *Plasmodium, Trypanosoma, Leishmania, Giardia*) with highly specialized genomes have been investigated in depth (GOLD, Genome Online Database; http://genomesonline.org/index2.htm). It is therefore likely that numerous sequences from environmental cDNA libraries will be incorrectly attributed to the most studied eukaryotic phylogenetic groups, which are fungi, animals, and green plants.

Functional annotation of 63 cDNAs encoding proteins of known functions was performed using the KOG classification scheme (Fig. 65.2). Sequences fell into 16 different KOG categories. The two main categories, J (translation, ribosomal structure and biogenesis, and post-translational modification) and O (protein turnover, chaperones), constituted one-third of the sequences. These two categories, which include for instance numerous ribosomal proteins, are of little value to understand the activities performed by the eukaryotes in soil. Genes of interest fall in categories which are less represented. This is in our case category Q (secondary metabolites biosynthesis, transport, and catabolism) or P (inorganic ion transport and metabolism). These preliminary data suggest that the metatranscriptomics analysis of eukaryotic microorganisms, as already reported for prokaryotes [Urich et al., 2008; see also Chapter 64, Vol. I; Frias-Lopez et al., 2008, Poretsky et al., 2009; Shrestha et al., 2009], will greatly benefit from high-throughput pyrosequencing technologies. Nevertheless, construction of cloned environmental cDNA libraries must not be considered as obsolete. It is indeed the most straightforward approach to isolate full-length coding sequences, whereas pyrosequencing, in most cases, only yields small gene fragments, which can be difficult to assemble.

An attractive possibility offered by metatranscriptomics is the selection of new genes from the environment by heterologous expression in a domesticated laboratory host. Numerous such examples have been reported for bacterial genes in the case of metagenomic libraries [e.g., see Chapters 47–53, Vol. II). Expression of environmental DNA in *E. coli* has been used to recover not only single genes coding enzymes of interest, but also different genes of a biosynthetic pathway responsible for the production of specific low molecular weight compounds (see Chapters 49, 50, 52, Vol. II). As a case study, we tested this approach by complementing yeast mutations that led to auxotrophies for histidine (HIS3 mutation)

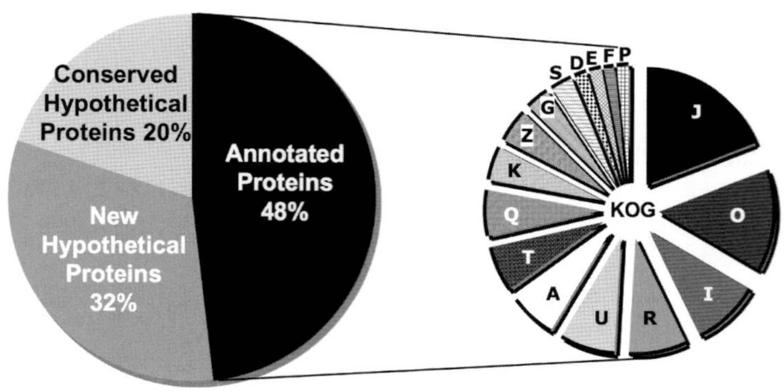

Figure 65.2 Annotation of the 119 sequenced cDNAs. "New hypothetical proteins" correspond to cDNAs with no homologues in sequence databases (Blastx searches). cDNAs encoding proteins of known functions were distributed into functional groups according to the KOG classification. A, RNA processing and modification; D, Cell cycle control, cell division, chromosome partitioning; E, Amino acid transport and metabolism, F, Nucleotide transport and metabolism; G, Carbohydrate transport and metabolism; I, Lipid transport and metabolism; J, Translation, ribosomal structure and biogenesis; K, Transcription; O, Posttranslational modification, protein turnover, chaperones; P, Inorganic ion transport and metabolism; Q, Secondary metabolites biosynthesis, transport, and catabolism; R, General function prediction only; S, Function unknown; T, Signal transduction mechanisms; U, Intracellular trafficking, secretion, and vesicular transport; Z, cytoskeleton.

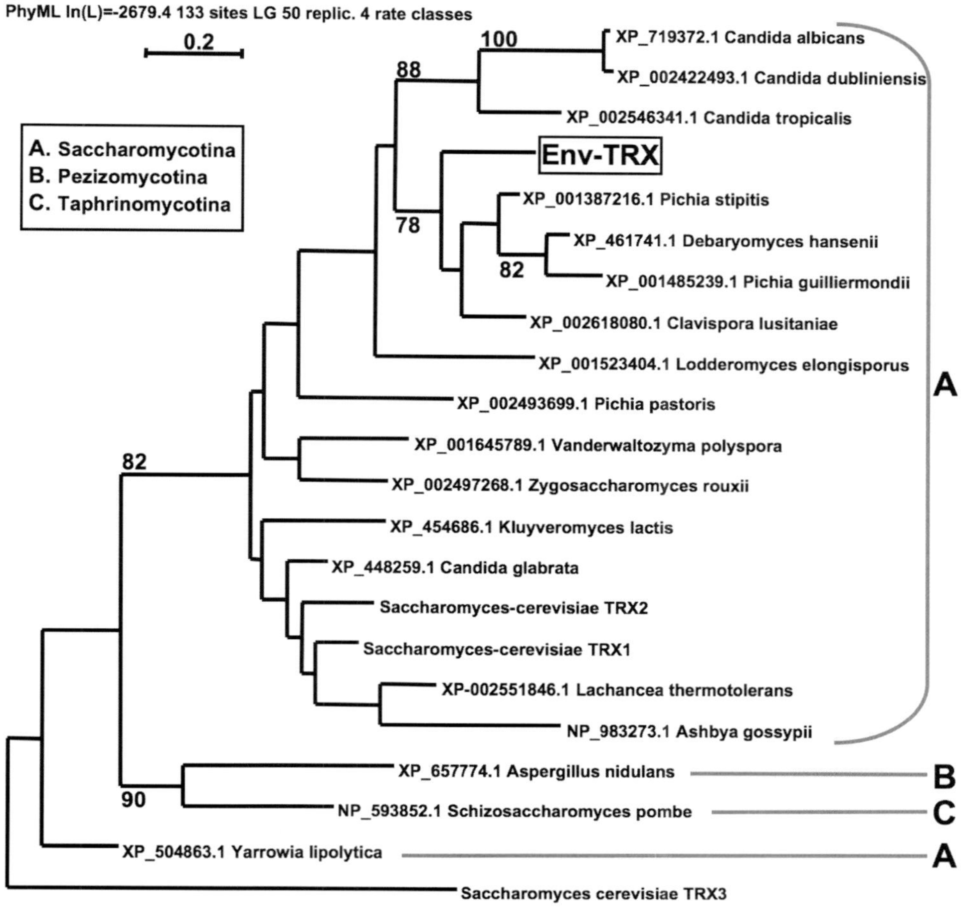

Figure 65.3 Phylogenetic affiliation (PhyML) of an environmental thioredoxin protein sequence (Env-TRX) which could complement the methionine auxotrophy of a *Saccharomyces cerevisiae* strain deleted of its two genes encoding cytoplasmic thioredoxins (TRX1 and 2). Bootstrap values (50 replicates) above 75% are indicated.

or methionine (combined TRX1 and TRX2 mutations leading to an absence of the two cytosolic thioredoxins). Among $\sim 3 \times 10^5$ cDNA screened, two and one complemented the *his*$^-$ and the *met*$^-$ phenotypes, respectively. All three genes were from fungi and encoded proteins homologous to either the His3p and Trx1/2p proteins. Interestingly, the environmental thioredoxin gene appears to derive from an hemiascomycete (Saccharomycotina) yeast species related to *S. cerevisiae* (**Fig. 65.3**), although this group of fungi is rather rare in soils.

These results demonstrate the potential of metatranscriptomics for the functional characterization of novel eukaryotic protein-coding genes. This clearly represents a step forward in biotechnology for the exploitation of the natural genetic diversity of complex communities of eukaryotic microorganisms which, for many of them, are probably unknown and yet uncultivable under standard laboratory conditions. In our study we targeted the model yeast *S. cerevisiae* as the recipient for expressing environmental cDNAs. In theory, prokaryotic hosts such as *E. coli* can also produce proteins from intronless cDNAs, although they will not be glycosylated. However, several alternatives to *S. cerevisiae* exist for producing proteins in eukaryotic cells. This is the case of other yeast species in the genera *Hansenula, Kluyveromyces*, or *Pichia* for which cloning and expression vectors are available [Wolf, 1996].

In conclusion, this study shows that metatranscriptomic analysis of microbial community can indeed be applied to eukaryotes.

INTERNET RESOURCES

Richness estimators: http://www.aslo.org/lomethods/free/2004/0114a.html

Functional annotation of eukaryotic genes (KOG): http://www.ncbi.nlm.nih.gov/COG/grace/kognitor.html

Eukaryote diversity: http://www.tolweb.org/tree/

Everything on *Saccharomyces cerevisiae*: http://www.yeastgenome.org/

Yeast general protocols: http://labs.fhcrc.org/gottschling/Yeast%20Protocols/index.html

Yeast transformation: http://www.umanitoba.ca/faculties/medicine/biochem/gietz/Trafo.html

Acknowledgments

We would like to thank Marie-Christine Verner for excellent technical assistance. Work on metatranscriptomics was supported by grants from the University Lyon 1 (BQR 2005), the PNETOX programme of the French Ministry for Ecology and Sustainable Development, and the ANR (programme Blanc 2006 Eumetatox and programme Biodiversité 2006 Fundiv).

REFERENCES

ADL SM, SIMPSON AGB, FARMER MA, ANDERSEN RA, ANDERSON OR, et al. 2005. The new higher level classification of eukaryotes with emphasis on the taxonomy of protists. *J. Eukaryot. Microbiol.* **52**:399–451.

BAILLY J, FRAISSINET-TACHET L, VERNER MC, DEBAUD JC, LEMAIRE M, et al. 2007. Soil eukaryotic functional diversity, a metatranscriptomic approach. *ISME J.* **1**:632–642.

DIEZ B, PEDROS-ALIO C, MASSANA R, et al. 2001. Study of genetic diversity of eukaryotic picoplankton in different oceanic regions by small-subunit rRNA gene cloning and sequencing. *Appl. Environ. Microbiol.* **67**:2932–2941.

FRIAS-LOPEZ J, SHI Y, TYSON GW, COLEMAN ML, SCHUSTER SC, et al. 2008. Microbial community gene expression in ocean surface waters. *Proc. Natl. Acad. Sci. USA* **105**:3805–3810.

GILBERT, FIELD D, HUANG Y, EDWARDS R, LI W, et al. 2008. Detection of large numbers of novel sequences in the metatranscriptomes of complex marine microbial communities. *PlosOne* **3**: e3042.

GRANT S, GRANT WD, COWAN DA, JONES BE, MA Y, et al. 2006. Identification of eukaryotic open reading frames in metagenomic cDNA libraries made from environmental samples. *Appl. Environ. Microbiol.* **72**:135–143.

KEMP PF, ALLER JY. 2004. Estimating prokaryotic diversity: When are 16S rDNA libraries large enough? *Limnol. Oceanogr. Methods* **2**:114–125.

LOPEZ-GARCIA P, MOREIRA D. 2008. Tracking microbial biodiversity through molecular and genomic ecology. *Res. Microbiol.* **159**:67–73.

MULLER EGD. 1991. Thioredoxin deficiency in yeast prolongs S phase and shortens the G1 interval of the cell cycle. *J. Biol. Chem.* **266**:9194–9202.

PORETSKY RS, HEWSON I, SUN S, ALLEN AE, ZEHR JP, MORAN MA. 2009. Comparative day/night metatranscriptomic analysis of microbial communities in the North Atlantic subtropical gyre. *Environ. Microbiol.* **11**:1358–1375.

ROSE MD, WINSTON F, HIETER P. 1990. *Methods in Yeast Genetics: A Laboratory Course Manual*. Cold Spring Harbor, NY: Cold Spring Harbor Press.

SHRESTHA PM, KUBE M, REINHARDT R, LIESACK W. 2009. Transcriptional activity of paddy soil bacterial communities. *Environ. Microbiol.* **11**:960–970.

TODAKA N, MORIYA S, SAITA K, HONDO T, KIUCHI I, et al. 2007. Environmental cDNA analysis of the genes involved in lignocellulose digestion in the symbiotic protist community of Reticulitermes speratus. *FEMS Microbiol. Ecol.* **59**:592–599.

URICH T, LANZÉN A, QI J, HUSON DH, SCHLEPER C, SCHUSTER SC. 2008. Simultaneous assessment of soil microbial community structure and function through analysis of the metatranscriptome. *PlosOne* **3**:e2527.

WOLF K, ed. 1996. *Nonconventional Yeasts in Biotechnology, A Handbook*. Berlin: Springer–Verlag.

Part 6C

Metaproteomics

Chapter 66

Proteomics for the Analysis of Environmental Stress Responses in Prokaryotes

KSENIA J. GROH, VICTOR J. NESATYY, AND MARC J.-F. SUTER

66.1 INTRODUCTION

Prokaryotic microorganisms constitute a large proportion of the inhabitants of aquatic and terrestrial ecosystems, greatly contributing to the maintenance of the natural nutrient cycles and the overall balance in their habitat. Having to adjust to ever-changing environmental conditions, bacteria and archaea have evolved sophisticated systems allowing them to respond and adjust to various stresses encountered, be it physical, chemical, or starvation-related. Some prokaryotes are able to tolerate extreme environments inaccessible for other organisms, such as those displaying temperature extremes, very high salinity, acidity, heavy metal contamination or other pollution, or nutrient limitation. This makes them interesting and useful research models for both investigating the mechanisms of stress response and searching for novel biomarkers of toxicity. In addition, microbes are well known for their outstanding catabolic versatility. They can, for instance, use many different electron donors and acceptors, and can drive dehalogenation, denitrification, sulfate reduction, and many other chemical reactions. Consequently, some species play an important role in bioremediation processes including detoxification and mineralization of contaminants [see Chapters 43–46, Vol. II], while others are used in a wide array of biotechnological applications (see also Chapters in Sections 6 and 7, Vol. II). All these features make the microorganisms important, interesting, and useful research models (see Sections 6 and 7, Vol. II).

In order to understand the mechanisms of stress response, find biomarkers of toxicity or evaluate the bioremediation potential offered by microorganisms, it is necessary to gain a detailed knowledge on the key physiological features and the molecular basis of the processes involved [Stenuit et al., 2008]. Among other conventional approaches, "omic" technologies are now widely used for characterization of the mechanisms defining specific states and metabolic capacities of an organism on the molecular level. In particular, transcriptomics, the study of the composition and abundance of mRNA transcripts in the specimen, mostly performed with the use of microarray technology, has delivered vast amounts of information on the gene expression patterns characteristic of normal development, as well as evolving in response to diverse stimuli [Steinberg et al., 2008. However, mRNA expression levels may be unreliable indicators of the abundance of the corresponding proteins, because they do not account for post-transcriptional events [Pradet-Balade et al., 2001]. By and large, proteins are the most relevant markers of biological function in comparison to all other biological molecules, since they reflect the actual activity and provide more direct links to the possible functions and capacities of the organism under investigation. In addition to their primary structure, encoded by genes and transmitted further by mRNA transcripts, many proteins carry post-translational modification (PTMs) that may occur either during or after the assembly of the polypeptide chain on a

Handbook of Molecular Microbial Ecology, Volume I: Metagenomics and Complementary Approaches, First Edition. Edited by Frans J. de Bruijn.

ribosome. Phosphorylation for signal transduction [Baker et al., 2006], ubiquitination for proteolysis, gaining particular importance during host–pathogen interactions [Munro et al., 2007], and glycosylation for cell/cell and cell/matrix adhesion [Abu-Qarn et al., 2008] are just a few examples of the most important protein PTMs encountered in prokaryotes. Modifications may drastically alter the physical and chemical properties of a protein, as well as alter its folding, stability, activity, and function, and thus it is crucial to monitor PTMs in order to make reliable predictions on the function of a specific gene product [Eichler and Adams, 2005]. The characterization of the total expressed proteome is addressed with proteomics, which relies on the use of several complementary methodologies. Proteomics may provide detailed information about protein abundances, stabilities, turnover rates, PTMs, and protein–protein interactions, delivering critical information on metabolic activity much more complete than that provided by transcriptomics. Going yet one step further, the field of metabolomics targets all low-molecular-weight (<1000 Da) primary and secondary metabolites present in and around cells growing under defined physiological conditions [Mashego et al., 2007]. This research area holds a promise to speed up the functional analysis of genes with unknown function, because a complementary use of proteomics and metabolomics may provide the best insights into the stress response of microorganisms [Singh, 2006; see also Chapters 71–74, Vol. I]. However, despite recent analytical advances, metabolome analysis may still require extensive methods development [Stenuit et al., 2008].

Among the "-omic" technologies described above, proteomics may at the moment be the most suitable one for the study of stress response in microorganisms, in terms of balancing technical feasibility and the ability to provide relevant information. In the next sections of this chapter we will first review the current proteomic methodologies and then discuss examples of applications of these methods to the characterization of the stress responses in microorganisms. We will deliberately avoid detailed description of metaproteomics work performed on microbial communities, such as the pioneering study of Wilmes and colleagues [Wilmes and Bond, 2004], because this topic is covered in Chapter 67 in this volume. Instead, we will focus on the proteomics studies performed on single, culturable microorganisms, which were aimed at deep characterization of stress response in one species, as well as at further development of proteomics methodologies. The initial application of proteomics to the study of single microorganisms is the necessary prerequisite for the establishment of robust and reliable techniques indispensable for future successes in metaproteomics.

66.2 EXPERIMENTAL METHODS AND TECHNOLOGIES IN PROTEOMICS

A proteome is defined as a total set of proteins expressed by a genome in a cell, tissue, or organism [Wilkins et al., 1996]. Its characterization allows associating individual proteins or groups of proteins with a specific physiological state. Comparison of protein patterns obtained under stressed conditions versus controls allows the detection of specific changes in protein expression [Blackstock and Weir, 1999] and an understanding of the underlying modes of action. Rather than targeting a particular family of proteins, this approach attempts to provide both qualitative and quantitative views on the whole proteome, which is essential for understanding the complex networks functioning in the organism.

The analysis of a proteome involves protein isolation and separation, data acquisition, and bioinformatics-driven identification of proteins based on available sequence data or de novo sequencing performed during the experiment. Among various techniques, mass spectrometry (MS) is the primary method for characterizing the presence and identities of proteins in native biological systems [Cravatt et al., 2007]. Information can be derived from the analysis of either an intact protein (top-down proteomics) or its peptides (bottom-up). Bottom-up proteomic experiments can be further separated into two major approaches: (i) gel-based, employing mainly two-dimensional gel electrophoresis (2D-GE) coupled with either peptide mass fingerprinting or tandem MS (MS/MS), and (ii) gel-free, using liquid chromatography (LC) coupled with MS/MS.

66.2.1 Top-Down Proteomics

The top-down analysis yields information about the molecular weight of the intact protein and its fragmentation ladders. This approach allows one to deduce the complete primary structure of the protein and reveals the majority of its modifications [Chait, 2006]. Conventional collision activation is unable to transfer the energy needed to produce extensive gas-phase fragmentation of protein larger than 50 kDa. To overcome this drawback, several alternative methods for protein fragmentation have been suggested, such as electron capture dissociation (ECD) [Ge et al., 2002] and electron-transfer dissociation (ETD) [Syka et al., 2004]. Recently, a novel prefolding dissociation method allowed the characterization of a protein as big as 229 kDa [Han et al., 2006]. Although the top-down proteomics was initially seen as a method more suitable for the analysis of single purified proteins, the future may see its wider application to the analysis of complex mixtures [Bunger et al., 2008].

66.2.2 Bottom-Up Gel-Based Proteomics: 2D-GE Combined with MALDI-TOF MS

Gel-based proteomics techniques involve the separation of complex protein samples by 2D-GE, which combines isoelectric focusing in the first dimension and denaturing sodium dodecyl sulfate polyacrylamide gel electrophoresis (SDS-PAGE) in the second dimension [Hanash, 2000]. This is still the most widely used separation methodology, performed with or without previous enrichment of sample with other techniques, such as organelle fractionation, chromatographic procedures, or equalizer technology. The separated proteins can then be quantified following suitable chemical staining and compared between different physiological states. Conventional protein staining methods used, such as Coomassie brilliant blue reagents or silver staining, suffer from low sensitivity and a problem of limited linear dynamic range and thus impede a real quantitative protein expression profiling. The intrinsic gel-to-gel variability of 2D-GE can mask the biological differences between the samples and compromise any quantitative protein comparison [Renzone et al., 2005]. This problem has been overcome with the development of a multiplexing technology for 2D-GE, the fluorescent two-dimensional difference gel electrophoresis (2D-DIGE) [Unlue et al., 1997]. This approach relies on the use of several fluorescent dyes that allow a simultaneous separation and independent visualization of two protein samples on a single gel. In addition to two different dyes that are used for independent labeling of each sample, the third dye is applied to label a pool of both samples, thus generating an internal standard [Wu, 2006]. The usage of a single gel for cross-sample comparison allows avoiding the artifacts introduced by gel-to-gel variability and thus increases the confidence in evaluating the real biological variations [van den Bergh and Arckens, 2004].

2D-GE offers the advantage of very high resolution power compared to other protein separation methods and a good visualization of the results. This technique may provide up to several thousands detectable protein spots in a single run [Gorg et al., 2004]. The proteins with PTMs can be detected as separate spots, and the presence of certain PTMs can be directly confirmed with antibody labeling [Ge et al., 2004]. It is also possible to deplete PTMs, compare the resulting 2D map with control samples, and identify the missing spots as PTM proteins [Yamagata et al., 2002]. Another interesting feature offered by gel-based proteomics is the ability to visualize the protein synthesis pattern in response to environmental stimuli, by labeling newly synthesized proteins with ^{35}S-methionine [Bernhardt et al., 1999]. Metabolic incorporation of stable isotopes such as ^{13}C or ^{15}N and subsequent electrophoretic resolving of labeled proteins provides another possibility for quantitative comparisons [Coute et al., 2007]. In the so-called In-gel Stable-Isotope Labeling (ISIL) technique, differences in visualized protein patterns serve as a preliminary quantification step, leading to a more accurate and sensitive MS-based relative quantification in the region of interest [Asara et al., 2006].

In theory, a spot separated by 2D-GE consists of an almost homogeneous protein, which can be identified following spot excision either by N-terminal sequencing [Edman, 1970; Shively, 2000; Tsugita, 1987] or, following the digestion with a sequence-specific protease, by MS procedures such as peptide mass fingerprinting (PMF), typically using matrix-assisted laser desorption ionization (MALDI)–time-of-flight (TOF) MS [Jungblut and Thiede, 1997], or tandem MS analysis (MS/MS) [Dongre et al., 1997]. PMFs are derived from the comparison of experimental masses of peptides produced during digestion by a specific protease against peptide masses produced by in silico digestion using the same protease [James et al., 1993; Pappin et al., 1993]. For protein identification based on sequence tags, MS/MS spectra are searched for obvious series of amino acid mass differences [Mann and Wilm, 1994; Wilm et al., 1996]. The latter technique is currently used most often, because it offers higher throughput and increased sensitivity compared to N-terminal sequencing and outperforms PMF in the analysis of spots containing multiple proteins, as well as in the identification of low-molecular-mass proteins that do not yield digests with a sufficient number of peptides to ensure unambiguous characterization by PMF. In addition, MS/MS sequencing of peptides proves extremely useful when working with organisms with incomplete genomic information.

The primary weakness of the use of 2D-GE for the study of a total proteome is its general bias toward higher abundance proteins. In addition, very acidic or basic proteins, as well as very large and/or hydrophobic proteins, may be absent or underrepresented on a 2D-GE map. In general, only proteins with a molecular weight of 10–100 kDa and a pI of 4–8 migrate well within 2D-GE gels [Gygi et al., 2000]. Consequently, although the coupling of 2D-GE with MS methodologies is a powerful tool for protein separation and identification, the above-mentioned limitations and the time necessary for extensive analysis, even with the use of automation, confirm the need for other proteomic approaches offering more information in a single step. This may be achieved by the gel-free techniques described below.

66.2.3 Bottom-Up Gel-Free Proteomics: LC-MS/MS

Non-gel-based approaches rely primarily on the separation of peptides rather than proteins and usually integrate online high-performance LC-MS/MS. In this case, a whole protein mixture, usually without a previous separation, is digested by a protease to produce a large collection of peptides. This mixture is then fractionated by LC and analyzed by MS using PMF or MS/MS approaches similar to those described above for the identification of peptides from digested 2D-GE protein spots. Usually, the proteins are identified with a high confidence based on the MS/MS spectrum of each corresponding peptide detected in the mixture. This method is also often referred to as "shotgun" proteomics.

For the samples containing a reduced number of proteins, one-dimensional chromatography may be sufficient, however, usually for the more complex samples, two-dimensional separation techniques are used (i.e., 2D-LC). Initially, the use of two orthogonal technologies (cation exchange and reversed-phase chromatography) was proposed to separate a complex peptide mixture in a single chromatographic run [Link et al., 1999]. This technique was then optimized by Washburn et al. [2001] and termed Multidimensional Protein Identification Technology (MudPIT).

The advantage of gel-free proteomics is that hydrophobicity, acidity, or molecular mass of intact proteins do not limit sample solubilization or separation, because the identification is performed on the peptide level. This method also provides a higher sensitivity compared to 2D-GE, allowing a better identification of lower abundance proteins. However, detection of PTMs by this method may not be as straightforward as with 2D-GE. In addition, direct quantitation of peptides with MS without internal standards may be relatively unreliable because of the difference in ionization efficiency or suppression effects. Therefore, the introduction of dedicated gel-free quantification techniques becomes necessary.

The enzymatic digestion of the whole protein sample that is carried out in "shotgun" proteomics produces tens of thousands of peptides. Direct MS screening for PTMs of all peptides in a complex digest is an inefficient and time-consuming process. To increase the efficiency of analysis, it is advisable to generate a subset of proteins/peptides carrying the PTMs of interest. This can be achieved with either chromatographic separation or affinity enrichment of the proteins/peptides that carry PTMs of interest using affinity tags that target specific amino acid residues [Kiernan, 2007; Zhao and Jensen, 2009]. Another useful method, developed for phosphoprotein enrichment, involves chemical replacement of phosphate moieties with affinity tags [Oda et al., 2001]. Furthermore, a variety of MS/MS techniques and in-source fragmentation methods that produce reporter ions characteristic for a particular type of PTM can be used to determine the presence and location of a PTM [Mann and Jensen, 2003].

As already mentioned before, lack of reproducibility during LC separation and variability of ion signals due to instable spray conditions and ion suppression effects make quantification on the peptide level a real challenge. Nevertheless, a number of protocols were developed recently that allow obtaining quantitative information within a single LC-MS run [Ong and Mann, 2005]. Most of these methods employ differential stable isotope labeling to create a specific mass tag that can be recognized by a mass spectrometer and at the same time provide the basis for quantification [Julka and Regnier, 2005]. In contrast, label-free quantification approaches aim to correlate the mass spectrometric signal of intact proteolytic peptides or the number of peptide hits for a given protein with the relative or absolute peptide quantity directly [Bantscheff et al., 2007].

Quantitative comparison of protein samples was first demonstrated with the Isotope-Coded Affinity Tag (ICAT) technique, which derivatizes Cys-containing peptides with reagents of a different isotopic composition that yield a pair of ions 8 Da apart [Gygi et al., 1999]. The problem of the biotin tag complicating the MS analysis was addressed with the introduction of the cleavable ICAT reagent (cICAT). A similar technique called Mass-Coded Abundance Tagging (MCAT) makes use of differential guanidylation of tryptic peptides with a C-terminal lysine, which produces pairs of ions 42 Da apart [Cagney and Emili, 2002]. The cICAT and MCAT techniques are obviously limited in a sense that they may miss those peptides lacking the derivatizable amino acid residues. To overcome this obstacle, the Global Internal Standard Technology (GIST) was developed that enables global labeling of peptides, irrespective of their amino acid composition or PTMs [Chakraborty and Regnier, 2002]. The GIST employs *N*-acetoxysuccinimide and its isotopic derivatives to differentially acetylate primary amino groups in peptides from case and control samples. However, all of the above-mentioned methods additionally suffer from the increased complexity introduced by differential tagging. This problem was addressed with the introduction of the tandem mass tag (TMT), which uses coding reagents of the same overall mass and physicochemical properties to ensure their chromatographic co-elution, but produces differential fragmentation patterns during MS/MS [Thompson et al., 2003]. An analogous approach, isobaric Tag for Relative and Absolute Quantification (iTRAQ), uses a multiplexed set of four reagents that label the N-terminus of all peptides generated by proteases and enables relative quantification based on MS/MS [Ross et al., 2004]. Generally, iTRAQ chemistry is expected to yield significantly better proteome coverage relative to the ICAT approach, as primary amines targeted by iTRAQ are more abundant than cysteins targeted by ICAT. As an example, an over threefold increase in coverage by iTRAQ compared to ICAT could be shown [Baliga et al., 2002; Whitehead et al., 2006]. However, a recent comparative study showed that, although more proteins are identified with iTRAQ, cICAT allowed identifying proteins with a much lower abundance, thus illustrating that the techniques based on specific and global labeling may be complementary to each other [de Souza et al., 2005].

Isotopic labeling can also be achieved by metabolic incorporation of stable isotopes present in the growth medium. In theory, this can lead to fully labeled amino acids and hence the complete quantification of every observable peptide. Commonly, stable isotopes are incorporated in newly synthesized proteins via ^{13}C-labeled arginine or lysine, a technique called Stable Isotope Labeling by Amino aCids (SILAC) [Ong et al., 2002]. In addition, SILAC-labeled cells can be used as an internal standard for relative and absolute quantification when comparing different samples (Culture-Derived Isotope Tag, CDIT) [Ishihama et al., 2005].

The labeling techniques described so far are mostly suited for relative quantification; however, absolute quantification technologies suitable for routine use may be needed for the future progress of proteomics. The shift from relative to absolute quantification will allow a more reliable comparison of different cases not only within a single experiment, but also between experiments conducted in different laboratories, even with a large variation in control samples. One of the absolute quantification methods, AQUA, relies on the use of internal standard peptides enriched in certain stable isotopes that mimic peptides produced during digestion of the target protein, and it allows identification and quantification with MS [Gerber et al., 2003; Kirkpatrick et al., 2005]. Another strategy uses artificial proteins (QconCATt; concatamers) composed of linearly assembled signature peptides from a protein set under investigation, which upon digestion in the protein mixture release labeled signature peptides suitable for quantification [Pratt et al., 2006]. Despite current progress, absolute quantification techniques still need further development, and it may take a while until they become established tools for general use.

The use of labeling-based methods for quantification has several disadvantages such as the expenses and high time demands, increased complexity of the peptide mixture and sample preparation procedures, and susceptibility to experimental artifacts. Some alternative "label-free" quantification methods are available, which either compare relative peptide peak intensities in unprocessed LC-MS datasets [Wong et al., 2008] or use spectral count sampling statistics to assess differential protein expression under multiple experimental conditions with a generalized G-test [Zhang et al., 2006]. The success of the label-free setting for comparative proteomics strongly depends on the temporal LC-MS alignment, which can be improved by using peptides detected within an LC-MS/MS analysis as references for realignment of highly complex samples that exhibit large variations [Fischer et al., 2006]. Despite recent improvements, significant method development is still needed to improve the reliability of the "label-free" quantification techniques and to promote their wider use. Meanwhile, isotopic labeling still remains the method of choice for performing quantification in a gel-free proteomics setup.

The above description of the protein characterization and quantification methodologies is intended as a brief introduction to the subject. For in-depth information on particular topics, the readers are invited to further consult a number of excellent reviews [Domon and Aebersold, 2006b; Leithner and Lindner, 2006; Nesatyy and Suter, 2008; Ong and Mann, 2005; Panchaud et al., 2008; see also Chapters 67–70, Vol. I].

66.2.4 Bioinformatics Solutions for MS-based Proteomics

Modern high-throughput proteomics is strongly dependent on the support and progress in bioinformatics, which in turn drives further development in MS instrumentation and experimental approaches. Dedicated software packages have been developed for the analysis and quantitative comparison of 2D-GE gels and differential LC-MS experiments. However, identification of the peptides either originating from the digestion of proteins spots separated by 2D-GE or detected during the LC-MS/MS experiment is one of the major tasks in bioinformatics for proteomics. Protein identification is based on the comparison of experimental MS (PMF approach) and/or MS/MS data against theoretical spectra generated from protein sequences stored in databases. Today the majority of proteomics data are produced through MS/MS-based identification. The prevalence of the latter method is due to its ability to work with complex peptide mixtures and the potential to obtain detailed information on protein sequence, location and type of PTMs, and amino acid mutations.

Database search algorithms may be either (a) heuristic such as the one used in SEQUEST [Eng et al., 1994] or X!Tandem [Craig and Beavis, 2003, 2004] or (b) probabilistic as in Mascot [Perkins et al., 1999], OMSSA [Geer et al., 2004], Phenyx [Colinge et al., 2003] or MASPIC [Narasimhan et al., 2005]. Heuristic algorithms match the theoretically and experimentally produced

MS/MS spectra based on the number of common peaks, and they calculate a score based on the similarity of the two spectra. In probabilistic algorithms, identification is based on modeling the peptide fragmentation process and calculating the probability of the appearance of the observed dissociation pattern. A recent evaluation and comparison of several publicly available MS/MS search algorithms [Kapp et al., 2005] revealed that some algorithms perform better in terms of sensitivity while others may be superior in terms of specificity. Consensus scoring may be used to increase confidence in protein identifications and to reduce the number of false-positive identifications—that is, to merge search results from multiple algorithms and to use only proteins identified in more than one search [Resing et al., 2004].

The techniques mentioned above rely on the availability of sequence information and cannot be used for the identification of proteins from unknown genomes. Although more than 550 prokaryotic genomes have already been sequenced (see http://cmr.jcvi.org/cgi-bin/CMR/shared/Genomes.cgi), this information for numerous other species is still lacking. Moreover, unaccounted alternative splicing events and post-translational and chemical modifications may not be included in the typical databases used for matching of MS/MS data. Due to these reasons, good-quality spectra are quite often left unexplained. An approach known as de novo sequencing offers the solution to these problems because it allows extracting the peptide sequence from the experimental tandem mass spectra on the basis of the masses of experimental peptide fragments. De novo sequencing algorithms search through all possible sequence combinations that could explain experimentally observed ions in a spectrum [Grossmann et al., 2005]. Due to several limitations such as the time needed to sift through all possible sequence combinations and the requirement for high-quality spectra with small fragment mass errors and relatively gap-free spectra, de novo sequencing is still used mainly as an independent validation tool for the results of database searches. However, the future will undoubtedly see the wider application of these algorithms—for example, when studying the recalcitrant unsequenced microorganisms from environmental microbial communities.

Diverse bioinformatics tools have been developed to facilitate further sophisticated proteome analysis, such as the characterization of PTMs, quantification, validation of obtained results, and systematic interpretation of the data. Due to the space constraint, they will not be discussed here. The reader is therefore referred to consult other comprehensive reviews [Bantscheff et al., 2007; Lisacek et al., 2006; Mueller et al., 2008; Nesatyy and Suter, 2008; Palagi et al., 2006; Panchaud et al., 2008].

66.3 APPLICATION OF PROTEOMICS TO THE ANALYSIS OF STRESS RESPONSE IN MICROORGANISMS

66.3.1 General and Specific Stress Responses in Prokaryotes

Facing the constantly changing sets of diverse environmental stimuli, bacteria and archaea have developed very complex adaptational networks, whose thorough characterization is crucial for understanding microbial physiology in general [Hecker et al., 2009]. The stress may be (a) physical, such as radiation, temperature, or pressure, and (b) chemical, including general, such as fluctuating pH or salinity, and specific stressors, such as heavy metals. Starvation for essential nutrients constitutes another growth-restricting stimulus in nature. By looking at the changes in the proteome, it becomes possible to link the stressor with its physiological consequences in the organism.

Table 66.1 provides some examples of stresses and organisms studied by diverse proteomics techniques, which will be reviewed below. To date, most of the proteomics studies of stress response in prokaryotes have used 2D-GE as a protein separation and quantification tool, coupled with MS and bioinformatics tools for protein identification. Because of their relatively low complexity, bacteria are well-suited for this kind of analysis, because a significant fraction of their proteome can be detected by gel-based techniques. Additional advantages of gel-based proteomics include the amenability for direct quantification of proteins originating from different samples in one gel (2D-DIGE), visualization of protein synthesis patterns, and resolving of proteins with different PTMs [Hecker et al., 2008]. Generation of 2D-maps of bacterial proteomes in many cases serves as a starting point for characterizing their physiology in normal and stressed conditions. These maps allow the detection of changes in response to stress, and they may also help in interspecies comparisons.

Considering the characterization depth and the range of stressors covered, gram-positive *Bacillus subtilis*, in addition to gram-negative *Escherichia coli*, is probably one of the most-studied prokaryotes so far, for which detailed 2D-GE proteome signatures of response to heat, cold, salt, oxidative, and starvation stresses have been defined [Antelmann et al., 2000; Bernhardt et al., 2003; Eymann et al., 1996; Graumann and Marahiel, 1999a; Hecker and Voelker, 2004; Mostertz et al., 2004], complemented with ^{35}S-methionine-labeling approach [Tam et al., 2006]. Cyanobacterium *Synechocystis* sp. PC 6803 has also been characterized in great detail [Los et al., 2008]. Other studies looked into the response to a

Table 66.1 Stress Responses in Prokaryotes: Selected Proteomics Studies

Stress	Technique Applied	Microorganism	Reference
High temperature	2D-GE	*Lactobacillus delbrueckii* subsp. *bulgaricus*	Gouesbet et al. [2002]
	2D-GE-MS	*Bacillus subtilis*	Tam et al. [2006]
		Thermoanaerobacter tengcongensis	Wang et al. [2007]
	2D-DIGE-MS	*Desulfovibrio vulgaris* Hildenborough	Chhabra et al. [2006]
		Synechocystis sp. PCC 6803	Slabas et al. [2006], Suzuki et al. [2006]
	1D-GE, iTRAQ, 2D-LC-MS	*B. subtilis*	Wolff et al. [2006]
Low temperature	2D-GE	*Lactococcus lactis* MG1363	Wouters et al. [1999a]
		Streptococcus thermophilus	Perrin et al. [1999], Wouters et al. [1999b]
	2D-GE-MS	*B. subtilis*	Brigulla et al. [2003], Graumann et al. [1996]
		Lc. lactis subsp. lactis	Wouters et al. [2000]
		Th. tengcongensis	Wang et al. [2007]
	2D-LC-MS	*Methanococcoides burtonii*	Nichols et al. [2004]
Starvation	2D-GE	*Escherichia coli*	Groat et al. [1986], Jenkins et al. [1990], Jenkins et al. [1988]
		Vibrio sp. S14	Nystroem et al. [1992]
	2D-GE-MS	*B. licheniformis*	Hoi et al. [2006]
		B. subtilis	Antelmann et al. [2000], Bernhardt et al. [2003]
Organic solvents	iTRAQ, LC-MS	*Sulfolobus solfataricus* P2	Chong et al. [2007]
Low pH	2D-GE	*Lb. delbrueckii* subsp. *bulgaricus*	Lim et al. [2000]
		Lb. sanfranciscensis	De Angelis et al. [2001]
	2D-GE-MS	*Bifidobacterium longum*	Sanchez et al. [2007]
		E. coli	Dukan et al. [1998]
		Salmonella SP1344	Adams et al. [2001]
		Syn. sp. PCC 6803	Kurian et al. [2006]
High pH	2D-DIGE-MS	*Syn. sp*. PCC 6803	Zhang et al. [2009]
Salinity	2D-GE-MS	*B. subtilis*	Hoeper et al. [2006]
		Bb. longum	Sanchez et al. [2005]
		Syn. sp. PCC 6803	Fulda et al. [2000] [2006], Huang et al., [2006]
	ICAT, 2D-LC-MS	*D. vulgaris* Hildenborough	Mukhopadhyay et al. [2006]
	iTRAQ, 1DE-LC-MS	*Euhalotece* sp. BAA*001* *Syn*. sp. PCC 6803	Pandhal et al. [2009]
Nitrate	iTRAQ, LC-MS	*D. vulgaris* Hildenborough	Redding et al. [2006]
Oxidants	2D-GE	*Francisella tularensis*	Lenco et al. [2005]
	2D-GE-MS	*B. subtilis*	Leichert et al. [2003], Mostertz et al. [2004]
		Staphylococcus aureus	Weber et al. [2004], Wolf et al. [2008]
Gamma radiation	2D-GE-MS	*Deinococcus radiodurans*	Ying et al. [2008], Zhang et al. [2005]
	iTRAQ, LC-MS	*Halobacterium NRC-1*	Whitehead et al. [2006]
UV	2D-GE-MS	*Anabaena doliolum*	Mishra et al. [2009]
		Syn. sp. PCC 6803	Gao et al. [2009]
High light	iTRAQ, LC-MS	*Prochlorococcus marinus* MED4	Pandhal et al. [2007]
High pressure	2D-GE	*E. coli*	Welch et al. [1993]
	2D-GE-MS	*Lb. sanfranciscensis*	Drews et al. [2002]
	2D-DIGE-MS	*B. cereus*	Martinez-Gomariz et al. [2009]

(*continued*)

Table 66.1 Stress Responses in Prokaryotes: Selected Proteomics Studies *(Continued)*

Stress	Technique Applied	Microorganism	Reference
Heavy metals	2D-GE-MS	*Campylobacter jejuni*	Kaakoush et al. [2008]
		Corynebacterium glutamicum	Fanous et al. [2008]
		E. coli	Bebien et al. [2002]
		Enterobacter liquefaciens	Marrero et al. [2004]
		Ferroplasma acidarmanus	Baker-Austin et al. [2005, 2007]
		Nitrosomonas europaea	Park and Ely [2008]
		Ralstonia metallidurans	Noeel-Georis et al. [2004]
		Stenotrophomonas maltophilia	Antonioli et al. [2007]
	2D-DIGE-MS	*Pseudomonas putida* UW4	Cheng et al. [2009]
Phytoestrogens	2D-GE-MS, LC-MS	*Bradyrhizobium japonicum*	Hempel et al. [2009]

wide range of physical and chemical stressors exhibited by different microorganisms significant for environment and biotechnology. Interestingly, the 2D-DIGE method, although being developed a relatively long time ago [Unlue et al., 1997], only recently saw a wider application in the studies of stress response in prokaryotes. In addition, various gel-free techniques are starting to be used more and more often (see Table 66.1).

Diverse studies of stress response have revealed that a defense system may be either specific, induced by a sublethal dose of stressor and permitting survival against an acute dose of the same agent, or a more general one, induced during the adaptation of cells to one condition and conferring concomitant resistance to other stressors [Hecker et al., 1996; Los et al., 2008]. The adaptive functions of the stress-specific regulons are to confer dedicated resistance mechanisms against the stress and thereby enable the cell to neutralize the stress or adapt to it and repair damages caused by the stressor [Hecker and Voelker, 2004; Tam et al., 2006]. In contrast, the general stress regulon encodes proteins with different functions conferring a multiple, nonspecific, and preventive stress resistance (collectively named general stress proteins, or GSPs), most strongly induced in nongrowing cells in a stationary phase [Gaidenko and Price, 1998; Hecker and Voelker, 2004]. Thus, induction of the general stress response by one stress also builds up a significant cross-protection against other stresses [Hecker and Voelker, 2004]. A common regulatory mechanism involves the modification of sigma factors whose primary role is to bind to core RNA polymerase conferring promoter specificity directing expression of specialty regulons [Abee and Wouters, 1999]. For instance, the major regulator of the general stress/starvation response has been identified in *E. coli* to be the product of the *katF* gene, which encodes a specific sigma factor named RpoS or σ^S [Hengge-Aronis et al., 1991; Lange and Hengge-Aronis, 1991]. The alternative factor sigma B (SigB, σ^B) is responsible for the induction of genes encoding GSPs in *B. subtilis* as well as in other gram-positive bacteria following heat, ethanol, salt, or acid stress, or during energy depletion [Hecker and Voelker, 2001].

66.3.2 Contribution of Proteomics to the Characterization of Stress Responses in Prokaryotes

The characterization of the SigB-controlled general stress regulon of *B. subtilis* presents an excellent example of the contribution of proteomics to the characterization of not only the regulon structure but also the regulon function, induction spectrum, and signal transduction mechanisms [Voelker and Hecker, 2005]. Starting from the initial observation of the induction of the SigB-dependent GSPs in response to a wide array of stressors, the concerted efforts have revealed that SigB controls one of the largest regulons of *B. subtilis* comprising about 150 genes [Hecker and Voelker, 2001]. As was shown mainly by transcriptomic and proteomic studies, at least three transduction pathways may activate SigB in response to environmental stresses (heat, cold, osmotic, acid, ethanol) and energy starvation [Brigulla et al., 2003]. The specific stress-response proteins are usually controlled by one single regulator, while the GSPs may be controlled either by single general regulator, by multiple general regulators, or by overlapping specific and general regulators [Tam et al., 2006]. Furthermore, some observed differences in the induction of individual members of SigB regulon might be explained by additional promoters, different promoter strength, and RNA or protein stability.

Below we will review selected proteomics studies describing the specific and general resistance systems in prokaryotes invoked in response to high or low temperature, starvation, pH, salinity, or oxidative stress imposed by various chemical and physical factors.

Exposure to high temperatures results in damage of membranes and nucleic acids, and in protein denaturation [Teixeira et al., 1997]. In addition, a decrease of the intracellular pH also occurs during the heat stress due to the disturbance of the transmembrane proton gradient [Piper et al., 1997]. The members of the two major families of heat shock proteins (HSPs), 70 kDa (DnaK) and 60 kDa (GroEL), are often found to be induced upon heat shock in diverse microorganisms [Chhabra et al., 2006; Gouesbet et al., 2002; Kovacs et al., 2001; Slabas et al., 2006; Uchiyama et al., 1999; Wang et al., 2007]. This induction of HSPs is the main response to heat stress on the gene expression level that is well-conserved among different organisms [Hecker et al., 1996]. HSPs promote the correct folding, degradation, and translocation of proteins and assist in the assembly of protein complexes [Bukau and Horwich, 1998]. It is generally accepted that HSPs belong to the group of GSPs, induced in response to a wide variety of other stress conditions. The upregulation of diverse Clp-ATPase proteins in response to heat shock has also been observed in several species [Suzuki et al., 2006; Tam et al., 2006]. In association with a serine protease ClpP, these proteins form a complex that degrades short peptides [Suzuki et al., 1997]. Upon heat shock, this activity is used for degradation of damaged protein that cannot be properly refolded.

The specific response of different species to the heat shock differs, depending on the genetic variation, the physiological status of the cells, and growth medium composition [Casadei et al., 2001]. For example, in heat-shocked *Syn*. sp. cells, an increase in the expression level of some proteases other than ClpP, enzymes involved in several metabolic processes such as protein synthesis, amino acid biosynthesis, and energy metabolism, as well as some PSII proteins, was observed [Slabas et al., 2006; Suzuki et al., 2006]. Strong heat sensitivity of specific enzymes involved in amino acid synthesis has also been observed in *B. subtilis*, revealed by gel-free proteomics [Wolff et al., 2006].

Exposure to low temperature causes a reduction in membrane fluidity and promotes the formation of stable secondary structures in DNA and RNA, leading to impairment of replication, transcription, and translation processes in the cell [Thieringer et al., 1998]. The response to these unfavorable conditions involves induction of so-called cold-inducible proteins (CIPs), which help to maintain the functions of the cellular machinery at lower temperatures [Graumann and Marahiel, 1999b; Phadtare et al., 1999], as well as promote an increase in the proportion of shorter and/or unsaturated fatty acids in the lipids in order to sustain membrane fluidity [Nichols et al., 2004; Paton et al., 1978]. Low-molecular-mass (approximately 7 kDa) CIPs are placed in a separate cold shock protein (CSP) family. These proteins contain the cold shock domain which consists of approximately 70 amino acids and exhibits nucleic acid binding motifs RNP-1 and RNP-2 [Weber et al., 2002]. In *B. subtilis* subjected to cold shock, the induction of several small acidic proteins was observed, which are homologous to previously characterized CspB [Willimsky et al., 1992]. These proteins were shown to affect the induction of several other CIPs [Graumann et al., 1996]. Putative CSP family members were also found to be induced in response to cold in other bacteria [Perrin et al., 1999; Wouters et al., 1999a,b]. Because induction of CSP family members also occurs in response to some other conditions, such as starvation or high-pressure stress, they may be viewed as GSPs [Graumann and Marahiel, 1999a]. Among other CIPs found in *B. subtilis*, those that function at various levels of cellular physiology, such as chemotaxis, sugar uptake, translation, protein folding, and general metabolism, were found [Graumann et al., 1996]. Similarly, CIPs identified in *Lactococcus lactis* subsp. *lactis* were found to be involved in the sugar metabolism, translation, chromosome structuring, and signal transduction [Wouters et al., 2000].

In addition to temperature fluctuations, a starvation stress is the one encountered probably most often in nature, because microorganisms have to adjust constantly to ever-changing concentrations of essential nutrients, such as carbohydrates (limiting for cell energy), phosphates (limiting for energy and DNA/RNA synthesis), and nitrogen (limiting for protein synthesis). Starvation stress limits the exponential growth of microorganisms, causing them to enter stationary phase. Transition to the stationary phase may also be triggered by diverse other environmental stressors. This resistant cell state allows survival for long periods under starvation conditions and multiple stresses [Jenkins et al., 1988, 1990; Nystroem et al., 1992] and is the most common among microorganisms encountered in the natural environment.

The adaptive response during growth usually involves the induction of specific groups of genes or regulons to cope with a specific stress condition, while the stress response during the stationary phase is mediated by numerous regulons that cope with many stress conditions. Adaptation during the stationary phase does not require exposure to stress to develop the response. For example, *Lc. lactis* in stationary phase developed a strong cross-protection against heat, ethanol, acid, osmotic, freezing, and oxidative stresses [Hartke et al., 1994; Wouters et al., 1999a]. Starved *Lactobacillus acidophilus* was found to be highly resistant to acidic stress [Lorca and Font de Valdez, 2001], while starved *E. coli* was found to be simultaneously protected against heat, oxidative and osmotic stress [Jenkins et al., 1988, 1990].

Distinct sets of proteins, essential for effective long-term survival and for multiple cross-protection, are induced during adaptation to starvation conditions [Groat et al., 1986; Hecker et al., 2009; Lorca and Font de Valdez, 2001; Nystroem et al., 1992; Schultz et al., 1988; Uchiyama et al., 1999]. By comparing the response to carbonate, phosphate, or nitrogen starvation, it was found that some induced proteins are unique to individual starvation condition, while others are common to one or more of these conditions [Groat et al., 1986; Nystroem et al., 1992]. Induced cross-protection is due to induction of GSPs, including HSPs, CSPs and antioxidant proteins [Graumann and Marahiel, 1999a; Hecker et al., 2009; Voigt et al., 2004]. Induction of CSPs upon entering the stationary phase or in response to cold shock may be triggered by a common signal, the inactivation of ribosomes, occurring under both conditions [Graumann and Marahiel, 1999b; VanBogelen and Neidhardt, 1990]. Interestingly, interspecies differences in response to starvation may exist. For example, phosphate starvation in *B. licheniformis* does not seem to induce the SigB-dependent general stress response observed in *B. subtilis* [Hoi et al., 2006].

Proteomics also allows investigation of the mechanisms of utilization of different carbohydrates as energy source and the stress caused by specific substrates. For example, the proteins found to be differentially expressed in the presence of l-malate in *Pseudoalteromonase haloplanktis* TAC125 hint at the presence of a functionally active two-component regulatory system in the upstream region of a specific outer-membrane porin gene [Papa et al., 2006]. As shown by iTRAQ shotgun proteomics, not only some specific alcohol dehydrogenases are induced during successful utilization of several alcohols and ketones by *Sulfolobus solfataricus* P2, but also some proteins with chaperone activities and implications in DNA repair, indicating that the cells were under stress and had an increased necessity for repair mechanisms to overcome damages caused by the addition of these substrates [Chong et al., 2007].

The homeostasis of pH during acid stress is regulated by the activity of proton-translocating ATPase [Bender and Marquis, 1987] and the overall proton permeability of the plasma membrane [Konings et al., 2002]. Another acid-stress response mechanism involves the arginine deiminase pathway, which allows the restoration of the optimal pH through arginine catabolism and NH_3 production [De Angelis et al., 2002]. Acid-tolerance response (ATR) allows preadaptation to low pH, leading to increased survival under strongly acidic conditions [Foster and Hall, 1991]. ATR may be associated with the induction of specific acid shock proteins (ASPs) [De Angelis et al., 2001; Dukan et al., 1998; Lim et al., 2000; O'Sullivan and Condon, 1999; Sanchez et al., 2007] or may occur independently of de novo protein synthesis [Hartke et al., 1996]. Some HSPs are often found among the induced proteins [De Angelis et al., 2001; Hartke et al., 1996; Lim et al., 2000]. Interestingly, the analysis of the cytoplasmic and periplasmic proteomes of *Syn*. sp cells under acidic stress failed to detect the induction of general stress response proteins [Kurian et al., 2006]. Proteomic analysis had revealed an acid-mediated decrease in flagellin expression and cell motility in *Salmonella* SL1344 cells, allowing the authors to hypothesize that flagellar repression at low pH conserves ATP for survival processes and helps to limit the influx of protons into the cytosol [Adams et al., 2001].

The analysis of plasma membranes of *Syn*. sp. under high pH stress allowed detection of significant perturbations in the proteome, affecting a large group of transport and substrate-binding proteins of ABC transporters, some photosynthetic and respiratory proteins, and some proteins with regulatory functions [Zhang et al., 2009]. In addition, differential expression of a phosphate transporter was observed, suggesting that the phosphate uptake system is highly susceptible to the elevated media pH [Zhang et al., 2009].

Increase in salinity (hyperosmotic stress) causes the water to move out of the cell, leading to a loss of cell turgor pressure, change of the cell volume, and variations in local ion distributions [Grant, 2004]. To retain water in the cell and thus maintain turgor pressure during hyperosmotic stress, many bacteria have systems for accumulating specific solutes that do not interfere with the physiology of the cell. Timely release of such solutes allows balancing the ionic strengths inside and outside of the cell. For example, in sulfate-reducing bacterium *Desulfovibrio vulgaris* Hildenbourough, it was demonstrated by the use of LC-MS proteomics and ICAT-based quantitation that import of osmoprotectans such as glycine-betaine and ectoine is the primary mechanism used by this organism to counter hyperionic stress. Several efflux systems and ATP synthesis pathway were upregulated, while flagellar biosynthesis, lactate uptake permeases and ABC transport systems were downregulated [Mukhopadhyay et al., 2006]. In the periplasmic protein fraction of *Syn*. sp., salt stress induced a group of enzymes involved in the alteration of the cell wall structure [Fulda et al., 2000]; in the total soluble protein fraction, it induced some enzymes involved in the synthesis of glucosylglycerol [Fulda et al., 2006], and in the plasma membranes fraction, it induced a large group of periplasmic-binding proteins of ABC-transporters, as well as some regulatory proteins [Huang et al., 2006]. Apart from proteins enhanced by salt stress, a number of proteins, mainly responsible for normal physiological processes in the cells, were reduced [Fulda et al., 2000; Huang et al., 2006]. In *B. subtilis*, many vegetative enzymes are initially repressed in response to salt stress, but they return back to usual levels after resumption of growth. Strongly induced proteins included the enzymes involved in sulfate assimilation and the formation of Fe–S clusters, suggesting an enhanced need for the assembly or repair of Fe–S clusters in response to salt stress [Hoeper et al., 2006]. Gene induction on the mRNA levels was not observed for 40% of the proteins that were accumulated in salt-acclimated *Syn*. sp. cells, indicating the potential involvement of posttranslational regulation in mechanisms of adaptation to salt stress [Castielli et al., 2009; Fulda et al., 2006], which was also observed in *D. vulgaris* Hildenborough [Mukhopadhyay et al., 2006].

iTRAQ proteomics was applied to investigate the response to nitrate in *D. vulgaris* [Redding et al., 2006]. Proteins involved in central metabolism and the sulfate reduction pathway were unperturbed, indicating that the applied stress was rather mild. The upregulation of transport systems for proline, glycine-betaine, and glutamate pointed out that the nitrate exposure simultaneously led to salt stress. In addition, upregulation of nitrate reduction systems, as well as of a large number of ABC transporters and Fe–S–cluster–containing proteins, was observed [Redding et al., 2006].

A comparative proteomics study of stress response to several salt concentrations in cyanobacteria had revealed many shared survival strategies between an extremely halotolerant *Euhalothece* sp. BAA001 and its moderately halotolerant relative *Syn*. sp. PCC6803 [Pandhal et al., 2009]. The method used was based on the concept that shared tryptic peptides from proteins which are encoded in different species are suitable for relative quantification across the species boundary [Snijders et al., 2007]. The relative abundance of proteins sharing tryptic peptides can be calculated by an MS analysis of the mixture of metabolically labeled (^{15}N) samples from one species and unlabeled ones from the other, thus allowing the cross-quantification for yet unsequenced species. In addition to in vivo metabolic labeling, an experiment using iTRAQ was also carried out, successfully demonstrating its applicability in cross-species proteomics [Pandhal et al., 2009]. Essential differences in cell behavior in three different salt concentrations could be documented based on the quantitative data for 340 unique proteins. Many salt adaptation strategies involving protein synthesis are shared by both cyanobacteria, and their quantitative expression affects acclimation. The results support the theory proposed for plants that many salt tolerance genes are present in both highly salt sensitive and highly salt tolerant species [Taji et al., 2004]. Such genes are "overexpressed" in the latter species under normal growth conditions, whereas they become induced only during salt stress in the former. The cross-species studies allow highlighting the level of control associated with salt adaptation strategies in different organisms [Pandhal et al., 2009].

In addition to induction of specific systems in response to salt stress, a strong upregulation of general stress response genes, including some HSPs, is usually observed. Furthermore, increased levels of superoxide dismutase and peroxiredoxin, pointing to the occurrence of oxidative stress after salt challenge, have been detected in many species [Fulda et al., 2006; Hoeper et al., 2006; Mukhopadhyay et al., 2006; Redding et al., 2006; Sanchez et al., 2005].

During oxidative stress, the overproduction of reactive oxygen species (ROS) leads to a damage of all major cellular components. Oxygen radicals may be reduced both by specific enzymes such as catalase or superoxide dismutase and by nonenzymatic compounds such as ascorbate, glutathione or tocopherols [Cabiscol et al., 2000; Dowds, 1994]. An induction of proteins typical for a complex oxidative stress response, such as antioxidant enzymes, chaperones, and proteases, was observed in *Francisella tularensis* [Lenco et al., 2005] and *B. subtilis* [Mostertz et al., 2004] in response to oxidative stress. Additionally, changes in iron metabolism were detected, such as enhanced expression of bacterioferritin in *F. tularensis* and upregulation of proteins contributing to a reduction of the free iron content in *B. subtilis*. The modulation of iron metabolism allows decreasing the overall toxicity of the peroxide by preventing excessive productive of hydroxyl radicals in a reaction with iron [Rabilloud et al., 2005]. The close analysis of *B. subtilis* responses to disulfide stress, induced by the specific thiol oxidant diamide, has revealed a complex combination of different regulatory networks, involving induction of antioxidants, HSPs, putative regulators of gene expression, and proteins protecting against toxic elements and heavy metals [Leichert et al., 2003]. The results of this proteomics study have shown that the response to disulfide stress is quite different from the "classic" oxidative stress response induced by hydrogen peroxide and the superoxide stress induced by the herbicide paraquat. Similarly, largely distinct responses to different oxidants could be observed in *Staphylococcus aureus*, where complex changes in the expression of proteins involved in detoxification, repair systems, and intermediary metabolism were revealed. Exposure to hydrogen peroxide caused a significant increase of proteins related to DNA metabolism, repair, and protection, while treatment with diamide produced stronger effects on the proteins involved in protein repair and degradation [Wolf et al., 2008]. In addition to monitoring the protein expression changes, proteomics also allows characterizing the proteins that are modified by oxidative stress, such as those containing cysteines oxidized into cysteic acid [Weber et al., 2004] or having formed disulfide bonds [Leichert et al., 2008].

Oxidative stress may be caused not only by direct oxidants such as hydrogen peroxide or other oxidizing chemicals, but also by heavy metals or physical stressors such as radiation or high pressure. The proteomics studies related to these stressors are reviewed below.

By causing DNA strand breaks and structural modification to nucleotide bases, γ-radiation induces severe DNA damage [Dianov et al., 2001; Hutchinson, 1985]. Most of the damage results from production of hydroxyl radicals via radiolysis of water, thereby imposing oxidative stress [Riley, 1994]. The proteins whose expression was altered after γ-radiation in *Deinococcus radiodurans* exhibited various cellular functions such as translation, transcription, signal transduction, carbohydrate transport and metabolism of carbohydrates, nucleotides, and inorganic ions, energy production and conversion, DNA replication, recombination, and repair, as well as some proteins involved in PTM processes and protein turnover and general chaperones [Zhang et al., 2005]. Extracellular proteins of *D. radiodurans* modified in response to radiation included those related to transport, metabolism, intracellular trafficking, and secretion, implying that the transmembrane transportation is involved in and contributes to the radioresistance in this organism [Ying et al., 2008]. Quantitative proteomics based on iTRAQ labeling were applied to study the response of *Halobacterium* NRC-1 to γ-radiation, revealing alterations in several cellular processes including increased DNA repair and protein turnover, shifts in metabolism favoring nucleotide biosynthesis, and an overall effort to repair or prevent oxidative damage, involving, for example, a decrease in components of electron transport chains [Whitehead et al., 2006].

UV radiation is known to impair various metabolic processes, including photosynthesis, pigmentation, CO_2 uptake, and nitrogen metabolism, resulting in decreased growth and survival [Rai et al., 1996; Sinha et al., 1996]. In addition, UV causes enhanced production of ROS, producing detrimental effects on all essential cellular macromolecules [He and Haeder, 2002]. Along with induction of specific protection systems, overexpression of HSPs and antioxidative enzymes allows cells to survive UV stress. Heat pretreatment associated with increased HSP levels alleviated UV toxicity in cyanobacterium *Anabaena doliolum* [Mishra et al., 2009]. UV-regulated proteins in *Syn*. sp. PCC 6803 included those involved in amino acid biosynthesis, photosynthesis and respiration, energy metabolism, protein biosynthesis and cell defense, with short-term response proteins being quite different from long-term response proteins, indicating the change in homeostatic mechanisms in response to UV stress [Gao et al., 2009].

Photosynthetic organisms need to be able to adapt quickly to varying light intensities in order to survive. Failure to do so may result in the production of ROS due to imbalanced energy supply. Adaptational response to high light includes reduction in light harvesting antennae and changes in energy transfer from the light harvesting complex to photosystem II, activation of energy dissipation pathways, and induction of repair mechanisms protecting from damage caused by ROS [Niyogi, 2000]. Analysis of proteins induced by light in dark-adapted cultures of *Syn*. showed induction of some chaperones related to general stress response [Choi et al., 2000]. Proteomic experiments using iTRAQ have been conducted to elucidate the light adaptation strategies of a globally abundant marine cyanobacterium *Prochlorococcus marinus* MED4 [Pandhal et al., 2007]. Photosystem-related proteins were downregulated in response to high light, allowing the organism to avoid the absorption of excess light energy, while several stress-related chaperones were upregulated, offering constant protection against possible detrimental effects [Pandhal et al., 2007].

High pressure causes protein denaturation and membrane damage, in severe cases leading to cell lysis [Malone et al., 2002; Pagan and Mackey, 2000]. Generally, a distinction between primary and secondary effects of high pressure can be made. The primary effect is the loss of concentration gradients across the cytoplasmic membrane due to increased permeability [Wouters et al., 1998]. The limited level of pH homeostasis is further influenced by the enhanced dissociation of weak organic acids caused by high pressure, thereby leading to acid stress. The resulting cell death may be considered a secondary effect promoted by pH and other environmental conditions [Molina-Guitierrez et al., 2002]. Several studies have looked into the differences in protein expression induced by high pressure. Interestingly, the induction of both HSPs and CSPs was observed in *E. coli* exposed to high pressure; some other induced proteins were unique to high-pressure conditions [Welch et al., 1993]. HSPs induction could arise from pressure-induced increase in the proportion of partially unfolded proteins, while both HSPs and CSPs could be induced by high pressure due to its inhibitory effects on the ribosome [Bartlett et al., 1995], which has been implicated to act as a sensor for both heat and cold shock response networks [VanBogelen and Neidhardt, 1990]. Induction of a CSP-homologue has also been found in pressure-treated *Lb. sanfranciscensis*, while some other pressure-induced proteins were putatively identified as ribokinase and elongation factor Tu [Drews et al., 2002]. The phosphorylation of ribose prevents leakage of the neutral sugar across the cell membrane, leading to its intracellular accumulation. This may protect from deleterious effects of high pressure, because carbohydrates are known to have a baroprotective function

[Saad-Nehme et al., 2001]. The induction of elongation factor Tu may also be beneficial in high-pressure conditions, because it promotes the protein biosynthesis, and may also protect against pressure-induced denaturation due to its chaperone-like function [Caldas et al., 1998]. Several mechanisms suggesting a reduction of virulence and protective response against oxidative stress could be identified in *B. cereus* exposed to high pressure [Martinez-Gomariz et al., 2009].

Heavy metals, often encountered by bacteria in soil and groundwater, may exert pronounced toxic effects on cellular metabolism. Heavy metal-induced toxicity is usually explained by oxidative tissue damage caused, however, additional factors such as alteration of enzyme activity may also play a role [Ercal et al., 2001].

The exposure to mercury (Hg) or cadmium (Cd) induced a set of antioxidant enzymes in soil bacterium *Corynebacterium glutamicum* [Fanous et al., 2008]. In addition, the induction of transporters conferring Hg elimination pathway was observed in response to Hg^{+} exposure, while Cd^{2+} caused additional increase in the levels of HSPs and proteins involved in energy metabolism and biosynthesis of lipoic acid and terpenoids [Fanous et al., 2008]. Similar responses have been observed in Cd-exposed *Campylobacter jejuni* [Kaakoush et al., 2008]. Overexpression of NirK (nitrite reductase) observed in *Nitrosomonas europaea* exposed to Cd^{2+} might support its involvement in the mechanisms of heavy metal tolerance in denitrifying bacteria [Holtan-Hartwig et al., 2002; Park and Ely, 2008]. A 2D-GE-MS analysis coupled with de novo sequencing revealed in a bioreactor microbial community exposed to Cd^{2+} the upregulation of several DNA repair proteins, oxidoreductases, some ATPases, and other ion transport proteins, while the downregulation of several metabolic enzymes was observed [Lacerda et al., 2007]. Exposure of *Enterobacter liquefaciens* to cobalt (Co) also induced a large group of proteins involved in cellular antioxidant defense mechanisms [Marrero et al., 2004].

Exposure to copper (Cu) or arsenic (As) led to induction of major HSPs in acidophilic archaeon "*Ferroplasma acidarmanus*" Fer1, in addition to proteins involved in DNA repair, transcription, and translation, and some enzymes related to central metabolic and biosynthetic pathways, as well as to protection from oxidative stress [Baker-Austin et al., 2005, 2007]. Analysis of *Ralstonia metallidurans* response to several heavy metals has shown an induction of CzcB, an element of the Co–Zn–Cd resistance operone coding for a membrane fusion protein that exports the heavy metal ions across inner and outer membranes without release in the periplasm. This was shown to be a specific response to zinc, but not to nickel or copper. On the contrary, the downregulation of the protein Cuf with yet unknown function seemed to occur due to a common regulation of this gene by divalent cations, because this response was triggered by both zinc and copper [Noeel-Georis et al., 2004]. Proteomic approaches have also been used to obtain important information on the mechanisms employed by cyanobacteria to maintain iron and copper homeostasis [De la Cerda et al., 2008].

Exposure of the plant growth-promoting bacterium *Pseudomonas putida* UW4 to nickel upregulated several pathways presumed to be involved in bacterial heavy metal detoxification, including general stress proteins, antioxidants, and heavy metal efflux proteins [Cheng et al., 2009]. Decreased activity of amino acid synthesis, protein synthesis and folding, DNA replication, cell division, and cell communication was also observed as expected [Cheng et al., 2009].

In *E. coli*, the toxicity of selenium (Se) was shown to be the result of superoxide production in the cells. This conclusion was based on the fact that increased superoxide dismutase levels are essential for Se tolerance [Bebien et al., 2002]. *Stenotrophomonas maltophilia* strain SeITEO2 is capable of resistance to high concentrations of selenite, reducing it to nontoxic elemental Se under aerobic conditions. Proteomic investigation of the response of this bacterium to selenite revealed proteins belonging to the following functional classes: nucleotide synthesis and metabolism, protein and amino acid metabolism, carbohydrate metabolism, and catabolism of damaged proteins, as well as some DNA-related proteins and proteins involved in cell division, oxidative stress, and cell wall synthesis [Antonioli et al., 2007].

Proteomics may also be used for investigating the effects of specific compounds of global significance, such as for example phytoestrogens, soil contaminants like polychlorinated biphenyls (PCBs), or antibiotics. Exposure to genistein induced three flagellar proteins in the secretome of soybean symbiont *Bradyrhizobium japonicum*, suggesting that this phytoestrogen influences the mobility of these bacteria, which in turn might affect their competitiveness in the rhizosphere [Hempel et al., 2009]. A comprehensive study investigated the catabolism of benzene, styrene, biphenyl, and ethylbenzene in *Rhodococcus jostii* RHA1, a soil bacterium with potent PCB-transforming properties, identifying the pathways essential for biodegradation of these compounds as well as oxidative stress response caused by increased uncoupling occurring during growth on these substrates [Patrauchan et al., 2008]. For more examples of other proteomics studies of the mechanisms of bioremediation by prokaryotes, the reader is referred to several recent reviews [Kim et al., 2007; Stenuit et al., 2008]. Likewise, proteomics-based investigations of bacterial response to antibiotics have not been discussed here due to space constraints, however, advanced proteomics techniques have also significantly contributed to the elucidation of the mode of action of drugs and the mechanisms of bacterial resistance. Proteomics also led to the generation of new testable hypotheses in the search for new antibiotics, as reviewed elsewhere [Bandow and Hecker, 2007].

66.4 CONCLUDING REMARKS

2D-GE has been the technique of choice for many scientists carrying out research in the field of stress response in prokaryotes, due to its ability to display significant fractions of a proteome and adequately quantify shifts in protein expression caused by changing conditions. However, despite the wide application seen today, gel-based techniques may need to be replaced, or at least complemented, by gel-free approaches, in order to overcome some inherent limitations such as (a) the inability to characterize the very hydrophobic and/or large proteins, leading to general omission of membrane proteins, as well as (b) the heavy bias toward proteins produced at higher concentrations and (c) difficulties in detecting low abundance proteins. The gel-free proteomic techniques, generally requiring less time and labor resources and therefore allowing higher throughput, in addition to deeper proteome coverage and characterization of theoretically any protein, regardless of its physicochemical properties [Delahunty and Yates, 2005], have significantly evolved during the last decade. Shotgun approaches generally result in a significantly higher number of protein identifications, with increased representation of basic and hydrophobic proteins [Arena et al., 2006; Barrios-Llerena et al., 2006; Gan et al., 2005; Wolff et al., 2006]. For example, in *B. subtilis*, to 745 protein identifications gained by 2D-GE [Eymann et al., 2004], 473 additional proteins could be added with use of shotgun proteomics [Wolff et al., 2006]. Furthermore, the quantitative data obtained by gel-based and gel-free technologies have been compared, demonstrating that the results obtained are generally similar [Choe et al., 2005; Wolff et al., 2006]. Thus, gel-free proteomics will probably be applied more and more in bacterial proteomics in the future, aiming to replace the 2D-GE technology. At the same time, the proteome data obtained by gel-based and gel-free approaches are complementary, in a way that usually no more than a half of the protein identifications obtained by either method overlap, as was recently demonstrated by several groups [Arena et al., 2006; Hempel et al., 2009; Renzone et al., 2005; Wolff et al., 2006]. Furthermore, correlation of the changes observed at the peptide level to individual protein isoforms is still a challenging task in gel-free proteomics, while it is easily accomplished in 2D-GE-based separation [Voelker and Hecker, 2005]. In general, to follow the fate and destination of single proteins from synthesis via targeting and modification to degradation and to discover the network of protein interactions on the large scale, an integrative approach using both gel-free and gel-based techniques may be necessary [Hecker et al., 2008].

In addition to diverse bottom-up proteomics methods, top-down proteomics carries the potential to offer a unique view on the prokaryotic proteome. Although the total number of proteins detected by top-down proteomics tends to be much lower than that detected by bottom-up approaches, this technique may be very useful in providing the view of intact protein structures, because it allows detecting PTMs and truncated protein forms that may be omitted in global shotgun analyses. In addition, robust quantification approaches based on stable isotope labeling, developed initially for shotgun proteomics, may also be suitable for top-down characterization and quantitation of intact proteins, as was recently shown for SILAC [Waanders et al., 2007]. Thus, although up to now this method has been used quite rarely for the characterization of total proteomes, it is slowly finding its way into the field. In a recently developed top-down proteomic workflow employing 2D-LC and a linear ion trap equipped with an ETD source, 322 different protein forms representing 174 proteins could be identified in the soluble protein extracts from *E. coli*, showing that top-down proteomics may provide a worthy complement to traditional bottom-up approaches [Bunger et al., 2008]. The complementarity of top-down and bottom-up proteomics has also been demonstrated in a study of ribosomal proteins of *Caulobacter crescentus*, which provided the basis for the characterization of ribosome structure and dependence of PTMs on growth conditions [Running et al., 2007]. Several groups have also shown the successful application of this method for the identification of protein biomarkers, such as those allowing to place unsequenced organisms in their correct phylogenetic context by homology analysis [Wynne et al., 2009]. The top-down analysis of the soluble subproteome of the obligate thermophile *Geobacillus thermoleovorans* T80 allowed identifying a variety of proteins known to play a role in stress response [Graham et al., 2006]. Even though, to the best of our knowledge, top-down proteomics has not yet been applied to the direct analysis of stress response, the above-mentioned studies provide the necessary basis and show that this method is feasible and sensitive enough to perform such investigations in the future.

It may be concluded that global proteomics has proved to be an invaluable technique for deciphering the mechanisms of general and specific stress responses in prokaryotes. This approach not only allows defining the specifically induced systems, but also serves as a starting point for finding sensitive biomarkers that might be used for routine screening for stress and toxic effects, and for studying the degradation pathways in microorganisms used for bioremediation [Stenuit et al., 2008]. Metaproteomics—that is, the study of the entire protein content of a given habitat [Wilmes and Bond, 2004]—is an emerging field carrying out functional analyses of whole microbial communities. One problem in analyzing the proteome of environmental microorganisms of interest may be the scarceness of sequence information available. However, the genome sequences of diverse microorganisms are being completed at an ever-increasing rate. Additionally, de novo sequencing approaches and sequence homologies across species can be used to identify proteins from yet unsequenced organisms [Pandhal et al., 2009]. Because the level and effects of stress imposed can only be defined in relation to the organism or population experiencing this particular stress [Bijlsma and Loeschcke, 2005], the differential response of subpopulations (such as multiple variants of the same species) has to be accounted for when analyzing the stress response in complex microbial communities.

One has to keep in mind that the identification of the proteins present in a sample is only the first step in the long process of understanding their function since it is now generally accepted that the diversity of biological processes is governed by intermolecular interactions in the cell [Alberts, 1998]. In a second step the horizontal approach (expression profiling) has to be complemented by a detailed analysis of regulons and the individual genes features (vertical approach) [Voelker and Hecker, 2005]. A successful model of protein function and regulation requires a broad understanding and thorough mapping of its molecular interactions with other proteins and ligands, for which the data acquired with transcriptomics, proteomics and metabolomics have to be integrated using dedicated bioinformatics tools [Nesatyy and Suter, 2007]. The addition of a temporal dimension to the static snapshots of the proteome at one particular point in time, currently acquired most often, will bring a much better understanding of biological processes [Blagoev et al., 2004], albeit at the expense of time, labor, and funds.

Although gel-free proteomics approaches usually provide a much deeper proteome coverage than the gel-based ones, substantial amounts of low abundance proteins still escape the detection with current shotgun techniques. A wide dynamic range of protein concentrations present in a complex biological sample has prompted the development of new strategies ranging from sample preparation to data acquisition. An increase in the sensitivity of the peptide detection can be achieved through the generation of inclusion lists produced as a result of prior experiments, the so-called "targeted" proteomics approach as opposed to global [Domon and Aebersold, 2006a]. Further, the use of selected reaction monitoring (SRM), based on specific fragmentation of chosen proteotypic peptides, may greatly enhance the sensitivity of MS detection, thus allowing identification and quantification for proteins of low copy counts with a high reproducibility [Lange et al., 2008b]. Multiple reaction monitoring (MRM) incorporates up to several hundred SRMs in a single MS detection run, sufficiently increasing the throughput for proteins of interest in multiple samples [Anderson and Hunter, 2006; Wolf-Yadlin et al., 2007]. Recently, MRM has been successfully applied to study the virulence factors in *Streptococcus pyogenes* [Lange et al., 2008a]. This technique has a high potential to deliver major breakthroughs in the proteomics studies of prokaryotes in the coming years.

MRM technology can also be used for identifying the proteins carrying specific PTMs [Cox et al., 2005; Wolf-Yadlin et al., 2007]. In addition, detection and quantification of PTMs can greatly benefit from sample preparation procedures designed to specifically isolate the peptides of interest. Subcellular targeted proteomic studies may provide relevant information to enhance the understanding of the different biological processes linked to the purified, or enriched, fractions for subsequent functional studies [Sanchez et al., 2008]. Further advances may be brought by the integration of de novo sequencing, neutral loss, and MS^3 scans into the experimental design [Nesatyy and Suter, 2007].

Undoubtedly, the future will see an even more rapid development of experimental and bioinformatics tools that will allow conducting high-throughput analyses of proteome both of single microorganisms as well as of complex communities sampled in the environment.

INTERNET RESOURCES

Comprehensive Microbial Resource, All genomes: http://cmr.jcvi.org/cgi-bin/CMR/shared/Genomes.cgi

Acknowledgments

KG would like to acknowledge the Swiss National Science Foundation for financial support. (project number 3100A0-118243)

REFERENCES

Abee T, Wouters JA. 1999. Microbial stress response in minimal processing. *Int. J. Food Microbiol*. **50**:65–91.

Abu-Qarn M, Eichler J, Sharon N. 2008. Not just for Eukarya anymore: Protein glycosylation in Bacteria and Archaea. *Curr. Opin. Struct. Biol*. **18**:544–550.

Adams P, Fowler R, Kinsella N, Howell G, Farris M, Coote P, O'Connor CD. 2001. Proteomic detection of PhoPQ- and acid-mediated repression of *Salmonella* motility. *Proteomics* **1**:597–607.

Alberts B. 1998. The cells as a collection of protein machines: Preparing the next generation of molecular biologists. *Cell* **92**:291–294.

Anderson L, Hunter CL. 2006. Quantitative mass spectrometric multiple reaction monitoring assays for major plasma proteins. *Mol. Cell. Proteom.* **5**:573–588.

Antelmann H, Scharf C, Hecker M. 2000. Phosphate starvation-inducible proteins of *Bacillus subtilis*: Proteomics and transcriptional analysis. *J. Bacteriol.* **182**:4478–4490.

Antonioli P, Lampis S, Chesini I. 2007. *Stenotrophomonas maltophilia* SeITE02, a new bacterial strain suitable for bioremediation of selenite-contaminated environmental matrices. *Appl. Environ. Microbiol.* **73**:6854–6863.

Arena S, D'Ambrosio C, Renzone G, Rullo R, Ledda L, Vitale F, Maglione G, Varcamonti M, Ferrara L, Scaloni A. 2006. A study of *Streptococcus thermophilus* proteome by integrated analytical procedures and differential expression investigations. *Proteomics* **6**:181–192.

Asara JM, Zhang X, Zheng B, Christofk HH, Wu N, Cantley LC. 2006. In-gel stable-isotope labeling (ISIL): A strategy for mass spectrometry-based relative quantification. *J. Proteome Res.* **5**:155–163.

Baker-Austin C, Dopson M, Wexler M. 2005. Molecular insight into extreme copper resistance in the extremophilic archaeon "*Ferroplasma acidarmanus*" Fer1. *Microbiology* **151**:2637–2646.

Baker-Austin C, Dopson M, Wexler M. 2007. Extreme arsenic resistance by the acidophilic archaeon "*Ferroplasma acidarmanus*" Fer1. *Extremophiles* **11**:425–434.

Baker MD, Wolanin PM, Stock JB. 2006. Signal transduction in bacterial chemotaxis. *BioEssays* **28**:9–22.

Baliga NS, Pan M, Goo YA, Yi EC, Goodlett DR, Dimitrov K, Shannon P, Aebersold R, Ng WV, Hood L. 2002. Coordinate regulation of energy transduction modules in *Halobacterium* sp. analyzed by a global systems approach. *Proc. Natl. Acad. Sci. USA* **99**:14913–14918.

Bandow JE, Hecker M. 2007. Proteomic profiling of cellular stresses in *Bacillus subtilis* reveals cellular networks and assists in elucidating antibiotic mechanisms of action. *Prog. Drug Res.* **64**:81–101.

Bantscheff M, Schirle M, Sweetman G, Rick J, Kuster B. 2007. Quantitative mass spectrometry in proteomics: A critical review. *Anal. Bioanal. Chem.* **389**:1017–1031.

Barrios-Llerena ME, Chong PK, Gan CS, Snijders AP, Reardon KF, Wright PC. 2006. Shotgun proteomics of cyanobacteria—Applications of experimental and data-mining techniques. *Brief. Funct. Genom. Proteom.* **5**:121–132.

Bartlett DH, Kato C, Horikoshi K. 1995. High pressure influences on gene and protein expression. *Res. Microbiol.* **146**:697–706.

Bebien M, Lagniel G, Garin J. 2002. Involvement of superoxide dismutases in the response of *Escherichia coli* to selenium oxides. *J. Bacteriol.* **184**:1556–1564.

Bender GR, Marquis RE. 1987. Membrane ATPases and acid tolerance of *Actinomyces viscosus* and *Lactobacillus casei*. *Appl. Environ. Microbiol.* **53**:2124–2128.

Bernhardt J, Buettner K, Scharf C, Hecker M. 1999. Dual channel imaging of two-dimensional electrophoregrams in *Bacillus subtilis*. *Electrophoresis* **20**:2225–2240.

Bernhardt J, Weibezahn J, Scharf C, Hecker M. 2003. *Bacillus subtilis* during feast and famine: Visualization of the overall regulation of protein synthesis during glucose starvation by proteome analysis. *Genome Res.* **13**:224–237.

Bijlsma R, Loeschcke V. 2005. Environmental stress, adaptation and evolution: An overview. *J. Evol. Biol.* **18**:744–749.

Blackstock WP, Weir MP. 1999. Proteomics: Quantitative and physical mapping of cellular proteins. *Trends Biotechnol.* **17**:121–127.

Blagoev B, Ong SE, Kratchmarova I, Mann M. 2004. Temporal analysis of phosphotyrosine-dependent signaling networks by quantitative proteomics. *Nat. Biotechnol.* **22**:1139–1145.

Brigulla M, Hoffmann T, Krisp A, Voelker A, Bremer E, Voelker U. 2003. Chill induction of the SigB-dependent general stress response in *Bacillus subtilis* and its contribution to low-temperature adaptation. *J. Bacteriol.* **185**:4305–4314.

Bukau B, Horwich AL. 1998. The Hsp70 and Hsp60 chaperone machines. *Cell* **92**:351–366.

Bunger MK, Cargile BJ, Ngunjiri A, Bundy JL, Stephenson JLJ. 2008. Automated proteomics of *Escherichia coli* via top-down electron-transfer dissociation mass spectrometry. *Anal. Chem.* **80**:1459–1467.

Cabiscol E, Tamarit J, Ros J. 2000. Oxidative stress in bacteria and protein damage by reactive oxygen species. *Int. Microbiol.* **3**:3–8.

Cagney G, Emili A. 2002. De novo peptide sequencing and quantitative profiling of complex protein mixtures using mass-coded abundance tagging. *Nat. Biotechnol.* **20**:163–170.

Caldas TD, El Yaagoubi A, Richarme G. 1998. Chaperone properties of bacterial elongation factor EF-Tu. *J. Biol. Chem.* **273**:11478–11482.

Casadei MA, Ingram R, Hitchings E, Archer J, Gaze JE. 2001. Heat resistance of *Bacillus cereus, Salmonella typhimurium* and *Lactobacillus delbrueckii* in relation to pH and ethanol. *Int. J. Food Microbiol.* **63**:125–134.

Castielli O, De la Cerda B, Navarro JA, Hervas M, De la Rosa MA. 2009. Proteomic analyses of the response of cyanobacteria to different stress conditions. *FEBS Lett.* **583**:1753–1758.

Chait BT. 2006. Mass spectrometry: Bottom-up or top-down ? *Science* **314**:65–66.

Chakraborty A, Regnier FE. 2002. Global internal standard technology for comparative proteomics. *J. Chromatogr. A* **949**:173–184.

Cheng Z, Wei Y-YC, Sung WWL, Glick BR, McConkey BJ. 2009. Proteomic analysis of the response to the plant growth-promoting bacterium *Pseudomonas putida* UW4 to nickel stress. *Proteome Sci.* **7**:1–8.

Chhabra SR, He Q, Huang KH, Gaucher SP, Alm EJ, et al. 2006. Global analysis of heat shock response in *Desulfovibrio vulgaris* Hildenborough. *J. Bacteriol.* **188**:1817–1828.

Choe LH, Aggarwal K, Franck Z, Lee KH. 2005. A comparison of the consistency of proteome quantitation using two-dimensional electrophoresis and shotgun isobaric tagging in *Escherichia coli* cells. *Electrophoresis* **26**:2437–2449.

Choi JS, Kim DS, Lee J, Kim SJ, Kim SI, Kim YH, Hong J, Yoo JS, Suh KH, Park YM. 2000. Proteome analysis of light-induced proteins in *Synechocystis* sp. PCC 6803: Identification of proteins separated by 2D-PAGE using N-terminal sequencing and MALDI-TOF MS. *Mol. Cells* **10**:705–711.

Chong PK, Burja AM, Radianingtyas H, Fazeli A, Wright PC. 2007. Translational and transcriptional analysis of *Sulfolobus solfataricus* P2 to provide insights into alcohol and ketone utilisation. *Proteomics* **7**:424–435.

Colinge J, Masselot A, Giron M, Dessingy T, Magnin J. 2003. OLAV: Towards high-throughput tandem mass spectrometry data identification. *Proteomics* **3**:1454–1463.

Coute Y, Hernandez C, Appel RD, Sanchez JC, Margolles A. 2007. Labeling of *Bifidobacterium longum* cells with 13C-substituted leucine for quantitative proteomic analyses. *Appl. Environ. Microbiol.* **73**:5653–5656.

Cox DM, Zhong F, Du M, Duchoslav E, Sakuma T, McDermott JC. 2005. Multiple reaction monitoring as a method for identifying protein posttranslational modifications. *J. Biomol. Tech.* **16**:83–90.

Craig R, Beavis RC. 2003. A method for reducing the time required to match protein sequences with tandem mass spectra. *Rapid Commun. Mass Spectrom.* **17**:2310–2316.

Craig R, Beavis RC. 2004. TANDEM: Matching proteins with tandem mass spectra. *Bioinformatics* **20**:1466–1467.

Cravatt BF, Simon GM, Yates Jr. 2007. The biological impact of mass-spectrometry-based proteomics. *Nature* **450**:991–1000.

De Angelis M, Bini L, Pallini V, Coccnocelli PS, Gobbetti M. 2001. The acid-stress response in *Lactobacillus sanfranciscensis* CB1. *Microbiology* **147**:1863–1873.

De Angelis M, Mariotti L, Rossi J, Servili M, Fox PF, Rollan G, Gobbetti M. 2002. Arginine catabolism by sourdough lactic acid bacteria: purification and characterization of the arginine deiminase pathway enzymes from *Lactobacillus sanfranciscensis* CB1. *Appl. Environ. Microbiol*. **68**:6193–6201.

De la Cerda B, Castielli O, Duran RV, Navarro JA, Hervas M, De la Rosa MA. 2008. A proteomic approach to iron and copper homeostasis in cyanobacteria. *Brief Funct. Genom. Proteom*. **6**:322–329.

de Souza L, Diehl G, Rodrigues MJ, Guo J, Romaschin AD, Colgan TJ, Siu KW. 2005. Search for cancer markers from endometrial tissues using differentially labeled tags iTRAQ and cICAT with multidimensional liquid chromatography and tandem mass spectrometry. *J. Proteome Res*. **4**:377–386.

Delahunty C, Yates JR. 2005. Protein identification using 2D-LC-MS/MS. *Methods* **35**:248–255.

Dianov GL, O'Neill P, Goodhead DT. 2001. Securing genome stability by orchestrating DNA repair: Removal of radiation-induced clustered lesions in DNA. *BioEssays* **23**:745–749.

Domon B, Aebersold R. 2006a. Challenges and opportunities in proteomics data analysis. *Mol. Cell. Proteom*. **5**:1921–1926.

Domon B, Aebersold R. 2006b. Mass spectrometry and protein analysis. *Science* **312**:212–217.

Dongre AR, Eng JK, Yates JR. 1997. Emerging tandem-mass-spectrometry techniques for the rapid identification of proteins. *Trends Biotechnol*. **15**:418–425.

Dowds BC. 1994. The oxidative stress response in *Bacillus subtilis*. *FEMS Microbiol*. **124**:155–264.

Drews O, Weiss W, Reil G, Parlar H, Wait R, Goerg A. 2002. High pressure effects stepwise altered protein expression in *Lactobacillus sanfranciscensis*. *Proteomics* **2**:765–774.

Dukan S, Turlin E, Biville F. 1998. Coupling 2D SDS-PAGE with CNBr cleavage and MALDI-TOFMS: A strategy applied to the identification of proteins induced by a hypochlorous acid stress in *Escherichia coli*. *Anal. Chem*. **70**:4433–4440.

Edman P. 1970. Sequence determination. *Mol. Biol. Biochem. Biophys*. **8**:211–255.

Eichler J, Adams MW. 2005. Posttranslational protein modification in Archaea. *Microbiol. Mol. Biol. Rev*. **69**:393–425.

Eng JK, McCormack AL, Yates JRI. 1994. An approach to correlate tandem mass spectral data of peptides with amino acid sequences in a protein database. *J. Am. Soc. Mass Spectrom*. **5**:976–989.

Ercal N, Gurer-Orhan H, Aykin-Burns N. 2001. Toxic metals and oxidative stress part I: Mechanisms involved in metal-induced oxidative damage. *Curr. Top. Med. Chem*. **1**:529–539.

Eymann C, Dreisbach A, Albrecht D, Bernhardt J, Becher D, et al. 2004. A comprehensive proteome map of growing *Bacillus subtilis* cells. *Proteomics* **4**:2849–2876.

Eymann C, Mach H, Harwood CR, Hecker M. 1996. Phosphate-starvation-inducible proteins in *Bacillus subtilis*: A two-dimensional gel electrophoresis study. *Microbiology* **142**:3163–3170.

Fanous A, Weiss W, Goerg A, Jacob F, Parlar H. 2008. A proteome analysis of the cadmium and mercury response in *Corynebacterium japonicum*. *Proteomics* **8**:4976–4986.

Fischer B, Grossmann J, Roth V, Gruissem W, Baginsky S, Buhmann JM. 2006. Semisupervised LC/MS alignment for differential proteomics. *Bioinformatics* **22**: e132–e140.

Foster JW, Hall HK. 1991. Inducible pH homeostasis and the acid tolerance response of *Salmonella typhimurium*. *J. Bacteriol*. **173**:5129–5135.

Fulda S, Huang F, Nilsson F, Hagemann M, Norling B. 2000. Proteomics of *Synechocystis* sp. strain PCC 6803. Identification of periplasmic protein in cells grown at low and high salt concentrations. *Eur. J. Biochem*. **267**:5900–5907.

Fulda S, Mikkat S, Huang F, Huckauf J, Marin K, Norling B, Hagemann M. 2006. Proteome analysis of salt stress response in the cyanobacterium *Synechocystis* sp. strain PCC 6803. *Proteomics* **6**:2733–2745.

Gaidenko TA, Price CW. 1998. General stress transcription factor σ^B and sporulation transcription factor σ^H each contribute to survival of *Bacillus subtilis* under extreme growth conditions. *J. Bacteriol*. **180**:3730–3733.

Gan CS, Reardon KF, Wright PC. 2005. Comparison of protein and peptide prefractionation methods for the shotgun proteomic analysis of *Synechocystis* sp. PCC 6803. *Proteomics* **5**:2468–2478.

Gao Y, Xiong W, Li XB, Gao CF, Zhang YL, Li H, Wu QY. 2009. Identification of the proteomic changes in *Synechocystis* sp. PCC 6803 following prolonged UV-B irradiation. *J. Exp. Bot*. **60**:1141–1154.

Ge Y, Lawhorn BG, ElNaggar M, Strauss E, Park J-H, Begley TP, McLafferty FW. 2002. Top down characterization of larger proteins (45kDa) by electron capture dissociation mass spectrometry. *J. Am. Chem. Soc*. **124**:672–678.

Ge Y, Rajkumar L, Guzman RC, Nandi S, Patton WF, Agnew BJ. 2004. Multiplexed fluorescence detection of phosphorylation, glycosylation, and total protein in the proteomic analysis of breast cancer refractoriness. *Proteomics* **4**:3464–3467.

Geer LY, Markey SP, Kowalak JA, Wagner L, Xu M, Maynard DM, Yang X, Shi W, Bryant SH. 2004. Open mass spectrometry search algorithm. *J. Proteome Res*. **3**:958–964.

Gerber SA, Rush J, Stemman O, Kirschner MW, Gygi SP. 2003. Absolute quantification of proteins and phosphoproteins from cell lysates by tandem MS. *Proc. Natl. Acad. Sci. USA* **100**:6940–6945.

Gorg A, Weiss W, Dunn MJ. 2004. Current two-dimensional electrophoresis technology for proteomics. *Proteomics* **4**:3665–3685.

Gouesbet G, Jan G, Boyaval P. 2002. Two-dimensional electrophoresis study of *Lactobacillus delbrueckii* subsp. *bulgaricus* thermotolerance. *Appl. Environ. Microbiol*. **68**:1055–1063.

Graham RLJ, Pollock CE, Ternan NG, McMullan G. 2006. Top-down proteomic analysis of the soluble sub-proteome of the obligate thermophile, *Geobacillus thermoleovorans* T80: Insights into the cellular processes. *J. Proteome Res*. **5**:822–828.

Grant WD. 2004. Life at low water activity. *Philos. Trans. R. Soc. Lond. B* **359**:1249–1266.

Graumann P, Schroeder K, Schmid R, Marahiel MA. 1996. Cold shock stress-induced proteins in *Bacillus subtilis*. *J. Bacteriol*. **178**:4611–4619.

Graumann PL, Marahiel MA. 1999a. Cold shock proteins CspB and CspC are major stationary-phase-induced proteins in *Bacillus subtilis*. *Arch. Microbiol*. **171**:135–138.

Graumann PL, Marahiel MA. 1999b. Cold shock response in *Bacillus subtilis*. *J. Mol. Microbiol. Biotechnol*. **1**:203–209.

Groat RG, Schultz JE, Zychlinsky E, Bockman A, Matin A. 1986. Starvation proteins in *Escherichia coli*: Kinetics of synthesis and role in starvation survival. *J. Bacteriol*. **168**:486–493.

Grossmann J, Roos FF, Cieliebak M, Liptak Z, Mathis LK, Muller M, Gruissem W, Baginsky S. 2005. AUDENS: A tool for automated peptide *de novo* sequencing. *J. Proteome Res*. **4**:1768–1774.

Gygi SP, Rist B, Gerber SA, Turacek F, Gelb MH, Aebersold R. 1999. Quantitative analysis of complex protein mixtures using isotope-coded affinity tags. *Nat. Biotechnol*. **17**:994–999.

Gygi SP, Corthals GL, Zhang Y, Rochon Y, Aebersold R. 2000. Evaluation of two-dimensional gel electrophoresis-based proteome analysis technology. *Proc. Natl. Acad. Sci. USA* **97**:9390–9395.

Han X, Jin M, Breuker K, McLafferty FW. 2006. Extending top-down mass spectrometry to proteins with masses greater than 200 kilodaltons. *Science* **314**:109–112.

Hanash SM. 2000. Biomedical applications of two-dimensional electrophoresis using immobilized pH gradients: Current status. *Electrophoresis* **21**:1202–1209.

Hartke A, Bouche S, Gansel X, Boutibonnes P, Auffray Y. 1994. Starvation-indued stress resistance in *Lactococcus lactis* subsp *lactis* IL1403. *Appl. Environ. Microbiol*. **60**:3474–3478.

Hartke A, Bouche S, Giard J-C, Benachour A, Boutibonnes P, Auffray Y. 1996. The lactic acid stress response of *Lactococcus lactis* subsp. *lactis*. *Curr. Microbiol*. **33**:194–199.

He YY, Haeder D. 2002. Reactive oxygen species and UV-B: Effect on cyanobacteria. *Photochem. Photobiol. Sci*. **1**:729–736.

Hecker M, Voelker U. 2001. General stress response of *Bacillus subtilis* and other bacteria. *Adv. Microb. Physiol*. **44**:35–91.

Hecker M, Voelker U. 2004. Towards a comprehensive understanding of *Bacillus subtilis* cell physiology by physiological proteomics. *Proteomics* **4**:3727–3750.

Hecker M, Schumann W, Voelker U. 1996. Heat-shock and general stress response in *Bacillus subtilis*. *Mol. Microbiol*. **19**:417–428.

Hecker M, Antelmann H, Buettner K, Bernhardt J. 2008. Gel-based proteomics of gram-positive bacteria: A powerful tool to address physiological questions. *Proteomics* **8**:4958–4975.

Hecker M, Reder A, Fuchs S, Pagels M, Engelmann S. 2009. Physiological proteomics and stress/starvation responses in *Bacillus subtilis* and *Staphylococcus aureus*. *Res. Microbiol*. **160**:245–258.

Hempel J, Zehner S, Goettfert M, Patschkowski T. 2009. Analysis of the secretome of the soybean symbiont *Bradyrhizobium japonicum*. *J. Biotechnol*. **140**:51–58.

Hengge-Aronis R, Klein W, Lange R, Rimmele M, Boos W. 1991. Trehalose synthesis genes are controlled by the putative sigma factor encoded by rpoS and are involved in stationary-phase thermotolerance in *Escherichia coli*. *J. Bacteriol*. **173**:7918–7924.

Hoeper D, Bernhardt J, Hecker M. 2006. Salt stress adaptation of *Bacillus subtilis*: A physiological proteomics approach. *Proteomics* **6**:1550–1562.

Hoi LT, Voigt B, Juergen B, Ehrenreich A, Gottschalk G, Evers S, Feesche J, Maurer KH, Hecker M, Schweder T. 2006. The phosphate-starvation response of *Bacillus licheniformis*. *Proteomics* **6**:3582–3601.

Holtan-Hartwig L, Bechmann M, Hoyas TR, Linjordeta R, Bakkenb LR. 2002. Heavy metals tolerance of soil denitrifying communities: N_2O dynamics. *Soil Biol. Biochem*. **34**:1181–1190.

Huang F, Fulda S, Hagemann M, Norling B. 2006. Proteomic screening of salt-stress-induced changes in plasma membranes of *Synechocystis* sp. strain PCC 6803. *Proteomics* **6**:910–920.

Hutchinson F. 1985. Chemical changes induced in DNA by ionizing radiation. *Prog. Nucleic Acid Res. Mol. Biol*. **32**:115–154.

Ishihama YST, Tabata t, Miyamoto N, Sagane K, Nagasu T, Oda Y. 2005. Quantitative mouse brain proteomics using culture-derived isotope tags as internal standards. *Nat. Biotechnol*. **23**:617–621.

James P, Quadroni M, Carafoli E, Gonnet G. 1993 Protein identification by mass profile fingerprinting. *Biochem. Biophys. Res. Commun*. **195**:58–64.

Jenkins DE, Schultz JE, Matin A. 1988. Starvation-induced cross protection against heat or H_2O_2 challenge in *Escherichia coli*. *J. Bacteriol*. **170**:3910–3914.

Jenkins DE, Chaisson SA, Matin A. 1990. Starvation-induced cross protection against osmotic challenge in *Escherichia coli*. *J. Bacteriol*. **172**:2779–2781.

Julka S, Regnier FE. 2005. Recent advancements in differential proteomics based on stable isotope coding. *Brief. Funct. Genom. Proteom*. **4**:158–177.

Jungblut P, Thiede B. 1997. Protein identification by 2D-E gels by MALDI mass spectrometry. *Mass Spectrom. Rev*. **16**:145–162.

Kaakoush NO, Raftery M, Mendz GL. 2008. Molecular response of *Campylobacter jejuni* to cadmium stress. *FEBS J*. **275**:5021–5033.

Kapp EA, Schuetz F, Connolly LM, Chakel JA, Meza JE, Miller, et al. 2005. An evaluation, comparison, and accurate benchmarking of several publicly available MS/MS search algorithms: Sensitivity and specificity analysis. *Proteomics* **5**:3475–3490.

Kiernan UA. 2007. Quantitation of target proteins and post-translational modifications in affinity-based proteomics approaches. *Expert Rev. Proteom*. **4**:421–428.

Kim SI, Choi JS, Kahng HY. 2007. A proteomics strategy for the analysis of bacterial biodegradation pathways. *OMICS* **11**:280–294.

Kirkpatrick DS, Gerber SA, Gygi SP. 2005. The absolute quantification strategy: A general procedure for the quantification of proteins and post-translational modifications. *Methods* **35**:265–273.

Konings WN, Albers S-V, Koning S, Driessen AJM. 2002. The cell membrane plays a crucial role in survival of bacteria and archaea in extreme environments. *Antonie van Leeuwenhoek* **81**:61–72.

Kovacs E, van der Vies SM, Glatz A, Toeroek Z, Varvasovszki V, Horvath I, Vigh L. 2001. The chaperonins of *Synechocystis* PCC 6803 differ in heat inducibility and chaperone activity. *Biochem. Biophys. Res. Commun*. **289**:908–915.

Kurian D, Phadwal K, Maeenpaeae P. 2006. Proteomic characterization of acid stress response in *Synechocystis* sp. PCC 6803. *Proteomics* **6**:3614–3624.

Lacerda CMR, Choe LH, Reardon KF. 2007. Metaproteomic analysis of a bacterial community response to cadmium exposure. *J. Proteome Res*. **6**:1145–1152.

Lange R, Hengge-Aronis R. 1991. Identification of a central regulator of stationary-phase gene expression in *Escherichia coli*. *Mol. Microbiol*. **5**:49–59.

Lange V, Malmstroem JA, Didion J, King NL, Johansson BP, et al. 2008a. Targeted quantitative analysis of *Streptococcus pyogenes* virulence factors by multiple reaction monitoring. *Mol. Cell. Proteom*. **7**:1489–1500.

Lange V, Picotti P, Domon B, Aebersold R. 2008b. Selected reaction monitoring for quantitative proteomics: A tutorial. *Mol. Syst. Biol*. **4**: 222.

Leichert LI, Scharf C, Hecker M. 2003. Global characterization of disulfide stress in *Bacillus subtilis*. *J. Bacteriol*. **185**:1967–1975.

Leichert LI, Gehrke F, Gudiseva HV, Blackwell T, Ilbert M, Walker AK, Strahler JR, Andrews PC, Jakob U. 2008. Quantifying changes in the thiol redox proteome upon oxidative stress in vivo. *Proc. Natl. Acad. Sci. USA* **105**:8197–8202.

Leithner A, Lindner W. 2006. Chemistry meets proteomics: The use of chemical tagging reactions for MS-based proteomics. *Proteomics* **6**:5418–5434.

Lenco J, Pavkova I, Hubalek M, Stulik J. 2005. Insights into the oxidative stress response in *Francisella tularensis* LVS and its mutant DeltaiglC1+2 by proteomic analysis. *FEMS Microbiol. Lett*. **246**:47–54.

Lim EM, Ehrlich SD, Maguin E. 2000. Identification of stress-inducible proteins in *Lactobacillus delbrueckii* subsp. *bulgaricus*. *Electrophoresis* **21**:2557–2561.

Link AJ, Eng J, Schieltz DM, Carmack E, Mize GJ, Morris DR, Garvik BM, Yates JR. 1999. Direct anlaysis of protein complexes using mass spectrometry. *Nat. Biotechnol*. **17**:676–682.

Lisacek F, Cohen-Boulakia S, Appel RD. 2006 Proteome informatics II: Bioinformatics for comparative proteomics. *Proteomics* **6**:5445–5466.

Lorca GL, Font de Valdez G. 2001. A low-pH-inducible, stationary-phase acid tolerance response in *Lactobacillus acidophilus* CRL 639. *Curr. Microbiol*. **42**:21–25.

Los DA, Suzuki I, Zinchenko VV, Murata N. 2008. Stress responses in *Synechocystis*: Regulated genes and regulatory systems. In Herrero A, Flores E, eds. *The Cyanobacteria: Molecular Biology, Genomics and Evolution*. Norfolk, VA: Caister Academic Press, pp. 117–157.

Malone AS, Shellhammer TH, Courtney PD. 2002. Effects of high pressure on the viability, morphology, lysis, and cell wall hydrolase activity of *Lactococcus lactis* subsp. *cremoris*. *Appl. Environ. Microbiol*. **68**:4357–4363.

Mann M, Jensen ON. 2003. Proteomic analysis of post-translational modifications. *Nat. Biotechnol*. **21**:255–261.

Mann M, Wilm M. 1994. Error-tolerant identification of peptides in sequence databases by peptide sequence tags. *Anal. Chem*. **66**:4390–4399.

Marrero J, Gonzalez LJ, Sanchez A, Ayala M, Paz-Lago D, Gonzalez W, Fallarero A, Castellanos-Serra L, Coto O. 2004. Effect of high concentration of Co (II) on *Enterobacter liquefaciens* strain C-1: A bacterium highly resistant to heavy metals with an unknown genome. *Proteomics* **4**:1265–1279.

Martinez-Gomariz M, Hernaez ML, Gutierrez D, Ximenez-Embun P, Prestamo G. 2009. Proteomic analysis by two-dimensional differential gel electrophoresis (2D DIGE) of a high-pressure effect in *Bacillus cereus*. J. Agric. Food Chem. **57**:3543–3549.

Mashego MR, Rumbold K, De Mey M, Vandamme E, Soetaert W, Heijnen JJ. 2007. Microbial metabolomics: Past, present and future methodologies. *Biotechnol. Lett*. **29**:1–16.

Mishra Y, Chaurasia N, Rai LC. 2009. Heat pretreatment alleviates UV-B toxicity in the cyanobacterium *Anabaena doliolum*: A proteomic analysis of cross tolerance. *Photochem. Photobiol*. **85**:824–833.

Molina-Guitierrez A, Stippl V, Delgado A, Gaenzle MG, Vogel RF. 2002. *In situ* determination of the intracellular pH of *Lactococcus lactis* and *Lactobacillus plantarum* during pressure treatment. *Appl. Environ. Microbiol*. **68**:4399–4406.

Mostertz J, Scharf C, Hecker M, Homuth G. 2004. Transcriptome and proteome analysis of *Bacillus subtilis* gene expression in response to superoxide and peroxide stress. *Microbiology* **150**:497–512.

Mueller LN, Brusniak M-Y, Mani DR, Aebersold R. 2008. An assessment of software solutions for the analysis of mass spectrometry based quantitative proteomics data. *J. Proteome Res*. **7**:51–61.

Mukhopadhyay A, He Z, Alm EJ, Arkin AP, Baidoo EE, et al. 2006. Salt stress in *Desulfovibrio vulgaris* Hildenborough: An integrated genomics approach. *J. Bacteriol*. **188**:4068–4078.

Munro P, Flatau G, Lemichez E. 2007. Bacteria and the ubiquitin pathway. *Curr. Opin. Microbiol*. **10**:39–46.

Narasimhan C, Tabb DL, Verberkmoes NC, Thompson MR, Hettich RL, Uberbacher EC. 2005. MASPIC: Intensity-based tandem mass spectrometry scoring scheme that improves peptide identification at high confidence. *Anal. Chem*. **77**:7581–7593.

Nesatyy VJ, Suter MJ-F. 2007. Proteomics for the analysis of environmental stress responses in organisms. *Environ. Sci. Technol*. **41**:6891–6900.

Nesatyy VJ, Suter MJ-F. 2008. Analysis of environmental stress response on the proteome level. *Mass Spectrom. Rev*. **27**:556–574.

Nichols DS, Miller MR, Davies NW, Goodchild A, Raftery M, Cavicchioli R. 2004. Cold adaptation in the antarctic archaeon *Methanococcoides burtonii* involves membrane lipid unsaturation. *J. Bacteriol*. **186**:8508–8515.

Niyogi KK. 2000. Safety valves for photosynthesis. *Curr. Opin. Plant Biol*. **3**:455–460.

Noeel-Georis I, Vallaeys T, Chauvaux R, Monchy S, Falmagne P, Mergeay M, Wattiez R. 2004. Global analysis of the *Ralstonia metallidurans* proteome: Prelude for the large-scale study of heavy metal response. *Proteomics* **4**:151–179.

Nystroem T, Olsson RM, Kjelleberg S. 1992. Survival, stress resistance, and alterations in protein expression in the marine *Vibrio* sp. strain S14 during starvation for different individual nutrients. *Appl. Environ. Microbiol*. **58**:55–65.

Oda Y, Nagasu T, Chait BT. 2001. Enrichment analysis of phosphorylated proteins as a tool for probing the phosphoproteome. *Nat. Biotechnol*. **19**:379–382.

Ong S-E, Blagoev B, Kratchmarova I, Bach Kristensen D, Steen H, Pandey A, Mann M. 2002. Stable isotope labeling by amino acids in cell culture, SILAC, is a simple and accurate approach to expression proteomics. *Mol. Cell. Proteom*. **1**:376–386.

Ong SE, Mann M. 2005. Mass spectrometry-based proteomics turns quantitative. *Nat. Chem. Biol*. **1**:252–262.

O'Sullivan E, Condon S. 1999. Relationship between acid tolerance, cytoplasmic pH, and ATP and H^+-ATPase levels in chemostat cultures of *Lactococcus lactis*. *Appl. Environ. Microbiol*. **65**:2287–2293.

Pagan R, Mackey B. 2000. Relationship between membrane damage and cell death in pressure-treated *Escherichia coli* cells: Differences between exponential- and stationary-phase cells and variation among strains. *Appl. Environ. Microbiol*. **66**:2829–2834.

Palagi PM, Hernandez P, Walther D, Appel RD. 2006. Proteome informatics I: Bioinformatics tools for processing experimental data. *Proteomics* **6**:5435–5444.

Panchaud A, Affolter M, Moreillon P, Kussmann M. 2008. Experimental and computational approaches to quantitative proteomics: Status quo and outlook. *J. Proteom*. **71**:19–33.

Pandhal J, Wright PC, Biggs CA. 2007. A quantitative proteomic analysis of light adaptation in a globally significant marine cyanobacterium *Prochlorococcus marinus* MED4. *J. Proteome Res*. **6**:996–1005.

Pandhal J, Ow SY, Wright PC, Biggs CA. 2009. Comparative proteomics study of salt tolerance between a nonsequenced extremely halotolerant Cyanobacterium and its mildly halotolerant relative using in vivo metabolic labeling and in vitro isobaric labeling. *J. Proteome Res*. **8**:818–828.

Papa R, Glagla S, Danchin A, Schweder T, Marino G, Duilio A. 2006. Proteomic identification of a two-component regulatory system in *Pseudoalteromonas haloplanktis* TAC125. *Extremophiles* **10**:483–491.

Pappin DJC, Hojrup P, Bleasby AJ. 1993. Rapid identification of proteins by peptide-mass fingerprinting. *Curr. Biol*. **3**:327–332.

Park S, Ely RL. 2008. Candidate stress genes of *Nitrosomonas europaea* for monitoring inhibition of nitrification by heavy metals. *Appl. Environ. Microbiol*. **74**:5475–5482.

Paton JC, McMurchie EJ, May BK, Elliott WH. 1978. Effect of growth temperature on membrane fatty acid composition and susceptibility to cold shock of *Bacillus amyloliquefaciens*. *J. Bacteriol*. **135**:754–759.

Patrauchan MA, Florizone C, Eapen S, Gomez-Gil L, Sethuraman B, Fukuda M, Davies J, Mohn WW, Eltis LD. 2008. Roles of ring-hydroxylating dioxygenases in styrene and benzene catabolism in *Rhodococcus jostii* RHA1. *J. Bacteriol*. **190**:37–47.

Perkins DN, Pappin DJ, Creasy DM, Cottrell JS. 1999. Probability-based protein identification by searching sequence databases using mass spectrometry data. *Electrophoresis* **20**:3551–3567.

Perrin C, Guimont c, Bracquart P, Gaillard JL. 1999. Expression of a new cold shock protein of 21.5kDa and of the major cold shock protein by *Streptococcus thermophilus* after cold shock. *Curr. Microbiol*. **39**:342–347.

Phadtare S, Alsina J, Inouye M. 1999. Cold-shock response and cold-shock proteins. *Curr. Opin. Microbiol*. **2**:175–180.

Piper PW, Ortiz-Calderon C, Holyoak C, Coote P, Cole M. 1997. Hsp30, the integral plasma membrane heat shock protein of *Saccharomyces cerevisiae*, is a stress-inducible regulator of plasma membrane H^+-ATPase. *Cell Stress Chaperones* **2**:12–24.

Pradet-Balade B, Boulme F, Beug H, Muellner EW, Garcia-Sanz JA. 2001. Translation control: Bridging the gap between genomics and proteomics. *Trends Biochem. Sci*. **26**:225–229.

Pratt JM, Simpson DM, Doherty MK, Rivers J, Gaskell SJ, Beynon RJ. 2006. Multiplexed absolute quantification for proteomics using concatenated signature peptides encoded by QconCAT genes. *Nat. Protoc*. **1**:1029–1043.

Rabilloud T, Chevallet M, Luche S, Leize-Wagner E. 2005. Oxidative stress response: A proteomic view. *Expert Rev. Proteomics* **2**:949–956.

Rai LC, Tyagi B, Mallick N. 1996. Alteration in photosynthetic characteristics of *Anabaena doliolum* following exposure to UV-B and Pb. *Photochem. Photobiol*. **64**:658–663.

Redding AM, Mukhopadhyay A, Joyner DC, Hazen TC, Keasling JD. 2006. Study of nitrate stress in *Desulfovibrio vulgaris* Hildenborough using iTRAQ proteomics. *Brief Funct. Genom. Proteom*. **5**:133–143.

Renzone G, D'Ambrosio c, Arena S, Rullo R, Ledda L, Ferrara L, Scaloni A. 2005. Differential proteomic analysis in the study of prokaryotes stress resistance. *Ann. Ist Super Sanita* **41**:459–468.

Resing KA, Meyer-Arendt K, Mendoza AM, Aveline-Wolf LD, et al. 2004. Improving reproducibility and sensitivity in identifying human proteins by shotgun proteomics. *Anal. Chem*. **76**:3556–3568.

Riley PA. 1994. Free radicals in biology: Oxidative stress and the effects of ionizing radiation. *Int. J. Radiat. Biol*. **65**:27–33.

Ross PL, Huang YLN, Marchese JN, Pappin DJC. 2004. Multiplexed protein quantification in *Saccharomyces cerevisiae* using amine-reactive isobaric tagging reagents. *Mol. Cell. Proteomics* **3**:1154–1169.

Running WE, Ravipaty S, Karty JA, Reilly JP. 2007. A top-down/bottom-up study of the ribosomal proteins of *Caulobacter crescentus*. *J. Proteome Res*. **6**:337–347.

Saad-Nehme J, Silva JL, Meyer-Fernandes Jr. 2001. Osmolytes protect mitochondrial F_0F_1-ATPase complex against pressure inactivation. *Biochim. Biophys. Acta* **1546**:164–170.

Sanchez A, Champomier-Verges MC, Anglade P, Baraige F, de Los Reyes-Gavilan CG, Margolles A, Zagorec M. 2005. Proteomic analysis of global changes in protein expression during bile salt exposure of *Bifidobacterium longum* NCIMB 8809. *J. Bacteriol*. **187**:5799–5808.

Sanchez B, Champomier-Verges MC, Collado Mdel C, Anglade P, Baraige F, Sanz Y, De los Reyes-Gavilan CG, Margolles A, Zagorec M. 2007. Low-pH adaptation and the acid tolerance response of *Bifidobacterium longum* biotype longum. *Appl. Environ. Microbiol*. **73**:6450–6459.

Sanchez B, Ruiz L, de los Reyes-Gavilan CG, Margolles A. 2008. Proteomics of stress response in *Bifidobacterium*. *Front. Biosci*. **13**:6905–6916.

Schultz JE, Latter GI, Matin A. 1988. Differential regulation by cyclic AMP of starvation protein synthesis in *Escherichia coli*. *J. Bacteriol*. **170**:3903–3909.

Shively JE. 2000. The chemistry of protein sequence analysis. *EXS* **88**:99–117.

Singh OV. 2006. Proteomics and metabolomics: The molecular make-up of toxic aromatic pollutant bioremediation. *Proteomics* **6**:5481–5492.

Sinha RP, Singh N, Kumar A, Kumar HD, Haeder M, Haeder DP. 1996. Effect of UV irradiation on certain physiological and biochemical processes in cyanobacterium. *J. Photochem. Photobiol. B* **32**:107–113.

Slabas AR, Suzuki I, Murata N, Simon WJ, Hall JJ. 2006. Proteomic analysis of the heat shock response in *Synechocystis* PCC6803 and a thermally tolerant knockout strain lacking the histidine kinase 34 gene. *Proteomics* **6**:845–864.

Snijders AP, de Koning B, Wright PC. 2007. Relative quantification of proteins across the species boundary through the use of shared peptides. *J. Proteome Res*. **6**:97–104.

Steinberg CE, Stuerzenbaum SR, Menzel R. 2008. Genes and environment—Striking the fine balance between sophisticated biomonitoring and true functional environmental genomics. *Sci. Total Environ*. **400**:142–161.

Stenuit B, Eyers L, Schuler L, Agathos SN, George I. 2008. Emerging high-throughput approaches to analyze bioremediation of sites contaminated with hazardous and/or recalcitrant waters. *Biotechnol. Adv*. **26**:561–575.

Suzuki CK, Rep M, Maarten van Dijl J, Suda K, Grivell LA, Schatz G. 1997. ATP-dependent proteases that also chaperone protein biogenesis. *Trends Biochem. Sci*. **22**:118–123.

Suzuki I, Simon WJ, Slabas AR. 2006. The heat shock response of *Synechocystis* sp. PCC 6803 analysed by transcriptomics and proteomics. *J. Exp. Bot*. **57**:1573–1578.

Syka JE, Coon JJ, Schroeder MJ, Shabanowitz J, Hunt DF. 2004. Peptide and protein sequence analysis by electron transfer dissociation mass spectrometry. *Proc. Natl. Acad. Sci. USA* **101**:9528–9533.

Taji T, Seki M, Satou M, Sakurai T, Kobayashi M, Ishiyama K, Narusaka Y, Narusaka M, Zhu JK, Shinozaki K. 2004. Comparative genomics in salt tolerance between *Arabidopsis* and *Arabidopsis*-related halophyte salt cress using *Arabidopsis* microarray. *Plant Physiol*. **135**:1697–1709.

Tam LT, Antelmann H, Eymann C, Albrecht D, Bernhardt J, Hecker M. 2006. Proteome signatures for stress and starvation in *Bacillus subtilis* as revealed by a 2-D gel image color coding approach. *Proteomics* **6**:4565–4585.

Teixeira P, Castro H, Mohacsi-Farkas C, Kirby R. 1997. Identification of sites of injury in *Lactobacillus bulgaricus* during heat stress. *J. Appl. Microbiol*. **83**:219–226.

Thieringer HA, Jones PG, Inouye M. 1998. Cold shock and adaptation. *BioEssays* **20**:49–57.

Thompson A, Schaefer J, Kuhn K, Kienle S, Schwarz J, Schmidt G, Neumann T, Hamon C. 2003. Tandem mass tags: A novel quantification strategy for comparative analysis of complex protein mixtures by MS/MS. *Anal. Chem*. **75**:1895–1904.

Tsugita A. 1987. Developments in protein microsequencing. *Adv. Biophys*. **23**:81–113.

Uchiyama H, Shinohara Y, Tomioka N, Kusakabe I. 1999. Induction and enhancement of stress proteins in a trichloroethylene-degrading methanotrophic bacterium, *Methylocystis* sp. M. *FEMS MIcrobiol. Lett*. **170**:125–130.

Unlue M, Morgan ME, Minden JS. 1997. Difference gel electrophoresis: a single gel method for detecting changes in protein extracts. *Electrophoresis* **18**:2071–2077.

van den Bergh G, Arckens L. 2004. Fluorescent two-dimensional difference gel electrophoresis unveils the potential of gel-based proteomics. *Curr. Opin. Biotechnol*. **15**:38–43.

VanBogelen RA, Neidhardt FC. 1990. Ribosomes as sensors of heat and cold shock in *Escherichia coli*. *Proc. Natl. Acad. Sci. USA* **87**:5589–5593.

Voelker U, Hecker M. 2005. From genomics via proteomics to cellular physiology of the Gram-positive model organism *Bacillus subtilis*. *Cell. Microbiol*. **7**:1077–1085.

Voigt B, Schweder T, Becher D, Ehrenreich A, Gottschalk G, Feesche J, Maurer KH, Hecker M. 2004. A proteomic view of cell physiology of *Bacillus licheniformis*. *Proteomics* **4**:1465–1490.

Waanders LF, Hanke S, Mann M. 2007. Top-down quantitation and characterization of SILAC-labeled proteins. *J. Am. Soc. Mass Spectrom*. **18**:2058–2064.

Wang J, Zhao CB, Meng B, Xie J, Zhou C, Chen X, Zhao K, Shao J, Xue Y, Xu N, Ma Y, Liu S. 2007. The proteomic alterations of *Thermoanaerobacter tengcongensis* cultured at different temperatures. *Proteomics* **7**:1409–1419.

Washburn MP, Wolters D, Yates JR. 2001. Large-scale analysis of the yeast proteome by multidimensional protein identification technology. *Nat. Biotechnol.* **19**:242–247.

Weber H, Engelmann S, Becher D, Hecker M. 2004. Oxidative stress triggers thiol oxidation in the glyceraldehyde-3-phosphate dehydrogenase of *Staphylococcus aureus*. *Mol. Microbiol.* **52**:133–140.

Weber MHW, Fricke I, Doll N, Marahiel MA. 2002. CSDBase: An interactive database for cold shock domain-containing proteins and the bacterial cold shock response. *Nucleic Acids Res.* **30**: 375–378.

Welch TJ, Farewell A, Neidhardt FC, Bartlett DH. 1993. Stress response of *Escherichia coli* to elevated hydrostatic pressure. *J. Bacteriol.* **175**:7170–7177.

Whitehead K, Kish A, Pan M, Kaur A, Reiss DJ, King N, Hohmann L, DiRuggiero J, Baliga NS. 2006. An integrated systems approach for understanding cellular responses to gamma radiation. *Mol. Syst. Biol.* **2**: 47.41–47.46.

Wilkins MR, Sanchez J-C, Williams KL, Hochstrasse DF. 1996. Current challenges and future applications for protein maps and post-translational vector maps in proteome projects. *Electrophoresis* **17**:830–838.

Willimsky G, Bang H, Fischer G, Marahiel MA. 1992. Characterization of cspB, a *Bacillus subtilis* inducible cold shock gene affecting cell viability at low temperatures. *J. Bacteriol.* **174**:6326–6335.

Wilm M, Shevchenko a, Houthaeve T, Breit S, Schweigerer L, Fotsis T, Mann M. 1996. Femtomole sequencing of proteins from polyacrlyamide gels by nano-electrospray mass spectrometry. *Nature* **379**:466–469.

Wilmes P, Bond PL. 2004. The application of two-dimensional polyacrylamide gel electrophoresis and downstream analyses to a mixed community of prokaryotic microorganisms. *Environ. Microbiol.* **6**:911–920.

Wolf-Yadlin A, Hautaniemi S, Lauffenburger DA, White FM. 2007. Multiple reaction monitoring for robust quantitative proteomic analysis of cellular signaling networks. *Proc. Natl. Acad. Sci. USA* **104**:5860–5865.

Wolf C, Hochgraefe F, Kusch H, Albrecht D, Hecker M, Engelmann S. 2008. Proteomic analysis of antioxidant strategies of *Staphylococcus aureus*: Diverse responses to different oxidants. *Proteomics* **8**:3139–3153.

Wolff S, Otto A, Albrecht D, Stahl Zeng J, Buettner K, Glueckmann M, Hecker M, Becher D. 2006. Gel-free and gel-based proteomics in *Bacillus subtilis*. *Mol. Cell. Proteom.* **5**:1183–1192.

Wong JW, Sullivan MJ, Cagney G. 2008. Computational methods for the comparative quantification of proteins in label-free LCn-MS experiments. *Brief Bioinform.* **9**:156–165.

Wouters PC, Glaasker E, Smelt JP. 1998. Effects of high pressure on inactivation kinetics and events related to proton efflux in *Lactobacillus plantarum*. *Appl. Environ. Microbiol.* **64**:509–514.

Wouters JA, Jeynov B, Rombouts FM, de Vos WM, Kuipers OP, Abee T. 1999a. Analysis of the role of 7kDa cold-shock proteins of *Lactococcus lactis* MG1363 in cryoprotection. *Microbiology* **145**:3185–3194.

Wouters JA, Rombouts FM, de Vos WM, Kuipers OP, Abee T. 1999b. Cold shock proteins and low-temperature response of *Streptococcus thermophilus* CNRZ302. *Appl. Environ. Microbiol.* **65**:4436–4442.

Wouters JA, Kamphuis HH, Hugenholtz J, Kuipers OP, De Vos WM, Abee T. 2000. Changes in glycolytic activity of *Lactococcus lactis* induced by low temperature. *Appl. Environ. Microbiol.* **66**:3686–3691.

Wu TL. 2006. Two-dimensional difference gel electrophoresis. *Methods Mol. Biol.* **328**:71–95.

Wynne C, Fenselau C, Demirev PA, Edwards N. 2009. Top-down identification of protein biomarkers in bacteria with unsequenced genomes. *Anal. Chem.* **81**:9633–9642.

Yamagata A, Kristensen DB, Takeda Y, Miyamoto Y, Okada K, Inamatsu M, Yoshizato K. 2002. Mapping of phosphorylated proteins on two-dimensional polyacrylamide gels using protein phosphatase. *Proteomics* **2**:1267–1276.

Ying N, Zheng Z, Xu H, Tian B, Hua Y. 2008. Extracellular proteome changes of *Deinococcus radiodurans* under gamma-irradiation stress conditions. *Protein Pept. Lett.* **15**:595–599.

Zhang B, VerBerkmoes NC, Langston MA, Uberbacher E, Hettich RL, Samatova NF. 2006. Detecting differential and correlated protein expression in label-free shotgun proteomics. *J. Proteome Res.* **5**:2909–2918.

Zhang C, Wei J, Zheng Z. 2005. Proteomic analysis of *Deinococcus radiodurans* recovering from gamma-irradiation. *Proteomics* **5**:138–143.

Zhang L-F, Yang H-M, Cui S-X, Hu J, Wang J, Kuang T-Y, Norling B, Huang F. 2009. Proteomic analysis of plasma membranes of cyanobacterium *Synechocystis* sp. strain PCC 6803 in response to high pH stress. *J. Proteome Res.* **8**:2892–2902.

Zhao Y, Jensen ON. 2009. Modification-specific proteomics: Strategies for characterization of post-translational modifications using enrichment techniques. *Proteomics* **9**:4632–4641.

Chapter 67

Microbial Community Proteomics

Paul Wilmes

67.1 INTRODUCTION

Within the biosphere, microorganisms represent the predominant mode of life. Microbial proteins thus constitute the primary catalytic entities that underlie the major biogeochemical cycles on our planet. Recent technological developments are facilitating the large-scale extraction, separation, and identification of proteins from various natural microbial communities. Most excitingly, microbial community proteomics is facilitating the high-resolution representation of community structure and function, a long-sought goal of microbial ecology.

Community genomic sequencing projects, which analyze genomic DNA directly from environmental samples, are providing opportunities to genetically characterize microbial ecosystems [Wilmes et al., 2009a]. These studies vastly expand our knowledge of the genetic diversity and the physiological potential within a rapidly increasing set of selected environments that include activated sludge [García-Martín et al., 2006], acid mine biofilms [Tyson et al., 2004], seawater [Venter et al., 2004], human guts [Gill et al., 2006], termite hindguts [Warnecke et al., 2007]; see also Vol. II. The exponentially growing DNA sequence data (genomic and metagenomic) forms the foundation for post-genomic approaches that provide functional insight into microbial ecosystems. Consequently, the field of microbial ecology is currently entering the era of Eco-Systems Biology with community proteomics occupying a central role.

The study of protein expression within environmental samples is not an entirely new concept [Ogunseitan, 1993; Ehlers and Cloete, 1999]. However, the present chapter discusses recent community proteomic studies (from 2004 onwards), which have developed on the back of vast accumulation of DNA sequences from a range of microbial habitats, improved protein separation techniques, and high-throughput protein identification by mass spectrometry.

67.2 COMMUNITY PROTEOMICS FOR NOVICES

The workflow for a community proteomic experiment is analogous to that utilized for a proteomic study of an axenic culture apart from the sample preparation step which needs to be more rigorous to ensure purity and representativeness. The procedural steps are briefly summarized as (i) sample preparation, (ii) protein extraction, (iii) separation of the proteins or peptides (usually in two dimensions), and (iv) mass spectrometry (MS) analysis followed by in silico spectral matching for identification of proteins (see **Fig. 67.1** for a detailed workflow).

The separations of proteins or peptides are typically performed either by classical single- or two-dimensional polyacrylamide gel electrophoresis (2DE), or by liquid chromatography (LC), respectively. 2DE provides a tangible and comprehensive visual representation of the protein complement after staining. The comparison of protein spot densities is convenient for detection of differential protein expression in response to changes in environmental conditions or community structure and function [Wilmes et al., 2008a; see also chapters 66, 68, 69 and 70, Vol I]. Chosen spots are excised and digested with a protease (typically trypsin is used) prior to MS analysis. A major limitation associated with the classical 2DE approach for community proteomics is the co-migration of distinct proteins into discrete spots hampering accurate quantification and identification. Large and hydrophobic proteins

Handbook of Molecular Microbial Ecology, Volume I: Metagenomics and Complementary Approaches, First Edition. Edited by Frans J. de Bruijn.

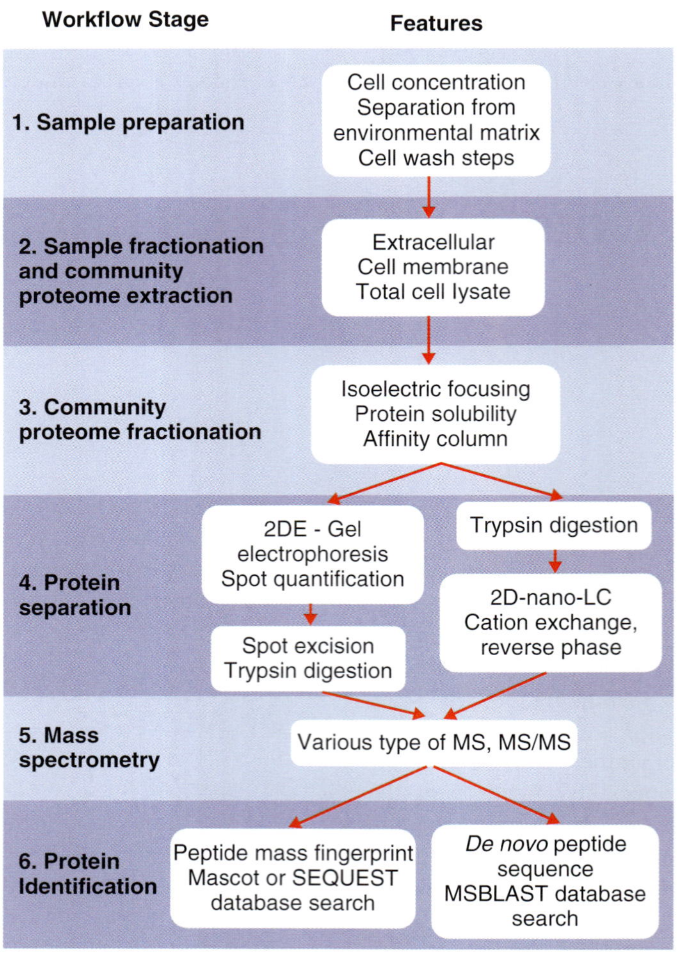

Figure 67.1 Community proteomics sample preparation, extraction, separation, and identification routes. The workflow for a community proteomic analysis may consist of six stages. Sample preparation may be required in stage 1. Cells may need to be concentrated or purified away from interfering substances—for example, humic acids in soil. Protein extraction is performed next (stage 2), and fractions of interest may be targeted—for example, extracellular, membrane, soluble and whole-cell fractions [Ram et al., 2005]. Cell lysis may involve French press lysis, sonication, chemical lysis, or bead beating [Wilmes and Bond, 2004]. The procedures in these stages must have minimal effect on the protein expression itself and sufficiently preserve the extracted proteins. To assist later separations, the extracted protein complement may be fractionated (stage 3)—for example, divided into fractions by preparative liquid isoelectric focusing or based on protein solubility. Protein or peptide separations may be performed by two-dimensional polyacrylamide gel electrophoresis (2DE) or by liquid chromatographic (LC) methods, respectively (stage 4). Following 2DE, gel images are analyzed and spots are quantified. Chosen spots are then excised and digested with a protease (trypsin) for mass spectrometry (MS) analysis [e.g., matrix-assisted laser desorption ionization time-of-flight (MALDI-ToF)]. For LC, the protein mixture is digested with trypsin prior to separation. Following LC, the separated peptides are directly introduced into the mass spectrometer (e.g., by electrospray ionization) and mass spectra acquired. MS analysis may also involve the fragmentation of the peptides and recording of the MS/MS spectra. If required, de novo protein sequence data can be determined from the MS/MS data. Algorithms such as Mascot [Perkins et al., 1999] and SEQUEST [Eng et al., 1994] enable the MS or MS/MS peptide mass fingerprint data to be searched against sequence databases for protein identification. Acronyms: 2DE, two-dimensional polyacrylamide gel electrophoresis; 2D-nano-LC, two-dimensional nano liquid chromatography. (Reproduced from Wilmes and Bond [2009] with permission from Elsevier.)

are also not separated well by 2DE (e.g., membrane proteins [Santoni et al., 2000]); the method suffers from limited automation potential, which, in turn, limits sample throughput. However, 2DE has good protein separation power for soluble proteins when combined with prefractionation procedures. Proteins of interest can be readily identified after proteolytic digestion, peptide mass fingerprints (PMF) generation by MS, and database searching. Although 2DE can be optimized substantially for community proteomics [Abram et al., 2009], LC-based approaches are rapidly gaining ground.

In the LC-based "shotgun" proteomic approach, complex protein samples are digested and the resulting peptides are separated, subjected to MS analysis, and PMFs are generated (Fig. 67.1). The approach circumvents many limitations of the 2DE approach and allows high-throughput analysis identifying some thousands of proteins within 1–2 days [VerBerkmoes et al., 2009a]. In particular, it facilitates the high-throughput analyses of insoluble membrane proteins [Wu et al., 2003]; hence, membrane fractions are typically generated in LC-based community proteomics [Ram et al., 2005]. The proteomic approach to separate and analyze at the peptide level, as opposed to protein separation, is referred to as "shotgun" or "bottom-up" proteomics. Different LC and MS systems are used. However, 2D nano-LC (strong cation exchange followed by reverse phase LC) coupled to an LTQ-Orbitrap mass spectrometer has proven well-suited for analysis of mixed microbial proteomes [Wilmes et al., 2008b; Denef et al., 2009; Verberkmoes et al., 2009b; Denef et al., 2010].

Parent proteins can be identified by sequence database matching of the PMFs using search algorithms such as Mascot [Perkins et al., 1999] and SEQUEST [Eng et al., 1994], and these are most easily achieved if representative genomic sequences are available. However, community proteomics is complicated by strain variation of protein species (**Fig. 67.2**) and precise identifications require high-mass accuracy peptide measurements and tandem MS (MS/MS) involving fragmentation of the peptides [Hunt et al., 1986].

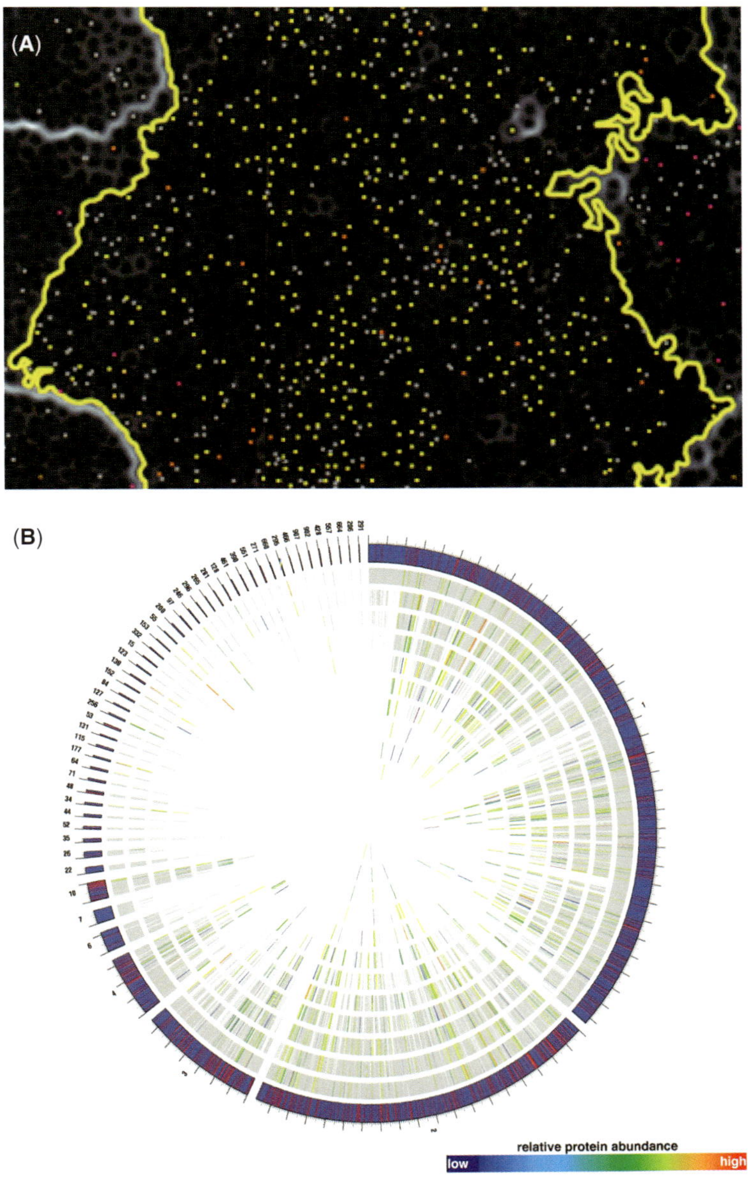

Figure 67.2 Strain-resolved proteomics in "*Candidatus Accumulibacter phosphatis*" (*A. phosphatis*)-dominated activated sludge. (**A**) Self-organizing map based on tetranucleotide frequency of genomic scaffold fragments with genomic fragments encoding identified proteins highlighted according to taxonomic groups (the *A. phosphatis* fragment cluster which includes numerous unclassified genomic fragments is delimited by a yellow line). Information in panel is color-coded according to concatenated metagenomic binning information [Martín et al., 2006, McHardy et al., 2007]: bright yellow, *A. phosphatis*; dark yellow, *Accumulibacter*; orange, Betaproteobacteria; medium purple, Gammaproteobacteria; bright green, Thiothrix; light blue, Actinobacteria; dark blue, Alphaproteobacteria; light brown, Sphingobacteria; dark pink, Deinococci; black, Deltaproteobacteria; dark green, Clostridia; medium pink, Planctomycetacia; dark brown, Spirochaetes; dark purple, Methanomicrobia; light purple, Chromadorea; light pink, Insecta; aquamarin, Mollicutes; grey, unclassified. (Reproduced from Wilmes et al. [2008b] with permission from Nature Publishing Group.) (**B**) Alignment of variant genomic fragments against the *A. phosphatis* composite genome (outer concentric ring in blue, USJ scaffold numbers indicated, and locations of inserted genes in the aligned genomic fragments highlighted in red). Relative *A. phosphatis* protein abundances highlighted according to each locus on the first inner concentric ring. Aligned variant genomic fragments with corresponding protein abundance in the following concentric rings. Gray color indicates no protein identification. The image was generated with Circos (M. Krzywinski, http://1477mkweb.bcgsc.ca/circos/). (Reproduced from Wilmes and Bond [2009] with permission from Elsevier.)

Importantly, the generation of characteristic tandem MS fragmentation patterns allows more precise spectral matching than from single MS alone (for a detailed description on strain-resolved proteomics using LC-MS/MS, see VerBerkmoes et al. [2009a]). Additionally, the fragmentation pattern can be employed to generate the de novo protein sequence and search for homologous sequences using MSBLAST [Shevchenko et al., 2001]. De novo peptide sequencing is especially useful for protein identification when corresponding sequence data are unavailable [Wilmes and Bond, 2004; Lacerda et al., 2007]. High-mass accuracy spectral matching can be used to positively identify proteins from closely related organisms, but the ability to do so decreases rapidly with amino acid sequence divergence [Denef et al., 2007]. Consequently, de novo peptide sequencing may prove indispensable in diverse samples in which the contribution of distinct microbial populations has to be assessed but for which full DNA sequence coverage is unavailable or unobtainable. Thus, with the advent of powerful de novo peptide sequencing algorithms, the implementation of the approach may become routine. For example, the de novo peptide sequencing approach was used to identify more than 100 proteins that were differentially expressed following exposure of bacterial communities to cadmium [Lacerda et al., 2007].

All community proteomic methods face challenges due to the large complexity of protein species and the large dynamic range of protein levels (see Chapters 66 and 70, Vol. I). This complexity increases for the LC approach as tryptic digestion produces some dozens of peptides per protein. A technique gaining popularity uses "off-gel" separation by isoelectric focusing (IEF) to divide the peptide mix into many fractions prior to LC-MS/MS [Ye et al., 2007]. This approach has additional benefit if the IEF is performed on whole proteins instead of peptides, and this approach has recently led to the identification of around 8000 proteins from a single microbial community sample [Wilmes et al., unpublished data]. This technique relies on the fact that the protein mixture is simplified in each fraction, and this increases the chance of protein identification by subsequent LC-MS/MS analysis.

67.3 THE STORY SO FAR

The application of community proteomics is expanding rapidly and recent studies directly examine community functional information in a phylogenetic context. As discussed, protein expression profiles based on gel separation provide overviews for the targeted identification of interesting proteins. Apart from activated sludge [Wilmes and Bond, 2004, 2006a; Wilmes et al., 2008a; Wexler et al., 2009], this approach has been applied to an estuary transect [Kan et al., 2005 discussed further in Chapter 68, Vol. I], contaminated soil and groundwater [Lacerda et al., 2007], *Riftia pachyptila* endosymbionts [Markert et al., 2007], infant faecal samples [Klaassens et al., 2007], freshwater samples following exposure to heavy metals [Maron et al., 2007a], lake water [Maron et al., 2007b], proteins associated with exopolysaccharides in activated sludge [Park and Helm, 2008; Park et al., 2008], sheep rumen [Toyoda et al., 2009], and an anaerobic wastewater digestor [Abram et al., 2009]. Other studies have employed LC or combinations of gel and LC separation. Proteins identified within dissolved organic matter (DOM) in soil and water have been used to determine the presence of broad taxonomic groups of microorganisms and highlight interesting functional differences between the microbial communities in forest soil (high abundance of peroxidases) and a peat bog (proteins involved in methanogenesis) [Schulze et al., 2005]. Similarly, proteins that form major constituents of DOM in seawater were identifiable due to the homology of identified peptides [Powell et al., 2005]. Due to a distinct lack of comprehensive genomic sequences in the majority of these studies, the numbers of proteins identified were rather limited. Albeit limited, all of the positive protein identifications did provide interesting new functional insights.

The availability of extensive genomic sequences from specific organisms or environments of interest greatly enhances both the number and quality of protein identifications. For example, proteins of pronounced biogeochemical importance (proteorhodopsins) were readily identifiable in environmental samples due to the availability of extensive gene sequences [Giovannoni et al., 2005]. A comprehensive community proteogenomic approach involving deep genomic and proteomic sampling was pioneered on acid mine biofilms [Ram et al., 2005; Lo et al., 2007; Denef et al., 2009, 2010] (see below for more information). This approach has since been applied to the enhanced biological phosphorus removal (EBPR) activated sludge system that is typically dominated by as yet uncultured organisms putatively named "*Candidatus* Accumulibacter phosphatis" [Wilmes et al., 2008b] (please see below for more information), to human fecal samples [Verberkmoes et al., 2009b], and to a *Geobacter*-dominated subsurface microbial community involved in uranium bioremediation [Wilkins et al., 2009]. An analogous shotgun proteomic approach has been applied to microbial communities inhabiting the euphotic zone of the Sargasso Sea and has led to the identification of several proteins linked to the dominant organisms SAR11, *Prochlorococcus*, and *Synechococcus* that are reflective of their lifestyle in this nutrient-limited environment [Sowell et al., 2009]. The power of combining genomics and proteomics on communities of immediate biotechnological interest (bioenergy) was highlighted by an integrative study of termite hindgut symbiotic bacteria involved in lignocellulose degradation [Warnecke et al., 2007; see also Chapter 22, Vol II]. Furthermore, an integrative high-throughput genomic and proteomic approach has recently been applied to natural bacterial phyllosphere communities and has highlighted specific proteomic traits essential for success and coexistence of two specific bacterial taxa (*Sphingomonas* and *Methylobacterium*) on plant leaves [Delmotte et al., 2009].

A full-cycle community proteomic investigation including comprehensive extraction, purification, separation, and identification following MS was first applied to a laboratory-scale activated sludge reactor operated in the United Kingdom [Wilmes and Bond, 2004]. In this study, comparisons of proteome profiles, generated by 2DE, were made to determine metabolic details of the EBPR wastewater treatment process. Protein expression was compared between the two operational stages (anaerobic and aerobic; **Fig. 67.3**) of EBPR. However, only minor differences in protein levels were detected using 2DE (Fig. 67.3a; Wilmes and Bond (2004); Wilmes and Bond (2006b)) which has recently been corroborated using a more sensitive ^{35}S met radiolabeling 2DE approach [Wexler et al., 2009] as well as a shotgun proteomics approach [Wilmes et al., 2008b] (Fig. 67.3b). Because of lack of differential protein expression, the initial community proteomic study focused on the identification of highly expressed proteins by de novo peptide sequencing [Wilmes and Bond, 2004]. When metagenomic sequences of EBPR sludges became available [García-Martín et al., 2006], the sequence data facilitated protein identification by analysis of PMF patterns and over 30% of highly expressed proteins chosen from 2DE gels were matched to the dominant organism—that is, *A. phosphatis* [Wilmes et al., 2008a]. Importantly, the majority of identified proteins is involved in major EBPR carbon transformations. In a further study, using the reference metagenomic sequences from the EBPR sludges cultured in the United States and Australia [García-Martín et al., 2006], the application of shotgun proteomics provided much deeper insight into

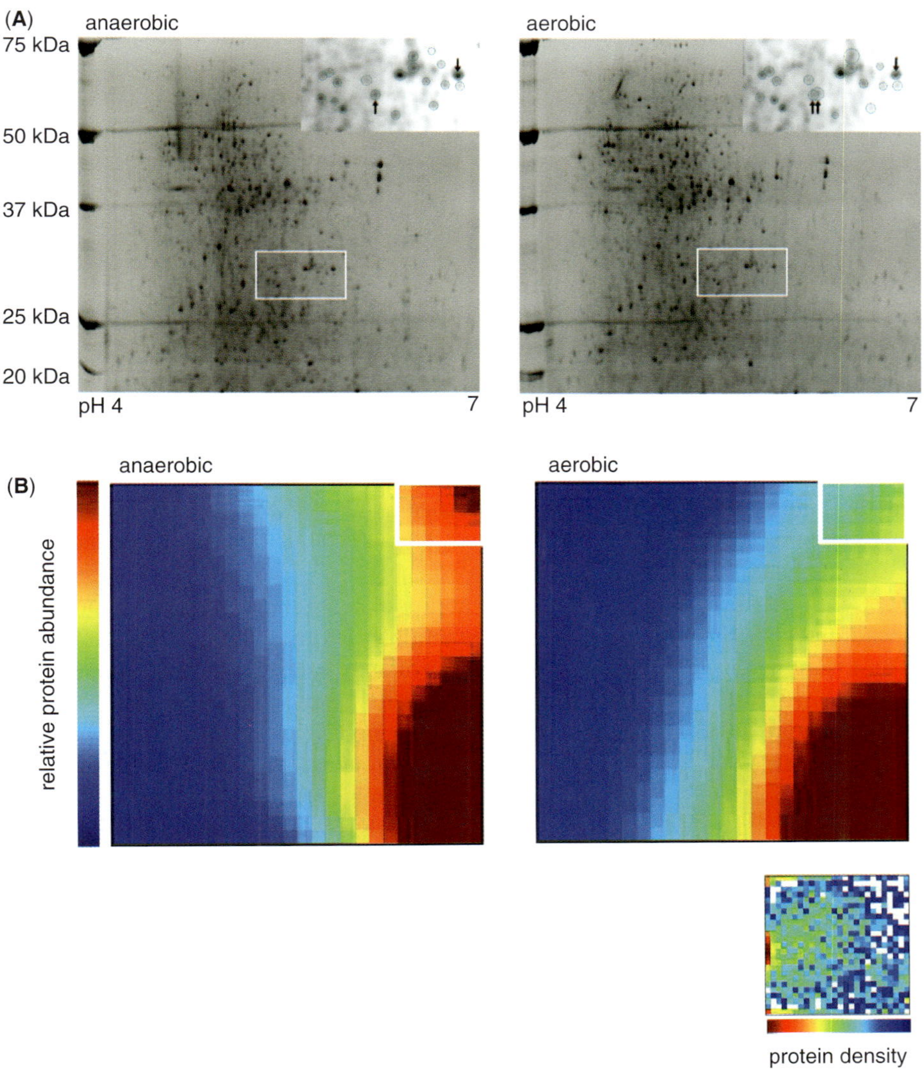

Figure 67.3 Time-resolved community proteomics highlighting subtle differences in protein abundance at the end of the anaerobic (120 min) and aerobic (330 min) phases within activated sludge performing enhanced biological phosphorus removal and dominated by *A. phosphatis*. (**A**) Two-dimensional polyacrylamide gel electrophoresis protein expression profiles. (**B**) Gene expression dynamics inspector [Eichler et al., 2003] profiles of LC-MS/MS normalized spectral abundance factor data. Interesting proteins that are slightly more abundant in the anaerobic phase are highlighted. Proteins in the highlighted cluster include a phosphate transport regulator (distant homologue of PhoU), an inorganic pyrophosphatase, an acyl-CoA dehydrogenases (fatty acid beta oxidation), and a nucleoside diphosphate kinase (governs the relative levels of GTP and ATP in cells). (Reproduced from Wilmes and Bond [2009] with permission from Elsevier.)

the metabolic transformations of *A. phosphatis* in the UK sludge (10% genomically encoded protein coverage [Wilmes et al., 2008b]). Interesting findings from placing identified proteins into metabolic context include the importance of denitrification, fatty acid cycling, and the glyoxylate shunt [Wilmes et al., 2008b]. This study also used strain-resolved proteomics to differentiate the expression of co-occurring protein variants within the EBPR sludge dominated by a *A. phosphatis* population [Wilmes et al., 2008b] (Fig. 67.2). This revealed that 59% of the most abundant protein variants derived from flanking *A. phosphatis* populations and not from the dominant *A. phosphatis* strain in the sequenced sludges. A significant subset of these was involved in core-metabolism and EBPR-specific pathways, suggesting an essential role for genetic diversity in maintaining the stable performance of microbial-mediated wastewater treatment. Importantly, Wexler et al. [2009] demonstrated that anaerobic versus aerobic tricarboxylic acid cycle protein expression differs according to *A. phosphatis* strain-level community composition. Consequently, this study highlights the importance of being able to functionally resolve by proteomics the extensive genetic heterogeneity that is a hallmark of natural microbial communities [Wilmes et al., 2009a].

The most extensive community proteomic analyses to date have been performed on acid mine biofilms that exhibit comparatively low diversity [Ram et al., 2005; Lo et al., 2007; Denef et al., 2009, 2010]. The Richmond Mine is characterized by low pH (~0.8) and microbially mediated iron oxidation that contributes to the acid mine drainage production. Here, an initial shotgun proteomic approach allowed the identification of over 2000 proteins [Ram et al., 2005]. High protein coverage (48%) was obtained for the dominant microorganism *(Leptospirillum* group II). One highly abundant protein, annotated as a hypothetical, was further investigated and found to be an iron-oxidizing cytochrome (Cyt_{579}), a key component of energy generation in the biofilms [Ram et al., 2005]. Thus, the proteomic results were instrumental in guiding the ensuing biochemical investigations [Jeans et al., 2008]. Lo et al. demonstrated that the high-resolution tandem mass spectrometry could differentiate between peptides originating from discrete populations within the mixed microbial community [Lo et al., 2007]. By assigning peptides to two different sequenced *Leptospirillum* group II populations, they inferred the genome architecture of a third unsequenced *Leptospirillum* group II population and demonstrated extensive interpopulation recombination [Lo et al., 2007]. This approach was expanded to conduct an extensive *Leptospirillum* group II proteomic genotype survey from 27 distinct biofilm samples [Denef et al., 2009]. The protein expression patterns suggest selection for particular recombinant types and revealed that recombination is a mechanism for fine-scale adaptation [Denef et al., 2009] and divergence [Denef et al., 2010] demonstrating the power of integrated genomics and proteomics to contribute extensively to our understanding of microbial ecology and evolution.

67.4 THE PATH AHEAD

Microbial community proteomics has already provided unprecedented functional insight into microbial communities with limited diversity and/or that are enriched in a particular organismal group. Detection limits for community proteomics suggest that each organism for which a protein is identified must be present at an abundance of at least a few percent of the total community. This represents a major limitation in systems that harbor extensive species richness—for example, 10^6 taxa in a gram of soil which may express 3×10^9 unique proteins (without considering strain-level variation; **Table 67.1**). Further complications include the unevenness of species distribution within samples and the fact that protein expression levels within a cell may differ by at least six orders of magnitude [Tyers and Mann, 2003]. Consequently, comprehensive proteomic analyses of mixed communities is challenging, and with current technology a typical analysis may only resolve $<<1\%$ of the protein complement within diverse samples [Wilmes and Bond, 2006b].

The range of detectable proteins will improve with future technical developments in proteomics especially advances in LC and MS. In addition, complex proteomes may be reduced in complexity prior to LC. Dividing protein complements into many fractions prior to LC, such as by IEF, holds promise to expand the dynamic range extensively [Ye et al., 2007]. Although DNA sequencing technology is currently advancing at an astonishing rate, the implementation of MS-based de novo peptide sequencing will diminish the requirement for comprehensive genomic (transcriptomic) foundations and will allow the identifications of proteins from low abundance community members for which no genomic sequences are available.

Due to the ability of being able to infer taxonomic and functional information from protein expression data, proteomics lends itself ideally to the monitoring of community structure and function over space and time (Fig. 67.3). For example, changes in protein abundance may be monitored between different natural conditions—for example, diurnal or anaerobic/aerobic cycles. Rapid changes within the environmental conditions may be manifested at the post-translational level [Mann and Jensen, 2003], and MS and bioinformatic methodologies need to be refined to resolve these. Another useful approach to detect rapid changes in protein expression is to observe incorporation

Table 67.1 Microbial Richness and Community Proteomic Expression. Modified and extended from [Wilmes and Bond, 2006b] with permission from Elsevier.

Ecosystem	Estimated Number of Unique Taxa[a]	Estimated Number of Unique Proteins[b]	Reference
Acid mine drainage biofilm	6	18,000	Tyson et al. [2004]
Activated sludge	17–268	51,000–800,000	Wagner and Loy [2002]
Ocean water	160	480,000	Curtis et al. [2002]
Sargasso Sea (combined sample, i.e., 1730 L)	1824–47,733	5,500,000–140,000,000	Venter et al. [2004]
Soil (1 g)	1,000,000	3,000,000,000	Gans et al. [2005]
Human			
Skin ($n = 10$)	9–67	27,000–201,000	Grice et al. [2009]
Oral cavity ($n = 3$)	500	1,500,000	Zaura et al. [2009]
Throat ($n = 6$)	204	612,000	Andersson et al. [2008]
Healthy stomach ($n = 3$)	375	1,125,000	Andersson et al. [2008]
Feces (gut; $n = 6$)	385	1,155,000	Andersson et al. [2008]

[a] As defined by author of referenced work.

[b] Number of unique proteins based on average environmental microbial genome size of 3 Mbp and 1 kbp of sequence coding for one gene. Uniqueness does not reflect strain-level variation.

of stable isotope- or radio-label into newly synthesized proteins [Wexler et al., 2009].

Although 2DE is labor-intensive and has limitations regarding separation of proteins, it remains convenient for expression quantification and comparative studies. For example, multiple samples differentiated by fluorescent tags (known as DIGE) can be run on the same gels [Lilley and Dupree, 2006], or metabolically active portions of communities can be detected by incorporation of a labeled substrate [Jehmlich et al., 2008].

Given the potential superiority of LC-MS/MS approaches, it is highly desirable to obtain quantitative information to detect systems level response to change. However, this is not so readily obtained from MS data, mainly due to the large variation in individual peptide chemistry. Furthermore, quantification may be confounded by the complexity of peptides such that only subsets of proteins may be identified from a sample [Bantscheff et al., 2007]. Nonetheless, emerging techniques for quantifying proteins from LC-MS/MS data include isotope-coded affinity tags (ICAT), metabolic labeling of proteins (using ^{13}C or ^{15}N), and isobaric tags for quantification (iTRAQ) [Ong and Mann, 2005]. These labeling techniques allow simultaneous analysis of multiple samples for comparison of the differentially tagged peptide abundances. Alternative methods to quantify MS data are commonly used, the so-called "label free" methods [Bantscheff et al., 2007]. A semiquantitative method that is commonly used in community proteomics, normalized spectral abundance factor (NSAF) calculation [Florens et al., 2006], relates the number of peptides identified from a protein, normalized to the protein size, to the overall protein abundance. An alternative approach uses the peptide MS peak signal intensity, which is collected during the chromatographic spread for each peptide [Ong and Mann, 2005]. The subsequent chromatogram peak area is then proportional to the peptide's abundance. These "label free" methods are reportedly not as accurate as labeling approaches for quantification, however, there is much interest to use these simple approaches and application will thus increase as statistical treatment continues to develop [Bantscheff et al., 2007; Choi et al., 2008].

So far, community proteomic studies have been carried out on bulk samples. However, microbial communities exhibit distinct organismal and functional organization [Wilmes et al., 2009b] and particular enzyme variants may be localized within distinct microniches [Wilmes et al., 2008b]. Hence, more fine-scale measurements will be necessary in the future to resolve the functional significance of protein localization within microbial communities. Mass spectrometry imaging techniques [Stoeckli et al., 2001] show great promise for resolving such fine-scale expression differences within microbial communities.

67.5 CONCLUSION

Community proteomics is providing unprecedented insight into genotypic and phenotypic traits within microbial consortia. In addition to other systems-level data that include genomics (discussed in Chapters 14–31), transcriptomics (discussed in Chapters 62–65,

Vol. I), and metabolomics (discussed in Chapters 71–74, Vol. I), proteomics is providing high-resolution molecular data that are allowing us to glean a more complete picture of microbial community composition, function, interaction, ecology, and evolution. Although numerous studies reviewed above have solely applied proteomics to specific mixed microbial communities of interest, the integration of different "omic" datasets from the same sample will become ever more important in the future to deconvolute and make sense of the complex metabolic processes and interactions that occur within complex microbial communities [Wilmes et al., 2010]. Such fundamental knowledge is essential for our understanding of the Earth's biogeochemical cycles, biotechnologies that rely on microbial communities as well as human health.

Acknowledgments

PW is supported by a Luxembourg National Research Fund ATTRACT grant (FNR/A09/03).

REFERENCES

ABRAM F, GUNNIGLE E, O'FLAHERTY V. 2009. Optimisation of protein extraction and 2-DE for metaproteomics of microbial communities from anaerobic wastewater treatment biofilms. *Electrophoresis* **30**:4149–4151.

ANDERSSON AF, LINDBERG M, JAKOBSSON H, BÄCKHED F, NYRÉN P, ENGSTRAND L. 2008. Comparative analysis of human gut microbiota by barcoded pyrosequencing. *PLoS ONE* **3**: e2836.

BANTSCHEFF M, SCHIRLE M, SWEETMAN G, RICK J, KUSTER B. 2007. Quantitative mass spectrometry in proteomics: A critical review. *Anal. Bioanal. Chem.* **389**:1017–1031.

CHOI H, FERMIN D, NESVIZHSKII AI. 2008. Significance analysis of spectral count data in label-free shotgun proteomics. *Mol. Cell Proteom.* **7**:2373–2385.

CURTIS TP, SLOAN WT, SCANNELL JW. 2002. Estimating prokaryotic diversity and its limits. *Proc. Natl. Acad. Sci. USA* **99**:10494–10499.

DELMOTTE N, KNIEF C, CHAFFRON S, INNEREBNER G, ROSCHITZKI B, SCHLAPBACH R, VON MERING C, VORHOLT JA. 2009. Community proteogenomics reveals insights into the physiology of phyllosphere bacteria. *Proc. Natl. Acad. Sci. USA* **106**:16428–16433.

DENEF VJ, SHAH MB, VERBERKMOES NC, HETTICH RL, BANFIELD JF. 2007. Implications of strain- and species-level sequence divergence for community and isolate shotgun proteomic analysis. *J. Proteome Res.* **6**:3152–3161.

DENEF VJ, VERBERKMOES NC, SHAH MB, ABRAHAM P, LEFSRUD M, HETTICH RL, BANFIELD JF. 2009. Proteomics-inferred genome typing (PIGT) demonstrates inter-population recombination as a strategy for environmental adaptation. *Environ. Microbiol.* **11**:313–325.

DENEF VJ, KALNEJAIS LH, MUELLER RS, WILMES P, BAKER BJ, THOMAS BC, VERBERKMOES NC, HETTICH RL, BANFIELD JF. 2010. Proteogenomic basis for ecological divergence of closely related bacteria in natural acidophilic microbial communities. *Proc. Natl. Acad. Sci. USA* **107**:2383–2390.

EHLERS M, CLOETE T. 1999. Protein profiles of phosphorus- and nitrate-removing activated sludge systems. *Water SA* **25**:351–356.

EICHLER GS, HUANG S, INGBER DE. 2003. Gene expression dynamics inspector (GEDI): For integrative analysis of expression profiles. *Bioinformatics* **19**:2321–2322.

ENG JK, MCCORMACK AL, YATES I, JOHN R. 1994. An approach to correlate tandem mass spectral data of peptides with amino acid sequences in a protein database. *J. Am. Soc. Mass Spectrom.* **5**:976–989.

FLORENS L, CAROZZA MJ, SWANSON SK, FOURNIER M, COLEMAN MK, WORKMAN JL, WASHBURN MP. 2006. Analyzing chromatin remodeling complexes using shotgun proteomics and normalized spectral abundance factors. *Methods* **40**:303–311.

GANS J, WOLINSKY M, DUNBAR J. 2005. Computational improvements reveal great bacterial diversity and high metal toxicity in soil. *Science* **309**:1387–1390.

GARCÍA-MARTÍN H, IVANOVA N, KUNIN V, WARNECKE F, BARRY KW, et al. 2006. Metagenomic analysis of two enhanced biological phosphorus removal (EBPR) sludge communities. *Nat. Biotechnol.* **24**:1263–1269.

GILL SR, POP M, DEBOY RT, ECKBURG PB, TURNBAUGH PJ, et al. 2006. Metagenomic analysis of the human distal gut microbiome. *Science* **312**:1355–1359.

GIOVANNONI SJ, BIBBS L, CHO J-C, STAPELS MD, DESIDERIO R, et al. 2005. Proteorhodopsin in the ubiquitous marine bacterium SAR11. *Nature* **438**:82–85.

GRICE EA, KONG HH, CONLAN S, DEMING CB, DAVIS J, et al. 2009. Topographical and temporal diversity of the human skin microbiome. *Science* **324**:1190–1192.

HUNT DF, YATES JR, SHABANOWITZ J, WINSTON S, HAUER CR. 1986. Protein sequencing by tandem mass spectrometry. *Proc. Natl. Acad. Sci. USA* **83**:6233–6237.

JEANS C, SINGER SW, CHAN CS, VERBERKMOES NC, SHAH M, HETTICH RL, BANFIELD JF, THELEN MP. 2008. Cytochrome 572 is a conspicuous membrane protein with iron oxidation activity purified directly from a natural acidophilic microbial community. *ISME J.* **2**:542–550.

JEHMLICH N, SCHMIDT F, VON BERGEN M, RICHNOW H-H, VOGT C. 2008. Protein-based stable isotope probing (Protein-SIP) reveals active species within anoxic mixed cultures. *ISME J.* **2**:1122–1133.

KAN J, HANSON TE, GINTER JM, WANG K, CHEN F. 2005. Metaproteomic analysis of Chesapeake Bay microbial communities. *Saline Systems* **1**: 7.

KLAASSENS ES, DE VOS WM, VAUGHAN EE. 2007. Metaproteomics approach to study the functionality of the microbiota in the human infant gastrointestinal tract. *Appl. Environ. Microbiol.* **73**:1388–1392.

LACERDA CMR, CHOE LH, REARDON KF. 2007. Metaproteomic analysis of a bacterial community response to cadmium exposure. *J. Proteome Res.* **6**:1145–1152.

LILLEY KS, DUPREE P. 2006. Methods of quantitative proteomics and their application to plant organelle characterization. *J. Exp. Bot.* **57**:1493–1499.

LO I, DENEF VJ, VERBERKMOES NC, SHAH MB, GOLTSMAN D, et al.. 2007. Strain-resolved community proteomics reveals recombining genomes of acidophilic bacteria. *Nature* **446**:537–541.

MANN M, JENSEN ON. 2003. Proteomic analysis of post-translational modifications. *Nat. Biotechnol.* **21**:255–261.

MARKERT S, ARNDT C, FELBECK H, BECHER D, SIEVERT SM, et al. 2007. Physiological proteomics of the uncultured endosymbiont of *Riftia pachyptila*. *Science* **315**:247–250.

MARON P-A, MOUGEL C, SIBLOT S, ABBAS H, LEMANCEAU P, RANJARD L. 2007a. Protein Extraction and Fingerprinting Optimization of Bacterial Communities in Natural Environment. *Microb. Ecol.* **53**:426–434.

MARON P-A, RANJARD L, MOUGEL C, LEMANCEAU P. 2007b. Metaproteomics: A new approach for studying functional microbial ecology. *Microb. Ecol.* **53**:486–493.

García-Martín H, Ivanova N, Kunin V, Warnecke F, Barry KW, McHardy AC, Yeates C, He S, Salamov AA, Szeto E, Dalin E, Putnam NH, Rigoutsos I, Kyrpides NC, Blackall LL, McMahon KD, Hugenholtz P. 2006. Metagenomic analysis of two enhanced biological phosphorus removal (EBPR) sludge communities. *Nature Biotechnology*. **24**:1263–1269.

McHardy AC, García-Martín HG, Tsirigos A, Hugenholtz P, Rigoutsos I. 2007. Accurate phylogenetic classification of variable-length DNA fragments. *Nat. Methods* **4**:63–72.

Ogunseitan OA. 1993. Direct extraction of proteins from environmental samples. *J. Microbiol. Methods* **17**:273–281.

Ong S-E, Mann M. 2005. Mass spectrometry-based proteomics turns quantitative. *Nat. Chem. Biol*. **1**:252–262.

Park C, Helm R. 2008. Application of metaproteomic analysis for studying extracellular polymeric substances (EPS) in activated sludge flocs and their fate in sludge digestion. *Water Sci. Technol*. **57**:2009–2015.

Park C, Novak JT, Helm RF, Ahn Y-O, Esen A. 2008. Evaluation of the extracellular proteins in full-scale activated sludges. *Water Res*. **42**:3879–3889.

Perkins DN, Pappin DJC, Creasy DM, Cottrell JS. 1999. Probability-based protein identification by searching sequence databases using mass spectrometry data. *Electrophoresis* **20**: 3551–3567.

Powell MJ, Sutton JN, Del Castillo CE, Timperman AT. 2005. Marine proteomics: Generation of sequence tags for dissolved proteins in seawater using tandem mass spectrometry. *Mar. Chem*. **95**:183–198.

Ram RJ, VerBerkmoes NC, Thelen MP, Tyson GW, Baker BJ, Blake RC, II, Shah M, Hettich RL, Banfield JF. 2005. Community proteomics of a natural microbial biofilm. *Science* **308**:1915–1920.

Santoni V, Molloy M, Rabilloud T. 2000. Membrane proteins and proteomics: Un amour impossible? *Electrophoresis* **21**:1054–1070.

Schulze W, Gleixner G, Kaiser K, Guggenberger G, Mann M, Schulze ED. 2005. A proteomic fingerprint of dissolved organic carbon and soil particles. *Oecologia* **142**:335–343.

Shevchenko A, Sunyaev S, Loboda A, Shevchenko A, Bork P, Ens W, Standing KG. 2001. Charting the proteomes of organisms with unsequenced genomes by MALDI-quadrupole time-of-flight mass spectrometry and BLAST homology searching. *Anal. Chem*. **73**:1917–1926.

Sowell SM, Wilhelm LJ, Norbeck AD, Lipton MS, Nicora CD, et al. 2009. Transport functions dominate the SAR11 metaproteome at low-nutrient extremes in the Sargasso Sea. *ISME J*. **3**:93–105.

Stoeckli M, Chaurand P, Hallahan DE, Caprioli RM. 2001. Imaging mass spectrometry: A new technology for the analysis of protein expression in mammalian tissues. *Nat. Med*. **7**:493–496.

Toyoda A, Iio W, Mitsumori M, Minato H. 2009. Isolation and identification of cellulose-binding proteins from sheep rumen contents. *Appl. Environ. Microbiol*. **75**:1667–1673.

Tyers M, Mann M. 2003. From genomics to proteomics. *Nat. Biotechnol*. **422**:193–197.

Tyson GW, Chapman J, Hugenholtz P, Allen EE, Ram RJ, et al. 2004. Community structure and metabolism through reconstruction of microbial genomes from the environment. *Nature* **428**:37–43.

Ultsch A, Moerchen F. 2005. ESOM-Maps: Tools for clustering, visualization, and classification with emergent SOM. In Technical Report of the Department of Mathematics and Computer Science, University of Marburg, Germany, Vol. 46, pp. 1–7.

Venter JC, Remington K, Heidelberg JF, Halpern AL, Rusch D, et al. 2004. Environmental genome shotgun sequencing of the Sargasso Sea. *Science* **304**:66–74.

VerBerkmoes NC, Denef VJ, Hettich RL, Banfield JF. 2009a. Systems biology: Functional analysis of natural microbial consortia using community proteomics. *Nat. Rev. Microbiol*. **7**:196–205.

Verberkmoes NC, Russell AL, Shah M, Godzik A, Rosenquist M, et al. 2009b. Shotgun metaproteomics of the human distal gut microbiota. *ISME J*. **3**:179–189.

Wagner M, Loy A. 2002. Bacterial community composition and function in sewage treatment systems. *Curr. Opin. Biotechnol*. **13**:218–227.

Warnecke F, Luginbuhl P, Ivanova N, Ghassemian M, Richardson TH, et al. 2007. Metagenomic and functional analysis of hindgut microbiota of a wood-feeding higher termite. *Nature* **450**: 560–565.

Wexler M, Richardson DJ, Bond PL. 2009. Radiolabelled proteomics to determine differential functioning of *Accumulibacter* during the anaerobic and aerobic phases of a bioreactor operating for enhanced biological phosphorus removal. *Environ. Microbiol*. **11**: 3029–3044.

Wilkins MJ, Verberkmoes NC, Williams KH, Callister SJ, Mouser PJ, et al. 2009. Proteogenomic monitoring of *Geobacter* physiology during stimulated uranium bioremediation. *Appl. Environ. Microbiol*. **75**:6591–6599.

Wilmes P, Bond PL. 2004. The application of two-dimensional polyacrylamide gel electrophoresis and downstream analyses to a mixed community of prokaryotic microorganisms. *Environ. Microbiol*. **6**:911–920.

Wilmes P, Bond PL. 2006a. Towards exposure of elusive metabolic mixed-culture processes: The application of metaproteomic analyses to activated sludge. *Water Sci. Technol*. **54**:217–226.

Wilmes P, Bond PL. 2006b. Metaproteomics: Studying functional gene expression in microbial ecosystems. *Trends Microbiol*. **14**:92–97.

Wilmes P, Bond PL. 2009. Microbial community proteomics: Elucidating the catalysts and metabolic mechanisms that drive the Earth's biogeochemical cycles. *Curr. Opin. Microbiol*. **12**:310–317.

Wilmes P, Wexler M, Bond PL. 2008a. Metaproteomics provides functional insight into activated sludge wastewater treatment. *PLoS ONE* **3**: e1778.

Wilmes P, Andersson AF, Lefsrud MG, Wexler M, Shah M, et al. 2008b. Community proteogenomics highlights microbial strain-variant protein expression within activated sludge performing enhanced biological phosphorus removal. *ISME J*. **2**: 853–864.

Wilmes P, Simmons SL, Denef VJ, Banfield JF. 2009a. The dynamic genetic repertoire of microbial communities. *FEMS Microbiol. Rev*. **33**:109–132.

Wilmes P, Remis JP, Hwang M, Auer M, Thelen MP, Banfield JF. 2009b. Natural acidophilic biofilm communities reflect distinct organismal and functional organization. *ISME J*. **3**:266–270.

Wu CC, MacCoss MJ, Howell KE, Yates JR. 2003. A method for the comprehensive proteomic analysis of membrane proteins. *Nat. Biotechnol*. **21**:532–538.

Ye M, Jiang X, Feng S, Tian R, Zou H. 2007. Advances in chromatographic techniques and methods in shotgun proteome analysis. *Trends Anal. Chem*. **26**:80–84.

Zaura E, Keijser B, Huse S, Crielaard W. 2009. Defining the healthy "core microbiome" of oral microbial communities. *BMC Microbiol*. **9**: 259.

Chapter 68

Synchronicity between Population Structure and Proteome Profiles: A Metaproteomic Analysis of Chesapeake Bay Bacterial Communities

JINJUN KAN, THOMAS E. HANSON, AND FENG CHEN

68.1 INTRODUCTION

Knowledge of environmental microbial complexity has been greatly advanced and improved by a progression of molecular approaches, including gene diversity, genomics, community-based genomics, transcriptomics, and proteomics. Proteomic techniques provide information on in situ protein abundances, thereby providing insight into enzymatic activities, stress responses, and cell structure remodeling that are integrated into cellular metabolism and ultimately functions of the population within an ecosystem (reviewed by Keller and Hettich [2009] and Wilmes and Bond [2009]). Proteomic techniques, namely 2D SDS-PAGE coupled with MS-based protein identification and/or shotgun LC-MS methods, originally used to analyze the responses of pure cultures in laboratory settings, are now being applied to mixed communities in natural environments—that is, metaproteomics. The environments and communities interrogated include activated sludge [Wilmes and Bond, 2004, 2006; Wilmes et al., 2008; Park and Helm, 2008], soil and groundwater [Schulze et al., 2005; Benndorf et al., 2007], freshwater [Maron et al., 2007; Pierre-Alain et al., 2007], estuarine water [Kan et al., 2005], Sargasso Sea [Sowell et al., 2009], acid mine drainage [Ram et al., 2005], bacterial endosymbionts of tubeworm [Markeert et al., 2007], and human intestinal bacterial consortia [Klaassens et al., 2007; see also Chapters 66–77, Vol. I]. These pioneering studies have demonstrated that the proteome profiles of environmental consortia correspond to microbial community structure and prevailing conditions of their local environments. For example, Wilmes and coworkers [Wilmes and Bond, 2004, 2006; Wilmes et al., 2008; see also Chapter 67, Vol. I] found that the proteome of active sludges employed in enhanced biological phosphorous removal reactors were dominated by proteins likely originating from a *Rhodocyclus*-type polyphosphate-accumulating organism that had been detected in the same samples by 16S rRNA studies. In addition, Lacerda et al. [2007] demonstrated that cadmium-shock proteins including ATPases, oxidoreductases, and transport proteins were found immediately after the exposure to cadmium in a continuous flow wastewater treatment reactor. Finally, the abundance of periplasmic substrate-binding proteins from SAR11, *Prochlorococcus*, and *Synechococcus* in Sargasso Sea metaproteome samples indicated community adaptations to the well-documented oligotrophic nature of this marine environment [Sowell et al., 2009]. Thus, metaproteomics as an approach appears to produce useful information about both the population composition and

Handbook of Molecular Microbial Ecology, Volume I: Metagenomics and Complementary Approaches, First Edition. Edited by Frans J. de Bruijn.

physiological responses of mixed bacterial communities to variable environmental conditions [Wilmes and Bond, 2009; see also Chapter 67, Vol. I].

The Chesapeake Bay is the largest estuarine ecosystem in the United States and displays strong environmental gradients including hydrographic regime, physical properties, and water chemistry. Our previous studies have shown that Chesapeake Bay water harbors both freshwater and marine microorganisms but also contains unique estuarine microbial phylotypes [Kan et al., 2008] and apparently endemic viral communities [Bench et al., 2007]. Previous DGGE-based studies (see Chapter 5, Vol. I) indicated that the planktonic communities exhibited stronger seasonal than spatial variations, and population structures in late spring and summer were distinct from those observed in the winter by both cultivation and molecular approaches [Kan et al., 2006b, 2007]. Furthermore, multiple-year monitoring of DGGE fingerprints revealed that repeatable and predictable patterns of bacterial community occurred in the Chesapeake Bay [Kan et al., 2006a].

To explore the potential functions and metabolic capability of these naturally occurring microbes, we developed and applied two-dimensional gel electrophoresis (2DGE)-based proteomics to study the gene expression of Chesapeake Bay microbial communities [Kan et al., 2005]. The microbial community proteomes from three stations along the salinity gradient showed a trend similar to the spatial pattern of microbial 16S ribosomal RNA fingerprints [Kan et al., 2005]. In this study, PCR-DGGE (denaturing gradient gel electrophoresis) and 2D SDS-PAGE fingerprints were used in parallel to compare the temporal variations of microbial composition and proteome profiles in the Chesapeake Bay in parallel with statistical techniques. It represents the first study of proteome dynamics for microbial communities, and it highlights the potential to link the population structure to gene expression of naturally occurring microbial complexes.

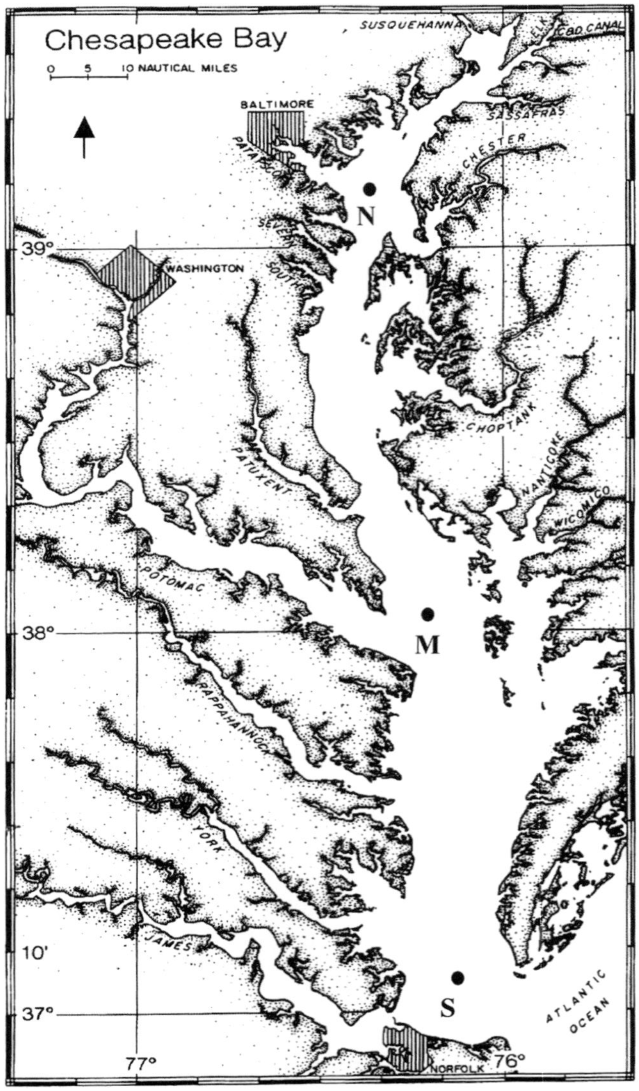

Figure 68.1 Chesapeake Bay map showing sampling stations. N, M, and S represent the northern, middle, and southern Bay, respectively.

68.2 MATERIALS AND METHODS

68.2.1 Microbial Community Collection

Planktonic microbial communities were collected following the optimized protocol as previously described [Kan et al., 2005]. Briefly, 20 L of surface water were prefiltered through 3-μm polycarbonate filters (142-mm diameter, Millipore, Bedford, MA) and concentrated with a tangential-flow ultrafiltration system (30,000 MW cutoff) at three sampling stations (N, northern; M, middle; S, southern) along the middle axis of the Chesapeake Bay (**Fig. 68.1**) in June 2004, August 2004, and February 2005. Microbial cells were pelleted using a GS-15R centrifuge (Beckman, Fullerton, CA) at 13,000g, 4°C for 10 min and stored frozen (−80°C) until further processing. Water temperature, salinity, and dissolved oxygen were recorded on board. Chl a, ammonia, nitrite and nitrate, phosphate, bacterial, and viral abundances were determined as previously described [Kan et al., 2006a].

68.2.2 DNA Extraction and PCR-DGGE Fingerprints

Bacterial genomic DNA extraction and PCR-DGGE fingerprints were performed as previously described [Kan et al., 2006b]. Bacterial population dynamics can be interpreted by banding patterns and by absence or presence of the representative bands.

68.2.3 Protein Extraction and Quantification, 2DGE and Image Analysis, and Protein Identification

Protein quantification, 2D SDS-PAGE, and downstream protein identification by mass spectrometry followed as previously described [Kan et al., 2005] except that (1) pH 5–8 immobilized pH gradient (IPG) strips were used in isoelectric focusing and (2) gel images were analyzed and quantitatively compared with ImageMaster 2D Platinum 6.0 (Amersham, GE healthcare) following the manufacturer's instructions. Briefly, the default ImageMaster algorithm was used to detect and annotate protein spots with collection of spot intensity, area, and volume data. All the gel images to be compared were grouped to a match set, and a master gel was used as a reference for matching and comparing for each season/site sample. Gel-to-gel comparisons were conducted by automatic matching with manual corrections. Downstream analyses were restricted to the spots with sufficient quality and intensity to permit subsequent attempts at protein identification via MALDI-TOF and LC-MS/MS [Kan et al., 2005].

68.2.4 MDS Analysis

Both 16S rRNA-based population structures (PCR-DGGE) and 2D SDS-PAGE proteomic profiles were analyzed by MDS (nonmetric multidimensional scaling) in SAS System software (SAS/STAT, 1992) [Kan et al., 2006a]. Distance matrices were calculated from intergel DGGE or 2D SDS-PAGE datasets by band/spot presence and absence. Two-dimensional MDS plots were then inspected for differences among the samples, and the stress values were extracted.

68.3 RESULTS AND DISCUSSION

68.3.1 Dynamics of Estuarine Environments and Bacterial Population Structure

The major stem of Chesapeake Bay is about 300 km long with a total tidal water area of 11,000 km^2 (www.gmu.edu/bios/bay). About half the volume of Chesapeake Bay water originates from more than 50 major tributaries, with the remainder supplied from the Atlantic Ocean. Salinity varied from 6.3 to 24.1 ppt from N-S in the bay (**Table 68.1**). Water temperature, dissolved oxygen, and Chl *a* varied significantly between late spring and summer (June/August; warm season) samples compared to fall and winter, while nutrients (i.e., ammonia, nitrite and nitrate, and phosphate) were not significantly different between seasons (Table 68.1). The total viral and bacterial abundance in the warm season were higher than those in the cold season, suggesting that microbial biomass and activities in these samples were primarily influenced by the physical variables and primary productivity, but not by mineral nutrients [Shiah and Ducklow, 1994; Schultz and Ducklow, 2000; Smith and Kemp, 2003].

Despite the geographical distance between the three sampling stations, DGGE fingerprints for the northern (N), middle (M) and southern (S) Bay showed similar microbial compositions for each sampling season (**Fig. 68.2**). Absence and presence of the bands indicated the distribution of bacterial groups in spatial and temporal manners. Bacterial groups in the summer (June and August 2004) were distinct from the winter (February 2005) (Fig. 68.2, band 8–10). Detailed annual population dynamics of Chesapeake Bay planktonic bacteria have been described previously [Kan et al., 2006b, 2007, 2008]. We demonstrated that the Chesapeake

Table 68.1 Environmental Parameters for Water Samples Taken from Northern (N), Middle (M), and Southern (S) Bay

	June 2004			August 2004			February 2005			*P* Value of *t*-test
Environmental Parameters	N	M	S	N	M	S	N	M	S	(Warm vs. Cold)
Water temperature (°C)	19.6	22.7	20.9	26.6	26.6	26.7	1.2	1.0	3.8	<0.0001[a]
Salinity (ppt)	6.3	10.1	19.8	7.4	13.7	24.1	9.6	10.5	15.4	0.706
Dissolved oxygen (mg/L)	7.1	3.9	3.5	7.2	6.2	6.5	10.3	11.8	7.9	0.01[a]
Chl a (μg/L)	10.8	15.8	10.5	13.8	6.3	8.0	5.3	5.0	4.8	0.028[a]
Ammonia (μM)	5.6	0.6	1.1	10.4	0.6	1.3	0.5	0.6	3.6	0.513
Nitrite and Nitrate (μM)	29.3	11.7	1.9	9.5	4.7	0.7	15.9	15.6	12.3	0.459
Phosphate (μM)	0.4	0.4	0.4	0.6	0.3	0.7	0.2	0.2	1.3	0.708
Bacterial abundance (10^6 cells/ml)	1.4	4.8	5.6	5.8	3.2	3.7	1.0	0.7	0.9	0.01[a]
Viral abundance (10^6 cells/ml)	10.8	13.5	10.5	25.2	39.0	30.8	3.1	2.6	1.0	0.029[a]

[a] Statistically significant at P value < 0.05.

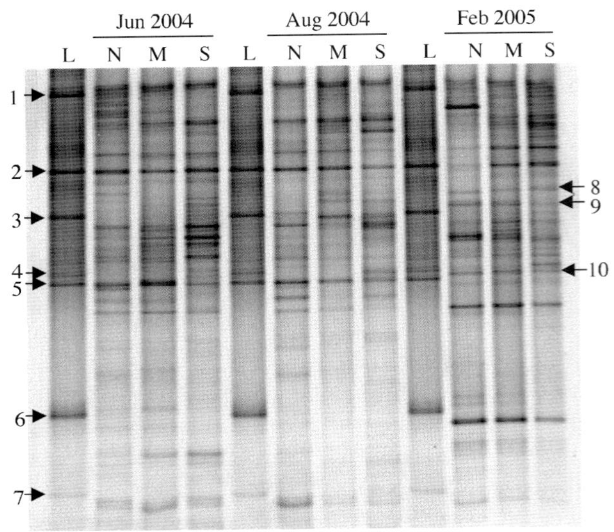

Figure 68.2 DGGE fingerprints of bacterial communities from the northern (N), middle (M), and southern (S) Chesapeake Bay stations in June 2004, August 2004, and February 2005. L, DGGE ladder with known bacterial phylotypes (1–7); 1, plastids; 2, Alphaproteobacterium (SAR11); 3, Roseobacter; 4, Alphaproteobacterium; 5, *Synechococcus*; 6, *Cytophaga* I; 7, *Cytophaga* II. Bands 8–10 represents different roseobacters present in cold season but absent in warm season [Kan et al., 2006b].

Bay microbial populations recur annually and can be predicted based on the environmental conditions [Kan et al., 2006a]. Similar temporal patterns have also been observed in Sargasso Sea [Morris et al., 2005], California Coastal Ocean [Fuhrman et al., 2006], and two independent rivers [Crump and Hobbie, 2005]. The spatial and seasonal variations of bacterial populations have important implications for understanding of the functional processes in aquatic ecosystems. Distinct bacterial groups dominated in different seasons suggested that these groups were likely involved in the microbial processes actively at certain times. An alternative explanation for this recurring pattern could be that summer and winter phylotypes are present at all times and that DGGE is only revealing the most abundant members of the community at a given time. An additional limitation to the extrapolation of ecosystem function from phylotype information is the lack of tight linkage between phylotype and physiology. While this is clear in some instances—that is, cyanobacterial phylotypes will all be oxygenic phototrophs—it is far less clear in the nitrogen cycle, for example, where nitrate reduction, nitrogen fixation, and ammonia oxidation are all widely phylogenetically disseminated.

68.3.2 Metaproteome Analysis

2D SDS-PAGE with first dimension of pH 5–8 revealed collections of proteins with apparent molecular masses between 10 and 80 kDa. More than 500 protein spots were detected for each gel in image analysis (**Fig. 68.3A–C**). Comparison of the proteomes from northern (N), middle (M), and southern (S) at three sampling seasons revealed both spatial and temporal variations. Qualitatively, the proteomes of samples collected at same time were more similar to one another than to samples collected in different months (Fig. 68.3A–C). Only a limited number of protein spots were maintained at the same site across seasons. Quantitatively, the similarity matrix also indicated that the samples collected at same time but from different sites shared higher level of similarity (highlighted in **Table 68.2**). Thus, similar to DGGE

Table 68.2 Similarity Matrix of 2DGE Proteome Patterns of Chesapeake Bay Microbial Communities[a]

Similarity (%)		June 2004			August 2004			February 2005		
		N	M	S	N	M	S	N	M	S
	N									
	M	45.8								
	S	41.6	53							
	N	21.7	13.4	18.6						
	M	28.8	17.5	15.1	55.7					
	S	28	29.6	23.5	46.7	45.8				
February 2005	N	10.8	13.8	8.1	21.5	28	9.8			
	M	16.4	9.2	12.5	29.9	42.1	12.9	70.3		
	S	18.9	10.8	14.8	29.3	32.7	14.5	66.4	72.3	

[a]Highlighted were similarities (%) among northern (N), middle (M), and southern (S) Bay samples from same sampling season.

fingerprints, proteome patterns showed more pronounced seasonal variations relative to spatial variations (Fig. 68.3 and Table 68.2).

68.3.3 MDS Pattern Analysis

MDS was used to separate both DGGE and 2D SDS-PAGE datasets independently. However, the parallel

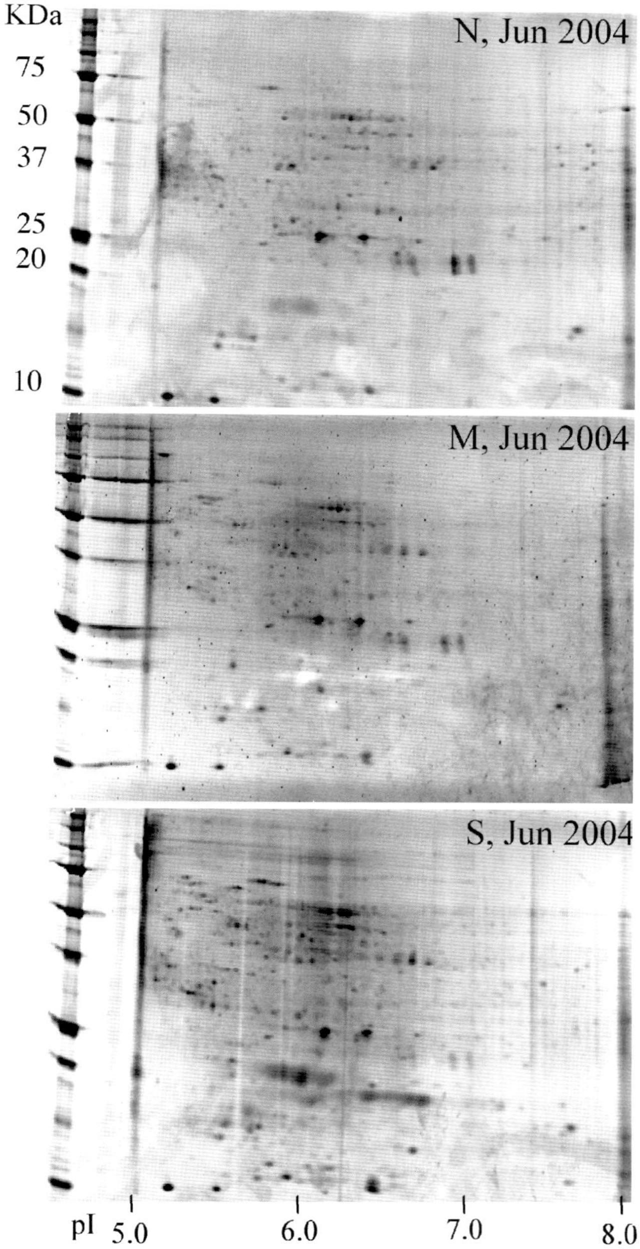

(A)

Figure 68.3 Two-dimensional gel electrophoresis (2DGE) images of metaproteomes obtained from the northern (N), middle (M), and southern (S) Chesapeake Bay stations in June 2004, August 2004, and February 2005.

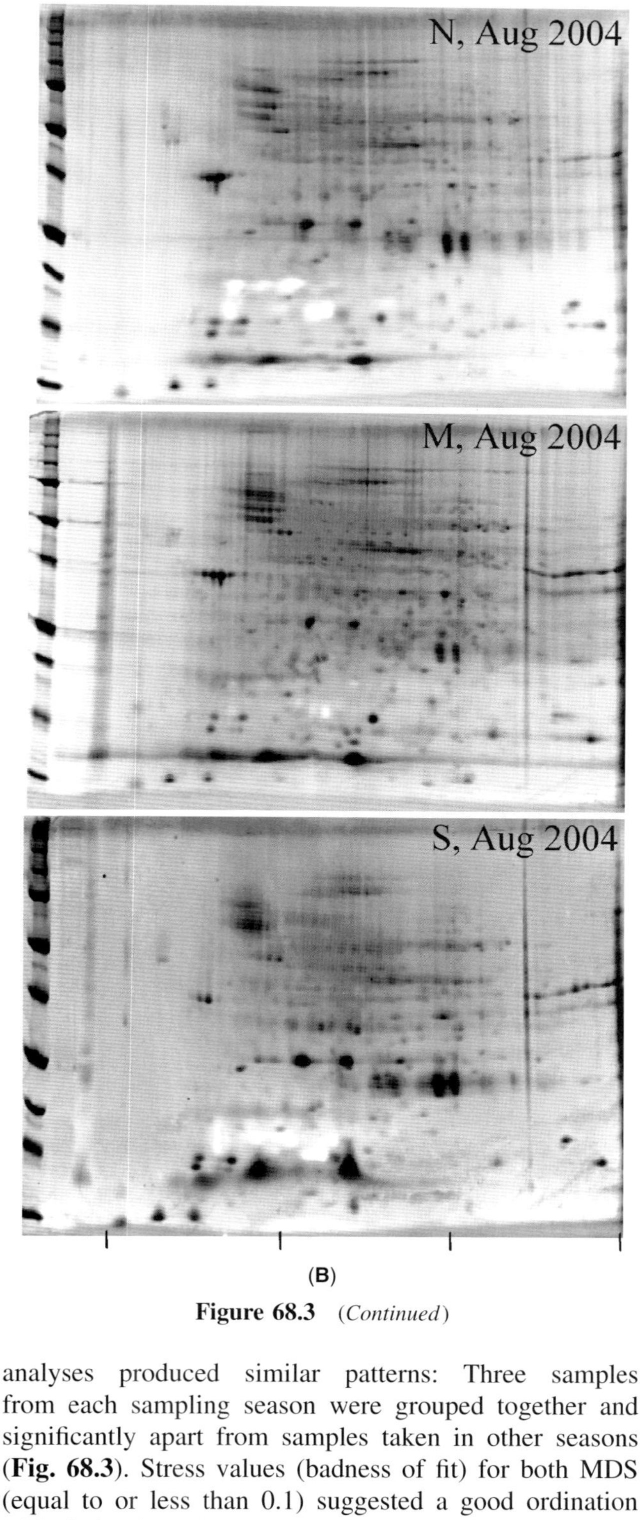

(B)

Figure 68.3 *(Continued)*

analyses produced similar patterns: Three samples from each sampling season were grouped together and significantly apart from samples taken in other seasons (**Fig. 68.3**). Stress values (badness of fit) for both MDS (equal to or less than 0.1) suggested a good ordination with little risk of misinterpretation of banding patterns [Clarke, 1993]. The seasonal synchronicity between microbial community structure and their protein profiles suggest that microbial community composition could strongly influence the metaproteomic signature and possible microbial processes.

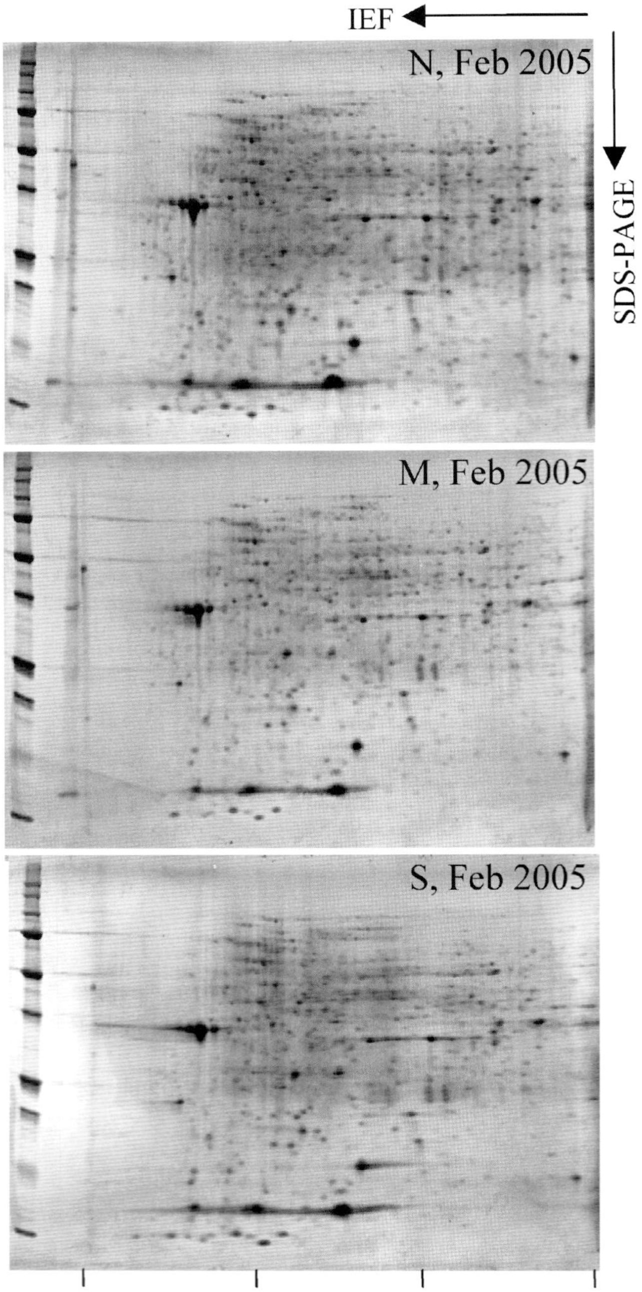

(C)

Figure 68.3 *(Continued)*

The similarity in sample aggregation by MDS between DGGE and 2D SDS-PAGE suggests that the proteome patterns are driven in these samples by changes in the microbial community structure across seasons [Kan et al., 2006b]. Metaproteomics has been proved to be a useful tool to study the response of a microbial community under environmental stresses even within a short period of time (e.g., cadmium exposure) [Lacerda et al., 2007; see also Chapter 70, Vol. I]. Therefore, the data here suggest a tight coupling between spatial and temporal proteome patterns of the Chesapeake Bay with microbial community structure shifts likely driven by seasonal fluctuations in hydrological regime, physical properties, and water chemistry.

68.3.4 Protein Identification

A total of 140 protein spots (41 from June 2004, 46 from August 2004, and 53 from February 2005) were selected and excised from 2DGE gels reflecting various molecular weights, charges, and relative abundance (for clarity, picked proteins were not shown in gel images). About 80% of the proteins yield interpretable MS profiles, but MALDI-TOF MS failed to provide identification for any of these samples (data not shown). However, MALDI-TOF peptide mass fingerprinting is incredibly sensitive to slight variations in protein sequence. De novo sequences obtained from MS/MS analyses indicated that nearly all proteins were bacterial origin, however, only a few proteins have been identified on the basis of two or more peptides [Kan et al., 2005]. Most of the de novo sequences resulted nonsignificant hits or identification conflicts among multiple peptides (data not shown).

This current study, in addition to others, realized that identification of proteins from natural aquatic samples is still facing challenges [Wilmes and Bond, 2009; Ram et al., 2005; Lacerda et al., 2007]. First, microbial communities in the aquatic environment are usually very diverse and produce dynamic levels of complex protein species. Second, community protein analysis requires a reliable metagenomic database. Currently, only very limited metagenome sequences are available for the Chesapeake Bay microbial communities, and these are mostly of viral origin, not bacterial [Bench et al., 2007]. Third, most of the metagenomic studies were based on a single sample collected at a certain place or in a certain season, which the collected DGGE data for the Chesapeake Bay suggest may be insufficient to capture the richness of microbial community dynamics in this system.

Unsatisfied protein identification leaves an open question about environmental proteomics: What proteins dominate metaproteomes, housekeeping genes, or ecologically important genes? One early concern about the proteomics is that it is biased toward the expression of the most dominant housekeeping genes and thus provides limited information on microbial functions [Keller and Hettich, 2009]. However, recent studies have identified functional gene products from relatively simple (acid mine drainage, Ram et al. [2005]) to high-complexity microbial communities (wastewater treatment bioreactor,

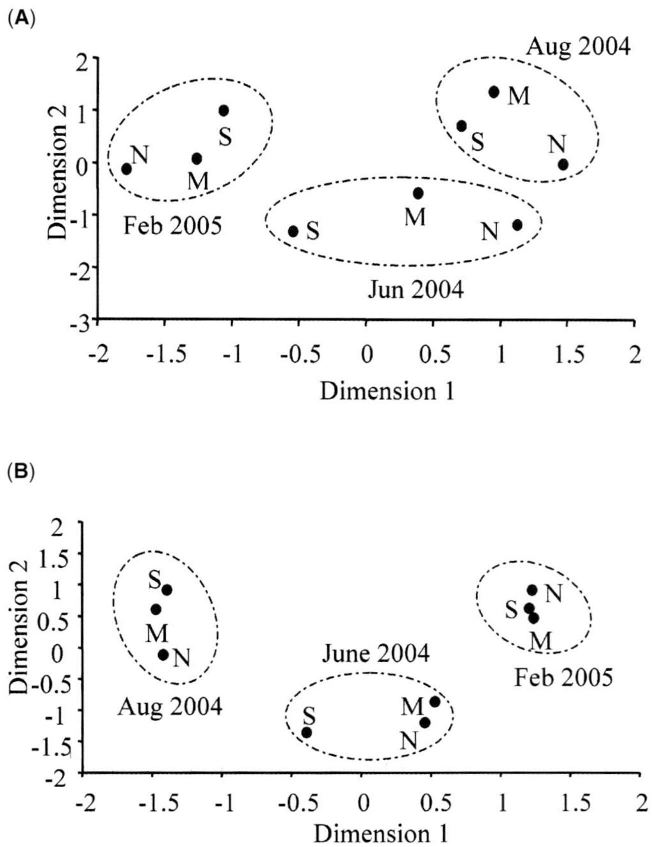

Figure 68.4 MDS plots of community structures (DGGE) and proteome patterns (2DGE). Badness of fit criterion for **A** and **B** were 0.1 and 0.06, respectively.

Lacerda et al. [2007]). For example, cadmium-shock proteins including ATPases, oxidoreductases and transport proteins have been consistently identified during cadmium exposures [Lacerda et al., 2007]. In addition, proteins facilitating substrate and nutrient transport were found dominant among the metaproteomes of surface water from Sargasso Sea [Sowell et al., 2009]. Therefore, the metaprotemes detected in the current study likely included proteins/enzymes that are related to the ecological functions under estuarine aquatic environments.

Chesapeake Bay is a eutrophic environment largely influenced by the nutrient input from its tributaries. Both bacterial composition and activities in the Bay were greatly affected by interactive environmental parameters [Shiah and Ducklow, 1994; Schulze and Ducklow, 2000; Smith and Kemp, 2003; Kan et al. 2006a]. In our previous study, a NADH:ubiquinone oxidoreductase and a predicted aminopeptidase were identified from Chesapeake Bay metaproteomes [Kan et al., 2005]. These proteins are involved in either electron transport or polypeptide/protein catalysis. Further characterization of Chesapeake Bay metaproteomes will improve our understanding of microbial activities and in situ functions in estuarine ecosystems.

68.4 CONCLUSIONS

This study first showed spatial and temporal metaproteome patterns of natural microbial communities in Chesapeake Bay. Similar to the community DNA fingerprint, the metaproteome fingerprints demonstrated strong seasonal variations with regard to spatial changes. The synchronicity between the population structures and proteome patterns suggested that species composition of microbes is a key factor determining community protein profile. We demonstrated that 2D SDS-PAGE-based proteomics was sensitive and reproducible to study functional dynamics for natural microbial communities. However, protein identification and linking a specific function with particular microbial taxa are the main challenges ahead. Downstream mass spectrometry analyses were not able to identify most of the selected proteins highlighting the need for dedicated, coupled metagenomic and metaproteomic approaches for each sample collected. While this is currently a major fiscal challenge for all but a few investigators, as sequencing and MS costs decrease, these studies will become more feasible for a larger group of investigators and therefore be applied to a larger number of systems. Nevertheless, although in its infancy, community-based proteomics (metaproteomics) provides a new strategy to monitor spatial and temporal variations of microbial diversity and activities in nature.

REFERENCES

Benndorf D, Balcke GU, Harms H, von Bergen M. 2007. Functional metaproteome analysis of protein extracts from contaminated soil and groundwater. *ISME J*. **1**:224–234.

Bench SR, Hanson TE, Williamson KE, Ghosh D, Radosovich M, Wang K, Wommack, KE. 2007. Metagenomic characterization of Chesapeake Bay virioplankton. *Appl. Environ. Microbiol*. **73**:7629–7641.

Clarke KR. 1993. Nonparametric multivariate analyses of changes in community structure. *Aust. J. Ecol*. **18**:117–143.

Crump BC, Hobbie JE. 2005. Synchrony and seasonality in bacterioplankton diversity of two temperate rivers. *Limnol. Oceanogr*. **50**:1718–1729.

Fuhrman, JA, Hewson I, Schwalbach MS, Steele JA, Brown MV, Naeem S. 2006. Annually reoccurring bacterial communities are predictable from ocean conditions. *Proc. Natl. Acad. Sci. USA* **103**:13104–13109.

Kan, J, Hanson TE, Ginter JM, Wang K, Chen F. 2005. Metaproteomic analysis of Chesapeake Bay microbial communities. *Saline Syst*. **1**:7.

Kan, J, Crump BC, Wang K, Chen F. 2006a. Bacterioplankton community in the Chesapeake Bay: Predictable or random assemblages. *Limnol. Oceanogr*. **51**:2157–2169.

Kan, J, Wang K, Chen F. 2006b. Temporal variation and detection limit of an estuarine bacterioplankton community analyzed by denaturing gradient gel electrophoresis (DGGE). *Aquat. Microb. Ecol*. **42**:7–18.

Kan, J, Suzuki MT, Wang K, Evans SE, Chen F. 2007. High temporal but low spatial heterogeneity of bacterioplankton in the Chesapeake Bay. *Appl. Environ. Microbiol*. **73**:6776–6789.

Kan, J, Evans SE, Chen F, Suzuki MT. 2008. Novel estuarine bacterioplankton in rRNA operon libraries from the Chesapeake Bay. *Aquat. Microb. Ecol*. **51**:55–66.

Keller, M, Hettich R. 2009. Environmental proteomics: a paradigm shift in characterizing microbial activities at the molecular level. *Microbiol. Mol. Biol. Rev*. **73**:62–70.

Klaassens, ES, de Vos WM, Vaughan EE. 2007. Metaproteomics approach to study the functionality of the microbiota in the human infant gastrointestinal tract. *Appl. Environ. Microbiol*. **73**:1388–1392

Lacerda, CMR, Choe LH, Reardon KF. 2007. Metaproteomic analysis of a bacterial community response to cadmium exposure. *J. Proteome Res*. **6**:1145–1152.

Markert, S, Arndt C, Felbeck H, Becher D, Sievert SM, Hugler M, et al. 2007. Physiological proteomics of the uncultured endosymbiont of *Riftia pachyptila*. *Science* **315**:247–250.

Maron PA, Ranjard L, Mougel C, Lemanceau P. 2007. Metaproteomics: A new approach for studying functional microbial ecology. *Microb. Ecol*. **53**:486–493.

Park C, Helm RF. 2008. Application of metaproteomic analysis for studying extracellular polymeric substances (EPS) in activated sludge flocs and their fate in sludge digestion. *Water Sci. Technol*. **57**:2009–2015.

Pierre-Alain M, Christophe M, Severine S, Houria A, Philippe L, Lionel R. 2007. Protein extraction and fingerprinting optimization of bacterial communities in natural environment. *Microb. Ecol*. **53**:426–434.

Ram, RJ, VerBerkmoes NC, Thelen MP, Tyson GW, Baker BJ, Blake II RC, et al. 2005. Community proteomics of a natural microbial biofilm. *Science* **308**:1915–1920.

SAS Institute Inc. 1992. *SAS User's Guide*. Cary, NC: SAS Institute Inc.

Schulze, GE, Ducklow HW. 2000. Changes in bacterioplankton metabolic capabilities along a salinity gradient in the York River estuary, Virginia, USA. *Aquat. Microb. Ecol*. **22**:163–174.

Schulze, WX, Gleixner G, Kaiser K, Guggenberger G, Mann M, Schulze ED. 2005. A proteomic fingerprint of dissolved organic carbon and of soil particles. *Oecologia* **142**:335–343.

Shiah, FK, Ducklow HW. 1994. Temperature regulation of heterotrophic bacterioplankton abundance, production, and specific growth rate. *Limnol. Oceanogr*. **39**:1243–1258.

Smith, EM, Kemp WM. 2003. Planktonic and bacterial respiration along an estuarine gradient: response to carbon and nutrient enrichment. *Aquat. Microb. Ecol*. **30**:251–261.

Sowell, SM, Wilhelm LJ, Norbeck AD, Lipton MS, Nicora CD, Barofsky DF, et al. 2009. Transport functions dominate the SAR11 metaproteome at low-nutrient extremes in the Sargasso Sea. *ISME J*. **3**:93–105.

Wilmes, P, Bond PL. 2004. The application of two-dimensional polyacrylamide gel electrophoresis and downstream analyses to a mixed community of prokaryotic microorganisms. *Environ. Microbiol*. **6**:911–920.

Wilmes, P, Bond PL. 2006. Metaproteomics: Studying functional gene expression in microbial ecosystems. *Trends. Microbiol*. **14**:92–97.

Wilmes, P, Bond PL. 2009. Microbial community proteomics: elucidating the catalysts and metabolic mechanisms that drive the Earth's biogeochemical cycles. *Curr. Opin. Microbiol*. **12**:310–317.

Wilmes, P, Wexleer M, Bond PL. 2008. Metaproteomics provides functional insight into activated sludge wastewater treatment. *PloS ONE* **3**:e1778.

Chapter 69

High-Throughput Cyanobacterial Proteomics: Systems-Level Proteome Identification and Quantitation

SAW YEN OW AND PHILLIP C. WRIGHT

69.1 INTRODUCTION

Cyanobacteria, oxygenic photoautotrophs, are one of Earth's oldest and most numerous groups of organisms [Whitton, 1992; Whitton and Potts, 2000]. They are thus interesting in understanding biological diversity and phylogeny [Stanier and Cohen-Bazire, 1977], but are also important in terms of the environment [Delpla et al., 2009] and climate change [Wiedner et al., 2007]. They have great potential for use for mankind in a wide variety of areas such as low-carbon fuels [Angermayr et al., 2009], agriculture [Dismukes et al., 2008], industrial products and therapeutics [Burja et al., 2003; Tamagnini et al., 2007; Wase and Wright 2008]. They have an interesting complement of genes that can be used for metabolic engineering/synthetic biology purposes either using native cyanobacteria themselves, or in a recombinant host chassis [Scaife et al., 2009].

Integrated systems biology has received widespread attention [Ideker et al., 2001; Kitano, 2002; Hood, 2003; Friboulet and Thomas, 2005], because it is clear that isolated reductionist observations are insufficient to properly describe biological systems and their inherent dynamics. In combination with advancements in basic biology, molecular biology, and computational and systems engineering and modeling, systems biology has iteratively co-motivated the invention of high-throughput techniques (HTT). The aim of HTT is to study whole biological systems. The success of the Human Genome Sequencing Project, and other high-profile genome projects are important landmarks in this enterprise [Venter et al., 2001].

Different HTT-based strategies have been explored in recent years. These advancements target key -omic levels, such as the transcriptome (see Chapters 62–65, Vol. I), proteome (see Chapters 66–70, Vol. I), and metabolome (see Chapters 71–74, Vol. I). DNA microarrays, for example, have been successfully developed to measure the transcriptome behavior, allowing for the possibility of global-scale understanding of transcription [Schena et al., 1995; see also Chapters 57, 58, 60, 61, Vol. I]. Similarly, global studies of proteins ("proteomics") have also been successfully carried out, despite some technical challenges (see Chapters 66–70, Vol. I). In combination with the recent emphasis on HTT-based mass spectrometry (MS), substantial advancements have been made [Souchelnytskyi, 2005; Smith and Figeys, 2006]. Currently, there are many choices of quantitative HTT workflows available to the research community. The depth and flexibility of these HTT technologies are rapidly improving, thus making it possible to obtain deeper insights into the dynamic functions and regulation of the proteome [Wu and MacCoss, 2002] and into how this integrates with other levels in the cell (see Chapter 66, Vol. I).

A great necessity for systems biology projects in general is the availability of a genome sequence. For the cyanobacteria, 35 projects are underway or competed (Kazusa Cyanobase, accessed January 24, 2010). The availability of genome-scale data importantly enables

Handbook of Molecular Microbial Ecology, Volume I: Metagenomics and Complementary Approaches, First Edition. Edited by Frans J. de Bruijn.

effective HTT workflows, which in turn drives modern research in the cyanobacteria.

Prior to the development of post-genomic HTT, earlier cyanobacterial protein research was primarily hypothesis-driven, and it focused on a limited number of specific proteins. Given the increased availability of HTT for proteomics, it is expected that cyanobacteriologists will be able to perform routine HT (quasi-)global measurements of the proteome [Barrios-Llerena et al., 2006]. As seen in **Table 69.1**, 15 HT studies have been carried out by a small number of groups and have been published from the first in 2004 until now in early 2010.

The purpose of this review is to summarize key technological changes arising from the transition from the low(er)-throughput gel-based proteomics workflows to the more modern and more rapid (usually gel-free) HTT. We focus on the trends that have changed since an earlier report [Burja et al., 2003]. We conclude with perspective on the potential for advancement of the proteomics field in general and the special opportunities and challenges for the cyanobacteria.

69.1.1 Summarizing the Gel Proteomics Era in Cyanobacteria

In common with all the proteomics fields for different biological systems, initial cyanobacterial proteomics relied (almost) exclusively on one- or two-dimensional gel electrophoresis [Barrios-Llerena et al., 2007]. The gel-based workflows deviated little in principle from the original 1975 report describing two-dimensional gel electrophoresis (2DGE) [O'Farrell, 1975]. The main differences between different biological systems tended to be in the development/adaptation of particular extraction methods [Barrios-Llerena et al., 2007]. These "tweaks" are still important, however, because they set the necessary foundations for HT proteomics in cyanobacteria [Barrios-Llerena et al., 2006], and extraction protocols are still an issue today. Extraction methods for cyanobacteria have been known to differ significantly from techniques employed for other bacteria, frequently relying on denaturants in order to gain access to the proteome (i.e. chaotrophs and detergents).

In common with many other bacterial systems, the major foci of cyanobacterial proteomics have tended to center on environmental changes/perturbations. Importantly, earlier investigations have demonstrated a multitude of integrated approaches that enabled targeted analysis via either narrow gradient isoelectric focusing, cellular enrichments, on protein cross-linking with gel electrophoresis [Razquin et al., 1996; Norling et al., 1998; Simon et al., 2002]. However, 2DGE in general and earlier studies were hampered by low-throughput and experimental designs required for high-speed large-scale analysis. A lack of a reliable protein identification technology (satisfying both high-throughput and confidence) further limits the investigation efficacy. Most studies prior to 2000 relied (almost) exclusively on mass spectrometry-based MALDI peptide mass fingerprinting (PMF), chemical-based N-terminal sequencing, and immunochemistry for detection [Shevchenko et al., 2000; Sazuka, 2001]. The shift in the detection and identification paradigm for the cyanobacteria did not come until 2004 or so, when confidence of protein detection and also its throughput was improved by the application of liquid chromatography and tandem mass spectrometers.

Despite a reliance on the combination of (usually) 2DGE and MALDI-TOF (later ion trap and TOF/TOF for gel-based analysis) MS, cyanobacterial proteomics research did have a solid degree of momentum, particularly, as mentioned above, for environmental perturbations. At the same time, there was a broadening interest in specialized areas such as improving the understanding of membrane proteins and protein localization. Other notable areas during the years 1999 to 2003 were (a) the analysis of the impact of increased concentrations of salt, (b) UV stress, and (c) the understanding of cellular differentiation in filamentous cyanobacteria [Pandhal et al., 2008d; Wase and Wright, 2008]. As noted earlier, reports in membrane proteomics during the years 1998 to 2004 demonstrated the extensive optimization of extraction and subcellular fractionation that were necessary to successfully measure hydrophobic proteins in the thykaloid membrane and other areas [Norling et al., 1998; Huang et al., 2002; Srivastava et al., 2005], while other important contributions include gel-based studies of the proteomes of the filamentous cyanobacteria *Nostoc* sp. PCC 7120 and *Nostoc punctiforme* ATCC 29133 [Kaneko et al., 2001; Hunsucker et al., 2004]. For more detailed discussions on this general area of cyanobacterial gel proteomics, the reader is directed elsewhere [Burja et al., 2003; Pandhal et al., 2008d].

69.2 ARRIVAL OF THE HIGH-THROUGHPUT PROTEOMICS ERA TO CYANOBACTERIA

The proteomics community began to realize the potential of shotgun proteomics or MudPIT after the publication of a key paper from the Yates group in San Diego. The rationale of MudPIT in 2001 was simple yet strategic, where a combination of a two-dimensional LC, ion-exchange plus reversed phase and an electrospray ion-trap MS was employed in tandem to profile the *Saccharomyces cerevisiae* proteome [Wolters et al., 2001]. The technical impact of the publication was also revolutionary, because the method opened up opportunities to perform fast (high-duty-cycle ion traps) and reliable identification (tandem

Table 69.1 Tabulation of Published Reports (up to January 2010) Applying a High-Throughput Proteomics Strategy in Cyanobacteria[a]

Organism of Interest (Reference)	Research Report and Novelty Focus	Cell Type	Nature of Experiments
Synechocystis sp. PCC 6803			
Gan et al. [2005]	Experimental focus: HT proteome profiling (qualitative) Main attraction: Technical development of prefractionation strategies for shotgun	Unicellular	ID-only shotgun
Chong et al. [2006]	Experimental focus: Quantitative HT proteome profiling Main attraction: Pilot introduction, technical assessments of methodology	Unicellular	iTRAQ shotgun
Gan et al. [2007]	Experimental focus: Quantitative HT proteome profiling Main attraction: Technical assessments of precision and experimental variation	Unicellular	iTRAQ shotgun
Kwon et al. [2009]	Experimental focus: HT proteome profiling (qualitative) Main attraction: Enrichment of membrane proteomes, subproteome profiling	Unicellular	ID-only shotgun (membrane)
Pandhal et al. [2008b]	Experimental focus: Quantitative HT proteome profiling Main attraction: Large-scale data integration, predictive analysis, salt tolerance in cyanobacteria	Unicellular	iTRAQ shotgun + predictive network analysis
Nostoc punctiforme ATCC 29133			
Anderson et al. [2006]	Experimental focus: HT proteome profiling (qualitative) Main attraction: Metabolic profiling of expressed proteome and speculations	Filamentous	ID-only shotgun
Ow et al. [2008a]	Experimental focus: Quantitative HT proteome profiling Main attraction: Detailed N_2 fixation directed heterocyst and vegetative cells metabolic mapping	Filamentous	iTRAQ shotgun
Nostoc sp. PCC 7120			
Stensjo et al. [2007]	Experimental focus: Quantitative HT proteome profiling Main attraction: First introduction to Nostoc, N_2 fixation and associated differential regulation	Filamentous	iTRAQ shotgun
Ow et al. [2008b]	Experimental focus: Quantitative HT proteome profiling Main attraction: N_2 fixation directed heterocyst and vegetative cells differential regulation	Filamentous	iTRAQ shotgun
Noirel et al. [2008][b]	Experimental focus: Bioinformatics Main attraction: Large-scale data integration to aid biological interpretation	Filamentous	In silico metabolic network analysis
Sanguinetti et al. [2008][b]	Experimental focus: Bioinformatics Main attraction: Statistical improvements to large-scale data integration	Filamentous	In silico metabolic network analysis
Euhalothece spp. BAA001			
Pandhal et al. [2008a]	Experimental focus: Quantitative HT proteome profiling Main attraction: Elemental N labeling, unsequenced genome, salt tolerance in cyanobacteria	Unicellular	Quantitative $^{15}N/^{14}N$ metabolic labeling
Pandhal et al. [2008b]	Experimental focus: Quantitative HT proteome profiling Main attraction: Complementary methodology, iTRAQ + elemental labeling, salt tolerance in cyanobacteria	Unicellular	Quantitative $^{15}N/^{14}N$ metabolic + iTRAQ shotgun

(continued)

Table 69.1 Tabulation of Published Reports (up to January 2010) Applying a High-Throughput Proteomics Strategy in Cyanobacteria[a] (*Continued*)

Organism of Interest (Reference)	Research Report and Novelty Focus	Cell Type	Nature of Experiments
Anabaena variabilis ATCC 29413			
Barrios-Llerena et al. [2006]	Experimental focus: HT proteome profiling (qualitative) Main attraction: Methods and technical review of shotgun for cyanobacteria	Filamentous	ID-only shotgun
Prochlorococcus marinus MED4			
Pandhal et al. [2007]	Experimental focus: Quantitative HT proteome profiling Main attraction: First introduction to MED4, light adaptation of Prochlorococcus	Unicellular	iTRAQ shotgun

[a] Aims of the analyses differ generally into two classes: I, address technical challenges; II, assess and study biological perturbations in the cyanobacteria. The focal attractions to the respective studies are described in the report focus.

[b] Analysis performed using data generated by *Nostoc* sp. PCC 7120 shotgun experiments by Stensjö et al. [2007] and Ow et al. [2008a].

Source: Modified from Ow et al. [2009].

MS sequencing) of proteins. Over the subsequent years since that 2001 paper, there has been a wide acceptance of high-performance liquid chromatography (HPLC) coupled to tandem mass spectrometers in online and offline combinations. The result was that key proteomics research areas that relied on gel-based workflows became increasingly displaced by these emerging HTT. The new HTT based on (rapid multiscale) multidimensional HPLC led to a new term, "gel-free" proteomics. This important technological paradigm shift led to the eventual effective HT analysis of complex proteomes [Wu et al., 2002] and also led to the term "shotgun proteomics," which arose from its quasi-stochastic detection approach. Shotgun proteomics requires that proteomes be (usually, though other variations also exists) directly digested and analyzed in situ using a combination of usually at least two-dimensional HPLC separation followed by detection with high-speed electrospray ionization (ESI) tandem mass spectrometers [Gevaert et al., 2004; Staes et al., 2004; Swanson and Washburn, 2005]. Similar approaches were also developed using offline approaches for MALDI-TOF/TOF tandem instruments to adapt shotgun proteomics [Zhen et al., 2004]. Shotgun proteomics approaches resulted in an increase of up to orders of magnitudes in speed over the older than gel-based proteomics workflows [Gan et al., 2005; Chong et al., 2007]. A new challenge arose from this increased number of protein/peptide detections though, an avalanche of data. This dramatic increase in data required the development of new algorithms/software, including proprietary and open-source tools [Elias et al., 2005; Lane, 2005]. Although the different software programs generally serve similar purposes, their philosophies and main applications can differ. A common identification strategy is to search a theoretical database. This process relies on one being able to match experimental MS/MS spectra of proteins/peptides to theoretical spectra generated by the software from data contained in the database. Key examples of this software include MASCOT and SEQUEST [Eng et al., 1994; Perkins et al., 1999]. It is clear that this method relies very heavily on the availability of a database from which to draw these theoretical spectra. The efficacy of the method decreases once the peptide sequences in the experimental system differ from those in the database by more than one or two amino acids (depending on stringency of substitution). This underlines the importance of genome sequences for proteomics. Nonetheless, it is possible for unsequenced organisms to be studied using this method, as long as the peptides being investigated are othologous to those in the database. Methods underlying these adaptations have been previously reported [Snijders et al., 2007; Pandhal et al., 2008a]. If this is not possible, then one can also apply BLAST-centric or de novo sequencing tools. There has been software developed for this, including (a) MS-based BLAST searches for MS/MS spectra (b) automated de novo sequencing software for MS/MS spectra (e.g., MS-BLAST and PEAKS) [Shevchenko et al., 1996; Han et al., 2004]. A combination of conventional sequencing methods and these latter tools are thus vital for successful investigation of organisms lacking a genome sequence. A review of such studies in the cyanobacteria has been published [Pandhal et al., 2008d].

For success, shotgun proteomics relies firstly on efficient orthologous separation of the peptide and protein pool. This is achieved either by Yates' original experimental template (an online two-dimensional system), or via a hybrid offline/online fractionation approach. The combined efficacy of these separation steps, and the duty cycle synchronization with the in situ interfaced tandem MS must be achieved to ensure efficient analysis of the large number of proteins. Hence, performance indicators such as separation resolution and MS duty cycle are commonly associated with shotgun proteomics and can directly dictate the effectiveness of the experiment [Swanson and Washburn, 2005]. **Figure 69.1** demonstrates an overview of the most commonly applied workflow templates for HT proteomics analysis in cyanobacteria. To maximize proteome coverage (depth) without excessive experimental time, both LC and MS parameters must be effectively balanced, and a pragmatic approach is also required. An example of such an approach in the cyanobacteria was first reported in 2003 as a technical benchmark of the efficiencies of diverse (pre-)fractionation methods (ion-exchange chromatography, PAGE-shotgun integration, and gel-based (IPG) or gel-free isoelectric focusing) for application in the cyanobacteria [Gan et al., 2005]. Subsequently, as summarized in Tables 69.1 and **69.2**, separate additional studies based on iterative developments from that work

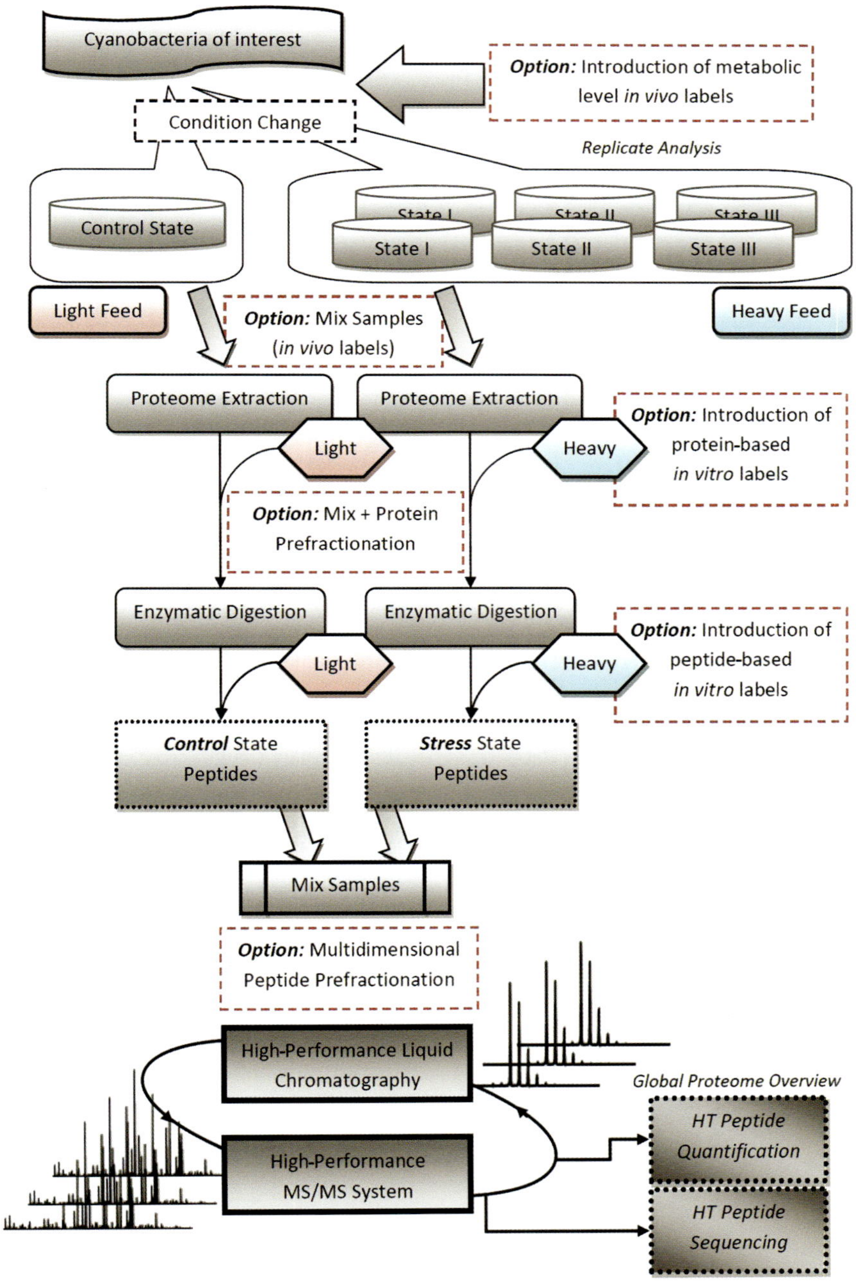

Figure 69.1 An overview of high-throughput shotgun proteomics as adapted from recent cyanobacterial analyses. Organisms of interest are usually subjected to a variety of extraction protocols. Label introduction stages are ad hoc to applications, and they also dictate analysis and buffering decisions for subsequent stages. Samples are commonly prefractionated using either gel-based or liquid chromatography, and they are finally analyzed using tandem mass spectrometry. Data generated from MS instruments are then paired together with the use of software interrogation to provide both identifications and quantification necessary for the abundance characterisation proteome population.

Table 69.2 Proteome Coverage Statistics of HT Experiments Made Using Variations of Shotgun Proteomics from 2003 until January 2010

Organism of Interest (Reference)	Report Focus	Proteins Identified	Proteins Quantified	Sequencing Method	Separation Method	Analysis Duty Cycle	Analysis Accuracy	Resolution Achievable
Synechocystis sp. PCC 6803								
Gan et al. [2005]	Qualitative shotgun	776	—	ESI-Qq-TOF	2D-LC + multiple modes of IEF and 1DGE	Intermediate	High	High
Chong et al. [2006]	Quantitative shotgun[a]	—	203	ESI-Qq-TOF	2D-LC with SCX and RP-LC	Intermediate	High	High
Kwon et al. [2009]	Qualitative shotgun (Membrane)	785	—	ESI-Q-IT-FT	2D-LC with protein/ peptide RP-LC	Very Fast (IT) Very Slow (FT)	Very high (on FT)	Very high (on FT)
Pandhal et al. [2008b]	Quantitative shotgun[a]	—	378	ESI-Qq-TOF	2D-LC with SCX and RP-LC	Intermediate	High	High
Nostoc punctiforme ATCC29133								
Anderson et al. [2006]	Qualitative shotgun	1575	—	ESI-Q-IT	3D-LC with SEC and online SCX-RP-LC	Very Fast	Intermediate	Low
Ow et al. [2008a]	Quantitative shotgun[a]	—	702	ESI-Qq-TOF	2D-LC + cellular enrichment with SCX and RP-LC on iTRAQ	Intermediate	High	High
Nostoc sp. PCC 7120								
Stensjo et al. [2007]	Quantitative shotgun[a]	—	486	ESI-Qq-TOF	2D-LC with SCX and RP-LC	Intermediate	High	High
Ow et al. [2008b]	Quantitative shotgun[a]	—	506	ESI-Qq-TOF	2D-LC + cellular enrichment	Intermediate	High	High
Euhalothece spp. BAA001								
Pandhal et al. [2008a]	Quantitative shotgun[b]	—	383	ESI-Qq-TOF	1DGE + RP-LC on $^{15}N/^{14}N$ Metabolic labeling	Intermediate	High	High
Pandhal et al. [2008b]	Quantitative shotgun[a,b]	—	207[a] 243[b]	ESI-Qq-TOF	1DGE + RP-LC & 2D-LC with SCX and RP-LC	Intermediate	High	High

were also carried out. In 2006, a qualitative shotgun study of *N. punctiforme* ATCC 29133 demonstrated the potential of an optimized workflow, by generating >1500 protein identifications in a single experiment [Anderson et al., 2006]. Their method was also the first to demonstrate size-exclusion chromatography of proteins in the shotgun proteomics investigation of a cyanobacterium.

To completely assess the implications of the work of Gan et al. and Anderson et al., it is important to consider the role of different mass spectrometers platforms play in the two studies. The report employed an ion trap (Q-IT) MS. It is well known that Q-IT-type instruments have advantages of high speed (short duty cycle) and high sensitivity. This is clearly a plus when one is trying to dig deeply into a complex proteome. Now if this were the only consideration, then one could leave it at that. However, in terms of systems biology, fast simultaneous measurements of identifications only is becoming of less importance when compared with the objective of obtaining quantitative data. Quadrupole time-of-flight (Q-TOF)-type mass spectrometers such as that employed by Gan et al. (usually) have lower duty cycles than equivalent Q-IT-type instruments. This would seem to be a disadvantage, except that they have substantially higher mass accuracy, typically an order magnitude higher, and resolution (i.e., will result in high-quality identifications) than Q-IT-type instruments; both aspects also dictate the confidence of the protein identification [March, 1997; Chernushevich et al., 2001]. Even more importantly, these features have an even greater impact for quantitation of proteins/peptides since greater resolution results in both the correct ion charge assignment and MS-based quantitation (see **Fig. 69.2**). In respect to applications so far, it is interesting to note that other major tandem MS instruments such as MALDI-TOF/TOF and Orbitrap FT-type instruments have to date not been applied to HT cyanobacterial proteomics. Only very recently has an FT-ion trap-based HT study been published, while MALDI-TOF/TOF instruments are still used primarily to sequence gel spots/bands [Kwon et al., 2009]. These early omissions are surprising, especially considering the similarity in operation of MALDI-TOF/TOF instruments compared to the MALDI-TOF [Medzihradszky et al., 2000] workhorses of the 1D and 2D-GE era. The LC-MALDI-type workflows (as noted above) necessarily provide complementary data to that obtained on ESI-instruments, albeit potentially requiring longer analysis time [Zhen et al., 2004].

In summary, research contributions of all early shotgun proteomics in cyanobacteria applications were of high value, and technical considerations raised in these studies led to several implications affecting later cyanobacterial shotgun proteomics studies. In a review, Barrios-Llerena et al., [2007] took these discussions further by critically evaluating these earlier investigations [Barrios-Llerena et al., 2006], especially with respect to the challenges involved in working with the high-abundance phycobiliproteins (up to 50% of the proteome) often found in cyanobacteria over a wide range of conditions [Gan et al., 2005]. It is well known in the proteomics community that the dynamic range of protein abundance in the cell has a significant impact on the ability to deeply mine the proteome. In the medical area, for example, implications of the serum proteome (in blood and tissue specimens) have been well known to express its proteome with dynamic range up to 10^{12} [Anderson et al., 2004], or even more.

In the cyanobacteria, the analog to the issues of actin and serum albumin for blood or tissue proteins is the abundance of the phycobiliproteins, in particular so far, in comparison to the concentrations of other proteins in the cyanobacterial cell. If there is a wide range of abundance in the cell, spanning over four orders of magnitude (or more), then this can lead to suppression and detection limitations in the MS (due to most instrument's dynamic range of less than four orders of magnitude) and other effects on separation efficiencies, sample loads, and detection sensitivities. Problems associated with phycobilisome abundance in several cyanobacterial proteomes have not been solved as yet. The cyanobacterial community does not have access to equivalent affinity separation/depletion columns that are available for human samples, for example. It also needs to be pointed out that the cyanobacteria are very diverse, and the problem is worse in some species than in others.

69.2.1 Developing Quantitative Approaches for HT Cyanobacteria Proteomics

Performing qualitative HT experiments, while impressive at their ability to simultaneously sequence and identify thousands of proteins in one experiment, does not directly relate quantitative information of the system. Apart from being a tool to provide a snapshot of the proteome, its immediate relevance to the actual metabolic and/or regulatory changes is often very limited. This led to the emergence and recent developments of quantitative aspects in HT proteomics.

There are currently a wide variety of tools available which have been developed to provide either relative or absolute quantification for MS-based proteomics experiments. These techniques can often be applied to on/off integration with shotgun techniques and can also provide very precise measurements of abundances [Righetti et al.,

Table 69.2 (*Continued*)

Organism of Interest (Reference)	Report Focus	Proteins Identified	Proteins Quantified	Sequencing Method	Separation Method	Analysis Duty Cycle	Analysis Accuracy	Resolution Achievable
Anabaena variabilis ATCC 29413								
Barrios-Llerena et al. [2006]	Qualitative shotgun	646	—	ESI-Qq-TOF	2D-LC with SCX and RP-LC	Intermediate	High	High
Prochlorococcus marinus MED4								
Pandhal et al. [2007]	Quantitative Shotgun[b]	—	184	ESI-Qq-TOF	2D-LC with SCX and RP-LC	Intermediate	High	High

[a] iTRAQ technique related,
[b] Other quantitative methods.
Source: Updated and modified from Ow et al. [2009].

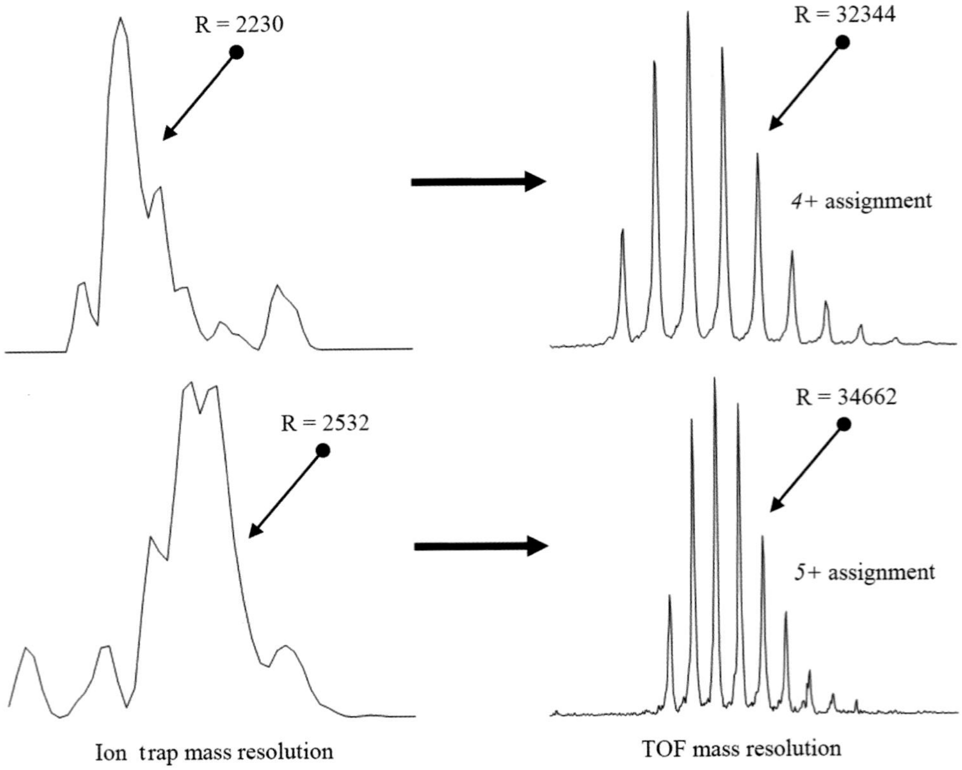

Figure 69.2 Benefits from increased mass resolution offered by increasingly more advanced MS instrumentation. The difference between intact peptide as reported by a next-generation time of flight (TOF) and an ion trap shows significantly elevated resolution (usually over an order of magnitude) that aids both identification and accurate quantification.

2004; Bantscheff et al., 2007]. Quantifications are made via the comparison of mass/charge (*m/z*) intensities of peptides or fragments of the peptides, where they can be compared between pairing peptides (with/without an *m/z* shift; see below) from control and condition phenotypes. State-of-the-art quantitative proteomics techniques also provide the option for multiplexing measurements, whereby simultaneous quantification can be made between several phenotypes in a single LC-MS run. This provides the benefits of improved information throughput and reduced total experiment time. Current methodologies often utilize the incorporation of stable isotopes (via conventional and silent techniques) to differentiate proteins between the different phenotypes [Bantscheff et al., 2007]. This principle is termed "labeling," because it allows effective segregation between the same peptide/protein species from different conditions. The consequence is a "mass shift" that is characteristic to the number of isotopes (and their mass difference) introduced to the proteins. Currently, two major classes of isotope incorporation exist; they can be introduced through either chemical labels (in vitro) via biochemical reactions after perturbation by a chemical label or in vivo via metabolic incorporation (such as SILAC [Ong et al., 2002], or elemental labeling [Snijders et al., 2005b]) before perturbation by using a "heavy isotope" feedstock change. Both variations produce similar characteristic "shifts" that can be compared as pairs (or multiplexes) in a mass spectrum to infer abundance changes during condition perturbations (see **Fig. 69.3**). Chemical-label-based techniques such as duplexing isotope-coded affinity tags (ICAT) were introduced in 1999 using chemicals modified by means of hydrogen (^{1}H) and deuterium (^{2}H) shifts [Gygi et al., 1999]. A later variation of chemical labeling (isobaric labels) also introduced multiplexing capabilities in 2003 (iTRAQ, octaplexing; or TMT, hexaplexing) using "silent isotope" labeling strategies [Ross et al., 2004]. This is achieved by the counterbalancing of masses between the incorporation of $^{15}N/^{14}N$ and $^{13}C/^{12}C$ isotopes between a reporter tag and a balance mass moiety [Gygi et al., 1999; Thompson et al., 2003; Ross et al., 2004]. Other variations of the ICAT and iTRAQ techniques also exist, including isotope-coded protein labels (ICPL) [Schmidt et al., 2005] and ExacTags [Liu et al., 2009], though they differ slightly in the chemistry and mechanism of quantification. Although principally all isotope-based techniques can also be used to provide absolute quantification (by introducing a known amount into a designated label), they are more commonly utilized to provide relative quantification in large-scale quantitative proteome profiling. The principle behind SILAC, ICAT, and iTRAQ will not be discussed further; however, readers are encouraged to refer to the relevant literature to better understand their respective applications [Gygi et al., 1999; Ross et al., 2004].

It is also important to understand the disparities, at least for mass-spectrometry-driven proteomics, between the data quality required for quantification. While the guidelines do vary between techniques, all methods require quantification to be accompanied by confident protein identification. It is therefore essential that the quality for both aspects is preserved. Thus, the most notable difference between qualitative and quantitative studies is to maintain not only qualitative confidence but also quantitative accuracy [Bantscheff et al., 2007]. This is one of the key criteria that must be carefully monitored because loosening the confidence threshold that dictates peptide identification could potentially affect both the precision and accuracy of the entire protein measurement. Insufficient stringency during identification also then leads to an increase in false discovery rate (FDR) [Elias and Gygi, 2007], and the significance of the quantification given the loss of precision as a result of outliers. Furthermore, in a number of quantitative techniques (such as iTRAQ), the relationship between optimal identification quality and quantitative accuracy are largely inversely proportional. Thus, achieving a pragmatic compromise between identifying and quantifying is a necessity. In other techniques (such as SILAC and other MS-level quantification techniques), quality of quantification can also depend on other factors such as mass resolution and accuracy (MS-level quantification techniques), and thus these factors are also interdependent. Therefore, as a general consensus, the FDR should always be kept low wherever possible—that is, at least by convention, no greater than 5%. In addition, the quality of quantification must also be insured by assessing the precision, and consistency of the measurements (biological replicate analysis and statistical tests). Fortunately, the majority of publications in the cyanobacteria thus far appear aware of this requirement and, in the authors' opinion, are also highly conservative with respect to the quality of data reported. Furthermore, publication guidelines are continually being reinforced, whereby stringent requirements are imposed by default in a number of key proteomics research journals [Tabb, 2008].

69.3 CHEMICAL LABELING-BASED METHODS IN CYANOBACTERIA HT STUDIES

From 2004 until the present, it can be observed that there are continued reports applying HT quantitative proteomics to the study of cyanobacteria. From 2004 to the end of 2009, there were eight reports that utilized chemical labels (also in vitro labeling) to perform HT proteomics analysis in cyanobacteria. In addition to two pilot experiments, six other reports studied the proteomic changes resulting from a number of environmental perturbations such as the N_2 fixation process and salt stress.

69.3.1 Best Technique for Cyanobacteria: Choices Are Dictated by Attachment Chemistries

With the large variation of chemical labelling strategies, it is often difficult to settle on a common workflow because each provides differing analytical/purification capabilities.

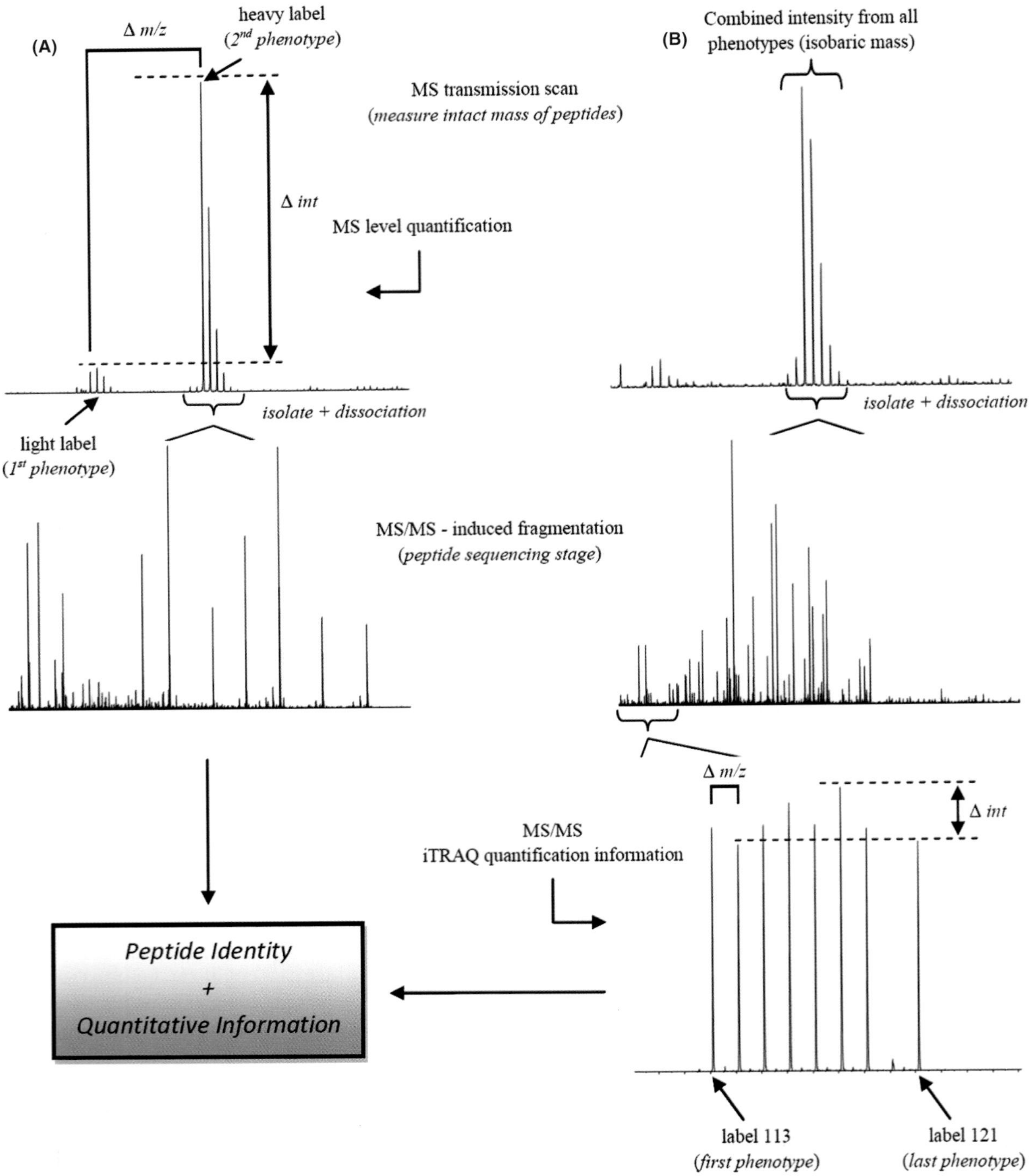

Figure 69.3 An illustration of the quantitative analysis MS cycle and its inferences. **(A)** Quantification as described using stable isotope labeling such as metabolic labeling and in vitro chemical nonisobaric techniques such as ICAT. Ions are quantified during the MS stage, where subsequent isolation of heavy/light moieties are made to dissociate and sequenced. **(B)** Quantification as described using 'silent' incorporation (isobaric mass) of stable isotopes such as iTRAQ and TMT; a compound peptide as a combination of signals from all phenotypes is isolated and dissociated; fragmentation provides ions that can be used to infer abundance differences. Symbols: Δ m/z denotes mass shifts as a consequence of mass difference between stable isotopes, Δ *int* denotes intensity difference as a consequence of abundance difference between phenotypes. In both cases, analysis returns similar but complementary information to infer the abundance differences of proteins in a complex/simple mixture.

For example, one of the major benefits of using ICAT labeling is its biotin tag and acid-cleavable chemistry. The biotin chemistry of its tags enables targeted purification of labeled samples by protein avidin tetramers, essentially simplifying sample complexity by selectively removing peptides other than those that are biotin labeled [Gygi et al., 1999]. This works as a method to simplify the proteome, enabling the segregation between quantifiable and nonquantifiable proteins. While the principles of ICAT are technically sound, its application to the cyanobacteria may be limited. ICAT chemistry was designed to target only cysteine residues, since they represent only a limited proportion of the amino acid population and are usually only prominent in eukaryotic proteomes. However, the presence of cysteines in cyanobacteria is extremely limited, and it typically represents only a limited region of their proteomes (for example, ~0.95% amino acids in *Nostoc punctiforme* ATCC 29133 and ~1% amino acids in *Synechocystis* sp. PCC 6803 are cysteines). Compared to other, more prominent amino acid residues such as lysines and arginines, ICAT's metabolic coverage in cyanobacteria is comparatively very limited. Not surprisingly, ICAT has not yet been applied to study cyanobacteria and represents only a method for directed analysis where the population of cysteines are significantly more prominent (e.g., electron-carrier enzymes where up to ~40% of the peptides contain cysteines). The iTRAQ method, in contrast, has been reported in all of the eight currently published peer-reviewed journal proteomics studies. Briefly, iTRAQ chemistry (also, tandem mass tags (TMTs); see later for details) has been developed to target primary/secondary amines using an NHS-ester reaction group, such as those from N-termini of peptides and lysines [Ross et al., 2004]. In principle, all peptides consequently have a freely reacting N-termini amine. Hence, one should expect that every single peptide (thus, protein) can be potentially labeled. Quantification is potentially more relevant because abundance is represented by an average of multiple peptides in a protein and not just one peptide species. The principle of abundance measurements using iTRAQ chemistry is also arguably more straightforward, given that the quantitative signals result from isolation plus fragmentation during tandem MS (MS/MS) and reside within a "quiet" region in the MS/MS spectrum. Collision-induced fragmentation of 8-plex iTRAQ labeled peptides releases an array of ions that report at multiple mass/charge positions 113.1–119.1 and 121.1 *m/z*, giving a reasonably clean, high signal-to-noise signal that infers phenotypical abundance differences [Aggarwal et al., 2006]. Conversely, the method of quantification employed by iTRAQ differs from those used by ICAT chemistry, where intensity difference is measured on the intact peptide ion itself and not on its constituent fragments. Intensity differences between the ions with the putative deuterium mass shift (+1 Da for each $^1H \rightarrow {}^2H$, eight H in total) are then compared to infer abundance differences. The advantages from obtaining relative abundances using the MS/MS fragment region or MS intact ion region is debatable, because the intact ion MS spectrum can be significantly noisier but is not subjected to precursor co-fragmentation and dynamic range suppression [Ow et al., 2008a]. An alternative isobaric mass tagging technique to iTRAQ, called tandem mass tags (TMTs) [Thompson et al., 2003], works with a similar principle, with labels in the MS/MS region 126 to 131, and has yet to be applied in the cyanobacteria.

69.3.2 Pilot iTRAQ Experiments in Cyanobacteria Systems

Prior to an actual biologically relevant study, there were two applications that investigated the compatibility of iTRAQ with cyanobacteria. Both papers were presented as technology manuscripts outlining the strengths and pitfalls of the technique [Chong et al., 2006; Gan et al., 2007]. From a pilot investigation standpoint, the results from these experiments were promising. Both investigations specifically assessed iTRAQ compatibility with high-throughput proteomics in three representatives biological systems, namely, the cyanobacterium *Synechocystis* sp. PCC 6803, the crenarchaeal strain *Sulfolobus sulfataricus* P2, and the model eukaryote *Saccharomyces cerevisiae*. The studies also described methods to improve and assess experimental reproducibility by estimating technical, experimental, and biological variation in iTRAQ measurements. In essence, both papers [Chong et al., 2006; Gan et al., 2007] demonstrated the applicability of the principles and are capable of producing highly precise and technically reproducible measurements. Gan et al. also demonstrated that biological variation should be assessed and monitored (average observed coefficient of variation <30%). Common ground established in both works suggested the need for multiple biological, experimental, and re-injection replicates in order to improve the quality of observations for the numbers of peptides identified and quantified for cross-validation in a large-scale analyses [Chong et al., 2006]. Following these foundations, it provided the necessary platform and implementation for other cyanobacterial iTRAQ studies. Below, we describe two of the most recent implementations that were derived from work established by these pilot experiments.

69.3.3 Application 1: Studying N_2 Fixation in Nostoc sp. PCC 7120 and *Nostoc punctiforme* ATCC 29133

A subsequent investigation in 2007 applied iTRAQ to study the proteome changes during N_2 fixation in cyanobacteria [Stensjo et al., 2007]. This study was the first iTRAQ investigation in cyanobacteria that investigated an actual biological perturbation. A total of 486 proteins were quantified from *Nostoc* sp. PCC 7120 during N_2 fixation. The investigation also mapped over 100 changes in protein abundances during the perturbation. Independently, the study provided the farthest-reaching quantitative overview for the proteome of *Nostoc* sp. PCC 7120 at the time of publication. These data also enabled the comparison between the HTT proteomics with existing HTT transcriptomics data in cyanobacteria. Metabolic changes observed proteomically largely corroborated findings from an earlier genome-wide microarray analysis that studied the same organism under comparable conditions [Ehira et al., 2003].

Two subsequent investigations were also conducted in *Nostoc* to delineate expression changes between differentiated and native vegetative cells. The reports, underlined by three separate iTRAQ experiments in two closely related strains, were designed to target cell-specific proteomes of the heterocysts and of the vegetative cells. The studies introduced the use of cell-level purification to respectively enrich N_2-fixing heterocyst cells and vegetative cells. As an improved understanding, the studies also attempted to address observations from both *Nostoc* sp. PCC 7120 and *Nostoc punctiforme* ATCC 29133 [Ow et al., 2008a,b]. The work led to an additional 57 targets that were potentially regulated during and after triggering of N_2 fixation. In both reports, the response differences between vegetative and N_2-fixing heterocyst cells were in agreement with the proposed relationship with the nitrogen-dependent NtcA transcription regulator. Moreover, the studies provided a generalized overview of *Nostoc*-specific proteome transitions during heterocyst differentiation. The impact of these HT proteomics investigations, together with earlier published DNA microarrays, was able to improve understanding in both physiology of cyanobacteria and industrial application of nitrogenase-mediated biohydrogen generation.

69.3.4 Application 2: Studying Light Intensity and Cellular Stresses in *Prochlorococcus marinus* MED4

There were a number of well-characterized model cyanobacteria that were receiving particular attention. More specifically, an important marine environmental strain, the unicellular *Prochlorococcus marinus* MED4, was one of the three earliest systems (including *Nostoc* and *Synechocystis*) to be studied using the iTRAQ workflow [Pandhal et al., 2007]. Oceanic, photoautotrophic, and a potential bioengineering strain, MED4, possesses one of the most compact naturally occurring genomes [Coleman et al., 2006]. At the time, there was significant interest in understanding the organisms' response to varying light intensities in its natural environment. Given the compactness of the genome, the iTRAQ study was able to profile over 11% of the proteome. The investigation, in turn, raised 15 potentially significant proteins and associated pathways that may be sensitive to light alteration and long-term adaptation.

69.4 EMERGENCE OF IN VIVO LABELING FOR CYANOBACTERIA HT PROTEOMICS

Apart from the more commonly applied chemical labeling techniques, in-vivo-based stable isotope methods such as elemental metabolic labeling and SILAC were also slowly emerging. Although techniques such as SILAC and elemental labeling [Conrads et al., 2001] share similar principles, the approach that enables isotope incorporation and associated cost is different. The most common form of elemental labeling is via use of a modified feedstock containing stable isotopes of nitrogen (^{15}N) and/or carbon (^{13}C) [Snijders et al., 2005a,b]. Often, one of the perceived limitations of metabolic labeling is the requirement for a feed-controlled "culture-to-harvest" environment. However, in a laboratory environment, these requirements are relatively easy to satisfy, though the level of incorporation must be frequently assessed [Pandhal et al., 2008c]. Given that feedstock constituents can be controlled, in vivo labeling becomes a viable and potentially more accurate technique. Thus, in principle, incorporating elemental changes (depending on efficiency) in N or C across different phenotypes results in predictable mass shifts of the proteins. However, unlike in vitro techniques, shifts are only semi-uniform and are dependent on the number of N/C in the protein. These differences in mass later appear "heavier" during MS analyses. Nevertheless, applying in vivo techniques to cyanobacteria has its limitations. Due to the autotrophic nature of cyanobacteria (ability to fix CO_2 and N_2), this makes both N- and C-based elemental labeling difficult, unless conditions for both liquid and gas environments are isolated and also isotopically modified. Despite this, there are scenarios where elemental labeling can be successfully applied, such as non-N_2 fixing strains or strains (such as *Synechocystis*) with silenced/absent N_2 fixing apparatus. The most common method of in vivo elemental labeling in cyanobacteria is through the incorporation of heavy nitrate (NO_3) or ammonia (NH_3). The method has been described in detail in a number of articles studying cyanobacteria and also other systems [Snijders et al., 2007; Pandhal et al., 2008b,c]. We detail below one of the most recent implementation of these in vivo methodologies.

69.4.1 Studying Salt Stress in an Unsequenced Halotolerant Cyanobacterium

As an alternative to the commonly applied in vitro iTRAQ technique, in vivo elemental labeling for proteomics was first reported for cyanobacteria in 2008 to investigate salt tolerance of an unsequenced cyanobacterium, *Euhalothece* sp. BAA001 [Pandhal et al., 2008c]. The influx of development for high throughput software at the time to support metabolic labelling allowed Pandhal et al. to innovate its primary experimental design and data mining methodologies to aid in the understanding of an unsequenced organism. The novelty of the study lies in an integration of a number of important data analysis methods (by applying principles of MS-BLAST and the cross-validation of light and heavy N labeling) to quantify proteins from *Euhalothece* using HTT. In essence, the availability of data from heavy and light pairing peptides and the availability of tools to provide homologous assignments provided greater confirmation of experimental observations that are otherwise very limited if dependent only on orthology-based analysis [Snijders et al., 2007]. This method was described in an earlier report intended to investigate *Sulfolobus* sp. [Snijders et al., 2007]. This, in turn, describes a systematic analysis that identified a population of highly orthologous sequenced genomes that act as identification "anchors" (among others, *Synechocystis* sp. PCC 6803) to match both heavy and light pairing peptides ($^{14}N/^{15}N$ isotopes), allowing cross-referencing of identifications.

A follow-up study reported shortly after by the same authors used a combination of in vivo and in vitro techniques to describe a HT comparative analysis [Pandhal et al., 2008a]. Relying on the previously described technique [Pandhal et al., 2008c], it demonstrated that a mixed population of cyanobacteria (*Euhalothece* BAA001 and *Synechocystis* sp. PCC 6803) can be combined and provide a mixed species analysis, given there were sufficient proteome orthology available between both cyanobacteria [Pandhal et al., 2008b]. The technique based its findings between unsequenced *Euhalothece* and its established orthology with a fully sequenced *Synechocystis* sp. and extended observations to assess their responses to increasing salt stress. The novelty of the study demonstrated the potential of HT quantitative analysis to compare not only an unsequenced organism, but a mixed population of sequenced/unsequenced organisms. The study also addressed the differences between quantitative information provided by iTRAQ and elemental labeling, suggesting that both techniques are essentially complementary.

69.5 IN PERSPECTIVE

While substantial effort has been directed toward improving HTT-proteomics-based understanding in the cyanobacteria, it is obvious that the current developments are pale in comparison to other biological systems. Certainly, financial constraints are fair considerations posed. Nonetheless, this scenario invites greater synergy between cyanobacterial proteomics laboratories. With a number of key laboratories possessing the required instruments, this presents an opportunity to work collaboratively, to develop defined scenario-specific methodologies, including the use/development of "top 5/10" antibodies strategies for affinity pull-down purifications commonly applied to serum samples.

There is still a wealth of subjects in cyanobacterial research that can be further investigated. New approaches are continually being developed, with these techniques covering both hypothesis-driven and discovery applications, and are expected to provide quantification. These methods, once matured and sufficiently optimized, will find their way into cyanobacterial proteomics laboratories. An established concern with iTRAQ (and also possibly TMT), is relative inflexibility with respect to the methods of prefractionation that can be employed [Wiese et al., 2007]. This offers the opportunity for other methods, such as QconCAT to target specific pathways [Beynon et al., 2005]. QconCAT introduces heterologously expressed signature 'heavy' peptides, unique to the parent protein, that can be spiked into samples in known amounts so as to facilitate absolute quantification [Pratt et al., 2006; Rivers et al., 2007; Mirzaei et al., 2008]. Other emerging methods also address the limitations of current techniques, include a protein level version of iTRAQ and ICPL (which has been lately applied to study glycoproteins in cyanobacteria [Kim et al., 2009]), and provide opportunities to improve compatibility between the different workflows. More recently, improvements in chromatography and MS acquisition software have resulted in workflows such as "label-free" quantitation. The principles of "label-free" techniques rely on the measurement of the proteins/peptides themselves without an explicit need for stable isotope manipulation. These techniques, however, are still in their infancy and, within the technique itself, have vastly different approaches in delivering the quantitative information (i.e., spectral counting [Old et al., 2005], molecular feature finding [Old et al., 2005], and more recently MS^E/AMRT [Silva et al., 2005]) Despite this, the absence of labels implies that protein samples will not have to be pretreated, thus simplifying the preparation biochemistry prior to sample analysis [Callister et al., 2006].

Interference from high-abundance phycobiliproteins in cyanobacteria has been a major concern in shotgun proteomics (as previously noted). Wide dynamic range is well known to detrimentally affect the detection range of the MS analysis, effectively "diluting" the presence of low-abundance species. Phycobilisome interference has been reported in a number of studies and has yet to be solved even with current state-of-the-art of shotgun proteomics [Gan et al., 2005; Barrios-Llerena et al., 2006]. Therefore, there is also need to effectively address this issue. Developments of high-efficiency ad hoc methods employing directed separation of these proteins are desperately needed. Examples, such as free-flowing-liquid electrophoresis (FFE), or other efficient techniques with similar principles, could be applied to segregate phycobiliproteins and other abundant species [Wang et al., 2004]. Other techniques such as high-efficiency affinity pull-down of these interfering proteins could also be devised as a method for downstream purification [Lindgren, 1994].

There are also opportunities to apply proteomics to understand other forms of cyanobacteria physiology. Transient differences such as circadian rhythm changes of the proteome may also be designed and monitored. Other forms of data integration, such as with other HTT systems-biology approaches, DNA microarrays, metabolome, or more generally flux data, can also be devised for improved modeling [Noirel et al., 2008; Sanguinetti et al., 2008]. Doing so also promotes the possibility to forward engineer industrially important strains for commercial uses. Likewise, global proteomics changes related to cellular differentiation have only been touched on [Ow et al., 2008a,b]. Other potential investigations, such as the proteomic changes in hormogonia, remain to be carried out [Meeks and Elhai, 2002]. There is also growing interest in the screening of the effects of genetic mutation on the host's proteome. With the increasing number of mutant strains in cyanobacteria, this is definitely a possibility. Other forms of environmental interactions such as symbiosis and marine mapping are some of the many general areas that should be further explored with HT quantitative proteomics [Wase and Wright 2008]. Finally, the foundations set today, and those tomorrow, can also be adapted to study and to understand potential post-translational modifications (PTM) in the cyanobacteria [Cox and Mann, 2007]. As yet, there are only limited reports of these PTM studies, and they have only very recently been reported with quantitative proteomics for discovering pattern changes of pilin glycan in *Synechocystis* sp. (using a targeted single protein analysis) [Kim et al., 2009]. A global examination of PTMs and their implications on cyanobacterial physiology has yet to be realized. These questions remain some of the most prominent investigations waiting to be undertaken.

From the outset, it is obvious that the future of HT proteomics for cyanobacteria will depend heavily on two synergistic factors. Foremost, it requires the willingness of other laboratories to explore the benefits offered by this field. The second key factor is the sustained effort to improve high-throughput proteomics methodologies and technologies by cyanobacterial proteomics laboratories. Developments to high-throughput proteomics, as a whole, are currently ubiquitous and will undoubtedly improve. Less importantly, the ingredients necessary to drive both factors are the aspirations from cyanobacteriologists and their desire to explore HT proteomics.

COMPUTER PROGRAMS PRESENTED

Matrix Science Mascot: http://www.matrixscience.com/

Genebio Phenyx: http://www.genebio.com/products/phenyx/

Thermo Sequest: http://fields.scripps.edu/sequest/

MS-BLAST: http://dove.embl-heidelberg.de/Blast2/msblast.html

BSI Peaks: http://www.bioinformaticssolutions.com/products/peaks/index.php

Acknowledgments

Dr. J. Pandhal and Dr. J. Noirel are thanked for the extensive discussions and constructive comments on the manuscript. We thank the EU FP6 NEST PATHFINDER BioModularH_2 project (contract #043340), and the EPSRC for provision of an Advanced Research Fellowship (GR/A11311/01) to PCW and for funding (EP/E036252/1) under the ChELSI initiative and high throughput technologies (GR/S84347/01).

REFERENCES

Aggarwal K, Choe LH, Lee KH. 2006. Shotgun proteomics using the iTRAQ isobaric tags. *Brief. Funct. Genomic Proteomic* **5**:112–120.

Anderson DC, Campbell EL, Meeks JC. 2006. A soluble 3D LC/MS/MS proteome of the filamentous cyanobacterium *Nostoc punctiforme*. *J. Proteome Res*. **5**:3096–3104.

Anderson NL, Polanski M, Pieper R, Gatlin T, Tirumalai RS, et al. 2004. The human plasma proteome: A nonredundant list developed by combination of four separate sources. *Mol. Cell. Proteomics* **3**:311–326.

Angermayr SA, Hellingwerf KJ, Lindblad P, de Mattos MJ. 2009. Energy biotechnology with cyanobacteria. *Curr. Opin. Biotechnol*. **20**:257–263.

Bantscheff M, Schirle M, Sweetman G, Rick J, Kuster B. 2007. Quantitative mass spectrometry in proteomics: A critical review. *Anal. Bioanal. Chem*. **389**:1017–1031.

Barrios-Llerena ME, Chong PK, Gan CS, Snijders AP, Reardon KF, et al. 2006. Shotgun proteomics of cyanobacteria—Applications of experimental and data-mining techniques. *Brief. Funct. Genomic Proteomic* **5**:121–132.

Barrios-Llerena ME, Reardon KF, Wright PC. 2007. 2-DE proteomic analysis of the model cyanobacterium *Anabaena variabilis*. *Electrophoresis* **28**:1624–1632.

Beynon RJ, Doherty MK, Pratt JM, Gaskell SJ. 2005. Multiplexed absolute quantification in proteomics using artificial QCAT proteins of concatenated. *Nat. Methods* **2**:587.

Burja AM, Dhamwichukorn S, Wright PC. 2003. Cyanobacterial postgenomic research and systems biology. *Trends Biotechnol*. **21**:504–511.

Callister SJ, Barry RC, Adkins JN, Johnson ET, Qian W, et al. 2006. Normalization approaches for removing systematic biases associated with mass spectrometry and label-free proteomics. *J. Proteome Res*. **5**:277.

Chernushevich IV, Loboda AV, Thomson BA. 2001. An introduction to quadrupole-time-of-flight mass spectrometry. *J. Mass Spectrom*. **36**:849–865.

Chong PK, Burja AM, Radianingtyas H, Fazeli A, Wright PC. 2007. Proteome analysis of *Sulfolobus solfataricus* P2 propanol metabolism. *J. Proteome Res*. **6**:1430–1439.

Chong PK, Gan CS, Pham TK, Wright PC. 2006. Isobaric tags for relative and absolute quantitation (iTRAQ) reproducibility: Implication of multiple injections. *J. Proteome Res*. **5**:1232–1240.

Coleman ML, Sullivan MB, Martiny AC, Steglich C, Barry K, et al. 2006. Genomic islands and the ecology and evolution of *Prochlorococcus*. *Science* **311**:1768–1770.

Conrads TP, Alving K, Veenstra TD, Belov ME, Anderson GA, et al. 2001. Quantitative analysis of bacterial and mammalian proteomes using a combination of cysteine affinity tags and 15N-metabolic labeling. *Anal. Chem*. **73**:2132–2139.

Cox J, Mann M. 2007. Is proteomics the new genomics? *Cell* **130**:395–398.

Delpla I, Jung AV, Baures E, Clement M, Thomas O. 2009. Impacts of climate change on surface water quality in relation to drinking water production. *Environ Int*. **35**:1225–1233.

Dismukes GC, Carrieri D, Bennette N, Ananyev GM, Posewitz MC. 2008. Aquatic phototrophs: Efficient alternatives to land-based crops for biofuels. *Curr. Opin. Biotechnol*. **19**:235–240.

Ehira S, Ohmori M, Sato N. 2003. Genome-wide expression analysis of the responses to nitrogen deprivation in the heterocyst-forming cyanobacterium *Anabaena* sp. strain PCC 7120. *DNA Res*. **10**:97–113.

Elias JE, Gygi SP. 2007. Target-decoy search strategy for increased confidence in large-scale protein identifications by mass spectrometry. *Nat. Methods* **4**:207–214.

Elias JE, Haas W, Faherty BK, Gygi SP. 2005. Comparative evaluation of mass spectrometry platforms used in large-scale proteomics investigations. *Nat. Methods* **2**:667–675.

Eng JK, McCormack AL, Yates Jr. 1994. An approach to correlate tandem mass spectral data of peptides with amino acid sequences in a protein database. *J. Am. Soc. Mass Spectrom*. **5**:976–989.

Friboulet A, Thomas D. 2005. Systems Biology-an interdisciplinary approach. *Biosens. Bioelectron*. **20**:2404–2407.

Gan CS, Reardon KF, Wright PC. 2005. Comparison of protein and peptide prefractionation methods for the shotgun proteomic analysis of *Synechocystis* sp. PCC 6803. *Proteomics* **5**:2468–2478.

Gan CS, Chong PK, Pham TK, Wright PC. 2007. Technical, experimental, and biological variations in isobaric tags for relative and absolute quantitation (iTRAQ). *J. Proteome Res*. **6**:821–827.

Gevaert K, Ghesquiere B, Staes A, Martens L, Van Damme J, et al. 2004. Reversible labeling of cysteine-containing peptides allows their specific chromatographic isolation for non-gel proteome studies. *Proteomics* **4**:897–908.

Gygi SP, Rist B, Gerber SA, Turecek F, Gelb MH, et al. 1999. Quantitative analysis of complex protein mixtures using isotope-coded affinity tags. *Nat. Biotechnol*. **17**:994–999.

Han Y, Ma B, Zhang K. 2004. SPIDER: Software for protein identification from sequence tags with de novo sequencing error. Computational Systems Bioinformatics Conference, 2004. CSB 2004. Proceedings. 2004 IEEE. 206–215.

Hood L. 2003. Systems biology: Integrating technology, biology, and computation. *Mech. Ageing Dev*. **124**:9–16.

Huang F, Parmryd I, Nilsson F, Persson AL, Pakrasi HB, et al. 2002. Proteomics of *Synechocystis* sp. strain PCC 6803: Identification of plasma membrane proteins. *Mol. Cell. Proteomics* **1**:956–966.

Hunsucker SW, Klage K, Slaughter SM, Potts M, Helm RF. 2004. A preliminary investigation of the *Nostoc punctiforme* proteome. *Biochem. Biophys. Res. Commun*. **317**:1121–1127.

Ideker T, Galitski T, Hood L. 2001. A new approach to decoding life: systems biology. *Annu. Rev. Genomics Hum. Genet*. **2**:343–372.

Kaneko T, Nakamura Y, Wolk CP, Kuritz T, Sasamoto S, et al. 2001. Complete genomic sequence of the filamentous nitrogen-fixing cyanobacterium *Anabaena* sp. strain PCC 7120. *DNA Res*. **8**:205–13; 227–253.

Kim YH, Kim JY, Kim SY, Lee JH, Lee JS, et al. 2009. Alteration in the glycan pattern of pilin in a nonmotile mutant of *Synechocystis* sp. PCC 6803. *Proteomics* **9**:1075–1086.

Kitano H. 2002. Systems biology: A brief overview. *Science* **295**:1662–1664.

Kwon J, Oh J, Park C, Cho K, Kim SI, et al. 2009. Systematic cyanobacterial membrane proteome analysis by combining acid hydrolysis and digestive enzymes with nano-liquid chromatography–Fourier transform mass spectrometry. *J. Chromatogr. A* **1217**(3):285–293.

Lane CS. 2005. Mass spectrometry-based proteomics in the life sciences. *Cell. Mol. Life Sci*. **62**:848–869.

Lindgren GE. 1994. Immobilized metal affinity chromatography (IMAC). *Am. Biotechnol. Lab*. **12**:36.

Liu Y, He J, Li C, Benitez R, Fu S, et al. 2009. Identification and confirmation of biomarkers using an integrated platform for quantitative analysis of glycoproteins and their glycosylations. *J. Proteome Res*. **9**(2):798–805.

March RE. 1997. An introduction to quadrupole ion trap mass spectrometry. *J. Mass Spectrom*. **32**:351É369.

Medzihradszky KF, Campbell JM, Baldwin MA, Falick AM, Juhasz P, et al. 2000. The characteristics of peptide collision-induced dissociation using a high-performance MALDI-TOF/TOF tandem mass spectrometer. *Anal. Chem*. **72**:552–558.

Meeks JC, Elhai J. 2002. Regulation of cellular differentiation in filamentous cyanobacteria in free-living and plant-associated symbiotic growth states. *Microbiol. Mol. Biol. Rev*. **66**:94–121; table of contents.

Mirzaei H, McBee JK, Watts J, Aebersold R. 2008. Comparative evaluation of current peptide production platforms used in absolute quantification in proteomics. *Mol. Cell. Proteomics* **7**:813–823.

Noirel J, Ow SY, Sanguinetti G, Jaramillo A, Wright PC. 2008. Automated extraction of meaningful pathways from quantitative proteomics data. *Brief. Funct. Genomic Proteomic* **7**(2):136–146.

Norling B, Zak E, Andersson B, Pakrasi H. 1998. 2D-isolation of pure plasma and thylakoid membranes from the cyanobacterium *Synechocystis* sp. PCC 6803. *FEBS Lett*. **436**:189–192.

O'Farrell PH. 1975. High resolution two-dimensional electrophoresis of proteins. *J. Biol. Chem*. **250**:4007–4021.

OLD WM, MEYER-ARENDT K, AVELINE-WOLF L, PIERCE KG, MENDOZA A, et al. 2005. Comparison of label-free methods for quantifying human proteins by shotgun proteomics. *Mol. Cell. Proteomics* **4**:1487–1502.

ONG SE, BLAGOEV B, KRATCHMAROVA I, KRISTENSEN DB, STEEN H, et al. 2002. Stable isotope labeling by amino acids in cell culture, SILAC, as a simple and accurate approach to expression proteomics. Mol. Cell. Proteomics **1**:376–386.

OW SY, CARDONA T, TATON A, MAGNUSON A, LINDBLAD P, et al. 2008a. Quantitative shotgun proteomics of enriched heterocysts from *Nostoc* sp. PCC 7120 using 8-plex isobaric peptide tags. *J. Proteome Res*. **7**(4):1615–1628.

OW SY, NOIREL J, CARDONA T, TATON A, LINDBLAD P, et al. 2008b. Quantitative overview of N_2 fixation in *Nostoc punctiforme* ATCC 29133 through cellular enrichments and iTRAQ shotgun proteomics. *J. Proteome Res*. **8**:187–198.

OW SY, SALIM M, NOIREL J, EVANS C, REHMAN I, et al. 2009. iTRAQ underestimation in simple and complex mixtures: "The good, the bad and the ugly". *J. Proteome Res*. **8**:5347–5355.

PANDHAL J, WRIGHT PC, BIGGS CA. 2007. A quantitative proteomic analysis of light adaptation in a globally significant marine cyanobacterium *Prochlorococcus marinus* MED4. *J. Proteome Res*. **6**:996–1005.

PANDHAL J, OW SY, WRIGHT PC, BIGGS CA. 2008a. Comparative proteomics study of salt tolerance between a Nonsequenced extremely halotolerant cyanobacterium and its mildly halotolerant relative using in vivo metabolic labeling and in vitro isobaric labeling. *J. Proteome Res*. **00**:000–000.

PANDHAL J, OW SY, WRIGHT PC, BIGGS CA. 2008b. Comparative proteomics study of salt tolerance between a nonsequenced extremely halotolerant cyanobacterium and its mildly halotolerant relative using in vivo metabolic labeling and in vitro isobaric labeling. *J Proteome Res*. ASAP Article.

PANDHAL J, SNIJDERS AP, WRIGHT PC, BIGGS CA. 2008c. A cross-species quantitative proteomic study of salt adaptation in a halotolerant environmental isolate using 15N metabolic labelling. *Proteomics* **8**:2266–2284.

PANDHAL J, WRIGHT PC, BIGGS CA. 2008d. Proteomics with a pinch of salt: A cyanobacterial perspective. *Saline Systems* **4**:1.

PERKINS DN, PAPPIN DJ, CREASY DM, COTTRELL JS. 1999. Probability-based protein identification by searching sequence databases using mass spectrometry data. *Electrophoresis* **20**:3551–3567.

PRATT JM, SIMPSON DM, DOHERTY MK, RIVERS J, GASKELL SJ, et al. 2006. Multiplexed absolute quantification for proteomics using concatenated signature peptides encoded by QconCAT genes. *Nat. Protoc*. **1**:1029–1043.

RAZQUIN P, FILLAT MF, SCHMITZ S, STRICKER O, BOHME H, et al. 1996. Expression of ferredoxin-NADP+ reductase in heterocysts from *Anabaena* sp. *Biochem. J*. **316**(Pt 1): 157–160.

RIGHETTI PG, CAMPOSTRINI N, PASCALI J, HAMDAN M, ASTNER H. 2004. Quantitative proteomics: A review of different methodologies. *Eur. J. Mass Spectrom. (Chichester, Eng.)* **10**:335–348.

RIVERS J, SIMPSON DM, ROBERTSON DH, GASKELL SJ, BEYNON RJ. 2007. Absolute multiplexed quantitative analysis of protein expression during muscle development using QconCAT. *Mol. Cell Proteomics* **6**:1416–27.

ROSS PL, HUANG YN, MARCHESE JN, WILLIAMSON B, PARKER K, et al. 2004. Multiplexed protein quantitation in *Saccharomyces cerevisiae* using amine-reactive isobaric tagging reagents. *Mol Cell Proteomics* **3**:1154–1169.

SANGUINETTI G, NOIREL J, WRIGHT PC. 2008. MMG: A probabilistic tool to identify submodules of metabolic pathways. *Bioinformatics* **24**(8):1078–1084.

SAZUKA T. 2001. Comparative proteomics of cyanobacteria: *Synechocystis* sp. *strain PCC 6803* and *Anabaena sp. strain PCC 7120*. Japanese *J. Electrophoresis*. **45**:123–128.

SCAIFE MA, BURJA AM, WRIGHT PC. 2009. Characterization of cyanobacterial beta-carotene ketolase and hydroxylase genes in *Escherichia coli*, and their application for astaxanthin biosynthesis. *Biotechnol. Bioeng*. **103**:944–955.

SCHENA M, SHALON D, DAVIS RW, BROWN PO. 1995. Quantitative monitoring of gene expression patterns with a complementary DNA microarray. *Science* **270**:467–470.

SCHMIDT A, KELLERMANN J, LOTTSPEICH F. 2005. A novel strategy for quantitative proteomics using isotope-coded protein labels. *Proteomics* **5**:4–15.

SHEVCHENKO A, SUNYAEV S, LOBODA A. 1996. Charting the proteomes of organisms with unsequenced genomes by MALDI-quadrupole time-of-flight mass spectrometry and BLAST homology searching. *Proc. Natl. Acad. Sci. USA* **93**:14440–14445.

SHEVCHENKO A, LOBODA A, SHEVCHENKO A, ENS W, STANDING KG. 2000. MALDI quadrupole time-of-flight mass spectrometry: a powerful tool for proteomic research. *Anal. Chem*. **72**:2132–2141.

SILVA JC, DENNY R, DORSCHEL CA, GORENSTEIN M, KASS IJ, et al. 2005. Quantitative proteomic analysis by accurate mass retention time pairs. *Anal. Chem*. **77**:2187–2200.

SIMON WJ, HALL JJ, SUZUKI I, MURATA N, SLABAS AR. 2002. Proteomic study of the soluble proteins from the unicellular cyanobacterium *Synechocystis* sp. PCC6803 using automated matrix-assisted laser desorption/ionization-time of flight peptide mass fingerprinting. *Proteomics* **2**:1735–1742.

SMITH JC, FIGEYS D. 2006. Proteomics technology in systems biology. *Mol. Biosyst*. **2**:364–370.

SNIJDERS AP, DE VOS MG, DE KONING B, WRIGHT PC. 2005a. A fast method for quantitative proteomics based on a combination between two-dimensional electrophoresis and 15N-metabolic labelling. *Electrophoresis* **26**:3191–3199.

SNIJDERS AP, DE VOS MG, WRIGHT PC. 2005b. Novel approach for peptide quantitation and sequencing based on 15N and 13C metabolic labeling. *J. Proteome Res*. **4**:578–585.

SNIJDERS AP, DE KONING B, WRIGHT PC. 2007. Relative quantification of proteins across the species boundary through the use of shared peptides. *J. Proteome Res*. **6**:97–104.

SOUCHELNYTSKYI S. 2005. Bridging proteomics and systems biology: What are the roads to be traveled ? *Proteomics* **5**:4123–4137.

SRIVASTAVA R, PISAREVA T, NORLING B. 2005. Proteomic studies of the thylakoid membrane of *Synechocystis* sp. PCC 6803. *Proteomics*. **5**:4905–4916.

STAES A, DEMOL H, VAN DAMME J, MARTENS L, VANDEKERCKHOVE J, et al. 2004. Global differential non-gel proteomics by quantitative and stable labeling of tryptic peptides with oxygen-18. *J. Proteome Res*. **3**:786–791.

STANIER RY, COHEN-BAZIRE G. 1977. Phototrophic prokaryotes: the cyanobacteria. *Annu. Rev. Microbiol*. **31**:225–274.

STENSJO K, OW SY, BARRIOS-LLERENA ME, LINDBLAD P, WRIGHT PC. 2007. An iTRAQ-based quantitative analysis to elaborate the proteomic response of Nostoc sp. PCC 7120 under N2 fixing conditions. *J. Proteome Res*. **6**:621–635.

SWANSON SK, WASHBURN MP. 2005. The continuing evolution of shotgun proteomics. *Drug Discovery Today* **10**:719–725.

TABB DL. 2008. What's driving false discovery rates ? J. *Proteome Res*. **7**:45–46.

TAMAGNINI P, LEITAO E, OLIVEIRA P, FERREIRA D, PINTO F, et al. 2007. Cyanobacterial hydrogenases: Diversity, regulation and applications. *FEMS Microbiol. Rev*. **31**:692–720.

THOMPSON A, SCHAFER J, KUHN K, KIENLE S, SCHWARZ J, et al. 2003. Tandem mass tags: A novel quantification strategy for comparative analysis of complex protein mixtures by MS/MS. *Anal. Chem*. **75**:1895–1904.

Venter JC, Adams MD, Myers EW, Li PW, Mural RJ, et al. 2001. The sequence of the human genome. *Science* **291**:1304–1351.

Wang Y, Hancock WS, Weber G, Eckerskorn C, Palmer-Toy D. 2004. Free flow electrophoresis coupled with liquid chromatography-mass spectrometry for a proteomic study of the human cell line (K562/CR3). *J. Chromatogr. A* **1053**:269–278.

Wase VN, Wright PC. 2008. Systems biology of cyanobacterial secondary metabolite production and its role in drug discovery. *Expert Opin. Drug Discovery* **3**:903–929.

Whitton BA. 1992. Diversity, ecology and taxonomy of the cyanobacteria. In: Mann NH and Carr NG, (eds) Photosynthetic Prokaryotes. Plenum Press, New York, 1–51.

Whitton BA, Potts M. 2000. Introduction to the cyanobacteria. In *The Ecology of Cyanobacteria: Their Diversity in Time and Space*, Kluwer Academic, Brian A Whitton, Malcolm Potts, Boston pp. 1–11.

Wiedner C, Rucker J, Bruggemann R, Nixdorf B. 2007. Climate change affects timing and size of populations of an invasive cyanobacterium in temperate regions. *Oecologia* **152**:473–484.

Wiese S, Reidegeld KA, Meyer HE, Warscheid B. 2007. Protein labeling by iTRAQ: A new tool for quantitative mass spectrometry in proteome research. *Proteomics* **7**:340–350.

Wolters DA, Washburn MP, Yates JR, 3rd. 2001. An automated multidimensional protein identification technology for shotgun proteomics. *Anal. Chem*. **73**:5683–5690.

Wu CC, MacCoss MJ. 2002. Shotgun proteomics: Tools for the analysis of complex biological systems. *Curr. Opin. Mol. Ther*. **4**:242–250.

Zhen Y, Xu N, Richardson B, Becklin R, Savage JR, et al. 2004. Development of an LC-MALDI method for the analysis of protein complexes. *J. Am. Soc. Mass Spectrom*. **15**:803–822.

Chapter 70

Protein Expression Profile of an Environmentally Important Bacterial Strain: The Chromate Response of *Arthrobacter* Species Strain FB24

KRISTENE L. HENNE, JOSHUA E. TURSE, CINDY H. NAKATSU, AND ALLAN E. KONOPKA

70.1 INTRODUCTION

Defining the complete proteome of a bacterial strain can be a daunting task, but is nonetheless a logical complement to genome annotation. Global, systematic approaches to analyzing organisms have developed sequentially from genomics, to transcriptomics, and more recently to proteomics. Similarly meta-analytical approaches have been applied to microbial communities to develop a systems-based model. More commonly metatranscriptomics and metaproteomics (see Chapters 62–65 and 66–69, Vol. I) are added to metagenomics analyses, permitting development of a holistic definition of microbial community lifestyle [Wilmes and Bond, 2006; Callister et al., 2008; VerBerkmoes et al., 2009]. Very few metaproteomic analyses of soil habitats have been completed. This is likely due to the technical hurdles that may accompany community-wide "omics" such as adequate protein extraction and accurate peptide identification [VerBerkmoes et al., 2009 (see Chapters 66, 67 Vol. I)]. Once these challenges are met, meta-based studies will undoubtedly provide valuable insight into microbial processes in mixed communities. These meta-analyses are informed by knowledge acquired about individual microbial taxa that are recognized as important components of communities. Proteomic studies at the strain level not only assist validation of genome annotations, but also can identify novel proteins responsible for ecologically important phenotypes [Graves and Haystead, 2002; de Souza et al., 2008, 2009].

We have applied [Henne et al., 2009b] both two-dimensional gel electrophoresis (2-DGE) and liquid chromatography coupled with tandem mass spectrometry (LC-MS/MS) to interrogate the global proteomic response of *Arthrobacter* sp. strain FB24 to chromate stress. *Arthrobacter* are high G + C gram-positive bacteria that are widespread in the environment [Jones and Keddie, 1999]. Of particular interest is their predominance in polluted soils [Crocker et al., 2000; Garten et al., 2000; Benyehuda et al., 2003; Fredrickson et al., 2004] and their ability to resist high levels of toxic metals [Margesin and Schinner, 1997; Crocker et al., 2000; Benyehuda et al., 2003]. *Arthrobacter* sp. strain FB24 was isolated from soils characterized by high levels of chromium, in addition to lead and organic pollutants [Beasley, 2004; Nakatsu et al., 2005]. Strain FB24 has the ability to tolerate chromate levels higher than most Proteobacteria [Camargo et al., 2005]. The *Arthrobacter* sp. strain FB24

Handbook of Molecular Microbial Ecology, Volume I: Metagenomics and Complementary Approaches, First Edition. Edited by Frans J. de Bruijn.

genome was found to contain a plasmid-borne chromate resistance determinant consisting of five genes similar to other chromate resistance genes, in addition to three novel genes that have not been associated with chromate resistance [Henne et al., 2009a]. The proteomic study was intended to examine both (a) the global response of strain FB24 to chromate and (b) the expression of this local cluster of genes at the protein level. This review highlights the major physiological changes of FB24 under chromate stress.

70.2 METHODS

This is a summary of the materials and methods. For full details, please see the original publication [Henne et al., 2009b].

70.2.1 Bacterial Culture Conditions

Log-phase cells of *Arthrobacter* sp. strain FB24 were harvested from three replicate cultures grown aerobically at 30°C in 500-ml batch cultures consisting of either 0.2X (for 2-DGE) or full-strength (for LC-MS/MS) nutrient broth (Difco) amended with 0 mM, 5 mM, or 20 mM potassium chromate [K_2CrO_4]. A stationary-phase, untreated control culture was included in the LC-MS/MS study.

70.2.2 Two-Dimensional Gel Electrophoresis, Image Analysis, and Protein Identification

Whole-cell lysates were prepared for 2-DGE analyses by sonicating cell pellets in solubilization buffer [9 M urea, 4% NP-40, 2% ampholytes (pH 3–10, Bio-Rad), 2% β-mercaptoethanol and protease inhibitors]. Protein concentration was determined by the modified Bradford protein assay [Ramagli, 1999]. Forty micrograms of protein sample was separated in the first dimension by isoelectric focusing (IEF) for 14,000 volt-hours (Vh) in tube gels containing a mixture of four parts pH 5–7 and one part pH 3–10 carrier ampholytes (Bio-Rad). Proteins were separated in the second dimension by SDS-PAGE in a 10–17% acrylamide gradient [O'Farrell, 1975; Anderson and Anderson, 1978a, 1978b] and visualized by silver staining [Giometti et al.,1991]. For preparative 2-DGE, 150 μg of protein samples were separated in the same manner and visualized by Coomassie staining [Giometti et al., 1988]. Spot detection and pattern matching were completed with the Progenesis software (Version 2003.03; Nonlinear USA). A representative 2-DGE image of proteins from lysates of cells grown without Cr(VI) was used as the reference pattern. Normalization of spot intensities was carried out as described previously [Ahrendt et al., 2007].

Protein spots showing significant ($p < 0.05$) differences in abundance between growth conditions were identified by peptide mass analysis by LC-MS/MS at Argonne National Laboratory and verified by independent analysis at the W.M. Keck Foundation Biotechnology Resource Laboratory, Yale University, New Haven, CT. Tryptic peptide analysis at Argonne National Laboratory was conducted as described previously [Ahrendt et al., 2007]. The MS/MS spectra were used in Mascot searches of the predicted protein sequence of the *Arthrobacter* sp. FB24 ORF database using the total genome sequence provided by the Department of Energy—Joint Genome Initiative (http://tinyurl.com/2tejrk). Protein identifications reported using the Mascot search algorithm had scores greater than 100, with at least two peptides matching the theoretical masses of tryptic peptides.

70.2.3 Preparation of Arthrobacter Digests for LC-MS/MS

Preparation of Arthrobacter digests and LC-MS/MS analysis was performed by the Environmental Molecular Sciences Laboratory (EMSL) at Pacific Northwest National Laboratory (PNNL), Richland, WA. Global, soluble and insoluble sample partitions were prepared as described previously [Lipton et al., 2002; Smith et al., 2002]. Samples were digested with trypsin and fractionated by gradient strong cation exchange chromatography (SCX) as described previously [Lipton et al., 2002]. Each SCX fraction was analyzed with an automated custom-built capillary HPLC system coupled online to an LTQ ion trap mass spectrometer (ThermoElectron) by using an electrospray ionization interface. The LC-MS/MS data were searched against a database of theoretical masses generated from the Arthrobacter sp. strain FB24 FastA (http://tinyurl.com/2zfpuw) and a sequence-reversed database using the SEQUEST algorithm [Eng et al., 1994]. The SEQUEST results were filtered using the following criteria: (1) Xcorr 1.9 for charge state +1 full tryptic peptides; (2) Xcorr 2.2 for charge state +2 full and partial tryptic peptides; and (3) Xcorr 3.75 for charge state +3 full tryptic and partial tryptic peptides. The delta correlation value (DelCn2) > 0.1 was used in all cases [Washburn et al., 2001]. The identified proteins were categorized based on their cellular locations and biological processes according to Gene Ontology (GO) information obtained from the European Bioinformatics Institute (http://tinyurl.com/2ckfsd) as well as by the Clusters of Orthologous Groups (COG) function categories available in The Integrated Microbial Genomes (IMG) system (http://tinyurl.com/27eyv3). Expression analysis was carried out using Agilent GeneSpring GX v 7.3.1. Data were analyzed using the 0 mM chromate/log phase growth condition as the basal state using the MS/MS spectral count information for the identified proteins as a measure of relative protein abundances [Zhu et al., 2004; Wang et al., 2006; Lu et al., 2007]. Comparison of relative protein abundances from bacteria in log-phase growth with no chromate stress to either chromate stress conditions or stationary-phase growth in the absence of chromate was used to obtain fold change in protein abundance. Clustering was performed using average linkage and Pearson correlation.

70.3 RESULTS AND DISCUSSION

70.3.1 Two-Dimensional Gel Electrophoresis

Arthrobacter sp. strain FB24 exhibited little change in growth rate and yield up to 5 mM chromate and tolerated chromate concentrations up to 200 mM. At Cr(VI) concentrations of 5 mM and greater, both the growth rate and yield decreased significantly [Henne et al., 2009b]. This is in stark contrast to other bacteria whose transcriptomes and proteomes have been interrogated in response to chromate stress such as *Escherichia coli*, *Shewanella oneidensis* MR-1, and *Caulobacter crescentus*. These three strains exhibited growth inhibition at Cr(VI) concentrations of 0.25 mM, 2 mM, and 0.05 mM, respectively [Hu et al., 2005; Ackerley et al., 2006; Thompson et al., 2007]. Two-dimensional gel electrophoresis (2-DGE) was applied to investigate the proteomic responses driving decreased growth rate and yield of *Arthrobacter* sp. strain FB24 at 5 mM and 20 mM Cr(VI). Two-dimensional gel electrophoresis detected approximately 1000 proteins representing 22% of the potential 4536 protein-coding ORFs within the genome. Twenty-six proteins exhibited significant ($p < 0.05$) differences in abundance in either 5 or 20 mM Cr(VI) compared to those found in the absence of chromate (**Fig. 70.1**). Nineteen of these proteins were identified by subsequent LC-MS/MS analysis (**Table 70.1**).

The proteins that showed changes in expression suggest alterations in central carbon metabolism, amino acid biosynthesis, and intracellular sulfur availability. Most notably, a threefold increase in cystathionine beta-synthase (Arth_3062) was observed under 20 mM Cr(VI). Increased abundance of proteins involved in cysteine synthesis in response to chromate stress is consistent with what has been shown in *E. coli* and *S. oneidensis* [Ackerley et al., 2006; Brown et al., 2006] and is likely due to the inhibition of sulfate transport induced by the structurally similar chromate ion [Ohtake et al., 1987; Cervantes and Silver, 1992], leading to limitations in sulfur-containing amino acids. Unlike *E. coli, S. oneidensis*, and *C. crescentus*, which also exhibited general stress and oxidative stress responses [Hu et al., 2005; Ackerley et al., 2006; Thompson et al., 2007], the 2-DGE data for FB24 suggest that the decreased growth at 5 and 20 mM Cr(VI) are due to other specific, metabolic changes. To obtain a more comprehensive look at the proteomic response of strain FB24 to chromate, global protein expression profiles of strain FB24 under different growth states were further determined using LC/LC-MS/MS.

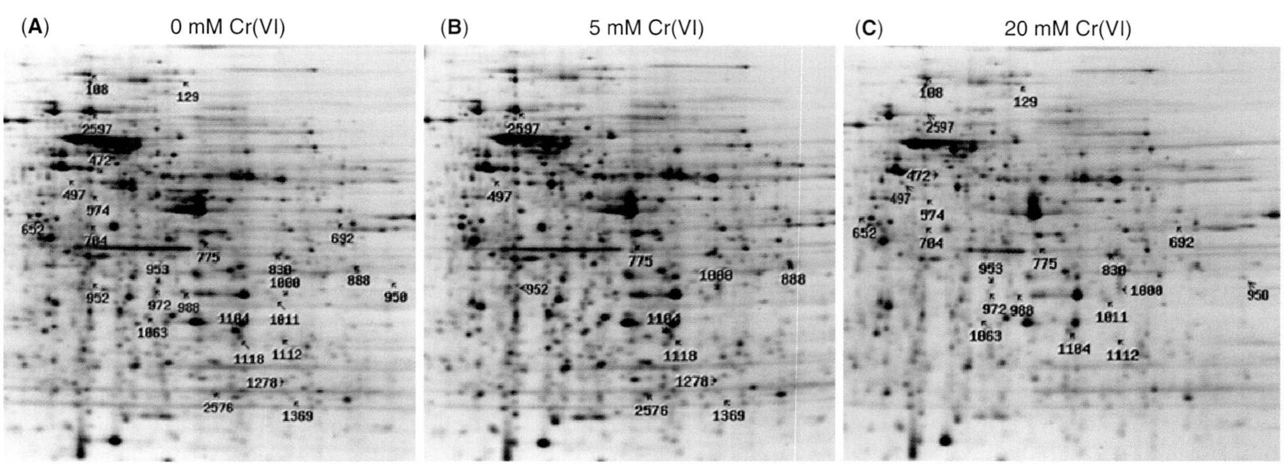

Figure 70.1 2-DGE proteomic profiles of *Arthrobacter* sp. strain FB24 in response to **(A)** 0 mM, **(B)** 5 mM, and **(C)** 20 mM Cr(VI). Proteins exhibiting a significant ($p < 0.05$) change in abundance in 5 or 20 mM Cr(VI) compared to the 0 mM Cr(VI) control state are marked in blue. (Reprinted with permission from Henne et al. [2009b]. Copyright 2009 American Chemical Society.)

Table 70.1 Identifications of Differentially Expressed Proteins from 2-DGE

Gel Spot Number	Locus	Relative Abundance Cr(VI)/No Cr(VI)	Student's *t*-Test (*p*-Value)	Mascot Score	Protein Identification
		Proteins Differentially Expressed Under 5 mM Cr(VI)			
888	3942	3.0	0.0003	673	Aldo/keto reductase
952	1058	2.3	0.0035	172	Oxidoreductase
1278	4418	0.53	0.0047	692	Iron (metal)-dependent repressor, DtxR family
1369	3395	0.42	0.028	759	Transcriptional regulator, Crp/Fnr family
2576	0093	2.1	0.016	689	LamB/YcsF family protein
		Proteins Differentially Expressed Under 20 mM Cr(VI)			
108	1434	2.0	0.0021	2667	Polyribonucleotide nucleotidyltransferase
129	2097	2.4	0.0071	1222	Transketolase
472	3062	3.3	0.0006	3113	Cystathionine beta-synthase
574	2096	0.60	0.0151	687	Transaldolase
652	3994	0.44	0.0032	735	Malate dehydrogenase (oxaloacetate-decarboxylating)
692	1499	2.2	0.0035	384	Acetylornithine/succinylornithine aminotransferases
704	0607	0.61	0.0081	1189	Extracellular solute-binding protein, family 1
830	1414	2.3	0.0026	621	NusA antitermination factor
950	1649	2.3	0.0070	78	Aldo-keto reductase
988	2087	0.49	0.0015	179	Glyceraldehyde-3-phosphate dehydrogenase, type 1
1011	2189	2.1	0.0003	516	Band 7 protein
		Proteins Differentially Expressed Under Both Conditions[a]			
497	2248	3.7	0.0001	862	Glycerol kinase
		3.4	0.0004		
775	3192	0.58	0.0018	948	Pyruvate dehydrogenase complex subunit
		0.55	0.0034		
1104	0627	0.44	0.0012	848	Nuclear export factor, GLE 1
		0.42	0.0005		
2597	0667	0.60	0.0011	1195	Phosphoenolpyruvate carboxykinase (GTP)
		0.18	0.0107		

[a]Top numbers for relative abundance and Student's *t*-test are for 5 mM Cr(VI) analysis. Bottom numbers are for 20 mM Cr(VI) analysis.

Source: Reprinted with permission from Henne et al. [2009b]. Copyright 2009 American Chemical Society.

70.3.2 LC/LC-MS/MS-Based Proteomics

The application of LC/LC-MS/MS based proteomics led to 71% proteome coverage of the 4536 predicted protein-coding ORFs within the *Arthrobacter* sp. strain FB24 genome, identifying 3220 proteins from 619,942 measurements of 56,920 high quality peptides. Abundance changes of proteins expressed in cells under chromate stress relative to log-phase cells grown in the absence of chromate revealed a twofold difference of 968 proteins in the global partition, 874 in the soluble partition and 1021 in the insoluble partition, with 247 proteins being present in each partition (**Figs. 70.2** and **70.3**). The predominant COG functional categories (**Fig. 70.4**) of the differentially expressed proteins were transcription (8–11%), amino acid transport and metabolism (12–15%), energy production and conversion (9%), and carbohydrate transport and metabolism (9–11%). Many of the changes observed were in proteins that centered on the pyruvate node of metabolism. This is an important hub in cellular physiology, connecting carbohydrate metabolism, energy production, and amino acid and lipid biosynthesis. These observations are consistent with the growth rate and biomass yield defects exhibited by *Arthrobacter* sp. strain FB24 when grown on 5 and 20 mM Cr(VI).

At least 20% or more of the differentially expressed proteins had no associated functional assignment. The FB24 genome contains 1036 genes annotated as "hypothetical protein," representing 23% of the predicted proteome (4536 predicted protein-encoding genes). Protein products for 589 of these genes were present

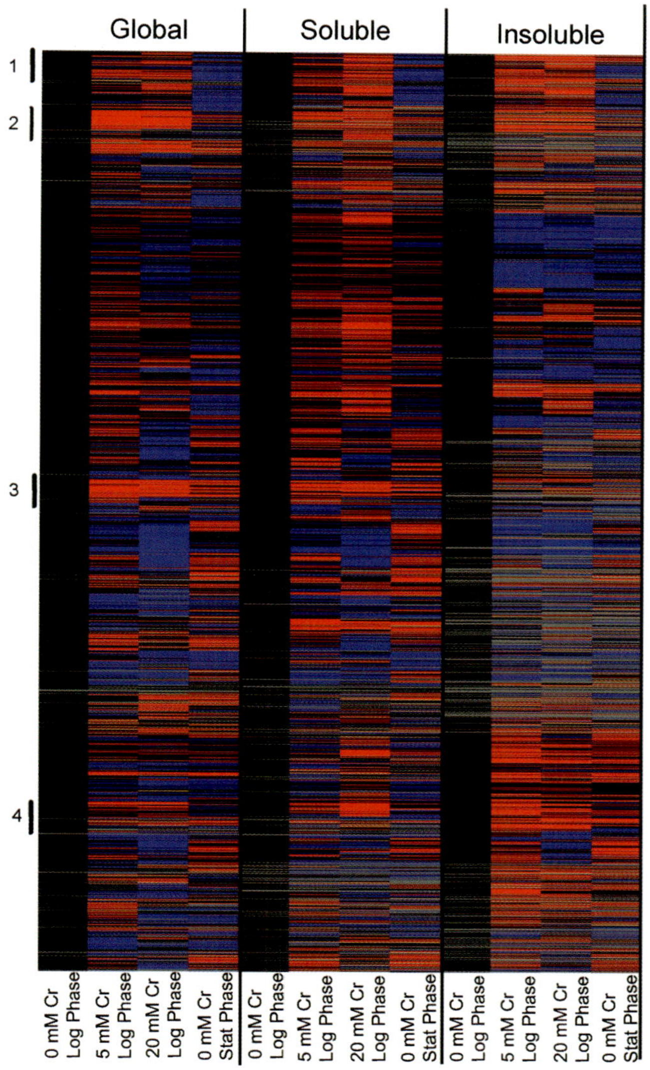

Figure 70.2 Heat map of differentially expressed proteins detected by LC/LC-MS/MS. Fold differences were calculated from comparing peptide count data for each experimental condition to the 0 mM Cr(VI) control condition (black columns). Red, increased abundance; blue, decreased abundance; black, no change; gray, no data. Proteins in the numbered sections of the heat map showed increased abundance in Cr(VI) in at least two of the three partitions and were chosen for further analysis. (Reprinted with permission from Henne et al. [2009b]. Copyright 2009 American Chemical Society.)

in the MS dataset. Of these proteins, 289 showed a twofold change in abundance under chromate conditions, with 28 present in all cellular partitions. Observations such as these will lead to improvements in gene annotation through the promotion of "hypothetical protein" annotations to at the very least "protein of unknown function." Additional data mining and biochemical analyses of proteins of interest—that is, exhibiting large fold-differences—should further aid in assigning functions to hypothetical genes and proteins. The need for functional assignment of poorly characterized proteins was emphasized by Callister et al. [2008], who, in attempting to define the core proteome as a set of conserved, essential proteins across bacterial species, found that a large number of the observed core proteins did not have complete functional assignments. This strongly suggests that there may still be important gaps in our knowledge of basic bacterial physiology.

Figure 70.3 Overlap of proteins from the global, soluble and insoluble partitions that exhibited a twofold difference in abundance in Cr(VI) conditions compared to the 0 mM Cr(VI) control. (Reprinted with permission from Henne et al. [2009b]. Copyright 2009 American Chemical Society.)

In this communication, we focus on proteins that are directly involved in chromate resistance (the chromate resistance determinant, or CRD), proteins with expression profiles similar to the CRD, and proteins suggestive of sulfur limitation conditions. A more comprehensive examination of the major functional categories of proteins, as well as candidate proteins exhibiting similar expression profiles and the highest changes in abundance in response to Cr(VI) exposure, can be found in Henne et al. [2009b].

70.3.3 Expression of the Chromate Resistance Determinant (CRD) Proteins

High-level chromate resistance in *Arthrobacter* sp. strain FB24 is likely due to chromate efflux through the ChrA transport protein, in addition to accessory genes that

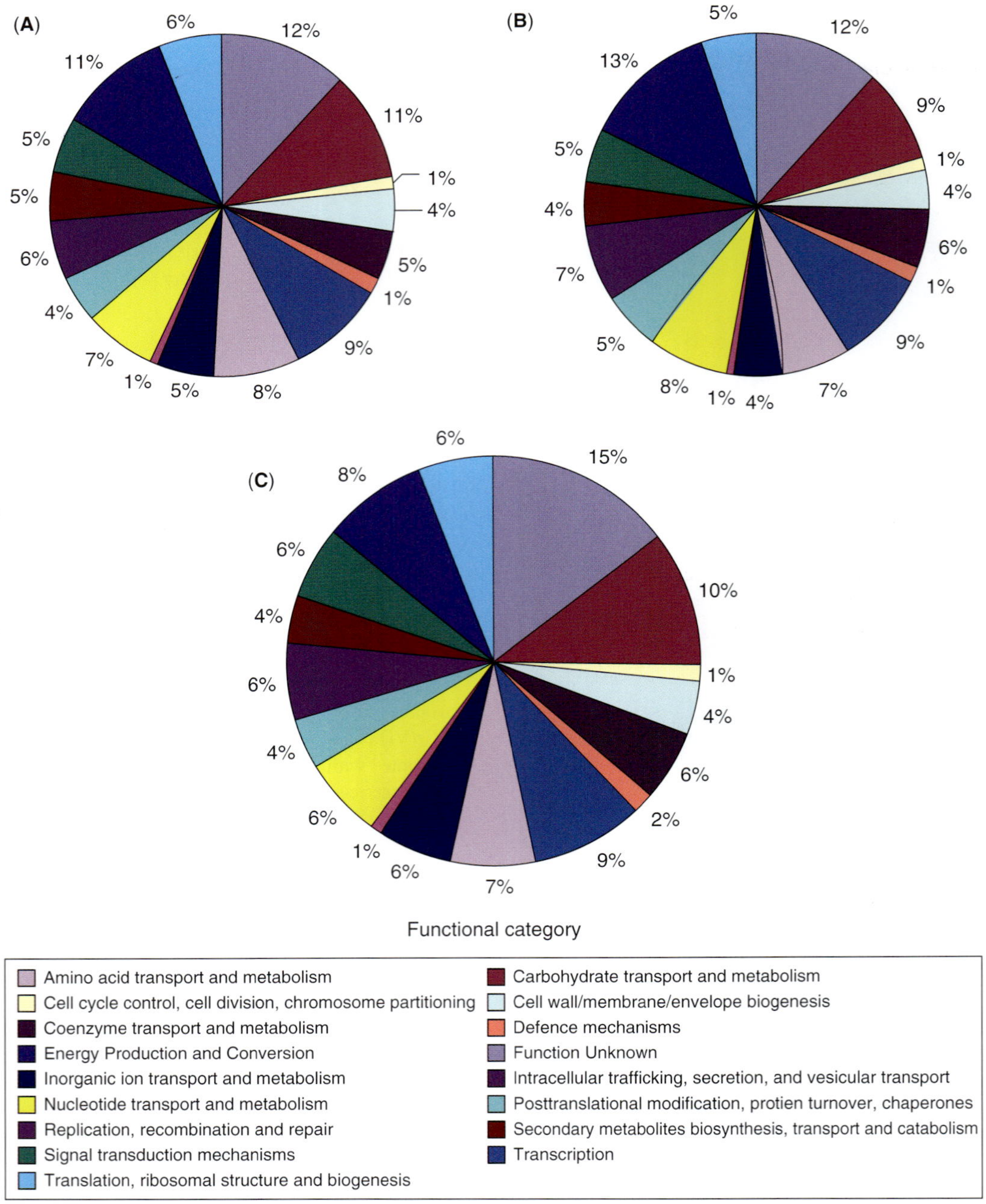

Figure 70.4 Major COG functional categories represented by the proteins in the **(A)** global, **(B)** soluble, and **(C)** insoluble partitions. (Reprinted with permission from Henne et al. [2009b]. Copyright 2009 American Chemical Society.)

have been shown to have significant roles in chromate resistance [Henne et al., 2009a]. The *chrA* gene and seven additional genes (Arth_4247-Arth_4255) comprise a plasmid-borne chromate resistance determinant (CRD). Each of the CRD genes exhibited a chromate-specific response at the transcript level and contributed to the recovery of chromate resistance in a chromate-sensitive derivative strain of FB24 [Henne et al., 2009a]. Proteomic analysis revealed large-fold differences (**Table 70.3**) for Arth_4247, 4252, 4255 and 4256, with the highest change observed for Arth_4247, a lipoprotein similar to the LppY/LpqO group in *Mycobacteria* [Sutcliffe and Harrington, 2004]. Peptides were observed for Arth_4248, Arth_4251 (*chrA* orthologs), and Arth_4252 (*chrB* ortholog) but failed to pass the stringent filtering process applied to the raw peptide data. This is likely due to technical difficulties in proteomic analysis of small (low molecular weight) proteins or those with many membrane-spanning domains rather than a lack of expression of these proteins. Chromate transport proteins have been classified into two main categories depending on their length and the number of conserved CHR domains present [Nies et al., 1998; Diaz-Perez et al., 2007; Ramirez-Diaz et al,. 2007]. Arth_4248 is a 450 amino acid protein with two conserved CHR domains and falls into the LCHR, or long ChrA, family, while Arth_4251 contains only 137 amino acids and one conserved CHR domain motif. Despite these features, the placement of Arth_4251 into the SCHR, or short ChrA, family is still a matter under scrutiny [Henne et al., 2009a]. The sequence of Arth_4251 is more similar to LCHR orthologs and there does not seem to be a "partner" gene for Arth_4251 in the FB24 genome, which is characteristic of most other SCHR genes [Diaz-Perez et al., 2007; Henne et al., 2009a]. The exact function of SCHR genes in chromate resistance awaits full characterization. This is, however, the first report of protein products detected for a SCHR gene.

Based on the putative functions of the remaining CRD proteins, it is expected that chromate resistance in strain FB24 is driven by chromate efflux and regulated by a signal transduction pathway. Regulatory roles have already been suggested for ChrB orthologs (Arth_4249, 4253, 4254) [Juhnke et al., 2002; Henne et al., 2009a], and conserved domains in Arth_4247 and Arth_4252 are present in other proteins that function in stress-responsive signal transduction pathways. As noted above, Arth_4247 is a lipoprotein, and although the involvement of lipoproteins in the response to anionic metal ions has not been demonstrated, they have been shown to function in resistance to divalent metal ions such as copper and lead [Borremans et al., 2001; Yamamoto and Ishihama, 2005]. For instance, the response of *E. coli* to copper involves the activation of the CpxAR stress responsive pathway by the lipoprotein NlpE [Yamamoto and Ishihama, 2005]. The Cpx stress response signal transduction pathway may also utilize an aminoglycoside phosphotransferase in a Ser/Thr protein [Zheng et al., 2007]. Interestingly, Arth_4256, which exhibited 25- and 30-fold increased abundance in 5 and 20 mM Cr(VI), respectively, is a putative aminoglycoside phosphotransferase with a Ser/Thr kinase domain and may prove to be integral to a signal cascade driving the expression of the CRD [Han and Zhang, 2001]. Likewise, Arth_4252, whose expression also increased in response to Cr(VI), contains a WD40 repeat domain, which is present in proteins involved in the response to environmental stimuli, such as envelope stress and initiating corresponding signal transduction pathways [Neer et al., 1994; Joshi et al., 2000].

The function of Arth_4255, a putative malate:quinine oxidoreductase, is unknown at this time, but its requirement in high-level chromate resistance in strain FB24 is supported by the protein expression data in response to (a) chromate presented here [Henne et al., 2009b] and (b) its apparent chromate-specific induction at the transcript level [Henne et al., 2009a]. Genetic complementation studies in a chromate-sensitive derivative of FB24 further showed that Arth_4255 is needed to recover the wild-type chromate-resistant phenotype [Henne et al., 2009a]. The basic function of malate:quinone oxidoreductases is the conversion of malate to oxaloacetate, with subsequent transfer of electrons to the quinone pool [Kather et al., 2000; Molenaar et al., 2000]. In terms of metal resistance mechanisms, NAD(P)H-dependent quinone oxidoreductases such as YieF from *E. coli* and ChrR from *P. putida* [Ackerley et al., 2004; Gonzalez et al., 2005] have been shown to reduce chromate. However, this is not likely in the case in *Arthrobacter* sp. strain FB24 given the lack of sufficient evidence of chromate reduction in this strain. In *Pseudomonas corrugata* strain 28, it was recently shown that disruption of a gene encoding a putative malic enzyme protein resulted in hypersensitivity to chromate compared to the wild-type strain. It was suggested that at least in *P. corrugata*, the malic enzyme may provide reducing powers to relieve oxidative stress induced by chromate [Decorosi et al., 2009]. Although further biochemical experiments are needed to confirm the function of Arth_4255, it is possible that it may serve a role similar to that of the inactivated malic enzyme in the *P. corrugata*.

In order to identify other proteins that may directly function in chromate resistance in *Arthrobacter* sp. strain FB24, the LC/LC-MS/MS data were queried to search for proteins exhibiting expression profiles similar to those from the genetically characterized CRD (**Table 70.2**). The findings for this focused group of proteins were similar to the overall findings for the entire dataset. Twenty-three proteins were identified as having at least a twofold increase in at least one partition under Cr(VI) conditions. These proteins fell into the COG functional categories of amino acid transport and metabolism, signal transduction mechanisms, and energy production and conversion, whereas 5 of the 23 proteins did not have any associated functional assignment. Consistent with the overarching theme of sulfur limitation due to chromate exposure, three proteins that may function in the uptake of cysteine and other sulfur-containing compounds showed large-fold increases under chromate stress conditions. These include a CBS (cystathionine beta-synthase) domain protein (Arth_2644), Arth_3901, which is classified as an extracellular solute-binding protein specific for polar amino acids and opines [Tam and Saier, 1993], and Arth_1300, which is similar to other proteins involved in the uptake of aliphatic sulfonates. These data also support those obtained with 2-DGE, which demonstrates consistency between the two proteomic approaches.

Table 70.2 Relative Fold Differences of CRD Proteins and Proteins with Similar Expression Profiles

Locus	Gene Description	Relative Fold Difference (Experimental/Control)[a]		
		5 mM Cr(VI)	20 mM Cr(VI)	No Cr(VI) Stat
	Proteins Encoded by Chromate Resistance Determinant			
Arth_4247	Protein of unknown function LppY and LpqO	85 G	21.6 I	2 g
Arth_4248	Chromate ion transporter (CHR) family	NO	NO	NO
Arth_4251	Hypothetical protein/chromate transport domain	NO	NO	NO
Arth_4252	40-residue YVTN family beta-propeller repeat protein	42 G	9 G	0.7 G
Arth_4253	Chromate resistance signal peptide, ChrB-Nterm	NO	2 G	NO
Arth_4254	Chromate resistance protein, ChrB-Cterm	NO	NO	NO
Arth_4255	Malate–quinone oxidoreductase	42 I	44 S	2.7 I
	Amino Acid Transport and Metabolism			
Arth_0642	Dehydrogenase	3 S	3 S	1 S
Arth_3746	ABC transporter related	9 I	4 I	1 I
Arth_3901	Extracellular solute-binding protein, family 3	82 G	42 G	NO
Arth_3977	Amino acid permease-associated region	2 G	1 G	NO
	Signal Transduction			
Arth_0657	Response regulator receiver protein	7 G	5.5 S	1 G
Arth_1303	Response regulator receiver protein	3 S	3 S	1 S
Arth_1376	Signal transduction histidine kinase, LytS	5 I	2.7 I	0.7 I
Arth_3901	Extracellular solute-binding protein, family 3	82 G	42 G	NO
Arth_4256	Aminoglycoside phosphotransferase	25 I	30 I	1.5 S
	Energy Production and Conversion			
Arth_1300	Putative nitrilotriacetate monooxygenase	50 S	35 S	1 S
Arth_1530	Sodium:dicarboxylate symporter	52 I	4 I	1 I
Arth_2628	Oxidoreductase FAD/NAD(P)-binding domain protein	4.5 I	2.5 I	1.5 I
Arth_3905	FMNH2-utilizing oxygenase	108 G	58 H	NO
	Proteins with no COG Function Assigned			
Arth_0044	Hypothetical protein	5 S	4 S	1 S
Arth_0514	Hypothetical protein	3 I	2 I	1 I
Arth_2155	Hypothetical protein	3 S	NO	1 I
Arth_3867	Hypothetical protein	3 I	2 I	NO

[a]Values reported reflect the highest expression level and the corresponding partition. G, global partition; S, soluble partition; I, Insoluble partition; NO, not observed.

Source: Reprinted with permission from Henne et al. [2009b]. Copyright 2009 American Chemical Society.

70.3.4 The Sulfur Limitation Response

Selected clusters of proteins showing increased abundance in at least two of the three cellular partitions (global, soluble, or insoluble) were further analyzed for any functional overlaps. The group of proteins with the highest-fold differences in abundance (up to 108-fold) under chromate stress point to an overall condition of cellular sulfur limitation (Table 70.3). The corresponding genes of several of these proteins are grouped into clusters and may represent sulfur-regulated operons. For instance, the clusters Arth_0857-0858 and Arth_3122-3125 are indicated in sulfate uptake and metabolism, Arth_3115-3117 in the uptake of alternative sulfur sources, while Arth_2446-2447 and Arth_3122-3125 are expected to be involved in cysteine synthesis. Biochemical evidence for the direct competition of the sulfate transporter by chromate has been established in other bacteria [Ohtake et al., 1987; Cervantes and Silver, 1992] and is considered the primary means for chromate to enter the cell. Transcriptomic and proteomic studies with other bacteria exposed to chromate stress have also demonstrated similar findings to those presented here [Hu et al., 2005; Ackerley et al., 2006; Brown

Table 70.3 Proteins from Selected Heat Map Clusters with Increased Abundance in Cr(VI)

Locus	Gene Description	Relative Fold Difference (Experimental/Control)[a]		
		5 mM Cr(VI)	20 mM Cr(VI)	No Cr(VI) Stat
	Replication, Recombination, and Repair			
Arth_0672	ATP-dependent DNA helicase, RecQ family	8.7 I	15.3 I	1.7 I
Arth_1211	Hydrolase, TatD family	2.0 S	4.7 S	2.0 S
Arth_1520	NUDIX hydrolase	3.0 S	6.0 S	1.0 S
Arth_1581	DNA polymerase III, alpha subunit	4.6 I	5.0 I	1.3 I
Arth_2063	DNA polymerase I	3.2 I	3.6 I	1.5 I
Arth_2074	DEAD/DEAH box helicase domain protein	4.0 G	10.0 G	2.5 S
Arth_2318	DNA polymerase III, alpha subunit	13.0 I	7.0 I	1.0 I
Arth_2756	UvrD/REP helicase	21.0 I	21.0 I	6.0 I
Arth_2757	UvrD/REP helicase	7.0 I	20.0 I	4.0 I
Arth_3798	SNF2-related protein	6.0 I	9.0 I	0.8 G
	Iron Sequestration/DNA Protection			
Arth_0334	Ferritin, Dps family protein	6.0 G	4.5 G	2.0 G
Arth_0761	Ferritin, Dps family protein	2.5 S	2.7 S	2.1 S
	General Function Prediction			
Arth_3714	Rieske (2Fe-2S) domain protein	13.0 G	18.0 G	2.0 G
Arth_4252	40-residue YVTN family beta-propeller repeat protein	42.0 G	9.0 G	0.7 G
Arth_4255	Malate–quinone oxidoreductase	42.0 I	44.0 S	2.7 I
Arth_4256	Aminoglycoside phosphotransferase	25.0 I	30.0 I	1.5 S
	Energy Production and Conversion			
Arth_0129	Nitroreductase	13.0 G	5.2 G	0.7 I
Arth_3948	Aldo/keto reductase	12.0 G	10.3 G	NO
	Inorganic Ion Transport/Putative Sulfonate Uptake Proteins			
Arth_0806	ABC transporter related	7.0 I	7.0 I	3.0 I
Arth_0857	ABC transporter related	4.0 S	9.0 S	3.0 G
Arth_0858	ABC transporter related	4.5 G	5.0 S	1.3 G
Arth_3115	Binding-protein-dependent transport systems inner membrane component	10 I	5 I	NO
Arth_3116	ABC transporter related	15.0 I	7.3 I	0.7 I
Arth_3117	NLPA lipoprotein	21.1 I	13.4 I	1.3 I
Arth_3122	Sulfate adenylyltransferase, large subunit	4.6 I	2.8 I	0.5 I
Arth_3123	Sulfate adenylyltransferase, small subunit	2.9 G	3.3 S	0.7 S
Arth_3124	Phosphoadenosine phosphosulfate reductase	2.3 G	2.0 I	0.7 G
Arth_3125	Ferredoxin–nitrite reductase	2.6 G	2.2 S	1.2 S
Arth_3905	FMNH2-utilizing oxygenase	108 G	58 G	NO

[a] Values reported reflect the highest expression level and the corresponding partition. G, global partition; S, soluble partition; I, insoluble partition; NO, not observed.

Source: Reprinted with permission from Henne et al. [2009b]. Copyright 2009 American Chemical Society.

et al., 2006; Thompson et al., 2007; Henne et al., 2009b]. In a separate study with *P. corrugata*, direct involvement of a putative regulatory gene, *oscA*, in activating a sulfur starvation response under chromate stress conditions was demonstrated [Viti et al., 2009]. The *oscA* gene is located directly upstream of a gene cluster encoding the sulfate transporter in *P. corrugata*. Both gene expression and complementation studies showed that *oscA* was expressed in response to chromate and could partially recover chromate resistance in a sensitive strain [Viti et al., 2009].

The proteomic results with *Arthrobacter* sp. strain FB24 and other bacteria referenced here, in addition to the physiological data obtained with strain FB24 and *P. corrugata*, strongly suggest that bacterial chromate resistance not only is driven by mechanisms such as chromate efflux, but also requires that cells take measures to sustain growth when faced with depletion of essential elements such as sulfur when exposed to chromate. Evidence in support of this hypothesis was obtained in separate investigation to determine the nutritional requirements of strain FB24. The addition of cysteine or cystine resulted in higher biomass yields when strain FB24 was cultured in minimal media supplemented with chromate compared to cultures without additional cysteine or cystine [Chen, 2007]. Additional experiments with other organic sulfur sources should provide further validation that chromate stress in strain FB24 can be alleviated through the induction of pathways leading to the synthesis or uptake of organosulfate compounds.

70.4 SUMMARY

The global proteomic response of *Arthrobacter* sp. strain FB24 to chromate was evaluated using both gel-based and mass spectrometry-based methods. In this case, two-dimensional gel electrophoresis provided initial insight into the cellular response of strain FB24 to chromate stress, resulting in 22% proteome coverage. A more comprehensive inventory of the FB24 proteome in response to chromate was obtained with liquid chromatography (LC/LC) coupled to tandem mass spectrometry (MS/MS). The MS approach achieved 71% proteome coverage. The pipeline of data analysis involved assigning peaks to peptides, peptides to proteins, and proteins to pathways in order to obtain a clear picture of the cellular response of strain FB24 to chromate stress. Initial clustering of the expression data resulted in a large matrix (Fig. 70.2). To simplify data interpretation, clusters of similar expression patterns from the heat map were analyzed on a local level and proteins grouped according to functional categories [Henne et al., 2009b]. The data provided protein-level support for genes annotated as "hypothetical proteins" and correlated well with transcript-level expression of CRD genes [Henne et al., 2009a]. Those proteins with the highest-fold increases in expression will also be likely candidates for future testing at the genetic and molecular levels to deduce function and direct roles in chromate resistance.

INTERNET RESOURCES AND SOFTWARE

DOE-JGI *Arthrobacter* sp. strain FB24 genome sequence (ftp://ftp.jgi_psf.org/pub/JGI_data/Microbial/arthrobacter_fb24/Finished2006/3634492.finished.fsa)

COG database (http://www.ncbi.nlm.nih.gov/COG/)

The Integrated Microbial Genomes (IMG) system (http://tinyurl.com/27eyv3)

GeneSpring GX v 7.3.1, Agilent Technologies, Santa Clara, CA

Progenesis software, Version 2003.03; Nonlinear USA, Inc, Durham, NC

Acknowledgments

We would like to thank Samuel Purvine and Angela Norbeck of Pacific Northwest National Laboratory and Gyorgy Babnigg and Carol Giometti from Argonne National Laboratory for thoughtful discussion and technical expertise. K. L. Henne would like to thank Walter Henne of Governors State University for editorial assistance.

REFERENCES

Ackerley DF, Gonzalez CF, Park CH, Blake 2nd R, Keyhan M, Matin, A. 2004. Chromate-reducing properties of soluble flavoproteins from *Pseudomonas putida* and *Escherichia coli*. *Appl. Environ. Microbiol*. 70:873–882.

Ackerley DF, Barak Y, Lynch SV, Curtin J, Matin A. 2006. Effect of chromate stress on *Escherichia coli* K-12. *J. Bacteriol*. **188**:3371–3381.

Ahrendt AJ, Tollaksen SL, Lindberg C, Zhu W, Yates 3rd JR, et al. 2007. Steady state protein levels in *Geobacter metallireducens* grown with iron (III) citrate or nitrate as terminal electron acceptor. *Proteomics* **7**:4148–4157.

Anderson NG, Anderson NL. 1978a. Analytical techniques for cell fractions. XXI. Two-dimensional analysis of serum and tissue proteins: multiple isoelectric focusing. *Anal. Biochem*. **85**:331–340.

Anderson NL, Anderson NG. 1978b. Analytical techniques for cell fractions. XXII. Two-dimensional analysis of serum and tissue proteins: multiple gradient-slab gel electrophoresis. *Anal. Biochem*. **85**:341–354.

Beasley FC. 2004. Characterization of Diversity, Chromate Resistance and Aromatic Hydrocarbon Degradation Among *Arthrobacter* Isolates from Mixed Waste Soil. Department of Agronomy, Purdue University, West Lafayette, IN, p. 219.

Benyehuda G, Coombs J, Ward PL, Balkwill D, Barkay T. 2003. Metal resistance among aerobic chemoheterotrophic bacteria from the deep terrestrial subsurface. *Can. J. Microbiol*. **49**:151–156.

Borremans B, Hobman JL, Provoost A, Brown NL, van Der Lelie D. 2001. Cloning and functional analysis of the *pbr* lead resistance determinant of *Ralstonia metallidurans* CH34. *J. Bacteriol*. **183**:5651–5658.

Brown SD, Thompson MR, Verberkmoes NC, Chourey K, Shah M, et al. 2006. Molecular dynamics of the *Shewanella oneidensis* response to chromate stress. *Mol. Cell. Proteomics* **5**:1054–1071.

Callister SJ, McCue LA, Turse JE, Monroe ME, Auberry KJ, et al. 2008. Comparative bacterial proteomics: Analysis of the core genome concept. *PLoS One* **3**:e1542.

Camargo FA, Okeke BC, Bento FM, Frankenberger WT. 2005. Diversity of chromium-resistant bacteria isolated from soils contaminated with dichromate. *Appl. Soil Ecol*. **29**:193–202.

Cervantes C, Silver S. 1992. Plasmid chromate resistance and chromate reduction. *Plasmid* **27**:65–71.

Chen, W. 2007. Characterization of Physiological Factors Associated with Chromate Resistance by *Arthrobacter* sp. FB24. Master of Science dissertation, Department of Agronomy, Purdue University, West Lafayette, IN, p. 87.

Crocker FH, Fredrickson JK, White DC, Ringelberg DB, Balkwill DL. 2000. Phylogenetic and physiological diversity of *Arthrobacter* strains isolated from unconsolidated subsurface sediments. *Microbiology* **146** (Pt 6):1295–1310.

de Souza GA, Malen H, Softeland T, Saelensminde G, Prasad S, et al. 2008. High accuracy mass spectrometry analysis as a tool to verify and improve gene annotation using *Mycobacterium tuberculosis* as an example. *BMC Genomics* **9**:316.

de Souza GA, Softeland T, Koehler CJ, Thiede B, Wiker HG. 2009. Validating divergent ORF annotation of the *Mycobacterium leprae* genome through a full translation data set and peptide identification by tandem mass spectrometry. *Proteomics* **9**:3233–3243.

Decorosi F, Tatti E, Mini A, Giovannetti L, Viti C. 2009. Characterization of two genes involved in chromate resistance in a Cr(VI)-hyper-resistant bacterium. *Extremophiles* **13**:917–923.

Diaz-Perez C, Cervantes C, Campos-Garcia J, Julian-Sanchez A, Riveros-Rosas H. 2007. Phylogenetic analysis of the chromate ion transporter (CHR) superfamily. *FEBS J*. **274**:6215–6227.

Eng JK, McCormack AL, Yates 3rd Jr. 1994. An approach to correlate tandem mass spectral data of peptides with amino acid sequences in a protein database. *J. Am. Soc. Mass Spectrom*. **5**:976–989.

Fredrickson JK, Zachara JM, Balkwill DL, Kennedy D, Li SM, et al. 2004. Geomicrobiology of high-level nuclear waste-contaminated vadose sediments at the Hanford site, Washington state. *Appl. Environ. Microbiol*. **70**:4230–4241.

Garten CT, Hamby DM Jr, Schreckhise RG. 2000. Radiocesium discharges and subsequent environmental transport at the major US weapons production facilities. *Sci. Total Environ*. **255**:55–73.

Giometti CS, Tollaksen SL, Gemmell MA, Burcham J, Peraino C. 1988. A heritable variant of mouse liver ornithine aminotransferase (EC 2.6.1.13) induced by ethylnitrosourea. *J. Biol. Chem*. **263**:15781–15784.

Giometti CS, Gemmell MA, Tollaksen SL, Taylor J. 1991. Quantitation of human leukocyte proteins after silver staining: A study with two-dimensional electrophoresis. *Electrophoresis* **12**:536–543.

Gonzalez CF, Ackerley DF, Lynch SV, Matin A. 2005. ChrR, a soluble quinone reductase of *Pseudomonas putida* that defends against H_2O_2. *J. Biol. Chem*. **280**:22590–22595.

Graves PR, Haystead TA. 2002. Molecular biologist's guide to proteomics. *Microbiol. Mol. Biol. Rev*. **66**:39–63.

Han G, Zhang CC. 2001. On the origin of Ser/Thr kinases in a prokaryote. *FEMS Microbiol. Lett*. **200**:79–84.

Henne KL, Nakatsu CH, Thompson DK, Konopka AE. 2009a. High-level chromate resistance in *Arthrobacter* sp. strain FB24 requires previously uncharacterized accessory genes. *BMC Microbiol*. **9**:199.

Henne KL, Turse JL, Nicora CD, Lipton MS, Tollaksen S, et al. 2009b. Global Proteomic Analysis of the Chromate Response in *Arthrobacter* sp. Strain FB24. J. *Proteome Res*. **8**:1704–1716.

Hu P, Brodie EL, Suzuki Y, McAdams HH, Andersen GL. 2005. Whole-genome transcriptional analysis of heavy metal stresses in *Caulobacter crescentus*. *J. Bacteriol*. **187**:8437–8449.

Jones D, Keddie RM. 1999. The genus *Arthrobacter*. In Dworkin E, et al., eds. The Prokaryotes: An Evolving Electronic Resource for the Microbiological Community. New York: Springer-Verlag.

Joshi B, Janda L, Stoytcheva Z, Tichy P. 2000. PkwA, a WD-repeat protein, is expressed in spore-derived mycelium of *Thermomonospora curvata* and phosphorylation of its WD domain could act as a molecular switch. *Microbiology* **146**:3259–3267.

Juhnke S, Peitzsch N, Hubener N, Grosse C, Nies DH. 2002. New genes involved in chromate resistance in *Ralstonia metallidurans* strain CH34. *Arch. Microbiol*. **179**:15–25.

Kather B, Stingl K, van der Rest ME, Altendorf K, Molenaar D. 2000. Another unusual type of citric acid cycle enzyme in *Helicobacter pylori*: The malate:quinone oxidoreductase. *J. Bacteriol*. **182**:3204–3209.

Lipton MS, Pasa-Tolic L, Anderson GA, Anderson DJ, Auberry DL, et al. 2002. From the Cover: Global analysis of the *Deinococcus radiodurans* proteome by using accurate mass tags. *Proc. Natl. Acad. Sci. USA* **99**:11049–11054.

Lipton MS, Romine MF, Monroe ME, Elias DA, Pasa-Tolic L, et al. 2006. AMT tag approach to proteomic characterization of *Deinococcus radiodurans* and *Shewanella oneidensis*. *Methods Biochem. Anal*. **49**:113–134.

Lu P, Vogel C, Wang R, Yao X, Marcotte EM. 2007. Absolute protein expression profiling estimates the relative contributions of transcriptional and translational regulation. *Nat. Biotechnol*. **25**:117–124.

Margesin R, Schinner F. 1997. Heavy metal resistant *Arthrobacter* sp.—A tool for studying conjugational plasmid transfer between gram-negative and gram-positive bacteria. *J. Basic Microbiol*. **37**:217–227.

Molenaar D, van der Rest MR, Drysch A, Yucel R. 2000. Functions of the membrane-associated and cytoplasmic malate dehydrogenases in the citric acid cycle of *Corynebacterium glutamicum*. *J. Bacteriol*. **182**:6884–6891.

Nakatsu CH, Carmosini N, Baldwin B, Beasley F, Kourtev P, Konopka A. 2005. Soil microbial community responses to additions of organic carbon substrates and heavy metals (Pb and Cr). *Appl. Environ. Microbiol*. **71**:7679–7689.

Neer E J, Schmidt CJ, Nambudripad R, Smith TF. 1994. The ancient regulatory-protein family of WD-repeat proteins. *Nature* **371**:297–300.

Nies DH, Koch S, Wachi S, Peitzsch N, Saier Jr MH. 1998. CHR, a novel family of prokaryotic proton motive force-driven transporters probably containing chromate/sulfate antiporters. *J. Bacteriol*. **180**:5799–5802.

O'Farrell PH. 1975. High resolution two-dimensional electrophoresis of proteins. *J. Biol. Chem*. **250**:4007–4021.

Ohtake H, Cervantes C, Silver S. 1987. Decreased chromate uptake in *Pseudomonas fluorescens* carrying a chromate resistance plasmid. *J. Bacteriol*. **169**:3853–3856.

Ramagli LS. 1999. Quantifying protein in 2-D PAGE solubilization buffers. *Methods Mol. Biol*. **112**:99–103.

Ramirez-Diaz MI, Diaz-Perez C, Vargas E, Riveros-Rosas H, Campos-Garcia J, Cervantes C. 2007. Mechanisms of bacterial resistance to chromium compounds. *Biometals* **21**:321–332.

Smith RD, Anderson GA, Lipton MS, Pasa-Tolic L, Shen Y, et al. 2002. An accurate mass tag strategy for quantitative and high-throughput proteome measurements. *Proteomics* **2**:513–523.

Sutcliffe IC, Harrington DJ. 2004. Lipoproteins of *Mycobacterium tuberculosis*: an abundant and functionally diverse class of cell envelope components. *FEMS Microbiol. Rev*. **28**:645–659.

Tam R, Saier MH Jr. 1993. Structural, functional, and dvolutionary relationships among extracellular solute-binding receptors of bacteria. *Microbiol. Rev*. **57**:320–346.

Thompson MR, Verberkmoes NC, Chourey K, Shah M, Thompson DK, Hettich RL. 2007. Dosage-dependent proteome response of *Shewanella oneidensis* MR-1 to acute chromate challenge. *J. Proteome Res*. **6**:1745–1757.

VerBerkmoes NC, Denef VJ, Hettich RL, Banfield JF. 2009. Systems biology: Functional analysis of natural microbial consortia using community proteomics. *Nat. Rev. Microbiol*. **7**:196–205.

Viti C, Decorosi F, Mini A, Tatti E, Giovannetti L. 2009. Involvement of the *oscA* gene in the sulphur starvation response and in Cr(VI) resistance in *Pseudomonas corrugata* 28. *Microbiology* **155**:95–105.

Wang H, Qian WJ, Chin MH, Petyuk VA, Barry RC, et al. 2006. Characterization of the mouse brain proteome using global proteomic analysis complemented with cysteinyl-peptide enrichment. *J. Proteome Res*. **5**:361–369.

Washburn MP, Wolters D, Yates JR 3rd. 2001. Large-scale analysis of the yeast proteome by multidimensional protein identification technology. *Nat. Biotechnol*. **19**:242–247.

Wilmes P, Bond PL. 2006. Metaproteomics: Studying functional gene expression in microbial ecosystems. *Trends Microbiol*. **14**:92–97.

Yamamoto K, Ishihama A. 2005. Transcriptional response of *Escherichia coli* to external copper. *Mol. Microbiol*. **56**:215–227.

Zheng J, He C, Singh VK, Martin NL, Jia Z. 2007. Crystal structure of a novel prokaryotic Ser/Thr kinase and its implication in the Cpx stress response pathway. *Mol. Microbiol*. **63**:1360–1371.

Zhu W, Reich CI, Olsen GJ, Giometti CS, Yates JR 3rd. 2004. Shotgun proteomics of *Methanococcus jannaschii* and insights into methanogenesis. *J. Proteome Res*. **3**:538–548.

Part 6D

Metabolomics

Chapter 71

The Small-Molecule Dimension: Mass-spectrometry-based Metabolomics, Enzyme Assays, and Imaging

Trent R. Northen

71.1 INTRODUCTION

Metabolomics is a rapidly growing field aimed at characterizing and often quantifying the small molecule composition of a biological sample [Fiehn, 2002]. The direct biochemical information obtained is complementary to other systems information (i.e., genomic, transcriptomic, and proteomic) in that metabolites are an ultimate downstream effector of gene expression (**Fig. 71.1**) and provide new insights into "omics" datasets described in these volumes. However, metabolomics is 'genome independent' and can therefore be used to study unsequenced communities i.e. to select interesting environmental samples for further characterization.

While small-molecule analysis has been performed using mass spectrometry for decades, recent advances in mass spectrometry instrumentation have led to rapid growth of the field. It is now routine to detect hundreds of metabolites using gas chromatography coupled to mass spectrometry (GC/MS) and liquid chromatography coupled to mass spectrometry (LC/MS). Detection in this context does not necessarily mean identification; moving from spectral detection to definitive identification of compounds is often a resource- and time-consuming process.

Mass spectrometry requires gas-phase ions (analytes) and there are a wide range of options for desorbing and ionizing molecules, the most used being GC/MS and LC/MS. For a detailed introduction, an excellent text is available [Siuzdak, 1996]. In GC/MS, volatile metabolites (typically by heating them to 200–300°C) are first separated on the GC and then ionized, usually by electron ionization (EI), a technique that often results in extensive molecular fragmentation. As one might expect, many metabolites have low vapor pressures even at these elevated temperatures (i.e., sugars, amino acids, etc.) and must be derivatized to increase their vapor pressure (typically by using a silane). However, this approach is not suited for all metabolites (i.e., thermally labile molecules, metabolites lacking a derivatizable group, etc.). That having been said, GC/MS is an excellent approach allowing research groups to immediately study metabolites with minimal investment of resources. The reader is referred to excellent protocols for detailed discussion of best practices for GC/MS metabolite analysis [Lisec et al., 2006].

The development of electrospray ionization (ESI) efficiently coupled to liquid chromatography (LC/MS) or capillary electrophoresis (CE/MS) has been a breakthrough for metabolomics (see Chapter 73, Vol. I). LC/MS methods have the advantage that the chromatography is performed on liquid-phase molecules relaxing the constraint on volatility. ESI is based on the drying of charged liquid droplets generated by a charged needle; as these droplets dry, the charges are concentrated until surface

Handbook of Molecular Microbial Ecology, Volume I: Metagenomics and Complementary Approaches, First Edition. Edited by Frans J. de Bruijn.

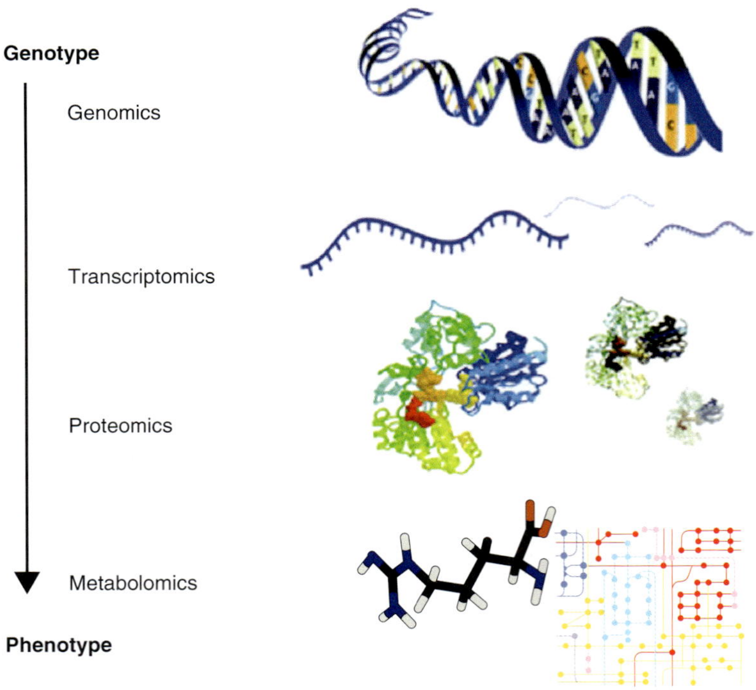

Figure 71.1 Metabolomic studies provide direct biochemical information on the existence of specific enzymatic transformations, making them an excellent complement to genomic, transcriptomic, and proteomic analysis.

charge repulsion exceeds surface tension causing droplets to breakup into smaller droplets—ultimately generating ions with minimal fragmentation. It should also be noted that ESI can also be coupled to take advantage of the excellent separation achieved by CE; however, this approach is not as widely used as LC/MS, due to signal suppression from the electrolyte (ampholyte) and the intensive effort required to efficiently couple with ESI.

Molecular fragmentation is a critical aspect of mass-spectrometry-based metabolomics. Matching experimental fragmentation patterns versus authentic standards and databases allows molecular structural confirmation. The existence of large spectral databases for GC/MS greatly facilities identifications, and the development of analogous LC/MS databases for collision-induced fragmentation (i.e., Metlin) is an important development [Smith et al., 2005; see also Chapter 74, Vol. I]. In cases where metabolites are not contained in such databases, it is critical to obtain the exact mass of the precursor (unfragmented) ion and the corresponding isotopic distribution to help identify the metabolite [Kind and Fiehn, 2007]. In this case, the softness of ESI is advantageous because the precursor ion often predominates, which greatly facilitates identification and simplifies the interpretation of co-eluting peaks.

71.1.1 Targeted and Global Approaches to Metabolomics

Researchers often divide metabolomic approaches into "targeted" and "untargeted" metabolomics. Just as the name suggests, targeted methods [Bennett et al., 2008] are designed to detect and often quantify specific metabolites within a sample. This approach has the advantage that methods can maximize the specificity and sensitivity of mass spectrometry by using metabolite standards. Untargeted methods, on the other hand, aim to maximize the coverage and number of metabolites, often with a loss of sensitivity.

Targeted approaches have been used for decades, a prime example of which is detecting clinically relevant diagnostic endogenous metabolites [Chace et al., 1993]. One area of particular interest for microbial functional genomics is that the use of targeted mass spectrometry approaches study enzymatic activities in vitro [Liesener and Karst, 2005]. Typically, development of targeted approaches requires analytical standards to define suitable methods for metabolite separation (i.e., GC or LC methods), determine unique metabolite fragments to assist in identification, and construction of standard curves for absolute quantification. Recently, Olszewski et al. [2009] used targeted approaches to study host–parasite (*Plasmodium falciparum*) interactions, which revealed modulations in metabolite abundance that correlated with parasite developmental stages. It also suggested that systemic arginine depletion by the parasite drives malarial hypoargininemia associated with cerebral malarial pathogenesis. Other examples of targeted approaches simultaneously monitoring multiple substrates and products include glycosylhydrolases/transferases [Yu et al., 2004] and sesquiterpene synthases [Yoshikuni et al.,

2006]. Such assays have also been used for the validation of enzymatic activities for putative enzymes [Fischbach et al., 2005; Lin et al., 2005], inhibitor screening [Greis, 2007], and identification of specific substrates for putative enzymes [Yu et al., 2004; Furuya et al., 2008].

Critical to all metabolomic experiments are the pre-processing steps required to prepare the biological materials for analysis. While these vary depending on the experimental goal, they typically include: a quenching step to stop metabolism, cell concentration to remove the medium, cell lysis to make metabolites solvent accessible, and finally an extraction step [Villas-Boas et al., 2005]. In the simplest protocol, all three are accomplished by treating cells with an organic solvent such as methanol [Want et al., 2004].

Untargeted metabolomic approaches using LC/MS generally require less up-front effort in comparison to targeted approaches but have much more complex outputs. The overall LC/MS metabolomic workflow is summarized in **Figure 71.2**, and there are detailed protocols to assist in implementation [De Vos et al., 2007]. Sample preparation, chromatographic conditions [Garcia et al., 2008], and MS ionization approaches are all selected to maximize the diversity of metabolites detected [Nordstrom et al., 2008]. Interpretation of the resulting hundreds or thousands of ions is extremely challenging and relies heavily on computational tools and existing metabolite databases (see Chapter 74, Vol. I).

The first step in the analysis of untargeted metabolomic data analysis is to define "features," which are the combination of the exact mass-to-charge ratio (*m/z*) and the corresponding LC retention time (RT) for the ion. These *m/z*, RT features are taken as an initial approximation of a metabolite identifier and can be quantitatively compared between samples to prioritize important features for further analysis. Since these features are dependent upon the experimental conditions it is critical to specify the exact experimental conditions under which they are detected. This issue is being addressed by the metabolomics community, which is working to standardize methods and data reporting [Jenkins et al., 2004; Sumner et al., 2004; Fiehn et al., 2008].

There are a wide range of algorithms for feature finding, and these have been the subject of a recent review [Werner et al., 2008]. The most widely used algorithms are XCMS [Smith et al., 2006], msInspect [Bellew et al., 2006], and mzMine [Katajamaa and Orešič, 2007], though there are many more including commercial options. Many of these tools also include statistical comparison between samples (fold change, *p*-values, etc.) to assist in selecting the most significant for further identification. It is typical to complement these results with other statistical and visualization tools to prioritize features further—for example, using principle component analysis (PCA) [Taylor et al., 2002], partial least-squares descrimant analysis (PLS-DA) [Jansson et al., 2009], and self-organizing networks [Patterson et al., 2008].

Feature identification relies heavily on searching exact masses against metabolite databases: Kegg [Kanehisa and Goto, 2000], Metlin [Smith et al., 2005], MetaCyc [Caspi et al., 2008], Golm [Kopka et al., 2005],

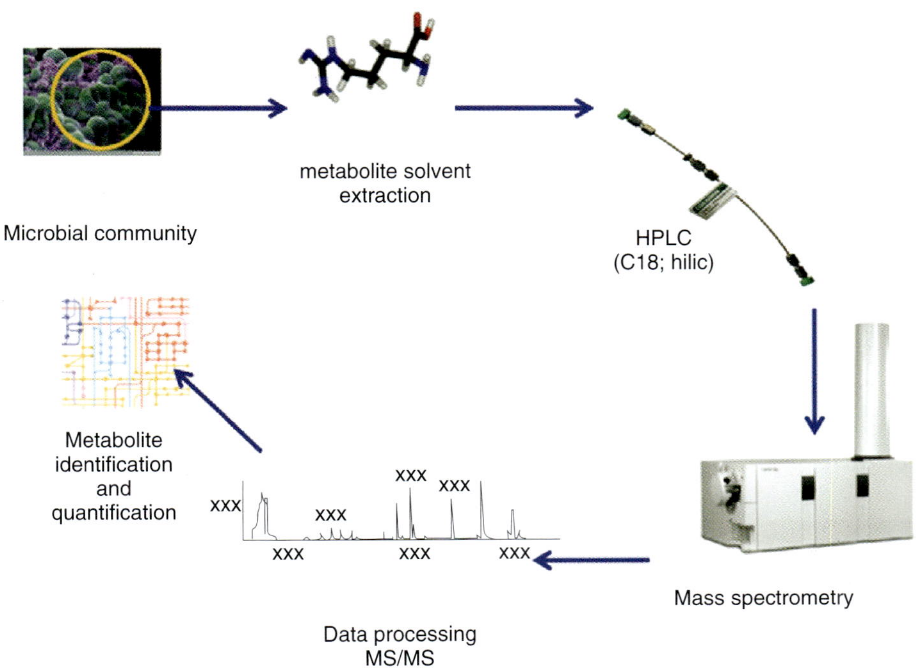

Figure 71.2 Conventional untargeted LC/MS metabolomics workflow. Samples are lysed and small-molecule metabolites are extracted, these are then separated using chromatographic steps, ionized and analyzed using mass spectrometry. The complex datasets are analyzed using a wide range of computational tools to identify ions for identification using database searching, authentic standards, and tandem mass spectrometry.

HMDB [Wishart et al., 2007], Lipidmaps [Fahy et al., 2005], and so on. It should be noted that organism specific databases are available through both Kegg and MetaCyc (BioCyc)—something that is useful for metabolite characterization in communities with known microbial compositions. In either case, database searching often results in several putative identifications for a particular ion. More definitive identification requires use of an authentic standard to define the molecule's retention time and fragmentation pattern. Unfortunately, only a small subset of metabolites can be obtained commercially. Hence, the identification of unknown ions is an extremely challenging process, often requiring either (a) enrichment to enable direct NMR studies or (b) chemical synthesis of libraries of related hypothetical molecules for comparison with unknowns using tandem mass spectrometry. The development of MS/MS spectral databases and search algorithms are an important goal for simplifying this process and predicting chemical structure based on MS/MS patterns [Benton et al., 2008]. These results can then be integrated into pathways and with other systems data, as discussed later in this volume by Davey (Chapter 74).

Several recent studies have used metabolomics to study gut microbiota including the chapter by Schmidt-Kopplin et al. (Chapter 72) in this volume. Wikoff et al. [2009] have used this approach to study the effect of the gut microbiome on blood metabolites by using untargeted mass spectrometry based on germ-free mice. This elegant study showed that production of indole-containing molecules (i.e., indol-3-propionic acid) was dependent on the action of the microbiome and that it could be established by colonization with *Clostridium sporogenes*. It was also shown that a number of glycine conjugates could be attributed to the action of the microbiome. In a related study (described in detail in this volume by Schmitt-Kopplin et al., Chapter 72), mass-spectrometry-based metabolomics was used to identify biomarkers of Crohn's disease by comparing fecal samples from 17 identical twins, including individuals with and without Crohn's disease [Jansson et al., 2009]. Interestingly, the similarity of metabolic profiles in monozygotic twin pairs suggests that genetics may be a major determinant of the gut environment and that tyrosine, phenylalanine, bile acid, and fatty acid biosynthesis may differentiate Crohn's disease etiologies.

Use of stable isotope-labeled metabolites is used to trace the fate of specific atoms within metabolic networks [Droste et al., 2008]. The resulting increased *m/z* for the resulting heavy metabolites are often easily detected and the specific labeling position identified using tandem mass spectrometry [Tang et al., 2007]. This information can be extremely useful in discriminating between alternative pathways and to estimate flux through metabolic networks [Sauer, 2006]. For example, researchers have used this to demonstrate that isoleucine is predominantly synthesized using an alternative pathway in *Geobacter sulfurreducens* [Risso et al., 2008].

MASS-SPECTROMETRY-BASED METABOLITE IMAGING

The loss of spatial information that results upon metabolite extraction can be a major limitation: it may be the local concentration of a metabolite, not the bulk average, which has biological significance. Therefore, metabolomic imaging technologies represent a critical new development within the field (**Fig. 71.3**). Such information is invaluable

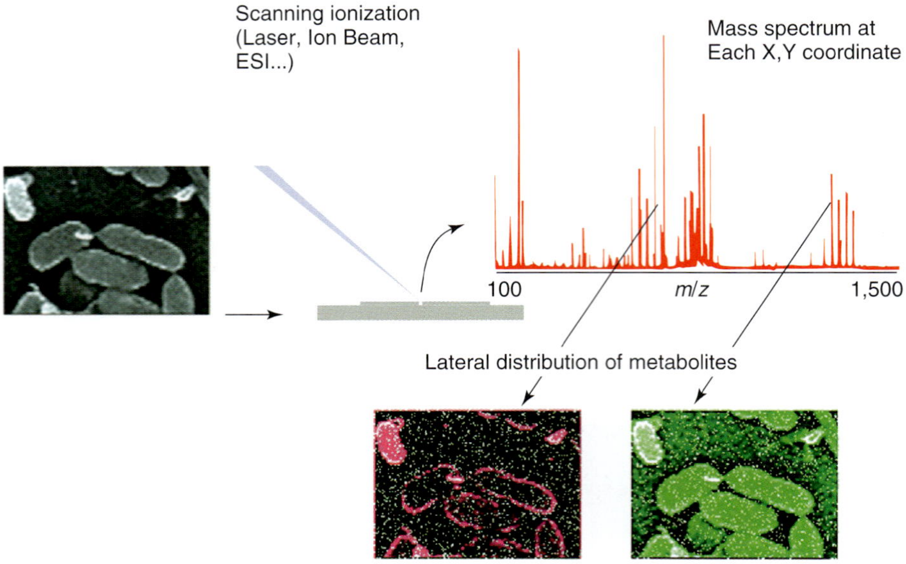

Figure 71.3 Mass spectrometry imaging approaches are used to study the lateral distribution of metabolites within biological samples. Mass spectra are collected by desorbing/ionizing metabolites from spatially defined regions of the sample, and the resulting datasets allow image reconstruction to define metabolite localization.

in understanding the species-specific metabolic processes occurring in complex communities.

The major mass spectrometry imaging approaches are matrix-assisted laser desorption ionization (MALDI) [Cornett et al., 2007] and secondary ion mass spectrometry (SIMS) [Winograd, 2005], both of which can be used to construct images. MALDI is a 'soft' desorption/ionization approach, and imaging is performed by spotting or spraying the sample with solvent containing "matrix," a light-absorbing molecule that desorbs and ionizes metabolites when irradiated. Matrix molecules also ionize, thereby producing background ions that make it difficult to detect small metabolites. In SIMS the untreated sample is sputtered with ions (i.e., gold, gallium, bismuth, etc.) that knock off secondary ions analyzed using time-of-flight mass analyzers (TOF-SIMS). This method is "matrix-free" and therefore does not have the background matrix ions and also has the advantage that ion beams can be focused to <100 nm. However, the energetic SIMS ionization process results in extensive molecular fragmentation, and typically only the most abundant ions can be detected at highest lateral resolution.

There are several new "soft" techniques that can be used for imaging: Desorption electrospray ionization (DESI), as the name implies, works by scanning the surface with an electrospray cone [Wiseman et al., 2006]. Nanostructure-initiator mass spectrometry (NIMS) is a "matrix-free" approach that uses a scanning laser system to generate ions from a liquid-filled nanostructured surface [Northen, et al., 2007]. In all cases, mass spectra are collected at defined $x-y$ coordinates across the surface, allowing one to define the $x-y$ distribution of particular metabolites within the microbial community. Ultra-high-resolution SIMS (NanoSIMS [Behrens et al., 2008]) has generated considerable interest amongst microbiologists. However, this approach is best reserved for studying the distribution of specific elements (not intact molecules). It has been used, for example, to image ion distributions including fluorine and bromine introduced into the cells via 16S rRNA-targeted probes within biofilms. MALDI imaging has been used to study metabolic exchange between *Bacillus subtilis* and *Streptomyces coelicolor* [Yang et al., 2009].

Mass-spectrometry-based meta–metabolomics provides direct biochemical information, making it an excellent complement to metagenomic, metatranscriptomic, and metaproteomic studies. This rapidly developing field provides insights into metabolic networks, specific enzymatic transformations, cellular dynamics, and the lateral distribution of metabolites within microbial communities. Targeted approaches allow metabolite quantification with the highest sensitivity and selectivity while more comprehensive approaches are used for discovering global changes. Computational tools and databases are used for metabolite identification and to link metabolites along pathways. The exact structure of these pathways can then be refined using stable isotopes of native compounds. While these methods are used to understand metabolic networks in detail, they do not resolve metabolite lateral distributions. Therefore, imaging mass spectrometry is a critical development to link metabolic pathways within the larger framework of the community.

Acknowledgments

I gratefully acknowledge support from the Department of Energy ENIGMA Scientific Focus Area [DE-AC02-05CH11231], and I thank F. Robles and B. Knierim for artwork and thank G. Siuzdak, R. Baran, W. Reindl, and S. Singer for useful comments.

REFERENCES

Behrens S, Losekann T, et al. 2008. Linking microbial phylogeny to metabolic activity at the single-cell level by using enhanced element labeling-catalyzed reporter deposition fluorescence in situ hybridization (EL-FISH) and NanoSIMS. *Appl. Environ. Microbiol*. **74**:3143–3150.

Bellew M, Coram M, et al. 2006. A suite of algorithms for the comprehensive analysis of complex protein mixtures using high-resolution LC-MS. *Bioinformatics* **22**:1902–1909.

Bennett BD, Yuan J, et al. 2008. Absolute quantitation of intracellular metabolite concentrations by an isotope ratio-based approach. *Nat. Protocols* **3**:1299–1311.

Benton H, Wong D, et al. 2008. XCMS2: processing tandem mass spectrometry data for metabolite identification and structural characterization. *Anal. Chem*. **80**:6382–6389.

Caspi R, Foerster H, et al. 2008. The MetaCyc database of metabolic pathways and enzymes and the BioCyc collection of pathway/genome databases. *Nucleic Acids Res*. **36**:D623.

Chace DH, Millington DS, et al. 1993. Rapid diagnosis of phenylketonuria by quantitative analysis for phenylalanine and tyrosine in neonatal blood spots by tandem mass spectrometry. *Clin. Chem*. **39**:66–71.

Cornett DS, Reyzer ML, et al. 2007. MALDI imaging mass spectrometry: molecular snapshots of biochemical systems. *Nat. Methods* **4**:828–833.

De Vos RCH, Moco S, et al. 2007. Untargeted large-scale plant metabolomics using liquid chromatography coupled to mass spectrometry. *Nat. Protocols* **2**:778–791.

Droste P, Weitzel M, et al. 2008. Visual exploration of isotope labeling networks in 3D. *Bioprocess Biosyst. Eng*. **31**:227–239.

Fahy E, Subramaniam S, et al. 2005. A comprehensive classification system for lipids. *J. Lipid. Res*. **46**:839–861.

Fiehn O. 2002. Metabolomics—The link between genotypes and phenotypes. *Plant Mol. Biol*. **48**:155–171.

Fiehn O, Wohlgemuth G, et al. 2008. Quality control for plant metabolomics: Reporting MSI-compliant studies. *Plant J*. **53**:691.

Fischbach MA, Lin H, et al. 2005. In vitro characterization of IroB, a pathogen-associated C-glycosyltransferase. *Proc. Natl. Acad. Sci. USA* **102**:571–576.

Furuya T, Nishi T, et al. 2008. Characterization of orphan monooxygenases by rapid substrate screening using FT-ICR mass spectrometry. *Chem. Biol*. **15**:563–572.

Garcia D, Baidoo E, et al. 2008. Separation and mass spectrometry in microbial metabolomics. *Curr. Opin. Microbiol.* **11**: 233–239.

Greis KD. 2007. Mass spectrometry for enzyme assays and inhibitor screening: An emerging application in pharmaceutical research. *Mass Spectrom. Rev.* **26**:324–339.

Jansson J, Willing B, et al. 2009. Metabolomics reveals metabolic biomarkers of Crohn's disease. *PLoS ONE* **4**:e6386.

Jenkins H, Hardy N, et al. 2004. A proposed framework for the description of plant metabolomics experiments and their results. *Nat. Biotechnol.* **22**:1601–1606.

Kanehisa M, Goto S. 2000. KEGG: Kyoto encyclopedia of genes and genomes. *Nucleic Acids Res.* **28**:27–30.

Katajamaa M and Orešič M. 2007. Data processing for mass spectrometry-based metabolomics. *J. Chrom. A* **1158**:318–328.

Kind T, Fiehn O. 2007. Seven golden rules for heuristic filtering of molecular formulas obtained by accurate mass spectrometry. *BMC Bioinfo.* **8**:105.

Kopka J, Schauer N, et al. 2005. GMD@CSB.DB: The Golm metabolome database. *Bioinformatics* **21**:1635–1638.

Liesener A, Karst U. 2005. Monitoring enzymatic conversions by mass spectrometry: A critical review. *Anal. Bioanal. Chem.* **382**:1451–1464.

Lin H, Fischbach MA, et al. 2005. In vitro characterization of salmochelin and enterobactin trilactone hydrolases IroD, IroE, and Fes. *J. Am. Chem. Soc.* **127**:11075–11084.

Lisec J, Schauer N, et al. 2006. Gas chromatography mass spectrometry-based metabolite profiling in plants. *Nat. Protocols* **1**:387–396.

Nordstrom A, Want E, et al. 2008. Multiple ionization mass spectrometry strategy used to reveal the complexity of metabolomics. *Anal. Chem.* **80**:421–429.

Northen TR, Yanes O, et al. 2007. Clathrate Nanostructures for Mass Spectrometry. *Nature.* **449**:1033–1036.

Olszewski KL, Morrisey JM, et al. 2009. Host-parasite interactions revealed by *Plasmodium falciparum* metabolomics. *Cell Host and Microbe* **5**:191–199.

Patterson AD, Li H, et al. 2008. UPLC-ESI-TOFMS-based metabolomics and gene expression dynamics inspector self-organizing metabolomic maps as tools for understanding the cellular response to ionizing radiation. *Anal. Chem.* **80**:665–674.

Risso C, Van Dien SJ, et al. 2008. Elucidation of an alternate isoleucine biosynthesis pathway in *Geobacter sulfurreducens*. *J. Bacteriol.* **190**:2266–2274.

Sauer U. 2006. Metabolic networks in motion: 13C-based flux analysis. *Mol. Syst. Biol.* **2**:62.

Siuzdak G. 1996. *Mass Spectrometry for Biotechnology*. San Diego, CA. Academic Press.

Smith CA, O'Maille G, et al. 2005. METLIN—A metabolite mass spectral database. *Ther. Drug Monit.* **27**:747–751.

Smith CA, Want EJ, et al. 2006. XCMS: Processing mass spectrometry data for metabolite profiling using Nonlinear peak alignment, matching, and identification. *Anal. Chem.* **78**:779–787.

Sumner L, Amberg A, et al. 2007. Proposed minimum reporting standards for chemical analysis. *Metabolomics* **3**:211–221.

Tang Y, Pingitore F, et al. 2007. Pathway confirmation and flux analysis of central metabolic pathways in *Desulfovibrio vulgaris* hildenborough using gas chromatography–mass spectrometry and Fourier transform-ion cyclotron resonance mass spectrometry. *J. Bacteriol.* **189**:940–949.

Taylor J, King R, et al. 2002. Application of metabolomics to plant genotype discrimination using statistics and machine learning. *Bioinformatics-Oxford* **18**:241–248.

Villas-Boas SG, Hojer-Pedersen J, et al. 2005. Global metabolite analysis of yeast: evaluation of sample preparation methods. *Yeast* **22**:1155–1169.

Want E, O'Maille G, et al. 2004. Solvent-dependent metabolite distribution, clustering, and protein extraction for serum profiling with mass spectrometry. *Cell. Biol.* **5**:763–769.

Werner E, Croixmarie V, et al. 2008. Mass spectrometry-based metabolomics: Accelerating the characterization of discriminating signals by combining statistical correlations and ultrahigh resolution. *Anal. Chem.* **80**:4918–4932.

Wikoff W, Anfora A, et al. 2009. Metabolomics analysis reveals large effects of gut microflora on mammalian blood metabolites. *Proc. Natl. Acad. Sci. USA* **106**:3698.

Winograd N. 2005. The magic of cluster SIMS. *Anal. Chem.* **77**:142A–149A.

Wiseman JM, Ifa DR, et al. 2006. Tissue imaging at atmospheric pressure using desorption electrospray ionization (DESI) mass spectrometry. *Angew. Chem. Int. Ed. Engl.* **45**:7188–7192.

Wishart D, Tzur D, et al. 2007. HMDB: The human metabolome database. *Nucleic Acids Res.* **35**:D521.

Yang Y, Xu Y, et al. 2009. Translating metabolic exchange with imaging mass spectrometry. *Nat. Chem. Biol.* **5**:885–887.

Yoshikuni Y, Ferrin T, et al. 2006. Designed divergent evolution of enzyme function. *Nature* **440**:1078–1082.

Yu Y, Ko K, et al. 2004. Discovery of the chemical function of glycosidases: Design, synthesis, and evaluation of mass-differentiated carbohydrate libraries. *Org. Lett.* **6**:2031–2033.

Chapter 72

Metabolomics: High-Resolution Tools Offer to Follow Bacterial Growth on a Molecular Level

LUCIO MARIANNA, AGNES FEKETE, MORITZ FROMMBERGER, AND PHILIPPE SCHMITT-KOPPLIN

72.1 INTRODUCTION

In a systems biology approach, metabolomics provides a functional readout of variations established by the genetic blueprint, protein abundance and modification, regulation, and environmental influence [Nicholson and Lindon, 2008; see also Chapter 71, Vol. I]. High-resolution molecular characterization techniques are particularly pertinent in biochemical and metabolome analysis because they enable faster and more accurate understanding of such complexity. The information about the actual cellular environment can consequently be retrieved looking more precisely and closely to the metabolome. The environment of the cell, cell associations, organs, or organisms is connected to many exogenous factors such as, for example, nutrition, drugs, pollutants, and environmental factors. Thus they can be used as tracers for human health predictions. In a similar manner, it is possible to follow environmental stresses (such as affected by pollutions or climate changes) in environment and biogeography metabolomic studies, allowing, for example, a mapping of metabolites as a result of organism adaptation strategies to particular pristine [Rossello-Mora et al., 2008] or polluted environments. The holistic view and interpretation of the metabolites present in a given system are governed by the technology employed. Here, the entire metabolomic approach starts with sample extraction, preparation, and concentration, and separation with spectrometric/spectrophotometric detection that has to be individually optimized and integrated into the total work flow [Harrigan and Goodacre, 2003]. The sample preparation is used to extract low-molecular-weight analytes, and thus the resolution power of the separation techniques applied—such as gas chromatography (GC), liquid chromatography (LC), and capillary electrophoresis (CE)—has to be as high as possible. Orthogonal, multidimensional systems were developed to enhance the chemical domain and the capacity of separable metabolites; GCxGC, LCxLC, and LCxCE have been cited as highly sophisticated approaches for a description of the molecular complexity of biological systems.

The main analytical tools used for metabolomics are so far based on nuclear magnetic resonance spectroscopy (NMR) and mass spectrometry (MS) techniques [Hollywood et al., 2006]. Although the NMR-based direct and quantitative mixture analyses in conjunction with mathematical data analysis and de novo structural elucidation were crucial supplements due its limitations in sensitivity, MS is promising and it more frequently applied technology in this field [Chen et al., 2008] either coupled to LC [Want et al., 2007], GC [Yuan et al., 2007], or CE [Yuan et al., 2007; Benavente et al., 2006; Baidoo et al., 2008; García-Villalba et al., 2008; Levandi et al., 2008] or in infusion mode as a stand-alone technique varying the types of atmospheric pressure ionization sources (ESI, APCI, APPI, APLI) resulting in increased structure-dependent ionization efficiencies of various metabolites [Hertkorn et al., 2007]. In this regard, the use

Handbook of Molecular Microbial Ecology, Volume I: Metagenomics and Complementary Approaches, First Edition. Edited by Frans J. de Bruijn.

of high-field Fourier transform-ion cyclotron resonance-mass spectrometry (ICR-FT/MS) enables the assignment of thousands of mass-to-charge ratio (m/z) of metabolites in complex mixtures by virtue of its unique mass accuracy (<100 ppb) and resolution (>500,000 in full scan mode at m/z 500) allowing the detection of fine-scale diversity in complex mixtures [Hertkorn et al., 2007]. The high mass accuracy allows the correspondence between masses to unique elementary compositions and annotation to potential metabolites giving as a result, for example, the generation of specific metabolic maps for each studied organism [Suhre and Schmitt-Kopplin, 2008] or the detection and structure confirmation of specific biomarkers [Rossello-Mora et al., 2008; Wang et al., 2008]. Additionally, metabolomics studies have also been possible thanks to the great advance in statistical and data evaluation tools [Eriksson et al., 1997; Crockford et al., 2006; see also Chapter 74, Vol. I]. Here we present the essential steps needed to process spectra coming from ICR-FT/MS. We illustrate the potential of this technique in the light of a specific example relative to the dynamic regulation of *N*-acyl-homoserine lacton (AHL) autoinducer derivative produced by *Pseudomonas putida* isoF. We recently found that the concentration of the AHLs was strongly dependent on the growing phases [Fekete et al., 2010]; [Li et al., 2006]; the target components did not show saturation in the function of the time but reached a maximum in the inflection point of the exponential phase and degraded completely until the beginning of the stationary phase. The nontargeted analysis presented here required the development of data evaluation tools of the mass spectra of the cell pellet and growing supernatant extracts to be capable of describing the dominant metabolisms in the different growing phases and thus describe the growth phenotypes of a bacterial batch culture.

72.2 MATERIALS AND METHODS

72.2.1 Experimental Condition

Cultures of *P. putida* IsoF [Steidle et al., 2001] were grown in ABC minimal medium containing citric acid as a carbon source. Cells were shaken at 30°C and 175 rpm in 500-ml baffled flasks with 200 ml of medium. Alternatively, a 5-L beaker was used containing 4 L of ABC medium, which was stirred by a magnetic stirrer at about 200 rpm. The medium was inoculated 1:1000 with an overnight culture and samples were taken right after inoculation and then every hour. The optical density was recorded by spectrophotometer at 436 nm, and an aliquot of cells was fixed with paraformaldehyde for subsequent cell counting under the microscope (10 randomly selected squares of 126 μm^2 each per time point). After taking the last sample, the ratio between living and dead cells was determined by a live/dead cell staining kit (Molecular Probes, Live/Dead BacLight Bacterial Viability Kit) and the percentage of living cells at the last time point was still above 90%.

The collected 200-ml samples were centrifuged at 15,000*g* and 4°C for 15 min and the supernatant was filtered through Millipore nitrocellulose filters (0.22 μm, type GSWP) with a vacuum pump. Then the samples were quenched by shock frosting with liquid nitrogen and stored at −25°C until measured. For analyzing bacterial cell pellets, the bacteria were dissolved in 2-ml ultrapure water, the same volume of methanol was added, and finally they were shock frosted with liquid nitrogen until their direct infusion to ICR/FT-MS.

A desalting procedure had to be applied for the growing supernatants before measurement ICR-FT/MS. The supernatants were purified by solid-phase extraction (SPE) and afterwards analyzed according to a previously published method [Fekete et al., 2006, 2007]. Briefly, after conditioning with methanol and water, the 200-ml sample extract was introduced onto a MegaBond Elute cartridge (Varian, Darmstadt, Germany). The cartridge was then washed with a methanol/water mixture and was subsequently dried under vacuum, and then the solutes were eluted with a 2-propanol/hexane mixture (85/15% v/v). The eluate was dried under a nitrogen stream, resolved with water containing 10% (v/v) acetonitrile, and filtered through a PTFE filter.

72.2.2 Instrument

Positive and negative ICR-FT mass spectra for qualitative analysis of AHLs were acquired on a Bruker Daltonics (Bremen, Germany) Apex Qe 12 T system equipped with an APOLLO II source and microspray infusion of 120 $\mu l\ h^{-1}$ with a scan number of 256. Prior to elemental formula calculation, all mass spectra needed to be calibrated to maximum accuracy externally and internally. External calibrations were done with a 10-ppm (part per million) arginine solution in positive and negative mode, while internal calibration was accomplished utilizing ubiquitary solvent impurities, either phthalate diesters (for positive ESI spectra) or fatty acids (for negative ESI spectra). For both external and internal calibration (where the further served as a positive control of proper system configuration), a maximum mass error of 100 ppb was accepted.

The raw data are processed with DataAnalysis 3.4 (Bruker Daltonik) software that is hard-coded in the instrument. An in-house software coded in Python aligns the spectra, and the chemical formulae are computed by a software developed in-house in Python and FORTRAN 95. Statistical analyses are done with SIMCA-P 11.5 (Umetrics, Umea, Sweden).

72.3 RESULTS

72.3.1 Growing Condition of *P. putida* isoF

For studying the growth of *P. putida* IsoF on a metabolome level, two different setups were used, namely, (i) a 5-L beaker from which 200 ml were taken at each time point and (ii) a set of 500-ml baffled flasks each containing sample volumes of 200 ml, allowing the comparison of growing rates in different culture conditions **Fig. 72.1**. The optical density and the cell counts were higher in the baffled flasks relative to the beakers, showing better growth conditions in the baffled flask presumably resulting mainly from better availability of dissolved oxygen. When a fresh medium is inoculated with given number of cells, a typical bacterial growth curve with four distinguishable phases could be observed: the lag, exponential (or log), stationary, and death phase. As shown in Figure 72.1, the growing phases were independent on the experimental setup; the lag phase was between 0 h and 7 h, the exponential phase lasted until 19 h, and the stationary phase was observed until the end of the study (34 h). Although several methods—such as direct microscopic count, viable cell count turbidity, or optical density measurement and cyclic voltammetry, describing the growth of the cells in time—have been developed, only limited information on metabolite production in function of growth phases has been found [Martins et al., 2004; Subrahmanyam et al., 2001]. Thus, direct infusion ICR/FT-MS was applied for the measurements of biomass and supernatant extracts to study the temporal metabolic processes and to investigate if the growth phases can be described as phenotypes of the bacterial species.

72.3.2 FT-ICR-MS Data and "Matrix Generator"

Several instrumental parameters were optimized in each ionization mode for nontargeted analysis applying direct infusion (see Chapter 73, Vol. I) ICR-FT/MS in order to get the maximum number of signals at highest intensity. In this study, electrospray ionization (ESI) was selected because it enables the detection of several middle apolar, polar and ionic substances. Positive and negative ESI modes generate different but complementary structural information. For the targeted determination of the autoinducers (putida), positive ESI was applied since AHLs showed an unstable and nonsensitive nature at negative ESI. However, by changing the mode to negative, more sample peaks were observed in the mass spectrum of the sample extracts, and thus it was applied for further nontargeted description of the bacterial growth on the metabolome level. A typical mass spectrum at broad mass range (*m/z* 150–1000) is shown in **Figure 72.2** using a dataset of 4 MW and 1 MW.

The mass accuracy did not change with resolution, but 4 MW certainly enables the additional differentiation of signals with narrow mass differences (i.e., m/z = 403.2337 in Fig. 72.2). Applying negative ESI and 4 MW, we observed 2600 and 4800 signals in the extract of bacterial pellet and growing medium, respectively.

To be able to compare different measurements, storing the data in a matrix is the first step of any type of experiments. In order to store properly all the data coming from an experiment, a reliable alignment algorithm has to be developed. The alignment starts with the mass spectra output of the ICR-FT/MS processed in this case with the Data Analysis Bruker Software exported in ASCII format. Each ASCII file represented a mass spectrum having on x axis the exact masses and on y axis the experimental intensities. Both of the acquired quantities were subjected to statistical investigations.

The extracted ICR-FT/MS spectra need to be analyzed as individual samples in the context of the population to which they belong, and this requires a preliminary data structure design. The union of the sample files is foreseen following the logical scheme visualized in the integration flux represented in **Figure 72.3**.

A first step required sorting the joined data according to the mass value, in order that each *m/z* value could be

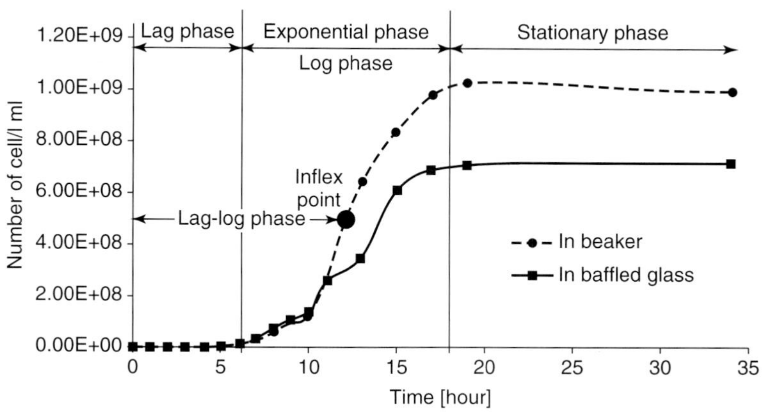

Figure 72.1 Cells number in function of time in beaker (– – –) and in baffled glass (——), and the growth phases according to microscopic the cell counting.

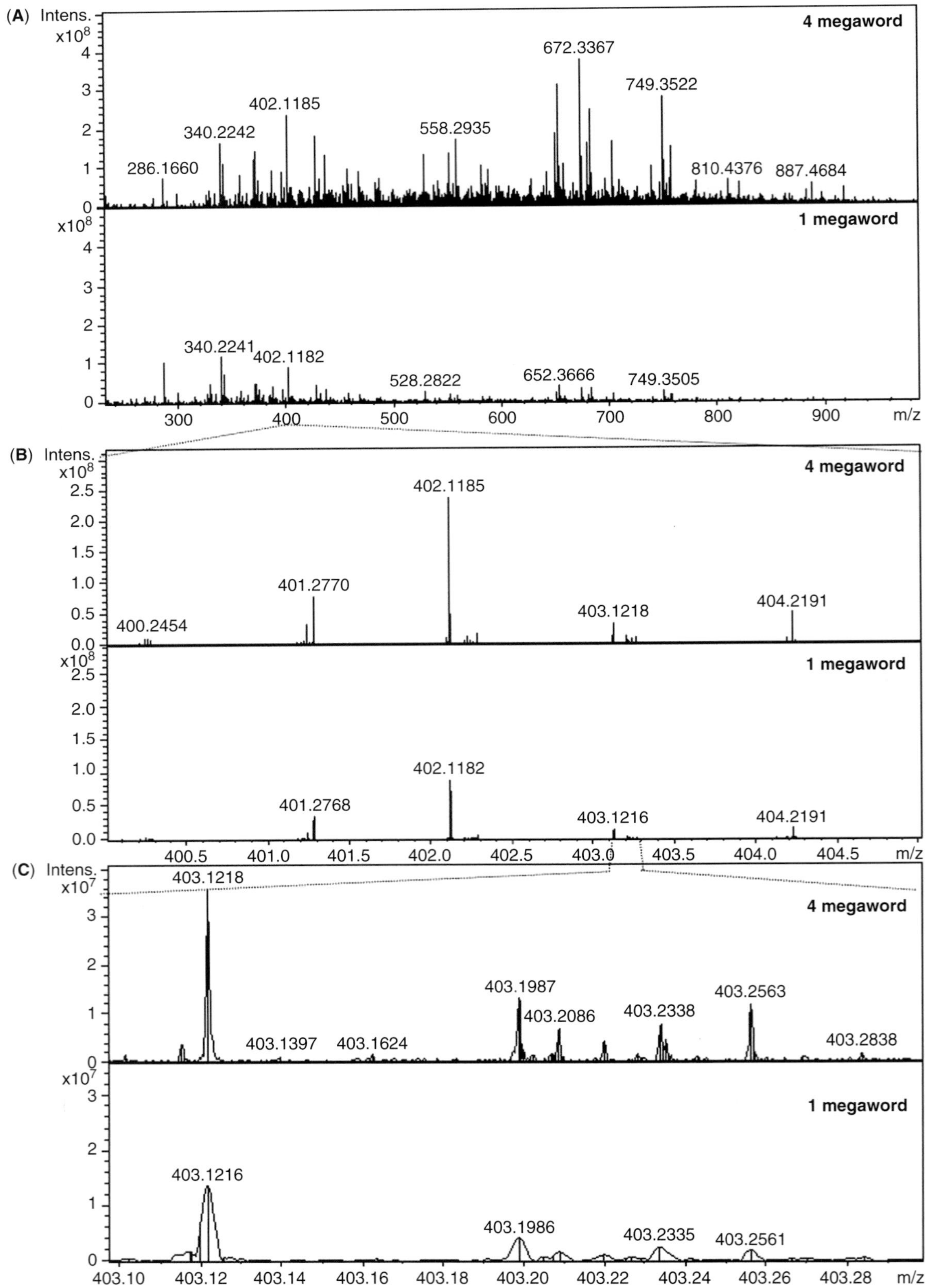

Figure 72.2 ESI-ICR-FTMS spectra of bacterial cultures between *m*/*z* of 250–1000 (**A**), 400–405 (**B**), and 403.1–403.3 at two different resolution setups (4 megawords and 1 megaword), with data measured in negative mode.

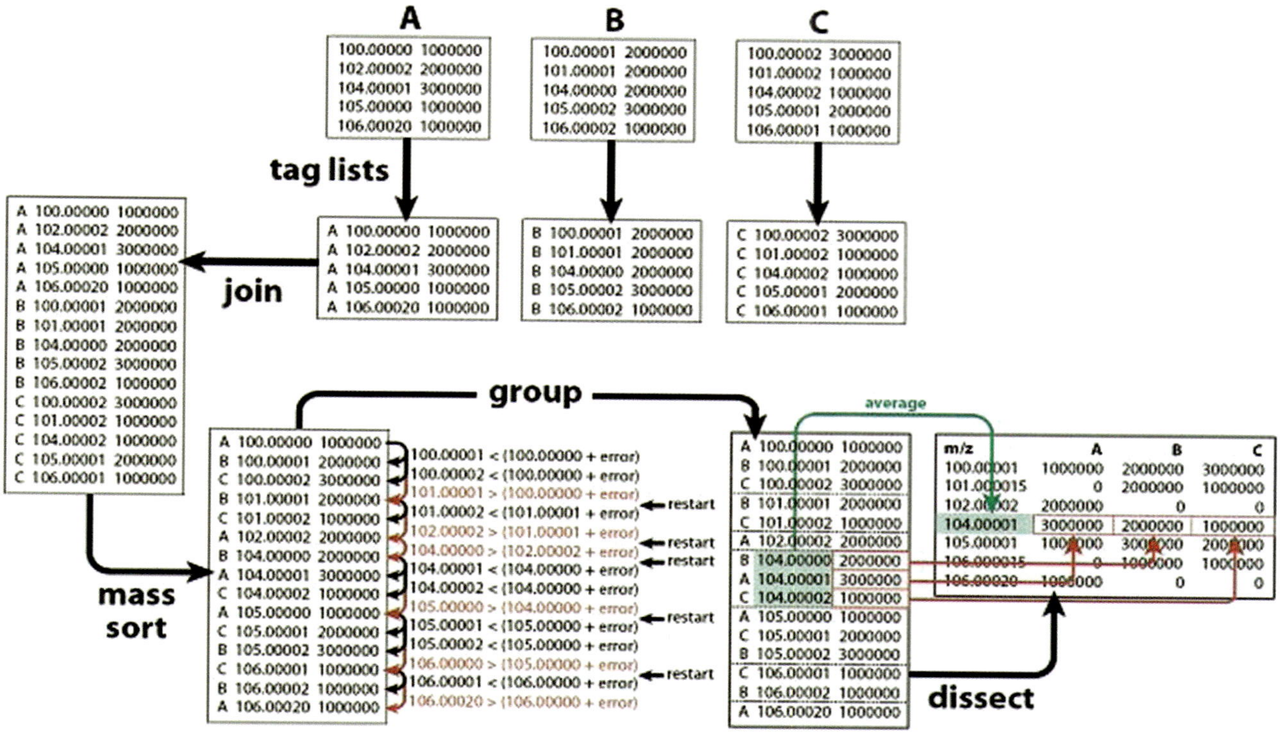

Figure 72.3 The input files are the spectra data (organized in: m/z, intensities); in this case **A**, **B** and **C**. They are tagged in order to be recognized during all the consecutive steps. After they have been joined and sorted. All the masses Within the window width chosen as a parameter are averaged. The intensities are reported in the formed matrix as the original intensity or 0, if there is no peak.

compared only with the nearest neighbor. To this end, the data belonging to the same sample are tagged. This way they were recognizable over the whole data processing pipeline. Once spectra data were tagged, joined, and sorted, the second step is grouping the next set of different masses. According to the mass error, moving within the range 0.1–1 ppm, a cross-correlation between the observed mass positions and error template was performed leading to averaged masses in case of positive correlation. The obtained masses are used in spectra assigning the respective measured intensities or the value 0 (zero) if the element associated to that mass is not present. As a result, we obtain a dataset composed of all the masses (measured and calculated) and their respective intensities (1 if present, otherwise 0), and then the peaks are aligned between samples so that one mass in a sample corresponds to the same compound in the next sample if present.

The first assumption in creating the matrix with this algorithm called "Matrix Generator" (in-house program written in Python) was a search for close peaks within each individual spectrum to avoid wrong comparison results between the different spectra. Whenever two peaks with a distance less than the defined minimum are detected within one spectrum, their mass values are averaged.

In **Figure 72.4** the number of peaks in the function of error is shown when the matrix was generated from 17 measurements. As was expected, a decreasing tendency was observed when the error level was increased. At low error value the same m/z-values in different spectrums may be recognized as distinct signals but at high error, very similar m/z-values having different elemental compositions might be considered together. A breaking point at 1 ppm was observed that was considered as an artifact-free value, permitted to reduce at best the presence of double peak events, and this did not build up any peak clusters from a single and unique m/z values. In alignment of the expectation, more charged masses were observed in the pellet extracts as in the growth medium applying 1 ppm for the matrix generation because an additional clean-up step for the medium had to be applied for desalting the samples causing components loss and also less excreted metabolites were expected.

72.3.3 Statistical Analysis

To set up a matrix is the start point to create a statistical model. It becomes useful to summarize and visualize a huge amount of data, to classify and to discriminate, and finally to find quantitative relationships among the

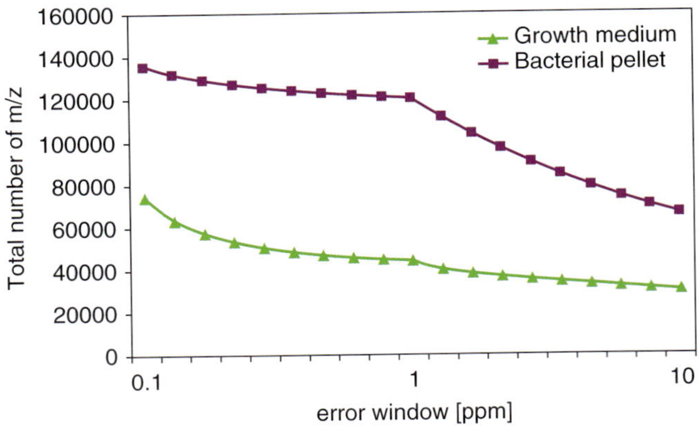

Figure 72.4 Total number of m/z peaks after spectra alignment of 17 samples of growth medium (▲) and pellet extract (■) at different values of error window.

variables with the possibility to make also a provision for new samples. In many cases the final aim, however, is obtain a plausible list of metabolites or biomarkers that explicate the phenomena being studied.

The large numbers of peaks in the spectra that are all potential biomarkers create modeling and validation challenges [Westerhuis et al., 2008]. A multivariate analysis based on projection methods represents a number of efficient and useful methods for the analysis and modeling of the complex data; this methods include principal component analysis (PCA) and partial least squares (PLS) [Trygg et al., 2006].

This modus operandi is robust and efficient for modeling and analysis of complicated chemical and biological data, which are characterized by noisy missing values and collinear data structure. However, it is difficult and is still widely discussed to determine when a model has become sufficiently detailed for its task and how confident one can be in its predictions. To fully realize the potential of this complexity science, much more investigation is needed to distinguish and understand the hidden information.

The objective in metabonomics is, first, to classify a spectrum based on identification of its inherent patterns of peaks and, second, to identify those spectral features responsible for the classification, which can be achieved via supervised and unsupervised pattern recognition technique [Coen et al., 2008].

A PLS model was applied to the extract of the growing medium grown in beaker and baffled flask to investigate the dominant metabolic pathways in the function of growing time [an increased biomass amount, this is considered to be a response matrix (Y)]. Before any statistical evaluation was applied, an orthogonal signal correction (OSC) to the data, which permits us to discard or retain the information noncorrelated (orthogonal) to Y [Bylesjö et al., 2006].

As shown in **Figure 72.5**, a border line was observed at 7 h for baffled flask and 8 h for beaker. This fitted to the time where the cells division accelerated to a constant growing rate differentiating the lag and exponential growing phase. The 1-h difference in the different experimental setup can be caused by the better growing condition in the baffled flask (Fig. 72.5). The time point 34 h was an observed outliner that belongs to the stationary phase according to the optical density measurement. Thus growing phases could be also differentiated by applying nontargeted analysis of the low-molecular-weight organics in the growing medium (excreted metabolites) by high-resolution MS along with direct microscopic count, viable cell count turbidity, or optical density measurement. But even more information can be extrapolated from the ICR-FT/MS measurements since the m/z values typical for the different phases can be annotated, thereby gaining more specific information on the division mechanism of the bacterial cell.

The mass spectra of the pellet extracts were also aligned and investigated with different multivariate statistical analysis tools. Biomass older than 5 h were possible to collect, extract, and measure by direct infusion ICR/FT-MS.

As shown in **Figure 72.6**, the first component differentiates the spectra at the inflection point of the exponential phase (11h and 15h shown in Fig. 72.1), reflecting the changing metabolic machinery to adapt to the competition for the dwindling nutrients that finally leads to zero growth rate. The second component differentiates the lag phase (5–7 h) from the front of the log phase (8–11 h) and also differentiates the end of the exponential (15–19 h) from the stationary phase (34 h) as shown in Figure 72.6. Interestingly, the lag and the beginning of the exponential phase showed more similarities applying the recent statistical model, and thus they were studied together and called a log-lag phase. Thus by investigating the metabolites in the bacterial biomass by nontargeted analysis approach, we could also precisely follow the actual state of cell division.

In order to validate the model, we applied internal validation; the significant components were calculated using sevenfold cross-validation, where the Q^2 and R^2 near to 1 denoted the quality of the resulting model (**Fig. 72.7**). Moreover, we validate the model using the so-called response permutation test (100 permutations) with $R^2 = 0.0$, 0.0349 and $Q^2 = 0.0$, -0.297. Eriksson et al. [2001] have shown that when $R^2 < 0.3$–0.4 and $Q^2 < 0.05$, the explanatory and predictive powers of the model are much higher than those obtained from randomly fitted Y data, and therefore the present model is valid.

When a statistical model is set up and validated, it is possible to extrapolate a list of masses characteristic for

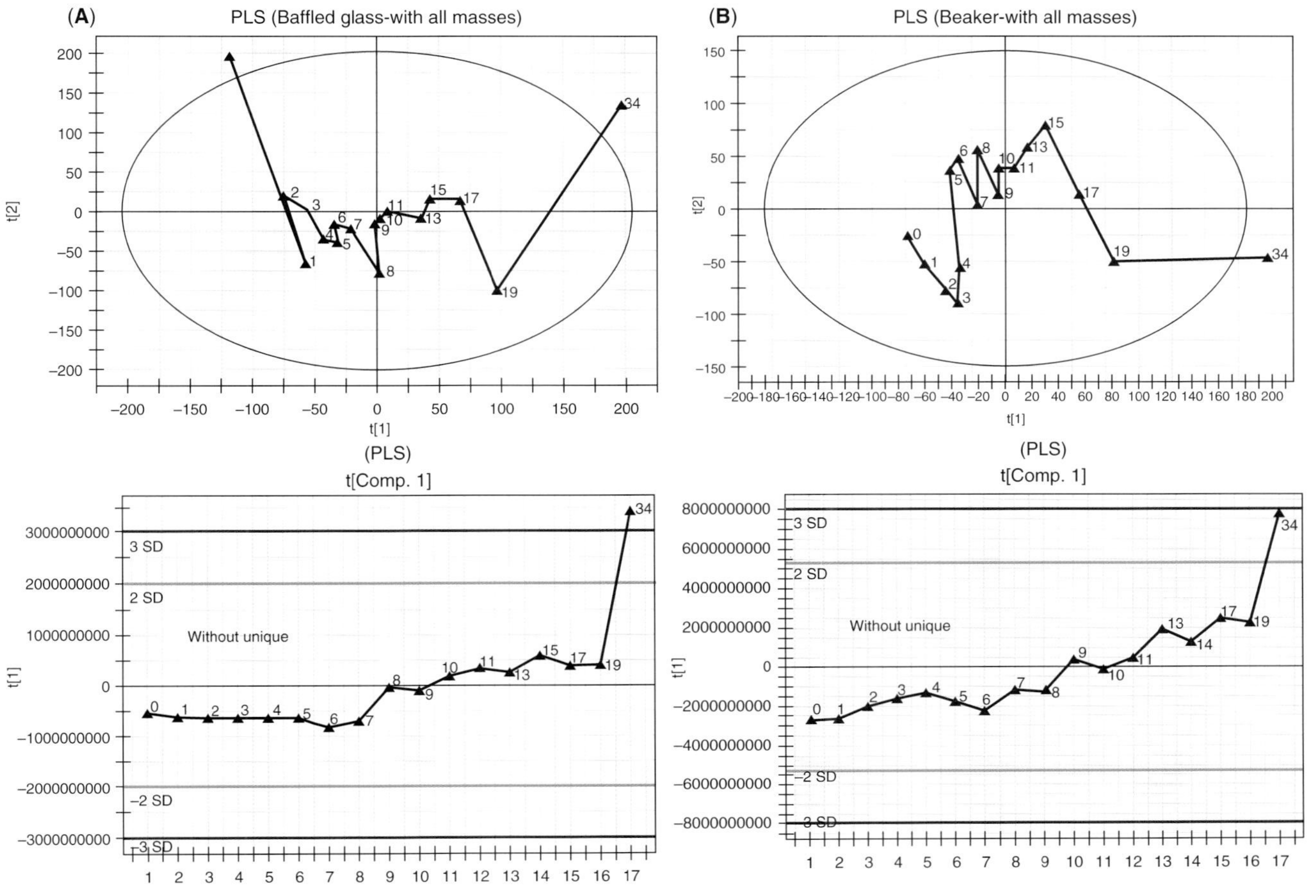

Figure 72.5 OSC-PLS score plots of the extracted mediums samples at different growth times. (T_X) grown in baffled flask and beaker. The log, exponential (or lag), and stationary phase were clearly separated after ignoring the unique masses. **(a)** Correlation between novelty and abundance of clones within each $OTU_{0.03}$ identified in KFS dataset. **(b)** Correlation between average percent sequence divergence and abundance of clones within KFS $OTU_{0.03}$ assignments.

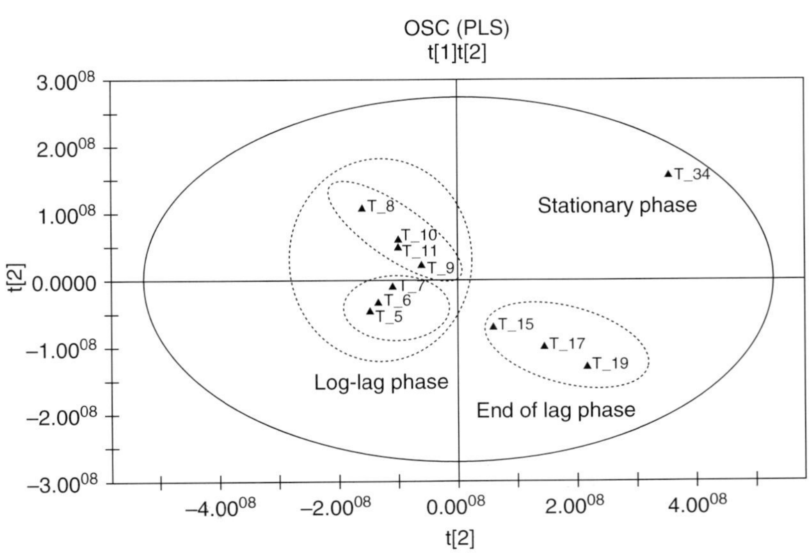

Figure 72.6 OSC-PLS score scatter plot of the pellet extracts spectra measured at different growth times (T_X). The log-lag phase is clearly separated from the end of the lag phase and the stationary phase.

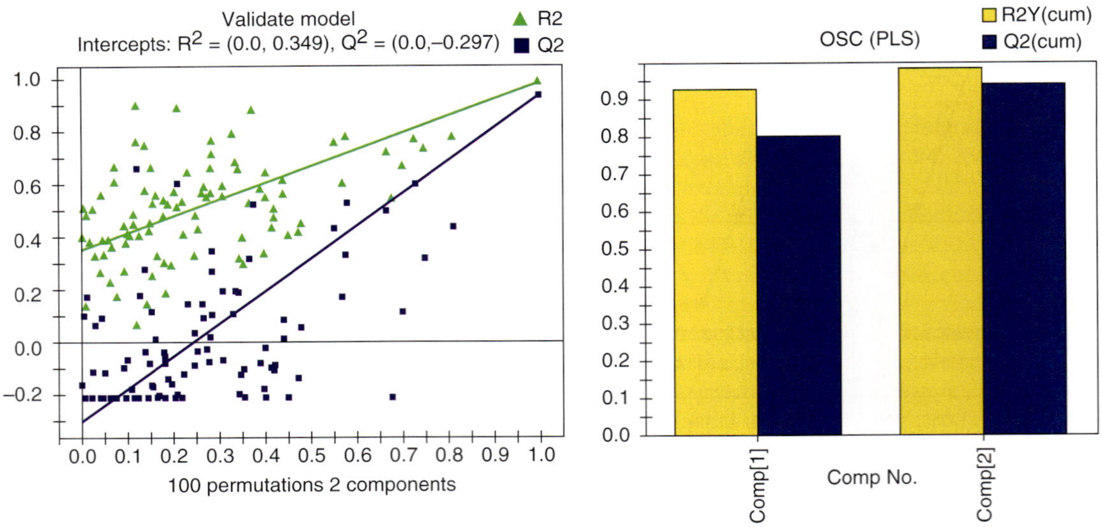

Figure 72.7 Internal validation. In the left-hand part the plot shows the different R^2 and Q^2 values and their intercepts obtained by a permutation test (100 permutations). In the right-hand part are plotted the Q^2(cum) and the $R^2\gamma$(cum) values obtained from the model with two valid components.

the different groups' object of study, and these masses are the possible biomarkers different tools that have been developed to annotate these *m/z* values. Through an in-house developed software, Formula Calculator (see paragraph 72.3.4), it was possible to calculate, starting from the accurate mass, the formula composition. The visualization of the hundreds of experimental elementary compositions calculated that way can be followed through van Krevelen diagram [Rossello-Mora et al., 2008]. Additionally, a metabolite–annotation web interface MassTRIX (http://www.masstrix.org) was developed that annotates the input experimental masses against all compounds in different databases (KEGG for a chosen organism, LipidMaps and Human Metabolome Database) [Suhre and Schmitt-Kopplin, 2008]. These visualization tools are described below.

72.3.4 Formula Calculator

A portable program written in Fortran 95 with a modular graphical user interface and processing front-end written in Python/TkInter was used for rapid batch calculation of possible elemental formulae. In principle, formula calculation was performed by subtracting (for positively charged peaks) or adding (for negatively charged peaks) an electron mass (0.000549 amu) from or to the measured *m/z* value. This formal neutralization was followed by a transformation of the mass value from a 12C-based IUPAC scale to a 1H-based hydrogen scale, where H = 1.000000 amu, C = 12.000000/1.007825 amu, O = 15.994915/1.007825 amu, and so on. The purpose of this transformation was to reduce the number of necessary calculation steps by a reduction of the number of possible elemental combinations that are needed to be considered for each mass. In this transformed mass, only the heavier elements contribute to the fractional part of the number, but not H, which normally is the most abundant element in biomolecules.

In parallel with this transformation, a database consisting of all possible combinations of elements (excluding H but analogously H-transformed) that were previously allowed, that fit in the mass window of interest, and that match chemical rules was generated (nitrogen rule, atomic oxygen to carbon ratio O/Cp1, element counts: hydrogen Hp(2+C2), carbon Cp100, oxygen Op80, nitrogen Np5, and sulfur Sp1), and only the masses in conjunction with their automated generated theoretical isotope pattern (existence of the 13C isotope) were taken into consideration [Hertkorn et al., 2007]; [Hertkorn et al., 2009].

In the next step, the fractional part of the transformed measured mass was compared to the fractional parts of the combinations in the database. Whenever a match within a previously defined error window (typically 0.5 ppm) was found, the integer part of the database mass was subtracted from the integer part of the ion mass which directly resulted in the number of hydrogens, because the H mass was previously scaled to 1.000000.

The formulae generated with this method were afterwards subjected to more advanced modular plausibility filtering steps—for example, involving a check for the nitrogen rule, minimum and maximum O/C and H/C ratio, presence of a 13C peak mass, or comparing to theoretical isotope patterns. The validity of the above-mentioned algorithm was extensively checked by comparing different known, but slower, methods of

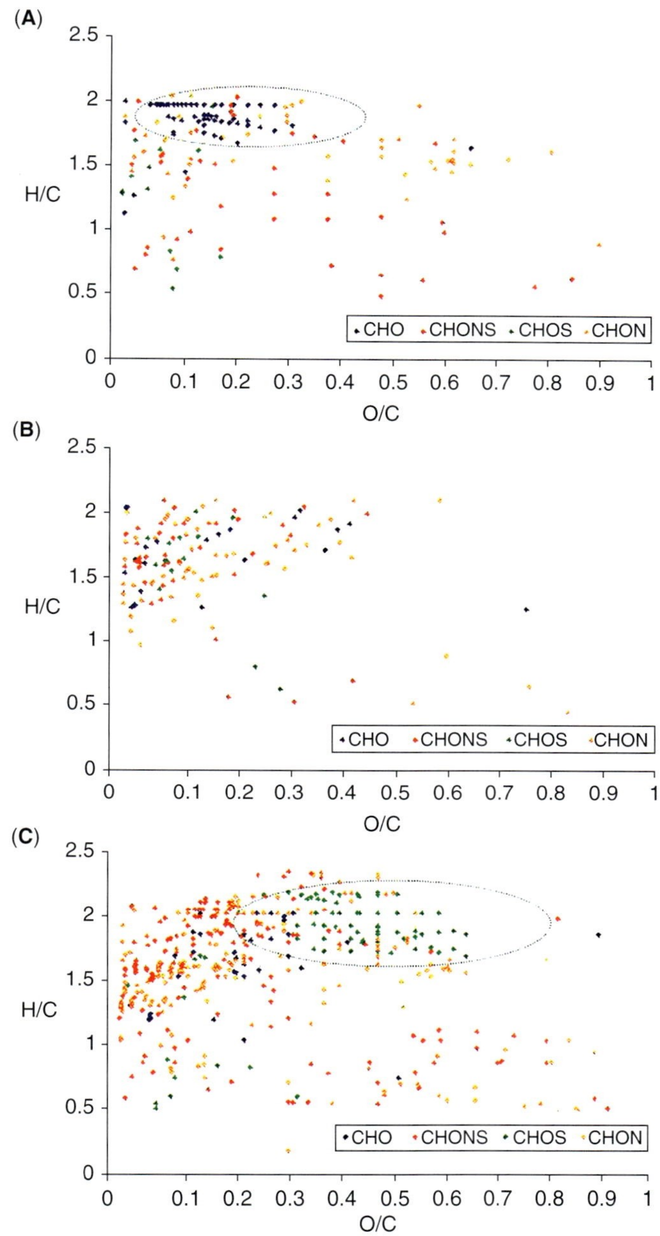

Figure 72.8 Chemical compositions of the discriminating *m/z* values of **(A)** the log-lag phase, **(B)** the end of the lag phase, **(C)** and the stationary phase visualized in a van Krevelen diagram..

elemental formula calculation. Across the comparison, no errors of imputation were found in the algorithm.

The elemental compositions of the *m/z* values describing the lag and the front of exponential phases (together called as lag-log phase), along with second part of lag and stationary phases in the extract of the pellet grown in the baffled flask, were determined applying the above-mentioned tool; 414, 520, and 912 elemental compositions described the lag-log phase, the end of the exponential phase, and the stationary phase, respectively. Although the calculated elemental compositions provide only limited information on the structure, it can be visualized with a van Krevelen diagram to be able to investigate the possible coherence in the chemical nature of the markers distinguishing the growing phases.

72.3.5 Data Interpretation and Visualization: van Krevelen Diagram

The chemical property of the list of candidate could be visualized with the van Krevelen diagram [Hertkorn et al., 2007]. It developed a graphical method to study the process in which the atomic hydrogen/carbon (H/C) ratio is plotted as a function of the atomic oxygen/carbon (O/C) ratio; similarly, other atomic ratios can be used Initially these diagrams were used to compare the bulk elementary composition of coal, but the methods were adapted for ICR-FT/MS datasets.

Although the van Krevelen diagram has been mostly applied to describe the quality of the complex natural organic matter (NOM) and in petroleomics, its practicability in metabolomics has been presented previously [Rossello-Mora et al., 2008; Leon et al., 2009]. The determined elemental compositions that discriminated the growth phases were visualized in the van Krevelen diagram and shown in **Figure 72.8** where different colors point to the presented elements (blue, CHO; red, CNONS; green, CHOS; and orange, CHON).

No typical pattern was recognized at the end of the lag phase (Fig. 72.8) until a pattern of C,H,O markers in the log-lag phase (Fig. 72.8A) and a pattern of saturated, oxygen-rich sulfur-containing components in the stationary phase were observed (Fig. 72.8C). In Fig. 72.8A, the data points at an H/C ratio of 2 reflect the saturated and O-depleted solutes like nonsubstituted fatty acids with carbon numbers between 10 and 16. However, at the same H/C ratio and higher O/C values, the number of oxygens increased, showing possible hydroxylation and dihydroxylation of the decanoyl and dodecanoyl derivatives. Additionally, four linear coherences with decreasing slope were observed within the CHO isomers as the markers of the growth before the inflection point of the growing curve (Fig. 72.1) that could be annotated as keto fatty acids, dicarboxylic acids, and nonsaturated carboxylic and dicarboxylic acids. Thus in the beginning of the growth, dominance of the fatty acid oxidation was concluded by applying statistical analysis of the broad-range mass spectra, which is a major source of the cellular ATP and thus operates the chemical energy transports within cells for metabolism. It was shown previously that although the growth of a bacterial strain was not caused by additional fatty acid derivatives; the high basal level of the oxidative enzymes is a consequence of an "internal induction" by the products of the fatty acid synthesizing system in the beginning of the bacterial growth [Weeks et al., 1969].

72.3.6 Data Visualization: MassTrix

Since metabolism of fatty acids showed dominance in the lag-log phase, we investigated which additional metabolism pathways play a role in the bacterial growth phase. Thus metabolite-annotation web interfaces, called MassTRIX (www.masstrix.org), were applied. MassTRIX processes the submitted mass peak list by comparing the input experimental masses against all compounds of the KEGG chemical compound database, additionally including ^{13}C, ^{15}N, and other isotopes and optionally adding selected lipids with variable fatty acid chain lengths. Raw input masses from ESI-MS can be corrected on-the-fly for the addition or the abstraction of a proton (and optionally a sodium ion in positive mode). To cope with the requirement of very low measurement errors (in the sub-ppm range), exact masses of all KEGG compounds have been recomputed from the corresponding chemical formula using high-precision atomic mass data [Wapstra et al., 2003]. MassTRIX then calls the KEGG/API (http://www.genome.jp/kegg/soap/) to generate pathway maps, where the identified compounds and genes are highlighted using different colors, thus differentiating between organism-specific and extra-organism items [Suhere and Schmitt-Kopplin, 2008].

The *m/z* values discriminating the three growth phases in the pellet extracts were submitted to MassTRIX. Since no *P. putida* isoF was found among the optional organisms, *P. putida* F was selected for the annotation since genetically they are close to each other. **Figure 72.9** shows the number of annotated metabolites within the pathways.

Metabolites in the citrate pathway were annotated as the discriminating factor in the lag-log phase, confirming the observation using a van Krevelen diagram since it is an important pathway for the oxidation of carbohydrates and fatty acids. Several common pathways were found assisting the changes of the growth rate, but the putative metabolite differs in the different growing phases. For example, discriminating metabolites in the fructose and mannose metabolism, in the fatty acid biosynthesis, and in the biosynthesis of fatty acids were observed in two phases, although a larger number of components within the pathways were found in the lag-log phase. The accurate mass measurement does not allow the differentiation of the isomers (molecules with identical elemental composition but different structure). Confirmation of the metabolite within a pathway is required with target analysis tools. Information on the contributed pathways of the different growth phase can be achieved, thereby enabling the differentiation at phenotypic level.

72.3.7 Summary and Perspectives

The use of FT/ICR-MS takes us to an unknown and not yet completed investigate world: Metabolomics, like all of the bio(geo)chemical analyses, generates data torrents that have to be interpreted in global and not aiming the targeted determination. Since lack of information was found on the data evaluation of FTICR mass spectra for nontargeted analysis of complex samples, a systematic approach was developed and applied for the characterization of the metabolites in function of the growth phases in batch culture (**Fig. 72.10**) A possible strategy of data investigation is presented here, and it

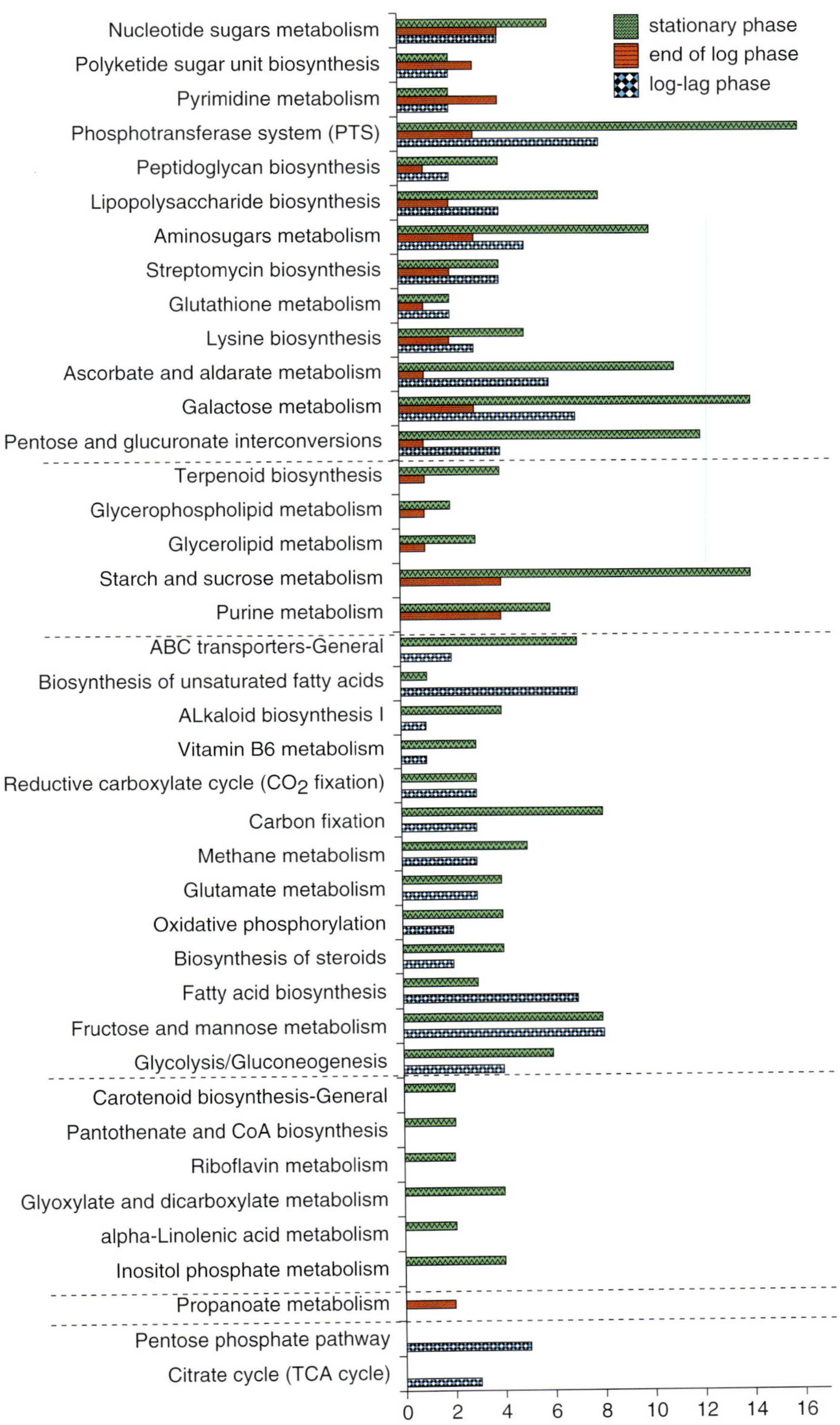

Figure 72.9 The *m/z* values discriminating the three growth phases in the pellet extracts were submitted to MassTRIX.

is a complete excursion starting from the raw data and ending with a list of biomarkers, with also the possibility to annotate part of them. The approach involves four parts; (i) optimization and characterization of the sample treatment and measurement to get *r* elevated number of signals at robust manner, (ii) data alignment enabling the comparison of the samples, (iii) statistical analysis to filter out the important peaks for the investigated phenomenon, and (iv) mass annotation applying database search and/or calculating the elemental composition from the accurate

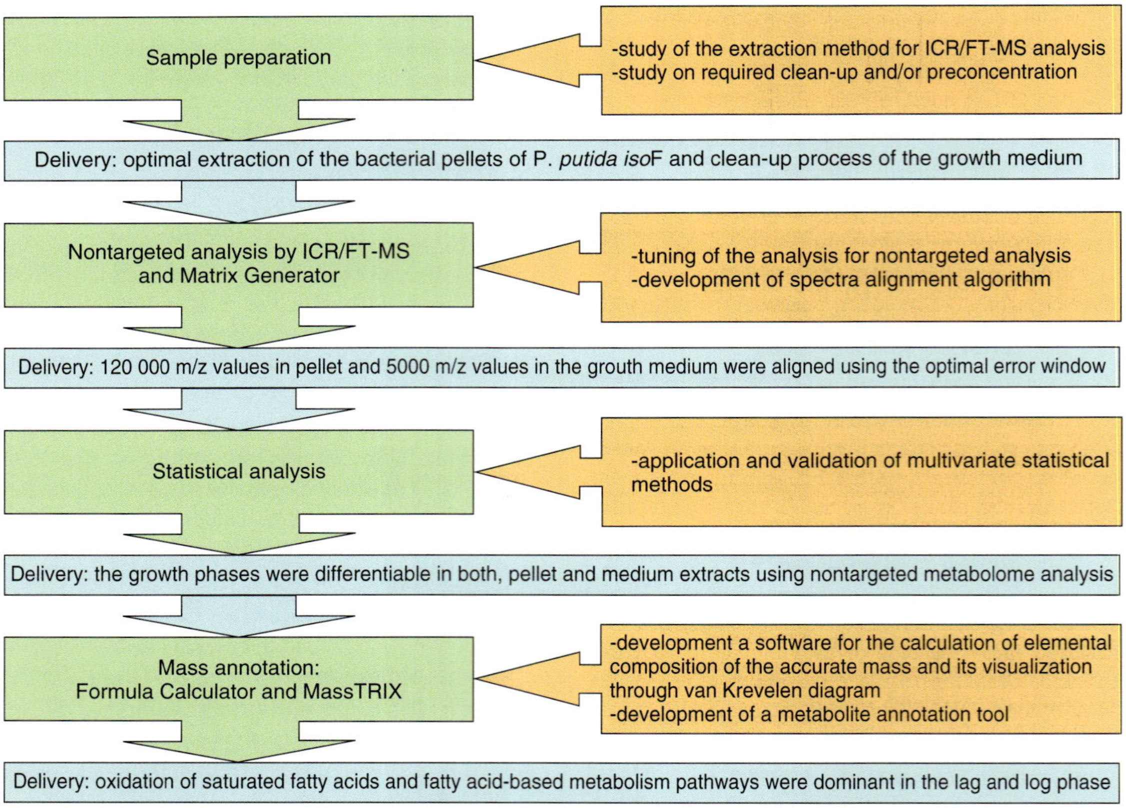

Figure 72.10 General scheme of the work.

charged mass. The processes are still technically complex, indeed there is a lack in terms of databases; for example, many of the biomarkers found are not yet classified.

All these aspects reflect the fact that metabolomics is a young field, especially with the new technologies (ICR-FT/MS 12T). The future is in the combination of different analytical techniques (separation/spectrometry/spectroscopy) that complement themselves to give a picture of the chemical space of the sample analyzed and hoping that its variability will be very challenging to a statistician. This needs to combine automatic sample preparation before ultrahigh-pressure liquid chromatographic and/or CE separation (CE) coupled to high-resolution tandem mass spectrometry via various atmospheric pressure ionization modes (ESI, APCI, APLI, APPI). The separated compounds can then be analyzed in a high dimensionality by adding a broad range of online and offline nuclear magnetic resonance spectroscopy (NMR) methods.

72.3.8 Software

The software used in order to develop the different tools were FORTRAN 95 and Python 2.6.4. The data analyses were done with DataAnalysis 3.4 (Bruker Daltonik), software that is hardcoded in the instrument (FT-ICR/MS). The analysis and elaboration were done in SIMCA-P 11.5 (Umetrics, Umea, Sweden).

Acknowledgments

The authors would like to thank Christina Kuttler and Michael Rothballer for their participation in the inoculation of the biomass, sampling, sample preparation, and cell count measurements. Special thanks to Brigitte Look for excellent technical assistance in the clean-up procedure.

REFERENCES

Baidoo EEK, Benket PI, Neususs C, Pelzing M, Kruppa G, et al. 2008. Capillary electrophoresis–Fourier transform ion cyclotron resonance mass Spectrometry for the identification of cationic metabolites via a pH-mediated stacking-transient isotachophoretic method. *Anal. Chem.* **80**:3112–3122.

Benavente F, Balaguer E, Balaguer E, Barbosa J, Sanz-Nebot V. 2006. Modelling migration behavior of peptide hormones in capillary electrophoresis–electrospray mass spectrometry. *J. Chromatogr. A* **1117**:94–102.

Bijlsma MF, Peppelenbosch MP, Speak CA. 2006. A dual role for 7-dehydrocholesterol reductase in regulating Hedgehog signalling? *Development* **133**:3951–3951.

Bylesjö M, Rantalainen M, Cloarec O, Nicholson JK, Holmes E, Trygg J. 2006. OPLS discriminant analysis: Combining the strengths of PLS-DA and SIMCA classificationy. *J. Chemometrics* **20**:341–351.

Chen J, Zhao X, Fritsche J, Yin P, Schmitt-Kopplin P, Wang W, et al. 2008. Practical approach for the identification and isomer elucidation of biomarkers detected in a metabonomic study for the discovery of individuals at risk for diabetes by integrating the chromatographic and mass spectrometric information. *Anal. Chem.* **80**:1280–1289.

Coen M, Holmes E, Lindon JC, Nicholson JK. 2008. NMR-based metabolic profiling and metabonomic approaches to problems in molecular toxicology. *Chem. Res. Toxicol.* **21**:9–27.

Crockford DJ, Holmes E, Lindon JC, Plumb RS, et al. 2006. Statistical heterospectroscopy, an approach to the integrated analysis of NMR and UPLC-MS data sets: Application in metabonomic toxicology studies. *Anal. Chem.* **78**:363–371.

Eriksson L, Johansson E, Wold S. 1997. *Quantitative Structure–Activity Relationships in Environmental Sciences—VII*. Pensacola, FL: SETAC.

Eriksson L, Johansson E, Kettaneh-Wold N, Wold S. 2001. *Multi- and Megavariate Data Analysis-Principles and Applications*. Umea, Sweden: Umetrics, ISBN: 94–97.

Fekete A, Frommberger M, Rothballer M, Li X, Englmann M, et al. 2007. Identification of bacterial *N*-acylhomoserine lactones (AHLs) with a combination of ultra-performance liquid chromatography (UPLC), ultra-high-resolution mass spectrometry, and in-situ biosensors. *Anal. Bioanal. Chem.* **387**:455–467.

Fekete A, Kuttler C, Rothballer M, Hense BA, et al. 2010. Dynamic regulation of *N*-acyl-homoserine lactone production and degradation in *Pseudomonas putida* IsoF, FEMS. *Microbiol. Ecol.* **00**:000–000.

Garcia-Villalba R, Leon C, Dinelli G, Segure-Carretero A, et al. 2008. Comparative metabolomic study of transgenic versus conventional soybean using capillary electrophoresis-time-of-flight mass spectrometry. *J. Chromatogr.* **1195**:164–173.

Harrigan G, Goodacre R. 2003. *Metabolic Profiling: Its Role in Biomarker Discovery and Gene Function Analysis*. Dordrecht: Kluwer Academic Publishers, ISBN 1-4020-7370-4.

Hertkorn N, Meringer M, Gugisch R, Ruecker C, Frommberger M, Perdue E. 2007. High-precision frequency measurements: Indispensable tools at the core of molecular-level analysis of complex systems. *Anal. Bioanal. Chem.* **389**:1311–1327.

Hertkorn N, Frommberger M, Schmitt-Kopplin P, Witt M, et al. 2008. Natural organic matter and the event horizon of mass spectrometry. *Anal. Chem.* **80**:8908–8919.

Hertkorn N, Schmitt-Kopplin P, Perdue EM. 2009. Natural organic matter and the event horizon of mass spectrometry. *Geochim. Cosmo. Acta* **73**:A526–A526.

Hollywood K, Brison DR, Goodacre R. 2006. Metabolomics: Current technologies and future trends. *Metabolomics* **6**:4716–4723.

Leon C, Rodriguez-Meizoso I, Lucio M, Garcia-Cañas V, Ibañez E, Schmitt-Kopplin P, Cifuentes A. 2009. Metabolomics of transgenic maize combining Fourier transform–ion cyclotron resonance–mass spectrometry, capillary electrophoresis–mass spectrometry and pressurized liquid extraction. *J. Chromatogr. A* **1216**:7314–7323.

Levandi T, Leon C, Kaljurand M, Garcia-Canas V, Cifuentes A. 2008. Capillary electrophoresis time-of-flight mass spectrometry for comparative metabolomics of transgenic versus conventional maize. *An. Chem.* **80**:6329–6335.

Li X, Fekete A, Englmann M, Götz C, Rothballer M, et al. 2006. Development and application of a method for the analysis of *N*-acylhomoserine lactones by solid-phase extraction and ultra high pressure liquid chromatography. *J. Chromatogr. A* **1134**: 186–193.

Martins AM, Camacho D, Shuman J, Sha W, et al. 2004. A systems biology study of two distinct growth phases of *Saccharomyces cerevisiae* cultures. *Curr. Genomics* **15**:649–663.

Nicholson JK, Lindon JC. 2008. Systems biology—Metabonomics. *Nature* **455**:1054–1056.

Rossello-Mora R, Lucio M, Pená A, Brito-Echeverría J, et al. 2008. Metabolic evidences of biogeographic isolation of the extremophilic bacterium *Salinibacter ruber*. *ISME* **2**:242–253.

Steidle K, Sigl K, Schuhegger R, Ihring A, et al. 2001. Visualization of *N*-acylhomoserine lactone-mediated cell-cell communication between bacteria colonizing the tomato rhizosphere. *Appl. Environ. Microbiol.* **67**:5761–5770.

Subrahmanyam S, Kodandapani N, Madhav VM, Murugesan M, et al. 2001. Cyclic voltammetric measurements of growth of *Aspergillus terreus*. *Analy. Sci.* **17**:481–484.

Suhre K, Schmitt-Kopplin P. 2008. MassTRIX: Mass translator into pathways. *Nucleic Acids Res.* **36**:W481–W484.

Trygg J, Gullberg J, Hohansson AI, Jonsson P, Moritz T. 2006. Chemometrics in metabolomics: An introduction. *Biotechnol*. Agriculture *Forestry* **57**:117–128.

Wang YY, Wang J, Yao M, Zhao X, Fritsche J, Schmitt-Kopplin P, et al. 2008. Metabonomics study on the effects of the ginsenoside Rg3 in a cyclodextrin-based formulation on tumor-bearing rats by a fully automatic hydrophilic interaction/reversed-phase column-switching HPLC-ESI-MS approach. *Anal. Chem.* **80**: 4680–4688.

Want E, Nordstro A, Morita H, Siuzdak G. 2007. From exogenous to endogenous: The inevitable imprint of mass spectrometry in metabolomics. *J. Proteome Res.* **6**:459–468.

Wapstra A, Audi G, Thibault C. 2003. The AME2003 atomic mass evaluation (I). Evaluation of input data, adjustment procedures. *Nuclear Phys. A* **729**:129–336.

Weeks G, Martin S, Burns RO, Wakil SJ. 1969. Control of fatty acid metabolism I. Induction of the enzymes of fatty acid oxidation in *Escherichia coli*. *J. Bacteriol.* **97**:827–836.

Westerhuis JA, Hoefsloot HCJ, Smit S, Vis DJ, et al. 2008. Assessment of PLSDA cross validation. *Metabolomics* **4**:81–89.

Yuan K, Kong H, Guan Y, Yang J, Xu G 2007. A GC-based metabonomics investigation of type 2 diabetes by organic acids metabolic profile. *Chromatrogr. B Anal. Tech. Biomed. Life Sci.* **850**: 236–240.

Chapter 73

Metabolic Profiling of Plant Tissues by Electrospray Mass Spectrometry

HEATHER WALKER

73.1 INTRODUCTION

Metabolic profiling is a fairly recent but rapidly expanding area of interest; and, as the use of mass spectrometry as an analysis tool has become more common, the number of publications on metabolite profiling has also increased. There appear to be several definitions of metabolic profiling [Fiehn, 2002; Dunn et al., 2005], but essentially it can be defined as an analysis to identify and quantify as many metabolites as possible. For nontargeted metabolic profiling, the objective is to analyze the maximum number of metabolites in a single run, from as many different chemical classes as possible, in a quantitative and nonbiased way, with the minimal loss of metabolites.

Metabolic profiling can be applied to both plants and bacteria because they are both very adaptable to environmental conditions, and it has become increasingly important to identify why and how they adapt (see Chapters 71–74, Vol. I).

Because there are estimated to be over 2600 metabolites in *Arabidopsis thaliana* (see Metabolites in *Arabidopsis thaliana* in Internet Resources) and over 1800 metabolites in *Escherichia coli* (see Metabolites in *Escherichia coli* in the Internet Resources section), any full metabolomic analysis needs to be an impartial process, and many analyses need to be carried out in order to gain the whole picture.

This particular chapter will focus on a nontargeted approach to metabolite profiling using direct infusion mass spectrometry (DIMS). This technique utilizes electrospray mass spectrometry whereby the sample is infused into the mass spectrometer at $\mu l\ min^{-1}$ flow rates without any prior separation by high-performance liquid chromatography. The electrospray ion source is a simple system and can produce many ions with few breakdown products or fragments; however, it can produce ions with multiple charges, which can make accurate measurement of large molecular masses such as proteins much easier. It is highly effective for the analysis of polar compounds and therefore very useful for the analysis of plant and bacterial material due to the abundance of polar compounds that make up their metabolic compositions. The example outlined in this chapter will focus on the metabolic profiling of tomatoes.

73.2 MATERIALS AND METHODS

73.2.1 Sample Preparation

A general extraction method needs to be employed in order to maximize the number of metabolites detected, and in fact the extraction procedure itself is often more important than the analysis because, if the extraction procedure is not uniform, the analysis will be meaningless.

The first important step, when extracting plant material, is to inhibit any enzymatic activity. This can be carried out by using methods such as freeze-clamping, rapid freezing in liquid nitrogen, or acidic treatments with perchloric or nitric acid. Acidic treatments tend to be avoided because they can cause problems for many analytical techniques further down the line. Freeze-clamping is not usually practical for large numbers of samples, so freezing in liquid nitrogen is usually the most convenient way of preserving the sample. Samples can then be stored at −80°C until such time as they are extracted. Tissue

Handbook of Molecular Microbial Ecology, Volume I: Metagenomics and Complementary Approaches, First Edition. Edited by Frans J. de Bruijn.

should not be used which has been stored at −80°C for longer than four weeks because there is evidence that considerable metabolic changes occur after this time (see Protocol for Plant Leaf Metabolite Profiling in the Internet Resources section). Freeze-drying is another method that can be utilized to preserve samples because enzymes are unable to work in the absence of water. Samples that have been freeze-dried do need to be kept in a desiccator in order to stop any hydrolyzation of the sample before it is extracted. Hydrolyzation of the sample can lead to the adsorption of metabolites onto cell walls and membranes, so particular care would be needed if metabolite levels in a specific plant area rather than in a whole sample were of interest.

The sample needs to be homogeneous, so the plant material should be ground in a pestle and mortar under liquid nitrogen, or alternatively by using a cooled ball-mill grinder. Once ground, the plant material needs to be extracted into a suitable polar medium for electrospray mass spectrometry. Many methods are based on a procedure used by Valle et al. [1998] utilizing a chloroform/methanol/water extraction whereby polar metabolites are extracted into the methanol/water layer and nonpolar metabolites are left behind in the chloroform layer. There are many variations on this theme such as using ethanol or heating the extraction mixture; however, no direct comparisons between similar methods have been made, and indeed heating the sample may be too harsh for metabolite analyses because it could lead to the breakdown of some metabolites.

Once extracted, the sample is centrifuged, which enables the two phases of the sample to be separated. These two extraction layers are treated differently. The nonpolar, chloroform phase is highly organic and therefore not particularly suitable for electrospray mass spectrometry analysis. This organic phase could be used for targeted analyses such as for chlorophyll, carbohydrate, or carotenoids. The polar, methanol/water, aqueous phase is used for the metabolite profiling analysis by electrospray ionization.

The samples, once extracted, should be stored at −80°C because it is generally assumed that no changes in metabolite levels or composition will occur at this temperature.

73.2.2 Introducing Samples to the Mass Spectrometer

Using a global metabolic profiling procedure, samples are introduced directly into the mass spectrometer for an "everything at once" analysis. Samples can be infused using a syringe pump, which is particularly effective at delivering a low, pulse-free flow or by the use of a liquid chromatograph. The use of a liquid chromatograph system allows for the analysis to be automated, whereas a syringe pump method does not.

There are, however, some matrix issues attached to this direct infusion method of sample introduction which must be taken into consideration. Ion suppression is a matrix effect that can negatively affect the efficiency of the ionization and can occur due to the large amount of metabolites that are infused into the mass spectrometer at the same time. The direct infusion method can also expose the instrument to very high concentrations of metabolites putting excess burden on the detector. It is therefore important to have metabolites at such a level that the larger peaks are monitored accurately without the loss of the smaller peaks, thereby maximizing the amount of information obtained from one run. For the analysis of samples of similar composition, matrix effects are thought to be minimal because they are assumed to be uniform for all samples within a set. When using direct infusion as a sample introduction method, the sample may need to be diluted further than the original extraction to try and minimize the effects of suppression. With care, the dilution can be carried out at a level that succeeds in lowering the level of ion suppression without the loss of many-low abundance compounds. The dilution of the sample also helps to minimize the buildup of contaminants upon the mass spectrometer sample introduction system.

The direct infusion method lends itself toward a qualitative approach rather than a fully quantitative one in that a comparison of peak area can be used for the metabolite analysis provided the sample reproducibility is good, or an internal standard can be used to compensate for any matrix effects before data comparison takes place.

73.2.3 Optimization of the Mass Spectrometer for the Analysis of Plant Material

Routine optimization of the mass spectrometer requires high sensitivity over a large mass range. For plant material the optimization needs to be across a wide range but starting at a lower mass range than is usual for other sample materials. This is because plant material contains a lot of low-molecular-weight compounds such as organic and amino acids. Conditions need to be set therefore, to allow analysis of metabolites down to around 50 atomic mass units up to around 800 atomic mass units. Above 800 atomic mass units there will be few polar metabolites in the plant extract which can be detected by electrospray mass spectrometry. To allow the analysis of the lower masses as well as a large mass range, the cone voltage of the instrument needs to be kept low. The cone voltage is the electric potential applied to attract ions into the mass spectrometer. By increasing the cone voltage, ions speed up and can be induced to fragment. If the cone voltage is reduced, the ions slow down and any collisions have insufficient energy for fragmentation. The reduced speed also allows the small ions to be focused into the mass spectrometer rather than being dispersed.

73.2.4 Quality Control of Large Batches

Biological samples always have some amount of natural variation that will lead to variability within the final dataset; however, other errors can occur due to the sample preparation and the instrumental analysis. For large batches of samples, in particular, a quality control check needs to be in place. Because it is not feasible to run all samples at the same time, it is imperative that analyses can be compared from one year's harvest, for instance, to the next.

Obtaining a reference material, useful for the quality control of the extraction procedure, is not always feasible for biological samples; however, some material can be purchased from companies such as the Laboratory of the Government Chemist (Teddington, UK). An in-house standard can be prepared, but this would need to be stored at -80°C for possibly long periods, which for raw, untreated sample is not always recommended. An extract could be used because this could be kept at -80°C storage for longer periods but evokes the question of how long to keep it, how much extract to make up, and what to do when the extract runs out. For long-term studies, it may be better to freeze-dry a large amount of material which can then be prepared along with every batch of samples extracted. This particular route will make it much easier for storage of samples, although the sample will have to be kept in a desiccator.

Instrumental quality control checks need to be in place for mass as well as intensity. An internal standard approach could be used whereby spikes of individual compounds can be added to the extracts or indeed added to the sample before extraction as a quality control measure. However, the compound needs to be selected carefully in that it must have a mass that occurs in a region of the spectra where there is an absence of sample peaks. It also needs to be easily and quantitatively extractable if it is undergoing the extraction procedure. This particular type of quality control, particularly for direct infusion methods, may not be the most suitable because it may be difficult to find an area free of peaks in order to choose a suitable mass. There is also the possible problem of the introduction of contaminants to a sample and indeed, the interaction of the compound introduced and any contaminants, with other peaks present in the spectrum. This could lead to the formation of other molecular species within the spectrum or lead to matrix effects. Some instruments on the market have the option of monitoring an external standard so, rather than adding an internal standard to a particular sample, the instrument monitors a standard that is introduced separately to the sample stream flow, but under the same conditions. An alternative approach for direct infusion mass spectrometry is to use a peak already present in the sample for the correction of mass error only. With plant material there are usually many peaks that are common even between different species—for instance, simple organic acids and amino acids which could be used for mass correction. For analyses over large mass ranges, more than one peak may have to be chosen. This is a quick and easy way to correct for mass without any adulteration of the sample.

Direct infusion samples are usually normalized to the largest peak in the spectrum or to the total ion count for the whole sample, in order to standardize large sets of data. The area of quality control and standardization of data, particularly for reporting purposes, where analyses are carried out by different research groups, is being looked at by several groups. Indeed an entire volume of the *Metabolomics Journal* was given over to this subject in 2007.

A standard procedure would involve the reporting of many parameters and the Metabolomics Standards Initiative (see Internet Resources section) is working toward developing guidelines for reporting metabolomics work utilizing working groups for Biological Context Metadata, Chemical Analysis, Data Processing, Ontology, and Exchange Format [Fiehn et al., 2007]. The proposed standard reporting guidelines need to encompass good analytical chemistry practices as well as allow for large datasets of metabolomic data to be compared electronically and across different instrumental techniques. The need is to obtain a minimum set of criteria for reporting which describe the experiments carried out so that any information published is useful across the wider community by standardizing any data submitted to academic journals. Frameworks and procedures for the standardization of reporting of metabolic experiments have been proposed by Jenkins et al. [2004] and Lindon et al. [2005]. The Metabolomics Society also, has begun a discussion forum on Standardization in Metabolomics Experiments encompassing drafting a document on best practice, reporting and data exchange (see Internet Resources section). Bino et al. [2004] suggested a checklist of information needed in order to publish metabolomic data but does not give a complete set of guidelines. Jenkins et al. [2004, 2005] also discussed the use of ArMet (an architecture for metabolomics) (see Internet Resources section) which is a complete and formal data description for plant metabolomics, supporting datasets, and experimental context information. The ArMet data model is freely available on the web, and it is hoped that it will lead to the long-term adoption of a standard model that will serve as a guideline for researchers, journals, funding, scientific and regulatory bodies, as well as vendors.

Data files are usually exported as text, thereby keeping the size of the files small and more manageable. Text file format is a universal one that allows the data to be introduced easily into many different applications—for instance, Excel spreadsheets as well as web- and grid-based systems.

The original spectra need not be kept because spectra data files can be very large, particularly if each individual spectrum is stored separately. Alternatively, the spectra can be added together so all scans within a run are stored as one "summed" spectrum, making the data file much smaller in size. Having the individual spectra, however, allows any discrepancies in the text format to be investigated easily so it can be invaluable, but only if storage facilities exist for these large data files.

The use of database software is becoming increasingly useful for the storage of spectra and other large files, and these can often be found published on the web. Metabolite and transcript data for *Arabidopsis thaliana* and tomato are already published on the web (see Internet Resources section).

73.2.5 Data Processing and Interpretation

The direct infusion of samples into a mass spectrometer for a global metabolic analysis produces very large quantities of data. These data need to be handled routinely, so any reduction in the volume of data is highly desired.

Several ways of reducing the data have been utilized, specifically a noise reduction and a binning procedure. These two procedures can be used separately or together.

Noise reduction is a way of removing small, noise peaks and ideally, needs to be carried out without placing a particular threshold value on the data. The use of a threshold value can mean that small peaks are lost where they fall below the cutoff value, and indeed there is the question of where to set this threshold value. A simple way to bypass this problem is to run all samples three times and then combine these data with the rule that only peaks seen within all three runs will be accepted as a real peak. If a peak appears in only one or two of the repeats, it will be excluded from the final dataset because it is assumed to be noise [Overy et al., 2005]. By using this method, the data can be significantly reduced without the loss of any of the small peaks.

The binning procedure involves binning data into set-sized atomic mass unit bins, thereby limiting the number of entries to deal with. For instance, a scan from 50 to 800 atomic mass units can easily produce over 4000 peaks, even after noise reduction has been carried out. By sorting these data into single atomic mass unit bins, the data are reduced to 750 entries for this particular scan range. Each bin contains metabolites from ± 0.5 atomic mass units around a whole atomic mass unit; for example, mass bin 191 contains all metabolites found between 190.5 and 191.5 atomic mass unit. This approach allows the dataset to be compacted, making it much easier to handle, and also provides a simple way to compare the profiles of samples where many masses are observed and where the masses obtained may not be exactly identical. Where differences in a particular bin are discovered, the original dataset and spectra can be looked at in order to investigate further and obtain an accurate mass for a specific peak within a bin. Different bin sizes other than one atomic mass unit can be used, although a minimum size is likely to be found whereby the binning procedure becomes less useful mainly due to the error in measuring the mass. The main disadvantage of using this binning data technique is that a bin may include more than one peak that may lead to a large peak within a bin masking a change in a smaller peak; however, the advantages of handling a much smaller dataset tend to outweigh any disadvantages.

Once a metabolite profile has been obtained for each sample within a set, the data are transferred to a software package to carry out a statistical analysis of the data. Some packages utilized for this purpose include SIMCA-P (Umetrics, Windsor, UK) and SPSS (SPSS UK Ltd., Woking, UK), which can run a principal component analysis. This analysis is a dimension reduction technique that is used to examine the relationships between a set of correlated variables. Its aim is to reduce the number of variables by creating a small number of principal components that explain most of the variation in the data. This is a standard way of looking at metabolite data, although other groups have investigated the use of other statistical techniques (see Chapter 74, Vol. I). Goodacre [2005] has investigated the use of evolutionary computation, Tokimatsu et al. [2005] have investigated a web-based analysis tool based on the use of vectors, and Fiehn [2001] and more recently Hageman et al. [2008] have looked at clustering techniques for analysis of the data.

These types of statistical analysis are particularly suited to the handling of data for a global metabolomic analysis where no identification of each peak is carried out and the nontargeted approach allows for an unbiased analysis. Any differences found using this approach can be investigated further using more targeted approaches. For instance, if a peak in mass bin 191 (in negative mode) was found to be changing, it could be identified using triple quadrupole mass spectrometry. If the peak was found to be citric acid and quantitation was needed, this could be achieved by using a standard high-performance liquid chromatography method. However, if a targeted approach had been used, this changing peak may not have been found, unless the analysis had specifically targeted organic acids.

73.2.6 Metabolic Profiling of Tomato

The example discussed will illustrate the metabolic profiling of tomatoes by electrospray time-of-flight mass spectrometry. Tomatoes are a commercially important crop, and tomato fruit yield and quality are regulated by a wide range of genetic and environmental factors. Therefore, the ability to enhance yield and quality is highly prioritized in current breeding programs because centuries of plant breeding and domestication have resulted in modern crops with a very narrow genetic base. Increased genetic diversity and agronomic value of cultivated crops can be gained naturally by cross breeding with wild relatives; however, in order to maximize and predict yield and quality, there must be an understanding of (a) the function of genes and (b) the relationship between a plant's genome and its metabolites. The tomato is currently in the process of having its entire genome sequenced, and the current status of this sequencing can be observed on the web (see Internet Resources section).

The common tomato, *Solanum lycopersicon* (formerly known as *Lycopersicon esculentum*), belongs to an extremely diverse and large family called the solanaceae. Often referred to as the Nightshade family, this family contains many commonly cultivated plants (potato, peppers, aubergine, tobacco, petunias) as well as various weeds (nightshades, jimson weed). The genus *Lycopersicon* consists of a vast reservoir of genetic variability, much of which remains underexploited, although some disease resistance in tomatoes has been derived from related species within the *Lycopersicon* genus. Other species within this genus may provide other potential benefits such as heat or cold tolerance, salt tolerance, or increase in nutrient amounts. Currently, tomatoes are of interest in the medical field due to the high levels of carotenoids they contain. Lycopene, a carotenoid present in tomatoes, is one of nature's most powerful antioxidants and has been found to be beneficial in preventing cancer, particularly prostate cancer [Gann and Khachik, 2003].

Eshed and Zamir and colleagues [Eshed et al., 1992; Eshed and Zamir, 1994] generated a population of introgressed lines (ILs) of the tomato species *Solanum lycopersicon* containing defined chromosomal introgressions from the wild species *Solanum pennellii* (formerly *Lycopersicon pennellii*). These introgression lines were produced by crossing a cultivated *Solanum lycopersicon* with a wild relative *Solanum pennellii* to create an F1 hybrid. This F1 hybrid was then backcrossed to the *Solanum lycopersicon* for six generations in order to successively reduce the proportion of the *Solanum pennellii* chromosome. Self-fertilization produced lines homozygous for the *Solanum pennellii* introgression, and each introgression line produced possesses a single chromosome fragment of the wild species in the cultivated *Solanum lycopersicon*. See **Figure 73.1** for the phenotypes of the two parents and the F1 hybrid tomato species.

The entire tomato genome in this set is represented by 50 introgression lines, with each line homozygous for a single introgressed region. See **Figure 73.2** for some examples of the phenotypes seen from the set of 50 introgression lines. Phenotypes for the other introgression lines in the set of samples did not differ significantly from the phenotype for the *Solanum lycopersicon* parent and thus are not reproduced here.

Subsequent analysis of these introgression lines can increase the knowledge of the genetics underlying certain given traits. Several quantitative trait loci (QTL) for beneficial agronomic traits such as fruit soluble solids, fruit weight, yield, and color have previously been identified [Eshed and Zamir, 1995; Eshed et al., 1996; Liu et al., 2003] as well as some unbeneficial traits such as undesirable volatiles [Tadmor et al., 2002] and leaf dissection [Holtan and Hake, 2003]. These traits have been identified using targeted approaches utilizing specific methods of sample preparation and analysis. The use of a metabolic profiling approach enables a global picture to be obtained for a sample after a standardized extraction procedure and a single instrumental analysis. The benefit of this approach

Figure 73.1 Phenotypes observed for parents in order from left to right *Solanum lycopersicon* parent, F1 hybrid, and *Solanum pennellii* parent.

Figure 73.2 Phenotypes observed for introgression lines 3-2 (LA4044), 4-4 (LA 3494), 2-5, (LA 4040) and *Solanum lycopersicon* parent (LA 3475).

is that it can then be used to target any specific traits that are identified.

73.2.7 Sample Preparation

The seeds for the tomatoes used in this study were obtained from the Tomato Genetic Resource Center, University of California, Davis. Eight plants of each line were grown in pots in a glasshouse under ambient conditions, using Levington M3 compost with supplemented nutrients (Osmocote, Scotts Ltd, UK). Plants were randomized throughout the glasshouse in order to minimize the effects of local environmental conditions. Plants were watered daily and fed weekly with a commercially available tomato food (Tomato liquid, LBS Horticulture, Lancashire, UK) after flowering. Samples were taken from the pericarp (the fruit wall) of ripe tomatoes and were immediately frozen in liquid nitrogen after harvesting. Ripeness was judged as being 7–10 days post breaker, with the breaker being classified as the first appearance of carotenoid coloration (i.e., turning red). Samples were extracted following a procedure developed by Overy et al. [2005] by taking 0.1g of the tomato pericarp and grinding it with 750 μl of an extraction medium mixture of 80 μl of water, 200 μl of chloroform, and 470 μl of methanol. This mixture was then left on ice for 30 minutes. The aqueous, polar phase was removed and the organic phase was then re-extracted twice more with 400 μl of water, and these polar phases were combined, centrifuged at 12,000*g* for two minutes, and retained for analysis by electrospray mass spectrometry. The nonpolar phase was retained and could be used for the analysis of nonpolar compounds such as chlorophyll or carotenoids, but was not used for this particular study. For metabolite profiling, samples were extracted from three fruits per plant and three plants per introgression line, thus giving nine replicates per line. Because there is always some degree of biological variation within a sample set, nine replicates gave enough data for a good statistical analysis. After extraction the extracts were stored at −80°C until analysis was carried out.

73.2.8 Analysis by Mass Spectrometry

All analysis was carried out using a Waters Micromass Liquid Chromatograph Time-of-Flight mass spectrometer (LCT) (Waters, Manchester, UK). Time-of-flight instruments work on the principal that ions of different m/z (mass to charge ratio) are accelerated by an electric field. These ions achieve velocities that are dependent upon their m/z values and the accelerating field to which they have been subjected. Among the advantages of time-of-flight measurement of mass spectra is the very short time needed to measure a mass spectrum, because only a few microseconds is needed for a mass range scan from 50 to 800 atomic mass units. Time-of-flight instruments also provide a better mass resolution than traditional quadrupole instruments [Dunn et al., 2005], and this higher resolution allows the detection of metabolites of different monoisotopic masses which have the same nominal mass. An accurate mass measurement can help with metabolite identification; however, for confirmation of metabolite identities, more specific analyses may need to be carried out because an accurate mass cannot distinguish between isomers. Accurate mass also becomes less useful at higher masses where the number of possible metabolites at a particular mass becomes much larger. The mass spectrometer was operated in both positive and negative modes in order to analyze as many metabolites as possible. See **Tables 73.1** and **73.2** for the conditions used.

Spectra were collected over the mass range 50–800 atomic mass units at a rate of one scan per 0.5 s with an interscan delay of 0.1 s. The RF lens value was set at 75 V to help optimize the detection of low mass ions as discussed earlier in the chapter.

A problem that occurs with the analysis of tomato fruit extracts is the buildup of material on the mass spectrometer inlet system, particularly the extraction cones. When analyzing plant material, the extracts produced are usually very clean and cause minimal problems to the sample inlet system apart from routine maintenance cleaning. With tomato fruit, however, a buildup of a dark, caramel-type deposit occurs after only a few days of running samples. This is probably caused by the large amount of sugars in the fruit samples but does mean that routine cleaning of the instrument must be carried out every week. It is possible that a different extraction procedure may help; however, as the purpose of this analysis is to try and monitor as many metabolites as possible within one standardized run, the extraction procedure needs to be as universal as possible without discrimination toward any particular class of compounds. For ease of analysis, one extraction procedure is used for both positive and negative analyses.

A Lockspray™ interface (Waters, Manchester, UK) was used which allowed a reference standard to run externally along with the samples to check for mass drift. All spectra were automatically corrected for mass using this external standard before data processing occurred.

The samples were run automatically using a liquid chromatograph system for the sample infusion, and a quality control standard was run after every 10 samples to check instrument performance and abundance. The liquid chromatograph system was used only as a flow injection system, to automate the analysis.

Table 73.1 Negative Ion Mode Electrospray Mass Spectrometer Conditions

Parameter	Setting
Resolution	4000 (full width at half height) at 300 *m/z*
Capillary Voltage	2800 V
Extraction Cone	1 V
Sample Cone	20 V
RF Lens	75 V
Source Temperature	110°C
Desolvation Temperature	120°C
Desolvation Gas Flow Rate	400 Lh^{-1}

Table 73.2 Positive Ion Mode Electrospray Mass Spectrometer Conditions

Parameter	Setting
Resolution	6000 (full width at half height) at 300 *m/z*
Capillary Voltage	3000 V
Extraction Cone	5 V
Sample Cone	20 V
RF Lens	75 V
Source Temperature	120°C
Desolvation Temperature	150°C
Desolvation Gas Flow Rate	400 Lh^{-1}

In order to try and minimize any matrix effects and contamination of the instrument, due to buildup of deposits on the sample introduction system, the samples were diluted 1:1 with 50:50 high-purity water:methanol, before analysis. All introgression lines and parents were run in both positive and negative modes.

Figure 73.3 shows two examples of mass spectra in negative ionization mode from the two parent fruit extracts of *Solanum lycopersicon* and *Solanum pennellii*. The differences between the two spectra can be observed immediately, and it can also be observed that most of the high abundance metabolites are observed at the lower-mass end of the spectra. This pattern is common for most aqueous plant extracts using electrospray mass spectrometry and is further enhanced by the fact that the instrument is optimized for these low masses.

The spectra obtained in positive mode are reproduced in **Figure 73.4** for the two parent species, and they are again observed to be very different from each other. There are fewer high-abundance peaks appearing in the low-mass region and more high-abundance peaks seen in the high-mass region for the *Solanum pennellii* parent as compared against the *Solanum lycopersicon* parent.

Higher-mass peaks tend to be seen more in the positive mode spectra rather than the negative mode spectra due to the increased stability for positive ions with increasing mass. Less identification has been possible for the metabolites seen in the positive ion mode as opposed to the negative ion mode. This is partly because of the more complex nature of the spectra, which can include more molecular species formation, but is also due to the increased number of high-mass peaks that appear. Identification of high-mass peaks, using the electrospray time-of-flight instrument, which relies on accurate mass without any structural identification, is very difficult due to the increasing number of isomers that are possible.

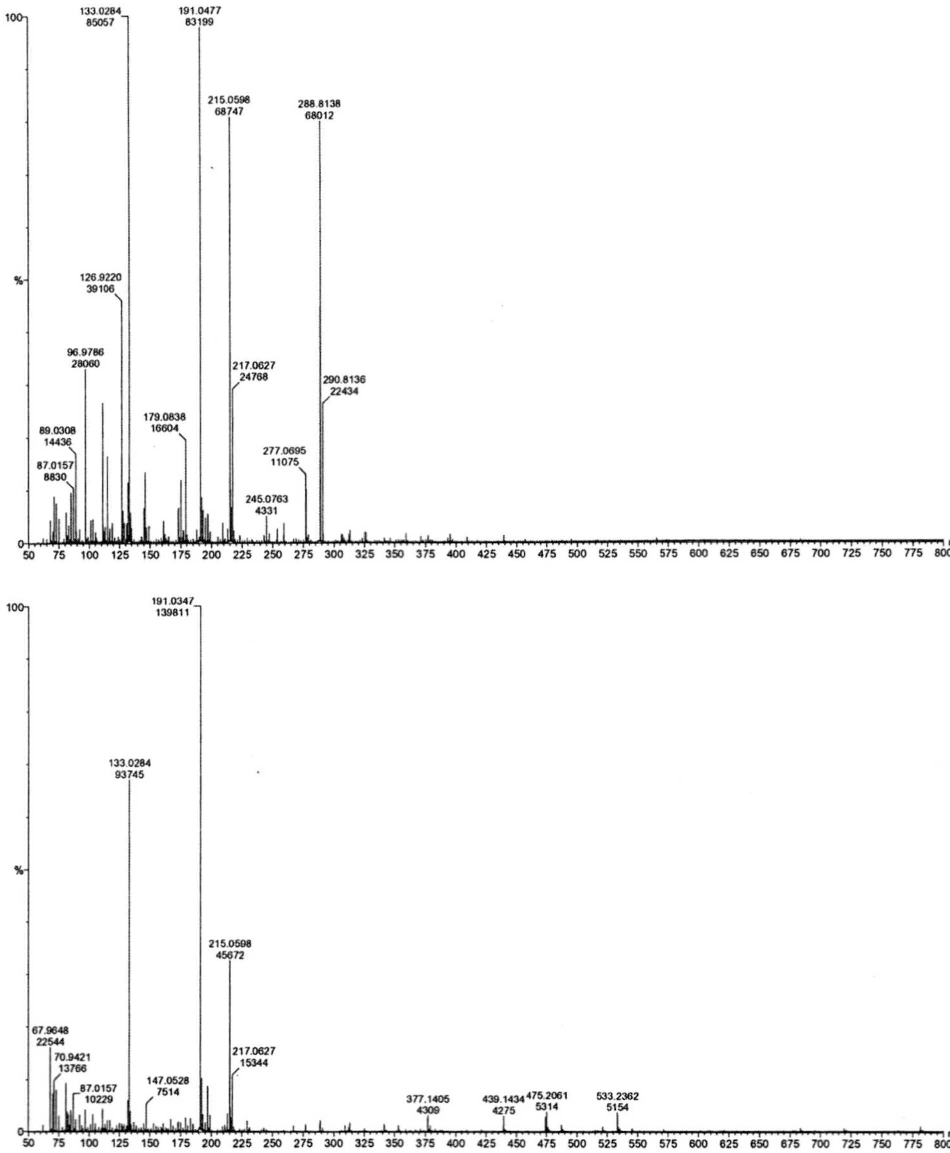

Figure 73.3 Raw metabolic profile mass spectra in negative ionisation mode for two parent ripe tomato fruit species extracts. The peaks in the spectra are annotated with m/z (top number) and intensity (bottom number). (**Top**) *Solanum lycopersicon*. (**Bottom**) *Solanum pennellii*.

73.2.9 Data Processing and Interpretation

Three replicate mass spectra of each individual sample were obtained in order to check the instrument reproducibility and enable noise reduction to take place. Noise reduction was carried out according to the procedure defined by Overy et al. [2005], which deems a true peak to be one that is present in all three instrumental replicate runs, as discussed earlier. This allows for all peaks regardless of intensity to be selected as true peaks and allows even the smallest peaks to be monitored. The method employs a linear equation to define the mass variance between the three replicate runs, because this was found to generate the maximum number of true peaks with the minimum false positives or non-true peaks. A different equation was used for both positive and negative modes. Once a peak was selected as a true peak, the mean of the three masses over the three replicate scans was used as the accurate mass, and this value along with the corresponding average intensity made up the metabolite profile. To minimize the variation between different samples and different analysis batches, the data for each sample were normalized to the total ion count for each replicate. Once these data transformations had taken place, the data were rounded into one atomic mass unit bins, as previously

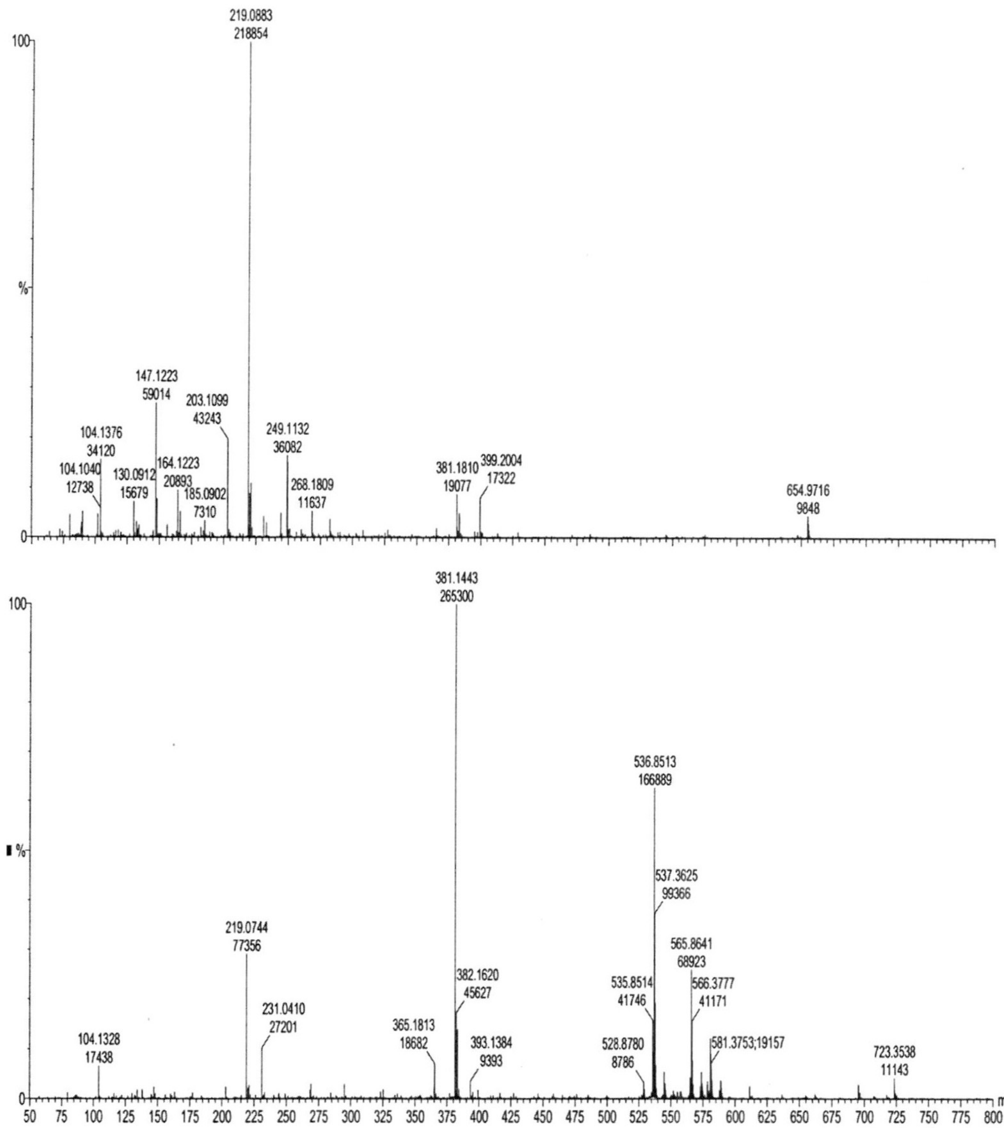

Figure 73.4 Raw metabolic profile mass spectra in positive ionization mode for two parent ripe tomato fruit species extracts. The peaks in the spectra are annotated with m/z (top number) and intensity (bottom number). (**Top**) *Solanum lycopersicon*. (**Bottom**) *Solanum pennellii*.

discussed, and the abundances within a bin were summed as a percentage of the total ion count. This binned dataset was saved as a metabolite profile in text file format.

Once a metabolite profile had been obtained for each sample, the data were transferred to a software package, SIMCA-P (Umetrics, Windsor, UK), to carry out a principal components analysis.

73.3 RESULTS AND DISCUSSION

In this particular example, using 50 introgression lines and two parents along with nine replicate samples per line, an enormous amount of data was generated. Hence, comparisons between the datasets rapidly become very complex and time-consuming to process. Also, the *Solanum pennellii* parent was observed to be very different from the *Solanum lycopersicon* parent, not only by the principal component analysis separation but also due to their very different phenotypes, seen in Figure 73.1. For these reasons, the 50 introgression lines were individually compared to the two parents, as well as to the other introgression lines within their chromosomes. Every introgression line showed clear principal component analysis discrimination between itself and the two parents. Not all principal component analysis graphical representations are reproduced here; a few are reproduced as an example of the type of plots expected. Two examples of principal component analysis plots are detailed below in **Figures 73.5** and **73.6**. Figure 73.5 reproduces the plot for introgression line 4-4 in negative mode, and Figure 73.6 reproduces the plot for introgression line 4-4 in positive mode. All introgression lines used in this study could be distinguished from the two parent lines within a principal component analysis, and all plots are published on the web (see Internet Resources section).

In many cases, the metabolites that were identified as changing were similar for all lines, and this is to be expected because genetically the difference between some of the introgression lines is small. Differences that are the most interesting, from a commercial point of view, are those that are associated with taste, smell, or color. For color comparison, the metabolites of interest are most likely found in the nonpolar, organic phase of the extract, so they would not be picked up by this particular analysis that has concentrated on the polar extracts. Also, compounds that contribute to smell may be too volatile to be analyzed by electrospray mass spectrometry and may be better suited to analysis by gas chromatography mass spectrometry. However, compounds that contribute to taste should be picked up by an analysis of the polar extract, because these include organic acids and sugars. These compounds can all be easily seen in negative mode, using electrospray mass spectrometry. Due to the lack of molecular species formation in negative mode, other than the $(M\text{-}H)^-$ molecular ion, it is possible to identify some of the small metabolites easily.

It is significant that different principal component analysis separations can be observed with data from the different modes, and often the separation seen can be better in one mode rather than the other. From the principal component analyses that were carried out, a list of the most changing metabolites, whether they be regulating up or down, could be obtained, and this leads us onto the next challenge. From this list, the search is for common metabolites as well as for major differences between the introgression lines. The spectra obtained in

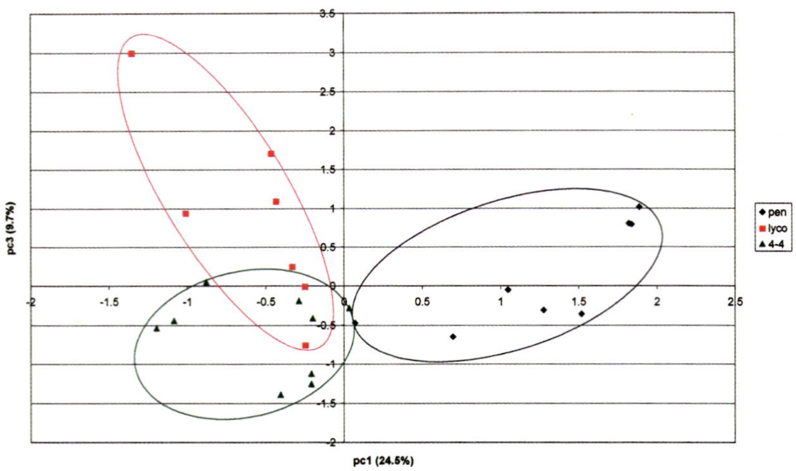

Figure 73.5 PCA plot for introgression line 4-4 (3494) in negative mode plotted against parents *Solanum lycopersicon* (lyco) and *Solanum pennellii* (pen).

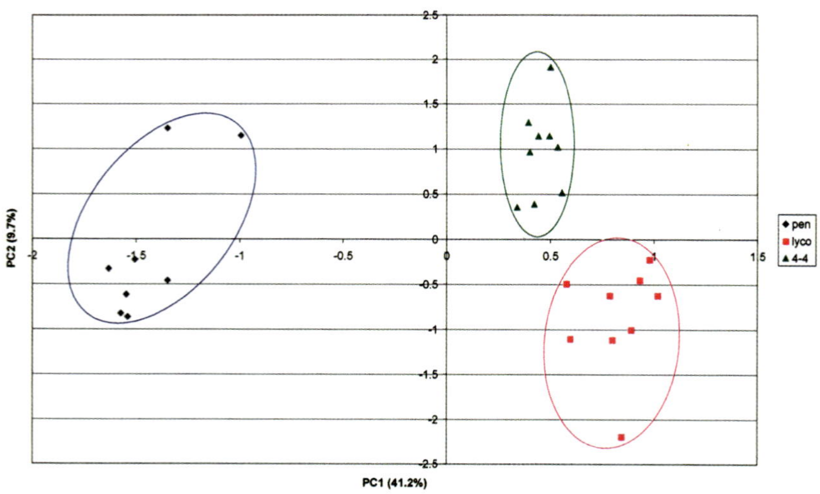

Figure 73.6 PCA plot for introgression line 4-4 (3494) in positive mode plotted against parents *Solanum lycopersicon* (lyco) and *Solanum pennellii* (pen).

negative mode were dominated by peaks for malic acid (peak at 133) and citric acid (peak at 191), as seen in Figure 73.3. These organic acids are very sensitive to electrospray mass spectrometry, particularly because the instrument is optimized for low mass, so it must be noted that peak intensity is not necessarily a reflection of the relative concentrations of these metabolites.

In the negative mode case, which has two very dominant peaks, differences are sometimes identified from the principal component analysis, which are due to the size of these large peaks rather than the changes between them. This can sometimes skew the principal component analysis slightly because small changes in the large peaks are not always significant. This is where scaled and more "educated" methods become increasingly useful, such as a discrimination function analysis as pioneered by Goodacre [2005]. This type of method is a little more intelligent in that it can look beyond the original analysis and mine the data for more specific changes by using a supervised and training approach to the analysis.

For the comparison of introgression lines within a chromosome, many different patterns were obtained because some introgression lines were very different from each other whereas others showed similarities. In this practical example the principal component analysis demonstrated that the *Solanum pennellii* parent was very different not only from the *Solanum lycopersicon* parent but also from all the other introgression lines. This is not unexpected, due to the inclusion of just one introgression from the *Solanum pennellii* parent into each introgression line.

Due to the small differences seen between some of the introgression lines, further investigation using a supervised data analysis technique, such as a discrimination function analysis, would be very useful in obtaining further information. The discriminant function analysis takes the data from the principal component analysis, discounts the data from the samples that have already separated out, and reanalyzes the data. This allows the analysis to concentrate on the differences between the remaining samples. In this case, using all the data, only the *Solanum pennellii* parent separates out from all the other introgression lines in the principal component analysis. The discriminant function analysis takes the data, excludes that obtained for the *Solanum pennellii*, and reprocesses. This is an iterative process because the discriminant function analysis can then be repeated several times.

The initial principal component analyses were investigated for differences between samples. Work has previously been carried out on the measurement of dissolved sugars in this set of tomato introgression lines [Eshed and Zamir, 1995], and this enables the comparison of results found using a metabolite profiling method with those observations obtained using more traditional methods.

The main purpose of this study was to identify any interesting differences using a global metabolic approach with the possibility to study them further using more targeted analyses. Therefore only minimal attempts were made to identify any metabolites found at this stage.

73.4 SUMMARY

Metabolic studies that utilize a direct infusion method can be used to detect concentrations down to micromole levels. Because electrospray ionization leads to minimal fragmentation of molecular ions, it produces a less complex mass spectrum than that of chemical or electron impact ionization methods. The spectra produced, however, can still be very complex, but the presence of molecular ions increases the ability to identify metabolites, without the use of chromatographic separation. Electrospray mass spectrometry is the technique of choice because it is also highly sensitive to many metabolites found routinely in plant and bacterial extracts.

For this particular case, metabolite profiling is invaluable in order to reduce the number of introgression lines of interest, down to a more manageable set. It also enables the identification of specific samples showing the largest or the most interesting differences, which can then be used for further metabolite profiling or for more targeted analysis.

Metabolite profiling can also be used with discriminative statistical methods in order to mine the data for further information. In this way the metabolic profile route gives a very quick and efficient way of targeting areas of interest within a large dataset.

The practical example outlined fulfills the criteria of a metabolic profile analysis by increasing knowledge of plant systems in a very quick and routine way, and this simple procedure can be used successfully with bacterial as well as plant material.

INTERNET RESOURCES

Metabolites in *Arabidopsis thaliana*: http://pmn.plantcyc.org/PLANT/class-instances?object=Compounds

Metabolites in *Escherichia coli*: http://biocyc.org/ECOLI/class-instances?object=Compounds

Protocol for Plant Leaf Metabolite Profiling: http://fiehnlab.ucdavis.edu/publications/Arabidopsis%20Protocols%202nd%20ed%20-%20Fiehn.pdf

Metabolomics Standards Initiative: http://msi-workgroups.sourceforge.net

Standardization in Metabolomics Experiments: http://129.128.185.121/metabolomics_society/standardization.html

ArMet: http://www.armet.org

Arabidopsis thaliana Transcript and Metabolite Data: http://kpv.kazusa.or.jp/kpv4/

Tomato Metabolite Data: http://ted.bti.cornell.edu/

Tomato Genomics: http://solgenomics.net/

PCA of all Tomato Introgression Lines: http://www.shef.ac.uk/aps/tomato/index.html

REFERENCES

Bino RJ, et al. 2004. Potential of metabolomics as a functional genomics tool. *Trends Plant Sci*. **9**(9):418–425.

Dunn W, Bailey NJC, Johnson HE. 2005. Measuring the metabolome: Current analytical technologies. *Analyst* **130**:606–625.

Dunn W, Overy S, Quick WP. 2005. Evaluation of automated electrospray–TOF mass spectrometry for metabolic fingerprinting of the plant metabolome. *Metabolomics* **1**(2):137–148.

Eshed Y, Zamir D. 1994. Introgressions from *Lycopersicon pennellii* can improve the soluble-solids yield of tomato hybrids. *Theor. Appl. Genet*. **88**:891–897.

Eshed Y, Zamir D. 1995. An introgression line population of *Lycopersicon pennellii* in the cultivated tomato enables the identification and fine mapping of yield-associated QTL. *Genetics* **141**: 1147–1162.

Eshed Y, Abu-Abied M, Saranga Y, Zamir D. 1992. *Lycopersicon esculentum* lines containing small overlapping introgressions from *L. pennellii*. *Theor. Appl. Genet*. **83**:1027–1034.

Eshed Y, Gera G, Zamir D. 1996. A genome-wide search for wild-species alleles that increase horticultural yield of processing tomatoes. *Theor. Appl. Genet*. **93**:877–886.

Fiehn O. 2001. Combining genomics, metabolome analysis, and biochemical modelling to understand metabolic networks. *Comp. Funct. Genomics* **2**:155–168.

Fiehn O. 2002. Metabolomics: the link between genotypes and phenotypes. *Plant Mol. Biol*. **48**:155–171.

Fiehn O, Robertson D, Griffin J, van der Werf M, Nikolau B, et al. 2007. The metabolomics standards initiative. *Metabolomics* **3**(3):175–178.

Gann PH, Khachik F. 2003. Tomatoes or lycopene versus prostate cancer: Is it evolution anti-reductionist? *J. Natl. Cancer Inst*. **95**:1563–1565.

Goodacre R. 2005. Making sense of the metabolome using evolutionary computation: Seeing the wood with the trees. *J. Exp. Bot*. **56**(410):245–254.

Hageman, JA, van der Berg RA, Westerhuis JA, van der Werf MJ, Smilde AK. 2008. Genetic algorithm based two-mode clustering of metabolomics data. *Metabolomics* **4**(2):141–149.

Holtan H, Hake S. 2003. Quantitative trait locus analysis of leaf dissection in tomato using *Lycopersicon pennellii* segmental introgression lines. *Genetics* **165**:1541–1550.

Jenkins H, Hardy N, Beckmann M, Draper J, Smith AR, et al. 2004. A proposed framework for the description of plant metabolomics experiments and their results. *Nat. Biotechnol*. **22**:1601–1606.

Jenkins H, Johnson H, Kular B, Wang T, Hardy N. 2005. Toward Supportive Data Collection Tools for Plant Metabolomics. *Plant Physiol*. **138**:67–77.

Lindon JC, Nicholson JK, Holmes E, Keun HC, Craig A, et al. 2005. Summary recommendations for standardization and reporting of metabolic analyses. *Nat. Biotechnol*. **23**:833–838.

Liu Y, Gur A, Ronen G, Causse M, Damidaux R, Buret M, Hirschberg J, Zamir, D. 2003. There is more to tomato fruit colour than candidate carotenoid genes. *Plant Biotechnol. J*. **1**:195–207.

Overy S, Walker HJ, Malone S, Howard TP, Baxter CJ, et al. 2005. Application of metabolite profiling to the identification of traits in a population of tomato introgression lines. *J. Exp. Bot*. **56**(410):287–298.

Tadmor Y, Fridman E, Gur A, Larkov O, Lastochkin E, Ravid U, Zamir D. 2002. Identification of malodorous, a wild species allele affecting tomato aroma that was selected against during domestication. *J. Agric. Food Chem*. **50**:2005–2009.

Tokimatsu T, Sakuri N, Suzuki H, Ohta H, Nishitani K, et al. 2005. KaPPA-View. A web-based analysis tool for integration of transcript and metabolite data on plant metabolic pathway maps. *Plant Physiol*. **138**:1289–1300.

Valle EM, Boggio SB, Heldt HW. 1998. Free amino acid composition of phloem sap and growing fruit of *Lycopersicon esculentum*. *Plant Cell Physiol*. **39**:458–4

Chapter 74

Metabolite Identification, Pathways, and Omic Integration Using Online Databases and Tools

MATTHEW P. DAVEY

74.1 INTRODUCTION

Metabolomics is the logical progression of the study of genes (genomics), transcripts (transcriptomics), and proteins (proteomics). The main intention of metabolomics research is to provide an unbiased assessment of metabolism across multiple pathways (see Chapters 71–73, Vol. I). Ideally, all metabolites should be identified and quantified at a time point of interest to the researcher. This assessment of metabolism is particularly important when we need to observe metabolic changes in response to an environmental perturbation or genetic alteration. By acquiring such knowledge and using either forward or reverse genetic approaches, we should amongst other benefits ultimately be able to assign functionality to genes [Bouché and Bouchez, 2001; Fiehn, 2002]. The theory is that with information obtained from mass spectrometry, we can detect major and minor changes in metabolism that traditionally may not be associated with the experimental treatment or genetically modified product [Weckwerth et al., 2004]. To date, no single research group has been successful in identifying and quantifying *all* compounds within an organism. Most research groups strive to reach this goal by using one of two methods [Shulaev, 2006]. The first is by untargeted metabolomic fingerprinting, which is a semiquantitative scan of predominantly soluble compounds [e.g., Dunn et al., 2005; Overy et al., 2005]. The second is targeted metabolite profiling, which can identify individual compounds and be fully quantitative [Broeckling et al., 2005; Catchpole et al., 2005]. These methods have been used in analytical biochemistry over the past few decades. However, the ease of application and software to analyze results, alongside the increase in interdisciplinary science, has opened up such technology to more research fields to answer a wider range of questions [Bundy et al., 2009; Davey et al., 2008, 2009].

Previous chapters in this book have described the analytical processes currently required for the acquisition of metabolomic information such as direct infusion/injection mass spectrometry (DIMS) (Chapters 71 and 72, Vol. I). A consequence of the current ease and speed of data acquisition is the sheer amount of targeted or nontargeted metabolite or mass information presented to the researcher.

The next step in the analytical process is the identification of metabolites from a list of masses obtained from a mass spectrometer. This identification process may require extra structural or mass information. There may also be a need to visualize the metabolite pathways associated with compounds of interest. This information may allow us to identify proteins and genes that are associated with the compounds or pathways of interest. Ultimately, this sequence of events should go some way to help the scientist understand the complexities and flux of processes within the organism of interest [Weckwerth, 2003]. This chapter briefly describes, within freely and publicly available web-based systems, the following: (i) how lists of masses (molecular ions) are searched within online databases for compound identification; (ii) the

Handbook of Molecular Microbial Ecology, Volume I: Metagenomics and Complementary Approaches, First Edition. Edited by Frans J. de Bruijn.

extra biochemical data required for metabolite confirmation; (iii) how data are visualized; and (iv) the putative gene and protein sequences associated with observed metabolic changes. Although most of the descriptions below are associated with plant material, other biological samples can be analyzed in a similar manner.

74.2 DATA PRESENTATION

The majority of data acquired from metabolomic analyses are usually presented in two forms. The first form is when the researcher is left with a comprehensive list of identified metabolites, all of which may be quantified. Such a list is usually acquired from nontargeted and targeted analyses, where as many compounds as possible or the compounds of a biochemical pathway of interest are analyzed in detail by separation techniques such as LC-MS and GC-MS. The unknown metabolites are usually identified by co-retention times, UV spectra, and mass and mass fragmentation in comparison with known standard compounds obtained from commercial suppliers. The second form of data is from nontargeted analyses where the user is presented with a list of molecular ions in the form of mass-to-charge ratios (*m/z* values) obtained from mass spectrometry. Depending on the selected mass range, thousands of *m/z* values are obtained from each analytical sample. Therefore, data and noise reduction techniques such as "binning" are commonplace. Binning is where the total ion counts from a certain mass range (e.g., from 89.5 Da to 90.5 Da) are combined to a single unit of data. In this case the bin value is 90 ± 0.5 Da. The term "binning" is sometimes called "bucketing," especially for NMR datasets. Usually, masses or bins are identified as biologically interesting after subjecting the data to multivariate chemometric analysis. Currently, the most common form of unsupervised data analysis is principal component analysis (PCA); a comprehensive review of current chemometric processes and interpretation is presented in Trygg et al. [2007]. These masses or bins are interesting to the researcher either because of the relatively high total ion count (TIC, the total number of times a metabolite ion hits the mass spectrometer detector) of a mass for a certain treatment or because of the fold change in TIC between treatments. Such masses of interest can be used for biomarker or substantial equivalence studies [Catchpole et al., 2005]. These fingerprints can also be used to produce metabolic phenotypes without knowing the exact identity of each mass [Kunin et al., 2009]. Overall changes in metabolism can be observed should the metabolic fingerprint, or phenotype, change in response to a genetic alteration or environmental treatment. Therefore, we have two starting points for the next sequence of data analyses. The first is that we already know the compound names, formulae, and possibly CAS (Chemical Abstracts Service) or IUPAC (International Union of Pure and Applied Chemistry) numbers. The second is that all we have are the raw *m/z* values obtained from the mass spectrometer. The next section describes how we assign putative compound identity by searching *m/z* values in publically available databases.

74.2.1 Compound Identification from Masses or Bins of Interest Obtained from Metabolic Fingerprinting and Chemometrics

Chemometric, or multivariate analyses, can be used to reduce data and to highlight the masses that are most likely to be discriminating sample groups [Trygg et al., 2007]. A free downloadable software package to carry out a variety of chemometric analysis can be found at the Pychem source forge site [Jarvis et al., 2006]. The first step toward compound identification from *m/z* values is to ensure that the mass of interest is likely to be the real parent molecular ion and not noise, an adduct, or a fragment ion. A check of the data within the raw chromatogram should inform us of the possibility of these events.

Most organic natural product masses will have a ^{13}C isotope ion at approximately 1 dalton, or thomson, larger than the mass of the parent compound. It is usually the case that if the mass of the parent compound has a high loading score from chemometric analysis such as PCA, then the isotope will also have a similar loading score pattern. This informs us that the mass is likely to be a real compound and that the identity of the daughter isotope mass can be tentatively dismissed.

The other effect in mass spectrometry is the presence of adducts, most commonly sodium and potassium adducts of the parent molecular ion or daughter fragment. These masses (mass + Na; mass + K, etc.) should be searched for, either in the list of raw masses from the mass spectrometer or in the list of bins. The presence of adducts mainly occurs in the positive ionization mode, rather than the negative ionization mode. However, care should be taken in identifying masses in the negative mode because hydroxyl groups can disassociate from the parent compound. The presence of these adducts for putatively identified compounds should inform the researcher that the mass of interest is likely to be a real compound and not an adduct itself.

Ion suppression is another factor to be aware of and is the main criticism of using a mass spectrometer for quantitative metabolomics, especially DIMS. This occurs when one compound takes most of the charge when simultaneously injecting multiple compounds into a mass spectrometer. This can cause low or no ionization of the other compounds and is detectable if a few masses from a metabolite fingerprint have very high TICs while the other masses have very low TIC. Therefore, the investigator should be aware that changes in metabolite phenotypes may be largely due to ion suppression caused by the increase in abundance of one or a few metabolites, rather than a supposed larger alteration in metabolism, especially if treatments or genotypes are compared.

Finally, there are many conjugates, such as glycones, that can occur on a single parent compound. Therefore, the mass of interest may consist of the parent and conjugate. The likelihood of conjugates forming or hydrolyzing can be dependent on the extraction procedure, such as the pH of the extractant. Enzyme or acid hydrolysis of conjugates prior to analysis is possible; however, details of such techniques are beyond the remit of this chapter, though interested readers are pointed to literature such as the classic text of Mabry et al. [1970].

Checking the raw data in this fashion will reduce the number of false positives in the dataset, and further reading is recommended to fully understand the complexities of ion formation in mass spectrometry. There are many books covering this area; for example, a comprehensive guide to mass spectrometry is provided by de Hoffmann and Stroobant [2001].

Once satisfied that the list of masses contains real metabolites and that the likelihood of false positives has been reduced, it is possible to start searching databases for putative mass identification. There can be thousands of metabolites that will match a mass range within a database. For example, in PubChem (**Table 74.1**), a search of the mass 342.0 at 1-Da resolution (molecular weight range 341.5 to 342.5) without any search filters returned 88,356 matched compounds in January 2010. Therefore, in order to reduce the number of matched masses, it is important to be aware of the bin or centroid resolution of the raw data. For example, the resolving power of peaks and centroiding parameters for data processed on software associated with the mass spectrometer should be known. Such resolution and parameters could potentially merge two or more compounds into one centroid mass unit. Therefore the search range of the masses of interest should not be smaller than the resolving power and centroiding parameters. The occurrence of merging compounds into one mass unit also increases with noise and data reduction functions such as binning. For example, it may be common practice to bin raw centroid values into 1-Da bins. However, it is possible to bin to a lower dalton range if the resolution of the mass spectrometer is known. The advantage of using binned data, rather than just using raw unbinned data, is that 'noise' peaks can be identified by selecting and deleting peaks that only occur in one or two analytical replicates (Chapter 73, Vol. I).

The ionization mode of the mass spectrometer should also be noted. Commonly, m/z values are charged such that M-H occurs in the negative ionization mode and M+H in the positive ionization mode. Therefore, the monoisotopic mass of hydrogen (1.007824) should be added or subtracted to the m/z of interest before the identity of the m/z value is searched within databases.

74.3 ONLINE DATABASES

The identity of the masses can now be searched for on a variety of publicly accessible online databases. These databases cover a range of metabolites for various organisms and classes of organic and inorganic compounds. A list of some of the databases commonly used for metabolite identification and information are presented in Table 74.1. This is not an exhaustive list of online databases, and new databases are frequently becoming online. There is a comprehensive review of databases in a special edition of Nucleic Acids Research [Cochrane and Galperin, 2010] which is revised each year. The URLs of databases and tools mentioned in this chapter are all presented in Table 74.1.

The KEGG resource is a collection of databases that primarily contain information about biological compounds, reactions, and genes [Goto et al., 2002; Kanehisa et al., 2006]. The MetaCyc [Caspi et al., 2006] and AraCyc [Rhee et al., 2003; Zhang et al., 2005] databases are part of the BioCyc family of databases and contain information about metabolites and their associated reactions, proteins and genes. The AraCyc database contains specific information for the plant *Arabidopsis thaliana* and is now part of the extensive databases that are accessible within the Plant Metabolic Network (PMN). The EcoCyc database contains information about *Escherichia coli*.

Databases such as the Chemical Abstract Service (CAS), KEGG, and MetaCyc are mainly data repositories so that searches are performed within their own datasets. However, databases such as the Plant Metabolic Network, BioMeta [Ott and Vriend, 2006], ChEBI (Chemical Entities of Biological Interest), Human Metabolome Project, PubChem, and Reactome [Joshi-Tope et al., 2005] perform some of their searches within the KEGG and MetaCyc databases. The PubChem system is particularly thorough because it searches many databases and results are easily exportable.

The dictionary of natural products (DNP) is one of the commercial databases from CHEMnetBASE. However, free basic search functions can be made on the DNP website and have other advantages such as searching accurate masses, and the database contains many metabolic conjugates. Another advantage of the DNP is that it only returns natural products whereas some other databases may also return inorganic or synthesized

Table 74.1 URLs that Provide Online or Downloadable Databases and/or Tools for Identifying Metabolites, Their Pathways and Related Genetic, Transcriptomic, and Proteomic Information

Database or Tool	URL
AraCyc. The Arabidopsis Information Resource (TAIR)	http://www.arabidopsis.org/biocyc/index.jsp
BioMeta	http://cheminf.cmbi.ru.nl/biometa/index.html
BioPathAt	http://www.wsu.edu/~lange-m/Research3_2.htm
BRENDA	http://www.brenda-enzymes.org/
ChemExper	http://www.chemexper.com/
Chemfinder	http://chemfinder.cambridgesoft.com
Chemical Abstracts Service (CAS)	http://www.cas.org
Chemical Entities of Biological Interest (ChEBI)	http://www.ebi.ac.uk/chebi/
ChemIDplus	http://chem.sis.nlm.nih.gov/chemidplus/
Chemspider	http://www.chemspider.com
Chlamycyc	http://chlamyto.mpimp-golm.mpg.de/chlamycyc/index.jsp
Cytoscape	http://www.cytoscape.org/
DatPAV	http://www.sdwa.nus.edu.sg/DATPAV/
Dictionary of Natural Products	http://www.chemnetbase.com
EcoCyc: Encyclopedia of *Escherichia coli* K-12 Genes and Metabolism	http://www.ecocyc.org/
ERGO	http://www.ergo-light.com/
ExPASy	http://www.expasy.ch/
Genevestigator	https://www.genevestigator.ethz.ch/
Golm Metabolome Database	http://csbdb.mpimp-golm.mpg.de/csbdb/gmd/msri/gmd_smq.html
Human Metabolome Project (HMP)	http://www.hmdb.ca/and http://www.metabolomics.ca/
KaPPA-View	http://kpv.kazusa.or.jp/kpv4/
International Union of Pure and Applied Chemistry (IUPAC)	http://www.iupac.org
KNApSAck	http://kanaya.naist.jp/KNApSAcK/
Kyoto Encyclopaedia of Genes and Genomes (KEGG) LIGAND database	http://www.genome.jp/kegg and http://www.kegg.org
MAPMAN	http://mapman.gabipd.org/web/guest
MetaCyc Encyclopaedia of Metabolic Pathways	http://metacyc.org
MetaGeneAlyse	http://metagenealyse.mpimp-golm.mpg.de/
MetaSHARK	http://bioinformatics.leeds.ac.uk/shark/
MetDAT: Metabolomics Data Analysis Tool	http://www.sdwa.nus.edu.sg/METDAT2/
MetExplore	http://metexplore.toulouse.inra.fr/metexplore/
METLIN	http://metlin.scripps.edu/index.php
National Centre for Biotechnology Information (NCBI)	http://www.ncbi.nlm.nih.gov/
National Institute of Standards and Technology (NIST)	http://webbook.nist.gov
National Cancer Institute (NCI)	http://129.43.27.140/ncidb2/?nsc=406942
Nomenclature Committee of the International Union of Biochemistry and Molecular Biology (NC-IUBMB)	http://www.chem.qmul.ac.uk/iubmb/enzyme/

Table 74.1 URLs that Provide Online or Downloadable Databases and/or Tools for Identifying Metabolites, Their Pathways and Related Genetic, Transcriptomic, and Proteomic Information (*Continued*)

Database or Tool	URL
ONDEX	http://pmn.plantcyc.org/ARA/expression.html
Pathway Tools Omics Viewer	http://www.arabidopsis.org:1555/expression.html
Pathway Hunter Ttool	http://pht.tu-bs.de/PHT/
Plant Metabolic Network (PMN)	http://plantcyc.org
PubChem	http://pubchem.ncbi.nlm.nih.gov
Pychem	http://sourceforge.net/projects/pychem
Reactome	http://www.reactome.org
Spectral Database for Organic Compounds SDBS	http://riodb01.ibase.aist.go.jp/sdbs/cgi-bin/cre_index.cgi?lang=eng
The Arabidopsis Information Resource (TAIR)	http://www.arabidopsis.org
The Pherobase: Database of insect pheromones and semiochemicals	http://pherobase.com/
3DMET: Three-dimensional-structure database of natural metabolites	http://www.3dmet.dna.affrc.go.jp/

and drug products. To distinguish between such products may be a cumbersome task to the non-chemist.

Most databases will hyperlink and reference search results to compound information in other databases. For example, KEGG will hyperlink compound information to information of the same compound in CAS, ChEBI, PubChem, and 3DMET. This hyperlink functionality is of particular use as individual databases hold related information that is not within the original queried database. It is also good practice to search for the identity of a mass in more than one database because errors in databases occasionally occur. Therefore, multiple database verification is advisable. Many of the freely available databases also allow the download of their data files by ftp (file transfer protocol). Masses can then be searched within the download database information files. These files can be used for in-house databases and are freely available for noncommercial uses. However, one disadvantage is that should error alterations, additions, and deletions occur in the main online database, then care must be taken to ensure that in-house files are regularly updated with newer versions. Such files can be downloaded at ftp://ftp.genome.jp/pub/kegg for KEGG files, http://bioinformatics.ai.sri.com/ecocyc/dist/flatfiles-52983746/for MetaCyc files, and ftp://ftp.arabidopsis.org/home/tair/Pathways/for AraCyc files.

It is very important to note at this point that in some databases, only the average molecular masses can be queried (compound molecular weight calculated using a weighted average of the natural isotopes of each element) [Rosman and Taylor, 1999; de Hoffmann and Stroobant, 2001; de Laeter et al., 2003]. The mass obtained from a mass spectrometer is generally considered the 'monoisotopic mass' (compound molecular weight calculated using the masses of the most naturally abundant isotopes of the elements). The mass difference between the two types of mass is small, for example, the average mass of the trisaccharide raffinose ($C_{18}H_32O_{16}$; CAS: 512-69-6) is 504.43708, but the monoisotopic mass is 504.16903. However, a database query on the monoisotopic mass may return a false identification if the database only holds average mass information or if the resolution of the search range is not large enough to incorporate average and monoisotopic mass discrepancies. The type(s) of mass(es) stored or searchable within a database should be clarified at the point of data entry and is a highly recommended action point for any database developer.

If the elemental composition of the metabolite has been obtained or computed, then this information may help to decrease the number of possible matches [Kind and Fiehn, 2006]. For example, in the PubChem database, only compounds containing selected elements will be returned.

It may also be useful to carry out a general metabolite identity search of the whole metabolite fingerprint before embarking on detailed mass identification. This can be carried out in the KNApSAck system developed by the Comparative Genomics Laboratory, Nara Institute of Science and Technology, Japan [Shinbo et al., 2006] by uploading the raw MS data, selecting the ionization mode and resolution (if necessary), and scrolling though the *m/z* values. A list of metabolites that match the *m/z* value from various species is automatically displayed. There is also an option to visualize the differences between the intensities of *m/z* values between chromatograms obtained from different experimental treatments.

Two widely used databases for compound information querying and to extract related omic information are KEGG and MetaCyc. An example of the search screenshots of the monoisotopic molecular ion *m/z* value of 504.169 (Raffinose; CAS: 512-69-6) in the KEGG LIGAND database is presented in **Figure 74.1**. A single mass is entered into the query box (Fig. 74.1A). The search returns all masses that fall within the inputted range. There were 13 matches returned (Fig. 74.1B). A table of basic information such as the metabolite entry identifier, structure, name, formula, and mass is provided about each returned metabolite. More information about a selected compound can be obtained from this point. This search process is similar for other databases such as AraCyc and ChEBI.

The list of metabolites returned per mass or mass range is usually between 10 and 200 matches on a 0.2-Da resolution. However, hundreds of matches can be returned for one single mass range. The number of returns and possible matches increases with a wider, less resolute, mass range. For example, searching the monoisotopic mass of sucrose (342.116; CAS: 57-50-1) at ±0.2-Da resolution in KEGG returned 45 possible compounds. The number of returned matches increases as the resolution decrease (Table 74.1). Even at "zero" error, KEGG returned 30 matches for the mass 342.116. This is because this mass is the same for other metabolites such as maltose, lactose, sucrose, and melibiose—that is, isomers.

With such a large number of putative metabolites, there is a requirement to reduce the number of possible matches. This can be done in a number of ways. The initial investigation of the metabolite list for a certain mass should tell you much information about the likelihood of that metabolite being a possible metabolite for that mass. There are a variety of questions that can be posed, such as: What is the likelihood of that metabolite being present in your species of interest? and Does your species have the genetic capacity to produce such metabolites? A certain amount of care should be taken here because even if there are no confirmed reports of a particular metabolite being present in an organism, it does not mean that the metabolite in question should be dismissed. It could just mean that it has not been actively measured and reported in that organism. Such compounds should certainly not be dismissed if a genetically modified organism is being analyzed. The polarity of the solvent and the ionization mode of the mass spectrometer should also be taken into account when assessing whether a metabolite is a possible candidate for the mass. For example, small polar metabolites such as sugars and organic acids ionize well in the negative mode, whereas amino acids ionize well in the positive mode. Lipids and carotenoids are usually present in the nonpolar organic phase. Another way to reduce the amount of returned possible identities for a mass is to obtain or compute the elemental composition of the metabolite [Kind and Fiehn, 2006]. Many mass spectrometer software packages have a function to calculate possible elemental compositions for a given mass. A function in the PubChem database only returns compounds containing selected user-elements. One disadvantage of using these databases in their current state is that only one mass, or one mass range, can be searched for. This can become a laborious task if hundreds or potentially thousands of masses of interest need identifying. Because the demand for multiple mass searching will no doubt increase, it is envisaged that automated multiple mass searching will become a common function in many databases. After searching the unidentified masses in a selection of databases as mentioned above, a "cleaned" metabolite list can now be produced for each unidentified mass from the original metabolite fingerprint data or directly from targeted analyses. It is important to note here that the positive identification of the mass should require extra independent and targeted analytical methods such as ms/ms^n. By using Q-ToF MS, basic quadruple mass spectrometry, or other similar technology, the mass of interest can be fragmented and the fragmentation pattern checked against published fragmentation patterns or the fragmentation pattern of a commercial standard. MS/MS fragmentation patterns can be searched within the Human Metabolome Project and METLIN databases. Similarly, targeted methods such as HPLC-MS or GC-MS could confirm the presence and quantify the abundance of the compounds in the extract.

74.4 WHAT BIOINFORMATIC TOOLS ARE AVAILABLE FOR METABOLIC PATHWAY VISUALIZATION?

Now that a list of metabolites that are associated with your organism of interest has been produced, either from the untargeted metabolite fingerprint or from the targeted metabolite profile, it is biologically and scientifically interesting to know the metabolic pathway or pathways that the compounds of interest are in. For the major, commonly known compounds, it may not be necessary to visualize the metabolic pathways in order to explore possible changes associated with the metabolites of interest. However, should the list of metabolites span a variety of metabolic pathways, and if the metabolites are outside the area of expertise to the researcher, then it should be beneficial to visualize where the compounds of interest are situated in metabolic maps.

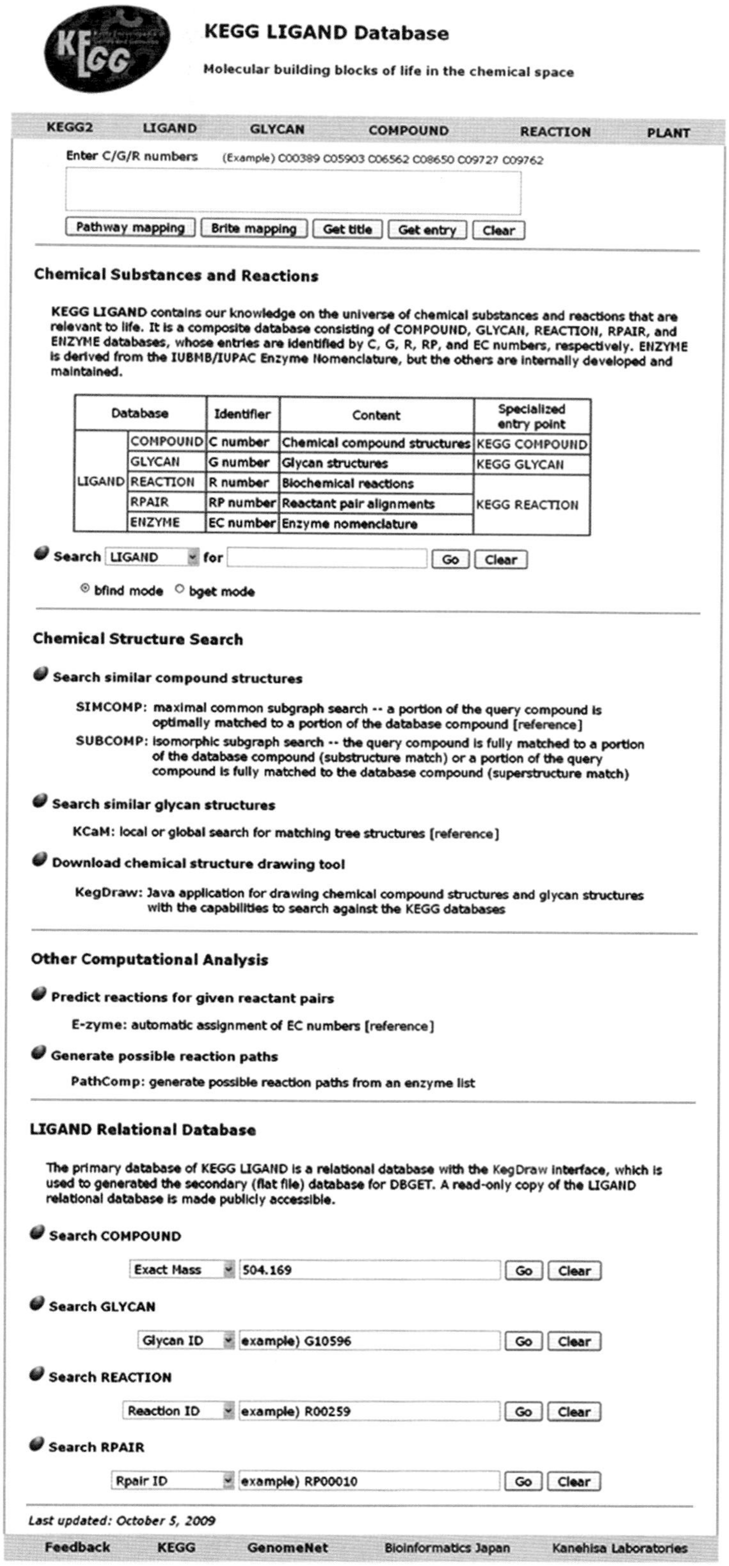

(A)

Figure 74.1 The search input of the monoisotopic molecular weight of 504.169 (**A**) in KEGG LIGAND (Release 52.1) returned 13 matched metabolites of the same molecular weight (**B**). (Screen shots reproduced with kind permission from KEGG http://www.genome.jp/kegg/ligand.html [Goto et al., 2002].)

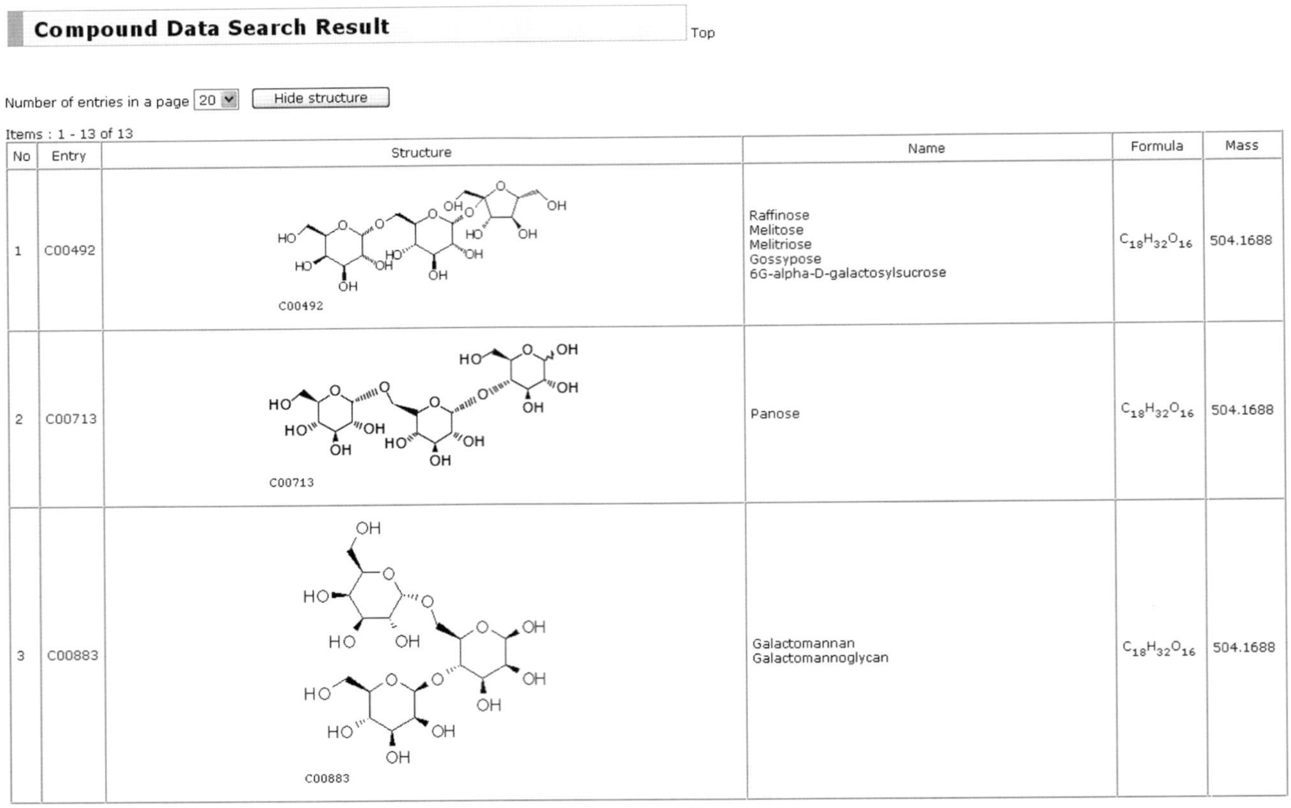

(B)

Figure 74.1 (*Continued*)

Table 74.2 Number of Matched and Returned Metabolites for the Monoisotopic Mass of Sucrose (342.116) in KEGG LIGAND (Release 52.1) at Decreasing Search Resolutions

Search Resolution in ± Da	0	0.1	0.2	0.4	0.8	1	2	5	10
Lower mass input	342.116	342.066	342.016	341.916	341.716	341.616	341.116	339.616	337.116
Higher mass input	342.116	342.166	342.216	342.316	342.516	342.616	343.116	344.616	347.116
Number of matched masses	30	39	45	50	51	51	66	198	370

There are a variety of publicly available bioinformatics tools to visualize metabolic pathways. Comprehensive publicly available pathway maps are found in the KEGG database: MetaCyc and AraCyc, where hundreds of metabolic pathways and reactions are stored; and Reactome, where metabolic maps of many species can be visualized. Of recent note is the MetExplore online system, which utilizes the Cytoscape software; of particular note for the *Arabidopsis* research community are: KaPPA-View [Tokimatsu et al., 2005], MAPMAN [Thimm et al., 2004], and BioPathAt [Lange and Ghassemian, 2005]. Another decent pathway tool is Pathway Hunter tool [Rahman et al., 2005], where the pathway between selected compounds can be retrieved. Other databases and systems mentioned in Table 74.1 also provide some function to view reactions and metabolic maps. In addition, Biocarta (http://www.biocarta.com/genes/allPathways.asp), ExPASy and Roche Applied Science biochemical pathways (http://www.expasy.org/tools/pathways/), the International Union of Biochemistry & Molecular Biology Metabolic Maps, Minimaps, and Animaps created by Donald Nicholson, University of Leeds, UK, and Sigma-Aldrich (http://www.sigmaaldrich.com/img/assets/4202/MetabolicPathways_6_17_04_.pdf) all provide visually impressive metabolic maps.

An example of the type of information obtained from these databases is provided below using pathway information for the polyhydric alcohol sorbitol ($C_6H_{14}O_6$; CAS: 50-70-4). In KEGG LIGAND, the metabolite of interest can be searched by compound name, formula, or a KEGG unique identifier code such as C00794 for sorbitol. Selecting the metabolite takes the user to a new screen where an overview of the information about sorbitol is given, such as formula, mass, reactions, pathways, enzymes, and links to other databases (**Fig. 74.2**).

If the metabolite is in a variety of pathways, then a list of maps is provided in the pathway box. By selecting a pathway—for example, fructose and mannose metabolism—it is possible to see the location of the metabolite (highlighted red in KEGG maps). An example of the type of visualization of the reaction and position in the metabolic pathway is provided in **Figures 74.3A** and 74.3B for KEGG and MetaCyc databases. The solid black arrows represent a molecular interaction or relation, and the open arrows denote a link to another map. A useful feature of KEGG maps is that if more information about your species is available, then such information can be automatically highlighted in the map. For example, green square sections in the KEGG pathway map provide an EC (Enzyme Commission) number that indicates the presence of genes and the completeness of the pathway of a specific organism. The location of compounds in multiple pathways can also be directly searched in KEGG by inserting the KEGG compound object code in their pathway search function (http://www.genome.jp/kegg/tool/search_pathway.html).

Much of the information given in the maps is hyperlinked to its original source either within the same database or in an external database. This allows the user to acquire more information regarding the metabolite and its neighboring metabolites.

In the AraCyc, MetDAT, MetExplore, and KaPPA-View web sites, it is possible to view the position and change of metabolites in an overview map by uploading multiple metabolites into the pathway tools omics viewer function. Although the number of metabolites that can be uploaded are limited by the data included in their metabolic maps, with some targeted analyses, this approach can provide a very powerful overview of metabolic responses in a plant system to experimental treatments. The relative abundance of the metabolites between treatments can also be visualized in a color

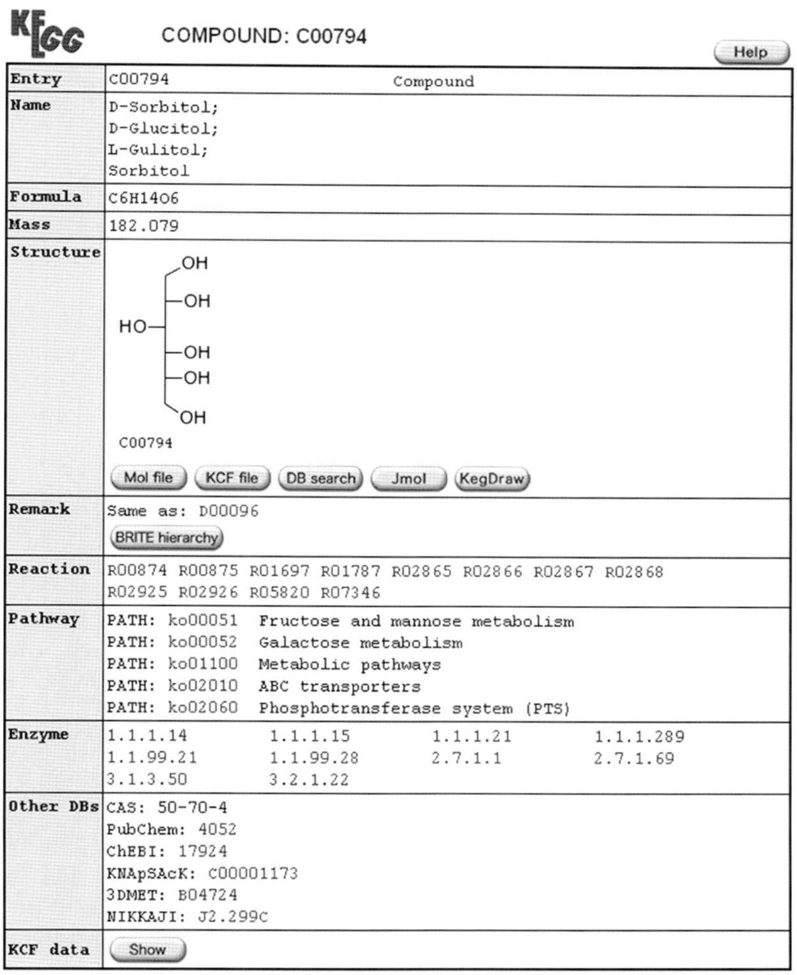

KEGG COMPOUND: C00794 Help

Entry	C00794 Compound
Name	D-Sorbitol; D-Glucitol; L-Gulitol; Sorbitol
Formula	C6H14O6
Mass	182.079
Structure	OH, OH, HO, OH, OH, OH C00794 Mol file KCF file DB search Jmol KegDraw
Remark	Same as: D00096 BRITE hierarchy
Reaction	R00874 R00875 R01697 R01787 R02865 R02866 R02867 R02868 R02925 R02926 R05820 R07346
Pathway	PATH: ko00051 Fructose and mannose metabolism PATH: ko00052 Galactose metabolism PATH: ko01100 Metabolic pathways PATH: ko02010 ABC transporters PATH: ko02060 Phosphotransferase system (PTS)
Enzyme	1.1.1.14 1.1.1.15 1.1.1.21 1.1.1.289 1.1.99.21 1.1.99.28 2.7.1.1 2.7.1.69 3.1.3.50 3.2.1.22
Other DBs	CAS: 50-70-4 PubChem: 4052 ChEBI: 17924 KNApSAcK: C00001173 3DMET: B04724 NIKKAJI: J2.299C
KCF data	Show

Figure 74.2 The metabolite information for the polyhydric alcohol sorbitol (MW 182.079; CAS: 50-70-4) in KEGG LIGAND (Release 52.1). (Screen shot reproduced with kind permission from KEGG [Goto et al., 2002].)

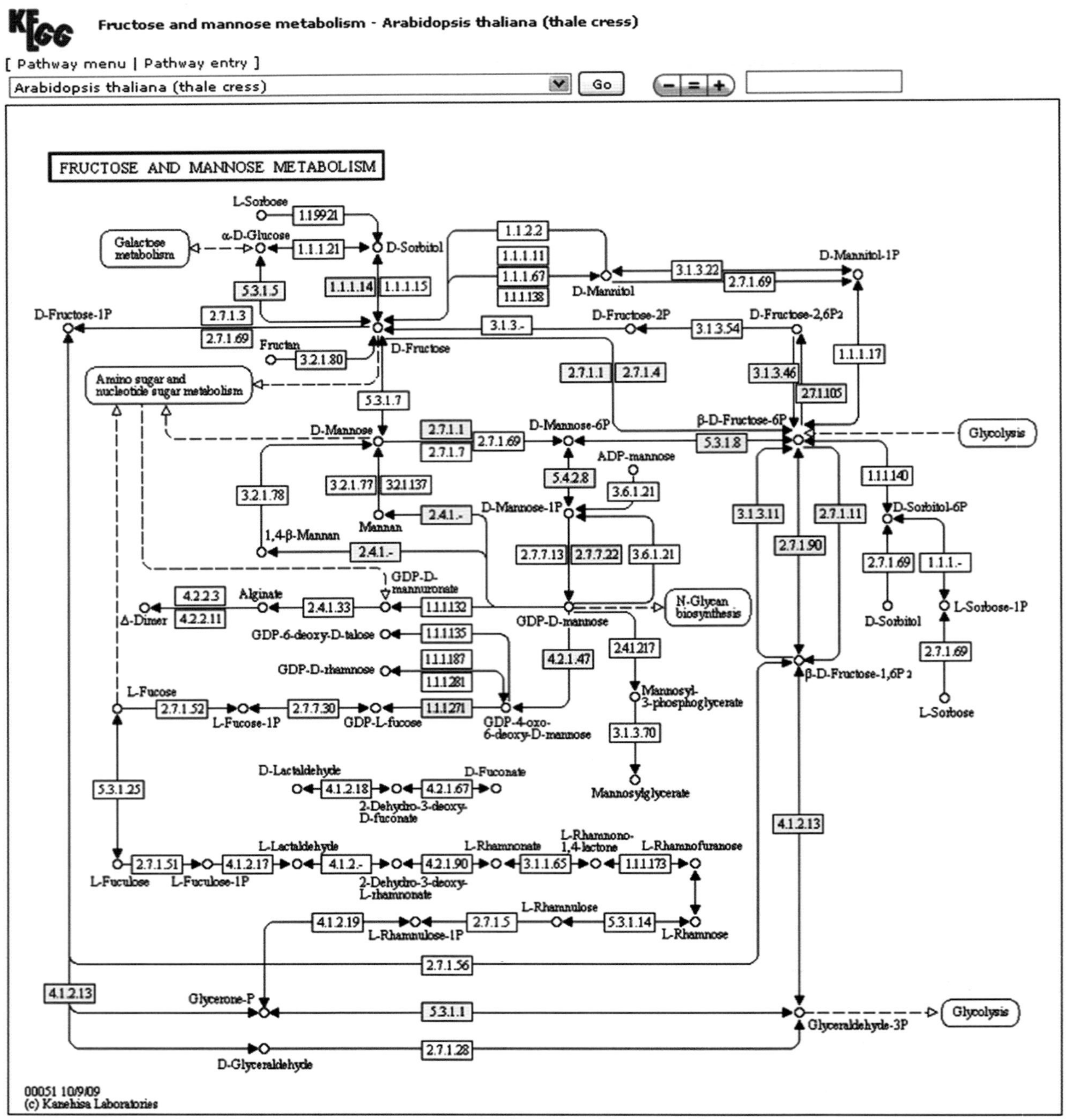

(A)

Figure 74.3 The location of the polyhydric alcohol sorbitol (MW 182.079; CAS: 50-70-4) in fructose and mannose metabolism in a KEGG LIGAND pathway map (Release 42.0) **(A)** and the reaction pathway of sorbitol biosynthesis in MetaCyc (Pathway tools Version 11.0) **(B)**. (Screen shots reproduced with kind permission from KEGG [Goto et al., 2002] and MetaCyc [Caspi et al., 2006].)

scale. Detailed information on pathways of interest can be viewed, and such results are hyperlinked to their original source. A screen shot of the KaPPA-View4 output is shown in **Figure 74.4**.

All of the pathways that the metabolites of interest are in should be inspected. Once a number of metabolic pathways have been identified, it may be the case that commonality in the pathways that the masses are associated with becomes more apparent. This can be indicative that genes and other post-translational factors are changing the whole pathway rather than single intermediates. The next section explores how such protein and

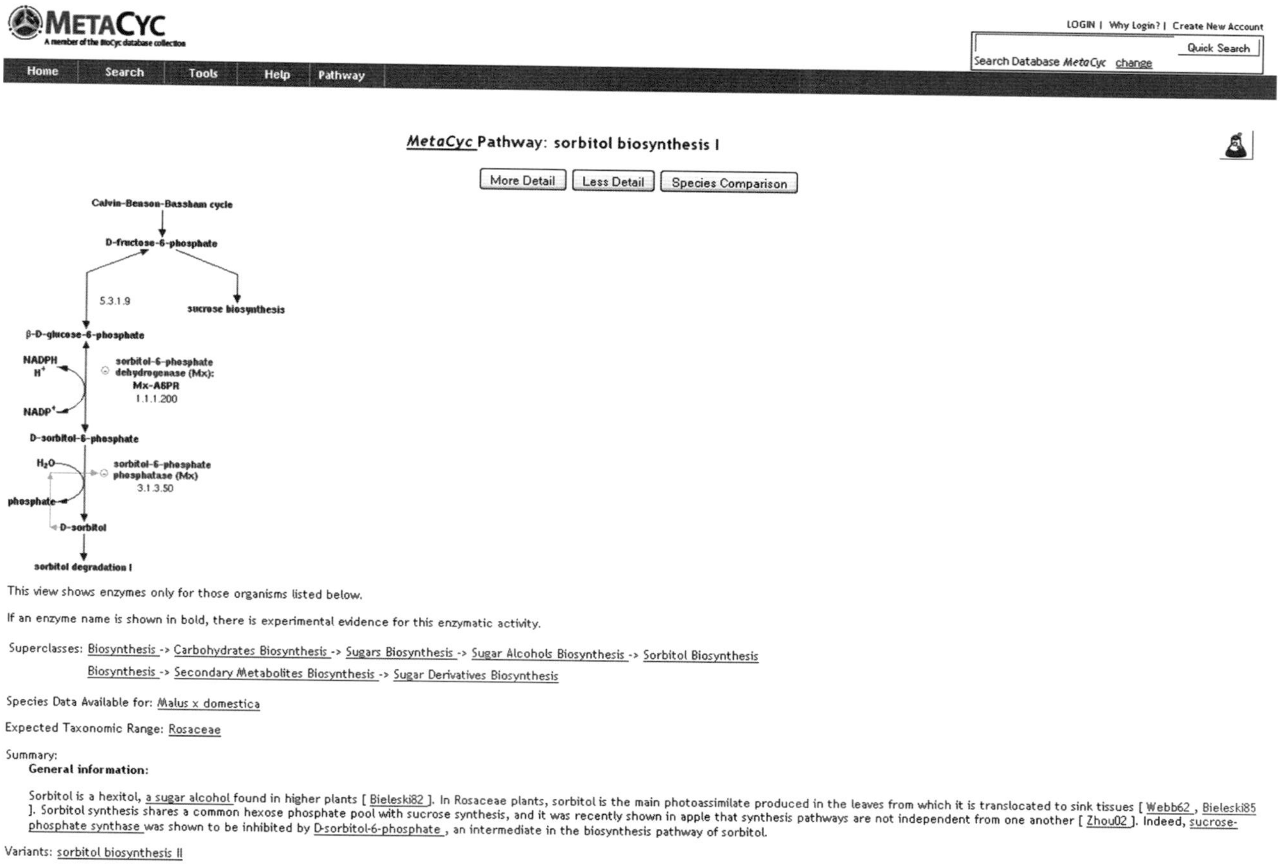

(B)

Figure 74.3 (*Continued*)

gene information around the pathways can be obtained from these online databases.

74.5 UPSTREAM INFORMATION FROM METABOLOMIC OUTPUTS: WHAT ELSE CAN THE DATA TELL US? – COMPOUND NEIGHBORS, ENZYMES, AND GENES

Once metabolic maps have been identified, then a whole suite of information can be acquired to build new hypotheses and experiments or relate to other acquired genomic or post-genomic data. The first step is to look at metabolite neighborhoods that are around the metabolite of interest. It is up to the researcher to decide how far upstream or downstream the pathway to investigate. If the researcher has metabolite fingerprints from DIMS, then the masses of neighboring metabolites can be searched within the raw mass profiles to identify whether it is plausible that such metabolites are present in the extract and organism. The TICs of the masses within the metabolic neighborhood can then be compared among samples and treatments. Once pathways of interest have been identified, it is be possible to develop targeted analyses such as LC-MS or GC-MS to separate and quantify each metabolite along the pathway. Such differences in abundance between metabolites in the pathway can imply that changes in enzyme activity or gene expression are also occurring. Such information surrounding metabolic pathways can be obtained from these online databases. The most useful sections of information from these databases are the hyperlinks to protein and genetic information. In its basic form, this information is usually in the form of EC codes for proteins and reactions and gene code for genetic loci. This information is usually hyperlinked to its original source database. For protein information, these data are usually linked to the Nomenclature Committee of the International Union of Biochemistry and Molecular Biology (NC-IUBMB); ExPASy, SwissProt, and UniProt; ERGO and/or BRENDA. The genetic codes are organism-specific, such as Atg numbers from the AGI coding system for the *Arabidopsis thaliana* genome

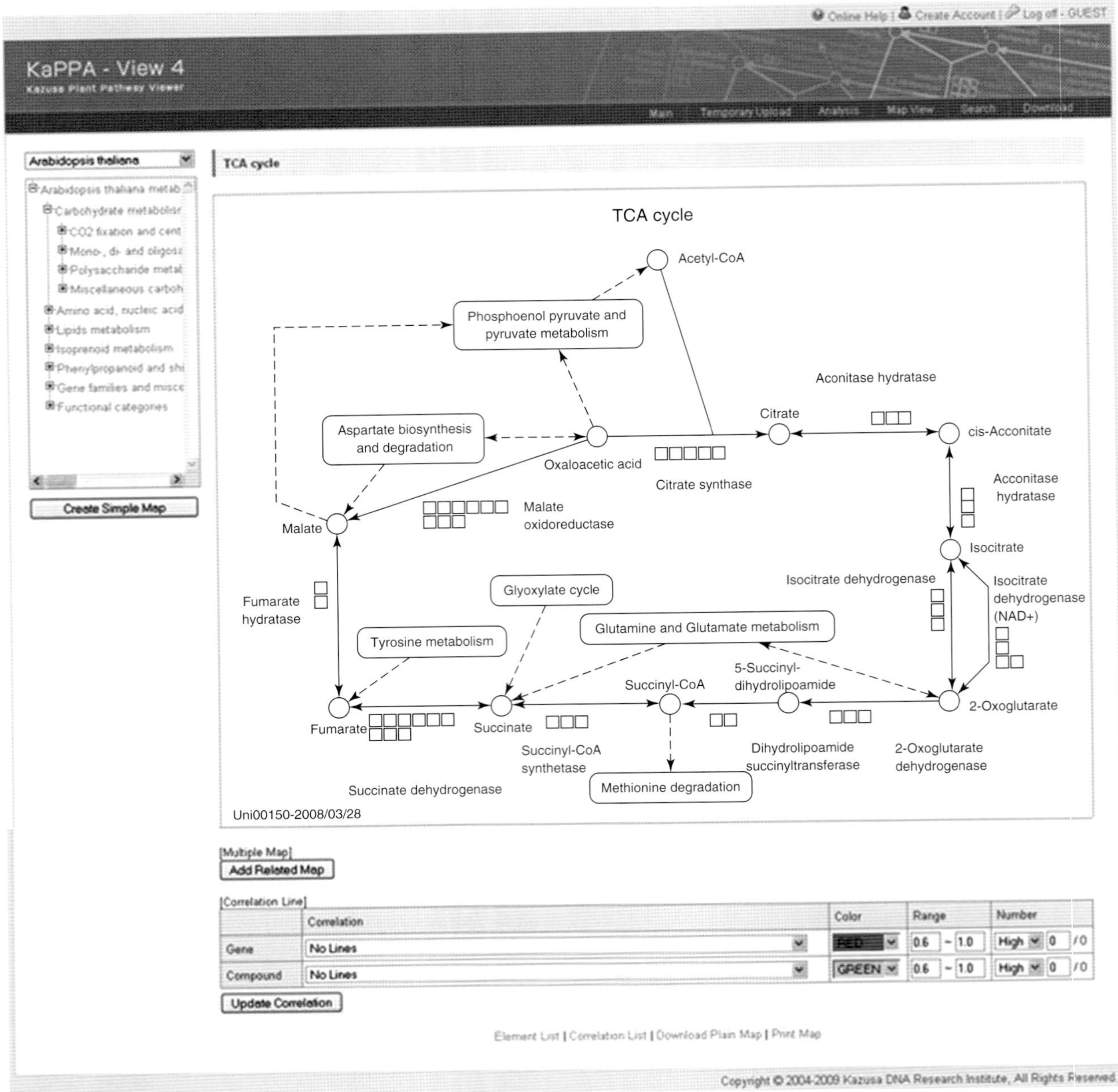

Figure 74.4 Screen shot of the KaPPA-View4 software. Metabolites and pathway details are listed on the left-hand side. (Screen shots reproduced with kind permission from KaPPA-View4 (http://kpv.kazusa.or.jp/kpv4/ [Tokimatsu et al., 2005]).)

[The Arabidopsis Genome Initiative, 2000]. A comprehensive list of codes and organisms is available in the KEGG Organisms section (http://www.genome.jp/dbget-bin/get_text?Organisms+-n). The genetic information is then hyperlinked to other sites such as The Arabidopsis Information Resource (TAIR) for *Arabidopsis thaliana* genetic information [Rhee et al., 2003; Swarbreck et al., 2008] or the Gene, GenBank, or Nucleotide databases in the National Centre for Biotechnology Information. There are usually also links to Gene Ontologies (GO terms) http://www.geneontology.org/[The Gene Ontology Consortium, 2000] for the genes of interest via the information returned in sites such as the KEGG Orthology (KO terms). The GO terms for genes can also be searched within TAIR. Alongside the main public databases, individual research groups have also developed software to study and model metabolic neighborhoods and networks and to integrate omic information—for example, the COPASI, DOME, and Gepasi programs at the Biochemical Networks Modelling Group at the Virginia

Bioinformatics Institute http://mendes.vbi.vt.edu/[Hoops et al., 2006] and MetaGeneAlyse at the Max Planck Institute for Molecular Plant Physiology [Daub et al., 2003].

Once we discover a list of genes that are potentially involved in the pathway of interest, it is possible to research other studies to see the type of reactions and pathways that these genes are associated with. One of the commonly used sites for gene investigation is Genevestigator® [Zimmermann et al., 2004]. This site will allow users to view the expression and regulation of selected genes, pathways, and networks in response to a variety of treatments. From this point in the analysis, there are two likely situations that the researcher will be in. The first is that from the new genetic and protein information obtained from the actions, new experiments can be designed and carried out to measure either gene expression or targeted protein activity. The second situation is that the researcher has already obtained other omic information from the same metabolomic experiment. One approach from the second situation is to cross check all hypothesized genetic and protein data obtained from the above databases with the expression and activity of the real data. Co-expression of omic data is also possible. However, it is important to remember here that changes in one omic may not necessarily coincide or be co-expressed with observed changes in another omic because temporal differences in expression do occur [Nicholson et al., 2002; Sauer, 2006; Wei et al., 2006]. There are a variety of bioinformatics tools to measure correlations between omic data. These tools are outlined in the next section.

74.6 WHAT SOFTWARE IS AVAILABLE FOR POLYOMIC, SYSTEMS BIOLOGY DATA?

It is becoming a common practice to measure more than one omic during an experiment. Alongside metabolomic information, the researcher may acquire transcriptomic and proteomic expression and abundance data. This information will allow us to identify proteins and genes that are associated with the compound or compounds of interest. Likewise, data from proteomics can inform us of possible alternations in the transcriptome and metabolome. Ultimately, this sequence of events should go some way to help the system biologist understand the complexities and flux of processes within an organism [Weckwerth, 2003].

There is a current requirement for bioinformatic tools to integrate and interrogate multiple omic datasets (Lange and Ghassemian, 2005]. Because of the singularity of each omic section (in terms of data format, assay, and experimental procedures), along with the lack of data standards and ontologies in several areas, integration is difficult at a computational level. Despite this difficultly, there are a few web-based programs available to integrate omic data. Under the BioCyc and Plant Metabolic Network systems, the Pathway Tools Omics Viewer can upload microarray expression data, proteomics, metabolomics, and reaction flux. The Kappa-View, MAPMAN, and the Comprehensive Systems Biology Database (CSB.DB) [Steinhauser et al., 2004] programs allow the upload of metabolomic and transcriptomic gene expression microarray data. The change in expression or abundance is usually shown as a change in a rainbow color scale on metabolic maps. The ONDEX [Köhler et al., 2006] and Cytoscape systems are one of the most comprehensive database integration and data analysis tools that is being developed. This system can integrate and produce visually impressive outputs of data from diverse omic inputs.

One of the limitations of viewing metabolomic and multiple omic data is that the results are viewed in static and predefined pathways. Of course, such two-dimensional pathways and compartmentalization do not exist in nature: The reality is four-dimensional. To overcome such visualization limits, the MetaSHARK (Metabolic Search and Reconstruction Kit) [Hyland et al., 2006] system allows the KEGG pathway maps to be viewed as a network without set pathway restrictions. Currently, only single compounds can be searched for within the system. Once the compound is found and selected, the network around a single compound can be explored. Protein names, GO terms, and EC numbers can also be viewed in the pathway. Gene expression data can also be uploaded, and the predicted metabolic pathways can be explored. Metabolic pathways and phenotypes can also be predicted from genetic sequence using programs such as Puma2 [Maltsev et al., 2006] and the GeNome Analysis and Research Environment (GNARE) server [Sulakhe et al., 2005].

74.7 CONCLUSIONS

Overall, this chapter has shown the wealth of bioinformatic tools that are now available to the scientific community. Information from mass spectrometry—either in its basic form, where the metabolite is unidentified, or targeted identified metabolite analysis—can be uploaded, explored, and integrated with other biological information in a way that until now has been very difficult to do for most research laboratories. With more biological information, agreed ontologies, increased computing power, and bioinformatic tools, the future for deciphering and understanding life systems and functions seems exciting, to say the least.

Acknowledgments

I thank KEGG, BioCyc/MetaCyc, and Kappa-View for their kind permission to use screen shots of the web front end of their databases and bioinformatic tools.

REFERENCES

Bouché N, Bouchez D. 2001. *Arabidopsis* gene knockout: phenotypes wanted. *Curr. Opin. Plant Biol*. **4**:111–117.

Broeckling CD, Huhman DV, Farag MA, Smith JT, May GC, Mendes P, Dixon RA, Sumner LW. 2005. Metabolic profiling of *Medicago truncatula* cell cultures reveals the effects of biotic and abiotic elicitors on metabolism. Making sense of the Metabolome Special Issue. *J. Exp. Bot*. **56**(410):323–336.

Bundy JG, Davey MP, Viant, MR. 2009. Environmental metabolomics: A critical review and future perspectives. *Metabolomics* **5**:3–21.

Caspi R, Foerster H, Fulcher CA, Hopkinson R, Ingraham J, Kaipa P, Krummenacker M, Paley S, Pick J, Rhee SY, Tissier C, Zhang P, Karp PD. 2006. MetaCyc: A multiorganism database of metabolic pathways and enzymes. *Nucleic Acids Res*. **34**:D511–D516.

Catchpole GS, Beckmann M, Enot DP, Mondhe M, Zywicki B, Taylor J, Hardy N, Smith A, Kinng RD, Kell DB, Fiehn O, Draper J. 2005. Hierarchical metabolomics demonstrates substantial compositional similarity between genetically modified and conventional potato crops. *Proc. Natl. Acad. Sci*. **102**:14458–14462.

Cochrane GR, Galperin MY. 2010. The 2010 Nucleic Acids Research Database Issue and online Database Collection: A community of data resources. *Nucleic Acids Res*. **38**(Database issue): D1–D4.

Daub CO, Kloska S, Selbig J. 2003. MetaGeneAlyse: Analysis of integrated transcriptional and metabolite data. *Bioinformatics* **19**:2332–2333.

Davey MP, Burrell MM, Woodward FI, Quick WP. 2008. Population specific metabolic phenotypes of *Arabidopsis lyrata* ssp. *petraea*. *New Phytol*. **177**(2):380–388.

Davey MP, Woodward FI and Quick WP. 2009. Intraspecific variation in cold-temperature metabolic phenotypes of *Arabidopsis lyrata* ssp. *petraea*. *Metabolomics* **5**:138–149.

de Hoffmann E, Stroobant V. 2001. *Mass Spectrometry: Principles and Applications*, 2nd ed. Chichester, UK: John Wiley & Sons.

de Laeter JR, Böhlke JK, de Bièvre P, Hidaka H, Peiser HS, Rosman KJR, Taylor PDP. 2003. Atomic weights of the elements. Review 2000 (IUPAC Technical Report). *Pure Appl. Chem*. **75**(6):683–799.

Dunn WB, Overy S, Quick WP. 2005. Evaluation of automated eletrospray–TOF mass spectrometry for metabolic fingerprinting of the plant metabolome. *Metabolomics* **1**:137–148.

Fiehn O. 2002. Metabolomics—The link between genotypes and phenotypes. *Plant Mol. Biol*. **48**:155–171.

Goto S, Okuno Y, Hattori M, Nishioka T, Kanehisa M. 2002. LIGAND: Database of chemical compounds and reactions in biological pathways. *Nucleic Acids Res*. **30**(1):402–404.

Hoops S, Sahle S, Gauges R, Lee C, Pahle J, Simus N, Singhal M, Xu L, Mendes P, Kummer U. 2006. COPASI–a COmplex PAthway Simulator. *Bioinformatics* **22**(24):3067–3074.

Hyland C, Pinney JW, McConkey GA, Westhead DR. 2006. metaSHARK: A WWW platform for interactive exploration of metabolic networks. *Nucleic Acids Res*. **34**:W504–W509.

Jarvis RM, Broadhurst D, Johnson H, O'Boyle NM, Goodacre R. 2006. PYCHEM: A multivariate analysis package for python. *Bioinformatics* **22**(20):2565–2566.

Joshi-Tope G, Gillespie M, Vastrik I, D'Eustachio P, Schmidt E, de Bono B, Jassal B, Gopinath GR, Wu GR, Matthews L, Lewis S, Birney E, Stein L. 2005. Reactome: A knowledgebase of biological pathways. *Nucleic Acids Res*. **33**:D428–D432.

Kanehisa M, Goto S, Hattori M, Aoki-Kinoshita KF, Itoh M, Kawashima S, Katayama T, Araki M, Hirakawa M. 2006. From genomics to chemical genomics: New developments in KEGG. *Nucleic Acids Res*. **34**:D354–D357.

Kind T, Fiehn O. 2006. Metabolomic database annotations via query of elemental compositions: Mass accuracy is insufficient even at less than 1ppm. *BMC Bioinform*. **7**:234–243.

Köhler J, Baumbach J, Taubert J, Specht M, Skusa A, Rüegg A, Rawlings C, Verrier P, Philippi S. 2006. Graph-based analysis and visualization of experimental results with ONDEX. *Bioinformatics* **22**(11):1383–1390.

Kunin WE, Vergeer P, Kenta T, Davey MP, Burke T, Woodward FI, Quick WP, Manerelli ME, Watson-Haigh NS, Butlin R. 2009. Variation at range margins across multiple spatial scales: Environmental temperature, population genetics and metabolomic phenotype. *Proc R. Soc B* **276**:1495–1506.

Lange BM, Ghassemian M. 2005. Comprehensive post-genomic data analysis approaches integrating biochemical pathway maps. *Phytochemistry* **66**:413–451.

Mabry TJ, Markham KR, Thomas MB. 1970. *The Systematic Identification of Flavonoids*. Berlin: Springer-Verlag.

Maltsev N, Glass E, Sulakhe D, Rodriguez A, Syed MH, Bompada T, Zhang Y, D'Souza M. 2006. PUMA2-grid-based high-throughput analysis of genomes and metabolic pathways. *Nucleic Acids Res*. **34**:D369–D372.

Nicholson JK, Connelly J, Lindon JC, Holmes E. 2002. Metabonomics: A platform for studying drug toxicity and gene function. *Nat Rev. Drug Discov*. **1**:153–162.

Ott MA, Vriend G. 2006. Correcting ligands, metabolites, and pathways. *BMC Bioinform*. **7**:517.

Overy SA, Walker HJ, Malone S, Howard TP, Baxter CJ, Sweetlove LJ, Hill SA, Quick WP. 2005. Application of metabolite profiling to the identification of traits in a population of tomato introgression lines. *J. Exp. Bot*. **56**:287–296.

Rahman SA, Advani P, Schunk R, Schrader R, Schomburg D. 2005 Metabolic pathway analysis web service (Pathway Hunter Tool at CUBIC). *Bioinformatics* **21**(7):1189–1193.

Rhee SY, Beavis W, Berardini TZ, Chen G, Dixon D, Doyle A, Garcia-Hernandez M, Huala E, Lander G, Montoya M, Miller N, Mueller LA, Mundodi S, Reiser L, Tacklind J, Weems DC, Wu Y, Xu I, Yoo D, Yoon J, Zhang P. 2003 The Arabidopsis Information Resource (TAIR): A model organism database providing a centralized, curated gateway to Arabidopsis biology, research materials and community. *Nucleic Acids Res*. **31**(1):224–228.

Rosman KHR, Taylor PDP. 1999. 1997 Report of the IUPAC Subcommittee for Isotopic Abundance Measurements. *Pure Appl. Chem*. **71**:1593–1607.

Sauer U. 2006. Metabolic networks in motion: ^{13}C-based flux analysis. *Mol. Syst. Biol*. **2**:62.

Shinbo Y, Sakaguchi S, Nakamura Y, Altaf-Ul-Amin M, Kurokawa K, Funatsu K, Kanaya S. 2006. Species-metabolite database (KNApSAcK): Elucidating diversity of flavonoids. [Special Issue: Fact Databases and Freewares] *J Comput. Aided Chem*. **7**:94–101.

Shulaev V. 2006 Metabolomics technology and bioinformatics. *Brief Bioinform*. **7**(2):128–139.

Steinhauser D, Usadel B, Luedemann A, Thimm O, Kopka J. 2004. CSB.DB: A comprehensive systems-biology database. *Bioinformatics* **20**:3647–3651.

SULAKHE D, RODRIGUEZ A, D'SOUZA M, WILDE M, NEFEDOVA V, FOSTER I, MALTSEV N. 2005. GNARE: Automated system for high-throughput genome analysis with grid computational backend. *J. Clin. Monitor Comp*. **19**:361–369.

SWARBRECK D, WILKS C, LAMESCH P, BERARDINI TZ, GARCIA-HERNANDEZ M, FOERSTER H, LI D, MEYER T, MULLER R, PLOETZ L, RADENBAUGH A, SINGH S, SWING V, TISSIER C, ZHANG P, HUALA E. 2008. The Arabidopsis Information Resource (TAIR): Gene structure and function annotation. *Nucleic Acids Res*. **36**:D1009–D1014.

The Arabidopsis Genome Initiative. 2000. Analysis of the genome sequence of the flowering plant *Arabidopsis thaliana*. *Nature* **408**(6814):796–815.

The Gene Ontology Consortium. 2000. Gene Ontology: tool for the unification of biology. *Nat. Genet*. **25**:25–29.

THIMM O, BLAESING O, GIBON Y, NAGEL A, MEYER S, KRÜGER P, SELBIG J, MÜLLER LA, RHEE SY, STITT M. 2004 MAPMAN: A user-driven tool to display genomics data sets onto diagrams of metabolic pathways and other biological processes. *Plant J*. **37**(6):914–939.

TOKIMATSU T, SAKURAI N, SUZUKI H, OHTA H, NISHITANI K, KOYAMA T, UMEZAWA T, MISAWA N, SAITO K, SHIBATA D. 2005. KaPPA-View. A web-based analysis tool for integration of transcript and metabolite data on plant metabolic pathway maps. *Plant Physiol*. **138**:1289–1300.

TRYGG J, HOLMES E, LUNDSTEDT T. 2007. Chemometrics in Metabonomics. *J Prot. Res*. **6**:469–479.

WECKWERTH W. 2003. Metabolomics in systems biology. *Annu. Rev. Plant Biol*. **54**:669–689.

WECKWERTH W, LOUREIRO ME, WENZEL K, FIEHN O. 2004 Differential metabolic networks silent plant phenotypes. *Proc. Natl. Acad. Sci. USA* **101**(20):7809–7814.

WEI H, PERSSON S, MEHTA T, SRINIVASASAINAGENDRA V, CHEN L, PAGE GP, SOMERVILLE C, LORAINE A. 2006 Transcriptional coordination of the metabolic network in Arabidopsis. *Plant Physiol*. **142**(2):762–774.

ZHANG P, FOERSTER H, TISSIER C, MUELLER L, PALEY S, KARP P, RHEE SY. 2005. MetaCyc and AraCyc. Metabolic pathway databases for plant research. *Plant Physiol*. **138**:27–37.

ZIMMERMANN P, HIRSCH-HOFFMANN M, HENNIG L, GRUISSEM W. 2004. GENEVESTIGATOR. Arabidopsis microarray database and analysis toolbox. *Plant Physiol*. **136**:2621–2632.

Part 6E

Single-Cell Analysis

Chapter 75

Application of Cytomics to Separate Natural Microbial Communities by their Physiological Properties

SUSANN MÜLLER AND DAVID R. JOHNSON

75.1 INTRODUCTION

Elucidating structure–function relationships in microbial communities is among the key goals of microbial ecology. Understanding these relationships is important not only for deriving general ecological principles that describe the behavior of microbial communities, but also for transforming microbial ecology into a predictive science that can be used to rationally manipulate microbial communities. Unraveling structure–function relationships, however, is challenging because of the dynamic behavior of microbial communities. Even seemingly simple tasks, such as assessing the metabolic activity and viability states of cells in changing environments, remain difficult, let alone assessing more cell-specific functional properties.

Cytomics is a relatively new discipline that originated from the medical field and seeks to address the need of describing the roles of individuals and subsets of cells within complex and dynamic cell networks. This discipline uses single cell analytics based on microscopic and especially flow cytometric techniques to unravel a cell's position, function, and physiological state within a community. FC interrogates single cells in solution by passing them one at a time through a focused beam of light [Shapiro, 2002]. Information on the individuals' characteristics is obtained from their behavior toward the incoming light or by labeling characteristic structures or functions with fluorescent dyes. This allows individual cells to be quickly assessed in a highly resolved manner and provides knowledge on coherence and confluence of joint microbial actions. FC is also able to measure the properties of a large number of cells in very short time periods, and it is therefore an effective tool to follow community dynamics.

In microbial ecology, FC can be used to interrogate the phylogenetic affiliations and physiological states of individual cells within communities. Phylogenetic information can be obtained using fluorescently labeled 16S rRNA probes, specific antibodies, or, to a less definitive degree, various lectins. These approaches all require knowledge on the queried species and, in some cases, require pure cultures to develop reliable techniques for their quantitative detection in natural environmental communities. Although such applications are widely accepted and used, they will not be the focus of this section. Instead, this chapter will focus on obtaining information on the physiological states of individual cells. These approaches use fluorescent probes that measure various physiological parameters, such as the content of DNA, metabolic activity, and rates of substrate utilization and uptake [Joux and Lebaron, 2000; Nebe-von-Caron et al., 2000; Hewitt and Nebe-von-Caron, 2004; Müller, 2007; Czechowska et al., 2008, Sträuber et al., 2010]. The power of these techniques is that they can be used to separate a microbial community into subsets of individuals with similar physiological properties and to follow the dynamics of different subsets in response to environmental changes, thus providing a quick tool for analyzing activity states of community members.

Handbook of Molecular Microbial Ecology, Volume I: Metagenomics and Complementary Approaches, First Edition. Edited by Frans J. de Bruijn.

Although they are powerful, cytomic techniques techniques also suffer several limitations. Currently available fluorescent dyes are limited in their specificity and the number of different physiological properties they can measure. There are also biophysical limitations, such as impermeable cell walls that prevent the uptake of specific high-molecular-weight dyes like labeled antibodies and enzymes. These limitations can be overcome by combining cytomic techniques with phylogenetic and omics approaches. The general idea is to first use FC to identify cells with specific physiological properties and to then use cell sorting instrumentation to physically separate the community into subcommunities and subpopulations based upon the target properties. After sorting, the genetic content of subcommunities and subpopulations can be independently analyzed. The cytomic separation of the community therefore provides a means to increase the resolution of community analyses.

Section 75.2.1 summarizes several FC staining techniques for identifying individuals with specific physiological properties, such as proliferation activity (Section 75.2.1.1) and viability (Section 75.2.1.2). Section 75.2.2 describes how to use FC to separate a community into subcommunities and subpopulations and to then assess the genetic content of these subsets. The chapter concludes with an outlook section (Section 75.3) that discusses future advances toward integrating FC and cell sorting with functional omics technologies, including transcriptomics (see Section 6B, Vol. I) and proteomics (see Section 6C, Vol. I). It is the authors' hope that this chapter not only provides an introduction to an exciting new set of strategies for analyzing microbial communities, but also inspires the reader to develop novel methodologies that combine these techniques.

75.2 APPROACH

75.2.1 Analysis of Individual Physiological States of Microorganisms

One of the major objectives in microbial ecology is to identify linkages between specific microorganisms and the functions they perform within communities. For example, which microorganisms within a community utilize a particular resource or are sensitive to a particular physical or chemical stressor? One approach for addressing these types of questions is to characterize the distinct physiological states of individual microorganisms within a community and under specific sets of environmental conditions. There are now numerous fluorescent probes that can be used in conjunction with FC to characterize different physiological states of individual cells, such as identifying proliferating and nonproliferating cells and viable and nonviable cells. These fluorescent probes can be used to cluster individual cells into subcommunities and subpopulations with shared physiological features.

A comprehensive survey of all the physiological states that can be characterized by FC is beyond the scope of this chapter and has been reviewed in detail elsewhere [Shapiro, 2002; Hewitt and Nebe-von-Caron, 2004; Müller and Nebe-von-Caron, 2010]. Instead, this section will focus on only two techniques; one technique identifies proliferating cells and subcommunities by quantifying chromosome contents (Section 75.2.1.1), while the other technique identifies viable and nonviable cells by measuring the integrity of the cell membrane (Section 75.2.1.2). This chapter focuses on these two techniques because, in the authors' opinion, they are among the most universally applicable across different microbial taxa and are therefore well-suited for interrogating uncharacterized microbial communities. Section 75.2.1 ends with a discussion of techniques for measuring species-specific functions within microbial communities that can be applied alone or in conjunction with the discussed proliferation and viability assays.

75.2.1.1 Identification of Proliferating Cells. One of the most effective methods for identifying proliferating cells within a community is to measure the amount of DNA within each cell. The premise of this approach is that cells have different numbers of chromosomes depending upon their genetic background, their state in the cell cycle, their velocity of growth, and the associated environmental conditions [Müller, 2007]. In general, rapidly growing cells often contain more copies of their chromosome than do slowly growing or dormant cells. Using a DNA specific dye, such as 4′-6-diamidino-2-phenylindole (DAPI), the DNA content of a cell can be stained and the fluorescence signal measured. As an example, changes in the DNA contents of *Escherichia coli* K12 as a function of growth conditions is shown in **Figure 75.1**. Even without being familiar with complex bacterial proliferation strategies, the increase in DNA contents and cell size during proliferation should be clear in Figure 75.1. Also, the more than linear increase in chromosome contents that is typical for bacterial 'uncoupled' DNA synthesis can be observed. The relative proliferation activity of cells within each subpopulation can be measured by tracking cell abundances over time and for different environmental conditions [Müller, 2007].

These patterns of DNA distributions can also be used to separate a microbial community into distinct subsets of cells based upon the number of chromosomes that they contain. If the quantity of cells within a cluster is changing, this movement of cells between the clusters points to the fact that distinct species are growing or outcompeting other species. Such distributions are shown in **Figure 75.2**,

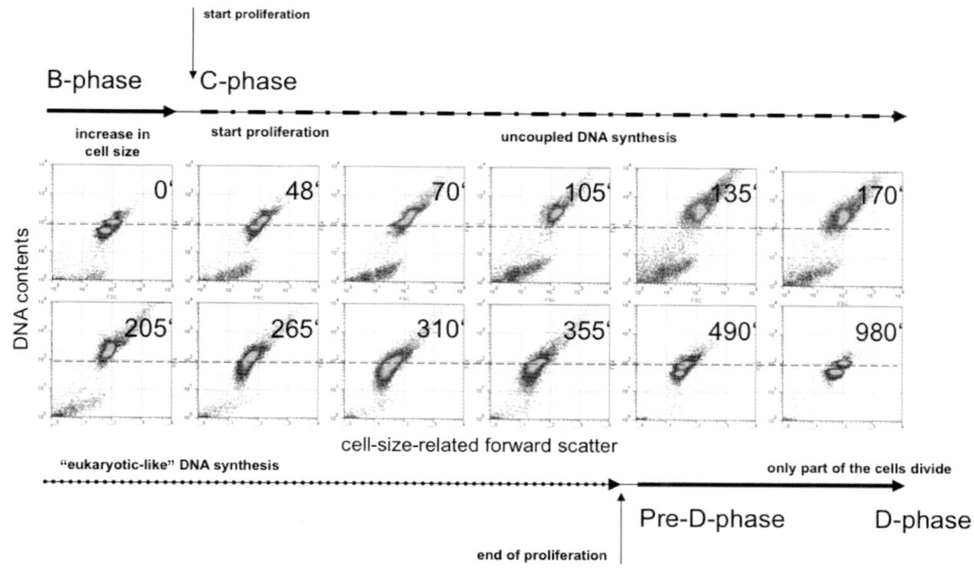

Figure 75.1 Cell cycling of *Echerichia coli* K12. *E. coli* K12 was grown in batch culture on LB medium at 37°C and 175 rpm and showed "uncoupled" DNA synthesis (C-phase) during early exponential growth and "eukaryotic-like" cycling (also C-phase) during mid-exponential growth. The cells do not divide completely at the end of cultivation. Therefore, some cells stay in the pre-D-phase (cells do not proliferate or divide; subpopulation above the dashed line) while others enter the D-phase (cell division; subpopulation below the dashed line).

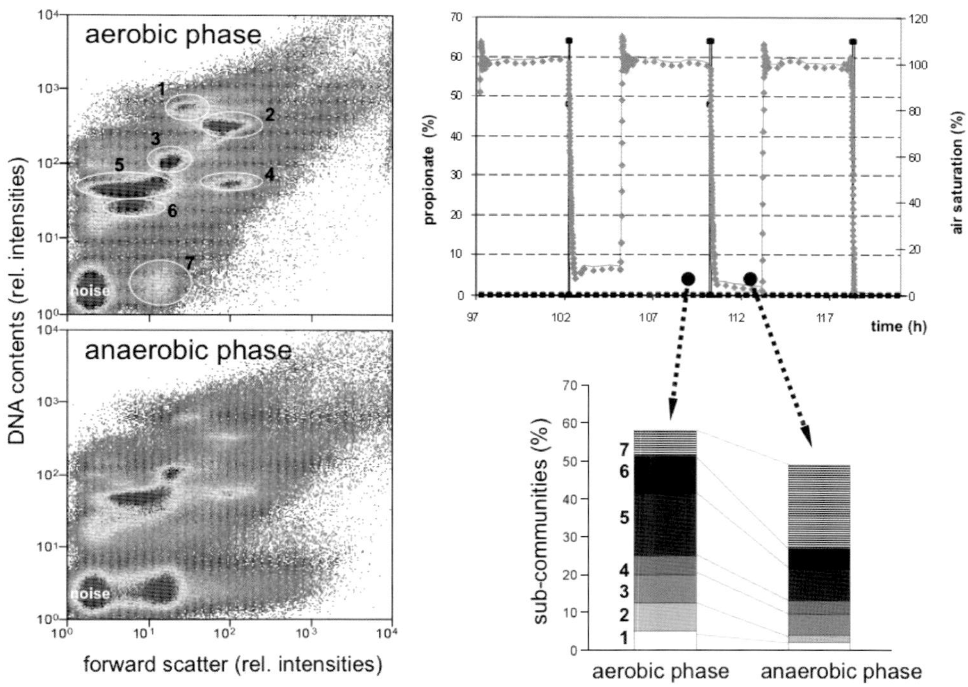

Figure 75.2 DNA distribution patterns for activated sludge subcommunities. Activated sludge from a wastewater treatment plant was obtained and cultivated in a 1-L sequencing batch reactor with aerated and nonaerated phases. The community was grown on synthetic wastewater medium with peptone, yeast extract and propionate as carbon and energy sources at 22°C, 150 rpm, and pH 7. The microorganisms were harvested every 30 min. After harvesting, cells were fixed, diluted, stained, and analyzed by FC according to Günther et al. [2009]. The DNA patterns versus forward light scatter distributions are shown for two samples from the aerobic and anaerobic phases of cultivation. Dominant subpopulations were exemplarily shown by gate settings, with the contained cell numbers determined quantitatively (black circles; the lowest left signals are related to machine-derived electronic noise). It can be observed that the general structure of the community is not varying in response to quick changes in oxygen availability; however, the numbers of cells in the respective subcommunities fluctuate.

where activated sludge was harvested after aerobic and anaerobic cultivation and subcommunities were identified based upon their DNA contents and their forward light scatter behavior. At least seven dominant subsets of cells emerged out of the community that retained their distribution but changed in their abundance according to oxygen availability. Other applications of this approach have identified proliferating subcommunities involved in the anaerobic degradation of toluene [Müller et al., 2009], as well as the aerobic degradation of 4-chlorosalicylate [Pawelczyk et al., 2008], crude oil components [Kleinsteuber et al., 2006], and complex substrates in wastewater treatment plants [Günther et al. 2009].

Although the DNA content approach is quite robust for microbial community analyses, there are several technical considerations that are important to consider. First, a fluorescent stain that specifically binds with DNA and not with other biological molecules within the cell, such as RNA, must be used. Nonspecific binding of the stain can obscure the cell-cycle and species-specific patterns of DNA content and lead to poor resolution of cell distributions [Müller, 2007]. Hoechst dyes and DAPI are well-suited for DNA content analyses because of their high specificity for DNA relative to other biological molecules [Müller, 2007]. Another consideration is how to preserve samples when there is only limited access to a flow cytometer [Günther et al., 2008]. Many conventional preservation protocols utilize alcohols and aldehydes [Shapiro, 2002], but these molecules can potentially cause problems for DNA content analyses. Alcohols often induce cell aggregation during storage [Schimenti and Jacobberger, 1992], which leads to an overestimation of per-event DNA content. Aldehydes can increase sample autofluorescence [Shapiro, 2002] and hinder proper staining of DNA [Cotner et al., 2001], which leads to an underestimation of the DNA content in preserved cells. To overcome these problems, fixation with 10% sodium azide, which inhibits energy production and respiration and preserves the cell constituents using the high osmolarity of the fixation solution, is typically sufficient [Günther et al., 2008]. For anaerobic communities, for which sodium azide is often ineffective, the addition of 5 mM each of barium chloride and nickel chloride is further recommended [Günther et al., 2008]. Using the latter approach, the authors found that samples could be safely stored for up to nine days at 4°C with no observed cell aggregation or changes in DNA staining patterns [Günther et al., 2008]. When using barium, RNA analyses cannot be performed afterwards because the metal changes the conformation of the RNA molecule. In that case, a fixation protocol using formaldehyde followed by suspension in ethanol and storage at −20°C is recommended [Wallner et al., 1993].

75.2.1.2 Identification of Viable and Nonviable Cells. Another challenge in microbial ecology is to identify viable and nonviable microorganisms within a microbial community. Natural communities contain cells that encompass a diverse range of viability states, including rapidly growing cells, dormant and resting cells, and cells that are in the process of death and turnover. Identifying viable microorganisms is important for systems where metabolically active organisms are desired, such as in biotechnological processes, but also important for systems where viable organisms are undesired, such as in drinking water resources. From both perspectives, it is of interest to quantify the fraction of a microbial community that maintains viability in a specific environment and to understand how viability is affected by environmental perturbations, such as by the addition of a stressor or disinfectant.

Although there is significant interest in identifying and quantifying viable and nonviable cells within a community, this task remains challenging. Among the numerous methods for assessing the viability states of single cells, staining with propidium iodide (PI) is often the technique of choice for microbial communities [Boulos et al., 1999; Gasol et al., 1999; Berney et al., 2007; Falcioni et al., 2008; Johnson et al., 2009]. PI is a red fluorescent molecule that readily penetrates across damaged cell membranes but cannot penetrate intact membranes. Thus, nonviable cells with damaged cell membranes will take up PI, while viable cells will not [Boulos et al., 1999]. PI can be used in conjunction with a second membrane-permeable fluorescent stain, such as SYTO9 or SYBR Green I. This second stain is useful for resolving all cells from the background irrespective of membrane integrity when the scatter characteristics of the cells cannot be sufficiently determined.

In a recent application, PI and SYBR Green I were used to quantify the fraction of viable cells within Lake Geneva, Switzerland, after exposure to the antibiotic triclosan [Johnson et al., 2009]. FC analyses revealed distinct subcommunities of PI-positive and PI-negative cells, with the fraction of PI positive cells increasing from 20% to 80% of the total population after triclosan exposure (see **Fig. 75.3** [Johnson et al., 2009]). The authors further showed that there are both triclosan-sensitive and triclosan-insensitive species in Lake Geneva and that as little as 10 nM of triclosan is sufficient to modify the composition of the community [Johnson et al., 2009]. In another example, PI and SYBR Green I were applied to quantify the effects of ultraviolet light irradiation on a freshwater microbial community [Berney et al., 2007]. The authors observed a clear differentiation between PI-positive and PI-negative subcommunities and showed that the fraction of PI-positive cells increased after UV

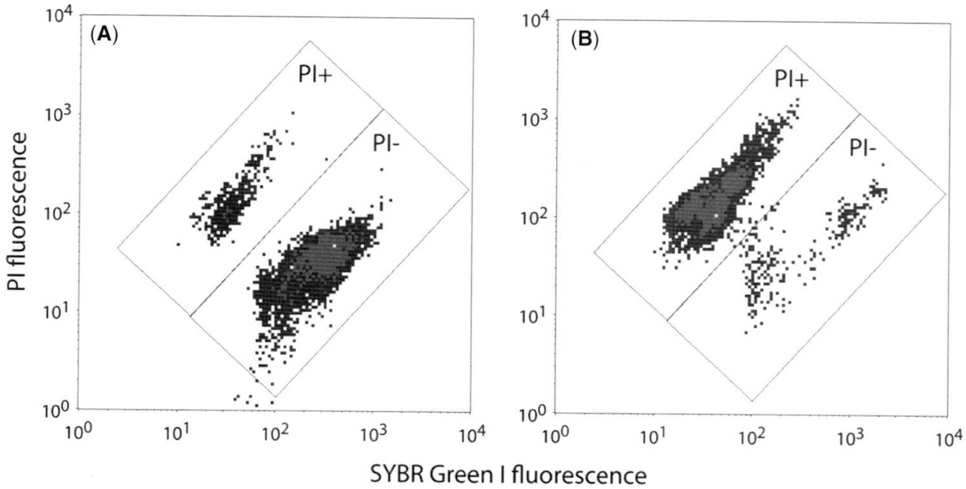

Figure 75.3 Effect of triclosan on the uptake of PI in the microbial community from Lake Geneva, Switzerland. Control cultures (**A**) and cultures exposed to 10 μM of triclosan for 18 h (**B**) were stained with a combination of PI and SYBR Green I and analyzed by FC. The PI− and PI+ regions were defined using viable and heat-killed control cultures, respectively. (Reprinted with permission from Johnson DR, Czechowska K, Chèvre N, van der Meer Jr. 2009. Toxicity of triclosan penconazole and metalaxyl on *Caulobacter crescentus* and a freshwater microbial community as assessed by flow cytometry. *Environ. Microbiol*. **11**:1682–1691. Copyright 2009, Society for Applied Microbiology and Blackwell Publishing Ltd).

irradiation [Berney et al., 2007]. Thus, the technique has potential application as a rapid and effective method for monitoring the efficiency of drinking water disinfection processes.

Although widely applied, caution should be taken when using PI to assess cell viability states because, in several cases, viable cells have been observed to take up significant levels of PI [Barbesti et al., 2000; Shi et al., 2007]. For example, studies with two environmental soil isolates revealed that up to 45% of cells from the early exponential phase were PI-positive [Shi et al., 2007]. Similar results have been observed in other environmental microbial communities. Johnson et al. [2009] found that 20% of cells in the microbial community in Lake Geneva, Switzerland, were PI-positive in the absence of triclosan (see Fig. 75.3). It remains unclear why some viable bacteria take up PI while others do not, but might be related to growth state-dependent differences in membrane composition and/or the rate of cell wall reconstruction [Barbesti et al., 2000; Shi et al., 2007]. Irrespective of the exact mechanism, it is clear that caution must be taken when interpreting results from PI-based viability investigations.

75.2.1.3 Measuring Species-Specific Functions in Microbial Communities. Within a microbial community, some species might perform a specific function while other species do not, which depends on both the genetic content of the species and the associated environmental conditions. FC staining techniques have been developed that allow researchers to identify microorganisms that perform specific functions. One example is an FC technique for identifying cells that store poly-ß-hydroxyalkanoate (PHA) granules. Several bacteria accumulate PHA under conditions of limiting concentrations of nutrients or electron acceptors and excess of carbon. To quantify and track PHA-accumulating bacteria within microbial communities, Nile Red can be used to stain this cellular component [Ackermann et al., 1995]. Another FC staining technique has been developed to identify polyphosphate granules in polyphosphate-accumulating organisms (PAOs), which have important roles in the removal of phosphate from wastewater [Günther et al., 2009]. The technique uses the fluorescent antibiotic tetracycline, which binds with the divalent counter-cations within intracellular polyphosphate granules [Hallett et al., 1972] and preferentially increases the fluorescence of PAOs. A further advantage of this method is that it can be combined with DNA-pattern analyses to reveal additional information on subcommunity proliferation and dynamics [Günther et al., 2009].

75.2.2 Analysis of FC-Separated Microbial Communities

FC is able to identify individual cells within a microbial community that share distinct physiological properties but is rarely able to directly link these properties with the genetic content of the individual cells. Such linkages, however, are critical for developing and testing hypotheses about structure–function relationships. This section will discuss how to sort cells based upon FC-measured physiological properties and to obtain information about the genetic content within sorted subcommunities.

75.2.2.1 Phylogenetic Analysis of Subcommunities. Molecular ecological techniques have revolutionized our ability to characterize the phylogenetic composition of microbial communities by targeting and analyzing the 16S rRNA genes of the contained individuals [Kowalchuk et al., 2004]. Although the analysis of these genes is well established (see also Chapter. 15, Vol. I), it remains a challenge to link different phylogenetic groups with specific biological functions within the community [Vogt et al., 2005]. Combining FC with cell sorting to physically separate physiologically distinct subcommunities followed by phylogenetic analyses of the sorted cells provides an additional and powerful means to make these important structure–function linkages. Such analyses are applicable to even low abundance subcommunities and populations that might otherwise be missed with conventional whole community approaches. The general approach is to first apply FC to identify individual cells with a specific physiological characteristic of interest, such as DNA content, viability, or a species-specific property like polyphosphate accumulation. Next, cells are clustered into subcommunities according to the target physiological properties and the subcommunities are quantified and physically separated from the whole community. Finally, conventional molecular ecological techniques are applied to characterize the phylogenetic composition within each subcommunity.

This general FC and sorting approach has proven useful for revealing novel phylogenetic structure–function relationships [Longnecker et al., 2005; Mou et al., 2005, 2007; Kleinsteuber et al., 2006; Günther et al., 2009; Müller et al., 2009]. For example, the DNA content approach discussed above (Section 75.2.1.1) was used to identify phylogenetic groups within subcommunities that proliferate during oil degradation [Kleinsteuber et al., 2006]. The authors showed that the dominant proliferating subcommunities were primarily composed of *Sphingomonas* spp. during low-salinity conditions but by *Ralstonia, Halomonas, Dietzia*, and *Alcanivorax* spp. during high salinity conditions [Kleinsteuber et al., 2006]. In another example using the same general approach, phylogenetic groups were identified within subpopulations that proliferate during the anaerobic degradation of toluene [Müller et al., 2009]. Strikingly, the authors found that a single *Desulfobulbaceae* phylotype comprised more than 80% of the actively proliferating subpopulations, suggesting that this phylogenetic group plays a key role in the anaerobic degradation of toluene [Müller et al., 2009].

75.3.1 Outlook to the future

75.3.1.1 Functional Omic Analysis of Subcommunities and Populations. Clearly, significant advances have been made toward developing approaches for genetic analyses of FC-separated subcommunities. Although these approaches allow researchers to separate a community based upon physiological properties and then obtain information about the structure and dynamics of subcommunities and subpopulations, these approaches provide limited information about the specific pathways that the organisms use under a given set of environmental conditions. For example, what pathways do cells use for the degradation of a specific substrate? What stress response systems does an individual use to survive exposure to a specific toxicant? Approaches that combine FC-based separation of communities with functional omics analyses of subcommunities would help answer these questions. In principle, this could include microarray (see also Chapters. 57–61, Vol. I) and sequencing-based metatranscriptomics (see also Chapters 62–65, Vol. I) [He et al., 2007; Parro et al., 2007; Frias-Lopez et al., 2008; Urich et al., 2008] and gel- or non-gel-based metaproteomics (see also Chapters. 66–70, Vol. I) [Wilmes and Bond, 2006; VerBerkmoes et al., 2009, Benndorf et al. 2007] that have been developed for total community analyses.

Although alluring, the application of functional omics approaches to FC-disintegrated subcommunities of microbial cells has not yet been accomplished due to a number of challenges beyond those typically encountered with total community analyses. The first challenge is how to obtain sufficient mRNA or protein from sorted cells. Typical transcriptomic and proteomic analyses require about 10^9 microbial cells. Even when using high-speed sorters with sorting rates greater than 20,000 cells per second, it remains difficult to obtain this required number of cells in a reasonable time frame. Although daunting, significant advances have been made with FC-sorted subpopulations of pure cultures. In the first reported example, proteomics was applied to analyze subpopulations from pure cultures of *Cupriavidus necator* after exposure to phenol [Wiacek et al., 2006]. In that study, the authors used 2D gel-based proteomics to identify subpopulation-specific proteins using single-cell one-mode sort purity (5000 events per second) [Wiacek et al., 2006]. A major problem in that study, however, was the dilution of the sample by the FC sheath stream, which led to a substantial loss of cells during the following centrifugation steps. A novel protocol is currently being developed to address this problem that includes a filtration step during the sorting procedure itself to concentrate cells [Jehmlich et al. 2010]. When combined with gel-free analysis of the proteins, this technique increases the proteome resolution by about 600-fold.

Another major challenge limiting the application of functional omics approaches to sorted cells is sample stability. The transcriptome and proteome are dynamic and might undergo significant changes during the sample collection process. This is especially problematic when analyzing transcriptomes, because mRNA is highly labile. Thus, care is needed to quickly and effectively preserve sorted cells. Although difficult, a previous study with pure cultures of *Saccharomyces cerevisiae* demonstrated that high-quality mRNA suitable for transcriptome analyses could be recovered from sorted subpopulations [Achilles et al., 2007]. The authors applied FC to sort viable cells based upon differences in the cells' affinity to glucose and their DNA content. To preserve transcriptomes during the sorting process, subpopulations were sorted into 50-ml tubes that were incubated on dry ice. Gel electrophoresis of isolated RNA revealed high levels of quality and integrity, with intact 18S and 26S rRNA bands [Achilles et al., 2007]. Liquid nitrogen could in principle also be used for preservation, but the close proximity to the sorting equipment combined with high rates of evaporation caused water to condense on the high-voltage electrostatic deflection plates and influenced sorting. Thus, the use of dry ice is recommended over liquid nitrogen [Achilles et al., 2007].

Although challenges remain, the acquisition of information at the gene, transcription, and protein levels are important additions to the concept of cytomics. Together, cytomics and functional omics promise to generate new breakthroughs in analyzing the regulation of pathways, assessing community functions, and revealing interactions between individual microbial cells within communities. For the integration of cytomics and omics to be successful, however, it is of paramount importance to continue extending the genomic, transcriptomic, and proteomic databases to biotechnologically and medically relevant microorganisms and to the massive amount of uncultivable microorganisms.

Acknowledgments

SM acknowledges the support of the Helmholtz Centre for Environmental Research (UFZ), the Deutsche Forschungsgemeinschaft (DFG), the European Union, and the Free State of Saxony. SM would like to thank Susanne Günther for performing the analyses for Figure 75.2 and greatly acknowledges the help of Thomas Hübschmann, Christine Süring, and Helga Engewald in the Leipzig Flow Cytometric Laboratory. DRJ acknowledges the support of the Swiss Federal Institute of Technology Zürich (ETHZ) and the Swiss Federal Institute of Aquatic Science and Technology (EAWAG).

REFERENCES

Achilles J, Stahl F, Harms H, Müller S. 2007. Isolation of intact RNA from cytometrically sorted *Saccharomyces cerevisiae* for the analysis of intrapopulation diversity of gene expression. *Nat. Protoc.* **2**:2203–2211.

Ackermann JU, Müller S, Lösche A, Bley T, Babel W. 1995. *Methylobacterium rhodesianum* cells tend to double the DNA content under growth limitations and accumulate PHB. *J. Biotechnol.* **39**:9–20.

Barbesti S, Citterio S, Labra M, Baroni MD, Neri MG, et al. 2000. Two- and three-color fluorescence flow cytometric analysis of immunoidentified viable bacteria. *Cytometry* **40**:214–218.

Benndorf D, Balcke GU, Harms H, von Bergen M. 2007. Functional metaproteome analysis of protein extracts from contaminated soil and groundwater. *ISME J*. **1**:224–234.

Berney M, Hammes F, Bosshard F, Weilenmann HU, Egli T. 2007. Assessment and interpretation of bacterial viability by using the LIVE/DEAD BacLight kit in combination with flow cytometry. *Appl. Environ. Microbiol*. **73**:3283–3290.

Boulos L, Prévost M, Barbeau B, Coallier J, Desjardins R. 1999. LIVE/DEAD BacLight: Application of a new rapid staining method for direct enumeration of viable and total bacteria in drinking water. *J. Microbiol. Methods* **37**:77–86.

Cotner JB, Ogdahl ML, Biddanda BA.. 2001. Double-stranded DNA measurement in lakes with the fluorescent stain PicoGreen and the application to bacterial bioassays. *Aquat. Microb. Ecol*. **25**:65–74.

Czechowska K, Johnson DR, van der Meer Jr. 2008. Use of flow cytometric methods for single-cell analysis in environmental microbiology. *Curr. Opin. Microbiol*. **11**:205–212.

Falcioni T, Papa S, Gasol JM. 2008. Evaluating the flow-cytometric nucleic acid double-staining protocol in realistic situations of planktonic bacterial death. *Appl. Environ. Microbiol*. **74**:1767–1779.

Frias-Lopez J, Shi Y, Tyson GW, Coleman ML, Schuster SC, et al. 2008. Microbial community gene expression in ocean surface waters. *Proc. Natl. Acad. Sci. USA* **105**:3805–3810.

Gasol JM, Zweifel UL, Peters F, Fuhrman JA, Hagström A. 1999. Significance of size and nucleic acid content heterogeneity as measured by flow cytometry in natural planktonic bacteria. *Appl. Environ. Microbiol*. **65**:4475–4483.

Günther S, Hübschmann T, Rudolf M, Eschenhagen M, Röske I, et al.. 2008. Fixation procedures for flow cytometric analysis of environmental bacteria. *J. Microbiol. Methods* **75**:127–134.

Günther S, Trutnau M, Kleinsteuber S, Hause G, Bley T, et al. 2009. Dynamics of polyphosphate-accumulating bacteria in wastewater treatment plant microbial communities detected via DAPI (4′,6′-diamidino-2-phenylindole) and tetracycline labeling. *Appl. Environ. Microbiol*. **75**:2111–2121.

Hallett M, Schneider AS, Carbone E. 1972. Tetracycline fluorescence as calcium-probe for nerve membrane with some model studies using erythrocyte ghosts. *J. Membr. Biol*. **10**:31–44.

He Z, Gentry TJ, Schadt CW, Wu L, Liebich J, et al. 2007. GeoChip: A comprehensive microarray for investigating biogeochemical, ecological and environmental processes. *ISME J*. **1**:67–77.

Hewitt CJ, Nebe-von-Caron G.. 2004. The application of multiparameter flow cytometry to monitor individual microbial cell physiological state. *Adv. Biochem. Eng. Biotechnol*. **89**:197–223.

Jehmlich N, Hübschmann T, Gesell Salazar M, Völker U, Benndorf D, Müller S, von Bergen M, Schmidt F, 2010. Advanced tool for characterization of microbial cultures by combining cytomics and proteomics. *Applied Microbiology and Biotechnology* **88**(2):575–584.

Johnson DR, Czechowska K, Chèvre N, van der Meer Jr. 2009. Toxicity of triclosan, penconazole, and metalaxyl on *Caulobacter crescentus* and a freshwater microbial community as assessed by flow cytometry. *Environ. Microbiol*. **11**:1682–1691.

Joux F, Lebaron P. 2000. Use of fluorescent probes to assess physiological functions of bacteria at single-cell level. *Microbes Infect*. **2**:1523–1535.

Kleinsteuber S, Riis V, Fetzer I, Harms H, Müller S. 2006. Population dynamics within a microbial consortium during growth on diesel fuel in saline environments. *Appl. Environ. Microbiol*. **72**:3531–3542.

Kowalchuk GA, de Bruijn FJ, Head IM, Akkermans AD, van Elsas JD. 2004. *Molecular Microbial Ecology Manual*, 2nd ed. London: Kluwer Academic Publishing.

Longnecker K, Sherr BF, Sherr EB. 2005. Activity and phylogenetic diversity of bacterial cells with high and low nucleic acid content and electron transport system activity in an upwelling ecosystem. *Appl. Environ. Microbiol*. **71**:7737–7749.

Mou X, Moran MA, Stepanauskas R, Gonzalez JM, Hodson E. 2005. Flow-cytometric cell sorting and subsequent molecular analyses for culture-independent identification of bacterioplankton involved in dimethylsulfoniopropionate transformations. *Appl. Environ. Microbiol*. **71**:1405–1416.

Mou X, Hodson RE, Moran MA. 2007. Bacterioplankton assemblages transforming dissolved organic compounds in coastal seawater. *Environ. Microbiol*. **9**:2025–2037.

Müller S. 2007. Modes of cytometric bacterial DNA pattern: A tool for pursuing growth. *Cell Prolif*. **40**:621–639.

Müller S, Vogt C, Laube M, Harms H, Kleinsteuber S. 2009. Community dynamics within a bacterial consortium during growth on toluene under sulfate-reducing conditions. *FEMS Microbiol. Ecol*. **70**(3):556–596.

Müller S, Nebe-von-Caron G. 2010. Functional single-cell analyses—flow cytometry and cell sorting of microbial populations and communities. *FEMS Microbial Rev*. **34**:554–587.

Nebe-von-Caron G, Stephens PJ, Hewitt CJ, Powell JR, Badley RA.. 2000. Analysis of bacterial function by multi-colour fluorescence flow cytometry and single cell sorting. *J. Microbiol. Methods* **42**:97–114.

Parro V, Moreno-Paz M, Gonzalez-Toril E. 2007. Analysis of environmental transcriptomes by DNA microarrays. *Environ. Microbiol*. **9**:453–464.

Pawelczyk S, Abraham WR, Harms H, Müller S. 2008. Community-based degradation of 4-chlorosalicylate tracked on the single cell level. *J. Microbiol. Methods* **75**:117–126.

Schimenti KJ, Jacobberger JW. 1992. Fixation of mammalian cells for flow cytometric evaluation of DNA content and nuclear immunofluorescence. *Cytometry* **13**:48–59.

Shapiro HM. 2002. *Practical Flow Cytometry*. 4th ed. New York: Wiley-Liss.

Shi L, Günther S, Hübschmann T, Wick LY, Harms H, et al. 2007. Limits of propidium iodide as a cell viability indicator for environmental bacteria. Cytometry A 71:592–598.

Sträuber H, Hübschmann T, Jehmlich N, Schmidt F, von Bergen M, et al. 2010. NBDT (3-(*N*-(7-nitrobenz-2-oxa-1,3-diazol-4-yl)amino)-3-toluene)-a novel fluorescent dye for studying mechanisms of toluene uptake into vital bacteria. *Cytometry Part A* **77**(A):113–120.

Urich T, Lanzén A, Qi J, Huson DH, Schleper C, et al. 2008. Simultaneous assessment of soil microbial community structure and function through analysis of the meta-transcriptome. *PLoS One* **3**: e2527.

VerBerkmoes NC, Denef VJ, Hettich RL, Banfield JF. 2009. Systems biology: Functional analysis of natural microbial consortia using community proteomics. *Nat. Rev. Microbiol*. **7**:196–205.

Vogt C, Lösche A, Kleinsteuber S, Müller S. 2005. Population profiles of a binary bacterial culture grown with toluene under sulphate reducing conditions. *Cytometry A* **66**:91–102.

Wallner G, Amann R, Beisker W. 1993. Optimizing fluorescent in situ hybridization with rRNA-targeted oligonucleotide probes for flow cytometric identification of *microorganisms*. *Cytometry* **14**: 136–143.

Wiacek C, Müller S, Benndorf D. 2006. A cytometric approach reveals population heterogeneity of *Cupriavidus necator* to harmful phenol concentrations. *Proteomics* **6**:5983–5994.

Wilmes P, Bond PL. 2006. Metaproteomics: Studying functional gene expression in microbial ecosystems. *Trends Microbiol*. **14**:92–97.

Chapter 76

Capturing Microbial Populations for Environmental Genomics

MARTHA SCHATTENHOFER AND ANNELIE WENDEBERG

76.1 INTRODUCTION

Culture-independent methodologies such as environmental metagenomic studies enable the study of novel microbial processes and interspecies relationships in situ and have expanded our understanding of the phylogenetic and metabolic diversity on Earth. However, metagenomic studies are limited by the ability to interpret the data in the context of organism identity, which becomes more difficult with increasing community complexity. While direct sequencing of complex environmental samples using high-throughput sequencing methods (shotgun cloning or clone-free pyrosequencing) yields important overviews of the metabolic potential of the entire community, information regarding specific groups of microorganisms and interspecies associations is often lacking. Methods such as microfluidics [Marcy et al., 2007], optical trapping [Ashkin et al., 1987], immunomagnetic capture [Yeung and Wittrup, 2002], or flow cytometry [Sekar et al., 2004, Kalyuzhnaya et al., 2006] enable the selective extraction of individual microorganisms from native microbial communities prior to genome sequencing. The ability to identify and characterize the metabolic potential of in situ syntrophic partnerships presents a greater challenge and requires new methodological strategies that are compatible with metagenomic analysis and can be applied in complex environmental samples, such as anoxic sediments, where these interspecies associations are likely to occur. In response to this need, Pernthaler and Orphan [2007] developed a culture-independent method dubbed "magneto-FISH" that allows for the population targeted capture microorganisms from complex environmental samples by combining CARD FISH and immunomagnetic cell capture. The capture of phylogenetically "tagged" microorganisms and physically associated microorganisms increases the probability of identifying potentially hidden microbial associations in situ and yields a low-complexity sample for environmental metagenomics. In a proof-of-concept study, magneto-FISH was used for targeted metagenomic investigations of in situ syntrophic assemblages involved in the anaerobic oxidation of methane [Pernthaler et al., 2008].

76.2 METHODS

For a step-by-step instruction see **Table 76.1** and **Figure 76.1**. Please also refer to the original publication [Pernthaler et al., 2008].

76.2.1 Fluorescence In Situ Hybridization

Mix formaldehyde (FA)-fixed sample (sediment or water sample) with 1× PBS [137 mM NaCl, 2.7 mM KCl, 10 mM Na_2HPO_4, and 2 mM KH_2PO_4 (pH 7.6)] containing 0.01 M sodium pyrophosphate. Sediment sample is sonicated on ice to dislodge cells and cell aggregates from sediment particles. Subsequently spin sample, discard supernatant, and place pellet on ice. For permeabilization of the target cells (and removal of external minerals associated with the cell aggregates), pellet is resuspended in 1× TE [10 mM TrisHCl and 1 mM EDTA (pH 9.0)] and treated in a histological microwave oven (Microwave Research & Applications, Inc., laurel, Md., USA) at

Handbook of Molecular Microbial Ecology, Volume I: Metagenomics and Complementary Approaches, First Edition. Edited by Frans J. de Bruijn.

Table 76.1 Magneto-FISH Protocol for Benthic and Planktonic Samples

Fluorescence in situ Hybridization	Mix formaldehyde fixed sediment sample with 50 ml of 1 × PBS, add 0.01M pyrophosphate. Sonicate on ice with two pulses of 10 s each (not necessary for water samples).
	Centrifuge at 4°C (sediment for 5 min at 1000 × g, water sample for 10–15 min at 16,000 × g), discard supernatant.
	Resuspend sample in 2 ml of 1 × TE (pH 9.0), microwave for 8 min at 65°C with power output 100%. Permeabilization may depend on target organism and/or sample (see Pernthaler et al., [2004]).
	Place sample on ice to cool down for a few minutes.
	Spin at 4°C, discard supernatant.
	Resuspend in 2 ml 1 × PBS containing 0.1% H_2O_2, vortex, keep at RT for 1 min.
	Spin at 4°C, discard supernatant.
	Wash sample twice with 1× TE by spinning and resuspending.
	Resuspend pellet in 2 ml hybridization buffer, and 0.5 ng μl^{-1} of HRP probe, microwave in a water bath at 46°C for 40 min with power output 50%.
	Resuspend in 50 ml of 1 × PBS, spin, discard supernatant.
	Resuspend in 50 ml of 1 × PBS and keep for 10 min at RT.
	Spin at 4°C, discard supernatant.
Catalyzed Reporter Deposition	Resuspend in 2 ml of amplification buffer containing 0.0015% H_2O_2 and 0.5 μg of each fluorescein and biotin-labeled tyramide, incubate for 20 min at 46°C in the dark.
	Mix sample with 50 ml of 1 × PBS, keep at RT for 5 min in the dark.
	Spin, discard supernatant.
	Resuspend in 2 ml of 1 × PBS containing 0.5% bovine serum albumin (BSA) and microwave at 40°C for 20 min at a power output of 50%.
	Spin, discard supernatant.
	Resuspend in 2 ml 1 × PBS containing 0.01 M sodium pyrophosphate
	Sonicate sample on ice for 5 s
Immunomagnetic Cell Capture	Mix 1 ml of CARD-FISH stained sample with 200 μl of 10% BSA and 100 μl of monoclonal mouse anti-fluorescein-antibody-labeled pan-mouse paramagnetic beads, incubate at RT for at least 1 h in the dark on a rotor
	Fill funnels A and B 1 × PBS, close valves. Place connector on top of funnel B, fill with 1 × PBS, and place ring magnet onto connector.
	Cut tip off a 1-ml pipette tip and gently pipette bead-sample mix into funnel A, mix gently by inversion. Attach funnel A and let buffer, beads, and sample flow slowly through connector and funnel B. Replace funnel A with a clean and PBS-filled funnel and let buffer flow through once again. Be careful not to let the buffer level sink below the magnetic field until the very end of the procedure.
	Drain 1 × PBS completely and resuspend bead-cell complex in 0.5 ml 1 × TE

65°C for 8 min at a power output of 100%. Subsequently the slurry is placed on ice to cool down for a few minutes and centrifuged at 4°C. To inactivate endogenous peroxidases, pellet is resuspended in 1 × PBS containing 0.1% H_2O_2, vortexed, kept at room temperature (RT) for 1 min, and centrifuged. The pellet is washed twice with 1 × TE by resuspension and centrifugation and subsequently resuspended in hybridization buffer (for reference see Pernthaler et al., [2004]), 0.5 ng μl^{-1} of horseradish peroxidase (HRP)-labeled oligonucleotide probe and pipetted into a 2-ml reaction tube, placed into a water bath, and heated in the microwave oven at 46°C for 40 min at a power output of 50%. Unbound probe is removed by mixing the hybridized sediment with 1× PBS and incubating at RT for 10 min, followed by centrifugation at 4°C. As a negative control, sample needs to be hybridized with an HRP-labeled nonsense-probe, NON338 [Manz et al., 1992].

76.2.2 Catalyzed Reporter Deposition

The washed pellet is resuspended in amplification buffer containing 0.0015% H_2O_2 and 0.5 μg of each

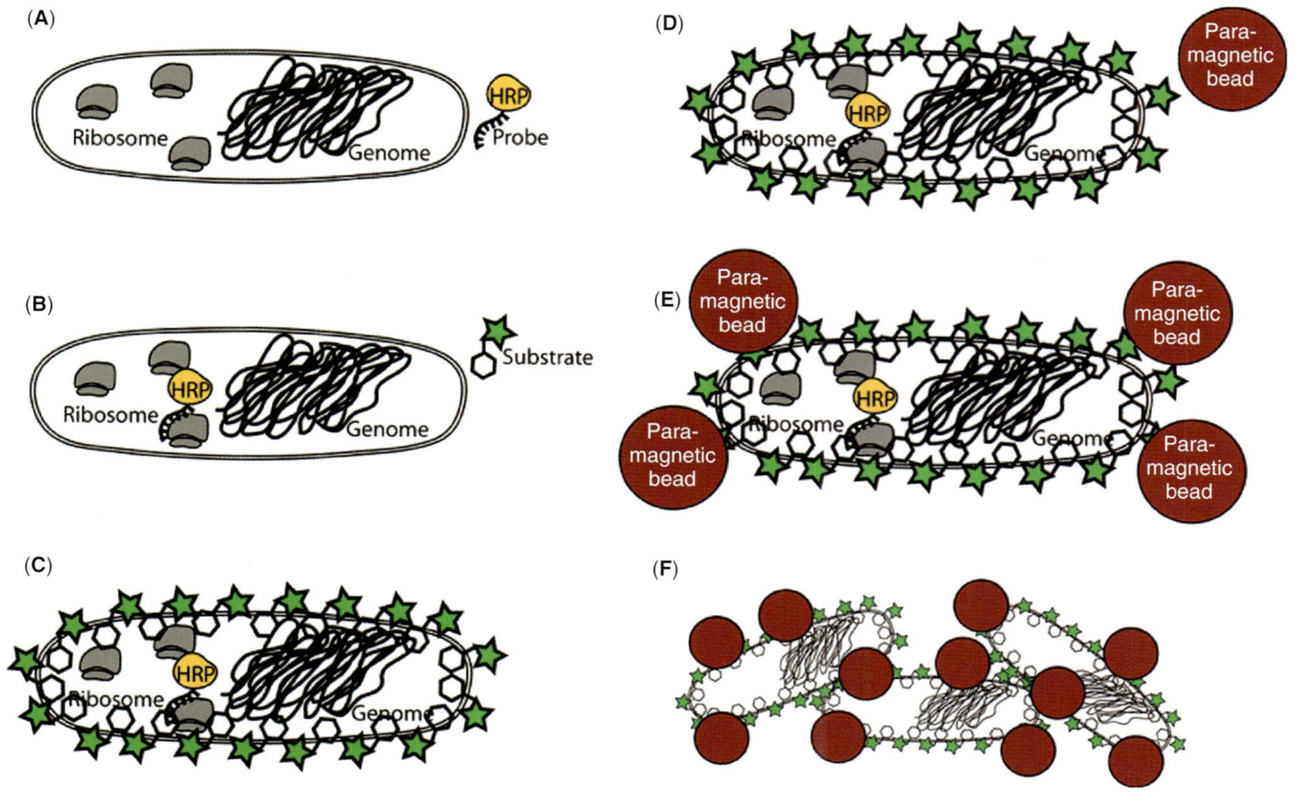

Figure 76.1 Cartoon of the steps of the magneto-FISH method. (**a**) Hybridization of target cells with a horseradish peroxidase (HRP)-labeled oligonucleotide probe (**b, c**) intracellular deposition of fluorescein-labeled tyramide by HRP activity. During the CARD-FISH procedure, fluorescein molecules bound to cellular proteins partially or fully protrude outside of the cell. (**d**) Addition of anti-fluorescein coated paramagnetic beads to hybridized sample suspension. (**e**) Binding of anti-fluorescein coated beads to fluorescein-labeled target cells. (**f**) Magnetic separation of labeled target cells bound to the paramagnetic beads from unlabeled microorganisms and other particles. (Modified from Pernthaler, et al., [2008], with permission of the publisher, copyright 2008, National Academy of Sciences, USA.

fluorescein and biotin-labeled tyramide (both custom labeled) according to Pernthaler et al. [2004] and incubated at 46°C for 20 min. The biotin-labeled tyramide serves to enhance the signal intensity resulting in a brighter and pH-independent Fluorescein-CARD-FISH signal [Pernthaler and Orphan, 2007]. The sample is then mixed with 1× PBS, centrifuged, resuspended in 1× PBS containing 0.5% Bovine Serum Albumin (BSA), and microwaved at 40°C for 20 min at a power output of 50% to remove unbound tyramide. Subsequently the sample is collected by centrifugation and resuspended in of 1× PBS containing 0.01 M sodium pyrophosphate. Sonicate sediment sample on ice for 5 s to dislodge cells and cell aggregates from sediment particles.

76.2.3 Cell Capture with Paramagnetic Beads

Add 200 μl of 10% BSA and 100 μl of monoclonal mouse anti-fluorescein-antibody (Molecular Probes)-labeled pan-mouse paramagnetic beads (5-μm diameter) (Dynal) to each 1 ml of CARD-FISH stained sample, and incubate mix at RT for at least 1 h on a rotator in the dark. To obtain a pure fraction of hybridized target cells free from particles and unlabeled cells, an apparatus consisting of two separation funnels with a connector is used (**Fig. 76.2**). Here, funnel B and the connector are filled with 1× PBS buffer and a strong Neodymium ring magnet placed around the connector. After incubation with paramagnetic beads, the slurry is gently pipetted into the PBS filled funnel A and gently mixed by inversion and placed on top of connector and funnel B. The connector between funnel A an B holds the magnet and provides easy access to the captured cell after magneto-FISH. Here, the cells captured by the paramagnetic beads are concentrated at the top of the connector with the ring magnet, whereas nonmagnetic residue consisting of nonhybridized cells or sediment is separated by gravity while the PBS-sample-bead mix slowly drains from funnel A. To gently wash away residual particulates adhering to the paramagnetic beads, a new and clean funnel A, filled with 1×PBS, is seated on top of the connector. The captured cells in the

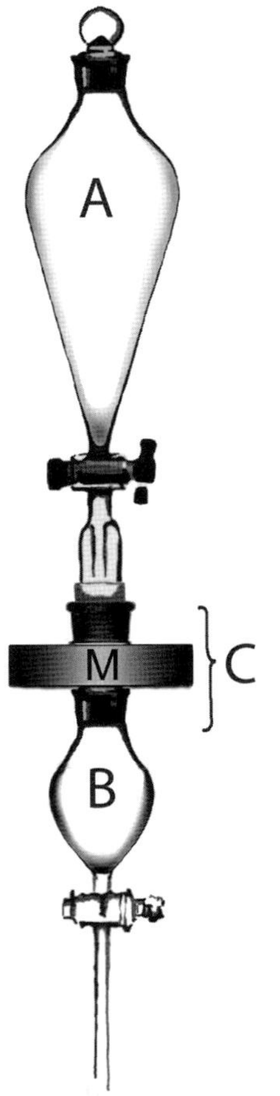

Figure 76.2 Setup for paramagnetic bead capture. Funnel A contains 1× PBS; the ring magnet (M) is seated around a small connector (C) between funnel A and B, to concentrate paramagnetic beads and captured cells. Funnel B functions as a reservoir for PBS; it maintains the liquid level above the bead-captured cells during the washing procedure, largely preventing loss of target cells. (Modified from Pernthaler et al. [2008], with permission of the publisher, copyright 2008, National Academy of Sciences, USA).

connector are always immersed in PBS, preventing disturbance and loss of the target cells, observed when the bead-cell complex had air contact. To maximize yield, noncaptured cells and sediment particles that washed out during this procedure can again be concentrated by centrifugation and incubated with fresh paramagnetic beads and captured as described above. In the last step, funnel A is disconnected from funnel B and the PBS is drained completely. To remove nonhybridized cells and particles adhered to the glass surface, the lower half of the connector has to be rinsed repeatedly with PBS while the magnet is still applied. To recover the paramagnetic beads and captured cells, the magnet is removed, the connector is tilted, and the captured cells are resuspended in a small volume of 1 × TE (0.5 ml), and pipette into a small reaction tube. In a second wash step with 0.5 ml of 1 × TE, the remaining cells in connector are recovered and pooled with the first.

76.2.4 DNA Extraction

The captured cells are incubated in a lysis solution [1 ml of Tris/EDTA buffer (pH 9.0) containing 1% SDS and 2 mg ml^{-1} of proteinase K] at 52°C for 2 h. The DNA can then be extracted with phenol–chloroform–isoamylalcohol, precipitated, and resuspended in 10 μl of MilliQ water.

76.2.5 DNA Amplification with Multiple Displacement Amplification (MDA)

Captured DNA template (1–2 μl) is subjected to multiple displacement amplification (MDA), using a commercial kit (GenomiPhi; GE Healthcare), following the manufacturer's instructions. The amplified DNA is then precipitated by using 0.5 M ammonium acetate and 95% ethanol for 3 h at −80°C, pelleted by centrifugation at 16,000 × g for 20 min at 4°C and rinsed twice with ice-cold absolute ethanol and once with cold 70% ethanol. The pellet is then resuspended in 20 μl of MilliQ water and stored in aliquots at −20°C.

76.3 RESULTS AND DISCUSSION

76.3.1 Validating the Specificity of the Magneto-FISH Capture

Magneto-FISH combines the widely used 16S rRNA-targeted fluorescence in situ hybridization and tyramide signal amplification (CARD-FISH) with the immunomagnetic capture of hybridized cells, using paramagnetic beads coated with an antibody targeting the fluorochrome applied in the CARD-FISH procedure (Fig. 76.1). Thus, the Magneto-FISH method enables both the selective extraction of select microorganisms and their physically associated partners directly from environmental samples and simplifies genomics by targeting the microorganisms of interest.

In the original publication by Pernthaler et al. [2008], the focus of the investigation was on a specific group of methane-oxidizing archaea known as the ANME-2c, using magneto-FISH to selectively capture intact ANME-2c consortia directly from deep-sea methane seep sediments from the Eel River Basin (ERB). The purity of the magneto-FISH captured ANME-2c consortia was evaluated by epifluorescence microscopy and DNA extraction with subsequent multiple displacement amplification of the genomic DNA and PCR-based screening of ribosomal and metabolic genes. Both examinations confirmed that all captured microorganisms were associated with fluorescently labeled ANME-2c aggregates (**Fig. 76.3**), representing 92% of the total archaeal 16S rRNA sequences recovered from the magneto-FISH captured fraction (as compared to 26% in the original sediment) [Pernthaler et al., 2008]. The capture of physically associated cells is a unique feature of the magneto-FISH technique. In the original publication [Pernthaler et al., 2008] we discovered so far unknown associations between the methane-oxidizing ANME-2c and at least three distinct bacterial lineages.

This technique has been developed and optimized for sediment samples and large cell aggregates, and each type of sample and target organism needs optimization of the magneto-FISH procedure. We are now optimizing magneto-FISH to capture single cells from a hydrocarbon contaminated aquifer in Leuna, Germany. In a zone where contaminant degradation coupled to sulfate reduction and methanogenesis occurs, we were particularly interested in members of the domains *Archaea* and of the class *Deltaproteobacteria*. We selectively extracted single cells of these organisms from groundwater samples with probes ARCH_915 [Stahl and Amann, 1991] and $DELTA_{495}a$ [Loy et al., 2002] (Fig. 76.3). While *Deltaproteobacteria* were enriched from 21% in the original groundwater sample to 94.5% in the captured fraction, *Archaea* where only enriched from 1.1% to 67.3%, respectively. During optimization of magneto-FISH, one has to consider whether abundance or purity of target cells is more important. Overstaining during the CARD-FISH procedure may result in higher cell numbers but also in a lower quality of the captured cell fraction. Repeated washing during the magnetic capture of the bead-cell complexes may result in target cell loss, but will also lead to a higher purity.

76.3.2 Summary

Magneto-FISH increases the probability of identifying potentially hidden microbial associations and simplifying environmental genomics by specifically targeting the microorganism(s) of interest. This method allows for the recovery of low-diversity metagenomic sequences that

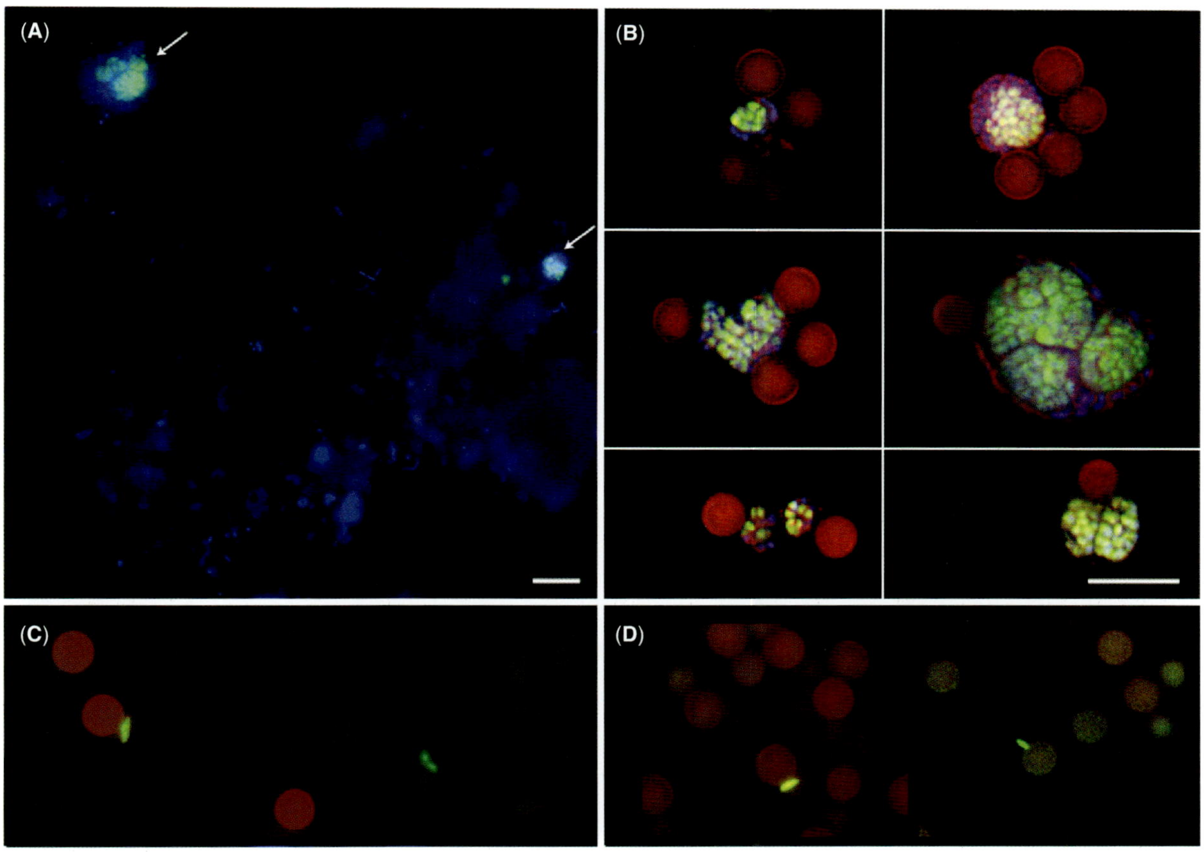

Figure 76.3 Magneto-FISH capture of aggregates (**a, b**), and single cells (**c, d**). Marine deep sea sediment from the Eel River Basin, hybridized with probe ANME-2c_760 (green) and a general DNA stain (DAPI, blue), before (**a**) and after (**b**) capture with paramagnetic beads (dark red, collage of six images, scale bars 10 μm). (Reproduced with permission of the publisher [Pernthaler et al., 2008], copyright 2008, National Academy of Sciences, USA) (**c, d**) Groundwater sample from a hydrocarbon contaminated aquifer, hybridized with (**c**) probe DELTA_495a (green) and (**d**) probe ARCH_915 (green) after capture with paramagnetic beads (collage of two images each).

can be attributed directly to the target microorganism and their physically associated partners within complex environments, as well as to free-living, single-celled organisms.

REFERENCES

Ashkin A, Dziedzic JM, Yamane T. 1987. Optical trapping and manipulation of single cells using infrared laser beams. *Nature* **330**:769–771.

Kalyuzhnaya MG, Zabinski R, Bowerman S, Baker DR, Lidstrom ME, Chistoserdova L. 2006. Fluorescence in situ hybridization-flow cytometry-cell sorting based method for separation and enrichment of type I and type II methanotroph populations. *Appl. Environ. Microbiol*. **72**(6):4293–4301.

Loy A, Lehner A, Lee N, Adamczyk J, Meier H, Ernst J, Schleifer K-H, Wagner M. 2002. Oligonucleotide Microarray for 16S rRNA Gene-Based Detection of All Recognized Lineages of Sulfate-Reducing Prokaryotes in the Environment. *Appl. Environ. Microbiol*. **68**(10):5064–5081.

Manz W, Amann R, Ludwig W, Wagner M, Schleifer K-H. 1992. Phylogenetic oligodeoxynucleotide probes for the major subclasses of *Proteobacteria*: Problems and solutions. *Syst. Appl. Microbiol*. **15**(4):593–600.

Marcy Y, Ouverney C, Bik EM, Lösekann T, Ivanova N, et al. 2007. Dissecting biological "dark matter" with single-cell genetic analysis of rare and uncultivated TM7 microbes from the human mouth. *Proc. Natl. Acad. Sci. USA* **104**(29): 11889–11894.

Pernthaler A, Orphan VJ. 2007. Process for seperating microorganisms. California Institute of Technology, United States of America, Patent 11/746,374.

Pernthaler A, Pernthaler J, Amann R. 2004. Sensitive multi-color fluorescence in situ hybridization for the identification of environmental microorganisms. In Kowalchuk G, de Bruijn FJ, Head IM, Akkermans ADL, van Elsas JD, eds. *Molecular Microbial Ecology Manual*. Dordrecht: Kluwer Academic Press, pp. 711–726.

Pernthaler A, Dekas AE, Brown CT, Goffredi SK, Embaye T, Orphan VJ. 2008. Diverse syntrophic partnerships from deep-sea methane vents revealed by direct cell capture and metagenomics. *Proc Natl Acad Sci USA* **105**(19):7052–7057.

Sekar R, Fuchs BM, Amann RI, Pernthaler J. 2004. Flow sorting of marine bacterioplankton after fluorescence in situ hybridization. *Appl. Environ. Microbiol*. **70**:6210–6219.

Stahl DA, Amann RI. 1991. Development and application of nucleic acid probes. In Stackebrandt E, Goodfellow M, eds. *Nucleic Acid Techniques in Bacterial Systematics*. Chichester, England: John Wiley & Sons., pp. 205–248.

Yeung YA, Wittrup KD. 2002. Quantitative screening of yeast surface-displayed polypeptide libraries by magnetic bead capture. *Biotechnol. Prog*. **18**:212–220.

Chapter 77

Microscopic Single-Cell Isolation and Multiple Displacement Amplification of Genomes from Uncultured Prokaryotes

PETER WESTERMANN AND THOMAS KVIST

77.1 INTRODUCTION

Direct metagenomic analysis of whole or enriched microbial populations can gain important knowledge of the functions and capabilities of complex microbial communities as pointed out in many of the preceding chapters. Despite the fact that availability of high-throughput sequencing has replaced much of the laborious establishment and sequencing of huge clone libraries, a reliable assembly of whole genomes derived from rich environments such as soils and sediments is almost impossible. Since many of these environments contain up to millions of different genomes [Gans et al., 2005; Schloss and Handelsman, 2005], a reduction in community complexity is necessary to manage the sequence information [Delong, 2005]. In projects where specific processes or specific taxonomic groups are investigated, it is even obvious to eliminate other members of the community by cell sorting such as FACS (fluorescent activated cell sorting) [Davey and Kell, 1996], dielectrophoresis [Hughes and Hoettges, 2008], elutriation [Macdonald, 1986], or similar methods. Another obvious approach is to pick the needle in the haystack by micromanipulation techniques. The general principle of these techniques is the manipulation of a single cell from a mixture of cells to a growth medium or a suitable buffer by means of mechanical, electromagnetic, or fluid forces. The two major principles exploited so far are laser forceps and mechanical micromanipulation. In the first technique a focused laser beam is used to capture and move a single cell from a population at one end of a microcapillary to the other end, which is then cut off and transferred to a tube containing culture medium [Mitchell et al., 1993; Huber, 1999]. The cell might subsequently be cultivated or genetically analyzed. The second principle is based upon a micro-injector developed for microscopic in vitro fertilization in combination with the precision of a servo-powered micromanipulator mounted on an inverted microscope [Frölich and König, 1999; Ishoy et al., 2006; Kvist et al., 2007]. Recently, single-cell isolation by means of microfluidics on a chip has been performed with success [Marcy et al., 2007]. Compared to FACS and other cell-sorting techniques, micromanipulation is laborious and the cell yield is much lower, but this might be outweighed by the value of microscopic observation during the process, which ensures that only single cells and only cells having the right morphology are isolated.

Although all prokaryotic cells in a sample can be isolated by micromanipulation, the technique does not solve the problem of uncultivable cells, although some authors have had successful cultivation of previously uncultivated cells after isolation [Mitchell et al., 1993]. If the isolated cell, however, can be lysed, multiple displacement amplification (MDA) can be used to obtain genomic information about the cell revealing the phylogenetic affiliation of the cell and also functional properties derived from

Handbook of Molecular Microbial Ecology, Volume I: Metagenomics and Complementary Approaches, First Edition. Edited by Frans J. de Bruijn.

sequencing the MDA product [Kvist et al., 2007; Hongoh et al., 2008].

In this chapter we demonstrate that a combination of conventional microbial ecology techniques with single-cell manipulation and multiple displacement amplification can give access to any prokaryote that can be detected in a 16S rRNA gene library. The isolation and subsequent cultivation of specific cells and thermophilic species was demonstrated using fluorescent *Escherichia coli* and *Pseudomonas putida* strains and sediments from a solfataric field [Ishoy et al., 2006]. The application of multiple displacement amplification on single cells was exerted on cells isolated from soil [Lasken et al., 2005]. As a proof of the whole concept, we detected, probed, microscopically isolated, and amplified the genome of a single soil archaeon belonging to the monophyletic clade C1b of the nonthermophilic *Crenarchaeota* [Kvist et al., 2007].

77.2 MATERIALS AND METHODS

In the following, a brief description of the setup and methods is given. For detailed descriptions, please refer to the original publications [Lasken et al., 2005; Ishoy et al., 2006; Kvist et al., 2007].

77.2.1 Soil Sample, Clone Library, and Sequence Analysis

Agricultural soil samples from a conventional farm were sieved and added to an outglowed mortar together with 50 ml liquid N_2. Immediately after the N_2 had evaporated, cells in the soil were carefully disrupted by grinding with a pistil. DNA was extracted from 1 g of the ground material after proteinase K addition and protein separation in chloroform:isoamylalcohol and finally precipitated in ethanol and dried overnight. The pellet was resuspended in sterile distilled water and purified using the Wizard genomic DNA clean-up system (Promega). Archaea-specific primers A21F [Delong, 1992] and Arch915r [Stahl and Amann, 1991] were used to amplify ≈900 bp of the 16S rRNA genes. The products were purified, and a clone library was established using the TOPO TA cloning kit for sequencing (Invitrogen). Plasmids were purified from positive transformants, and inserts were sequenced at MWG Biotech (Germany). Sequences were analyzed for chimeras and matched to related sequences using "Chimera check" and "Sequence match" in the Ribosomal Database Project II [Maidak et al., 1999]. Sequences were aligned with the selected sequences in Clustal W within the Bio Edit Sequence alignment editor [Hall, 1999], and a phylogenetic tree was constructed using TREECON for Windows [van de Peer and De Wachter, 1997]. The tree was validated using the Ribosomal Database Project Phylip interface and FastDNA mlk within the Bio Edit platform,

77.2.2 Cell extraction and FISH Probing

To obtain an enriched cell suspension with fewer soil particles, a density gradient centrifugation was carried out using the Nycodenz extraction protocol described by Burmølle et al., [2003]. The cell fraction was then fixed in ethanol/PBS and mounted on a cover slide. The FISH procedure was carried out as described by Amann et al. [1996] using two Archaea-specific probes (Arch 915 [Stahl and Amann, 1991], SoilCren 750 [Kvist et al., 2005]) both attaching to 16S rRNA to maximize the signal. Cells were counterstained using DAPI.

77.2.3 Microscope and Micromanipulation

The equipment used for isolation of single cells (**Fig. 77.1**) consists of a Leica DMIRBE inverted microscope equipped with a mercury lamp for fluorescent microscopy. A servo-powered micromanipulator (Multicontrol 2000, ITK Lohnau, Germany) was mounted on the cross table, and a microinjector (Eppendorf Transjector 5246) was connected to a custom made capillary, which had been heated and pulled to an inner diameter of 2–10 μm [Lasken et al., 2005]. The components were connected to a computer system and operated by a joystick to avoid vibrations and increase control. A Kappa DX20H camera was mounted on the microscope to ensure that all operations could be carried out without touching the microscope.

For the isolation of viable cells, cell suspensions or environmental samples were diluted to achieve an appropriate cell density and loaded in a 20-mm-wide hole drilled in a 2-mm-thick microscope slide. The bottom of the chamber was covered by a coverslip tightened to the edge of the hole with petroleum jelly. For genomic analysis, FISH-probed cells were fixed to a coverslip, which was then mounted with the cells facing the chamber of the microscope slide. Prior to isolation of a single cell, adjacent cells were released from the cover slide by use of the micromanipulator and sucked into a capillary. A sterile microcapillary was then mounted, and the desired cell was removed and transferred to 13.5 μl of sterile TE buffer (pH8) in PCR tubes.

77.2.4 Cell Lysis and Genomic Amplification

Isolated cells suspended in 13.5 μl of TE buffer were lysed by five serial freeze/thaw cycles at −80°C and room

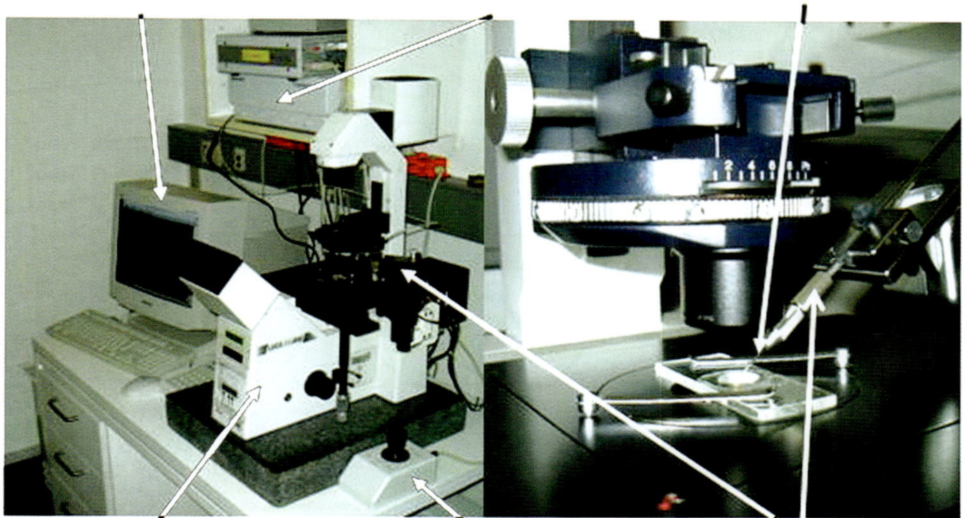

Figure 77.1 The microscopic micromanipulation and isolation setup. The cover slide with probed cells facing upward is mounted underneath the thick microscope slide. This allows observation from beneath and isolation of single cells by the microcapillary mounted on the micromanipulator from above.

temperature. Whole genome amplification from lysed cells was carried out using the Phi 29 DNA polymerase containing Repli-g kit from Quiagen as described by Kvist et al. [2007]. For subsequent PCR of specific genes, 0.2 μl from the genomic amplification was used as template. In the described experiment, partial 16S rRNA genes from isolated archaeal cells were amplified using the primers described above.

77.3 RESULTS

The concept is outlined in **Figure 77.2** and has the following components:

1. DNA extraction from an environmental sample and establishment of a 16S rRNA gene clone library and a phylogenetic tree of a specific taxonomic group addressed by specific primers.
2. Extraction of cells from the same environment by Nycodenz gradient centrifugation and labeling of specific cells with fluorescent probes constructed from sequence information attained from the clone library.
3. Microscopic isolation of labeled cells by micromanipulation and whole genome amplification by MDA.
4. Amplification and sequencing of the 16S rRNA gene(s) from the MDA product to confirm the identity of the isolated cell.
5. Shotgun-cloning and sequencing or direct amplification/sequencing and analysis of the MDA product.

The full procedure described above was used to investigate uncultured soil crenarchaeota. Using the Archaea-specific primers, a clone library of 100 16S rRNA gene sequences was established and grouped into 14 distinct patterns by restriction analysis. Between 1 and 3 clones from each group was sequenced and aligned with related sequences. In **Figure 77.3**, a phylogenetic analysis showed that the clones all belonged to the C1b clade of the non-thermophilic Crenarchaeota and fell within the cluster of crenarchaeotal soil clones. Except for one, all clones were found to match the two FISH probes (Arch 915 and Soil-Cren 750). Two probes were used to ensure a strong signal during the microscopic isolation and to minimize the number of false positive cells (see below). Probing of a soil sample from the same environment yielded clear signals from irregular cocci growing either as filaments or as single cells (Fig. 77.3). Eight distinct fluorescent cells were isolated by micromanipulation and subjected to MDA. Subsequent PCR testing of the MDA product with Archaea-specific primers yielded two positive cells, both of which fell nicely within the agricultural crenarchaotal soil cluster (Fig. 77.3).

From a shotgun library derived from one of the cells, 97 clones were sequenced yielding a 130-kbp genomic sequence. Based upon G+C content and alignments with Genbank archaeal sequences, 21 kbp containing 20 open

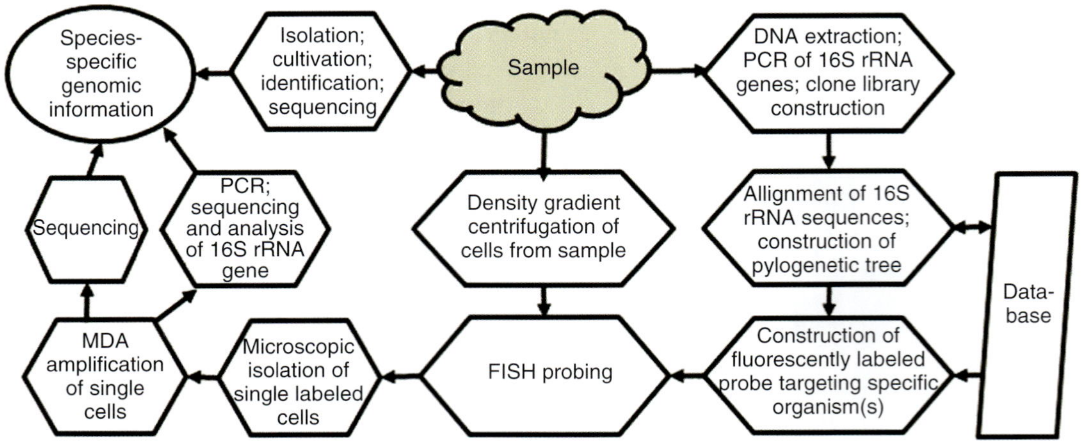

Figure 77.2 Schematic diagram of the single-cell isolation and amplification process. Based upon a clone library of 16SrRNA genes from the environment, FISH probe(s) are constructed and used for labeling of cells of interest. The labeled cells are isolated one by one using the micromanipulation system, and the genome is amplified using MDA. The identity of the cell is confirmed by PCR of the 16S rRNA gene followed by sequencing of the product.

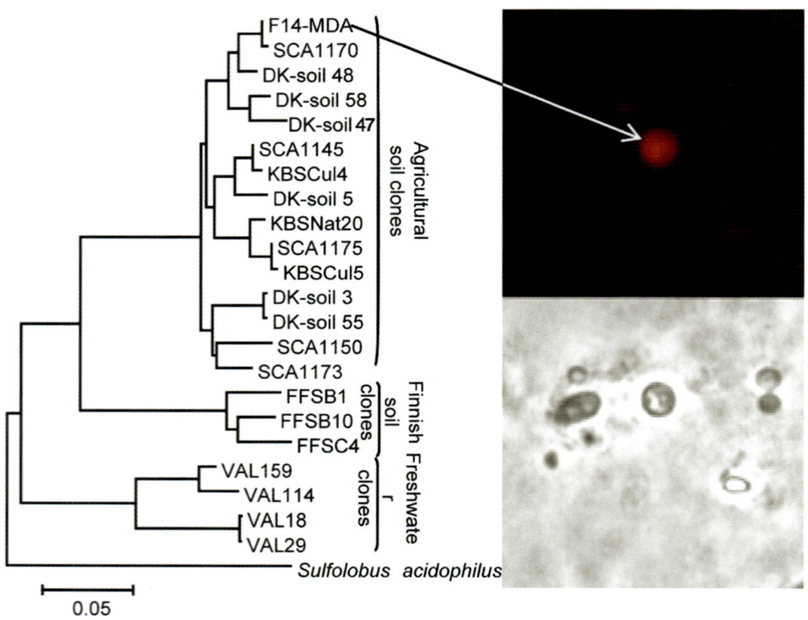

Figure 77.3 Phylogenetic tree of the cloned 16S rRNA genes and their closest relatives from the soil sample. The error bar indicates 5 substitutions per 100 nucleotides. The sequence F14-MDA is derived from the FISH-probed fluorescent cell. The same cell is viewed as a bright-field image below the epifluorescent microscopic image.

reading frames were affiliated with the isolated cell while the remaining sequences had a more uncertain affiliation [Kvist et al., 2007].

In a similar experiment, cells were only labeled with one probe (Arch 915) and one rod-shaped cell was picked, lysed, and exposed to MDA. Analysis of the 16S rRNA gene amplified from the product revealed surprisingly that this cell was most closely related to *Bacillus alcalophilus* [Lasken et al., 2005]. The Arch 915 probe also matches some gram-positive bacteria [Stahl and Amann, 1991] and almost completely matched the 16S rRNA gene sequence of the isolated cell.

77.4 DISCUSSION

The establishment of axenic cultures derived from a single cell isolated from an environmental sample by differential streaking on solid substrate surfaces has been a fundamental requirement for physiological studies of microorganisms for more than a century. The single cell isolation procedure outlined in this chapter follows this principle and has been demonstrated to be very efficient for isolation and subsequent cultivation of mesophilic, thermophilic, and hyperthermophilic bacteria [Ishoy et al., 2006]. In the genomic amplification, process, the genome of a single cell is propagated by multiple displacement amplification leaving the opportunity to identify the cell by 16S rRNA gene sequencing and to suggest physiological capabilities from genome analysis paralleling the propagation of a single cell to an axenic colony. The FISH probing and microscopic isolation yields the morphological characteristics of the cell, leaving specific growth-related knowledge as the major information deficiency compared to classical isolation and analysis of microorganisms. More than 90% of the prokaryotic biodiversity found in soil has been considered uncultivable [Amann et al., 1995; Torsvik et al., 1990]. The cultivation-free approach implied by the single cell isolation and genomic amplification technique yields the possibility to gain important genetic insight into specific uncultivated prokaryotes.

There are two major problems in single-cell genomic amplification: (1) production of unspecific high-molecular-weight DNA from contaminating DNA and primer dimers and (2) incomplete coverage and uneven representation of the amplified genome.

Contaminating DNA from lysed cells, enzyme preparations, and so on, will be amplified to the same extent as the target DNA due to the lack of specificity of the multiple displacement amplification process. Therefore, it is important to exert the same care as exercised when setting up PCR reactions with low amounts of template. This implies a clean working environment during the microscopic isolation and the use of DNA-free instruments and containers. All solutions should be prepared in PCR-grade water. Setting up control reactions using specific PCR primers is necessary to exclude false-positive reactions containing high-molecular-weight DNA produced in the multiple displacement reaction from primer dimers. Knowledge on the identity of enzyme producing strains used in commercial molecular kits used to open cells, purify DNA, and amplify the genome can be helpful to be able to deduct false sequences amplified from DNA impurities in the multiple displacement amplification.

Incomplete coverage and uneven representation is a problem when only a single genome of an unknown organism is amplified by multiple displacement amplification. During the cell lysis and handling of the genome, it is sheared into an unknown number of fragments. The resulting gaps will terminate the amplification process at these loci. It was previously shown by Raghunathan et al., [2005] that a specific locus only was represented in 74% of 66 parallel multiple displacement reactions using single *E. coli* cells as template. If identical cells can be obtained, then parallel amplification followed by 16S rRNA gene similarity verification and pooling of DNA from identical cells before sequencing can circumvent this problem.

The incubation volume of 50 μl is huge compared to the volume of a single prokaryotic cell ($\sim 10^{16}$ times) for the replication of a single genome, and an increased specificity has been achieved by reducing the volume of the reaction to 0.6 μl [Hutchison et al., 2005]. To improve this specificity further, a single-cell isolation and genome amplification microfluidic chip was constructed by Marcy et al. [2007]. On this arrangement a single cell could be isolated and lysed, and a multiple displacement reaction was performed in a 60-nl reaction chamber, which led to a further reduction in the amplification bias. Even smaller amplification reactors of 450 pl have been constructed [Marcus et al., 2006] but have to our knowledge not been tested in whole-genome amplification experiments.

The rapid and ongoing development of sequencing technologies achieved by next-generation sequencing machines has led to a dramatic increase in throughput at a much lower price compared to classical Sanger techniques (see Chapter 18, Vol. I). A major difference between the Sanger technique used in a capillary sequencer and the new technologies is the much shorter reads achieved [Pop and Salzberg, 2008], which complicates a reliable assembly of full microbial genomes. In metagenomics studies where the purpose is to study the diversity of specific genes or/and metabolic pathways or to find genes that code for functions of interest, this problem is manageable. However, the high diversity of closely related strains of microbes in soil such as seen for the crenarchaeotes in Fig. 77.3 will probably render a full genomic assembly impossible without previous isolation. The single-cell genomic approach also leaves the possibility to associate highly variable sequences with the individual strains, which would have been impossible in a mixture of genomes. Finally, metagenomics combined with single-cell genomics can elucidate the boundaries between functionally related species with high genomic similarity.

Acknowledgments

Roger Lasken and Thomas Ishoey are acknowledged for the collaboration leading to the development of the described methods. Parts of the described study was financed by the The Danish Natural Science Council and performed within the Danish Archaea Centre.

REFERENCES

Amann R, Ludwig W, Schleifer KH. 1995. Phylogenetic identification and in situ detection of individual microbial cells without cultivation. *Microbiol. Rev*. **59**:143–169.

Amann R, Snaidr J, Wagner M, Ludwig W, Schleifer KH. 1996. In situ visualization of high genetic diversity in a natural microbial community. *J. Bacteriol*. **178**:3496–3500.

Burmolle M, Hansen LH, Oregaard G, Sorensen SJ. 2003. Presence of N-acyl homoserine lactones in soil detected by a whole-cell biosensor and flow cytometry. *Microb. Ecol*. **45**:226–236.

Davey HM, Kell DB. 1996. Flow cytometry and cell sorting of heterogeneous microbial populations: The importance of single-cell analyses. *Microbiol. Rev*. **60**:641–696.

DeLong EF. 1992. Archaea in coastal marine environments. *Proc. Natl. Acad. Sci. USA* **89**:5685–5689.

DeLong EF. 2005. Microbial community genomics in the ocean. *Nature Rev. Microbiol.* **3**:459–469.

Frölich J, König H. 1999. Rapid isolation of single microbial cells from mixed culture and laboratory populations with the aid of a micromanipulator. *Syst. Appl. Microbiol*. **22**:249–257.

Gans J, Wolinsky M, Dunbar J. 2005. Computational improvements reveal great bacterial diversity and high metal toxicity in soil. *Science* **309**:1387–1390.

Hall TA. 1999. BioEdit: A user-friendly biological sequence alignment editor and analysis program for Windows 95/98/NT. *Nucleic Acids. Symp. Ser*. **41**:95–98.

Hongoh Y, Sharma VK, Prakash T, Noda S, Taylor TD, et al. 2008. Complete genome of the uncultured Termite Group 1 bacteria in a single host protist cell. *Proc. Natl. Acad. Sci. USA* **105**:5555–5560.

Huber R. 1999. Die Laserpinzette als Basis für Einzelzellkultivierungen. *BIOspektrum* **5**:289–291.

Hughes MP, Hoettges KH. 2008. Bacterial concentration, separation and analysis by dielectrophoresis. In Zourob M, Elwary S, Turner A, eds. *Principles of Bacterial Detection: Biosensors, Recognition Receptors and Microsystems*. Springer, New York pp. 895–907.

Hutchison CA 3rd, Smith HO, Pfannkoch C, Venter JC. 2005 Cell-free cloning using Φ 29 DNA polymerase. *Proc. Natl. Acad. Sci. USA* **102**:17332–17336.

Ishoy T, Kvist T, Westermann P, Ahring, BK. 2006. An improved method for single cell isolation of prokaryotes from meso-, thermo- and hyperthermophilic environments using micromanipulation. *Appl. Microbiol. Biotechnol*. **69**:510–514.

Kvist T, Mengewein A, Manzei S, Ahring BK, Westermann P. 2005. Diversity of thermophilic and nonthermophilic crenarchaeota at 80 degrees C. *FEMS Microbiol. Lett*. **244**:61–68.

Kvist T, Ahring BK, Lasken R, Westermann P. 2007. Specific single-cell isolation and genomic amplification of uncultured microorganisms. *Appl. Microbiol. Biotechnol*. **74**:926–935.

Lasken R, Raghunathan A, Kvist T, Ishoy T, Westermann P, et al. 2005. Multiple displacement amplification from single bacterial cells. In Hughes S, Lasken RS, eds. *Whole Genome Amplification*. Oxford, UK: Methods Express Scion Publishing, pp. 119–147.

Macdonald RM. 1986. Sampling soil microfloras: Dispersion of soil by ion exchange and extraction of specific microorganisms from suspension by elutriation. *Soil Biol. Biochem*. **18**:394–406.

Maidak BL, Cole JR, Parker CT, Jr, Garrity GM, Larsen N, Li B, Lilburn TG, McCaughey MJ, Olsen GJ, Overbeek R, Pramanik S, Schmidt TM, Tiedje JM, Woese CR, 1999. A new version of the RDP (Ribosomal Database Project). *Nucleic. Acids. Res.* **27**:171–173.

Marcus JS, Anderson WF, Quake Sr. 2006. Parallel picoliter RT-PCR assays using microfluidics. *Anal. Chem*. **78**:956–958.

Marcy Y, Ishoey T, Lasken RS, Stockwell TB, Walenz BP, et al. 2007. Nanoliter reactors improve multiple displacement amplification of genomes from single cells. *PLOS Genet*. **3**:1702–1708.

Mitchell JG, Weller R, Beconi M, Sell J, Holland J. 1993. A practical optical trap for manipulating and isolating bacteria from complex microbial communities. *Microb. Ecol*. **25**:113–119.

Pop M, Salzberg L. 2008.

Raghunathan A, Ferguson HR, Jr. Bornarth CJ, Song W, Driscoll M, Lasken RS. 2005. Genomic DNA amplification from a single bacterium. *Appl. Environ. Microbiol*. **71**:3342–3347.

Schloss PD, Handelsman J. 2005. Metagenomics for studying unculturable microorganisms: cutting the Gordian knot. *Genome Biol*. **6**:229–233.

Stahl DA, Amann R. 1991. Development and application of nucleic acid probes In Stackebrandt E, Goodfellow M. eds. *Nucleic Acid Techniques in Bacterial Systematic*. Chichester, UK: John Wiley & Sons, pp. 205–248.

Torsvik V, Goksoyr J, Daae FL. 1990. High diversity in DNA of soil bacteria. *Appl. Environ. Microbiol*. **56**:782–787.

Torsvik V. 1995. Cell extraction method. In Akkermans ADL, van Elsas JD, De Bruijn FJ, eds. *Molecular Microbial Ecology Manual*. Dordrecht: Kluwer Academic Publishers, Netherlands, pp. 1–15.

van de Peer Y, De Wachter R. 1997. Construction of evolutionary distance trees with TREECON for Windows: Accounting for variation in nucleotide substitution rate among sites. *Comput. Appl. Biosci*. **13**:227–230.

Index

^{12}C-DNA, 492, *493*
^{13}C-DNA, 492, *493*, 494
^{13}C isotope ion, 711
βAMO primers, 42, *43*, 46
β-proteobacteria species, 271, 283, 510, 545
γ-proteobacteria, 283, 545
2D SDS-PAGE, 637, 638, 639, 640, 641, 642, 643
3DMET, 713
4–6-diamidino-2-phenylindole (DAPI), 728
5S rRNA gene, 30, 31
16S PCR primer empirical testing, 135
in silico versus in vitro validation and, 138
nonspecific amplification effect on real-time qPCR and clone libraries and, 138–139
primer design and in silico testing and, 135–136
primer optimization and in vitro testing and, 136–137
primer specificity confirmation in complex environment and, 138
real-time PCR assays and, 137
16S rDNA sequences, 163–167, *200*, 201
16S rRNA, 41, 42, 60, *64*, 72, 77, 95–96, 102, 116–117, 118, 120, 123, 130, 140, 143, 144, 145–146, 163, 205, 206, 209, 228, 229, 235, 238, 266, 268, *270*, 311, 316, 334, 336, 353, 370, 376, 409, 437, 461, 462, 466, 493, 494, 497, 502, 512, 524, *527*, 528, *529*, 543, 545, 546, 552, 553, 557, *561*, 580
analytical tools, 124–125
case studies, 125–126
challenges, 126
clones, *124*
community profiling and, 589
data generation, 123–124
gene libraries, 576
gene sequence data analysis, 131–132
materials and methods, 123
see also ribosomal RNA (rRNA) operon and genomic repetitive sequences
18S rDNA, 598–599, *599*
23S rRNA genes, 17, *19*, *23*, 30, 31, 387, 388
2° structures comparison and, 18
annotation of, 18
diversity of, in primary structure, 18
gene database, 18
intervening sequences (IVS), *21*, 22
and point mutations, 22–23
small, 22
transposon-mediated insertion of, 22
intragenomic diversity analysis in, 18
secondary structure effect and, 18, 20–22
454 pyrosequencing technique, 552

Accumulibacter. phosphatis, *629*, 631–632
acid mine drainage (AMD), 512
Acidobacteria species, 228, 270, 283, 558
Acidobacterium capsulatum, 136
Acidovorax facilis, 136
Acidovorax species, 545
acid shock proteins (ASPs), 614
acid-tolerance response (ATR), 614
Acinetobacter baumannii, 252, 258, *259*, *260*, 261
Acinetobacter johnsonii, 544
Acinetobacter lwoffii, 557
Acinetobacter radioresistens, 543
Acinetobacter species, 60, 258, 511, 545
ACLAME database, 326
Actinobacteriales demonstration dataset, 442
Actinobacteria species, 200, 201, 231, 270, 283, 447
activated sludge subcommunities, DNA distribution patterns for, 728, *729*, 730
acyl carrier protein (ACP) domain, 424, 428
acyl transferase (AT) domain, 424
ad hoc committee reports, for species definition reevaluation, 99
bacterial systematics and, 100–101
DNA–DNA reassociation and gene sequencing and, 100
DNA–DNA similarities and rRNA gene similarities, relationship between, 101–102
in bacteriology, 102–103
Aeromonas species, 101
Affimetrix
GeneChip, 522
PhyloChip, 554, 555, 556, 558, *559–560*
Afipia species, 545
Agilent GeneSpring GX v 7.3.1, 665
Agrobacterium species, 79, 187, 558
Agrobacterium tumefaciens, 22, 283
Alcaligenes species, 187
Alcanivorax species, 732
Alicyclobacillus acidocaldarius (AB089859), *34*
alignment mask, 315
Alphaproteobacteria species, 492
Altered Schaedler's Flora (ASF), 528
Alteromondales species, 209

Handbook of Molecular Microbial Ecology, Volume I: Metagenomics and Complementary Approaches, First Edition. Edited by Frans J. de Bruijn.

Alu gene, 498
*Alu*I sequences, 63
Aminobacter species, 52
ammonia oxidizers, 41, 42
marine, 44
soil, 43–44
ammonia-oxidizing Archaea (AOA), 303
ammonia-oxidizing bacteria (AOB), 41, 42, 44, *45*, 303
amoA genes, 42, 510, 593
amoC gene, 593, 594
amplified fragment-length polymorphism (AFLP) analyses, 50
and fingerprinting, 54
amplified rDNA restriction analysis (ARDRA) analysis, 50, *51*, 54 ,59, 67, 198, *199*, 551
experimental procedures
choice of electrophoresis conditions and reagents, 61
choice of primers and restriction enzymes, 60
functional genes, 61
in-silico analysis of enzymes and primers, 61
materials, 62
PCR, 61, 62
restriction endonucleases (REs) digestion, 62
result analysis and data storage, 62
specific taxonomic groups profiling, 60
phylogenetic analysis of profiles of, 63
primers used for, *61*
profile-based identification of clones and, 63
theoretical considerations, 63–64
Anabaena doliolum, 616
Anaeromyxobacter species, 524
anf gene, 510
AQUA method, 609
Aquaspirillum, 543
Aquifex species, 213
Arabidopsis thaliana, 509, 697, 700, 711, 719–720
AraCyc database, 711, 716, 717
ARB software, 132, 393, 399, 554, 588
availability and training, 405
central database of, 400–401
data access and visualization of, 401
main window, 399–400
phylogenetic treeing and, 403–404
positional tree server, 404
probe design and evaluation, 404–405
profiles, masks, and filters and, 402–403
sequence alignment and quality checks and, 404
sequence editors and, 401–402
SSU and LSU Rrna databases for, 395
tools, 405
arbuscular mycorrhizal fungi (AMF), 302, 303
Archeoglobus fulgidus, 362
arg6 site, 260
ArMet (anarchitecture for metabolomics) data model, 699–700
ArrayOme, 252, 253, 254
arthrobacter species, 187, 663
bacterial culture conditions and, 664
chromate resistance determinant (CRD) proteins expression, 667–670, *670*
digest preparation, 664–665
and identification, 664
LC/LC-MS/MS-based proteomics, 666–667
sulfur limitation response, 670–672
two-dimensional gel electrophoresis (2-DGE), 665–666
Arthromitus species, 528
asn34 site, 260
Aspergillus fumigatus, 281, 382–383
assembly, testing, and launch operations (ATLO), 544
Astragalus adsurgens, 56
Astragalus ernestii, 49, 50, 56
Astragalus luteolus, 49, 50, 53, 56
Astragalus sinicus, 56
Astragalus species, 49, 52, 53, 54, 56
attC gene, 217, 218, 220, 221
attI gene, 217, 218
Aureobasidium, 543
automated pipeline for phylogenomic analysis (AMPHORA), 206, 207, 211, *213*
additional markers and reference genomes and, 213–214
automated sequence alignment and trimming, 207
bacteria genome tree construction, 207
bacterial genome trees and genome tree construction, 209
genome-based microbial taxonomy, 209, 211
high-quality and highly reproducible sequence alignments, 208–209
metagenomic phylotyping and global ocean survey, 211–212
phylogeny-based and similarity-based phylotyping comparison and, 213
phylotyping by phylogenetic analyses, 208
phylotyping simulation study, 208
protein phylogenetic marker database, 206–208
speed, 209
autotrophs, 115
average molecular masses, 713
AWE (Argonne Work Engine), 326
Azorhizobium caulinodans, 240

Bacillus alcalophilus, 744
Bacillus amyloliquefaciens, 35, 36
Bacillus anthracis, 35, 78, 84, 85, 525
Bacillus atrophaeus, 35
Bacillus cereus, 31, 35, 36, 38, 85, 617
Bacillus licheniformis, 35, 36, 544, 614
Bacillus megterium, 544
Bacillus mycoides, 35
Bacillus pseudomycoides, 35
Bacillus pumilus, 35, 36, 542, 544
bacillus species, 60, 101, 187, 283, 543
identification approaches, 32–34
Bacillus sphaericus, 36
Bacillus subtilis, 35, 36, 175, 278, 281, 362, 500, 537, 610, 612, 613, 614, 615, 618, 681
Bacillus thuringiensis, 35
Bacillus weihenstephanensis, 35
bacterial artificial chromosomes (BAC), 117, 174–175, 235, 280, 283, 333, 497
bacterial genome
complexity of, *476*
genome-based microbial taxonomy, 209, 211
genome tree construction, 209
bacterial species and metagenomic insights, 89, *90*
biogeography and sequence-discrete population and, 94–95
complex microbial communities and, 92–93
material and methods and, 90–91
population cohesion forces and, 93–94
species definition implications and, 95–96
bacterial taxonomy, 29, *30*
bacteriophage genomes, 229–231
Bacteroides species, 241
Bacteroides stercoris, 241
Bacteroides thetaiotaomicron, 241
Bacteroidetes species, 270, 558
Bam gene, 198
Barcrawl utility program, 412
bartab utility program, 413
Basic Local Alignment Search Tool (BLAST), 132, 219, 326, 337, 344, 354, 359, 369, 374, 376, 388, 424, 572, 588
Bdellovibrio species, 60, 114
Bergey's Manual of Determinative/Systematic Bacteriology, 99, 100, 165, 395
Bergey's Taxonomic Outline of Prokaryotes, 209, 395

Big Dye Terminator v 1.1 cycle sequencing kit (Applied Biosystem), 219
binning procedure, *160*, 700, 710
compositional, 226–227
composition-based supervised, 357
composition-based unsupervised, 357
GC content and read coverage-based
homology-based, 356
see also seeded growing self-organizing map (GSOM)
Biocarta, 716
BioCyc database, 711
biodivprep utility program, 414
biodiv utility program, 414
biofilm formation in soil particle and tools development, 303
bioinformatics, in metagenomics, 118
tools and web resources for, *482–483*
bioinformatic tools, for metabolic pathway visualization, 714–719
biological pathway reconstruction, 453
benchmarking experiments and, 456
for genomes
human genome, 457
KEGG pathway, 457
metagenomes, 457–458
performance overview of MinPath, 457
viral metagenomes, 458–459
implementation details, 456
integer programming algorithm and, 455–456
minimal pathway reconstruction problem and, 455
naïve mapping approach and, 455
protein function and function annotation and, 456
biological species concept (BSC), 80, 114
BioMeta, 711
Bionumerics, 4.0 131
BioNumerics software, 68, 70
BioPathAt, 716
biphenyl dioxygenase, 237, 239
bisbenzimidazole, 184–185, 186
black box, microorganisms as, 5
BLASTN program, 164, *165*, 166, 167, 198, 268, 381, 382, 383, 384, 388, 389, 576, *590*
BlastN v2.2.20 algorithm, 380
BLASTX, 239, 577, 588
BLOSOM64, 238
Bordetella species, 230
BOX sequences, 32, *36*, 38
Bradyrhizobium canariense, 56
Bradyrhizobium japonicum, 56, 136, 617
Bradyrhizobium yunmingense, 240
BRENDA, 719
Brevibacillus species, 60
Brevundimonas diminuta, 544
Brevundimonas species, 545
Brucella melitensis, 84
Bruker Daltonics Apex Qe 12 T system, 684
*Bsg*I sites, 270
Buchnera aphidicola, 362
buk gene, 240, 241
Burkholderiaceae species, 227, 231
Burkholderia cepacia, 240
Burkholderia graminis, 283
Burkholderiales species, 271
Burkholderia pseudomallei, 362
Burkholderia phytofirmans, 240
burkholderia species, 302–303, 354, 557
Burkholderia xenovorans, *240*
but gene, 240, 241

CalcOligo software, 554
Calvin cycle, 454
Campylobacter jejuni, 537, 617
Campylobacter species, 68, *71*
Candida albicans, 511
Candidatus Arthromitus, 528
Candidatus Desulforudis audaxviator, 541
Candidatus Liberibacter species, 558, 561
canonical correspondence analysis (CCA), 514
capillary electrophoresis, 677
Carboxydothermus hydrogenoformans, 18, 20, 25
CARMA, 337
Carsonella ruddii species, 247
catalyzed reporter deposition, 736–737, *736*
catalyzed reporter deposition fluorescence in situ hybridization (CARD-FISH), 129, 551, *737*, 738, 739
Caulobacter crescentus, 618, 665
Caulobacter vibrioides, 283
cca package, 523
cDNA sequence annotation, 576–577, 577–578
Celera assembler, 336
cell capture with paramagnetic beads, 737–738
cells lysis stringency fractionation, 291–292
cellulolytic bacteria analysis, 302
Cenarchaeum symbiosum, 594
Chateaurenard *Fusarium*, 279
ChEBI (Chemical Entities of Biological Interest), 711, 713
ChemAxon, 426
chemical abstract service (CAS), 711, 713
CHEMnetBASE, 711
chemolithotrophs, 115
chemoorganotrophs, 115
chemotrophs, 115
Chesapeake bay bacterial communities, 637
DNA extraction and PCR-DGGE fingerprints, 638–639
estuarine dynamics environment and bacterial population structure, 639–640
MDS analysis, 639
of patterns, 641–642
metaproteome analysis, 640–641
microbial community collection, 638
protein extraction and quantification and identification, 639
protein identification, 642–643
ChipChecker program, 554, 555
Chlamydia pneumoniae, 84
Chlamydia species, 247
Chlamydomonas eugametos, 246
Chlamydophila species, 247
Chlorobium tepidum, 362
Chloroflexi species, 247
chrA gene, 668
chrB gene, 668
Citrobacter rodentium, 529
Cladophora species, 68
CLONTECH, 498, *499*
closed habitat system, 545
closed pan-genome, *84*, 85
Clostridia species, 231
Clostridium acetobutylicum, 437
Clostridium difficile, 241
Clostridium perfringens, 18, *20*, 25, 186, *187*, 241
Clostridium species, 240, 501, 502
Clostridium sporogenes, 680
Clostridium tetani, 557
Clostridium thermocellum, 500
ClustalW, 404
CLUSTALW program, *165*, 166, 206, 209
cluster analysis (CA), 514
clustering-based peak alignment algorithm, 67, *69*, 71–72
comparison with BioNumerics software and, 70
data preparation and, 68
dataset, 68
metagenomics and, 72–73
sample-specific DGGE bands identification, 70–71
statistical analyses and, 70
clustering percentage (CP) determination, 371, *373*, 375
Clusters of Orthologous Groups (COG), 578, 669
ClustScan, 423–425
analysis steps, 426–427
cluster definition, 427–429
future developments of, 430–431
materials and methods and, 426
metagenomic data limitations and, 431
product chemistry prediction, 429

ClustScan (continued)
as productivity tool for metagenomics, 429–430
CODEHOP algorithm, 442, 445–446
codon and dicodon usage and Orphelia, 360–361
codon usage table (CUT), 442, 444, 445
cold-inducible proteins (CIPs), 613
cold shock protein (CSP), 613, 614, 616
collision-induced fragmentation, 678
Comamonadaceae species, 227, 231
Comamonas testosterone, 45
comM gene, 261
CommOligo software, 512
common oligo reference standard (CORS), 512
Community Cyberinfrastructure for Advanced Marine Microbial Ecology Research and Analysis (CAMERA), 236, 577
community genome arrays (CGAs), 510, 552
comparative day/night metatranscriptomic analysis, of microbial communities, 575
16S rRNA gene libraries and, 576
biogeochemically relevant genes and, *583*
cDNA sequencing and annotation and, 576–577, 577–578
community composition and transcript taxonomic origin, 580
eukaryotic sequence annotation and, 577
predicted highly expressed genes and, 577
qPCR verifications, 577
quality analysis, 578–579
RNA and DNA preparation and, 576
and samples, 576, 581–584
statistical analysis, 577
comparative genome hybridization (CGH), 84, 252, 253
Comparative Genomics Laboratory, 713
compositional binning, 226–227
composition-based supervised binning, 357
composition-based unsupervised binning, 357
Comprehensive Systems Biology Database (CSB.DB), 721
Constrained K-means, 371
COPASI, 720–721
COP K-means, 371
Coprococcus comes, 241
Coprococcus eutactus, 241
CopyControl Fosmid Library Production Kit, 493
core genome, *84*, 86
core genome hypothesis, 79–80
Corticoviridae species, 230
Corynebacterium glutamicum, 617
Corynebacterium jeikeium, 362
Corynebacterium species, 188, 444, 448
Coxiella burnetii, 247, 535–536
Cramer–von Mises statistics, 239
Crenarchaea, 91, 94, 95
Crenarchaeota, 742
CS Analyzer version 3 (ATTO), 68
CTO primers, 42, 43, 44, 46
culture-independent methodologies, 735
cyanobacteria, 645
best technique for, 653–655
gel proteomics era in, 646
high-throughput proteomics arrival to, 646–653
Cyanobacteria species, 228
Cymbastela concentrica, 247
cytomics, applications of, 727
FC-separated microbial communities, analysis of, 731
subcommunities, phylogenetic analysis of, 732
future
subcommunities and populations, functional omic analysis of, 732–733
individual physiological states of microorganisms, analysis of, 728
proliferating cells, identification of, 728–730
species-specific functions measuring, in microbial communities, 731
viable and nonviable cells, identification of, 730–731
Cytoscape systems, 716, 721

DataAnalysis 3.4 (Bruker Daltonik) software, 684
*Dde*I sequences, 63
dehydratase domain (DH), 424
Deinococcus radiodurans, 22, *24*, 25, 208, 616
Delftia acidovorans, 545
Deltaproteobacteria species, 739
denaturing gradient gel electrophoresis (DGGE), 42, 43, *44*, *45*, 46, 67, 68, 70–71, 72, 191–192, 131, 492, *493*, 494, 551
de novo sequencing, 610
Department of Homeland Security BioWatch program, 525
desorption electrospray ionization (DESI), 681
Desulfitobacterium hafniense, 502
Desulfomicrobium species, 557
Desulfovibrio vulgaris, 614, 615
detrended correspondence analysis (DCA), 514
Deutsche Sammlung von Mikroorganismen und Zellkulturen (DSMZ), 314
dictionary of natural products (DNP), 711–712
Dictyostelium discoideum, 426, *428*
Dietzia species, 732
dimethylsulfoniopropionate (DMSP), 241
Dinoroseobacter shibae, 580
directed homogeneity test, 348, 350
direct infusion mass spectrometry (DIMS), 697
dispensable genome, *84*, 86
dissolved organic matter (DOM), 630
dmdA gene, 241
DNA, 5, 116
content analyses, 728–730
–DNA reassociation and gene sequencing, 100, 114
–DNA similarities and rRNA gene similarities, relationship between, 101–102
and evolution and development of diversity, 5
and exploration, 5–6
extraction, 7, 226, 279, 738
fingerprinting, *32*, *33*, *35*, 67–73
genetic diversity estimation and, 8–10
melting of metagenomic, 7–8
microbial biodiversity and metagenome diversity and, 6–7
pyrosequencing, 144, 149–154
reassociation applications
diversity and composition of microbial isolates and total community, 10–11
metagenomics along salinity gradient, 12–14
perturbation impact assessment and microbial diversity pollution, 11–12
reassociation kinetics, 6–7, *11*
reassociation of metagenomic, 8
recovering strategy
cells lysis stringency fractionation, 291–292
molecular DNA weight separation, 292–293
prokaryote cells extraction, 290–291
prokaryote cells separation, 291
soil fractionation, 290
recovery and isopycnic centrifugation, 226
sequencing and assembly, 226
stable isotope probing (DNA-SIP), 225, 282, 491, 494, 495
see also individual entries
DNA DataBank of Japan (DDBJ), 337
DNADIST v3.6 program, 166
dNTP, 144
DOME, 720–721

dot blot hybridization, 499–500
DOTUR, 200, 201, 462, 465, 466
double-RNA approach, 587
analysis pipeline and, 589
bioinformatic analysis and, 588
cDNA synthesis and pyrosequencing and, 588
community profiling and, 589–592
data deposition and, 588–589
low-abundant uncultured group metabolism probing and, 592–594
nitrifier community and, 594
putative mRNA-tags global functional analysis and, 592
site description and soil sample processing and, 588
taxonomic binning critical assessment and, 592
dsrAB gene, 557
dsr gene, 240

EBI-EMBL database, 394, 395
EcoCyc database, 711
ecological control analysis (ECA), 486
ecological diversity, 6
ECOSIM analysis, 126
effective genome size (EGS), 334
EF-Tu gene, 205
electron capture dissociation (ECD), 607
electron ionization (EI), 677
electron-transfer dissociation (ETD), 607
electrospray ionization (ESI), 677, 685
electrospray ionization (ESI) tandem mass spectrometers, 648
electrospray mass spectrometry, for metabolic profiling
data processing and interpretation and, 700–701
large batches, quality control of, 699–700
mass spectrometer, optimization of, 698–699
sample preparation and, 697–698
samples introduction to mass spectrometer and, 698
tomato, metabolic profiling of, 701–702
analysis, 702–703
data processing and interpretation, 704–705
sample preparation, 702
emPCR, 149, 153
EMVEC database, 394
endomycorrhizae analysis, 303–304
enhanced biological phosphorus removal (EBPR), 486, 630, 631
enoyl reductase (ER) domain, 424
Enterobacter agglomerans, 278
Enterobacteriaceae species, 511
enterobacterial repetitive intergenic consensus (ERIC) sequences, 32
polymerase chain reaction (ERIC-PCR), *37*, 38
Enterobacter liquefaciens, 617
Enterobacter species, 511
Enterococcus faecalis, 281
Enterococcus faecium, 511
Enterococcus species, 117, 118, 511
environmental genomics, microbial populations for, 735
catalyzed reporter deposition and, 736–737
cell capture with paramagnetic beads and, 737–738
DNA extraction and, 738
fluorescence in situ hybridization and, 735–736
magneto-FISH capture, specificity validation of, 738–739
multiple displacement amplification (MDA), DNA amplification with, 738
Environmental Microbial Genomics Group, 189
Environmental Molecular Sciences Laboratory (EMSL), 664
environmental shotgun sequencing (ESS), 157, 206
comparative metagenomics and, 161
methods to study microbes in environment, *158*
microbe cataloging and, 158–159
top down and bottom up approaches and, 159–161
environmental stress response analysis in prokaryotes, 605
experimental methods and technologies in proteomics and bioinformatics, 609–610
bottom-up gel-based proteomics, 607
bottom-up gel-free proteomics, 607–609
top-down proteomics, 606
proteomic applications,
contribution to stress characterization in prokaryotes, 612–618
general and specific stress response in prokaryotes, 610–612
Equitability index, 9
ERGO, 719
erythromycin biosynthesis, *425*
Escherichia coli, 6, 8, 13, 24, 25, 38, 78, 83, 85, 86, 114, 117, 120, 143, 145, 146, 150, 154, 159, 171, 174, 175, 186, *187*, 201, 247, 248, 251, 279, 280, 281, 282, 283, 325, 362, 402, 492, 494, 499, 502, 511, 537, 569, 570, 571, 572, 591, 598, 600, 610, 612, 613, 616, 617, 618, 665, 669, 697, 728, *729*, 742, 745
ESI-ICR-FTMS spectra of bacterial cultures, *686*
ESPIRIT program, 268, 461
air sample experiments, 466
benchmark study of, 464–465
efficiency of, 470
low-quality reads removal and, 462
pairwise and multiple sequence alignment, 466, 469–470
pairwise distance computation and, 463
seawater sample experiments, 465–466
sequence assignment to OTUs and, 463–464
statistical inference of species richness and, 464
Eubacteria genomes, 171
Eucommia ulmoides, 201
Euhalothece species, 615, 657
eukaryotic sequence annotation, 577
EULER assembly method, 483
European Bioinformatics Institute (EBI) metagenomics, 333
16S rRNA analysis, 334
bioinformatic data analysis and, 336
data archiving and retrieval, 337
metagenome archive, integration, and analysis resource, 338–339
nucleotide sequence archiving, 337–338
protein archiving, 338
functional elucidation, 337
metagenomic resource summary and, *338*
minimum information standard, 333–334
open reading frame (ORF) prediction, 336
preanalysis and assembly and, 335–336
sample and metadata collection, 334
sample diversity analysis and, 336–337
sample diversity and sequence coverage, 334
sample sequencing and, 335
Sequence Read Archive (EBI-SRA), 337
European Molecular Biology Open Software Suite (EMBOSS), 63
European Nucleotide Archive (ENA), 334, 337
European Space Agency
BioPan, 542
European Technology Exposure Platform and Experiment Facility (EuTEF), 542
ExPASy, 716, 719

fa2fq utility program, 415
fa2nw utility program, 415
false discovery rate (FDR), 653

Fasta, 315, 337, 354, 442, 447
FC analyses, 727
FC-separated microbial communities, analysis of, 731
subcommunities, phylogenetic analysis of, 732
Ferroplasma acidarmanus, 617
Fibrobacteres species, 357
fic gene, 260
FIG protein family, 453, 456
Finite Supragenome Model, 85
Firmicutes species, 201, 270
Flavobacterium species, 101, 187, 188
flowgram, 149
fluorescence-activated cell sorting (FACS), 248
fluorescence cell sorting (FACS), 177
fluorescence in situ hybridization (FISH), 41, 129, 380, 551, 557, 735–736
probing and microscopic isolation, 745
Fluorinert®, 186
focused metagenomics approach, 491–495
formatdb utility program, 414
fosmids, 280
Fourier transform-ion cyclotron resonance-mass spectrometry (ICR-FT/MS), 684
fq2fa utility program, 415
fractionated centrifugation method, 6
Francisella species, 525
Francisella tularensis, 247, 525, 615
free-flowing-liquid electrophoresis (FFE), 658
Freeze-drying, 698
FT-ICR-MS data and "Matrix Generator", 685
ftp (file transfer protocol), 713
functional gene arrays (FGAs), 510, 511, 512, 552
functional gene pipeline/repository (FGPR), 236
functional screening, metagenomic libraries for, 171, *172*
based on induced gene expression, 177–178
cloning strategies and, 175–177
environmental DNA isolation and target genes enrichment, 171–173
by growth selection, 177
sequence-based, 178
by specific phenotype detection, 177
and vectors, 174, *176*
host range of, 175
for large-insert metagenomic libraries, 174–175
for small-insert metagenomic libraries, 174
Fusarium graminarum, 537
Fusarium oxysporum, 281
gacA gene, 302
gas chromatography coupled to mass spectrometry (GC/MS), 677, 714
GC content and read coverage-based binning, 357
gcd gene, 303
GC fractionation, 183–185, 187–188
comparative statistical analysis of community profiles and, 192–194
DGGE analysis of, 191–192
functional gene probing of, 190–191
microbial community DNA purification and, 185–186
cesium chloride-bisbenzimidazole gradients, 186
DNA stock solutions, 186
downstream analyses, 186
sequence libraries creation and, 188–190
GDE package, 206
Geaomannomyces graminis var. *tritici*, 278
GELCOMPAR 4.0 program, 50
gel-free proteomics, 648
gel proteomics era in cyanobacteria, 646
GenBank, 22, 23, 30, 51, 85, 146, 239, 315, 337, 365, 366, 388, 442, 576, 720
GeneAmp PCR System 2400 thermal cycler, 498
Gene database, 720
Gene Expression Omnibus (GEO) database, NCBI, 268
GeneMark, 360, 362, 364
GeneMark-PS, 426
GeneMarkS, 438
gene ontology (GO), 343, 346–347, 348, 720
general stress proteins (GSPs), 612, 613
Genesis mission, 543
gene-targeted -(GT) metagenomics, 235
biphenyl dioxygenase diversity in soils and, 239
deep sea functional gene analysis, 241
flowchart of, *236*
functional gene primers design and testing and, 240–241
nitrogen fixation investigation in soils and, 239–240
postsequencing and
clustering and operational taxonomic units, 238
diversaity estimation and community comparison, 238
sequence alignment, 237–238
presequencing and
collection of previously known gene sequence, 236–237
primer specificity, 237
region selection for primers, 237
Genevestigator®, 721
GenFrac program, 420
Genome Sequencer FLX System, 239, 241
Genomic Encyclopedia of Bacteria and Archaea (GEBA), 209, 337, 546
Genomics Standards Consortium, 126
Geobacillus thermoleovorans, 618
Geobacter species, 524
Geobacter sulfurreducens, 680
GeoChip, 509
data analysis, 514
design, 512
examples, 514–517
functional gene array development and, 510
group-specific arrays, 510–511
pathogen detection, 511
specific environments, 512
microbial community studies based on, *515–516*
posthybridization analysis and, 513
sample preparation and hybridization and, 512–513
Geothrix fermentans, 524
Gepasi programs, 720–721
GFX DNA and Gel Band Purification kit (GE Healthcare), 219
Giardia species, 600
Glimmer, 426
Glimmer3, 438
Gliocladium species, 278
glnA genes, 51, 52, 54, *55*, 56
glnII genes, 51, 52, 54, *55*
Global Alignment for Sequence Taxonomy (GAST), 322
Global Internal Standard Technology (GIST), 608
Global Ocean Sampling (GOS), 336, 359, 546, 577
global-scale gene mapping (GGM), 284
glomalin-related soil proteins (GRSP), 302
glycol granular activated sludge (GAS), 190
glycosylhydrolases/transferases, 678
Good Agricultural Practices Certification, 300
Great Plate Count Anomaly, 277
green fluorescent protein (GFP), 177, 178
GREENGENES, 326
growing SOM (GSOM) methods, 370, *373*
GS20 next generation sequencing, 143
GS-FLX pyrosequencing technology, 119, 149, 150, 152, 153, 154
GTG_5 polytrinucleotide sequences, 32, *36*, 38
guild, 114

*Hae*III sequences, 63
Haemophilus influenzae, 85, 251, 373
Haemophilus somnus, 373

Haemophilus species, 79
Haloarcula marismortui, 18, 20, 25
Halobacterium salinarum, 535
Halobacterium species, 616
Halomonas species, 732
Haloquadratum walsbyi, 12, 13
Hansenula species, 601
Heaps' law, 85
heat shock proteins (HSPs), 613, 614, 616, 617
Helicobacter pylori, 362, 536–537, *538*
Helicos, 144
Herbaspirillum species, 71
heterotrophs, 115
*Hha*I sequences, 63
hidden Markov models (HMMs), 206, *207*, 208–209, 236, 237, 314, 387, 388, 426
Hierarchy Browser, RDP, 315, 317
Hierarchy View, RDP, 315
highmolecular-weight (HMW) metagenomic DNA, 173, 279, 283
high-performance liquid chromatography (HPLC), 648
  -mass spectrometry (HPLC-MS), 714
high-resolution metagenomics, 225
  bacteriophage genomes and, 229–231
  community composition shifts and methylotroph DNA enrichment, 227–228
    methylotenera mobilis composite genome, 228–229
  compositional binning and, 226–227
  DNA extraction and, 226
  DNA sequencing and assembly and, 226
  isopycnic centrifugation and DNA recovery and, 226
  protein recruitment and, 227
  sampling and stable isotope probing and, 226
high-resolution molecular characterization techniques, 683
  experimental condition, 684
  instrument, 684
high-throughput cyanobacterial proteomics, 645
  chemical labeling-based methods in, 653
    best technique for cyanobacteria, 653–655
    pilot iTRAQ experiments, 655
    Prochlorococcus marinus MED4, studying light intensity and cellular stresses in, 656
    studying N_2 fixation in *Nostoc* sp. PCC 7120 and *Nostoc* punctiforme ATCC, 29133 655–656
  developing quantitative approaches for, 652–653, *654*
  emergence of in vivo labeling for, 656
    unsequenced halotolerant cyanobacterium, studying salt stress in, 657
  gel proteomics era in cyanobacteria, 646
  high-throughput proteomics arrival to cyanobacteria, 646–653, *647–648*, *649*, *650*, *652*
  in perspective, 657–658
Hinf I sequences, 63
HIS3 gene deletion, 598
HiSeq2000, 144
HITChip, 558
HMMER, 426, 512
holistic approach, 5
homing endonuclease restriction and marker insertion (HERMI), 246, 247, 248
homology-based binning, 356
homopolymer, 149, 151, 153, 242, 419
horizontal gene transfer (HGT), 79
horizontal gene transfer and recombination shape mesorhizobial populations, 49
  AFLP analyses, 50
  bacterial strains sampling and symbiotic status and, 50
  data treatment and sequence analysis and, 50
  genetic diversity and, 52
  glnA and *glnll* sequence phylogenetic analysis and, 52
  nifH gene sequence phylogenetic analysis, 52–53
  nucleotide sequencing and, 50
    accession numbers, 51
  PCR amplification and, 50
  phylogenetic analysis and, 51
  RFLP and, 50
  rrs phylogeny and, 52
HSP70 gene, 205
Human Gut Microbiome Initiative, 307
Human Metabolome Project, 711
Human Metabolome Project and METLIN databases, 714
human microbiome project, 307
  body sites sampled in, *311*
  and disease, 311
  funded projects of, *308–309*
  normal, 311
  reference genome catalog, 310–311
HyTher software, 554

*I-Ceu*I homing endonuclease, 246–247
*I-Cre*I gene, 248
Illumina (Solexa), 521, 552
  Gallx platform, 144, 387
Illumina Genome Analyzer, 242
ImageMaster algorithm, 639
immunomagnetic cell capture, *736*
Infernal Stochastic Context-Free Grammar (SCFG)-based aligner, 314
inferred contigs (ICs), 252
inferred gaps (IGs), 254
INSDC, 313, 314, 315, 337, 338
insurance hypothesis, 3
integer programming (IP), 456
Integrated Microbial Genomes (IMG), 236, 664
  with Microbiome Samples (IMG/M), 227, 228, 229
integron first gene cassettes, 217
  amplified fragments screening, 220
  analysis, 219
  and application to metagenome, 220–221
  clone screening and, 219
  cloning and, 219
  DNA samples and, 218
  library construction, 219–220
  PCR amplifications of, 218–219
Internal Active Thermal Coolant System (IATCS), 545
internal transcribed spacer (ITS), 31
  polymerase chain reaction (ITS-PCR), 31, 34, 35, 36
International Committee for Systematic Bacteriology (ICSB), 100
International Human Microbiome Consortium, 312
International Journal of Systematic and Evolutionary Microbiology, 102
International Journal of Systematic Bacteriology, 101
International Nucleotide Sequence Database Collaboration, 394
International Space Station (ISS), 542, 545
International Space Station Internal Active Thermal Coolant System, 545–546
International Space Station Water System, 545
International Union of Biochemistry & Molecular Biology Metabolic Maps, Minimaps, and Animaps, 716
InterPro database, 338
InterProScan, 339
intervening sequences (IVS), *21*, 22
  and point mutations, 22–23
  small, 22
  transposon-mediated insertion of, 22
intI gene, 217, *218*, 219, 220, 221
intragenomic heterogeneity, 17, 18, *19*
inverted repeat lacking clade (IRLC), 56
isobaric Tag for Relative and Absolute Quantification (iTRAQ), 608–609, 615, 616, 633, 655, 657
isoelectric focusing (IEF), 119, 630
isotope-coded affinity tags (ICAT), 608, 609, 614, 653, 655

isotope-coded protein labels (ICPL), 653, 657
iterative reads, 379
assembly, 380–381
consensus genome and, 383–384
data, 380
depth and coverage increase and, 381–382
initial mapping, to reference genome, 381
quality check, 382–383
quasispecies diversity and, 383

Jannaschia species, 580
JAZZ assembler, 336, 374
Jetflex Genomic DNA Purification Kit (GENOMED, 600100) 130
Jmol window, *424*, 426
Joint Genome Institute—Production Genomics Facility (JGI-PGF), 226, 228
Jukes–Cantor model, 418

KAAS pathway, 453, 455
KaPPA-View4 software, *720*
KaPPA-View program, 716, 717, 721
katF gene, 612
KEGG database, 711, 713, 714, 716, 717
KEGG Orthology (KO), 720, 456
KEGG pathway, 453, 455, 456, 457, 458, *459*, 578, 581, *582*
Kellogg Biological Station Long-Term Ecological Research site (KBS-LTER), 135, 138
Kennedy Space Center
Payload Hazardous Servicing Facility (PHSF), 544
ketoreductase domain (KR), 424
ketosynthase domain (KS) transfer, 424
Kitasatospora species, 556
Klebsiella pneumoniae, 260–261
Klebsiella species, 511
Kluyveromyces species, 601
k-mer, 369, 374
Krebs cycle, 454
k-tuple, 463

label-free techniques, 657
labeling principle, 652–653
Laboratory of the Government Chemist, 699
Lactobacillus acidophilus, 613
Lactobacillus casei, 252, 253, *254*, *255*, *256*
Lactobacillus murinus, 528, *529*
Lactobacillus paracasei, 252, 255, *256*
Lactobacillus sanfranciscensis, 616
Lactobacillus species, 60, 502
Lactococcus lactis, 613
Lawrence Berkeley National Laboratory (LBNL), 522, 554
Leishmania species, 600
Leptospirillum group II organism, 120, 632
Library Compare, RDP, 316
Life Technologies (Applied Biosystems), 143, 144
linear programming (LP), 455–456
liquid chromatograph time-of-flight mass spectrometer (LCT), 702
liquid chromatography coupled to mass spectrometry (LC/MS), 677, 678
metabolomic workflow, 679, *679*
Listeria monocytogenes, 537
Lockspray™ interface, 703
lowest common ancestor (LCA) algorithm, 344
lycopene, 701

MAFFT program, 206, 461
magneto-FISH capture, 735
for benthic and planktonic samples, *736*
catalyzed reporter deposition and, 736–737
cell capture with paramagnetic beads and, 737–738
DNA extraction and, 738
fluorescence in situ hybridization and, 735–736
multiple displacement amplification (MDA), DNA amplification with, 738
specificity validation of, 738–739
step-by-step instruction, *737*
MALDI-TOF, 642
Mallotus nudiflorus, 197, 198, 199, 200
MAPMAN program, 721
Maq mapping algorithm, 381, 384
marine ammonia oxidizer communities analysis, 44
Mars Exploration Rovers Mission, 544
Mars Odyssey Orbiter Mission, 543
Mars Reconnaissance Orbiter Mission, 544
Mascot version2.2 (Matrix Science Inc., USA), of metaproteomic sequencing, 380
Mass-Coded Abundance Tagging (MCAT), 608
mass spectrometry (MS), 10, 119, 606, 677
based metabolite imaging, 680–681, *680*
based meta-metabolomics, 681
MassTRIX, 692
MasterPure Gram-Positive DNA Purification Kit (Epicentre MGP04100), 130
matrix-assisted laser desorption ionization (MALDI), 681
Matrix Generator, 685
*Mbo*I sequences, 63
MEGAN (MEta Genome Analyzer), 118, 213, 343, 420, 483, 485, 577, 580, *590*, 592
dataset comparison and
of charts, 348
of functional content, 348
of taxonomical content, 348, 350
visual, 347–348
functional analysis and, 346–347
similarity analysis of multiple datasets and, 350
taxonomical analysis and, 344–346
Mesorhizobium amorphae, 52
Mesorhizobium ciceri, 52
Mesorhizobium huakuii, 52, 54
Mesorhizobium loti, 52, 54
Mesorhizobium plurifarum, 52
Mesorhizobium septentrionale, 52, 56
Mesorhizobium species, 49, 52, *53*, *54*, 56, 240
Mesorhizobium temperatum, 56
MetaBioME, 486
metabolic flux analysis (^{13}C-MFA), 486
metabolic profiling of plant tissues, by electrospray mass spectrometry, 697
data processing and interpretation and, 700–701
large batches, quality control of, 699–700
mass spectrometer, optimization of, 698–699
sample preparation and, 697–698
samples introduction to mass spectrometer and, 698
tomato, metabolic profiling of, 701–702
analysis by mass spectrometry, 702–703
data processing and interpretation, 704–705
sample preparation, 702
metabolite identification, pathways, and omic integration, 709
bioinformatics tools and, 714–719
data presentation, 710
compound identification from masses/bins of interest obtained from metabolic fingerprinting and chemometrics, 710–711
metabolomic outputs, upstream information and, 719–721
online databases and, 711–714
polyomic, systems biology data, software and, 721
metabolite information for polyhydric alcohol sorbitol, 716–717, *718–719*
metabolite-related expression (METREX), 178
metabolomics, 120–121, 709

high-resolution molecular characterization techniques, 683
experimental condition, 684
instrument, 684
outputs, upstream information from, 719–721
small-molecule dimension, 677
mass-spectrometry-based metabolite imaging, 680–681, *680*
targeted and global approaches to metabolomics, 678–680
targeted and global approaches to, 678–680
Metacontrol, 277, 278
MetaCyc database, 711, 714, 716, 717
MetaGene, 360, 362, 363, 365, 366, 433–434
basic architecture of, 434
codon frequencies and
estimation by G%, 434
regression analysis, 435
dataset, 434
length and distance distributions of, 434
Mycoplasma gene and, 439
prediction performance on complete genomes and, 437
prediction performance on short genomic fragments and, 437–438
ribosomal binding site (RBS) model, *436*
construction, 434–435
species-specific patterns, 437
MetaGeneAlyse, 721
MetaGeneAnnotator, 362, 364, 365, 366
metagenomic array (MGA), 178, 510, 553
Metagenomics RAST server (MG-RAST), 118, 325, *328*, 388, 456, 458, 500, 588, *593*
annotations, 327
comparative heat maps
metabolic, 328
phylogenetic, 328–329
comparative metagenomics, 327–328
data types, 326
implementation, 326–327
metabolic reconstructions and models and, 330
recruitment plot, 329–330
user registration and management and, 326
metagenomic whole genome shotgun (mWGS) sequencing, 144, 146, 310, 311
metaproteomics, 119–120
MetaSHARK (Metabolic Search and Reconstruction Kit), 721
MetaSim, 362, 417
population sampler and, 418
read sampling and, 418–419
read simulation and empirical error models and, 420
Sanger sequencing simulation and, 419
simulation of sequencing by synthesis and, 419–420
species profiles configuration and, 418
metatranscriptomics, 119
MetDAT, 717
MetExplore online system, 716, 717
Methanobacterium formicicum, 502
Methanobrevibacter ruminantium, 502
Methanococcus jannaschii, 362
Methanomicrobium mobile, 502
Methanosarcina acetivorans, 502
Methanosarcina barkeri, 502, 503
Methanothermobacter thermautotrophicus, 500, 503
Methods in Molecular Biology, 498
Methylobacillus flagellatus, 229, *230*
Methylobacter. tundripaludum, 227, 228, 231
Methylobacterium radiotolerans, 545
Methylobacterium species, 631
Methylobacter species, 227
Methylocaldum species, 511
Methylococcaceae species, 227
Methylococcus. capsulatus, 227, 231
Methylocystis species, 492, 493, 511
Methylomirabilis oxyfera, 380
Methylomonas species, 228
Methylophaga species, 492
Methylophilaceae species, 227
Methylotenera mobilis, 228–229, *230*, 231, 232
Methylotenera species, 227, 231
methylotrophy, 225
Methyloversatilis universalis, 228
Mfold software, 554
mGenomeSubtractor, 252, 258, *259*, *260*, 261, 262
microarray technology, 118
microarray types, in microbial ecology, 552–553
see also phylogenetic oligonucelotide microarray
microarray-visualised genome (MVG), 252, 253, *254*, *255*
microautoradiography (MAR), 129
MICROB*Express*™, 569, 572, 573
microbial biodiversity and metagenome diversity, 6–7
microbial community proteomics, 627
history and significance of, 630–632
for novices, 627–630
microbial ecology, 113
historical background of, 113
metabolic diversity and distribution patterns and, 115
metagenomics and impact on, 115–116
metabolomics, 120–121
metaproteomics, 119–120
metatranscriptomics, 119
narrow-sense metagenomics, 116–119
microbes in, 113–114
scales in, 114
technical and methodological developments and, 114–115
microbial persistance, 541
closed habitat system and, 545
and extreme conditions
cleanroom with low-biomass, 542–543
with low biomass, 541–542
with low microbial diversity, 541
Genesis mission and, 543
International Space Station Internal Active Thermal Coolant System and, 545–546
International Space Station Water System and, 545
low-biomass environment challenges, 546, 548
Mars Exploration Rovers mission and, 544
Mars Odyssey Orbiter mission and, 543
Mars Reconnaissance Orbiter mission and, 544
microbial diversity of spacecraft assembly cleanrooms and, 543
Phoenix mission and, 544–545
microbial species identification, metagenomic approaches for, 105–107
cultivation-independent multilocus sequence analysis and, 107
metagenomic assembly and species populations, 107–108
Synechococcus species population, theory-based prediction of, 107
Micrococcus lysodeikticus, 186, *187*
Micromonas species, 580
microorganisms, individual physiological states of, 728
microscopic single-cell isolation and multiple displacement amplification, 741
cell extraction and FISH probing, 742
cell lysis and genomic amplification, 742–743
microscope and micromanipulation, 742
soil sample, clone library, and sequence analysis, 742
minerna utility program, 414–415
MiniElute Gel Extraction Kit, 268
Minimum Information about a (Meta)Genome Sequence (MIGS/MIMS), 326, *327*

Minimum Information about an Marker Gene Sequence (MIMARKS), 396
MinPath, 454, 456, 457, *459*
mmoX gene, 494
mobile genetic elements (MGE), 217, 218
mobile integrons, 218, 220
mobilome discovery, arrayOme-and tRNAcc-facilitated, 251
  databases and, 252
  Lactobacillus casei and *Lactobacillus paracasei* genomes analysis and, 252–258
  mGenomeSubtractor usage and, 258
  mobilomeFINDER-enabled pipeline and, 258, 260
  Shigella resistance locus island family comparative mapping and, 261
  TnAbaR1-like resistance mega-transposons, 261
  tools and, 252
  tRIP PCR-based discovery of *Klebsiella pneumoniae* niche-adaptation islands and, 260–261
mobilomeFINDER, 252, *253*, 255
  -enabled pipeline, 258, 260
molecular fragmentation, 678
monoisotopic mass, 713
  sucrose, matched and returned metabolites for, *716*
MOTHUR, 462
mRNA isolation, 569
  cDNA synthesis, cloning, and sequence analysis and, 570–571
  RNA extraction optimization from different environmental microbial communities and, 571
  sequence analysis of transcripts from environmental microbial communities and, 572
  from total RNA, 571–572
  total RNA isolation and, 570
mRNA-tags functional analysis, 592, *593*
MS/MS fragmentation patterns, 714
MSBLAST, 630
*Msp*I sequences, 63
multilocus sequence analysis (MLSA), 49, 102
  cultivation-independent, 107
multi-locus sequence typing (MLST), 78, 86, 102
multiple displacement amplification (MDA), 492, 494, 546, 570, 738, 741
multiple reaction monitoring (MRM), 619
multiple sequence alignment (MSA), 388, 461, 466, 469–470
MUSCLE program, 209, 238, 462, 465, 466
mvpart package, 523
mxaF gene, 493, 494
mxaFJGIRSAC genes, 494
Mycobacterium abscessus, 449
Mycobacterium avium, 449
Mycobacterium chelonae, 449
Mycobacterium fortuitum, 449
Mycobacterium intracellulare, 449
Mycobacterium kansasii, 449
Mycobacterium leprae, 449
Mycobacterium mucogenicum, 449
Mycobacterium phocaicum, 449
Mycobacterium scrophulaceum, 449
Mycobacterium species, 60, 441, 444, 447, 448, 449, 450, 668
Mycobacterium tuberculosis, 449
Mycobacterium ulcerans, 449
Mycobacterium vanbalenii, 449
Mycoplasma agalactiae, 63
Mycoplasma anatis, 63, 64
Mycoplasma bovis, 63
Mycoplasma collis, 63
Mycoplasma columborale, 63
Mycoplasma cricetuli, 63
Mycoplasma gallopavonis, 63
Mycoplasma mustelae, 63, 64
Mycoplasma oxoniensis, 63, 64
Mycoplasma pneumoniae, 537
Mycoplasma species, 60, 63– 64, *64*, *438*, 439

N_2 fixation in *Nostoc* sp. PCC 7120 and *Nostoc* punctiforme ATCC, 29133 655–656
N-acyl-homoserine lacton (AHL), 684
Naïve Bayesian Classifier, RDP, 316
naïve mapping approach, 455
nanostructure-initiator mass spectrometry (NIMS), 681
*Nap*A, 303
narrow-sense metagenomics
  bioinformatics, 118
  microarrays, 118–119
  random but comprehensive metagenome sequencing, 118
  single-gene approach, 116–117
  targeted partial metagenome sequencing, 117–118
NASA, 548
  Long Duration Exposure Facility, 542
  missions, 543–545, *547*
NAST software, 412
National Center for Biotechnology Information, 77
National Human Genome Research Institute (NHGRI), 307
National Institute of Allergy and Infectious Diseases, 307
National Institute of Dental and Craniofacial Research, 307
National Institutes of Health (NIH), 307
Natronomonas pharaonis, 362
natural microbial communities
  characterization, PCR primer-based strategies for, 41
  marine ammonia oxidizer communities analysis, 44
  methods, 42–43
  soil ammonia oxidizer communities analysis, 43–44
nearest-neighbor interchange (NNI), 403, 404
Needleman–Wunsch algorithm, 463
NEIGHBOR v3.6 program, 166
Neisseria gonorrhoeae, 79
Neisseria lactamica, 79
Neisseria meningitides, 79, 83
Neisseria meningitidis, 83, 84
Neisseria meningitis, 83
Neisseria species, 78
neural network analysis (NNA), 514
Neurospora crassa, 281
next-generation sequencing (NGS), 143, 333, 335, 336, 417
  application types of
    16S rRNA sequencing, 145–146
    metagenomic whole genome shotgun sequencing, 146
    reference genome sequencing, 146–147
  computational challenges in, 147
  instrumentation and reactions and, 143–144
    Illumina (Solexa) Gallx platform, 144
    Roche-454 platform, 144
  platform features of, 144–145
nifH gene, 52–53, *55*, 303, 510, 511
NIH Roadmap Human Microbiome Project (NIH HMP), 146, 147
Nile Red, 31
nirK gene, 510
nirS gene, 510
nitrogen cycle genes and phosphorous mobilization genes and metagenomic studies, 303
Nitrosococcus species, 42
Nitrosomonas communis, 42
Nitrosomonas europea, 190, *191*, 617
Nitrosomonas oligotropha, 42
Nitrosomonas species, 42, 44, *45*
Nitrosospira briensis, 43
Nitrosospira species, 42, 43, 44, *45*
Nitrospirae species, 228, 231
nm2nw utility program, 415
Nocardia farcinica, 18, *20*, 25
Nocardia species, 448
nodC gene, 512
Nomenclature Committee of the International Union of Biochemistry

and Molecular Biology (NC-IUBMB), 719
nonribosomal peptide synthetases (NRPS), 424, 425, 426, *428*, 429, 430, 431
Nostoc sp. PCC, 7120 646
studying N_2 fixation in, 655–656
Nostoc species, 22, *24*
nucleotide database, 720
nucleotide sequencing, 50
accession numbers, 51
archiving, 337–338

Oceanospirillales species, 209
Olavius algarvensis, 357
OligoCheck software, 554
oligonucleotides, *52*, 108, 510
omics datasets, 677
ONDEX systems, 721
online databases, 711–714
open pan-genome, *84*, 85
open reading frames (ORFs), 219, 221, 283, 336, 339, 359, 361, 362, 363, 365, 381, 383, 433, 434, 483, 493, 509, 553
putative genes corresponding to, *495*
operational protein families (OPFs), 238
operational taxonomic units (OTUs), 62, 72, 102, 124–125, 238, 268, 334, 405, 413–414, 461, 465, *524*
oprF gene, 302
Orphelia, gene prediction in metagenomic fragments with, 359
accuracy evaluation and, 362–363
accuracy in reads from different species and, 363
accuracy in reads with sequencing errors, 363–364
accuracy on different read lengths and, 363
codon and dicodon usage and, 360–361
final candidate selection and, 362
GC content and, 361
length feature and, 361
neural network and, 361–362
open reading frame (ORF) and, 360
tool, 363
translation initiation site and, 361
orthogonal signal correction (OSC), 688
Oryza longistaminata, 510
Ostreococcus species, 580

Pacific Northwest National Laboratory (PNNL), 664
Paenibacillus borealis, 240
Pandoraea species, 60
pangenomes, 83, 84–85
applications of, 86
classical reverse vaccinology for vaccine development, 83–84
recent developments of, 85
parallel analysis of ribosomal sequence tags (PARST), 271
partial least squares (PLS) model, 688
partial least-squares descrimant analysis (PLS-DA), 679
particulate methane monooxygenase (pMMO), 492
PathwayHuntertool, 716
PCR-Select™Bacterial Genome Subtraction Kit, 498, *499*
Pedobacter heparinus, 136
Pelagibacter (α-*Proteobacteria*), 94, 201, 283, 580
peptide mass fingerprinting (PMF), 607, 629
peptidyl carrier protein domain (PCP), 425
phenotype microarrays (PMs), 533
applications
cell growth in culture, 534–536
community level physiological profiles reconsideration, 537–538
gene and genome annotation improvement, 536
metabolic modeling improvement, 536–537
secondary metabolism and chemical resistance analysis, 537
significance of, 533–534
phenotypic diversity, 6
pheV site, 260
Phi29 DNA polymerase, 120
*Pho*C, 303
Phoenix Mission, 544–545
phospholipid fatty acid (PLFA), 592
Photobacterium damselae species, 31
Photobacterium profundum, 22
Photorhabdus luminescens, 22
phototrophs, 115
*Phy*A gene, 303
PhylArray algorithm, 554
PHYLIP software, 132, 404, 442
PhylochipAnalyzer, 555
PhyloChips, 124, 125, 126, 509, 521, 543, 554
data analysis, 523
high-density oligonucleotide microarray application, 523–524
high-density phylogenetic oligonucleotide microarray development and, 522
intestinal Th17 cell induction by filamentous bacteria and, 525, 527–530
sample preparation and, 522
urban aerosols and, 524–525
phylogenetic binning, of metagenome sequence samples, 353
case studies
composition-based supervised binning, 357
composition-based unsupervised binning, 357
GC content and read coverage-based binning, 357
homology-based binning, 356
recent developments and challenges and, 357–358
sequence composition-based approaches and, 354, 356
sequence homology-based approaches and, 353–354
phylogenetic oligonucelotide microarray (POA), 551
application of, 556–558
case study, 558, 561
development and analytical performance
data analysis, 554–555
marker gene and probe length choice, 553
probe design, 553
validation, 554
technology challenges and
cost, 556
quantification, 556
sensitivity, 555–556
specificity, 555
Phylogenetic Profiler tool, 227
phylogenetic screening, of metagenomic libraries, 245
approach, 246
homing endonuclease restriction and marker insertion and, 247
specificity and selectivity, 247
kanamycin cassette construction with *I-Ceu*I recognition sequences and, 246–247
phylogenetic species evenness (PSE), 125
phylogenetic species variability (PSV), 125
phylogenetic test, 239
PhyloPythia algorithm, 226, 245, 371, 374
PhyloTrac, Fast UniFrac in, 555
phyloware project, 407
goals, 408
implementation
barcoded primer design, 412
batch processing, 410
biodiversity measures, 414
multiplexed sequencing output deconvolution, 413
overview, 409
parallel computing, 410
parse rRNA sequences from GenBank files, 414–415
sequence clustering, 413–414
sequence polishing, 413
third-party software integration, 412
utility programs, 415

phyloware project (*continued*)
XplorSeq, 409–410, *410*, *411–412*, 413
metagenomic sequence analysis best practices and, 408–409
problem statement, 407–408
PHYML algorithm, 207, 404, 447, *601*
Pichia species, 601
Pintail software, 394
Pinus pinaster, 598
Planctomycetes species, 228, 231, 247, 270, 558
plant metabolic network (PMN), 711
plant microbiota enrichment, 197
assessment, by 16S rDNA-based techniques, 198
enhancement, 199–201
metagenomic library construction and evaluation and, 198
prokaryotic inserts and, 201
by SDS and NaCl coupling, 198–199
with stem barks, 198
plant pathogen-suppressive soil metagenomics, 277, *280*
disease-suppressive soils and
antibiotic production, 278
chitinolytic activity, 278
functional screening, 281
genetic screening, 281–282
metagenomic libraries, 279–281
soil DNA extraction and processing, 279
target genes hit rates, 282
hit rates improvement and
deliberate bias in sampled communities and, 282
direct pyrosequencing, 283–284
improved metagenomic library host searching, 282–283
improved screening methods, 283
metagenomic library vector, 283
Plasmodiophora brassicae, 279
Plasmodium species, 600
pmoA gene, 494, 510
polychlorinated biphenyls (PCBs), 617
polyketide synthases (PKS), 424, 425, 426, *428*, 429, 430, 431
polymerase chain reaction (PCR), 29, 31, 237, 241, 441, 444
amplification, 42, 46, 50, 218, *219*, 266, 498–499, *499*, 509, 510, 511
denaturing gradient gel electrophoresis (DGGE), 638–639
genetic fingerprinting techniques and, 31–32
primers, 41–44, *52*, 117, 123, *220*, 239, 442, 462, 510
product digestion and primer removal and, 266–267
strategies, *43*, 44, *45*
template, 61
polymerase chain reaction (PCR)-based rRNA gene sequence analysis, 129
16S rRNA gene sequence data analysis, 131–132
cell lysis and extraction of nucleic acids and, 130–131
chimeric structure formation, 131
denturating gel electrophoresis patterns, statistical analysis of, 131
polyomic, systems biology data, software for, 721
polyphosphate-accumulating organisms (PAOs), 731
poly-ß-hydroxyalkanoate (PHA), 731
population genomics, 77
core genome hypothesis and, 79–80
data, 78–79
information, 78
Porphyromonas gingivalis, 84
Postgres relational database, 326
post-translational modification (PTMs), 605–606, 607, 608, 618, 619
PowerSoil™, 570
pqq gene, 303
precision recall curve (PRC), 361
predicted highly expressed (PHX) genes, 577
PRIMER 5 software, 131
primers4clades server, 441, *443*, 554
Actinobacteriales demonstration dataset and, 442
analysis, 447
customization of, 444
Mycobacterium species and rpoB and, 447–448
results returned to user by, 444–447
standard and extended CODEHOP primer design strategies, 442, 444
statistical evaluation and, 449
PRIMROSE software, 135, *137*, 138, 554
principal component analysis (PCA), 70, 291, 294, *295*, 514, 688, 710
principal coordinates analysis (PCoA), 125
principle component analysis (PCA), 679
prnD copy number, 278
ProbeBase, 557
Probe Match, RDP, 317
Prochlorococcus (*Cyanobacteria*), 94
Prochlorococcus marinus, 213, 362, *579*, 580, 581, 584, 616, 656
Prochlorococcus species, 577, 630, 637
Prochlorococcus ubique, 580, 581
proliferating cells, identification of, 728–730
propidium iodide (PI), 730
Propionibacterium species, 545
protein archiving, 338
protein coding genes (PCGs), 359, 360, 366, 441
protein encoding genes (PEGs), 326
Proteobacteria species, 200, 201, 201, 228, 270, 558
proteorhodopsin, 117
Proteus species, 511
psaA gene, 579
psaB gene, 579
Pseudoalteromonase haloplanktis, 614
Pseudomonadales species, 209
pseudomona fluorescens, 302
Pseudomonas aeruginosa, 125, 281, 362, 511, 537
Pseudomonas corrugata, 669, 671, 672
Pseudomonas fluorescens, 278
pseudomona species, 302
Pseudomonas putida, 136, 175, 283, 617, 669, 742
Pseudomonas putida isoF, 684, 692
growing condition of, 685
Pseudomonas species, 79, 101, 187, 536, 545
Pseudomonas syringae, 510, 537
Psychrobacter species, 545
PubChem database, 711, 713, 714
pulsed field gel electrophoresis (PFGE), 292, 493, *494*
*Pvu*II gene, 198, 199
Pyrolobus fumarii, 115
pyrosequencing, 119, 144, 150–154, 271, 521, 588
direct, 283–284
error rate calculations and, 150
V6 Amplicon libraries generation and sequencing and, 150

QconCAT, 657
QIAquick PCR Purification Kit (QIAGEN), 138, 266, 268, 499
Q-ToFMS, 714
quadrupole time-of-flight (Q-TOF)-type mass spectrometers, 651
quantitative real-time PCR(qPCR), 137, 138–149
verifications, 577
quickdist algorithm, 463

Ralstonia eutropha, 227, 228, 231
Ralstonia metallidurans, 283, 617
Ralstonia species, 60, 543, 545, 732
random amplified polymorphic DNAPCR (RAPD), 31
random but comprehensive metagenome sequencing, 118
Random Primers DNA Labeling System (Invitrogen), 499

random sequence reads (RSRs) versus 16S rDNA sequences, 163–164
  comparison from human gut metagenomic libraries, 166–167
  computational pipeline validation, 166
  database choice, 165–166
  gut microbiota biodiversity, 166
  Sargasso sea shotgun sequences
    dataset analyses, 167–168
    taxonomy assignment pipeline (TAP), 164–165
  sequence datasets, 165
RAxML program, 132, 208, 404
reactive oxygen species (ROS), 615
reactome, 711
recA gene, 205
reductive carboxylate cycle, 454
reference genome sequencing, 146–147
Regenerative Enclosed Life Support Module Simulator (REMS), 545
repetitive extragenic palindromic (REP) sequences, 32
repetitive sequence-based polymerase chain reaction (rep-PCR), 31, 32, 34, 35, 36, 67, 68, *71*
restriction endonucleases (REs), 59, 62
restriction fragment length polymorphism (RFLP) patterns, 50, 51, 67
reverse vaccinology, for vaccine development, 83–84
rhizobia, 49, 50, 53, 56
Rhizobium galegae, 56
Rhizobium gallicum, 56
Rhizobium mongolense, 56
Rhizobium species, 52, *53*, *54*, 56, 59
Rhizobium yanglingense, 56
Rhizoctonia solani, 278, 279
Rhodococcus jostii, 617
Rhodococcus species, 448
Rhodocyclaceae species, 227, 231
Rhodocyclales species, 557
Rhodopseudomonas species, 545
rhodopsin family, 117–118
ribosomal database project (RDP), 313
  analysis tools, 315
    Assignment Generator, 318
    Hierarchy Browser, 315, 317
    Hierarchy View, 315
    Library Compare, 316
    Naïve Bayesian Classifier, 316
    Probe Match, 317
    SeqCart, 315, 319
    Sequence Match, 316–317
    Taxomatic, 318
    Tree Builder, 317–318
    web service interfaces, 318
  *my*RDP personal account system
    Library Run, 320
    *my*RDP pipeline, 319–320
    Research Buddies, 319
    sequence groups, 318–319
  public sequence growth, *314*
  pyro pipeline, 320, *321*
    experimental choices, 322
    initial process, 320
    taxonomy-based analysis, 320–321
    taxonomy-independent analysis, 321–322
  sequence data, 314
    quality testing, 314
    sequence alignments, 314–315
  version, 9.60 63
Ribosomal Database Project II, 30
ribosomal RNA (rRNA) operon and genomic repetitive sequences, 29, *31*, 100, 102
  bacillus species identification approaches
    materials and methods, 32–34
  bacterial taxonomy based on, 30–31
  –PCR, 158–159
  PCR-based genetic fingerprinting techniques, 31–32
  see also 23S rRNA genes
ribosomal RNA identification, 387
  algorithm descriptions and, 388
  algorithm evaluation and, 388
  algorithm performance and, 389
  application on real datasets and, 389
  computation speed and, 389
ribosomal RNA sequencing, 126
ribosomal rRNA intergenic spacer analysis (RISA), 60, 67, 130, 290, 291
Riftia pachyptila, 630
RNase away™, 570
Roadmap Initiative in the Human Microbiome Project, 17
Roche-454 platform, 144, 149
Roche Applied Science biochemical pathways, 716
Roche GS analysis software, 150, 153
Roseiflexus species, 247
Roseobacter denitrificans, 580
rpoB gene, 205, 448, 449
rrnA23S, 22, *24*
rrnB23S, 22, 23
rrnC23S, 22, 23
rrnC operon, 25
rrnD23S, 22, 23
rrnE23S, 22
rrnJ23S, 22
rrs genes, 49, *51*, 54, 56
  phylogeny, 52, 54
*Rsa*I sequences, 63
Rubrivivax gelatinosus, 45
Rumen metagenomes, rRNA analysis from, 502
Saccharomyces cerevisiae, 120, 598, 646, 655, 733
  affiliation of environmental thioredoxin protein sequence, *601*
Saccharopolyspora species, 448
Salinibacter genus, 12
Salinibacter ruber, 13
Salinispora species, 448
Salmonella enterica, 78, 511
Salmonella Paratyphi A, 260
Salmonella species, 511, 614
Salmonella typhimurium, 18, 20, 22, 25
Sanger (dideoxy) sequencing, 149
Sargasso sea shotgun sequences
  dataset analyses, 167–168
  taxonomy assignment pipeline (TAP), 164–165
SASRT method, 270
secondary ion mass spectrometry (SIMS), 681
SEED database, 325, 326, 453, 455, 456, 457, 458, 459, 588, 592
seeded growing self-organizing map (GSOM), 369, 370–372, 374–375
  binning fidelity comparison and, 374
  clustering percentage determination and, 371
  datasets of binning fidelity comparison and, 369–370
  datasets of clustering performance comparison and, 369
  growing SOM (GSOM) methods and, 370
  seeded growing self-organizing map (S-GSOM) method and, 370–371
  seeds and metagenomic fragments preparation and, 370
  self-organizing map (SOM) and, 370
  semisupervised algorithms for binning comparison and, 371–374
  training features and, 371
Seeded K-means, 371
segmented filamentous bacterium (SFB), 528–530
selected reaction monitoring (SRM), 619
self-organizing map (SOM), 370
  see also seeded growing self-organizing map (GSOM)
semisupervised algorithms for binning comparison, 371–374
SeqCart, RDP, 315, 319
SEQ MATCH program, 268
sequence associated information (SAIs), 402, 403
sequence composition-based approaches, 354, 356
sequence homology-based approaches, 353–354
Sequence Match, RDP, 316–317

Sequence Read Archive (SRA), 268
SEQUEST, 609, 629, 664
serial analysis of ribosomal sequence tags (SARST), 265
applications of, 270–271
cloned RST concatamers sequencing and phylogenetic analysis and, 268–269
gel sizing, end repair, and concatemers cloning and, 267–268
individual RST concatenation and, 267
PCR amplification and, 266
PCR product digestion and primer removal and, 266–267
potential bias and limitations of, 269–270
primers and linkers used in, *268*
Serratia marcescens, 278
Serratia species, 545
sesquiterpene synthases, 678
Shannon entropy, *240*
Shannon–Weaver index, 9
Shewanella oneidensis, 18, 20, 25, 665
Shewanella species, 354
Shigella flexneri, 261
Shigella species, 261
shotgun proteomics, 648
si2fa utility program, 415
silent isotope labeling strategies, 653
Silicibacter pomeroyi, 580
SILVA database, 393, 403, 405, 588
aligner, 394
alignment, 396
anomaly check and, 394–395
availability/website, 395–396
data retrieval and processing and, 396
nomenclature and rDNAs from genome projects, 395
quality checks and, 394
sequence data retrieval and rRNA extraction and, 394
SSU and LSU Rrna databases for ARB, 395
taxonomy and type strain information and, 395
SIMCA-P, 700
single-gene approach, 116–117
single-nucleotide polymorphisms (SNP), 383
single-strand conformation polymorphism (SSCP), 42
Sinorhizobium species, 52, *53*, *54*
SMILES, 429, *430*
sodium dodecyl sulfate polyacrylamide gel electrophoresis (SDS-PAGE), 119, 607
soil ammonia oxidizer communities analysis, 43–44
Soil Biology and Sustainable Agricultural Production (BIOSPAS) Consortium, 299
certified agriculture and new challenges and, 300
complementary approaches and, 301
biochemical analyses of soil samples, 302
biofilm formation in soil particle and tools development, 303
burkholderia bacteria group analysis, 302–303
cellulolytic bacteria analysis, 302
endomycorrhizae analysis, 303–304
glomalin-related soil proteins (GRSP), 302
molecular analysis of microbial diversity in soil and microorganisms, 303
nitrogen cycle genes and phosphorous mobilization genes and metagenomic studies, 303
organic matter fraction analysis, 302
pathogen incidence and isolation within studied environments, 304
pseudomona bacteria group analysis, 302
soil chemical composition, structure, and physical behavior, 301–302
soil meso-and macro fauna studies, 304
new paradigms in Argentine Agriculture and, 299–300
priority research areas in Argentina and creation of, 300–301
soil eukaryotic diversity, 597–601
soil meso-and macro fauna studies, 304
soil metagenomics, 288
advances, 288–289
comparisons, 293–296
DNA recovering strategy
cells lysis stringency fractionation, 291–292
molecular DNA weight separation, 292–293
prokaryote cells extraction, 290–291
prokaryote cells separation, 291
soil fractionation, 290
international partnerships to sequence environments and, 289–290
limits, 288
members of, *301*
metasoil project and, 290
Solanum lycopersicon, 701, 703, 705
Solanum pennellii, 701, 703, 705
Solexa Genome Analyzer and Applied Biosystems (SOLiD) *see* Illumina (Solexa) Gallx platform
sortx utility program, 413–414
species abundance distribution (SAD), 14
species-specific functions measuring, in microbial communities, 731
Sphingomonas species, 60, 545, 631, 732
Sphingomonas yanoikuyae, 545
SPSS, 700
SQLite, 326
Stable Isotope Labeling by Amino aCids (SILAC), 609, 618, 656
Standardization in Metabolomics Experiments, 699
Staphylococcus aureus, 85, 281, 511, 615
Staphylococcus epidermidis, 150, 151, 154, 511
Staphylococcus species, 511
Stenotrophomonas maltophilia, 278, 617
Stenotrophomonas species, 511
Streptococcus agalactiae, 84–85, 86
Streptococcus pneumoniae, 32, 84, 85
Streptococcus pyogenes, 18, 20, 25, 619
Streptococcus species, 78, 79, 187, 511, 545, 557
Streptomyces coelicolor, 681
Streptomyces griseus, 136
Streptomyces lividans, 175
Streptomyces species, 101, 201, 280, 281, 535, 556
strong cation exchange chromatography (SCX), 664
subcommunities, phylogenetic analysis of, 732
subcommunities and populations, functional omic analysis of, 732–733
substrate-induced gene expression screening (SIGEX), 177
subtractive hybridization, in silico, *259*
subunit ribosomal RNA (SSU rRNA) *see* 16S rRNA
sulfate-reducing prokaryotes (SRPs), 557
Sulfolobus solfataricus, 614, 655
Sulfolobus species, 657
summed spectrum, 700
super integrons, 218
support vector machine (SVM), 226, 356, 369
suppressive subtractive hybridization (SSH), 497, 498–499
clone libraries domain composition analysis and, 500
dot blot hybridization and, 499–500
grain-specific SSH clone libraries analysis and, 502
hay-specific clone libraries analysis and, 500–501
rRNA analysis from Rumen metagenomes and, 502
sample and collection and total DNA extraction and, 498
sequencing and data analysis and, 500

SuSE distribution, 405
SwissProt, 719
SYBRGreenI fluorescence, 730, *731*
Synechococcus lividus, 105
Synechococcus species, 105–108, *106*, 630, 637
Synechocystis species, 610, 613, 614, 615, 616, 655, 657
systems biology, 475, 721
computational approaches in, 479–481
computer programs for metagenomic data analysis and, 481–486
databases related to metagenomic applications and, *484*
molecular components of complex, *478*
movement to metagenomes, 477–479
omics approaches in, *476*, 486
SYTO9, 730

TACOA algorithm, 245
TaKaRa *ExTaq* PCR kit (TaKaRa), 498
tandem mass tag (TMT), 608, 655
*Taq*I sequences, 63
Taq polymerase, 131
targeted partial metagenome sequencing, 117–118
taxa-area relationships (TARs), 514
Taxomatic, RDP, 318
taxonomy assignment pipeline (TAP), 164–165, *164*, 166, *167*, *168*
tDNA, 32
PCR, *35*, 36, 38
temperature gradient gel electrophoresis (TGGE), 42
terminal-restriction fragment length polymorphism (T-RFLP), 42, 67, 509, 551
tet gene, 511
TETRA, 369, 481
The Arabidopsis Information Resource (TAIR), 720
Thermoanaerobacter tengcongensis, 18, *21*, 22, *23*, *24*, 25
Thermus thermophilus, *21*
thioesterase (TE), 424
time-of-flight instruments, 702
time-of-flight mass analyzers (TOF-SIMS), 681
Titanium FLX, 143, 144
tmRNA site, 260
TnAbaR1-like resistance mega-transposons, 261
tomato, metabolic profiling of, 701–702
data processing and interpretation and, 704–705
mass spectrometry analysis and, 702–703
sample preparation and, 702
total ion count (TIC), 710
traditional reductionist approach, 5
Transductive Support Vector Machine (TSVM), 371
Tree Builder, RDP, 317–318
TREE-PUZZLE program, 404
Treponema species, 357
Trewia nudiflora see *Mallotus nudiflorus*
Trichoderma species, 278
Tris–acetate–EDTA (TAE) buffer, 61
Tris–borate–EDTA (TBE) buffer, 61
tRNA gene, 31, 251
truly unique genes (TUG), 85
TRX1 gene, 598
TRX2 gene, 598
Trypanosoma species, 600
Tsukamurella species, 448
two-dimensional difference gel electrophoresis (2D-DIGE), 607, 610
two-dimensional gel electrophoresis (2-DGE), 119, 607, 608, 618, 638, *640*, *641*, *643*, 646, 665–666

Ubuntu Linux distribution, 405
UGA codon and MetaGene, 439
ultra-high-resolution SIMS, 681
UniFrac, 125, 239, 523
UniProt, 719
UniProt Archive (UniParc), 338, 339
UniVec database, 394
unsequenced halotolerant cyanobacterium, studying salt stress in, 657
unweighted pair group method using averages (UPGMA), 34, *37*, 44, *45*, 50, 70, 131
V6 Amplicon libraries generation and sequencing, 150
van Krevelen Diagram, 690, 691–692
Variovorax species, 43, 44, 46, 543
vegan package, 523
ventilator-associated pneumonia (VAP), 125
Vernicia montana, 201
Verrucomicrobia species, 228, 231, 232, 270, 558
viable and nonviable cells, identification of, 730–731
Vibrio splendidus, 92
Vicia nodules, 59
virulence and marker gene (VMG), 511

Weighbor Weighted Neighbor Joining program, 317
weighted UniFrac, 125
whole community genome/RNA amplification (WCGA/WCRA), 513
whole-genome amplification (WGA), 545, 546
wholegenome shotgun (WGS) sequences, *91*, 92, 146, 225
Windshield splatter analysis, 333
Wise2 package, 237, 242
Wolbachia species, 362

Xanthomonas campestris, 218, 219, *220*
*Xho*I gene, 198
XplorSeq, 409–410, *410*, 413
functionality, *411–412*
Toolkit, 412
xsdstcmp utility program, 415
xsel utility program, 415
xsgap utility program, 415
xsrep utility program, 415
xstrim utility program, 413
xsub utility program, 415

Yersinia pseudotuberulosis, 535, *536*, 537
Yule–Harding model, 418

ZORRO algorithm, 206, 209